STUDENT'S SOLUTIONS MANUAL

C TRIMBLE & ASSOCIATES

INTERMEDIATE ALGEBRA
THIRD EDITION

Michael Sullivan, III
Joliet Junior College

Katherine Struve
Columbus State Community College

PEARSON

Boston   Columbus   Indianapolis   New York   San Francisco   Upper Saddle River
Amsterdam   Cape Town   Dubai   London   Madrid   Milan   Munich   Paris   Montreal   Toronto
Delhi   Mexico City   São Paulo   Sydney   Hong Kong   Seoul   Singapore   Taipei   Tokyo

Copyright © 2014, 2010, 2007 Pearson Education, Inc.
Publishing as Pearson, 75 Arlington Street, Boston, MA 02116.

ISBN-13: 978-0-321-88135-9
ISBN-10: 0-321-88135-4

1 2 3 4 5 6 EBM 16 15 14 13

www.pearsonhighered.com

PEARSON

Contents

Chapter R

Section R.1

R.1 Exercises

Answers will vary.

Section R.2

Section R.2 Quick Checks

1. A <u>set</u> is a well-defined collection of objects.

2. The objects in a set are called <u>elements</u>.

3. We will let D represent the set of all digits that are less than 5.
   set-builder: $D = \{x \mid x \text{ is a digit less than } 5\}$
   roster: $\{0, 1, 2, 3, 4\}$

4. We will let G represent the set of all digits that are greater than or equal to 6.
   set-builder:
   $G = \{x \mid x \text{ is a digit greater than or equal to } 6\}$
   roster: $\{6, 7, 8, 9\}$

5. The statement is true because the order in which elements are listed in a set does not matter.

6. If every element of a set A is also an element of a set B, then we say that A is a <u>subset</u> of B and write $A \subseteq B$.

7. False. The empty set can be denoted as { } or $\varnothing$, but not as $\{\varnothing\}$.

8. The statement $B \subseteq A$ is true because all the elements that are in B are also in A.

9. The statement $B = C$ is false because there are elements in B (a and b) that are not in C. There is also an element in C, namely d, that is not in B.

10. The statement $B \subset D$ is false. In order for B to be a proper subset of D, it must be the case that all the elements that are in B are also in D. In addition, there must be at least one element in D that is not in B. Since $B = D$, this is not the case.

11. The statement $\varnothing \subseteq A$ is true. The empty set is a subset of every set.

12. The statement $5 \in \{0, 1, 2, 3, 4, 5\}$ is true because 5 is an element of the set.

13. The statement Michigan $\notin$ {Illinois, Indiana, Michigan, Wisconsin} is false because Michigan **is** an element of the set.

14. The statement
    $\frac{8}{3} \in \{x \mid x = \frac{p}{q} \text{ where } p \text{ and } q \text{ are digits}, q \neq 0\}$
    is true because $\frac{8}{3}$ is of the form $\frac{p}{q}$ where $p = 8$ and $q = 3$.

15. The <u>whole</u> numbers are numbers in the set $\{0, 1, 2, 3, 4, ...\}$.

16. The numbers in the set
    $\left\{ x \mid x = \frac{p}{q}, \text{ where } p \text{ and } q \text{ are integers}, q \neq 0 \right\}$
    are called <u>rational</u> numbers.

17. The set of <u>irrational</u> numbers have decimal representations that neither terminate nor repeat.

18. False. If a number is rational, it cannot be irrational and vice-versa.

19. False. Decimals that terminate or repeat represent rational numbers, but decimals that do not repeat and do not terminate represent irrational numbers.

20. True. The set of rational numbers is a subset of the set of real numbers.

21. True. The set of rational numbers and the set of irrational numbers do not have any numbers in common.

22. 10 and $\frac{12}{4} = 3$ are the only natural numbers.

23. 10, $\frac{0}{3} = 0$, and $\frac{12}{4} = 3$ are the whole numbers.

24. -9, 10, $\frac{0}{3} = 0$, and $\frac{12}{4} = 3$ are the integers.

25. $\dfrac{7}{3}$, -9, 10, $4.\overline{56}$, $\dfrac{0}{3}$, $-\dfrac{4}{7}$, and $\dfrac{12}{4}$ are the rational numbers.

26. $5.7377377737777\ldots$ and π and $\sqrt{11}$ are the irrational numbers. Note that $5.7377377737777\ldots$ is irrational because the decimal portion is non-terminating and non-repeating.

27. All numbers listed are real numbers.

28. a. To truncate, we remove all digits after the third decimal place. So, 5.694392 truncated to three decimal places is 5.694.

    b. To round to three decimal places, we first examine the digit in the fourth decimal place and then truncate. Since the fourth decimal place contains a 3, we do not change the digit in the third decimal place. Thus, 5.694392 rounded to three decimal places is 5.694.

29. a. To truncate, we remove all digits after the second decimal place. So, -4.9369102 truncated to two decimal places is -4.93.

    b. To round to two decimal places, we first examine the digit in the third decimal place and then truncate. Since the third decimal place contains a 6, we increase the digit in the second decimal place by 1. Thus, -4.9369102 rounded to two decimal places is -4.94.

30. $\left\{-2, 0, \dfrac{1}{2}, 3, 3.5\right\}$

31. $3 < 6$ because 3 is further to the left on a number line.

32. $-3 < -2$ because -3 is further to the left on a number line.

33. $\dfrac{2}{3} = 0.\overline{6}$ and $\dfrac{1}{2} = 0.5$

    $\dfrac{2}{3} > \dfrac{1}{2}$ because $\dfrac{2}{3}$ is further to the right on a number line.

34. $\dfrac{5}{7} \approx 0.714$

    $\dfrac{5}{7} > 0.7$ because $\dfrac{5}{7}$ is further to the right on a number line.

35. $\dfrac{2}{3} = 0.\overline{6}$ and $\dfrac{10}{15} = 0.\overline{6}$

    $\dfrac{2}{3} = \dfrac{10}{15}$ because the numbers are the same.

36. $\pi \approx 3.14159$

    $\pi > 3.14$ because π is further to the right on a number line.

R.2 Exercises

37. $\{0, 1, 2, 3, 4, 5\}$

39. $\{-2, -1, 0, 1, 2, 3, 4\}$

41. $\varnothing$ or $\{\ \}$

43. The statement $B \subseteq C$ is true since all the elements of B are also elements of C.

45. In order for B to be a proper subset of D, it must be the case that all the elements that are in B are also elements in D. In addition, there must be at least one element that is in D that is not in B. Because $B = D$, this statement is false.

47. The statement $B = D$ is true because sets B and D have the same elements.

49. The statement $\varnothing \subset C$ is true because the empty set is a subset of every set, and the set C is not empty.

51. $\dfrac{1}{2} \notin \{x \mid x \text{ is an integer}\}$

53. $\pi \in \{x \mid x \text{ is a real number}\}$

55. a. 4

    b. $-5, 4$

    c. $-5, 4, \dfrac{4}{3}, -\dfrac{7}{5}, 5.\overline{1}$

    d. π

   e. $-5, 4, \dfrac{4}{3}, -\dfrac{7}{5}, 5.\bar{1}, \pi$

57. a.   100

    b. $100, -64$

    c. $100, -5.423, \dfrac{8}{7}, -64$

    d. $\sqrt{2} + 4$

    e. $100, -5.423, \dfrac{8}{7}, \sqrt{2} + 4, -64$

59. a.   To truncate, we remove all digits after the fourth decimal place. So, 19.93483 truncated to four decimal places is 19.9348.

    b.   To round to four decimal places, we first examine the digit in the fifth decimal place and then truncate. Since the fifth decimal place contains a 3, we do not change the digit in the fourth decimal place. Thus, 19.93483 rounded to four decimal places is 19.9348.

61. a.   To truncate, we remove all digits after the first decimal place. So, 0.06345 truncated to one decimal place is 0.0.

    b.   To round to one decimal place, we first examine the digit in the second decimal place and then truncate. Since the second decimal place contains a 6, we increase the digit in the first decimal place by 1. Thus, 0.06345 rounded to one decimal place is 0.1.

63.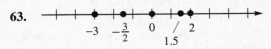

65. $-5 < -3$ since -5 is further to the left on a real number line.

67. $\dfrac{3}{2} = 1.5$ since $\dfrac{3}{2}$ and 1.5 are at the same point on a real number line.

69. $\dfrac{1}{3} > 0.3$ since $\dfrac{1}{3}$ is further to the right on a real number line.

71. Since sea level is an elevation of 0 ft, Death Valley has an elevation of -282 feet.

73. $-\$3.37$
The value of Best Buy common stock fell $3.37 during 2012.

75. -6
A score of 66 is 6 under par.

77. Answers will vary.

79. Answers will vary.

81. No; if a number is rational, it cannot be irrational and vice-versa.
No; every real number, when written in decimal form, will either terminate (rational), have a repeating block (rational), or neither terminate nor repeat (irrational).

83. Answers will vary.

85. If *A* is a subset of *B*, then all the elements that are in set *A* are also in set *B*. Thus *B* is a subset of *B*. If *A* is a proper subset of *B*, then all the elements that are in set *A* are also elements in set *B*, but there are elements in *B* that are not in set *A*. Thus, *B* is not a proper subset of *B*.

87. If the digit *after* the specified final digit is 4 or less, then truncating and rounding will yield the same decimal approximation.

89. Rounded: $\dfrac{8}{7} \approx 1.143$

    Truncated: $\dfrac{8}{7} \approx 1.142$

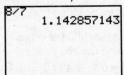

Section R.3

Section R.3 Quick Checks

  1. In the expression $a \cdot b$, the expressions *a* and *b* are called <u>factors</u>.

  2. False. The absolute value of 0 is 0, and 0 is neither positive nor negative.

3. $|6| = 6$ because the distance from 0 to 6 on the real number line is 6 units.

4. $|-10| = 10$ because the distance from 0 to -10 on the real number line is 10 units.

5. $18 + (-6)$

One number is positive, while the other is negative. We determine the absolute value of each number: $|18| = 18$ and $|-6| = 6$. We now subtract the smaller absolute value from the larger and obtain $18 - 6 = 12$. Because the number with the larger absolute value is positive, the sum will be positive. So, $18 + (-6) = 12$.

6. $-21 + 10$

One number is positive, while the other is negative. We determine the absolute value of each number: $|-21| = 21$ and $|10| = 10$. We now subtract the smaller absolute value from the larger and obtain $21 - 10 = 11$. Because the number with the larger absolute value is negative, the sum will be negative. So, $-21 + 10 = -11$.

7. $-5.4 + (-1.2)$

Both numbers are negative, so we first determine their absolute values: $|-5.4| = 5.4$ and $|-1.2| = 1.2$. We now add the absolute values and obtain $5.4 + 1.2 = 6.6$. Because both numbers to be added are negative, the sum is also negative. So, $-5.4 + (-1.2) = -6.6$.

8. $-6.5 + 4.3$

One number is positive, while the other is negative. We determine the absolute value of each number: $|-6.5| = 6.5$ and $|4.3| = 4.3$. We now subtract the smaller absolute value from the larger and obtain $6.5 - 4.3 = 2.2$. Because the number with the larger absolute value is negative, the sum will be negative. So, $-6.5 + 4.3 = -2.2$.

9. $-9 + 9$

One number is positive, while the other is negative. We determine the absolute value of each number: $|-9| = 9$ and $|9| = 9$. Since the absolute values are the same, the difference will be 0. So, $-9 + 9 = 0$.

10. For any real number a, there is a real number $-a$, called the <u>additive inverse</u>, or <u>opposite</u>, of a such that $a + (-a) = -a + a = \underline{0}$.

11. For any real number a, $-(-a) = \underline{a}$.

12. The additive inverse of 5 is -5 because $5 + (-5) = 0$.

13. The additive inverse of $\frac{4}{5}$ is $-\frac{4}{5}$ because $\frac{4}{5} + \left(-\frac{4}{5}\right) = 0$.

14. The additive inverse of -12 is $-(-12) = 12$ because $-12 + 12 = 0$.

15. The additive inverse of $-\frac{5}{3}$ is $-\left(-\frac{5}{3}\right) = \frac{5}{3}$ because $-\frac{5}{3} + \frac{5}{3} = 0$.

16. The additive inverse of 0 is 0 because $0 + 0 = 0$.

17. $6 - 2 = 6 + (-2) = 4$

18. $4 - 13 = 4 + (-13) = -9$

19. $-3 - 8 = -3 + (-8) = -11$

20. $12.5 - 3.4 = 12.5 + (-3.4) = 9.1$

21. $-8.5 - (-3.4) = -8.5 + 3.4 = -5.1$

22. $-6.9 - 9.2 = -6.9 + (-9.2) = -16.1$

23. True; the product of two negative real numbers is positive.

24. True. The commutative property of addition states $a + b = b + a$, and the commutative property of multiplication states $a \cdot b = b \cdot a$.

25. $-6 \cdot (8) = -(6 \cdot 8) = -48$

26. $12 \cdot (-5) = -(12 \cdot 5) = -60$

27. $4 \cdot 14 = 56$

28. $-7 \cdot (-15) = 7 \cdot 15 = 105$

29. $-1.9 \cdot (-2.7) = 1.9 \cdot 2.7 = 5.13$

30. The additive inverse of a, $-a$, is also called the <u>opposite</u> of a. The multiplicative inverse of a, $\dfrac{1}{a}$, is also called the <u>reciprocal</u> of a.

31. The multiplicative inverse of 10 is $\dfrac{1}{10}$.

32. The multiplicative inverse of -8 is $-\dfrac{1}{8}$.

33. The multiplicative inverse of $\dfrac{2}{5}$ is $\dfrac{5}{2}$.

34. The multiplicative inverse of $-\dfrac{1}{5}$ is $-\dfrac{5}{1} = -5$.

35. A fraction is written in <u>lowest terms</u> if the numerator and denominator share no common factor other than 1.

36. $\dfrac{2 \cdot 6}{2 \cdot 5} = \dfrac{\cancel{2} \cdot 6}{\cancel{2} \cdot 5} = \dfrac{6}{5}$

37. $\dfrac{33}{24} = \dfrac{\cancel{3} \cdot 11}{\cancel{3} \cdot 8} = \dfrac{11}{8}$

38. $-\dfrac{24}{20} = -\dfrac{\cancel{4} \cdot 6}{\cancel{4} \cdot 5} = -\dfrac{6}{5}$

39. $\dfrac{4}{9}$ is in lowest terms.

40. $\dfrac{5}{3} \cdot \dfrac{12}{25} = \dfrac{5 \cdot 12}{3 \cdot 25}$

$= \dfrac{5 \cdot 3 \cdot 4}{3 \cdot 5 \cdot 5}$

$= \dfrac{\cancel{5} \cdot \cancel{3} \cdot 4}{\cancel{3} \cdot \cancel{5} \cdot 5}$

$= \dfrac{4}{5}$

41. $\dfrac{2}{3} \cdot \left(-\dfrac{5}{4}\right) = -\dfrac{2 \cdot 5}{3 \cdot 4}$

$= -\dfrac{2 \cdot 5}{3 \cdot 2 \cdot 2}$

$= -\dfrac{\cancel{2} \cdot 5}{3 \cdot \cancel{2} \cdot 2}$

$= -\dfrac{5}{3 \cdot 2}$

$= -\dfrac{5}{6}$

42. $\dfrac{4}{3} \div \dfrac{8}{3} = \dfrac{4}{3} \cdot \dfrac{3}{8}$

$= \dfrac{4 \cdot 3}{3 \cdot 8}$

$= \dfrac{\cancel{4} \cdot \cancel{3}}{\cancel{3} \cdot \cancel{4} \cdot 2}$

$= \dfrac{1}{2}$

43. $\dfrac{\frac{10}{3}}{\frac{5}{12}} = \dfrac{10}{3} \cdot \dfrac{12}{5}$

$= \dfrac{10 \cdot 12}{3 \cdot 5}$

$= \dfrac{2 \cdot \cancel{5} \cdot \cancel{3} \cdot 4}{\cancel{3} \cdot \cancel{5}}$

$= 8$

44. $\dfrac{3}{11} + \dfrac{2}{11} = \dfrac{3+2}{11} = \dfrac{5}{11}$

45. $\dfrac{8}{15} - \dfrac{13}{15} = \dfrac{8-13}{15} = \dfrac{-5}{15} = -\dfrac{\cancel{5}}{3 \cdot \cancel{5}} = -\dfrac{1}{3}$

46. $\dfrac{3}{7} + \dfrac{11}{7} = \dfrac{3+11}{7} = \dfrac{14}{7} = \dfrac{2 \cdot \cancel{7}}{\cancel{7}} = 2$

47. $\dfrac{8}{5} - \dfrac{3}{5} = \dfrac{8-3}{5} = \dfrac{5}{5} = 1$

48. The <u>least common denominator</u> is the smallest number that each denominator has as a common multiple.

49. $20 = 2 \cdot 2 \cdot 5$
$15 = 3 \cdot 5$
The common factor is 5. The remaining factors are 2, 2, and 3.
$LCD = 2 \cdot 2 \cdot 3 \cdot 5 = 60$
$$\frac{3}{20} = \frac{3}{20} \cdot \frac{3}{3} = \frac{3 \cdot 3}{20 \cdot 3} = \frac{9}{60}$$
$$\frac{2}{15} = \frac{2}{15} \cdot \frac{4}{4} = \frac{2 \cdot 4}{15 \cdot 4} = \frac{8}{60}$$

50. $18 = 2 \cdot 3 \cdot 3$
$45 = 3 \cdot 3 \cdot 5$
The common factors are 3 and 3. The remaining factors are 2 and 5.
$LCD = 3 \cdot 3 \cdot 2 \cdot 5 = 90$
$$\frac{5}{18} = \frac{5}{18} \cdot \frac{5}{5} = \frac{5 \cdot 5}{18 \cdot 5} = \frac{25}{90}$$
$$-\frac{1}{45} = -\frac{1}{45} \cdot \frac{2}{2} = -\frac{1 \cdot 2}{45 \cdot 2} = -\frac{2}{90}$$

51. $LCD = 2 \cdot 2 \cdot 5 \cdot 3 = 60$
$$\frac{3}{20} + \frac{2}{15} = \frac{3}{20} \cdot \frac{3}{3} + \frac{2}{15} \cdot \frac{4}{4}$$
$$= \frac{9}{60} + \frac{8}{60}$$
$$= \frac{9 + 8}{60}$$
$$= \frac{17}{60}$$

52. $LCD = 2 \cdot 7 \cdot 3 = 42$
$$\frac{5}{14} - \frac{11}{21} = \frac{5}{14} + \left(\frac{-11}{21}\right)$$
$$= \frac{5}{14} \cdot \frac{3}{3} + \left(\frac{-11}{21} \cdot \frac{2}{2}\right)$$
$$= \frac{15}{42} + \left(\frac{-22}{42}\right)$$
$$= \frac{15 + (-22)}{42}$$
$$= \frac{-7}{42} = \frac{-1 \cdot 7}{6 \cdot 7}$$
$$= -\frac{1}{6}$$

53. $LCD = 5 \cdot 5 \cdot 6 = 150$
$$-\frac{4}{25} - \frac{7}{30} = \frac{-4}{25} + \left(\frac{-7}{30}\right)$$
$$= \frac{-4}{25} \cdot \frac{6}{6} + \left(\frac{-7}{30} \cdot \frac{5}{5}\right)$$
$$= \frac{-24}{150} + \left(\frac{-35}{150}\right)$$
$$= \frac{-24 + (-35)}{150}$$
$$= -\frac{59}{150}$$

54. $LCD = 2 \cdot 3 \cdot 3 \cdot 5 = 90$
$$-\frac{5}{18} - \frac{1}{45} = \frac{-5}{18} + \left(\frac{-1}{45}\right)$$
$$= \frac{-5}{18} \cdot \frac{5}{5} + \left(\frac{-1}{45} \cdot \frac{2}{2}\right)$$
$$= \frac{-25}{90} + \left(\frac{-2}{90}\right)$$
$$= \frac{-25 + (-2)}{90}$$
$$= \frac{-27}{90}$$
$$= -\frac{3 \cdot 9}{10 \cdot 9}$$
$$= -\frac{3}{10}$$

55. The Distributive Property states that
$a \cdot (b + c) = \underline{ab} + \underline{ac}$.

56. $-(a + b) = \underline{-a - b}$

57. $5(x + 3) = 5 \cdot x + 5 \cdot 3$
$ = 5x + 15$

58. $-6(x + 1) = -6 \cdot x + (-6) \cdot 1$
$ = -6x - 6$

59. $-4(z - 8) = -4 \cdot z - (-4) \cdot 8$
$ = -4z + 32$

60. $\frac{1}{3}(6x + 9) = \frac{1}{3} \cdot 6x + \frac{1}{3} \cdot 9$
$$= \frac{6x}{3} + \frac{9}{3}$$
$$= 2x + 3$$

61. $-(11p + 8) = -11p + (-8) = -11p - 8$

62. $-(-9 - 4t) = -(-9) - (-4t) = 9 + 4t$

R.3 Exercises

63. $\left|\dfrac{2}{3}\right| = \dfrac{2}{3}$

65. $\left|-\dfrac{8}{3}\right| = \dfrac{8}{3}$

67. $-13 + 4 = (-13) + 4 = -9$

69. $12 - 5 = 12 + (-5) = 7$

71. $4.3 - 6.8 = 4.3 + (-6.8) = -2.5$

73. $4 \cdot (-8) = -(4 \cdot 8) = -32$

75. $-6 \cdot (-14) = 6 \cdot 14 = 84$

77. $-4.3 \cdot (8.5) = -(4.3 \cdot 8.5) = -36.55$

79. $\dfrac{28}{12} = \dfrac{\cancel{4} \cdot 7}{\cancel{4} \cdot 3} = \dfrac{7}{3}$

81. $\dfrac{30}{25} = \dfrac{5 \cdot 6}{5 \cdot 5} = \dfrac{\cancel{5} \cdot 6}{\cancel{5} \cdot 5} = \dfrac{6}{5}$

83. $\dfrac{3}{4} \cdot \dfrac{20}{9} = \dfrac{3 \cdot 20}{4 \cdot 9} = \dfrac{3 \cdot 4 \cdot 5}{4 \cdot 3 \cdot 3} = \dfrac{\cancel{3} \cdot \cancel{4} \cdot 5}{\cancel{4} \cdot \cancel{3} \cdot 3} = \dfrac{5}{3}$

85. $\dfrac{2}{3} \cdot \left(-\dfrac{9}{14}\right) = -\dfrac{2}{3} \cdot \dfrac{9}{14} = -\dfrac{2 \cdot 3 \cdot 3}{3 \cdot 2 \cdot 7} = -\dfrac{\cancel{2} \cdot \cancel{3} \cdot 3}{\cancel{3} \cdot \cancel{2} \cdot 7} = -\dfrac{3}{7}$

87. $\dfrac{7}{4} \div \dfrac{21}{8} = \dfrac{7}{4} \cdot \dfrac{8}{21} = \dfrac{7 \cdot 8}{4 \cdot 21} = \dfrac{7 \cdot 2 \cdot 4}{4 \cdot 3 \cdot 7} = \dfrac{\cancel{7} \cdot 2 \cdot \cancel{4}}{\cancel{4} \cdot 3 \cdot \cancel{7}} = \dfrac{2}{3}$

89. $\dfrac{\tfrac{2}{5}}{\tfrac{8}{25}} = \dfrac{2}{5} \cdot \dfrac{25}{8} = \dfrac{2 \cdot 25}{5 \cdot 8} = \dfrac{2 \cdot 5 \cdot 5}{5 \cdot 4 \cdot 2} = \dfrac{\cancel{2} \cdot \cancel{5} \cdot 5}{\cancel{5} \cdot 4 \cdot \cancel{2}} = \dfrac{5}{4}$

91. $-\dfrac{5}{12} + \dfrac{7}{12} = \dfrac{-5 + 7}{12} = \dfrac{2}{12} = \dfrac{2}{2 \cdot 6} = \dfrac{\cancel{2}}{\cancel{2} \cdot 6} = \dfrac{1}{6}$

93. LCD $= 2 \cdot 2 \cdot 3 \cdot 5$

$\dfrac{7}{15} + \dfrac{9}{20} = \dfrac{7}{15} \cdot \dfrac{4}{4} + \dfrac{9}{20} \cdot \dfrac{3}{3}$

$= \dfrac{28}{60} + \dfrac{27}{60}$

$= \dfrac{28 + 27}{60}$

$= \dfrac{55}{60} = \dfrac{11 \cdot \cancel{5}}{12 \cdot \cancel{5}}$

$= \dfrac{11}{12}$

95. $2(x + 4) = 2 \cdot x + 2 \cdot 4 = 2x + 8$

97. $-3(z - 2) = -3 \cdot z - (-3) \cdot 2 = -3z + 6$

99. $(x - 10) \cdot 3 = x \cdot 3 = -10 \cdot 3 = 3x - 30$

101. $\dfrac{3}{4}(8x - 12) = \dfrac{3}{4} \cdot 8x - \dfrac{3}{4} \cdot 12 = \dfrac{24x}{4} - \dfrac{36}{4} = 6x - 9$

103. $-(5z + 17) = -5z + (-17) = -5z - 17$

105. $|13 - 16| = |13 + (-16)| = |-3| = 3$

107. $\dfrac{51}{42} = \dfrac{17 \cdot \cancel{3}}{14 \cdot \cancel{3}} = \dfrac{17}{14}$

LCD $= 2 \cdot 5 \cdot 7 = 70$

$\dfrac{51}{42} - \left(-\dfrac{8}{35}\right) = \dfrac{17}{14} - \left(-\dfrac{8}{35}\right)$

$= \dfrac{17}{14} + \dfrac{8}{35}$

$= \dfrac{17}{14} \cdot \dfrac{5}{5} + \dfrac{8}{35} \cdot \dfrac{2}{2}$

$= \dfrac{85}{70} + \dfrac{16}{70}$

$= \dfrac{85 + 16}{70}$

$= \dfrac{101}{70}$

109. $5 \div \dfrac{15}{4} = \dfrac{5}{1} \cdot \dfrac{4}{15} = \dfrac{5 \cdot 4}{15} = \dfrac{5 \cdot 4}{5 \cdot 3} = \dfrac{\cancel{5} \cdot 4}{\cancel{5} \cdot 3} = \dfrac{4}{3}$

111. $|6.2 - 9.5| = |6.2 + (-9.5)| = |-3.3| = 3.3$

113. $-|-8 \cdot (4)| = -|-32| = -(32) = -32$

115. $\left|-\dfrac{1}{2} - \dfrac{4}{5}\right| = \left|-\left(\dfrac{1}{2} + \dfrac{4}{5}\right)\right| = \dfrac{1}{2} + \dfrac{4}{5}$

$\qquad = \dfrac{1}{2}\cdot\dfrac{5}{5} + \dfrac{4}{5}\cdot\dfrac{2}{2} = \dfrac{5}{10} + \dfrac{8}{10}$

$\qquad = \dfrac{5+8}{10} = \dfrac{13}{10}$

117. $\left|-\dfrac{5}{6} - \dfrac{3}{10}\right| = \left|-\left(\dfrac{5}{6} + \dfrac{3}{10}\right)\right| = \dfrac{5}{6} + \dfrac{3}{10}$

$\qquad = \dfrac{5}{6}\cdot\dfrac{5}{5} + \dfrac{3}{10}\cdot\dfrac{3}{3}$

$\qquad = \dfrac{25}{30} + \dfrac{9}{30} = \dfrac{34}{30} = \dfrac{17\cdot 2}{15\cdot 2}$

$\qquad = \dfrac{17\cdot\cancel{2}}{15\cdot\cancel{2}} = \dfrac{17}{15}$

119. $\dfrac{21}{32} + (-5) = \dfrac{21}{32} + \left(\dfrac{-5}{1}\cdot\dfrac{32}{32}\right) = \dfrac{21}{32} + \left(\dfrac{-160}{32}\right)$

$\qquad = \dfrac{21+(-160)}{32} = \dfrac{-139}{32} = -\dfrac{139}{32}$

121. $\dfrac{\dfrac{2}{3}}{6} = \dfrac{2}{3}\cdot\dfrac{1}{6} = \dfrac{2}{3\cdot 6} = \dfrac{2}{3\cdot 2\cdot 3} = \dfrac{\cancel{2}}{3\cdot\cancel{2}\cdot 3} = \dfrac{1}{9}$

123. $\dfrac{18}{0}$ is undefined.

125. $\dfrac{0}{20} = 0$

127. $5\cdot 3 = 3\cdot 5$

Commutative Property of Multiplication

129. $5\cdot\dfrac{1}{5} = 1$

Multiplicative Inverse Property

131. $\dfrac{42}{10} = \dfrac{21}{5}$

Reduction Property

133. $3+(4+5) = (3+4)+5$

Associative Property of Addition

135. $69 - 42 = 69 + (-42) = 27$

The difference in age of the oldest and youngest president at the time of inauguration is 27 years.

137. $(4+(-3))+8 = 1+8 = 9$

The Bears gained a total of 9 yards in the first 3 plays. Since 9 is less than 10, they did not obtain a first down.

139. $20{,}320 - (-282) = 20{,}320 + 282 = 20{,}602$

The difference between the highest and lowest elevation is 20,602 feet.

141. Multiplying two numbers, say a and b, means that we add the number a a total of b times. On a number line, we would start at 0 and move to the right a units (since a is positive) a total of b times. Since we keep moving to the right, the end result will be positive. For example, $5\cdot 3$ means we would add the number 5 three times. That is, $5\cdot 3 = 5+5+5 = 15$

On a number line we would get:

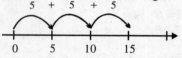

143. $d(P,Q) = |10-(-4)| = |10+4| = |14| = 14$

145. $d(P,Q) = |7.2-(-3.2)| = |7.2+3.2| = |10.4| = 10.4$

147. $d(P,Q) = \left|\dfrac{6}{5} - \left(-\dfrac{10}{3}\right)\right| = \left|\dfrac{6}{5} + \dfrac{10}{3}\right|$

$\qquad = \left|\dfrac{6}{5}\cdot\dfrac{3}{3} + \dfrac{10}{3}\cdot\dfrac{5}{5}\right| = \left|\dfrac{18}{15} + \dfrac{50}{15}\right|$

$\qquad = \left|\dfrac{18+50}{15}\right| = \left|\dfrac{68}{15}\right|$

$\qquad = \dfrac{68}{15}$

149. a. $a\cdot 0 = 0$

b. $\qquad a\cdot 0 = 0$

$a\cdot(b+(-b)) = 0$

c. $a\cdot(b+(-b)) = 0$

$ab + a(-b) = 0$

d. If $a < 0$ and $b > 0$, then $ab < 0$ since the product of a negative and a positive is negative. In addition, $-b < 0$. Now, if $ab < 0$ then $a(-b)$ must be positive so that the sum $ab + a(-b)$ is 0. Since $a < 0$ and $-b < 0$, we have that the product of two negative numbers is a positive number.

151. Zero does not have a multiplicative inverse because division by zero is not defined.

153. The Reduction Property only applies to dividing out factors.

155. No. For example,
$$4 - (5 - 3) \neq (4 - 5) - 3$$
$$4 - 2 \neq -1 - 3$$
$$2 \neq -4$$

157. No. For example,
$$16 \div (8 \div 2) \neq (16 \div 8) \div 2$$
$$16 \div 4 \neq 2 \div 2$$
$$4 \neq 1$$

159.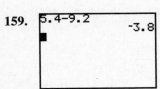

161.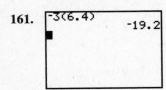

163.

165.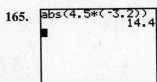

167.

Section R.4

Section R.4 Quick Checks

1. In the expression 5^6, the number 5 is the <u>base</u> and 6 is the <u>exponent</u> or <u>power</u>.

2. False. $-7^4 = -(7 \cdot 7 \cdot 7 \cdot 7) = -7 \cdot 7 \cdot 7 \cdot 7$

3. $4^3 = 4 \cdot 4 \cdot 4 = 64$

4. $(-7)^2 = (-7) \cdot (-7) = 49$

5. $(-10)^3 = (-10) \cdot (-10) \cdot (-10) = -1000$

6. $\left(\dfrac{2}{3}\right)^3 = \dfrac{2}{3} \cdot \dfrac{2}{3} \cdot \dfrac{2}{3} = \dfrac{8}{27}$

7. $-8^2 = -(8 \cdot 8) = -64$

8. $-(-5)^3 = -\left[(-5) \cdot (-5) \cdot (-5)\right] = -(-125) = 125$

9. $5 \cdot 2 + 6 = 10 + 6 = 16$

10. $3 \cdot 2 + 5 \cdot 6 = 6 + 30 = 36$

11. $4 \cdot (5 + 3) = 4 \cdot 8 = 32$

12. $8 \cdot (9 - 3) = 8 \cdot 6 = 48$

13. $(12 - 4) \cdot (18 - 13) = 8 \cdot 5 = 40$

14. $(4 + 9) \cdot (6 - 4) = 13 \cdot 2 = 26$

15. $\dfrac{3 + 7}{4 + 9} = \dfrac{10}{13}$

16. $1 - 4 + 8 \cdot 2 + 5 = 1 - 4 + 16 + 5$
$$= 22 - 4$$
$$= 18$$

17. $25 \cdot \left[2(8-3)-8\right] = 25 \cdot \left[2(5)-8\right]$
$= 25 \cdot \left[10-8\right]$
$= 25 \cdot \left[2\right]$
$= 50$

18. $\dfrac{3+5}{2 \cdot (9-4)} = \dfrac{8}{2 \cdot (5)} = \dfrac{8}{10} = \dfrac{4}{5}$

19. The order of operations is (1) <u>parentheses</u>, (2) <u>exponents</u>, (3) <u>multiplication</u> and <u>division</u>, (4) <u>addition</u> and <u>subtraction</u>.

20. $6+5 \cdot 2 = 6+10 = 16$

21. $(3+9) \cdot 4 = 12 \cdot 4 = 48$

22. $\dfrac{7+5}{4+10} = \dfrac{12}{14} = \dfrac{6}{7}$

23. $4 + \left[(8-3) \cdot 2\right] = 4 + \left[5 \cdot 2\right]$
$= 4 + 10$
$= 14$

24. $\left[4 \cdot (6-2)-9\right] = 4 \cdot 4 - 9$
$= 16 - 9$
$= 7$

25. $-8 + 2 \cdot 5^2 = -8 + 2 \cdot 25$
$= -8 + 50$
$= 42$

26. $5 \cdot 3 - 3 \cdot 2^3 = 5 \cdot 3 - 3 \cdot 8$
$= 15 - 24$
$= -9$

27. $5 \cdot (10-8)^2 = 5 \cdot 2^2$
$= 5 \cdot 4$
$= 20$

28. $3 \cdot (-2)^2 + 6 \cdot 3 - 3 \cdot 4^2 = 3 \cdot 4 + 6 \cdot 3 - 3 \cdot 16$
$= 12 + 18 - 48$
$= 30 - 48$
$= -18$

29. $\dfrac{4+6^2}{2 \cdot 3 + 2} = \dfrac{4+36}{2 \cdot 3 + 2}$
$= \dfrac{4+36}{6+2}$
$= \dfrac{40}{8}$
$= 5$

30. $-3 \cdot \left|7^2 - 2 \cdot (8-5)^3\right| = -3 \cdot \left|7^2 - 2 \cdot 3^3\right|$
$= -3 \cdot \left|49 - 2 \cdot 27\right|$
$= -3 \cdot \left|49 - 54\right|$
$= -3 \cdot \left|-5\right|$
$= -3 \cdot 5$
$= -15$

R.4 Exercises

31. $(-3)^4 = (-3) \cdot (-3) \cdot (-3) \cdot (-3) = 81$

33. $-5^4 = -(5 \cdot 5 \cdot 5 \cdot 5) = -625$

35. $-(-2)^3 = -\left[(-2) \cdot (-2) \cdot (-2)\right] = -(-8) = 8$

37. $-\left(\dfrac{2}{3}\right)^2 = -\left(\dfrac{2}{3} \cdot \dfrac{2}{3}\right) = -\dfrac{4}{9}$

39. $4^2 - 3^2 = 16 - 9 = 7$

41. $3 \cdot 2 + 9 = 6 + 9 = 15$

43. $4 + 2 \cdot (6-2) = 4 + 2 \cdot (4) = 4 + 8 = 12$

45. $-2\left[10 - (3-7)\right] = -2\left[10 - (-4)\right]$
$= -2\left[10 + 4\right]$
$= -2\left[14\right]$
$= -28$

47. $\dfrac{5-(-7)}{4} = \dfrac{5+7}{4} = \dfrac{12}{4} = \dfrac{\cancel{4} \cdot 3}{\cancel{4} \cdot 1} = 3$

49. $\left|3 \cdot 2 - 4 \cdot 5\right| = \left|6 - 20\right| = \left|-14\right| = 14$

51. $12 \cdot \dfrac{2}{3} - 5 \cdot 2 = \dfrac{24}{3} - 10 = 8 - 10 = -2$

53. $3\cdot\left[2+3\cdot(1+5)\right]=3\cdot\left[2+3\cdot(6)\right]=3\cdot\left[2+18\right]$
$=3\cdot\left[20\right]=60$

55. $\left(3^2-3\right)\left(3-(-3)^3\right)=(9-3)(3-(-27))$
$=(6)(3+27)$
$=(6)(30)=180$

57. $\left|3\cdot\left(6-3^2\right)\right|=\left|3\cdot(6-9)\right|=\left|3\cdot(-3)\right|=\left|-9\right|=9$

59. $\left|4\cdot\left[2\cdot5+(-3)\cdot4\right]\right|=\left|4\cdot\left[10+(-12)\right]\right|$
$=\left|4\cdot\left[-2\right]\right|=\left|-8\right|=8$

61. $\dfrac{2\cdot5+15}{2^2+3\cdot2}=\dfrac{2\cdot5+15}{4+3\cdot2}=\dfrac{10+15}{4+6}=\dfrac{25}{10}=\dfrac{\cancel{5}\cdot5}{\cancel{5}\cdot2}=\dfrac{5}{2}$

63. $\dfrac{2\cdot(4+8)}{3+3^2}=\dfrac{2\cdot(12)}{3+3^2}=\dfrac{2(12)}{3+9}=\dfrac{24}{3+9}$
$=\dfrac{24}{12}=\dfrac{\cancel{12}\cdot2}{\cancel{12}\cdot1}=2$

65. $\dfrac{6\cdot\left[12-3\cdot(5-2)\right]}{5\cdot\left[21-2\cdot(4+5)\right]}=\dfrac{6\cdot\left[12-3\cdot3\right]}{5\cdot\left[21-2\cdot9\right]}=\dfrac{6\cdot\left[12-9\right]}{5\cdot\left[21-18\right]}$
$=\dfrac{6\cdot3}{5\cdot3}=\dfrac{6\cdot\cancel{3}}{5\cdot\cancel{3}}=\dfrac{6}{5}$

67. $\left(\dfrac{2}{3}\right)^2\cdot\left(\dfrac{1+2^3}{2^3-2}\right)=\left(\dfrac{4}{9}\right)\left(\dfrac{1+8}{8-2}\right)=\left(\dfrac{4}{9}\right)\left(\dfrac{9}{6}\right)$
$=\dfrac{\cancel{2}\cdot2\cdot\cancel{9}}{\cancel{9}\cdot3\cdot\cancel{2}}=\dfrac{2}{3}$

69. $\dfrac{3^3-2^4\cdot3}{4\cdot\left(3^2-2\cdot3\right)}=\dfrac{27-16\cdot3}{4\cdot(9-2\cdot3)}=\dfrac{27-48}{4\cdot(9-6)}$
$=\dfrac{-21}{4\cdot3}=-\dfrac{21}{12}=-\dfrac{\cancel{3}\cdot7}{\cancel{3}\cdot4}=-\dfrac{7}{4}$

71. $\dfrac{2\cdot4-5}{4^2+(-2)^3}+\dfrac{3^2}{2^3}=\dfrac{2\cdot4-5}{16-8}+\dfrac{9}{8}=\dfrac{8-5}{16-8}+\dfrac{9}{8}$
$=\dfrac{3}{8}+\dfrac{9}{8}=\dfrac{3+9}{8}=\dfrac{12}{8}=\dfrac{\cancel{4}\cdot3}{\cancel{4}\cdot2}=\dfrac{3}{2}$

73. $\dfrac{\dfrac{2\cdot3+3^2}{2\cdot5-8}+\dfrac{4}{3}}{7\cdot(5-3)}{14-2^3}=\dfrac{\dfrac{2\cdot3+9}{2\cdot5-8}+\dfrac{4}{3}}{\dfrac{7\cdot(5-3)}{14-8}}=\dfrac{\dfrac{2\cdot3+9}{2\cdot5-8}+\dfrac{4}{3}}{\dfrac{7\cdot(2)}{14-8}}$

$=\dfrac{\dfrac{6+9}{10-8}+\dfrac{4}{3}}{\dfrac{14}{14-8}}=\dfrac{\dfrac{15}{2}+\dfrac{4}{3}}{\dfrac{14}{6}}$

$=\dfrac{\dfrac{15}{2}\cdot\dfrac{3}{3}+\dfrac{4}{3}\cdot\dfrac{2}{2}}{\dfrac{14}{6}}=\dfrac{\dfrac{45}{6}+\dfrac{8}{6}}{\dfrac{14}{6}}=\dfrac{\dfrac{45+8}{6}}{\dfrac{14}{6}}$

$=\dfrac{\dfrac{53}{6}}{\dfrac{14}{6}}=\dfrac{53}{6}\cdot\dfrac{6}{14}=\dfrac{53\cdot\cancel{6}}{14\cdot\cancel{6}}=\dfrac{53}{14}$

75. $3\cdot(7-2)=15$

77. $3+5\cdot(6-3)=18$

79. $2\cdot3.1416\cdot5^2+2\cdot3.1416\cdot5\cdot12$
$=3.1416\left(2\cdot5^2+2\cdot5\cdot12\right)$
$=3.1416(2\cdot25+2\cdot5\cdot12)$
$=3.1416(50+120)$
$=3.1416(170)$
≈534.07
The surface area of the cylinder is about 534.07 square inches.

81. $-16\cdot3^2+50\cdot3=-16\cdot9+50\cdot3$
$=-144+150$
$=6$
After 3 seconds, the height of the ball is 6 feet.

83. $\dfrac{105+80+115+95+105}{5}=\dfrac{500}{5}=\dfrac{\cancel{5}\cdot100}{\cancel{5}}=100$

85. Answers will vary.

87. Answers will vary.

89. $\dfrac{4}{5}-\left(\dfrac{2}{3}\right)^2=\dfrac{4}{5}-\dfrac{4}{9}=\dfrac{36}{45}-\dfrac{20}{45}=\dfrac{16}{45}$

```
(4/5)-(2/3)²▶Fra
c
            16/45
■
```

91. $\dfrac{4^2+1}{13} = \dfrac{16+1}{13} = \dfrac{17}{13}$

```
(4²+1)/13►Frac
                17/13
■
```

93. $\dfrac{3.5^3}{1.3^2} - 6.2^3 = \dfrac{42.875}{1.69} - 238.328 \approx -212.96$

```
(3.5³/1.3²)-6.2³
►Frac
          -212.9581775
■
```

95. $4.3\left[9.3^2 - 4(34.2+18.5)\right]$

$= 4.3\left[86.49 - 4(52.7)\right]$

$= 4.3\left[86.49 - 210.8\right]$

$= 4.3\left[-124.31\right] = -534.533 \approx -534.53$

```
4.3(9.3²-4(34.2+
18.5))►Frac
         -534533/1000
-534533/1000
           -534.533
■
```

Section R.5

Section R.5 Quick Checks

1. A <u>variable</u> is a letter used to represent any number from a given set of numbers.

2. A <u>constant</u> is either a fixed number or a letter used to represent a fixed (possibly unspecified) number.

3. The word *sum* indicates addition so we write $3+11$.

4. The word *product* indicates multiplication so we write $6 \cdot 7$.

5. The word *quotient* indicates division so we write $\dfrac{y}{4}$.

6. The word *difference* indicates subtraction so we write $3-z$.

7. *Twice* means to multiply by 2 and *difference* indicates subtraction.
$2(x-3)$

8. The word *twice* indicates multiplication and the word *difference* indicates subtraction.
$2x - 3$

9. To <u>evaluate</u> an algebraic expression, substitute a numerical value for each variable in the expression, and simplify the result.

10. $-5x+3$
$-5(2)+3 = -10+3 = -7$

11. $y^2 - 6y + 1$
$(-4)^2 - 6(-4) + 1 = 16 + 24 + 1 = 40 + 1 = 41$

12. $\dfrac{w+8}{3w}$
$\dfrac{(4)+8}{3(4)} = \dfrac{12}{12} = 1$

13. $|4x-5|$
$\left|4\left(\dfrac{1}{2}\right)-5\right| = |2-5| = |-3| = 3$

14. $80x$
$x = 100: 80(100) = 8000$ Yen
$x = 1000: 80(1000) = 80,000$ Yen
$x = 10,000: 80(10,000) = 800,000$ Yen

15. $\dfrac{5}{9}(x-32)$

$x = 32: \dfrac{5}{9}(32-32) = \dfrac{5}{9}(0) = 0° \text{ C}$

$x = 86: \dfrac{5}{9}(86-32) = \dfrac{5}{9}(54) = 30° \text{ C}$

$x = 212: \dfrac{5}{9}(212-32) = \dfrac{5}{9}(180) = 100° \text{ C}$

16. Terms that have the same variable(s) and the same exponent(s) on the variables are called <u>like terms</u>.

17. The coefficient of the term $-mn$ is <u>−1</u>.

18. $4x - 9x = (4-9)x = -5x$

19. $-2x^2 + 13x^2 = (-2+13)x^2 = 11x^2$

20. $-5x - 3x + 6 - 3 = (-5-3)x + (6-3)$
$= -8x + 3$

21. $6x - 10x - 4y + 12y = (6-10)x + (-4+12)y$
$= -4x + 8y$

22. $10y - 3 + 5y + 2 = 10y + 5y - 3 + 2$
$= (10+5)y + (-3+2)$
$= 15y - 1$

23. $0.5x^2 + 1.3 + 1.8x^2 - 0.4$
$= 0.5x^2 + 1.8x^2 + 1.3 - 0.4$
$= (0.5 + 1.8)x^2 + (1.3 - 0.4)$
$= 2.3x^2 + 0.9$

24. $4z + 6 - 8z - 3 - 2z$
$= 4z - 8z - 2z + 6 - 3$
$= (4 - 8 - 2)z + (6 - 3)$
$= -6z + 3$

25. $3(x-2) + x = 3 \cdot x - 3 \cdot 2 + x$
$= 3x - 6 + x$
$= 3x + x - 6$
$= 4x - 6$

26. $5(y+3) - 10y - 4 = 5 \cdot y + 5 \cdot 3 - 10y - 4$
$= 5y + 15 - 10y - 4$
$= 5y - 10y + 15 - 4$
$= -5y + 11$

27. $3(z+4) - 2(3z+1)$
$= 3 \cdot z + 3 \cdot 4 - 2 \cdot 3z - 2 \cdot 1$
$= 3z + 12 - 6z - 2$
$= 3z - 6z + 12 - 2$
$= -3z + 10$

28. $-4(x-2) - (2x+4)$
$= -4 \cdot x - (-4) \cdot 2 - 2x - 4$
$= -4x + 8 - 2x - 4$
$= -4x - 2x + 8 - 4$
$= -6x + 4$

29. $\dfrac{1}{2}(6x+4) - \dfrac{15x+5}{5} = \dfrac{1}{2} \cdot 6x + \dfrac{1}{2} \cdot 4 - \dfrac{15}{5}x - \dfrac{5}{5}$
$= 3x + 2 - 3x - 1$
$= 3x - 3x + 2 - 1$
$= 1$

30. $\dfrac{5x-1}{3} + \dfrac{5x+9}{2} = \dfrac{5}{3}x - \dfrac{1}{3} + \dfrac{5}{2}x + \dfrac{9}{2}$
$= \dfrac{5}{3}x + \dfrac{5}{2}x - \dfrac{1}{3} + \dfrac{9}{2}$
$= \dfrac{10}{6}x + \dfrac{15}{6}x - \dfrac{2}{6} + \dfrac{27}{6}$
$= \dfrac{25}{6}x + \dfrac{25}{6}$
$= \dfrac{25x+25}{6}$ or $\dfrac{25(x+1)}{6}$

31. The set of values that a variable may assume is called the <u>domain</u> of the variable.

32. We need to determine whether the value of the variable causes division by 0. That is, we need to determine if the value makes $x - 4 = 0$.

    a. When $x = 2$, we have that
$x - 4 = 2 - 4 = -2$, so 2 is in the domain of the variable.

    b. When $x = 0$, we have that
$x - 4 = 0 - 4 = -4$, so 0 is in the domain of the variable.

    c. When $x = 4$, we have that $x - 4 = 4 - 4 = 0$, so 4 is **not** in the domain of the variable.

    d. When $x = -3$, we have that
$x - 4 = -3 - 4 = -7$, so -3 is in the domain of the variable.

33. We need to determine whether the value of the variable causes division by 0. That is, we need to determine if the value makes $x + 3 = 0$.

    a. When $x = 2$, we have that $x + 3 = 2 + 3 = 5$, so 2 is in the domain of the variable.

    b. When $x = 0$, we have that $x + 3 = 0 + 3 = 3$, so 0 is in the domain of the variable.

    c. When $x = 4$, we have that $x + 3 = 4 + 3 = 7$, so 4 is in the domain of the variable.

    d. When $x = -3$, we have that
$x + 3 = -3 + 3 = 0$, so -3 is **not** in the domain of the variable.

34. We need to determine whether the value of the variable causes division by 0. That is, we need to determine if the value makes $x^2 + x - 6 = 0$.

a. When $x = 2$, we have that

$x^2 + x - 6 = (2)^2 + (2) - 6 = 4 + 2 - 6 = 0$, so 2 is **not** in the domain of the variable.

b. When $x = 0$, we have that

$x^2 + x - 6 = (0)^2 + (0) - 6 = -6$, so 0 is in the domain of the variable.

c. When $x = 4$, we have that

$x^2 + x - 6 = (4)^2 + (4) - 6 = 14$, so 4 is in the domain of the variable.

d. When $x = -3$, we have that

$x^2 + x - 6 = (-3)^2 + (-3) - 6 = 9 - 3 - 6 = 0$, so -3 is **not** in the domain of the variable.

R.5 Exercises

35. $5 + x$

37. $4z$

39. $y - 7$

41. $2(t + 4)$

43. $5x - 3$

45. $\dfrac{y}{3} + 6x$

47. $4x + 3$

$4(2) + 3 = 8 + 3 = 11$

49. $x^2 + 5x - 3$

$(-2)^2 + 5(-2) - 3 = 4 - 10 - 3 = -6 - 3 = -9$

51. $4 - z^2$

$4 - (-5)^2 = 4 - 25$

53. $\dfrac{2w}{w^2 + 2w + 1}$

$\dfrac{2(3)}{(3)^2 + 2(3) + 1} = \dfrac{6}{9 + 6 + 1} = \dfrac{6}{16} = \dfrac{3 \cdot \cancel{2}}{8 \cdot \cancel{2}} = \dfrac{3}{8}$

55. $\dfrac{v^2 + 2v + 1}{v^2 + 3v + 2}$

$\dfrac{(5)^2 + 2(5) + 1}{(5)^2 + 3(5) + 2} = \dfrac{25 + 10 + 1}{25 + 15 + 2} = \dfrac{36}{42} = \dfrac{\cancel{6} \cdot 6}{\cancel{6} \cdot 7} = \dfrac{6}{7}$

57. $|5x - 4|$

$|5(-5) - 4| = |-25 - 4| = |-29| = 29$

59. $(x + 2y)^2$

$(3 + 2(-4))^2 = (3 + (-8))^2 = (-5)^2 = 25$

61. $\dfrac{(x + 2)^2}{|4x - 10|}$

$\dfrac{((1) + 2)^2}{|4(1) - 10|} = \dfrac{3^2}{|4 - 10|} = \dfrac{9}{|-6|} = \dfrac{9}{6} = \dfrac{\cancel{3} \cdot 3}{\cancel{3} \cdot 2} = \dfrac{3}{2}$

63. $3x - 2x = (3 - 2)x = 1x = x$

65. $-4z - 2z + 3 = (-4 - 2)z + 3 = -6z + 3$

67. $13z + 2 - 14z - 7 = 13z - 14z + 2 - 7$

$= (13 - 14)z + (2 - 7)$

$= -1z + (-5) = -z - 5$

69. $\dfrac{3}{4}x + \dfrac{1}{6}x = \left(\dfrac{3}{4} + \dfrac{1}{6}\right)x = \left(\dfrac{9}{12} + \dfrac{2}{12}\right)x = \dfrac{11}{12}x$

71. $2x + 3x^2 - 5x + x^2 = 3x^2 + x^2 + 2x - 5x$

$= (3 + 1)x^2 + (2 - 5)x$

$= 4x^2 - 3x$

73. $-1.3x - 3.4 + 2.9x + 3.4 = -1.3x + 2.9x - 3.4 + 3.4$

$= (-1.3 + 2.9)x + 0$

$= 1.6x$

75. $3x - 2 - x + 3 - 5x = 3x - x - 5x - 2 + 3$

$= (3 - 1 - 5)x + (-2 + 3)$

$= -3x + 1$

77. $-2(5x - 4) - (4x + 1) = -10x + 8 - 4x - 1$

$= -10x - 4x + 8 - 1$

$= (-10 - 4)x + (8 - 1)$

$= -14x + 7$

79. $5(z+2)-6z=5z+10-6z$
$$=5z-6z+10$$
$$=(5-6)z+10$$
$$=-z+10$$

81. $\dfrac{2}{5}(5x-10)+\dfrac{1}{4}(8x+4)$
$$=\frac{2}{5}\cdot 5x-\frac{2}{5}\cdot 10+\frac{1}{4}\cdot 8x+\frac{1}{4}\cdot 4$$
$$=2x-4+2x+1=2x+2x-4+1$$
$$=(2+2)x+(-4+1)$$
$$=4x-3$$

83. $2(v-3)+5(2v-1)=2v-6+10v-5$
$$=2v+10v-6-5$$
$$=(2+10)v+(-6-5)$$
$$=12v-11$$

85. $\dfrac{3}{5}(10x+4)-\dfrac{8x+3}{2}=\dfrac{3}{5}(10x+4)-\dfrac{1}{2}(8x+3)$
$$=\frac{3}{5}\cdot 10x+\frac{3}{5}\cdot 4-\frac{1}{2}\cdot 8x-\frac{1}{2}\cdot 3$$
$$=6x+\frac{12}{5}-4x-\frac{3}{2}=6x-4x+\frac{12}{5}-\frac{3}{2}$$
$$=(6-4)x+\left(\frac{12}{5}-\frac{3}{2}\right)=2x+\left(\frac{24}{10}-\frac{15}{10}\right)$$
$$=2x+\frac{9}{10}$$

87. $\dfrac{5}{6}\left(\dfrac{3}{10}x-\dfrac{2}{5}\right)+\dfrac{2}{3}\left(\dfrac{1}{6}x+\dfrac{1}{2}\right)$
$$=\frac{5}{6}\cdot\frac{3}{10}x-\frac{5}{6}\cdot\frac{2}{5}+\frac{2}{3}\cdot\frac{1}{6}x+\frac{2}{3}\cdot\frac{1}{2}$$
$$=\frac{\cancel{5}\cdot\cancel{3}}{2\cdot\cancel{3}\cdot\cancel{5}\cdot 2}x-\frac{\cancel{5}\cdot\cancel{2}}{3\cdot\cancel{2}\cdot\cancel{5}}+\frac{\cancel{2}}{3\cdot\cancel{2}\cdot 3}x+\frac{\cancel{2}}{3\cdot\cancel{2}}$$
$$=\frac{1}{4}x-\frac{1}{3}+\frac{1}{9}x+\frac{1}{3}$$
$$=\frac{1}{4}x+\frac{1}{9}x-\frac{1}{3}+\frac{1}{3}$$
$$=\left(\frac{1}{4}+\frac{1}{9}\right)x=\left(\frac{9}{36}+\frac{4}{36}\right)x$$
$$=\frac{13}{36}x$$

89. $4.3(1.2x-2.3)+9.3x-5.6$
$$=5.16x-9.89+9.3x-5.6$$
$$=5.16x+9.3x-9.89-5.6$$
$$=(5.16+9.3)x+(-9.89-5.6)$$
$$=14.46x-15.49$$

91. $6.2(x-1.4)-5.4(3.2x-0.6)$
$$=6.2x-8.68-17.28x+3.24$$
$$=6.2x-17.28x-8.68+3.24$$
$$=(6.2-17.28)x+(-8.68+3.24)$$
$$=-11.08x-5.44$$

93. We need to determine whether the value of the variable causes division by 0. That is, we need to determine if the value makes $x-5=0$.

  a. When $x=5$, we have $x-5=5-5=0$, so 5 is **not** in the domain of the variable.

  b. When $x=-1$, we have $x-5=-1-5=-6$, so -1 is in the domain of the variable.

  c. When $x=-4$, we have $x-5=-4-5=-9$, so -4 is in the domain of the variable.

  d. When $x=0$, we have $x-5=0-5=-5$, so 0 is in the domain of the variable.

95. We need to determine whether the value of the variable causes division by 0. That is, we need to determine if the value makes $x+5=0$.

  a. When $x=5$, we have $x+5=5+5=10$, so 5 is in the domain of the variable.

  b. When $x=-1$, we have $x+5=-1+5=4$, so -1 is in the domain of the variable.

  c. When $x=-4$, we have $x+5=-4+5=1$, so -4 is in the domain of the variable.

  d. When $x=0$, we have $x+5=0+5=5$, so 0 is in the domain of the variable.

97. We need to determine whether the value of the variable causes division by 0. That is, we need to determine if the value makes $x^2-5x=0$.

  a. When $x=5$, we have
$$x^2-5x=(5)^2-5(5)=25-25=0,$$ so 5 is **not** in the domain of the variable.

  b. When $x=-1$, we have
$$x^2-5x=(-1)^2-5(-1)=1+5=6,$$ so -1 is in the domain of the variable.

c. When $x = -4$, we have
$$x^2 - 5x = (-4)^2 - 5(-4) = 16 + 20 = 36, \text{ so}$$
-4 is in the domain of the variable.

d. When $x = 0$, we have
$$x^2 - 5x = (0)^2 - 5(0) = 0 - 0 = 0, \text{ so } 0 \text{ is}$$
not in the domain of the variable.

99. For $s = 1$ inch, $s^3 = (1)^3 = 1$ cubic inch.

For $s = 2$ inches, $s^3 = (2)^3 = 8$ cubic inches.

For $s = 3$ inches, $s^3 = (3)^3 = 27$ cubic inches.

For $s = 4$ inches, $s^3 = (4)^3 = 64$ cubic inches.

101. a. For $t = 0$:
$$-16(0)^2 + 75(0) = 0 + 0 = 0 \text{ ft}$$
For $t = 1$:
$$-16(1)^2 + 75(1) = -16 + 75 = 59 \text{ ft}$$
For $t = 2$:
$$-16(2)^2 + 75(2) = -16(4) + 75(2)$$
$$= -64 + 150 = 86 \text{ ft}$$
For $t = 3$:
$$-16(3)^2 + 75(3) = -16(9) + 75(3)$$
$$= -144 + 225 = 81 \text{ ft}$$
For $t = 4$:
$$-16(4)^2 + 75(4) = -16(16) + 75(4)$$
$$= -256 + 300 = 44 \text{ ft}$$

b. The ball begins on the ground. When it is hit, it rises in the air (for somewhere around 2 seconds) and then begins to fall back towards the ground.

103. Let $x =$ Bob's age in years.

Tony's age is expressed as $x + 5$.
When $x = 13$, $x + 5 = 13 + 5 = 18$. When Bob is 13 years old, Tony is 18 years old.

105. Let $p =$ the original price in dollars.

"half off regular price" is $p - \dfrac{1}{2}p = \dfrac{1}{2}p$

The discounted price can be expressed as $\dfrac{1}{2}p$.

When $p = 900$, $\dfrac{1}{2}p = \dfrac{1}{2}(900) = 450$. When the original price is \$900, the discount price is \$450.

107. $\dfrac{X - \mu}{\sigma}$

When $X = 120, \mu = 100,$ and $\sigma = 15$, we have
$$\frac{X - \mu}{\sigma} = \frac{120 - 100}{15} = \frac{20}{15} = \frac{\cancel{5} \cdot 4}{\cancel{5} \cdot 3} = \frac{4}{3}$$

109. Answers may vary. One possible answer is: *"Twice a number z decreased by 5"*

111. Answers may vary. One possible answer is: *"Twice the difference of a number z and 5"*

113. Answers may vary. One possible answer is: *"One-half the sum of a number z and 3"*

115. Answers will vary. A *variable* is a letter used to represent any number from a given set of numbers. A *constant* is a letter used to represent a fixed (though possibly unknown) value.

117. Answers will vary. Like terms are terms that have the same variable(s) and the same corresponding exponents on the variables. The Distributive Property is used to first remove parentheses and then later 'in reverse' to combine the coefficients of like terms.

119. a. $-4x + 3$ when $x = 0$.

b. $-4x + 3$ when $x = -3$.

121. a. $4x^2 - 8x + 3$ when $x = 5$.

b. $4x^2 - 8x + 3$ when $x = -2$.

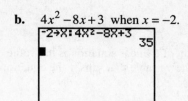

123. a. $\dfrac{3z-1}{z^2+1}$ when $z = -2$.

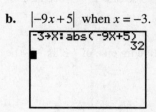

b. $\dfrac{3z-1}{z^2+1}$ when $z = 8$.

```
8→Z:(3Z-1)/(Z²+1
)►Frac
                  23/65
```

125. a. $\left| -9x + 5 \right|$ when $x = 8$.

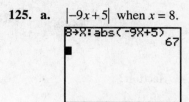

b. $\left| -9x + 5 \right|$ when $x = -3$.

```
-3→X:abs(-9X+5)
                    32
```

127. The calculator displays an error message because $x = 5$ makes the denominator equal to 0.

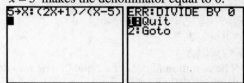

Chapter 1

Section 1.1

Are You Ready for This Section?

R1. The additive inverse of 5 is -5 because
$5 + (-5) = 0$.

R2. The multiplicative inverse of -3 is $-\dfrac{1}{3}$ because

$$(-3) \cdot \left(-\frac{1}{3}\right) = 1.$$

R3. $\dfrac{1}{5} \cdot 5x = \dfrac{5 \cdot x}{5} = \dfrac{\cancel{5} \cdot x}{\cancel{5}} = x$

R4. $8 = 2 \cdot 2 \cdot 2$
$12 = 2 \cdot 2 \cdot 3$
The common factors are 2 and 2. The remaining
factors are 2 and 3.
$\text{LCD} = 2 \cdot 2 \cdot 2 \cdot 3 = 24$

R5. $6(z-2) = 6 \cdot z - 6 \cdot 2 = 6z - 12$

R6. The coefficient of $-4x$ is -4.

R7. $4(y-2) - y + 5 = 4y - 8 - y + 5$
$\qquad\qquad\qquad\quad = 4y - y - 8 + 5$
$\qquad\qquad\qquad\quad = 3y - 3$

R8. $-5(x+3) - 8$
$-5((-2)+3) - 8 = -5(1) - 8$
$\qquad\qquad\qquad\quad = -5 - 8$
$\qquad\qquad\qquad\quad = -13$

R9. Since division by 0 is undefined, we need to
determine if the value for the variable makes
$x + 3 = 0$.
When $x = 3$, we have $x + 3 = (3) + 3 = 6$, so 3 is
in the domain of the variable.
When $x = -3$, we have $x + 3 = (-3) + 3 = 0$, so
-3 is **not** in the domain of the variable.

Section 1.1 Quick Checks

1. The equation $3x + 5 = 2x - 3$ is a <u>linear</u> equation
in one variable. The expressions $3x + 5$ and
$2x - 3$ are called <u>sides</u> of the equation.

2. The <u>solution</u> of a linear equation is the value or
values of the variable that satisfy the equation.

3. $-5x + 3 = -2$
Let $x = -2$ in the equation.
$-5(-2) + 3 \overset{?}{=} -2$
$\qquad 10 + 3 \overset{?}{=} -2$
$\qquad\qquad 13 \neq -2$
$x = -2$ is not a solution to the equation.

Let $x = 1$ in the equation.
$-5(1) + 3 \overset{?}{=} -2$
$\quad -5 + 3 \overset{?}{=} -2$
$\qquad\quad -2 = -2 \ \text{T}$
$x = 1$ is a solution to the equation.

Let $x = 3$ in the equation.
$-5(3) + 3 \overset{?}{=} -2$
$\quad -15 + 3 \overset{?}{=} -2$
$\qquad\quad -12 \neq -2$
$x = 3$ is not a solution to the equation.

4. $3x + 2 = 2x - 5$
Let $x = 0$ in the equation.
$3(0) + 2 \overset{?}{=} 2(0) - 5$
$\quad 0 + 2 \overset{?}{=} 0 - 5$
$\qquad\quad 2 \neq -5$
$x = 0$ is not a solution to the equation.

Let $x = 6$ in the equation.
$3(6) + 2 \overset{?}{=} 2(6) - 5$
$\quad 18 + 2 \overset{?}{=} 12 - 5$
$\qquad\quad 20 \neq 7$
$x = 6$ is not a solution to the equation.

Let $x = -7$ in the equation.
$3(-7) + 2 \overset{?}{=} 2(-7) - 5$
$\quad -21 + 2 \overset{?}{=} -14 - 5$
$\qquad\quad -19 = -19 \ \text{T}$
$x = -7$ is a solution to the equation.

5. $-3(z+2) = 4z + 1$
Let $z = -3$ in the equation.
$-3((-3)+2) \overset{?}{=} 4(-3) + 1$
$\quad -3(-1) \overset{?}{=} -12 + 1$
$\qquad\quad 3 \neq -11$
$z = -3$ is not a solution to the equation.

Let $z = -1$ in the equation.
$$-3\big((-1)+2\big) \overset{?}{=} 4(-1)+1$$
$$-3(1) \overset{?}{=} -4+1$$
$$-3 = -3 \text{ T}$$
$z = -1$ is a solution to the equation.

Let $z = 2$ in the equation.
$$-3\big((2)+2\big) \overset{?}{=} 4(2)+1$$
$$-3(4) \overset{?}{=} 8+1$$
$$-12 \ne 9$$
$z = 2$ is not a solution to the equation.

6. True

7. To <u>isolate</u> the variable means to get the variable by itself with a coefficient of 1.

8.
$$3x+8=17$$
$$3x+8-8=17-8$$
$$3x=9$$
$$\frac{3x}{3}=\frac{9}{3}$$
$$x=3$$
Check:
$$3(3)+8 \overset{?}{=} 17$$
$$9+8 \overset{?}{=} 17$$
$$17=17 \text{ T}$$
Solution set: $\{3\}$

9.
$$-4a-7=1$$
$$-4a-7+7=1+7$$
$$-4a=8$$
$$\frac{-4a}{-4}=\frac{8}{-4}$$
$$a=-2$$
Check:
$$-4(-2)-7 \overset{?}{=} 1$$
$$8-7 \overset{?}{=} 1$$
$$1=1 \text{ T}$$
Solution set: $\{-2\}$

10.
$$5y+1=2$$
$$5y+1-1=2-1$$
$$5y=1$$
$$\frac{5y}{5}=\frac{1}{5}$$
$$y=\frac{1}{5}$$

Check:
$$5\left(\frac{1}{5}\right)+1 \overset{?}{=} 2$$
$$1+1 \overset{?}{=} 2$$
$$2=2 \text{ T}$$
Solution set: $\left\{\dfrac{1}{5}\right\}$

11.
$$2x+3+5x+1=4x+10$$
$$7x+4=4x+10$$
$$7x+4-4x=4x+10-4x$$
$$3x+4=10$$
$$3x+4-4=10-4$$
$$3x=6$$
$$\frac{3x}{3}=\frac{6}{3}$$
$$x=2$$
Check:
$$2(2)+3+5(2)+1 \overset{?}{=} 4(2)+10$$
$$4+3+10+1 \overset{?}{=} 8+10$$
$$18=18 \text{ T}$$
Solution set: $\{2\}$

12.
$$4b+3-b-8-5b=2b-1-b-1$$
$$-2b-5=b-2$$
$$-2b-5-b=b-2-b$$
$$-3b-5=-2$$
$$-3b-5+5=-2+5$$
$$-3b=3$$
$$\frac{-3b}{-3}=\frac{3}{-3}$$
$$b=-1$$
Check:
$$4(-1)+3-(-1)-8-5(-1) \overset{?}{=} 2(-1)-1-(-1)-1$$
$$-4+3+1-8+5 \overset{?}{=} -2-1+1-1$$
$$-3=-3 \text{ T}$$
Solution set: $\{-1\}$

13.
$$2w+8-7w+1=3w-1+2w-5$$
$$-5w+9=5w-6$$
$$-5w+9-5w=5w-6-5w$$
$$-10w+9=-6$$
$$-10w+9-9=-6-9$$
$$-10w=-15$$
$$\frac{-10w}{-10}=\frac{-15}{-10}$$
$$w=\frac{3}{2}$$

Check:

$$2\left(\frac{3}{2}\right)+8-7\left(\frac{3}{2}\right)+1 \overset{?}{=} 3\left(\frac{3}{2}\right)-1+2\left(\frac{3}{2}\right)-5$$

$$3+8-\frac{21}{2}+1 \overset{?}{=} \frac{9}{2}-1+3-5$$

$$\frac{3}{2}=\frac{3}{2} \text{ T}$$

Solution set: $\left\{\dfrac{3}{2}\right\}$

14. $4(x-1)=12$

$$4x-4=12$$
$$4x-4+4=12+4$$
$$4x=16$$
$$\frac{4x}{4}=\frac{16}{4}$$
$$x=4$$

Check:

$$4(4-1) \overset{?}{=} 12$$
$$4(3) \overset{?}{=} 12$$
$$12=12 \text{ T}$$

Solution set: $\{4\}$

15. $-2(x-4)-6=3(x+6)+4$

$$-2x+8-6=3x+18+4$$
$$-2x+2=3x+22$$
$$-2x+2-3x=3x+22-3x$$
$$-5x+2=22$$
$$-5x+2-2=22-2$$
$$-5x=20$$
$$\frac{-5x}{-5}=\frac{20}{-5}$$
$$x=-4$$

Check:

$$-2(-4-4)-6 \overset{?}{=} 3(-4+6)+4$$
$$-2(-8)-6 \overset{?}{=} 3(2)+4$$
$$16-6 \overset{?}{=} 6+4$$
$$10=10 \text{ T}$$

Solution set: $\{-4\}$

16. $4(x+3)-8x=3(x+2)+x$

$$4x+12-8x=3x+6+x$$
$$-4x+12=4x+6$$
$$-4x+12-4x=4x+6-4x$$
$$-8x+12=6$$
$$-8x+12-12=6-12$$
$$-8x=-6$$
$$\frac{-8x}{-8}=\frac{-6}{-8}$$
$$x=\frac{3}{4}$$

Check:

$$4\left(\frac{3}{4}+3\right)-8\left(\frac{3}{4}\right) \overset{?}{=} 3\left(\frac{3}{4}+2\right)+\frac{3}{4}$$

$$4\left(\frac{15}{4}\right)-6 \overset{?}{=} 3\left(\frac{11}{4}\right)+\frac{3}{4}$$

$$15-6 \overset{?}{=} \frac{33}{4}+\frac{3}{4}$$

$$9=9 \text{ T}$$

Solution set: $\left\{\dfrac{3}{4}\right\}$

17. $5(x-3)+3(x+3)=2x-3$

$$5x-15+3x+9=2x-3$$
$$8x-6=2x-3$$
$$8x-6-2x=2x-3-2x$$
$$6x-6=-3$$
$$6x-6+6=-3+6$$
$$6x=3$$
$$\frac{6x}{6}=\frac{3}{6}$$
$$x=\frac{1}{2}$$

Check:

$$5\left(\frac{1}{2}-3\right)+3\left(\frac{1}{2}+3\right) \overset{?}{=} 2\left(\frac{1}{2}\right)-3$$

$$5\left(-\frac{5}{2}\right)+3\left(\frac{7}{2}\right) \overset{?}{=} 1-3$$

$$-\frac{25}{2}+\frac{21}{2} \overset{?}{=} -2$$

$$-2=-2 \text{ T}$$

Solution set: $\left\{\dfrac{1}{2}\right\}$

18. To solve a linear equation containing fractions, we can multiply each side of the equation by the <u>least common denominator</u> to clear the fractions.

19.
$$\frac{3y}{2} + \frac{y}{6} = \frac{10}{3}$$
$$6\left(\frac{3y}{2} + \frac{y}{6}\right) = 6\left(\frac{10}{3}\right)$$
$$6 \cdot \frac{3y}{2} + 6 \cdot \frac{y}{6} = 6 \cdot \frac{10}{3}$$
$$9y + y = 20$$
$$10y = 20$$
$$\frac{10y}{10} = \frac{20}{10}$$
$$y = 2$$

Check:
$$\frac{3(2)}{2} + \frac{2}{6} \stackrel{?}{=} \frac{10}{3}$$
$$3 + \frac{1}{3} \stackrel{?}{=} \frac{10}{3}$$
$$\frac{10}{3} = \frac{10}{3} \ \text{T}$$
Solution set: $\{2\}$

20.
$$\frac{3x}{4} - \frac{5}{12} = \frac{5x}{6}$$
$$12\left(\frac{3x}{4} - \frac{5}{12}\right) = 12\left(\frac{5x}{6}\right)$$
$$12 \cdot \frac{3x}{4} - 12 \cdot \frac{5}{12} = 12 \cdot \frac{5x}{6}$$
$$9x - 5 = 10x$$
$$9x - 5 - 9x = 10x - 9x$$
$$-5 = x$$

Check:
$$\frac{3(-5)}{4} - \frac{5}{12} \stackrel{?}{=} \frac{5(-5)}{6}$$
$$\frac{-15}{4} - \frac{5}{12} \stackrel{?}{=} -\frac{25}{6}$$
$$\frac{-45}{12} - \frac{5}{12} \stackrel{?}{=} -\frac{25}{6}$$
$$-\frac{50}{12} \stackrel{?}{=} -\frac{25}{6}$$
$$-\frac{25}{6} = -\frac{25}{6} \ \text{T}$$
Solution set: $\{-5\}$

21.
$$\frac{x+2}{6} + 2 = \frac{5}{3}$$
$$6\left(\frac{x+2}{6} + 2\right) = 6\left(\frac{5}{3}\right)$$
$$6 \cdot \frac{(x+2)}{6} + 6 \cdot 2 = 6 \cdot \frac{5}{3}$$
$$x + 2 + 12 = 10$$
$$x + 14 = 10$$
$$x + 14 - 14 = 10 - 14$$
$$x = -4$$

Check:
$$\frac{-4+2}{6} + 2 \stackrel{?}{=} \frac{5}{3}$$
$$\frac{-2}{6} + 2 \stackrel{?}{=} \frac{5}{3}$$
$$-\frac{1}{3} + \frac{6}{3} \stackrel{?}{=} \frac{5}{3}$$
$$\frac{5}{3} = \frac{5}{3} \ \text{T}$$
Solution set: $\{-4\}$

22.
$$\frac{4x+3}{9} - \frac{2x+1}{2} = \frac{1}{6}$$
$$18\left(\frac{4x+3}{9} - \frac{2x+1}{2}\right) = 18\left(\frac{1}{6}\right)$$
$$18 \cdot \frac{(4x+3)}{9} - 18 \cdot \frac{(2x+1)}{2} = 18 \cdot \frac{1}{6}$$
$$2(4x+3) - 9(2x+1) = 3$$
$$8x + 6 - 18x - 9 = 3$$
$$-10x - 3 = 3$$
$$-10x - 3 + 3 = 3 + 3$$
$$-10x = 6$$
$$\frac{-10x}{-10} = \frac{6}{-10}$$
$$x = -\frac{3}{5}$$

Check:

$$\frac{4\left(-\frac{3}{5}\right)+3}{9}-\frac{2\left(-\frac{3}{5}\right)+1}{2}\overset{?}{=}\frac{1}{6}$$

$$\frac{\frac{-12}{5}+\frac{15}{5}}{9}-\frac{\frac{-6}{5}+\frac{5}{5}}{2}\overset{?}{=}\frac{1}{6}$$

$$\frac{\frac{3}{5}}{9}-\frac{\frac{-1}{5}}{2}\overset{?}{=}\frac{1}{6}$$

$$\frac{3}{5}\cdot\frac{1}{9}+\frac{1}{5}\cdot\frac{1}{2}\overset{?}{=}\frac{1}{6}$$

$$\frac{1}{15}+\frac{1}{10}\overset{?}{=}\frac{1}{6}$$

$$\frac{2}{30}+\frac{3}{30}\overset{?}{=}\frac{5}{30}$$

$$\frac{5}{30}=\frac{5}{30}\ \text{T}$$

Solution set: $\left\{-\dfrac{3}{5}\right\}$

23. Our first step is to multiply both sides of the equation by 100 to eliminate the decimals.

$$100\left(0.07x-1.3\right)=100\left(0.05x-1.1\right)$$
$$7x-130=5x-110$$
$$7x-130-5x=5x-110-5x$$
$$2x-130=-110$$
$$2x-130+130=-110+130$$
$$2x=20$$
$$\frac{2x}{2}=\frac{20}{2}$$
$$x=10$$

Check:
$$0.07\left(10\right)-1.3\overset{?}{=}0.05\left(10\right)-1.1$$
$$0.7-1.3\overset{?}{=}0.5-1.1$$
$$-0.6=-0.6\ \text{T}$$

Solution set: $\{10\}$

24. Our first step is to multiply both sides of the equation by 10 to eliminate the decimals.

$$10\left[0.4\left(y+3\right)\right]=10\left[0.5\left(y-4\right)\right]$$
$$4\left(y+3\right)=5\left(y-4\right)$$
$$4y+12=5y-20$$
$$4y+12-5y=5y-20-5y$$
$$-y+12=-20$$
$$-y+12-12=-20-12$$
$$-y=-32$$
$$\frac{-y}{-1}=\frac{-32}{-1}$$
$$y=32$$

Check:
$$0.4\left(32+3\right)\overset{?}{=}0.5\left(32-4\right)$$
$$0.4\left(35\right)\overset{?}{=}0.5\left(28\right)$$
$$14=14\ \text{T}$$

Solution set: $\{32\}$

25. A <u>conditional equation</u> is an equation that is true for some values of the variable and false for others.

26. A <u>contradiction</u> is an equation that is false for every value of the variable. An <u>identity</u> is an equation that is satisfied by every allowed choice of the variable.

27. $$4\left(x+2\right)=4x+2$$
$$4x+8=4x+2$$
$$4x+8-4x=4x+2-4x$$
$$8=2$$

The last statement is a contradiction. The equation has no solution.

Solution set: $\{\ \}$ or $\varnothing$

28. $$3\left(x-2\right)=2x-6+x$$
$$3x-6=3x-6$$
$$3x-6-3x=3x-6-3x$$
$$-6=-6$$

The last statement is an identity. All real numbers are solutions.

Solution set: $\mathbb{R}$

29. $$-4x+2+x+1=-4\left(x+2\right)+11$$
$$-3x+3=-4x-8+11$$
$$-3x+3=-4x+3$$
$$-3x+3+4x=-4x+3+4x$$
$$x+3=3$$
$$x+3-3=3-3$$
$$x=0$$

This is a conditional equation.

Solution set: $\{0\}$

30. $$-3\left(z+1\right)+2\left(z-3\right)=z+6-2z-15$$
$$-3z-3+2z-6=-z-9$$
$$-z-9=-z-9$$
$$-z-9+z=-z-9+z$$
$$-9=-9$$

The last statement is an identity. All real numbers are solutions.

Solution set: $\mathbb{R}$

1.1 Exercises

31. $8x - 10 = 6$
Let $x = -2$ in the equation.
$8(-2) - 10 \stackrel{?}{=} 6$
$-16 - 10 \stackrel{?}{=} 6$
$-26 \neq 6$
$x = -2$ is **not** a solution to the equation.
Let $x = 1$ in the equation.
$8(1) - 10 \stackrel{?}{=} 6$
$8 - 10 \stackrel{?}{=} 6$
$-2 \neq 6$
$x = 1$ is **not** a solution to the equation.
Let $x = 2$ in the equation.
$8(2) - 10 \stackrel{?}{=} 6$
$16 - 10 \stackrel{?}{=} 6$
$6 = 6$ T
$x = 2$ is a solution to the equation.

33. $5m - 3 = -3m + 5$
Let $m = -2$ in the equation.
$5(-2) - 3 \stackrel{?}{=} -3(-2) + 5$
$-10 - 3 \stackrel{?}{=} 6 + 5$
$-13 \neq 11$
$m = -2$ is **not** a solution to the equation.
Let $m = 1$ in the equation.
$5(1) - 3 \stackrel{?}{=} -3(1) + 5$
$5 - 3 \stackrel{?}{=} -3 + 5$
$2 = 2$ T
$m = 1$ is a solution to the equation.
Let $m = 3$ in the equation.
$5(3) - 3 \stackrel{?}{=} -3(3) + 5$
$15 - 3 \stackrel{?}{=} -9 + 5$
$12 \neq -4$
$m = 3$ is **not** a solution to the equation.

35. $4(x - 1) = 3x + 1$
Let $x = -1$ in the equation.
$4((-1) - 1) \stackrel{?}{=} 3(-1) + 1$
$4(-2) \stackrel{?}{=} -3 + 1$
$-8 \neq -2$
$x = -1$ is **not** a solution to the equation.
Let $x = 2$ in the equation.
$4((2) - 1) \stackrel{?}{=} 3(2) + 1$
$4(1) \stackrel{?}{=} 6 + 1$
$4 \neq 7$

$x = 2$ is **not** a solution to the equation.
Let $x = 5$ in the equation.
$4((5) - 1) \stackrel{?}{=} 3(5) + 1$
$4(4) \stackrel{?}{=} 15 + 1$
$16 = 16$ T
$x = 5$ is a solution to the equation.

37. $3x + 1 = 7$
$3x + 1 - 1 = 7 - 1$
$3x = 6$
$\dfrac{3x}{3} = \dfrac{6}{3}$
$x = 2$
Check:
$3(2) + 1 \stackrel{?}{=} 7$
$6 + 1 \stackrel{?}{=} 7$
$7 = 7$ T
Solution set: $\{2\}$

39. $4z + 3 = 2$
$4z + 3 - 3 = 2 - 3$
$4z = -1$
$\dfrac{4z}{4} = \dfrac{-1}{4}$
$z = -\dfrac{1}{4}$
Check:
$4\left(-\dfrac{1}{4}\right) + 3 \stackrel{?}{=} 2$
$-1 + 3 \stackrel{?}{=} 2$
$2 = 2$ T
Solution set: $\left\{-\dfrac{1}{4}\right\}$

41. $-3w + 2w + 5 = -4$
$-w + 5 = -4$
$-w + 5 - 5 = -4 - 5$
$-w = -9$
$-1(-w) = -1(-9)$
$w = 9$
Check:
$-3(9) + 2(9) + 5 \stackrel{?}{=} -4$
$-27 + 18 + 5 \stackrel{?}{=} -4$
$-4 = -4$ T
Solution set: $\{9\}$

43.
$$5x + 2 - 2x + 3 = 7x + 2 - x + 5$$
$$5x - 2x + 2 + 3 = 7x - x + 2 + 5$$
$$3x + 5 = 6x + 7$$
$$3x + 5 - 5 = 6x + 7 - 5$$
$$3x = 6x + 2$$
$$3x - 6x = 6x + 2 - 6x$$
$$-3x = 2$$
$$\frac{-3x}{-3} = \frac{2}{-3}$$
$$x = -\frac{2}{3}$$

Check: $5\left(-\dfrac{2}{3}\right) + 2 - 2\left(-\dfrac{2}{3}\right) + 3 \overset{?}{=} 7\left(-\dfrac{2}{3}\right) + 2 - \left(-\dfrac{2}{3}\right) + 5$

$$-\frac{10}{3} + 2 + \frac{4}{3} + 3 \overset{?}{=} -\frac{14}{3} + 2 + \frac{2}{3} + 5$$
$$-\frac{6}{3} + 5 \overset{?}{=} -\frac{12}{3} + 7$$
$$-2 + 5 \overset{?}{=} -4 + 7$$
$$3 = 3 \text{ T}$$

Solution set: $\left\{-\dfrac{2}{3}\right\}$

45.
$$3(x + 2) = -6$$
$$3x + 6 = -6$$
$$3x + 6 - 6 = -6 - 6$$
$$3x = -12$$
$$\frac{3x}{3} = \frac{-12}{3}$$
$$x = -4$$

Check:
$$3(-4 + 2) \overset{?}{=} -6$$
$$3(-2) \overset{?}{=} -6$$
$$-6 = -6 \text{ T}$$

Solution set: $\{-4\}$

47.
$$\frac{4y}{5} - \frac{14}{15} = \frac{y}{3}$$
$$15\left(\frac{4y}{5} - \frac{14}{15}\right) = 15\left(\frac{y}{3}\right)$$
$$15 \cdot \frac{4y}{5} - 15 \cdot \frac{14}{15} = 15 \cdot \frac{y}{3}$$
$$12y - 14 = 5y$$
$$12y - 14 - 12y = 5y - 12y$$
$$-14 = -7y$$
$$\frac{-14}{-7} = \frac{-7y}{-7}$$
$$2 = y \quad \text{or} \quad y = 2$$

Check:

$$\frac{4(2)}{5} - \frac{14}{15} \stackrel{?}{=} \frac{(2)}{3}$$

$$\frac{8}{5} - \frac{14}{15} \stackrel{?}{=} \frac{2}{3}$$

$$\frac{24}{15} - \frac{14}{15} \stackrel{?}{=} \frac{2}{3}$$

$$\frac{10}{15} \stackrel{?}{=} \frac{2}{3}$$

$$\frac{2}{3} = \frac{2}{3} \quad \text{T}$$

Solution set: $\{2\}$

49.
$$\frac{4x+3}{9} - \frac{2x+1}{2} = \frac{1}{6}$$

$$18\left(\frac{4x+3}{9}\right) - 18\left(\frac{2x+1}{2}\right) = 18\left(\frac{1}{6}\right)$$

$$2(4x+3) - 9(2x+1) = 3$$

$$8x+6-18x-9 = 3$$

$$-10x-3 = 3$$

$$-10x-3+3 = 3+3$$

$$-10x = 6$$

$$\frac{-10x}{-10} = \frac{6}{-10}$$

$$x = -\frac{3}{5}$$

Check:
$$\frac{4\left(-\frac{3}{5}\right)+3}{9} - \frac{2\left(-\frac{3}{5}\right)+1}{2} \stackrel{?}{=} \frac{1}{6}$$

$$\frac{-\frac{12}{5}+\frac{15}{5}}{9} - \frac{-\frac{6}{5}+\frac{5}{5}}{2} \stackrel{?}{=} \frac{1}{6}$$

$$\frac{\frac{3}{5}}{9} + \frac{\frac{1}{5}}{2} \stackrel{?}{=} \frac{1}{6}$$

$$\frac{3}{5} \cdot \frac{1}{9} + \frac{1}{5} \cdot \frac{1}{2} \stackrel{?}{=} \frac{1}{6}$$

$$\frac{1}{15} + \frac{1}{10} \stackrel{?}{=} \frac{1}{6}$$

$$\frac{2}{30} + \frac{3}{30} \stackrel{?}{=} \frac{1}{6}$$

$$\frac{5}{30} \stackrel{?}{=} \frac{1}{6}$$

$$\frac{1}{6} = \frac{1}{6} \quad \text{T}$$

Solution set: $\left\{-\frac{3}{5}\right\}$

51.
$$\frac{y}{10} + 6 = \frac{y}{4} + 12$$

$$20\left(\frac{y}{10} + 6\right) = 20\left(\frac{y}{4} + 12\right)$$

$$20 \cdot \frac{y}{10} + 20(6) = 20 \cdot \frac{y}{4} + 20(12)$$

$$2y + 120 = 5y + 240$$

$$2y + 120 - 2y = 5y + 240 - 2y$$

$$120 = 3y + 240$$

$$120 - 240 = 3y + 240 - 240$$

$$-120 = 3y$$

$$\frac{-120}{3} = \frac{3y}{3}$$

$$-40 = y$$

Check: $\frac{-40}{10} + 6 \stackrel{?}{=} \frac{-40}{4} + 12$

$$-4 + 6 \stackrel{?}{=} -10 + 2$$

$$2 = 2 \quad \text{True}$$

Solution set: $\{-40\}$

53.
$$0.5x - 3.2 = -1.7$$

$$10(0.5x - 3.2) = 10(-1.7)$$

$$5x - 32 = -17$$

$$5x - 32 + 32 = -17 + 32$$

$$5x = 15$$

$$\frac{5x}{5} = \frac{15}{5}$$

$$x = 3$$

Check:
$$0.5(3) - 3.2 \stackrel{?}{=} -1.7$$

$$1.5 - 3.2 \stackrel{?}{=} -1.7$$

$$-1.7 = -1.7 \quad \text{T}$$

Solution set: $\{3\}$

55.
$$0.14x + 2.23 = 0.09x + 1.98$$

$$100(0.14x + 2.23) = 100(0.09x + 1.98)$$

$$14x + 223 = 9x + 198$$

$$14x + 223 - 223 = 9x + 198 - 223$$

$$14x = 9x - 25$$

$$14x - 9x = 9x - 25 - 9x$$

$$5x = -25$$

$$\frac{5x}{5} = \frac{-25}{5}$$

$$x = -5$$

Check: $0.14(-5) + 2.23 \stackrel{?}{=} 0.09(-5) + 1.98$

$$-0.7 + 2.23 \stackrel{?}{=} -0.45 + 1.98$$

$$1.53 = 1.53 \quad \text{T}$$

Solution set: $\{-5\}$

57.
$$4(x+1) = 4x$$
$$4x + 4 = 4x$$
$$4x - 4x + 4 = 4x - 4x$$
$$4 = 0$$
This statement is a contradiction. The equation has no solution.

Solution set: $\{\ \}$ or $\varnothing$

59.
$$4m + 1 - 6m = 2(m+3) - 4m$$
$$-2m + 1 = 2m + 6 - 4m$$
$$-2m + 1 = -2m + 6$$
$$-2m + 1 + 2m = -2m + 6 + 2m$$
$$1 = 6$$
This statement is a contradiction. The equation has no solution.

Solution set: $\{\ \}$ or $\varnothing$

61.
$$2(y+1) - 3(y-2) = 5y + 8 - 6y$$
$$2y + 2 - 3y + 6 = 8 - y$$
$$8 - y = 8 - y$$
$$8 - y + y = 8 - y + y$$
$$8 = 8$$
This statement is an identity. All real numbers are solutions.

Solution set: $\{y \mid y \text{ is a real number}\}$ or $\mathbb{R}$

63.
$$\frac{x}{4} + \frac{3x}{10} = -\frac{33}{20}$$
$$20\left(\frac{x}{4} + \frac{3x}{10}\right) = 20\left(-\frac{33}{20}\right)$$
$$20 \cdot \frac{x}{4} + 20 \cdot \frac{3x}{10} = -33$$
$$5x + 6x = -33$$
$$11x = -33$$
$$\frac{11x}{11} = \frac{-33}{11}$$
$$x = -3$$
Check:
$$\frac{(-3)}{4} + \frac{3(-3)}{10} \stackrel{?}{=} -\frac{33}{20}$$
$$-\frac{3}{4} - \frac{9}{10} \stackrel{?}{=} -\frac{33}{20}$$
$$-\frac{15}{20} - \frac{18}{20} \stackrel{?}{=} -\frac{33}{20}$$
$$-\frac{33}{20} = -\frac{33}{20} \quad \text{T}$$
This is a conditional equation.

Solution set: $\{-3\}$

65.
$$3p - \frac{p}{4} = \frac{11p}{4} + 1$$
$$4\left(3p - \frac{p}{4}\right) = 4\left(\frac{11p}{4} + 1\right)$$
$$4 \cdot 3p - 4 \cdot \frac{p}{4} = 4 \cdot \frac{11p}{4} + 4 \cdot 1$$
$$12p - p = 11p + 4$$
$$11p = 11p + 4$$
$$11p - 11p = 11p + 4 - 11p$$
$$0 = 4$$
This statement is a contradiction. The equation has no solution.

Solution set: $\{\ \}$ or $\varnothing$

67.
$$\frac{2x+1}{2} - \frac{x+1}{5} = \frac{23}{10}$$
$$10\left(\frac{2x+1}{2} - \frac{x+1}{5}\right) = 10 \cdot \frac{23}{10}$$
$$5(2x+1) - 2(x+1) = 23$$
$$10x + 5 - 2x - 2 = 23$$
$$8x + 3 = 23$$
$$8x + 3 - 3 = 23 - 3$$
$$8x = 20$$
$$\frac{8x}{8} = \frac{20}{8}$$
$$x = \frac{5}{2}$$
Check:
$$\frac{2\left(\frac{5}{2}\right)+1}{2} - \frac{\left(\frac{5}{2}\right)+1}{5} \stackrel{?}{=} \frac{23}{10}$$
$$\frac{5+1}{2} - \frac{\frac{7}{2}}{5} \stackrel{?}{=} \frac{23}{10}$$
$$\frac{6}{2} - \frac{7}{10} \stackrel{?}{=} \frac{23}{10}$$
$$3 - \frac{7}{10} \stackrel{?}{=} \frac{23}{10}$$
$$\frac{30}{10} - \frac{7}{10} \stackrel{?}{=} \frac{23}{10}$$
$$\frac{23}{10} = \frac{23}{10} \quad \text{T}$$
This is a conditional equation.

Solution set: $\left\{\frac{5}{2}\right\}$

69. $0.4(z+1)-0.7z = -0.1z+0.7-0.2z-0.3$

$0.4z+0.4-0.7z = -0.3z+0.4$

$-0.3z+0.4 = -0.3z+0.4$

$-0.3z+0.4-0.4 = -0.3z+0.4-0.4$

$-0.3z = -0.3z$

$-0.3z+0.3z = -0.3z+0.3z$

$0 = 0$

This is an identity.

Solution set: $\{z \mid z \text{ is a real number}\}$ or $\mathbb{R}$

71. $\frac{1}{3}(2x-3)+2 = \frac{5}{6}(x+3)-\frac{11}{12}$

$12\left[\frac{1}{3}(2x-3)+2\right] = 12\left[\frac{5}{6}(x+3)-\frac{11}{12}\right]$

$4(2x-3)+24 = 10(x+3)-11$

$8x-12+24 = 10x+30-11$

$8x+12 = 10x+19$

$8x+12-12 = 10x+19-12$

$8x = 10x+7$

$8x-10x = 10x+7-10x$

$-2x = 7$

$\frac{-2x}{-2} = \frac{7}{-2}$

$x = -\frac{7}{2}$

Check:

$\frac{1}{3}\left(2\left(-\frac{7}{2}\right)-3\right)+2 \stackrel{?}{=} \frac{5}{6}\left(-\frac{7}{2}+3\right)-\frac{11}{12}$

$\frac{1}{3}(-7-3)+2 \stackrel{?}{=} \frac{5}{6}\left(-\frac{1}{2}\right)-\frac{11}{12}$

$\frac{1}{3}(-10)+2 \stackrel{?}{=} -\frac{5}{12}-\frac{11}{12}$

$-\frac{10}{3}+\frac{6}{3} \stackrel{?}{=} -\frac{16}{12}$

$-\frac{4}{3} = -\frac{4}{3}$ T

This is a conditional equation.

Solution set: $\left\{-\frac{7}{2}\right\}$

73. $7y-8 = -7$

$7y-8+8 = -7+8$

$7y = 1$

$\frac{7y}{7} = \frac{1}{7}$

$y = \frac{1}{7}$

Check: $7\left(\frac{1}{7}\right)-8 \stackrel{?}{=} -7$

$1-8 \stackrel{?}{=} -7$

$-7 = -7$ T

This is a conditional equation.

Solution set: $\left\{\frac{1}{7}\right\}$

75. $4a+3-2a+4 = 5a-7+a$

$2a+7 = 6a-7$

$2a+7-7 = 6a-7-7$

$2a = 6a-14$

$2a-6a = 6a-14-6a$

$-4a = -14$

$\frac{-4a}{-4} = \frac{-14}{-4}$

$a = \frac{7}{2}$

Check:

$4\left(\frac{7}{2}\right)+3-2\left(\frac{7}{2}\right)+4 \stackrel{?}{=} 5\left(\frac{7}{2}\right)-7+\frac{7}{2}$

$14+3-7+4 \stackrel{?}{=} \frac{35}{2}-7+\frac{7}{2}$

$21-7 \stackrel{?}{=} \frac{42}{2}-7$

$14 \stackrel{?}{=} 21-7$

$14 = 14$ T

This is a conditional equation.

Solution set: $\left\{\frac{7}{2}\right\}$

77. $4(p+3) = 3(p-2)+p+18$

$4p+12 = 3p-6+p+18$

$4p+12 = 4p+12$

$4p+12-4p = 4p+12-4p$

$12 = 12$

This statement is an identity. The solution set is all real numbers.

Solution set: $\{p \mid p \text{ is any real number}\}$ or $\mathbb{R}$

79. $4b-3(b+1)-b = 5(b-1)-5b$

$4b-3b-3-b = 5b-5-5b$

$-3 = -5$

This statement is a contradiction. The equation has no solution.

Solution set: $\{\ \}$ or $\varnothing$

81.
$$8(4x+6)=11-(x+7)$$
$$32x+48=11-x-7$$
$$32x+48=4-x$$
$$32x+48+x=4-x+x$$
$$33x+48=4$$
$$33x+48-48=4-48$$
$$33x=-44$$
$$\frac{33x}{33}=\frac{-44}{33}$$
$$x=-\frac{4}{3}$$

Check: $8\left[4\left(-\frac{4}{3}\right)+6\right]\overset{?}{=}11-\left(-\frac{4}{3}+7\right)$

$$8\left(\frac{-16}{3}+\frac{18}{3}\right)\overset{?}{=}11-\left(-\frac{4}{3}+\frac{21}{3}\right)$$
$$8\left(\frac{2}{3}\right)\overset{?}{=}\frac{33}{3}-\frac{17}{3}$$
$$\frac{16}{3}=\frac{16}{3}\ \text{True}$$

This is a conditional equation.

Solution set: $\left\{-\frac{4}{3}\right\}$

83.
$$\frac{m+1}{4}+\frac{5}{6}=\frac{2m-1}{12}$$
$$12\left(\frac{m+1}{4}+\frac{5}{6}\right)=12\left(\frac{2m-1}{12}\right)$$
$$3(m+1)+10=2m-1$$
$$3m+3+10=2m-1$$
$$3m+13=2m-1$$
$$3m+13-13=2m-1-13$$
$$3m=2m-14$$
$$3m-2m=2m-14-2m$$
$$m=-14$$

Check:
$$\frac{-14+1}{4}+\frac{5}{6}=\frac{2(-14)-1}{12}$$
$$\frac{-13}{4}+\frac{5}{6}=\frac{-28-1}{12}$$
$$\frac{-39}{12}+\frac{10}{12}=\frac{-29}{12}$$
$$\frac{-29}{12}=\frac{-29}{12}\ \text{T}$$

This is a conditional equation.

Solution set: $\{-14\}$

85.
$$0.3x-1.3=0.5x-0.7$$
$$0.3x-1.3+1.3=0.5x-0.7+1.3$$
$$0.3x=0.5x+0.6$$
$$0.3x-0.5x=0.5x+0.6-0.5x$$
$$-0.2x=0.6$$
$$\frac{-0.2x}{-0.2}=\frac{0.6}{-0.2}$$
$$x=-3$$

Check:
$$0.3(-3)-1.3\overset{?}{=}0.5(-3)-0.7$$
$$-0.9-1.3\overset{?}{=}-1.5-0.7$$
$$-2.2=-2.2\ \text{T}$$

This is a conditional equation.

Solution set: $\{-3\}$

(Note: Alternatively, we could have started by multiplying both sides of the equation by 10 to clear the decimals.)

87.
$$-0.8(x+1)=0.2(x+4)$$
$$-0.8x-0.8=0.2x+0.8$$
$$-0.8x-0.8+0.8=0.2x+0.8+0.8$$
$$-0.8x=0.2x+1.6$$
$$-0.8x-0.2x=0.2x+1.6-0.2x$$
$$-1.0x=1.6$$
$$\frac{-1.0x}{-1.0}=\frac{1.6}{-1.0}$$
$$x=-1.6$$

Check: $-0.8(-1.6+1)\overset{?}{=}0.2(-1.6+4)$
$$-0.8(-0.6)\overset{?}{=}0.2(2.4)$$
$$0.48=0.48\ \text{T}$$

This is a conditional equation.

Solution set: $\{-1.6\}$

(Note: Alternatively, we could have started by multiplying both sides of the equation by 10 to clear the decimals.)

89. $\dfrac{1}{4}(x-4)+3=\dfrac{1}{3}(2x+6)-\dfrac{5}{6}$

$12\left[\dfrac{1}{4}(x-4)+3\right]=12\left[\dfrac{1}{3}(2x+6)-\dfrac{5}{6}\right]$

$\qquad 3(x-4)+36=4(2x+6)-10$

$\qquad 3x-12+36=8x+24-10$

$\qquad\qquad 3x+24=8x+14$

$\qquad 3x+24-24=8x+14-24$

$\qquad\qquad\quad 3x=8x-10$

$\qquad\quad 3x-8x=8x-10-8x$

$\qquad\qquad\quad -5x=-10$

$\qquad\qquad\dfrac{-5x}{-5}=\dfrac{-10}{-5}$

$\qquad\qquad\qquad x=2$

Check: $\dfrac{1}{4}(2-4)+3\overset{?}{=}\dfrac{1}{3}(2(2)+6)-\dfrac{5}{6}$

$\qquad\qquad \dfrac{1}{4}(-2)+3\overset{?}{=}\dfrac{1}{3}(10)-\dfrac{5}{6}$

$\qquad\qquad\quad -\dfrac{1}{2}+3\overset{?}{=}\dfrac{1}{3}(10)-\dfrac{5}{6}$

$\qquad\qquad -\dfrac{1}{2}+\dfrac{6}{2}\overset{?}{=}\dfrac{20}{6}-\dfrac{5}{6}$

$\qquad\qquad\qquad\quad \dfrac{5}{2}\overset{?}{=}\dfrac{15}{6}$

$\qquad\qquad\qquad\quad \dfrac{5}{2}=\dfrac{5}{2}\ \text{T}$

This is a conditional equation.
Solution set: $\{2\}$

91. $ax+3=15$

Let $x=-3:\qquad a(-3)+3=15$

$\qquad\qquad\qquad\qquad -3a+3=15$

$\qquad\qquad\qquad -3a+3-3=15-3$

$\qquad\qquad\qquad\qquad\quad -3a=12$

$\qquad\qquad\qquad\qquad\dfrac{-3a}{-3}=\dfrac{12}{-3}$

$\qquad\qquad\qquad\qquad\qquad a=-4$

Check: $\qquad -4x+3=15$

$\qquad\quad -4x+3-3=15-3$

$\qquad\qquad\qquad -4x=12$

$\qquad\qquad\quad \dfrac{-4x}{-4}=\dfrac{12}{-4}$

$\qquad\qquad\qquad\quad x=-3$

93. $a(x-1)=3(x-1)$

To have the solution set be all real numbers, we need both sides of the linear equation to be identical. This can be achieved by letting $a=3$.

95. We need to set the denominator equal to 0 and solve the resulting equation.

$\qquad\quad 2x+1=0$

$\quad 2x+1-1=0-1$

$\qquad\qquad 2x=-1$

$\qquad\qquad \dfrac{2x}{2}=\dfrac{-1}{2}$

$\qquad\qquad\quad x=-\dfrac{1}{2}$

Check:

$2\left(-\dfrac{1}{2}\right)+1\overset{?}{=}0$

$\qquad\quad -1+1\overset{?}{=}0$

$\qquad\qquad\quad 0=0\ \text{T}$

We must exclude $x=-\dfrac{1}{2}$ from the domain.

97. We need to set the denominator equal to 0 and solve the resulting equation.

$\qquad\quad 4x-3=0$

$\quad 4x-3+3=0+3$

$\qquad\qquad 4x=3$

$\qquad\qquad \dfrac{4x}{4}=\dfrac{3}{4}$

$\qquad\qquad\quad x=\dfrac{3}{4}$

Check:

$4\left(\dfrac{3}{4}\right)-3\overset{?}{=}0$

$\qquad\quad 3-3\overset{?}{=}0$

$\qquad\qquad 0=0\ \text{T}$

We must exclude $x=\dfrac{3}{4}$ from the domain.

99. We need to set the denominator equal to 0 and solve the resulting equation.

$\quad 3(x+1)-6=0$

$\qquad 3x+3-6=0$

$\qquad\qquad 3x-3=0$

$\quad 3x-3+3=0+3$

$\qquad\qquad\quad 3x=3$

$\qquad\qquad\quad \dfrac{3x}{3}=\dfrac{3}{3}$

$\qquad\qquad\qquad x=1$

Check:

$3(1+1)-6\overset{?}{=}0$

$\qquad 3(2)-6\overset{?}{=}0$

$\qquad\quad 6-6\overset{?}{=}0$

$\qquad\qquad\quad 0=0\ \text{T}$

We must exclude $x=1$ from the domain.

101.
$$25 = \frac{2000}{12} \cdot r$$
$$\frac{12}{2000} \cdot 25 = \frac{12}{2000} \cdot \frac{2000}{12} \cdot r$$
$$\frac{300}{2000} = r$$
$$0.15 = r$$
The card's annual interest rate is 0.15 or 15%.

103.
$$4412.5 = 0.15(I - 8500) + 850$$
$$4412.5 = 0.15I - 1275 + 850$$
$$4412.5 = 0.15I - 425$$
$$4412.5 + 425 = 0.15I - 425 + 425$$
$$4837.5 = 0.15I$$
$$\frac{4837.5}{0.15} = \frac{0.15I}{0.15}$$
$$32,250 = I$$
Your adjusted gross income was $32,250 in 2011.

105. $4(x+1) - 2$ is an expression and $4(x+1) = 2$ is an equation. In general, an equation can be distinguished from an algebraic expression by the presence of an equal sign.

107. Answers will vary. One possibility:
One solution: $3x + 2 = 2x - 7$
No solution: $4x - 3 = 4x + 5$
Identity: $3x + 2 = 3x + 2$
In both the no solution case and the identity case, the variable terms will cancel out leaving a constant on each side of the equation. If the constants are the same, then you have an identity. If they differ, then there is no solution. If the variable terms do not cancel when put on the same side, then there will be one solution.

Section 1.2

Are You Ready for This Section?

R1. $12 + z$
The keyword 'sum' indicates addition.

R2. $4x$
The keyword 'product' indicates multiplication.

R3. $y - 87$
The keywords 'decreased by' indicate subtraction. Since order matters here, we subtract the quantity after the keywords.

R4. $\dfrac{z}{12}$
The keyword 'quotient' indicates division. The first quantity is the numerator and the second quantity is the denominator.

R5. $4(x+7)$
The keyword 'times' indicates multiplication and the keyword 'sum' indicates addition. Because the statement says 'times the sum' we multiply 4 by the entire sum, not just the x.

R6. $4x + 7$
The keyword 'times' indicates multiplication and the keyword 'sum' indicates addition. After the keywords 'sum of', we look for the two quantities to add. These are separated by the keyword 'and'. Thus, the first addend is 'four times a number x' while the second addend is the number '7'.

Section 1.2 Quick Checks

1. Mathematical statements are represented symbolically as <u>equations</u>.

2. $x + 7 = 12$
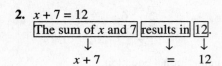

3. $3y = 21$
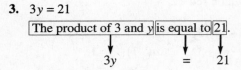

4. $2(n + 3) = 5$
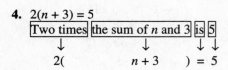

5. $x - 10 = \dfrac{x}{2}$
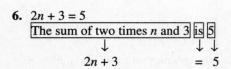

6. $2n + 3 = 5$

The sum of two times n and 3 is 5

7. False; if n represents the first of two consecutive odd integers, $n + 2$ represents the next consecutive odd integer.

8. Let x = the first of the consecutive even integers.
Second: $x + 2$
Third: $x + 4$
$$x + (x + 2) + (x + 4) = 60$$
$$x + x + 2 + x + 4 = 60$$
$$3x + 6 = 60$$
$$3x + 6 - 6 = 60 - 6$$
$$3x = 54$$
$$\frac{3x}{3} = \frac{54}{3}$$
$$x = 18$$
The three consecutive even integers are 18, 20, and 22.
Check:
$$18 + (18 + 2) + (18 + 4) = 18 + 20 + 22 = 60$$

9. Let x = the first of the consecutive integers.
Second: $x + 1$
Third: $x + 2$
$$x + (x + 1) + (x + 2) = 78$$
$$x + x + 1 + x + 2 = 78$$
$$3x + 3 = 78$$
$$3x + 3 - 3 = 78 - 3$$
$$3x = 75$$
$$\frac{3x}{3} = \frac{75}{3}$$
$$x = 25$$
The three consecutive integers are 25, 26, and 27.
Check:
$$25 + (25 + 1) + (25 + 2) = 25 + 26 + 27 = 78$$

10. Let w = Melody's hourly wage.
Overtime rate: $1.5w$
$$40w + 6(1.5w) = 735$$
$$40w + 9w = 735$$
$$49w = 735$$
$$\frac{49w}{49} = \frac{735}{49}$$
$$w = 15$$
Melody makes $15 per hour.
Check:
$$40(15) + 6[1.5(15)] = 600 + 6(22.5)$$
$$= 600 + 135$$
$$= 735$$

11. Let w = Jim's regular hourly rate.
Saturday rate: $1.5w$
Sunday rate: $2w$
$$30w + 6(1.5w) + 4(2w) = 564$$
$$30w + 9w + 8w = 564$$
$$47w = 564$$
$$\frac{47w}{47} = \frac{564}{47}$$
$$w = 12$$
Jim's regular hourly rate is $12 per hour.
Check:
$$30(12) + 6[1.5(12)] + 4[2(12)]$$
$$= 360 + 6(18) + 4(24)$$
$$= 360 + 108 + 96$$
$$= 564$$

12. Let m = the number of miles.
EZ-Rental cost = Do It Yourself cost
$$35 + 0.15m = 20 + 0.25m$$
$$0.15m = -15 + 0.25m$$
$$-0.1m = -15$$
$$\frac{-0.1m}{-0.1} = \frac{-15}{-0.1}$$
$$m = 150$$
The rental costs will be the same for 150 miles.
Check:
$$35 + 0.15(150) = 35 + 22.5 = \$57.50$$
$$20 + 0.25(150) = 20 + 37.5 = \$57.50$$

13. Let m = the number of minutes used.
Company A cost = Company B cost
$$12 + 0.1m = 0.15m$$
$$12 = 0.05m$$
$$\frac{12}{0.05} = \frac{0.05m}{0.05}$$
$$240 = m$$
The monthly costs will be the same for 240 minutes.
Check:
$$12 + 0.1(240) = 12 + 24 = 36$$
$$0.15(240) = 36$$

14. Percent means "divided by 100."

15. Let x = the desired result.
$$x = 0.4(100) = 40$$
40% of 100 is 40.

16. Let x = the desired value.
$$8 = 0.05(x)$$
$$\frac{8}{0.05} = \frac{0.05x}{0.05}$$
$$160 = x$$
8 is 5% of 160.

17. Let p = the desired percent (as a decimal).
$$15 = p \cdot 20$$
$$15 = 20p$$
$$\frac{15}{20} = \frac{20p}{20}$$
$$0.75 = p$$
15 is 75% of 20.

18. Let p = the original price.
Discount: $0.3p$
$$p - 0.3p = 21$$
$$0.7p = 21$$
$$\frac{0.7p}{0.7} = \frac{21}{0.7}$$
$$p = 30$$
The original price of the shirt was $30.

19. Let c = Milex's cost for a spark plug.
Mark up: $0.35c$
$$c + 0.35c = 1.62$$
$$1.35c = 1.62$$
$$\frac{1.35c}{1.35} = \frac{1.62}{1.35}$$
$$c = 1.2$$
Milex's cost for each spark plug is $1.20.

20. Interest is money paid for the use of money. The total amount borrowed is called the principal.

21. Note that 1 month is $\frac{1}{12}$ year.
$$I = Prt$$
$$I = (6500)(0.06)\left(\frac{1}{12}\right) = 32.5$$
The interest charge on Dave's car loan after 1 month is $32.50.

22. Note that 6 months is $\frac{1}{2}$ year.
$$I = Prt$$
$$I = (1400)(0.015)\left(\frac{1}{2}\right) = 10.50$$
The interest paid after 6 months would be $10.50.
The balance is the original amount in the account

plus the accrued interest.
$$1400 + 10.50 = 1410.50$$
The account will have a balance of $1410.50.

23. Let x = the amount invested in Aaa-rated bonds.
B-rated bonds: $90,000 - x$
Int. from Aaa-rated + Int. from B-rated = 5400
$$0.05x + 0.09(90,000 - x) = 5400$$
$$0.05x + 8100 - 0.09x = 5400$$
$$8100 - 0.04x = 5400$$
$$-0.04x = -2700$$
$$\frac{-0.04x}{-0.04} = \frac{-2700}{-0.04}$$
$$x = 67,500$$
Sophia should place $67,500 in Aaa-rated bonds and $90,000 - \$67,500 = \$22,500$ in B-rated bonds.

24. Let x = the amount invested in a 5-yr CD.
Corporate bonds: $25,000 - x$
CD interest + bond interest = total interest
$$0.04x + 0.09(25,000 - x) = 0.08(25,000)$$
$$0.04x + 2250 - 0.09x = 2000$$
$$2250 - 0.05x = 2000$$
$$-0.05x = -250$$
$$\frac{-0.05x}{-0.05} = \frac{-250}{-0.05}$$
$$x = 5000$$
Steve should invest $5000 in the CD and $25,000 - \$5000 = \$20,000$ in corporate bonds.

25. Let x = pounds of Tea A.
Tea B pounds: $10 - x$

	Tea A	Tea B	Mix
price per pound	4	2.75	3.50
pounds	x	$10 - x$	10
revenue	$4x$	$2.75(10 - x)$	$3.5(10)$

We want the individual revenues to be the same as the revenue from the mix. Therefore, we get
$$4x + 2.75(10 - x) = 3.5(10)$$
$$4x + 27.5 - 2.75x = 35$$
$$1.25x + 27.5 = 35$$
$$1.25x = 7.5$$
$$\frac{1.25x}{1.25} = \frac{7.5}{1.25}$$
$$x = 6$$
You should blend 6 pounds of Tea A and $10 - 6 = 4$ pounds of Tea B.

26. Let x = pounds of cashews.
Peanuts: $30 - x$

	Cashews	Peanuts	Mix
price per pound	6	1.5	3
pounds	x	$30-x$	30
revenue	$6x$	$1.5(30-x)$	$3(30)$

We want the individual revenues to be the same as the revenue from the mix. Therefore, we get
$$6x + 1.5(30-x) = 3(30)$$
$$6x + 45 - 1.5x = 90$$
$$4.5x + 45 = 90$$
$$4.5x = 45$$
$$\frac{4.5x}{4.5} = \frac{45}{4.5}$$
$$x = 10$$
You should blend 10 pounds of cashews and $30 - 10 = 20$ pounds of peanuts.

27. Objects that move at a constant velocity are said to be in <u>uniform motion</u>.

28. Another term for speed is <u>rate</u>.

29. We want to know when the distance traveled by both cars will be the same.
Let t = the time of travel by the Cruze.
Then the time of travel for the Dodge will be $t - 2$ since it left two hours later.
     Cruze distance = Dodge distance
$$(\text{rate})(\text{time})_{\text{Cruze}} = (\text{rate})(\text{time})_{\text{Dodge}}$$
$$40t = 60(t-2)$$
$$40t = 60t - 120$$
$$-20t = -120$$
$$t = 6$$
The Dodge will catch the Cruze after the Cruze has been traveling for 6 hours (4 hours for the Dodge). When the Dodge catches the Cruze, both cars will have traveled $40(6) = 240$ miles.

30. We want to know when the distance traveled by both the train and the helicopter will be the same.
Let t = the time of travel by the train.
Then the time of travel for the helicopter will be $t - 4$ since it left four hours later.
     train distance = helicopter distance
$$(\text{rate})(\text{time})_{\text{train}} = (\text{rate})(\text{time})_{\text{helicopter}}$$
$$50t = 90(t-4)$$
$$50t = 90t - 360$$
$$-40t = -360$$
$$t = 9$$

The helicopter will catch the train after the train has been traveling for 9 hours (5 hours for the helicopter). When the helicopter catches the train, both the helicopter and the train will have traveled $50(9) = 450$ miles.

1.2 Exercises

31. Let x = desired result.
$$x = 0.25(40) = 10$$
25% of 40 is 10.

33. Let x = original amount.
$$12 = 0.3x$$
$$\frac{12}{0.3} = \frac{0.3x}{0.3}$$
$$40 = x$$
12 is 30% of 40.

35. Let x = percent as a decimal.
$$30 = x \cdot 80$$
$$30 = 80x$$
$$\frac{30}{80} = \frac{80x}{80}$$
$$0.375 = x$$
30 is 37.5% of 80.

37.
$$x + 12 = 20$$
$$x + 12 - 12 = 20 - 12$$
$$x = 8$$

39.
$$2(y+3) = 16$$
$$2y + 6 = 16$$
$$2y = 10$$
$$\frac{2y}{2} = \frac{10}{2}$$
$$y = 5$$

41.
$$w - 22 = 3w$$
$$-22 = 2w$$
$$\frac{-22}{2} = \frac{2w}{2}$$
$$-11 = w$$

43.
$$4x = 2x + 14$$
$$2x = 14$$
$$\frac{2x}{2} = \frac{14}{2}$$
$$x = 7$$

45. $0.8x = x + 5$
$$-0.2x = 5$$
$$\frac{-0.2x}{-0.2} = \frac{5}{-0.2}$$
$$x = -25$$

47. First number: x
Second number: $2x$
$$x + 2x = 39$$
$$3x = 39$$
$$\frac{3x}{3} = \frac{39}{3}$$
$$x = 13$$
Check:
$$13 + 2(13) = 13 + 26 = 39$$
The two numbers are 13 and 26.

49. First of the consecutive integers: x
Second consecutive integer: $x + 1$
Third consecutive integer: $x + 2$
$$x + (x + 1) + (x + 2) = 75$$
$$x + x + 1 + x + 2 = 75$$
$$3x + 3 = 75$$
$$3x + 3 - 3 = 75 - 3$$
$$3x = 72$$
$$\frac{3x}{3} = \frac{72}{3}$$
$$x = 24$$
Check:
$$24 + (24 + 1) + (24 + 2) = 24 + 25 + 26 = 75$$
The integers are 24, 25, and 26.

51. Let x = score on Kendra's final exam.
$$\frac{84 + 78 + 64 + 88 + 2x}{6} = 80$$
$$\frac{314 + 2x}{6} = 80$$
$$6 \cdot \frac{314 + 2x}{6} = 6 \cdot 80$$
$$314 + 2x = 480$$
$$314 + 2x - 314 = 480 - 314$$
$$2x = 166$$
$$\frac{2x}{2} = \frac{166}{2}$$
$$x = 83$$
Check:
$$\frac{84 + 78 + 64 + 88 + 2(83)}{6} = \frac{314 + 166}{6} = \frac{480}{6} = 80$$
Kendra needs an 83 on her final exam to have an average of 80.

53. Let x = number of pages printed.
HP printer cost: $180 + 0.03x$
Brother printer cost: $230 + 0.01x$
$$180 + 0.03x = 230 + 0.01x$$
$$180 + 0.03x - 180 = 230 + 0.01x - 180$$
$$0.03x = 0.01x + 50$$
$$0.03x - 0.01x = 0.01x + 50 - 0.01x$$
$$0.02x = 50$$
$$\frac{0.02x}{0.02} = \frac{50}{0.02}$$
$$x = 2500$$
Check:
$$180 + 0.03(2500) \stackrel{?}{=} 230 + 0.01(2500)$$
$$180 + 75 \stackrel{?}{=} 230 + 25$$
$$255 = 255 \text{ T}$$
Jacob would need to print 2500 pages for the cost to be the same for the two printers.

55. Let x = amount that Connor receives.

Olivia: $\dfrac{3}{4}x$

Avery: $\dfrac{1}{4}x$

$$x + \frac{3}{4}x + \frac{1}{4}x = 800,000$$
$$2x = 800,000$$
$$\frac{2x}{2} = \frac{800,000}{2}$$
$$x = 400,000$$
Check:
$$400,000 + \frac{3}{4}(400,000) + \frac{1}{4}(400,000) \stackrel{?}{=} 800,000$$
$$400,000 + 300,000 + 100,00 \stackrel{?}{=} 800,000$$
$$800,000 = 800,000 \text{ T}$$
Connor: \$400,000; Olivia: \$300,000; Avery: \$100,000.

57. Let x = total of final bill.
$$599 + 0.0635 \cdot 599 = x$$
$$599 + 38.0365 = x$$
$$637.0365 = x$$
The final bill will be \$637.04.

59. Let x = dealer's cost.
$$x + 0.15x = 23,950$$
$$1.15x = 23,950$$
$$\frac{1.15x}{1.15} = \frac{23,950}{1.15}$$
$$x \approx 20,826.09$$

Check:
$$20,826.09 + 0.15(20,826.09)$$
$$= 20,826.09 + 3123.9135$$
$$= 23,950.0035$$
The dealer's cost is about $20,826.09.

61. Let x = original price of flash drive.
Discount price: $x - 0.60x = 0.4x$
$$0.4x = 27.55$$
$$\frac{0.4x}{0.4} = \frac{27.55}{0.4}$$
$$x = 68.88$$
Check:
$$0.4(68.88) \stackrel{?}{=} 27.55$$
$$27.55 = 27.55 \text{ True}$$
The flash drive originally cost $68.88.

63. Let x = weight in pounds of Nissan Altima
Mazda 6s: $x + 136$
Honda Accord EX: $x + 119$
$$x + (x+136) + (x+119) = 9834$$
$$x + x + 136 + x + 119 = 9834$$
$$3x + 255 = 9834$$
$$3x + 255 - 255 = 9834 - 255$$
$$3x = 9579$$
$$\frac{3x}{3} = \frac{9579}{3}$$
$$x = 3193$$
The Nissan Altima weighs 3193 pounds, the
Mazda 6s weighs 3329 pounds, and the Honda
Accord EX weighs 3312 pounds.
Check: $3193 + 3329 + 3312 = 9834$

65. Let x = amount that Adam will receive.
Krissy: $x + 3000$
$$x + (x + 3000) = 20,000$$
$$x + x + 3000 = 20,000$$
$$2x + 3000 = 20,000$$
$$2x + 3000 - 3000 = 20,000 - 3000$$
$$2x = 17,000$$
$$\frac{2x}{2} = \frac{17,000}{2}$$
$$x = 8500$$
Check:
$$8500 + (8500 + 3000) = 8500 + 11,500 = 20,000$$
Adam will get $8500 and Krissy will get
$11,500.

67. Let x = amount invested in stocks.
Bonds: $\frac{3}{5}x$
$$x + \frac{3}{5}x = 24,000$$
$$\frac{8}{5}x = 24,000$$
$$\frac{5}{8} \cdot \frac{8}{5}x = \frac{5}{8} \cdot 24,000$$
$$x = 15,000$$
Check:
$$15,000 + \frac{3}{5}(15,000) = 15,000 + 9000 = 24,000$$
You should invest $15,000 in stocks and $9000
in bonds.

69. Let I = interest charge after one month.
$$I = Prt = (2500)(0.14)\left(\frac{1}{12}\right) \approx 29.17$$
The interest charge after one month will be
$29.17.
Check:
$$\left(\frac{29.17}{2500}\right) \cdot 12 = (0.011668) \cdot 12 = 0.140016$$

71. Let x = amount loaned at 6%.
Amount at 11%: $500,000 - x$
$$0.06x + 0.11(500,000 - x) = 43,750$$
$$0.06x + 55,000 - 0.11x = 43,750$$
$$-0.05x + 55,000 = 43,750$$
$$-0.05x = -11,250$$
$$\frac{-0.05x}{-0.05} = \frac{-11,250}{-0.05}$$
$$x = 225,000$$
The bank loaned $225,000 at 6% interest.

73. Let x = the amount invested in bonds.
Amount in stocks: $25,000 - x$
Here we need the simple interest formula
$I = P \cdot r \cdot t$. We need to add the interest from
each investment after $t = 1$ year.
$$0.05x + 0.09(25,000 - x) = 1875$$
$$0.05x + 2250 - 0.09x = 1875$$
$$2250 - 0.04x = 1875$$
$$2250 - 0.04x - 2250 = 1875 - 2250$$
$$-0.04x = -375$$
$$\frac{-0.04x}{-0.04} = \frac{-375}{-0.04}$$
$$x = 9375$$
Pedro should invest $9375 in the 5% bond and
$25,000 - 9,375 = \$15,625$ in the 9% stock fund.

Check:
$$0.05(9375)+0.09(15,625) \stackrel{?}{=} 1875$$
$$468.75+1406.25 \stackrel{?}{=} 1875$$
$$1875 = 1875 \text{ T}$$

75. Let x = pounds of coffee A. Since there will be a total of 50 pounds of coffee, you must add $50-x$ pounds of coffee B.

77. Let d = number of dimes.
Quarters: $47-d$
$$0.1(d)+0.25(47-d) = 9.50$$
$$0.1d+11.75-0.25d = 9.50$$
$$-0.15d+11.75 = 9.50$$
$$-0.15d = -2.25$$
$$\frac{-0.15d}{-0.15} = \frac{-2.25}{-0.15}$$
$$d = 15$$
Check:
$$0.1(15)+0.25(32) = 1.5+8 = 9.5$$
Bobby has 15 dimes and 32 quarters saved.

79. Let x = grams of pure gold.
12 karat: $72-x$
$$\text{18k gold} = \text{12k gold} + \text{pure gold}$$
$$(\%) \cdot (\text{grams}) = (\%) \cdot (\text{grams}) + (\%) \cdot (\text{grams})$$
$$\frac{18}{24}(72) = \frac{12}{24}(72-x) + \frac{24}{24}(x)$$
$$54 = 36 - \frac{1}{2}x + x$$
$$54 = 36 + \frac{1}{2}x$$
$$18 = \frac{1}{2}x$$
$$2 \cdot 18 = 2 \cdot \frac{1}{2}x$$
$$36 = x$$
36 grams of pure gold should be mixed with 36 grams of 12 karat gold.
Check:
$$\frac{12}{24}(72-36) + \frac{24}{24}(36) = \frac{1}{2}(36) + 1(36)$$
$$= 18 + 36$$
$$= 54$$
$$= \frac{18}{24}(72)$$

81. Let x = time spent running (hours).
Bicycle: $4-x$
For this problem we will need the distance traveled formula: $d = r \cdot t$.
tot. dist. = dist. run + dist. biked
$$\text{tot. dist.} = (\text{run rate})(\text{time}) + (\text{bike rate})(\text{time})$$
$$62 = 8(x) + 20(4-x)$$
$$62 = 8x + 80 - 20x$$
$$62 = 80 - 12x$$
$$-18 = -12x$$
$$\frac{-18}{-12} = \frac{-12x}{-12}$$
$$1.5 = x$$
Check:
$$8(1.5) + 20(4-1.5) = 12 + 20(2.5) = 12 + 50 = 62$$
We now know that the time spent running is 1.5 hours and the time spent biking is 2.5 hours. However, the question asked for distances. We can use the distance traveled formula here.
Run: distance $= 8(1.5) = 12$ miles

Bike: distance $= 20(2.5) = 50$ miles

The race consists of running for 12 miles and biking for 50 miles.

83. Let x = speed of slower car (mph).
Faster car: $10+x$
tot. dist. = slow dist. + fast dist.
$$455 = x(3.5) + (10+x)(3.5)$$
$$455 = 3.5x + 35 + 3.5x$$
$$455 = 7x + 35$$
$$420 = 7x$$
$$\frac{420}{7} = \frac{7x}{7}$$
$$60 = x$$
Check:
$$60(3.5) + 70(3.5) = 210 + 245 = 455$$
The slow car is traveling at 60 mph while the faster car travels at 70 mph.

85. Let t = time in hours.
Speed of slow boat: 25 mph
Speed of fast boat: 25 + 12 = 37 mph
tot. dist. = slow boat dist. + fast boat dist.
$$155 = 25t + 37t$$
$$155 = 62t$$
$$\frac{155}{62} = t$$
$$2.5 = t$$
Check: $25(2.5) + 37(2.5) = 62.5 + 92.5 = 155$

The boats will be 155 miles apart after 2.5 hours.

87. Let x = speed of slower person (mph).

Speed of faster person: $2 + x$

tot. dist. = slow dist. + fast dist.

$$15 = x(1.5) + (2 + x)(1.5)$$
$$15 = 1.5x + 3 + 1.5x$$
$$15 = 3x + 3$$
$$12 = 3x$$
$$\frac{12}{3} = x$$
$$4 = x$$

Check: $4(1.5) + 6(1.5) = 6 + 9 = 15$

One person is walking at a rate of 4 miles per hour and the other is walking at a rate of 6 miles per hour.

89. Written answers will vary.

Algebraic:

Let x = distance to Florida (miles).

When thinking about this problem, we are assuming that the distance to Florida is the same as the distance back. We will again need the distance traveled formula: $d = r \cdot t$.

Also note that $t = \dfrac{d}{r}$ (this is used to express the times of travel to and from Florida). This gives us the following:

$$\text{avg. speed} = \frac{\text{total distance traveled}}{\text{total time of travel}}$$
$$= \frac{\text{dist. to} + \text{dist. from}}{\text{time to} + \text{time from}}$$
$$= \frac{x + x}{\frac{x}{50} + \frac{x}{60}} = \frac{2x}{\frac{6x}{300} + \frac{5x}{300}} = \frac{2x}{\frac{11x}{300}}$$
$$= \frac{2x}{1} \cdot \frac{300}{11x} = \frac{600}{11} \approx 54.55$$

The average speed of the trip to Florida and back is roughly 54.55 miles per hour.

91. Answers will vary. One possibility:

"Payton has DSL service and gets a 5% discount for bundling this service with her local phone service. If her discounted monthly rate for DSL is $60, how much would she pay without the discount?"

93. Start by noting that you and the train travel for the same amount of time. We also need the distance traveled formula: $d = r \cdot t$.

Distance traveled by train:

$$\frac{20 \text{ miles}}{1 \text{ hr}} \cdot 1 \text{ min} \cdot \frac{1 \text{ hr}}{60 \text{ min}} = \frac{1}{3} \text{ mile}$$

Distance traveled by you:

$$\frac{2 \text{ miles}}{1 \text{ hr}} \cdot 1 \text{ min} \cdot \frac{1 \text{ hr}}{60 \text{ min}} = \frac{1}{30} \text{ mile}$$

Since you went from the front of the train to the back, the difference in the distances traveled is the length of the train.

$$\frac{1}{3} - \frac{1}{30} = \frac{10}{30} - \frac{1}{30} = \frac{9}{30} = \frac{3}{10} = 0.3$$

The train is 0.3 mile long, or 1584 feet.

From a physical point of view, the situation would be the same as if you were standing still and the train was traveling at 18 mph ($20 - 2$). We know the rate (18 mph) and we know the time (1 min = $\dfrac{1}{60}$ hr). Using the distance traveled formula we get:

$$d = r \cdot t = 18\left(\frac{1}{60}\right) = 0.3 \text{ miles}$$

The train is 0.3 mile long, or 1584 feet.

95. Answers will vary. Mathematical modeling is related to problem solving in that it requires us to use information, tools (mathematical formulas), and our own skills to solve math related problems.

97. Categories: Direct Translation, Mixtures, Geometry, Uniform Motion, and Work Problems.

Two types of mixture problems are interest and blends.

Section 1.3

Are You Ready for This Section?

R1. a. To round to three decimal places, we first examine the digit in the fourth decimal place and then truncate. Since the fourth decimal place contains a 4, we do not change the digit in the third decimal place. Thus, 3.00343 rounded to three decimal places is 3.003.

b. To truncate, we remove all digits after the third decimal place. So, 3.00343 truncated to three decimal places is 3.003.

R2. a. To round to two decimal places, we first examine the digit in the third decimal place and then truncate. Since the third decimal place contains a 7, we increase the digit in the second decimal place by 1. Thus, 14.957 rounded to two decimal places is 14.96.

b. To truncate, we remove all digits after the second decimal place. So, 14.957 truncated to two decimal places is 14.95.

Section 1.3 Quick Checks

1. A <u>formula</u> is an equation that describes how two or more variables are related.

2. $A = \pi r^2$

3. $V = \pi r^2 h$

4. $C = 175x + 7000$

5. $s = \dfrac{1}{2} g t^2$

6. The <u>area</u> is the amount of space enclosed by a two-dimensional figure and is measured in square units, whereas <u>volume</u> is the amount of space occupied by a three-dimensional figure and is measured in cubic units.

7. False; the units are feet, not square feet.

8. **a.**
$$A = \frac{1}{2} bh$$
$$2 \cdot A = 2 \cdot \frac{1}{2} bh$$
$$2A = bh$$
$$\frac{2A}{b} = \frac{bh}{b}$$
$$\frac{2A}{b} = h \quad \text{or} \quad h = \frac{2A}{b}$$

   b.
$$h = \frac{2A}{b} = \frac{2(10)}{4} = 5$$
   A triangle whose area is 10 square inches and whose base is 4 inches has a height of 5 inches.

9. **a.**
$$P = 2a + 2b$$
$$P - 2a = 2a + 2b - 2a$$
$$P - 2a = 2b$$
$$\frac{P - 2a}{2} = \frac{2b}{2}$$
$$\frac{P - 2a}{2} = b \quad \text{or} \quad b = \frac{P - 2a}{2}$$

   b.
$$b = \frac{P - 2a}{2}$$
$$b = \frac{60 - 2(20)}{2} = \frac{60 - 40}{2} = \frac{20}{2} = 10$$
   A parallelogram whose perimeter is 60 cm and whose adjacent length is 20 cm has a length of 10 cm.

10.
$$I = P \cdot r \cdot t$$
$$\frac{I}{r \cdot t} = \frac{P \cdot r \cdot t}{r \cdot t}$$
$$\frac{I}{rt} = P \quad \text{or} \quad P = \frac{I}{rt}$$

11.
$$Ax + By = C$$
$$Ax + By - Ax = C - Ax$$
$$By = C - Ax$$
$$\frac{By}{B} = \frac{C - Ax}{B}$$
$$y = \frac{C - Ax}{B}$$

12.
$$2xh - 4x = 3h - 3$$
$$2xh - 4x - 3h = 3h - 3 - 3h$$
$$2xh - 4x - 3h = -3$$
$$2xh - 4x - 3h + 4x = -3 + 4x$$
$$2xh - 3h = 4x - 3$$
$$h(2x - 3) = 4x - 3$$
$$\frac{h(2x - 3)}{2x - 3} = \frac{4x - 3}{2x - 3}$$
$$h = \frac{4x - 3}{2x - 3}$$

13.
$$S = na + (n - 1)d$$
$$S = na + nd - d$$
$$S + d = na + nd - d + d$$
$$S + d = na + nd$$
$$S + d = n(a + d)$$
$$\frac{S + d}{a + d} = \frac{n(a + d)}{a + d}$$
$$\frac{S + d}{a + d} = n \quad \text{or} \quad n = \frac{S + d}{a + d}$$

14. Let w = the width of the pool in feet. Then the
length is $l = w + 10$.

$$P = 2l + 2w$$
$$180 = 2(w + 10) + 2w$$
$$180 = 2w + 20 + 2w$$
$$180 = 4w + 20$$
$$160 = 4w$$
$$40 = w$$

The pool is 40 feet wide and $40 + 10 = 50$ feet
long.

15. Since the bookcase is rectangular, we can
consider the height as the length in our formula.
Let w = the width of the bookcase in inches.
Then the height is $h = w + 32$.

$$P = 2(\text{length}) + 2(\text{width})$$
$$224 = 2(w + 32) + 2(w)$$
$$224 = 2w + 64 + 2w$$
$$224 = 4w + 64$$
$$160 = 4w$$
$$40 = w$$

The opening of the bookcase has a width of 40
inches and a height of $40 + 32 = 72$ inches.

16.
$$A = 2\pi r^2 + 2\pi rh$$
$$51.8 = 2\pi(1.5)^2 + 2\pi(1.5)h$$
$$51.8 - 2\pi(1.5)^2 = 2\pi(1.5)h$$
$$\frac{51.8 - 2\pi(1.5)^2}{2\pi(1.5)} = h$$

```
(51.8-2π(1.5)²)/
(2π(1.5))
         3.996150701
```

The height of the can is roughly 4.00 inches.

1.3 Exercises

17. $F = ma$

19. $V = \dfrac{4}{3}\pi r^3$

21. $d = rt$
$$\frac{d}{t} = \frac{rt}{t}$$
$$\frac{d}{t} = r$$

23.
$$y - y_1 = m(x - x_1)$$
$$\frac{y - y_1}{x - x_1} = \frac{m(x - x_1)}{(x - x_1)}$$
$$\frac{y - y_1}{x - x_1} = m$$

25.
$$Z = \frac{x - \mu}{\sigma}$$
$$\sigma \cdot Z = \sigma \cdot \frac{x - \mu}{\sigma}$$
$$\sigma Z = x - \mu$$
$$\sigma Z + \mu = x - \mu + \mu$$
$$\mu + \sigma Z = x$$

27.
$$F = G\frac{m_1 m_2}{r^2}$$
$$r^2 \cdot F = r^2 \cdot G\frac{m_1 m_2}{r^2}$$
$$r^2 F = Gm_1 m_2$$
$$\frac{r^2 F}{Gm_2} = \frac{Gm_1 m_2}{Gm_2}$$
$$\frac{r^2 F}{Gm_2} = m_1$$

29.
$$A = P + Prt$$
$$A = P(1 + rt)$$
$$\frac{A}{1 + rt} = \frac{P(1 + rt)}{1 + rt}$$
$$\frac{A}{1 + rt} = P$$

31.
$$C = \frac{5}{9}(F - 32)$$
$$\frac{9}{5} \cdot C = \frac{9}{5} \cdot \frac{5}{9}(F - 32)$$
$$\frac{9}{5}C = F - 32$$
$$\frac{9}{5}C + 32 = F - 32 + 32$$
$$\frac{9}{5}C + 32 = F$$

33.
$$2x + y = 13$$
$$2x + y - 2x = 13 - 2x$$
$$y = 13 - 2x \text{ or } y = -2x + 13$$

35.
$$9x - 3y = 15$$
$$9x - 3y - 9x = 15 - 9x$$
$$-3y = 15 - 9x$$
$$\frac{-3y}{-3} = \frac{15 - 9x}{-3}$$
$$y = -5 + 3x \quad \text{or} \quad y = 3x - 5$$

37.
$$4x + 3y = 13$$
$$4x + 3y - 4x = 13 - 4x$$
$$3y = 13 - 4x$$
$$\frac{3y}{3} = \frac{13 - 4x}{3}$$
$$y = \frac{13}{3} - \frac{4}{3}x \quad \text{or} \quad y = -\frac{4}{3}x + \frac{13}{3}$$

39.
$$\frac{1}{2}x + \frac{1}{6}y = 2$$
$$6\left(\frac{1}{2}x + \frac{1}{6}y\right) = 6(2)$$
$$3x + y = 12$$
$$3x + y - 3x = 12 - 3x$$
$$y = 12 - 3x \quad \text{or} \quad y = -3x + 12$$

41. a.
$$V = \pi r^2 h$$
$$\frac{V}{\pi r^2} = \frac{\pi r^2 h}{\pi r^2}$$
$$\frac{V}{\pi r^2} = h$$

    b. $h = \dfrac{V}{\pi r^2} = \dfrac{32\pi}{\pi(2)^2} = \dfrac{32}{4} = 8$ inches

    The height of the cylinder is 8 inches.

43. a.
$$M = -0.711A + 206.3$$
$$M - 206.3 = -0.711A + 206.3 - 206.3$$
$$M - 206.3 = -0.711A$$
$$\frac{M - 206.3}{-0.711} = \frac{-0.711A}{-0.711}$$
$$\frac{206.3 - M}{0.711} = A$$

    b. $A = \dfrac{206.3 - M}{0.711}$

$$= \frac{206.3 - 160}{0.711}$$
$$= \frac{46.3}{0.711} \approx 65.12$$

An individual whose maximum heart rate is 160 should be about 65 years old.

45. a.
$$A = P(1 + r)^t$$
$$\frac{A}{(1 + r)^t} = \frac{P(1 + r)^t}{(1 + r)^t}$$
$$\frac{A}{(1 + r)^t} = P$$

    b.

$$P = \frac{A}{(1 + r)^t} = \frac{5000}{(1 + 0.04)^5} = \frac{5000}{1.04^5} \approx 4109.64$$

Roughly \$4109.64 should be deposited today to have \$5000 in 5 years in an account that pays 4% annual interest.

47. Let x = measure of smaller angle (degrees) .
Larger angle: $x + 30$
$$x + (x + 30) = 180$$
$$x + x + 30 = 180$$
$$2x + 30 = 180$$
$$2x + 30 - 30 = 180 - 30$$
$$2x = 150$$
$$\frac{2x}{2} = \frac{150}{2}$$
$$x = 75$$
The smaller angle measures $75°$ and its supplement measures $105°$.

49. Let y = measure of the smaller angle (degrees).
Larger angle: $10 + 3y$
$$y + (10 + 3y) = 90$$
$$y + 10 + 3y = 90$$
$$4y + 10 = 90$$
$$4y + 10 - 10 = 90 - 10$$
$$4y = 80$$
$$\frac{4y}{4} = \frac{80}{4}$$
$$y = 20$$
The smaller angle measures $20°$ and its complement measures $70°$.

51. Let l = length of the window (feet).
width: $w = l + 3$
$$P = 2l + 2w$$
$$26 = 2l + 2(l + 3)$$
$$26 = 2l + 2l + 6$$
$$26 = 4l + 6$$
$$26 - 6 = 4l + 6 - 6$$
$$20 = 4l$$
$$\frac{20}{4} = \frac{4l}{4}$$
$$5 = l$$

The window is 5 feet long and 8 feet wide.

53. We can start by finding the length, s, of one side of the square given that $P = 40$.
$$\frac{P}{4} = s$$
$$\frac{40}{4} = s$$
$$10 \text{ inches} = s$$

The circle with the largest area will have a diameter equal to the length of one side of the square. Since the diameter is twice the radius, r, of the circle, we have
$$2r = 10 \quad \text{or} \quad r = 5$$
The radius of the desired circle is 5 inches.
$$A = \pi r^2 = \pi(5)^2 = 25\pi \approx 78.54 \text{ in.}^2$$

The area of the circle is 25π square inches (roughly 78.54 square inches).

55. Let x = measure of the first angle (degrees).
Second angle: $x + 15$
Third angle: $x + 45$
$$x + (x + 15) + (x + 45) = 180$$
$$x + x + 15 + x + 45 = 180$$
$$3x + 60 = 180$$
$$3x + 60 - 60 = 180 - 60$$
$$3x = 120$$
$$\frac{3x}{3} = \frac{120}{3}$$
$$x = 40$$

The first angle measures $40°$, the second measures $55°$, and the third measures $85°$.

57. a. Let w = width of patio in feet.
Length: $l = w + 5$
$$P = 2l + 2w$$
$$80 = 2(w + 5) + 2w$$
$$80 = 2w + 10 + 2w$$
$$80 = 4w + 10$$
$$80 - 10 = 4w + 10 - 10$$
$$70 = 4w$$
$$\frac{70}{4} = \frac{4w}{4}$$
$$17.5 = w$$

The patio is 17.5 ft wide and 22.5 ft long.

b. $h = 4 \; \cancel{\text{in}} \cdot \dfrac{1 \text{ ft}}{12 \; \cancel{\text{in}}} = \dfrac{1}{3} \text{ ft}$

$$V = l \cdot w \cdot h$$
$$V = (22.5)(17.5)\left(\frac{1}{3}\right)$$
$$V = 131.25 \text{ cubic feet}$$

You would need to purchase 131.25 cubic feet of cement.

59. a. Notice that the edge of the deck forms a large circle and the pool forms a smaller circle that sits inside the larger circle. We can subtract the area covered by the pool from the area of the larger circle to obtain the area of the deck. Since the diameter of the pool is 25 ft, the radius must be 12.5 ft.
Area of pool:
$$A = \pi r^2 = \pi(12.5)^2 = 156.25\pi \text{ ft}^2$$
Area of large circle:
$$A = \pi(12.5 + 3)^2$$
$$= \pi(15.5)^2$$
$$= 240.25\pi \text{ ft}^2$$
Area of deck:
$$A = 240.25\pi - 156.25\pi$$
$$= 84\pi \approx 263.89 \text{ ft}^2$$
The deck has an area of 84π square feet (roughly 264 square feet).

b. The fence would go on the perimeter of the deck so we need the circumference of the large circle.
$$C = 2\pi r = 2\pi(15.5) = 31\pi \text{ feet} \approx 97.39 \text{ feet}$$

It would require roughly 97.39 feet of fencing to encircle the pool.

c. $31\pi(25) = 775\pi \approx 2434.73$

The fence would cost about $2434.73.

61. No, the area would increase by a factor of 4.
Let r = original radius and $R = 2r$ be the new
radius.

$$\pi R^2 = \pi(2r)^2 = \pi \cdot 4r^2 = 4 \cdot \pi r^2$$

If the length of a side of a cube is doubled, the
volume increases by a factor of 8.
Let s = original length of a side and $S = 2s$ be
the new length.

$$V = S^3 = (2s)^3 = 8s^3$$

Section 1.4

Are You Ready for This Section?

R1. $3 < 6$ because 6 is further to the right on a real
number line.

R2. $-3 > -6$ because -3 is further to the right on a
real number line.

R3. $\dfrac{1}{2} = 0.5$ because both numbers are at the same

point on a real number line.

R4. $\dfrac{2}{3} = \dfrac{10}{15} > \dfrac{9}{15} = \dfrac{3}{5}$ because $\dfrac{2}{3}$ is further to the

right on a real number line.

R5. False; strict inequalities are either < or >.

R6. $\{x \mid x \text{ is a digit that is divisible by } 3\}$

Section 1.4 Quick Checks

1. A <u>closed interval</u>, denoted $[a, b]$ consists of all
real numbers x for which $a \le x \le b$.

2. In the interval (a, b), a is called the <u>left endpoint</u>
and b is called the <u>right endpoint</u> of the interval.

3. False; the interval notation is $(-\infty, -11]$.

4. $-3 \le x \le 2$
Interval: $[-3, 2]$
Graph:
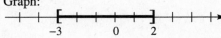

5. $3 \le x < 6$
Interval: $[3, 6)$
Graph:
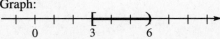

6. $x \le 3$
Interval: $(-\infty, 3]$
Graph:

7. $\dfrac{1}{2} < x < \dfrac{7}{2}$

Interval: $\left(\dfrac{1}{2}, \dfrac{7}{2}\right)$

Graph:

8. $(0, 5]$
Inequality: $0 < x \le 5$
Graph:

9. $(-6, 0)$
Inequality: $-6 < x < 0$
Graph:

10. $(5, \infty)$
Inequality: $x > 5$
Graph:
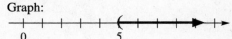

11. $\left(-\infty, \dfrac{8}{3}\right]$

Inequality: $x \le \dfrac{8}{3}$

Graph:
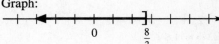

12. $\quad 4 < 7$
$\quad 4 + 5 < 7 + 5$
$\quad\quad 9 < 12$
Addition Property of Inequalities

13. $\quad x + 3 > -6$
$\quad x + 3 - 3 > -6 - 3$
$\quad\quad\quad x > -9$
Addition Property of Inequalities

14.
$$2 > 8$$
$$\tfrac{1}{2} \cdot 2 > \tfrac{1}{2} \cdot 8$$
$$1 > 4$$
Multiplication Property of Inequalities

15. $-6 < 9$
$$\frac{-6}{-3} > \frac{9}{-3}$$
$$2 > -3$$
Multiplication Property of Inequalities

16. $5x < 30$
$$\frac{5x}{5} < \frac{30}{5}$$
$$x < 6$$
Multiplication Property of Inequalities

17. $x + 3 > 5$
$$x + 3 - 3 > 5 - 3$$
$$x > 2$$
Interval: $(2, \infty)$
Set-builder: $\{x \mid x > 2\}$
Graph:
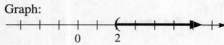

18. $\dfrac{1}{3}x \le 2$
$$3 \cdot \frac{1}{3}x \le 3 \cdot 2$$
$$x \le 6$$
Interval: $(-\infty, 6]$
Set-builder: $\{x \mid x \le 6\}$
Graph:

19. $-2x + 1 \le 13$
$$-2x + 1 - 1 \le 13 - 1$$
$$-2x \le 12$$
$$\frac{-2x}{-2} \ge \frac{12}{-2}$$
$$x \ge -6$$
Interval: $[-6, \infty)$
Set-builder: $\{x \mid x \ge -6\}$
Graph:
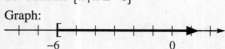

20. $3x + 1 > x - 5$
$$3x + 1 - 1 > x - 5 - 1$$
$$3x > x - 6$$
$$3x - x > x - 6 - x$$
$$2x > -6$$
$$\frac{2x}{2} > \frac{-6}{2}$$
$$x > -3$$
Interval: $(-3, \infty)$
Set-builder: $\{x \mid x > -3\}$
Graph:

21. $-2x + 1 \le 3x + 11$
$$-2x + 1 - 1 \le 3x + 11 - 1$$
$$-2x \le 3x + 10$$
$$-2x - 3x \le 3x + 10 - 3x$$
$$-5x \le 10$$
$$\frac{-5x}{-5} \ge \frac{10}{-5}$$
$$x \ge -2$$
Interval: $[-2, \infty)$
Set-builder: $\{x \mid x \ge -2\}$
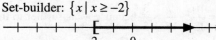

22. $-5x + 12 < x - 3$
$$-5x + 12 - 12 < x - 3 - 12$$
$$-5x < x - 15$$
$$-5x - x < x - 15 - x$$
$$-6x < -15$$
$$\frac{-6x}{-6} > \frac{-15}{-6}$$
$$x > \frac{5}{2}$$
Interval: $\left(\dfrac{5}{2}, \infty\right)$

Set-builder: $\left\{x \mid x > \dfrac{5}{2}\right\}$

23. $4(x-2) < 3x-4$
$$4x-8 < 3x-4$$
$$4x-8+8 < 3x-4+8$$
$$4x < 3x+4$$
$$4x-3x < 3x+4-3x$$
$$x < 4$$
Interval: $(-\infty, 4)$
Set-builder: $\{x \mid x < 4\}$

24. $-2(x+1) \geq 4(x+3)$
$$-2x-2 \geq 4x+12$$
$$-2x-2+2 \geq 4x+12+2$$
$$-2x \geq 4x+14$$
$$-2x-4x \geq 4x+14-4x$$
$$-6x \geq 14$$
$$\frac{-6x}{-6} \leq \frac{14}{-6}$$
$$x \leq -\frac{7}{3}$$
Interval: $\left(-\infty, -\frac{7}{3}\right]$

Set-builder: $\left\{x \mid x \leq -\frac{7}{3}\right\}$

25. $7-2(x+1) \leq 3(x-5)$
$$7-2x-2 \leq 3x-15$$
$$-2x+5 \leq 3x-15$$
$$-2x+5-5 \leq 3x-15-5$$
$$-2x \leq 3x-20$$
$$-2x-3x \leq 3x-20-3x$$
$$-5x \leq -20$$
$$\frac{-5x}{-5} \geq \frac{-20}{-5}$$
$$x \geq 4$$
Interval: $[4, \infty)$
Set-builder: $\{x \mid x \geq 4\}$

26. $\dfrac{3x+1}{5} \geq 2$
$$5 \cdot \frac{3x+1}{5} \geq 5 \cdot 2$$
$$3x+1 \geq 10$$
$$3x+1-1 \geq 10-1$$
$$3x \geq 9$$
$$\frac{3x}{3} \geq \frac{9}{3}$$
$$x \geq 3$$
Interval: $[3, \infty)$
Set-builder: $\{x \mid x \geq 3\}$

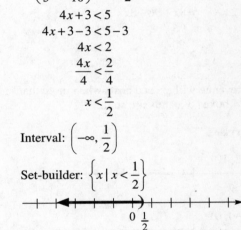

27. $\dfrac{2}{5}x + \dfrac{3}{10} < \dfrac{1}{2}$
$$10\left(\frac{2}{5}x + \frac{3}{10}\right) < 10 \cdot \frac{1}{2}$$
$$4x+3 < 5$$
$$4x+3-3 < 5-3$$
$$4x < 2$$
$$\frac{4x}{4} < \frac{2}{4}$$
$$x < \frac{1}{2}$$
Interval: $\left(-\infty, \frac{1}{2}\right)$

Set-builder: $\left\{x \mid x < \frac{1}{2}\right\}$

28. $\dfrac{1}{2}(x+3) > \dfrac{1}{3}(x-4)$
$$6 \cdot \frac{1}{2}(x+3) > 6 \cdot \frac{1}{3}(x-4)$$
$$3(x+3) > 2(x-4)$$
$$3x+9 > 2x-8$$
$$3x+9-9 > 2x-8-9$$
$$3x > 2x-17$$
$$3x-2x > 2x-17-2x$$
$$x > -17$$
Interval: $(-17, \infty)$
Set-builder: $\{x \mid x > -17\}$

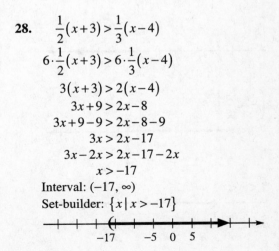

29. Let b = annual balance in dollars.
The annual cost is the sum of the annual fee and the interest charge. To find the interest charge, we use the simple interest formula $I = Prt$ and let $t = 1$ year.

$$\text{Bank A cost} < \text{Bank B cost}$$
$$\text{fee}_A + \text{interest}_A < \text{fee}_B + \text{interest}_B$$
$$24 + 0.0995b < 0 + 0.1495b$$
$$24 + 0.0995b < 0.1495b$$
$$24 + 0.0995b - 0.0995b < 0.1495b - 0.0995b$$
$$24 < 0.05b$$
$$\frac{24}{0.05} < \frac{0.05b}{0.05}$$
$$480 < b \quad \text{or} \quad b > 480$$

The card from Bank A will cost less if the annual balance is more than \$480.

30. Let x = the number of boxes of candy.
$$R > C$$
$$12x > 8x + 96$$
$$12x - 8x > 8x + 96 - 8x$$
$$4x > 96$$
$$\frac{4x}{4} > \frac{96}{4}$$
$$x > 24$$

Revenue will exceed costs when more than 24 boxes of candy are sold.

1.4 Exercises

31. $[2, 10]$

32. $[-4, 0)$

35. $[6, \infty)$

37. $\left(-\infty, \dfrac{3}{2}\right)$

39. $1 < x < 8$

41. $-5 < x \le 1$

43. $x < 5$

45. $x \ge 3$

47. $<$; Addition Property of Inequalities

49. $>$; Multiplication Property of Inequalities

51. $\le$; Addition Property of Inequalities

53. $\le$; Multiplication Property of Inequalities

55. $$x - 4 \le 2$$
$$x - 4 + 4 \le 2 + 4$$
$$x \le 6$$
$$\{x \mid x \le 6\}; \ (-\infty, 6]$$

57. $$6x < 24$$
$$\frac{6x}{6} < \frac{24}{6}$$
$$x < 4$$
$$\{x \mid x < 4\}; \ (-\infty, 4)$$

59. $$-7x < 21$$
$$\frac{-7x}{-7} > \frac{21}{-7}$$
$$x > -3$$
$$\{x \mid x > -3\}; \ (-3, \infty)$$

61. $$\frac{4}{15}x > \frac{8}{5}$$
$$\frac{15}{4} \cdot \frac{4}{15}x > \frac{15}{4} \cdot \frac{8}{5}$$
$$x > 6$$
$$\{x \mid x > 6\}; \ (6, \infty)$$

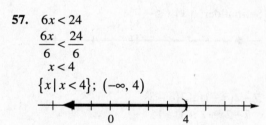

63.
$$3x + 2 > 11$$
$$3x + 2 - 2 > 11 - 2$$
$$3x > 9$$
$$\frac{3x}{3} > \frac{9}{3}$$
$$x > 3$$
$$\{x \mid x > 3\}; \ (3, \infty)$$

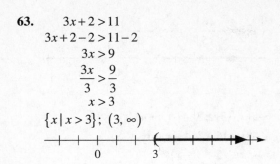

65.
$$-3x + 1 > 13$$
$$-3x + 1 - 1 > 13 - 1$$
$$-3x > 12$$
$$\frac{-3x}{-3} < \frac{12}{-3}$$
$$x < -4$$
$$\{x \mid x < -4\}; \ (-\infty, -4)$$

67.
$$6x + 5 \le 3x + 2$$
$$6x + 5 - 5 \le 3x + 2 - 5$$
$$6x \le 3x - 3$$
$$6x - 3x \le 3x - 3 - 3x$$
$$3x \le -3$$
$$\frac{3x}{3} \le \frac{-3}{3}$$
$$x \le -1$$
$$\{x \mid x \le -1\}; \ (-\infty, -1]$$

69.
$$-3x + 1 < 2x + 11$$
$$-3x + 1 - 1 < 2x + 11 - 1$$
$$-3x < 2x + 10$$
$$-3x - 2x < 2x + 10 - 2x$$
$$-5x < 10$$
$$\frac{-5x}{-5} > \frac{10}{-5}$$
$$x > -2$$
$$\{x \mid x > -2\}; \ (-2, \infty)$$

71.
$$3(x - 3) < 2(x + 4)$$
$$3x - 9 < 2x + 8$$
$$3x - 9 + 9 < 2x + 8 + 9$$
$$3x < 2x + 17$$
$$3x - 2x < 2x + 17 - 2x$$
$$x < 17$$
$$\{x \mid x < 17\}; \ (-\infty, 17)$$

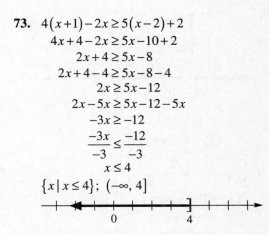

73.
$$4(x + 1) - 2x \ge 5(x - 2) + 2$$
$$4x + 4 - 2x \ge 5x - 10 + 2$$
$$2x + 4 \ge 5x - 8$$
$$2x + 4 - 4 \ge 5x - 8 - 4$$
$$2x \ge 5x - 12$$
$$2x - 5x \ge 5x - 12 - 5x$$
$$-3x \ge -12$$
$$\frac{-3x}{-3} \le \frac{-12}{-3}$$
$$x \le 4$$
$$\{x \mid x \le 4\}; \ (-\infty, 4]$$

75.
$$0.5x + 4 \le 0.2x - 5$$
$$0.5x + 4 - 4 \le 0.2x - 5 - 4$$
$$0.5x \le 0.2x - 9$$
$$0.5x - 0.2x \le 0.2x - 9 - 0.2x$$
$$0.3x \le -9$$
$$\frac{0.3x}{0.3} \le \frac{-9}{0.3}$$
$$x \le -30$$
$$\{x \mid x \le -30\}; \ (-\infty, -30]$$

77.
$$\frac{3x+1}{4} < \frac{1}{2}$$
$$4\left(\frac{3x+1}{4}\right) < 4\left(\frac{1}{2}\right)$$
$$3x+1 < 2$$
$$3x+1-1 < 2-1$$
$$3x < 1$$
$$\frac{3x}{3} < \frac{1}{3}$$
$$x < \frac{1}{3}$$
$$\left\{x \mid x < \frac{1}{3}\right\}; \ \left(-\infty, \frac{1}{3}\right)$$

79.
$$\frac{1}{2}(x-4) > \frac{3}{4}(2x+1)$$
$$4 \cdot \frac{1}{2}(x-4) > 4 \cdot \frac{3}{4}(2x+1)$$
$$2(x-4) > 3(2x+1)$$
$$2x-8 > 6x+3$$
$$2x > 6x+11$$
$$-4x > 11$$
$$\frac{-4x}{-4} < \frac{11}{-4}$$
$$x < -\frac{11}{4}$$
$$\left\{x \mid x < -\frac{11}{4}\right\}; \ \left(-\infty, -\frac{11}{4}\right)$$

81.
$$\frac{3}{5} - x > \frac{5}{3}$$
$$\frac{3}{5} - x - \frac{3}{5} > \frac{5}{3} - \frac{3}{5}$$
$$-x > \frac{25}{15} - \frac{9}{15}$$
$$-x > \frac{16}{15}$$
$$-1(-x) < -1\left(\frac{16}{15}\right)$$
$$x < -\frac{16}{15}$$

$$\left\{x \mid x < -\frac{16}{15}\right\}; \ \left(-\infty, -\frac{16}{15}\right)$$

83.
$$-5(x-3) \geq 3\big[4-(x+4)\big]$$
$$-5x+15 \geq 3\big[4-x-4\big]$$
$$-5x+15 \geq 3(-x)$$
$$-5x+15 \geq -3x$$
$$-5x \geq -3x-15$$
$$-2x \geq -15$$
$$\frac{-2x}{-2} \leq \frac{-15}{-2}$$
$$x \leq \frac{15}{2}$$
$$\left\{x \mid x \leq \frac{15}{2}\right\}; \ \left(-\infty, \frac{15}{2}\right]$$

85.
$$4(3x-1)-5(x+4) \geq 3\big[2-(x+3)\big]-6x$$
$$12x-4-5x-20 \geq 3\big[2-x-3\big]-6x$$
$$7x-24 \geq 3\big[-1-x\big]-6x$$
$$7x-24 \geq -3-3x-6x$$
$$7x-24 \geq -3-9x$$
$$7x \geq 21-9x$$
$$16x \geq 21$$
$$\frac{16x}{16} \geq \frac{21}{16}$$
$$x \geq \frac{21}{16}$$
$$\left\{x \mid x \geq \frac{21}{16}\right\}; \ \left[\frac{21}{16}, \infty\right)$$

87. $\dfrac{2}{3}(4x-1)-\dfrac{4}{9}(x-4)>\dfrac{5}{12}(2x+3)$

$36\cdot\dfrac{2}{3}(4x-1)-36\cdot\dfrac{4}{9}(x-4)>36\cdot\dfrac{5}{12}(2x+3)$

$24(4x-1)-16(x-4)>15(2x+3)$

$96x-24-16x+64>30x+45$

$80x+40>30x+45$

$80x>30x+5$

$50x>5$

$\dfrac{50x}{50}>\dfrac{5}{50}$

$x>\dfrac{1}{10}$

$\left\{x\,\middle|\,x>\dfrac{1}{10}\right\};\ \left(\dfrac{1}{10},\infty\right)$

89. $\dfrac{4x-3}{3}<3$

$3\cdot\dfrac{4x-3}{3}<3\cdot3$

$4x-3<9$

$4x<12$

$\dfrac{4x}{4}<\dfrac{12}{4}$

$x<3$

$\{x\mid x<3\};\ (-\infty,3)$

91. $\dfrac{2}{3}x<\dfrac{1}{4}(2x+3)$

$12\cdot\dfrac{2}{3}x<12\cdot\dfrac{1}{4}(2x+3)$

$8x<3(2x+3)$

$8x<6x+9$

$2x<9$

$\dfrac{2x}{2}<\dfrac{9}{2}$

$x<\dfrac{9}{2}$

$\left\{x\,\middle|\,x<\dfrac{9}{2}\right\};\ \left(-\infty,\dfrac{9}{2}\right)$

93. $y+8>-7$

$y+8-8>-7-8$

$y>-15$

$\{y\mid y>-15\};\ (-15,\infty)$

95. $-3a\le-21$

$\dfrac{-3a}{-3}\ge\dfrac{-21}{-3}$

$a\ge7$

$\{a\mid a\ge7\};\ [7,\infty)$

97. $13x-5>10x-6$

$13x-5+5>10x-6+5$

$13x>10x-1$

$13x-10x>10x-1-10x$

$3x>-1$

$\dfrac{3x}{3}>\dfrac{-1}{3}$

$x>-\dfrac{1}{3}$

$\left\{x\,\middle|\,x>-\dfrac{1}{3}\right\};\ \left(-\dfrac{1}{3},\infty\right)$

99. $4(x+2)\le3(x-2)$

$4x+8\le3x-6$

$4x+8-8\le3x-6-8$

$4x\le3x-14$

$4x-3x\le3x-14-3x$

$x\le-14$

$\{x\mid x\le-14\};\ (-\infty,-14]$

101.
$$3(4-3x) > 6-5x$$
$$12-9x > 6-5x$$
$$12-9x-12 > 6-5x-12$$
$$-9x > -5x-6$$
$$-9x+5x > -5x-6+5x$$
$$-4x > -6$$
$$\frac{-4x}{-4} < \frac{-6}{-4}$$
$$x < \frac{3}{2}$$
$$\left\{x \mid x < \frac{3}{2}\right\};\ \left(-\infty, \frac{3}{2}\right)$$

103.
$$2\left[4-3(x+1)\right] \le -4x+8$$
$$2\left[4-3x-3\right] \le -4x+8$$
$$2\left[1-3x\right] \le -4x+8$$
$$2-6x \le -4x+8$$
$$-6x \le -4x+6$$
$$-2x \le 6$$
$$\frac{-2x}{-2} \ge \frac{6}{-2}$$
$$x \ge -3$$
$$\{x \mid x \ge -3\};\ [-3, \infty)$$

105.
$$\frac{x}{2}+\frac{3}{4} \ge \frac{3}{8}$$
$$8\cdot\frac{x}{2}+8\cdot\frac{3}{4} \ge 8\cdot\frac{3}{8}$$
$$4x+6 \ge 3$$
$$4x \ge -3$$
$$\frac{4x}{4} \ge \frac{-3}{4}$$
$$x \ge -\frac{3}{4}$$
$$\left\{x \mid x \ge -\frac{3}{4}\right\};\ \left[-\frac{3}{4}, \infty\right)$$

107.
$$2x+5 \ge 13$$
$$2x+5-5 \ge 13-5$$
$$2x \ge 8$$
$$\frac{2x}{2} \ge \frac{8}{2}$$
$$x \ge 4$$
$$\{x \mid x \ge 4\};\ [4, \infty)$$

109.
$$4z-3 \le 9$$
$$4z-3+3 \le 9+3$$
$$4z \le 12$$
$$\frac{4z}{4} \le \frac{12}{4}$$
$$z \le 3$$
$$\{z \mid z \le 3\};\ (-\infty, 3]$$

111. Let x = score on final exam.
$$90+83+95+90+x \ge 540$$
$$358+x \ge 540$$
$$358+x-358 \ge 540-358$$
$$x \ge 182$$
Jackie must earn at least 182 points on the final exam to earn an A in Mr. Ruffatto's class.

113. Let x = number of hamburgers.
$$19+21+9x \le 67$$
$$40+9x \le 67$$
$$40+9x-40 \le 67-40$$
$$9x \le 27$$
$$\frac{9x}{9} \le \frac{27}{9}$$
You can order no more than 3 hamburgers to keep the fat content to no more than 67 grams.

115. Let x = cargo weight.
$$150(179)+x \le 45,686$$
$$26,850+x \le 45,686$$
$$26,850+x-26,850 \le 45,686-26,850$$
$$x \le 18,836$$
The plane can carry up to 18,836 pounds of luggage and cargo.

117.
$$19.25t+585.72 > 1000$$
$$19.25t+585.72-585.72 > 1000-585.72$$
$$19.25t > 414.28$$
$$\frac{19.25t}{19.25} > \frac{414.28}{19.25}$$
$$t > \frac{414.28}{19.25} \approx 21.52$$
The monthly benefit will exceed $1000 in 2012.

119. Let x = value of computer systems sold.

$$34,000 + 0.012x \geq 100,000$$
$$34,000 + 0.012x - 34,000 \geq 100,000 - 34,000$$
$$0.012x \geq 66,000$$
$$\frac{0.012x}{0.012} \geq \frac{66,000}{0.012}$$
$$x \geq \frac{66,000}{0.012} = 5,500,000$$

Susan will need to sell at least $5,500,000$ in computer systems to earn $100,000$.

121.
$$S > D$$
$$-200 + 10p > 1000 - 20p$$
$$-200 + 10p - 1000 > 1000 - 20p - 1000$$
$$-1200 + 10p > -20p$$
$$-1200 + 10p - 10p > -20p - 10p$$
$$-1200 > -30p$$
$$\frac{-1200}{-30} < \frac{-30p}{-30}$$
$$40 < p$$

Supply will exceed demand when the price is greater than 40.

123.
$$3(x+2) + 2x > 5(x+1)$$
$$3x + 6 + 2x > 5x + 5$$
$$5x + 6 > 5x + 5$$
$$5x + 6 - 6 > 5x + 5 - 6$$
$$5x > 5x - 1$$
$$5x - 5x > 5x - 1 - 5x$$
$$0 > -1$$

This statement is an identity. All real numbers are solutions.

125. Answers may vary. The direction, or sense, of an inequality changes if you multiply or divide both sides by a negative number, or if you interchange both sides of the inequality.

127. Answers will vary. In the inequality $4 < x > 7$, we are stating that x is greater than 4 *and* x is greater than 7. Any number that is greater than 7 must be greater than 4. A similar statement can be made if the inequalities in the example are reversed. When inequalities are mixed in such a fashion, only one is truly necessary.

Putting the Concepts Together (Sections 1.1–1.4)

1. a. Let $x = -3$ in the equation.
$$5(2(-3) - 3) + 1 \stackrel{?}{=} 2(-3) - 6$$
$$5(-6 - 3) + 1 \stackrel{?}{=} -6 - 6$$
$$5(-9) + 1 \stackrel{?}{=} -12$$
$$-45 + 1 \stackrel{?}{=} -12$$
$$-44 \neq -12$$

$x = -3$ is **not** a solution to the equation.

b. Let $x = 1$ in the equation.
$$5(2(1) - 3) + 1 \stackrel{?}{=} 2(1) - 6$$
$$5(2 - 3) + 1 \stackrel{?}{=} 2 - 6$$
$$5(-1) + 1 \stackrel{?}{=} -4$$
$$-5 + 1 \stackrel{?}{=} -4$$
$$-4 = -4$$

$x = 1$ is a solution to the equation.

2.
$$3(2x - 1) + 6 = 5x - 2$$
$$6x - 3 + 6 = 5x - 2$$
$$6x + 3 = 5x - 2$$
$$6x + 3 - 5x = 5x - 2 - 5x$$
$$x + 3 = -2$$
$$x + 3 - 3 = -2 - 3$$
$$x = -5$$

Solution set: $\{-5\}$

3.
$$\frac{7}{3}x + \frac{4}{5} = \frac{5x + 12}{15}$$
$$15\left(\frac{7}{3}x + \frac{4}{5}\right) = 15\left(\frac{5x + 12}{15}\right)$$
$$35x + 12 = 5x + 12$$
$$35x + 12 - 12 = 5x + 12 - 12$$
$$35x = 5x$$
$$35x - 5x = 5x - 5x$$
$$30x = 0$$
$$\frac{30x}{30} = \frac{0}{30}$$
$$x = 0$$

Solution set: $\{0\}$

4. $5 - 2(x+1) + 4x = 6(x+1) - (3+4x)$
$$5 - 2x - 2 + 4x = 6x + 6 - 3 - 4x$$
$$3 + 2x = 2x + 3$$

These expressions are equivalent, so the statement is an identity.

5. Let x represent the number. $x - 3 = \frac{1}{2}x + 2$

6. Let x represent the number.

$$\frac{x}{2} < x + 5$$

7. Let x = liters of 20% solution.
Amount of 40% solution: $16 - x$ liters
    tot. acid = acid from 20% + acid from 40%
$$(\%)(\text{liters}) = (\%)(\text{liters}) + (\%)(\text{liters})$$
$$0.35(16) = 0.2(x) + 0.4(16 - x)$$
$$5.6 = 0.2x + 6.4 - 0.4x$$
$$5.6 = 6.4 - 0.2x$$
$$5.6 - 6.4 = 6.4 - 0.2x - 6.4$$
$$-0.8 = -0.2x$$
$$\frac{-0.8}{-0.2} = \frac{-0.2x}{-0.2}$$
$$4 = x$$
The chemist needs to mix 4 liters of the 20% solution with 12 liters of the 40% solution.

8. Let t = time of travel in hours.
$d = rt$
total distance = dist. for car 1 + dist. for car 2
$$255 = 30(t) + 45(t)$$
$$255 = 75t$$
$$\frac{255}{75} = \frac{75t}{75}$$
$$3.4 = t$$
After 3.4 hours, the two cars will be 255 miles apart.

9.
$$3x - 2y = 4$$
$$3x - 2y - 3x = 4 - 3x$$
$$-2y = -3x + 4$$
$$\frac{-2y}{-2} = \frac{-3x + 4}{-2}$$
$$y = \frac{-3x}{-2} + \frac{4}{-2}$$
$$y = \frac{3}{2}x - 2$$

10.
$$A = P + Prt$$
$$A - P = P + Prt - P$$
$$A - P = Prt$$
$$\frac{A - P}{Pt} = \frac{Prt}{Pt}$$
$$\frac{A - P}{Pt} = r \quad \text{or} \quad r = \frac{A - P}{Pt}$$

11. a.
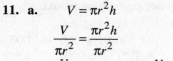
$$V = \pi r^2 h$$
$$\frac{V}{\pi r^2} = \frac{\pi r^2 h}{\pi r^2}$$
$$\frac{V}{\pi r^2} = h \quad \text{or} \quad h = \frac{V}{\pi r^2}$$

b. $h = \dfrac{V}{\pi r^2} = \dfrac{294\pi}{\pi(7)^2} = \dfrac{294}{49} = 6$ inches

12. a. Interval: $(-3, \infty)$
Graph:
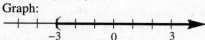

b. Interval: $(2, 5]$
Graph:
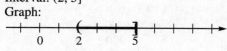

13. a. Inequality: $x \le -1.5$
Graph:
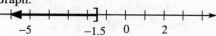

b. Inequality: $-3 < x \le 1$
Graph:
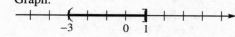

14.
$$2x + 3 \le 4x - 9$$
$$2x + 3 - 3 \le 4x - 9 - 3$$
$$2x \le 4x - 12$$
$$2x - 4x \le 4x - 12 - 4x$$
$$-2x \le -12$$
$$\frac{-2x}{-2} \ge \frac{-12}{-2}$$
$$x \ge 6$$
Interval: $[6, \infty)$
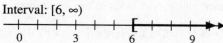

15.
$$-3 > 3x - (x + 5)$$
$$-3 > 3x - x - 5$$
$$-3 > 2x - 5$$
$$-3 + 5 > 2x - 5 + 5$$
$$2 > 2x$$
$$\frac{2}{2} > \frac{2x}{2}$$
$$1 > x \quad \text{or} \quad x < 1$$
Interval: $(-\infty, 1)$
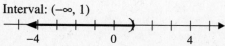

16.
$$x - 9 \le x + 3(2 - x)$$
$$x - 9 \le x + 6 - 3x$$
$$x - 9 \le -2x + 6$$
$$x - 9 + 2x \le -2x + 6 + 2x$$
$$3x - 9 \le 6$$
$$3x - 9 + 9 \le 6 + 9$$
$$3x \le 15$$
$$\frac{3x}{3} \le \frac{15}{3}$$
$$x \le 5$$

Interval: $(-\infty, 5]$

17. Let x = number of children at the party.
$$75 + 5x \le 125$$
$$75 + 5x - 75 \le 125 - 75$$
$$5x \le 50$$
$$\frac{5x}{5} \le \frac{50}{5}$$
$$x \le 10$$

Including Logan, there can be at most 10 children at the party. Therefore, Logan can invite at most 9 children to the party.

Section 1.5

Are You Ready for This Section?

R1.

R2. $3x - 5(x + 2) = 4$

    a. Let $x = 0$ in the equation.
$$3(0) - 5(0 + 2) \stackrel{?}{=} 4$$
$$0 - 5(2) \stackrel{?}{=} 4$$
$$-10 = 4 \text{ False}$$
$x = 0$ is not a solution to the equation.

    b. Let $x = -3$ in the equation.
$$3(-3) - 5(-3 + 2) \stackrel{?}{=} 4$$
$$-9 - 5(-1) \stackrel{?}{=} 4$$
$$-9 + 5 \stackrel{?}{=} 4$$
$$-4 = 4 \text{ False}$$
$x = -3$ is not a solution to the equation.

    c. Let $x = -7$ in the equation.
$$3(-7) - 5(-7 + 2) \stackrel{?}{=} 4$$
$$-21 - 5(-5) \stackrel{?}{=} 4$$
$$-21 + 25 \stackrel{?}{=} 4$$
$$4 = 4 \text{ True}$$
$x = -7$ is a solution to the equation.

R3. **a.** Let $x = 0$:
$$2x^2 - 3x + 1 = 2(0)^2 - 3(0) + 1$$
$$= 0 - 0 + 1$$
$$= 1$$

    b. Let $x = 2$:
$$2x^2 - 3x + 1 = 2(2)^2 - 3(2) + 1$$
$$= 2(4) - 6 + 1$$
$$= 8 - 6 + 1$$
$$= 3$$

    c. Let $x = -3$:
$$2x^2 - 3x + 1 = 2(-3)^2 - 3(-3) + 1$$
$$= 2(9) + 9 + 1$$
$$= 18 + 9 + 1$$
$$= 28$$

R4.
$$3x + 2y = 8$$
$$3x + 2y - 3x = 8 - 3x$$
$$2y = 8 - 3x$$
$$\frac{2y}{2} = \frac{8 - 3x}{2}$$
$$y = \frac{8}{2} - \frac{3x}{2}$$
$$y = 4 - \frac{3}{2}x \quad \text{or} \quad y = -\frac{3}{2}x + 4$$

R5. $|-4| = 4$ because -4 is 4 units away from 0 on a real number line.

Section 1.5 Quick Checks

1. The point where the x-axis and the y-axis intersect in the Cartesian coordinate system is called the <u>origin</u>.

2. True

3. A : quadrant I
    B : quadrant IV
    C : y-axis
    D : quadrant III

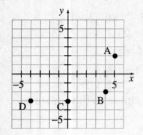

4. A: quadrant II
   B: x-axis
   C: quadrant IV
   D: quadrant I

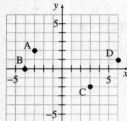

5. True

6. a. Let $x = 2$ and $y = -3$.
$$2(2) - 4(-3) \stackrel{?}{=} 12$$
$$4 + 12 \stackrel{?}{=} 12$$
$$16 = 12 \text{ False}$$
The point $(2, -3)$ is not on the graph.

b. Let $x = 2$ and $y = -2$.
$$2(2) - 4(-2) \stackrel{?}{=} 12$$
$$4 + 8 \stackrel{?}{=} 12$$
$$12 = 12 \text{ T}$$
The point $(2, -2)$ is on the graph.

c. Let $x = \dfrac{3}{2}$ and $y = -\dfrac{9}{4}$.
$$2\left(\frac{3}{2}\right) - 4\left(-\frac{9}{4}\right) \stackrel{?}{=} 12$$
$$3 + 9 \stackrel{?}{=} 12$$
$$12 = 12 \text{ T}$$
The point $\left(\dfrac{3}{2}, -\dfrac{9}{4}\right)$ is on the graph.

7. a. Let $x = 1$ and $y = 4$.
$$(4) \stackrel{?}{=} (1)^2 + 3$$
$$4 \stackrel{?}{=} 1 + 3$$
$$4 = 4 \text{ T}$$
The point $(1, 4)$ is on the graph.

b. Let $x = -2$ and $y = -1$.
$$(-1) \stackrel{?}{=} (-2)^2 + 3$$
$$-1 \stackrel{?}{=} 4 + 3$$
$$-1 = 7 \text{ False}$$
The point $(-2, -1)$ is not on the graph.

c. Let $x = -3$ and $y = 12$.
$$(12) \stackrel{?}{=} (-3)^2 + 3$$
$$12 \stackrel{?}{=} 9 + 3$$
$$12 = 12 \text{ T}$$
The point $(-3, 12)$ is on the graph.

8. $y = 3x + 1$

x	$y = 3x + 1$	(x, y)
-2	$y = 3(-2) + 1 = -5$	$(-2, -5)$
-1	$y = 3(-1) + 1 = -2$	$(-1, -2)$
0	$y = 3(0) + 1 = 1$	$(0, 1)$
1	$y = 3(1) + 1 = 4$	$(1, 4)$
2	$y = 3(2) + 1 = 7$	$(2, 7)$

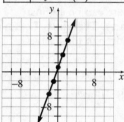

9. $2x + 3y = 8$
$$3y = -2x + 8$$
$$y = -\frac{2}{3}x + \frac{8}{3}$$

x	$y = -\frac{2}{3}x + \frac{8}{3}$	(x, y)
-5	$y = -\frac{2}{3}(-5) + \frac{8}{3} = 6$	$(-5, 6)$
-2	$y = -\frac{2}{3}(-2) + \frac{8}{3} = 4$	$(-2, 4)$
1	$y = -\frac{2}{3}(1) + \frac{8}{3} = 2$	$(1, 2)$
4	$y = -\frac{2}{3}(4) + \frac{8}{3} = 0$	$(4, 0)$
7	$y = -\frac{2}{3}(7) + \frac{8}{3} = -2$	$(7, -2)$

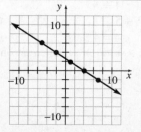

10. $y = x^2 + 3$

x	$y = x^2 + 3$	(x, y)
-2	$y = (-2)^2 + 3 = 7$	$(-2, 7)$
-1	$y = (-1)^2 + 3 = 4$	$(-1, 4)$
0	$y = (0)^2 + 3 = 3$	$(0, 3)$
1	$y = (1)^2 + 3 = 4$	$(1, 4)$
2	$y = (2)^2 + 3 = 7$	$(2, 7)$

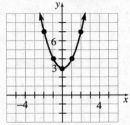

11. $x = y^2 + 2$

y	$x = y^2 + 2$	(x, y)
-2	$x = (-2)^2 + 2 = 6$	$(6, -2)$
-1	$x = (-1)^2 + 2 = 3$	$(3, -1)$
0	$x = (0)^2 + 2 = 2$	$(2, 0)$
1	$x = (1)^2 + 2 = 3$	$(3, 1)$
2	$x = (2)^2 + 2 = 6$	$(6, 2)$

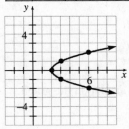

12. $x = (y - 1)^2$

y	$x = (y - 1)^2$	(x, y)
-1	$x = (-1 - 1)^2 = 4$	$(4, -1)$
0	$x = (0 - 1)^2 = 1$	$(1, 0)$
1	$x = (1 - 1)^2 = 0$	$(0, 1)$
2	$x = (2 - 1)^2 = 1$	$(1, 2)$
3	$x = (3 - 1)^2 = 4$	$(4, 3)$

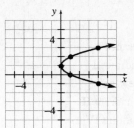

13. The points, if any, at which a graph crosses or touches a coordinate axis are called <u>intercepts</u>.

14. False

15. To find the intercepts, we look for the points where the graph crosses or touches either coordinate axis. From the graph, we see that the intercepts are $(-5, 0)$, $(0, -0.9)$, $(1, 0)$, and $(6.7, 0)$.

The x-intercepts are the points where the graph crosses or touches the x-axis. From the graph we see that the x-intercepts are $(-5, 0)$, $(1, 0)$, and $(6.7, 0)$.

The y-intercept is the point where the graph crosses the y-axis. From the graph we see that the y-intercept is $(0, -0.9)$.

16. a. Locate the value 250 along the x-axis, go up to the graph, and then read the corresponding value on the y-axis. According to the graph, the cost of refining 250 thousand gallons of gasoline per hour is $200 thousand.

    b. Locate the value 400 along the x-axis, go up to the graph, and then read the corresponding value on the y-axis. According to the graph, the cost of refining 400 thousand gallons of gasoline per hour is $350 thousand.

    c. Since the horizontal axis represents the number of gallons per hour that can be refined, the graph ending at 700 thousand gallons per hour represents the capacity of the refinery per hour.

    d. The intercept is $(0, 100)$. This represents the fixed costs of operating the refinery. That is, refining 0 gallons per hour costs $100 thousand.

1.5 Exercises

17. $A : (2, 3)$; in quadrant I

   $B : (-5, 2)$; in quadrant II

   $C : (0, -2)$; on the y-axis

   $D : (-4, -3)$; in quadrant III

   $E : (3, -4)$; in quadrant IV

   $F : (4, 0)$; on the x-axis

19. A: quadrant I; B: quadrant III; C: x-axis; D: quadrant IV; E: y-axis; F: quadrant II

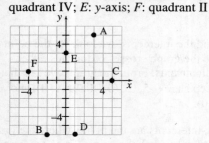

21. **a.** Let $x = 1$ and $y = 2$.

$$2(1) + 5(2) \overset{?}{=} 12$$
$$2 + 10 \overset{?}{=} 12$$
$$12 = 12 \text{ True}$$

The point $(1, 2)$ is on the graph.

   b. Let $x = -2$ and $y = 3$.

$$2(-2) + 5(3) \overset{?}{=} 12$$
$$-4 + 15 \overset{?}{=} 12$$
$$11 = 12 \text{ False}$$

The point $(-2, 3)$ is not on the graph.

   c. Let $x = -4$ and $y = 4$.

$$2(-4) + 5(4) \overset{?}{=} 12$$
$$-8 + 20 \overset{?}{=} 12$$
$$12 = 12 \text{ True}$$

The point $(-4, 4)$ is on the graph.

   d. Let $x = -\dfrac{3}{2}$ and $y = 3$.

$$2\left(-\frac{3}{2}\right) + 5(3) \overset{?}{=} 12$$
$$-3 + 15 \overset{?}{=} 12$$
$$12 = 12 \text{ True}$$

The point $\left(-\dfrac{3}{2}, 3\right)$ is on the graph.

23. **a.** Let $x = -2$ and $y = -15$.

$$(-15) \overset{?}{=} -2(-2)^2 + 3(-2) - 1$$
$$-15 \overset{?}{=} -2(4) - 6 - 1$$
$$-15 \overset{?}{=} -8 - 6 - 1$$
$$-15 = -15 \text{ True}$$

The point $(-2, -15)$ is on the graph.

   b. Let $x = 3$ and $y = 10$.

$$(10) \overset{?}{=} -2(3)^2 + 3(3) - 1$$
$$10 \overset{?}{=} -2(9) + 9 - 1$$
$$10 \overset{?}{=} -18 + 9 - 1$$
$$10 = -10 \text{ False}$$

The point $(3, 10)$ is not on the graph.

   c. Let $x = 0$ and $y = 1$.

$$(1) \overset{?}{=} -2(0)^2 + 3(0) - 1$$
$$1 \overset{?}{=} 0 + 0 - 1$$
$$1 = -1 \text{ False}$$

The point $(0, 1)$ is not on the graph.

   d. Let $x = 2$ and $y = -3$.

$$(-3) \overset{?}{=} -2(2)^2 + 3(2) - 1$$
$$-3 \overset{?}{=} -2(4) + 6 - 1$$
$$-3 \overset{?}{=} -8 + 6 - 1$$
$$-3 = -3 \text{ True}$$

The point $(2, -3)$ is on the graph.

25. **a.** Let $x = 1$ and $y = 4$.

$$(4) \overset{?}{=} |1 - 3|$$
$$4 \overset{?}{=} |-2|$$
$$4 = 2 \text{ False}$$

The point $(1, 4)$ is not on the graph.

   b. Let $x = 4$ and $y = 1$.

$$(1) \overset{?}{=} |4 - 3|$$
$$1 \overset{?}{=} |1|$$
$$1 = 1 \text{ True}$$

The point $(4, 1)$ is on the graph.

   c. Let $x = -6$ and $y = 9$.

$$(9) \overset{?}{=} |-6 - 3|$$
$$9 \overset{?}{=} |-9|$$
$$9 = 9 \text{ True}$$

The point $(-6, 9)$ is on the graph.

d. Let $x = 0$ and $y = 3$.

$$(3) \overset{?}{=} |0 - 3|$$

$$3 \overset{?}{=} |-3|$$

$$3 = 3 \text{ True}$$

The point $(0, 3)$ is on the graph.

27. $y = 4x$

x	$y = 4x$	(x, y)
-2	$y = 4(-2) = -8$	$(-2, -8)$
-1	$y = 4(-1) = -4$	$(-1, -4)$
0	$y = 4(0) = 0$	$(0, 0)$
1	$y = 4(1) = 4$	$(1, 4)$
2	$y = 4(2) = 8$	$(2, 8)$

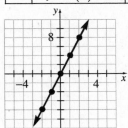

29. $y = -\dfrac{1}{2}x$

x	$y = -\dfrac{1}{2}x$	(x, y)
-4	$y = -\dfrac{1}{2}(-4) = 2$	$(-4, 2)$
-2	$y = -\dfrac{1}{2}(-2) = 1$	$(-2, 1)$
0	$y = -\dfrac{1}{2}(0) = 0$	$(0, 0)$
2	$y = -\dfrac{1}{2}(2) = -1$	$(2, -1)$
4	$y = -\dfrac{1}{2}(4) = -2$	$(4, -2)$

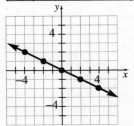

31. $y = x + 3$

x	$y = x + 3$	(x, y)
-4	$y = (-4) + 3 = -1$	$(-4, -1)$
-2	$y = (-2) + 3 = 1$	$(-2, 1)$
0	$y = (0) + 3 = 3$	$(0, 3)$
2	$y = (2) + 3 = 5$	$(2, 5)$
4	$y = (4) + 3 = 7$	$(4, 7)$

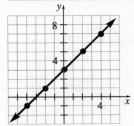

33. $y = -3x + 1$

x	$y = -3x + 1$	(x, y)
-2	$y = -3(-2) + 1 = 7$	$(-2, 7)$
-1	$y = -3(-1) + 1 = 4$	$(-1, 4)$
0	$y = -3(0) + 1 = 1$	$(0, 1)$
1	$y = -3(1) + 1 = -2$	$(1, -2)$
2	$y = -3(2) + 1 = -5$	$(2, -5)$

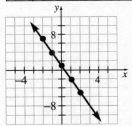

35. $y = \dfrac{1}{2}x - 4$

x	$y = \dfrac{1}{2}x - 4$	(x, y)
-4	$y = \dfrac{1}{2}(-4) - 4 = -6$	$(-4, -6)$
-2	$y = \dfrac{1}{2}(-2) - 4 = -5$	$(-2, -5)$
0	$y = \dfrac{1}{2}(0) - 4 = -4$	$(0, -4)$
2	$y = \dfrac{1}{2}(2) - 4 = -3$	$(2, -3)$
4	$y = \dfrac{1}{2}(4) - 4 = -2$	$(4, -2)$

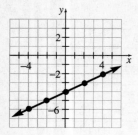

37. $2x + y = 7$
$y = -2x + 7$

x	$y = -2x + 7$	(x, y)
-5	$y = -2(-5) + 7 = 17$	$(-5, 17)$
-3	$y = -2(-3) + 7 = 13$	$(-3, 13)$
0	$y = -2(0) + 7 = 7$	$(0, 7)$
3	$y = -2(3) + 7 = 1$	$(3, 1)$
5	$y = -2(5) + 7 = -3$	$(5, -3)$

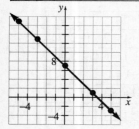

39. $y = -x^2$

x	$y = -x^2$	(x, y)
-3	$y = -(-3)^2 = -9$	$(-3, -9)$
-2	$y = -(-2)^2 = -4$	$(-2, -4)$
0	$y = -(0)^2 = 0$	$(0, 0)$
2	$y = -(2)^2 = -4$	$(2, -4)$
3	$y = -(3)^2 = -9$	$(3, -9)$

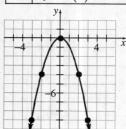

41. $y = 2x^2 - 8$

x	$y = 2x^2 - 8$	(x, y)
-3	$y = 2(-3)^2 - 8 = 10$	$(-3, 10)$
-2	$y = 2(-2)^2 - 8 = 0$	$(-2, 0)$
-1	$y = 2(-1)^2 - 8 = -6$	$(-1, -6)$
0	$y = 2(0)^2 - 8 = -8$	$(0, -8)$
1	$y = 2(1)^2 - 8 = -6$	$(1, -6)$
2	$y = 2(2)^2 - 8 = 0$	$(2, 0)$
3	$y = 2(3)^2 - 8 = 10$	$(3, 10)$

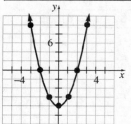

43. $y = |x|$

x	$y = |x|$	(x, y)
-4	$y = |-4| = 4$	$(-4, 4)$
-2	$y = |-2| = 2$	$(-2, 2)$
0	$y = |0| = 0$	$(0, 0)$
2	$y = |2| = 2$	$(2, 2)$
4	$y = |4| = 4$	$(4, 4)$

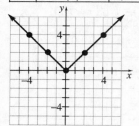

45. $y = |x - 1|$

x	$y = |x - 1|$	(x, y)
-4	$y = |-4 - 1| = 5$	$(-4, 5)$
-1	$y = |-1 - 1| = 2$	$(-1, 2)$
0	$y = |0 - 1| = 1$	$(0, 1)$
1	$y = |1 - 1| = 0$	$(1, 0)$
3	$y = |3 - 1| = 2$	$(3, 2)$
5	$y = |5 - 1| = 4$	$(5, 4)$

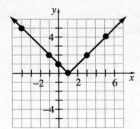

47. $y = x^3$

x	$y = x^3$	(x, y)
-3	$y = (-3)^3 = -27$	$(-3, -27)$
-2	$y = (-2)^3 = -8$	$(-2, -8)$
-1	$y = (-1)^3 = -1$	$(-1, -1)$
0	$y = (0)^3 = 0$	$(0, 0)$
1	$y = (1)^3 = 1$	$(1, 1)$
2	$y = (2)^3 = 8$	$(2, 8)$
3	$y = (3)^3 = 27$	$(3, 27)$

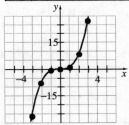

49. $y = x^3 + 1$

x	$y = x^3 + 1$	(x, y)
-2	$y = (-2)^3 + 1 = -7$	$(-2, -7)$
-1	$y = (-1)^3 + 1 = 0$	$(-1, 0)$
0	$y = (0)^3 + 1 = 1$	$(0, 1)$
1	$y = (1)^3 + 1 = 2$	$(1, 2)$
2	$y = (2)^3 + 1 = 9$	$(2, 9)$

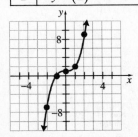

51. $x^2 - y = 4$

$x^2 = y + 4$

$x^2 - 4 = y$

$y = x^2 - 4$

x	$y = x^2 - 4$	(x, y)
-3	$y = (-3)^2 - 4 = 5$	$(-3, 5)$
-2	$y = (-2)^2 - 4 = 0$	$(-2, 0)$
0	$y = (0)^2 - 4 = -4$	$(0, -4)$
2	$y = (2)^2 - 4 = 0$	$(2, 0)$
3	$y = (3)^2 - 4 = 5$	$(3, 5)$

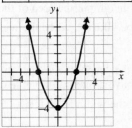

53. $x = y^2 - 1$

y	$x = y^2 - 1$	(x, y)
-2	$x = (-2)^2 - 1 = 3$	$(3, -2)$
-1	$x = (-1)^2 - 1 = 0$	$(0, -1)$
0	$x = (0)^2 - 1 = -1$	$(-1, 0)$
1	$x = (1)^2 - 1 = 0$	$(0, 1)$
2	$x = (2)^2 - 1 = 3$	$(3, 2)$

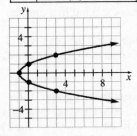

55. The intercepts are $(-2, 0)$ and $(0, 3)$.

57. The intercepts are $(-2, 0)$, $(1, 0)$, and $(0, -4)$.

59. If $(a, 4)$ is a point on the graph, then we have

$$4 = 4(a) - 3$$
$$4 = 4a - 3$$
$$7 = 4a$$
$$\frac{7}{4} = a$$

We need $a = \dfrac{7}{4}$.

61. If $(3, b)$ is a point on the graph, then we have

$$b = (3)^2 - 2(3) + 1 = 9 - 6 + 1 = 4$$

We need $b = 4$.

63. a. According to the graph, the area of the opening will be 400 ft^2 when the width is 10 feet.

   b. According to the graph, the area of the opening will have a maximum value of 625 ft^2 when the width is 25 feet.

   c. The x-intercepts are $(0, 0)$ and $(50, 0)$. These indicate that the window opening will have 0 area when the width is 0 feet or 50 feet. The x-coordinates form the bounds for the width of the opening.
   The y-intercept is $(0, 0)$. This indicates that the area of the opening will be 0 ft^2 when the width is 0 feet.

65. a. According to the graph, the Sprint PCS 2000-minute plan would cost $100 for 200 minutes of talking in a month and $100 for 500 minutes of talking in a month.

   b. According to the graph the plan would cost $1600 for 8000 minutes of talking in a month.

   c. The only intercept for the graph is the y-intercept, $(0, 100)$. This indicates that the monthly cost will be $100 if no minutes are used. In other words, the base charge for the plan is $100 per month.

67. The set of all points of the form $(4, y)$ form a vertical line with an x-intercept of $(4, 0)$.

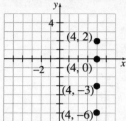

69. Answers will vary. One possible graph is below.

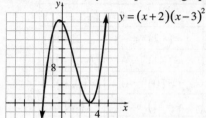

71. Answers will vary. One possibility:
All three points have a y-coordinate of 0. Therefore, they all lie on the horizontal line with equation $y = 0$.

73. A graph is considered to be complete when enough is shown to indicate main features and allow the viewer to 'see' the rest of the graph as an obvious continuation of what is shown.

75. The point-plotting method is when we choose values for one variable and find the corresponding value of the other variable. The resulting ordered pairs form points that lie on the graph of the equation.

77. $y = 3x - 9$

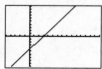

79. $y = -x^2 + 8$

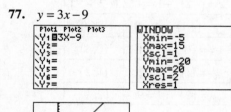

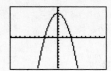

81. $y + 2x^2 = 13$

$y = -2x^2 + 13$

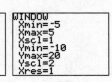

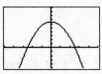

83. $y = x^3 - 6x + 1$

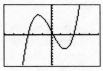
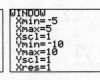

Section 1.6

Are You Ready for This Section?

R1. $3x + 12 = 0$

$3x + 12 - 12 = 0 - 12$

$3x = -12$

$\dfrac{3x}{3} = \dfrac{-12}{3}$

$x = -4$

The solution set is $\{-4\}$.

R2. $2x + 5 = 13$

$2x + 5 - 5 = 13 - 5$

$2x = 8$

$\dfrac{2x}{2} = \dfrac{8}{2}$

$x = 4$

The solution set is $\{4\}$.

R3. $3x - 2y = 10$

$3x - 2y - 3x = 10 - 3x$

$-2y = 10 - 3x$

$\dfrac{-2y}{-2} = \dfrac{10 - 3x}{-2}$

$y = \dfrac{3x - 10}{2}$ or $y = \dfrac{3}{2}x - 5$

R4. $\dfrac{5 - 2}{-2 - 4} = \dfrac{3}{-6} = -\dfrac{1}{2}$

R5. $\dfrac{-7 - 2}{-5 - (-2)} = \dfrac{-9}{-5 + 2} = \dfrac{-9}{-3} = 3$

R6. $-2(x + 3) = (-2) \cdot x + (-2) \cdot 3$

$= -2x - 6$

Section 1.6 Quick Checks

1. A <u>linear equation</u> is an equation of the form $Ax + By = C$, where A, B, and C are real numbers, A and B cannot both be 0.

2. The graph of a linear equation is a <u>line</u>.

3. $y = 2x - 3$

Let $x = -1$, 0, 1, and 2.

$x = -1:\quad y = 2(-1) - 3$

$y = -2 - 3$

$y = -5$

$x = 0:\quad y = 2(0) - 3$

$y = 0 - 3$

$y = -3$

$x = 1:\quad y = 2(1) - 3$

$y = 2 - 3$

$y = -1$

$x = 2:\quad y = 2(2) - 3$

$y = 4 - 3$

$y = 1$

Thus, the points $(-1, -5)$, $(0, -3)$, $(1, -1)$, and $(2, 1)$ are on the graph.

61

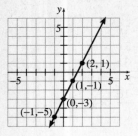

4. $\frac{1}{2}x + y = 2$

Let $x = -2, 0, 2,$ and 4.

$x = -2:$ $\quad \frac{1}{2}(-2) + y = 2$
$$-1 + y = 2$$
$$y = 3$$

$x = 0:$ $\quad \frac{1}{2}(0) + y = 2$
$$0 + y = 2$$
$$y = 2$$

$x = 2:$ $\quad \frac{1}{2}(2) + y = 2$
$$1 + y = 2$$
$$y = 1$$

$x = 4:$ $\quad \frac{1}{2}(4) + y = 2$
$$2 + y = 2$$
$$y = 0$$

Thus, the points $(-2, 3)$, $(0, 2)$, $(2, 1)$, and $(4, 0)$ are on the graph.

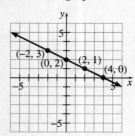

5. $-6x + 3y = 12$

Let $x = -2, -1, 0,$ and 1.

$x = -2:$ $\quad -6(-2) + 3y = 12$
$$12 + 3y = 12$$
$$3y = 0$$
$$y = 0$$

$x = -1:$ $\quad -6(-1) + 3y = 12$
$$6 + 3y = 12$$
$$3y = 6$$
$$y = 2$$

$x = 0:$ $\quad -6(0) + 3y = 12$
$$0 + 3y = 12$$
$$3y = 12$$
$$y = 4$$

$x = 1:$ $\quad -6(1) + 3y = 12$
$$-6 + 3y = 12$$
$$3y = 18$$
$$y = 6$$

Thus, the points $(-2, 0)$, $(-1, 2)$, $(0, 4)$, and $(1, 6)$ are on the graph.

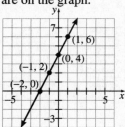

6. True

7. $x + y = 4$

To find the y-intercept, let $x = 0$ and solve for y.
$x = 0:$ $\quad 0 + y = 4$
$$y = 4$$
The y-intercept is $(0, 4)$. To find the x-intercept, let $y = 0$ and solve for x.
$y = 0:$ $\quad x + 0 = 4$
$$x = 4$$
The x-intercept is $(4, 0)$.

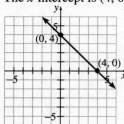

8. $4x - 5y = 20$

To find the y-intercept, let $x = 0$ and solve for y.
$x = 0:$ $\quad 4(0) - 5y = 20$
$$-5y = 20$$
$$y = -4$$

The y-intercept is $(0, -4)$. To find the x-intercept, let $y = 0$ and solve for x.

$$y = 0: \quad 4x - 5(0) = 20$$
$$4x = 20$$
$$x = 5$$

The x-intercept is $(5, 0)$.

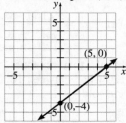

9. $3x - 2y = 0$

To find the y-intercept, let $x = 0$ and solve for y.
$$x = 0: \quad 3(0) - 2y = 0$$
$$-2y = 0$$
$$y = 0$$

The y-intercept is $(0, 0)$. To find the x-intercept, let $y = 0$ and solve for x.
$$y = 0: \quad 3x - 2(0) = 0$$
$$3x = 0$$
$$x = 0$$

The x-intercept is $(0, 0)$. Since the x- and y-intercepts both result in the origin, we must find other points on the graph. Let $x = -2$ and 2, and solve for y.
$$x = -2: \quad 3(-2) - 2y = 0$$
$$-6 - 2y = 0$$
$$-2y = 6$$
$$y = -3$$
$$x = 2: \quad 3(2) - 2y = 0$$
$$6 - 2y = 0$$
$$-2y = -6$$
$$y = 3$$

Thus, the points $(-2, -3)$ and $(2, 3)$ are also on the graph.

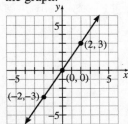

10. The equation of a vertical line is of the form $\underline{x = a}$. The equation of a horizontal line is of the form $\underline{y = b}$.

11. The equation $x = 5$ will be a vertical line with x-intercept $(5, 0)$.

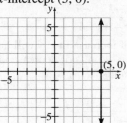

12. The equation $y = -4$ will be a horizontal line with y-intercept $(0, -4)$.

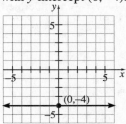

13. $$-3x + 4 = 1$$
$$-3x + 4 - 4 = 1 - 4$$
$$-3x = -3$$
$$\frac{-3x}{-3} = \frac{-3}{-3}$$
$$x = 1$$

The equation $-3x + 4 = 1$ will be a vertical line with x-intercept $(1, 0)$.

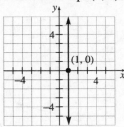

14. If a line is vertical, then its slope is <u>undefined</u>.

15. $\dfrac{\text{Rise}}{\text{Run}} = \dfrac{4}{10} = \dfrac{2}{5}$

16. False; the slope is $m = \dfrac{y_2 - y_1}{x_2 - x_1}, \ x_1 \neq x_2$

17. True

18. $m = \dfrac{y_2 - y_1}{x_2 - x_1} = \dfrac{12 - 3}{3 - 0} = \dfrac{9}{3} = 3$

For every 1 unit of increase in x, y will increase by 3 units.

19. $m = \dfrac{y_2 - y_1}{x_2 - x_1} = \dfrac{-4 - 3}{3 - (-1)} = \dfrac{-7}{4} = -\dfrac{7}{4}$

For every 4 units of increase in x, y will decrease by 7 units. For every 4 units of decrease in x, y will increase by 7 units.

20. $m = \dfrac{y_2 - y_1}{x_2 - x_1} = \dfrac{2 - 2}{-3 - 3} = \dfrac{0}{-6} = 0$

The line is horizontal.

21. $m = \dfrac{y_2 - y_1}{x_2 - x_1} = \dfrac{-1 - 4}{-2 - (-2)} = \dfrac{-5}{0} = $ undefined

The line is vertical.

22. True

23. A horizontal line has slope = $\underline{0}$.

24. $m_1 = \dfrac{4 - 3}{6 - 1} = \dfrac{1}{5}$; $m_2 = \dfrac{8 - 3}{1 - 1} = \dfrac{5}{0} = $ undefined;

$m_3 = \dfrac{7 - 3}{-3 - 1} = \dfrac{4}{-4} = -1$; $m_4 = \dfrac{3 - 3}{-4 - 1} = \dfrac{0}{-5} = 0$

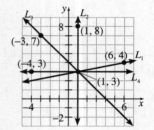

25. a.

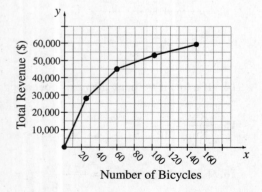

Number of Bicycles

b. Average rate of change $= \dfrac{28{,}000 - 0}{25 - 0}$
$= \$1120$ per bicycle

For each bicycle sold, total revenue increased by $1120 when between 0 and 25 bicycles were sold.

c. Average rate of change $= \dfrac{59{,}160 - 53{,}400}{150 - 102}$
$= \$120$ per bicycle

For each bicycle sold, total revenue increased by $120 when between 102 and 150 bicycles were sold.

d. No, the revenue does not grow linearly. The average rate of change (slope) is not constant.

26. a. Start at $(-1,\ 3)$. Because $m = \dfrac{1}{3}$, move 3 units to the right and 1 unit up to find point $(2,\ 4)$.

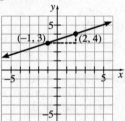

b. Start at $(-1,\ 3)$. Because $m = -4 = -\dfrac{4}{1}$, move 1 unit to the right and 4 units down to find point $(0,\ -1)$.

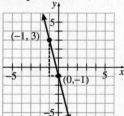

c. Because $m = 0$, the line is horizontal passing through the point $(-1, 3)$.

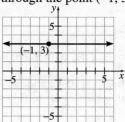

27. An equation in the form $y - y_1 = m(x - x_1)$ is said to be in <u>point-slope</u> form.

28. $y - y_1 = m(x - x_1)$
$y - 5 = 2(x - 3)$

To graph the line, start at $(3, 5)$. Because

$m = 2 = \dfrac{2}{1}$, move 1 unit to the right and 2 units

up to find point $(4, 7)$.

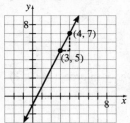

29. $y - y_1 = m(x - x_1)$
$y - 3 = -4(x - (-2))$
$y - 3 = -4(x + 2)$

To graph the line, start at $(-2, 3)$. Because

$m = -4 = -\dfrac{4}{1}$, move 1 unit to the right and

4 units down to find point $(-1, -1)$.

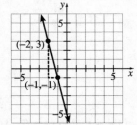

30. $y - y_1 = m(x - x_1)$
$y - (-4) = \dfrac{1}{3}(x - 3)$
$y + 4 = \dfrac{1}{3}(x - 3)$

To graph the line, start at $(3, -4)$. Because

$m = \dfrac{1}{3}$, move 3 units to the right and 1 unit up to

find point $(6, -3)$.

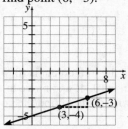

31. $y - y_1 = m(x - x_1)$
$y - (-2) = 0(x - 4)$
$y + 2 = 0$
$y = -2$

To graph the line, we draw a horizontal line that passes through the point $(4, -2)$.

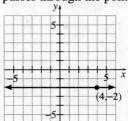

32. An equation in the form $y = mx + b$ is said to be in <u>slope-intercept</u> form.

33. $3x - y = 2$
$-y = -3x + 2$
$y = \dfrac{-3x + 2}{-1}$
$y = 3x - 2$

The slope is 3 and the y-intercept is $(0, -2)$. To graph the line, begin at $(0, -2)$ and move to the right 1 unit and up 3 units to find the point $(1, 1)$.

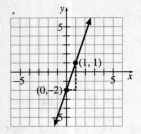

34. $6x + 2y = 8$
$$2y = -6x + 8$$
$$y = \frac{-6x + 8}{2}$$
$$y = -3x + 4$$

The slope is -3 and the y-intercept is $(0, 4)$. To graph the line, begin at $(0, 4)$ and move to the right 1 unit and down 3 units to find the point $(1, 1)$.

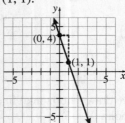

35. $3x - 2y = 7$
$$-2y = -3x + 7$$
$$y = \frac{-3x + 7}{-2}$$
$$y = \frac{3}{2}x - \frac{7}{2}$$

The slope is $\frac{3}{2}$ and the y-intercept is $\left(0, -\frac{7}{2}\right)$.

To graph the line, begin at $\left(0, -\frac{7}{2}\right)$ and move to the right 2 units and up 3 units to find the point $\left(2, -\frac{1}{2}\right)$.

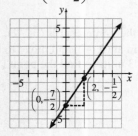

36. $7x + 3y = 0$
$$3y = -7x$$
$$y = -\frac{7}{3}x$$

The slope is $-\frac{7}{3}$ and the y-intercept is $(0, 0)$. To graph the line, begin at $(0, 0)$ and move to the right 3 units and down 7 units to find point $(3, -7)$.

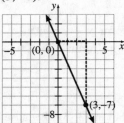

37. $m = \dfrac{y_2 - y_1}{x_2 - x_1} = \dfrac{9 - 3}{4 - 1} = \dfrac{6}{3} = 2$
$$y - y_1 = m(x - x_1)$$
$$y - 3 = 2(x - 1)$$
$$y - 3 = 2x - 2$$
$$y = 2x + 1$$

The slope is 2 and the y-intercept is $(0, 1)$.

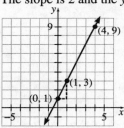

38. $m = \dfrac{y_2 - y_1}{x_2 - x_1} = \dfrac{2 - 4}{2 - (-2)} = \dfrac{-2}{4} = -\dfrac{1}{2}$
$$y - y_1 = m(x - x_1)$$
$$y - 4 = -\frac{1}{2}(x - (-2))$$
$$y - 4 = -\frac{1}{2}x - 1$$
$$y = -\frac{1}{2}x + 3$$

The slope is $-\dfrac{1}{2}$ and the y-intercept is $(0, 3)$.

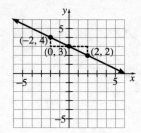

39. $m = \dfrac{y_2 - y_1}{x_2 - x_1} = \dfrac{6-6}{3-(-4)} = \dfrac{0}{7} = 0$

The slope is 0, so the line is horizontal. The equation of the line is $y = 6$.

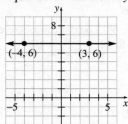

40. $m = \dfrac{y_2 - y_1}{x_2 - x_1} = \dfrac{-4-2}{3-3} = \dfrac{-6}{0} =$ undefined

The slope is undefined, so the line is vertical. The equation of the line is $x = 3$.

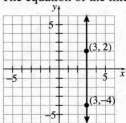

1.6 Exercises

41. $x - 2y = 6$

Let $x = -2,\ 0,\ 2,$ and 4.

$x = -2:\quad -2 - 2y = 6$
$\qquad\qquad -2y = 8$
$\qquad\qquad\ \ y = -4$

$x = 0:\quad 0 - 2y = 6$
$\qquad\qquad -2y = 6$
$\qquad\qquad\ \ y = -3$

$x = 2:\quad 2 - 2y = 6 \qquad x = 4:\quad 4 - 2y = 6$
$\qquad\qquad -2y = 4 \qquad\qquad\qquad -2y = 2$
$\qquad\qquad\ \ y = -2 \qquad\qquad\qquad\ \ y = -1$

Thus, the points $(-2, -4)$, $(0, -3)$, $(2, -2)$, and $(4, -1)$ are on the graph.

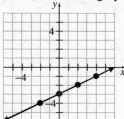

43. $3x + 2y = 12$

Let $x = -2,\ 0,\ 2,$ and 4.

$x = -2:\quad 3(-2) + 2y = 12$
$\qquad\qquad\ \ -6 + 2y = 12$
$\qquad\qquad\qquad\ 2y = 18$
$\qquad\qquad\qquad\ \ y = 9$

$x = 0:\quad 3(0) + 2y = 12$
$\qquad\qquad\qquad 2y = 12$
$\qquad\qquad\qquad\ y = 6$

$x = 2:\quad 3(2) + 2y = 12$
$\qquad\qquad\ \ 6 + 2y = 12$
$\qquad\qquad\qquad 2y = 6$
$\qquad\qquad\qquad\ y = 3$

$x = 4:\quad 3(4) + 2y = 12$
$\qquad\qquad 12 + 2y = 12$
$\qquad\qquad\qquad\ 2y = 0$
$\qquad\qquad\qquad\ \ y = 0$

Thus, the points $(-2, 9)$, $(0, 6)$, $(2, 3)$, and $(4, 0)$ are on the graph.

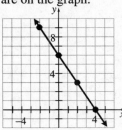

45. $\dfrac{2}{3}x + y = 6$

Let $x = -3,\ 0,\ 3,$ and 6.

$x = -3:\quad \dfrac{2}{3}(-3) + y = 6$
$\qquad\qquad\qquad -2 + y = 6$
$\qquad\qquad\qquad\qquad y = 8$

$x = 0: \quad \frac{2}{3}(0) + y = 6$
$$y = 6$$

$x = 3: \quad \frac{2}{3}(3) + y = 6$
$$2 + y = 6$$
$$y = 4$$

$x = 6: \quad \frac{2}{3}(6) + y = 6$
$$4 + y = 6$$
$$y = 2$$

Thus, the points $(-3, 8)$, $(0, 6)$, $(3, 4)$, and $(6, 2)$ are on the graph.

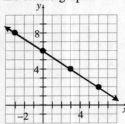

47. $5x - 3y = 6$

Let $x = -3,\ 0,\ 3,$ and 6.

$x = -3: \quad 5(-3) - 3y = 6$
$$-15 - 3y = 6$$
$$-3y = 21$$
$$y = -7$$

$x = 0: \quad 5(0) - 3y = 6$
$$-3y = 6$$
$$y = -2$$

$x = 3: \quad 5(3) - 3y = 6$
$$15 - 3y = 6$$
$$-3y = -9$$
$$y = 3$$

$x = 6: \quad 5(6) - 3y = 6$
$$30 - 3y = 6$$
$$-3y = -24$$
$$y = 8$$

Thus, the points $(-3, -7)$, $(0, -2)$, $(3, 3)$, and $(6, 8)$ are on the graph.

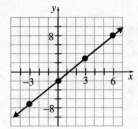

49. $3x + y = 6$

To find the *y*-intercept, let $x = 0$ and solve for *y*.

$x = 0: \quad 3(0) + y = 6$
$$0 + y = 6$$
$$y = 6$$

The *y*-intercept is $(0, 6)$. To find the *x*-intercept, let $y = 0$ and solve for *x*.

$y = 0: \quad 3x + 0 = 6$
$$3x = 6$$
$$x = 2$$

The *x*-intercept is $(2, 0)$.

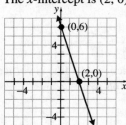

51. $5x - 3y = 15$

To find the *y*-intercept, let $x = 0$ and solve for *y*.

$x = 0: \quad 5(0) - 3y = 15$
$$-3y = 15$$
$$y = -5$$

The *y*-intercept is $(0, -5)$. To find the *x*-intercept, let $y = 0$ and solve for *x*.

$y = 0: \quad 5x - 3(0) = 15$
$$5x = 15$$
$$x = 3$$

The *x*-intercept is $(3, 0)$.

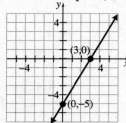

53. $\frac{1}{3}x - \frac{1}{2}y = 1$

To find the *y*-intercept, let $x = 0$ and solve for *y*.

$x = 0:$ $\frac{1}{3}(0) - \frac{1}{2}y = 1$

$-\frac{1}{2}y = 1$

$y = -2$

The *y*-intercept is (0, −2). To find the *x*-intercept, let $y = 0$ and solve for *x*.

$y = 0:$ $\frac{1}{3}x - \frac{1}{2}(0) = 1$

$\frac{1}{3}x = 1$

$x = 3$

The *x*-intercept is (3, 0).

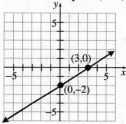

55. $2x + y = 0$

To find the *y*-intercept, let $x = 0$ and solve for *y*.

$x = 0:$ $2(0) + y = 0$

$y = 0$

The *y*-intercept is (0, 0). To find the *x*-intercept, let $y = 0$ and solve for *x*.

$y = 0:$ $2x + 0 = 0$

$2x = 0$

$x = 0$

The *x*-intercept is (0, 0). Since the *x*- and *y*-intercepts both result in the origin, we must find other points on the graph. Let $x = -1$ and 1, and solve for *y*.

$x = -1:$ $2(-1) + y = 0$

$-2 + y = 0$

$y = 2$

$x = 1:$ $2(1) + y = 0$

$2 + y = 0$

$y = -2$

Thus, the points (−1, 2) and (1, −2) are also on the graph.

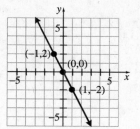

57. $\frac{2}{3}x - \frac{1}{2}y = 0$

To find the *y*-intercept, let $x = 0$ and solve for *y*.

$x = 0:$ $\frac{2}{3}(0) - \frac{1}{2}y = 0$

$-\frac{1}{2}y = 0$

$y = 0$

The *y*-intercept is (0, 0). To find the *x*-intercept, let $y = 0$ and solve for *x*.

$y = 0:$ $\frac{2}{3}x - \frac{1}{2}(0) = 0$

$\frac{2}{3}x = 0$

$x = 0$

The *x*-intercept is (0, 0). Since the *x*- and *y*-intercepts both result in the origin, we must find other points on the graph. Let $x = -3$ and 3, and solve for *y*.

$x = -3:$ $\frac{2}{3}(-3) - \frac{1}{2}y = 0$

$-2 - \frac{1}{2}y = 0$

$-\frac{1}{2}y = 2$

$y = -4$

$x = 3:$ $\frac{2}{3}(3) - \frac{1}{2}y = 0$

$2 - \frac{1}{2}y = 0$

$-\frac{1}{2}y = -2$

$y = 4$

Thus, the points (−3, −4) and (3, 4) are also on the graph.

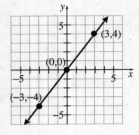

59. The equation $x = -5$ will be a vertical line with x-intercept $(-5, 0)$.

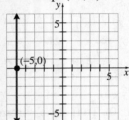

61. The equation $y = 1$ will be a horizontal line with y-intercept $(0, 1)$.

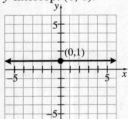

63. $2y + 8 = -6$
$\quad\quad 2y = -14$
$\quad\quad\quad y = -7$

The equation $y = -7$ will be a horizontal line with y-intercept $(0, -7)$.

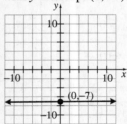

65. a. $m = \dfrac{y_2 - y_1}{x_2 - x_1} = \dfrac{4 - 0}{3 - 0} = \dfrac{4}{3}$

    b. For every 3 units of increase in x, y will increase by 4 units.

67. a. $m = \dfrac{y_2 - y_1}{x_2 - x_1} = \dfrac{-4 - 4}{2 - (-1)} = \dfrac{-8}{3} = -\dfrac{8}{3}$

    b. For every 3 units of increase in x, y will decrease by 8 units. For every 3 units of decrease in x, y will increase by 8 units.

69. $m = \dfrac{y_2 - y_1}{x_2 - x_1} = \dfrac{5 - 0}{1 - 0} = \dfrac{5}{1} = 5$

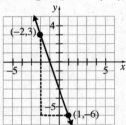

71. $m = \dfrac{y_2 - y_1}{x_2 - x_1} = \dfrac{-6 - 3}{1 - (-2)} = \dfrac{-9}{3} = -3$

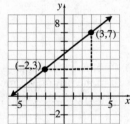

73. $m = \dfrac{y_2 - y_1}{x_2 - x_1} = \dfrac{7 - 3}{3 - (-2)} = \dfrac{4}{5}$

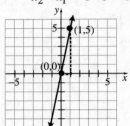

75. $m = \dfrac{y_2 - y_1}{x_2 - x_1} = \dfrac{2 - 2}{4 - (-3)} = \dfrac{0}{7} = 0$

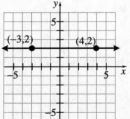

77. $m = \dfrac{y_2 - y_1}{x_2 - x_1} = \dfrac{-3 - 2}{10 - 10} = \dfrac{-5}{0} = \text{undefined}$

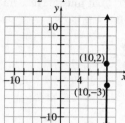

79. $m = \dfrac{y_2 - y_1}{x_2 - x_1} = \dfrac{\frac{11}{6} - \frac{5}{3}}{\frac{9}{4} - \frac{1}{2}} = \dfrac{\frac{11}{6} - \frac{10}{6}}{\frac{9}{4} - \frac{2}{4}} = \dfrac{\frac{1}{6}}{\frac{7}{4}} = \dfrac{1}{6} \cdot \dfrac{4}{7} = \dfrac{2}{21}$

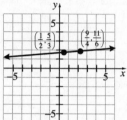

81. Start at (1, 2). Because $m = 3 = \dfrac{3}{1}$, move 1 unit to the right and 3 units up to find point (2, 5).

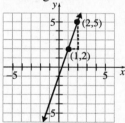

83. Start at (−3, 1). Because $m = -2 = -\dfrac{2}{1}$, move 1 unit to the right and 2 units down to find point (−2, −1). We can also move 1 unit to the left and 2 units up to find point (−4, 3).

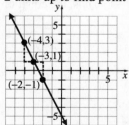

85. Start at (−3, 4). Because $m = \dfrac{1}{3}$, move 3 units to the right and 1 unit up to find point (0, 5).

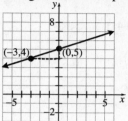

87. Start at (2, 5). Because $m = -\dfrac{3}{2}$, move 2 units to the right and 3 units down to find point (4, 2). We can also move 2 units to the left and 3 units up to find point (0, 8).

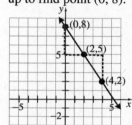

89. Because $m = 0$, the line is horizontal passing through the point (1, 2).

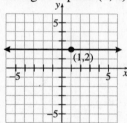

91. Answers will vary. One possible answer follows. We know $m = \dfrac{5}{2}$. Beginning at the point (−2, 3), and using a change in x of 2 and a change of y of 5, $-2 + 2 = 0$ and $3 + 5 = 8$, so (0, 8) is on the line. Next, beginning at (0, 8), $0 + 2 = 2$ and $8 + 5 = 13$, so (2, 13) is on the line. Now, beginning at (2, 13), $2 + 2 = 4$ and $13 + 5 = 18$, so (4, 18) is on the line. In summary, the points (0, 8), (2, 13), and (4, 18) will all be on the line.

93. $m = \dfrac{y_2 - y_1}{x_2 - x_1} = \dfrac{1-(-3)}{-4-5} = \dfrac{4}{-9} = -\dfrac{4}{9}$

$\quad y - y_1 = m(x - x_1)$

$\quad\quad y - 1 = -\dfrac{4}{9}\left(x-(-4)\right)$

$\quad\quad y - 1 = -\dfrac{4}{9}(x+4)$

$\quad\quad y - 1 = -\dfrac{4}{9}x - \dfrac{16}{9}$

$\quad\quad\quad y = -\dfrac{4}{9}x - \dfrac{7}{9}$

95. $m = \dfrac{y_2 - y_1}{x_2 - x_1} = \dfrac{3-3}{-4-1} = \dfrac{0}{-5} = 0$

$\quad y - y_1 = m(x - x_1)$

$\quad\quad y - 3 = 0(x-1)$

$\quad\quad y - 3 = 0$

$\quad\quad\quad y = 3$

97. $y - y_1 = m(x - x_1)$

$\quad\quad y - 0 = 2(x-0)$

$\quad\quad\quad y = 2x$

99. $y - y_1 = m(x - x_1)$

$\quad\quad y - 1 = -3\left(x-(-1)\right)$

$\quad\quad y - 1 = -3(x+1)$

$\quad\quad y - 1 = -3x - 3$

$\quad\quad\quad y = -3x - 2$

101. $y - y_1 = m(x - x_1)$

$\quad\quad y - 2 = \dfrac{4}{3}(x-3)$

$\quad\quad y - 2 = \dfrac{4}{3}x - 4$

$\quad\quad\quad y = \dfrac{4}{3}x - 2$

103. $y - y_1 = m(x - x_1)$

$\quad\quad y - 4 = -\dfrac{5}{4}\left(x-(-2)\right)$

$\quad\quad y - 4 = -\dfrac{5}{4}(x+2)$

$\quad\quad y - 4 = -\dfrac{5}{4}x - \dfrac{5}{2}$

$\quad\quad\quad y = -\dfrac{5}{4}x + \dfrac{3}{2}$

105. The slope is undefined, so the line is vertical and the equation is of the form $x = a$. Since the graph contains the point $(6, 1)$, it must have x-intercept 6. Thus, the equation is $x = 6$.

107. $m = \dfrac{y_2 - y_1}{x_2 - x_1} = \dfrac{7-0}{5-0} = \dfrac{7}{5}$

$\quad y - y_1 = m(x - x_1)$

$\quad\quad y - 0 = \dfrac{7}{5}(x-0)$

$\quad\quad\quad y = \dfrac{7}{5}x$

109. $m = \dfrac{y_2 - y_1}{x_2 - x_1} = \dfrac{7-2}{4-3} = \dfrac{5}{1} = 5$

$\quad y - y_1 = m(x - x_1)$

$\quad\quad y - 2 = 5(x-3)$

$\quad\quad y - 2 = 5x - 15$

$\quad\quad\quad y = 5x - 13$

111. $m = \dfrac{y_2 - y_1}{x_2 - x_1} = \dfrac{-2-1}{5-(-2)} = \dfrac{-3}{7} = -\dfrac{3}{7}$

$\quad y - y_1 = m(x - x_1)$

$\quad\quad y - 1 = -\dfrac{3}{7}\left(x-(-2)\right)$

$\quad\quad y - 1 = -\dfrac{3}{7}(x+2)$

$\quad\quad y - 1 = -\dfrac{3}{7}x - \dfrac{6}{7}$

$\quad\quad\quad y = -\dfrac{3}{7}x + \dfrac{1}{7}$

113. $m = \dfrac{y_2 - y_1}{x_2 - x_1} = \dfrac{5-(-3)}{-1-(-1)} = \dfrac{8}{0} = $ undefined

Since the slope is undefined, the line is vertical and the equation is of the form $x = a$. Since the graph contains $(-1, -3)$ and $(-1, 5)$, the x-intercept must be -1. Thus, the equation is $x = -1$.

115. $m = \dfrac{y_2 - y_1}{x_2 - x_1} = \dfrac{-7 - 3}{-3 - 1} = \dfrac{-10}{-4} = \dfrac{5}{2}$

$$y - y_1 = m(x - x_1)$$

$$y - 3 = \frac{5}{2}(x - 1)$$

$$y - 3 = \frac{5}{2}x - \frac{5}{2}$$

$$y = \frac{5}{2}x + \frac{1}{2}$$

117. $m = \dfrac{y_2 - y_1}{x_2 - x_1} = \dfrac{4 - 4}{-4 - 2} = \dfrac{0}{-6} = 0$

$$y - y_1 = m(x - x_1)$$

$$y - 4 = 0(x - 2)$$

$$y - 4 = 0$$

$$y = 4$$

119. For $y = 2x - 1$, the slope is 2 and the y-intercept is $(0, -1)$. Begin at $(0, -1)$ and move to the right 1 unit and up 2 units to find point $(1, 1)$.

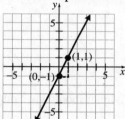

121. For $y = -4x$, the slope is -4 and the y-intercept is $(0, 0)$. Begin at $(0, 0)$ and move to the right 1 unit and down 4 units to find point $(1, -4)$. We can also move 1 unit to the left and 4 units up to find point $(-1, 4)$.

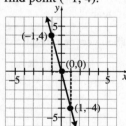

123. $2x + y = 3$

$$y = -2x + 3$$

The slope is -2 and the y-intercept is $(0, 3)$. Begin at $(0, 3)$ and move to the right 1 unit and down 2 units to find the point $(1, 1)$. We can also move 1 unit to the left and 2 units up to find the point $(-1, 5)$.

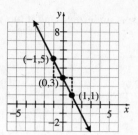

125. $4x + 2y = 8$

$$2y = -4x + 8$$

$$y = \frac{-4x + 8}{2}$$

$$y = -2x + 4$$

The slope is -2 and the y-intercept is $(0, 4)$. Begin at $(0, 4)$ and move to the right 1 unit and down 2 units to find the point $(1, 2)$. We can also move 1 unit to the left and 2 units up to find the point $(-1, 6)$.

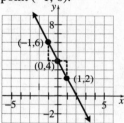

127. $x - 4y - 2 = 0$

$$-4y = -x + 2$$

$$y = \frac{-x + 2}{-4}$$

$$y = \frac{1}{4}x - \frac{1}{2}$$

The slope is $\dfrac{1}{4}$ and the y-intercept is $\left(0, -\dfrac{1}{2}\right)$.

Begin at $\left(0, -\dfrac{1}{2}\right)$ and move to the right 4 units

and up 1 unit to find the point $\left(4, \dfrac{1}{2}\right)$.

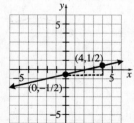

129. For $x = 3$, the slope is undefined and it has no y-intercept. It is a vertical line with x-intercept $(3, 0)$.

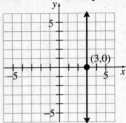

131. The x-axis is a horizontal line passing through the origin. Thus, the slope is 0 and the y-intercept is $(0, 0)$. So, the equation of the x-axis is $y = 0x + 0$ or simply $y = 0$.

133. a.

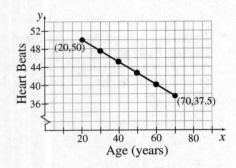

b. Average rate of change
$$= \frac{47.5 - 50}{30 - 20}$$
$= -0.25$ heartbeats per year

Between ages 20 and 30, the maximum number of heartbeats (that a healthy individual should have during a 15-second time interval while exercising) decreases at a rate of 0.25 heartbeat per year.

c. Average rate of change
$$= \frac{40 - 42.5}{60 - 50}$$
$= -0.25$ heartbeats per year

Between ages 50 and 60, the maximum number of heartbeats (that a healthy individual should have during a 15-second time interval while exercising) decreases at a rate of 0.25 heartbeat per year.

d. Yes. The maximum number of heart beats appears to be linearly related to age. The average rate of change (slope) is constant for the data provided.

135. a.

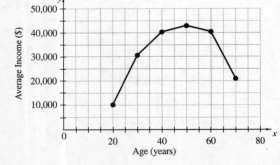

b.   Average rate of change $= \dfrac{31,161 - 10,469}{30 - 20}$
$= \$2069.20$ per year

Between ages 20 and 30, the average individual's income increases at a rate of $2069.20 per year.

c.   Average rate of change $= \dfrac{40,654 - 43,627}{60 - 50}$
$= -\$297.30$ per year

Between ages 50 and 60, the average individual's income decreases at a rate of $297.30 per year.

d.   No. The average income is not linearly related to age. The average rate of change (slope) is not constant.

137. Let F represent the Fahrenheit temperature and C represent the Celsius temperature.

$$m = \frac{100 - 0}{212 - 32} = \frac{100}{180} = \frac{5}{9}$$

$$C - 0 = \frac{5}{9}(F - 32)$$

$$C = \frac{5}{9}(F - 32)$$

Finally, when $F = 60$, $C = \dfrac{5}{9}(60 - 32) \approx 15.6$.

Thus, 60°F is equivalent to 15.6°C.

139. The graph shown has both a positive slope and a positive y-intercept. The only given equation that meets both of these conditions is the one in part (c), $y = 2x + 3$.

141. Standard form: $Ax + By = C$

Slope-intercept form: $y = mx + b$

Point-slope form: $y - y_1 = m(x - x_1)$

Horizontal line: $y = b$

Vertical line: $x = a$

143. A horizontal line of the form $y = b$ where $b \neq 0$ has one y-intercept, but no x-intercept.

145. No. Answers may vary. One possibility follows: Every line must have at least one intercept. Vertical lines have one x-intercept, but no y-intercept. Horizontal lines have one y-intercept, but no x-intercept. Lines that are neither vertical nor horizontal will have one x-intercept and one y-intercept.

147. The graphs of the three lines are shown below.

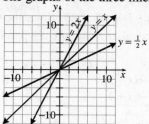

Answers will vary. All three lines all pass through the origin and slant upward from left to right. The graph of $y = x$ slants more steeply than the graph of $y = \dfrac{1}{2}x$. The graph of $y = 2x$ slants even more steeply than that of $y = x$. The graph of $y = ax$, with $a > 0$, will be a line that passes through the origin and slants upward from left to right.

149.

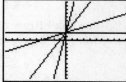

Answers may vary. When $m \geq 0$, the line is either horizontal or rises from left to right.

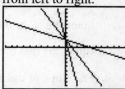

Answers may vary. When $m < 0$, the line falls from left to right.

Section 1.7

Are You Ready for This Section?

R1. The reciprocal of 3 is $\dfrac{1}{3}$ because $3 \cdot \dfrac{1}{3} = 1$.

R2. The reciprocal of $-\dfrac{3}{5}$ is $-\dfrac{5}{3}$ because
$$\left(-\frac{3}{5}\right)\left(-\frac{5}{3}\right) = 1.$$

Section 1.7 Quick Checks

1. Two lines are parallel if and only if they have the same <u>slope</u> and different <u>y-intercepts</u>.

2. The slope of $L_1 : y = 3x + 1$ is $m_1 = 3$. The slope of $L_2 : y = -3x - 3$ is $m_2 = -3$. Since the two slopes are not equal, the lines are not parallel.

3.
$$L_1 : 6x + 3y = 3 \qquad\qquad L_2 : -8x - 4y = 12$$
$$3y = -6x + 3 \qquad\qquad -4y = 8x + 12$$
$$y = \frac{-6x + 3}{3} \qquad\qquad y = \frac{8x + 12}{-4}$$
$$y = -2x + 1 \qquad\qquad y = -2x - 3$$

Since the slopes $m_1 = -2$ and $m_2 = -2$ are equal but the y-intercepts are different, the two lines are parallel.

4.
$$L_1 : -3x + 5y = 10$$
$$5y = 3x + 10$$
$$y = \frac{3x + 10}{5}$$
$$y = \frac{3}{5}x + 2$$

$$L_2 : 6x + 10y = 10$$
$$10y = -6x + 10$$
$$y = \frac{-6x + 10}{10}$$
$$y = -\frac{3}{5}x + 1$$

Since the slopes $m_1 = \frac{3}{5}$ and $m_2 = -\frac{3}{5}$ are not equal, the two lines are not parallel.

5. The slope of the line we seek is $m = 3$, the same as the slope of $y = 3x + 1$. The equation of the line is
$$y - y_1 = m(x - x_1)$$
$$y - 8 = 3(x - 5)$$
$$y - 8 = 3x - 15$$
$$y = 3x - 7$$

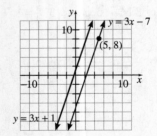

6. The slope of the line we seek is $m = -\frac{3}{2}$, the same as the slope of the line
$$3x + 2y = 10$$
$$2y = -3x + 10$$
$$y = \frac{-3x + 10}{2}$$
$$y = -\frac{3}{2}x + 5$$

Thus, the equation of the line we seek is
$$y - y_1 = m(x - x_1)$$
$$y - 4 = -\frac{3}{2}\left(x - (-2)\right)$$
$$y - 4 = -\frac{3}{2}(x + 2)$$
$$y - 4 = -\frac{3}{2}x - 3$$
$$y = -\frac{3}{2}x + 1$$

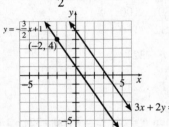

7. Two lines are perpendicular if and only if the product of their slopes is -1.

8. The negative reciprocal of -3 is $-\dfrac{1}{-3} = \dfrac{1}{3}$. Any line whose slope is $\dfrac{1}{3}$ will be perpendicular to a line whose slope is -3.

9. The slope of $L_1 : y = 5x - 3$ is $m_1 = 5$. The slope of $L_2 : y = -\dfrac{1}{5}x - 4$ is $m_2 = -\dfrac{1}{5}$. Since the two slopes are negative reciprocals (i.e., since $5\left(-\dfrac{1}{5}\right) = -1$), the lines are perpendicular.

10. $L_1 : 4x - y = 3$ $L_2 : x - 4y = 2$

$$-y = -4x + 3 \qquad\qquad -4y = -x + 2$$

$$y = \frac{-4x + 3}{-1} \qquad\qquad y = \frac{-x + 2}{-4}$$

$$y = 4x - 3 \qquad\qquad y = \frac{1}{4}x - \frac{1}{2}$$

Since the slopes $m_1 = 4$ and $m_2 = \frac{1}{4}$ are not

negative reciprocals (i.e., since $4 \cdot \frac{1}{4} = 1 \ne -1$),

the two lines are not perpendicular.

11. $L_1 : 2y + 4 = 0$ $L_2 : 3x - 6 = 0$

$$2y = -4 \qquad\qquad 3x = 6$$

$$\frac{2y}{2} = \frac{-4}{2} \qquad\qquad \frac{3x}{3} = \frac{6}{3}$$

$$y = -2 \qquad\qquad x = 2$$

Since L_1 is horizontal with slope 0 and L_2 is
vertical with an undefined slope, the two lines
are perpendicular.

12. The slope of the line we seek is $m = -\frac{1}{2}$, the

negative reciprocal of the slope of the line
$y = 2x + 1$. Thus, the equation of the line we seek
is

$$y - y_1 = m(x - x_1)$$

$$y - 2 = -\frac{1}{2}(x - (-4))$$

$$y - 2 = -\frac{1}{2}(x + 4)$$

$$y - 2 = -\frac{1}{2}x - 2$$

$$y = -\frac{1}{2}x$$

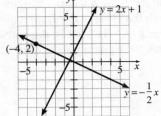

13. The slope of the line we seek is $m = -\frac{4}{3}$, the

negative reciprocal of the slope of the line
$3x - 4y = 8$

$$-4y = -3x + 8$$

$$y = \frac{-3x + 8}{-4}$$

$$y = \frac{3}{4}x - 2$$

Thus, the equation of the line we seek is

$$y - y_1 = m(x - x_1)$$

$$y - (-4) = -\frac{4}{3}(x - (-3))$$

$$y + 4 = -\frac{4}{3}x - 4$$

$$y = -\frac{4}{3}x - 8$$

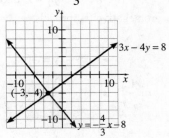

14. The line we seek is a horizontal line, because the
line we are given is a vertical line. Since the line
we seek must pass through the point (3, −2), its
equation is $y = -2$.

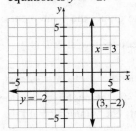

1.7 Exercises

15. a. Slope of parallel line: $m = 5$.

     b. Slope of perpendicular line: $m = -1/5$.

17. a. Slope of parallel line: $m = -5/6$.

     b. Slope of perpendicular line: $m = 6/5$.

19. Parallel. The two lines have equal slopes, $m = 5$,
but different y-intercepts.

21. $8x + y = 12$ $2x - 8y = 3$
$$y = -8x + 12 \qquad -8y = -2x + 3$$
$$y = \frac{-2x + 3}{-8}$$
$$y = \frac{1}{4}x - \frac{3}{8}$$

Neither. The two slopes, $m_1 = -8$ and $m_2 = 1/4$, are not equal and are not negative reciprocals.

23. $-4x + 2y = 12$ $x + 2y = 6$
$$2y = 4x + 12 \qquad 2y = -x + 6$$
$$y = \frac{4x + 12}{2} \qquad y = \frac{-x + 6}{2}$$
$$y = 2x + 6 \qquad\qquad y = -\frac{1}{2}x + 3$$

Perpendicular. The two slopes, $m_1 = 2$ and $m_2 = -1/2$, are negative reciprocals.

25. $-x + \frac{1}{3}y = \frac{1}{3}$ $x - \frac{1}{3}y = \frac{5}{3}$
$$3\left(-x + \frac{1}{3}y\right) = 3\left(\frac{1}{3}\right) \quad -3\left(x - \frac{1}{3}y\right) = -3\left(\frac{5}{3}\right)$$
$$-3x + y = 1 \qquad\qquad -3x + y = -5$$
$$y = 3x + 1 \qquad\qquad y = 3x - 5$$

Parallel. The two lines have equal slopes, $m = 3$, but different y-intercepts.

27. Since line L is parallel to $y = \frac{3}{2}x$, the slope of L is $m = 3/2$. From the graph, the y-intercept of L is $(0, -3)$. Thus, the equation of line L is $y = \frac{3}{2}x - 3$.

29. Since line L is perpendicular to $y = -2x - 1$, the slope of L is $m = 1/2$, the negative reciprocal of -2. From the graph, the y-intercept of L is $(0, 2)$. Thus, the equation of line L is
$$y = \frac{1}{2}x + 2.$$

31. Since line L is perpendicular to $x = -2$ and since $x = -2$ is a vertical line, then L must be a horizontal line. From the graph, the y-intercept of L is $(0, 2)$. Thus, the equation of line L is $y = 2$.

33. The slope of the line we seek is $m = 2$, the same as the slope of $y = 2x + 3$. The equation of the line is
$$y - y_1 = m(x - x_1)$$
$$y - 1 = 2(x - 3)$$
$$y - 1 = 2x - 6$$
$$y = 2x - 5.$$

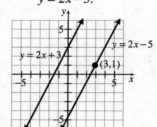

35. The slope of the line we seek is $m = 1/2$, the negative reciprocal of the slope of $y = -2x + 1$. The equation of the line is
$$y - y_1 = m(x - x_1)$$
$$y - 3 = \frac{1}{2}(x - 2)$$
$$y - 3 = \frac{1}{2}x - 1$$
$$y = \frac{1}{2}x + 2.$$

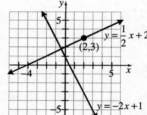

37. The line we seek is horizontal, parallel to the horizontal line $y = 1$. Now a horizontal line passing through the point $(-1, -3)$ must also pass through the point $(0, -3)$. That is, the y-intercept is $(0, -3)$. Thus, the equation of the line is $y = -3$.

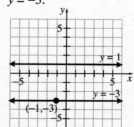

39. The line we seek is horizontal, perpendicular to the vertical line $x = 1$. Now a horizontal line passing through the point (1, 3) must also pass through the point (0, 3). That is, the y-intercept is (0, 3). Thus, the equation of the line is $y = 3$.

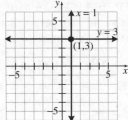

41. The slope of the line we seek is $m = 3$, the same as the slope of the line
$$3x - y = 2$$
$$-y = -3x + 2$$
$$y = 3x - 2.$$
Thus, the equation of the line we seek is
$$y - y_1 = m(x - x_1)$$
$$y - 5 = 3(x - 1)$$
$$y - 5 = 3x - 3$$
$$y = 3x + 2.$$

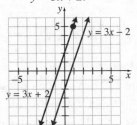

43. The slope of the line we seek is $m = 3/4$, the negative reciprocal of the slope of the line
$$4x + 3y - 1 = 0$$
$$3y = -4x + 1$$
$$y = \frac{-4x + 1}{3}$$
$$y = -\frac{4}{3}x + \frac{1}{3}.$$
Thus, the equation of the line we seek is

$$y - y_1 = m(x - x_1)$$
$$y - 1 = \frac{3}{4}(x - (-4))$$
$$y - 1 = \frac{3}{4}(x + 4)$$
$$y - 1 = \frac{3}{4}x + 3$$
$$y = \frac{3}{4}x + 4.$$

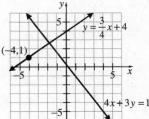

45. The slope of the line we seek is $m = -5/2$, the same as the slope of the line
$$5x + 2y = 1$$
$$2y = -5x + 1$$
$$y = \frac{-5x + 1}{2}$$
$$y = -\frac{5}{2}x + \frac{1}{2}.$$
Thus, the equation of the line we seek is
$$y - y_1 = m(x - x_1)$$
$$y - (-3) = -\frac{5}{2}(x - (-2))$$
$$y + 3 = -\frac{5}{2}(x + 2)$$
$$y + 3 = -\frac{5}{2}x - 5$$
$$y = -\frac{5}{2}x - 8.$$

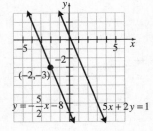

47. $m_1 = \dfrac{5-2}{6-1} = \dfrac{3}{5}$; $m_2 = \dfrac{-2-3}{1-(-2)} = \dfrac{-5}{3} = -\dfrac{5}{3}$

Perpendicular. The two slopes, $m_1 = 3/5$ and $m_2 = -5/3$, are negative reciprocals.

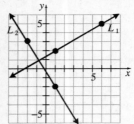

49. $m_1 = \dfrac{-3-4}{1-(-2)} = \dfrac{-7}{3} = -\dfrac{7}{3}$;

$m_2 = \dfrac{1-(-6)}{-4-(-1)} = \dfrac{7}{-3} = -\dfrac{7}{3}$

Parallel. The two lines have equal slopes, but different y-intercepts.

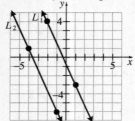

51. $m_1 = \dfrac{3-5}{1-0} = \dfrac{-2}{1} = -2$; $m_2 = \dfrac{14-6}{0-(-4)} = \dfrac{8}{4} = 2$

Neither. The two slopes are not equal and are not negative reciprocals.

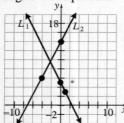

53. a. $A = (1, 1)$, $B = (4, 3)$, $C = (2, 6)$

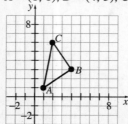

b. Slope of $\overline{AB} = \dfrac{3-1}{4-1} = \dfrac{2}{3}$

Slope of $\overline{BC} = \dfrac{6-3}{2-4} = \dfrac{3}{-2} = -\dfrac{3}{2}$

Because the slopes are negative reciprocals, segments $\overline{AB}$ and $\overline{BC}$ are perpendicular. Thus, triangle ABC is a right triangle.

55. a. $A = (2, 2)$, $B = (7, 3)$, $C = (8, 6)$, $D = (3, 5)$

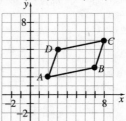

b. Slope of $\overline{AB} = \dfrac{3-2}{7-2} = \dfrac{1}{5}$

Slope of $\overline{BC} = \dfrac{6-3}{8-7} = \dfrac{3}{1} = 3$

Slope of $\overline{CD} = \dfrac{5-6}{3-8} = \dfrac{-1}{-5} = \dfrac{1}{5}$

Slope of $\overline{DA} = \dfrac{2-5}{2-3} = \dfrac{-3}{-1} = 3$

Because the slopes of $\overline{AB}$ and $\overline{CD}$ are equal, $\overline{AB}$ and $\overline{CD}$ are parallel. Because the slopes of $\overline{BC}$ and $\overline{DA}$ are equal, $\overline{BC}$ and $\overline{DA}$ are parallel. Thus, quadrilateral $ABCD$ is a parallelogram.

57. The line
$$4x + y = 3$$
$$y = -4x + 3$$
has a slope of -4. The line
$$Ax + 4y = 12$$
$$4y = -Ax + 12$$
$$y = \dfrac{-Ax + 12}{4}$$
$$y = -\dfrac{A}{4}x + 3$$

has a slope of $-A/4$. For the two lines to be perpendicular, the product of their slopes must be -1. Use this fact to determine the value of A:
$$-4 \cdot \left(-\dfrac{A}{4}\right) = -1$$
$$A = -1$$

59. In the graph, the two parallel lines have positive slopes. In addition, one of the lines has a positive y-intercept while the other has a negative y-intercept. The only pair of equations that meets these criteria is the pair in part (c). (Note: The graphs of each pair of equations in parts (b) and (d) are parallel, but have all positive y-intercepts. The pair of equations in part (a) are perpendicular, not parallel. The pair of equations in part (e) represents the same line.)

61. No. If the two nonvertical lines have the same x-intercept but different y-intercepts, then their slopes cannot be equal. Therefore, they cannot be parallel lines.

Section 1.8

Are You Ready for This Section?

R1. $3(4)+1 \overset{?}{\geq} 7$
$12+1 \overset{?}{\geq} 7$
$13 \geq 7 \leftarrow$ True
Yes, $x = 4$ does satisfy the inequality $3x + 1 \geq 7$.

R2. $\quad -4x-3 > 9$
$-4x-3+3 > 9+3$
$-4x > 12$
$\dfrac{-4x}{-4} < \dfrac{12}{-4}$
$x < -3$
The solution set is $\{x \,|\, x < -3\}$ or, using interval notation, $(-\infty, -3)$.

R3. True

Section 1.8 Quick Checks

1. If we replace the inequality symbol in $Ax + By > C$ with an equal sign, we obtain the equation of a line, $Ax + By = C$. The line separates the xy-plane into two regions called <u>half-planes</u>.

2. a. $x=4, \; y=1 \quad -2(4)+3(1) \overset{?}{\geq} 3$
$-8+3 \overset{?}{\geq} 3$
$-5 \geq 3 \;$ False
So, $(4, 1)$ is not a solution to $-2x + 3y \geq 3$.

b. $x=-1, \; y=2 \quad -2(-1)+3(2) \overset{?}{\geq} 3$
$2+6 \overset{?}{\geq} 3$
$8 \geq 3 \;$ True
So, $(-1, \, 2)$ is a solution to $-2x + 3y \geq 3$.

c. $x=2, \; y=3 \quad -2(2)+3(3) \overset{?}{\geq} 3$
$-4+9 \overset{?}{\geq} 3$
$5 \geq 3 \;$ True
So, $(2, \, 3)$ is a solution to $-2x + 3y \geq 3$.

d. $x=0, \; y=1 \quad -2(0)+3(1) \overset{?}{\geq} 3$
$0+3 \overset{?}{\geq} 3$
$3 \geq 3 \;$ True
So, $(0, \, 1)$ is a solution to $-2x + 3y \geq 3$.

3. False; it also consists of a half-plane.

4. True

5. Replace the inequality symbol with an equal sign to obtain $y = -2x + 3$. Because the inequality is strict, graph $y = -2x + 3$ using a dashed line.

Test Point: $(0,0)$: $\quad 0 \overset{?}{<} -2(0)+3$
$0 \overset{?}{<} 0+3$
$0 \overset{?}{<} 3 \quad$ True

Therefore, $(0, 0)$ is a solution to $y < -2x + 3$. Shade the half-plane that contains $(0, 0)$.

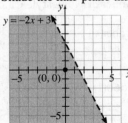

6. Replace the inequality symbol with an equal sign to obtain $6x - 3y = 15$. Because the inequality is non-strict, graph $6x - 3y = 15$ $(y = 2x - 5)$ using a solid line.

Test Point: $(0,0)$: $6(0)-3(0)\overset{?}{\leq}15$

$$0\leq15 \quad \text{True}$$

Therefore, $(0, 0)$ is solution to $6x - 3y \leq 15$.
Shade the half-plane that contains $(0, 0)$.

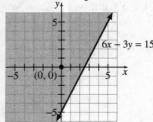

7. Replace the inequality symbol with an equal sign
    to obtain $2x + y = 0$. Because the inequality is
    strict, graph $2x + y = 0$ $(y = -2x)$ using a
    dashed line.

    Test Point: $(1, 1)$: $2(1)+1\overset{?}{<}0$

    $$2+1\overset{?}{<}0$$

    $$3<0 \quad \text{False}$$

    Therefore, $(1, 1)$ is not a solution to $2x + y < 0$.
    Shade the half-plane that does not contain
    $(1, 1)$.

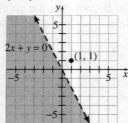

8. a. Let x = the number of Homestyle Chicken
        filets.
        Let y = the number of Frosties.
        $560x + 300y \leq 900$

    b. $x = 1, y = 1$: $560(1)+300(1)\overset{?}{\leq}900$

        $$560+300\overset{?}{\leq}900$$

        $$860\leq900 \quad \text{True}$$

        Yes, Avery can stay within his allotment of
        calories by eating one Homestyle Chicken
        filet and one Frosty.

    c. $x = 2, y = 1$: $560(2)+300(1)\overset{?}{\leq}900$

        $$1120+300\overset{?}{\leq}900$$

        $$1420\leq900 \quad \text{False}$$

    No, Avery cannot stay within his allotment
    of calories by eating two Homestyle
    Chicken filets and one Frosty.

1.8 Exercises

9. a. $x = 0, y=1$ $\quad 0+3(1)\overset{?}{<}6$

        $$3<6 \quad \text{True}$$

        So, $(0, 1)$ is a solution to $x + 3y < 6$.

    b. $x = -2, y=4$ $\quad -2+3(4)\overset{?}{<}6$

        $$-2+12\overset{?}{<}6$$

        $$10<6 \quad \text{False}$$

        So, $(-2, 4)$ is not a solution to $x + 3y < 6$.

    c. $x = 8, y=-1$ $\quad 8+3(-1)\overset{?}{<}6$

        $$8-3\overset{?}{<}6$$

        $$5<6 \quad \text{True}$$

        So, $(8, -1)$ is a solution to $x + 3y < 6$.

11. a. $x = -4, y=2$ $\quad -3(-4)+4(2)\overset{?}{\geq}12$

        $$12+8\overset{?}{\geq}12$$

        $$20\geq12 \quad \text{True}$$

        So, $(-4, 2)$ is a solution to $-3x + 4y \geq 12$.

    b. $x = 0, y = 2$ $\quad -3(0)+4(2)\overset{?}{\geq}12$

        $$0+8\overset{?}{\geq}12$$

        $$8\geq12 \quad \text{False}$$

        So, $(0, 2)$ is not a solution to $-3x + 4y \geq 12$.

    c. $x = 0, y = 3$ $\quad -3(0)+4(3)\overset{?}{\geq}12$

        $$0+12\overset{?}{\geq}12$$

        $$12\geq12 \quad \text{True}$$

        So, $(0, 3)$ is a solution to $-3x + 4y \geq 12$.

13. Replace the inequality symbol with an equal sign to obtain $y = 3$. Because the inequality is strict, graph $y = 3$ using a dashed line.

Test Point: $(0,0)$: $0 \overset{?}{>} 3$ False

Therefore, (0, 0) is not a solution to $y > 3$.
Shade the half-plane that does not contain (0, 0).

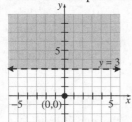

15. Replace the inequality symbol with an equal sign to obtain $x = -2$. Because the inequality is non-strict, graph $x = -2$ using a solid line.

Test Point: $(0,0)$: $0 \overset{?}{\geq} -2$ True

Therefore, (0, 0) is a solution to $x \geq -2$. Shade the half-plane that contains (0, 0).

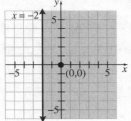

17. Replace the inequality symbol with an equal sign to obtain $y = 5x$. Because the inequality is strict, graph $y = 5x$ using a dashed line.

Test Point: $(1, 1)$: $1 \overset{?}{<} 5(1)$

$1 \overset{?}{<} 5$ True

Therefore, (1, 1) is a solution to $y < 5x$. Shade the half-plane that contains (1, 1).

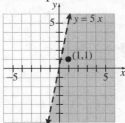

19. Replace the inequality symbol with an equal sign to obtain $y = 2x + 3$. Because the inequality is strict, graph $y = 2x + 3$ using a dashed line.

Test Point: $(0,0)$: $0 \overset{?}{>} 2(0) + 3$

$0 \overset{?}{>} 3$ False

Therefore, (0, 0) is not a solution to $y > 2x + 3$.
Shade the half-plane that does not contain (0, 0).

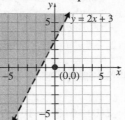

21. Replace the inequality symbol with an equal sign to obtain $y = \frac{1}{2}x - 5$. Because the inequality is non-strict, graph $y = \frac{1}{2}x - 5$ using a solid line.

Test Point: $(0,0)$: $0 \overset{?}{\leq} \frac{1}{2}(0) - 5$

$0 \overset{?}{\leq} -5$ False

Therefore, (0, 0) is not a solution to $y \leq \frac{1}{2}x - 5$.
Shade the half-plane that does not contain (0, 0).

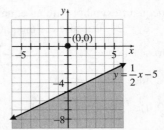

23. Replace the inequality symbol with an equal sign to obtain $3x + y = 4$. Because the inequality is non-strict, graph $3x + y = 4$ $(y = -3x + 4)$ using a solid line.

Test Point: $(0,0)$: $3(0) + 0 \overset{?}{\leq} 4$

$0 \overset{?}{\leq} 4$ True

Therefore, (0, 0) is solution to $3x + y \leq 4$. Shade the half-plane that contains (0, 0).

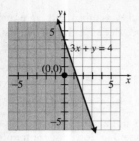

25. Replace the inequality symbol with an equal sign to obtain $2x + 5y = -10$. Because the inequality is non-strict, graph $2x + 5y = -10$

$\left(y = -\dfrac{2}{5}x - 2 \right)$ using a solid line.

Test Point: $(0,0)$: $2(0) + 5(0) \overset{?}{\le} -10$

$\qquad\qquad\qquad\qquad 0 \overset{?}{\le} -10 \qquad$ False

Therefore, $(0, 0)$ is not a solution to $2x + 5y \le -10$. Shade the half-plane that does not contain $(0, 0)$.

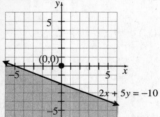

27. Replace the inequality symbol with an equal sign to obtain $-4x + 6y = 24$. Because the inequality is strict, graph $-4x + 6y = 24$ $\left(y = \dfrac{2}{3}x + 4 \right)$

using a dashed line.

Test Point: $(0,0)$: $-4(0) + 6(0) \overset{?}{>} 24$

$\qquad\qquad\qquad\qquad 0 \overset{?}{>} 24 \qquad$ False

Therefore, $(0, 0)$ is not a solution to $-4x + 6y > 24$. Shade the half-plane that does not contain $(0, 0)$.

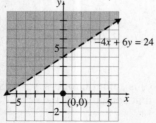

29. Replace the inequality symbol with an equal sign to obtain $\dfrac{x}{2} + \dfrac{y}{3} = 1$. Because the inequality is strict, graph $\dfrac{x}{2} + \dfrac{y}{3} = 1$ $\left(y = -\dfrac{3}{2}x + 3 \right)$ using a dashed line.

Test Point: $(0,0)$: $\dfrac{0}{2} + \dfrac{0}{3} \overset{?}{<} 1$

$\qquad\qquad\qquad\qquad 0 \overset{?}{<} 1 \qquad$ True

Therefore, $(0, 0)$ is a solution to $\dfrac{x}{2} + \dfrac{y}{3} < 1$.

Shade the half-plane that contains $(0, 0)$.

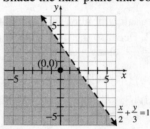

31. Replace the inequality symbol with an equal sign to obtain $\dfrac{2}{3}x - \dfrac{3}{2}y = 2$. Because the inequality is non-strict, graph $\dfrac{2}{3}x - \dfrac{3}{2}y = 2$ $\left(y = \dfrac{4}{9}x - \dfrac{4}{3} \right)$ using a solid line.

Test Point: $(0,0)$: $\dfrac{2}{3}(0) - \dfrac{3}{2}(0) \overset{?}{\ge} 2$

$\qquad\qquad\qquad\qquad 0 \overset{?}{\ge} 2 \qquad$ False

Therefore, $(0, 0)$ is not a solution to $\dfrac{2}{3}x - \dfrac{3}{2}y \ge 2$. Shade the half-plane that does not contain $(0, 0)$.

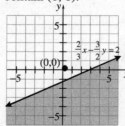

33. a. Let x = the number of Filet-o-Fish.
Let y = the number of orders of fries.
$39x + 29y \le 125$

b. $x = 2$, $y = 1$: $39(2) + 29(1) \overset{?}{\le} 125$

$$107 \overset{?}{\le} 125 \quad \text{True}$$

Sammy can stay below his limit of 125 grams of carbohydrates by eating two Filet-o-Fish and one order of fries.

c. $x = 3$, $y = 1$: $39(3) + 29(1) \overset{?}{\le} 125$

$$146 \overset{?}{\le} 125 \quad \text{False}$$

Sammy cannot stay below his limit of 125 grams of carbohydrates by eating three Filet-o-Fish and one order of fries.

35. a. Let x = the number of Switch A assemblies. Let y = the number of Switch B assemblies.
$2x + 1.5y \le 80$

b. $x = 24$, $y = 41$:

$$2(24) + 1.5(41) \overset{?}{\le} 80$$

$$109.5 \overset{?}{\le} 80 \quad \text{False}$$

Acme cannot manufacture 24 of Switch A and 41 of Switch B in a week (80 hours).

c. $x = 16$, $y = 45$: $2(16) + 1.5(45) \overset{?}{\le} 80$

$$99.5 \overset{?}{\le} 80 \quad \text{False}$$

Acme cannot manufacture 16 of Switch A and 45 of Switch B in a week (80 hours).

37. The graph contains a line passing through the points $(0, -2)$ and $(2, 2)$. The slope of the line is $m = \dfrac{2 - (-2)}{2 - 0} = \dfrac{4}{2} = 2$. The y-intercept is $(0, -2)$.
Thus, the equation of the line is $y = 2x - 2$.
Because the half-plane that is shaded is above the line, our inequality will contain a "greater than." Because the line is dashed, this means that the inequality is strict. That is, our inequality will consist only of "greater than." Thus, the linear inequality is $y > 2x - 2$.

39. The graph contains a line passing through the points $(-2, 3)$ and $(2, -1)$. The slope of the line is $m = \dfrac{-1 - 3}{2 - (-2)} = \dfrac{-4}{4} = -1$. The equation of the line is

$$y - (-1) = -1(x - 2)$$
$$y + 1 = -x + 2$$
$$y = -x + 1.$$

Because the half-plane that is shaded is below the line, our inequality will contain a "less than." Because the line is solid, this means that the inequality is non-strict. That is, our inequality will consist of "less than or equal to." Thus, the linear inequality is $y \le -x + 1$.

41. Answers may vary. The solution of a linear inequality is a half-plane. In each half plane, either all of the points satisfy the inequality or no points satisfy the inequality. If a test point does not satisfy the inequality, then it must lie in the half-plane in which no points satisfy the inequality. The opposite half-plane must then contain the solutions.

43. $y > 3$

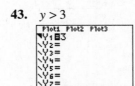 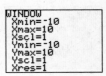

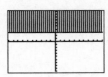

45. $y < 5x$

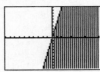 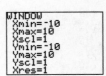

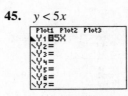

47. $y > 2x + 3$

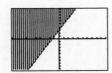 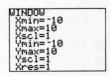

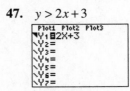

49. $y \le \dfrac{1}{2}x - 5$

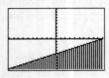

51. $3x + y \le 4$

$y \le -3x + 4$

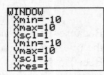
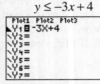

53. $2x + 5y \le -10$

$5y \le -2x - 10$

$y \le \dfrac{-2x - 10}{5}$

$y \le -\dfrac{2}{5}x - 2$

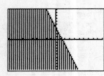

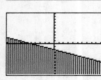

Chapter 1 Review

1. $3x - 4 = 6 + x$

Let $x = 5$ in the equation.

$3(5) - 4 \stackrel{?}{=} 6 + (5)$

$15 - 4 \stackrel{?}{=} 11$

$11 = 11$ T

$x = 5$ is a solution to the equation.

Let $x = 6$ in the equation.

$3(6) - 4 \stackrel{?}{=} 6 + (6)$

$18 - 4 \stackrel{?}{=} 12$

$14 \ne 12$

$x = 6$ is **not** a solution to the equation.

2. $-1 - 4x = 2(3 - 2x) - 7$

Let $x = -2$ in the equation.

$-1 - 4(-2) \stackrel{?}{=} 2(3 - 2(-2)) - 7$

$-1 + 8 \stackrel{?}{=} 2(3 + 4) - 7$

$7 \stackrel{?}{=} 2(7) - 7$

$7 \stackrel{?}{=} 14 - 7$

$7 = 7$ T

$x = -2$ is a solution to the equation.

Let $x = -1$ in the equation.

$-1 - 4(-1) \stackrel{?}{=} 2(3 - 2(-1)) - 7$

$-1 + 4 \stackrel{?}{=} 2(3 + 2) - 7$

$3 \stackrel{?}{=} 2(5) - 7$

$3 \stackrel{?}{=} 10 - 7$

$3 = 3$ T

$x = -1$ is a solution to the equation.

3. $4y - (1 - y) + 5 = -6 - 2(3y - 5) - 2y$

Let $y = -2$ in the equation.

$4(-2) - (1 - (-2)) + 5 \stackrel{?}{=} -6 - 2(3(-2) - 5) - 2(-2)$

$-8 - (3) + 5 \stackrel{?}{=} -6 - 2(-6 - 5) + 4$

$-11 + 5 \stackrel{?}{=} -2 - 2(-11)$

$-6 \ne 20$

$y = -2$ is **not** a solution to the equation.

Let $y = 0$ in the equation.

$4(0) - (1 - 0) + 5 \stackrel{?}{=} -6 - 2(3(0) - 5) - 2(0)$

$0 - 1 + 5 \stackrel{?}{=} -6 - 2(-5) - 0$

$4 \stackrel{?}{=} -6 + 10$

$4 = 4$ T

$y = 0$ is a solution to the equation.

4. $\dfrac{w-7}{3} - \dfrac{w}{4} = -\dfrac{7}{6}$

Let $w = -14$ in the equation.

$$\dfrac{-14-7}{3} - \dfrac{-14}{4} \overset{?}{=} -\dfrac{7}{6}$$

$$\dfrac{-21}{3} + \dfrac{14}{4} \overset{?}{=} -\dfrac{7}{6}$$

$$-7 + \dfrac{7}{2} \overset{?}{=} -\dfrac{7}{6}$$

$$-\dfrac{7}{2} \neq -\dfrac{7}{6}$$

$w = -14$ is **not** a solution to the equation.

Let $w = 7$ in the equation.

$$\dfrac{7-7}{3} - \dfrac{7}{4} \overset{?}{=} -\dfrac{7}{6}$$

$$0 - \dfrac{7}{4} \overset{?}{=} -\dfrac{7}{6}$$

$$-\dfrac{7}{4} \neq -\dfrac{7}{6}$$

$w = 7$ is **not** a solution to the equation.

5. $\quad 2w + 9 = 15$

$$2w + 9 - 9 = 15 - 9$$

$$2w = 6$$

$$\dfrac{2w}{2} = \dfrac{6}{2}$$

$$w = 3$$

Check: $\quad 2(3) + 9 \overset{?}{=} 15$

$$6 + 9 \overset{?}{=} 15$$

$$15 = 15 \ \text{T}$$

This is a conditional equation. Solution set: $\{3\}$

6. $\quad -4 = 8 - 3y$

$$-4 - 8 = 8 - 3y - 8$$

$$-12 = -3y$$

$$\dfrac{-12}{-3} = \dfrac{-3y}{-3}$$

$$4 = y$$

Check: $\quad -4 \overset{?}{=} 8 - 3(4)$

$$-4 \overset{?}{=} 8 - 12$$

$$-4 = -4 \ \text{T}$$

This is a conditional equation. Solution set: $\{4\}$

7. $\quad 2x + 5x - 1 = 20$

$$7x - 1 = 20$$

$$7x - 1 + 1 = 20 + 1$$

$$7x = 21$$

$$x = 3$$

Check: $\quad 2(3) + 5(3) - 1 \overset{?}{=} 20$

$$6 + 15 - 1 \overset{?}{=} 20$$

$$20 = 20 \ \text{T}$$

This is a conditional equation. Solution set: $\{3\}$

8. $\quad 7x + 5 - 8x = 13$

$$-x + 5 = 13$$

$$-x + 5 - 5 = 13 - 5$$

$$-x = 8$$

$$x = -8$$

Check: $\quad 7(-8) + 5 - 8(-8) \overset{?}{=} 13$

$$-56 + 5 + 64 \overset{?}{=} 13$$

$$13 = 13 \ \text{T}$$

This is a conditional equation.
Solution set: $\{-8\}$

9. $\quad -2(x - 4) = 8 - 2x$

$$-2x + 8 = 8 - 2x$$

$$-2x + 8 - 8 = 8 - 2x - 8$$

$$-2x = -2x$$

$$-2x + 2x = -2x + 2x$$

$$0 = 0$$

This statement is an identity. The equation is true for all real numbers. Solution set:
$\{x \mid x \text{ is a real number}\}$, or $\mathbb{R}$

10. $\quad 3(2r + 1) - 5 = 9(r - 1) - 3r$

$$6r + 3 - 5 = 9r - 9 - 3r$$

$$6r - 2 = 6r - 9$$

$$6r - 2 + 2 = 6r - 9 + 2$$

$$6r = 6r - 7$$

$$6r - 6r = 6r - 7 - 6r$$

$$0 = -7$$

This statement is a contradiction. The equation has no solution. Solution set: $\{ \ \}$ or $\varnothing$

11. $\quad \dfrac{2y + 3}{4} - \dfrac{y}{2} = 5$

$$4\left(\dfrac{2y + 3}{4}\right) - 4\left(\dfrac{y}{2}\right) = 4(5)$$

$$2y + 3 - 2y = 20$$

$$3 = 20$$

This statement is a contradiction. The equation has no solution. Solution set: $\{ \ \}$ or $\varnothing$

12.
$$\frac{x}{3}+\frac{2x}{5}=\frac{x-20}{15}$$

$$15\left(\frac{x}{3}\right)+15\left(\frac{2x}{5}\right)=15\left(\frac{x-20}{15}\right)$$

$$5x+6x=x-20$$

$$11x=x-20$$

$$11x-x=x-20-x$$

$$10x=-20$$

$$\frac{10x}{10}=\frac{-20}{10}$$

$$x=-2$$

Check:

$$\frac{-2}{3}+\frac{2(-2)}{5}\overset{?}{=}\frac{(-2)-20}{15}$$

$$-\frac{2}{3}-\frac{4}{5}\overset{?}{=}-\frac{22}{15}$$

$$-\frac{10}{15}-\frac{12}{15}\overset{?}{=}-\frac{22}{15}$$

$$-\frac{22}{15}=-\frac{22}{15}\ \ \text{T}$$

This is a conditional equation.
Solution set: $\{-2\}$

13.
$$0.2(x-6)+1.75=4.25+0.1(3x+10)$$

$$0.2x-1.2+1.75=4.25+0.3x+1$$

$$0.2x+0.55=5.25+0.3x$$

$$0.2x-0.3x+0.55=5.25+0.3x-0.3x$$

$$-0.1x+0.55=5.25$$

$$-0.1x+0.55-0.55=5.25-0.55$$

$$-0.1x=4.7$$

$$\frac{-0.1x}{-0.1}=\frac{4.7}{-0.1}$$

$$x=-47$$

Check:

$$0.2(-47-6)+1.75\overset{?}{=}4.25+0.1(3(-47)+10)$$

$$0.2(-53)+1.75\overset{?}{=}4.25+0.1(-141+10)$$

$$-10.6+1.75\overset{?}{=}4.25+0.1(-131)$$

$$-8.85\overset{?}{=}4.25-13.1$$

$$-8.85=-8.85\ \ \text{T}$$

This is a conditional equation.
Solution set: $\{-47\}$

(Note: Alternatively, we could have started by multiplying both sides of the equation by 100 to clear the decimals.)

14.
$$2.1w-3(2.4-0.2w)=0.9(3w-5)-2.7$$

$$2.1w-7.2+0.6w=2.7w-4.5-2.7$$

$$2.7w-7.2=2.7w-7.2$$

$$0=0$$

This statement is an identity. The equation is true for all real numbers. Solution set:
$\{w\,|\,w\text{ is a real number}\}$, or $\mathbb{R}$

(Note: Alternatively, we could have started by multiplying both sides of the equation by 10 to clear the decimals.)

15. We need to set the denominator equal to 0 and solve the resulting equation.

$$2x+3=0$$

$$2x+3-3=0-3$$

$$2x=-3$$

$$\frac{2x}{2}=\frac{-3}{2}$$

$$x=-\frac{3}{2}$$

Thus, $x=-\dfrac{3}{2}$ must be excluded from the domain.

16. We need to set the denominator equal to 0 and solve the resulting equation.

$$6(x-1)+3=0$$

$$6x-6+3=0$$

$$6x-3=0$$

$$6x-3+3=0+3$$

$$6x=3$$

$$\frac{6x}{6}=\frac{3}{6}$$

$$x=\frac{1}{2}$$

Thus, $x=\dfrac{1}{2}$ must be excluded from the domain.

17.
$$2370=0.06(x-9000)+315$$

$$2370=0.06x-540+315$$

$$2370=0.06x-225$$

$$2370+225=0.06x-225+225$$

$$2595=0.06x$$

$$\frac{2595}{0.06}=\frac{0.06x}{0.06}$$

$$43,250=x$$

Her Missouri taxable income was $43,250.

18.
$$x + 4(x - 10) = 69.75$$
$$x + 4x - 40 = 69.75$$
$$5x - 40 = 69.75$$
$$5x - 40 + 40 = 69.75 + 40$$
$$5x = 109.75$$
$$\frac{5x}{5} = \frac{109.75}{5}$$
$$x = 21.95$$
The regular club price for a DVD is $21.95.

19. Let x = the number.
$$3x + 7 = 22$$

20. Let x = the number.
$$x - 3 = \frac{x}{2}$$

21. Let x = the number.
$$0.2x = x - 12$$

22. Let x = the number.
$$6x = 2x - 4$$

23. Let x = Payton's age.
Shawn's age: $x + 8$
$$x + (x + 8) = 18$$
$$2x + 8 = 18$$
$$2x + 8 - 8 = 18 - 8$$
$$2x = 10$$
$$\frac{2x}{2} = \frac{10}{2}$$
$$x = 5$$
Payton is 5 years old and Shawn is 13 years old.

24. Let x = the first odd integer.
Second odd integer: $x + 2$
Third odd integer: $x + 4$
Fourth odd integer: $x + 6$
Fifth odd integer: $x + 8$
$$x + (x + 2) + (x + 4) + (x + 6) + (x + 8) = 125$$
$$5x + 20 = 125$$
$$5x + 20 - 20 = 125 - 20$$
$$5x = 105$$
$$\frac{5x}{5} = \frac{105}{5}$$
$$x = 21$$
The five odd integers are 21, 23, 25, 27, and 29.

25. Let x = the final exam score.
$$\frac{85 + 81 + 84 + 77 + 2x}{6} = 80$$
$$\frac{327 + 2x}{6} = 80$$
$$6 \cdot \frac{327 + 2x}{6} = 6 \cdot 80$$
$$327 + 2x = 480$$
$$327 + 2x - 327 = 480 - 327$$
$$2x = 153$$
$$\frac{2x}{2} = \frac{153}{2}$$
$$x = 76.5$$
Logan needs to get a score of 76.5 on the final exam to have an average of 80.

26. Here we can use the simple interest formula $I = Prt$. Remember that t is in years and that the interest rate should be written as a decimal.
$$I = 15,000(0.0549)\left(\frac{1}{12}\right)$$
$$= 68.625$$
After 1 month, Cherie will accrue $68.63 in interest.

27. Let x = the original price.
Discount: $0.3x$
$$\text{sale price} = \text{original price} - \text{discount}$$
$$94.50 = x - 0.3x$$
$$94.50 = 0.7x$$
$$\frac{94.50}{0.7} = \frac{0.7x}{0.7}$$
$$135 = x$$
The original price of the sleeping bag was $135.00.

28. Let x = the federal minimum wage.
Increase: $0.107x$
$$7.25 = x + 0.107x$$
$$7.25 = 1.107x$$
$$\frac{7.25}{1.107} = \frac{1.107x}{1.107}$$
$$6.55 \approx x$$
The federal minimum wage was $6.55 (per hour).

29. Let x = pounds of blueberries.
Strawberries: $12 - x$

tot. rev. = rev. from b.b. + rev. from s.b.
$$(\text{price})(\text{lbs}) = (\text{price})(\text{lbs}) + (\text{price})(\text{lbs})$$
$$(12.95)(12) = (10.95)(x) + (13.95)(12 - x)$$
$$155.4 = 10.95x + 167.4 - 13.95x$$
$$155.4 = 167.4 - 3x$$
$$155.4 - 167.4 = 167.4 - 3x - 167.4$$
$$-12 = -3x$$
$$\frac{-12}{-3} = \frac{-3x}{-3}$$
$$4 = x$$

The store should mix 4 pounds of chocolate covered blueberries with 8 pounds of chocolate covered strawberries.

30. Let x = pounds of baseball gum.
Soccer gum: $10 - x$

tot. rev. = rev. from b.b. + rev. from s.b.
$$(\text{price})(\text{lbs}) = (\text{price})(\text{lbs}) + (\text{price})(\text{lbs})$$
$$(3.75)(10) = (3.50)(x) + (4.50)(10 - x)$$
$$37.5 = 3.5x + 45 - 4.5x$$
$$37.5 = 45 - x$$
$$37.5 - 45 = 45 - x - 45$$
$$-7.5 = -x$$
$$7.5 = x$$

The store should mix 7.5 pounds of baseball gumballs with 2.5 pounds of soccer gumballs.

31. Let x = amount invested at 8%.
Amount at 18%: $8000 - x$

tot. return = ret. from 8% + ret. from 18%
$$(\%)(\$) = (\%)(\$) + (\%)(\$)$$
$$(0.12)(8000) = (0.08)(x) + (0.18)(8000 - x)$$
$$960 = 0.08x + 1440 - 0.18x$$
$$960 = 1440 - 0.1x$$
$$960 - 1440 = 1440 - 0.1x - 1440$$
$$-480 = -0.1x$$
$$\frac{-480}{-0.1} = \frac{-0.1x}{-0.1}$$
$$4800 = x$$

Angie should invest $4800 at 8% and $3200 at 18% to achieve a 12% return.

32. Let x = quarts drained = quarts added.
We can write the following equation for the amount of antifreeze:

final amt = amt now - amt drain + amt add
$$(\%)(\text{L}) = (\%)(\text{L}) - (\%)(\text{L}) + (\%)(\text{L})$$
$$(0.5)(7.5) = (0.3)(7.5) - (0.3)(x) + (1.0)(x)$$
$$3.75 = 2.25 - 0.3x + x$$
$$3.75 = 2.25 + 0.7x$$
$$3.75 - 2.25 = 2.25 + 0.7x - 2.25$$
$$1.5 = 0.7x$$
$$\frac{1.5}{0.7} = \frac{0.7x}{0.7}$$
$$2.14 \approx x$$

About 2.14 quarts would need to be drained and replaced with pure antifreeze.

33. For this problem, we will need the distance traveled formula: $d = rt$.
Let x = miles driven at 60 mph.
Miles at 70 mph: $300 - x$

tot. time = time at 60 mph + time at 70 mph
$$4.5 = \frac{x}{60} + \frac{300 - x}{70}$$
$$420(4.5) = 420\left(\frac{x}{60} + \frac{300 - x}{70}\right)$$
$$1890 = 7x + 1800 - 6x$$
$$1890 = x + 1800$$
$$1890 - 1800 = x + 1800 - 1800$$
$$90 = x$$

Josh drove 90 miles at 60 miles per hour and 210 miles at 70 miles per hour.

34. For this problem, we will need the distance traveled formula: $d = rt$

We also need to note that 50 minutes is $\dfrac{5}{6}$ of an hour.

Let x = speed of the F14.
Speed of F15: $x + 200$.

tot. dist. = dist. of F14 + dist. of F15
$$2200 = \frac{5}{6}x + \frac{5}{6}(x + 200)$$
$$2200 = \frac{5}{6}x + \frac{5}{6}x + \frac{500}{3}$$
$$2200 = \frac{5}{3}x + \frac{500}{3}$$
$$3(2200) = 3\left(\frac{5}{3}x + \frac{500}{3}\right)$$
$$6600 = 5x + 500$$
$$6600 - 500 = 5x + 500 - 500$$
$$6100 = 5x$$
$$\frac{6100}{5} = \frac{5x}{5}$$
$$1220 = x$$

The F14 is traveling at a speed of 1220 miles per hour and the F15 is traveling at a speed of 1420 miles per hour.

35.
$$y = \frac{k}{x}$$
$$y \cdot x = \frac{k}{x} \cdot x$$
$$xy = k$$
$$\frac{xy}{y} = \frac{k}{y}$$
$$x = \frac{k}{y}$$

36.
$$F = \frac{9}{5}C + 32$$
$$F - 32 = \frac{9}{5}C + 32 - 32$$
$$F - 32 = \frac{9}{5}C$$
$$\frac{5}{9}(F - 32) = \frac{5}{9}\left(\frac{9}{5}C\right)$$
$$\frac{5}{9}(F - 32) = C$$

37.
$$P = 2L + 2W$$
$$P - 2L = 2L + 2W - 2L$$
$$P - 2L = 2W$$
$$\frac{P - 2L}{2} = \frac{2W}{2}$$
$$\frac{P - 2L}{2} = W$$

38.
$$\rho = m_1 v_1 + m_2 v_2$$
$$\rho - m_1 v_1 = m_1 v_1 + m_2 v_2 - m_1 v_1$$
$$\rho - m_1 v_1 = m_2 v_2$$
$$\frac{\rho - m_1 v_1}{v_2} = \frac{m_2 v_2}{v_2}$$
$$\frac{\rho - m_1 v_1}{v_2} = m_2$$

39.
$$PV = nRT$$
$$\frac{PV}{nR} = \frac{nRT}{nR}$$
$$\frac{PV}{nR} = T$$

40.
$$S = 2LW + 2LH + 2WH$$
$$S - 2LH = 2LW + 2LH + 2WH - 2LH$$
$$S - 2LH = 2LW + 2WH$$
$$S - 2LH = W(2L + 2H)$$
$$\frac{S - 2LH}{2L + 2H} = \frac{W(2L + 2H)}{2L + 2H}$$
$$\frac{S - 2LH}{2L + 2H} = W$$

41.
$$3x + 4y = 2$$
$$3x + 4y - 3x = 2 - 3x$$
$$4y = -3x + 2$$
$$\frac{4y}{4} = \frac{-3x + 2}{4}$$
$$y = -\frac{3}{4}x + \frac{1}{2}$$

42.
$$-5x + 4y = 10$$
$$-5x + 4y + 5x = 10 + 5x$$
$$4y = 5x + 10$$
$$\frac{4y}{4} = \frac{5x + 10}{4}$$
$$y = \frac{5}{4}x + \frac{5}{2}$$

43.
$$48x - 12y = 60$$
$$48x - 12y - 48x = 60 - 48x$$
$$-12y = -48x + 60$$
$$\frac{-12y}{-12} = \frac{-48x + 60}{-12}$$
$$y = 4x - 5$$

44.
$$\frac{2}{5}x + \frac{1}{3}y = 8$$
$$\frac{2}{5}x + \frac{1}{3}y - \frac{2}{5}x = 8 - \frac{2}{5}x$$
$$\frac{1}{3}y = -\frac{2}{5}x + 8$$
$$3\left(\frac{1}{3}y\right) = 3\left(-\frac{2}{5}x + 8\right)$$
$$y = -\frac{6}{5}x + 24$$

45.
$$C = \frac{5}{9}(3221.6 - 32)$$
$$C = \frac{5}{9}(3189.6)$$
$$C = 1772$$
The melting point of platinum is $1772°\text{C}$.

46. Let x = the measure of the congruent angles.
Third angle: $x - 30$
$$x + x + (x - 30) = 180$$
$$3x - 30 = 180$$
$$3x - 30 + 30 = 180 + 30$$
$$3x = 210$$
$$\frac{3x}{3} = \frac{210}{3}$$
$$x = 70$$
The angles measure 70°, 70°, and 40°.

47. Let w = width of the window in feet.
Length: $l = w + 8$
$$2l + 2w = P$$
$$2(w + 8) + 2w = 76$$
$$2w + 16 + 2w = 76$$
$$4w + 16 = 76$$
$$4w + 16 - 16 = 76 - 16$$
$$4w = 60$$
$$\frac{4w}{4} = \frac{60}{4}$$
$$w = 15$$
$l = w + 8 = 15 + 8 = 23$
The window measures 15 feet by 23 feet.

48. a.
$$C = 2.95 + 0.04x$$
$$C - 2.95 = 2.95 + 0.04x - 2.95$$
$$C - 2.95 = 0.04x$$
$$\frac{C - 2.95}{0.04} = \frac{0.04x}{0.04}$$
$$25C - 73.75 = x \quad \text{or} \quad x = 25C - 73.75$$

b.
$$x = 25(20) - 73.75$$
$$x = 500 - 73.75$$
$$x = 426.25$$
Debbie can talk for 426 minutes in one month and not spend more than $20 on long distance.

49. Let x = thickness in feet.
$$V = lwh$$
$$80 = 12 \cdot 18 \cdot x$$
$$80 = 216x$$
$$\frac{80}{216} = \frac{216x}{216}$$
$$\frac{10}{27} = x \quad \text{or} \quad x \approx 0.37$$
The patio will be $\frac{10}{27}$ of a foot thick (i.e. about 4.44 inches).

50. a.
$$A = \pi s (R + r)$$
$$\frac{A}{\pi s} = \frac{\pi s (R + r)}{\pi s}$$
$$\frac{A}{\pi s} = R + r$$
$$\frac{A}{\pi s} - R = r \quad \text{or} \quad r = \frac{A - \pi R s}{\pi s}$$

b.
$$r = \frac{10\pi - \pi(3)(2)}{\pi(2)}$$
$$r = \frac{10 - 6}{2}$$
$$r = \frac{4}{2} = 2$$
The radius of the top of the frustum is 2 feet.

51. a.
$$C = 7.48 + 0.08674x$$
$$C - 7.48 = 7.48 + 0.08674x - 7.48$$
$$C - 7.48 = 0.08674x$$
$$\frac{C - 7.48}{0.08674} = \frac{0.08674x}{0.08674}$$
$$x = \frac{C - 7.48}{0.08674}$$

b.
$$x = \frac{206.03 - 7.48}{0.08674}$$
$$= \frac{198.55}{0.08674}$$
$$\approx 2289$$
Approximately 2289 kwh were used.

52. Let x = the measure of the angle.
Complement: $90 - x$
Supplement: $180 - x$
$$(90 - x) + (180 - x) = 150$$
$$90 - x + 180 - x = 150$$
$$270 - 2x = 150$$
$$270 - 2x - 270 = 150 - 270$$
$$-2x = -120$$
$$\frac{-2x}{-2} = \frac{-120}{-2}$$
$$x = 60$$
The angle measures 60°.

53. $2 < x \le 7$
Interval: (2, 7]

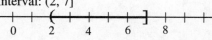

54. $x > -2$

Interval: $(-2, \infty)$

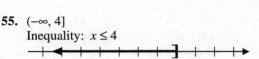

55. $(-\infty, 4]$

Inequality: $x \le 4$

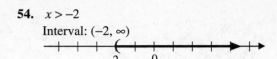

56. $[-1, 3)$

Inequality: $-1 \le x < 3$

57.
$$5 \le x \le 9$$
$$2 \cdot 5 \le 2 \cdot x \le 2 \cdot 9$$
$$10 \le 2x \le 18$$
$$10 - 3 \le 2x - 3 \le 18 - 3$$
$$7 \le 2x - 3 \le 15$$

Therefore, $a = 7$ and $b = 15$.

58.
$$-2 < x < 0$$
$$3(-2) < 3(x) < 3(0)$$
$$-6 < 3x < 0$$
$$-6 + 5 < 3x + 5 < 0 + 5$$
$$-1 < 3x + 5 < 5$$

Therefore, $a = -1$ and $b = 5$.

59.
$$3x + 12 \le 0$$
$$3x + 12 - 12 \le 0 - 12$$
$$3x \le -12$$
$$\frac{3x}{3} \le \frac{-12}{3}$$
$$x \le -4$$

Solution set: $\{x \mid x \le -4\}$

Interval: $(-\infty, -4]$

Graph:

60.
$$2 < 1 - 3x$$
$$2 - 1 < 1 - 3x - 1$$
$$1 < -3x$$
$$\frac{1}{-3} > \frac{-3x}{-3}$$
$$-\frac{1}{3} > x \text{ or } x < -\frac{1}{3}$$

Solution set: $\left\{ x \mid x < -\frac{1}{3} \right\}$

Interval: $\left(-\infty, -\frac{1}{3} \right)$

Graph:

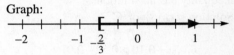

61.
$$-7 \le 3(h + 1) - 8$$
$$-7 \le 3h + 3 - 8$$
$$-7 \le 3h - 5$$
$$-7 + 5 \le 3h - 5 + 5$$
$$-2 \le 3h$$
$$\frac{-2}{3} \le \frac{3h}{3}$$
$$-\frac{2}{3} \le h \text{ or } h \ge -\frac{2}{3}$$

Solution set: $\left\{ h \mid h \ge -\frac{2}{3} \right\}$

Interval: $\left[-\frac{2}{3}, \infty \right)$

Graph:

62.
$$-7x - 8 < -22$$
$$-7x - 8 + 8 < -22 + 8$$
$$-7x < -14$$
$$\frac{-7x}{-7} > \frac{-14}{-7}$$
$$x > 2$$

Solution set: $\{x \mid x > 2\}$

Interval: $(2, \infty)$

Graph:

63. $3(p - 2) + (5 - p) > 2 - (p - 3)$
$$3p - 6 + 5 - p > 2 - p + 3$$
$$2p - 1 > 5 - p$$
$$2p + p - 1 > 5 - p + p$$
$$3p - 1 > 5$$
$$3p - 1 + 1 > 5 + 1$$
$$3p > 6$$
$$\frac{3p}{3} > \frac{6}{3}$$
$$p > 2$$

Solution set: $\{p \mid p > 2\}$

Interval: $(2, \infty)$

Graph:

64. $2(x+1)+1 > 2(x-2)$

$2x+2+1 > 2x-4$

$2x+3 > 2x-4$

$2x+3-2x > 2x-4-2x$

$3 > -4$

This statement is true for all real numbers.

Solution set: $\{x \mid x \text{ is any real number}\}$

Interval: $(-\infty, \infty)$

Graph:

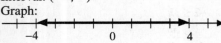

65. $5(x-1)-7x > 2(2-x)$

$5x-5-7x > 4-2x$

$-2x-5 > 4-2x$

$-2x-5+2x > 4-2x+2x$

$-5 > 4$ false

This statement is false for all real numbers. The inequality has no solutions.

Solution set: $\{\ \}$ or $\varnothing$

66. $0.03x+0.10 > 0.52-0.07x$

$0.03x+0.10-0.10 > 0.52-0.07x-0.10$

$0.03x > 0.42-0.07x$

$0.03x+0.07x > 0.42-0.07x+0.07x$

$0.1x > 0.42$

$\dfrac{0.1x}{0.1} > \dfrac{0.42}{0.1}$

$x > 4.2$

Solution set: $\{x \mid x > 4.2\}$

Interval: $(4.2, \infty)$

Graph:

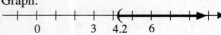

67. $-\dfrac{4}{9}w+\dfrac{7}{12} < \dfrac{5}{36}$

$36\left(-\dfrac{4}{9}w+\dfrac{7}{12}\right) < 36\left(\dfrac{5}{36}\right)$

$-16w+21 < 5$

$-16w+21-21 < 5-21$

$-16w < -16$

$\dfrac{-16w}{-16} > \dfrac{-16}{-16}$

$w > 1$

Solution set: $\{w \mid w > 1\}$

Interval: $(1, \infty)$

Graph:

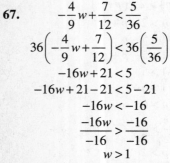

68. $\dfrac{2}{5}y-20 > \dfrac{2}{3}y+12$

$15\left(\dfrac{2}{5}y-20\right) > 15\left(\dfrac{2}{3}y+12\right)$

$6y-300 > 10y+180$

$6y-300-10y > 10y+180-10y$

$-4y-300 > 180$

$-4y-300+300 > 180+300$

$-4y > 480$

$\dfrac{-4y}{-4} < \dfrac{480}{-4}$

$y < -120$

Solution set: $\{y \mid y < -120\}$

Interval: $(-\infty, -120)$

Graph:

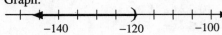

69. Let x = number of people attending.

$7.5x+150 \le 600$

$7.5x+150-150 \le 600-150$

$7.5x \le 450$

$\dfrac{7.5x}{7.5} \le \dfrac{450}{7.5}$

$x \le 60$

To stay within budget, no more than 60 people can attend the banquet.

70. Let x = number of miles driven daily (average)

Miles charged: $x-150$

$43.46+0.25(x-150) \le 60$

$43.46+0.25x-37.5 \le 60$

$0.25x+5.96 \le 60$

$0.25x+5.96-5.96 \le 60-5.96$

$0.25x \le 54.04$

$\dfrac{0.25x}{0.25} \le \dfrac{54.04}{0.25}$

$x \le 216.16$

To stay within budget, you can drive an average of 216 miles per day.

71. Let x = number of candy bars sold.

$1.0x > 50+0.6x$

$1.0x-0.6x > 50+0.6x-0.6x$

$0.4x > 50$

$\dfrac{0.4x}{0.4} > \dfrac{50}{0.4}$

$x > 125$

The band must sell more than 125 candy bars to make a profit.

72. Let x = number of DVDs purchased.

$$24.95 + 9.95(x-1) \le 72$$
$$24.95 + 9.95x - 9.95 \le 72$$
$$15 + 9.95x \le 72$$
$$15 + 9.95x - 15 \le 72 - 15$$
$$9.95x \le 57$$
$$\frac{9.95x}{9.95} \le \frac{57}{9.95}$$
$$x \le \frac{57}{9.95} \approx 5.73$$

You can purchase up to 5 DVDs and still be within budget.

73. A: quadrant IV; B: quadrant III; C: y-axis; D: quadrant II; E: x-axis; F: quadrant I

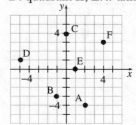

74. A: x-axis; B: quadrant I; C: quadrant III; D: quadrant II; E: quadrant IV; F: y-axis

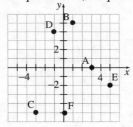

75. a. Let $x = 3$ and $y = 1$.

$$3(3) - 2(1) \overset{?}{=} 7$$
$$9 - 2 \overset{?}{=} 7$$
$$7 = 7 \text{ True}$$

The point (3, 1) is on the graph.

b. Let $x = 2$ and $y = -1$.

$$3(2) - 2(-1) \overset{?}{=} 7$$
$$6 + 2 \overset{?}{=} 7$$
$$8 = 7 \text{ False}$$

The point (2, −1) is not on the graph.

c. Let $x = 4$ and $y = 0$.

$$3(4) - 2(0) \overset{?}{=} 7$$
$$12 - 0 \overset{?}{=} 7$$
$$12 = 7 \text{ False}$$

The point (4, 0) is not on the graph.

d. Let $x = \dfrac{1}{3}$ and $y = -3$.

$$3\left(\frac{1}{3}\right) - 2(-3) \overset{?}{=} 7$$
$$1 + 6 \overset{?}{=} 7$$
$$7 = 7 \text{ True}$$

The point $\left(\dfrac{1}{3}, -3\right)$ is on the graph.

76. a. Let $x = -1$ and $y = 3$.

$$3 \overset{?}{=} 2(-1)^2 - 3(-1) + 2$$
$$3 \overset{?}{=} 2(1) + 3 + 2$$
$$3 \overset{?}{=} 2 + 3 + 2$$
$$3 = 7 \text{ False}$$

The point (−1, 3) is not on the graph.

b. Let $x = 1$ and $y = 1$.

$$1 \overset{?}{=} 2(1)^2 - 3(1) + 2$$
$$1 \overset{?}{=} 2(1) - 3 + 2$$
$$1 \overset{?}{=} 2 - 3 + 2$$
$$1 = 1 \text{ True}$$

The point (1, 1) is on the graph.

c. Let $x = -2$ and $y = 16$.

$$16 \overset{?}{=} 2(-2)^2 - 3(-2) + 2$$
$$16 \overset{?}{=} 2(4) + 6 + 2$$
$$16 \overset{?}{=} 8 + 6 + 2$$
$$16 = 16 \text{ True}$$

The point (−2, 16) is on the graph.

d. Let $x = \dfrac{1}{2}$ and $y = \dfrac{3}{2}$.

$$\frac{3}{2} \overset{?}{=} 2\left(\frac{1}{2}\right)^2 - 3\left(\frac{1}{2}\right) + 2$$
$$\frac{3}{2} \overset{?}{=} 2\left(\frac{1}{4}\right) - \frac{3}{2} + 2$$
$$\frac{3}{2} \overset{?}{=} \frac{1}{2} - \frac{3}{2} + 2$$
$$\frac{3}{2} = 1 \text{ False}$$

The point $\left(\dfrac{1}{2}, \dfrac{3}{2}\right)$ is not on the graph.

77. $y = x + 2$

x	$y = x + 2$	(x, y)
-3	$y = (-3) + 2 = -1$	$(-3, -1)$
-1	$y = (-1) + 2 = 1$	$(-1, 1)$
0	$y = (0) + 2 = 2$	$(0, 2)$
1	$y = (1) + 2 = 3$	$(1, 3)$
3	$y = (3) + 2 = 5$	$(3, 5)$

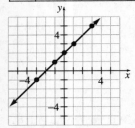

78. $2x + y = 3$
$y = -2x + 3$

x	$y = -2x + 3$	(x, y)
-3	$y = -2(-3) + 3 = 9$	$(-3, 9)$
-1	$y = -2(-1) + 3 = 5$	$(-1, 5)$
0	$y = -2(0) + 3 = 3$	$(0, 3)$
1	$y = -2(1) + 3 = 1$	$(1, 1)$
3	$y = -2(3) + 3 = -3$	$(3, -3)$

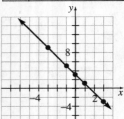

79. $y = -x^2 + 4$

x	$y = -x^2 + 4$	(x, y)
-3	$y = -(-3)^2 + 4 = -5$	$(-3, -5)$
-1	$y = -(-1)^2 + 4 = 3$	$(-1, 3)$
0	$y = -(0)^2 + 4 = 4$	$(0, 4)$
1	$y = -(1)^2 + 4 = 3$	$(1, 3)$
3	$y = -(3)^2 + 4 = -5$	$(3, -5)$

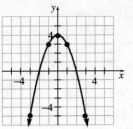

80. $y = |x + 2| - 1$

x	$y = |x + 2| - 1$	(x, y)
-5	$y = |-5 + 2| - 1 = 2$	$(-5, 2)$
-3	$y = |-3 + 2| - 1 = 0$	$(-3, 0)$
-1	$y = |-1 + 2| - 1 = 0$	$(-1, 0)$
0	$y = |0 + 2| - 1 = 1$	$(0, 1)$
1	$y = |1 + 2| - 1 = 2$	$(1, 2)$

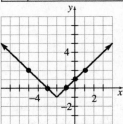

81. $y = x^3 + 2$

x	$y = x^3 + 2$	(x, y)
-2	$y = (-2)^3 + 2 = -6$	$(-2, -6)$
-1	$y = (-1)^3 + 2 = 1$	$(-1, 1)$
0	$y = (0)^3 + 2 = 2$	$(0, 2)$
1	$y = (1)^3 + 2 = 3$	$(1, 3)$
2	$y = (2)^3 + 2 = 10$	$(2, 10)$

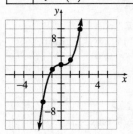

82. $x = y^2 + 1$

y	$x = y^2 + 1$	$(x,\ y)$
-2	$x = (-2)^2 + 1 = 5$	$(5, -2)$
-1	$x = (-1)^2 + 1 = 2$	$(2, -1)$
0	$x = (0)^2 + 1 = 1$	$(1, 0)$
1	$x = (1)^2 + 1 = 2$	$(2, 1)$
2	$x = (2)^2 + 1 = 5$	$(5, 2)$

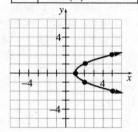

83. The intercepts are $(-3, 0)$, $(0, -1)$, and $(0, 3)$.

84. a. Since 2250 is less than 3000, the monthly bill will still be $40. Using the graph, we find that when $x = 2.25$ (2250 minutes) the y-value is 40, or $40.

   b. For 12,000 minutes, we have $x = 12$. According to the graph, the monthly bill for 12,000 minutes will be about $500.

85. Let $x = -2, -1, 0,$ and 1.

$$x = -2: \quad -2 + y = 7 \qquad x = -1: \quad -1 + y = 7$$
$$y = 9 \qquad\qquad\qquad\qquad y = 8$$

$$x = 0: \quad 0 + y = 7 \qquad x = 1: \quad 1 + y = 7$$
$$y = 7 \qquad\qquad\qquad\qquad y = 6$$

Thus, the points $(-2, 9)$, $(-1, 8)$, $(0, 7)$, and $(1, 6)$ are on the graph.

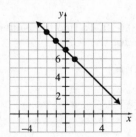

86. Let $x = -2, -1, 0,$ and 1.

$$x = -2: \quad -2 - y = -4 \qquad x = -1: \quad -1 - y = -4$$
$$-y = -2 \qquad\qquad\qquad\qquad -y = -3$$
$$y = 2 \qquad\qquad\qquad\qquad\quad y = 3$$

$$x = 0: \quad 0 - y = -4 \qquad x = 1: \quad 1 - y = -4$$
$$-y = -4 \qquad\qquad\qquad\qquad -y = -5$$
$$y = 4 \qquad\qquad\qquad\qquad\quad y = 5$$

Thus, the points $(-2, 2)$, $(-1, 3)$, $(0, 4)$, and $(1, 5)$ are on the graph.

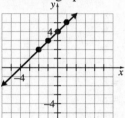

87. Let $x = -2, 0, 2,$ and 4.

$$x = -2: \quad 5(-2) - 2y = 6$$
$$-10 - 2y = 6$$
$$-2y = 16$$
$$y = -8$$

$$x = 0: \quad 5(0) - 2y = 6$$
$$-2y = 6$$
$$y = -3$$

$$x = 2: \quad 5(2) - 2y = 6$$
$$10 - 2y = 6$$
$$-2y = -4$$
$$y = 2$$

$$x = 4: \quad 5(4) - 2y = 6$$
$$20 - 2y = 6$$
$$-2y = -14$$
$$y = 7$$

Thus, the points $(-2, -8)$, $(0, -3)$, $(2, 2)$, and $(4, 7)$ are on the graph.

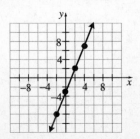

88. Let $x = -4, -2, 0,$ and 2.

$x = -4:$ $\quad -3(-4) + 2y = 8$
$\qquad\qquad 12 + 2y = 8$
$\qquad\qquad\quad 2y = -4$
$\qquad\qquad\quad\; y = -2$

$x = -2:$ $\quad -3(-2) + 2y = 8$
$\qquad\qquad 6 + 2y = 8$
$\qquad\qquad\; 2y = 2$
$\qquad\qquad\;\; y = 1$

$x = 0:$ $\quad -3(0) + 2y = 8$
$\qquad\qquad 2y = 8$
$\qquad\qquad\; y = 4$

$x = 2:$ $\quad -3(2) + 2y = 8$
$\qquad\qquad -6 + 2y = 8$
$\qquad\qquad\;\; 2y = 14$
$\qquad\qquad\;\;\; y = 7$

Thus, the points $(-4, -2), (-2, 1), (0, 4),$ and $(2, 7)$ are on the graph.

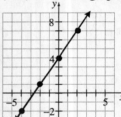

89. To find the *y*-intercept, let $x = 0$ and solve for *y*.

$x = 0:$ $\quad 5(0) + 3y = 30$
$\qquad\qquad 3y = 30$
$\qquad\qquad\; y = 10$

The *y*-intercept is (0, 10). To find the *x*-intercept, let $y = 0$ and solve for *x*.

$y = 0:$ $\quad 5x + 3(0) = 30$
$\qquad\qquad 5x = 30$
$\qquad\qquad\; x = 6$

The *x*-intercept is (6, 0).

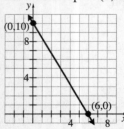

90. To find the *y*-intercept, let $x = 0$ and solve for *y*.

$x = 0:$ $\quad 4(0) + 3y = 0$
$\qquad\qquad 4y = 0$
$\qquad\qquad\; y = 0$

The *y*-intercept is (0, 0). To find the *x*-intercept, let $y = 0$ and solve for *x*.

$y = 0:$ $\quad 4x + 3(0) = 0$
$\qquad\qquad 4x = 0$
$\qquad\qquad\; x = 0$

The *x*-intercept is (0, 0). Since the *x*- and *y*-intercepts both result in the origin, we must find other points on the graph. Let $x = -3$ and 3 and solve for *y*.

$x = -3:$ $\quad 4(-3) + 3y = 0$
$\qquad\qquad -12 + 3y = 0$
$\qquad\qquad\quad\; 3y = 12$
$\qquad\qquad\qquad y = 4$

$x = 3:$ $\quad 4(3) + 3y = 0$
$\qquad\qquad 12 + 3y = 0$
$\qquad\qquad\;\; 3y = -12$
$\qquad\qquad\qquad y = -4$

Thus, the points $(-3, 4)$ and $(3, -4)$ are also on the graph.

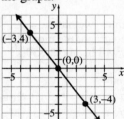

91. To find the *y*-intercept, let $x = 0$ and solve for *y*.

$x = 0:$ $\quad \dfrac{3}{4}(0) - \dfrac{1}{2}y = 1$
$\qquad\qquad\quad -\dfrac{1}{2}y = 1$
$\qquad\qquad\qquad\;\; y = -2$

The *y*-intercept is (0, −2). To find the *x*-intercept, let $y = 0$ and solve for *x*.

$y = 0:$ $\quad \dfrac{3}{4}x - \dfrac{1}{2}(0) = 1$
$\qquad\qquad\quad \dfrac{3}{4}x = 1$
$\qquad\qquad\qquad x = \dfrac{4}{3}$

The *x*-intercept is $\left(\dfrac{4}{3}, 0\right)$.

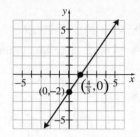

92. To find the *y*-intercept, let $x = 0$ and solve for *y*.

$x = 0$: $4(0) + y = 8$
$\qquad\qquad\qquad y = 8$

The *y*-intercept is (0, 8). To find the *x*-intercept, let $y = 0$ and solve for *x*.

$y = 0$: $4x + 0 = 8$
$\qquad\qquad 4x = 8$
$\qquad\qquad\quad x = 2$

The *x*-intercept is (2, 0).

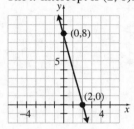

93. The equation $x = 4$ will be a vertical line with *x*-intercept (4, 0).

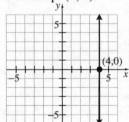

94. The equation $y = -8$ will be a horizontal line with *y*-intercept (0, −8).

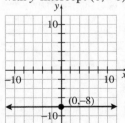

95. $3x + 5 = -1$
$\qquad\quad 3x = -6$
$\qquad\quad\ \ x = -2$

The equation $x = -2$ will be a vertical line with *x*-intercept (−2, 0).

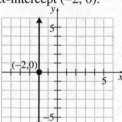

96. a. $m = \dfrac{y_2 - y_1}{x_2 - x_1} = \dfrac{4 - (-1)}{2 - (-2)} = \dfrac{5}{4}$

    b. For every 4 units of increase in *x*, *y* will increase by 5 units.

97. a. $m = \dfrac{y_2 - y_1}{x_2 - x_1} = \dfrac{-1 - 1}{5 - (-3)} = \dfrac{-2}{8} = -\dfrac{1}{4}$

    b. For every 4 units of increase in *x*, *y* will decrease by 1 unit. For every 4 units of decrease in *x*, *y* will increase by 1 unit.

98. $m = \dfrac{y_2 - y_1}{x_2 - x_1} = \dfrac{-1 - 5}{2 - (-1)} = \dfrac{-6}{3} = -2$

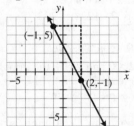

99. $m = \dfrac{y_2 - y_1}{x_2 - x_1} = \dfrac{-1 - 5}{0 - 4} = \dfrac{-6}{-4} = \dfrac{3}{2}$

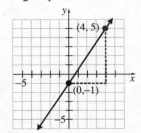

100. a.

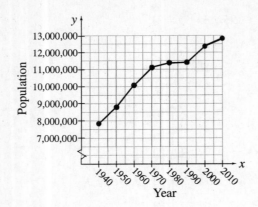

b. Average rate of change
$$= \frac{8,712,176 - 7,897,241}{1950 - 1940}$$
$= 81,493.5$ people per year

Between 1940 and 1950, the population of Illinois increased at a rate of 81,493.5 people per year.

c. Average rate of change
$$= \frac{11,430,602 - 11,427,409}{1990 - 1980}$$
$= 319.3$ people per year

Between 1980 and 1990, the population of Illinois increased at a rate of 319.3 people per year.

d. Average rate of change
$$= \frac{12,830,632 - 12,419,293}{2010 - 2000}$$
$= 41,133.9$ people per year

Between 2000 and 2010, the population of Illinois increased at a rate of 41,133.9 people per year.

e. No. The population of Illinois is not linearly related to the year. The average rate of change (slope) is not constant.

101. Start at $(-1, -5)$. Because $m = 4 = \frac{4}{1}$, move 1 unit to the right and 4 units up to find point $(0, -1)$.

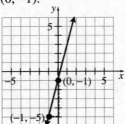

102. Start at $(3, 2)$. Because $m = -\frac{2}{3}$, move 3 units to the right and 2 units down to find point $(6, 0)$. We can also move 3 units to the left and 2 units up to find point $(0, 4)$.

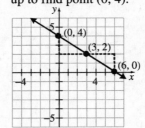

103. Because m is undefined, the line is vertical passing through the point $(2, -4)$.

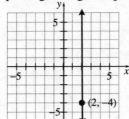

104. Because $m = 0$, the line is horizontal passing through the point $(-3, 1)$.

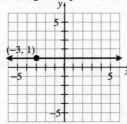

105. $m = \dfrac{y_2 - y_1}{x_2 - x_1} = \dfrac{-1-3}{6-(-2)} = \dfrac{-4}{8} = -\dfrac{1}{2}$

$y - y_1 = m(x - x_1)$

$y - (-1) = -\dfrac{1}{2}(x - 6)$

$y + 1 = -\dfrac{1}{2}x + 3$

$y = -\dfrac{1}{2}x + 2 \ \text{ or } \ x + 2y = 4$

106. $m = \dfrac{y_2 - y_1}{x_2 - x_1} = \dfrac{6-(-3)}{2-(-1)} = \dfrac{9}{3} = 3$

$y - y_1 = m(x - x_1)$

$y - 6 = 3(x - 2)$

$y - 6 = 3x - 6$

$y = 3x \ \text{ or } \ 3x - y = 0$

107. $y - y_1 = m(x - x_1)$

$y - 2 = -1(x - 3)$

$y - 2 = -x + 3$

$y = -x + 5 \ \text{ or } \ x + y = 5$

108. $y - y_1 = m(x - x_1)$

$y - (-4) = \dfrac{3}{5}(x - (-10))$

$y + 4 = \dfrac{3}{5}(x + 10)$

$y + 4 = \dfrac{3}{5}x + 6$

$y = \dfrac{3}{5}x + 2 \ \text{ or } \ 3x - 5y = -10$

109. $m = \dfrac{y_2 - y_1}{x_2 - x_1} = \dfrac{5-2}{-3-6} = \dfrac{3}{-9} = -\dfrac{1}{3}$

$y - y_1 = m(x - x_1)$

$y - 2 = -\dfrac{1}{3}(x - 6)$

$y - 2 = -\dfrac{1}{3}x + 2$

$y = -\dfrac{1}{3}x + 4 \ \text{ or } \ x + 3y = 12$

110. $m = \dfrac{y_2 - y_1}{x_2 - x_1} = \dfrac{3-3}{4-(-2)} = \dfrac{0}{6} = 0$

$y - y_1 = m(x - x_1)$

$y - 3 = 0(x - 4)$

$y - 3 = 0$

$y = 3$

111. $m = \dfrac{y_2 - y_1}{x_2 - x_1} = \dfrac{-7-(-1)}{1-4} = \dfrac{-6}{-3} = 2$

$y - y_1 = m(x - x_1)$

$y - (-7) = 2(x - 1)$

$y + 7 = 2x - 2$

$y = 2x - 9 \ \text{ or } \ 2x - y = 9$

112. $m = \dfrac{y_2 - y_1}{x_2 - x_1} = \dfrac{-1-2}{8-(-1)} = \dfrac{-3}{9} = -\dfrac{1}{3}$

$y - y_1 = m(x - x_1)$

$y - (-1) = -\dfrac{1}{3}(x - 8)$

$y + 1 = -\dfrac{1}{3}x + \dfrac{8}{3}$

$y = -\dfrac{1}{3}x + \dfrac{5}{3} \ \text{ or } \ x + 3y = 5$

113. For $y = 4x - 6$, the slope is 4 and the y-intercept is $(0, -6)$. Begin at $(0, -6)$ and move to the right 1 unit and up 4 units to find point $(1, -2)$.

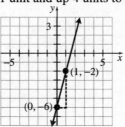

114. $2x + 3y = 12$

$3y = -2x + 12$

$y = \dfrac{-2x + 12}{3}$

$y = -\dfrac{2}{3}x + 4$

The slope is $-\dfrac{2}{3}$ and the y-intercept is $(0, 4)$.

Begin at $(0, 4)$ and move to the right 3 units and down 2 units to find the point $(3, 2)$. We can also move 3 units to the left and 2 units up to find the point $(-3, 6)$.

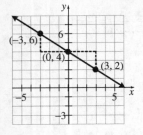

115. A line parallel to L will have a slope equal to the slope of L. Thus, the slope of the line parallel to

L is $-\dfrac{3}{8}$.

116. A line perpendicular to L will have a slope that is the negative reciprocal to the slope of L. Thus, the slope of the line perpendicular to L is

$$\frac{-1}{-\frac{3}{8}} = -1 \cdot -\frac{8}{3} = \frac{8}{3}.$$

117.
$$
\begin{aligned}
x - 3y &= 9 & 9x + 3y &= -3 \\
-3y &= -x + 9 & 3y &= -9x - 3 \\
y &= \frac{-x+9}{-3} & y &= \frac{-9x-3}{3} \\
y &= \frac{1}{3}x - 3 & y &= -3x - 1
\end{aligned}
$$

Perpendicular. The two slopes, $m_1 = \dfrac{1}{3}$ and

$m_2 = -3$, are negative reciprocals.

118.
$$
\begin{aligned}
6x - 8y &= 16 & 3x + 4y &= 28 \\
-8y &= -6x + 16 & 4y &= -3x + 28 \\
y &= \frac{-6x+16}{-8} & y &= \frac{-3x+28}{4} \\
y &= \frac{3}{4}x - 2 & y &= -\frac{3}{4}x + 7
\end{aligned}
$$

Neither. The two slopes are $m_1 = \dfrac{3}{4}$ and

$m_2 = -\dfrac{3}{4}$. The two slopes are not equal and are not negative reciprocals.

119.
$$
\begin{aligned}
2x - y &= 3 & -6x + 3y &= 0 \\
-y &= -2x + 3 & 3y &= 6x \\
y &= \frac{-2x+3}{-1} & y &= \frac{6x}{3} \\
y &= 2x - 3 & y &= 2x
\end{aligned}
$$

Parallel. The two lines have equal slopes, $m = 2$, but different y-intercepts.

120. Perpendicular. The line $x = 2$ is vertical, and the line $y = 2$ is horizontal. Vertical lines are perpendicular to horizontal lines.

121. The slope of the line we seek is $m = -2$, the same as the slope of $y = -2x - 5$. The equation of the line is:

$$
\begin{aligned}
y - 2 &= -2(x - 1) \\
y - 2 &= -2x + 2 \\
y &= -2x + 4
\end{aligned}
$$

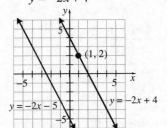

122. The slope of the line we seek is $m = \dfrac{5}{2}$, the same as the slope of the line
$$
\begin{aligned}
5x - 2y &= 8 \\
-2y &= -5x + 8 \\
y &= \frac{-5x+8}{-2} \\
y &= \frac{5}{2}x - 4
\end{aligned}
$$

Thus, the equation of the line we seek is:

$$
\begin{aligned}
y - 3 &= \frac{5}{2}(x - 4) \\
y - 3 &= \frac{5}{2}x - 10 \\
y &= \frac{5}{2}x - 7
\end{aligned}
$$

123. The line we seek is vertical, parallel to the
vertical line $x = -3$. Now a vertical line passing
through the point $(1, -4)$ must also pass through
the point $(1, 0)$. That is, the x-intercept is $(1, 0)$.
Thus, the equation of the line is $x = 1$.

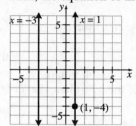

124. The slope of the line we seek is $m = -\dfrac{1}{3}$, the
negative reciprocal of the slope of $y = 3x + 7$.
The equation of the line is:

$$y - 2 = -\frac{1}{3}(x - 6)$$

$$y - 2 = -\frac{1}{3}x + 2$$

$$y = -\frac{1}{3}x + 4$$

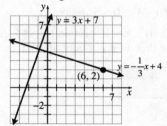

125. The slope of the line we seek is $m = \dfrac{4}{3}$, the
negative reciprocal of the line
$$3x + 4y = 6$$
$$4y = -3x + 6$$
$$y = \frac{-3x + 6}{4}$$
$$y = -\frac{3}{4}x + \frac{3}{2}$$
Thus, the equation of the line we seek is:

$$y - (-2) = \frac{4}{3}(x - (-3))$$

$$y + 2 = \frac{4}{3}x + 4$$

$$y = \frac{4}{3}x + 2$$

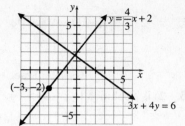

126. The line we seek is horizontal, perpendicular to
the vertical line $x = 2$. Now a horizontal line
passing through the point $(5, -4)$ must also pass
through the point $(0, -4)$. That is, the y-intercept
is $(0, -4)$. Thus, the equation of the line is
$y = -4$.

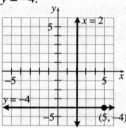

127. a. $x = 4,\ y = -2$ $5(4) + 3(-2) \overset{?}{\le} 15$

                             $20 - 6 \overset{?}{\le} 15$

                             $14 \overset{?}{\le} 15$ True

So, $(4, -2)$ is a solution to $5x + 3y \le 15$.

b. $x = -6,\ y = 15$ $5(-6) + 3(15) \overset{?}{\le} 15$

                            $-30 + 45 \overset{?}{\le} 15$

                            $15 \overset{?}{\le} 15$ True

So, $(-6, 15)$ is a solution to $5x + 3y \le 15$.

c. $x = 5,\ y = -1$ $5(5) + 3(-1) \overset{?}{\le} 15$

                            $25 - 3 \overset{?}{\le} 15$

                            $22 \overset{?}{\le} 15$ False

So, $(5, -1)$ is not a solution to $5x + 3y \le 15$.

128. a. $x = 2,\ y = 3$ $2 - 2(3) \overset{?}{>} -4$

                            $2 - 6 \overset{?}{>} -4$

                            $-4 \overset{?}{>} -4$ False

So, $(2, 3)$ is not a solution to $x - 2y > -4$.

b. $x = 5, \ y = -2 \quad 5 - 2(-2) \overset{?}{>} -4$

$$5 + 4 \overset{?}{>} -4$$

$$9 \overset{?}{>} -4 \quad \text{True}$$

So, $(5, -2)$ is a solution to $x - 2y > -4$.

c. $x = -1, \ y = 3 \quad (-1) - 2(3) \overset{?}{>} -4$

$$-1 - 6 \overset{?}{>} -4$$

$$-7 \overset{?}{>} -4 \quad \text{False}$$

So, $(-1, 3)$ is not a solution to $x - 2y > -4$.

129. Replace the inequality symbol with an equal sign to obtain $y = 3x - 2$. Because the inequality is strict, graph $y = 3x - 2$ using a dashed line.

Test Point: $(0,0)$: $\quad 0 \overset{?}{<} 3(0) - 2$

$$0 \overset{?}{<} -2 \quad \text{False}$$

Therefore, $(0, 0)$ is not a solution to $y < 3x - 2$. Shade the half-plane that does not contain $(0, 0)$.

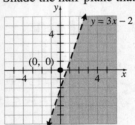

130. Replace the inequality symbol with an equal sign to obtain $2x - 4y = 12$. Because the inequality is non-strict, graph $2x - 4y = 12$ $\left(y = \dfrac{1}{2}x - 3 \right)$ using a solid line.

Test Point: $(0,0)$: $\quad 2(0) - 4(0) \overset{?}{\le} 12$

$$0 \overset{?}{\le} 12 \quad \text{True}$$

Therefore, $(0, 0)$ is a solution to $2x - 4y \le 12$. Shade the half-plane that contains $(0, 0)$.

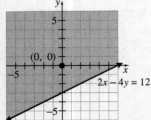

131. Replace the inequality symbol with an equal sign to obtain $3x + 4y = 20$. Because the inequality is strict, graph $3x + 4y = 20$ $\left(y = -\dfrac{3}{4}x + 5 \right)$ using a dashed line.

Test Point: $(0, 0)$: $\quad 3(0) + 4(0) \overset{?}{>} 20$

$$0 \overset{?}{>} 20 \quad \text{False}$$

Therefore, $(0, 0)$ is not a solution to $3x + 4y > 20$. Shade the half-plane that does not contain $(0, 0)$.

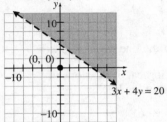

132. Replace the inequality symbol with an equal sign to obtain $y = 5$. Because the inequality is non-strict, graph $y = 5$ using a solid line.

Test Point: $(0, 0)$: $\quad 0 \overset{?}{>} 5 \quad \text{False}$

Therefore, $(0, 0)$ is not a solution to $y > 5$. Shade the half-plane that does not contain $(0, 0)$.

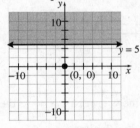

133. Replace the inequality symbol with an equal sign to obtain $2x + 3y = 0$. Because the inequality is strict, graph $2x + 3y = 0$ $\left(y = -\dfrac{2}{3}x \right)$ using a dashed line.

Test Point: $(1, 1)$: $\quad 2(1) + 3(1) \overset{?}{<} 0$

$$2 + 3 \overset{?}{<} 0$$

$$5 \overset{?}{<} 0 \quad \text{False}$$

Therefore, $(1, 1)$ is not a solution to $2x + 3x < 0$. Shade the half-plane that does not contain $(1, 1)$.

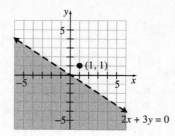

134. Replace the inequality symbol with an equal sign to obtain $x = -8$. Because the inequality is strict, graph $x = -8$ using a dashed line.

Test Point: $(0, 0)$: $\quad 0 \overset{?}{>} -8 \quad$ True

Therefore, $(0, 0)$ is a solution to $x > -8$. Shade the half-plane that contains $(0, 0)$.

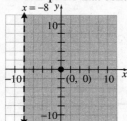

135. a. Let x = the number of movie tickets.
Let y = the number of music downloads.
$7.50x + 2y \le 60$

   b. $x = 5, y = 15$:

$$7.50(5) + 2(15) \overset{?}{\le} 60$$
$$37.50 + 30 \overset{?}{\le} 60$$
$$67.50 \le 60 \qquad \text{False}$$

No, Ethan cannot buy 5 movie tickets and 15 music downloads on his \$60 budget.

   c. $x = 6, y = 5$: $\quad 7.50(6) + 2(5) \overset{?}{\le} 60$
$$45 + 10 \overset{?}{\le} 60$$
$$55 \le 60$$

Yes, Ethan can buy 6 movie tickets and 5 music downloads on his \$60 budget.

136. a. Let x = the number of candy bars.
Let y = the number of candles.
$0.50x + 2y \ge 1000$

   b. $x = 500, y = 350$:

$$0.50(500) + 2(350) \overset{?}{\ge} 1000$$
$$250 + 700 \overset{?}{\ge} 1000$$
$$950 \ge 1000 \qquad \text{False}$$

No, the math club will not earn enough money for the field trip by selling 500 candy bars and 350 candles.

   c. $x = 600, y = 400$:

$$0.50(600) + 2(400) \overset{?}{\ge} 1000$$
$$300 + 800 \overset{?}{\ge} 1000$$
$$1100 \ge 1000 \qquad \text{True}$$

Yes, the math club will earn enough money for the field trip by selling 600 candy bars and 400 candles.

Chapter 1 Test

1. a. Let $x = 6$ in the equation.
$$3(6-7) + 5 \overset{?}{=} 6 - 4$$
$$3(-1) + 5 \overset{?}{=} 2$$
$$-3 + 5 \overset{?}{=} 2$$
$$2 = 2 \quad \text{True}$$
$x = 6$ is a solution to the equation.

   b. Let $x = -2$ in the equation.
$$3(-2-7) + 5 \overset{?}{=} -2 - 4$$
$$3(-9) + 5 \overset{?}{=} -6$$
$$-27 + 5 \overset{?}{=} -6$$
$$-22 \ne -6$$
$x = -2$ is **not** a solution to the equation.

2. a. Interval: $(-4, \infty)$

   b. Interval: $(3, 7]$

3. Let x represent the number.
$3x - 8 = x + 4$

4. Let x represent the number.
$$\frac{2}{3}x + 2(x-5) > 7$$

5. $5x - (x - 2) = 6 + 2x$

$$5x - x + 2 = 6 + 2x$$
$$4x + 2 = 6 + 2x$$
$$4x + 2 - 2 = 6 + 2x - 2$$
$$4x = 4 + 2x$$
$$4x - 2x = 4 + 2x - 2x$$
$$2x = 4$$
$$\frac{2x}{2} = \frac{4}{2}$$
$$x = 2$$

Solution set: $\{2\}$

This is a conditional equation.

6. $7 + x - 3 = 3(x + 1) - 2x$

$$7 + x - 3 = 3x + 3 - 2x$$
$$4 + x = x + 3$$
$$4 + x - x = x + 3 - x$$
$$4 = 3 \quad \text{False}$$

This statement is a contradiction. The equation has no solution. Solution set: $\{\ \}$ or $\varnothing$

7. $x + 2 \leq 3x - 4$

$$x + 2 - 3x \leq 3x - 4 - 3x$$
$$-2x + 2 \leq -4$$
$$-2x + 2 - 2 \leq -4 - 2$$
$$-2x \leq -6$$
$$\frac{-2x}{-2} \geq \frac{-6}{-2}$$
$$x \geq 3$$

Solution set: $\{x \mid x \geq 3\}$

Interval: $[3, \infty)$

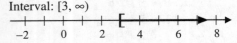

8. $4x + 7 > 2x - 3(x - 2)$

$$4x + 7 > 2x - 3x + 6$$
$$4x + 7 > -x + 6$$
$$4x + 7 + x > -x + 6 + x$$
$$5x + 7 > 6$$
$$5x + 7 - 7 > 6 - 7$$
$$5x > -1$$
$$\frac{5x}{5} > \frac{-1}{5}$$
$$x > -\frac{1}{5}$$

Solution set: $\left\{x \mid x > -\dfrac{1}{5}\right\}$

Interval: $\left(-\dfrac{1}{5}, \infty\right)$

9. $-x + 4 \leq x + 3$

$$-x + 4 - x \leq x + 3 - x$$
$$-2x + 4 \leq 3$$
$$-2x + 4 - 4 \leq 3 - 4$$
$$-2x \leq -1$$
$$\frac{-2x}{-2} \geq \frac{-1}{-2}$$
$$x \geq \frac{1}{2}$$

Solution set: $\left\{x \mid x \geq \dfrac{1}{2}\right\}$

Interval: $\left[\dfrac{1}{2}, \infty\right)$

10. $7x + 4y = 3$

$$7x + 4y - 7x = 3 - 7x$$
$$4y = -7x + 3$$
$$\frac{4y}{4} = \frac{-7x + 3}{4}$$
$$y = -\frac{7}{4}x + \frac{3}{4}$$

11. Let $x =$ Glen's weekly sales.

$$400 + 0.08x \geq 750$$
$$400 + 0.08x - 400 \geq 750 - 400$$
$$0.08x \geq 350$$
$$\frac{0.08x}{0.08} \geq \frac{350}{0.08}$$
$$x \geq 4375$$

Glen's weekly sales must be at least \$4375 for him to earn at least \$750.

12. Let $x =$ number of children at the party.

$$75 + 5x = 145$$
$$75 + 5x - 75 = 145 - 75$$
$$5x = 70$$
$$\frac{5x}{5} = \frac{70}{5}$$
$$x = 14$$

There were 14 children at Payton's party.

13. Let x = the width of the sandbox.
length: $x + 2$

$$P = 2L + 2W$$
$$20 = 2(x+2) + 2(x)$$
$$20 = 2x + 4 + 2x$$
$$20 = 4x + 4$$
$$20 - 4 = 4x + 4 - 4$$
$$16 = 4x$$
$$\frac{16}{4} = \frac{4x}{4}$$
$$4 = x$$

The sandbox has a width of 4 feet and a length of 6 feet.

14. Let x = liters of 10% solution.
Amount of 40% solution: $12 - x$ liters

tot. acid = acid from 10% + acid from 40%
$$(\%)(\text{liters}) = (\%)(\text{liters}) + (\%)(\text{liters})$$
$$0.2(12) = 0.1(x) + 0.4(12 - x)$$
$$2.4 = 0.1x + 4.8 - 0.4x$$
$$2.4 = 4.8 - 0.3x$$
$$2.4 - 4.8 = 4.8 - 0.3x - 4.8$$
$$-2.4 = -0.3x$$
$$\frac{-2.4}{-0.3} = \frac{-0.3x}{-0.3}$$
$$8 = x$$

The chemist needs to mix 8 liters of the 10% solution with 4 liters of the 40% solution.

15. Let t = time to catch up in hours.
Running time for A: $t + 0.5$
Running time for B: t
The total distance traveled by both runners will be the same when they meet. Therefore, we have
distance A = distance B
$$(\text{rate})(\text{time}) = (\text{rate})(\text{time})$$
$$8(t + 0.5) = 10(t)$$
$$8t + 4 = 10t$$
$$8t + 4 - 8t = 10t - 8t$$
$$4 = 2t$$
$$\frac{4}{2} = \frac{2t}{2}$$
$$2 = t$$

It will take contestant B two hours to catch up to contestant A.

16. A: quadrant IV; B: y-axis; C: x-axis; D: quadrant I; E: quadrant III; F: quadrant II

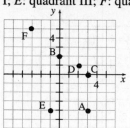

17. a. Let $x = -2$ and $y = 4$.

$$4 \overset{?}{=} 3(-2)^2 + (-2) - 5$$
$$4 \overset{?}{=} 3(4) - 7$$
$$4 \overset{?}{=} 12 - 7$$
$$4 = 5 \text{ False}$$

The point $(-2, 4)$ is not on the graph.

b. Let $x = -1$ and $y = -3$.

$$-3 \overset{?}{=} 3(-1)^2 + (-1) - 5$$
$$-3 \overset{?}{=} 3(1) - 6$$
$$-3 \overset{?}{=} 3 - 6$$
$$-3 = -3 \text{ True}$$

The point $(-1, -3)$ is on the graph.

c. Let $x = 2$ and $y = 9$.

$$9 \overset{?}{=} 3(2)^2 + (2) - 5$$
$$9 \overset{?}{=} 3(4) - 3$$
$$9 \overset{?}{=} 12 - 3$$
$$9 = 9 \text{ True}$$

The point $(2, 9)$ is on the graph.

18. $y = 4x - 1$

x	$y = 4x - 1$	(x, y)
-1	$y = 4(-1) - 1 = -5$	$(-1, -5)$
0	$y = 4(0) - 1 = -1$	$(0, -1)$
1	$y = 4(1) - 1 = 3$	$(1, 3)$

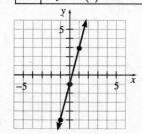

19. $y = 4x^2$

x	$y = 4x^2$	(x, y)
-1	$y = 4(-1)^2 = 4$	$(-1, 4)$
$-\frac{1}{2}$	$y = 4\left(-\frac{1}{2}\right)^2 = 1$	$\left(-\frac{1}{2}, 1\right)$
0	$y = 4(0)^2 = 0$	$(0, 0)$
$\frac{1}{2}$	$y = 4\left(\frac{1}{2}\right)^2 = 1$	$\left(\frac{1}{2}, 1\right)$
1	$y = 4(1)^2 = 4$	$(1, 4)$

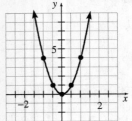

20. The intercepts are $(-3, 0)$, $(0, 1)$, and $(0, 3)$. The x-intercept is $(-3, 0)$ and the y-intercepts are $(0, 1)$ and $(0, 3)$.

21. **a.** 30 miles per hour

b. The graph passes through the origin and the point $(32, 0)$. This means the car was stopped (its speed was 0 mph) at 0 seconds and at 32 seconds.

22. Find the y-intercept by letting $x = 0$ and solving for y.
$$x = 0: \quad 0 - y = 8$$
$$-y = 8$$
$$y = -8$$
The y-intercept is $(0, -8)$. Find the x-intercept by letting $y = 0$ and solving for x.
$$y = 0: \quad x - 0 = 8$$
$$x = 8$$
The x-intercept is $(8, 0)$.

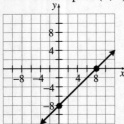

23. Let $x = -5$, 0, 5, and 10.
$$x = -5: \quad 3(-5) + 5y = 0$$
$$-15 + 5y = 0$$
$$5y = 15$$
$$y = 3$$

$$x = 0: \quad 3(0) + 5y = 0$$
$$0 + 5y = 0$$
$$y = 0$$

$$x = 5: \quad 3(5) + 5y = 0$$
$$15 + 5y = 0$$
$$5y = -15$$
$$y = -3$$

$$x = 10: \quad 3(10) + 5y = 0$$
$$30 + 5y = 0$$
$$5y = -30$$
$$y = -6$$

Thus, the points $(-5, 3)$, $(0, 0)$, $(5, -3)$, and $(10, -6)$ are on the graph.

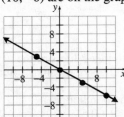

24. Find the y-intercept by letting $x = 0$ and solving for y.
$$x = 0: \quad 3(0) + 2y = 12$$
$$2y = 12$$
$$y = 6$$
The y-intercept is $(0, 6)$. Find the x-intercept by letting $y = 0$ and solving for x.
$$y = 0: \quad 3x + 2(0) = 12$$
$$3x = 12$$
$$x = 4$$
The x-intercept is $(4, 0)$.

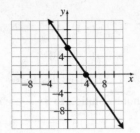

25. Let $x = -1,\ 0,\ 1,$ and 2.

$x = -1:$ $\dfrac{3}{2}(-1) - \dfrac{1}{4}y = 1$

$-\dfrac{3}{2} - \dfrac{1}{4}y = 1$

$-4\left(-\dfrac{3}{2} - \dfrac{1}{4}y\right) = -4(1)$

$6 + y = -4$

$y = -10$

$x = 0:$ $\dfrac{3}{2}(0) - \dfrac{1}{4}y = 1$

$-\dfrac{1}{4}y = 1$

$-4\left(-\dfrac{1}{4}y\right) = -4(1)$

$y = -4$

$x = 1:$ $\dfrac{3}{2}(1) - \dfrac{1}{4}y = 1$

$\dfrac{3}{2} - \dfrac{1}{4}y = 1$

$-4\left(\dfrac{3}{2} - \dfrac{1}{4}y\right) = -4(1)$

$-6 + y = -4$

$y = 2$

$x = 2:$ $\dfrac{3}{2}(2) - \dfrac{1}{4}y = 1$

$3 - \dfrac{1}{4}y = 1$

$-\dfrac{1}{4}y = -2$

$-4\left(-\dfrac{1}{4}y\right) = -4(-2)$

$y = 8$

Thus, the points $(-1, -10)$, $(0, -4)$, $(1, 2)$, and $(2, 8)$ are on the graph.

26. The equation $x = -7$ will be a vertical line with x-intercept $(-7, 0)$.

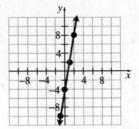

27. $m = \dfrac{y_2 - y_1}{x_2 - x_1} = \dfrac{6 - (-2)}{-1 - 5} = \dfrac{8}{-6} = -\dfrac{4}{3};$

For every 3 units of increase in x, y will decrease by 4 units. For every 3 units of decrease in x, y will increase by 4 units.

28. Start at $(2, -4)$. Because $m = -\dfrac{3}{5}$, move 5 units to the right and 3 units down to find point $(7, -7)$. We can also move 5 units to the left and 3 units up to find point $(-3, -1)$.

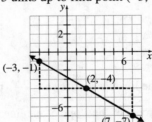

29. $8x - 2y = 1$ $x + 4y = -2$

$-2y = -8x + 1$ $4y = -x - 2$

$y = \dfrac{-8x + 1}{-2}$ $y = \dfrac{-x - 2}{4}$

$y = 4x - \dfrac{1}{2}$ $y = -\dfrac{1}{4}x - \dfrac{1}{2}$

Perpendicular. The two lines have slopes that are negative reciprocals.

30. $y - y_1 = m(x - x_1)$

$\quad y - 1 = 4(x - (-3))$

$\quad y - 1 = 4(x + 3)$

$\quad y - 1 = 4x + 12$

$\quad\quad y = 4x + 13$ or $4x - y = -13$

31. $m = \dfrac{y_2 - y_1}{x_2 - x_1} = \dfrac{7 - 1}{-3 - 6} = \dfrac{6}{-9} = -\dfrac{2}{3}$

$\quad y - y_1 = m(x - x_1)$

$\quad y - 1 = -\dfrac{2}{3}(x - 6)$

$\quad y - 1 = -\dfrac{2}{3}x + 4$

$\quad\quad y = -\dfrac{2}{3}x + 5$ or $2x + 3y = 15$

32. The slope of the line we seek is $m = \dfrac{1}{5}$, the same

as the slope of the line
$x - 5y = 15$

$\quad -5y = -x + 15$

$\quad\quad y = \dfrac{-x + 15}{-5}$

$\quad\quad y = \dfrac{1}{5}x - 3$

Thus, the equation of the line we seek is:

$y - (-1) = \dfrac{1}{5}(x - 10)$

$\quad y + 1 = \dfrac{1}{5}x - 2$

$\quad\quad y = \dfrac{1}{5}x - 3$ or $x - 5y = 15$

33. The slope of the line we seek is $m = -\dfrac{1}{3}$, the

negative reciprocal of the slope of the line
$3x - y = 4$

$\quad -y = -3x + 4$

$\quad\quad y = 3x - 4$

Thus, the equation of the line we seek is:

$y - 2 = -\dfrac{1}{3}(x - 6)$

$\quad y - 2 = -\dfrac{1}{3}x + 2$

$\quad\quad y = -\dfrac{1}{3}x + 4$ or $x + 3y = 12$

34. a. $x = 3, y = -1 \quad 3(3) - (-1) \overset{?}{>} 10$

$\quad\quad\quad\quad\quad\quad 9 + 1 \overset{?}{>} 10$

$\quad\quad\quad\quad\quad\quad 10 \overset{?}{>} 10 \quad$ False

So, $(3, -1)$ is not a solution to $3x - y > 10$.

b. $x = 4, y = 5 \quad 3(4) - 5 \overset{?}{>} 10$

$\quad\quad\quad\quad\quad\quad 12 - 5 \overset{?}{>} 10$

$\quad\quad\quad\quad\quad\quad 7 \overset{?}{>} 10 \quad$ False

So, $(4, 5)$ is not a solution to $3x - y > 10$.

c. $x = 5, y = 3 \quad 3(5) - 3 \overset{?}{>} 10$

$\quad\quad\quad\quad\quad\quad 15 - 3 \overset{?}{>} 10$

$\quad\quad\quad\quad\quad\quad 12 \overset{?}{>} 10 \quad$ True

So, $(5, 3)$ is a solution to $3x - y > 10$.

35. Replace the inequality symbol with an equal sign
to obtain $y = -2x + 1$. Because the inequality is
non-strict, graph $y = -2x + 1$ using a solid line.

Test Point: $(0, 0)$: $\quad 0 \overset{?}{\le} -2(0) + 1$

$\quad\quad\quad\quad\quad\quad 0 \overset{?}{\le} 1 \quad$ True

Therefore, $(0, 0)$ is a solution to $y \le -2x + 1$.
Shade the half-plane that contains $(0, 0)$.

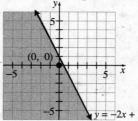

36. Replace the inequality symbol with an equal sign
to obtain $5x - 2y = 0$. Because the inequality is

strict, graph $5x - 2y = 0$ $\left(y = \dfrac{5}{2}x \right)$ using a

dashed line.

Test Point: $(1, 1)$: $\quad 5(1) - 2(1) \overset{?}{<} 0$

$\quad\quad\quad\quad\quad\quad 5 - 2 \overset{?}{<} 0$

$\quad\quad\quad\quad\quad\quad 3 \overset{?}{<} 0 \quad$ False

Therefore, $(1, 1)$ is not a solution to $5x - 2y < 0$.
Shade the half-plane that does not contain $(1, 1)$.

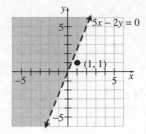

37. a.

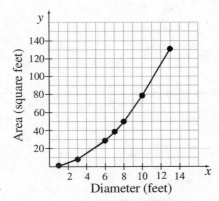

b. Average rate of change

$$= \frac{7.07 - 0.79}{3 - 1}$$

$$= 3.14 \text{ square feet per foot}$$

Between diameters of 1 foot and 3 feet, the area of the circle increases at a rate of 3.14 square feet per foot.

c. Average rate of change

$$= \frac{132.73 - 78.54}{13 - 10}$$

$$\approx 18.06 \text{ square feet per foot}$$

Between diameters of 10 feet and 13 feet, the area of the circle increases at a rate of approximately 18.06 square feet per foot.

d. No. The area of circles is not linearly related to the diameter. The average rate of change (slope) is not constant.

Cumulative Review Chapters R–1

1. a. (i) 27.235
(ii) 27.236

b. (i) 1.0
(ii) 1.1

2.

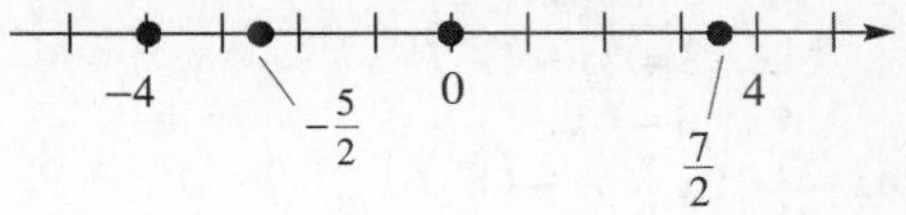

<!-- number line image at top right -->

3. $-|-14| = -(14) = -14$

4. $-3 + 4 - 7 = 1 - 7 = -6$

5. $\dfrac{-3(12)}{-6} = \dfrac{-36}{-6} = 6$

6. $(-3)^4 = (-3)(-3)(-3)(-3) = 81$

7. $5 - 2(1-4)^3 + 5 \cdot 3 = 5 - 2(-3)^3 + 5 \cdot 3$
$= 5 + 54 + 15 = 74$

8. $\dfrac{2}{3} + \dfrac{1}{2} - \dfrac{1}{4} = \dfrac{8}{12} + \dfrac{6}{12} - \dfrac{3}{12} = \dfrac{8 + 6 - 3}{12} = \dfrac{11}{12}$

9. $3(2)^2 + 2(2) - 7 = 3(4) + 4 - 7 = 12 - 3 = 9$

10. $4a^2 - 6a + a^2 - 12 + 2a - 1$
$= 4a^2 + a^2 - 6a + 2a - 12 - 1$
$= 5a^2 - 4a - 13$

11. a. Evaluate the denominator when $x = -2$.
$(-2)^2 + (-2) - 2 = 4 - 2 - 2 = 0$
Since $x = -2$ makes the denominator 0, it is not in the domain.

b. Evaluate the denominator when $x = 0$.
$(0)^2 + (0) - 2 = -2$
Since $x = 0$ does not make the denominator equal 0, it is in the domain.

12. $3(x+2) - 4(2x-1) + 8 = 3x + 6 - 8x + 4 + 8$
$= 3x - 8x + 6 + 4 + 8$
$= -5x + 18$

13. Let $x = 3$ in the equation.
$3 - (2(3) + 3) \overset{?}{=} 5(3) - 1$
$3 - (6 + 3) \overset{?}{=} 15 - 1$
$3 - 9 \overset{?}{=} 14$
$-6 \neq 14$
$x = 3$ is **not** a solution to the equation.

14.
$$4x - 3 = 2(3x - 2) - 7$$
$$4x - 3 = 6x - 4 - 7$$
$$4x - 3 = 6x - 11$$
$$4x - 3 - 6x = 6x - 11 - 6x$$
$$-2x - 3 = -11$$
$$-2x - 3 + 3 = -11 + 3$$
$$-2x = -8$$
$$\frac{-2x}{-2} = \frac{-8}{-2}$$
$$x = 4$$
Solution set: $\{4\}$

15.
$$\frac{x+1}{3} = x - 4$$
$$3\left(\frac{x+1}{3}\right) = 3(x - 4)$$
$$x + 1 = 3x - 12$$
$$x + 1 - 3x = 3x - 12 - 3x$$
$$-2x + 1 = -12$$

$$-2x + 1 - 1 = -12 - 1$$
$$-2x = -13$$
$$\frac{-2x}{-2} = \frac{-13}{-2}$$
$$x = \frac{13}{2}$$
Solution set: $\left\{\dfrac{13}{2}\right\}$

16.
$$2x - 5y = 6$$
$$2x - 5y - 2x = 6 - 2x$$
$$-5y = -2x + 6$$
$$\frac{-5y}{-5} = \frac{-2x + 6}{-5}$$
$$y = \frac{2}{5}x - \frac{6}{5}$$

17.
$$\frac{x+3}{2} \le \frac{3x-1}{4}$$
$$4\left(\frac{x+3}{2}\right) \le 4\left(\frac{3x-1}{4}\right)$$
$$2x + 6 \le 3x - 1$$
$$2x + 6 - 3x \le 3x - 1 - 3x$$
$$-x + 6 \le -1$$
$$-x + 6 - 6 \le -1 - 6$$
$$-x \le -7$$
$$x \ge 7$$

Solution set: $\{x \mid x \ge 7\}$

Interval: $[7, \infty)$

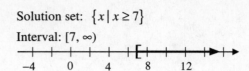

18.
$$5(x - 3) \ge 7(x - 4) + 3$$
$$5x - 15 \ge 7x - 28 + 3$$
$$5x - 15 \ge 7x - 25$$
$$5x \ge 7x - 10$$
$$-2x \ge -10$$
$$x \le 5$$
$$\{x \mid x \le 5\} \text{ or } (-\infty, 5]$$

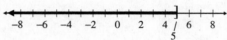

19.

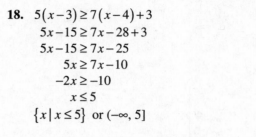

20. For $y = -\dfrac{1}{2}x + 4$, the slope is $-\dfrac{1}{2}$ and the y-intercept is (0, 4). Begin at (0, 4) and move to the right 2 units and down 1 unit to find the point (2, 3). We can also move 2 unit to the left and 1 unit up to find the point (−2, 5).

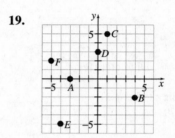

21.
$$4x - 5y = 15$$
$$-5y = -4x + 15$$
$$y = \frac{-4x + 15}{-5}$$
$$y = \frac{4}{5}x - 3$$

The slope is $\dfrac{4}{5}$ and the y-intercept is (0, −3).

Begin at the point (0, −3) and move to the right 5 units and up 4 units to find the point (5, 1).

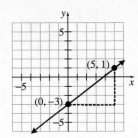

22. $m = \dfrac{y_2 - y_1}{x_2 - x_1} = \dfrac{10 - (-2)}{-6 - 3} = \dfrac{12}{-9} = -\dfrac{4}{3}$

$y - y_1 = m(x - x_1)$

$y - 10 = -\dfrac{4}{3}\big(x - (-6)\big)$

$y - 10 = -\dfrac{4}{3}(x + 6)$

$y - 10 = -\dfrac{4}{3}x - 8$

$y = -\dfrac{4}{3}x + 2 \ \text{ or } \ 4x + 3y = 6$

23. The slope of the line we seek is $m = -3$, the same as the slope of the line $y = -3x + 10$

Thus, the equation of the line we seek is:

$y - 7 = -3\big(x - (-5)\big)$

$y - 7 = -3(x + 5)$

$y - 7 = -3x - 15$

$y = -3x - 8 \ \text{ or } \ 3x + y = -8$

24. Replace the inequality symbol with an equal sign to obtain $x - 3y = 12$. Because the inequality is strict, graph $x - 3y = 12$ $\left(y = \dfrac{1}{3}x - 4\right)$ using a dashed line.

Test Point: $(0, 0)$: $\quad 0 - 3(0) \overset{?}{>} 12$

$\qquad\qquad\qquad\qquad\quad 0 \overset{?}{>} 12 \qquad$ False

Therefore, $(0, 0)$ is a not a solution to $x - 3y > 12$. Shade the half-plane that does not contain $(0, 0)$.

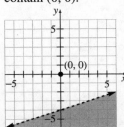

25. Let x = score on final exam.

$93 \le \dfrac{94 + 95 + 90 + 97 + 2x}{6} \le 100$

$93 \le \dfrac{376 + 2x}{6} \le 100$

$6(93) \le 6\left(\dfrac{376 + 2x}{6}\right) \le 6(100)$

$558 \le 376 + 2x \le 600$

$558 - 376 \le 376 + 2x - 376 \le 600 - 376$

$182 \le 2x \le 224$

$\dfrac{182}{2} \le \dfrac{2x}{2} \le \dfrac{224}{2}$

$91 \le x \le 112$

Shawn needs to score at least 91 on the final exam to earn an A (assuming the maximum score on the exam is 100).

26. Let x = weight in pounds.

$0.2x - 2 \ge 30$

$0.2x - 2 + 2 \ge 30 + 2$

$0.2x \ge 32$

$\dfrac{0.2x}{0.2} \ge \dfrac{32}{0.2}$

$x \ge 160$

A person 62 inches tall would be considered obese if they weighed 160 pounds or more.

27. Let x = measure of the smaller angle.
Larger angle: $15 + 2x$

$x + (15 + 2x) = 180$

$3x + 15 = 180$

$3x + 15 - 15 = 180 - 15$

$3x = 165$

$\dfrac{3x}{3} = \dfrac{165}{3}$

$x = 55$

The angles measure $55°$ and $125°$.

28. Let h = height of cylinder in inches.

$S = 2\pi r^2 + 2\pi r h$

$100 = 2\pi (2)^2 + 2\pi (2) h$

$100 = 8\pi + 4\pi h$

$100 - 8\pi = 4\pi h$

$h = \dfrac{100 - 8\pi}{4\pi} \approx 5.96$

The cylinder should be about 5.96 inches tall.

29. Let x = first even integer.
Second even integer: $x + 2$
Third even integer: $x + 4$

$$x + (x + 2) = 22 + (x + 4)$$
$$2x + 2 = x + 26$$
$$2x + 2 - x = x + 26 - x$$
$$x + 2 = 26$$
$$x + 2 - 2 = 26 - 2$$
$$x = 24$$

The three consecutive even integers are 24, 26, and 28.

Chapter 2

Section 2.1

Are You Ready for This Section?

R1. Inequality: $-4 \leq x \leq 4$
Interval: $[-4, 4]$
We use square brackets in interval notation because the inequalities are not strict.

R2. Interval: $[2, \infty)$
Inequality: $x \geq 2$
The square bracket indicates a non-strict inequality.

R3.

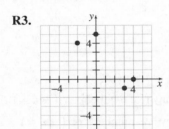

R4. $2x + 5y = 10$

Let $x = 0$: $2(0) + 5y = 10$
$$0 + 5y = 10$$
$$5y = 10$$
$$y = 2$$

y-intercept is 2.
Let $y = 0$: $2x + 5(0) = 10$
$$2x + 0 = 10$$
$$2x = 10$$
$$x = 5$$

x-intercept is 5.

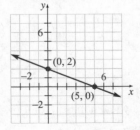

R5. $y = x^2 - 3$

x	$y = x^2 - 3$	(x, y)
-2	$y = (-2)^2 - 3 = 1$	$(-2, 1)$
-1	$y = (-1)^2 - 3 = -2$	$(-1, -2)$
0	$y = (0)^2 - 3 = -3$	$(0, -3)$
1	$y = (1)^2 - 3 = -2$	$(1, -2)$
2	$y = (2)^2 - 3 = 1$	$(2, 1)$

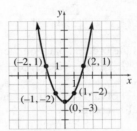

Section 2.1 Quick Checks

1. If a relation exists between x and y, then we say that x <u>corresponds</u> to y or that y <u>depends</u> on x, and we write $x \rightarrow y$.

2. The first element of the ordered pair comes from the set 'Friend' and the second element is the corresponding element from the set 'Birthday'.
{(Max, November 8), (Alesia, January 20), (Trent, March 3), (Yolanda, November 8), (Wanda, July 6), (Elvis, January 8)}

3. The first elements of the ordered pairs make up the first step and the second elements make up the second set.

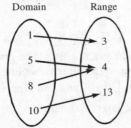

4. The <u>domain</u> of a relation is the set of all inputs of the relation. The <u>range</u> is the set of all outputs of the relation.

5. The domain is the set of all inputs and the range is the set of all outputs. The inputs are the elements in the set 'Friend' and the outputs are the elements in the set 'Birthday'.
   <u>Domain:</u>
   {Max, Alesia, Trent, Yolanda, Wanda, Elvis}

   <u>Range:</u>
   {January 20, March 3, July 6, November 8, January 8}

6. The domain is the set of all inputs and the range is the set of all outputs. The inputs are the first elements in the ordered pairs and the outputs are the second elements in the ordered pairs.
   <u>Domain:</u>   <u>Range:</u>
   {1, 5, 8, 10}   {3, 4, 13}

7. First we notice that the ordered pairs on the graph are (−2, 0), (−1, 2), (−1, −2), (2, 3), (3, 0), and (4, −3).
   The domain is the set of all x-coordinates and the range is the set of all y-coordinates.
   <u>Domain:</u>   <u>Range:</u>
   {−2, −1, 2, 3, 4}   {−3, −2, 0, 2, 3}

8. True

9. False

10. To find the domain, we first determine the x-values for which the graph exists. The graph exists for all x-values between −2 and 4, inclusive. Thus, the domain is $\{x \mid -2 \le x \le 4\}$, or [−2, 4] if we use interval notation.

    To find the range, we first determine the y-values for which the graph exists. The graph exists for all y-values between −2 and 2, inclusive. Thus, the range is $\{y \mid -2 \le y \le 2\}$, or [−2, 2] if we use interval notation.

11. To find the domain, we first determine the x-values for which the graph exists. The graph exists for all x-values on a real number line. Thus, the domain is $\{x \mid x \text{ is any real number}\}$, or (−∞, ∞) if we use interval notation.

    To find the range, we first determine the y-values for which the graph exists. The graph exists for

all y-values on a real number line. Thus, the range is $\{y \mid y \text{ is any real number}\}$, or (−∞, ∞) if we use interval notation.

12. $y = 3x - 8$

x	$y = 3x - 8$	(x, y)
−1	$y = 3(-1) - 8 = -11$	$(-1, -11)$
0	$y = 3(0) - 8 = -8$	$(0, -8)$
1	$y = 3(1) - 8 = -5$	$(1, -5)$
2	$y = 3(2) - 8 = -2$	$(2, -2)$
3	$y = 3(3) - 8 = 1$	$(3, 1)$

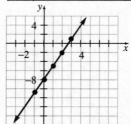

Domain: $\{x \mid x \text{ is any real number}\}$ or $(-\infty, \infty)$

Range: $\{y \mid y \text{ is any real number}\}$ or $(-\infty, \infty)$

13. $y = x^2 - 8$

x	$y = x^2 - 8$	(x, y)
−3	$y = (-3)^2 - 8 = 1$	$(-3, 1)$
−2	$y = (-2)^2 - 8 = -4$	$(-2, -4)$
0	$y = (0)^2 - 8 = -8$	$(0, -8)$
2	$y = (2)^2 - 8 = -4$	$(2, -4)$
3	$y = (3)^2 - 8 = 1$	$(3, 1)$

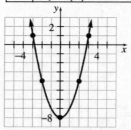

Domain: $\{x \mid x \text{ is any real number}\}$ or $(-\infty, \infty)$

Range: $\{y \mid y \ge -8\}$ or $[-8, \infty)$

14. $x = y^2 + 1$

y	$x = y^2 + 1$	(x, y)
-2	$x = (-2)^2 + 1 = 5$	$(5, -2)$
-1	$x = (-1)^2 + 1 = 2$	$(2, -1)$
0	$x = (0)^2 + 1 = 1$	$(1, 0)$
1	$x = (1)^2 + 1 = 2$	$(2, 1)$
2	$x = (2)^2 + 1 = 5$	$(5, 2)$

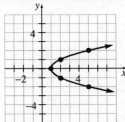

Domain: $\{x \mid x \geq 1\}$ or $[1, \infty)$

Range: $\{y \mid y \text{ is any real number}\}$ or $(-\infty, \infty)$

2.1 Exercises

15. {(USA Today, 1.83), (Wall Street Journal, 2.12),
(New York Times, 0.92),
(Los Angeles Times, 0.61),
(Washington Post, 0.55)}

Domain: {USA Today, Wall Street Journal,
New York Times, Los Angeles Times,
Washington Post}

Range: {0.55, 0.61, 0.92, 1.83, 2.12}

17. {(Less than 9th Grade, $13,992),
(9th-12th Grade – No Diploma, $14,460),
(High School Graduate, $23,520),
(Associate's Degree, $36,012),
(Bachelor's Degree, $51,108)}

Domain: {Less than 9th Grade,
9th-12th Grade – No Diploma,
High School Graduate, Associate's Degree,
Bachelor's Degree}

Range:
{$13,992, $14,460, $23,520, $36,012, $51,108}

19.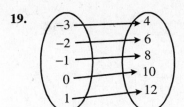

Domain: $\{-3, -2, -1, 0, 1\}$
Range: $\{4, 6, 8, 10, 12\}$

21.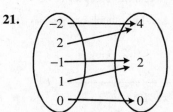

Domain: $\{-2, -1, 0, 1, 2\}$
Range: $\{0, 2, 4\}$

23.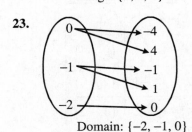

Domain: $\{-2, -1, 0\}$
Range: $\{-4, -1, 0, 1, 4\}$

25. Domain: $\{-3, -2, 0, 2, 3\}$
Range: $\{-3, -1, 2, 3\}$

27. Domain: $\{x \mid -4 \leq x \leq 4\}$ or $[-4, 4]$
Range: $\{y \mid -2 \leq y \leq 2\}$ or $[-2, 2]$

29. Domain: $\{x \mid -1 \leq x \leq 3\}$ or $[-1, 3]$
Range: $\{y \mid 0 \leq y \leq 4\}$ or $[0, 4]$

31. Domain: $\{x \mid x \text{ is a real number}\}$ or $(-\infty, \infty)$
Range: $\{y \mid y \geq -3\}$ or $[-3, \infty)$

33. $y = -3x + 1$

x	$y = -3x + 1$	(x, y)
-2	$y = -3(-2) + 1 = 7$	$(-2, 7)$
-1	$y = -3(-1) + 1 = 4$	$(-1, 4)$
0	$y = -3(0) + 1 = 1$	$(0, 1)$
1	$y = -3(1) + 1 = -2$	$(1, -2)$
2	$y = -3(2) + 1 = -5$	$(2, -5)$

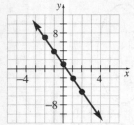

Domain: $\{x \mid x \text{ is a real number}\}$ or $(-\infty, \infty)$

Range: $\{y \mid y \text{ is a real number}\}$ or $(-\infty, \infty)$

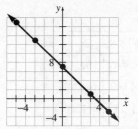

Domain: $\{x \mid x \text{ is a real number}\}$ or $(-\infty, \infty)$

Range: $\{y \mid y \text{ is a real number}\}$ or $(-\infty, \infty)$

35. $y = \dfrac{1}{2}x - 4$

x	$y = \dfrac{1}{2}x - 4$	(x, y)
-4	$y = \dfrac{1}{2}(-4) - 4 = -6$	$(-4, -6)$
-2	$y = \dfrac{1}{2}(-2) - 4 = -5$	$(-2, -5)$
0	$y = \dfrac{1}{2}(0) - 4 = -4$	$(0, -4)$
2	$y = \dfrac{1}{2}(2) - 4 = -3$	$(2, -3)$
4	$y = \dfrac{1}{2}(4) - 4 = -2$	$(4, -2)$

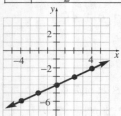

Domain: $\{x \mid x \text{ is a real number}\}$ or $(-\infty, \infty)$

Range: $\{y \mid y \text{ is a real number}\}$ or $(-\infty, \infty)$

37. $2x + y = 7$
$y = -2x + 7$

x	$y = -2x + 7$	(x, y)
-5	$y = -2(-5) + 7 = 17$	$(-5, 17)$
-3	$y = -2(-3) + 7 = 13$	$(-3, 13)$
0	$y = -2(0) + 7 = 7$	$(0, 7)$
3	$y = -2(3) + 7 = 1$	$(3, 1)$
5	$y = -2(5) + 7 = -3$	$(5, -3)$

39. $y = -x^2$

x	$y = -x^2$	(x, y)
-3	$y = -(-3)^2 = -9$	$(-3, -9)$
-2	$y = -(-2)^2 = -4$	$(-2, -4)$
0	$y = -(0)^2 = 0$	$(0, 0)$
2	$y = -(2)^2 = -4$	$(2, -4)$
3	$y = -(3)^2 = -9$	$(3, -9)$

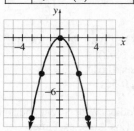

Domain: $\{x \mid x \text{ is a real number}\}$ or $(-\infty, \infty)$

Range: $\{y \mid y \le 0\}$ or $(-\infty, 0]$

41. $y = 2x^2 - 8$

x	$y = 2x^2 - 8$	(x, y)
-3	$y = 2(-3)^2 - 8 = 10$	$(-3, 10)$
-2	$y = 2(-2)^2 - 8 = 0$	$(-2, 0)$
-1	$y = 2(-1)^2 - 8 = -6$	$(-1, -6)$
0	$y = 2(0)^2 - 8 = -8$	$(0, -8)$
1	$y = 2(1)^2 - 8 = -6$	$(1, -6)$
2	$y = 2(2)^2 - 8 = 0$	$(2, 0)$
3	$y = 2(3)^2 - 8 = 10$	$(3, 10)$

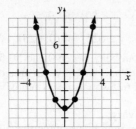

Domain: $\{x \mid x \text{ is a real number}\}$ or $(-\infty, \infty)$

Range: $\{y \mid y \geq -8\}$ or $[-8, \infty)$

43. $y = |x|$

x	$y = |x|$	(x, y)
-4	$y = |-4| = 4$	$(-4, 4)$
-2	$y = |-2| = 2$	$(-2, 2)$
0	$y = |0| = 0$	$(0, 0)$
2	$y = |2| = 2$	$(2, 2)$
4	$y = |4| = 4$	$(4, 4)$

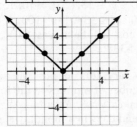

Domain: $\{x \mid x \text{ is a real number}\}$ or $(-\infty, \infty)$

Range: $\{y \mid y \geq 0\}$ or $[0, \infty)$

45. $y = |x - 1|$

x	$y = |x - 1|$	(x, y)
-4	$y = |-4 - 1| = 5$	$(-4, 5)$
-1	$y = |-1 - 1| = 2$	$(-1, 2)$
0	$y = |0 - 1| = 1$	$(0, 1)$
1	$y = |1 - 1| = 0$	$(1, 0)$
3	$y = |3 - 1| = 2$	$(3, 2)$
5	$y = |5 - 1| = 4$	$(5, 4)$

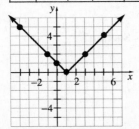

Domain: $\{x \mid x \text{ is a real number}\}$ or $(-\infty, \infty)$

Range: $\{y \mid y \geq 0\}$ or $[0, \infty)$

47. $y = x^3$

x	$y = x^3$	(x, y)
-3	$y = (-3)^3 = -27$	$(-3, -27)$
-2	$y = (-2)^3 = -8$	$(-2, -8)$
-1	$y = (-1)^3 = -1$	$(-1, -1)$
0	$y = (0)^3 = 0$	$(0, 0)$
1	$y = (1)^3 = 1$	$(1, 1)$
2	$y = (2)^3 = 8$	$(2, 8)$
3	$y = (3)^3 = 27$	$(3, 27)$

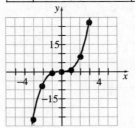

Domain: $\{x \mid x \text{ is a real number}\}$ or $(-\infty, \infty)$

Range: $\{y \mid y \text{ is a real number}\}$ or $(-\infty, \infty)$

49. $y = x^3 + 1$

x	$y = x^3 + 1$	(x, y)
-2	$y = (-2)^3 + 1 = -7$	$(-2, -7)$
-1	$y = (-1)^3 + 1 = 0$	$(-1, 0)$
0	$y = (0)^3 + 1 = 1$	$(0, 1)$
1	$y = (1)^3 + 1 = 2$	$(1, 2)$
2	$y = (2)^3 + 1 = 9$	$(2, 9)$

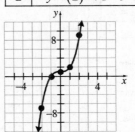

Domain: $\{x \mid x \text{ is a real number}\}$ or $(-\infty, \infty)$

Range: $\{y \mid y \text{ is a real number}\}$ or $(-\infty, \infty)$

51. $x^2 - y = 4$

$$x^2 = y + 4$$
$$x^2 - 4 = y$$
$$y = x^2 - 4$$

x	$y = x^2 - 4$	(x, y)
-3	$y = (-3)^2 - 4 = 5$	$(-3, 5)$
-2	$y = (-2)^2 - 4 = 0$	$(-2, 0)$
0	$y = (0)^2 - 4 = -4$	$(0, -4)$
2	$y = (2)^2 - 4 = 0$	$(2, 0)$
3	$y = (3)^2 - 4 = 5$	$(3, 5)$

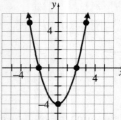

Domain: $\{x \mid x \text{ is a real number}\}$ or $(-\infty, \infty)$

Range: $\{y \mid y \geq -4\}$ or $[-4, \infty)$

53. $x = y^2 - 1$

y	$x = y^2 - 1$	(x, y)
-2	$x = (-2)^2 - 1 = 3$	$(3, -2)$
-1	$x = (-1)^2 - 1 = 0$	$(0, -1)$
0	$x = (0)^2 - 1 = -1$	$(-1, 0)$
1	$x = (1)^2 - 1 = 0$	$(0, 1)$
2	$x = (2)^2 - 1 = 3$	$(3, 2)$

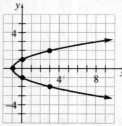

Domain: $\{x \mid x \geq -1\}$ or $[-1, \infty)$

Range: $\{y \mid y \text{ is a real number}\}$ or $(-\infty, \infty)$

55. a. Domain: $\{x \mid 0 < x < 50\}$ or $(0, 50)$

Range: $\{y \mid 0 < y \leq 625\}$ or $(0, 625]$

The window can be between 0 and 50 feet wide (inclusive) and will have an area of between 0 and 625 square feet (inclusive).

b. Answers may vary;
Assuming the window is rectangular, the perimeter is given by the equation
$P = 2l + 2w$
where l is the length and w is the width. Since the window cannot have a negative width, it is reasonable for the domain to begin with 0. If the width were greater than 50, this would require a negative value for the length in order to have a perimeter of 100 in the above equation. Since length cannot be negative, it is reasonable for the domain to stop at 50.

57. a. Domain: $\{m \mid 0 \leq m \leq 15,120\}$ or $[0, 15,120]$
Range: $\{c \mid 100 \leq c \leq 3380\}$ or $[100, 3380]$
The monthly cost will be at least \$100, but no more than \$3380.

b. Answers may vary. It is not possible to talk a negative number of minutes so the lower bound of 0 makes sense. Assuming there are 21 non-weekend days in a month, we get:

$$\frac{21 \text{ days}}{1} \cdot \frac{12 \text{ hrs}}{1 \text{ day}} \cdot \frac{60 \text{ min}}{1 \text{ hr}} = 15,120$$

That is, in 21 non-weekend days there are 15,120 'anytime' minutes so this is the most someone could use in a month.

59. Actual graphs will vary but each graph should be a horizontal line.

61. Answers will vary. A relation could be thought of as a rule that tells us what to do to a value in the domain (input value) to obtain its corresponding value in the range (output value).

Section 2.2

Are You Ready for This Section?

R1. a. Let $x = 1$:
$$2x^2 - 5x = 2(1)^2 - 5(1) = 2 - 5 = -3$$

b. Let $x = 4$:
$$2x^2 - 5x = 2(4)^2 - 5(4)$$
$$= 2(16) - 20$$
$$= 32 - 20$$
$$= 12$$

c. Let $x = -3$:
$$2x^2 - 5x = 2(-3)^2 - 5(-3)$$
$$= 2(9) + 15$$
$$= 18 + 15$$
$$= 33$$

R2. $\dfrac{3}{2x + 1}$

$\dfrac{3}{2\left(-\frac{1}{2}\right) + 1} = \dfrac{3}{-1 + 1} = \dfrac{3}{0}$ is undefined.

R3. Inequality: $x \leq 5$
Interval: $(-\infty, 5]$

R4. Interval: $(2, \infty)$
Set notation: $\{x \mid x > 2\}$
The inequality is strict since the parenthesis was used instead of a square bracket.

Section 2.2 Quick Checks

1. A <u>function</u> is a relation in which each element in the domain of the relation corresponds to exactly one element in the range of the relation.

2. False

3. The relation is a function because each element in the domain (Friend) corresponds to exactly one element in the range (Birthday).
Domain: {Max, Alesia, Trent, Yolanda, Wanda, Elvis}
Range: {January 20, March 3, July 6, November 8, January 8}

4. The relation is not a function because there is an element in the domain, 210, that corresponds to more than one element in the range. If 210 is selected from the domain, a single sugar content cannot be determined.

5. The relation is a function because there are no ordered pairs with the same first coordinate but different second coordinates.
Domain: {−3, −2, −1, 0, 1}
Range: {0, 1, 2, 3}

6. The relation is not a function because there are two ordered pairs, (−3, 2) and (−3, 6), with the same first coordinate but different second coordinates.

7. $y = -2x + 5$
The relation is a function since there is only one output than can result for each input.

8. $y = \pm 3x$
The relation is not a function since a single input for x will yield two output values for y. For example, if $x = 1$, then $y = \pm 3$.

9. $y = x^2 + 5x$
The relation is a function since there is only one output than can result for each input.

10. True

11. The graph is that of a function because every vertical line will cross the graph in at most one point.

12. The graph is not that of a function because a vertical line can cross the graph in more than one point.

13. $f(x) = 3x + 2$
$f(x) = 3(4) + 2$
$\qquad = 12 + 2$
$\qquad = 14$

14. $f(x) = 3x + 2$
$f(-2) = 3(-2) + 2 = -6 + 2 = -4$

15. $g(x) = -2x^2 + x + 3$
$g(-3) = -2(-3)^2 + (-3) + 3$
$\qquad\quad = -2(9) - 3 + 3$
$\qquad\quad = -18 - 3 + 3$
$\qquad\quad = -18$

16. $g(x) = -2x^2 + x + 3$
$g(1) = -2(1)^2 + 1 + 3$
$\qquad = -2(1) + 1 + 3$
$\qquad = -2 + 1 + 3$
$\qquad = 2$

17. In the function $H(q) = 2q^2 - 5q + 1$, H is called the <u>dependent</u> variable, and q is called the <u>independent</u> variable or <u>argument</u>.

18. $f(x) = 2x - 5$

$f(x - 2) = 2(x - 2) - 5$

$\qquad = 2x - 4 - 5$

$\qquad = 2x - 9$

19. $f(x) - f(2) = [2x - 5] - [2(2) - 5]$

$\qquad\qquad = 2x - 5 - (-1)$

$\qquad\qquad = 2x - 5 + 1$

$\qquad\qquad = 2x - 4$

20. When only the equation of a function f is given, we agree that the <u>domain</u> of f is the set of real numbers x for which $f(x)$ is a real number.

21. $f(x) = 3x^2 + 2$

The function tells us to square a number x, multiply by 3, and then add 2. Since these operations can be performed on any real number, the domain of f is the set of all real numbers. The domain can be written as $\{x \mid x \text{ is any real number}\}$, or $(-\infty, \infty)$ in interval notation.

22. $h(x) = \dfrac{x + 1}{x - 3}$

The function tells us to divide $x + 1$ by $x - 3$. Since division by 0 is not defined, the denominator $x - 3$ can never be 0. Therefore, x can never equal 3. The domain can be written as $\{x \mid x \neq 3\}$.

23. $A(r) = \pi r^2$

Since r represents the radius of the circle, it must take on positive values. Therefore, the domain is $\{r \mid r > 0\}$, or $(0, \infty)$ in interval notation.

24. a. Independent variable: t (number of days)
Dependent variable: A (square miles)

   b. $A(t) = 0.25\pi t^2$

$A(30) = 0.25\pi(30)^2 \approx 706.86$ sq. miles

After oil has been leaking for 30 days, the circular oil slick will cover about 706.86 square miles.

2.2 Exercises

25. Function. Each state corresponds to exactly one number of Representatives.
Domain: {Virginia, Nevada, Arkansas, Tennessee, Texas}
Range: {4, 9, 11, 36}

27. Not a function. The domain element 174 for horsepower corresponds to two different top speeds in the range.
Domain: {150, 174, 180|
Range: {118, 130, 140}

29. Function. There are no ordered pairs that have the same first coordinate but different second coordinates.
Domain: {0, 1, 2, 3}
Range: {3, 4, 5, 6}

31. Function. There are no ordered pairs that have the same first coordinate but different second coordinates.
Domain: {−3, 1, 4, 7}
Range: {5}

33. Not a function. There are two ordered pairs that have the same first coordinate but different second coordinates.
Domain: {−10, −5, 0}
Range: {1, 2, 3, 4}

35. $y = 2x + 9$

Since there is only one output y that can result from any given input x, this relation is a function.

37. $2x + y = 10$

$\qquad y = -2x + 10$

Since there is only one output y that can result from any given input x, this relation is a function.

39. $y = \pm 5x$

Since a given input x can result in more than one output y, this relation is not a function.

41. $y = x^2 + 2$

Since there is only one output y that can result from any given input x, this relation is a function.

43. $x + y^2 = 10$

$\qquad y^2 = 10 - x$

Since a given input x can result in more than one output y, this relation is not a function. For example, if $x = 1$ then $y = 3$ or $y = -3$.

45. Function. The graph passes the vertical line test so it is the graph of a function.

47. Not a function. The graph fails the vertical line test so it is not the graph of a function.

49. Function. The graph passes the vertical line test so it is the graph of a function.

51. Function. The graph passes the vertical line test so it is the graph of a function.

53. a. $f(0) = 2(0) + 3 = 0 + 3 = 3$

    b. $f(3) = 2(3) + 3 = 6 + 3 = 9$

    c. $f(-2) = 2(-2) + 3 = -4 + 3 = -1$

55. a. $f(0) = -5(0) + 2 = 0 + 2 = 2$

    b. $f(3) = -5(3) + 2 = -15 + 2 = -13$

    c. $f(-2) = -5(-2) + 2 = 10 + 2 = 12$

57. a. $f(0) = (0)^2 - 3(0) = 0 - 0 = 0$

    b. $f(3) = (3)^2 - 3(3) = 9 - 9 = 0$

    c. $f(-2) = (-2)^2 - 3(-2) = 4 + 6 = 10$

59. a. $f(0) = -(0)^2 + (0) + 3 = 0 + 0 + 3 = 3$

    b. $f(3) = -(3)^2 + (3) + 3 = -9 + 3 + 3 = -3$

    c. $f(-2) = -(-2)^2 + (-2) + 3 = -4 - 2 + 3 = -3$

61. a. $f(-x) = 2(-x) - 5 = -2x - 5$

    b. $f(x + 2) = 2(x + 2) - 5 = 2x + 4 - 5 = 2x - 1$

    c. $f(2x) = 2(2x) - 5 = 4x - 5$

    d. $-f(x) = -(2x - 5) = -2x + 5$

    e. $f(x + h) = 2(x + h) - 5 = 2x + 2h - 5$

63. a. $f(-x) = 7 - 5(-x) = 7 + 5x$

    b. $f(x + 2) = 7 - 5(x + 2)$
$$= 7 - 5x - 10$$
$$= -3 - 5x$$

    c. $f(2x) = 7 - 5(2x) = 7 - 10x$

    d. $-f(x) = -(7 - 5x) = -7 + 5x$

    e. $f(x + h) = 7 - 5(x + h) = 7 - 5x - 5h$

65. $f(x) = x^2 + 3$
$$f(2) = (2)^2 + 3 = 4 + 3 = 7$$

67. $s(t) = -t^3 - 4t$
$$s(-2) = -(-2)^3 - 4(-2)$$
$$= -(-8) + 8$$
$$= 8 + 8$$
$$= 16$$

69. $F(x) = |x - 2|$
$$F(-3) = |(-3) - 2| = |-5| = 5$$

71. $F(z) = \dfrac{z + 2}{z - 5}$
$$F(4) = \dfrac{(4) + 2}{(4) - 5} = \dfrac{6}{-1} = -6$$

73. $f(x) = 4x + 7$
Since each operation in the function can be performed for any real number, the domain of the function is all real numbers.
Domain: $\{x \mid x \text{ is a real number}\}$ or $(-\infty, \infty)$

75. $F(z) = \dfrac{2z + 1}{z - 5}$
The function tells us to divide $2z + 1$ by $z - 5$. Since division by 0 is not defined, the denominator can never equal 0. Thus, z can never equal 5.
Domain: $\{z \mid z \neq 5\}$

77. $f(x) = 3x^4 - 2x^2$
Since each operation in the function can be performed for any real number, the domain of the function is all real numbers.
Domain: $\{x \mid x \text{ is a real number}\}$ or $(-\infty, \infty)$

79. $G(x) = \dfrac{3x - 5}{3x + 1}$
The function tells us to divide $3x - 5$ by $3x + 1$. Since division by 0 is not defined, the denominator can never equal 0.

$$3x+1=0$$
$$3x=-1$$
$$x=-\frac{1}{3}$$

Thus, x can never equal $-\frac{1}{3}$.

Domain: $\left\{x \mid x \neq -\frac{1}{3}\right\}$

81. $f(x)=3x^2-x+C; f(3)=18$
$$18=3(3)^2-(3)+C$$
$$18=3(9)-3+C$$
$$18=27-3+C$$
$$18=24+C$$
$$-6=C$$

83. $f(x)=\dfrac{2x+5}{x-A}; f(0)=-1$
$$-1=\frac{2(0)+5}{0-A}$$
$$-1=\frac{0+5}{0-A}$$
$$-1=\frac{5}{-A}$$
$$A=5$$

85. $A(r)=\pi r^2$
$$A(4)=\pi(4)^2=16\pi \approx 50.27$$
The area is roughly 50.27 square inches.

87. $h=$ number of hours worked
$G=$ gross salary
$$G(h)=15h$$
$$G(25)=15(25)=375$$
Jackie's gross salary for 25 hours is \$375.

89. a. The dependent variable is the population, P, and the independent variable is the age, a.

b. $P(20)=18.75(20)^2-5309.62(20)+321,783.32$
$$=18.75(400)-106,192.4+321,783.32$$
$$=7500-106,192.4+321,783.32$$
$$=223,090.92$$

The population of the U.S. that were 20 years of age or older in 2007 was roughly 223,091 thousand (223,091,000).

    c. $P(0) = 18.75(0)^2 - 5309.62(0) + 321,783.32$

             $= 321,783.32$

      $P(0)$ represents the entire population of the U.S. since every member of the population is at least 0 years of age. The population of the U.S. in 2007 was roughly 321,783 thousand (321,783,000).

91. a. The dependent variable is revenue, R, and the independent variable is price, p.

    b. $R(50) = -(50)^2 + 200(50)$

             $= -2500 + 10,000$

             $= 7500$

      Selling MP3 players for \$50 will yield a daily revenue of \$7500 for the company.

    c. $R(120) = -(120)^2 + 200(120)$

               $= -14,400 + 24,000$

               $= 9600$

      Selling MP3 players for \$120 will yield a daily revenue of \$9600 for the company.

93. $V(r) = \dfrac{4}{3}\pi r^3$

    Since the radius must have a positive length, the domain is all positive real numbers.

    Domain: $\{r \mid r > 0\}$ or $(0, \infty)$

95. $G(h) = 22.5h$

    Since Jackie cannot work a negative number of hours and she can work a maximum of 60 hours in a week, the domain is all real numbers between 0 and 60 (inclusive).

    Domain: $\{h \mid 0 \le h \le 60\}$ or $[0, 60]$

97. $D(p) = 1200 - 10p$

    A graph of the function would indicate that the domain is all real numbers. However, the context of the problem needs to be considered. It is not feasible that the price would be negative, nor that the demand for hot dogs would be negative. Thus, the domain is all real numbers between 0 and 120, inclusive.

    Domain: $\{p \mid 0 \le p \le 120\}$ or $[0, 120]$

99. a. **i.** $3 \ge 0$ so use $f(x) = -2x + 1$

          $f(3) = -2(3) + 1 = -6 + 1 = -5$

          $f(3) = -5$

      ii. $-2 < 0$ so use $f(x) = x + 3$

          $f(-2) = -2 + 3 = 1$

          $f(-2) = 1$

      iii. $0 \ge 0$ so use $f(x) = -2x + 1$

          $f(0) = -2(0) + 1 = 0 + 1 = 1$

          $f(0) = 1$

    b. **i.** $-4 < -2$ so use $f(x) = -3x + 1$

          $f(-4) = -3(-4) + 1 = 12 + 1 = 13$

          $f(-4) = 13$

ii. $2 \geq -2$ so use $f(x) = x^2$

$$f(2) = (2)^2 = 4$$
$$f(2) = 4$$

iii $-2 \geq -2$ so use $f(x) = x^2$

$$f(-2) = (-2)^2 = 4$$
$$f(-2) = 4$$

101. Answers will vary.

103. Answers will vary. A function could be thought of as a relation that tells us what to do to a value in the domain (input value) to obtain exactly one corresponding value in the range (output value).

105. The four forms of functions presented in this section are: maps, ordered pairs, equations, and graphs.

107. $f(x) = x^2 + 3$

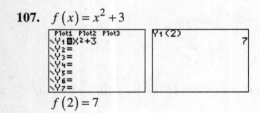

$$f(2) = 7$$

109. $F(x) = |x - 2|$

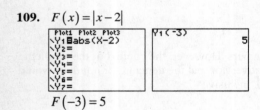

$$F(-3) = 5$$

111. $H(x) = \sqrt{4x - 3}$

$$H(7) = 5$$

113. $F(z) = \dfrac{z+2}{z-5}$

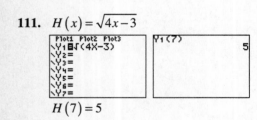

$$F(4) = -6$$

Section 2.3

Are You Ready for This Section?

R1.
$$3x - 12 = 0$$
$$3x - 12 + 12 = 0 + 12$$
$$3x = 12$$
$$\frac{3x}{3} = \frac{12}{3}$$
$$x = 4$$

The solution set is $\{4\}$.

R2. $y = x^2$

x	$y = x^2$	(x, y)
-2	$y = (-2)^2 = 4$	$(-2, 4)$
-1	$y = (-1)^2 = 1$	$(-1, 1)$
0	$y = (0)^2 = 0$	$(0, 0)$
1	$y = (1)^2 = 1$	$(1, 1)$
2	$y = (2)^2 = 4$	$(2, 4)$

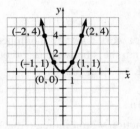

Section 2.3 Quick Checks

1. When a function is defined by an equation in x and y, the <u>graph</u> of the <u>function</u> is the set of all ordered pairs (x, y) such that $y = f(x)$.

2. If $f(4) = -7$, then the point whose ordered pair is $(\underline{4}, \underline{-7})$ is on the graph of $y = f(x)$.

3. $f(x) = -2x + 9$

x	$y = f(x) = -2x + 9$	(x, y)
-2	$f(-2) = -2(-2) + 9 = 13$	$(-2, 13)$
0	$f(0) = -2(0) + 9 = 9$	$(0, 9)$
2	$f(2) = -2(2) + 9 = 5$	$(2, 5)$
4	$f(4) = -2(4) + 9 = 1$	$(4, 1)$
6	$f(6) = -2(6) + 9 = -3$	$(6, -3)$

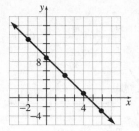

4. $f(x) = x^2 + 2$

x	$y = f(x) = x^2 + 2$	(x, y)
-3	$f(-3) = (-3)^2 + 2 = 11$	$(-3, 11)$
-1	$f(-1) = (-1)^2 + 2 = 3$	$(-1, 3)$
0	$f(0) = (0)^2 + 2 = 2$	$(0, 2)$
1	$f(1) = (1)^2 + 2 = 3$	$(1, 3)$
3	$f(3) = (3)^2 + 2 = 11$	$(3, 11)$

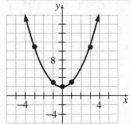

5. $f(x) = |x - 2|$

x	$y = f(x) = |x - 2|$	(x, y)
-2	$f(-2) = |-2 - 2| = 4$	$(-2, 4)$
0	$f(0) = |0 - 2| = 2$	$(0, 2)$
2	$f(2) = |2 - 2| = 0$	$(2, 0)$
4	$f(4) = |4 - 2| = 2$	$(4, 2)$
6	$f(6) = |6 - 2| = 4$	$(6, 4)$

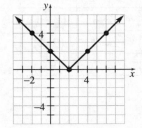

6. a. The arrows on the ends of the graph indicate that the graph continues indefinitely. Therefore, the domain is $\{x \mid x \text{ is any real number}\}$, or $(-\infty, \infty)$ in interval notation.
The function reaches a maximum value of 2, but has no minimum value. Therefore, the range is $\{y \mid y \le 2\}$, or $(-\infty, 2]$ in interval notation.

   b. The intercepts are $(-2, 0)$, $(0, 2)$, and $(2, 0)$. The x-intercepts are $(-2, 0)$ and $(2, 0)$, and the y-intercept is $(0, 2)$.

7. If the point $(3, 8)$ is on the graph of a function f, then $f(\underline{3}) = \underline{8}$. $f(-2) = 4$, then $(\underline{-2}, \underline{4})$ is a point on the graph of g.

8. a. Since $(-3, -15)$ and $(1, -3)$ are on the graph of f, then $f(-3) = -15$ and $f(1) = -3$.

   b. To determine the domain, notice that the graph exists for all real numbers. Thus, the domain is $\{x \mid x \text{ is any real number}\}$, or $(-\infty, \infty)$ in interval notation.

   c. To determine the range, notice that the function can assume any real number. Thus, the range is $\{y \mid y \text{ is any real number}\}$, or $(-\infty, \infty)$ in interval notation.

   d. The intercepts are $(-2, 0)$, $(0, 0)$, and $(2, 0)$. The x-intercepts are $(-2, 0)$, $(0, 0)$, and $(2, 0)$. The y-intercept is $(0, 0)$.

   e. Since $(3, 15)$ is the only point on the graph where $y = f(x) = 15$, the solution set to $f(x) = 15$ is $\{3\}$.

9. a. When $x = -2$, then
$$f(x) = -3x + 7$$
$$f(-2) = -3(-2) + 7$$
$$= 6 + 7$$
$$= 13$$
Since $f(-2) = 13$, the point $(-2, 13)$ is on the graph. This means the point $(-2, 1)$ is **not** on the graph.

b. If $x = 3$, then
$$f(x) = -3x + 7$$
$$f(3) = -3(3) + 7$$
$$= -9 + 7$$
$$= -2$$
The point $(3, -2)$ is on the graph.

c. If $f(x) = -8$, then
$$f(x) = -8$$
$$-3x + 7 = -8$$
$$-3x = -15$$
$$x = 5$$
If $f(x) = -8$, then $x = 5$. The point $(5, -8)$ is on the graph.

10. $f(x) = 2x + 6$
$$f(-3) = 2(-3) + 6 = -6 + 6 = 0$$
Yes, -3 is a zero of f.

11. $g(x) = x^2 - 2x - 3$
$$g(1) = (1)^2 - 2(1) - 3 = 1 - 2 - 3 = -4$$
No, 1 is not a zero of g.

12. $h(z) = -z^3 + 4z$
$$h(2) = -(2)^3 + 4(2) = -8 + 8 = 0$$
Yes, 2 is a zero of h.

13. The zeros of the function are the *x*-intercepts: -2 and 2.

14. Maria's distance from home is a function of time so we put time (in minutes) on the horizontal axis and distance (in blocks) on the vertical axis. Starting at the origin $(0, 0)$, we draw a straight line to the point $(5, 5)$. The ordered pair $(5, 5)$ represents Maria being 5 blocks from home after 5 minutes. From the point $(5, 5)$, we draw a straight line to the point $(7, 0)$ that represents her trip back home. The ordered pair $(7, 0)$ represents Maria being back at home after 7 minutes. Draw a line segment from $(7, 0)$ to $(8, 0)$ to represent the time it takes Maria to find her keys and lock the door. Next, draw a line segment from $(8, 0)$ to $(13, 8)$ that represents her 8 block run in 5 minutes. Then draw a line segment from $(13, 8)$ to $(14, 11)$ that represents her 3 block run in 1 minute. Now draw a horizontal line from $(14, 11)$ to $(16, 11)$ that represents Maria's resting period. Finally, draw a line segment from $(16, 11)$ to $(26, 0)$ that represents her walk home.

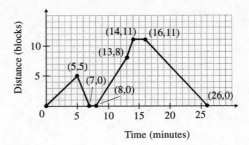

2.3 Exercises

15. $f(x) = 4x - 6$

x	$y = f(x) = 4x - 6$	(x, y)
-2	$f(-2) = 4(-2) - 6 = -14$	$(-2, -14)$
-1	$f(-1) = 4(-1) - 6 = -10$	$(-1, -10)$
0	$f(0) = 4(0) - 6 = -6$	$(0, -6)$
1	$f(1) = 4(1) - 6 = -2$	$(1, -2)$
2	$f(2) = 4(2) - 6 = 2$	$(2, 2)$

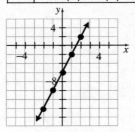

17. $h(x) = x^2 - 2$

x	$y = h(x) = x^2 - 2$	(x, y)
-3	$h(-3) = (-3)^2 - 2 = 7$	$(-3, 7)$
-2	$h(-2) = (-2)^2 - 2 = 2$	$(-2, 2)$
-1	$h(-1) = (-1)^2 - 2 = -1$	$(-1, -1)$
0	$h(0) = (0)^2 - 2 = -2$	$(0, -2)$
1	$h(1) = (1)^2 - 2 = -1$	$(1, -1)$
2	$h(2) = (2)^2 - 2 = 2$	$(2, 2)$
3	$h(3) = (3)^2 - 2 = 7$	$(3, 7)$

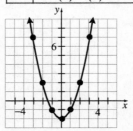

19. $G(x) = |x-1|$

x	$y = G(x) = |x-1|$	(x, y)
-3	$G(-3) = |-3-1| = 4$	$(-3, 4)$
-1	$G(-1) = |-1-1| = 2$	$(-1, 2)$
1	$G(1) = |1-1| = 0$	$(1, 0)$
3	$G(3) = |3-1| = 2$	$(3, 2)$
5	$G(5) = |5-1| = 4$	$(5, 4)$

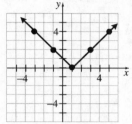

21. $g(x) = x^3$

x	$y = g(x) = x^3$	(x, y)
-3	$y = (-3)^3 = -27$	$(-3, -27)$
-2	$y = (-2)^3 = -8$	$(-2, -8)$
-1	$y = (-1)^3 = -1$	$(-1, -1)$
0	$y = (0)^3 = 0$	$(0, 0)$
1	$y = (1)^3 = 1$	$(1, 1)$
2	$y = (2)^3 = 8$	$(2, 8)$
3	$y = (3)^3 = 27$	$(3, 27)$

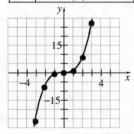

23. a. Domain: $\{x \mid x \text{ is a real number}\}$ or $(-\infty, \infty)$
Range: $\{y \mid y \text{ is a real number}\}$ or $(-\infty, \infty)$

   b. The intercepts are $(0, 2)$ and $(1, 0)$. The x-intercept is $(1, 0)$ and the y-intercept is $(0, 2)$.

   c. Zero: 1

25. a. Domain: $\{x \mid x \text{ is a real number}\}$ or $(-\infty, \infty)$
Range: $\{y \mid y \geq -2.25\}$ or $[-2.25, \infty)$

   b. The intercepts are $(-2, 0)$, $(4, 0)$, and $(0, -2)$. The x-intercepts are $(-2, 0)$ and $(4, 0)$, and the y-intercept is $(0, -2)$.

   c. Zeros: -2, 4

27. a. Domain: $\{x \mid x \text{ is a real number}\}$ or $(-\infty, \infty)$
Range: $\{y \mid y \text{ is a real number}\}$ or $(-\infty, \infty)$

   b. The intercepts are $(-3, 0)$, $(-1, 0)$, $(2, 0)$, and $(0, -3)$. The x-intercepts are $(-3, 0)$, $(-1, 0)$, and $(2, 0)$, and the y-intercept is $(0, -3)$.

   c. Zeros: -3, -1, 2

29. a. Domain: $\{x \mid x \text{ is a real number}\}$ or $(-\infty, \infty)$
Range: $\{y \mid y \geq 0\}$ or $[0, \infty)$

   b. The intercepts are $(-3, 0)$, $(3, 0)$, and $(0, 9)$. The x-intercepts are $(-3, 0)$ and $(3, 0)$, and the y-intercept is $(0, 9)$.

   c. Zeros: -3, 3

31. a. Domain: $\{x \mid x \leq 4\}$ or $(-\infty, 4]$
Range: $\{y \mid y \leq 3\}$ or $(-\infty, 3]$

   b. The intercepts are $(-2, 0)$ and $(0, 2)$. The x-intercept is $(-2, 0)$ and the y-intercept is $(0, 2)$.

   c. Zero: -2

33. a. $f(-7) = -2$

   b. $f(-3) = 3$

   c. $f(6) = 2$

   d. negative

   e. $f(x) = 0$ for $\{-6, -1, 4\}$

   f. Domain: $\{x \mid -7 \leq x \leq 6\}$ or $[-7, 6]$

   g. Range: $\{y \mid -2 \leq y \leq 3\}$ or $[-2, 3]$

   h. The x-intercepts are $(-6, 0)$, $(-1, 0)$, and $(4, 0)$.

   i.   The y-intercept is $(0, -1)$.

   j. $f(x) = -2$ for $\{-7, 2\}$.

   k. $f(x) = 3$ for $\{-3\}$.

   l.   The zeros are $-6, -1$, and 4.

35. a.   From the table, when $x = -2$ the value of the function is 3. Therefore,
$$F(-2) = 3$$

   b.   From the table, when $x = 3$ the value of the function is -6. Therefore,
$$F(3) = -6$$

   c.   From the table, $F(x) = 5$ when $x = -1$.

   d.   The x-intercept is the point for which the function value is 0. From the table, $F(x) = 0$ when $x = -4$. Therefore, the x-intercept is $(-4, 0)$.

   e.   The y-intercept is the point for which $x = 0$. From the table, when $x = 0$ the value of the function is 2. Therefore, the y-intercept is $(0, 2)$.

37. a. $f(2) = 4(2) - 9 = 8 - 9 = -1$

      Since $f(2) = -1$, the point $(2, 1)$ is not on the graph of the function.

   b. $f(3) = 4(3) - 9 = 12 - 9 = 3$
      The point $(3, 3)$ is on the graph.

   c. $4x - 9 = 7$
$$4x = 16$$
$$x = 4$$
      The point $(4, 7)$ is on the graph.

   d. $f(2) = 4(2) - 9 = 8 - 9 = -1$
      2 is not a zero of f.

39. a. $g(4) = -\dfrac{1}{2}(4) + 4 = -2 + 4 = 2$

      Since $g(4) = 2$, the point $(4, 2)$ is on the graph of the function.

   b. $g(6) = -\dfrac{1}{2}(6) + 4 = -3 + 4 = 1$
      The point $(6, 1)$ is on the graph.

   c. $-\dfrac{1}{2}x + 4 = 10$
$$-\dfrac{1}{2}x = 6$$
$$x = -12$$
      The point $(-12, 10)$ is on the graph.

   d. $g(8) = -\dfrac{1}{2}(8) + 4 = -4 + 4 = 0$
      8 is a zero of g.

41.   Square function, (c)

43.   Absolute value function, (a)

45.   Cube function, (d)

47. $f(x) = x^2$

x	$y = f(x) = x^2$	(x, y)
-3	$y = (-3)^2 = 9$	$(-3, 9)$
-2	$y = (-2)^2 = 4$	$(-2, 4)$
0	$y = (0)^2 = 0$	$(0, 0)$
2	$y = (2)^2 = 4$	$(2, 4)$
3	$y = (3)^2 = 9$	$(3, 9)$

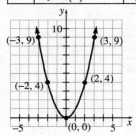

49. $f(x) = |x|$

x	$y = f(x) = |x|$	(x, y)
-5	$y = |-5| = 5$	$(-5, 5)$
-3	$y = |-3| = 3$	$(-3, 3)$
0	$y = |0| = 0$	$(0, 0)$
3	$y = |3| = 3$	$(3, 3)$
5	$y = |5| = 5$	$(5, 5)$

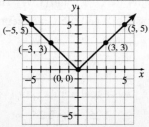

51. a. Graph (III). The height should fluctuate up and down as the person jumps. Thus, we expect to see a graph that oscillates.

b. Graph (I). Billed phone calls are generally charged per minute so the cost will continue to increase the longer you talk. Thus, we expect to see a graph that continually increases.

c. Graph (IV). As a person grows, his or her height increases until they reach adulthood at which time the height tends to stabilize. Thus, we expect to see a graph that starts increasing but then levels off.

d. Graph (V). As price increases, we expect that revenues will increase to a point. If prices get too high, few people will be able to afford the cars and little or no revenue will be generated. We would look for a graph that starts increasing but then begins to decrease.

e. Graph (II). Starting at some initial value, we expect the value to decrease over time until it reaches 0. We expect to see a graph that is constantly decreasing but stops when it hits 0.

53.

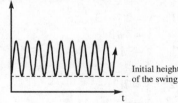

55. Answers will vary. One possibility:

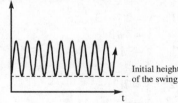

Initial height of the swing

57. Answers will vary. One possibility: Starting from the person's birth weight, the individual gained weight as they grew to adulthood. As an adult, the individual's weight fluctuated a little and then started to decline slightly after middle-age.

59. Answers will vary. One possibility:

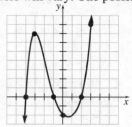

61. If a graph has more than one y-intercept, then a single input, $x = 0$, would yield more than one output (the y-intercepts). Thus, the graph could not be of a function.

63. Answers will vary. The range is the set of all possible output values.

Putting the Concepts Together (Sections 2.1–2.3)

1. The relation is a function because each element in the domain corresponds to exactly one element in the range.
$\{(-2, 1), (-1, 0), (0, 1), (1, 2), (2, 3)\}$

2. a. $y = x^3 - 4x$ is a function because any specific value of x (input) yields exactly one value of y (output).

b. $y = \pm 4x + 3$ is not a function because with the exception of 0, any value of x can yield two values of y. For instance, if $x = 1$, then $y = 7$ or $y = -1$.

3. Yes, the graph represents a function.
Domain: $\{-4, -1, 0, 3, 6\}$
Range: $\{-3, -2, 2, 6\}$

4. This relation is a function because it passes the vertical line test.
$f(5) = -6$

5. The zero is 4.

6. a. $f(4) = -5(4) + 3 = -20 + 3 = -17$

b. $g(-3) = -2(-3)^2 + 5(-3) - 1$
$= -2(9) - 15 - 1$
$= -18 - 15 - 1$
$= -34$

c. $f(x) - f(4) = [-5x+3] - [-17]$
$$= -5x + 3 + 17$$
$$= -5x + 20$$

d. $f(x-4) = -5(x-4) + 3$
$$= (-5)x - (-5)4 + 3$$
$$= -5x + 20 + 3$$
$$= -5x + 23$$

7. a. Domain: $\{h \mid h \text{ is a real number}\}$ or $(-\infty, \infty)$

b. Since we cannot divide by zero, we must find the values of w which make the denominator equal to zero.
$$3w + 1 = 0$$
$$3w + 1 - 1 = 0 - 1$$
$$3w = -1$$
$$\frac{3w}{3} = \frac{-1}{3}$$
$$w = -\frac{1}{3}$$
Domain: $\left\{ w \mid w \neq -\frac{1}{3} \right\}$

8. $y = |x| - 2$

x	$y = |x| - 2$	(x, y)
-4	$y = |-4| - 2 = 2$	$(-4, 2)$
-2	$y = |-2| - 2 = 0$	$(-2, 0)$
0	$y = |0| - 2 = -2$	$(0, -2)$
2	$y = |2| - 2 = 0$	$(2, 0)$
4	$y = |4| - 2 = 2$	$(4, 2)$

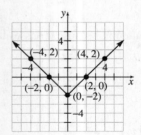

9. a. $h(2.5) = 80$
The ball is 80 feet high after 2.5 seconds.

b. $[0, 3.8]$

c. $[0, 105]$

d. 1.25 seconds

10. a. $f(3) = 5(3) - 2 = 15 - 2 = 13$
Since the point $(3, 13)$ is on the graph, the point $(3, 12)$ is not on the graph of the function.

b. $f(-2) = 5(-2) - 2 = -10 - 2 = -12$
The point $(-2, -12)$ is on the graph of the function.

c.
$$f(x) = -22$$
$$5x - 2 = -22$$
$$5x - 2 + 2 = -22 + 2$$
$$5x = -20$$
$$\frac{5x}{5} = \frac{-20}{5}$$
$$x = -4$$
The point $(-4, -22)$ is on the graph of f.

d. $f\left(\frac{2}{5}\right) = 5\left(\frac{2}{5}\right) - 2 = 2 - 2 = 0$

$\frac{2}{5}$ is a zero of f.

Section 2.4

Are You Ready for This Section?

R1. $y = 2x - 3$
Let $x = -1, 0, 1$, and 2.

$x = -1$: $y = 2(-1) - 3$
$$y = -2 - 3$$
$$y = -5$$

$x = 0$: $y = 2(0) - 3$
$$y = 0 - 3$$
$$y = -3$$

$x = 1$: $y = 2(1) - 3$
$$y = 2 - 3$$
$$y = -1$$

$x = 2$: $y = 2(2) - 3$
$$y = 4 - 3$$
$$y = 1$$

Thus, the points $(-1, -5)$, $(0, -3)$, $(1, -1)$, and $(2, 1)$ are on the graph.

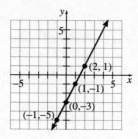

R2. $\frac{1}{2}x + y = 2$

Let $x = -2$, 0, 2, and 4.

$x = -2:\quad \frac{1}{2}(-2) + y = 2$
$$-1 + y = 2$$
$$y = 3$$

$x = 0:\quad \frac{1}{2}(0) + y = 2$
$$0 + y = 2$$
$$y = 2$$

$x = 2:\quad \frac{1}{2}(2) + y = 2$
$$1 + y = 2$$
$$y = 1$$

$x = 4:\quad \frac{1}{2}(4) + y = 2$
$$2 + y = 2$$
$$y = 0$$

Thus, the points $(-2, 3)$, $(0, 2)$, $(2, 1)$, and $(4, 0)$ are on the graph.

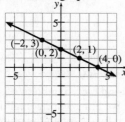

R3. The graph of $y = -4$ is a horizontal line with y-intercept -4.

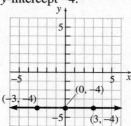

R4. The graph of $x = 5$ is a vertical line with x-intercept 5. It consists of all ordered pairs whose x-coordinate is 5.

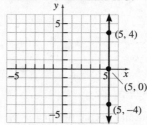

R5. $m = \dfrac{-4 - 3}{3 - (-1)} = \dfrac{-7}{4} = -\dfrac{7}{4}$

Using $m = \dfrac{-7}{4}$ we would interpret the slope as saying that y will decrease 7 units if we increase x by 4 units. We could also say $m = \dfrac{7}{-4}$ in which case we would interpret the slope as saying that y will increase by 7 units if we decrease x by 4 units. In either case, the slope is the average rate of change of y with respect to x.

R6. We start by finding the slope of the line using the two given points.
$$m = \frac{9 - 3}{4 - 1} = \frac{6}{3} = 2$$
Now we use the point-slope form of the equation of a line:
$$y - y_1 = m(x - x_1)$$
$$y - 3 = 2(x - 1)$$
$$y - 3 = 2x - 2$$
$$y = 2x + 1$$
The equation of the line is $y = 2x + 1$.

R7.
$$0.5(x-40)+100=84$$
$$(0.5)x-(0.5)40+100=84$$
$$0.5x-20+100=84$$
$$0.5x+80=84$$
$$0.5x+80-80=84-80$$
$$0.5x=4$$
$$\frac{0.5x}{0.5}=\frac{4}{0.5}$$
$$x=8$$

R8.
$$4x+20\geq32$$
$$4x+20-20\geq32-20$$
$$4x\geq12$$
$$\frac{4x}{4}\geq\frac{12}{4}$$
$$x\geq3$$
$$\{x\,|\,x\geq3\}\text{ or }[3,\infty)$$

Section 2.4 Quick Checks

1. For the graph of a linear function $f(x)=mx+b$, m is the <u>slope</u> and $(0, b)$ is the <u>y-intercept</u>.

2. The graph of a linear function is called a <u>line</u>.

3. False

4. For the linear function $G(x)=-2x+3$, the slope is <u>−2</u> and the y-intercept is <u>(0, 3)</u>.

5. Comparing $f(x)=2x-3$ to $f(x)=mx+b$, we see that the slope m is 2 and the y-intercept b is −3. We begin by plotting the point $(0,-3)$. Because $m=2=\dfrac{2}{1}=\dfrac{\Delta y}{\Delta x}=\dfrac{\text{Rise}}{\text{Run}}$, from the point $(0,-3)$ we go up 2 units and to the right 1 unit and end up at $(1,-1)$. We draw a line through these points and obtain the graph of $f(x)=2x-3$.

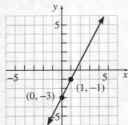

6. Comparing $G(x)=-5x+4$ to $G(x)=mx+b$, we see that the slope m is −5 and the y-intercept b is 4. We begin by plotting the point $(0, 4)$. Because $m=-5=\dfrac{-5}{1}=\dfrac{\Delta y}{\Delta x}=\dfrac{\text{Rise}}{\text{Run}}$, from the point $(0, 4)$ we go down 5 units and to the right 1 unit and end up at $(1, -1)$. We draw a line through these points and obtain the graph of $G(x)=-5x+4$.

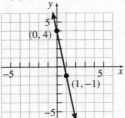

7. Comparing $h(x)=\dfrac{3}{2}x+1$ to $h(x)=mx+b$, we see that the slope m is $\dfrac{3}{2}$ and the y-intercept b is 1. We begin by plotting the point $(0, 1)$. Because $m=\dfrac{3}{2}=\dfrac{\Delta y}{\Delta x}=\dfrac{\text{Rise}}{\text{Run}}$, from the point $(0, 1)$ we go up 3 units and to the right 2 units and end up at $(2, 4)$. We draw a line through these points and obtain the graph of $h(x)=\dfrac{3}{2}x+1$.

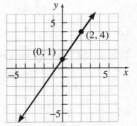

8. Comparing $f(x)=4$ to $f(x)=mx+b$, we see that the slope m is 0 and the y-intercept b is 4. Since the slope is 0, we have a horizontal line. We draw a horizontal line through the point $(0, 4)$ to obtain the graph of $f(x)=4$.

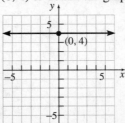

9. $f(x) = 0$
$$3x - 15 = 0$$
$$3x = 15$$
$$x = 5$$
5 is the zero.

10. $G(x) = 0$
$$\frac{1}{2}x + 4 = 0$$
$$\frac{1}{2}x = -4$$
$$x = -8$$
−8 is the zero.

11. $F(p) = 0$
$$-\frac{2}{3}p + 8 = 0$$
$$-\frac{2}{3}p = -8$$
$$-2p = -24$$
$$p = 12$$
12 is the zero.

12. a.   The independent variable is the number of miles driven, *x.* It does not make sense to drive a negative number of miles, we have that the domain of the function is $\{x \mid x \geq 0\}$ or, using interval notation, $[0, \infty)$.

   b.   To determine the *C*-intercept, we find $C(0) = 0.35(0) + 40 = 40$. The *C*-intercept is $(0, 40)$.

   c. $C(80) = 0.35(80) + 40 = 28 + 40 = 68$. If the truck is driven 80 miles, the rental cost will be $68.

   d.   We solve $C(x) = 85.50$:
$$0.35x + 40 = 85.50$$
$$0.35x = 45.50$$
$$x = 130$$
If the rental cost is $85.50, then the truck was driven 130 miles.

   e.   We plot the independent variable, *number of miles driven*, on the horizontal axis and the dependent variable, *rental cost*, on the vertical axis. From parts (b) and (c), we have that the points (0, 40) and (80, 68) are on the graph. We find one more point by evaluating the function for $x = 200$:
$C(200) = 0.35(200) + 40 = 70 + 40 = 110$.
The point (200, 110) is also on the graph.

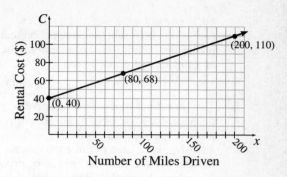

   f.   We solve $C(x) \leq 127.50$:
$$0.35x + 40 \leq 127.50$$
$$0.35x \leq 87.50$$
$$x \leq 250$$
You can drive up to 250 miles if you can spend up to $127.50.

13. a.   From Example 4, the daily fixed costs were $2000 with a variable cost of $80 per bicycle. The tax of $1 per bicycle changes the variable cost to $81 per bicycle. Thus, the cost function is $C(x) = 81x + 2000$.

   b. $C(5) = 81(5) + 2000 = 2405$
So, the cost of manufacturing 5 bicycles in a day is $2405.

   c. $$C(x) = 2810$$
$$81x + 2000 = 2810$$
$$81x = 810$$
$$x = 10$$
So, 10 bicycles can be manufactured for a cost of $2810.

   d.   Label the horizontal axis *x* and the vertical axis *C.*

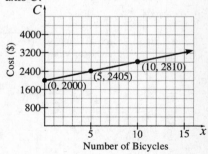

14. a.   We let $C(x)$ represent the monthly cost of operating the car after driving *x* miles, so $C(x) = mx + b$. The monthly cost before the car is driven is $250, so $C(0) = 250$. The *C*-intercept of the linear function is 250. Because the maintenance and gas cost is

$0.18 per mile, the slope of the linear function is 0.18. The linear function that relates the monthly cost of operating the car as a function of miles driven is
$C(x) = 0.18x + 250.$

b. The car cannot be driven a negative distance, the number of miles driven, x, must be greater than or equal to zero. In addition, there is no definite maximum number of miles that the car can be driven. Therefore, the implied domain of the function is $\{x \mid x \geq 0\}$, or using interval notation $[0, \infty)$.

c. $C(320) = 0.18(320) + 250 = 307.6$
So, the monthly cost of driving 320 miles is $307.60.

d.
$$C(x) = 282.40$$
$$0.18x + 250 = 282.40$$
$$0.18x = 32.40$$
$$x = 180$$
So, Roberta can drive 180 miles each month for the monthly cost of $282.40.

e. Label the horizontal axis x and the vertical axis C. From part (a) we know $C(0) = 250$, and from part (c) we know $C(320) = 307.6$, so $(0, 250)$ and $(320, 307.60)$ are on the graph.

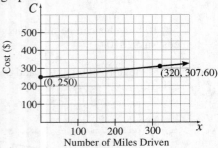

15. a.

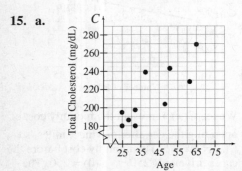

b. From the scatter diagram, we can see that as the age increases, the total cholesterol also increases.

16. Nonlinear

17. Linear with a positive slope.

18. a. Answers will vary. We will use the points $(25, 180)$ and $(65, 269)$.
$$m = \frac{269 - 180}{65 - 25} = \frac{89}{40} = 2.225$$
$$y - 180 = 2.225(x - 25)$$
$$y - 180 = 2.225x - 55.625$$
$$y = f(x) = 2.225x + 124.375$$

b.

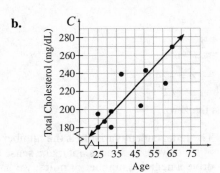

c. $f(39) = 2.225(39) + 124.375 = 211.15$
We predict that the total cholesterol of a 39-year-old male will be approximately 211 mg/dL.

d. The slope of the linear function is 2.225. This means that, for males, the total cholesterol increases by 2.225 mg/dL for each one-year increase in age. The y-intercept, 124.375, would represent the total cholesterol of a male who is 0 years old. Thus, it does not make sense to interpret this y-intercept.

2.4 Exercises

19. Comparing $F(x) = 5x - 2$ to $F(x) = mx + b$, we see that the slope m is 5 and b is -2. We begin by plotting the point $(0, -2)$. Because
$$m = 5 = \frac{5}{1} = \frac{\Delta y}{\Delta x} = \frac{\text{Rise}}{\text{Run}}, \text{ from the point } (0, -2)$$
we go up 5 units and to the right 1 unit and end up at $(1, 3)$. We draw a line through these points and obtain the graph of $F(x) = 5x - 2$.

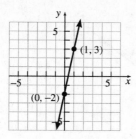

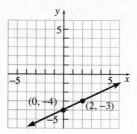

21. Comparing $G(x) = -3x + 7$ to $G(x) = mx + b$,

we see that the slope m is -3 and the b is 7. We begin by plotting the point (0, 7). Because

$$m = -3 = \frac{-3}{1} = \frac{\Delta y}{\Delta x} = \frac{\text{Rise}}{\text{Run}}, \text{ from the point } (0, 7)$$

we go down 3 units and to the right 1 unit and end up at (1, 4). We draw a line through these points and obtain the graph of $G(x) = -3x + 7$.

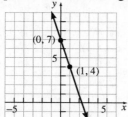

23. Comparing $H(x) = -2$ to $H(x) = mx + b$, we see that the slope m is 0 and b is -2. The graph is a horizontal line through the point (0, -2). We draw a horizontal line through this point and obtain the graph of $H(x) = -2$.

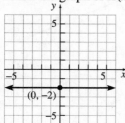

25. Comparing $f(x) = \frac{1}{2}x - 4$ to $f(x) = mx + b$, we

see that the slope m is $\frac{1}{2}$ and b is -4. We begin

by plotting the point (0, -4). Because

$$m = \frac{1}{2} = \frac{\Delta y}{\Delta x} = \frac{\text{Rise}}{\text{Run}}, \text{ from the point } (0, -4) \text{ we}$$

go up 1 unit and to the right 2 units and end up at (2, -3). We draw a line through these points and

obtain the graph of $f(x) = \frac{1}{2}x - 4$.

27. Comparing $F(x) = -\frac{5}{2}x + 5$ to $F(x) = mx + b$,

we see that the slope m is $-\frac{5}{2}$ and b is 5. We

begin by plotting the point (0, 5). Because

$$m = -\frac{5}{2} = \frac{-5}{2} = \frac{\Delta y}{\Delta x} = \frac{\text{Rise}}{\text{Run}}, \text{ from the point } (0, 5)$$

we go down 5 units and to the right 2 units and end up at (2, 0). We draw a line through these

points and obtain the graph of $F(x) = -\frac{5}{2}x + 5$.

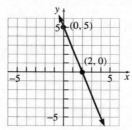

29. Comparing $G(x) = -\frac{3}{2}x$ to $G(x) = mx + b$, we

see that the slope m is $-\frac{3}{2}$ and b is 0. We begin

by plotting the point (0, 0). Because

$$m = -\frac{3}{2} = \frac{-3}{2} = \frac{\Delta y}{\Delta x} = \frac{\text{Rise}}{\text{Run}}, \text{ from the point } (0, 0)$$

we go down 3 units and to the right 2 units and end up at (2, -3). We draw a line through these

points and obtain the graph of $G(x) = -\frac{3}{2}x$.

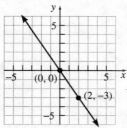

31. $f(x) = 0$
$2x + 10 = 0$
$2x = -10$
$x = -5$
-5 is the zero.

33. $G(x) = 0$
$-5x + 40 = 0$
$-5x = -40$
$x = 8$
8 is the zero.

35. $s(t) = 0$
$\frac{1}{2}t - 3 = 0$
$\frac{1}{2}t = 3$
$t = 6$
6 is the zero.

37. $P(z) = 0$
$-\frac{4}{3}z + 12 = 0$
$-\frac{4}{3}z = -12$
$-4z = -36$
$z = 9$
9 is the zero.

39. Nonlinear

41. Linear with positive slope

43. a.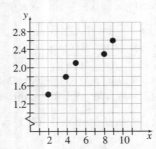

    b. Answers will vary. We will use the points (4, 1.8) and (9, 2.6).
$$m = \frac{2.6 - 1.8}{9 - 4} = \frac{0.8}{5} = 0.16$$
$$y - 1.8 = 0.16(x - 4)$$
$$y - 1.8 = 0.16x - 0.64$$
$$y = 0.16x + 1.16$$

c.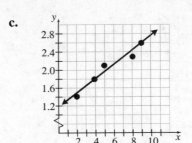

45. a.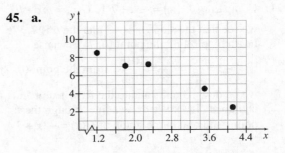

    b. Answers will vary. We will use the points (1.2, 8.4) and (4.1, 2.4).
$$m = \frac{2.4 - 8.4}{4.1 - 1.2} = \frac{-6.0}{2.9} \approx -2.1$$
$$y - 8.4 = -2.1(x - 1.2)$$
$$y - 8.4 = -2.1x + 2.52$$
$$y = -2.1x + 10.92$$

    c.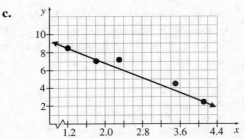

47. a. 3

    b. (0, 2)

    c. $f(x) = 0$
$3x + 2 = 0$
$3x = -2$
$x = -\frac{2}{3}$
$-\frac{2}{3}$ is the zero.

d. $f(x) = 5$

$3x + 2 = 5$

$3x = 3$

$x = 1$

The point (1, 5) is on the graph of *f*.

e. $f(x) \leq -1$

$3x + 2 \leq -1$

$3x \leq -3$

$x \leq -1$

$\{x \mid x \leq -1\}$ or $(-\infty, -1]$

f.

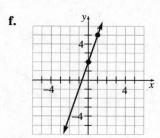

49. a. $f(x) = g(x)$

$x - 5 = -3x + 7$

$x + 3x = 7 + 5$

$4x = 12$

$x = 3$

$f(3) = (3) - 5 = -2$

(3, −2) is on the graph of *f*(x) and *g*(x).

b. $f(x) > g(x)$

$x - 5 > -3x + 7$

$x + 3x > 7 + 5$

$4x > 12$

$x > 3$

$\{x \mid x > 3\}$ or $(3, \infty)$

c.

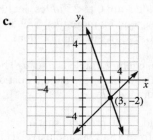

51. Since $f(2) = 6$ and $f(5) = 12$, the points (2, 6) and (5, 12) are on the graph of *f*. Thus,

$$m = \frac{y_2 - y_1}{x_2 - x_1} = \frac{12 - 6}{5 - 2} = \frac{6}{3} = 2$$

$$y - y_1 = m(x - x_1)$$

$$y - 6 = 2(x - 2)$$

$$y - 6 = 2x - 4$$

$$y = 2x + 2 \quad \text{or} \quad f(x) = 2x + 2$$

Finally, $f(-2) = 2(-2) + 2 = -4 + 2 = -2.$

53. Since $h(3) = 7$ and $h(-1) = 14$, the points (3, 7) and (−1, 14) are on the graph of *h*. Thus,

$$m = \frac{y_2 - y_1}{x_2 - x_1} = \frac{14 - 7}{-1 - 3} = \frac{7}{-4} = -\frac{7}{4}$$

$$y - y_1 = m(x - x_1)$$

$$y - 7 = -\frac{7}{4}(x - 3)$$

$$y - 7 = -\frac{7}{4}x + \frac{21}{4}$$

$$y = -\frac{7}{4}x + \frac{49}{4} \quad \text{or} \quad h(x) = -\frac{7}{4}x + \frac{49}{4}$$

Finally, $h\left(\frac{1}{2}\right) = -\frac{7}{4}\left(\frac{1}{2}\right) + \frac{49}{4} = -\frac{7}{8} + \frac{49}{4} = \frac{91}{8}.$

55. a. The point (3, 1) is on the graph of *f*, so $f(3) = 1$. Thus, the solution of $f(x) = 1$ is $x = 3$.

b. The point (−1, −3) is on the graph of *f*, so $f(-1) = -3$. Thus, the solution of $f(x) = -3$ is $x = -1$.

c. The point (4, 2) is on the graph of *f*, so $f(4) = 2$.

d. The intercepts of $y = f(x)$ are (0, −2) and (2, 0). The *y*-intercept is (0, −2) and the *x*-intercept is (2, 0).

e. Use any two points to determine the slope. Here we use (3, 1) and (4, 2):

$$m = \frac{2 - 1}{4 - 3} = \frac{1}{1} = 1$$

From part (d), we know the *y*-intercept is −2, so the equation of the function is $f(x) = x - 2$.

57. a. We are told that the tax function T is for adjusted gross incomes x between \$10,850 and \$37,500, inclusive. Thus the domain is $\{x \mid 10{,}850 \le x \le 37{,}500\}$ or, using interval notation, [10,850, 37,500].

b. Evaluate T at $x = 20{,}000$.
$$T(20{,}000) = 0.15(20{,}000 - 10{,}850) + 870$$
$$= 2242.5$$
A single filer will pay \$2242.50 in taxes if his or her adjusted gross income is \$20,000.

c. The independent variable is adjusted gross income, x. The dependent variable is the tax bill, T.

d. Evaluate T at $x = 10{,}850$ and $37{,}500$.
$$T(10{,}850) = 0.15(10{,}850 - 10{,}850) + 870$$
$$= 870$$
$$T(37{,}500) = 0.15(37{,}500 - 10{,}850) + 870$$
$$= 4867.5$$
Thus the points (10,850, 870), (37,500, 4867.50), and (20,000, 2242.50) (from part b) are on the graph.

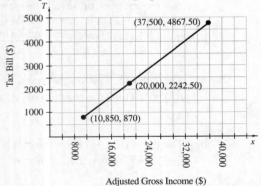

e. We must solve $T(x) = 2996.25$.
$$0.15(x - 10{,}850) + 870 = 2996.25$$
$$0.15x - 1627.5 + 870 = 2996.25$$
$$0.15x - 757.5 = 2996.25$$
$$0.15x = 3753.75$$
$$x = 25{,}025$$
A single filer with an adjusted gross income of \$25,025 will have a tax bill of \$2996.25.

59. a. The independent variable is the number of miles traveled, m. It would not make sense to travel a negative number of miles. Thus, the domain of C is $\{m \mid m \ge 0\}$ or, using interval notation, $[0, \infty)$.

b. $C(0) = 1.5(0) + 2 = 2$
The base fare is \$2.00 before any distance is driven.

c. Evaluate C at $m = 5$.
$$C(5) = 1.5(5) + 2 = 9.5$$
The cab fare for a 5-mile ride is \$9.50.

d. Evaluate C at $m = 0$, 10, and 15.
$$C(0) = 1.5(0) + 2 = 2$$
$$C(10) = 1.5(10) + 2 = 17$$
$$C(15) = 1.5(15) + 2 = 24.5$$
Thus, the points (0, 2), (10, 17), and (15, 24.5) are on the graph.

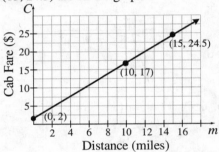

e. We must solve $C(m) = 13.25$.
$$1.50m + 2.00 = 13.25$$
$$1.50m = 11.25$$
$$m = 7.5$$
A person can travel 7.5 miles in a cab for \$13.25.

f. $C(m) \le 39.50$
$$1.5m + 2 \le 39.50$$
$$1.5m \le 37.50$$
$$m \le 25$$
You can ride from 0 to 25 miles, inclusive, if you can spend no more than \$39.50, which is [0, 25] in interval notation.

61. a. The independent variable is age, a. The dependent variable is the annual cost of health insurance, H.

b. We are told in the problem that a is restricted from 15 to 90, inclusive. Thus, the domain of H is $\{a \mid 15 \le a \le 90\}$ or, using interval notation, [15, 90].

c. Evaluate H at $a = 30$.
$$H(30) = 22.8(30) - 117.5 = 566.5$$
The health insurance premium of a 30 year old is $566.50.

d. Evaluate H at $a = 15,\ 50,$ and 90.
$$H(15) = 22.8(15) - 117.5 = 224.5$$
$$H(50) = 22.8(50) - 117.5 = 1022.5$$
$$H(90) = 22.8(90) - 117.5 = 1934.5$$
Thus, the points (15, 224.5), (50, 1022.5), and (90, 1934.5) are on the graph.

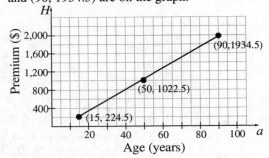

e. We must solve $H(a) = 976.9$.
$$22.8a - 117.5 = 976.9$$
$$22.8a = 1094.4$$
$$a = 48$$
A 48-year-old individual will have a health insurance premium of $976.90.

63. a. $B(m) = 0.05m + 5.95$

b. The number of minutes, m, is the independent variable. The amount of the bill, B, is the dependent variable.

c. Because the number of minutes can not be negative, m must be greater than or equal to zero. The implied domain is $\{m \mid m \geq 0\}$, or using interval notation $[0, \infty)$.

d. $B(300) = 0.05(300) + 5.95 = 20.95$
If 300 minutes are used, the monthly bill will be $20.95.

e. $0.05m + 5.95 = 17.95$
$0.05m = 12$
$m = 240$
If the monthly bill is $17.95, then 240 minutes were used.

f.

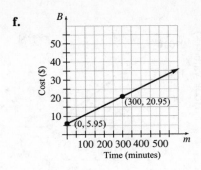

g. $0.05m + 5.95 \leq 18.45$
$0.05m \leq 12.50$
$m \leq 250$
You can speak from 0 to 250 minutes, included, if you don't want to spend more than $18.45. Or using interval notation, [0, 250].

65. a. The computer will depreciate by
$$\frac{\$2700}{3} = \$900 \text{ per year. Thus, the slope is}$$
-900. The y-intercept will be $2700, the initial value of the computer. The linear function that represents book value, V, of the computer after x years is
$V(x) = -900x + 2700$.

b. Because the computer cannot have a negative age, the age, x, must be greater than or equal to 0. After 3 years, the book value will be $V(3) = -900(3) + 2700 = 0$, and the book value cannot be negative. Therefore the implied domain of function is $\{x \mid 0 \leq x \leq 3\}$, or using interval notation [0, 3].

c. $V(1) = -900(1) + 2700 = 1800$
After one year, the book value of the computer will be $1800.

d. The intercepts are (0, 2700) and (3, 0). The V-intercept is (0, 2700) and the x-intercept is (3, 0).

e. $-900x + 2700 = 900$
$-900x = -1800$
$x = 2$
After two years, the book value of the computer will be $900.

f.

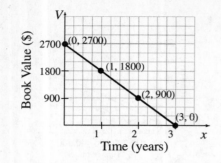

67. a. Let x represent the weight of the diamond and C represent the cost.

$$m = \frac{4378 - 3543}{0.8 - 0.7} = \$8350 \text{ per carat}$$

$$C - 3543 = 8350(x - 0.7)$$
$$C - 3543 = 8350x - 5845$$
$$C = 8350x - 2302$$

Using function notation,
$$C(x) = 8350x - 2302.$$

b. $C(0.77) = 8350(0.77) - 2302 = 4127.5$

The price of a 0.77 carat diamond would be $4127.50.

c. The slope indicates that the cost of diamonds increases at a rate of $8350 per carat.

d. $8350x - 2302 = 5300$
$$8350x = 7602$$
$$x = \frac{7602}{8350} \approx 0.91$$

A diamond weighing approximately 0.91 carat should cost $5300.

69. a. Let x represent disposable income (in billions of dollars) and C represent personal consumption expenditures (in billions of dollars).

$$m = \frac{10{,}349 - 9323}{11{,}380 - 9916} \approx 0.701$$

$$C - 9323 = 0.701x - 9916$$
$$C - 9323 \approx 0.701x - 6951.116$$
$$C \approx 0.701x + 2371.884$$

In function notation, and replacing the approximately equals sign, we have
$$C(x) = 0.701x + 2371.884.$$

b. $C(9742) = 0.701(9742) + 2371.884$
$$\approx 9201.0$$

In 2007, personal consumption expenditures were $9201.0 billion.

c. The slope indicates that the personal consumption expenditures are increasing at a rate of approximately $0.70 for every $1.00 increase in personal disposable income.

d. $0.701x + 2371.884 = 9520$
$$0.701x = 7148.116$$
$$x = 10{,}197$$

Personal disposable income would be approximately $10,197 billion if personal consumption expenditures were $9520 billion.

71. a.

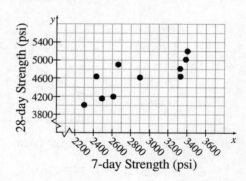

b. Linear

c. Answers will vary. We will use the points (2300, 4070) and (3390, 5220).

$$m = \frac{5220 - 4070}{3390 - 2300} = \frac{1150}{1090} \approx 1.055$$

$$y - 4070 = 1.06(x - 2300)$$
$$y - 4070 = 1.06x - 2438$$
$$y = 1.06x + 1632$$

d.

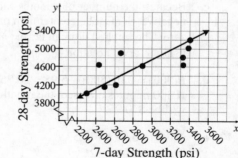

e. $x = 3000$: $y = 1.06(3000) + 1632 = 4812$
We predict that the 28-day strength will be 4812 psi if the 7-day strength is 3000 psi.

f. The slope of the line found is 1.06. This means that if the 7-day strength is increased by 1 psi, then the 28-day strength will increase by 1.06 psi.

73. a. No, the relation does not represent a function. Several w-coordinates are paired with multiple N-coordinates. For example, the w-coordinate 42.3 is paired with the two different N-coordinates 87 and 82.

b.

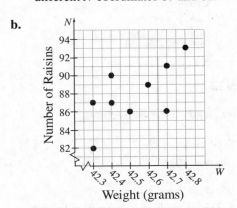

c. Answers will vary. We will use the points (42.3, 82) and (42.8, 93).

$$m = \frac{93 - 82}{42.8 - 42.3} = \frac{11}{0.5} = 22$$
$$N - 82 = 22(w - 42.3)$$
$$N - 82 = 22w - 930.6$$
$$N = 22w - 848.6$$

d.

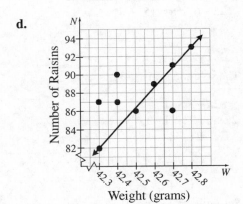

e. Let N represent the number of raisins in the box, and let w represent the weight (in grams) of the box of raisins.
$$N(w) = 22w - 848.6$$

f. $N(42.5) = 22(42.5) - 848.6 = 86.4$

We predict that approximately 86 raisins will be in a box weighing 42.5 grams.

g. The slope of the line found is 22 raisins per gram. This means that if the weight is to be increased by one gram, then the number of raisins must be increased by 22 raisins.

75. a.

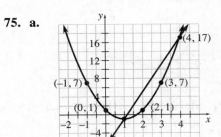

b. $f(1) = -1$
$f(4) = 17$
$$m = \frac{17 - (-1)}{4 - 1} = \frac{18}{3} = 6$$
The slope is 6.

c. $y - 17 = 6(x - 4)$
$y - 17 = 6x - 24$
$y = 6x - 7$

d.

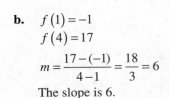

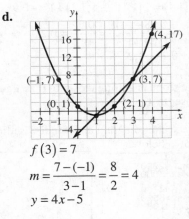

$f(3) = 7$
$$m = \frac{7 - (-1)}{3 - 1} = \frac{8}{2} = 4$$
$$y = 4x - 5$$

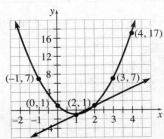

$f(2) = 1$

$m = \dfrac{1-(-1)}{2-1} = \dfrac{2}{1} = 2$

$y = 2x - 3$

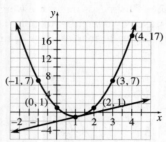

$f(1.5) = 2(1.5)^2 - 4(1.5) + 1 = 4.5 - 6 + 1 = -0.5$

$m = \dfrac{-0.5-(-1)}{1.5-1} = \dfrac{0.5}{0.5} = 1$

$y = x - 2$

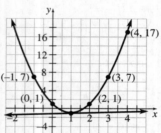

$f(1.1) = 2(1.1)^2 - 4(1.1) + 1 = -0.98$

$m = \dfrac{-0.98-(-1)}{1.1-1} = \dfrac{0.02}{0.1} = 0.2$

$y = 0.2x - 1.2$

   e. The slope is getting closer to 0 as x gets closer to 1.

77. a. The scatter diagram and window settings are shown below.

 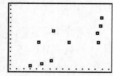

   b. As shown below the line of best fit is approximately $y = 0.676x + 2675.562$.

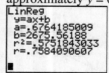

79. a. The scatter diagram and window settings are shown below.

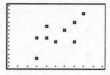 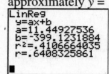

   b. As shown below the line of best fit is approximately $y = 11.449x - 399.123$.

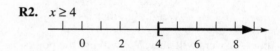

Section 2.5

Are You Ready for This Section?

R1. Set-builder: $\{x \mid -2 \le x \le 5\}$

    Interval: $[2, 5]$

R2. $x \ge 4$

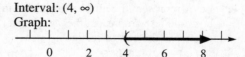

R3. The parenthesis indicates that we do *not* include -1 in our interval, while the square bracket indicates that we *do* include 3. Our interval is $(-1, 3]$.

R4. $2(x+3) - 5x = 15$

      $2x + 6 - 5x = 15$

         $-3x + 6 = 15$

            $-3x = 9$

               $x = -3$

    The solution set is $\{-3\}$.

R5. $2x + 3 > 11$

    $2x + 3 - 3 > 11 - 3$

         $2x > 8$

        $\dfrac{2x}{2} > \dfrac{8}{2}$

          $x > 4$

    Set-builder: $\{x \mid x > 4\}$

    Interval: $(4, \infty)$

    Graph:

R6.
$$x+8 \geq 4(x-1)-x$$
$$x+8 \geq 4x-4-x$$
$$x+8 \geq 3x-4$$
$$x+8-x \geq 3x-4-x$$
$$8 \geq 2x-4$$
$$8+4 \geq 2x-4+4$$
$$12 \geq 2x$$
$$\frac{12}{2} \geq \frac{2x}{2}$$
$$6 \geq x \quad \text{or} \quad x \leq 6$$

Set-builder: $\{x \mid x \leq 6\}$

Interval: $(-\infty, 6]$

Graph:

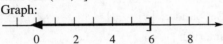

Section 2.5 Quick Checks

1. The <u>intersection</u> of two sets A and B, denoted $A \cap B$, is the set of all elements that belong to both set A and set B.

2. The word <u>and</u> implies intersection. The word <u>or</u> implies union.

3. True. If the two sets have no elements in common, the intersection will be the empty set.

4. False. The symbol for the union of two sets is $\cup$ while the symbol for intersection is $\cap$.

5. $A \cap B = \{1, 3, 5\}$

6. $A \cap C = \{2, 4, 6\}$

7. $A \cup B = \{1, 2, 3, 4, 5, 6, 7\}$

8. $A \cup C = \{1, 2, 3, 4, 5, 6, 8\}$

9. $B \cap C = \{\ \ \}$ or $\varnothing$

10. $B \cup C = \{1, 2, 3, 4, 5, 6, 7, 8\}$

11. $A \cap B$ is the set of all real numbers that are greater than 2 and less than 7.

Set-builder: $\{x \mid 2 < x < 7\}$

Interval: $(2, 7)$

12. $A \cup C$ is the set of real numbers that are greater than 2 or less than or equal to -3.

Set-builder: $\{x \mid x \leq -3 \text{ or } x > 2\}$

Interval: $(-\infty, -3] \cup (2, \infty)$

13.
$$2x+1 \geq 5 \quad \text{and} \quad -3x+2 < 5$$
$$2x \geq 4 \phantom{\text{and}\quad} -3x < 3$$
$$x \geq 2 \phantom{\text{and}\quad aaa} x > -1$$
We need $x \geq 2$ and $x > -1$.

Set-builder: $\{x \mid x \geq 2\}$

Interval: $[2, \infty)$

14.
$$4x-5 < 7 \quad \text{and} \quad 3x-1 > -10$$
$$4x < 12 \phantom{\text{and}\quad} 3x > -9$$
$$x < 3 \phantom{\text{and}\quad aaa} x > -3$$
We need $x < 3$ and $x > -3$.

Set-builder: $\{x \mid -3 < x < 3\}$

Interval: $(-3, 3)$

15.
$$-8x+3 < -5 \quad \text{and} \quad \frac{2}{3}x+1 < 3$$
$$-8x < -8 $$
$$x > 1 \frac{2}{3}x < 2$$
$$ x < 3$$
We need $x > 1$ and $x < 3$.

Set-builder: $\{x \mid 1 < x < 3\}$

Interval: $(1, 3)$

16.
$$3x-5 < -8 \quad \text{and} \quad 2x+1 > 5$$
$$3x < -3 \phantom{\text{and}\quad} 2x > 4$$
$$x < -1 \phantom{\text{and}\quad aa} x > 2$$
We need $x < -1$ and $x > 2$. Looking at the graphs of the inequalities separately we see that there are no such numbers that satisfy both inequalities. Therefore, the solution set is empty.

Solution set: $\{\ \ \}$ or $\varnothing$

17.
$$5x+1 \leq 6 \quad \text{and} \quad 3x+2 \geq 5$$
$$5x \leq 5 \phantom{\text{and}\quad} 3x \geq 3$$
$$x \leq 1 \phantom{\text{and}\quad aa} x \geq 1$$
We need $x \leq 1$ and $x \geq 1$. Looking at the graphs of the inequalities separately we see that the only number that is both less than or equal to 1 and greater than or equal to 1 is the number 1.

Solution set: $\{1\}$

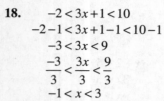

18.
$$-2 < 3x+1 < 10$$
$$-2-1 < 3x+1-1 < 10-1$$
$$-3 < 3x < 9$$
$$\frac{-3}{3} < \frac{3x}{3} < \frac{9}{3}$$
$$-1 < x < 3$$

Set-builder: $\{x \mid -1 < x < 3\}$

Interval: $(-1, 3)$

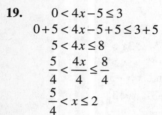

19.
$$0 < 4x-5 \le 3$$
$$0+5 < 4x-5+5 \le 3+5$$
$$5 < 4x \le 8$$
$$\frac{5}{4} < \frac{4x}{4} \le \frac{8}{4}$$
$$\frac{5}{4} < x \le 2$$

Set-builder: $\left\{ x \mid \frac{5}{4} < x \le 2 \right\}$

Interval: $\left(\frac{5}{4}, 2 \right]$

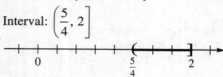

20.
$$3 \le -2x-1 \le 11$$
$$3+1 \le -2x-1+1 \le 11+1$$
$$4 \le -2x \le 12$$
$$\frac{4}{-2} \ge \frac{-2x}{-2} \ge \frac{12}{-2}$$
$$-2 \ge x \ge -6 \quad \text{or} \quad -6 \le x \le -2$$

Set-builder: $\{x \mid -6 \le x \le -2\}$

Interval: $[-6, -2]$

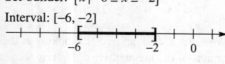

21. $x+3 < 1 \quad$ or $\quad x-2 > 3$
$\qquad x < -2 \qquad\qquad x > 5$
We need $x < -2$ or $x > 5$.
Set-builder: $\{x \mid x < -2 \text{ or } x > 5\}$
Interval: $(-\infty, -2) \cup (5, \infty)$

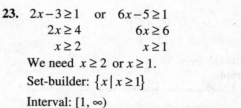

22. $3x+1 \le 7 \quad$ or $\quad 2x-3 > 9$
$\qquad 3x \le 6 \qquad\qquad 2x > 12$
$\qquad x \le 2 \qquad\qquad x > 6$
We need $x \le 2$ or $x > 6$.
Set-builder: $\{x \mid x \le 2 \text{ or } x > 6\}$

Interval: $(-\infty, 2] \cup (6, \infty)$

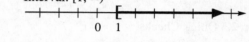

23. $2x-3 \ge 1 \quad$ or $\quad 6x-5 \ge 1$
$\qquad 2x \ge 4 \qquad\qquad 6x \ge 6$
$\qquad x \ge 2 \qquad\qquad x \ge 1$
We need $x \ge 2$ or $x \ge 1$.
Set-builder: $\{x \mid x \ge 1\}$

Interval: $[1, \infty)$

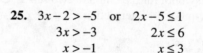

24. $\dfrac{3}{4}(x+4) < 6 \quad$ or $\quad \dfrac{3}{2}(x+1) > 15$
$\qquad x+4 < 8 \qquad\qquad x+1 > 10$
$\qquad x < 4 \qquad\qquad x > 9$
We need $x < 4$ or $x > 9$.
Set-builder: $\{x \mid x < 4 \text{ or } x > 9\}$

Interval: $(-\infty, 4) \cup (9, \infty)$

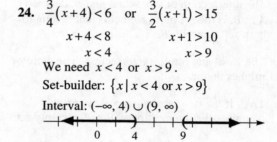

25. $3x-2 > -5 \quad$ or $\quad 2x-5 \le 1$
$\qquad 3x > -3 \qquad\qquad 2x \le 6$
$\qquad x > -1 \qquad\qquad x \le 3$
If we look at the graphs of the inequalities separately, we see that the union of their solution sets is the set of real numbers.
Set-builder: $\{x \mid x \text{ is any real number}\}$

Interval: $(-\infty, \infty)$

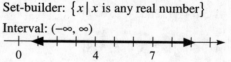

26. $-5x-2 \le 3 \quad$ or $\quad 7x-9 > 5$
$\qquad -5x \le 5 \qquad\qquad 7x > 14$
$\qquad x \ge -1 \qquad\qquad x > 2$
We need $x \ge -1$ or $x > 2$. Since the solution set of the inequality $x > 2$ is a subset of the solution set for the inequality $x \ge -1$, we only need to consider $x \ge -1$.
Set-builder: $\{x \mid x \ge -1\}$

Interval: $[-1, \infty)$

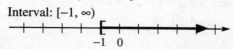

27. Let x = taxable income (in dollars). The federal income tax in the 25% bracket is $4867.50 plus 25% of the amount over $35,350. In general, the income tax for the 25% bracket is given by
$$4867.50 + 0.25(x - 35,350)$$
Because the federal income tax is between $4867.50 and $17,442.50, we have
$$4867.50 \le 4867.50 + 0.25(x - 35,350) \le 17,442.50$$
$$4867.50 \le 4867.50 + 0.25x - 8837.5 \le 17,442.50$$
$$4867.50 \le 0.25x - 3970 \le 17,442.50$$
$$8837.50 \le 0.25x \le 21,412.50$$
$$35,350 \le x \le 85,650$$
To be in the 25% tax bracket, an individual would have an income between $35,350 and $85,650.

28. Let x = number of minutes. The long distance plan charges $2.00 per month and $0.10 per minute. In general, the charge is given by $2.00 + 0.10x$. Sophia's charges ranged from $6.50 to $26.50, so we get the inequality
$$6.50 \le 2.00 + 0.10x \le 26.50$$
$$4.50 \le 0.10x \le 24.50$$
$$45 \le x \le 245$$
Over the course of the year, Sophia's monthly minutes were between 45 minutes and 245 minutes.

2.5 Exercises

29. $A \cup B = \{1, 4, 5, 6, 7, 8, 9\}$

31. $A \cap B = \{5, 7, 9\}$

33. $B \cap C = \varnothing$ or $\{\ \}$

35. a. $A \cap B = \{x \mid -2 < x \le 5\}$

Interval: $(-2, 5]$

b. $A \cup B = \{x \mid x \text{ is a real number}\}$

Interval: $(-\infty, \infty)$

37. a. $E \cap F = \varnothing$ or $\{\ \}$

b. $E \cup F = \{x \mid x < -1 \text{ or } x > 3\}$

Interval: $(-\infty, -1) \cup (3, \infty)$

39. a. We need the set of all values for x such that $2x - 1 \le 3$ and $2x - 1 \ge -5$. On the graph, we look for where the graph of $g(x) = 2x - 1$ is between the horizontal lines $f(x) = -5$ and $h(x) = 3$ (inclusive).

Set-builder: $\{x \mid -2 \le x \le 2\}$

Interval: $[-2, 2]$

b. We need the set of all values for x such that $2x - 1 < -5$ or $2x - 1 > 3$. On the graph, we look for where the graph of $g(x) = 2x - 1$ is below the horizontal line $f(x) = -5$ or above the horizontal line $h(x) = 3$.

Set-builder: $\{x \mid x < -2 \text{ or } x > 2\}$

Interval: $(-\infty, -2) \cup (2, \infty)$

41. a. We need the set of all values for x such that $-\dfrac{5}{3}x + 1 < 6$ and $-\dfrac{5}{3}x + 1 > -4$. On the graph, we look for where the graph of $g(x) = -\dfrac{5}{3}x + 1$ is between the horizontal lines $f(x) = -4$ and $h(x) = 6$, not inclusive.

Set-builder: $\{x \mid -3 < x < 3\}$

Interval: $(-3, 3)$

b. We need the set of all values for x such that $-\dfrac{5}{3}x + 1 \le -4$ or $-\dfrac{53}{x+1} \ge 6$. On the graph, we look for where the graph of $g(x) = -\dfrac{5}{3}x + 1$ is below the horizontal line $f(x) = -4$ or above the horizontal line $h(x) = 6$ (inclusive).

Set-builder: $\{x \mid x \le -3 \text{ or } x \ge 3\}$

Interval: $(-\infty, -3] \cup [3, \infty)$

43. $x < 3$ and $x \geq -2$

Set-builder: $\{x \mid -2 \leq x < 3\}$

Interval: $[-2, 3)$

45. $4x - 4 < 0$ and $-5x + 1 \leq -9$

$\quad 4x < 4 \qquad\qquad -5x \leq -10$

$\quad\; x < 1 \qquad\qquad\;\; x \geq 2$

We need $x < 1$ and $x \geq 2$.

Solution set: $\varnothing$ or $\{\ \}$

47. $4x - 3 < 5$ and $-5x + 3 > 13$

$\quad 4x < 8 \qquad\qquad -5x > 10$

$\quad\; x < 2 \qquad\qquad\;\; x < -2$

We need $x < 2$ and $x < -2$.

Set-builder: $\{x \mid x < -2\}$

Interval: $(-\infty, -2)$

49. $7x + 2 \geq 9$ and $4x + 3 \leq 7$

$\quad 7x \geq 7 \qquad\qquad 4x \leq 4$

$\quad\; x \geq 1 \qquad\qquad\; x \leq 1$

We need $x \geq 1$ and $x \leq 1$.

Solution set: $\{1\}$

51. $-3 \leq 5x + 2 < 17$

$-3 - 2 \leq 5x + 2 - 2 < 17 - 2$

$\quad -5 \leq 5x < 15$

$\quad \dfrac{-5}{5} \leq \dfrac{5x}{5} < \dfrac{15}{5}$

$\quad -1 \leq x < 3$

Set-builder: $\{x \mid -1 \leq x < 3\}$

Interval: $[-1, 3)$

53. $-3 \leq 6x + 1 \leq 10$

$-3 - 1 \leq 6x + 1 - 1 \leq 10 - 1$

$\quad -4 \leq 6x \leq 9$

$\quad \dfrac{-4}{6} \leq \dfrac{6x}{6} \leq \dfrac{9}{6}$

$\quad -\dfrac{2}{3} \leq x \leq \dfrac{3}{2}$

Set-builder: $\left\{x \mid -\dfrac{2}{3} \leq x \leq \dfrac{3}{2}\right\}$

Interval: $\left[-\dfrac{2}{3}, \dfrac{3}{2}\right]$

55. $\quad 3 \leq -5x + 7 < 12$

$\quad 3 - 7 \leq -5x + 7 - 7 < 12 - 7$

$\quad\; -4 \leq -5x < 5$

$\quad \dfrac{-4}{-5} \geq \dfrac{-5x}{-5} > \dfrac{5}{-5}$

$\quad\; \dfrac{4}{5} \geq x > -1 \;$ or $\; -1 < x \leq \dfrac{4}{5}$

Set-builder: $\left\{x \mid -1 < x \leq \dfrac{4}{5}\right\}$

Interval: $\left(-1, \dfrac{4}{5}\right]$

57. $\quad -1 \leq \dfrac{1}{2}x - 1 \leq 3$

$\quad -1 + 1 \leq \dfrac{1}{2}x - 1 + 1 \leq 3 + 1$

$\quad\quad 0 \leq \dfrac{1}{2}x \leq 4$

$\quad 2(0) \leq 2 \cdot \dfrac{1}{2}x \leq 2 \cdot 4$

$\quad\quad 0 \leq x \leq 8$

Set-builder: $\{x \mid 0 \leq x \leq 8\}$

Interval: $[0, 8]$

59. $\quad 3 \leq -2x - 1 \leq 11$

$\quad 3 + 1 \leq -2x - 1 + 1 \leq 11 + 1$

$\quad\; 4 \leq -2x \leq 12$

$\quad \dfrac{4}{-2} \geq \dfrac{-2x}{-2} \geq \dfrac{12}{-2}$

$\quad -2 \geq x \geq -6 \;$ or $\; -6 \leq x \leq -2$

Set-builder: $\{x \mid -6 \leq x \leq -2\}$

Interval: $[-6, -2]$

61. $\dfrac{2}{3}x + \dfrac{1}{2} < \dfrac{5}{6}$ and $-\dfrac{1}{5}x + 1 < \dfrac{3}{10}$

$6\left(\dfrac{2}{3}x + \dfrac{1}{2}\right) < 6\left(\dfrac{5}{6}\right)$ $10\left(-\dfrac{1}{5}x + 1\right) < 10\left(\dfrac{3}{10}\right)$

$\qquad 4x + 3 < 5 \qquad\qquad\qquad -2x + 10 < 3$

$\qquad\quad 4x < 2 \qquad\qquad\qquad\qquad -2x < -7$

$\qquad\qquad x < \dfrac{1}{2} \qquad\qquad\qquad\qquad x > \dfrac{7}{2}$

We need $x < \dfrac{1}{2}$ and $x > \dfrac{7}{2}$.

Solution set: $\varnothing$ or $\{\ \}$

63. $-2 < \dfrac{3x+1}{2} \le 8$

$2(-2) < 2\left(\dfrac{3x+1}{2}\right) \le 2(8)$

$\quad -4 < 3x + 1 \le 16$

$-4 - 1 < 3x + 1 - 1 \le 16 - 1$

$\qquad -5 < 3x \le 15$

$\qquad \dfrac{-5}{3} < \dfrac{3x}{3} \le \dfrac{15}{3}$

$\qquad -\dfrac{5}{3} < x \le 5$

Set-builder: $\left\{ x \,\middle|\, -\dfrac{5}{3} < x \le 5 \right\}$

Interval: $\left(-\dfrac{5}{3}, 5\right]$

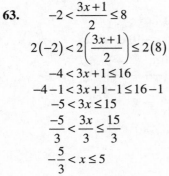

65. $-8 \le -2(x+1) < 6$

$\quad -8 \le -2x - 2 < 6$

$-8 + 2 \le -2x - 2 + 2 < 6 + 2$

$\quad -6 \le -2x < 8$

$\quad \dfrac{-6}{-2} \ge \dfrac{-2x}{-2} > \dfrac{8}{-2}$

$\qquad 3 \ge x > -4$ or $-4 < x \le 3$

Set-builder: $\{ x \,|\, -4 < x \le 3 \}$

Interval: $(-4, 3]$

67. $x < -2$ or $x > 3$

Solution set: $\{ x \,|\, x < -2 \text{ or } x > 3 \}$

Interval: $(-\infty, -2) \cup (3, \infty)$

69. $x - 2 < -4$ or $x + 3 > 8$

$\quad x < -2 \qquad\qquad x > 5$

We need $x < -2$ or $x > 5$.

Set-builder: $\{ x \,|\, x < -2 \text{ or } x > 5 \}$

Interval: $(-\infty, -2) \cup (5, \infty)$

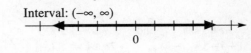

71. $6(x-2) < 12$ or $4(x+3) > 12$

$\quad 6x - 12 < 12 \qquad\qquad 4x + 12 > 12$

$\qquad 6x < 24 \qquad\qquad\qquad 4x > 0$

$\qquad\quad x < 4 \qquad\qquad\qquad\quad x > 0$

We need $x < 4$ or $x > 0$.

Set-builder: $\{ x \,|\, x \text{ is a real number} \}$

Interval: $(-\infty, \infty)$

73. $-8x + 6x - 2 > 0$ or $5x > 3x + 8$

$\quad -2x - 2 > 0 \qquad\qquad 2x > 8$

$\qquad -2x > 2 \qquad\qquad\quad x > 4$

$\qquad\quad x < -1$

We need $x < -1$ or $x > 4$.

Set-builder: $\{ x \,|\, x < -1 \text{ or } x > 4 \}$

Interval: $(-\infty, -1) \cup (4, \infty)$

75. $2x + 5 \le -1$ or $\dfrac{4}{3}x - 3 > 5$

$\quad 2x \le -6 \qquad\qquad\qquad \dfrac{4}{3}x > 8$

$\quad\; x \le -3 \qquad\qquad\qquad\quad 4x > 24$

$\qquad\qquad\qquad\qquad\qquad\qquad x > 6$

We need $x \le -3$ or $x > 6$.

Set-builder: $\{ x \,|\, x \le -3 \text{ or } x > 6 \}$

Interval: $(-\infty, -3] \cup (6, \infty)$

77. $\dfrac{1}{2}x < 3$ or $\dfrac{3x-1}{2} > 4$

$\quad x < 6 \qquad\qquad 3x - 1 > 8$

$\qquad\qquad\qquad\qquad 3x > 9$

$\qquad\qquad\qquad\qquad\; x > 3$

We need $x < 6$ or $x > 3$.

Set-builder: $\{ x \,|\, x \text{ is a real number} \}$

Interval: $(-\infty, \infty)$

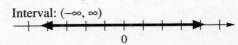

79. $3(x-1)+5<2$ or $-2(x-3)<1$
$3x-3+5<2$ $-2x+6<1$
$3x+2<2$ $-2x<-5$
$3x<0$ $x>\dfrac{5}{2}$
$x<0$

We need $x<0$ or $x>\dfrac{5}{2}$.

Set-builder: $\left\{x\,\middle|\,x<0 \text{ or } x>\dfrac{5}{2}\right\}$

Interval: $(-\infty,\,0)\cup\left(\dfrac{5}{2},\,\infty\right)$

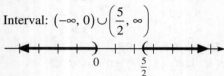

81. $3a+5<5$ and $-2a+1\le7$
$3a+5-5<5-5$ $-2a+1-1\le7-1$
$3a<0$ $-2a\le6$
$\dfrac{3a}{3}<\dfrac{0}{3}$ $\dfrac{-2a}{-2}\ge\dfrac{6}{-2}$
$a<0$ $a\ge-3$

We need $a<0$ and $a\ge-3$.

Set-builder: $\{a\,|\,-3\le a<0\}$

Interval: $[-3,\,0)$

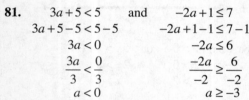

83. $5(x+2)<20$ or $4(x-4)>-20$
$\dfrac{5(x+2)}{5}<\dfrac{20}{5}$ $\dfrac{4(x-4)}{4}>\dfrac{-20}{4}$
$x+2<4$ $x-4>-5$
$x+2-2<4-2$ $x-4+4>-5+4$
$x<2$ $x>-1$

We need $x<2$ or $x>-1$.

Set-builder: $\{x\,|\,x \text{ is a real number}\}$

Interval: $(-\infty,\,\infty)$

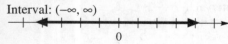

85. $-4\le3x+2\le10$
$-4-2\le3x+2-2\le10-2$
$-6\le3x\le8$
$\dfrac{-6}{3}\le\dfrac{3x}{3}\le\dfrac{8}{3}$
$-2\le x\le\dfrac{8}{3}$

Set-builder: $\left\{x\,\middle|\,-2\le x\le\dfrac{8}{3}\right\}$

Interval: $\left[-2,\,\dfrac{8}{3}\right]$

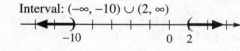

87. $2x+7<-13$ or $5x-3>7$
$2x+7-7<-13-7$ $5x-3+3>7+3$
$2x<-20$ $5x>10$
$\dfrac{2x}{2}<\dfrac{-20}{2}$ $\dfrac{5x}{5}>\dfrac{10}{5}$
$x<-10$ $x>2$

We need $x<-10$ or $x>2$.

Set-builder: $\{x\,|\,x<-10 \text{ or } x>2\}$

Interval: $(-\infty,\,-10)\cup(2,\,\infty)$

89. $5<3x-1<14$
$5+1<3x-1+1<14+1$
$6<3x<15$
$\dfrac{6}{3}<\dfrac{3x}{3}<\dfrac{15}{3}$
$2<x<5$

Set-builder: $\{x\,|\,2<x<5\}$

Interval: $(2,\,5)$

91. $\dfrac{x}{3}\le-1$ or $\dfrac{4x-1}{2}>7$
$3\cdot\dfrac{x}{3}\le3\cdot(-1)$ $2\cdot\dfrac{4x-1}{2}>2\cdot7$
$x\le-3$ $4x-1>14$
                    $4x-1+1>14+1$
                    $4x>15$
                    $\dfrac{4x}{4}>\dfrac{15}{4}$
                    $x>\dfrac{15}{4}$

We need $x\le-3$ or $x>\dfrac{15}{4}$.

Set-builder: $\left\{x\,\middle|\,x\le-3 \text{ or } x>\dfrac{15}{4}\right\}$

Interval: $(-\infty,\,-3]\cup\left(\dfrac{15}{4},\,\infty\right)$

93.
$$-3 \le -2(x+1) < 8$$
$$-3 \le -2x - 2 < 8$$
$$-3 + 2 \le -2x - 2 + 2 < 8 + 2$$
$$-1 \le -2x < 10$$
$$\frac{-1}{-2} \ge \frac{-2x}{-2} > \frac{10}{-2}$$
$$\frac{1}{2} \ge x > -5 \quad \text{or} \quad -5 < x \le \frac{1}{2}$$

Set-builder: $\left\{ x \mid -5 < x \le \frac{1}{2} \right\}$

Interval: $\left(-5, \frac{1}{2} \right]$

95.
$$-3 < x < 4$$
$$-3 + 4 < x + 4 < 4 + 4$$
$$1 < x + 4 < 8$$
We need $a = 1$ and $b = 8$.

97.
$$4 < x < 10$$
$$3 \cdot 4 < 3 \cdot x < 3 \cdot 10$$
$$12 < 3x < 30$$
We need $a = 12$ and $b = 30$.

99.
$$-2 < x < 6$$
$$3 \cdot (-2) < 3 \cdot x < 3 \cdot 6$$
$$-6 < 3x < 18$$
$$-6 + 5 < 3x + 5 < 18 + 5$$
$$-1 < 3x + 5 < 23$$
We need $a = -1$ and $b = 23$.

101. Let x = systolic blood pressure.
$$90 < x < 140$$

103. Let x = final exam score.
$$80 \le \frac{74 + 86 + 77 + 89 + 2x}{6} \le 89$$
$$80 \le \frac{326 + 2x}{6} \le 89$$
$$6 \cdot 80 \le 6 \cdot \frac{326 + 2x}{6} \le 6 \cdot 89$$
$$480 \le 326 + 2x \le 534$$
$$480 - 326 \le 326 + 2x - 326 \le 534 - 326$$
$$154 \le 2x \le 208$$
$$\frac{154}{2} \le \frac{2x}{2} \le \frac{208}{2}$$
$$77 \le x \le 104$$
Joanna needs to score at least a 77 on the final.
That is, $77 \le x \le 100$ (assuming 100 is the max score, otherwise $77 \le x \le 104$).

105. Let x = weekly wages.
$$800 \le x \le 900$$
$$800 - 721 \le x - 721 \le 900 - 721$$
$$79 \le x - 721 \le 179$$
$$0.25(79) \le 0.25(x - 721) \le 0.25(179)$$
$$19.75 \le 0.25(x - 721) \le 44.75$$
$$19.75 + 93.60 \le 0.25(x - 721) + 93.60 \le 44.75 + 93.60$$
$$113.35 \le 0.25(x - 721) + 93.60 \le 138.35$$
The amount withheld ranges between \$113.35 and \$138.35, inclusive.

107. Let x = total sales.
$$3000 \le 2500 + 0.01x \le 5000$$
$$500 \le 0.01x \le 2500$$
$$50{,}000 \le x \le 250{,}000$$
To earn between \$3000 and \$5000 per month, total sales must be between \$50,000 and \$250,000.

109. Let x = number of kwh.
$$88.86 \le 0.092897(x - 350) + 42.41 \le 137.16$$
$$46.45 \le 0.092897(x - 350) \le 94.75$$
$$\frac{46.45}{0.092897} \le x - 350 \le \frac{94.75}{0.092897}$$
$$\frac{46.45}{0.092897} + 350 \le x \le \frac{94.75}{0.092897} + 350$$
$$850 \le x \le 1370 \quad \text{(approx.)}$$
The electricity usage ranged from 850 to 1370 kwh.

111. Step 1:
$$a < b$$
$$a + a < a + b$$
$$2a < a + b$$
$$\frac{2a}{2} < \frac{a + b}{2}$$
$$a < \frac{a + b}{2}$$

Step 2:
$$a < b$$
$$a + b < b + b$$
$$a + b < 2b$$
$$\frac{a + b}{2} < \frac{2b}{2}$$
$$\frac{a + b}{2} < b$$

Step 3:

Since $a < \dfrac{a+b}{2}$ and $\dfrac{a+b}{2} < b$, it follows that

$a < \dfrac{a+b}{2} < b$.

113. $2x+1 \le 5x+7 \le x-5$

We can rewrite the inequality as

$2x+1 \le 5x+7$ and $5x+7 \le x-5$

$\begin{array}{ll} 2x+1 \le 5x+7 & 5x+7 \le x-5 \\ 2x+1-5x \le 5x+7-5x & 5x+7-x \le x-5-x \\ -3x+1 \le 7 & 4x+7 \le -5 \\ -3x+1-1 \le 7-1 & 4x+7-7 \le -5-7 \\ -3x \le 6 & 4x \le -12 \\ \dfrac{-3x}{-3} \ge \dfrac{6}{-3} & \dfrac{4x}{4} \le \dfrac{-12}{4} \\ x \ge -2 & x \le -3 \end{array}$

Therefore, we need $x \ge -2$ AND $x \le -3$. Since $-2 > -3$, there are no values for x that can satisfy both inequalities. Thus, the solution set is $\{\ \}$ or $\varnothing$.

115. $\quad 4x+1 > 2(2x+1)$

$\quad\quad 4x+1 > 4x+2$

$4x+1-4x > 4x+2-4x$

$\quad\quad\quad\quad 1 > 2$

This is a contradiction. There is no solution. If, during simplification, the variable terms all cancel out and a contradiction results, then there is no solution to the inequality.

117. If $x < 2$ then $x-2 < 2-2 \;\Rightarrow\; x-2 < 0$. When multiplying both sides of the inequality by $x-2$ in the second step, the direction of the inequality must switch.

Section 2.6

Are You Ready for This Section?

R1. $|3| = 3$ because the distance from 0 to 3 on a real number line is 3 units.

R2. $|-4| = 4$ because the distance from 0 to -4 on a real number line is 4 units.

R3. $|-1.6| = 1.6$ because the distance from 0 to -1.6 on a real number line is 1.6 units.

R4. $|0| = 0$ because the distance from 0 to 0 on a real number line is 0 units.

R5. The distance between 0 and 5 on a real number line can be expressed as $|5|$.

R6. The distance between 0 and -8 on a real number line can be expressed as $|-8|$.

R7. $\quad\quad 4x+5 = -9$

$\quad\quad 4x+5-5 = -9-5$

$\quad\quad\quad\quad 4x = -14$

$\quad\quad\quad \dfrac{4x}{4} = \dfrac{-14}{4}$

$\quad\quad\quad\quad x = -\dfrac{7}{2}$

The solution set is $\left\{ -\dfrac{7}{2} \right\}$.

R8. $\quad\quad -2x+1 > 5$

$\quad -2x+1-1 > 5-1$

$\quad\quad\quad -2x > 4$

$\quad\quad\quad \dfrac{-2x}{-2} < \dfrac{4}{-2}$

$\quad\quad\quad\quad x < -2$

Set-builder: $\{x \mid x < -2\}$

Interval: $(-\infty, -2)$

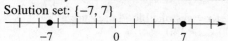

Section 2.6 Quick Checks

1. $|x| = 7$

$x = 7$ or $x = -7$ because both numbers are 7 units away from 0 on a real number line.
Solution set: $\{-7, 7\}$

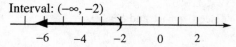

2. $|z| = 1$

$z = 1$ or $z = -1$ because both numbers are 1 unit away from 0 on a real number line.
Solution set: $\{-1, 1\}$

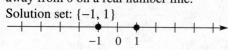

3. $|u| = a$ is equivalent to $u = \underline{a}$ or $u = \underline{-a}$.

4. $|2x + 3| = 5$ is equivalent to $2x + 3 = 5$ or $\underline{2x + 3 = -5}$.

5. $|2x-3| = 7$

$\begin{array}{ll} 2x-3=7 & \text{or} \quad 2x-3=-7 \\ 2x=10 & 2x=-4 \\ x=5 & x=-2 \end{array}$

Check:

Let $x = 5$:	Let $x = -2$:
$\lvert 2(5) - 3 \rvert \overset{?}{=} 7$	$\lvert 2(-2) - 3 \rvert \overset{?}{=} 7$
$\lvert 10 - 3 \rvert \overset{?}{=} 7$	$\lvert -4 - 3 \rvert \overset{?}{=} 7$
$7 = 7$ T	$7 = 7$ T

Solution set: $\{-2, 5\}$

6. $\lvert 3x - 2 \rvert + 3 = 10$

$\lvert 3x - 2 \rvert = 7$

$3x - 2 = 7$ or $3x - 2 = -7$

$ 3x = 9 \phantom{\text{or}} 3x = -5$

$ x = 3 \phantom{\text{or 3x}} x = -\dfrac{5}{3}$

Check:

Let $x = 3$:   Let $x = -\dfrac{5}{3}$:

$\lvert 3(3) - 2 \rvert + 3 \overset{?}{=} 10$ $\left\lvert 3\left(-\dfrac{5}{3}\right) - 2 \right\rvert + 3 \overset{?}{=} 10$

$\lvert 9 - 2 \rvert + 3 \overset{?}{=} 10$ $\lvert -5 - 2 \rvert + 3 \overset{?}{=} 10$

$7 + 3 \overset{?}{=} 10$ $7 + 3 \overset{?}{=} 10$

$10 = 10$ T $10 = 10$ T

Solution set: $\left\{ -\dfrac{5}{3}, 3 \right\}$

7. $\lvert -5x + 2 \rvert - 2 = 5$

$\lvert -5x + 2 \rvert = 7$

$-5x + 2 = 7$ or $-5x + 2 = -7$

$ -5x = 5 \phantom{\text{or}} -5x = -9$

$ x = -1 \phantom{\text{or } -5x} x = \dfrac{9}{5}$

Check:

Let $x = -1$:   Let $x = \dfrac{9}{5}$:

$\lvert -5(-1) + 2 \rvert - 2 \overset{?}{=} 5$ $\left\lvert -5\left(\dfrac{9}{5}\right) + 2 \right\rvert - 2 \overset{?}{=} 5$

$\lvert 5 + 2 \rvert - 2 \overset{?}{=} 5$ $\lvert -9 + 2 \rvert - 2 \overset{?}{=} 5$

$7 - 2 \overset{?}{=} 5$ $7 - 2 \overset{?}{=} 5$

$5 = 5$ T $5 = 5$ T

Solution set: $\left\{ -1, \dfrac{9}{5} \right\}$

8. $3 \lvert x + 2 \rvert - 4 = 5$

$3 \lvert x + 2 \rvert = 9$

$\lvert x + 2 \rvert = 3$

$x + 2 = 3$ or $x + 2 = -3$

$ x = 1 \phantom{\text{or } x +} x = -5$

Check:

Let $x = 1$:   Let $x = -5$:

$3 \lvert 1 + 2 \rvert - 4 \overset{?}{=} 5$ $3 \lvert -5 + 2 \rvert - 4 \overset{?}{=} 5$

$3(3) - 4 \overset{?}{=} 5$ $3(3) - 4 \overset{?}{=} 5$

$9 - 4 \overset{?}{=} 5$ $9 - 4 \overset{?}{=} 5$

$5 = 5$ T $5 = 5$ T

Solution set: $\{-5, 1\}$

9. True. The absolute value of a number represents the distance of the number from 0 on a real number line. Since distance is never negative, absolute value is never negative.

10. $\lvert 5x + 3 \rvert = -2$

Since absolute values are never negative, this equation has no solution.

Solution set: $\{\ \}$ or $\varnothing$

11. $\lvert 2x + 5 \rvert + 7 = 3$

$\lvert 2x + 5 \rvert = -4$

Since absolute values are never negative, this equation has no solution.

Solution set: $\{\ \}$ or $\varnothing$

12. $\lvert x + 1 \rvert + 3 = 3$

$\lvert x + 1 \rvert = 0$

$x + 1 = 0$

$x = -1$

Check:

Let $x = -1$:

$\lvert -1 + 1 \rvert + 3 \overset{?}{=} 3$

$0 + 3 \overset{?}{=} 3$

$3 \overset{?}{=} 3$ T

Solution set: $\{-1\}$

13. $\lvert u \rvert = \lvert v \rvert$ is equivalent to $\underline{u = v}$ or $\underline{u = -v}$.

14. $\lvert x - 3 \rvert = \lvert 2x + 5 \rvert$

$x - 3 = 2x + 5$ or $x - 3 = -(2x + 5)$

$ x = 2x + 8 \phantom{\text{or }} x - 3 = -2x - 5$

$ -x = 8 \phantom{\text{or } x - 3 =} x = -2x - 2$

$ x = -8 \phantom{\text{or } x - 3 =} 3x = -2$

$\phantom{x - 3 = \text{or } x - 3 =3} x = -\dfrac{2}{3}$

Check:

Let $x = -8$:   Let $x = -\dfrac{2}{3}$:

$\big|-8-3\big| \overset{?}{=} \big|2(-8)+5\big|$ $\left|-\dfrac{2}{3}-3\right| \overset{?}{=} \left|2\left(-\dfrac{2}{3}\right)+5\right|$

$\qquad\big|-11\big| \overset{?}{=} \big|-16+5\big|$ $\left|-\dfrac{11}{3}\right| \overset{?}{=} \left|-\dfrac{4}{3}+5\right|$

$\qquad\qquad 11 = 11 \ \ \text{T}$

$\qquad\qquad\qquad\qquad\qquad \dfrac{11}{3} = \dfrac{11}{3} \ \ \text{T}$

Solution set: $\left\{-8, \ -\dfrac{2}{3}\right\}$

15. $\big|8z+11\big| = \big|6z+17\big|$

$8z+11 = 6z+17 \quad \text{or} \quad 8z+11 = -(6z+17)$
$\qquad 8z = 6z+6 \qquad\qquad 8z+11 = -6z-17$
$\qquad 2z = 6 \qquad\qquad\qquad 8z = -6z-28$
$\qquad\ z = 3 \qquad\qquad\qquad 14z = -28$
$\qquad\qquad\qquad\qquad\qquad\qquad\ z = -2$

Check:

Let $z = -2$:
$\big|8(-2)+11\big| \overset{?}{=} \big|6(-2)+17\big|$
$\quad \big|-16+11\big| \overset{?}{=} \big|-12+17\big|$
$\qquad\qquad\qquad 5 = 5 \ \ \text{T}$

Let $z = 3$:
$\big|8(3)+11\big| \overset{?}{=} \big|6(3)+17\big|$
$\quad \big|24+11\big| \overset{?}{=} \big|18+17\big|$
$\qquad\qquad 35 = 35 \ \ \text{T}$

Solution set: $\{-2, 3\}$

16. $\big|3-2y\big| = \big|4y+3\big|$

$3-2y = 4y+3 \quad \text{or} \quad 3-2y = -(4y+3)$
$\quad -2y = 4y \qquad\qquad 3-2y = -4y-3$
$\quad -6y = 0 \qquad\qquad\ \ -2y = -4y-6$
$\qquad y = 0 \qquad\qquad\qquad 2y = -6$
$\qquad\qquad\qquad\qquad\qquad\quad y = -3$

Check:

Let $y = 0$:   Let $y = -3$:
$\big|3-2(0)\big| \overset{?}{=} \big|4(0)+3\big|$ $\big|3-2(-3)\big| \overset{?}{=} \big|4(-3)+3\big|$
$\qquad \big|3\big| \overset{?}{=} \big|3\big|$ $\big|3+6\big| \overset{?}{=} \big|-12+3\big|$
$\qquad 3 = 3 \ \ \text{T}$ $9 = 9 \ \ \text{T}$

Solution set: $\{-3, 0\}$

17. $\big|2x-3\big| = \big|5-2x\big|$

$2x-3 = 5-2x \quad \text{or} \quad 2x-3 = -(5-2x)$
$\quad 2x = 8-2x \qquad\qquad 2x-3 = -5+2x$
$\quad 4x = 8 \qquad\qquad\qquad 2x = -2+2x$
$\qquad x = 2 \qquad\qquad\qquad\ 0 = -2 \ \ \text{false}$

The second equation leads to a contradiction. Therefore, the only solution is $x = 2$.

Check:

Let $x = 2$:
$\big|2(2)-3\big| \overset{?}{=} \big|5-2(2)\big|$
$\quad \big|4-3\big| \overset{?}{=} \big|5-4\big|$
$\qquad\qquad 1 = 1 \ \ \text{T}$

Solution set: $\{2\}$

18. If $a > 0$, then $|u| < a$ is equivalent to $\underline{-a < u < a}$.

19. To solve $|3x+4| < 10$, solve $\underline{-10} < 3x+10 < \underline{10}$.

20. $|x| \le 5$

$-5 \le x \le 5$

Set-builder: $\{x \mid -5 \le x \le 5\}$

Interval: $[-5, 5]$

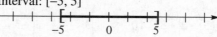

21. $|x| < \dfrac{3}{2}$

$-\dfrac{3}{2} < x < \dfrac{3}{2}$

Set-builder: $\left\{x \mid -\dfrac{3}{2} < x < \dfrac{3}{2}\right\}$

Interval: $\left(-\dfrac{3}{2}, \dfrac{3}{2}\right)$

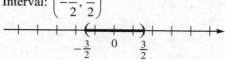

22. $|x+3| < 5$

$-5 < x+3 < 5$
$-5-3 < x+3-3 < 5-3$
$\quad -8 < x < 2$

Set-builder: $\{x \mid -8 < x < 2\}$

Interval: $(-8, 2)$

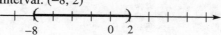

23. $|2x - 3| \leq 7$

$$-7 \leq 2x - 3 \leq 7$$
$$-7 + 3 \leq 2x - 3 + 3 \leq 7 + 3$$
$$-4 \leq 2x \leq 10$$
$$\frac{-4}{2} \leq \frac{2x}{2} \leq \frac{10}{2}$$
$$-2 \leq x \leq 5$$

Set-builder: $\{x \mid -2 \leq x \leq 5\}$

Interval: $[-2, 5]$

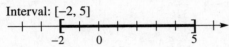

24. $|7x + 2| < -3$

Since absolute values are never negative, this inequality has no solution.

Solution set: $\{\ \}$ or $\varnothing$

25. True

26. $|x| + 4 < 6$

$$|x| < 2$$
$$-2 < x < 2$$

Set-builder: $\{x \mid -2 < x < 2\}$; Interval: $(-2, 2)$

27. $|x - 3| + 4 \leq 8$

$$|x - 3| \leq 4$$
$$-4 \leq x - 3 \leq 4$$
$$-4 + 3 \leq x - 3 + 3 \leq 4 + 3$$
$$-1 \leq x \leq 7$$

Set-builder: $\{x \mid -1 \leq x \leq 7\}$

Interval: $[-1, 7]$

28. $3|2x + 1| \leq 9$

$$|2x + 1| \leq 3$$
$$-3 \leq 2x + 1 \leq 3$$
$$-3 - 1 \leq 2x + 1 - 1 \leq 3 - 1$$
$$-4 \leq 2x \leq 2$$
$$\frac{-4}{2} \leq \frac{2x}{2} \leq \frac{2}{2}$$
$$-2 \leq x \leq 1$$

Set-builder: $\{x \mid -2 \leq x \leq 1\}$

Interval: $[-2, 1]$

29. $|-3x + 1| - 5 < 3$

$$|-3x + 1| < 8$$
$$-8 < -3x + 1 < 8$$
$$-8 - 1 < -3x + 1 - 1 < 8 - 1$$
$$-9 < -3x < 7$$
$$\frac{-9}{-3} > \frac{-3x}{-3} > \frac{7}{-3}$$
$$3 > x > -\frac{7}{3} \quad \text{or} \quad -\frac{7}{3} < x < 3$$

Set-builder: $\left\{x \mid -\frac{7}{3} < x < 3\right\}$

Interval: $\left(-\frac{7}{3}, 3\right)$

30. $|u| > a$ is equivalent to $\underline{u < -a}$ or $\underline{u > a}$.

31. $|5x - 2| \geq 7$ is equivalent to $5x - 2 \leq \underline{-7}$ or $5x - 2 \geq \underline{7}$.

32. $|x| \geq 6$

$$x \leq -6 \quad \text{or} \quad x \geq 6$$

Set-builder: $\{x \mid x \leq -6 \ \text{or} \ x \geq 6\}$

Interval: $(-\infty, -6] \cup [6, \infty)$

33. $|x| > \frac{5}{2}$

$$x < -\frac{5}{2} \quad \text{or} \quad x > \frac{5}{2}$$

Set-builder: $\left\{x \mid x < -\frac{5}{2} \ \text{or} \ x > \frac{5}{2}\right\}$

Interval: $\left(-\infty, -\frac{5}{2}\right) \cup \left(\frac{5}{2}, \infty\right)$

34. $|x - 9| > 6$ is equivalent to $x - 9 \geq 6$ or $x - 9 \leq -6$.

35. $|x + 3| > 4$

$$x + 3 < -4 \quad \text{or} \quad x + 3 > 4$$
$$x < -7 \qquad\qquad x > 1$$

Set-builder: $\{x \mid x < -7 \ \text{or} \ x > 1\}$

Interval: $(-\infty, -7) \cup (1, \infty)$

36. $|4x-3| \geq 5$

$$4x - 3 \leq -5 \quad \text{or} \quad 4x - 3 \geq 5$$
$$4x \leq -2 \qquad\qquad 4x \geq 8$$
$$x \leq -\frac{1}{2} \qquad\qquad x \geq 2$$

Set-builder: $\left\{ x \,\middle|\, x \leq -\frac{1}{2} \text{ or } x \geq 2 \right\}$

Interval: $\left(-\infty, -\frac{1}{2} \right] \cup [2, \infty)$

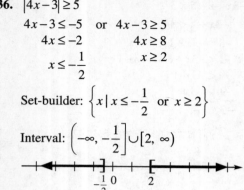

37. $|-3x+2| > 7$

$$-3x + 2 < -7 \quad \text{or} \quad -3x + 2 > 7$$
$$-3x < -9 \qquad\qquad -3x > 5$$
$$x > 3 \qquad\qquad\qquad x < -\frac{5}{3}$$

Set-builder: $\left\{ x \,\middle|\, x < -\frac{5}{3} \text{ or } x > 3 \right\}$

Interval: $\left(-\infty, -\frac{5}{3} \right) \cup (3, \infty)$

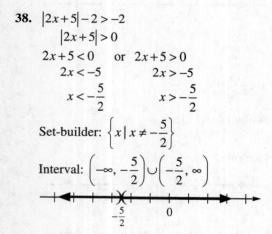

38. $|2x+5| - 2 > -2$

$$|2x+5| > 0$$

$$2x + 5 < 0 \quad \text{or} \quad 2x + 5 > 0$$
$$2x < -5 \qquad\qquad 2x > -5$$
$$x < -\frac{5}{2} \qquad\qquad x > -\frac{5}{2}$$

Set-builder: $\left\{ x \,\middle|\, x \neq -\frac{5}{2} \right\}$

Interval: $\left(-\infty, -\frac{5}{2} \right) \cup \left(-\frac{5}{2}, \infty \right)$

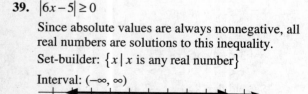

39. $|6x-5| \geq 0$

Since absolute values are always nonnegative, all real numbers are solutions to this inequality.

Set-builder: $\left\{ x \,\middle|\, x \text{ is any real number} \right\}$

Interval: $(-\infty, \infty)$

40. $|2x+1| > -3$

Since absolute values are always nonnegative, all real numbers are solutions to this inequality.

Set-builder: $\left\{ x \,\middle|\, x \text{ is any real number} \right\}$

Interval: $(-\infty, \infty)$

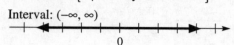

41. $|x-4| \leq \frac{1}{32}$

$$-\frac{1}{32} \leq x - 4 \leq \frac{1}{32}$$
$$-\frac{1}{32} + 4 \leq x - 4 + 4 \leq \frac{1}{32} + 4$$
$$-\frac{1}{32} + \frac{128}{32} \leq x \leq \frac{1}{32} + \frac{128}{32}$$
$$\frac{127}{32} \leq x \leq \frac{129}{32}$$

The acceptable belt widths are between $\frac{127}{32}$ inches and $\frac{129}{32}$ inches.

42. $|p-9| \leq 1.7$

$$-1.7 \leq p - 9 \leq 1.7$$
$$-1.7 + 9 \leq p - 9 + 9 \leq 1.7 + 9$$
$$7.3 \leq p \leq 10.7$$

The percentage of people that have been shot at is between 7.3 percent and 10.7 percent.

2.6 Exercises

43. $|x| = 10$

$$x = 10 \text{ or } x = -10$$

Solution set: $\{-10, 10\}$

45. $|y-3| = 4$

$$y - 3 = 4 \quad \text{or} \quad y - 3 = -4$$
$$y = 7 \quad \text{or} \qquad y = -1$$

Solution set: $\{-1, 7\}$

47. $|-3x+5| = 8$

$$-3x + 5 = 8 \quad \text{or} \quad -3x + 5 = -8$$
$$-3x = 3 \quad \text{or} \qquad -3x = -13$$
$$x = -1 \quad \text{or} \qquad x = \frac{13}{3}$$

Solution set: $\left\{ -1, \frac{13}{3} \right\}$

49. $|y| - 7 = -2$

$\quad |y| = 5$

$\quad\quad y = 5 \text{ or } y = -5$

Solution set: $\{-5, 5\}$

51. $|2x + 3| - 5 = 3$

$\quad |2x + 3| = 8$

$\quad\quad 2x + 3 = 8 \text{ or } 2x + 3 = -8$

$\quad\quad\quad 2x = 5 \text{ or } \quad 2x = -11$

$\quad\quad\quad\quad x = \dfrac{5}{2} \text{ or } \quad x = -\dfrac{11}{2}$

Solution set: $\left\{ -\dfrac{11}{2}, \dfrac{5}{2} \right\}$

53. $-2|x - 3| + 10 = -4$

$\quad\quad -2|x - 3| = -14$

$\quad\quad\quad |x - 3| = 7$

$\quad\quad\quad\quad x - 3 = 7 \text{ or } x - 3 = -7$

$\quad\quad\quad\quad\quad\quad x = 10 \text{ or } \quad x = -4$

Solution set: $\{-4, 10\}$

55. $|-3x| - 5 = -5$

$\quad |-3x| = 0$

$\quad\quad -3x = 0$

$\quad\quad\quad x = 0$

Solution set: $\{0\}$

57. $\left| \dfrac{3x - 1}{4} \right| = 2$

$\quad \dfrac{3x - 1}{4} = 2 \text{ or } \dfrac{3x - 1}{4} = -2$

$\quad 3x - 1 = 8 \text{ or } 3x - 1 = -8$

$\quad\quad 3x = 9 \text{ or } \quad 3x = -7$

$\quad\quad\quad x = 3 \text{ or } \quad\quad x = -\dfrac{7}{3}$

Solution set: $\left\{ -\dfrac{7}{3}, 3 \right\}$

59. $|3x + 2| = |2x - 5|$

$\quad 3x + 2 = 2x - 5 \text{ or } 3x + 2 = -(2x - 5)$

$\quad 3x = 2x - 7 \text{ or } 3x + 2 = -2x + 5$

$\quad\quad x = -7 \quad\quad \text{ or } \quad 3x = -2x + 3$

$\quad\quad x = -7 \quad\quad \text{ or } \quad 5x = 3$

$\quad\quad x = -7 \quad\quad \text{ or } \quad\quad x = \dfrac{3}{5}$

Solution set: $\left\{ -7, \dfrac{3}{5} \right\}$

61. $|8 - 3x| = |2x - 7|$

$\quad 8 - 3x = 2x - 7 \text{ or } 8 - 3x = -(2x - 7)$

$\quad\quad -3x = 2x - 15 \text{ or } 8 - 3x = -2x + 7$

$\quad\quad -5x = -15 \quad\quad \text{ or } \quad -3x = -2x - 1$

$\quad\quad\quad x = 3 \quad\quad\quad \text{ or } \quad\quad -x = -1$

$\quad\quad\quad x = 3 \quad\quad\quad \text{ or } \quad\quad\quad x = 1$

Solution set: $\{1, 3\}$

63. $|4y - 7| = |9 - 4y|$

$\quad 4y - 7 = 9 - 4y \text{ or } 4y - 7 = -(9 - 4y)$

$\quad 4y = 16 - 4y \text{ or } 4y - 7 = -9 + 4y$

$\quad\quad 8y = 16 \quad\quad \text{ or } \quad\quad 4y = -2 + 4y$

$\quad\quad\quad y = 2 \quad\quad\quad \text{ or } \quad\quad\quad 0 = -2$

Solution set: $\{2\}$

65. $|x| < 9$

$\quad -9 < x < 9$

Set-builder: $\{x \mid -9 < x < 9\}$

Interval: $(-9, 9)$

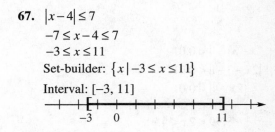

67. $|x - 4| \le 7$

$\quad -7 \le x - 4 \le 7$

$\quad -3 \le x \le 11$

Set-builder: $\{x \mid -3 \le x \le 11\}$

Interval: $[-3, 11]$

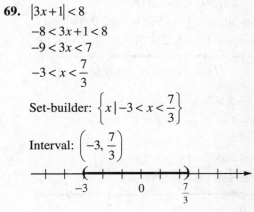

69. $|3x + 1| < 8$

$\quad -8 < 3x + 1 < 8$

$\quad -9 < 3x < 7$

$\quad -3 < x < \dfrac{7}{3}$

Set-builder: $\left\{ x \mid -3 < x < \dfrac{7}{3} \right\}$

Interval: $\left(-3, \dfrac{7}{3} \right)$

71. $|6x + 5| < -1$

No solution. Absolute value is never negative.

Solution set: $\varnothing$ or $\{ \ \}$

73. $2|x-3|+3<9$
$\qquad 2|x-3|<6$
$\qquad |x-3|<3$
$\qquad\quad -3<x-3<3$
$\qquad\qquad 0<x<6$
Set-builder: $\{x\mid 0<x<6\}$
Interval: $(0, 6)$

75. $|2-5x|+3<10$
$\qquad |2-5x|<7$
$-7<2-5x<7$
$-9<-5x<5$
$\dfrac{9}{5}>x>-1$ or $-1<x<\dfrac{9}{5}$
Set-builder: $\left\{x\mid -1<x<\dfrac{9}{5}\right\}$
Interval: $\left(-1,\dfrac{9}{5}\right)$
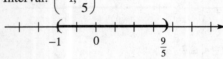

77. $|(2x-3)-1|<0.01$
$\qquad |2x-3-1|<0.01$
$\qquad\quad |2x-4|<0.01$
$\qquad -0.01<2x-4<0.01$
$\qquad\quad 3.99<2x<4.01$
$\qquad\quad 1.995<x<2.005$
Set-builder: $\{x\mid 1.995<x<2.005\}$
Interval: $(1.995, 2.005)$
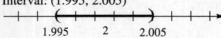

79. $|y-5|>2$
$\quad y-5<-2$ or $y-5>2$
$\qquad y<3$ or $\quad y>7$
Set-builder: $\{y\mid y<3 \text{ or } y>7\}$
Interval: $(-\infty, 3)\cup(7, \infty)$
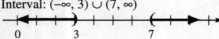

81. $|-4x-3|\geq 5$
$-4x-3\leq -5$ or $-4x-3\geq 5$
$\quad -4x\leq -2$ or $\quad -4x\geq 8$
$\qquad x\geq\dfrac{1}{2}$ or $\qquad x\leq -2$
Set-builder: $\left\{x\mid x\leq -2 \text{ or } x\geq\dfrac{1}{2}\right\}$
Interval: $(-\infty, -2]\cup\left[\dfrac{1}{2}, \infty\right)$
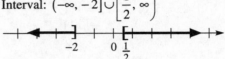

83. $2|y|+3>1$
$\quad 2|y|>-2$
$\quad |y|>-1$
Since $|y|\geq 0>-1$ for all y, all real numbers are solutions.
Set-builder: $\{y\mid y \text{ is a real number}\}$
Interval: $(-\infty, \infty)$
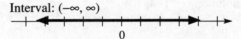

85. $|-5x-3|-7>0$
$\qquad |-5x-3|>7$
$\quad -5x-3<-7$ or $-5x-3>7$
$\qquad -5x<-4$ or $\quad -5x>10$
$\qquad\quad x>\dfrac{4}{5}$ or $\qquad x<-2$
Set-builder: $\left\{x\mid x<-2 \text{ or } x>\dfrac{4}{5}\right\}$
Interval: $(-\infty, -2)\cup\left(\dfrac{4}{5}, \infty\right)$
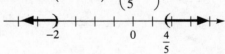

87. $4|-2x+1|>4$
$\quad |-2x+1|>1$
$\quad -2x+1<-1$ or $-2x+1>1$
$\qquad -2x<-2$ or $\quad -2x>0$
$\qquad\quad x>1$ or $\qquad x<0$
Set-builder: $\{x\mid x<0 \text{ or } x>1\}$
Interval: $(-\infty, 0)\cup(1, \infty)$
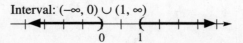

89. $|1-2x| \geq |-5|$

$|1-2x| \geq 5$

$1-2x \leq -5$ or $1-2x \geq 5$

$-2x \leq -6$ or $-2x \geq 4$

$x \geq 3$ or $x \leq -2$

Set-builder: $\{x \mid x \leq -2 \text{ or } x \geq 3\}$

Interval: $(-\infty, -2] \cup [3, \infty)$

91. a. $f(x) = g(x)$ when $x = -5$ and $x = 5$.

The solution set is $\{-5, 5\}$.

b. $f(x) \leq g(x)$ when $-5 \leq x \leq 5$.

Set-builder: $\{x \mid -5 \leq x \leq 5\}$

Interval: $[-5, 5]$

c. $f(x) > g(x)$ for $x < -5$ or $x > 5$.

Set-builder: $\{x \mid x < -5 \text{ or } x > 5\}$

Interval: $(-\infty, -5) \cup (5, \infty)$

93. a. $f(x) = g(x)$ when $x = -5$ and $x = 1$.

The solution set is $\{-5, 1\}$.

b. $f(x) < g(x)$ when $-5 < x < 1$.

Set-builder: $\{x \mid -5 < x < 1\}$

Interval: $(-5, 1)$

c. $f(x) \geq g(x)$ for $x \leq -5$ or $x \geq 1$.

Set-builder: $\{x \mid x \leq -5 \text{ or } x \geq 1\}$

Interval: $(-\infty, -5] \cup [1, \infty)$

95. $|x| > 5$

$x < -5$ or $x > 5$

Set-builder: $\{x \mid x < -5 \text{ or } x > 5\}$

Interval: $(-\infty, -5) \cup (5, \infty)$

97. $|2x+5| = 3$

$2x+5 = 3$ or $2x+5 = -3$

$2x = -2$ or $2x = -8$

$x = -1$ or $x = -4$

Solution set: $\{-4, -1\}$

99. $7|x| = 35$

$|x| = 5$

$x = 5$ or $x = -5$

Solution set: $\{-5, 5\}$

101. $|5x+2| \leq 8$

$-8 \leq 5x+2 \leq 8$

$-10 \leq 5x \leq 6$

$-2 \leq x \leq \dfrac{6}{5}$

Set-builder: $\left\{ x \mid -2 \leq x \leq \dfrac{6}{5} \right\}$

Interval: $\left[-2, \dfrac{6}{5} \right]$

103. $|-2x+3| = -4$

No solution. Absolute value is never negative.

Solution set: $\varnothing$ or $\{\ \}$

105. $|3x+2| \geq 5$

$3x+2 \leq -5$ or $3x+2 \geq 5$

$3x \leq -7$ or $3x \geq 3$

$x \leq -\dfrac{7}{3}$ or $x \geq 1$

Set-builder: $\left\{ x \mid x \leq -\dfrac{7}{3} \text{ or } x \geq 1 \right\}$

Interval: $\left(-\infty, -\dfrac{7}{3} \right] \cup [1, \infty)$

107. $|3x-2| + 7 > 9$

$|3x-2| > 2$

$3x-2 < -2$ or $3x-2 > 2$

$3x < 0$ or $3x > 4$

$x < 0$ or $x > \dfrac{4}{3}$

Set-builder: $\left\{ x \mid x < 0 \text{ or } x > \dfrac{4}{3} \right\}$

Interval: $(-\infty, 0) \cup \left(\dfrac{4}{3}, \infty \right)$

109. $|5x+3| = |3x+5|$

$5x+3 = 3x+5$ or $5x+3 = -(3x+5)$

$\quad 5x = 3x+2$ or $\quad 5x+3 = -3x-5$

$\quad\ 2x = 2$ or $\qquad 8x = -8$

$\qquad x = 1$ or $\qquad\ x = -1$

Solution set: $\{-1, 1\}$

111. $|4x+7| + 6 < 5$

$\quad |4x+7| < -1$

Since absolute value is always nonnegative, that is ≥ 0, this inequality has no solution.

Solution set: $\varnothing$ or $\{\ \}$

113. $\left|\dfrac{x-2}{4}\right| = \left|\dfrac{2x+1}{6}\right|$

$\dfrac{x-2}{4} = \dfrac{2x+1}{6}$ or $\dfrac{x-2}{4} = -\dfrac{2x+1}{6}$

$12 \cdot \dfrac{x-2}{4} = 12 \cdot \dfrac{2x+1}{6}$ or $12 \cdot \dfrac{x-2}{4} = 12 \cdot \dfrac{-2x-1}{6}$

$\quad 3x-6 = 4x+2$ or $3x-6 = -4x-2$

$\qquad 3x = 4x+8$ or $\qquad 3x = -4x+4$

$\qquad\ -x = 8$ or $\qquad\ 7x = 4$

$\qquad\ x = -8$ or $\qquad\ x = \dfrac{4}{7}$

Solution set: $\left\{-8, \dfrac{4}{7}\right\}$

115. $|x-5| < 3$

$-3 < x-5 < 3$

$\ \ 2 < x < 8$

Set-builder: $\{x \mid 2 < x < 8\}$

Interval: $(2, 8)$

117. $|2x-(-6)| > 3$

$|2x+6| > 3$

$2x+6 < -3$ or $2x+6 > 3$

$\ \ 2x < -9$ or $\quad 2x > -3$

$\quad x < -\dfrac{9}{2}$ or $\qquad x > -\dfrac{3}{2}$

Set-builder: $\left\{x \mid x < -\dfrac{9}{2} \text{ or } x > -\dfrac{3}{2}\right\}$

Interval: $\left(-\infty, -\dfrac{9}{2}\right) \cup \left(-\dfrac{3}{2}, \infty\right)$

119. $|x-5.7| \leq 0.0005$

$-0.0005 \leq x-5.7 \leq 0.0005$

$\ \ 5.6995 \leq x \leq 5.7005$

The acceptable rod lengths are between 5.6995 inches and 5.7005 inches, inclusive.

121. $\left|\dfrac{x-100}{15}\right| > 1.96$

$\dfrac{x-100}{15} < -1.96$ or $\dfrac{x-100}{15} > 1.96$

$x-100 < -29.4$ or $x-100 > 29.4$

$\quad x < 70.6$ or $\qquad x > 129.4$

An unusual IQ score would be less than 70.6 or greater than 129.4.

123. $|x| - x = 5$

$|x| = x+5$

$x = x+5$ or $x = -(x+5)$

$0 = 5$ or $x = -x-5$

$\qquad\qquad\ 2x = -5$

$\qquad\qquad\ x = -\dfrac{5}{2}$

Check: $-\dfrac{5}{2}$:

$\left|-\dfrac{5}{2}\right| - \left(-\dfrac{5}{2}\right) \overset{?}{=} 5$

$\dfrac{5}{2} + \dfrac{5}{2} \overset{?}{=} 5$

$5 = 5$ T

Solution set: $\left\{-\dfrac{5}{2}\right\}$

125. $z + |-z| = 4$

$|-z| = 4-z$

$-z = 4-z$ or $-z = -(4-z)$

$0 = 4$ or $-z = -4+z$

$\qquad\qquad -2z = -4$

$\qquad\qquad\quad z = 2$

Check 2: $2 + |-(2)| \overset{?}{=} 4$

$2 + 2 \overset{?}{=} 4$

$4 = 4$ T

Solution set: $\{2\}$

127. $|4x+1| = x-2$

$4x+1 = x-2$ or $4x+1 = -(x-2)$

$\quad 4x = x-3$ or $\quad 4x+1 = -x+2$

$\quad 3x = -3$ or $\qquad 4x = -x+1$

$\qquad x = -1$ or $\qquad 5x = 1$

$\qquad x = -1$ or $\qquad\ x = \dfrac{1}{5}$

Check:

$$\left|4(-1)+1\right| \overset{?}{=} (-1)-2 \qquad \left|4\left(\tfrac{1}{5}\right)+1\right| \overset{?}{=} \tfrac{1}{5}-2$$

$$\left|-4+1\right| \overset{?}{=} -3$$

$$\left|-3\right| \overset{?}{=} -3 \qquad\qquad \left|\tfrac{4}{5}+1\right| \overset{?}{=} -\tfrac{9}{5}$$

$$3 \neq -3 \qquad\qquad\qquad \left|\tfrac{9}{5}\right| \overset{?}{=} -\tfrac{9}{5}$$

$$\tfrac{9}{5} \neq -\tfrac{9}{5}$$

Solution set: $\varnothing$ or $\{\ \}$

129. $\left|x+5\right| = -(x+5)$

Since we have $|u| = -u$, we need $u \leq 0$ so the absolute value will not be negative. Thus, we can say that $x + 5 \leq 0$ or $x \leq -5$.

Set-builder: $\{x \mid x \leq -5\}$

Interval: $(-\infty, -5]$

131. $|2x - 3| + 1 = 0$ has no solution because the absolute value, when isolated, is equal to a negative number which is not possible.

133. $|4x + 3| + 3 < 0$ has the empty set as the solution set because the absolute value, when isolated, is less than a negative number (-3). Since absolute values are always nonnegative, this is not possible.

Chapter 2 Review

1. {(Cent, 2.500), (Nickel, 5.000), (Dime, 2.268), (Quarter, 5.670), (Half Dollar, 11.340), (Dollar, 8.100)}
Domain:
{Cent, Nickel, Dime, Quarter, Half Dollar, Dollar}
Range:
{2.268, 2.500, 5.000, 5.670, 8.100, 11.340}

2. {(16, $12.99), (28, $14.99), (30, $14.99), (59, $24.99), (85, $29.99)}
Domain:
{16, 28, 30, 59, 85}
Range:
{$12.99, $14.99, $24.99, $29.99}

3. Domain: $\{-4, -2, 2, 3, 6\}$
Range: $\{-9, -1, 5, 7, 8\}$

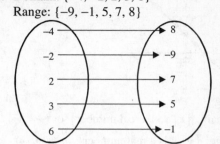

4. Domain: $\{-2, 1, 3, 5\}$
Range: $\{1, 4, 7, 8\}$

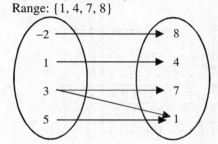

5. Domain: $\{x \mid x \text{ is a real number}\}$ or $(-\infty, \infty)$
Range: $\{y \mid y \text{ is a real number}\}$ or $(-\infty, \infty)$

6. Domain: $\{x \mid -6 \leq x \leq 4\}$ or $[-6, 4]$
Range: $\{y \mid -4 \leq y \leq 6\}$ or $[-4, 6]$

7. Domain: $\{2\}$
Range: $\{y \mid y \text{ is a real number}\}$ or $(-\infty, \infty)$

8. Domain: $\{x \mid x \geq -1\}$ or $[-1, \infty)$
Range: $\{y \mid y \geq -2\}$ or $[-2, \infty)$

9. $y = x + 2$

x	$y = x + 2$	(x, y)
-3	$y = (-3) + 2 = -1$	$(-3, -1)$
-1	$y = (-1) + 2 = 1$	$(-1, 1)$
0	$y = (0) + 2 = 2$	$(0, 2)$
1	$y = (1) + 2 = 3$	$(1, 3)$
3	$y = (3) + 2 = 5$	$(3, 5)$

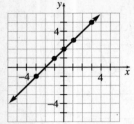

Domain: $\{x \mid x$ is a real number$\}$ or $(-\infty, \infty)$

Range: $\{y \mid y$ is a real number$\}$ or $(-\infty, \infty)$

10. $2x + y = 3$
$y = -2x + 3$

x	$y = -2x + 3$	(x, y)
-3	$y = -2(-3) + 3 = 9$	$(-3, 9)$
-1	$y = -2(-1) + 3 = 5$	$(-1, 5)$
0	$y = -2(0) + 3 = 3$	$(0, 3)$
1	$y = -2(1) + 3 = 1$	$(1, 1)$
3	$y = -2(3) + 3 = -3$	$(3, -3)$

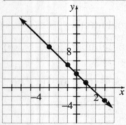

Domain: $\{x \mid x$ is a real number$\}$ or $(-\infty, \infty)$

Range: $\{y \mid y$ is a real number$\}$ or $(-\infty, \infty)$

11. $y = -x^2 + 4$

x	$y = -x^2 + 4$	(x, y)
-3	$y = -(-3)^2 + 4 = -5$	$(-3, -5)$
-1	$y = -(-1)^2 + 4 = 3$	$(-1, 3)$
0	$y = -(0)^2 + 4 = 4$	$(0, 4)$
1	$y = -(1)^2 + 4 = 3$	$(1, 3)$
3	$y = -(3)^2 + 4 = -5$	$(3, -5)$

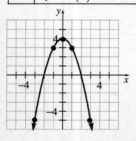

Domain: $\{x \mid x$ is a real number$\}$ or $(-\infty, \infty)$

Range: $\{y \mid y \le 4\}$ or $(-\infty, 4]$

12. $y = |x + 2| - 1$

x	$y = |x + 2| - 1$	(x, y)
-5	$y = |-5 + 2| - 1 = 2$	$(-5, 2)$
-3	$y = |-3 + 2| - 1 = 0$	$(-3, 0)$
-1	$y = |-1 + 2| - 1 = 0$	$(-1, 0)$
0	$y = |0 + 2| - 1 = 1$	$(0, 1)$
1	$y = |1 + 2| - 1 = 2$	$(1, 2)$

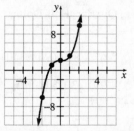

Domain: $\{x \mid x$ is a real number$\}$ or $(-\infty, \infty)$

Range: $\{y \mid y \ge -1\}$ or $[-1, \infty)$

13. $y = x^3 + 2$

x	$y = x^3 + 2$	(x, y)
-2	$y = (-2)^3 + 2 = -6$	$(-2, -6)$
-1	$y = (-1)^3 + 2 = 1$	$(-1, 1)$
0	$y = (0)^3 + 2 = 2$	$(0, 2)$
1	$y = (1)^3 + 2 = 3$	$(1, 3)$
2	$y = (2)^3 + 2 = 10$	$(2, 10)$

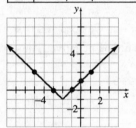

Domain: $\{x \mid x$ is a real number$\}$ or $(-\infty, \infty)$

Range: $\{y \mid y$ is a real number$\}$ or $(-\infty, \infty)$

14. $x = y^2 + 1$

y	$x = y^2 + 1$	(x, y)
-2	$x = (-2)^2 + 1 = 5$	$(5, -2)$
-1	$x = (-1)^2 + 1 = 2$	$(2, -1)$
0	$x = (0)^2 + 1 = 1$	$(1, 0)$
1	$x = (1)^2 + 1 = 2$	$(2, 1)$
2	$x = (2)^2 + 1 = 5$	$(5, 2)$

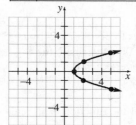

Domain: $\{x \mid x \geq 1\}$ or $[1, \infty)$

Range: $\{y \mid y \text{ is a real number}\}$ or $(-\infty, \infty)$

15. a. Domain: $\{x \mid 0 \leq x \leq 44.64\}$ or $[0, 44.64]$

Range: $\{y \mid 40 \leq y \leq 2122\}$ or $[40, 2122]$

The monthly cost will be at least \$40 but no more than \$2122. The number of minutes used must be between 0 and 44,640.

b. Answers may vary. There are at most 31 days in a month. Since 31 days is equivalent to 44,640 minutes, this must be the largest value in the domain. It is not possible to talk for a negative number of minutes, so the domain should begin at 0.

16. Domain: $\{t \mid 0 \leq t \leq 4\}$ or $[0, 4]$

Range: $\{y \mid 0 \leq y \leq 121\}$ or $[0, 121]$

The ball will be in the air from 0 to 4 seconds and will reach heights from 0 feet up to a maximum of 121 feet.

17. a. Not a function. The domain element -1 corresponds to two different values in the range.
Domain: $\{-1, 5, 7, 9\}$
Range: $\{-2, 0, 2, 3, 4\}$

b. Function. Each animal corresponds to exactly one typical lifespan.
Domain: {Camel, Macaw, Deer, Fox, Tiger, Crocodile}
Range: {14, 22, 35, 45, 50}

18. a. Function. There are no ordered pairs that have the same first coordinate but different second coordinates.
Domain: $\{-3, -2, 2, 4, 5\}$
Range: $\{-1, 3, 4, 7\}$

b. Not a function. The domain element 'Blue' corresponds to two different types of cars in the range.
Domain: {Red, Blue, Green, Black}
Range: {Camry, Taurus, Windstar, Durango}

19. $3x - 5y = 18$
$$-5y = -3x + 18$$
$$y = \frac{-3x + 18}{-5}$$
$$y = \frac{3}{5}x - \frac{18}{5}$$

Since there is only one output y that can result from any given input x, this relation is a function.

20. $x^2 + y^2 = 81$
$$y^2 = 81 - x^2$$

Since a given input x can result in more than one output y, this relation is not a function. For example, if $x = 0$ then $y = 9$ or $y = -9$.

21. $y = \pm 10x$

Since a given input x can result in more than one output y, this relation is not a function.

22. $y = x^2 - 14$

Since there is only one output y that can result from any given input x, this relation is a function.

23. Not a function. The graph fails the vertical line test so it is not the graph of a function.

24. Function. The graph passes the vertical line test so it is the graph of a function.

25. Function. The graph passes the vertical line test so it is the graph of a function.

26. Not a function. The graph fails the vertical line test so it is not the graph of a function.

27. a. $f(-2) = (-2)^2 + 2(-2) - 5$
$= 4 - 4 - 5$
$= -5$

b. $f(3) = (3)^2 + 2(3) - 5$
$= 9 + 6 - 5$
$= 10$

28. a. $g(0) = \dfrac{2(0)+1}{(0)-3}$
$= \dfrac{0+1}{-3}$
$= -\dfrac{1}{3}$

b. $g(2) = \dfrac{2(2)+1}{(2)-3}$
$= \dfrac{4+1}{-1}$
$= -5$

29. a. $F(5) = -2(5) + 7$
$= -10 + 7$
$= -3$

b. $F(-x) = -2(-x) + 7$
$= 2x + 7$

30. a. $G(7) = 2(7) + 1$
$= 14 + 1$
$= 15$

b. $G(x+h) = 2(x+h) + 1$
$= 2x + 2h + 1$

31. $f(x) = -\dfrac{3}{2}x + 5$

Since each operation in the function can be performed for any real number, the domain of the function is all real numbers.
Domain: $\{x \mid x \text{ is a real number}\}$ or $(-\infty, \infty)$

32. $g(w) = \dfrac{w-9}{2w+5}$

The function tells us to divide $w-9$ by $2w+5$. Since division by 0 is not defined, the denominator can never equal 0.

$2w + 5 = 0$
$2w = -5$
$w = -\dfrac{5}{2}$

Thus, the domain is all real numbers except $-\dfrac{5}{2}$.

Domain: $\left\{ w \mid w \neq -\dfrac{5}{2} \right\}$

33. $h(t) = \dfrac{t+2}{t-5}$

The function tells us to divide $t+2$ by $t-5$. Since division by 0 is not defined, the denominator can never equal 0.
$t - 5 = 0$
$t = 5$
Thus, the domain of the function is all real numbers except 5.
Domain: $\{t \mid t \neq 5\}$

34. $G(t) = 3t^2 + 4t - 9$

Since each operation in the function can be performed for any real number, the domain of the function is all real numbers.
Domain: $\{t \mid t \text{ is a real number}\}$ or $(-\infty, \infty)$

35. a. The dependent variable is the population, P, and the independent variable is the number of years after 1900, t.

b. $P(120)$
$= 0.132(120)^2 - 4.782(120) + 43.947$
$= 1370.907$
According to the model, the population of Orange County will be roughly 1,370,907 in 2020.

c. $P(-70)$
$= 0.132(-70)^2 - 4.782(-70) + 43.947$
$= 1025.487$
According to the model, the population of Orange County was roughly 1,025,487 in 1830. This is not reasonable. (The population of the entire Florida territory was roughly 35,000 in 1830.)

36. a. The dependent variable is the percent of the population with an advanced degree, P, and the independent variable is the age of the population, a.

b. $P(30) = -0.0064(30)^2 + 0.6826(30) - 6.82$
$= 7.898$
Approximately 7.9% of 30 year olds have
an advanced degree.

37. $f(x) = 2x - 5$

x	$y = f(x) = 2x - 5$	(x, y)
-1	$f(-1) = 2(-1) - 5 = -7$	$(-1, -7)$
0	$f(0) = 2(0) - 5 = -5$	$(0, -5)$
1	$f(1) = 2(1) - 5 = -3$	$(1, -3)$
2	$f(2) = 2(2) - 5 = -1$	$(2, -1)$
3	$f(3) = 2(3) - 5 = 1$	$(3, 1)$

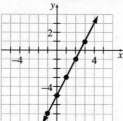

38. $g(x) = x^2 - 3x + 2$

x	$y = g(x) = x^2 - 3x + 2$	(x, y)
-1	$g(-1) = (-1)^2 - 3(-1) + 2 = 6$	$(-1, 6)$
0	$g(0) = (0)^2 - 3(0) + 2 = 2$	$(0, 2)$
1	$g(1) = (1)^2 - 3(1) + 2 = 0$	$(1, 0)$
2	$g(2) = (2)^2 - 3(2) + 2 = 0$	$(2, 0)$
3	$g(3) = (3)^2 - 3(3) + 2 = 2$	$(3, 2)$
4	$g(4) = (4)^2 - 3(4) + 2 = 6$	$(4, 6)$

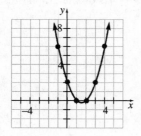

39. $h(x) = (x-1)^3 - 3$

x	$y = h(x) = (x-1)^3 - 3$	(x, y)
-1	$h(-1) = (-1-1)^3 - 3 = -11$	$(-1, -11)$
0	$h(0) = (0-1)^3 - 3 = -4$	$(0, -4)$
1	$h(1) = (1-1)^3 - 3 = -3$	$(1, -3)$
2	$h(2) = (2-1)^3 - 3 = -2$	$(2, -2)$
3	$h(3) = (3-1)^3 - 3 = 5$	$(3, 5)$

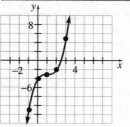

40. $f(x) = |x+1| - 4$

x	$y = f(x) = |x+1| - 4$	(x, y)
-5	$f(-5) = |-5+1| - 4 = 0$	$(-5, 0)$
-3	$f(-3) = |-3+1| - 4 = -2$	$(-3, -2)$
-1	$f(-1) = |-1+1| - 4 = -4$	$(-1, -4)$
1	$f(1) = |1+1| - 4 = -2$	$(1, -2)$
3	$f(3) = |3+1| - 4 = 0$	$(3, 0)$

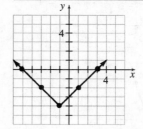

41. a. Domain: $\{x \mid x \text{ is a real number}\}$ or $(-\infty, \infty)$

Range: $\{y \mid y \text{ is a real number}\}$ or $(-\infty, \infty)$

b. The intercepts are $(0, 2)$ and $(4, 0)$. The
x-intercept is 4 and the y-intercept is 2.

42. a. Domain: $\{x \mid x \text{ is a real number}\}$ or $(-\infty, \infty)$

Range: $\{y \mid y \geq -3\}$ or $[-3, \infty)$

b. The intercepts are $(-2, 0)$, $(2, 0)$, and
$(0, -3)$. The x-intercepts are -2 and 2, and
the y-intercept is -3.

43. a. Domain: $\{x \mid x \text{ is a real number}\}$ or $(-\infty, \infty)$

Range: $\{y \mid y \text{ is a real number}\}$ or $(-\infty, \infty)$

b. The intercepts are $(0, 0)$ and $(2, 0)$. The x-intercepts are 0 and 2; the y-intercept is 0.

44. a. Domain: $\{x \mid x \geq -3\}$ or $[-3, \infty)$

Range: $\{y \mid y \geq 1\}$ or $[1, \infty)$

b. The only intercept is $(0, 3)$. There are no x-intercepts, but there is a y-intercept of 3.

45. a. Since the point $(-3, 4)$ is on the graph, $f(-3) = 4$.

b. Since the point $(1, -4)$ is on the graph, when $x = 1, f(x) = 4$.

c. Since the x-intercepts are -1 and 3, the zeroes of f are -1 and 3.

46. a. $y = x^2$

x	$y = x^2$	(x, y)
-2	$y = (-2)^2 = 4$	$(-2, 4)$
-1	$y = (-1)^2 = 1$	$(-1, 1)$
0	$y = (0)^2 = 0$	$(0, 0)$
1	$y = (1)^2 = 1$	$(1, 1)$
2	$y = (2)^2 = 4$	$(2, 4)$

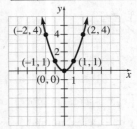

b. $y = x$

x	$y = x$	(x, y)
-2	$y = -2$	$(-2, -2)$
0	$y = 0$	$(0, 0)$
2	$y = 2$	$(2, 2)$

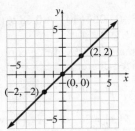

47. a. $h(3) = 2(3) - 7 = 6 - 7 = -1$

Since $h(3) = -1$, the point $(3, -1)$ is on the graph of the function.

b. $h(-2) = 2(-2) - 7 = -4 - 7 = -11$

The point $(-2, -11)$ is on the graph of the function.

c. $h(x) = 4$

$2x - 7 = 4$

$2x = 11$

$x = \dfrac{11}{2}$

The point $\left(\dfrac{11}{2}, 4\right)$ is on the graph of h.

48. a. $g(-5) = \dfrac{3}{5}(-5) + 4 = -3 + 4 = 1$

Since $g(-5) = 1$, the point $(-5, 2)$ is not on the graph of the function.

b. $g(3) = \dfrac{3}{5}(3) + 4 = \dfrac{9}{5} + 4 = \dfrac{29}{5}$

The point $\left(3, \dfrac{29}{5}\right)$ is on the graph of the function.

c. $g(x) = -2$

$\dfrac{3}{5}x + 4 = -2$

$\dfrac{3}{5}x = -6$

$x = -10$

The point $(-10, -2)$ is on the graph of g.

49.

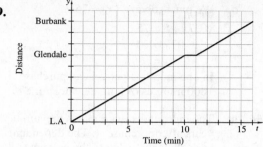

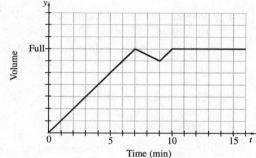

50.

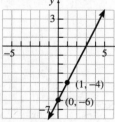

51. Comparing $g(x) = 2x - 6$ to $g(x) = mx + b$, we see that the slope m is 2 and the y-intercept b is -6. We begin by plotting the point $(0, -6)$.

Because $m = 2 = \dfrac{2}{1} = \dfrac{\Delta y}{\Delta x} = \dfrac{\text{Rise}}{\text{Run}}$, from the point $(0, -6)$ we go up 2 units and to the right 1 unit and end up at $(1, -4)$. We draw a line through these points and obtain the graph of $g(x) = 2x - 6$.

$f(x) = 0$
$2x - 6 = 0$
$\quad 2x = 6$
$\quad\ \ x = 3$
The zero of f is 3.

52. Comparing $H(x) = -\dfrac{4}{3}x + 5$ to $H(x) = mx + b$, we see that the slope m is $-\dfrac{4}{3}$ and the y-intercept b is 5. We begin by plotting the point

(0, 5). Because $m = -\dfrac{4}{3} = \dfrac{-4}{3} = \dfrac{\Delta y}{\Delta x} = \dfrac{\text{Rise}}{\text{Run}}$,

from the point $(0, 5)$ we go down 4 units and to the right 3 units and end up at $(3, 1)$. We draw a line through these points and obtain the graph of

$H(x) = -\dfrac{4}{3}x + 5.$

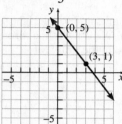

$H(x) = 0$

$-\dfrac{4}{3}x + 5 = 0$

$\quad -\dfrac{4}{3}x = -5$

$\qquad\ \ x = \dfrac{15}{4}$

The zero of H is $\dfrac{15}{4}$.

53. Comparing $F(x) = -x - 3$ to $F(x) = mx + b$, we see that the slope m is -1 and the y-intercept b is -3. We begin by plotting the point $(0, -3)$.

Because $m = -1 = \dfrac{-1}{1} = \dfrac{\Delta y}{\Delta x} = \dfrac{\text{Rise}}{\text{Run}}$, from the

point $(0, -3)$ we go down 1 unit and to the right 1 unit and end up at $(1, -4)$. We draw a line through these points and obtain the graph of $F(x) = -x - 3.$

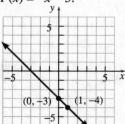

$F(x) = 0$
$-x - 3 = 0$
$\quad -x = 3$
$\quad\ \ x = -3$
The zero of F is -3.

54. Comparing $f(x) = \dfrac{3}{4}x - 3$ to $f(x) = mx + b$, we

see that the slope m is $\dfrac{3}{4}$ and the y-intercept b is

-3. We begin by plotting the point $(0, -3)$.

Because $m = \dfrac{3}{4} = \dfrac{\Delta y}{\Delta x} = \dfrac{\text{Rise}}{\text{Run}}$, from the point

$(0, -3)$ we go up 3 units and to the right 4 units
and end up at $(4, 0)$. We draw a line through
these points and obtain the graph of

$f(x) = \dfrac{3}{4}x - 3$.

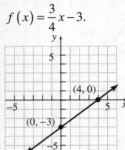

From the graph, we see that the x-intercept is 4,
so the zero of f is 4.

55. a. The independent variable is the number of
long-distance minutes used, x. It would not
make sense to talk for a negative number of
minutes. Thus, the domain of C is
$\{x \mid x \geq 0\}$ or, using interval notation, $[0, \infty)$.

b. $C(235) = 0.07(235) + 5 = 21.45$

The cost for 235 minutes of long-distance
calls during one month is $21.45.

c. Evaluate C at $x = 0$, 100, and 500.

$C(0) = 0.07(0) + 5 = 5$

$C(100) = 0.07(100) + 5 = 12$

$C(500) = 0.07(500) + 5 = 40$

Thus, the points $(0, 5)$, $(100, 12)$, and
$(500, 40)$ are on the graph.

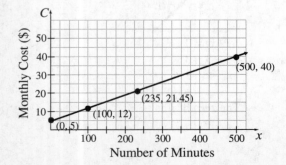

d. We must solve $C(x) = 75$.

$0.07x + 5 = 75$

$0.07x = 70$

$x = 1000$

In one month, a person can purchase
1000 minutes of long distance for $75.

56. a. The independent variable is the number of
years after purchase, x. The dependent
variable is the value of the computer, V.

b. We are told that the value function V is for 0
to 5 years, inclusive. Thus, the domain is
$\{x \mid 0 \leq x \leq 5\}$ or, using interval notation,
$[0, 5]$.

c. The initial value of the computer will be the
value at $x = 0$ years.

$V(0) = 1800 - 360(0) = 1800$

The initial value of the computer is $1800.

d. $V(2) = 1800 - 360(2) = 1080$

After 2 years, the value of the computer is
$1080.

e. Evaluate V at $x = 3$, 4, and 5.

$V(3) = 1800 - 360(3) = 720$

$V(4) = 1800 - 360(4) = 360$

$V(5) = 1800 - 360(5) = 0$

Thus, the points $(3, 720)$, $(4, 360)$, and $(5, 0)$
are on the graph.

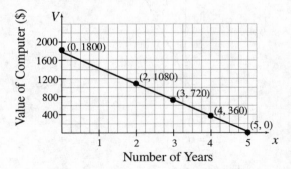

f. We must solve $V(x) = 0$.

$1800 - 360x = 0$

$-360x = -1800$

$x = 5$

After 5 years, the computer's value will be
$0.

57. a. Let x = FICO score. Let L = loan rate.

$$m = \frac{5-9}{750-675} = \frac{-4}{75}$$

$$L - 5 = -\frac{4}{75}(x-750)$$

$$L - 5 = -\frac{4}{75}x + 40$$

$$L = -\frac{4}{75}x + 45$$

In function notation,

$$L(x) = -\frac{4}{75}x + 45.$$

b. $L(710) = -\frac{4}{75}(710) + 45 = 7$

With a FICO credit score of 710, the auto loan rate should be approximately 7%.

c. The slope, $-\frac{4}{75} \approx -0.053$, indicates that the loan rate decreases approximately 0.05% for every 1-unit increase in the FICO score.

d. $-\frac{4}{75}x + 45 = 6.5$

$$-\frac{4}{75}x = -38.5$$

$$x \approx 722$$

A bank will offer a rate of 6.5% for a FICO score of 722.

58. a. Let x represent the age of men and H represent the maximum recommended heart rate for men under stress.

$$m = \frac{160 - 200}{60 - 20}$$

$$= -1 \text{ beat per minute per year}$$

$$H - 200 = -1(x - 20)$$

$$H - 200 = -x + 20$$

$$H = -x + 220$$

In function notation, $H(x) = -x + 220$.

b. $H(45) = -(45) + 220 = 175$

The maximum recommended heart rate for a 45 year old man under stress is 175 beats per minute.

c. The slope (−1) indicates that the maximum recommended heart rate for men under stress decreases at a rate of 1 beat per minute per year.

d. $-x + 220 = 168$

$$-x = -52$$

$$x = 52$$

The maximum recommended heart rate under stress is 168 beats per minute for 52-year-old men.

59. a. $C(m) = 0.12m + 35$

b. The number of miles driven, m, is the independent variable. The rental cost, C, is the dependent variable.

c. Because the number of miles cannot be negative, the it must be greater than or equal to zero. That is, the implied domain is $\{m \mid m \geq 0\}$ or, using interval notation, $[0, \infty)$.

d. $C(124) = 0.12(124) + 35 = 49.88$

If 124 miles are driven during a one-day rental, the charge will be $49.88.

e. $0.12m + 35 = 67.16$

$$0.12m = 32.16$$

$$m = 268$$

If the charge for a one-day rental is $67.16, then 268 miles were driven.

f.

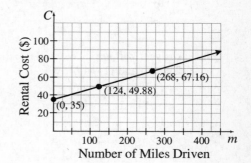

60. a. $B(x) = 3.50x + 33.99$

b. The number of pay-per-view movies watched, x, is the independent variable. The monthly bill, B, is the dependent variable.

c. Because the number pay-per-view movies watched cannot be negative, it must be greater than or equal to zero. That is, the implied domain is $\{x \mid x \geq 0\}$ or, using interval notation, $[0, \infty)$.

d. $B(5) = 3.50(5) + 33.99 = 51.49$

If 5 pay-per-view movies are watched one month, the bill will be $51.49.

e. $3.50x + 33.99 = 58.49$
$$3.50x = 24.50$$
$$x = 7$$

If the bill one month is $58.49, then 7 pay-per-view movies were watched.

f.

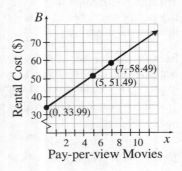

61. a.

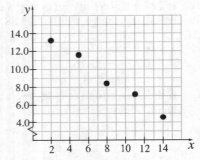

b.   Answers will vary. We will use the points (2, 13.3) and (14, 4.6).

$$m = \frac{4.6 - 13.3}{14 - 2} = \frac{-8.7}{12} = -0.725$$
$$y - 13.3 = -0.725(x - 2)$$
$$y - 13.3 = -0.725x + 1.45$$
$$y = -0.725x + 14.75$$

c.

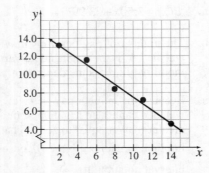

62. a.

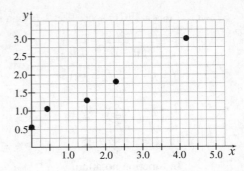

b.   Answers will vary. We will use the points (0, 0.6) and (4.2, 3.0).

$$m = \frac{3.0 - 0.6}{4.2 - 0} = \frac{2.4}{4.2} = \frac{4}{7}$$
$$y - 0.6 = \frac{4}{7}(x - 0)$$
$$y - 0.6 = \frac{4}{7}x$$
$$y = \frac{4}{7}x + 0.6$$

c.

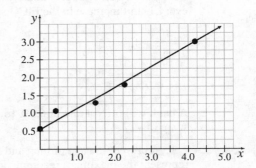

63. a.

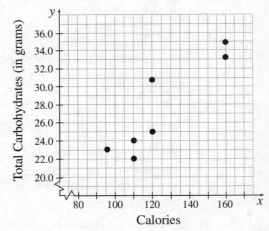

b.   Approximately linear

c. Answers will vary. We will use the points (96, 23.2) and (160, 33.3).

$$m = \frac{33.3 - 23.2}{160 - 96} = \frac{10.1}{64} \approx 0.158$$

$$y - 23.2 = 0.158(x - 96)$$

$$y - 23.2 = 0.158x - 15.168$$

$$y = 0.158x + 8.032$$

d.

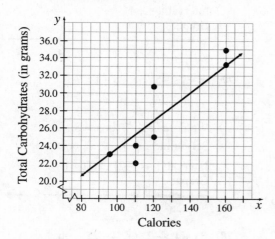

e. $x = 140$: $y = 0.158(140) + 8.032$

$$= 30.152$$

We predict that a one-cup serving of cereal having 140 calories will have approximately 30.2 grams of total carbohydrates.

f. The slope of the line found is 0.158. This means that, in a one-cup serving of cereal, total carbohydrates will increase by 0.158 gram for each one-calorie increase.

64. a.

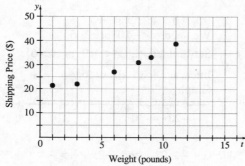

b. Approximately linear

c. Answers will vary. We will use the point (1, 21.09) and (11, 37.79).

$$m = \frac{37.79 - 21.09}{11 - 1} = \frac{16.7}{10} = 1.67$$

$$y - 21.09 = 1.67(x - 1)$$

$$y - 21.09 = 1.67x - 1.67$$

$$y = 1.67x + 19.42$$

d.

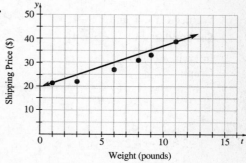

e. $x = 5$; $y = 1.67(5) + 19.42$

$$= 27.77$$

We predict that the FedEx 2Day ® price for shipping a 5-pound package will be $27.77.

f. The slope of the line found is 1.67. This means that the FedEx 2Day ® shipping price increases by $1.67 for each one-pound increase in the weight of a package.

65. $A \cup B = \{-1, 0, 1, 2, 3, 4, 6, 8\}$

66. $A \cap C = \{2, 4\}$

67. $B \cap C = \{1, 2, 3, 4\}$

68. $A \cup C = \{1, 2, 3, 4, 6, 8\}$

69. a. $A \cap B = \{x \mid 2 < x \le 4\}$

Interval: (2, 4]

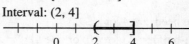

b. $A \cup B = \{x \mid x \text{ is a real number}\}$

Interval: $(-\infty, \infty)$

70. a. $E \cap F = \{\ \}$ or $\varnothing$

  b. $E \cup F = \{x \mid x < -2 \text{ or } x \geq 3\}$

   Interval: $(-\infty, -2) \cup [3, \infty)$

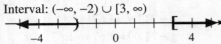

71. $x < 4$ and $x + 3 > 2$

   $\qquad\qquad x > -1$

   We need $x < 4$ and $x > -1$.

   Set-builder: $\{x \mid -1 < x < 4\}$

   Interval: $(-1, 4)$

   Graph:

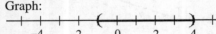

72. $\qquad 3 < 2 - x < 7$

   $3 - 2 < 2 - x - 2 < 7 - 2$

   $\qquad 1 < -x < 5$

   $-1(1) > -1(-x) > -1(5)$

   $\qquad -1 > x > -5$

   $\qquad -5 < x < -1$

   Set-builder: $\{x \mid -5 < x < -1\}$

   Interval: $(-5, -1)$

   Graph:

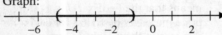

73. $x + 3 < 1$ or $x > 2$

   $\quad x < -2$

   We need $x < -2$ or $x > 2$.

   Set-builder: $\{x \mid x < -2 \text{ or } x > 2\}$

   Interval: $(-\infty, -2) \cup (2, \infty)$

   Graph:

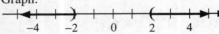

74. $x + 6 \geq 10$ or $x \leq 0$

   $\quad x \geq 4$

   We need $x \geq 4$ or $x \leq 0$.

   Set-builder: $\{x \mid x \leq 0 \text{ or } x \geq 4\}$

   Interval: $(-\infty, 0] \cup [4, \infty)$

   Graph:

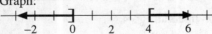

75. $3x + 2 \leq 5$ and $-4x + 2 \leq -10$

   $\quad 3x \leq 3 \qquad\qquad -4x \leq -12$

   $\quad x \leq 1 \qquad\qquad\quad x \geq 3$

   We need $x \leq 1$ and $x \geq 3$.

   Solution set: $\{\ \}$ or $\varnothing$

76. $\qquad 1 \leq 2x + 5 < 13$

   $1 - 5 \leq 2x + 5 - 5 < 13 - 5$

   $\qquad -4 \leq 2x < 8$

   $\qquad \dfrac{-4}{2} \leq \dfrac{2x}{2} < \dfrac{8}{2}$

   $\qquad -2 \leq x < 4$

   Set-builder: $\{x \mid -2 \leq x < 4\}$

   Interval: $[-2, 4)$

   Graph:

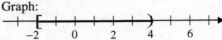

77. $x - 3 \leq -5$ or $2x + 1 > 7$

   $\quad x \leq -2 \qquad\qquad 2x > 6$

   $\qquad\qquad\qquad\qquad x > 3$

   We need $x \leq -2$ or $x > 3$.

   Set-builder: $\{x \mid x \leq -2 \text{ or } x > 3\}$

   Interval: $(-\infty, -2] \cup (3, \infty)$

   Graph:

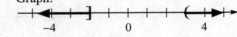

78. $3x + 4 > -2$ or $4 - 2x \geq -6$

   $\quad 3x > -6 \qquad\qquad -2x \geq -10$

   $\quad x > -2 \qquad\qquad\quad x \leq 5$

   We need $x > -2$ or $x \leq 5$.

   Set-builder: $\{x \mid x \text{ is any real number}\}$

   Interval: $(-\infty, \infty)$

   Graph:

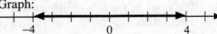

79. $\dfrac{1}{3}x > 2$ or $\dfrac{2}{5}x < -4$

   $\quad x > 6 \qquad\qquad x < -10$

   We need $x < -10$ or $x > 6$.

   Set-builder: $\{x \mid x < -10 \text{ or } x > 6\}$

   Interval: $(-\infty, -10) \cup (6, \infty)$

   Graph:

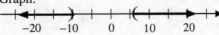

80. $x + \dfrac{3}{2} \geq 0$ and $-2x + \dfrac{3}{2} > \dfrac{1}{4}$

   $\qquad x \geq -\dfrac{3}{2} \qquad\qquad -2x > -\dfrac{5}{4}$

   $\qquad\qquad\qquad\qquad\qquad\quad x < \dfrac{5}{8}$

   We need $x \geq -\dfrac{3}{2}$ and $x < \dfrac{5}{8}$.

Set-builder: $\left\{ x \mid -\dfrac{3}{2} \le x < \dfrac{5}{8} \right\}$

Interval: $\left[-\dfrac{3}{2}, \dfrac{5}{8} \right)$

Graph:

81. $70 \le x \le 75$

82. Let x = number of kilowatt-hours. Then, the number of kilowatt-hours for usage above 800 Kwh is given by the expression $x - 800$. Therefore, we need to solve the following inequality:

$$52.62 \le 43.56 + 0.038752(x - 800) \le 88.22$$
$$9.06 \le 0.038752(x - 800) \le 44.66$$
$$\dfrac{9.06}{0.038752} \le x - 800 \le \dfrac{44.66}{0.038752}$$
$$\dfrac{9.06}{0.038752} + 800 \le x \le \dfrac{44.66}{0.038752} + 800$$
$$1033.8 \le x \le 1952.5 \text{ (approx.)}$$

The electric usage varied from roughly 1033.8 kilowatt-hours up to roughly 1952.5 kilowatt-hours.

83. $|x| = 4$

$x = 4$ or $x = -4$
Solution set: $\{-4, 4\}$

84. $|3x - 5| = 4$

$3x - 5 = 4$ or $3x - 5 = -4$
$3x = 9$ $3x = 1$
$x = 3$ $x = \dfrac{1}{3}$

Solution set: $\left\{ \dfrac{1}{3}, 3 \right\}$

85. $|-y + 4| = 9$

$-y + 4 = 9$ or $-y + 4 = -9$
$-y = 5$ $-y = -13$
$y = -5$ $y = 13$

Solution set: $\{-5, 13\}$

86. $-3|x + 2| - 5 = -8$

$-3|x + 2| = -3$
$|x + 2| = 1$
$x + 2 = 1$ or $x + 2 = -1$
$x = -1$ $x = -3$

Solution set: $\{-3, -1\}$

87. $|2w - 7| = -3$

This equation has no solution since an absolute value can never yield a negative result.
Solution set: $\{ \ \}$ or $\varnothing$

88. $|x + 3| = |3x - 1|$

$x + 3 = 3x - 1$ or $x + 3 = -(3x - 1)$
$-2x = -4$ $x + 3 = -3x + 1$
$x = 2$ $4x = -2$
                      $x = -\dfrac{1}{2}$

Solution set: $\left\{ -\dfrac{1}{2}, 2 \right\}$

89. $|x| < 2$

$-2 < x < 2$
Set-builder: $\{ x \mid -2 < x < 2 \}$
Interval: $(-2, 2)$

90. $|x| \ge \dfrac{7}{2}$

$x \le -\dfrac{7}{2}$ or $x \ge \dfrac{7}{2}$

Set-builder: $\left\{ x \mid x \le -\dfrac{7}{2} \text{ or } x \ge \dfrac{7}{2} \right\}$

Interval: $\left(-\infty, -\dfrac{7}{2} \right] \cup \left[\dfrac{7}{2}, \infty \right)$

91. $|x + 2| \le 3$

$-3 \le x + 2 \le 3$
$-3 - 2 \le x + 2 - 2 \le 3 - 2$
$-5 \le x \le 1$
Set-builder: $\{ x \mid -5 \le x \le 1 \}$
Interval: $[-5, 1]$

92. $|4x - 3| \ge 1$

$4x - 3 \le -1$ or $4x - 3 \ge 1$
$4x \le 2$ $4x \ge 4$
$x \le \dfrac{1}{2}$ $x \ge 1$

Set-builder: $\left\{ x \mid x \le \dfrac{1}{2} \text{ or } x \ge 1 \right\}$

Interval: $\left(-\infty, \frac{1}{2}\right] \cup [1, \infty)$

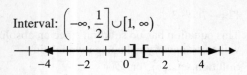

93. $3|x| + 6 \ge 1$

$3|x| \ge -5$

$|x| \ge -\dfrac{5}{3}$

Since the result of an absolute value is always nonnegative, any real number is a solution to this inequality.

Set-builder: $\{x \mid x \text{ is a real number}\}$

Interval: $(-\infty, \infty)$

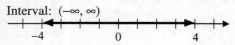

94. $|7x + 5| + 4 < 3$

$|7x + 5| < -1$

Since the result of an absolute value is never negative, this inequality has no solutions.

Solution set: $\{\ \}$ or $\varnothing$

95. $|(x - 3) - 2| \le 0.01$

$|x - 3 - 2| \le 0.01$

$|x - 5| \le 0.01$

$-0.01 \le x - 5 \le 0.01$

$-0.01 + 5 \le x - 5 + 5 \le 0.01 + 5$

$4.99 \le x \le 5.01$

Set-builder: $\{x \mid 4.99 \le x \le 5.01\}$

Interval: $[4.99, 5.01]$

96. $\left|\dfrac{2x - 3}{4}\right| > 1$

$\dfrac{2x - 3}{4} < -1 \quad \text{or} \quad \dfrac{2x - 3}{4} > 1$

$2x - 3 < -4 \qquad\qquad 2x - 3 > 4$

$2x < -1 \qquad\qquad\quad 2x > 7$

$x < -\dfrac{1}{2} \qquad\qquad\quad x > \dfrac{7}{2}$

Solution set: $\left\{x \mid x < -\dfrac{1}{2} \text{ or } x > \dfrac{7}{2}\right\}$

Interval: $\left(-\infty, -\dfrac{1}{2}\right) \cup \left(\dfrac{7}{2}, \infty\right)$

97. $|x - 0.503| \le 0.001$

$-0.001 \le x - 0.503 \le 0.001$

$0.502 \le x \le 0.504$

The acceptable diameters of the bearing are between 0.502 inch and 0.504 inch, inclusive.

98. $\left|\dfrac{x - 40}{2}\right| > 1.96$

$\dfrac{x - 40}{2} < -1.96 \quad \text{or} \quad \dfrac{x - 40}{2} > 1.96$

$x - 40 < -3.92 \qquad\qquad x - 40 > 3.92$

$x < 36.08 \qquad\qquad\qquad x > 43.92$

Tensile strengths below 36.08 lb/in.^{2} or above 43.92 lb/in.^{2} would be considered unusual.

Chapter 2 Test

1. Domain: $\{-4, 2, 5, 7\}$

Range: $\{-7, -2, -1, 3, 8, 12\}$

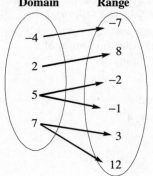

2. Domain: $\left\{x \mid -\dfrac{5\pi}{2} \le x \le \dfrac{5\pi}{2}\right\}$ or $\left[-\dfrac{5\pi}{2}, \dfrac{5\pi}{2}\right]$

Range: $\{y \mid 1 \le y \le 5\}$ or $[1, 5]$

3. $y = x^2 - 3$

x	$y = x^2 - 3$	(x, y)
-2	$y = (-2)^2 - 3 = 1$	$(-2, 1)$
-1	$y = (-1)^2 - 3 = -2$	$(-1, -2)$
0	$y = (0)^2 - 3 = -3$	$(0, -3)$
1	$y = (1)^2 - 3 = -2$	$(1, -2)$
2	$y = (2)^2 - 3 = 1$	$(2, 1)$

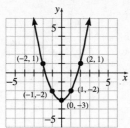

Domain: $\{x \mid x \text{ is a real number}\}$ or $(-\infty, \infty)$

Range: $\{y \mid y \geq -3\}$ or $[-3, \infty)$

4. Function. Each element in the domain corresponds to exactly one element in the range.
   Domain: $\{-5, -3, 0, 2\}$
   Range: $\{3, 7\}$

5. Not a function. The graph fails the vertical line test so it is not the graph of a function.
   Domain: $\{x \mid x \leq 3\}$ or $(-\infty, 3]$
   Range: $\{y \mid y \text{ is a real number}\}$ or $(-\infty, \infty)$

6. No, $y = \pm 5x$ is not a function because a single input, x, can yield two different outputs. For example, if $x = 1$ then $y = -5$ or $y = 5$.

7. $f(x+h) = -3(x+h) + 11 = -3x - 3h + 11$

8. **a.** $g(-2) = 2(-2)^2 + (-2) - 1 = 2(4) - 3$
   $= 8 - 3 = 5$

   b. $g(0) = 2(0)^2 + (0) - 1 = 0 + 0 - 1 = -1$

   c. $g(3) = 2(3)^2 + (3) - 1$
   $= 2(9) + 2$
   $= 18 + 2$
   $= 20$

9. $f(x) = x^2 + 3$

x	$y = f(x) = x^2 + 3$	(x, y)
-2	$f(-2) = (-2)^2 + 3 = 7$	$(-2, 7)$
-1	$f(-1) = (-1)^2 + 3 = 4$	$(-1, 4)$
0	$f(0) = (0)^2 + 3 = 3$	$(0, 3)$
1	$f(1) = (1)^2 + 3 = 4$	$(1, 4)$
2	$f(2) = (2)^2 + 3 = 7$	$(2, 7)$

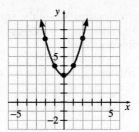

10. **a.** The dependent variable is the ticket price, P, and the independent variable is the number of years after 1996, x.

    b. $P(14) = 0.24(14) + 4.39 = 7.7$
    According to the model, the average ticket price in 2010 ($x = 14$) was $7.70.

    c. $10 = 0.24x + 4.34$
    $5.66 = 0.24x$
    $x \approx 24$
    According to the model, the average movie ticket price will be $10.00 in 2020 ($x = 24$).

11. The function tells us to divide -15 by $x + 2$. Since we can't divide by zero, we need $x \neq -2$.
    Domain: $\{x \mid x \neq -2\}$

12. **a.** $h(2) = -5(2) + 12$
    $= -10 + 12$
    $= 2$
    Since $h(2) = 2$, the point $(2, 2)$ is on the graph of the function.

    b. $h(3) = -5(3) + 12$
    $= -15 + 12$
    $= -3$
    Since $h(3) = -3$, the point $(3, -3)$ is on the graph of the function.

    c. $h(x) = 27$
    $-5x + 12 = 27$
    $-5x = 15$
    $x = -3$
    The point $(-3, 27)$ is on the graph of h.

    d. $h(x) = 0$
    $-5x + 12 = 0$
    $-5x = -12$
    $x = \dfrac{12}{5}$
    $\dfrac{12}{5}$ is the zero of h.

13. a. The car stops accelerating when the speed stops increasing. Thus, the car stops accelerating after 6 seconds.

    b. The car has a constant speed when the graph is horizontal. Thus, the car maintains a constant speed for 18 seconds.

14. a. The profit is $\$30 - \$12 = \$18$ times the number of shelves sold x, minus the $\$100$ for renting the display. Thus, the function is $P(x) = 18x - 100$.

    b. The independent variable is the number of shelves sold, x. Henry could not sell a negative number of shelves. Thus, the domain of P is $\{x \mid x \geq 0\}$ or, using interval notation, $[0, \infty)$.

    c. $P(34) = 18(34) - 100$
$$= 512$$
If Henry sells 34 shelves, his profit will be $\$512$.

    d. Evaluate P at $x = 0$, 20, and 50.
$$P(0) = 18(0) - 100$$
$$= -100$$
$$P(20) = 18(20) - 100$$
$$= 260$$
$$P(50) = 18(50) - 100$$
$$= 800$$
Thus, the points $(0, -100)$, $(20, 260)$, and $(50, 800)$ are on the graph.

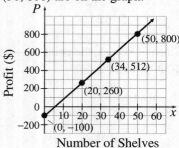

Number of Shelves

    e. We must solve $P(x) = 764$.
$$18x - 100 = 764$$
$$18x = 864$$
$$x = 48$$
If Henry sells 48 shelves, his profit will be $\$764$.

15. a.

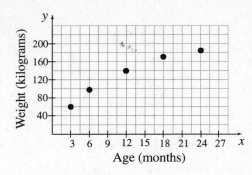

    b. Approximately linear

    c. Answers will vary. We will use the points $(6, 95)$ and $(18, 170)$.
$$m = \frac{170 - 95}{18 - 6} = \frac{75}{12} = 6.25$$
$$y - 95 = 6.25(x - 6)$$
$$y - 95 = 6.25x - 37.5$$
$$y = 6.25x + 57.5$$

    d.

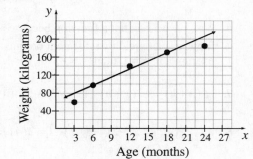

    e. $x = 9$: $\quad y = 6.25(9) + 57.5$
$$= 113.75$$
We predict that a 9-month-old Shetland pony will weigh 113.75 kilograms.

    f. The slope of the line found is 6.25. This means that a Shetland pony's weight will increase by 6.25 kilograms for each one-month increase in age.

16. $|2x + 5| - 3 = 0$
$$|2x + 5| = 3$$
$$2x + 5 = -3 \quad \text{or} \quad 2x + 5 = 3$$
$$2x = -8 \qquad\qquad 2x = -2$$
$$x = -4 \qquad\qquad x = -1$$
Solution set: $\{-4, -1\}$

17. $x + 2 < 8$ and $2x + 5 \geq 1$

$\quad\quad x < 6 \quad\quad\quad\quad 2x \geq -4$

$\quad\quad\quad\quad\quad\quad\quad\quad\quad x \geq -2$

We need $x \geq -2$ and $x < 6$.

Set-builder: $\{x \mid -2 \leq x < 6\}$

Interval: $[-2, 6)$

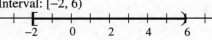

18. $x > 4$ or $2(x - 1) + 3 < -2$

$\quad\quad\quad\quad\quad\quad 2x - 2 + 3 < -2$

$\quad\quad\quad\quad\quad\quad 2x + 1 < -2$

$\quad\quad\quad\quad\quad\quad 2x + 1 - 1 < -2 - 1$

$\quad\quad\quad\quad\quad\quad 2x < -3$

$\quad\quad\quad\quad\quad\quad \dfrac{2x}{2} < \dfrac{-3}{2}$

$\quad\quad\quad\quad\quad\quad x < -\dfrac{3}{2}$

We need $x > 4$ or $x < -\dfrac{3}{2}$.

Set-builder: $\left\{ x \mid x < -\dfrac{3}{2} \text{ or } x > 4 \right\}$

Interval: $\left(-\infty, -\dfrac{3}{2} \right) \cup (4, \infty)$

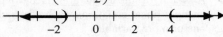

19. $2|x - 5| + 1 < 7$

$\quad\quad 2|x - 5| < 6$

$\quad\quad |x - 5| < 3$

$\quad -3 < x - 5 < 3$

$\quad\quad 2 < x < 8$

Set-builder: $\{x \mid 2 < x < 8\}$

Interval: $(2, 8)$

20. $|-2x + 1| \geq 5$

$\quad -2x + 1 \leq -5$ or $-2x + 1 \geq 5$

$\quad\quad -2x \leq -6 \quad\quad\quad -2x \geq 4$

$\quad\quad\quad x \geq 3 \quad\quad\quad\quad\quad x \leq -2$

Set-builder: $\{x \mid x \leq -2 \text{ or } x \geq 3\}$

Interval: $(-\infty, -2] \cup [3, \infty)$

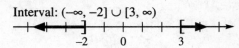

Chapter 3

Section 3.1

Are You Ready for This Section?

R1. Substitute $x = 5$ and $y = 4$, and simplify:
$$2x - 3y = 2(5) - 3(4)$$
$$= 10 - 12$$
$$= -2$$

R2. $2(4) - 3(-1) \stackrel{?}{=} 11$
$$8 + 3 \stackrel{?}{=} 11$$
$$11 = 11 \quad \leftarrow \text{True}$$
Yes, $(4, -1)$ is on the graph of $2x - 3y = 11$.

R3. $y = 3x - 7$
Let $x = 0$, 1, and 2.

$x = 0: \quad y = 3(0) - 7$
$$y = 0 - 7$$
$$y = -7$$

$x = 1: \quad y = 3(1) - 7$
$$y = 3 - 7$$
$$y = -4$$

$x = 2: \quad y = 3(2) - 7$
$$y = 6 - 7$$
$$y = -1$$

Thus, the points $(0, -7)$, $(1, -4)$, and $(2, -1)$ are on the graph.

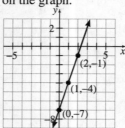

R4. The slope of the line we seek is $m = -3$, the same as the slope of $y = -3x + 1$. The equation of the line is
$$y - y_1 = m(x - x_1)$$
$$y - 3 = -3(x - 2)$$
$$y - 3 = -3x + 6$$
$$y = -3x + 9$$

R5. $4x - 3y = 15$
$$-3y = -4x + 15$$
$$y = \frac{-4x + 15}{-3}$$
$$y = \frac{4}{3}x - 5$$
The slope is $\frac{4}{3}$ and the y-intercept is -5.

R6. The additive inverse of 4 is -4 because $4 + (-4) = 0$.

R7. $2x - 3(-3x + 1) = -36$
$$2x + 9x - 3 = -36$$
$$11x - 3 = -36$$
$$11x = -33$$
$$x = \frac{-33}{11} = -3$$
The solution set is $\{-3\}$.

Section 3.1 Quick Checks

1. A <u>system of linear equations</u> is a grouping of two or more linear equations, each of which contains one or more variables.

2. A <u>solution</u> of a system of equations consists of values of the variables that satisfy each equation of the system.

3. $\begin{cases} 2x + 3y = 7 & (1) \\ 3x + y = -7 & (2) \end{cases}$

   a. Let $x = 2$ and $y = 1$ in both equations (1) and (2).
Equation (1): $2(2) + 3(1) \stackrel{?}{=} 7$
$$4 + 3 \stackrel{?}{=} 7$$
$$7 = 7$$
Equation (2): $3(2) + (1) \stackrel{?}{=} -7$
$$6 + 1 \stackrel{?}{=} -7$$
$$7 \neq -7$$
Although these values satisfy equation (1), they do not satisfy equation (2). Therefore, the ordered pair $(2, 1)$ is not a solution.

   b. Let $x = -4$ and $y = 5$ in both equations (1) and (2).
Equation (1): $2(-4) + 3(5) \stackrel{?}{=} 7$
$$-8 + 15 \stackrel{?}{=} 7$$
$$7 = 7$$

Equation (2): $3(-4)+5 \overset{?}{=} -7$

$$-12+5 \overset{?}{=} -7$$

$$-7 = -7$$

These values satisfy both equations, so the ordered pair $(-4,\ 5)$ is a solution.

c.   Let $x = -2$ and $y = -1$ in both equations (1) and (2).

Equation (1): $2(-2)+3(-1) \overset{?}{=} 7$

$$-4-3 \overset{?}{=} 7$$

$$-7 \neq 7$$

Equation (2): $3(-2)+(-1) \overset{?}{=} -7$

$$-6-1 \overset{?}{=} -7$$

$$-7 = -7$$

Although these values satisfy equation (2), they do not satisfy equation (1). Therefore, the ordered pair $(-2, -1)$ is not a solution.

4.   If a system of equations has no solution, it is said to be <u>inconsistent</u>.

5.   If a system of equations has infinitely many solutions, the system is said to be <u>consistent</u> and the equations are <u>dependent</u>.

6.   False. If the graphs of the equations are parallel lines, then the system will have no solution.

7.   True

8.   True

9. $\begin{cases} y = -3x + 10 & (1) \\ y = 2x - 5 & (2) \end{cases}$

The two equations are in slope-intercept form. Graph each equation and find the point of intersection.

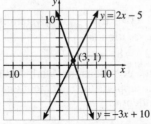

The solution is the ordered pair $(3,\ 1)$.

10. $\begin{cases} 2x + y = -1 & (1) \\ -2x + 2y = 10 & (2) \end{cases}$

Equation (1) in slope-intercept form is $y = -2x - 1$. Equation (2) in slope-intercept form is $y = x + 5$. Graph each equation and find the point of intersection.

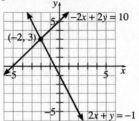

The solution is the ordered pair $(-2,\ 3)$.

11. $\begin{cases} y = -3x - 5 & (1) \\ 5x + 3y = 1 & (2) \end{cases}$

Substituting $-3x - 5$ for y in equation (2), we obtain

$$5x + 3(-3x - 5) = 1$$

$$5x - 9x - 15 = 1$$

$$-4x - 15 = 1$$

$$-4x = 16$$

$$x = -4$$

Substituting -4 for x in equation (1), we obtain $y = -3(-4) - 5 = 12 - 5 = 7$.

The solution is the ordered pair $(-4, 7)$.

12. $\begin{cases} 2x + y = -2 & (1) \\ -3x - 2y = -2 & (2) \end{cases}$

Equation (1) solved for y is $y = -2x - 2$. Substituting $-2x - 2$ for y in equation (2), we obtain

$$-3x - 2(-2x - 2) = -2$$

$$-3x + 4x + 4 = -2$$

$$x + 4 = -2$$

$$x = -6$$

Substituting -6 for y in equation (1), we obtain

$$2(-6) + y = -2$$

$$-12 + y = -2$$

$$y = 10$$

The solution is the ordered pair $(-6, 10)$.

13.   The basic idea in using the elimination method is to get the coefficients of one of the variables to be <u>additive inverses</u>, such as 5 and −5.

14. $\begin{cases} 2x - 3y = -6 & (1) \\ -8x + 3y = 3 & (2) \end{cases}$

There is no need to multiply either equation by any constant. Add the two equations.

$$2x - 3y = -6$$
$$\underline{-8x + 3y = 3}$$
$$-6x \quad\quad = -3$$
$$x = \frac{1}{2}$$

Substituting $\frac{1}{2}$ for x in equation (1), we obtain

$$2\left(\frac{1}{2}\right) - 3y = -6$$
$$1 - 3y = -6$$
$$-3y = -7$$
$$y = \frac{7}{3}$$

The solution is the ordered pair $\left(\frac{1}{2}, \frac{7}{3}\right)$.

15. $\begin{cases} -2x + y = 4 & (1) \\ -5x + 3y = 7 & (2) \end{cases}$

Multiply both sides of equation (1) by -3 and add the result to equation (2).

$$6x - 3y = -12$$
$$\underline{-5x + 3y = 7}$$
$$x \quad\quad = -5$$

Substituting -5 for x in equation (1), we obtain

$$-2(-5) + y = 4$$
$$10 + y = 4$$
$$y = -6$$

The solution is the ordered pair $(-5, -6)$.

16. $\begin{cases} -3x + 2y = 3 & (1) \\ 4x - 3y = -6 & (2) \end{cases}$

Multiply both sides of equation (1) by 4, multiply both sides of equation (2) by 3, and add the results.

$$-12x + 8y = 12$$
$$\underline{12x - 9y = -18}$$
$$-y = -6$$
$$y = 6$$

Substituting 6 for y in equation (1), we obtain

$$-3x + 2(6) = 3$$
$$-3x + 12 = 3$$
$$-3x = -9$$
$$x = 3$$

The solution is the ordered pair $(3, 6)$.

17. While solving a system of equations, you have eliminated the variable, and a false statement, such as $-8 = 0$, results. This means that the solution of the system is $\varnothing$ or { }.

18. $\begin{cases} -3x + y = 2 & (1) \\ 6x - 2y = 1 & (2) \end{cases}$

Multiply both sides of equation (1) by 2 and add the result to equation (2).

$$-6x + 2y = 4$$
$$\underline{6x - 2y = 1}$$
$$0 = 5$$

The equation $0 = 5$ is false, so the system has no solution. The solution set is $\varnothing$ or { }. The system is inconsistent.

The graphs of the equations (shown below) are parallel, which supports the statement that the system has no solution.

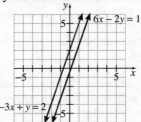

19. When solving a system of equations, you have eliminated the variable, and a true statement, such as $11 = 11$, results. This means that the equations are <u>dependent</u>.

20. $\begin{cases} -3x + 2y = 8 & (1) \\ 6x - 4y = -16 & (2) \end{cases}$

Multiply both sides of equation (1) by 2 and add the results.

$$-6x + 4y = 16$$
$$\underline{6x - 4y = -16}$$
$$0 = 0$$

The equation $0 = 0$ is true, so the system is dependent. The solution is
$$\{(x, y) \mid -3x + 2y = 8\}.$$

The graphs of the equations (shown below) coincide, which supports the statement that the system is dependent.

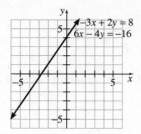

21. $\begin{cases} 2x - 3y = -16 & (1) \\ -3x + 2y = 19 & (2) \end{cases}$

Multiply both sides of equation (1) by 3, multiply both sides of equation (2) by 2, and add the results.

$$6x - 9y = -48$$
$$\underline{-6x + 4y = 38}$$
$$-5y = -10$$
$$y = 2$$

Substituting 2 for y in equation (1), we obtain

$$2x - 3(2) = -16$$
$$2x - 6 = -16$$
$$2x = -10$$
$$x = -5$$

The solution is the ordered pair $(-5, 2)$.
The graphs of the equations (shown below) intersect at the point $(-5, 2)$, which supports the statement that the solution of the system is $(-5, 2)$.

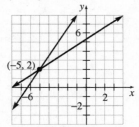

22. $\begin{cases} \dfrac{1}{3}x + \dfrac{1}{5}y = 7 & (1) \\ \dfrac{1}{6}x - \dfrac{2}{5}y = -4 & (2) \end{cases}$

Multiply both sides of equation (1) by 60, multiply both sides of equation (2) by 30, and add the results.

$$20x + 12y = 420$$
$$\underline{5x - 12y = -120}$$
$$25x = 300$$
$$x = 12$$

Substituting 12 for x in equation (1), we obtain

$$\frac{1}{3}(12) + \frac{1}{5}y = 7$$
$$4 + \frac{1}{5}y = 7$$
$$\frac{1}{5}y = 3$$
$$y = 15$$

The solution is the ordered pair $(12, 15)$.

3.1 Exercises

23. a. Let $x = 5$, $y = 3$ in both equations (1) and (2).
$$\begin{cases} 2(5) + 3 = 10 + 3 = 13 \\ -5(5) + 3(3) = -25 + 9 = -16 \neq 6 \end{cases}$$

Although these values satisfy equation (1), they do not satisfy equation (2). Therefore, the ordered pair $(5, 3)$ is not a solution of the system.

b. Let $x = 3$, $y = 7$ in both equations (1) and (2).
$$\begin{cases} 2(3) + 7 = 6 + 7 = 13 \\ -5(3) + 3(7) = -15 + 21 = 6 \end{cases}$$

Because these values satisfy both equations (1) and (2), the ordered pair $(3, 7)$ is a solution of the system.

25. a. Let $x = 1$, $y = 2$ in both equations (1) and (2).
$$\begin{cases} 5(1) + 2(2) = 5 + 4 = 9 \\ -10(1) - 4(2) = -10 - 8 = -18 \end{cases}$$

Because these values satisfy both equations (1) and (2), the ordered pair $(1, 2)$ is a solution of the system.

b. Let $x = 2$, $y = -1/2$ in both equations (1) and (2).
$$\begin{cases} 5(2) + 2(-1/2) = 10 + (-1) = 9 \\ -10(2) - 4(-1/2) = -20 + 2 = -18 \end{cases}$$

Because these values satisfy both equations (1) and (2), the ordered pair $(2, -1/2)$ is a solution of the system.

27. consistent; independent

29. inconsistent

31. $\begin{cases} y = 3x & (1) \\ y = -2x + 5 & (2) \end{cases}$

The two equations are in slope-intercept form. Graph each equation and find the point of intersection.

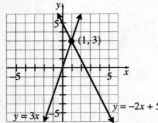

The solution is the ordered pair (1, 3).

33. $\begin{cases} 2x + y = 2 & (1) \\ x + 3y = -9 & (2) \end{cases}$

Equation (1) in slope-intercept form is $y = -2x + 2$. Equation (2) in slope-intercept form is $y = -\frac{1}{3}x - 3$. Graph each equation and find the point of intersection.

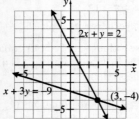

The solution is the ordered pair (3, –4).

35. $\begin{cases} y = -\dfrac{1}{2}x + 1 & (1) \\ y + 2x = 10 & (2) \end{cases}$

Substituting $-\dfrac{1}{2}x + 1$ for y in equation (2), we get

$$-\frac{1}{2}x + 1 + 2x = 10$$
$$2\left(-\frac{1}{2}x + 1 + 2x\right) = 2(10)$$
$$-x + 2 + 4x = 20$$
$$3x = 18$$
$$x = 6$$

Substituting 6 for x in equation (2), we obtain $y = -2(6) + 10 = -12 + 10 = -2$.

The solution is the ordered pair (6, –2).

37. $\begin{cases} x = \dfrac{2}{3}y & (1) \\ 3x - y = -3 & (2) \end{cases}$

Substituting $\dfrac{2}{3}y$ for x in equation (2), we obtain

$$3\left(\frac{2}{3}y\right) - y = -3$$
$$2y - y = -3$$
$$y = -3$$

Substituting –3 for y in equation (1), we obtain

$$x = \frac{2}{3}(-3) = -2 .$$

The solution is the ordered pair $(-2, -3)$.

39. $\begin{cases} 2x - 4y = 2 & (1) \\ x + 2y = 0 & (2) \end{cases}$

Equation (2) solved for x is $x = -2y$. Substituting $-2y$ for x in equation (1), we obtain

$$2(-2y) - 4y = 2$$
$$-4y - 4y = 2$$
$$-8y = 2$$
$$y = \frac{2}{-8} = -\frac{1}{4}$$

Substituting $-\dfrac{1}{4}$ for y in equation (2), we obtain

$$x + 2\left(-\frac{1}{4}\right) = 0$$
$$x - \frac{1}{2} = 0$$
$$x = \frac{1}{2}$$

The solution is the ordered pair $\left(\dfrac{1}{2}, -\dfrac{1}{4}\right)$.

41. $\begin{cases} x + y = 10,000 & (1) \\ 0.05x + 0.07y = 650 & (2) \end{cases}$

Equation (1) solved for x is $x = 10,000 - y$. Substituting $10,000 - y$ for x in equation (2), we obtain

$$0.05(10,000 - y) + 0.07y = 650$$
$$500 - 0.05y + 0.07y = 650$$
$$0.02y + 500 = 650$$
$$0.02y = 150$$
$$y = 7500$$

Substituting 7500 for y in equation (1), we obtain

$$x + 7500 = 10{,}000$$
$$x = 2500$$
The solution is the ordered pair (2500, 7500).

43. $\begin{cases} x + y = -5 & (1) \\ -x + 2y = 14 & (2) \end{cases}$

Add equations (1) and (2).

$$\begin{cases} \;\;\;x + y = -5 \\ -x + 2y = 14 \end{cases}$$
$$\overline{3y = 9}$$
$$y = 3$$

Substituting 3 for y in equation (1), we obtain
$$x + 3 = -5$$
$$x = -8$$
The solution is the ordered pair (–8, 3).

45. $\begin{cases} x + 2y = -5 & (1) \\ 3x + 3y = 9 & (2) \end{cases}$

Multiply both sides of equation (1) by –3, and add the result to equation (2).

$$\begin{cases} -3x - 6y = 15 \\ \;\;\;3x + 3y = 9 \end{cases}$$
$$\overline{-3y = 24}$$
$$y = -8$$

Substituting –8 for y in equation (1), we obtain
$$x + 2(-8) = -5$$
$$x - 16 = -5$$
$$x = 11$$
The solution is the ordered pair (11, –8).

47. $\begin{cases} 2x + 5y = -3 & (1) \\ x + \dfrac{5}{4}y = -\dfrac{1}{2} & (2) \end{cases}$

Multiply both sides of equation (2) by –4 and add the result to equation (1).

$$\begin{cases} 2x + 5y = -3 \\ -4x - 5y = 2 \end{cases}$$
$$\overline{-2x \quad\;\; = -1}$$
$$x = \frac{1}{2}$$

Substituting $1/2$ for x in equation (1), we obtain

$$2\left(\frac{1}{2}\right) + 5y = -3$$
$$1 + 5y = -3$$
$$5y = -4$$
$$y = -\frac{4}{5}$$

The solution is the ordered pair $\left(\dfrac{1}{2}, -\dfrac{4}{5}\right)$.

49. $\begin{cases} 0.05x + 0.1y = 5.25 & (1) \\ 0.08x - 0.02y = 1.2 & (2) \end{cases}$

Multiply both sides of equation (1) by 0.2, and add the result to equation (2).

$$\begin{cases} 0.01x + 0.02y = 1.05 \\ 0.08x - 0.02y = 1.2 \end{cases}$$
$$\overline{0.09x = 2.25}$$
$$x = 25$$

Substituting 25 for x in equation (1), we obtain
$$0.05(25) + 0.1y = 5.25$$
$$1.25 + 0.1y = 5.25$$
$$0.1y = 4$$
$$y = 40$$
The solution is the ordered pair (25, 40).

51. $\begin{cases} 3x + y = 1 & (1) \\ -6x - 2y = -4 & (2) \end{cases}$

Equation (1) solved for y is $y = -3x + 1$.
Substituting $-3x + 1$ for y in equation (2), we obtain

$$-6x - 2(-3x + 1) = -4$$
$$-6x + 6x - 2 = -4$$
$$-2 = -4$$

The system has no solution. The solution set is $\varnothing$ or $\{\ \}$. The system is inconsistent.

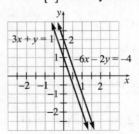

53. $\begin{cases} 5x - 2y = 2 & (1) \\ -10x + 4y = 3 & (2) \end{cases}$

Multiply both sides of equation (1) by 2 and add the result to equation (2).

$$\begin{cases} 10x - 4y = 4 \\ \underline{-10x + 4y = 3} \\ 0 = 7 \end{cases}$$

The system has no solution. The solution set is
$\varnothing$ or $\{\ \}$. The system is inconsistent.

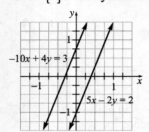

55. $\begin{cases} y = \dfrac{1}{2}x + 1 & (1) \\ 2x - 4y = -4 & (2) \end{cases}$

Substituting $\dfrac{1}{2}x + 1$ for y in equation (2), we

obtain

$$2x - 4\left(\dfrac{1}{2}x + 1\right) = -4$$
$$2x - 2x - 4 = -4$$
$$-4 = -4$$

The system is dependent. The solution is
$\{(x, y) \mid 2x - 4y = -4\}$.

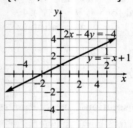

57. $\begin{cases} x + 3y = 6 & (1) \\ -\dfrac{x}{3} - y = -2 & (2) \end{cases}$

Equation (2) solved for y is $y = -\dfrac{x}{3} + 2$.

Substituting $-\dfrac{x}{3} + 2$ for y in equation (1), we

obtain

$$x + 3\left(-\dfrac{x}{3} + 2\right) = 6$$
$$x - x + 6 = 6$$
$$6 = 6$$

The system is dependent. The solution is

$\{(x, y) \mid x + 3y = 6\}$.

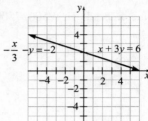

59. $\begin{cases} \dfrac{1}{3}x - 2y = 6 & (1) \\ -\dfrac{1}{2}x + 3y = -9 & (2) \end{cases}$

Multiply both sides of equation (1) by 3,
multiply both sides of equation (2) by 2, and add
the results.

$$\begin{array}{r} x - 6y = 18 \\ \underline{-x + 6y = -18} \\ 0 = 0 \end{array}$$

The system is dependent. The solution is

$\left\{(x, y) \mid \dfrac{1}{3}x - 2y = 6\right\}$.

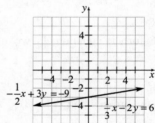

61. $\begin{cases} x + 3y = 0 & (1) \\ -2x + 4y = 30 & (2) \end{cases}$

Because the coefficient of x in equation (1) is 1,
we use substitution to solve the system.
Equation (1) solved for x is $x = -3y$.
Substituting $-3y$ for x in equation (2), we

obtain

$$-2(-3y) + 4y = 30$$
$$6y + 4y = 30$$
$$10y = 30$$
$$y = 3$$

Substituting 3 for y in equation (1), we obtain
$$x + 3(3) = 0$$
$$x + 9 = 0 \quad .$$
$$x = -9$$

The solution is the ordered pair $(-9, 3)$.

Copyright © 2014 Pearson Education, Inc.

63. $\begin{cases} x = 5y - 3 & (1) \\ -3x + 15y = 9 & (2) \end{cases}$

Because equation (1) is already solved for x, we use substitution to solve the system.

Substituting $5y - 3$ for x in equation (2), we obtain

$$-3(5y - 3) + 15y = 9$$
$$-15y + 9 + 15y = 9$$
$$9 = 9$$

The system is dependent. The solution is
$\{(x, y) \mid x = 5y - 3\}$.

65. $\begin{cases} 2x - 4y = 18 & (1) \\ 3x + 5y = -3 & (2) \end{cases}$

Because none of variables have a coefficient of 1, we use elimination to solve the system. Multiply both sides of equation (1) by 3, multiply both sides of equation (2) by -2, and add the results.

$$\begin{cases} 6x - 12y = 54 \\ -6x - 10y = 6 \end{cases}$$
$$-22y = 60$$
$$y = \frac{60}{-22} = -\frac{30}{11}$$

Substituting $-\dfrac{30}{11}$ for y in equation (1), we obtain

$$2x - 4\left(-\frac{30}{11}\right) = 18$$
$$2x + \frac{120}{11} = 18$$
$$11\left(2x + \frac{120}{11}\right) = 11(18)$$
$$22x + 120 = 198$$
$$22x = 78$$
$$x = \frac{78}{22} = \frac{39}{11}$$

The solution is the ordered pair $\left(\dfrac{39}{11}, -\dfrac{30}{11}\right)$.

67. $\begin{cases} \dfrac{5}{6}x - \dfrac{1}{3}y = -5 & (1) \\ -x + \dfrac{2}{5}y = 1 & (2) \end{cases}$

Because none of the variables have a coefficient of 1, we use elimination to solve the system. Multiply both sides of equation (1) by 6, multiply both sides of equation (2) by 5, and add the results.

$$\begin{cases} 5x - 2y = -30 \\ -5x + 2y = 5 \end{cases}$$
$$0 = -25$$

The system has no solution. The solution set is $\varnothing$ or $\{\ \}$. The system is inconsistent.

69. $\begin{cases} 2x + y = -5 & (1) \\ 5x + 3y = 1 & (2) \end{cases}$

Write each equation in slope-intercept form.

$$\begin{array}{ll} 2x + y = -5 & 5x + 3y = 1 \\ y = -2x - 5 & 3y = -5x + 1 \\ & y = \dfrac{-5x + 1}{3} \\ & y = -\dfrac{5}{3}x + \dfrac{1}{3} \end{array}$$

Since the equations have different slopes, the lines are neither parallel nor coincident. Thus, the system must have exactly one solution.

71. $\begin{cases} 3x - 2y = -2 & (1) \\ -6x + 4y = 4 & (2) \end{cases}$

Write each equation in slope-intercept form.

$$\begin{array}{ll} 3x - 2y = -2 & -6x + 4y = 4 \\ -2y = -3x - 2 & 4y = 6x + 4 \\ y = \dfrac{-3x - 2}{-2} & y = \dfrac{6x + 4}{4} \\ y = \dfrac{3}{2}x + 1 & y = \dfrac{3}{2}x + 1 \end{array}$$

Because both equations have the same slope-intercept form, the two lines are coincident. Thus, the system is dependent and has an infinite number of solutions.

73. a. Equation of line through $(-1, 3)$ and $(3, 1)$:

$$m = \frac{1 - 3}{3 - (-1)} = \frac{-2}{4} = -\frac{1}{2}$$
$$y - 1 = -\frac{1}{2}(x - 3)$$
$$y - 1 = -\frac{1}{2}x + \frac{3}{2}$$
$$y = -\frac{1}{2}x + \frac{5}{2}$$

Equation of line through $(-2, -1)$ and $(4, 5)$:

$$m = \frac{y_2 - y_1}{x_2 - x_1} = \frac{5 - (-1)}{4 - (-2)} = \frac{6}{6} = 1$$
$$y - 5 = 1(x - 4)$$
$$y - 5 = x - 4$$
$$y = x + 1$$

b. Solve the system formed by the equations found in part a.

$$\begin{cases} y = x+1 & (1) \\ y = -\dfrac{1}{2}x + \dfrac{5}{2} & (2) \end{cases}$$

Substituting $x+1$ for y in equation (2), we obtain

$$x+1 = -\frac{1}{2}x + \frac{5}{2}$$

$$2(x+1) = 2\left(-\frac{1}{2}x + \frac{5}{2}\right)$$

$$2x+2 = -x+5$$

$$3x = 3$$

$$x = 1$$

Substituting 1 for x in equation (1), we obtain $y = 1 + 1 = 2$.

The solution is the ordered pair (1, 2).

75. The two lines intersect in the second quadrant. Of the choices provided, the ordered pairs $(-3, 1)$ and $(-1, 3)$ are the only ones that are in the second quadrant. Therefore, the answers are (c) and (f).

77. $\begin{cases} Ax + 3By = 2 \\ -3Ax + By = -11 \end{cases}$

Substitute 3 for x and 1 for y in the system and solve for A and B.

$$\begin{cases} A(3) + 3B(1) = 2 & \text{or} \quad 3A + 3B = 2 & (1) \\ -3A(3) + B(1) = -11 & \text{or} \quad -9A + B = -11 & (2) \end{cases}$$

Multiply equation (1) by 3 and add the result to equation (2).

$$9A + 9B = 6$$
$$-9A + B = -11$$
$$10B = -5$$

$$B = \frac{-5}{10} = -\frac{1}{2}$$

Substituting $-\dfrac{1}{2}$ for B in equation (2), we obtain

$$-9A + \left(-\frac{1}{2}\right) = -11$$

$$-2\left(-9A - \frac{1}{2}\right) = -2(-11)$$

$$18A + 1 = 22$$

$$18A = 21$$

$$A = \frac{21}{18} = \frac{7}{6}$$

Thus, for the system to have $x = 3$, $y = 1$ as a solution, then $A = \dfrac{7}{6}$ and $B = -\dfrac{1}{2}$.

79. Answers will vary. One possibility follows:

$$\begin{cases} x + y = 3 \\ x - y = -5 \end{cases}$$

81. $\begin{cases} 3x + y = 5 & (1) \\ x + y = 3 & (2) \\ x + 3y = 7 & (3) \end{cases}$

Multiply both sides of equation (2) by -1 and add the result to equation (1).

$$\begin{cases} 3x + y = 5 \\ -x - y = -3 \end{cases}$$
$$\overline{ \quad 2x = 2}$$
$$x = 1$$

Substituting 1 for x in equation (1), we obtain

$$3(1) + y = 5$$
$$3 + y = 5$$
$$y = 2$$

The point (1, 2) is a solution to the first two equations. Check to see if it is a solution to equation (3).

$$x + 3y = 7$$
$$(1) + 3(2) \overset{?}{=} 7$$
$$1 + 6 = 7 \checkmark$$

The solution of the system is the ordered pair (1, 2).

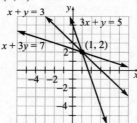

83. $\begin{cases} y = \dfrac{2}{3}x - 5 & (1) \\ 4x - 6y = 30 & (2) \\ x - 5y = 11 & (3) \end{cases}$

Substituting $\dfrac{2}{3}x - 5$ for y in equation (2), we obtain

$$4x - 6\left(\frac{2}{3}x - 5\right) = 30$$

$$4x - 4x + 30 = 30$$

$$30 = 30$$

These two equations form a dependent system.

Substituting $\dfrac{2}{3}x - 5$ for y in equation (3), we obtain

$$x - 5\left(\frac{2}{3}x - 5\right) = 11$$

$$x - \frac{10}{3}x + 25 = 11$$

$$-\frac{7}{3}x = -14$$

$$x = 6$$

Substituting 6 for x in equation (1), we obtain

$$y = \frac{2}{3}(6) - 5$$

$$y = 4 - 5$$

$$y = -1$$

The solution of the system is the ordered pair $(6, -1)$.

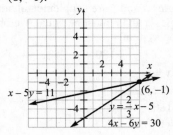

85. Yes. The method of elimination is preferred over the method of substitution if substitution leads to fractions.

87. The lines intersect at the point $(3, -2)$.

89. $\begin{cases} y = 3x - 1 & (1) \\ y = -2x + 5 & (2) \end{cases}$

The solution is the ordered pair $(1.2, \ 2.6)$.

91. $\begin{cases} 3x - y = -1 & (1) \\ -4x + y = -3 & (2) \end{cases}$

Writing each equation in slope-intercept form, we obtain $y = 3x + 1$ and $y = 4x - 3$.

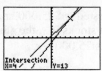

The solution is the ordered pair $(4, 13)$.

93. $\begin{cases} 4x - 3y = 1 & (1) \\ -8x + 6y = -2 & (2) \end{cases}$

Writing each equation in slope-intercept form, we obtain the same equation for both,

$$y = \frac{4}{3}x - \frac{1}{3}.$$

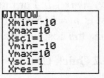

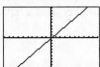

The two lines are coincident. The system is dependent. The solution is $\{(x, y) \mid 4x - 3y = 1\}$.

95. $\begin{cases} 2x - 3y = 12 & (1) \\ 5x + y = -2 & (2) \end{cases}$

Writing each equation in slope-intercept form,

we obtain $y = \frac{2}{3}x - 4$ and $y = -5x - 2$.

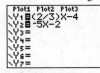

The solution is approximately the ordered pair $(0.35, -3.76)$.

Section 3.2

Are You Ready for This Section?

R1. Step 1: Identify what you are looking for.
Step 2: Give names to the unknowns.
Step 3: Translate the problem into the language of mathematics.
Step 4: Solve the equation(s) found in Step 3.
Step 5: Check the reasonableness of your answer.
Step 6: Answer the question.

R2. Since the total to be invested is $25,000 and the amount to be invested in stocks is s, the amount to be invested in bonds is $25,000 - s$.

R3. The total interest for the year would be $0.125(\$3500) = \$437.50$. Since there are 12 months in a year, the interest that would be paid after one month would be

$$\frac{\$437.50}{12} \approx \$36.46.$$

R4. A linear cost function will be of the form $C(x) = mx + b$. For this problem, the variable cost is $m = \$15$ and the fixed cost is $b = \$500$. Thus, the cost function is $C(x) = 15x + 500$.

Section 3.2 Quick Checks

1. Let x represent the first number, and y represent the second number.

$$\begin{cases} 2x + 3y = 9 & (1) \\ x - y = 22 & (2) \end{cases}$$

Multiply equation (2) by 3 and add the result to equation (1).

$$\begin{aligned} 2x + 3y &= 9 \\ 3x - 3y &= 66 \\ \hline 5x &= 75 \\ x &= 15 \end{aligned}$$

Substituting 15 for x in equation (2), we obtain

$$\begin{aligned} x - y &= 22 \\ 15 - y &= 22 \\ y &= -7 \end{aligned}$$

The two numbers are 15 and -7.

2. Let c represent the cost of a cheeseburger and let s represent the cost of a medium shake.

$$\begin{cases} 4c + 2s = 10.10 & (1) \\ 3c + 3s = 10.35 & (2) \end{cases}$$

Multiply both sides of equation (1) by 3, multiply both sides of equation (2) by -2, and add the results.

$$\begin{aligned} 12c + 6s &= 30.30 \\ -6c - 6s &= -20.70 \\ \hline 6c &= 9.60 \\ c &= 1.60 \end{aligned}$$

Substituting 1.60 for c in equation (1), we obtain

$$\begin{aligned} 4(1.60) + 2s &= 10.10 \\ 6.40 + 2s &= 10.10 \\ 2s &= 3.70 \\ s &= 1.85 \end{aligned}$$

A cheeseburger costs $1.60. A medium shake costs $1.85.

3. Let x and y represent the length and width of the rectangle, respectively.

$$\begin{cases} 2x + 2y = 360 & (1) \\ x = 2y & (2) \end{cases}$$

Substituting $2y$ for x in equation (1), we obtain

$$\begin{aligned} 2(2y) + 2y &= 360 \\ 4y + 2y &= 360 \\ 6y &= 360 \\ y &= 60 \end{aligned}$$

Substituting 60 for y in equation (2), we obtain $x = 2(60) = 120$.
The length is 120 yards, and the width is 60 yards.

4. Because angles a and c are supplemental, and because angles a and e must be equal, we obtain the following system

$$\begin{cases} (x + 3y) + (3x + y) = 180 \\ x + 3y = 5x + y \end{cases}$$

Simplify each equation.

$$\begin{cases} 4x + 4y = 180 & (1) \\ -4x + 2y = 0 & (2) \end{cases}$$

Add the two equations.

$$\begin{aligned} 4x + 4y &= 180 \\ -4x + 2y &= 0 \\ \hline 6y &= 180 \\ y &= 30 \end{aligned}$$

Substituting 30 for y in equation (2), we obtain

$$\begin{aligned} -4x + 2(30) &= 0 \\ -4x + 60 &= 0 \\ -4x &= -60 \\ x &= 15 \end{aligned}$$

Thus, $x = 15$ and $y = 30$.

5. Let a represent the amount invested in the Aaa-rated bond, and let b represent the amount invested in the B-rated bond. We organize the information in the table below, using the interest formula: Interest = Principal × rate × time.

	Principal $	Rate %	Time Yr	Interest $
Aaa-rated bond	a	0.05	1	0.05a
B-rated bond	b	0.10	1	0.10b
Total	120,000			7200

Using the columns for Principal and Interest, we obtain the following system.

$$\begin{cases} a + b = 120,000 & (1) \\ 0.05a + 0.10b = 7200 & (2) \end{cases}$$

Solving equation (1) for b, we obtain $b = 120,000 - a$. Substituting $120,000 - a$ for b in equation (2), we obtain

$$0.05a + 0.10(120,000 - a) = 7200$$
$$0.05a + 12000 - 0.10a = 7200$$
$$-0.05a + 12000 = 7200$$
$$-0.05a = -4800$$
$$a = 96,000$$

Substituting 96,000 for a into equation (1), we obtain $96,000 + b = 120,000$

$$b = 24,000$$

Maria should invest $96,000 in the Aaa-rated bond and $24,000 in the B-rated bond.

6. Let c represent the weight of cashews to be used and let p represent the weight of peanuts to be used. We organize the information in the table below, using the formula
Price per pound × Number of pounds = Revenue.

	Price per pound	Number of pounds	Revenue
Cashews	7.00	c	$7c$
Peanuts	2.50	p	$2.5p$
Trail mix	4.00	30	$4(30) = 120$

Using the columns for Number of pounds and Revenue, we obtain the following system.

$$\begin{cases} c + p = 30 & (1) \\ 7c + 2.5p = 120 & (2) \end{cases}$$

Solving equation (1) for c yields $c = 30 - p$. Substituting $30 - p$ for c in equation (2), we obtain

$$7(30 - p) + 2.5p = 120$$
$$210 - 7p + 2.5p = 120$$
$$210 - 4.5p = 120$$
$$-4.5p = -90$$
$$p = 20$$

Substituting 20 for p in equation (1) yields

$$c + 20 = 30$$
$$c = 10$$

The trail mix requires 10 pounds of cashews and 20 pounds of peanuts.

7. Let a represent the airspeed of the plane, and let w represent the impact of wind resistance on the plane. We organize the information in the table below.

	Distance	Rate	Time
Against Wind (West)	1200	$a - w$	4
With Wind (East)	1200	$a + w$	3

Using the formula Distance = Rate × Time, we obtain the following system.

$$\begin{cases} 1200 = 4(a - w) \\ 1200 = 3(a + w) \end{cases}$$

Simplify each equation.

$$\begin{cases} 4a - 4w = 1200 & (1) \\ 3a + 3w = 1200 & (2) \end{cases}$$

Multiply both sides of equation (1) by 3, multiply both sides of equation (2) by 4, and add the results

$$12a - 12w = 3600$$
$$12a + 12w = 4800$$
$$24a \qquad\quad = 8400$$
$$a = 350$$

Substituting 350 for a in equation (2) yields

$$3(350) + 3w = 1200$$
$$1050 + 3w = 1200$$
$$3w = 150$$
$$w = 50$$

The airspeed of the plain is 350 miles per hour. The impact of wind resistance on the plane is 50 miles per hour.

8. a. If one Austrian Pine tree is sold, revenue will be $230. If two trees are sold, revenue will be $230(2) = \$460$. If x trees are sold, revenue will be $230x. The revenue function is $R(x) = 230x$.

   b. The cost function is $C(x) = ax + b$ where a is the variable cost and b is the fixed cost. With $a = 160$ and $b = 2100$, the cost function is $C(x) = 160x + 2100$.

c.

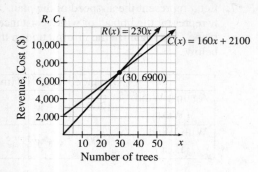

d. $R(x) = C(x)$

$230x = 160x + 2100$

$70x = 2100$

$x = 30$

$R(30) = 230(30) = 6900$

$C(30) = 160(30) + 2100 = 6900$

The break-even point is $(30, \ 6900)$.

If 30 Austrian Pine trees are sold, the cost and revenue will be even at $6900.

3.2 Exercises

9. Let x and y represent the two numbers.

$$\begin{cases} x + y = 18 & (1) \\ x - y = -2 & (2) \end{cases}$$

Adding equations (1) and (2) result in

$2x = 16$

$x = 8$

Substituting 8 for x in equation (1), we obtain

$8 + y = 18$

$y = 10$

The two numbers are 8 and 10.

11. Let x represent the first number and let y represent the second number.

$$\begin{cases} 2x + y = 47 & (1) \\ x + 3y = 81 & (2) \end{cases}$$

Multiply both sides of equation (2) by -2 and add the result to equation (1).

$$\begin{cases} 2x + y = 47 \\ \underline{-2x - 6y = -162} \end{cases}$$

$-5y = -115$

$y = 23$

Substituting 23 for y in equation (2), we obtain

$x + 3(23) = 81$

$x + 69 = 81$

$x = 12$

The 1st number is 12 and the 2nd number is 23.

13. Let x represent the flat set-up fee and let y represent the hourly rental rate.

$$\begin{cases} x + 5y = 235 & (1) \\ x + 3y = 165 & (2) \end{cases}$$

Multiply both sides of equation (2) by -1 and add the result to equation (1).

$$\begin{cases} x + 5y = 235 \\ \underline{-x - 3y = -165} \end{cases}$$

$2y = 70$

$y = 35$

Substituting 35 for y in equation (2), we obtain

$x + 3(35) = 165$

$x + 105 = 165$

$x = 60$

The flat set-up fee is $60 and the rental rate is $35 per hour.

15. Let x and y represent the length and width of the rectangle, respectively.

$$\begin{cases} 2x + 2y = 120 & (1) \\ x = y + 20 & (2) \end{cases}$$

Substituting $y + 20$ for x in equation (1), we obtain

$2(y + 20) + 2y = 120$

$2y + 40 + 2y = 120$

$4y + 40 = 120$

$4y = 80$

$y = 20$

Substituting 20 for y in equation (2), we obtain

$x = 20 + 20 = 40$.

The dimensions of the rectangle are 40 meters by 20 meters.

17. Because angles 1 and 2 are supplemental, and because angles 2 and 3 must be equal, we obtain the following system.

$$\begin{cases} (10x + 6y) + 4y = 180 & (1) \\ 4y = 7x + 2y & (2) \end{cases}$$

Simplify each equation.

$$\begin{cases} 10x + 10y = 180 & (1) \\ 7x - 2y = 0 & (2) \end{cases}$$

Divide both sides of equation (1) by 5 and add the result to equation (2).

$$\begin{cases} 2x + 2y = 36 \\ \underline{7x - 2y = 0} \end{cases}$$

$9x = 36$

$x = 4$

Substituting 4 for x in equation (2), we obtain
$$7(4)-2y=0$$
$$28-2y=0$$
$$-2y=-28$$
$$y=14$$
Thus, $x=4$ and $y=14$.

19. Let s represent the amount to be invested in stocks and let b represent the amount to be invested in bonds.
$$\begin{cases} s+b=36{,}000 & (1) \\ 5b=3s & (2) \end{cases}$$
Write equation (2) in standard form.
$$\begin{cases} s+b=36{,}000 & (1) \\ -3s+5b=0 & (2) \end{cases}$$
Multiply both sides of equation (1) by 3 and add the result to equation (2).
$$\begin{cases} 3s+3b=108{,}000 \\ -3s+5b=0 \end{cases}$$
$$8b=108{,}000$$
$$b=13{,}500$$
Substituting 13,500 for b in equation (1), we obtain
$$s+13{,}500=36{,}000$$
$$s=22{,}500$$
Invest \$22,500 in stocks and \$13,500 in bonds.

21. Let a represent the weight of Arabica beans to be used and let r represent the weight of African Robustas beans to be used. We organize the information in the table below, using the formula Price per pound $\times$ number of pounds = Revenue.

	Price per pound	Number of pounds	Revenue
Arabica	9.00	a	$9a$
African Robustas	11.50	r	$11.5r$
Blend	10.00	100	$10(100)=1000$

Using the columns for Number of pounds and Revenue, we obtain the following system.
$$\begin{cases} a+r=100 & (1) \\ 9a+11.5r=1000 & (2) \end{cases}$$
Solving equation (1) for a yields $a=100-r$. Substituting $100-r$ for a in equation (2), we

obtain $9(100-r)+11.5r=1000$
$$900-9r+11.5r=1000$$
$$900+2.5r=1000$$
$$2.5r=100$$
$$r=40$$
Substituting 40 for r in equation (1) yields
$$a+40=100$$
$$a=60$$
The blend requires 60 pounds of Arabica beans and 40 pounds of African Robustas beans.

23. Let x represent the speed at which Jonathon and Samantha can row in still water and let y represent the speed of the current. The rate at which the pair will travel downstream will be $x+y$, and the rate upstream will be $x-y$. Using the formula rate $\times$ time = distance, we obtain the following system:
$$\begin{cases} (x+y)\cdot 2=26 & (1) \\ (x-y)\cdot 3=9 & (2) \end{cases}$$
Divide both sides of equation (1) by 2, divide both sides of equation (2) by 3, and add the results.
$$x+y=13$$
$$x-y=3$$
$$2x=16$$
$$x=8$$
Jonathon and Samantha can row at a rate of 8 miles per hour in still water.

25. Let t represent the time for both cars. Let d represent the distance traveled by the Lincoln. Then the distance traveled by the Infiniti is $d+100$. We obtain the following system by utilizing the formula rate $\times$ time = distance :
$$\begin{cases} 40t=d & (1) \\ 50t=d+100 & (2) \end{cases}$$
Substituting $40t$ for d in equation (2) yields
$$50t=40t+100$$
$$10t=100$$
$$t=10$$
Substituting 10 for t in equation (1), we obtain
$$40(10)=d$$
$$400=d$$
The distance traveled by the Lincoln was 400 miles. The distance traveled by the Infiniti is $400+100=500$ miles. The time for both trips was 10 hours each.

27. a. $\begin{cases} R(x) = 12x \\ C(x) = 5.5x + 9880 \end{cases}$

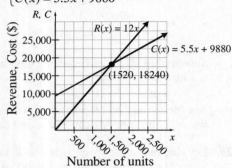

b. $R(x) = C(x)$

$12x = 5.5x + 9880$

$6.5x = 9880$

$x = 1520$

$R(1520) = 12(1520) = 18,240$

$C(1520) = 5.5(1520) + 9880 = 18,240$

The break-even point is (1520, 18,240).

29. Let x represent the length of one of the congruent sides and let y represent the length of the third side.

$\begin{cases} 2x + y = 35 & (1) \\ x = 3y & (2) \end{cases}$

Substituting $3y$ for x in equation (1), we obtain

$2(3y) + y = 35$

$6y + y = 35$

$7y = 35$

$y = 5$

Substituting 5 for y in equation (2), we obtain

$x = 3(5) = 15.$

The lengths of the three sides of the triangle are 15 cm, 15 cm, and 5 cm.

31. Let x represent the amount loaned at 5% per year and let y represent the amount loaned at 8% per year. We organize the information in the table below, using the interest formula:

Interest = Principal × rate × time or $I = Prt$.

	P	r	t	I
5% loan	x	0.05	1	$0.05x$
8% loan	y	0.08	1	$0.08y$
Total	750,000		1	52,500

Using the columns for Principal and Interest, we obtain the following system.

$\begin{cases} x + y = 750,000 & (1) \\ 0.05x + 0.08y = 52,500 & (2) \end{cases}$

Solving equation (1) for y, we obtain

$y = 750,000 - x$. Substituting $750,000 - x$ for y in equation (2), we obtain

$0.05x + 0.08(750,000 - x) = 52,500$

$0.05x + 60,000 - 0.08x = 52,500$

$-0.03x + 60,000 = 52,500$

$-0.03x = -7500$

$x = 250,000$

The bank loaned $250,000 at 5%.

33. Let x represent the first number and let y represent the second number.

$\begin{cases} 2x + 3y = 81 & (1) \\ 3x - y = 17 & (2) \end{cases}$

Multiply both sides of equation (2) by 3 and add the result to equation (1).

$\begin{cases} 2x + 3y = 81 \\ 9x - 3y = 51 \end{cases}$

$11x \quad = 132$

$x = 12$

Substituting 12 for x in equation (2), we obtain

$3(12) - y = 17$

$36 - y = 17$

$-y = -19$

$y = 19$

The 1st number is 12 and the 2nd number is 19.

35. Let r_n represent the normal walking speed of Drenell. Let r_w represent the speed of the walkway.

	d	r	t
with the walkway	126	$r_n + r_w$	18
against the walkway	126	$r_n - r_w$	63

$\begin{cases} (r_n + r_w) \cdot 18 = 126 \\ (r_n - r_w) \cdot 63 = 126 \end{cases}$

Expanding these equations yields the system to be solved.

$\begin{cases} 18r_n + 18r_w = 126 & (1) \\ 63r_n - 63r_w = 126 & (2) \end{cases}$

Multiply equation (1) by 7 and multiply equation (2) by 2. Add the results.

$$126r_n + 126r_w = 882$$
$$\underline{126r_n - 126r_w = 252}$$
$$252r_n \qquad\quad = 1134$$
$$r_n = 4.5$$

Note the units of measurement for the rate is feet per second. Drenell's normal walking speed is 4.5 fps.

37. Because angles 1, 2, and 3 add up to $180°$, and because angles 2 and 4 are supplemental, we obtain the following system.
$$\begin{cases} (4x+2y)+(10x+5y)+(7x+5y)=180 & (1) \\ (10x+5y)+(9x+8y)=180 & (2) \end{cases}$$

Simplify each equation.
$$\begin{cases} 21x+12y=180 & (1) \\ 19x+13y=180 & (2) \end{cases}$$

Multiply both sides of equation (1) by -19, multiply both sides of equation (2) by 21, and add the results.
$$\begin{cases} -399x-228y=-3420 \\ 399x+273y=3780 \end{cases}$$
$$\overline{45y=360}$$
$$y=8$$

Substituting 8 for y in equation (2), we obtain
$$19x+13(8)=180$$
$$19x+104=180$$
$$19x=76$$
$$x=4$$
Thus, $x=4$ and $y=8$.

39. Let x represent the number of milligrams of the first liquid and let y represent the number of milligrams of the second liquid to be used. Then the amount of vitamin C in the two liquids is $0.20x+0.50y$. The amount of vitamin D in the two liquids is $0.30x+0.25y$. This results in the following system:
$$\begin{cases} 0.20x+0.50y=40 & (1) \\ 0.30x+0.25y=30 & (2) \end{cases}$$

Multiply both sides of equation (1) by 2, multiply both sides of equation (2) by -4, and add the results.
$$\begin{cases} 0.4x+y=80 \\ -1.2x-y=-120 \end{cases}$$
$$\overline{-0.8x \qquad\; = -40}$$
$$x=50$$
Substituting 50 for x in equation (1), we obtain

$$0.20(50)+0.50y=40$$
$$10+0.50y=40$$
$$0.50y=30$$
$$y=60$$

To fill the prescription, 50 milligrams of the first liquid and 60 milligrams of the second liquid must be mixed.

41. Let h represent the number of calories in an order of chicken nuggets and let c represent the number of calories in a medium Coke.
$$\begin{cases} 2h+c=770 & (1) \\ 3h+2c=1260 & (2) \end{cases}$$

Multiply both sides of equation (1) by -2 and add the result to equation (2).
$$\begin{cases} -4h-2c=-1540 \\ 3h+2c=1260 \end{cases}$$
$$\overline{-h \qquad\quad =-280}$$
$$h=280$$

Substituting 280 for h in equation (1), we obtain
$$2(280)+c=770$$
$$560+c=770$$
$$c=210$$
An order of chicken nuggets has 280 calories. A medium Coke has 210 calories.

43. Let a represent the amount of the 10% alloy and b represent the amount of the 25% alloy. We organize the information in the table below, using the formula

% titanium × amount alloy = amount titanium

	%	Amt alloy	Amt titanium
10% alloy	0.10	a	$0.1a$
25% alloy	0.25	b	$0.25b$
Blend (16%)	0.16	100 g	$0.16\cdot100=16$

Using the columns for the amount of alloy and amount of titanium, we obtain the following system.
$$\begin{cases} a+b=100 & (1) \\ 0.1a+0.25b=16 & (2) \end{cases}$$

Solving equation (1) for a yields $a=100-b$. Substituting $100-b$ for a in equation (2), we obtain
$$0.1(100-b)+0.25b=16$$
$$10-0.1b+0.25b=16$$
$$0.15b=6$$
$$b=40$$

Substituting 40 for b in equation (1) yields
$$a + 40 = 100$$
$$a = 60$$
The blend requires 60 grams of the 10% alloy and 40 grams of the 25% alloy.

45. a.
$$\begin{cases} S(p) = -2000 + 3000p \\ D(p) = 10,000 - 1000p \end{cases}$$

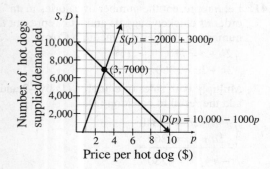

b.
$$S(p) = D(p)$$
$$-2000 + 3000p = 10,000 - 1000p$$
$$-2000 + 4000p = 10,000$$
$$4000p = 12,000$$
$$p = 3$$

$$S(3) = -2000 + 3000(3) = 7000$$
$$D(3) = 10,000 - 1000(3) = 7000$$

The equilibrium price is $3. The equilibrium quantity is 7000 hot dogs.

47. a.
$$\begin{cases} M(t) = 0.25t + 14.4 \\ F(t) = 0.29t + 12 \end{cases}$$

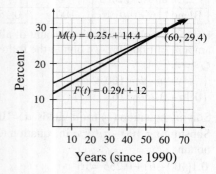

b.
$$M(t) = F(t)$$
$$0.25t + 14.4 = 0.29t + 12$$
$$2.4 = 0.04t$$
$$60 = t$$
The percent of male and female college graduates will be equal approximately

60 years after 1990; that is in approximately 2050.
$$M(60) = 0.25(60) + 14.4 = 29.4$$
$$F(60) = 0.29(60) + 12 = 29.4$$
The percent will be approximately 29%.

49. a. Let x represent the amount of sales, let A represent the pay from option A, and let B represent the pay from option B.
$$\begin{cases} A(x) = 15,000 + 0.01x \\ B(x) = 25,000 + 0.0075x \end{cases}$$

b.
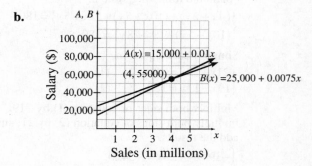

c.
$$A(x) = B(x)$$
$$15,000 + 0.01x = 25,000 + 0.0075x$$
$$0.0025x = 10,000$$
$$x = 4,000,000$$
$4,000,000 of sales is required for the two options to result in the same salary is.

$$A(4,000,000) = 15,000 + 0.01(4,000,000)$$
$$= 55,000$$
$$B(4,000,000) = 25,000 + 0.0075(4,000,000)$$
$$= 55,000$$
The annual salary would be $55,000.

51. a. $R(x) = 60x$

b. $C(x) = 3500 + 35x$

c.
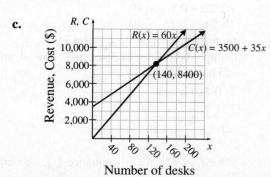

   d. $R(x) = C(x)$

$$60x = 3500 + 35x$$
$$25x = 3500$$
$$x = 140$$

The number of desks at which the revenue equals cost is 140 desks.

$$R(140) = 60(140) = 8400$$
$$C(140) = 3500 + 35(140) = 8400$$

The cost and revenue for producing and selling 140 desks will both be $8400.

53. a. and **b.**   Let x represent the number of years since 1968.

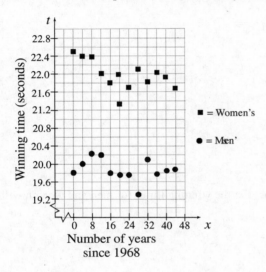

Number of years
since 1968

   c.   Answers may vary. We use the points $(4, 20.00)$ and $(36, 19.79)$.

$$m = \frac{19.79 - 20.00}{36 - 4} = \frac{-0.21}{32} = -0.0065625$$
$$y - 20.00 = -0.0065625(x - 4)$$
$$y - 20.00 = -0.0065625x + 0.02625$$
$$y = -0.0065625x + 20.02625$$

See graph below part d.

   d.   Answers may vary. We use the points $(0, 22.50)$ and $(32, 21.84)$.

$$m = \frac{21.84 - 22.50}{32 - 0} = \frac{-0.66}{32} = -0.020625$$
$$y - 22.50 = -0.020625(x - 0)$$
$$y - 22.50 = -0.020625x$$
$$y = -0.020625x + 22.50$$

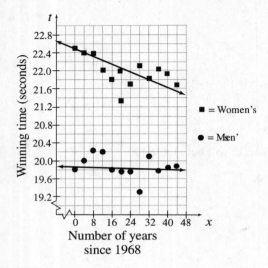

Number of years
since 1968

e. $-0.0065625x + 20.02625 = -0.020625x + 22.50$
$$0.0140625x + 20.02625 = 22.50$$
$$0.0140625x = 2.47375$$
$$x \approx 176$$

Now, $1968 + 176 = 2144$ will be the year in which we predict the winning times for men and women will be equal.

Men: $y = -0.0065625(176) + 20.02625 \approx 18.87$

Women: $y = -0.020625(176) + 22.50 \approx 18.87$

We predict the common winning times will be approximately 18.87 seconds.

f. Answers will vary. Our prediction in part (e) is not very reliable. Neither data set is very linear, so the linear models that we found in parts (c) and (d) are suspect.

g. Answers will vary. The winning times for the 2012 Olympic games are:
men: 19.32 seconds
women: 21.69 seconds

Section 3.3

Are You Ready for This Section?

R1. Substitute $x = 1$, $y = -2$, and $z = 3$:
$$3x - 2y + 4z = 3(1) - 2(-2) + 4(3)$$
$$= 3 + 4 + 12$$
$$= 19$$

Section 3.3 Quick Checks

1. If a system of equations has no solution, it is said to be <u>inconsistent</u>. If a system of equations has infinitely many solutions, the system is said to be <u>consistent</u> and the equations are <u>dependent</u>.

2. A <u>solution</u> to a system of equations consists of values for the variables that are solutions of each equation of the system.

3. False. For example, if the graphs of the equations are parallel planes, then the system will have no solution.

4. True

5. a. Substitute $x = 3$, $y = 2$, and $z = -2$ into all three equations.
Equation (1): $3 + 2 + (-2) \stackrel{?}{=} 3$
$$3 = 3$$

Equation (2): $3(3) + 2 - 2(-2) \stackrel{?}{=} -23$
$$9 + 2 + 4 \stackrel{?}{=} -23$$
$$15 \neq -23$$

Equation (3): $-2(3) - 3(2) + 2(-2) \stackrel{?}{=} 17$
$$-6 - 6 - 4 \stackrel{?}{=} 17$$
$$-16 \neq 17$$

Although these values satisfy equation (1), they do not satisfy equations (2) and (3). Therefore, the ordered triple $(3, 2, -2)$ is not a solution of the system.

b. Substitute $x = -4$, $y = 1$, and $z = 6$ into all three equations.
Equation (1): $-4 + 1 + 6 \stackrel{?}{=} 3$
$$3 = 3$$

Equation (2): $3(-4) + 1 - 2(6) \stackrel{?}{=} -23$
$$-12 + 1 - 12 \stackrel{?}{=} -23$$
$$-23 = -23$$

Equation (3): $-2(-4) - 3(1) + 2(6) \stackrel{?}{=} 17$
$$8 - 3 + 12 \stackrel{?}{=} 17$$
$$17 = 17$$

Because these values satisfy all three equations, the ordered triple $(-4, 1, 6)$ is a solution of the system.

6. $\begin{cases} x + y + z = -3 & (1) \\ 2x - 2y - z = -7 & (2) \\ -3x + y + 5z = 5 & (3) \end{cases}$
Multiply both sides of equation (1) by -2 and add the result to equation (2).
$$-2x - 2y - 2z = 6$$
$$2x - 2y - z = -7$$
$$-4y - 3z = -1$$

Multiply both sides of equation (1) by 3 and add the result to equation (3).
$$3x + 3y + 3z = -9$$
$$-3x + y + 5z = 5$$
$$4y + 8z = -4$$

Rewriting the system, we have
$\begin{cases} x + y + z = -3 & (1) \\ -4y - 3z = -1 & (2) \\ 4y + 8z = -4 & (3) \end{cases}$

Add equations (2) and (3).
$$-4y - 3z = -1$$
$$4y + 8z = -4$$
$$5z = -5$$

Once again rewriting the system, we have
$\begin{cases} x + y + z = -3 & (1) \\ -4y - 3z = -1 & (2) \\ 5z = -5 & (3) \end{cases}$

Solving equation (3) for z, we obtain
$$5z = -5$$
$$z = -1$$

Back-substituting -1 for z in equation (2), we obtain
$$-4y - 3(-1) = -1$$
$$-4y + 3 = -1$$
$$-4y = -4$$
$$y = 1$$

Back-substituting 1 for y and -1 for z in equation (1), we obtain
$$x + 1 + (-1) = -3$$
$$x = -3$$
The solution is the ordered triple $(-3, 1, -1)$.

7. $\begin{cases} 2x \quad\;\; - 4z = -7 & (1) \\ x + 6y \quad\;\; = 5 & (2) \\ 2y - z = 2 & (3) \end{cases}$
Multiply both sides of equation (2) by -2 and add the result to equation (1).
$$-2x - 12y \quad\quad\; = -10$$
$$2x \quad\quad\; - 4z = -7$$
$$-12y - 4z = -17$$

Rewriting the system, we have
$$\begin{cases} 2x \quad\ -4z = -7 & (1) \\ -12y - 4z = -17 & (2) \\ \quad\ 2y - z = 2 & (3) \end{cases}$$

Multiply both sides of equation (3) by 6 and add the result to equation (2).

$$\begin{aligned} 12y - 6z &= 12 \\ -12y - 4z &= -17 \\ \hline -10z &= -5 \end{aligned}$$

Once again rewriting the system, we have
$$\begin{cases} 2x \quad\ -4z = -7 & (1) \\ -12y - 4z = -17 & (2) \\ \quad\ -10z = -5 & (3) \end{cases}$$

Solving equation (3) for z, we obtain
$$-10z = -5$$
$$z = \frac{-5}{-10} = \frac{1}{2}$$

Back-substituting $\frac{1}{2}$ for z in equation (2), we obtain
$$-12y - 4\left(\frac{1}{2}\right) = -17$$
$$-12y - 2 = -17$$
$$-12y = -15$$
$$y = \frac{-15}{-12} = \frac{5}{4}$$

Back-substituting $\frac{1}{2}$ for z in equation (1), we obtain
$$2x - 4\left(\frac{1}{2}\right) = -7$$
$$2x - 2 = -7$$
$$2x = -5$$
$$x = -\frac{5}{2}$$

The solution is the ordered triple $\left(-\frac{5}{2}, \frac{5}{4}, \frac{1}{2}\right)$.

8. $\begin{cases} x - y + 2z = -7 & (1) \\ -2x + y - 3z = 5 & (2) \\ x - 2y + 3z = 2 & (3) \end{cases}$

Multiply both sides of equation (1) by 2 and add the result to equation (2).

$$\begin{aligned} 2x - 2y + 4z &= -14 \\ -2x + y - 3z &= 5 \\ \hline -y + z &= -9 \end{aligned}$$

Multiply both sides of equation (1) by -1 and add the result to equation (3).
$$\begin{aligned} -x + y - 2z &= 7 \\ x - 2y + 3z &= 2 \\ \hline -y + z &= 9 \end{aligned}$$

Rewriting the system, we have
$$\begin{cases} x - y + 2z = -7 & (1) \\ \quad -y + z = -9 & (2) \\ \quad -y + z = 9 & (3) \end{cases}$$

Multiply both sides of equation (2) by -1 and add the result to equation (3).

$$\begin{aligned} y - z &= 9 \\ -y + z &= 9 \\ \hline 0 &= 18 \end{aligned}$$

Once again rewriting the system, we have
$$\begin{cases} x - y + 2z = -7 & (1) \\ \quad -y + z = -9 & (2) \\ \quad 0 = 18 & (3) \quad \text{False} \end{cases}$$

Equation (3) is a false statement (contradiction). Therefore, the system is inconsistent. The solution set is $\varnothing$ or { }.

9. $\begin{cases} x - y + 3z = 2 & (1) \\ -x + 2y - 5z = -3 & (2) \\ 2x - y + 4z = 3 & (3) \end{cases}$

Add equations (1) and (2).
$$\begin{aligned} x - y + 3z &= 2 \\ -x + 2y - 5z &= -3 \\ \hline y - 2z &= -1 \end{aligned}$$

Multiply both sides of equation (1) by -2 and add the result to equation (3).
$$\begin{aligned} -2x + 2y - 6z &= -4 \\ 2x - y + 4z &= 3 \\ \hline y - 2z &= -1 \end{aligned}$$

Rewriting the system, we have
$$\begin{cases} x - y + 3z = 2 & (1) \\ \quad y - 2z = -1 & (2) \\ \quad y - 2z = -1 & (3) \end{cases}$$

Multiply both sides of equation (2) by -1 and add the result to equation (3).

$$-y + 2z = 1$$
$$\underline{y - 2z = -1}$$
$$0 = 0$$

Once again rewriting the system, we have

$$\begin{cases} x - y + 3z = 2 & (1) \\ y - 2z = -1 & (2) \\ 0 = 0 & (3) \quad \text{True} \end{cases}$$

Thus, the system is dependent and has an infinite number of solutions.

Solve equation (2) for y.

$$y = 2z - 1$$

Substituting $2z - 1$ for y in equation (1), we obtain

$$x - (2z - 1) + 3z = 2$$
$$x - 2z + 1 + 3z = 2$$
$$x + z + 1 = 2$$
$$x + z = 1$$
$$x = -z + 1$$

The solution to the system is

$$\{(x, y, z) \mid x = -z + 1, y = 2z - 1, z \text{ is any real number}\}$$

10. Let x represent the number of 21-inch mowers, let y represent the number of 24-inch mowers, and let z represent the number of 40-inch mowers.

$$\begin{cases} 2x + 3y + 4z = 81 & (1) \\ 3x + 3y + 4z = 95 & (2) \\ x + y + 2z = 35 & (3) \end{cases}$$

Multiply both sides of equation (3) by -2 and add the result to equation (1).

$$-2x - 2y - 4z = -70$$
$$\underline{2x + 3y + 4z = 81}$$
$$y \qquad\quad = 11$$

Multiply both sides of equation (3) by -3 and add the result to equation (2).

$$-3x - 3y - 6z = -105$$
$$\underline{3x + 3y + 4z = 95}$$
$$-2z = -10$$

Rewriting the system, we have

$$\begin{cases} 2x + 3y + 4z = 81 & (1) \\ y \quad = 11 & (2) \\ -2z = -10 & (3) \end{cases}$$

From equation (2), we know $y = 11$. Solving equation (3) for z, we obtain

$$-2z = -10$$
$$z = 5$$

Substituting 11 for y and 5 for z and 11 in equation (1), we obtain

$$2x + 3(11) + 4(5) = 81$$
$$2x + 33 + 20 = 81$$
$$2x + 53 = 81$$
$$2x = 28$$
$$x = 14$$

The company can manufacture 14 twenty-one-inch mowers, 11 twenty-four-inch mowers, and 5 forty-inch mowers.

3.3 Exercises

11. a. Substitute $x = 6$, $y = 2$, $z = -1$ into all three equations.

$$\begin{cases} 6 + 2 + 2(-1) = 6 + 2 - 2 = 6 \\ -2(6) - 3(2) + 5(-1) = -12 - 6 - 5 = -23 \neq 1 \\ 2(6) + 2 + 3(-1) = 12 + 2 - 3 = 11 \neq 5 \end{cases}$$

Although these values satisfy equation (1), they do not satisfy equations (2) and (3). Therefore, the ordered triple $(6, 2, -1)$ is not a solution of the system.

b. Substitute $x = -3$, $y = 5$, $z = 2$ into all three equations).

$$\begin{cases} -3 + 5 + 2(2) = -3 + 5 + 4 = 6 \\ -2(-3) - 3(5) + 5(2) = 6 - 15 + 10 = 1 \\ 2(-3) + 5 + 3(2) = -6 + 5 + 6 = 5 \end{cases}$$

Because these values satisfy all three equations, the ordered triple $(-3, 5, 2)$ is a solution of the system.

13. $$\begin{cases} x + y + z = 5 & (1) \\ -2x - 3y + 2z = 8 & (2) \\ 3x - y - 2z = 3 & (3) \end{cases}$$

Multiply both sides of equation (1) by 2 and add the result to equation (2).

$$2x + 2y + 2z = 10$$
$$\underline{-2x - 3y + 2z = 8}$$
$$-y + 4z = 18 \qquad (4)$$

Multiply both sides of equation (1) by –3 and add the result to equation (3).

$$-3x - 3y - 3z = -15$$
$$\underline{3x - y - 2z = 3}$$
$$-4y - 5z = -12 \qquad (5)$$

Multiply both sides of equation (4) by –4 and add the result to equation (5).

$$4y - 16z = -72$$
$$\underline{-4y - 5z = -12}$$
$$-21z = -84$$
$$z = 4$$

Substituting 4 for z in equation (4), we obtain

$$-y + 4(4) = 18$$
$$-y + 16 = 18$$
$$-y = 2$$
$$y = -2$$

Substituting –2 for y and 4 for z in equation (1), we obtain

$$x + (-2) + 4 = 5$$
$$x + 2 = 5$$
$$x = 3$$

The solution is the ordered triple $(3, -2, 4)$.

15. $\begin{cases} x - 3y + z = 13 & (1) \\ 3x + y - 4z = 13 & (2) \\ -4x - 4y + 2z = 0 & (3) \end{cases}$

Multiply equation (1) by -3 and add the result to equation (2).

$$-3x + 9y - 3z = -39$$
$$\underline{3x + y - 4z = 13}$$
$$10y - 7z = -26 \qquad (4)$$

Multiply both sides of equation (1) by 4 and add the result to equation (3).

$$4x - 12y + 4z = 52$$
$$\underline{-4x - 4y + 2z = 0}$$
$$-16y + 6z = 52 \qquad (5)$$

Multiply equation (4) by 8, multiply equation (5) by 5, and add the results.

$$80y - 56z = -208$$
$$\underline{-80y + 30z = 260}$$
$$-26z = 52$$
$$z = -2$$

Substituting –2 for z in equation (4) and solving for y, we obtain

$$10y - 7(-2) = -26$$
$$10y + 14 = -26$$
$$10y = -40$$
$$y = -4$$

Substituting –2 for z and –4 for y in equation (1), we obtain

$$x - 3(-4) + (-2) = 13$$
$$x + 12 - 2 = 13$$
$$x + 10 = 13$$
$$x = 3$$

The solution is the ordered triple $(3, -4, -2)$.

17. $\begin{cases} x - 4y + z = 5 & (1) \\ 4x + 2y + z = 2 & (2) \\ -4x + y - 3z = -8 & (3) \end{cases}$

Multiply both sides of equation (1) by -4 and add the result to equation (2).

$$-4x + 16y - 4z = -20$$
$$\underline{4x + 2y + z = 2}$$
$$18y - 3z = -18 \qquad (4)$$

Add equations (2) and (3).

$$4x + 2y + z = 2$$
$$\underline{-4x + y - 3z = -8}$$
$$3y - 2z = -6 \qquad (5)$$

Multiply both sides of equation (5) by -6 and add the result to equation (4).

$$-18y + 12z = 36$$
$$\underline{18y - 3z = -18}$$
$$9z = 18$$
$$z = 2$$

Substituting 2 for z in equation (5), we obtain

$$3y - 2(2) = -6$$
$$3y - 4 = -6$$
$$3y = -2$$
$$y = -\frac{2}{3}$$

Substituting $-\frac{2}{3}$ for y and 2 for z in equation (1), we obtain

$$x-4\left(-\frac{2}{3}\right)+2=5$$
$$x+\frac{8}{3}+2=5$$
$$3\left(x+\frac{8}{3}+2\right)=3(5)$$
$$3x+8+6=15$$
$$3x=1$$
$$x=\frac{1}{3}$$

The solution is the ordered triple $\left(\frac{1}{3}, -\frac{2}{3}, 2\right)$.

19. $\begin{cases} x-3y=12 & (1) \\ 2y-3z=-9 & (2) \\ 2x+z=7 & (3) \end{cases}$

Multiply both sides of equation (1) by -2 and add the result to equation (3).
$$-2x+6y=-24$$
$$\underline{2x+z=7}$$
$$6y+z=-17 \qquad (4)$$

Multiply both sides of equation (4) by 3 and add the result to equation (2).
$$18y+3z=-51$$
$$\underline{2y-3z=-9}$$
$$20y=-60$$
$$y=-3$$

Substituting -3 for y in equation (4), we obtain
$$6(-3)+z=-17$$
$$-18+z=-17$$
$$z=1$$
Substituting -3 for y in equation (1), we obtain
$$x-3(-3)=12$$
$$x+9=12$$
$$x=3$$
The solution is the ordered triple $(3, -3, 1)$.

21. $\begin{cases} x+y-2z=6 & (1) \\ -2x-3y+z=12 & (2) \\ -3x-4y+3z=2 & (3) \end{cases}$

Multiply both sides of equation (1) by 3, and add the result to equation (2).
$$3x+3y-6z=18$$
$$\underline{-2x-3y+z=12}$$
$$x-5z=30 \qquad (4)$$
Multiply both sides of equation (1) by 4, and add the result to equation (3).

$$4x+4y-8z=24$$
$$\underline{-3x-4y+3z=2}$$
$$x-5z=26 \qquad (5)$$
Multiply both sides of equation (4) by -1, and add the result to equation (5).
$$-x+5z=-30$$
$$\underline{x-5z=26}$$
$$0=-4$$

Because we have arrived at a contradiction, this system of equations has no solution.

23. $\begin{cases} x+y+z=4 & (1) \\ -2x-y+2z=6 & (2) \\ x+2y+5z=18 & (3) \end{cases}$

Add equations (1) and (2).
$$x+y+z=4$$
$$\underline{-2x-y+2z=6}$$
$$-x+3z=10 \qquad (4)$$
Multiply both sides of equation (2) by 2, and add the result to equation (3).
$$-4x-2y+4z=12$$
$$\underline{x+2y+5z=18}$$
$$-3x+9z=30 \qquad (5)$$
Multiply both sides of equation (4) by -3, and add the result to equation (5).
$$3x-9z=-30$$
$$\underline{-3x+9z=30}$$
$$0=0 \qquad \text{True}$$
Thus, the system is dependent and has an infinite number of solutions.
Solve equation (4) for x.
$$-x+3z=10$$
$$x=3z-10$$
Substituting $3z-10$ for x in equation (1), we obtain
$$(3z-10)+y+z=4$$
$$y+4z-10=4$$
$$y=-4z+14$$
The solution to the system is
$$\{(x,y,z)\,|\,x=3z-10, y=-4z+14,$$
$$z \text{ is any real number}\}$$

25. $\begin{cases} x+3z=5 & (1) \\ -2x+y=1 & (2) \\ y+6z=11 & (3) \end{cases}$

Multiply both sides of equation (1) by 2, and add the result to equation (2).

$$2x \quad\ +6z =10$$
$$\underline{-2x+y \quad\ \ =1}$$
$$\qquad\ \ y+6z =11 \qquad (4)$$

Multiply both sides of equation (3) by −1, and add the result to equation (4).

$$-y-6z =-11$$
$$\underline{\ \ y+6z =\ \ 11}$$
$$\qquad\ 0 =\quad 0 \qquad \text{True}$$

Thus, the system is dependent and has an infinite number of solutions.
Solve equation (4) for y.

$$y+6z =11$$
$$y =-6z +11$$

Substituting $-6z+11$ for y in equation (2), we obtain

$$-2x+(-6z+11)=1$$
$$-2x-6z+11=1$$
$$\qquad\ -2x =6z-10$$
$$\qquad\quad\ x =-3z+5$$

The solution to the system is
$$\{(x, y, z)\,|\, x =-3z +5, y =-6z +11,$$
$$z \text{ is any real number}\}$$

27. $\begin{cases} 2x-y+2z =1 & (1) \\ -2x+3y-2z =3 & (2) \\ 4x-y+6z =7 & (3) \end{cases}$

Add equations (1) and (2).

$$2x-y+2z =1$$
$$\underline{-2x+3y-2z =3}$$
$$\quad\ 2y \qquad\ =4$$
$$\qquad\quad\ y =2$$

Multiply both sides of equation (2) by 2 and add the result to equation (3).

$$-4x+6y-4z =6$$
$$\underline{\ \ 4x-y+6z =7}$$
$$\qquad 5y+2z =13 \qquad (4)$$

Substituting 2 for y in equation (4), we obtain

$$5(2)+2z =13$$
$$10+2z =13$$
$$\quad\ 2z =3$$
$$\qquad z =\frac{3}{2}$$

Substituting 2 for y and $\frac{3}{2}$ for z in equation (1), we obtain

$$2x-2+2\left(\frac{3}{2}\right)=1$$
$$2x-2+3 =1$$
$$\quad\ 2x =0$$
$$\qquad x =0$$

The solution is the ordered triple $\left(0,\ 2,\ \dfrac{3}{2}\right)$.

29. $\begin{cases} x-y+z =5 & (1) \\ -2x+y-z =2 & (2) \\ x-2y+2z =1 & (3) \end{cases}$

Multiply both sides of equation (1) by 2 and add the result to equation (2).

$$2x-2y+2z =10$$
$$\underline{-2x+\ y-\ z =2}$$
$$\qquad\quad -y+z =12 \qquad (4)$$

Multiply both sides of equation (3) by 2 and add the result to equation (2).

$$2x-4y+4z =2$$
$$\underline{-2x+\ y-\ z =2}$$
$$\qquad -3y+3z =4 \qquad (5)$$

Multiply both sides of equation (4) by −3 and add the result to equation (5).

$$3y-3z =-36$$
$$\underline{-3y+3z =4}$$
$$\qquad 0 =-32 \qquad \text{False}$$

The system has no solution. The solution set is $\varnothing$ or $\{\ \}$. The system is inconsistent.

31. $\begin{cases} 2y-z =-3 & (1) \\ -2x+3y =10 & (2) \\ 4x \quad\ +3z =-11 & (3) \end{cases}$

Multiply both sides of equation (2) by 2 and add the result to equation (3).

$$-4x+6y \quad\ =20$$
$$\underline{\ \ 4x \quad\ +3z =-11}$$
$$\qquad 6y+3z =9 \qquad (4)$$

Multiply both sides of equation (1) by 3 and add the result to equation (4).

$$6y-3z =-9$$
$$\underline{6y+3z =9}$$
$$\quad 12y =0$$
$$\qquad y =0$$

Substituting 0 for y in equation (1), we obtain
$$2(0) - z = -3$$
$$-z = -3$$
$$z = 3$$

Substituting 0 for y in equation (2), we obtain
$$-2x + 3(0) = 10$$
$$-2x = 10$$
$$x = -5$$
The solution is the ordered triple $(-5, 0, 3)$.

33. $\begin{cases} x - 2y + z = 5 & (1) \\ -2x + y - z = 2 & (2) \\ x - 5y - 4z = 8 & (3) \end{cases}$

Add equations (1) and (2).
$$\begin{array}{r} x - 2y + z = 5 \\ \underline{-2x + y - z = 2} \\ -x - y \quad\;\; = 7 \;\; (4) \end{array}$$

Multiply equation (1) by 4 and add to equation (3).
$$\begin{array}{r} 4x - 8y + 4z = 20 \\ \underline{x - 5y - 4z = 8} \\ 5x - 13y \qquad = 28 \;\; (5) \end{array}$$

Multiply equation (4) by 5 and add to equation (5).
$$\begin{array}{r} -5x - 5y = 35 \\ \underline{5x - 13y = 28} \\ -18y = 63 \end{array}$$
$$y = \frac{63}{-18} = -\frac{7}{2}$$

Substituting $-\frac{7}{2}$ for y in equation (4), we obtain
$$-2 - \left(-\frac{7}{2}\right) = 7$$
$$-x + \frac{7}{2} = 7$$
$$-x = \frac{7}{2}$$
$$x = -\frac{7}{2}$$

Substituting $-\frac{7}{2}$ for x and $-\frac{7}{2}$ for y in equation (1), we obtain

$$-\frac{7}{2} - 2\left(-\frac{7}{2}\right) + z = 5$$
$$\frac{7}{2} + z = 5$$
$$z = \frac{3}{2}$$

The solution is the ordered triple $\left(-\frac{7}{2}, -\frac{7}{2}, \frac{3}{2}\right)$.

35. $\begin{cases} x + 2y - z = 1 & (1) \\ 2x + 7y + 4z = 11 & (2) \\ x + 3y + z = 4 & (3) \end{cases}$

Multiply both sides of equation (1) by -2 and add the result to equation (2).
$$\begin{array}{r} -2x - 4y + 2z = -2 \\ \underline{2x + 7y + 4z = 11} \\ 3y + 6z = 9 \quad (4) \end{array}$$

Multiply both sides of equation (1) by -1 and add the result to equation (3).
$$\begin{array}{r} -x - 2y + z = -1 \\ \underline{x + 3y + z = 4} \\ y + 2z = 3 \qquad (5) \end{array}$$

Multiply both sides of equation (5) by -3 and add the result to equation (4).
$$\begin{array}{r} -3y - 6z = -9 \\ \underline{3y + 6z = 9} \\ 0 = 0 \quad \text{True} \end{array}$$

Thus, the system is dependent and has an infinite number of solutions.

Solve equation (5) for y.
$$y = -2z + 3$$
Substituting $-2z + 3$ for y in equation (1), we obtain
$$x + 2(-2z + 3) - z = 1$$
$$x - 4z + 6 - z = 1$$
$$x - 5z + 6 = 1$$
$$x - 5z = -5$$
$$x = 5z - 5$$

The solution to the system is
$$\{(x, y, z) \mid x = 5z - 5,$$
$$y = -2z + 3, z \text{ is any real number}\}.$$

37. $\begin{cases} x+y+z=5 & (1) \\ 3x+4y+z=16 & (2) \\ -x-4y+z=-6 & (3) \end{cases}$

Add equations (1) and (3).

$\begin{aligned} x+y+z &= 5 \\ -x-4y+z &= -6 \\ \hline -3y+2z &= -1 \qquad (4) \end{aligned}$

Multiply both sides of equation (3) by 3 and add the result to equation (2).

$\begin{aligned} -3x-12y+3z &= -18 \\ 3x+\;4y+\;z &= 16 \\ \hline -8y+4z &= -2 \qquad (5) \end{aligned}$

Multiply both sides of equation (4) by –2 and add the result to equation (5).

$\begin{aligned} 6y-4z &= 2 \\ -8y+4z &= -2 \\ \hline -2y &= 0 \\ y &= 0 \end{aligned}$

Substituting 0 for y in equation (4), we obtain

$\begin{aligned} -3(0)+2z &= -1 \\ 2z &= -1 \\ z &= -\frac{1}{2} \end{aligned}$

Substituting 0 for y and $-\frac{1}{2}$ for z in equation (1), we obtain

$\begin{aligned} x+0+\left(-\frac{1}{2}\right) &= 5 \\ x &= \frac{11}{2} \end{aligned}$

The solution is the ordered triple $\left(\frac{11}{2},\ 0,-\frac{1}{2}\right)$.

39. $\begin{cases} x+y+z=3 & (1) \\ -x+\dfrac{1}{2}y+z=\dfrac{1}{2} & (2) \\ -x+2y+3z=4 & (3) \end{cases}$

Add equations (1) and (3).

$\begin{aligned} x+y+z &= 3 \\ -x+2y+3z &= 4 \\ \hline 3y+4z &= 7 \qquad (4) \end{aligned}$

Multiply both sides of equation (1) by 2, multiply both sides of equation (2) by 2, and add the results.

$\begin{aligned} 2x+2y+2z &= 6 \\ -2x+y+2z &= 1 \\ \hline 3y+4z &= 7 \qquad (5) \end{aligned}$

Multiply both sides of equation (4) by –1 and add the result to equation (5).

$\begin{aligned} -3y-4z &= -7 \\ 3y+4z &= 7 \\ \hline 0 &= 0 \qquad \text{True} \end{aligned}$

Thus, the system is dependent and has an infinite number of solutions.

Solve equation (5) for y.

$\begin{aligned} 3y+4z &= 7 \\ 3y &= -4z+7 \\ y &= -\frac{4}{3}z+\frac{7}{3} \end{aligned}$

Substituting $-\frac{4}{3}z+\frac{7}{3}$ for y in equation (1), we obtain

$\begin{aligned} x+\left(-\frac{4}{3}z+\frac{7}{3}\right)+z &= 3 \\ x-\frac{1}{3}z+\frac{7}{3} &= 3 \\ x &= \frac{1}{3}z+\frac{2}{3} \end{aligned}$

The solution to the system is

$\left\{(x,y,z)\,\middle|\, x=\frac{1}{3}z+\frac{2}{3},\right.$

$\left. y=-\frac{4}{3}z+\frac{7}{3}, z \text{ is any real number}\right\}.$

41. Answers will vary. One possibility follows.
For the ordered triple $(2,-1,\ 3)$, evaluate the expressions $x+y+z,\ x-y+z,$ and $x+y-z$:

$x+y+z = 2+(-1)+3 = 4$

$x-y+z = 2-(-1)+3 = 6$

$x+y-z = 2+(-1)-3 = -2$

So, the system below has the solution $(2,-1,3)$.

$\begin{cases} x+y+z=4 \\ x-y+z=6 \\ x+y-z=-2 \end{cases}$

43. a. If $f(-1) = -6$, then $a(-1)^2 + b(-1) + c = -6$
or $a - b + c = -6$.
If $f(2) = 3$, then $a(2)^2 + b(2) + c = 3$ or
$4a + 2b + c = 3$.

b. To find a, b, and c, we must solve the
following system:

$$\begin{cases} a + b + c = 4 & (1) \\ a - b + c = -6 & (2) \\ 4a + 2b + c = 3 & (3) \end{cases}$$

Add equations (1) and (2).
$a + b + c = 4$
$\underline{a - b + c = -6}$
$2a \quad\;\; + 2c = -2 \qquad (4)$

Multiply both sides of equation (2) by 2 and
add the result to equation (3).
$2a - 2b + 2c = -12$
$\underline{4a + 2b + c = 3}$
$6a \qquad + 3c = -9 \qquad (5)$

Divide both sides of equation (4) by 2,
divide equation (5) by -3, and add the
results.
$a + c = -1$
$\underline{-2a - c = 3}$
$\qquad -a = 2$
$\qquad\;\; a = -2$

Substituting -2 for a in equation (4), we
obtain
$2(-2) \;\; + 2c = -2$
$\qquad -4 + 2c = -2$
$\qquad\qquad 2c = 2$
$\qquad\qquad\; c = 1$

Substituting -2 for a and 1 for c in equation
(1), we obtain

$-2 + b + 1 = 4$
$\qquad\;\; b = 5$

Thus, the quadratic equation that contains
$(-1, -6)$, $(1, 4)$, and $(2, 3)$ is
$f(x) = -2x^2 + 5x + 1$.

45. Rewrite the system with each equation in
standard form.

$$\begin{cases} i_1 - i_2 + i_3 = 0 & (1) \\ -3i_1 \quad\;\; + 2i_3 = 3 & (2) \\ \quad\;\; 4i_2 + 2i_3 = 22 & (3) \end{cases}$$

Multiply both sides of equation (1) by 3 and add
the result to equation (2).
$3i_1 - 3i_2 + 3i_3 = 0$
$\underline{-3i_1 \qquad\;\; + 2i_3 = 3}$
$\quad\;\; -3i_2 + 5i_3 = 3 \qquad (4)$

Multiply both sides of equation (3) by 3,
multiply both sides of equation (4) by 4, and add
the results.
$12i_2 + 6i_3 = 66$
$\underline{-12i_2 + 20i_3 = 12}$
$\qquad\quad 26i_3 = 78$
$\qquad\qquad i_3 = 3$

Substituting 3 for i_3 in equation (2), we obtain
$-3i_1 + 2(3) = 3$
$\quad -3i_1 + 6 = 3$
$\qquad\;\; -3i_1 = -3$
$\qquad\qquad i_1 = 1$

Substituting 3 for i_3 and 1 for i_1 in equation (1),
we obtain
$1 - i_2 + (3) = 0$
$\quad\; -i_2 + 4 = 0$
$\qquad\;\; -i_2 = -4$
$\qquad\quad i_2 = 4$

The currents are $i_1 = 1$, $i_2 = 4$, and $i_3 = 3$.

47. Let b represent the number of box seats, let r
represent the number of reserved seats, and let l
represent the number of lawn seats.

$$\begin{cases} b + r + l = 4100 & (1) \\ 9b + 7r + 5l = 28,400 & (2) \\ \frac{1}{2}(9b) + \frac{1}{2}(7r) + 5l = 18,300 & (3) \end{cases}$$

Simplify equation (3) and rewrite the system.
$\frac{1}{2}(9b) + \frac{1}{2}(7r) + 5l = 18,300$
$4.5b + 3.5r + 5l = 18,300$

Rewriting the system, we have

$$\begin{cases} b + r + l = 4100 & (1) \\ 9b + 7r + 5l = 28,400 & (2) \\ 4.5b + 3.5r + 5l = 18,300 & (3) \end{cases}$$

Multiply both sides of equation (1) by –9 and add the result to equation (2).

$$-9b - 9r - 9l = -36,900$$
$$\underline{9b + 7r + 5l = 28,400}$$
$$-2r - 4l = -8500 \qquad (4)$$

Multiply both sides of equation (3) by –2 and add the result to equation (2).
$$-9b - 7r - 10l = -36,600$$
$$\underline{9b + 7r + 5l = 28,400}$$
$$-5l = -8200$$
$$l = 1640$$

Substituting 1640 for l in equation (4), we obtain
$$-2r - 4(1640) = -8500$$
$$-2r - 6460 = -8500$$
$$-2r = -1940$$
$$r = 970$$

Substituting 1640 for l and 970 for r in equation (1), we obtain
$$b + 970 + 1640 = 4100$$
$$b + 2610 = 4100$$
$$b = 1490$$
There are 1490 box seats, 970 reserve seats, and 1640 lawn seats in the stadium.

49. Let c represent the number servings of Chex® cereal, let m represent the number of servings of 2% milk, and let j represent the number of servings of orange juice.
$$\begin{cases} 220c + 125m = 470 & \text{(sodium)} & (1) \\ 26c + 12m + 26j = 89 & \text{(carbs)} & (2) \\ c + 8m + 2j = 20 & \text{(protein)} & (3) \end{cases}$$
Multiply both sides of equation (3) by –13 and add the result to equation (2).
$$-13c - 104m - 26j = -260$$
$$\underline{26c + 12m + 26j = 89}$$
$$13c - 92m = -171 \qquad (4)$$

Multiply both sides of equation (1) by 13, multiply both sides of equation (4) by –220, and add the results.

$$2860c + 1625m = 6110$$
$$\underline{-2860c + 20,240m = 37,620}$$
$$21,865m = 43,730$$
$$m = 2$$

Substituting 2 for m in equation (1), we obtain
$$220c + 125(2) = 470$$
$$220c + 250 = 470$$
$$220c = 220$$
$$c = 1$$

Substituting 2 for m and 1 for c in equation (3), we obtain
$$1 + 8(2) + 2j = 20$$
$$17 + 2j = 20$$
$$2j = 3$$
$$j = 1.5$$
Nancy needs 1 serving of Chex® cereal, 2 servings of 2% milk, and 1.5 servings of orange juice.

51. Let t represent the amount to be invested in Treasury bills, let m represent the amount to be invested in municipal bonds, and let c represent the amount to be invested in corporate bonds.
$$\begin{cases} t + m + c = 25,000 & (1) \\ 0.03t + 0.05m + 0.09c = 1210 & (2) \\ t - c = 7000 & (3) \end{cases}$$

Multiply both sides of equation (1) by –0.05 and add the result to equation (2).
$$-0.05t - 0.05m - 0.05c = -1250$$
$$\underline{0.03t + 0.05m + 0.09c = 1210}$$
$$-0.02t + 0.04c = -40 \qquad (4)$$

Multiply both sides of equation (3) by 0.02 and add the result to equation (4).
$$0.02t - 0.02c = 140$$
$$\underline{-0.02t + 0.04c = -40}$$
$$0.02c = 100$$
$$c = 5000$$

Substituting 5000 for c in equation (3), we obtain
$$t - 5000 = 7000$$
$$t = 12,000$$

Substituting 5000 for c and 12,000 for t in equation (1), we obtain
$$12,000 + m + 5000 = 25,000$$
$$m + 17,000 = 25,000$$
$$m = 8000$$

Sachi should invest $12,000 in Treasury bills, $8000 in municipal bonds, and $5000 in corporate bonds.

53. Let x represent the lengths of $\overline{AO}$ and $\overline{AM}$, let y represent the lengths of $\overline{OC}$ and $\overline{NC}$, and let z represent the lengths of $\overline{BM}$ and $\overline{BN}$.
Because $\overline{AO} + \overline{OC} = \overline{AC}$, $\overline{BN} + \overline{NC} = \overline{BC}$, and $\overline{AM} + \overline{MB} = \overline{AB}$, we obtain the following system.

$$\begin{cases} x + y \quad\;\; = 14 & (1) \\ x \quad\;\; + z = 6 & (2) \\ \quad\; y + z = 12 & (3) \end{cases}$$

Multiply both sides of equation (1) by -1 and add the result to equation (2).

$$\begin{aligned} -x - y \quad\;\; &= -14 \\ \underline{x \quad\;\; + z} \; &= 6 \\ -y + z &= -8 \qquad (4) \end{aligned}$$

Add equations (3) and (4).

$$\begin{aligned} y + z &= 12 \\ \underline{-y + z} &= -8 \\ 2z &= 4 \\ z &= 2 \end{aligned}$$

Substituting 2 for z in equation (3), we obtain

$$\begin{aligned} y + 2 &= 12 \\ y &= 10 \end{aligned}$$

Substituting 2 for z in equation (2), we obtain

$$\begin{aligned} x + 2 &= 6 \\ x &= 4 \end{aligned}$$

Therefore, $\overline{AM} = 4$, $\overline{BN} = 2$, and $\overline{OC} = 10$.

55. $\begin{cases} \dfrac{2}{5}x + \dfrac{1}{2}y - \dfrac{1}{3}z = 0 & (1) \\[2mm] \dfrac{3}{5}x - \dfrac{1}{4}y + \dfrac{1}{2}z = 10 & (2) \\[2mm] -\dfrac{1}{5}x + \dfrac{1}{4}y - \dfrac{1}{6}z = -4 & (3) \end{cases}$

Simplify each equation and rewrite the system.

(1) $\quad 30\left(\dfrac{2}{5}x + \dfrac{1}{2}y - \dfrac{1}{3}z\right) = 30(0)$
$\qquad\qquad 12x + 15y - 10z = 0$

(2) $\quad 20\left(\dfrac{3}{5}x - \dfrac{1}{4}y + \dfrac{1}{2}z\right) = 20(10)$
$\qquad\qquad 12x - 5y + 10z = 200$

(3) $\quad 60\left(-\dfrac{1}{5}x + \dfrac{1}{4}y - \dfrac{1}{6}z\right) = 60(-4)$
$\qquad\qquad -12x + 15y - 10z = -240$

Rewriting the system, we have

$$\begin{cases} 12x + 15y - 10z = 0 & (1) \\ 12x - 5y + 10z = 200 & (2) \\ -12x + 15y - 10z = -240 & (3) \end{cases}$$

Add equations (1) and (3).

$$\begin{aligned} 12x + 15y - 10z &= 0 \\ \underline{-12x + 15y - 10z} &= -240 \\ 30y - 20z &= -240 \qquad (4) \end{aligned}$$

Add equations (2) and (3).

$$\begin{aligned} 12x - 5y + 10z &= 200 \\ \underline{-12x + 15y - 10z} &= -240 \\ 10y \quad\quad\; &= -40 \\ y &= -4 \end{aligned}$$

Substituting -4 for y in equation (4), we obtain

$$\begin{aligned} 30(-4) - 20z &= -240 \\ -120 - 20z &= -240 \\ -20z &= -120 \\ z &= 6 \end{aligned}$$

Substituting -4 for y and 6 for z in equation (1), we obtain

$$\begin{aligned} 12x + 15(-4) - 10(6) &= 0 \\ 12x - 60 - 60 &= 0 \\ 12x &= 120 \\ x &= 10 \end{aligned}$$

The solution is the ordered triple $(10, -4, 6)$.

57. $\begin{cases} x + y + z + w = 3 & (1) \\ -2x - y + 3z - w = -1 & (2) \\ 2x + 2y - 2z + w = 2 & (3) \\ -x + 2y - 3z + 2w = 12 & (4) \end{cases}$

Add equations (1) and (4).

$$\begin{aligned} x + y + z + w &= 3 \\ \underline{-x + 2y - 3z + 2w} &= 12 \\ 3y - 2z + 3w &= 15 \qquad (5) \end{aligned}$$

Add equations (2) and (3).

$$\begin{aligned} -2x - y + 3z - w &= -1 \\ \underline{2x + 2y - 2z + w} &= 2 \\ y + z \quad\quad &= 1 \qquad (6) \end{aligned}$$

Multiply both sides of equation (1) by 2 and add the result to equation (2).

$$\begin{aligned} 2x + 2y + 2z + 2w &= 6 \\ \underline{-2x - y + 3z - w} &= -1 \\ y + 5z + w &= 5 \qquad (7) \end{aligned}$$

207

Multiply both sides of equation (7) by –3 and add the result to equation (5).

$$-3y-15z-3w=-15$$
$$\underline{3y-2z+3w=15}$$
$$-17z\qquad\;\;\;=0$$
$$z=0$$

Substituting 0 for z in equation (6), we obtain

$$y+0=1$$
$$y=1$$

Substituting 0 for z and 1 for y equation (7), we obtain

$$1+5(0)+w=5$$
$$1+w=5$$
$$w=4$$

Substituting 0 for z and 1 for y and 4 for w in equation (1), we obtain

$$x+1+0+4=3$$
$$x+5=3$$
$$x=-2$$

The solution is $(-2,\,1,\,0,\,4)$.

59. Answers will vary. One possibility follows: We must eliminate the same variable in the first step so that we will obtain a system of two equations containing two variables, which we can then solve using the methods of Section 3.1. Once we solve for the two variables, we can then back-substitute in order to find the third variable.

Putting the Concepts Together (Sections 3.1–3.3)

1. The graphs of the linear equations intersect at the point $(-3, 5)$. Thus, the solution is $(-3, 5)$.

2. $\begin{cases} 3x+\;\;y=7 & (1) \\ -2x+3y=-12 & (2) \end{cases}$

Equation (1) in slope-intercept form is $y=-3x+7$. Equation (2) in slope-intercept form is $y=\dfrac{2}{3}x-4$. Graph each equation and find the point of intersection.

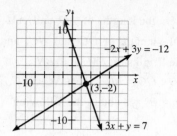

The solution is the ordered pair $(3, -2)$.

3. $\begin{cases} x=2-3y & (1) \\ 3x+10y=5 & (2) \end{cases}$

Because equation (1) is solved for x, we use substitution to solve the system.
Substituting $2-3y$ for x in equation (2), we obtain

$$3(2-3y)+10y=5$$
$$6-9y+10y=5$$
$$6+y=5$$
$$y=-1$$

Substituting -1 for y in equation (1), we obtain $x = 2 - 3(-1) = 2 + 3 = 5$.
The solution is the ordered pair $(5, -1)$.

4. $\begin{cases} 4x+3y=-1 & (1) \\ 2x-y=3 & (2) \end{cases}$

Because the coefficient of y in equation (2) is -1, we use substitution to solve the system.
Equation (2) solved for y is $y=2x-3$.
Substituting $2x-3$ for y in equation (1), we obtain

$$4x+3(2x-3)=-1$$
$$4x+6x-9=-1$$
$$10x-9=-1$$
$$10x=8$$
$$x=\frac{8}{10}=\frac{4}{5}$$

Substituting $\dfrac{4}{5}$ for x in equation (2), we obtain

$$2\left(\frac{4}{5}\right) - y = 3$$

$$\frac{8}{5} - y = \frac{15}{5}$$

$$-y = \frac{7}{5}$$

$$y = -\frac{7}{5}$$

The solution is the ordered pair $\left(\frac{4}{5},\ -\frac{7}{5}\right)$.

5. $\begin{cases} x + 3y = 8 & (1) \\ \dfrac{1}{5}x + \dfrac{1}{2}y = 1 & (2) \end{cases}$

Because equation (2) contains fractions, we choose to use the elimination method to solve the system.
Multiply both sides of equation (1) by 2, multiply both side of equation (2) by -10, and add the results.

$$2x + 6y = 16$$
$$\underline{-2x - 5y = -10}$$
$$y = 6$$

Substituting 6 for y in equation (1), we obtain

$$x + 3(6) = 8$$
$$x + 18 = 8$$
$$x = -10$$

The solution is the ordered pair $(-10, 6)$.

6. $\begin{cases} 8x - 4y = 12 & (1) \\ -10x + 5y = -15 & (2) \end{cases}$

Because none of variables have a coefficient of 1, we use elimination to solve the system.
Multiply both sides of equation (1) by 5, multiply both sides of equation (2) by 4, and add the results.

$$40x - 20y = 60$$
$$\underline{-40x + 20y = -60}$$
$$0 = 0 \qquad \text{True}$$

The system is dependent. The solution is $\{(x,\ y) \mid 8x - 4y = 12\}$.

7. $\begin{cases} 2x + y + 3z = 10 & (1) \\ x - 2y + z = 10 & (2) \\ -4x + 3y + 2z = 5 & (3) \end{cases}$

Multiply both sides of equation (2) by -2 and add the result to equation (1).

$$-2x + 4y - 2z = -20$$
$$\underline{2x + y + 3z = 10}$$
$$5y + z = -10 \qquad (4)$$

Multiply both sides of equation (1) by 2 and add the result to equation (3).

$$4x + 2y + 6z = 20$$
$$\underline{-4x + 3y + 2z = 5}$$
$$5y + 8z = 25 \qquad (5)$$

Multiply both sides of equation (4) by -1 and add the result to equation (5).

$$-5y - z = 10$$
$$\underline{5y + 8z = 25}$$
$$7z = 35$$
$$z = 5$$

Substituting 5 for z in equation (4), we obtain

$$5y + 5 = -10$$
$$5y = -15$$
$$y = -3$$

Substituting 5 for z and -3 for y in equation (1), we obtain

$$2x + (-3) + 3(5) = 10$$
$$2x - 3 + 15 = 10$$
$$2x + 12 = 10$$
$$2x = -2$$
$$x = -1$$

The solution is the ordered triple $(-1, -3, 5)$.

8. $\begin{cases} x + 2y - 2z = 3 & (1) \\ x + 3y - 4z = 6 & (2) \\ 4x + 5y - 2z = 6 & (3) \end{cases}$

Multiply both sides of equation (1) by -1 and add the result to equation (2).

$$-x - 2y + 2z = -3$$
$$\underline{x + 3y - 4z = 6}$$
$$y - 2z = 3 \qquad (4)$$

Multiply both sides of equation (1) by -4 and add the result to equation (3).

$$-4x - 8y + 8z = -12$$
$$\underline{4x + 5y - 2z = 6}$$
$$-3y + 6z = -6 \qquad (5)$$

Multiply both sides of equation (4) by 3 and add the result to equation (5).

$$3y - 6z = 9$$
$$\underline{-3y + 6z = -6}$$
$$0 = 3 \quad \text{False}$$

The system has no solution. The solution set is $\varnothing$ or $\{\ \}$. The system is inconsistent.

9. Let a represent the number of adult tickets and let y represent the number of youth tickets sold.

$$\begin{cases} 9a + 5y = 5925 & (1) \\ a + y = 825 & (2) \end{cases}$$

Multiply both sides of equation (2) by -5 and add the result to equation (1).

$$-5a - 5y = -4125$$
$$\underline{9a + 5y = 5925}$$
$$4a = 1800$$
$$a = 450$$

Substituting 450 for a in equation (2), we obtain

$$450 + y = 825$$
$$y = 375$$

The museum sold 450 adult tickets and 375 youth tickets.

10. Let r represent the number of orchestra seats, let m represent the number of mezzanine seats, and let b represent the number of balcony seats.

$$\begin{cases} r + m + b = 1200 & (1) \\ 65r + 48m + 35b = 55,640 & (2) \\ b = m + 150 & (3) \end{cases}$$

Writing the equations in the system in standard form, we obtain

$$\begin{cases} r + m + b = 1200 & (1) \\ 65r + 48m + 35b = 55,640 & (2) \\ m - b = -150 & (3) \end{cases}$$

Multiply equation (1) by -65 and add the result to equation (2).

$$-65r - 65m - 65b = -78,000$$
$$\underline{65r + 48m + 35b = 55,640}$$
$$-17m - 30b = -22,360 \quad (4)$$

Multiply equation (3) by 17 and add the result to equation (4).

$$17m - 17b = -2,550$$
$$\underline{-17m - 30b = -22,360}$$
$$-47b = -24,910$$
$$b = 530$$

Substituting 530 for b in equation (3), we obtain

$$m - 530 = -150$$
$$m = 380$$

Substituting 530 for b and 380 for m in equation (1), we obtain

$$r + 380 + 530 = 1200$$
$$r + 910 = 1200$$
$$r = 290$$

The theatre contains 290 orchestra seats, 380 mezzanine seats, and 530 balcony seats.

Section 3.4

Are You Ready for This Section?

R1. For the expression $4x - 2y + z$, the coefficients are 4, -2, and 1.

R2.
$$x - 4y = 3$$
$$x - 4y + 4y = 3 + 4y$$
$$x = 4y + 3$$

R3. Substitute $x = 1$, $y = -3$, and $z = 2$:
$$3x - 2y + z = 3(1) - 2(-3) + 2$$
$$= 3 + 6 + 2$$
$$= 11$$

Section 3.4 Quick Checks

1. An m by n rectangular array of numbers is called a <u>matrix</u>.

2. The matrix used to represent a system of linear equations is called an <u>augmented</u> matrix.

3. A 4×3 matrix has <u>4</u> rows and <u>3</u> columns.

4. False. The augmented matrix of a system of two equations containing two unknowns has 2 rows and 3 columns.

5. In the augmented matrix, the first column represents the coefficients on the x variable. The second column represents the coefficients on the y variable. The vertical line signifies the equal signs. The third column represents the constants to the right of the equal sign.

$$\begin{bmatrix} 3 & -1 & | & -10 \\ -5 & 2 & | & 0 \end{bmatrix}$$

6. The system
$$\begin{cases} x+2y-2z=11 \\ -x-2z=4 \\ 4x-y+z-3=0 \end{cases}$$
gets rearranged as
$$\begin{cases} x+2y-2z=11 \\ -x+0y-2z=4 \\ 4x-y+z=3 \end{cases}$$
Thus, the augmented matrix is
$$\begin{bmatrix} 1 & 2 & -2 & | & 11 \\ -1 & 0 & -2 & | & 4 \\ 4 & -1 & 1 & | & 3 \end{bmatrix}.$$

7. Since the augmented matrix has two rows, it represents a system of two equations. Because there are two columns to the left of the vertical bar, the system has two variables. If we call the variables x and y, the system of equations is
$$\begin{cases} x-3y=7 \\ -2x+5y=-3 \end{cases}$$

8. Since the augmented matrix has three rows, it represents a system of three equations. Because there are three columns to the left of the vertical bar, the system has three variables. If we call the variables x, y, and z, the system of equations is
$$\begin{cases} x-3y+2z=4 \\ 3x \quad -z=-1 \\ -x+4y \quad =0 \end{cases}$$

9. $\begin{bmatrix} 1 & -2 & | & 5 \\ -4 & 5 & | & -11 \end{bmatrix}$ $(R_2 = 4r_1 + r_2)$

$= \begin{bmatrix} 1 & -2 & | & 5 \\ 4(1)+(-4) & 4(-2)+5 & | & 4(5)+(-11) \end{bmatrix}$

$= \begin{bmatrix} 1 & -2 & | & 5 \\ 0 & -3 & | & 9 \end{bmatrix}$

10. We want a 0 in row 1, column 2. We accomplish this by multiplying row 2 by -5 and adding the result to row 1. That is, we apply the row operation $R_1 = -5r_2 + r_1$.

$\begin{bmatrix} 1 & 5 & | & 13 \\ 0 & 1 & | & 2 \end{bmatrix}$ $(R_1 = -5r_2 + r_1)$

$= \begin{bmatrix} -5(0)+1 & -5(1)+5 & | & -5(2)+13 \\ 0 & 1 & | & 2 \end{bmatrix}$

$= \begin{bmatrix} 1 & 0 & | & 3 \\ 0 & 1 & | & 2 \end{bmatrix}$

11. True

12. Write the augmented matrix of the system and then put it in row echelon form.

$\begin{bmatrix} 2 & -4 & | & 20 \\ 3 & 1 & | & 16 \end{bmatrix}$ $\left(R_1 = \dfrac{1}{2}r_1\right)$

$= \begin{bmatrix} 1 & -2 & | & 10 \\ 3 & 1 & | & 16 \end{bmatrix}$ $(R_2 = -3r_1 + r_2)$

$= \begin{bmatrix} 1 & -2 & | & 10 \\ 0 & 7 & | & -14 \end{bmatrix}$ $\left(R_2 = \dfrac{1}{7}r_2\right)$

$= \begin{bmatrix} 1 & -2 & | & 10 \\ 0 & 1 & | & -2 \end{bmatrix}$

From row 2, we have that $y = -2$. Row 1 represents the equation $x - 2y = 10$. Back-substitute -2 for y and solve for x.
$$x - 2(-2) = 10$$
$$x + 4 = 10$$
$$x = 6$$
The solution is the ordered pair $(6, -2)$.

13. Write the augmented matrix of the system and then put it in row echelon form.

$\begin{bmatrix} 1 & -1 & 2 & | & 7 \\ 2 & -2 & 1 & | & 11 \\ -3 & 1 & -3 & | & -14 \end{bmatrix}$ $\left(\begin{matrix} R_2 = -2r_1 + r_2 \\ R_3 = 3r_1 + r_3 \end{matrix}\right)$

$= \begin{bmatrix} 1 & -1 & 2 & | & 7 \\ 0 & 0 & -3 & | & -3 \\ 0 & -2 & 3 & | & 7 \end{bmatrix}$ (Interchange r_1 and r_2)

$= \begin{bmatrix} 1 & -1 & 2 & | & 7 \\ 0 & -2 & 3 & | & 7 \\ 0 & 0 & -3 & | & -3 \end{bmatrix}$ $\left(\begin{matrix} R_2 = -\dfrac{1}{2}r_2 \\ R_3 = -\dfrac{1}{3}r_3 \end{matrix}\right)$

$= \begin{bmatrix} 1 & -1 & 2 & | & 7 \\ 0 & 1 & -\dfrac{3}{2} & | & -\dfrac{7}{2} \\ 0 & 0 & 1 & | & 1 \end{bmatrix}$

Write the system of equations that corresponds to the row-echelon matrix
$$\begin{cases} x-y+2z=7 & (1) \\ y-\dfrac{3}{2}z=-\dfrac{7}{2} & (2) \\ z=1 & (3) \end{cases}$$

Substituting 1 for z in equation (2), we obtain
$$y - \frac{3}{2}(1) = -\frac{7}{2}$$
$$y - \frac{3}{2} = -\frac{7}{2}$$
$$y = -\frac{4}{2} = -2$$

Substituting 1 for z and -2 for y in equation (1), we obtain

$$x-(-2)+2(1)=7$$
$$x+2+2=7$$
$$x+4=7$$
$$x=3$$

The solution is the ordered triple $(3, -2, 1)$.

14. Write the augmented matrix of the system and then put it in row echelon form.

$$\begin{bmatrix} 2 & 3 & -2 & | & 1 \\ 2 & 0 & -4 & | & -9 \\ 4 & 6 & -1 & | & 14 \end{bmatrix} \quad \left(R_1 = \tfrac{1}{2}r_1\right)$$

$$=\begin{bmatrix} 1 & \tfrac{3}{2} & -1 & | & \tfrac{1}{2} \\ 2 & 0 & -4 & | & -9 \\ 4 & 6 & -1 & | & 14 \end{bmatrix} \quad \begin{pmatrix} R_2=-2r_1+r_2 \\ R_3=-4r_1+r_3 \end{pmatrix}$$

$$=\begin{bmatrix} 1 & \tfrac{3}{2} & -1 & | & \tfrac{1}{2} \\ 0 & -3 & -2 & | & -10 \\ 0 & 0 & 3 & | & 12 \end{bmatrix} \quad \begin{pmatrix} R_2=-\tfrac{1}{3}r_2 \\ R_3=\tfrac{1}{3}r_3 \end{pmatrix}$$

$$=\begin{bmatrix} 1 & \tfrac{3}{2} & -1 & | & \tfrac{1}{2} \\ 0 & 1 & \tfrac{2}{3} & | & \tfrac{10}{3} \\ 0 & 0 & 1 & | & 4 \end{bmatrix}$$

Write the system of equations that corresponds to the row-echelon matrix

$$\begin{cases} x+\tfrac{3}{2}y-z=\tfrac{1}{2} & (1) \\ y+\tfrac{2}{3}z=\tfrac{10}{3} & (2) \\ z=4 & (3) \end{cases}$$

Substituting 4 for z in equation (2), we obtain

$$y+\tfrac{2}{3}(4)=\tfrac{10}{3}$$
$$y+\tfrac{8}{3}=\tfrac{10}{3}$$
$$y=\tfrac{2}{3}$$

Substituting 4 for z and $\tfrac{2}{3}$ for y in equation (1), we obtain

$$x+\tfrac{3}{2}\left(\tfrac{2}{3}\right)-(4)=\tfrac{1}{2}$$
$$x+1-4=\tfrac{1}{2}$$
$$x=\tfrac{1}{2}+3=\tfrac{7}{2}$$

The solution is the ordered triple $\left(\tfrac{7}{2},\tfrac{2}{3},4\right)$.

15. Write the augmented matrix of the system and then put it in row echelon form.

$$\begin{bmatrix} 2 & 5 & | & -6 \\ -6 & -15 & | & 18 \end{bmatrix} \quad \left(R_1=\tfrac{1}{2}r_1\right)$$

$$=\begin{bmatrix} 1 & \tfrac{5}{2} & | & -3 \\ -6 & -15 & | & 18 \end{bmatrix} \quad (R_2=6r_1+r_2)$$

$$=\begin{bmatrix} 1 & \tfrac{5}{2} & | & -3 \\ 0 & 0 & | & 0 \end{bmatrix}$$

Write the system of equations that corresponds to the row-echelon matrix.

$$\begin{cases} x+\tfrac{5}{2}y=-3 & (1) \\ 0=0 & (2) \end{cases}$$

The statement $0=0$ in equation (2) indicates that the system is dependent and has an infinite number of solutions.
The solution to the system is
$$\{(x,y)\,|\,2x+5y=-6\}.$$

16. Write the augmented matrix of the system and then put it in row echelon form.

$$\begin{bmatrix} 1 & 1 & -3 & | & 8 \\ 2 & 3 & -10 & | & 19 \\ -1 & -2 & 7 & | & -11 \end{bmatrix} \quad \begin{pmatrix} R_2=-2r_1+r_2 \\ R_3=r_1+r_3 \end{pmatrix}$$

$$=\begin{bmatrix} 1 & 1 & -3 & | & 8 \\ 0 & 1 & -4 & | & 3 \\ 0 & -1 & 4 & | & -3 \end{bmatrix} \quad (R_3=r_2+r_3)$$

$$=\begin{bmatrix} 1 & 1 & -3 & | & 8 \\ 0 & 1 & -4 & | & 3 \\ 0 & 0 & 0 & | & 0 \end{bmatrix}$$

Write the system of equations that corresponds to the row-echelon matrix.

$$\begin{cases} x+y-3z=8 & (1) \\ y-4z=3 & (2) \\ 0=0 & (3) \end{cases}$$

The statement $0 = 0$ in equation (3) indicates that the system is dependent and has an infinite number of solutions.

Solve equation (2) for y.
$$y - 4z = 3$$
$$y = 4z + 3$$

Substituting $4z + 3$ for y in equation (1), we obtain
$$x + (4z + 3) - 3z = 8$$
$$x + z + 3 = 8$$
$$x = -z + 5$$

The solution to the system is
$\{(x, y, z) | x = -z + 5,\ y + 4z + 3,\ z$ is any real number$\}$

17. Write the augmented matrix of the system and then put it in row echelon form.

$$\begin{bmatrix} -2 & 3 & | & 4 \\ 10 & -15 & | & 2 \end{bmatrix} \quad \left(R_1 = -\frac{1}{2} r_1 \right)$$

$$= \begin{bmatrix} 1 & -\dfrac{3}{2} & | & -2 \\ 10 & -15 & | & 2 \end{bmatrix} \quad \left(R_2 = -10 r_1 + r_2 \right)$$

$$= \begin{bmatrix} 1 & -\dfrac{3}{2} & | & -2 \\ 0 & 0 & | & 22 \end{bmatrix}$$

Write the system of equations that corresponds to the row-echelon matrix.

$$\begin{cases} x - \dfrac{3}{2} y = -2 & (1) \\ 0 = 22 & (2) \end{cases}$$

The statement $0 = 22$ in equation (2) indicates that the system is inconsistent. The system has no solution. The solution set is $\varnothing$ or { }.

18. Write the augmented matrix of the system and then put it in row echelon form.

$$\begin{bmatrix} -1 & 2 & -1 & | & 5 \\ 2 & 1 & 4 & | & 3 \\ 3 & -1 & 5 & | & 0 \end{bmatrix} \quad \left(R_1 = -1 \cdot r_1 \right)$$

$$= \begin{bmatrix} 1 & -2 & 1 & | & -5 \\ 2 & 1 & 4 & | & 3 \\ 3 & -1 & 5 & | & 0 \end{bmatrix} \quad \left(\begin{matrix} R_2 = -2 r_1 + r_2 \\ R_3 = -3 r_1 + r_3 \end{matrix} \right)$$

$$= \begin{bmatrix} 1 & -2 & 1 & | & -5 \\ 0 & 5 & 2 & | & 13 \\ 0 & 5 & 2 & | & 15 \end{bmatrix} \quad \left(R_2 = \frac{1}{5} r_2 \right)$$

$$= \begin{bmatrix} 1 & -2 & 1 & | & -5 \\ 0 & 1 & \dfrac{2}{5} & | & \dfrac{13}{5} \\ 0 & 5 & 2 & | & 15 \end{bmatrix} \quad \left(R_3 = -5 r_2 + r_3 \right)$$

$$= \begin{bmatrix} 1 & -2 & 1 & | & -5 \\ 0 & 1 & \dfrac{2}{5} & | & \dfrac{13}{5} \\ 0 & 0 & 0 & | & 2 \end{bmatrix}$$

Write the system of equations that corresponds to the row-echelon matrix.

$$\begin{cases} x - 2y + z = -5 & (1) \\ y + \dfrac{2}{5} z = \dfrac{13}{5} & (2) \\ 0 = 2 & (3) \end{cases}$$

The statement $0 = 2$ in equation (3) indicates that the system is inconsistent. The system has no solution. The solution set is $\varnothing$ or { }.

3.4 Exercises

19. $\begin{bmatrix} 1 & -3 & | & 2 \\ 2 & 5 & | & 1 \end{bmatrix}$

21. $\begin{bmatrix} 1 & 1 & 1 & | & 3 \\ 2 & -1 & 3 & | & 1 \\ -4 & 2 & -5 & | & -3 \end{bmatrix}$

23. Write each equation in the system in standard form.
$$\begin{cases} -x + y = 2 \\ 5x + y = -5 \end{cases}$$
Thus, the augmented matrix is
$$\begin{bmatrix} -1 & 1 & | & 2 \\ 5 & 1 & | & -5 \end{bmatrix}$$

25. Write each equation in the system in standard form.
$$\begin{cases} x \quad\ \ + z = 2 \\ 2x + y \quad\ = 13 \\ x - y + 4z = -4 \end{cases}$$
Thus, the augmented matrix is
$$\begin{bmatrix} 1 & 0 & 1 & | & 2 \\ 2 & 1 & 0 & | & 13 \\ 1 & -1 & 4 & | & -4 \end{bmatrix}$$

27. $\begin{cases} 2x + 5y = 3 \\ -4x + y = 10 \end{cases}$

29. $\begin{cases} x + 5y - 3z = 2 \\ \quad\quad 3y - z = -5 \\ 4x \quad\quad + 8z = 6 \end{cases}$

31. $\begin{cases} x - 2y + 9z = 2 \\ \quad\quad y - 5z = 8 \\ \quad\quad\quad z = \dfrac{4}{3} \end{cases}$

33. a. $\begin{bmatrix} 1 & -3 & | & 2 \\ -2 & 5 & | & 1 \end{bmatrix}$ $\quad (R_2 = 2r_1 + r_2)$

$= \begin{bmatrix} 1 & -3 & | & 2 \\ 2(1) + (-2) & 2(-3) + 5 & | & 2(2) + 1 \end{bmatrix}$

$= \begin{bmatrix} 1 & -3 & | & 2 \\ 0 & -1 & | & 5 \end{bmatrix}$

b. $\begin{bmatrix} 1 & -3 & | & 2 \\ 0 & -1 & | & 5 \end{bmatrix}$ $\quad (R_2 = -1 \cdot r_2)$

$= \begin{bmatrix} 1 & -3 & | & 2 \\ -1(0) & -1(-1) & | & -1(5) \end{bmatrix}$

$= \begin{bmatrix} 1 & -3 & | & 2 \\ 0 & 1 & | & -5 \end{bmatrix}$

35. a. $\begin{bmatrix} 1 & 1 & -1 & | & 4 \\ 2 & 5 & 3 & | & -3 \\ -1 & -3 & 2 & | & 1 \end{bmatrix}$ $\quad (R_2 = -2r_1 + r_2)$

$= \begin{bmatrix} 1 & 1 & -1 & | & 4 \\ -2(1) + 2 & -2(1) + 5 & -2(-1) + 3 & | & -2(4) + (-3) \\ -1 & -3 & 2 & | & 1 \end{bmatrix}$

$= \begin{bmatrix} 1 & 1 & -1 & | & 4 \\ 0 & 3 & 5 & | & -11 \\ -1 & -3 & 2 & | & 1 \end{bmatrix}$

b. $\begin{bmatrix} 1 & 1 & -1 & | & 4 \\ 0 & 3 & 5 & | & -11 \\ -1 & -3 & 2 & | & 1 \end{bmatrix}$ $\quad (R_3 = r_1 + r_3)$

$= \begin{bmatrix} 1 & 1 & -1 & | & 4 \\ 0 & 3 & 5 & | & -11 \\ 1 + (-1) & 1 + (-3) & -1 + 2 & | & 4 + 1 \end{bmatrix}$

$= \begin{bmatrix} 1 & 1 & -1 & | & 4 \\ 0 & 3 & 5 & | & -11 \\ 0 & -2 & 1 & | & 5 \end{bmatrix}$

37. a. $\begin{bmatrix} 1 & 1 & 1 & | & 4 \\ 0 & 5 & 3 & | & -3 \\ 0 & -4 & 2 & | & 8 \end{bmatrix}$ $\quad (R_2 = r_3 + r_2)$

$= \begin{bmatrix} 1 & 1 & 1 & | & 4 \\ 0+0 & -4+5 & 2+3 & | & 8+(-3) \\ 0 & -4 & 2 & | & 8 \end{bmatrix}$

$= \begin{bmatrix} 1 & 1 & 1 & | & 4 \\ 0 & 1 & 5 & | & 5 \\ 0 & -4 & 2 & | & 8 \end{bmatrix}$

b. $\begin{bmatrix} 1 & 1 & 1 & | & 4 \\ 0 & 1 & 5 & | & 5 \\ 0 & -4 & 2 & | & 8 \end{bmatrix}$ $\quad \left(R_3 = \dfrac{1}{2} r_3\right)$

$= \begin{bmatrix} 1 & 1 & 1 & | & 4 \\ 0 & 1 & 5 & | & 5 \\ \dfrac{1}{2}(0) & \dfrac{1}{2}(-4) & \dfrac{1}{2}(2) & | & \dfrac{1}{2}(8) \end{bmatrix}$

$= \begin{bmatrix} 1 & 1 & 1 & | & 4 \\ 0 & 1 & 5 & | & 5 \\ 0 & -2 & 1 & | & 4 \end{bmatrix}$

39. Write the augmented matrix of the system and then put it in row echelon form.

$\begin{bmatrix} 2 & 3 & | & 1 \\ -1 & 4 & | & -28 \end{bmatrix}$ $\quad \left(R_1 = \dfrac{1}{2} r_1\right)$

$= \begin{bmatrix} 1 & \dfrac{3}{2} & | & \dfrac{1}{2} \\ -1 & 4 & | & -28 \end{bmatrix}$ $\quad (R_2 = r_1 + r_2)$

$= \begin{bmatrix} 1 & \dfrac{3}{2} & | & \dfrac{1}{2} \\ 0 & \dfrac{11}{2} & | & -\dfrac{55}{2} \end{bmatrix}$ $\quad \left(R_2 = \dfrac{2}{11} r_2\right)$

$= \begin{bmatrix} 1 & \dfrac{3}{2} & | & \dfrac{1}{2} \\ 0 & 1 & | & -5 \end{bmatrix}$

Write the system of equations that corresponds to the row-echelon matrix.

$\begin{cases} x + \dfrac{3}{2} y = \dfrac{1}{2} & (1) \\ y = -5 & (2) \end{cases}$

Substituting -5 for y in equation (1), we obtain

$x + \dfrac{3}{2}(-5) = \dfrac{1}{2}$

$x - \dfrac{15}{2} = \dfrac{1}{2}$

$x = \dfrac{16}{2} = 8$

The solution is the ordered pair $(8, -5)$.

41. Write the augmented matrix of the system and then put it in row echelon form.

$\begin{bmatrix} 1 & 5 & -2 & | & 23 \\ -2 & -3 & 5 & | & -11 \\ 3 & 2 & 1 & | & 4 \end{bmatrix}$ $\quad \begin{pmatrix} R_2 = 2r_1 + r_2 \\ R_3 = -3r_1 + r_3 \end{pmatrix}$

$= \begin{bmatrix} 1 & 5 & -2 & | & 23 \\ 0 & 7 & 1 & | & 35 \\ 0 & -13 & 7 & | & -65 \end{bmatrix}$ $\quad \left(R_2 = \dfrac{1}{7} r_2\right)$

$= \begin{bmatrix} 1 & 5 & -2 & | & 23 \\ 0 & 1 & \dfrac{1}{7} & | & 5 \\ 0 & -13 & 7 & | & -65 \end{bmatrix}$ $\quad (R_3 = 13r_2 + r_3)$

$= \begin{bmatrix} 1 & 5 & -2 & | & 23 \\ 0 & 1 & \dfrac{1}{7} & | & 5 \\ 0 & 0 & \dfrac{62}{7} & | & 0 \end{bmatrix}$ $\quad \left(R_3 = \dfrac{7}{62} r_3\right)$

$= \begin{bmatrix} 1 & 5 & -2 & | & 23 \\ 0 & 1 & \dfrac{1}{7} & | & 5 \\ 0 & 0 & 1 & | & 0 \end{bmatrix}$

Write the system of equations that corresponds to the row-echelon matrix.

$\begin{cases} x + 5y - 2z = 23 & (1) \\ y + \dfrac{1}{7} z = 5 & (2) \\ z = 0 & (3) \end{cases}$

By equation (3), we know $z = 0$. Substituting 0 for z in equation (2), we obtain

$y + \dfrac{1}{7}(0) = 5$

$y = 5$

Substituting 0 for z and 5 for y in equation (1), we obtain

$x + 5(5) - 2(0) = 23$

$x + 25 = 23$

$x = -2$

The solution is the ordered triple $(-2, 5, 0)$.

43. Write the augmented matrix of the system and then put it in row echelon form.

$$\begin{bmatrix} 4 & 5 & 0 & | & 0 \\ -8 & 10 & -2 & | & -13 \\ 0 & 15 & 4 & | & -7 \end{bmatrix} \quad (R_2 = 2r_1 + r_2)$$

$$= \begin{bmatrix} 4 & 5 & 0 & | & 0 \\ 0 & 20 & -2 & | & -13 \\ 0 & 15 & 4 & | & -7 \end{bmatrix} \left(R_2 = \frac{1}{20} r_2\right)$$

$$= \begin{bmatrix} 4 & 5 & 0 & | & 0 \\ 0 & 1 & -\frac{1}{10} & | & -\frac{13}{20} \\ 0 & 15 & 4 & | & -7 \end{bmatrix} \quad (R_3 = -15 r_2 + r_3)$$

$$= \begin{bmatrix} 4 & 5 & 0 & | & 0 \\ 0 & 1 & -\frac{1}{10} & | & -\frac{13}{20} \\ 0 & 0 & \frac{11}{2} & | & \frac{11}{4} \end{bmatrix} \left(R_3 = \frac{2}{11} r_3\right)$$

$$= \begin{bmatrix} 4 & 5 & 0 & | & 0 \\ 0 & 1 & -\frac{1}{10} & | & -\frac{13}{20} \\ 0 & 0 & 1 & | & \frac{1}{2} \end{bmatrix}$$

Write the system of equations that corresponds to the row-echelon matrix.

$$\begin{cases} 4x + 5y = 0 & (1) \\ y - \frac{1}{10} z = -\frac{13}{20} & (2) \\ z = \frac{1}{2} & (3) \end{cases}$$

Substituting $\frac{1}{2}$ for z in equation (2), we obtain

$$y - \frac{1}{10}\left(\frac{1}{2}\right) = -\frac{13}{20}$$

$$y - \frac{1}{20} = -\frac{13}{20}$$

$$y = -\frac{12}{20} = -\frac{3}{5}$$

Substituting $-\frac{3}{5}$ for y in equation (1), we obtain

$$4x + 5\left(-\frac{3}{5}\right) = 0$$

$$4x - 3 = 0$$

$$4x = 3$$

$$x = \frac{3}{4}$$

The solution is the ordered triple $\left(\frac{3}{4}, -\frac{3}{5}, \frac{1}{2}\right)$.

45. Write the augmented matrix of the system and then put it in row echelon form.

$$\begin{bmatrix} 1 & -3 & | & 3 \\ -2 & 6 & | & -6 \end{bmatrix} \quad (R_2 = 2r_1 + r_2)$$

$$= \begin{bmatrix} 1 & -3 & | & 3 \\ 0 & 0 & | & 0 \end{bmatrix}$$

Write the system of equations that corresponds to the row-echelon matrix.

$$\begin{cases} x - 3y = 3 & (1) \\ 0 = 0 & (2) \end{cases}$$

The statement $0 = 0$ in equation (2) indicates that the system is dependent and has an infinite number of solutions.
The solution to the system is

$$\{(x, y) \mid x - 3y = 3\}.$$

47. Write the augmented matrix of the system and then put it in row echelon form.

$$\begin{bmatrix} 1 & -2 & 1 & | & 3 \\ -2 & 4 & -1 & | & -5 \\ -8 & 16 & 1 & | & -21 \end{bmatrix} \begin{pmatrix} R_2 = 2r_1 + r_2 \\ R_3 = 8r_1 + r_3 \end{pmatrix}$$

$$= \begin{bmatrix} 1 & -2 & 1 & | & 3 \\ 0 & 0 & 1 & | & 1 \\ 0 & 0 & 9 & | & 3 \end{bmatrix} \quad (R_3 = -9r_2 + r_3)$$

$$= \begin{bmatrix} 1 & -2 & 1 & | & 3 \\ 0 & 0 & 1 & | & 1 \\ 0 & 0 & 0 & | & -6 \end{bmatrix}$$

Write the system of equations that corresponds to the row-echelon matrix.

$$\begin{cases} x - 2y + z = 3 & (1) \\ z = 1 & (2) \\ 0 = -6 & (3) \end{cases}$$

The statement $0 = -6$ in equation (3) indicates that the system is inconsistent. The system has no solution. The solution set is $\varnothing$ or { }.

49. Write the augmented matrix of the system and then put it in row echelon form.

$$\begin{bmatrix} 1 & 1 & -2 & | & 4 \\ -4 & 0 & 3 & | & -4 \\ -2 & 2 & -1 & | & 4 \end{bmatrix} \begin{pmatrix} R_2 = 4r_1 + r_2 \\ R_3 = 2r_1 + r_3 \end{pmatrix}$$

$$= \begin{bmatrix} 1 & 1 & -2 & | & 4 \\ 0 & 4 & -5 & | & 12 \\ 0 & 4 & -5 & | & 12 \end{bmatrix} \left(R_2 = \frac{1}{4} r_2\right)$$

$$= \begin{bmatrix} 1 & 1 & -2 & | & 4 \\ 0 & 1 & -\dfrac{5}{4} & | & 3 \\ 0 & 4 & -5 & | & 12 \end{bmatrix} \quad (R_3 = -4r_2 + r_3)$$

$$= \begin{bmatrix} 1 & 1 & -2 & | & 4 \\ 0 & 1 & -\dfrac{5}{4} & | & 3 \\ 0 & 0 & 0 & | & 0 \end{bmatrix}$$

Write the system of equations that corresponds to the row-echelon matrix.

$$\begin{cases} x + y - 2z = 4 & (1) \\ y - \dfrac{4}{5}z = 3 & (2) \\ 0 = 0 & (3) \end{cases}$$

The statement $0 = 0$ in equation (3) indicates that the system is dependent and has an infinite number of solutions.
Solve equation (2) for y.

$$y = \frac{5}{4}z + 3$$

Substituting $\dfrac{5}{4}z + 3$ for y in equation (1), we obtain

$$x + \left(\frac{5}{4}z + 3 \right) - 2z = 4$$

$$x + \frac{5}{4}z - \frac{8}{4}z = 1$$

$$x = \frac{3}{4}z + 1$$

The solution is the ordered triple

$$\left\{ (x, y, z) \mid x = \frac{3}{4}z + 1, y = \frac{5}{4}z + 3, \right.$$

$$\left. z \text{ is any real number} \right\}$$

51. $\begin{cases} x + 4y = -5 & (1) \\ y = -2 & (2) \end{cases}$

This system is consistent and independent.
Substituting -2 for y in equation (1), we obtain
$$x + 4(-2) = -5$$
$$x - 8 = -5$$
$$x = 3$$
The solution is the ordered pair $(3, -2)$.

53. $\begin{cases} x + 3y - 2z = 6 & (1) \\ y + 5z = -2 & (2) \\ 0 = 4 & (3) \end{cases}$

This system is inconsistent. The solution is $\varnothing$ or $\{ \ \}$.

55. $\begin{cases} x - 2y - z = 3 & (1) \\ y - 2z = -8 & (2) \\ z = 5 & (3) \end{cases}$

This system is consistent and independent.
Substituting 5 for z in equation (2), we obtain
$$y - 2(5) = -8$$
$$y - 10 = -8$$
$$y = 2$$
Substituting 5 for z and 2 for y in equation (1), we obtain
$$x - 2(2) - 5 = 3$$
$$x - 4 - 5 = 3$$
$$x - 9 = 3$$
$$x = 12$$
The solution is the ordered triple $(12, \ 2, \ 5)$.

57. Write the augmented matrix of the system and then put it in row echelon form.

$$\begin{bmatrix} 1 & -3 & | & 18 \\ 2 & 1 & | & 1 \end{bmatrix} \quad (R_2 = -2r_1 + r_2)$$

$$= \begin{bmatrix} 1 & -3 & | & 18 \\ 0 & 7 & | & -35 \end{bmatrix} \quad \left(R_2 = \frac{1}{7}r_2 \right)$$

$$= \begin{bmatrix} 1 & -3 & | & 18 \\ 0 & 1 & | & -5 \end{bmatrix}$$

Write the system of equations that corresponds to the row-echelon matrix

$$\begin{cases} x - 3y = 18 & (1) \\ y = -5 & (2) \end{cases}$$

This system is consistent and independent.
Substituting -5 for y in equation (1), we obtain
$$x - 3(-5) = 18$$
$$x + 15 = 18$$
$$x = 3$$
The solution is the ordered pair $(3, -5)$.

59. Write the augmented matrix of the system and then put it in row echelon form.

$$\begin{bmatrix} 2 & 4 & | & 10 \\ 1 & 2 & | & 3 \end{bmatrix} \quad (\text{Interchange } r_1 \text{ and } r_2)$$

$$= \begin{bmatrix} 1 & 2 & | & 3 \\ 2 & 4 & | & 10 \end{bmatrix} \quad (R_2 = -2r_1 + r_2)$$

$$= \begin{bmatrix} 1 & 2 & | & 3 \\ 0 & 0 & | & 4 \end{bmatrix}$$

The system is inconsistent. The system has no solution. The solution set is $\varnothing$ or $\{ \ \}$.

61. Write the augmented matrix of the system and then put it in row echelon form.

$$\begin{bmatrix} 1 & -6 & | & 8 \\ 2 & 8 & | & -9 \end{bmatrix} \qquad (R_2 = -2r_1 + r_2)$$

$$= \begin{bmatrix} 1 & -6 & | & 8 \\ 0 & 20 & | & -25 \end{bmatrix} \qquad \left(R_2 = \frac{1}{20}r_2\right)$$

$$= \begin{bmatrix} 1 & -6 & | & 8 \\ 0 & 1 & | & -\frac{5}{4} \end{bmatrix}$$

Write the system of equations that corresponds to the row-echelon matrix

$$\begin{cases} x - 6y = 8 & (1) \\ \quad\ y = -\dfrac{5}{4} & (2) \end{cases}$$

This system is consistent and independent.

Substituting $-\dfrac{5}{4}$ for y in equation (1), we obtain

$$x - 6\left(-\frac{5}{4}\right) = 8$$

$$x + \frac{15}{2} = 8$$

$$x = \frac{1}{2}$$

The solution is the ordered pair $\left(\dfrac{1}{2}, -\dfrac{5}{4}\right)$.

63. Write the augmented matrix of the system and then put it in row echelon form.

$$\begin{bmatrix} 4 & -1 & | & 8 \\ 2 & -\frac{1}{2} & | & 4 \end{bmatrix} \qquad \left(R_1 = \frac{1}{4}r_1\right)$$

$$= \begin{bmatrix} 1 & -\frac{1}{4} & | & 2 \\ 2 & -\frac{1}{2} & | & 4 \end{bmatrix} \qquad (R_2 = -2r_1 + r_2)$$

$$= \begin{bmatrix} 1 & -\frac{1}{4} & | & 2 \\ 0 & 0 & | & 0 \end{bmatrix}$$

The system is dependent. The solution to the system is $\{(x, y)\mid 4x - y = 8\}$.

65. Write the augmented matrix of the system and then put it in row echelon form.

$$\begin{bmatrix} 1 & 1 & 1 & | & 0 \\ 2 & -3 & 1 & | & 19 \\ -3 & 1 & -2 & | & -15 \end{bmatrix} \qquad \left(\begin{matrix} R_2 = -2r_1 + r_2 \\ R_3 = 3r_1 + r_3 \end{matrix}\right)$$

$$= \begin{bmatrix} 1 & 1 & 1 & | & 0 \\ 0 & -5 & -1 & | & 19 \\ 0 & 4 & 1 & | & -15 \end{bmatrix} \qquad \left(R_2 = -\frac{1}{5}r_2\right)$$

$$= \begin{bmatrix} 1 & 1 & 1 & | & 0 \\ 0 & 1 & \frac{1}{5} & | & -\frac{19}{5} \\ 0 & 4 & 1 & | & -15 \end{bmatrix} \qquad (R_3 = -4r_2 + r_3)$$

$$= \begin{bmatrix} 1 & 1 & 1 & | & 0 \\ 0 & 1 & \frac{1}{5} & | & -\frac{19}{5} \\ 0 & 0 & \frac{1}{5} & | & \frac{1}{5} \end{bmatrix} \qquad (R_3 = 5r_3)$$

$$= \begin{bmatrix} 1 & 1 & 1 & | & 0 \\ 0 & 1 & \frac{1}{5} & | & -\frac{19}{5} \\ 0 & 0 & 1 & | & 1 \end{bmatrix}$$

Write the system of equations that corresponds to the row-echelon matrix

$$\begin{cases} x + y + z = 0 & (1) \\ \quad\ y + \dfrac{1}{5}z = -\dfrac{19}{5} & (2) \\ \qquad\qquad z = 1 & (3) \end{cases}$$

This system is consistent and independent.
Substituting 1 for z in equation (2), we obtain

$$y + \frac{1}{5}(1) = -\frac{19}{5}$$

$$y + \frac{1}{5} = -\frac{19}{5}$$

$$y = -\frac{20}{4} = -4$$

Substituting 1 for z and -4 for y in equation (1), we obtain

$$x + (-4) + 1 = 0$$

$$x - 3 = 0$$

$$x = 3$$

The solution is the ordered triple $(3, -4, 1)$.

67. Write the augmented matrix of the system and then put it in row echelon form.

$$\begin{bmatrix} 2 & 1 & -1 & | & 13 \\ -1 & -3 & 2 & | & -14 \\ -3 & 2 & -3 & | & 3 \end{bmatrix} \quad \text{(Interchange } r_1 \text{ and } r_2\text{)}$$

$$= \begin{bmatrix} -1 & -3 & 2 & | & -14 \\ 2 & 1 & -1 & | & 13 \\ -3 & 2 & -3 & | & 3 \end{bmatrix} \quad (R_1 = -1 \cdot r_1)$$

$$= \begin{bmatrix} 1 & 3 & -2 & | & 14 \\ 2 & 1 & -1 & | & 13 \\ -3 & 2 & -3 & | & 3 \end{bmatrix} \quad \begin{pmatrix} R_2 = -2r_1 + r_2 \\ R_3 = 3r_1 + r_3 \end{pmatrix}$$

$$= \begin{bmatrix} 1 & 3 & -2 & | & 14 \\ 0 & -5 & 3 & | & -15 \\ 0 & 11 & -9 & | & 45 \end{bmatrix} \quad \left(R_2 = -\frac{1}{5}r_2\right)$$

$$= \begin{bmatrix} 1 & 3 & -2 & | & 14 \\ 0 & 1 & -\frac{3}{5} & | & 3 \\ 0 & 11 & -9 & | & 45 \end{bmatrix} \quad (R_3 = -11r_2 + r_3)$$

$$= \begin{bmatrix} 1 & 3 & -2 & | & 14 \\ 0 & 1 & -\frac{3}{5} & | & 3 \\ 0 & 0 & -\frac{12}{5} & | & 12 \end{bmatrix} \quad \left(R_3 = -\frac{5}{12}r_3\right)$$

$$= \begin{bmatrix} 1 & 3 & -2 & | & 14 \\ 0 & 1 & -\frac{3}{5} & | & 3 \\ 0 & 0 & 1 & | & -5 \end{bmatrix}$$

Write the system of equations that corresponds to the row-echelon matrix

$$\begin{cases} x + 3y - 2z = 14 & (1) \\ y - \dfrac{3}{5}z = 3 & (2) \\ z = -5 & (3) \end{cases}$$

This system is consistent and independent. Substituting –5 for z in equation (2), we obtain

$$y - \frac{3}{5}(-5) = 3$$
$$y + 3 = 3$$
$$y = 0$$

Substituting –5 for z and 0 for y in equation (1), we obtain

$$x + 3(0) - 2(-5) = 14$$
$$x + 10 = 14$$
$$x = 4$$

The solution is the ordered triple $(4, 0, -5)$.

69. Write the augmented matrix of the system and then put it in row echelon form.

$$\begin{bmatrix} 2 & -1 & 3 & | & 1 \\ -1 & 3 & 1 & | & -4 \\ 3 & 1 & 7 & | & -2 \end{bmatrix} \quad \text{(Interchange } r_1 \text{ and } r_2\text{)}$$

$$= \begin{bmatrix} -1 & 3 & 1 & | & -4 \\ 2 & -1 & 3 & | & 1 \\ 3 & 1 & 7 & | & -2 \end{bmatrix} \quad (R_1 = -1 \cdot r_1)$$

$$= \begin{bmatrix} 1 & -3 & -1 & | & 4 \\ 2 & -1 & 3 & | & 1 \\ 3 & 1 & 7 & | & -2 \end{bmatrix} \quad \begin{pmatrix} R_2 = -2r_1 + r_2 \\ R_3 = -3r_1 + r_3 \end{pmatrix}$$

$$= \begin{bmatrix} 1 & -3 & -1 & | & 4 \\ 0 & 5 & 5 & | & -7 \\ 0 & 10 & 10 & | & -14 \end{bmatrix} \quad (R_3 = -2r_2 + r_3)$$

$$= \begin{bmatrix} 1 & -3 & -1 & | & 4 \\ 0 & 5 & 5 & | & -7 \\ 0 & 0 & 0 & | & 0 \end{bmatrix}$$

The system is dependent and has an infinite number of solutions.

Write the system of equations that corresponds to the row-echelon matrix

$$\begin{cases} x - 3y - z = 4 & (1) \\ 5y + 5z = -7 & (2) \\ 0 = 0 & (3) \end{cases}$$

Solve equation (2) for y.

$$5y + 5z = -7$$
$$5y = -5z - 7$$
$$y = -z - \frac{7}{5} = -z - 1.4$$

Substituting $-z - 1.4$ for y in equation (1), we obtain

$$x - 3(-z - 1.4) - z = 4$$
$$x + 3z + 4.2 - z = 4$$
$$x + 2z + 4.2 = 4$$
$$x = -2z - 0.2$$

The solution to the system is
$$\{(x, y, z) \mid x = -2z - 0.2,$$
$$y = -z - 1.4, \ z \text{ is any real number}\}.$$

71. Write the augmented matrix of the system and then put it in row echelon form.

$$\begin{bmatrix} 3 & 1 & -4 & | & 0 \\ -2 & -3 & 1 & | & 5 \\ -1 & -5 & -2 & | & 3 \end{bmatrix} \quad (\text{Interchange } r_1 \text{ and } r_3)$$

$$= \begin{bmatrix} -1 & -5 & -2 & | & 3 \\ -2 & -3 & 1 & | & 5 \\ 3 & 1 & -4 & | & 0 \end{bmatrix} \quad (R_1 = -1 \cdot r_1)$$

$$= \begin{bmatrix} 1 & 5 & 2 & | & -3 \\ -2 & -3 & 1 & | & 5 \\ 3 & 1 & -4 & | & 0 \end{bmatrix} \quad \begin{pmatrix} R_2 = 2r_1 + r_2 \\ R_3 = -3r_1 + r_3 \end{pmatrix}$$

$$= \begin{bmatrix} 1 & 5 & 2 & | & -3 \\ 0 & 7 & 5 & | & -1 \\ 0 & -14 & -10 & | & 9 \end{bmatrix} \quad (R_3 = 2r_2 + r_3)$$

$$= \begin{bmatrix} 1 & 5 & 2 & | & -3 \\ 0 & 7 & 5 & | & -1 \\ 0 & 0 & 0 & | & 7 \end{bmatrix}$$

The system is inconsistent. The system has no solution. The solution set is $\varnothing$ or $\{\ \}$.

73. Write the augmented matrix of the system and then put it in row echelon form.

$$\begin{bmatrix} 2 & -1 & 3 & | & -1 \\ 3 & 1 & -4 & | & 3 \\ 1 & 7 & -2 & | & 2 \end{bmatrix} \quad (\text{Interchange } r_1 \text{ and } r_3)$$

$$= \begin{bmatrix} 1 & 7 & -2 & | & 2 \\ 3 & 1 & -4 & | & 3 \\ 2 & -1 & 3 & | & -1 \end{bmatrix} \quad \begin{pmatrix} R_2 = -3r_1 + r_2 \\ R_3 = -2r_1 + r_3 \end{pmatrix}$$

$$= \begin{bmatrix} 1 & 7 & -2 & | & 2 \\ 0 & -20 & 2 & | & -3 \\ 0 & -15 & 7 & | & -5 \end{bmatrix} \quad \left(R_2 = -\frac{1}{20}r_2\right)$$

$$= \begin{bmatrix} 1 & 7 & -2 & | & 2 \\ 0 & 1 & -\frac{1}{10} & | & \frac{3}{20} \\ 0 & -15 & 7 & | & -5 \end{bmatrix} \quad (R_3 = 15r_2 + r_3)$$

$$= \begin{bmatrix} 1 & 7 & -2 & | & 2 \\ 0 & 1 & -\frac{1}{10} & | & \frac{3}{20} \\ 0 & 0 & \frac{11}{2} & | & -\frac{11}{4} \end{bmatrix} \quad \left(R_3 = \frac{2}{11}r_3\right)$$

$$= \begin{bmatrix} 1 & 7 & -2 & | & 2 \\ 0 & 1 & -\frac{1}{10} & | & \frac{3}{20} \\ 0 & 0 & 1 & | & -\frac{1}{2} \end{bmatrix}$$

Write the system of equations that corresponds to the row-echelon matrix

$$\begin{cases} x + 7y - 2z = 2 & (1) \\ y - \dfrac{1}{10}z = \dfrac{3}{20} & (2) \\ z = -\dfrac{1}{2} & (3) \end{cases}$$

This system is consistent and independent.

Substituting $-\dfrac{1}{2}$ for z in equation (2), we obtain

$$y - \frac{1}{10}\left(-\frac{1}{2}\right) = \frac{3}{20}$$

$$y + \frac{1}{20} = \frac{3}{20}$$

$$y = \frac{2}{20} = \frac{1}{10}$$

Substituting $-\dfrac{1}{2}$ for z and $\dfrac{1}{10}$ for y in equation (1), we obtain

$$x + 7\left(\frac{1}{10}\right) - 2\left(-\frac{1}{2}\right) = 2$$

$$x + \frac{7}{10} + 1 = 2$$

$$x + \frac{17}{10} = 2$$

$$x = \frac{3}{10}$$

The solution is the ordered triple $\left(\dfrac{3}{10}, \dfrac{1}{10}, -\dfrac{1}{2}\right)$.

75. Write the augmented matrix of the system and then put it in row echelon form.

$$\begin{bmatrix} 3 & 5 & 2 & | & 6 \\ 0 & 10 & -2 & | & 5 \\ 6 & 0 & 4 & | & 8 \end{bmatrix} \quad \left(R_1 = \frac{1}{3} r_1 \right)$$

$$= \begin{bmatrix} 1 & \frac{5}{3} & \frac{2}{3} & | & 2 \\ 0 & 10 & -2 & | & 5 \\ 6 & 0 & 4 & | & 8 \end{bmatrix} \quad \left(R_3 = -6 r_1 + r_3 \right)$$

$$= \begin{bmatrix} 1 & \frac{5}{3} & \frac{2}{3} & | & 2 \\ 0 & 10 & -2 & | & 5 \\ 0 & -10 & 0 & | & -4 \end{bmatrix} \quad \left(R_3 = -\frac{1}{10} r_3 \right)$$

$$= \begin{bmatrix} 1 & \frac{5}{3} & \frac{2}{3} & | & 2 \\ 0 & 10 & -2 & | & 5 \\ 0 & 1 & 0 & | & \frac{2}{5} \end{bmatrix} \quad (\text{Interchange } r_2 \text{ and } r_3)$$

$$= \begin{bmatrix} 1 & \frac{5}{3} & \frac{2}{3} & | & 2 \\ 0 & 1 & 0 & | & \frac{2}{5} \\ 0 & 10 & -2 & | & 5 \end{bmatrix} \quad \left(R_3 = -10 r_2 + r_3 \right)$$

$$= \begin{bmatrix} 1 & \frac{5}{3} & \frac{2}{3} & | & 2 \\ 0 & 1 & 0 & | & \frac{2}{5} \\ 0 & 0 & -2 & | & 1 \end{bmatrix} \quad \left(R_3 = -\frac{1}{2} r_3 \right)$$

$$= \begin{bmatrix} 1 & \frac{5}{3} & \frac{2}{3} & | & 2 \\ 0 & 1 & 0 & | & \frac{2}{5} \\ 0 & 0 & 1 & | & -\frac{1}{2} \end{bmatrix}$$

Write the system of equations that corresponds to the row-echelon matrix

$$\begin{cases} x + \frac{5}{3}y + \frac{2}{3}z = 2 & (1) \\ \quad\quad y = \frac{2}{5} & (2) \\ \quad\quad\quad z = -\frac{1}{2} & (3) \end{cases}$$

This system is consistent and independent.

We have $y = \frac{2}{5}$ and $z = -\frac{1}{2}$. Substituting $\frac{2}{5}$ for

y and $-\frac{1}{2}$ for z in equation (1), we obtain

$$x + \frac{5}{3}\left(\frac{2}{5}\right) + \frac{2}{3}\left(-\frac{1}{2}\right) = 2$$

$$x + \frac{2}{3} - \frac{1}{3} = 2$$

$$x + \frac{1}{3} = 2$$

$$x = \frac{5}{3}$$

The solution is the ordered triple $\left(\frac{5}{3}, \frac{2}{5}, -\frac{1}{2} \right)$.

77. Write the augmented matrix of the system and then put it in row echelon form.

$$\begin{bmatrix} 1 & 0 & -1 & | & 3 \\ 2 & 1 & 0 & | & -3 \\ 0 & 2 & -1 & | & 7 \end{bmatrix} \quad \left(R_2 = -2 r_1 + r_2 \right)$$

$$= \begin{bmatrix} 1 & 0 & -1 & | & 3 \\ 0 & 1 & 2 & | & -9 \\ 0 & 2 & -1 & | & 7 \end{bmatrix} \quad \left(R_3 = -2 r_2 + r_3 \right)$$

$$= \begin{bmatrix} 1 & 0 & -1 & | & 3 \\ 0 & 1 & 2 & | & -9 \\ 0 & 0 & -5 & | & 25 \end{bmatrix} \quad \left(R_3 = -\frac{1}{5} r_3 \right)$$

$$= \begin{bmatrix} 1 & 0 & -1 & | & 3 \\ 0 & 1 & 2 & | & -9 \\ 0 & 0 & 1 & | & -5 \end{bmatrix}$$

Write the system of equations that corresponds to the row-echelon matrix

$$\begin{cases} x \quad\quad - z = 3 & (1) \\ \quad y + 2z = -9 & (2) \\ \quad\quad\quad z = -5 & (3) \end{cases}$$

This system is consistent and independent. Substituting -5 for z in equation (2), we obtain

$$y + 2(-5) = -9$$
$$y - 10 = -9$$
$$y = 1$$

Substituting -5 for z in equation (1), we obtain

$$x - (-5) = 3$$
$$x + 5 = 3$$
$$x = -2$$

The solution is the ordered triple $(-2, 1, -5)$.

79. a. If $f(1) = 0$, then $a(1)^2 + b(1) + c = 0$ or $a + b + c = 0$.

If $f(2) = 3$, then $a(2)^2 + b(2) + c = 3$ or $4a + 2b + c = 3$.

b. To find a, b, and c, we must solve the following system:

$$\begin{cases} a-b+c=6 & (1) \\ a+b+c=0 & (2) \\ 4a+2b+c=3 & (3) \end{cases}$$

Write the augmented matrix of the system and then put it in row echelon form.

$$\begin{bmatrix} 1 & -1 & 1 & | & 6 \\ 1 & 1 & 1 & | & 0 \\ 4 & 2 & 1 & | & 3 \end{bmatrix} \quad \begin{pmatrix} R_2 = -1r_1 + r_2 \\ R_3 = -4r_1 + r_3 \end{pmatrix}$$

$$= \begin{bmatrix} 1 & -1 & 1 & | & 6 \\ 0 & 2 & 0 & | & -6 \\ 0 & 6 & -3 & | & -21 \end{bmatrix} \quad \left(R_2 = \frac{1}{2}r_2 \right)$$

$$= \begin{bmatrix} 1 & -1 & 1 & | & 6 \\ 0 & 1 & 0 & | & -3 \\ 0 & 6 & -3 & | & -21 \end{bmatrix} \quad (R_3 = -6r_2 + r_3)$$

$$= \begin{bmatrix} 1 & -1 & 1 & | & 6 \\ 0 & 1 & 0 & | & -3 \\ 0 & 0 & -3 & | & -3 \end{bmatrix} \quad \left(R_3 = -\frac{1}{3}r_3 \right)$$

$$= \begin{bmatrix} 1 & -1 & 1 & | & 6 \\ 0 & 1 & 0 & | & -3 \\ 0 & 0 & 1 & | & 1 \end{bmatrix}$$

Write the system of equations that corresponds to the row-echelon matrix

$$\begin{cases} a-b+c=6 & (1) \\ b=-3 & (2) \\ c=1 & (3) \end{cases}$$

This system is consistent and independent. We have $b=-3$ and $c=1$. Substituting -3 for b and 1 for c in equation (1), we obtain

$$a-(-3)+1=6$$
$$a+4=6$$
$$a=2$$

Thus, the quadratic equation that contains $(-1, 6)$, $(1, 0)$, and $(2, 3)$ is

$$f(x)=2x^2-3x+1.$$

81. Let t represent the amount to be invested in Treasury bills, let m represent the amount to be invested in municipal bonds, and let c represent the amount to be invested in corporate bonds.

$$\begin{cases} t+m+c=20,000 & (1) \\ 0.04t+0.05m+0.08c=1070 & (2) \\ t-c=3000 & (3) \end{cases}$$

Write the augmented matrix of the system and then put it in row echelon form.

$$\begin{bmatrix} 1 & 1 & 1 & | & 20,000 \\ 0.04 & 0.05 & 0.08 & | & 1070 \\ 1 & 0 & -1 & | & 3000 \end{bmatrix} \quad \begin{pmatrix} R_2 = -0.04r_1 + r_2 \\ R_3 = -1r_1 + r_3 \end{pmatrix}$$

$$= \begin{bmatrix} 1 & 1 & 1 & | & 20,000 \\ 0 & 0.01 & 0.04 & | & 270 \\ 0 & -1 & -2 & | & -17,000 \end{bmatrix} \quad (R_2 = 100r_2)$$

$$= \begin{bmatrix} 1 & 1 & 1 & | & 20,000 \\ 0 & 1 & 4 & | & 27,000 \\ 0 & -1 & -2 & | & -17,000 \end{bmatrix} \quad (R_3 = r_2 + r_3)$$

$$= \begin{bmatrix} 1 & 1 & 1 & | & 20,000 \\ 0 & 1 & 4 & | & 27,000 \\ 0 & 0 & 2 & | & 10,000 \end{bmatrix} \quad \left(R_3 = \frac{1}{2}r_3 \right)$$

$$= \begin{bmatrix} 1 & 1 & 1 & | & 20,000 \\ 0 & 1 & 4 & | & 27,000 \\ 0 & 0 & 1 & | & 5000 \end{bmatrix}$$

Write the system of equations that corresponds to the row-echelon matrix

$$\begin{cases} t+m+c=20,000 & (1) \\ m+4c=27,000 & (2) \\ c=5000 & (3) \end{cases}$$

Substituting 5000 for c in equation (2), we obtain

$$m+4(5000)=27,000$$
$$m+20,000=27,000$$
$$m=7000$$

Substituting 5000 for c and 7000 for m in equation (1), we obtain

$$t+7000+5000=20,000$$
$$t+12,000=20,000$$
$$t=8000$$

Therefore, Carissa should invest $8000 in Treasury bills, $7000 in municipal bonds, and $5000 in corporate bonds.

83. Write the augmented matrix of the system and then put it in reduced row echelon form.

$$\begin{bmatrix} 2 & 1 & | & 1 \\ -3 & -2 & | & -5 \end{bmatrix} \quad (R_1 = 2r_1 + r_2)$$

$$= \begin{bmatrix} 1 & 0 & | & -3 \\ -3 & -2 & | & -5 \end{bmatrix} \quad (R_2 = 3r_1 + r_2)$$

$$= \begin{bmatrix} 1 & 0 & | & -3 \\ 0 & -2 & | & -14 \end{bmatrix} \quad \left(R_2 = -\frac{1}{2}r_2 \right)$$

$$= \begin{bmatrix} 1 & 0 & | & -3 \\ 0 & 1 & | & 7 \end{bmatrix}$$

Write the system of equations that corresponds to the reduced row echelon matrix

$$\begin{cases} x = -3 \\ y = 7 \end{cases}$$

The solution is the ordered pair $(-3,\ 7)$.

85. Write the augmented matrix of the system and then put it in reduced row echelon form.

$$\begin{bmatrix} 1 & 1 & 1 & 3 \\ 2 & 1 & -4 & 25 \\ -3 & 2 & 1 & 0 \end{bmatrix} \quad \begin{pmatrix} R_2 = -2r_1 + r_2 \\ R_3 = 3r_1 + r_3 \end{pmatrix}$$

$$= \begin{bmatrix} 1 & 1 & 1 & 3 \\ 0 & -1 & -6 & 19 \\ 0 & 5 & 4 & 9 \end{bmatrix} \quad \begin{pmatrix} R_1 = r_1 + r_2 \\ R_3 = 5r_2 + r_3 \end{pmatrix}$$

$$= \begin{bmatrix} 1 & 0 & -5 & 22 \\ 0 & -1 & -6 & 19 \\ 0 & 0 & -26 & 104 \end{bmatrix} \quad \left(R_3 = -\dfrac{1}{26} r_3 \right)$$

$$= \begin{bmatrix} 1 & 0 & -5 & 22 \\ 0 & -1 & -6 & 19 \\ 0 & 0 & 1 & -4 \end{bmatrix} \quad \begin{pmatrix} R_1 = 5r_3 + r_1 \\ R_2 = 6r_3 + r_2 \end{pmatrix}$$

$$= \begin{bmatrix} 1 & 0 & 0 & 2 \\ 0 & -1 & 0 & -5 \\ 0 & 0 & 1 & -4 \end{bmatrix} \quad (R_2 = -1 \cdot r_2)$$

$$= \begin{bmatrix} 1 & 0 & 0 & 2 \\ 0 & 1 & 0 & 5 \\ 0 & 0 & 1 & -4 \end{bmatrix}$$

Write the system of equations that corresponds to the reduced row echelon matrix

$$\begin{cases} x = 2 \\ y = 5 \\ z = -4 \end{cases}$$

The solution is the ordered triple $(2,\ 5,\ -4)$.

87. Answers will vary. One possibility follows: First, perform row operations so that the entry in row 1, column 1 is 1. Second, perform row operations so that all the entries below the 1 in row 1, column 1 are 0's. Third, perform row operations so that the entry in row 2, column 2 is 1. Make sure that the entries in column 1 remain unchanged. If it is impossible to place a 1 in row 2, column 2, then use row operations to place a 1 in row 2, column 3. (If any row with all 0's occurs, then place it in the last row of the matrix.) Once a 1 is in place, perform row operations to place 0's below it. Continue this process until the augmented matrix is in row echelon form.

89. Multiply each entry in row 2 by $\dfrac{1}{5}$ (or divide each entry of row 2 by 5). That is, use the row operation $R_2 = \dfrac{1}{5} r_2$.

91. Write the augmented matrix of the system.

$$\begin{bmatrix} 2 & 3 & 1 \\ -3 & -4 & -3 \end{bmatrix}$$

Enter the system into a 2 by 3 matrix, [A]. Then, use the **ref(** command along with the ▸ **frac** command to write the matrix in row echelon form with the entries in fractional form.

Thus, the row echelon matrix is

$$\begin{bmatrix} 1 & \frac{4}{3} & 1 \\ 0 & 1 & -3 \end{bmatrix}$$

Write the system of equations that corresponds to the row echelon matrix.

$$\begin{cases} x + \dfrac{4}{3} y = 1 & (1) \\ y = -3 & (2) \end{cases}$$

Substituting -3 for y in equation (1), we obtain

$$x + \dfrac{4}{3}(-3) = 1$$
$$x - 4 = 1$$
$$x = 5$$

The solution is the ordered pair $(5, -3)$.

93. Write the augmented matrix of the system.

$$\begin{bmatrix} 2 & 3 & -2 & -12 \\ -3 & 1 & 2 & 0 \\ 4 & 3 & -1 & 3 \end{bmatrix}$$

Enter the system into a 3 by 4 matrix, [A]. Then, use the **ref(** command along with the ▸ **frac** command to write the matrix in row echelon form with the entries in fractional form. Since the entire matrix does not fit on the screen, we need to scroll right to see the rest of it.

```
[A]
 [[2   3  -2  -12]
  [-3  1  2   0 ]
  [4   3  -1  3   ]]
```

```
ref([A])▶Frac
[[1  3/4  -1/4  3/…
 [0  1    5/13  9/…
 [0  0    1     7  …
```

```
ref([A])▶Frac
…3/4  -1/4  3/4  ]
…1    5/13  9/13]
…0    1     7    ]]
```

Thus, the row echelon matrix is

$$\left[\begin{array}{ccc|c} 1 & \frac{3}{4} & -\frac{1}{4} & \frac{3}{4} \\ 0 & 1 & \frac{5}{13} & \frac{9}{13} \\ 0 & 0 & 1 & 7 \end{array}\right]$$

Write the system of equations that corresponds to the row-echelon matrix

$$\begin{cases} x + \dfrac{3}{4}y - \dfrac{1}{4}z = \dfrac{3}{4} & \quad (1) \\[2mm] \qquad y + \dfrac{5}{13}z = \dfrac{9}{13} & \quad (2) \\[2mm] \qquad\qquad z = 7 & \quad (3) \end{cases}$$

Substituting 7 for z in equation (2), we obtain

$$y + \frac{5}{13}(7) = \frac{9}{13}$$
$$y + \frac{35}{13} = \frac{9}{13}$$
$$y = -\frac{26}{13} = -2$$

Substituting 7 for z and -2 for y in equation (1), we obtain

$$x + \frac{3}{4}(-2) - \frac{1}{4}(7) = \frac{3}{4}$$
$$x - \frac{3}{2} - \frac{7}{4} = \frac{3}{4}$$
$$x - \frac{13}{4} = \frac{3}{4}$$
$$x = \frac{16}{4} = 4$$

The solution is the ordered triple $(4, -2, 7)$.

Section 3.5

Are You Ready for This Section?

R1. $4 \cdot 2 - 3 \cdot (-3) = 8 + 9 = 17$

R2. $\dfrac{18}{6} = 3$

Section 3.5 Quick Checks

1. $ad - bc$

2. A matrix is <u>square</u> if the number of rows and the number of columns are equal.

3. $\begin{vmatrix} 5 & 3 \\ 4 & 6 \end{vmatrix} = 5(6) - 4(3) = 30 - 12 = 18$

4. $\begin{vmatrix} -2 & -5 \\ 1 & 7 \end{vmatrix} = -2(7) - 1(-5) = -14 + 5 = -9$

5. $D = \begin{vmatrix} 3 & 2 \\ -2 & -1 \end{vmatrix} = 3(-1) - (-2)(2) = -3 + 4 = 1$

   $D_x = \begin{vmatrix} 1 & 2 \\ 1 & -1 \end{vmatrix} = 1(-1) - 1(2) = -1 - 2 = -3$

   $D_y = \begin{vmatrix} 3 & 1 \\ -2 & 1 \end{vmatrix} = 3(1) - (-2)(1) = 3 + 2 = 5$

   $x = \dfrac{D_x}{D} = \dfrac{-3}{1} = -3$; $y = \dfrac{D_y}{D} = \dfrac{5}{1} = 5$

   Thus, the solution is the ordered pair $(-3,\ 5)$.

6. $D = \begin{vmatrix} 4 & -2 \\ -6 & 3 \end{vmatrix} = 4(3) - (-6)(-2) = 12 - 12 = 0$

   Since $D = 0$, Cramer's Rule does not apply.

7. $\begin{vmatrix} 2 & -3 & 5 \\ 0 & 4 & -1 \\ 3 & 8 & -7 \end{vmatrix} = 2\begin{vmatrix} 4 & -1 \\ 8 & -7 \end{vmatrix} - (-3)\begin{vmatrix} 0 & -1 \\ 3 & -7 \end{vmatrix} + 5\begin{vmatrix} 0 & 4 \\ 3 & 8 \end{vmatrix}$

   $= 2\big[4(-7) - 8(-1)\big] + 3\big[0(-7) - 3(-1)\big] + 5\big[0(8) - 3(4)\big]$

   $= 2(-28 + 8) + 3(0 + 3) + 5(0 - 12)$

   $= 2(-20) + 3(3) + 5(-12)$

   $= -40 + 9 - 60$

   $= -91$

8. $\begin{vmatrix} 3 & 2 & 1 \\ 1 & 1 & -3 \\ -5 & -1 & -5 \end{vmatrix} = 3\begin{vmatrix} 1 & -3 \\ -1 & -5 \end{vmatrix} - 2\begin{vmatrix} 1 & -3 \\ -5 & -5 \end{vmatrix} + 1\begin{vmatrix} 1 & 1 \\ -5 & -1 \end{vmatrix}$

   $= 3[1(-5) - (-1)(-3)] - 2[1(-5) - (-5)(-3)] + 1[1(-1) - (-5)(1)]$

   $= 3(-5 - 3) - 2(-5 - 15) + 1(-1 + 5)$

   $= 3(-8) - 2(-20) + 1(4)$

   $= -24 + 40 + 4$

   $= 20$

9. $x = \dfrac{D_x}{D} = \dfrac{2}{4} = \dfrac{1}{2}$

$y = \dfrac{D_y}{D} = \dfrac{-8}{4} = -2$

$z = \dfrac{D_z}{D} = \dfrac{-4}{4} = -1$

Thus, the solution is the ordered triple $\left(\dfrac{1}{2}, -2, -1\right)$.

10. $D = \begin{vmatrix} 1 & -1 & 3 \\ 4 & 3 & 1 \\ -2 & 0 & 5 \end{vmatrix}$

$= 1\begin{vmatrix} 3 & 1 \\ 0 & 5 \end{vmatrix} - (-1)\begin{vmatrix} 4 & 1 \\ -2 & 5 \end{vmatrix} + 3\begin{vmatrix} 4 & 3 \\ -2 & 0 \end{vmatrix}$

$= 1\big[3(5) - 0(1)\big] - (-1)\big[4(5) - (-2)(1)\big] + 3\big[4(0) - (-2)(3)\big]$

$= 1(15 - 0) - (-1)(20 + 2) + 3(0 + 6)$

$= 1(15) - (-1)(22) + 3(6)$

$= 15 + 22 + 18$

$= 55$

$D_x = \begin{vmatrix} -2 & -1 & 3 \\ 9 & 3 & 1 \\ 7 & 0 & 5 \end{vmatrix}$

$= -2\begin{vmatrix} 3 & 1 \\ 0 & 5 \end{vmatrix} - (-1)\begin{vmatrix} 9 & 1 \\ 7 & 5 \end{vmatrix} + 3\begin{vmatrix} 9 & 3 \\ 7 & 0 \end{vmatrix}$

$= -2\big[3(5) - 0(1)\big] - (-1)\big[9(5) - 7(1)\big] + 3\big[9(0) - 7(3)\big]$

$= -2(15 - 0) - (-1)(45 - 7) + 3(0 - 21)$

$= -2(15) - (-1)(38) + 3(-21)$

$= -30 + 38 - 63$

$= -55$

$D_y = \begin{vmatrix} 1 & -2 & 3 \\ 4 & 9 & 1 \\ -2 & 7 & 5 \end{vmatrix}$

$= 1\begin{vmatrix} 9 & 1 \\ 7 & 5 \end{vmatrix} - (-2)\begin{vmatrix} 4 & 1 \\ -2 & 5 \end{vmatrix} + 3\begin{vmatrix} 4 & 9 \\ -2 & 7 \end{vmatrix}$

$= 1\big[9(5) - 7(1)\big] - (-2)\big[4(5) - (-2)(1)\big] + 3\big[4(7) - (-2)(9)\big]$

$= 1(45 - 7) - (-2)(20 + 2) + 3(28 + 18)$

$= 1(38) - (-2)(22) + 3(46)$

$= 38 + 44 + 138$

$= 220$

$$D_z = \begin{vmatrix} 1 & -1 & -2 \\ 4 & 3 & 9 \\ -2 & 0 & 7 \end{vmatrix}$$

$$= 1\begin{vmatrix} 3 & 9 \\ 0 & 7 \end{vmatrix} - (-1)\begin{vmatrix} 4 & 9 \\ -2 & 7 \end{vmatrix} + (-2)\begin{vmatrix} 4 & 3 \\ -2 & 0 \end{vmatrix}$$

$$= 1\big[3(7) - 0(9)\big] - (-1)\big[4(7) - (-2)(9)\big] + (-2)\big[4(0) - (-2)(3)\big]$$

$$= 1(21 - 0) - (-1)(28 + 18) + (-2)(0 + 6)$$

$$= 1(21) - (-1)(46) + (-2)(6)$$

$$= 21 + 46 - 12$$

$$= 55$$

$$x = \frac{D_x}{D} = \frac{-55}{55} = -1\,; \ \ y = \frac{D_y}{D} = \frac{220}{55} = 4\,; \ \ z = \frac{D_z}{D} = \frac{55}{55} = 1$$

Thus, the solution is the ordered triple $(-1,\ 4,\ 1)$.

3.5 Exercises

11. $\begin{vmatrix} 4 & 2 \\ 1 & 3 \end{vmatrix} = 4(3) - 1(2) = 12 - 2 = 10$

13. $\begin{vmatrix} -2 & -4 \\ 1 & 3 \end{vmatrix} = -2(3) - 1(-4) = -6 + 4 = -2$

15. $D = \begin{vmatrix} 1 & 1 \\ 1 & -1 \end{vmatrix} = 1(-1) - 1(1) = -1 - 1 = -2$

$$D_x = \begin{vmatrix} -4 & 1 \\ -12 & -1 \end{vmatrix}$$
$$= -4(-1) - (-12)(1) = 4 + 12 = 16$$

$$D_y = \begin{vmatrix} 1 & -4 \\ 1 & -12 \end{vmatrix} = 1(-12) - 1(-4) = -12 + 4 = -8$$

$$x = \frac{D_x}{D} = \frac{16}{-2} = -8\,; \ \ y = \frac{D_y}{D} = \frac{-8}{-2} = 4$$

Thus, the solution is the ordered pair $(-8,\ 4)$.

17. $D = \begin{vmatrix} 2 & 3 \\ -3 & 1 \end{vmatrix} = 2(1) - (-3)(3) = 2 + 9 = 11$

$$D_x = \begin{vmatrix} 3 & 3 \\ -10 & 1 \end{vmatrix} = 3(1) - (-10)(3) = 3 + 30 = 33$$

$$D_y = \begin{vmatrix} 2 & 3 \\ -3 & -10 \end{vmatrix}$$
$$= 2(-10) - (-3)(3) = -20 + 9 = -11$$

$$x = \frac{D_x}{D} = \frac{33}{11} = 3\,; \ \ y = \frac{D_y}{D} = \frac{-11}{11} = -1$$

Thus, the solution is the ordered pair $(3, -1)$.

19. $D = \begin{vmatrix} 3 & 4 \\ -6 & 8 \end{vmatrix} = 3(8) - (-6)(4) = 24 + 24 = 48$

$D_x = \begin{vmatrix} 1 & 4 \\ 4 & 8 \end{vmatrix} = 1(8) - 4(4) = 8 - 16 = -8$

$D_y = \begin{vmatrix} 3 & 1 \\ -6 & 4 \end{vmatrix} = 3(4) - (-6)(1) = 12 + 6 = 18$

$x = \dfrac{D_x}{D} = \dfrac{-8}{48} = -\dfrac{1}{6}; \quad y = \dfrac{D_y}{D} = \dfrac{18}{48} = \dfrac{3}{8}$

Thus, the solution is the ordered pair $\left(-\dfrac{1}{6}, \dfrac{3}{8}\right)$.

21. The system in standard form is: $\begin{cases} 2x - 6y = 12 \\ 3x - 5y = 11 \end{cases}$

$D = \begin{vmatrix} 2 & -6 \\ 3 & -5 \end{vmatrix} = 2(-5) - 3(-6) = -10 + 18 = 8$

$\begin{aligned} D_x &= \begin{vmatrix} 12 & -6 \\ 11 & -5 \end{vmatrix} \\ &= 12(-5) - 11(-6) \\ &= -60 + 66 \\ &= 6 \end{aligned}$

$D_y = \begin{vmatrix} 2 & 12 \\ 3 & 11 \end{vmatrix} = 2(11) - 3(12) = 22 - 36 = -14$

$x = \dfrac{D_x}{D} = \dfrac{6}{8} = \dfrac{3}{4}; \quad y = \dfrac{D_y}{D} = \dfrac{-14}{8} = -\dfrac{7}{4}$

Thus, the solution is the ordered pair $\left(\dfrac{3}{4}, -\dfrac{7}{4}\right)$.

23. $\begin{vmatrix} 2 & 0 & -1 \\ 3 & 8 & -3 \\ 1 & 5 & -2 \end{vmatrix}$

$= 2\begin{vmatrix} 8 & -3 \\ 5 & -2 \end{vmatrix} - 0\begin{vmatrix} 3 & -3 \\ 1 & -2 \end{vmatrix} + (-1)\begin{vmatrix} 3 & 8 \\ 1 & 5 \end{vmatrix}$

$= 2[8(-2) - 5(-3)] - 0[3(-2) - 1(-3)]$
$\qquad\qquad + (-1)[3(5) - 1(8)]$

$= 2(-16 + 15) - 0(-6 + 3) + (-1)(15 - 8)$

$= 2(-1) - 0(-3) + (-1)(7)$

$= -2 + 0 - 7$

$= -9$

25. $\begin{vmatrix} -3 & 2 & 3 \\ 0 & 5 & -2 \\ 1 & 4 & 8 \end{vmatrix}$

$= (-3)\begin{vmatrix} 5 & -2 \\ 4 & 8 \end{vmatrix} - 2\begin{vmatrix} 0 & -2 \\ 1 & 8 \end{vmatrix} + 3\begin{vmatrix} 0 & 5 \\ 1 & 4 \end{vmatrix}$

$= -3[5(8) - 4(-2)] - 2[0(8) - 1(-2)]$
$\qquad\qquad\qquad + 3[0(4) - 1(5)]$

$= -3(40 + 8) - 2(0 + 2) + 3(0 - 5)$

$= -3(48) - 2(2) + 3(-5)$

$= -144 - 4 - 15$

$= -163$

27. $\begin{vmatrix} 0 & 2 & 1 \\ 1 & -6 & -4 \\ -3 & 4 & 5 \end{vmatrix}$

$= 0\begin{vmatrix} -6 & -4 \\ 4 & 5 \end{vmatrix} - 2\begin{vmatrix} 1 & -4 \\ -3 & 5 \end{vmatrix} + 1\begin{vmatrix} 1 & -6 \\ -3 & 4 \end{vmatrix}$

$= 0[-6(5) - 4(-4)] - 2[1(5) - (-3)(-4)]$
$\qquad\qquad\qquad + 1[1(4) - (-3)(-6)]$

$= 0(-30 + 16) - 2(5 - 12) + 1(4 - 18)$

$= 0(-14) - 2(-7) + 1(-14)$

$= 0 + 14 - 14$

$= 0$

29. $D = \begin{vmatrix} 1 & -1 & 1 \\ 1 & 2 & -1 \\ 2 & 1 & 2 \end{vmatrix}$

$= 1\begin{vmatrix} 2 & -1 \\ 1 & 2 \end{vmatrix} - (-1)\begin{vmatrix} 1 & -1 \\ 2 & 2 \end{vmatrix} + 1\begin{vmatrix} 1 & 2 \\ 2 & 1 \end{vmatrix}$

$= 1[2(2) - 1(-1)] - (-1)[1(2) - 2(-1)]$
$\qquad\qquad\qquad + 1[1(1) - 2(2)]$

$= 1(4 + 1) - (-1)(2 + 2) + 1(1 - 4)$

$= 1(5) - (-1)(4) + 1(-3)$

$= 5 + 4 - 3$

$= 6$

$D_x = \begin{vmatrix} -4 & -1 & 1 \\ 1 & 2 & -1 \\ -5 & 1 & 2 \end{vmatrix}$

$= -4\begin{vmatrix} 2 & -1 \\ 1 & 2 \end{vmatrix} - (-1)\begin{vmatrix} 1 & -1 \\ -5 & 2 \end{vmatrix} + 1\begin{vmatrix} 1 & 2 \\ -5 & 1 \end{vmatrix}$

$= -4[2(2) - 1(-1)] - (-1)[1(2) - (-5)(-1)]$
$\qquad\qquad\qquad + 1[1(1) - (-5)(2)]$

$= -4(4 + 1) - (-1)(2 - 5) + 1(1 + 10)$

$= -4(5) - (-1)(-3) + 1(11)$

$= -20 - 3 + 11$

$= -12$

$$D_y = \begin{vmatrix} 1 & -4 & 1 \\ 1 & 1 & -1 \\ 2 & -5 & 2 \end{vmatrix}$$

$$= 1\begin{vmatrix} 1 & -1 \\ -5 & 2 \end{vmatrix} - (-4)\begin{vmatrix} 1 & -1 \\ 2 & 2 \end{vmatrix} + 1\begin{vmatrix} 1 & 1 \\ 2 & -5 \end{vmatrix}$$

$$= 1\big[1(2) - (-5)(-1)\big] - (-4)\big[1(2) - 2(-1)\big]$$
$$+ 1\big[1(-5) - 2(1)\big]$$

$$= 1(2-5) - (-4)(2+2) + 1(-5-2)$$

$$= 1(-3) - (-4)(4) + 1(-7)$$

$$= -3 + 16 - 7$$

$$= 6$$

$$D_z = \begin{vmatrix} 1 & -1 & -4 \\ 1 & 2 & 1 \\ 2 & 1 & -5 \end{vmatrix}$$

$$= 1\begin{vmatrix} 2 & 1 \\ 1 & -5 \end{vmatrix} - (-1)\begin{vmatrix} 1 & 1 \\ 2 & -5 \end{vmatrix} + (-4)\begin{vmatrix} 1 & 2 \\ 2 & 1 \end{vmatrix}$$

$$= 1\big[2(-5) - 1(1)\big] - (-1)\big[1(-5) - 2(1)\big]$$
$$+ (-4)\big[1(1) - 2(2)\big]$$

$$= 1(-10-1) - (-1)(-5-2) + (-4)(1-4)$$

$$= 1(-11) - (-1)(-7) + (-4)(-3)$$

$$= -11 - 7 + 12$$

$$= -6$$

$$x = \frac{D_x}{D} = \frac{-12}{6} = -2; \quad y = \frac{D_y}{D} = \frac{6}{6} = 1;$$

$$z = \frac{D_z}{D} = \frac{-6}{6} = -1$$

Thus, the solution is the ordered triple $(-2, 1, -1)$.

31. $$D = \begin{vmatrix} 1 & 1 & 1 \\ 5 & 2 & -3 \\ 2 & -1 & -1 \end{vmatrix}$$

$$= 1\begin{vmatrix} 2 & -3 \\ -1 & -1 \end{vmatrix} - 1\begin{vmatrix} 5 & -3 \\ 2 & -1 \end{vmatrix} + 1\begin{vmatrix} 5 & 2 \\ 2 & -1 \end{vmatrix}$$

$$= 1\big[2(-1) - (-1)(-3)\big] - 1\big[5(-1) - 2(-3)\big]$$
$$+ 1\big[5(-1) - 2(2)\big]$$

$$= 1(-2-3) - 1(-5+6) + 1(-5-4)$$

$$= 1(-5) - 1(1) + 1(-9)$$

$$= -5 - 1 - 9$$

$$= -15$$

$$D_x = \begin{vmatrix} 4 & 1 & 1 \\ 7 & 2 & -3 \\ 5 & -1 & -1 \end{vmatrix}$$

$$= 4\begin{vmatrix} 2 & -3 \\ -1 & -1 \end{vmatrix} - 1\begin{vmatrix} 7 & -3 \\ 5 & -1 \end{vmatrix} + 1\begin{vmatrix} 7 & 2 \\ 5 & -1 \end{vmatrix}$$

$$= 4\big[2(-1) - (-1)(-3)\big] - 1\big[7(-1) - 5(-3)\big]$$
$$+ 1\big[7(-1) - 5(2)\big]$$

$$= 4(-2-3) - 1(-7+15) + 1(-7-10)$$

$$= 4(-5) - 1(8) + 1(-17)$$

$$= -20 - 8 - 17$$

$$= -45$$

$$D_y = \begin{vmatrix} 1 & 4 & 1 \\ 5 & 7 & -3 \\ 2 & 5 & -1 \end{vmatrix}$$

$$= 1\begin{vmatrix} 7 & -3 \\ 5 & -1 \end{vmatrix} - 4\begin{vmatrix} 5 & -3 \\ 2 & -1 \end{vmatrix} + 1\begin{vmatrix} 5 & 7 \\ 2 & 5 \end{vmatrix}$$

$$= 1\big[7(-1) - 5(-3)\big] - 4\big[5(-1) - 2(-3)\big]$$
$$+ 1\big[5(5) - 2(7)\big]$$

$$= 1(-7+15) - 4(-5+6) + 1(25-14)$$

$$= 1(8) - 4(1) + 1(11)$$

$$= 8 - 4 + 11$$

$$= 15$$

$$D_z = \begin{vmatrix} 1 & 1 & 4 \\ 5 & 2 & 7 \\ 2 & -1 & 5 \end{vmatrix}$$

$$= 1\begin{vmatrix} 2 & 7 \\ -1 & 5 \end{vmatrix} - 1\begin{vmatrix} 5 & 7 \\ 2 & 5 \end{vmatrix} + 4\begin{vmatrix} 5 & 2 \\ 2 & -1 \end{vmatrix}$$

$$= 1\big[2(5) - (-1)(7)\big] - 1\big[5(5) - 2(7)\big]$$
$$+ 4\big[5(-1) - 2(2)\big]$$

$$= 1(10+7) - 1(25-14) + 4(-5-4)$$

$$= 1(17) - 1(11) + 4(-9)$$

$$= 17 - 11 - 36$$

$$= -30$$

$$x = \frac{D_x}{D} = \frac{-45}{-15} = 3; \quad y = \frac{D_y}{D} = \frac{15}{-15} = -1;$$

$$z = \frac{D_z}{D} = \frac{-30}{-15} = 2$$

Thus, the solution is the ordered triple $(3, -1, 2)$.

33. $D = \begin{vmatrix} 2 & 1 & -1 \\ -1 & 2 & 2 \\ 5 & 5 & -1 \end{vmatrix}$

$= 2 \begin{vmatrix} 2 & 2 \\ 5 & -1 \end{vmatrix} - 1 \begin{vmatrix} -1 & 2 \\ 5 & -1 \end{vmatrix} + (-1) \begin{vmatrix} -1 & 2 \\ 5 & 5 \end{vmatrix}$

$= 2 \big[2(-1) - 5(2) \big] - 1 \big[(-1)(-1) - 5(2) \big]$
$\qquad\qquad + (-1) \big[(-1)(5) - 5(2) \big]$

$= 2(-2 - 10) - 1(1 - 10) + (-1)(-5 - 10)$

$= 2(-12) - 1(-9) + (-1)(-15)$

$= -24 + 9 + 15$

$= 0$

Since $D = 0$, we know that Cramer's Rule does not apply and that this is not a consistent and independent system.

35. $D = \begin{vmatrix} 3 & 1 & 1 \\ 1 & 1 & -3 \\ -5 & -1 & -5 \end{vmatrix}$

$= 3 \begin{vmatrix} 1 & -3 \\ -1 & -5 \end{vmatrix} - 1 \begin{vmatrix} 1 & -3 \\ -5 & -5 \end{vmatrix} + 1 \begin{vmatrix} 1 & 1 \\ -5 & -1 \end{vmatrix}$

$= 3 \big[1(-5) - (-1)(-3) \big] - 1 \big[1(-5) - (-5)(-3) \big]$
$\qquad\qquad + 1 \big[1(-1) - 1(-5) \big]$

$= 3(-5 - 3) - 1(-5 - 15) + 1(-1 + 5)$

$= 3(-8) - 1(-20) + 1(4)$

$= -24 + 20 + 4$

$= 0$

Since $D = 0$, Cramer's Rule does not apply.

37. $D = \begin{vmatrix} 2 & 0 & 1 \\ -1 & -3 & 0 \\ 1 & -2 & 1 \end{vmatrix}$

$= 2 \begin{vmatrix} -3 & 0 \\ -2 & 1 \end{vmatrix} - 0 \begin{vmatrix} -1 & 0 \\ 1 & 1 \end{vmatrix} + 1 \begin{vmatrix} -1 & -3 \\ 1 & -2 \end{vmatrix}$

$= 2 \big[-3(1) - (-2)(0) \big] - 0 \big[-1(1) - 1(0) \big]$
$\qquad\qquad + 1 \big[-1(-2) - 1(-3) \big]$

$= 2(-3 - 0) - 0(-1 - 0) + 1(2 + 3)$

$= 2(-3) - 0(-1) + 1(5)$

$= -6 - 0 + 5$

$= -1$

$D_x = \begin{vmatrix} 27 & 0 & 1 \\ 6 & -3 & 0 \\ 27 & -2 & 1 \end{vmatrix}$

$= 27 \begin{vmatrix} -3 & 0 \\ -2 & 1 \end{vmatrix} - 0 \begin{vmatrix} 6 & 0 \\ 27 & 1 \end{vmatrix} + 1 \begin{vmatrix} 6 & -3 \\ 27 & -2 \end{vmatrix}$

$= 27 \big[-3(1) - (-2)(0) \big] - 0 \big[6(1) - 27(0) \big]$
$\qquad\qquad + 1 \big[6(-2) - 27(-3) \big]$

$= 27(-3 - 0) - 0(6 - 0) + 1(-12 + 81)$

$= 27(-3) - 0(6) + 1(69)$

$= -81 - 0 + 69$

$= -12$

$D_y = \begin{vmatrix} 2 & 27 & 1 \\ -1 & 6 & 0 \\ 1 & 27 & 1 \end{vmatrix}$

$= 2 \begin{vmatrix} 6 & 0 \\ 27 & 1 \end{vmatrix} - 27 \begin{vmatrix} -1 & 0 \\ 1 & 1 \end{vmatrix} + 1 \begin{vmatrix} -1 & 6 \\ 1 & 27 \end{vmatrix}$

$= 2 \big[6(1) - 27(0) \big] - 27 \big[-1(1) - 1(0) \big]$
$\qquad\qquad + 1 \big[(-1)(27) - 1(6) \big]$

$= 2(6 - 0) - 27(-1 - 0) + 1(-27 - 6)$

$= 2(6) - 27(-1) + 1(-33)$

$= 12 + 27 - 33$

$= 6$

$D_z = \begin{vmatrix} 2 & 0 & 27 \\ -1 & -3 & 6 \\ 1 & -2 & 27 \end{vmatrix}$

$= 2 \begin{vmatrix} -3 & 6 \\ -2 & 27 \end{vmatrix} - 0 \begin{vmatrix} -1 & 6 \\ 1 & 27 \end{vmatrix} + 27 \begin{vmatrix} -1 & -3 \\ 1 & -2 \end{vmatrix}$

$= 2 \big[-3(27) - (-2)(6) \big] - 0 \big[-1(27) - 1(6) \big]$
$\qquad\qquad + 27 \big[-1(-2) - 1(-3) \big]$

$= 2(-81 + 12) - 0(-27 - 6) + 27(2 + 3)$

$= 2(-69) - 0(-33) + 27(5)$

$= -138 - 0 + 135$

$= -3$

$x = \dfrac{D_x}{D} = \dfrac{-12}{-1} = 12; \quad y = \dfrac{D_y}{D} = \dfrac{6}{-1} = -6;$

$z = \dfrac{D_z}{D} = \dfrac{-3}{-1} = 3$

The solution is the ordered triple $(12, -6, 3)$.

39. $D = \begin{vmatrix} 5 & 3 & 0 \\ -10 & 0 & 3 \\ 0 & 1 & -2 \end{vmatrix}$

$= 5\begin{vmatrix} 0 & 3 \\ 1 & -2 \end{vmatrix} - 3\begin{vmatrix} -10 & 3 \\ 0 & -2 \end{vmatrix} + 0\begin{vmatrix} -10 & 0 \\ 0 & 1 \end{vmatrix}$

$= 5\left[0(-2)-1(3)\right] - 3\left[-10(-2)-0(3)\right] + 0\left[-10(1)-0(0)\right]$

$= 5(0-3) - 3(20-0) + 0(-10-0)$

$= 5(-3) - 3(20) + 0(-10)$

$= -15 - 60 + 0$

$= -75$

$D_x = \begin{vmatrix} 2 & 3 & 0 \\ -3 & 0 & 3 \\ -9 & 1 & -2 \end{vmatrix}$

$= 2\begin{vmatrix} 0 & 3 \\ 1 & -2 \end{vmatrix} - 3\begin{vmatrix} -3 & 3 \\ -9 & -2 \end{vmatrix} + 0\begin{vmatrix} -3 & 0 \\ -9 & 1 \end{vmatrix}$

$= 2\left[0(-2)-1(3)\right] - 3\left[-3(-2)-(-9)(3)\right] + 0\left[-3(1)-(-9)(0)\right]$

$= 2(0-3) - 3(6+27) + 0(-3+0)$

$= 2(-3) - 3(33) + 0(-3)$

$= -6 - 99 + 0$

$= -105$

$D_y = \begin{vmatrix} 5 & 2 & 0 \\ -10 & -3 & 3 \\ 0 & -9 & -2 \end{vmatrix}$

$= 5\begin{vmatrix} -3 & 3 \\ -9 & -2 \end{vmatrix} - 2\begin{vmatrix} -10 & 3 \\ 0 & -2 \end{vmatrix} + 0\begin{vmatrix} -10 & -3 \\ 0 & -9 \end{vmatrix}$

$= 5\left[-3(-2)-(-9)(3)\right] - 2\left[-10(-2)-0(3)\right] + 0\left[-10(-9)-0(-3)\right]$

$= 5(6+27) - 2(20-0) + 0(90+0)$

$= 5(33) - 2(20) + 0(90)$

$= 165 - 40 + 0$

$= 125$

$D_z = \begin{vmatrix} 5 & 3 & 2 \\ -10 & 0 & -3 \\ 0 & 1 & -9 \end{vmatrix}$

$= 5\begin{vmatrix} 0 & -3 \\ 1 & -9 \end{vmatrix} - 3\begin{vmatrix} -10 & -3 \\ 0 & -9 \end{vmatrix} + 2\begin{vmatrix} -10 & 0 \\ 0 & 1 \end{vmatrix}$

$= 5\left[0(-9)-1(-3)\right] - 3\left[-10(-9)-0(-3)\right] + 2\left[-10(1)-0(0)\right]$

$= 5(0+3) - 3(90+0) + 2(-10-0)$

$= 5(3) - 3(90) + 2(-10)$

$= 15 - 270 + -20$

$= -275$

$x = \dfrac{D_x}{D} = \dfrac{-105}{-75} = \dfrac{7}{5}; \quad y = \dfrac{D_y}{D} = \dfrac{125}{-75} = -\dfrac{5}{3}; \quad z = \dfrac{D_z}{D} = \dfrac{-275}{-75} = \dfrac{11}{3}$

The solution is the ordered triple $\left(\dfrac{7}{5}, -\dfrac{5}{3}, \dfrac{11}{3}\right)$.

41. $D = \begin{vmatrix} 1 & 1 & 1 \\ 0 & -1 & 2 \\ -1 & 0 & 1 \end{vmatrix}$

$= 1 \begin{vmatrix} -1 & 2 \\ 0 & 1 \end{vmatrix} - 1 \begin{vmatrix} 0 & 2 \\ -1 & 1 \end{vmatrix} + 1 \begin{vmatrix} 0 & -1 \\ -1 & 0 \end{vmatrix}$

$= 1[-1(1) - 0(2)] - 1[0(1) - (-1)(2)] + 1[0(0) - (-1)(-1)]$

$= 1(-1 - 0) - 1(0 + 2) + 1(0 - 1)$

$= 1(-1) - 1(2) + 1(-1)$

$= -1 - 2 - 1$

$= -4$

$D_x = \begin{vmatrix} 3 & 1 & 1 \\ 1 & -1 & 2 \\ 0 & 0 & 1 \end{vmatrix}$

$= 3 \begin{vmatrix} -1 & 2 \\ 0 & 1 \end{vmatrix} - 1 \begin{vmatrix} 1 & 2 \\ 0 & 1 \end{vmatrix} + 1 \begin{vmatrix} 1 & -1 \\ 0 & 0 \end{vmatrix}$

$= 3[(-1)(1) - 0(2)] - 1[1(1) - 0(2)] + 1[1(0) - 0(-1)]$

$= 3(-1 - 0) - 1(1 - 0) + 1(0 - 0)$

$= -3 - 1 + 0$

$= -4$

$D_y = \begin{vmatrix} 1 & 3 & 1 \\ 0 & 1 & 2 \\ -1 & 0 & 1 \end{vmatrix}$

$= 1 \begin{vmatrix} 1 & 2 \\ 0 & 1 \end{vmatrix} - 3 \begin{vmatrix} 0 & 2 \\ -1 & 1 \end{vmatrix} + 1 \begin{vmatrix} 0 & 1 \\ -1 & 0 \end{vmatrix}$

$= 1[1(1) - 0(2)] - 3[0(1) - (-1)(2)] + 1[0(0) - (-1)(1)]$

$= 1(1 - 0) - 3(0 + 2) + 1(0 + 1)$

$= 1 - 6 + 1$

$= -4$

$D_z = \begin{vmatrix} 1 & 1 & 3 \\ 0 & -1 & 1 \\ -1 & 0 & 0 \end{vmatrix}$

$= 1 \begin{vmatrix} -1 & 1 \\ 0 & 0 \end{vmatrix} - 1 \begin{vmatrix} 0 & 1 \\ -1 & 0 \end{vmatrix} + 3 \begin{vmatrix} 0 & -1 \\ -1 & 0 \end{vmatrix}$

$= 1[-1(0) - 0(1)] - 1[0(0) - (-1)(1)] + 3[0(0) - (-1)(-1)]$

$= 1(0 - 0) - 1(0 + 1) + 3(0 - 1)$

$= 0 - 1 + (-3)$

$= -4$

$x = \dfrac{D_x}{D} = \dfrac{-4}{-4} = 1, \quad y = \dfrac{D_y}{D} = \dfrac{-4}{-4} = 1, \quad z = \dfrac{D_z}{D} = \dfrac{-4}{-4} = 1$

The solution is the ordered triple (1, 1, 1).

43. $\begin{vmatrix} x & 3 \\ 1 & 2 \end{vmatrix} = 7$

$x(2) - 1(3) = 7$

$2x - 3 = 7$

$2x = 10$

$x = 5$

45.

$$\begin{vmatrix} x & -1 & -2 \\ 1 & 0 & 4 \\ 3 & 2 & 5 \end{vmatrix} = 5$$

$$x\begin{vmatrix} 0 & 4 \\ 2 & 5 \end{vmatrix} - (-1)\begin{vmatrix} 1 & 4 \\ 3 & 5 \end{vmatrix} + (-2)\begin{vmatrix} 1 & 0 \\ 3 & 2 \end{vmatrix} = 5$$

$$x\left[0(5) - 2(4)\right] - (-1)\left[1(5) - 3(4)\right] + (-2)\left[1(2) - 3(0)\right] = 5$$

$$x(-8) - (-1)(-7) + (-2)(2) = 5$$

$$-8x - 7 - 4 = 5$$

$$-8x - 11 = 5$$

$$-8x = 16$$

$$x = -2$$

47. a. Triangle *ABC* is graphed below.

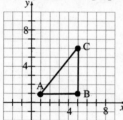

b. $D = \dfrac{1}{2}\begin{vmatrix} 1 & 5 & 5 \\ 1 & 1 & 6 \\ 1 & 1 & 1 \end{vmatrix}$

$$= \frac{1}{2}\left(1\begin{vmatrix} 1 & 6 \\ 1 & 1 \end{vmatrix} - 5\begin{vmatrix} 1 & 6 \\ 1 & 1 \end{vmatrix} + 5\begin{vmatrix} 1 & 1 \\ 1 & 1 \end{vmatrix}\right)$$

$$= \frac{1}{2}\left[1(1 \cdot 1 - 1 \cdot 6) - 5(1 \cdot 1 - 1 \cdot 6) + 5(1 \cdot 1 - 1 \cdot 1)\right]$$

$$= \frac{1}{2}\left[1(-5) - 5(-5) + 5(0)\right]$$

$$= \frac{1}{2}(-5 + 25 + 0)$$

$$= \frac{1}{2}(20)$$

$$= 10$$

Thus, the area of triangle *ABC* is $|10| = 10$.

49. a. Parallelogram *ABCD* is graphed below.

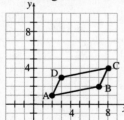

b. Triangle *ABC* is formed by the points (2, 1), (7, 2), and (8, 4).

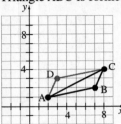

$$D = \frac{1}{2}\begin{vmatrix} 2 & 7 & 8 \\ 1 & 2 & 4 \\ 1 & 1 & 1 \end{vmatrix}$$

$$= \frac{1}{2}\left[2\begin{vmatrix} 2 & 4 \\ 1 & 1 \end{vmatrix} - 7\begin{vmatrix} 1 & 4 \\ 1 & 1 \end{vmatrix} + 8\begin{vmatrix} 1 & 2 \\ 1 & 1 \end{vmatrix} \right]$$

$$= \frac{1}{2}\left[2(2\cdot1 - 1\cdot4) - 7(1\cdot1 - 1\cdot4) + 8(1\cdot1 - 1\cdot2) \right]$$

$$= \frac{1}{2}\left[2(-2) - 7(-3) + 8(-1) \right]$$

$$= \frac{1}{2}(-4 + 21 - 8)$$

$$= \frac{1}{2}(9)$$

$$= 4.5$$

Thus, the area of triangle *ABC* is $|4.5| = 4.5$.

c. Triangle *ADC* is formed by the points (2, 1), (3, 3), and (8, 4).

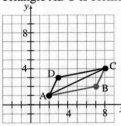

$$D = \frac{1}{2}\begin{vmatrix} 2 & 3 & 8 \\ 1 & 3 & 4 \\ 1 & 1 & 1 \end{vmatrix}$$

$$= \frac{1}{2}\left[2\begin{vmatrix} 3 & 4 \\ 1 & 1 \end{vmatrix} - 3\begin{vmatrix} 1 & 4 \\ 1 & 1 \end{vmatrix} + 8\begin{vmatrix} 1 & 3 \\ 1 & 1 \end{vmatrix} \right]$$

$$= \frac{1}{2}\left[2(3\cdot1 - 1\cdot4) - 3(1\cdot1 - 1\cdot4) + 8(1\cdot1 - 1\cdot3) \right]$$

$$= \frac{1}{2}\left[2(-1) - 3(-3) + 8(-2) \right]$$

$$= \frac{1}{2}(-2 + 9 - 16)$$

$$= \frac{1}{2}(-9)$$

$$= -4.5$$

Thus, the area of triangle *ADC* is $|-4.5| = 4.5$.

d. The areas of triangle *ABC* and triangle *ADC* are equal. (That is, the diagonal of the parallelogram forms two triangles of equal area.) Thus, the area of parallelogram *ABCD* is $4.5 + 4.5 = 9$.

51. a.
$$\begin{vmatrix} x & y & 1 \\ 3 & 2 & 1 \\ 5 & 1 & 1 \end{vmatrix} = 0$$

$$x\begin{vmatrix} 2 & 1 \\ 1 & 1 \end{vmatrix} - y\begin{vmatrix} 3 & 1 \\ 5 & 1 \end{vmatrix} + 1\begin{vmatrix} 3 & 2 \\ 5 & 1 \end{vmatrix} = 0$$

$$x\left[2(1) - 1(1)\right] - y\left[3(1) - 5(1)\right] + 1\left[3(1) - 5(2)\right] = 0$$

$$x(1) - y(-2) + 1(-7) = 0$$

$$x + 2y - 7 = 0$$

$$x + 2y = 7$$

b. $m = \dfrac{y_2 - y_1}{x_2 - x_1} = \dfrac{1-2}{5-3} = \dfrac{-1}{2} = -\dfrac{1}{2}$

$$y - y_1 = m(x - x_2)$$

$$y - 2 = -\frac{1}{2}(x - 3)$$

$$y - 2 = -\frac{1}{2}x + \frac{3}{2}$$

$$2y - 4 = -x + 3$$

$$x + 2y = 7$$

53. $\begin{vmatrix} 3 & -2 \\ 1 & 4 \end{vmatrix} = 3(4) - 1(-2)$

$$= 12 + 2$$

$$= 14$$

Interchanging rows 1 and 2 and recomputing the determinant, we obtain

$$\begin{vmatrix} 1 & 4 \\ 3 & -2 \end{vmatrix} = 1(-2) - 3(4)$$

$$= -2 - 12$$

$$= -14$$

The two determinants are opposites.
Answers may vary. In general, it is true that the value of a determinant will change signs if any two rows (or any two columns) are interchanged.

55. Using $A = D$, $B = D_x$, and $C = D_y$.

```
[A]
        [[1  1 ]
         [1 -1]]
[B]
        [[-4  1 ]
         [-12 -1]]
```

```
[C]
        [[1  -4 ]
         [1 -12]]
```

```
det([B])/det([A]
)▶Frac
                  -8
det([C])/det([A]
)▶Frac
                   4
```

The solution is the ordered pair $(-8, 4)$.

57. Using $A = D$, $B = D_x$, and $C = D_y$.

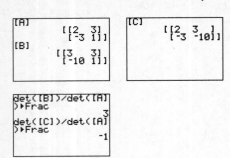

The solution is the ordered pair $(3, -1)$.

59. Using $A = D$, $B = D_x$, $C = D_y$, and $D = D_z$.

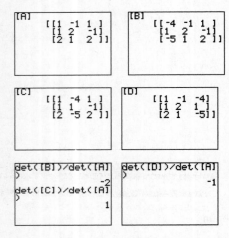

The solution is the ordered triple $(-2, 1, -1)$.

Section 3.6

Are You Ready for This Section?

R1. $\overset{?}{-3(-2)+2 \geq 7}$
$$\overset{?}{6+2 \geq 7}$$
$$8 \geq 7 \leftarrow \text{True}$$
Yes, $x = -2$ does satisfy the inequality
$-3x + 2 \geq 7$.

R2. $-2x + 1 > 7$
$$-2x + 1 - 1 > 7 - 1$$
$$-2x > 6$$
$$\frac{-2x}{-2} < \frac{6}{-2}$$
$$x < -3$$
The solution set is $\{x \mid x < -3\}$ or, using interval notation, $(-\infty, -3)$.

R3. Replace the inequality symbol with an equal sign to obtain $3x + 2y = -6$. Because the inequality is strict, graph $3x + 2y = -6$ $\left(y = -\dfrac{3}{2}x - 3\right)$ using a dashed line.

Test Point: $(0,0)$: $\quad \overset{?}{3(0) + 2(0) > -6}$
$$\overset{?}{0 + 0 > -6}$$
$$\overset{?}{0 > -6} \quad \text{True}$$
Therefore, $(0, 0)$ is solution to $3x + 2y > -6$.
Shade the half-plane that contains $(0, 0)$.

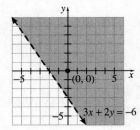

Section 3.6 Quick Checks

1. An ordered pair <u>satisfies</u> a system of linear inequalities if it makes each inequality in the system a true statement.

2. $\begin{cases} -4x + y < -5 & (1) \\ 2x - 5y < 10 & (2) \end{cases}$

   a. Let $x = 1$ and $y = 2$ in both inequalities (1) and (2).

       Inequality (1): $\overset{?}{-4(1) + 2 < -5}$
$$\overset{?}{-4 + 2 < -5}$$
$$-2 \not< -5$$

Inequality (2): $2(1) - 5(2) \overset{?}{<} 10$

$$2 - 10 \overset{?}{<} 10$$

$$-8 < 10$$

The inequality $-4x + y < -5$ is not true when $x = 1$ and $y = 2$, so $(1, 2)$ is not a solution.

b. Let $x = 3$ and $y = 1$ in both inequalities (1) and (2).

Inequality (1): $-4(3) + 1 \overset{?}{<} -5$

$$-12 + 1 \overset{?}{<} -5$$

$$-11 < -5$$

Inequality (2): $2(3) - 5(1) \overset{?}{<} 10$

$$6 - 5 \overset{?}{<} 10$$

$$1 < 10$$

These values satisfy both inequalities, so the ordered pair $(3, 1)$ is a solution.

c. Let $x = 1$ and $y = -2$ in both inequalities (1) and (2).

Inequality (1): $-4(1) + (-2) \overset{?}{<} -5$

$$-4 - 2 \overset{?}{<} -5$$

$$-6 < -5$$

Inequality 2: $2(1) - 5(-2) \overset{?}{<} 10$

$$2 + 10 \overset{?}{<} 10$$

$$12 \not< 10$$

The inequality $2x - 5y < 10$ is not true when $x = 1$ and $y = -2$, so $(1, -2)$ is not a solution.

3. $\begin{cases} 2x + y \le 5 \\ -x + y \ge -4 \end{cases}$

First, graph the inequality $2x + y \le 5$. To do so, replace the inequality symbol with an equal sign to obtain $2x + y = 5$. Because the inequality is non-strict, graph $2x + y = 5$ $(y = -2x + 5)$ using a solid line.

Test Point: $(0,0)$: $2(0) + 0 \overset{?}{\le} 5$

$$0 \le 5$$

Therefore, the half-plane containing $(0, 0)$ is the solution set of $2x + y \le 5$.

Second, graph the inequality $-x + y \ge -4$. To do so, replace the inequality symbol with an equal sign to obtain $-x + y = -4$. Because the inequality is non-strict, graph $-x + y = -4$ $(y = x - 4)$ using a solid line.

Test Point: $(0,0)$: $-(0) + 0 \overset{?}{\ge} -4$

$$0 \ge -4$$

Therefore, the half-plane containing $(0, 0)$ is the solution set of $-x + y \ge -4$.

The overlapping shaded region (that is, the shaded region in the graph below) is the solution to the system of linear inequalities.

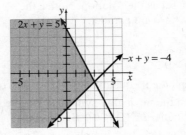

4. $\begin{cases} 3x + y > -2 \\ 2x + 3y < 3 \end{cases}$

First, graph the inequality $3x + y > -2$. To do so, replace the inequality symbol with an equal sign to obtain $3x + y = -2$. Because the inequality is strict, graph $3x + y = -2$ $(y = -3x - 2)$ using a dashed line.

Test Point: $(0,0)$: $3(0) + 0 \overset{?}{>} -2$

$$0 > -2$$

Therefore, the half-plane containing $(0, 0)$ is the solution set of $3x + y > -2$.

Second, graph the inequality $2x + 3y < 3$. To do so, replace the inequality symbol with an equal sign to obtain $2x + 3y = 3$. Because the inequality is strict, graph $2x + 3y = 3$ $\left(y = -\dfrac{2}{3}x + 1 \right)$ using a dashed line.

Test Point: $(0,0)$: $-2(0) + 3(0) \overset{?}{<} 3$

$$0 < 3$$

Therefore, the half-plane containing $(0, 0)$ is the solution set of $2x + 3y < 3$.

The overlapping shaded region (that is, the shaded region in the graph below) is the solution to the system of linear inequalities.

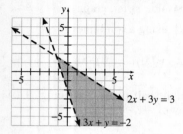

5. False. For example, the system $\begin{cases} x+y>5 \\ x+y<-1 \end{cases}$ has the empty set as the solution. Notice the boundary lines are parallel.

6. $\begin{cases} -2x+3y \geq 9 \\ 6x-9y \geq 9 \end{cases}$

   First, graph the inequality $-2x+3y \geq 9$. To do so, replace the inequality symbol with an equal sign to obtain $-2x+3y=9$. Because the inequality is non-strict, graph $-2x+3y \geq 9$ $\left(y=\dfrac{2}{3}x+3 \right)$ using a solid line.

   Test Point: $(0,0)$: $-2(0)+3(0) \overset{?}{\geq} 9$
   $$0 \ngeq 9$$

   Therefore, the half-plane not containing $(0, 0)$ is the solution set of $-2x+3y \geq 9$.

   Second, graph the inequality $6x-9y \geq 9$. To do so, replace the inequality symbol with an equal sign to obtain $6x-9y=9$. Because the inequality is non-strict, graph $6x-9y=9$ $\left(y=\dfrac{2}{3}x-1 \right)$ using a solid line.

   Test Point: $(0,0)$: $6(0)-9(0) \overset{?}{\geq} 9$
   $$0 \ngeq 9$$

   Therefore, the half-plane not containing $(0, 0)$ is the solution set of $6x-9y \geq 9$.

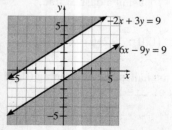

Because no overlapping region results, there are no points in the Cartesian plane that satisfy both inequalities. The system has no solution, so the solution set is $\varnothing$ or $\{ \ \}$.

7. The points of intersection between two boundary lines in a system of linear inequalities are called <u>corner points</u>.

8. If the solution set of a system o linear inequalities cannot be contained within a circle, then the system is <u>unbounded</u>.

9. $\begin{cases} x+y \leq 6 \\ 2x+y \leq 10 \\ x \geq 0 \\ y \geq 0 \end{cases}$

   First, graph the inequality $x+y \leq 6$. To do so, replace the inequality symbol with an equal sign to obtain $x+y=6$. Because the inequality is non-strict, graph $x+y=6$ $(y=-x+6)$ using a solid line.

   Test Point: $(0,0)$: $0+0 \overset{?}{\leq} 6$
   $$0 \leq 6$$

   Therefore, the half-plane containing $(0, 0)$ is the solution set of $x+y \leq 6$.

   Next, graph the inequality $2x+y \leq 10$. To do so, replace the inequality symbol with an equal sign to obtain $2x+y=10$. Because the inequality is non-strict, graph $2x+y=10$ $(y=-2x+10)$ using a solid line.

   Test Point: $(0,0)$: $2(0)+0 \overset{?}{\leq} 10$
   $$0 \leq 10$$

   Therefore, the half-plane containing $(0, 0)$ is the solution set of $2x+y \leq 10$.

   The two inequalities $x \geq 0$ and $y \geq 0$ require that the graph be in quadrant I.

The overlapping shaded region (that is, the shaded region in the graph below) is the solution to the system of linear inequalities.

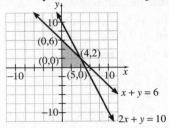

The graph is bounded. The four corner points of the bounded region are (0, 0), (5, 0), (4, 2), and (0, 6).

10. Let x represent the amount to be invested in Treasury notes and let y represent the amount to be invested in corporate bonds. Now, Jack and Mary have at most $25,000 to invest, so $x + y \leq 25,000$. The amount invested in Treasury notes must be at least $10,000, so $x \geq 10,000$. The amount to be invested in corporate bonds must be no more than $15,000, so $y \leq 15,000$. They cannot invest a negative amount in corporate bonds, so we have the non-negativity constraint $y \geq 0$. Thus, we have the following system:

$$\begin{cases} x + y \leq 25,000 \\ x \geq 10,000 \\ y \leq 15,000 \\ y \geq 0 \end{cases}$$

To graph the system, begin with the inequality $x + y \leq 25,000$. To do so, replace the inequality symbol with an equal sign to obtain $x + y = 25,000$. Because the inequality is non-strict, graph $x + y = 25,000$ $\left(y = -x + 25,000\right)$ using a solid line.

Test Point: $(0,0)$: $0 + 0 \overset{?}{\leq} 25,000$

$0 \leq 25,000$

Thus, the solution set of $x + y \leq 25,000$ is the half-plane containing (0, 0).

The inequality $x \geq 10,000$ requires that the graph be to the right of the vertical line $x = 10,000$.

The inequality $y \leq 15,000$ requires that the graph be below the horizontal line $y = 15,000$.

The inequality $y \geq 0$ requires that the graph be above the x-axis.

The overlapping shaded region (that is, the shaded region in the graph below) is the solution to the system of linear inequalities.

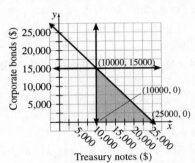

The corner points of the bounded region are (10,000, 15,000), (10,000, 0), and (25,000, 0).

3.6 Exercises

11. **a.** Let $x = 3$ and $y = 2$ in each inequality.

$$\begin{cases} x + y = 3 + 2 = 5 \not\leq 4 \\ 3x + y = 3(3) + 2 = 11 \geq 10 \end{cases}$$

The inequality $x + y \leq 4$ is not true when $x = 3$ and $y = 2$, so $(3, \ 2)$ is not a solution.

b. Let $x = 5$ and $y = -1$ in each inequality.

$$\begin{cases} x + y = 5 + (-1) = 4 \leq 4 \\ 3x + y = 3(5) + (-1) = 14 \geq 10 \end{cases}$$

Both inequalities are true when $x = 5$ and $y = -1$, so $(5, -1)$ is a solution.

13. **a.** Let $x = 3$ and $y = 2$ in each inequality.

$$\begin{cases} 3x + 2y = 3(3) + 2(2) = 13 \not< 12 \\ -2x + 3y = -2(3) + 3(2) = 0 \not> 12 \end{cases}$$

Neither inequality is true when $x = 3$ and $y = 2$, so $(3, \ 2)$ is not a solution.

b. Let $x = 1$ and $y = 4$ in each inequality.

$$\begin{cases} 3x + 2y = 3(1) + 2(4) = 11 < 12 \\ -2x + 3y = -2(1) + 3(4) = 10 \not> 12 \end{cases}$$

The inequality $-2x + 3y > 12$ is not true when $x = 1$ and $y = 4$, so $(1, \ 4)$ is not a solution.

15. a. Let $x = 4$ and $y = 2$ in each inequality.

$$\begin{cases} x + y = 4 + 2 = 6 \le 8 \\ 3x + y = 3(4) + 2 = 14 \not\le 12 \\ x = 4 \ge 0 \\ y = 2 \ge 0 \end{cases}$$

The inequality $3x + y \le 12$ is not true when

$x = 4$ and $y = 2$, so $(4,\ 2)$ is not a solution.

b. Let $x = 2$ and $y = 5$ in each inequality.

$$\begin{cases} x + y = 2 + 5 = 7 \le 8 \\ 3x + y = 3(2) + 5 = 11 \le 12 \\ x = 2 \ge 0 \\ y = 5 \ge 0 \end{cases}$$

All four inequalities are true when $x = 2$

and $y = 5$, so $(2,\ 5)$ is a solution.

17. $\begin{cases} x + y \le 4 \\ 3x + y \ge 10 \end{cases}$

First, graph the inequality $x + y \le 4$. To do so,

replace the inequality symbol with an equal sign

to obtain $x + y = 4$. Because the inequality is

non-strict, graph $x + y = 4$ $(y = -x + 4)$ using a

solid line.

Test Point: $(0,0)$: $0 + 0 = 0 \le 4$

Therefore, the half-plane containing (0, 0) is the

solution set of $x + y \le 4$.

Second, graph the inequality $3x + y \ge 10$. To do

so, replace the inequality symbol with an equal

sign to obtain $3x + y = 10$. Because the

inequality is non-strict, graph $3x + y = 10$

$(y = -3x + 10)$ using a solid line.

Test Point: $(0,0)$: $3(0) + 0 = 0 \not\ge 10$

Therefore, the half-plane not containing (0, 0) is

the solution set of $3x + y \ge 10$.

The overlapping shaded region (that is, the

shaded region in the graph below) is the solution

to the system of linear inequalities.

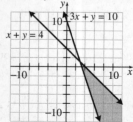

19. $\begin{cases} 3x + 2y < 12 \\ -2x + 3y > 12 \end{cases}$

First, graph the inequality $3x + 2y < 12$. To do

so, replace the inequality symbol with an equal

sign to obtain $3x + 2y = 12$. Because the

inequality is strict, graph $3x + 2y = 12$

$\left(y = -\dfrac{3}{2}x + 6 \right)$ using a dashed line.

Test Point: $(0,0)$: $3(0) + 2(0) = 0 < 12$

Therefore, the half-plane containing (0, 0) is the

solution set of $3x + 2y < 12$.

Second, graph the inequality $-2x + 3y > 12$. To

do so, replace the inequality symbol with an

equal sign to obtain $-2x + 3y = 12$. Because the

inequality is strict, graph $-2x + 3y = 12$

$\left(y = \dfrac{2}{3}x + 4 \right)$ using a dashed line.

Test Point: $(0,0)$: $-2(0) + 3(0) = 0 \not> 12$

Therefore, the half-plane not containing (0, 0) is

the solution set of $-2x + 3y > 12$.

The overlapping shaded region (that is, the

shaded region in the graph below) is the solution

to the system of linear inequalities.

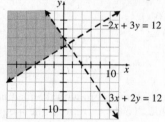

21. $\begin{cases} x - \dfrac{1}{2}y \le 3 \\ \dfrac{3}{2}x + y > 3 \end{cases}$

First, graph the inequality $x - \dfrac{1}{2}y \le 3$. To do so,

replace the inequality symbol with an equal sign

to obtain $x - \dfrac{1}{2}y = 3$. Because the inequality is

non-strict, graph $x - \dfrac{1}{2}y = 3$ $(y = 2x - 6)$ using

a solid line.

Test Point: $(0,0)$: $0 - \dfrac{1}{2}(0) = 0 \le 3$

Therefore, the half-plane containing $(0,0)$ is the solution set of $x - \frac{1}{2}y \le 3$.

Second, graph the inequality $\frac{3}{2}x + y > 3$. To do so, replace the inequality symbol with an equal sign to obtain $\frac{3}{2}x + y = 3$. Because the inequality is strict, graph $\frac{3}{2}x + y = 3$ $\left(y = -\frac{3}{2}x + 3 \right)$ using a dashed line.

Test Point: $(0,0)$: $\frac{3}{2}(0) + 0 = 0 \not> 3$

Therefore, the half-plane not containing $(0, 0)$ is the solution set of $\frac{3}{2}x + y > 3$.

The overlapping shaded region (that is, the shaded region in the graph below) is the solution to the system of linear inequalities.

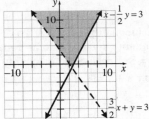

23. $\begin{cases} -2x + y < -8 \\ 2x - y > 8 \end{cases}$

First, graph the inequality $-2x + y < -8$. To do so, replace the inequality symbol with an equal sign to obtain $-2x + y = -8$. Because the inequality is strict, graph $-2x + y = -8$ $(y = 2x - 8)$ using a dashed line.

Test Point: $(0,0)$: $-2(0) + 0 = 0 \not< -8$

Therefore, the half-plane not containing $(0, 0)$ is the solution set of $-2x + y < -8$.

Second, graph the inequality $2x - y > 8$. To do so, replace the inequality symbol with an equal sign to obtain $2x - y = 8$. Because the inequality is strict, graph $2x - y = 8$ $(y = 2x - 8)$ using a dashed line.

Test Point: $(0,0)$: $2(0) - 0 = 0 \not> 8$

Therefore, the half-plane not containing $(0, 0)$ is the solution set of $2x - y > 8$.

The overlapping shaded region (that is, the shaded region in the graph below) is the solution to the system of linear inequalities. In this case, notice that the two inequalities that make up the system both have the exact same solution set. Therefore, the overlapping shaded region is the solution set of both individual inequalities.

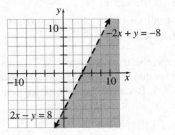

25. $\begin{cases} -5x + 3y < 12 \\ 5x - 3y < 9 \end{cases}$

First, graph the inequality $-5x + 3y < 12$. To do so, replace the inequality symbol with an equal sign to obtain $-5x + 3y = 12$. Because the inequality is strict, graph $-5x + 3y = 12$ $\left(y = \frac{5}{3}x + 4 \right)$ using a dashed line.

Test Point: $(0,0)$: $-5(0) + 3(0) = 0 < 12$

Therefore, the half-plane containing $(0, 0)$ is the solution set of $-5x + 3y < 12$.

Second, graph the inequality $5x - 3y < 9$. To do so, replace the inequality symbol with an equal sign to obtain $5x - 3y = 9$. Because the inequality is strict, graph $5x - 3y = 9$ $\left(y = \frac{5}{3}x - 3 \right)$ using a dashed line.

Test Point: $(0,0)$: $5(0) - 3(0) = 0 < 9$

Therefore, the half-plane containing $(0, 0)$ is the solution set of $5x - 3y < 9$.

The overlapping shaded region (that is, the shaded region in the graph below) is the solution to the system of linear inequalities.

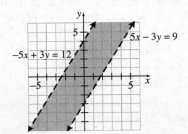

27. $\begin{cases} y \le 8 \\ x \ge 3 \end{cases}$

First, graph the inequality $y \le 8$. To do so, replace the inequality symbol with an equal sign to obtain $y = 8$. Because the inequality is non-strict, graph $y = 8$ using a solid line.

Test point: $(0,0)$: $0 \le 8$

Therefore, the half-plane containing $(0, 0)$ is the solution set of $y \le 8$.

Second, graph the inequality $x \ge 3$. To do so, replace the inequality symbol with an equal sign to obtain $x = 3$. Because the inequality is non-strict, graph $x = 3$ using a solid line.

Test point: $(0,0)$: $0 \not\ge 3$

Therefore, the half-plane not containing $(0, 0)$ is the solution set of $x \ge 3$.

The overlapping shaded region (that is, the shaded region in the graph below) is the solution to the system of linear inequalities.

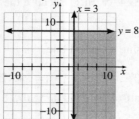

29. $\begin{cases} x + y \le 6 \\ 3x + y \le 12 \\ x \ge 0 \\ y \ge 0 \end{cases}$

First, graph the inequality $x + y \le 6$. To do so, replace the inequality symbol with an equal sign to obtain $x + y = 6$. Because the inequality is non-strict, graph $x + y = 6$ $(y = -x + 6)$ using a solid line.

Test Point: $(0,0)$: $0 + 0 = 0 \le 6$

Therefore, the half-plane containing $(0, 0)$ is the solution set of $x + y \le 6$.

Next, graph the inequality $3x + y \le 12$. To do so, replace the inequality symbol with an equal sign to obtain $3x + y = 12$. Because the inequality is non-strict, graph $3x + y = 12$ $(y = -3x + 12)$ using a solid line.

Test Point: $(0,0)$: $-3(0) + 0 = 0 \le 12$

Therefore, the half-plane containing $(0, 0)$ is the solution set of $3x + y \le 12$.

The two inequalities $x \ge 0$ and $y \ge 0$ require that the graph be in quadrant I.

The overlapping shaded region (that is, the shaded region in the graph below) is the solution to the system of linear inequalities.

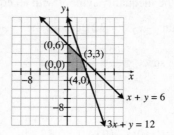

The graph is bounded. The four corner points of the bounded region are $(0, 0)$, $(4, 0)$, $(3, 3)$, and $(0, 6)$.

31. $\begin{cases} x + y \ge 8 \\ 4x + 2y \ge 28 \\ x \ge 0 \\ y \ge 0 \end{cases}$

First, graph the inequality $x + y \ge 8$. To do so, replace the inequality symbol with an equal sign to obtain $x + y = 8$. Because the inequality is non-strict, graph $x + y = 8$ $(y = -x + 8)$ using a solid line.

Test Point: $(0,0)$: $0 + 0 = 0 \not\ge 8$

Therefore, the half-plane not containing $(0, 0)$ is the solution set of $x + y \ge 8$.

Next, graph the inequality $4x + 2y \ge 28$. To do so, replace the inequality symbol with an equal sign to obtain $4x + 2y = 28$. Because the inequality is non-strict, graph $4x + 2y = 28$ $(y = -2x + 14)$ using a solid line.

Test Point: $(0,0)$: $4(0) + 2(0) = 0 \not\ge 28$

Therefore, the half-plane not containing $(0, 0)$ is the solution set of $4x + 2y \ge 28$.

The two inequalities $x \ge 0$ and $y \ge 0$ require that the graph be in quadrant I.

The overlapping shaded region (that is, the shaded region in the graph below) is the solution to the system of linear inequalities.

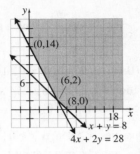

The graph is unbounded. The three corner points of the unbounded region are (0, 14), (6, 2), and (8, 0).

33. $\begin{cases} 2x + 3y \leq 30 \\ 3x + 2y \leq 25 \\ 5x + 2y \leq 35 \\ x \geq 0 \\ y \geq 0 \end{cases}$

First, graph the inequality $2x + 3y \leq 30$. To do so, replace the inequality symbol with an equal sign to obtain $2x + 3y = 30$. Because the inequality is non-strict, graph $2x + 3y = 30$

$\left(y = -\dfrac{2}{3}x + 10 \right)$ using a solid line.

Test Point: $(0,0)$: $2(0) + 3(0) = 0 \leq 30$

Therefore, the half-plane containing (0, 0) is the solution set of $2x + 3y \leq 30$.

Second, graph the inequality $3x + 2y \leq 25$. To do so, replace the inequality symbol with an equal sign to obtain $3x + 2y = 25$. Because the inequality is non-strict, graph $3x + 2y = 25$

$\left(y = -\dfrac{3}{2}x + \dfrac{25}{2} \right)$ using a solid line.

Test Point: $(0,0)$: $3(0) + 2(0) = 0 \leq 25$

Therefore, the half-plane containing (0, 0) is the solution set of $3x + 2y \leq 25$.

Third, graph the inequality $5x + 2y \leq 35$. To do so, replace the inequality symbol with an equal sign to obtain $5x + 2y = 35$. Because the inequality is non-strict, graph $5x + 2y = 35$

$\left(y = -\dfrac{5}{2}x + \dfrac{35}{2} \right)$ using a solid line.

Test Point: $(0,0)$: $5(0) + 2(0) = 0 \leq 35$

Therefore, the half-plane containing (0, 0) is the solution set of $5x + 2y \leq 35$.

The two inequalities $x \geq 0$ and $y \geq 0$ require that the graph be in quadrant I.

The overlapping shaded region (that is, the shaded region in the graph below) is the solution to the system of linear inequalities.

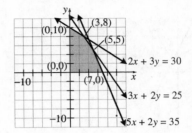

The graph is bounded. The five corner points of the bounded region are (0, 0), (7, 0), (5, 5), (3, 8), and (0, 10).

35. $\begin{cases} 2x + y \geq 13 \\ x + 2y \geq 11 \\ x \geq 4 \\ y \geq 0 \end{cases}$

First, graph the inequality $2x + y \geq 13$. To do so, replace the inequality symbol with an equal sign to obtain $2x + y = 13$. Because the inequality is non-strict, graph $2x + y = 13$ $\left(y = -2x + 13 \right)$ using a solid line.

Test Point: $(0,0)$: $2(0) + 0 = 0 \ngeq 13$

Therefore, the half-plane not containing (0, 0) is the solution set of $2x + y \geq 13$.

Second, graph the inequality $x + 2y \geq 11$. To do so, replace the inequality symbol with an equal sign to obtain $x + 2y = 11$. Because the inequality is non-strict, graph $x + 2y = 11$

$\left(y = -\dfrac{1}{2}x + \dfrac{11}{2} \right)$ using a solid line.

Test Point: $(0,0)$: $0 + 2(0) = 0 \ngeq 11$

Therefore, the half-plane not containing (0, 0) is the solution set of $x + 2y \geq 11$.

Third, graph the inequality $x \geq 4$. To do so, replace the inequality symbol with an equal sign to obtain $x = 4$. Because the inequality is non-strict, graph $x = 4$ using a solid line.

Test Point: $(0,0)$: $0 \ngeq 4$

Therefore, the half-plane not containing (0, 0) is the solution set of $x \geq 4$.

The inequality $y \geq 0$ requires that the graph be above the x-axis.

The overlapping shaded region (that is, the shaded region in the graph below) is the solution to the system of linear inequalities.

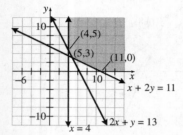

The graph is unbounded. The three corner points of the unbounded region are (4, 5), (5, 3), and (11, 0).

37. a. Let x represent the number of servings of orange juice and let y represent the number of servings of cereal. Now, the amount of potassium Daria consumes must be at least 500 mg, so $450x + 50y \geq 500$. The amount of protein Daria consumes must be at least 14 grams, so $2x + 6y \geq 14$. Finally, Daria cannot consume negative servings of orange juice or cereal, so we have the non-negativity constraints $x \geq 0$ and $y \geq 0$. Thus, we have the following system:

$$\begin{cases} 450x + 50y \geq 500 \\ 2x + 6y \geq 14 \\ x \geq 0 \\ y \geq 0 \end{cases}$$

b. First, graph the inequality $450x + 50y \geq 500$. To do so, replace the inequality symbol with an equal sign to obtain $450x + 50y = 500$. Because the inequality is non-strict, graph $450x + 50y = 500$ $(y = -9x + 10)$ using a solid line.

Test Point: $(0,0)$: $450(0) + 50(0) = 0 \not\geq 500$

Thus, the solution set of $450x + 50y \geq 500$ is the half-plane not containing (0, 0).

Second, graph the inequality $2x + 6y \geq 14$. To do so, replace the inequality symbol with an equal sign to obtain $2x + 6y = 14$. Because the inequality is non-strict, graph

$2x + 6y = 14$ $\left(y = -\dfrac{1}{3}x + \dfrac{7}{3}\right)$ using a solid line.

Test Point: $(0,0)$: $2(0) + 6(0) = 0 \not\geq 14$

Thus, the solution set of $2x + 6y \geq 14$ is the half-plane not containing (0, 0).

The two inequalities $x \geq 0$ and $y \geq 0$ require that the graph be in quadrant I.

The overlapping shaded region (that is, the shaded region in the graph below) is the solution to the system of linear inequalities.

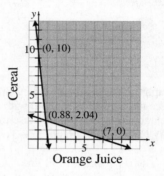

The three corner points of the unbounded region are (0, 10), $\approx (0.88, 2.04)$ and (7, 0).

39. a. Let x represent the amount to be invested in Treasury Notes and let y represent the amount to be invested in corporate bonds. Now, Jack and Mary have at most $25,000 to invest, so $x + y \leq 25,000$. They want to earn at least $1,400 per year in interest, so $0.05x + 0.08y \geq 1,400$. The amount invested in Treasury Notes must be at least $5,000, so $x \geq 5,000$. The amount to be invested in corporate bonds must be no more than $15,000, so $y \leq 15,000$. They cannot invest a negative amount in corporate bonds, so we have the non-negativity constraint $y \geq 0$. Thus, we have the following system:

$$\begin{cases} x + y \leq 25,000 \\ 0.05x + 0.08y \geq 1,400 \\ x \geq 5,000 \\ y \leq 15,000 \\ y \geq 0 \end{cases}$$

b. First, graph the inequality $x + y \le 25{,}000$. To do so, replace the inequality symbol with an equal sign to obtain $x + y = 25{,}000$. Because the inequality is non-strict, graph
$x + y = 25{,}000 \ \left(y = -x + 25{,}000 \right)$ using a solid line.

Test Point: $(0, 0)$: $0 + 0 = 0 \le 25{,}000$

Thus, the solution set of $x + y \le 25{,}000$ is the half-plane containing (0, 0).

Second, graph the inequality
$0.05x + 0.08y \ge 1{,}400$. To do so, replace the inequality symbol with an equal sign to obtain $0.05x + 0.08y = 1{,}400$. Because the inequality is non-strict, graph
$0.05x + 0.08y = 1{,}400 \ \left(y = -\dfrac{5}{8}x + 17{,}500 \right)$
using a solid line.

Test Point: $(0, 0)$:

$0.05(0) + 0.08(0) = 0 \not\ge 1{,}400$

Thus, the solution set of
$0.05x + 0.08y \ge 1{,}400$ is the half-plane not containing (0, 0).

The inequality $x \ge 5{,}000$ requires that the graph be to the right of the vertical line
$x = 5{,}000$.

The inequality $y \le 15{,}000$ requires that the graph be below the horizontal line
$y = 15{,}000$.

The inequality $y \ge 0$ requires that the graph be above the x-axis.

The overlapping shaded region (that is, the shaded region in the graph below) is the solution to the system of linear inequalities.

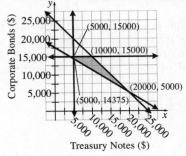

The four corner points of the bounded region are $(5000, 15000)$, $(5000, 14375)$, $(20000, 5000)$, and $(10000, 15000)$.

41. The graph contains a line passing through the points $(0, 4)$ and $(2, 6)$. The slope of the line is
$m = \dfrac{6 - 4}{2 - 0} = \dfrac{2}{2} = 1$. The y-intercept is 4. Thus, the equation of the line is $y = x + 4$. Because the shaded region is above this line, our inequality will contain a "greater than." Because the line is solid, this means that the inequality is non-strict. That is, our inequality will consist of "greater than or equal to." Thus, the linear inequality is $y \ge x + 4$.

The second line in the graph passes through the points (0, 10) and (2, 6). The slope of the line is
$m = \dfrac{6 - 10}{2 - 0} = \dfrac{-4}{2} = -2$. The y-intercept is 10.
Thus, the equation of the line is $y = -2x + 10$. Because the shaded region is above this line, our inequality will contain a "greater than." Because the line is solid, this means that the inequality is non-strict. That is, our inequality will consist of "greater than or equal to." Thus, the linear inequality is $y \ge -2x + 10$.

Thus, the system of linear inequalities is
$$\begin{cases} y \ge x + 4 \\ y \ge -2x + 10 \end{cases}$$

43. The graph is restricted to the first quadrant. Therefore, the system will contain the inequalities $x \ge 0$ and $y \ge 0$.

The graph contains a line passing through the points (0, 12) and (4, 8). The slope of the line is
$m = \dfrac{8 - 12}{4 - 0} = \dfrac{-4}{4} = -1$. The y-intercept is 12.
Thus, the equation of the line is $y = -x + 12$. Because the shaded region is below this line, our inequality will contain a "less than." Because the line is solid, this means that the inequality is non-strict. That is, our inequality will consist of "less than or equal to." Thus, the linear inequality is $y \le -x + 12$.

The second line in the graph passes through the points (4, 8) and (8, 0). The slope of the line is
$m = \dfrac{0 - 8}{8 - 4} = \dfrac{-8}{4} = -2$. Using the point-slope form, the equation of the second line is
$$\begin{aligned} y - y_1 &= m(x - x_1) \\ y - 0 &= -2(x - 8) \\ y &= -2x + 16 \end{aligned}$$

Because the shaded region is below this line, our inequality will contain a "less than." Because the line is solid, this means that the inequality is non-strict. That is, our inequality will consist of "less than or equal to." Thus, the linear inequality is $y \leq -2x + 16$.

Thus, the system of linear inequalities is

$$\begin{cases} y \leq -x + 12 \\ y \leq -2x + 16 \\ x \geq 0 \\ y \geq 0 \end{cases}$$

45. a.
$$\begin{cases} x + y \leq 8 \\ 2x + y \leq 10 \\ x \geq 0 \\ y \geq 0 \end{cases}$$

First, graph the inequality $x + y \leq 8$. To do so, replace the inequality symbol with an equal sign to obtain $x + y = 8$. Because the inequality is non-strict, graph $x + y = 8$

$(y = -x + 8)$ using a solid line.

Test Point: $(0,0): 0 + 0 = 0 \leq 8$

Therefore, the half-plane containing $(0, 0)$ is the solution set of $x + y \leq 8$.
Next, graph the inequality $2x + y \leq 10$. To do so, replace the inequality symbol with an equal sign to obtain $2x + y = 10$. Because the inequality is non-strict, graph $2x + y = 10$

$(y = -2x + 10)$ using a solid line.

Test Point: $(0,0): 2 \cdot 0 + 0 = 0 \leq 10$

Therefore, the half-plane containing $(0, 0)$ is the solution set of $2x + y \leq 10$.
The two inequalities $x \geq 0$ and $y \geq 0$ require that the graph be in quadrant I. The overlapping shaded region (that is, the shaded region in the graph below) is the solution to the system of linear inequalities. In this context, it is called the feasible region.

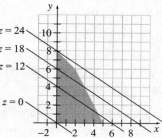

b. $z = 2x + 3y$

$z = 0: 2x + 3y = 0$
$$3y = -2x$$
$$y = -\frac{2}{3}x$$

$z = 12: 2x + 3y = 12$
$$3y = -2x + 12$$
$$y = -\frac{2}{3}x + 4$$

$z = 18: 2x + 3y = 18$
$$3y = -2x + 18$$
$$y = -\frac{2}{3}x + 6$$

$z = 24: 2x + 3y = 24$
$$3y = -2x + 24$$
$$y = -\frac{2}{3}x + 8$$

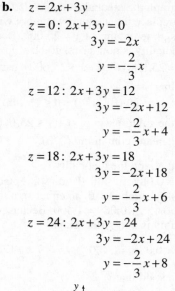

c. The largest possible value of z while still remaining in the feasible region is 24. The values $x = 0$ and $y = 8$ correspond to this value.

47. No, the corner point would not be a solution to the system. Solutions must satisfy all of the inequalities in the system. However, the corner point would only satisfy the non-strict inequality. Since it would not satisfy the strict inequality, it would not be a solution to the system.

49. Yes, it is possible for a system to have a straight line as the solution set. Examples may vary. One possibility follows: The solution to the following system of linear inequalities is the line $x + y = 1$:
$$\begin{cases} x + y \leq 1 \\ x + y \geq 1 \end{cases}$$

51. $\begin{cases} x+y \le 4 & (y \le -x+4) \\ 3x+y \ge 10 & (y \ge -3x+10) \end{cases}$

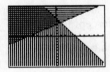

53. $\begin{cases} 3x+2y < 12 & \left(y < -\dfrac{3}{2}x+6\right) \\ -2x+3y > 12 & \left(y > \dfrac{2}{3}x+4\right) \end{cases}$

55. $\begin{cases} x+y \le 6 & (y \le -x+6) \\ 3x+y \le 12 & (y \le -3x+12) \\ x \ge 0 \\ y \ge 0 \end{cases}$

The inequalities $x \ge 0$ and $y \ge 0$ restrict the graph to the first quadrant.

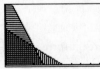

Chapter 3 Review

1. a. Let $x=3$, $y=-1$ in both equations (1) and (2).

$$\begin{cases} 3+3(-1) = 3-3 = 0 \ne -2 \\ 2(3)-(-1) = 6+1 = 7 \ne 10 \end{cases}$$

These values do not satisfy either equation. Therefore, the ordered pair $(3, -1)$ is not a solution of the system.

b. Let $x=4$, $y=-2$ in both equations (1) and (2).

$$\begin{cases} 4+3(-2) = 4-6 = -2 \\ 2(4)-(-2) = 8+2 = 10 \end{cases}$$

Because these values satisfy both equations, the ordered pair $(4, -2)$ is a solution of the system.

2. a. Let $x=\dfrac{1}{2}$, $y=1$ in both equations (1) and (2).

$$\begin{cases} -2\left(\dfrac{1}{2}\right)+5(1) = -1+5 = 4 \\ 4\left(\dfrac{1}{2}\right)-5(1) = 2-5 = -3 \end{cases}$$

Because these values satisfy both equations, the ordered pair $\left(\dfrac{1}{2},\, 1\right)$ is a solution of the system.

b. Let $x=-1$, $y=\dfrac{1}{3}$ in both equations (1) and (2).

$$\begin{cases} -2(-1)+5\left(\dfrac{1}{3}\right) = 2+\dfrac{5}{3} = \dfrac{11}{3} \ne 4 \\ 4(-1)-5\left(\dfrac{1}{3}\right) = -4-\dfrac{5}{3} = -\dfrac{17}{3} \ne -3 \end{cases}$$

These values do not satisfy either equation. Therefore, the ordered pair $\left(-1,\, \dfrac{1}{3}\right)$ is not a solution of the system.

3. a. Let $x=3$, $y=6$ in both equation (1) and (2).

$$\begin{cases} 6(3)-5(6) = 18-30 = -12 \\ 3-6 = -3 \end{cases}$$

These values satisfy both equations. Therefore, the ordered pair $(3, 6)$ is a solution of the system.

b. Let $x = 1$, $y = 4$ in both equations (1) and (2).

$$\begin{cases} 6(1) - 5(4) = 6 - 20 = -14 \neq -12 \\ 1 - 4 = -3 \end{cases}$$

Although these values satisfy equation (2), they do not satisfy the equation (1). Therefore, the ordered pair (1, 4) is not a solution of the system.

4. a. Let $x = 3$, $y = 0$ in both equations (1) and (2).

$$\begin{cases} 3(3) - 0 = 9 - 0 = 9 \\ 8(3) + 3(0) = 24 + 0 = 24 \neq 7 \end{cases}$$

Although these values satisfy equation (1), they do not satisfy the equation (2). Therefore, the ordered pair (3, 0) is not a solution of the system.

b. Let $x = 2$, $y = -3$ in both equations (1) and (2).

$$\begin{cases} 3(2) - (-3) = 6 + 3 = 9 \\ 8(2) + 3(-3) = 16 - 9 = 7 \end{cases}$$

Because these values satisfy both equations, the ordered pair (2, –3) is a solution of the system.

5. The graphs of the linear equations intersect at the point (4, 2). Thus, the solution is (4, 2).

6. The graphs of the linear equations intersect at the point (2, –1). Thus, the solution is (2, –1).

7. $\begin{cases} y = -3x + 1 & (1) \\ y = \dfrac{1}{2}x - 6 & (2) \end{cases}$

The two equations are in slope-intercept form. Graph each equation and find the point of intersection.

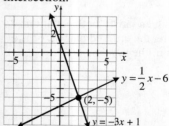

The solution is the ordered pair (2, –5).

8. $\begin{cases} -2x + 3y = -9 & (1) \\ 3x + y = 8 & (2) \end{cases}$

Equation (1) in slope-intercept form is

$y = \dfrac{2}{3}x - 3$. Equation (2) in slope-intercept form

is $y = -3x + 8$. Graph each equation and find the point of intersection.

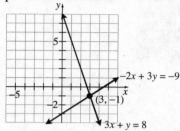

The solution is the ordered pair (3, –1).

9. $\begin{cases} y = -\dfrac{1}{3}x + 2 & (1) \\ x + 3y = -9 & (2) \end{cases}$

Equation (1) is already in slope-intercept form. Equation (2) in slope-intercept form is

$y = -\dfrac{1}{3}x - 3$. Graph each equation.

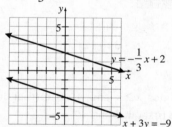

The two lines are parallel. The system is inconsistent. The system has no solution. That is, $\varnothing$ or { }.

10. $\begin{cases} 2x - 3y = 0 & (1) \\ 2x - y = -4 & (2) \end{cases}$

Equation (1) in slope-intercept form is $y = \dfrac{2}{3}x$.

Equation (2) in slope-intercept form is $y = 2x + 4$. Graph each equation and find the point of intersection.

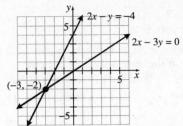

The solution is the ordered pair $(-3, -2)$.

11. $\begin{cases} y = -\dfrac{1}{4}x + 2 & (1) \\ y = 4x - 32 & (2) \end{cases}$

Substituting $-\dfrac{1}{4}x + 2$ for y in equation (2), we

obtain

$$-\frac{1}{4}x + 2 = 4x - 32$$

$$-4\left(-\frac{1}{4}x + 2\right) = -4(4x - 32)$$

$$x - 8 = -16x + 128$$

$$17x = 136$$

$$x = 8$$

Substituting 8 for x in equation (2), we obtain

$y = 4(8) - 32 = 32 - 32 = 0$.

The solution is the ordered pair $(8, 0)$.

12. $\begin{cases} y = -\dfrac{3}{4}x + 2 & (1) \\ 3x + 4y = 8 & (2) \end{cases}$

Substituting $-\dfrac{3}{4}x + 2$ for y in equation (2), we

obtain

$$3x + 4\left(-\frac{3}{4}x + 2\right) = 8$$

$$3x - 3x + 8 = 8$$

$$8 = 8$$

The system is dependent. The solution is

$$\left\{(x, y) \;\middle|\; y = -\frac{3}{4}x + 2\right\}.$$

13. $\begin{cases} y = 3x - 9 & (1) \\ 4x + 3y = -1 & (2) \end{cases}$

Substituting $3x - 9$ for y in equation (2), we

obtain

$$4x + 3(3x - 9) = -1$$

$$4x + 9x - 27 = -1$$

$$13x - 27 = -1$$

$$13x = 26$$

$$x = 2$$

Substituting 2 for x in equation (1), we obtain

$y = 3(2) - 9 = 6 - 9 = -3$.

The solution is the ordered pair $(2, -3)$.

14. $\begin{cases} x - 2y = 7 & (1) \\ 3x - y = -4 & (2) \end{cases}$

Equation (1) solved for x is $x = 2y + 7$.

Substituting $2y + 7$ for x in equation (2), we

obtain

$$3(2y + 7) - y = -4$$

$$6y + 21 - y = -4$$

$$5y + 21 = -4$$

$$5y = -25$$

$$y = -5$$

Substituting -5 for y in equation (1), we obtain

$$x - 2(-5) = 7$$

$$x + 10 = 7$$

$$x = -3$$

The solution is the ordered pair $(-3, -5)$.

15. $\begin{cases} 2x - y = 9 & (1) \\ 3x + y = 11 & (2) \end{cases}$

Add equations (1) and (2).

$$2x - y = 9$$

$$3x + y = 11$$

$$5x = 20$$

$$x = 4$$

Substituting 4 for x in equation (2), we obtain

$$3(4) + y = 11$$

$$12 + y = 11$$

$$y = -1$$

The solution is the ordered pair $(4, -1)$.

16. $\begin{cases} -x + 3y = 4 & (1) \\ 3x - 4y = -2 & (2) \end{cases}$

Multiply both sides of equation (1) by 3 and add

the result to equation (2).

$$-3x + 9y = 12$$

$$3x - 4y = -2$$

$$5y = 10$$

$$y = 2$$

Substituting 2 for y in equation (1), we obtain

$$-x + 3(2) = 4$$
$$-x + 6 = 4$$
$$2 = x$$

The solution is the ordered pair (2, 2).

17. $\begin{cases} 2x - 4y = 8 & (1) \\ -3x + 6y = -12 & (2) \end{cases}$

Multiply both sides of equation (1) by 3, multiply both sides of equation (2) by 2, and add the results.

$$6x - 12y = 24$$
$$-6x + 12y = -24$$
$$0 = 0$$

The system is dependent. The solution is $\{(x,\ y) \mid 2x - 4y = 8\}$.

18. $\begin{cases} 3x - 4y = -11 & (1) \\ 2x - 3y = -7 & (2) \end{cases}$

Multiply both sides of equation (1) by 2, multiply both sides of equation (2) by –3, and add the results.

$$6x - 8y = -22$$
$$-6x + 9y = 21$$
$$y = -1$$

Substituting –1 for y in equation (1), we obtain

$$3x - 4(-1) = -11$$
$$3x + 4 = -11$$
$$3x = -15$$
$$x = -5$$

The solution is the ordered pair (–5, –1).

19. $\begin{cases} x + y = -4 & (1) \\ 2x - 3y = 12 & (2) \end{cases}$

Because the coefficients of x and y in equation (1) are both 1, we use substitution to solve the system.

Equation (1) solved for x is $x = -y - 4$.

Substituting $-y - 4$ for x in equation (2), we obtain

$$2(-y - 4) - 3y = 12$$
$$-2y - 8 - 3y = 12$$
$$-5y - 8 = 12$$
$$-5y = 20$$
$$y = -4$$

Substituting –4 for y in equation (1), we obtain

$$x + (-4) = -4$$
$$x - 4 = -4$$
$$x = 0$$

The solution is the ordered pair (0, –4).

20. $\begin{cases} 5x - 3y = 2 & (1) \\ x + 2y = -10 & (2) \end{cases}$

Because the coefficient of x in equation (2) is 1, we use substitution to solve the system.

Equation (2) solved for x is $x = -2y - 10$.

Substituting $-2y - 10$ for x in equation (1), we obtain

$$5(-2y - 10) - 3y = 2$$
$$-10y - 50 - 3y = 2$$
$$-13y - 50 = 2$$
$$-13y = 52$$
$$y = -4$$

Substituting –4 for y in equation (2), we obtain

$$x + 2(-4) = -10$$
$$x - 8 = -10$$
$$x = -2$$

The solution is the ordered pair (–2, –4).

21. $\begin{cases} 3x - 2y = 5 & (1) \\ 4x - 5y = 9 & (2) \end{cases}$

Because none of variables have a coefficient of 1, we use elimination to solve the system. Multiply both sides of equation (1) by 4, multiply both sides of equation (2) by –3, and add the results.

$$12x - 8y = 20$$
$$-12x + 15y = -27$$
$$7y = -7$$
$$y = -1$$

Substituting –1 for y in equation (1), we obtain

$$3x - 2(-1) = 5$$
$$3x + 2 = 5$$
$$3x = 3$$
$$x = 1$$

The solution is the ordered pair (1, –1).

22. $\begin{cases} 12x + 20y = 21 & (1) \\ 3x - 2y = 0 & (2) \end{cases}$

Because none of variables have a coefficient of 1, we use elimination to solve the system. Multiply both sides of equation (2) by –4 and add the result to equation (1).

$$12x + 20y = 21$$
$$-12x + 8y = 0$$
$$28y = 21$$
$$y = \frac{21}{28} = \frac{3}{4}$$

Substituting $\dfrac{3}{4}$ for y in equation (1), we obtain

$$12x + 20\left(\dfrac{3}{4}\right) = 21$$
$$12x + 15 = 21$$
$$12x = 6$$
$$x = \dfrac{6}{12} = \dfrac{1}{2}$$

The solution is the ordered pair $\left(\dfrac{1}{2}, \dfrac{3}{4}\right)$.

23. $\begin{cases} 6x + 9y = -3 & (1) \\ 8x + 12y = 7 & (2) \end{cases}$

Because none of variables have a coefficient of 1, we use elimination to solve the system. Multiply both sides of equation (1) by 4, multiply both sides of equation (2) by –3, and add the results.

$$24x + 36y = -12$$
$$-24x - 36y = -21$$
$$0 = -33$$

The system has no solution. The solution set is $\varnothing$ or $\{\ \}$. The system is inconsistent.

24. $\begin{cases} 6x + 11y = 2 & (1) \\ 5x + 8y = -3 & (2) \end{cases}$

Because none of variables have a coefficient of 1, we use elimination to solve the system. Multiply both sides of equation (1) by 5, multiply both sides of equation (2) by –6, and add the results.

$$30x + 55y = 10$$
$$-30x - 48y = 18$$
$$7y = 28$$
$$y = 4$$

Substituting 4 for y in equation (1), we obtain

$$6x + 11(4) = 2$$
$$6x + 44 = 2$$
$$6x = -42$$
$$x = -7$$

The solution is the ordered pair (–7, 4).

25. Let x and y represent the two numbers.
$\begin{cases} x + y = 56 & (1) \\ x - y = 14 & (2) \end{cases}$
Adding equations (1) and (2) results in
$$2x = 70$$
$$x = 35$$
Substituting 35 for x in equation (1), we obtain
$$35 + y = 56$$
$$y = 21$$
The two numbers are 35 and 21.

26. Let m represent the number of males to be inducted and f represent the number of females to be inducted.
$\begin{cases} m + f = 73 & (1) \\ f = m + 11 & (2) \end{cases}$
Substituting $m + 11$ for f in equation (1), we obtain
$$m + (m + 11) = 73$$
$$2m + 11 = 73$$
$$2m = 62$$
$$m = 31$$
Substituting 31 for m in equation (2), we obtain
$$f = 31 + 11 = 42$$
The local chapter of Phi Theta Kappa will induct 31 males and 42 females.

27. Let p represent the number of calories in a slice of pepperoni pizza and s represent the number of calories in a slice of Italian sausage pizza.
$\begin{cases} 3p + 5s = 2600 & (1) \\ 4p + 2s = 1880 & (2) \end{cases}$
Multiply both sides of equation (1) by 4, multiply both sides of equation (2) by -3, and add the results.
$$12p + 20s = 10,400$$
$$-12p - 6s = -5640$$
$$14s = 4760$$
$$s = 340$$
Substituting 340 for s in equation (1), we obtain
$$3p + 5(340) = 2600$$
$$3p + 1700 = 2600$$
$$3p = 900$$
$$p = 300$$
There are 300 calories in a slice of pepperoni pizza and 340 calories in a slice of Italian sausage pizza.

28. Let l represent the length of the rectangle and w represent the width of the rectangle.

$$\begin{cases} l = 2w - 5 & (1) \\ 2l + 2w = 68 & (2) \end{cases}$$

Substituting $2w - 5$ for l in equation (2), we obtain

$$2(2w - 5) + 2w = 68$$
$$4w - 10 + 2w = 68$$
$$6w - 10 = 68$$
$$6w = 78$$
$$w = 13$$

Substituting 13 for w in equation (1), we obtain

$$l = 2(13) - 5 = 26 - 5 = 21$$

The dimensions of the rectangle are 21 inches by 13 inches.

29. Because the sum of the measures of the three angles in a triangle must equal 180°, we obtain the following system.

$$\begin{cases} x + y + 90 = 180 & (1) \\ y = 2x & (2) \end{cases}$$

Substituting $2x$ for y in equation (1), we obtain

$$x + 2x + 90 = 180$$
$$3x + 90 = 180$$
$$3x = 90$$
$$x = 30$$

Substituting 30 for x in equation (2), we obtain $y = 2(30) = 60$.

Thus, angle $x = 30°$ and angle $y = 60°$.

30. Because angles 1 and 2 are supplemental, and because angles 1 and 3 must be equal, we obtain the following system.

$$\begin{cases} (x + 4y) + (8x + 5y) = 180 & (1) \\ x + 4y = 2x + y & (2) \end{cases}$$

Simplify each equation.

$$\begin{cases} 9x + 9y = 180 & (1) \\ -x + 3y = 0 & (2) \end{cases}$$

Divide both sides of equation (1) by 9 and add the result to equation (2).

$$x + y = 20$$
$$-x + 3y = 0$$
$$\overline{ 4y = 20}$$
$$y = 5$$

Substituting 5 for y in equation (2), we obtain

$$-x + 3(5) = 0$$
$$-x + 15 = 0$$
$$15 = x$$

Thus, $x = 15$ and $y = 5$.

31. Let n represent the number of nickels and q represent the number of quarters.

	Number of Coins	Value per Coin	Total Value
Nickels	n	0.05	$0.05n$
Quarters	q	0.25	$0.25q$
Total	40		4.40

Using the columns for Number of Coins and Total Value, we obtain the following system.

$$\begin{cases} n + q = 40 & (1) \\ 0.05n + 0.25q = 4.40 & (2) \end{cases}$$

Multiply both sides of equation (1) by -0.05 and add the result to equation (2).

$$-0.05n - 0.05q = -2$$
$$0.05n + 0.25q = 4.40$$
$$\overline{ 0.20q = 2.40}$$
$$q = 12$$

Substituting 12 for q in equation (1), we obtain

$$n + 12 = 40$$
$$n = 28$$

Jerome's piggy bank contains 28 nickels and 12 quarters.

32. Let x represent the number of liters of the 25%-hydrochloric-acid (HCl) solution to be used, and let y represent the number of liters of the 40%-HCl solution to be used.

	Liters of Solution	Percent HCl	Liters of HCl
25%-HCl Solution	x	25% = 0.25	$0.25x$
40%-HCl Solution	y	40% = 0.40	$0.40y$
30%-HCl Solution	12	30% = 0.30	$0.30(12)$ $= 3.6$

Using the columns for Liters of Solution and Liters of HCl, we obtain the following system.

$$\begin{cases} x + y = 12 & (1) \\ 0.25x + 0.40y = 3.6 & (2) \end{cases}$$

Multiply both sides of equation (1) by -0.25 and add the result to equation (2).

$$-0.25x - .25y = -3$$
$$0.25x + 0.40y = 3.6$$
$$\overline{ 0.15y = 0.6}$$
$$y = 4$$

Substituting 4 for y in equation (1), we obtain
$$x + 4 = 12$$
$$x = 8$$
The chemist should mix 8 liters of the 25%-hydrochloric-acid solution with 4 liters of the 40%-hydrochloric-acid solution.

33. Let s represent the amount invested in stocks, and let b represent the amount invested in bonds. We organize the information in the table below, using the interest formula:
Interest = Principal $\times$ rate $\times$ time or $I = Prt$.

	P	r	t	I
Stocks	s	0.065	1	$0.065s$
Bonds	b	0.0425	1	$0.0425b$
Total	10,000		1	582.50

Using the columns for Principal and Interest, we obtain the following system.
$$\begin{cases} s + b = 10,000 & (1) \\ 0.065s + 0.0425b = 582.50 & (2) \end{cases}$$
Solving equation (1) for b, we obtain
$b = 10,000 - s$. Substituting $10,000 - s$ for b in equation (2), we obtain
$$0.065s + 0.0425(10,000 - s) = 582.50$$
$$0.065s + 425 - 0.0425s = 582.50$$
$$0.0225s + 425 = 582.50$$
$$0.0225s = 157.50$$
$$s = 7000$$
Substituting 7000 for s in equation (1), we obtain
$$7000 + b = 10,000$$
$$b = 3000$$
Verna invested $7000 in stocks and $3000 in bonds.

34. Let p represent the speed of the plane and let w represent the effect of the wind resistance.
$$\begin{cases} p + w = 160 & (1) \\ p - w = 112 & (2) \end{cases}$$
Adding equations (1) and (2) results in
$$\begin{array}{l} p + w = 160 \\ p - w = 112 \\ \hline 2p \phantom{{}+ w} = 272 \\ p = 136 \end{array}$$
Substituting 136 for p in equation (1), we obtain
$$136 + w = 160$$
$$w = 24$$
The plane's speed is 136 miles per hour and the effect of the wind resistance is 24 miles per hour.

35. Let d represent the distance that both cars have traveled on the turnpike at the point of catch up. Let t represent the Mustang's time on the turnpike at the point of catch up. Then $t + \dfrac{1}{2}$ represents the Grand Am's time on the turnpike at that point. We obtain the following system by utilizing the formula rate $\times$ time = distance.
$$\begin{cases} 80t = d & (1) \\ 50\left(t + \dfrac{1}{2}\right) = d & (2) \end{cases}$$
Substituting $80t$ for d in equation (2) yields
$$50\left(t + \frac{1}{2}\right) = 80t$$
$$50t + 25 = 80t$$
$$-30t + 25 = 0$$
$$-30t = -25$$
$$t = \frac{-25}{-30} = \frac{5}{6}$$
Substituting $\dfrac{5}{6}$ for t in equation (1), we obtain
$$80\left(\frac{5}{6}\right) = d$$
$$d = \frac{200}{3} = 66\frac{2}{3}$$
It will take $\dfrac{5}{6}$ hour (that is, 50 minutes) for the Mustang to catch up to the Grand Am. At that time, the two cars will have traveled $66\dfrac{2}{3}$ miles on the turnpike.

36. Let b represent the speed of the boat in still water and let c represent the speed of the current. The rate at which the boat will travel downstream will be $b + c$, and the rate upstream will be $b - c$. Using the formula rate $\times$ time = distance, we obtain the following system:
$$\begin{cases} (b - c) \cdot 1.5 = 30 & (1) \\ (b + c) \cdot 1 = 30 & (2) \end{cases}$$
Divide both sides of equation (1) by 1.5, simplify equation (2), and add the results.
$$\begin{array}{l} b - c = 20 \\ b + c = 30 \\ \hline 2b \phantom{{}+ c} = 50 \\ b = 25 \end{array}$$

Substituting 25 for b in equation (2), we obtain
$$(25+c)\cdot 1 = 30$$
$$25+c = 30$$
$$c = 5$$
The speed of the boat in still water is 25 miles per hour and the speed of the current is 5 miles per hour.

37. a. $\begin{cases} M(t) = 0.056t + 3.354 \\ F(t) = 0.103t + 1.842 \end{cases}$

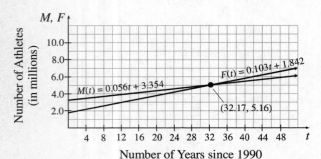

Number of Years since 1990

b. $M(t) = F(t)$
$$0.056t + 3.354 = 0.103t + 1.842$$
$$-0.047t + 3.354 = 1.842$$
$$-0.047t = -1.512$$
$$t \approx 32.17$$
Male and female participation will be the same approximately 32 years after 1990, which is the year 2022.

38. a. $R(x) = 15x$

b. $C(x) = 1200 + 2.50x$

c.

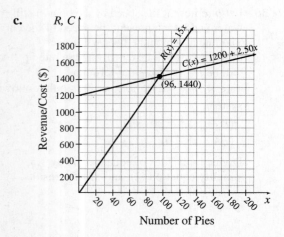

Number of Pies

d. $R(x) = C(x)$
$$15x = 1200 + 2.50x$$
$$12.50x = 1200$$
$$x = 96$$
$$R(96) = 15(96) = 1440$$
$$C(96) = 1200 + 2.50(96) = 1440$$
The break-even number of pies is 96 pies. The revenue and cost of 96 pies is $1440.

39. $\begin{cases} x + y - z = 1 & (1) \\ x - y + z = 7 & (2) \\ x + 2y + z = 1 & (3) \end{cases}$

Multiply both sides of equation (1) by −1 and add the result to equation (2).
$$-x - y + z = -1$$
$$x - y + z = 7$$
$$-2y + 2z = 6 \qquad (4)$$

Multiply both sides of equation (1) by −1 and add the result to equation (3).
$$-x - y + z = -1$$
$$x + 2y + z = 1$$
$$y + 2z = 0 \qquad (5)$$

Divide both sides of equation (4) by 2 and add the result to equation (5).

$$-y + z = 3$$
$$y + 2z = 0$$
$$3z = 3$$
$$z = 1$$
Substituting 1 for z in equation (4), we obtain
$$-2y + 2(1) = 6$$
$$-2y + 2 = 6$$
$$-2y = 4$$
$$y = -2$$
Substituting −2 for y and 1 for z in equation (1), we obtain
$$x + (-2) - 1 = 1$$
$$x - 3 = 1$$
$$x = 4$$
The solution is the ordered triple (4, −2, 1).

40. $\begin{cases} 2x - 2y + z = -10 & (1) \\ 3x + y - 2z = 4 & (2) \\ 5x + 2y - 3z = 7 & (3) \end{cases}$

Multiply both sides of equation (1) by 2 and add the result to equation (2).

$$4x - 4y + 2z = -20$$
$$3x + y - 2z = 4$$
$$7x - 3y \quad = -16 \quad (4)$$

Multiply both sides of equation (1) by 3 and add the result to equation (3).
$$6x - 6y + 3z = -30$$
$$5x + 2y - 3z = 7$$
$$11x - 4y \quad = -23 \quad (5)$$

Multiply both sides of equation (4) by 4, multiply both sides of equation (5) by −3, and add the results.
$$28x - 12y = -64$$
$$-33x + 12y = 69$$
$$-5x \quad\quad = 5$$
$$x = -1$$

Substituting −1 for x in equation (4), we obtain
$$7(-1) - 3y = -16$$
$$-7 - 3y = -16$$
$$-3y = -9$$
$$y = 3$$

Substituting −1 for x and 3 for y in equation (1), we obtain
$$2(-1) - 2(3) + z = -10$$
$$-2 - 6 + z = -10$$
$$z = -2$$

The solution is the ordered triple (−1, 3, −2).

41. $\begin{cases} x + 2y \quad\quad = -1 & (1) \\ \quad 3y + 4z = 7 & (2) \\ 2x \quad\quad - z = 6 & (3) \end{cases}$

Multiply both sides of equation (1) by −2 and add the result to equation (3).
$$-2x - 4y \quad = 2$$
$$2x \quad\quad - z = 6$$
$$-4y - z = 8 \quad (4)$$

Multiply both sides of equation (4) by 4 and add the result to equation (2).
$$-16y - 4z = 32$$
$$3y + 4z = 7$$
$$-13y \quad = 39$$
$$y = -3$$

Substituting −3 for y in equation (4), we obtain
$$-4(-3) - z = 8$$
$$12 - z = 8$$
$$-z = -4$$
$$z = 4$$

Substituting −3 for y in equation (1), we obtain
$$x + 2(-3) = -1$$
$$x - 6 = -1$$
$$x = 5$$
The solution is the ordered triple (5, −3, 4).

42. $\begin{cases} x + 2y - 3z = -4 & (1) \\ x + y + 3z = 5 & (2) \\ 3x + 4y + 3z = 7 & (3) \end{cases}$

Add equations (1) and (2).
$$x + 2y - 3z = -4$$
$$x + y + 3z = 5$$
$$2x + 3y \quad = 1 \quad (4)$$

Add equations (1) and (3).
$$x + 2y - 3z = -4$$
$$3x + 4y + 3z = 7$$
$$4x + 6y \quad = 3 \quad (5)$$

Multiply both sides of equation (4) by −2 and add the result to equation (5).
$$-4x - 6y = -2$$
$$4x + 6y = 3$$
$$0 = 1 \quad \text{False}$$
The system has no solution. The solution set is $\varnothing$ or $\{ \ \}$. The system is inconsistent.

43. $\begin{cases} 3x + y - 2z = 6 & (1) \\ x + y - z = -2 & (2) \\ -x - 3y + 2z = 14 & (3) \end{cases}$

Multiply both sides of equation (3) by 3 and add the result to equation (1).
$$-3x - 9y + 6z = 42$$
$$3x + y - 2z = 6$$
$$-8y + 4z = 48 \quad (4)$$

Add equations (2) and (3).
$$x + y - z = -2$$
$$-x - 3y + 2z = 14$$
$$-2y + z = 12 \quad (5)$$

Multiply both sides of equation (5) by −4 and add the result to equation (4).
$$8y - 4z = -48$$
$$-8y + 4z = 48$$
$$0 = 0 \quad \text{True}$$
Thus, the system is dependent and has an infinite number of solutions.

Solve equation (5) for y.

$$-2y + z = 12$$
$$-2y = -z + 12$$
$$y = \frac{-z + 12}{-2}$$
$$y = \frac{1}{2}z - 6$$

Substituting $\frac{1}{2}z - 6$ for y in equation (2), we obtain

$$x + \left(\frac{1}{2}z - 6\right) - z = -2$$
$$x - \frac{1}{2}z - 6 = -2$$
$$x = \frac{1}{2}z + 4$$

The solution to the system is

$$\left\{(x, y, z) \middle| x = \frac{1}{2}z + 4,\right.$$
$$\left. y = \frac{1}{2}z - 6, \; z \text{ is any real number}\right\}.$$

44. $\begin{cases} 9x - y + 2z = -5 & (1) \\ -3x - 4y + 4z = 3 & (2) \\ 15x + 3y - 2z = -10 & (3) \end{cases}$

Multiply both sides of equation (2) by 3 and add the result to equation (1).

$$-9x - 12y + 12z = 9$$
$$\underline{9x - y + 2z = -5}$$
$$-13y + 14z = 4 \qquad (4)$$

Multiply both sides of equation (2) by 5 and add the result to equation (3).

$$-15x - 20y + 20z = 15$$
$$\underline{15x + 3y - 2z = -10}$$
$$-17y + 18z = 5 \qquad (5)$$

Multiply both sides of equation (4) by 17, multiply both sides of equation (5) by -13, and add the results.

$$-221y + 238z = 68$$
$$\underline{221y - 234z = -65}$$
$$4z = 3$$
$$z = \frac{3}{4}$$

Substituting $\frac{3}{4}$ for z in equation (4), we obtain

$$-13y + 14\left(\frac{3}{4}\right) = 4$$
$$-13y + \frac{21}{2} = 4$$
$$2\left(-13y + \frac{21}{2}\right) = 2(4)$$
$$-26y + 21 = 8$$
$$-26y = -13$$
$$y = \frac{-13}{-26} = \frac{1}{2}$$

Substituting $\frac{1}{2}$ for y and $\frac{3}{4}$ for z in equation (1), we obtain

$$9x - \frac{1}{2} + 2\left(\frac{3}{4}\right) = -5$$
$$9x - \frac{1}{2} + \frac{3}{2} = -5$$
$$9x + 1 = -5$$
$$9x = -6$$
$$x = \frac{-6}{9} = -\frac{2}{3}$$

The solution is the ordered triple $\left(-\frac{2}{3}, \frac{1}{2}, \frac{3}{4}\right)$.

45. $\begin{cases} 4x - 5y + 2z = -8 & (1) \\ 3x + 7y - 3z = 21 & (2) \\ 7x - 4y + 2z = -5 & (3) \end{cases}$

Multiply both sides of equation (1) by 3, multiply both sides of equation (2) by -4, and add the results.

$$12x - 15y + 6z = -24$$
$$\underline{-12x - 28y + 12z = -84}$$
$$-43y + 18z = -108 \qquad (4)$$

Multiply both sides of equation (2) by 7, multiply both sides of equation (3) by -3, and add the results.

$$21x + 49y - 21z = 147$$
$$\underline{-21x + 12y - 6z = 15}$$
$$61y - 27z = 162 \qquad (5)$$

Multiply both sides of equation (4) by 3, multiply both sides of equation (5) by 2, and add the results.

$$-129y + 54z = -324$$
$$122y - 54z = 324$$
$$-7y = 0$$
$$y = 0$$

Substituting 0 for y in equation (4), we obtain
$$-43(0) + 18z = -108$$
$$18z = -108$$
$$z = -6$$

Substituting 0 for y and -6 for z in equation (1), we obtain
$$4x - 5(0) + 2(-6) = -8$$
$$4x - 0 - 12 = -8$$
$$4x = 4$$
$$x = 1$$

The solution is the ordered triple (1, 0, −6).

46. $\begin{cases} 3x - 2y + 5z = -7 & (1) \\ 4x + y + 3z = -2 & (2) \\ 2x - 3y + 7z = -4 & (3) \end{cases}$

Multiply both sides of equation (1) by 4, multiply both sides of equation (2) by −3, and add the results.
$$12x - 8y + 20z = -28$$
$$-12x - 3y - 9z = 6$$
$$ -11y + 11z = -22 \quad (4)$$

Multiply both sides of equation (3) by −2 and add the result to equation (2).
$$-4x + 6y - 14z = 8$$
$$4x + y + 3z = -2$$
$$ 7y - 11z = 6 \quad (5)$$

Add equations (4) and (5).

$$-11y + 11z = -22$$
$$7y - 11z = 6$$
$$-4y = -16$$
$$y = 4$$

Substituting 4 for y in equation (4), we obtain
$$-11(4) + 11z = -22$$
$$-44 + 11z = -22$$
$$11z = 22$$
$$z = 2$$

Substituting 4 for y and 2 for z in equation (1), we obtain
$$3x - 2(4) + 5(2) = -7$$
$$3x - 8 + 10 = -7$$
$$3x + 2 = -7$$
$$3x = -9$$
$$x = -3$$

The solution is the ordered triple (−3, 4, 2).

47. Because the sum of the measures of the three angles in a triangle must equal 180°, we obtain the following system.
$$\begin{cases} x + y + z = 180 & (1) \\ z = 2y & (2) \\ y = 3x - 10 & (3) \end{cases}$$

Writing the equations in the system in standard form, we obtain
$$\begin{cases} x + y + z = 180 & (1) \\ 2y - z = 0 & (2) \\ 3x - y = 10 & (3) \end{cases}$$

Adding equations (1) and (2), we obtain
$$x + y + z = 180$$
$$2y - z = 0$$
$$x + 3y = 180 \quad (4)$$

Multiply equation (3) by 3 and add the result to equation (4).
$$9x - 3y = 30$$
$$x + 3y = 180$$
$$10x = 210$$
$$x = 21$$

Substituting 21 for x in equation (3), we obtain
$$y = 3(21) - 10 = 63 - 10 = 53$$

Substituting 53 for y in equation (2), we obtain
$$z = 2(53) = 106$$

Thus, angle $x = 21°$, angle $y = 53°$, and angle $z = 106°$.

48. Let b represent the number of calories in a cheeseburger, let f represent the number of calories in a medium order of fries, and let c represent the number of calories in a medium Coke.
$$\begin{cases} b + f + c = 1000 & (1) \\ 2b + f + c = 1300 & (2) \\ 2b + f + 2c = 1590 & (3) \end{cases}$$

Multiply equation (1) by −1 and add the result to equation (3).
$$-b - f - c = -1000$$
$$2b + f + 2c = 1590$$
$$\overline{b + c = 590} \quad (4)$$

Multiply equation (1) by −1 and add the result to equation (2).

$$-b - f - c = -1000$$
$$\underline{2b + f + c = 1300}$$
$$b \qquad\qquad = 300$$

Substituting 300 for b in equation (4), we obtain
$$300 + c = 590$$
$$c = 290$$

Substituting 300 for b and 290 for c in equation (1), we obtain
$$300 + f + 290 = 1000$$
$$f + 590 = 1000$$
$$f = 410$$

A cheeseburger contains 300 calories, a medium order of fries contains 410 calories, and a medium Coke contains 290 calories.

49. $\begin{bmatrix} 3 & 1 & | & 7 \\ 2 & 5 & | & 9 \end{bmatrix}$

50. $\begin{bmatrix} 1 & -5 & | & 14 \\ -1 & 1 & | & -3 \end{bmatrix}$

51. $\begin{bmatrix} 5 & -1 & 4 & | & 6 \\ -3 & 0 & -3 & | & -1 \\ 1 & -2 & 0 & | & 0 \end{bmatrix}$

52. $\begin{bmatrix} 8 & -1 & 3 & | & 14 \\ -3 & 5 & -6 & | & -18 \\ 7 & -4 & 5 & | & 21 \end{bmatrix}$

53. $\begin{cases} x + 2y = 12 \\ \quad\;\; 3y = 15 \end{cases}$

54. $\begin{cases} 3x - 4y = -5 \\ -x + 2y = 7 \end{cases}$

55. $\begin{cases} x + 3y + 4z = 20 \\ \quad\;\; y - 2z = -16 \\ \qquad\qquad z = 7 \end{cases}$

56. $\begin{cases} -3x + 7y + 9z = 1 \\ 4x + 10y + 7z = 5 \\ 2x - 5y - 6z = -8 \end{cases}$

57. a. $\begin{bmatrix} 1 & -5 & | & 22 \\ -2 & 9 & | & -40 \end{bmatrix} \qquad (R_2 = 2r_1 + r_2)$

$= \begin{bmatrix} 1 & -5 & | & 22 \\ 2(1) + (-2) & 2(-5) + 9 & | & 2(22) + (-40) \end{bmatrix}$

$= \begin{bmatrix} 1 & -5 & | & 22 \\ 0 & -1 & | & 4 \end{bmatrix}$

b. $\begin{bmatrix} 1 & -5 & | & 22 \\ 0 & -1 & | & 4 \end{bmatrix}$ $\left(R_2 = -1 \cdot r_2 \right)$

$= \begin{bmatrix} 1 & -5 & | & 22 \\ -1(0) & -1(-1) & | & -1(4) \end{bmatrix}$

$= \begin{bmatrix} 1 & -5 & | & 22 \\ 0 & 1 & | & -4 \end{bmatrix}$

58. a. $\begin{bmatrix} 1 & -4 & | & 7 \\ 3 & -7 & | & 6 \end{bmatrix}$ $\left(R_2 = -3r_1 + r_2 \right)$

$= \begin{bmatrix} 1 & -4 & | & 7 \\ -3(1)+3 & -3(-4)+(-7) & | & -3(7)+6 \end{bmatrix}$

$= \begin{bmatrix} 1 & -4 & | & 7 \\ 0 & 5 & | & -15 \end{bmatrix}$

b. $\begin{bmatrix} 1 & -4 & | & 7 \\ 0 & 5 & | & -15 \end{bmatrix}$ $\left(R_2 = \dfrac{1}{5} r_2 \right)$

$= \begin{bmatrix} 1 & -4 & | & 7 \\ \dfrac{1}{5}(0) & \dfrac{1}{5}(5) & | & \dfrac{1}{5}(-15) \end{bmatrix}$

$= \begin{bmatrix} 1 & -4 & | & 7 \\ 0 & 1 & | & -3 \end{bmatrix}$

59. a. $\begin{bmatrix} -1 & 2 & 1 & | & 1 \\ 2 & -1 & 3 & | & -3 \\ -1 & 5 & 6 & | & 2 \end{bmatrix}$ $\left(R_2 = 2r_1 + r_2 \right)$

$= \begin{bmatrix} -1 & 2 & 1 & | & 1 \\ 2(-1)+2 & 2(2)+(-1) & 2(1)+3 & | & 2(1)+(-3) \\ -1 & 5 & 6 & | & 2 \end{bmatrix}$

$= \begin{bmatrix} -1 & 2 & 1 & | & 1 \\ 0 & 3 & 5 & | & -1 \\ -1 & 5 & 6 & | & 2 \end{bmatrix}$

b. $\begin{bmatrix} -1 & 2 & 1 & | & 1 \\ 0 & 3 & 5 & | & -1 \\ -1 & 5 & 6 & | & 2 \end{bmatrix}$ $\left(R_3 = -1 \cdot r_1 + r_3 \right)$

$= \begin{bmatrix} -1 & 2 & 1 & | & 1 \\ 0 & 3 & 5 & | & -1 \\ -1(-1)+(-1) & -1(2)+5 & -1(1)+6 & | & -1(1)+2 \end{bmatrix}$

$= \begin{bmatrix} -1 & 2 & 1 & | & 1 \\ 0 & 3 & 5 & | & -1 \\ 0 & 3 & 5 & | & 1 \end{bmatrix}$

60. a.
$$\begin{bmatrix} 1 & 3 & 4 & | & 4 \\ 0 & 5 & 10 & | & -15 \\ 0 & -4 & -7 & | & 7 \end{bmatrix} \qquad \left(R_2 = -\frac{1}{5} \cdot r_2\right)$$

$$= \begin{bmatrix} 1 & 3 & 4 & | & 4 \\ \frac{1}{5}(0) & \frac{1}{5}(5) & \frac{1}{5}(10) & | & \frac{1}{5}(-15) \\ 0 & -4 & -7 & | & 7 \end{bmatrix}$$

$$= \begin{bmatrix} 1 & 3 & 4 & | & 4 \\ 0 & 1 & 2 & | & -3 \\ 0 & -4 & -7 & | & 7 \end{bmatrix}$$

b.
$$\begin{bmatrix} 1 & 3 & 4 & | & 4 \\ 0 & 1 & 2 & | & -3 \\ 0 & -4 & -7 & | & 7 \end{bmatrix} \qquad (R_3 = 4r_2 + r_3)$$

$$= \begin{bmatrix} 1 & 3 & 4 & | & 4 \\ 0 & 1 & 2 & | & -3 \\ 4(0)+0 & 4(1)+(-4) & 4(2)+(-7) & | & 4(-3)+7 \end{bmatrix}$$

$$= \begin{bmatrix} 1 & 3 & 4 & | & 4 \\ 0 & 1 & 2 & | & -3 \\ 0 & 0 & 1 & | & -5 \end{bmatrix}$$

61. Write the augmented matrix of the system and then put it in row echelon form.
$$\begin{bmatrix} 1 & 2 & | & 1 \\ 1 & 3 & | & -2 \end{bmatrix} \qquad (R_2 = -1 \cdot r_1 + r_2)$$

$$= \begin{bmatrix} 1 & 2 & | & 1 \\ 0 & 1 & | & -3 \end{bmatrix}$$

Write the system of equations that corresponds to the row-echelon matrix
$$\begin{cases} x + 2y = 1 & (1) \\ \quad\quad y = -3 & (2) \end{cases}$$
This system is consistent and independent.
Substituting -3 for y in equation (1), we obtain
$$x + 2(-3) = 1$$
$$x - 6 = 1$$
$$x = 7$$
The solution is the ordered pair $(7, -3)$.

62. Write the augmented matrix of the system and then put it in row echelon form.
$$\begin{bmatrix} 6 & -2 & | & -7 \\ 4 & 3 & | & 17 \end{bmatrix} \qquad \left(R_1 = \frac{1}{6}r_1\right)$$

$$= \begin{bmatrix} 1 & -\frac{1}{3} & | & -\frac{7}{6} \\ 4 & 3 & | & 17 \end{bmatrix} \qquad (R_2 = -4r_1 + r_2)$$

$$= \begin{bmatrix} 1 & -\frac{1}{3} & | & -\frac{7}{6} \\ 0 & \frac{13}{3} & | & \frac{65}{3} \end{bmatrix} \qquad \left(R_2 = \frac{3}{13}r_2\right)$$

$$= \begin{bmatrix} 1 & -\frac{1}{3} & | & -\frac{7}{6} \\ 0 & 1 & | & 5 \end{bmatrix}$$

Write the system of equations that corresponds to the row-echelon matrix

$$\begin{cases} x - \dfrac{1}{3}y = -\dfrac{7}{6} & (1) \\ \quad\quad y = 5 & (2) \end{cases}$$

This system is consistent and independent. Substituting 5 for y in equation (1), we obtain

$$x - \frac{1}{3}(5) = -\frac{7}{6}$$
$$x - \frac{5}{3} = -\frac{7}{6}$$
$$x = \frac{3}{6} = \frac{1}{2}$$

The solution is the ordered pair $\left(\dfrac{1}{2},\ 5\right)$.

63. Write the augmented matrix of the system and then put it in row echelon form.

$$\begin{bmatrix} 3 & 2 & \big| & -10 \\ 2 & -1 & \big| & -9 \end{bmatrix} \quad \left(R_1 = \frac{1}{3}r_1\right)$$

$$= \begin{bmatrix} 1 & \frac{2}{3} & \big| & -\frac{10}{3} \\ 2 & -1 & \big| & -9 \end{bmatrix} \quad \left(R_2 = -2r_1 + r_2\right)$$

$$= \begin{bmatrix} 1 & \frac{2}{3} & \big| & -\frac{10}{3} \\ 0 & -\frac{7}{3} & \big| & -\frac{7}{3} \end{bmatrix} \quad \left(R_2 = -\frac{3}{7}r_2\right)$$

$$= \begin{bmatrix} 1 & \frac{2}{3} & \big| & -\frac{10}{3} \\ 0 & 1 & \big| & 1 \end{bmatrix}$$

Write the system of equations that corresponds to the row-echelon matrix

$$\begin{cases} x + \dfrac{2}{3}y = -\dfrac{10}{3} & (1) \\ \quad\quad y = 1 & (2) \end{cases}$$

This system is consistent and independent. Substituting 1 for y in equation (1), we obtain

$$x + \frac{2}{3}(1) = -\frac{10}{3}$$
$$x + \frac{2}{3} = -\frac{10}{3}$$
$$x = -\frac{12}{3} = -4$$

The solution is the ordered pair $(-4, 1)$.

64. Write the augmented matrix of the system and then put it in row echelon form.

$$\begin{bmatrix} -3 & 9 & \big| & 15 \\ 5 & -15 & \big| & 11 \end{bmatrix} \quad \left(R_1 = -\frac{1}{3}r_1\right)$$

$$= \begin{bmatrix} 1 & -3 & \big| & -5 \\ 5 & -15 & \big| & 11 \end{bmatrix} \quad \left(R_2 = -5r_1 + r_2\right)$$

$$= \begin{bmatrix} 1 & -3 & \big| & -5 \\ 0 & 0 & \big| & 36 \end{bmatrix}$$

The system is inconsistent. The system has no solution. The solution set is $\varnothing$ or $\{\ \}$.

65. Write the augmented matrix of the system and then put it in row echelon form.

$$\begin{bmatrix} 4 & -2 & \big| & 6 \\ 6 & -3 & \big| & 9 \end{bmatrix} \quad \left(R_1 = \frac{1}{4}r_1\right)$$

$$= \begin{bmatrix} 1 & -\frac{1}{2} & \big| & \frac{3}{2} \\ 6 & -3 & \big| & 9 \end{bmatrix} \quad \left(R_2 = -6r_1 + r_2\right)$$

$$= \begin{bmatrix} 1 & -\frac{1}{2} & \big| & \frac{3}{2} \\ 0 & 0 & \big| & 0 \end{bmatrix}$$

The system is dependent. The solution to the system is $\{(x,\ y)\ |\ 4x - 2y = 6\}$.

66.
$$\begin{bmatrix} 1 & 4 & \big| & 4 \\ 4 & -8 & \big| & 7 \end{bmatrix} \quad \left(R_2 = -4r_1 + r_2\right)$$

$$= \begin{bmatrix} 1 & 4 & \big| & 4 \\ 0 & -24 & \big| & -9 \end{bmatrix} \quad \left(R_2 = -\frac{1}{24}r_2\right)$$

$$= \begin{bmatrix} 1 & 4 & \big| & 4 \\ 0 & 1 & \big| & \frac{3}{8} \end{bmatrix}$$

Write the system of equations that corresponds to the row-echelon matrix

$$\begin{cases} x + 4y = 4 & (1) \\ \quad\quad y = \dfrac{3}{8} & (2) \end{cases}$$

This system is consistent and independent.

Substituting $\dfrac{3}{8}$ for y in equation (1), we obtain

$$x + 4\left(\frac{3}{8}\right) = 4$$
$$x + \frac{3}{2} = 4$$
$$x = \frac{5}{2}$$

The solution is the ordered pair $\left(\dfrac{5}{2},\ \dfrac{3}{8}\right)$.

67. Write the augmented matrix of the system and then put it in row echelon form.

$$\begin{bmatrix} 1 & 2 & -2 & | & -11 \\ 2 & 0 & 1 & | & -6 \\ 0 & 5 & -3 & | & -7 \end{bmatrix} \quad (R_2 = -2r_1 + r_2)$$

$$=\begin{bmatrix} 1 & 2 & -2 & | & -11 \\ 0 & -4 & 5 & | & 16 \\ 0 & 5 & -3 & | & -7 \end{bmatrix} \quad \left(R_2 = -\frac{1}{4}r_2\right)$$

$$=\begin{bmatrix} 1 & 2 & -2 & | & -11 \\ 0 & 1 & -\frac{5}{4} & | & -4 \\ 0 & 5 & -3 & | & -7 \end{bmatrix} \quad (R_3 = -5r_2 + r_3)$$

$$=\begin{bmatrix} 1 & 2 & -2 & | & -11 \\ 0 & 1 & -\frac{5}{4} & | & -4 \\ 0 & 0 & \frac{13}{4} & | & 13 \end{bmatrix} \quad \left(R_3 = \frac{4}{13}r_3\right)$$

$$=\begin{bmatrix} 1 & 2 & -2 & | & -11 \\ 0 & 1 & -\frac{5}{4} & | & -4 \\ 0 & 0 & 1 & | & 4 \end{bmatrix}$$

Write the system of equations that corresponds to the row-echelon matrix

$$\begin{cases} x + 2y - 2z = -11 & (1) \\ y - \frac{5}{4}z = -4 & (2) \\ z = 4 & (3) \end{cases}$$

This system is consistent and independent. Substituting 4 for z in equation (2), we obtain

$$y - \frac{5}{4}(4) = -4$$
$$y - 5 = -4$$
$$y = 1$$

Substituting 4 for z and 1 for y in equation (1), we obtain

$$x + 2(1) - 2(4) = -11$$
$$x + 2 - 8 = -11$$
$$x - 6 = -11$$
$$x = -5$$

The solution is the ordered triple $(-5, 1, 4)$.

68. Write the augmented matrix of the system and then put it in row echelon form.

$$\begin{bmatrix} 2 & -5 & 2 & | & 9 \\ -1 & 1 & -2 & | & -2 \\ -1 & -2 & -4 & | & 8 \end{bmatrix} \quad (\text{Interchange } r_1 \text{ and } r_3)$$

$$=\begin{bmatrix} -1 & -2 & -4 & | & 8 \\ -1 & 1 & -2 & | & -2 \\ 2 & -5 & 2 & | & 9 \end{bmatrix} \quad (R_1 = -1 \cdot r_1)$$

$$=\begin{bmatrix} 1 & 2 & 4 & | & -8 \\ -1 & 1 & -2 & | & -2 \\ 2 & -5 & 2 & | & 9 \end{bmatrix} \quad \begin{pmatrix} R_2 = r_1 + r_2 \\ R_3 = 2r_2 + r_3 \end{pmatrix}$$

$$=\begin{bmatrix} 1 & 2 & 4 & | & -8 \\ 0 & 3 & 2 & | & -10 \\ 0 & -3 & -2 & | & 5 \end{bmatrix} \quad (R_3 = r_2 + r_3)$$

$$=\begin{bmatrix} 1 & 2 & 4 & | & -8 \\ 0 & 3 & 2 & | & -10 \\ 0 & 0 & 0 & | & -5 \end{bmatrix}$$

The system is inconsistent. The system has no solution. The solution set is $\varnothing$ or $\{\ \}$.

69. Write the augmented matrix of the system and then put it in row echelon form.

$$\begin{bmatrix} 2 & -7 & 11 & | & -5 \\ 4 & -2 & 6 & | & 2 \\ -2 & 19 & -27 & | & 17 \end{bmatrix} \quad \left(R_1 = \frac{1}{2} \cdot r_1\right)$$

$$=\begin{bmatrix} 1 & -\frac{7}{2} & \frac{11}{2} & | & -\frac{5}{2} \\ 4 & -2 & 6 & | & 2 \\ -2 & 19 & -27 & | & 17 \end{bmatrix} \quad \begin{pmatrix} R_2 = -4r_1 + r_2 \\ R_3 = 2r_1 + r_3 \end{pmatrix}$$

$$=\begin{bmatrix} 1 & -\frac{7}{2} & \frac{11}{2} & | & -\frac{5}{2} \\ 0 & 12 & -16 & | & 12 \\ 0 & 12 & -16 & | & 12 \end{bmatrix} \quad \left(R_2 = \frac{1}{12}r_2\right)$$

$$=\begin{bmatrix} 1 & -\frac{7}{2} & \frac{11}{2} & | & -\frac{5}{2} \\ 0 & 1 & -\frac{4}{3} & | & 1 \\ 0 & 12 & -16 & | & 12 \end{bmatrix} \quad (R_3 = -12r_2 + r_3)$$

$$=\begin{bmatrix} 1 & -\frac{7}{2} & \frac{11}{2} & | & -\frac{5}{2} \\ 0 & 1 & -\frac{4}{3} & | & 1 \\ 0 & 0 & 0 & | & 0 \end{bmatrix}$$

The system is dependent and has an infinite number of solutions.

Write the system of equations that corresponds to the row-echelon matrix

$$\begin{cases} x - \dfrac{7}{2}y + \dfrac{11}{2}z = -\dfrac{5}{2} & (1) \\ \qquad y - \dfrac{4}{3}z = 1 & (2) \\ \qquad\qquad 0 = 0 & (3) \end{cases}$$

Solve equation (2) for y.

$$y - \frac{4}{3}z = 1$$
$$y = \frac{4}{3}z + 1$$

Substituting $\dfrac{4}{3}z + 1$ for y in equation (1), we obtain

$$x - \frac{7}{2}\left(\frac{4}{3}z + 1\right) + \frac{11}{2}z = -\frac{5}{2}$$
$$x - \frac{14}{3}z - \frac{7}{2} + \frac{11}{2}z = -\frac{5}{2}$$
$$6\left(x - \frac{14}{3}z - \frac{7}{2} + \frac{11}{2}z\right) = 6\left(-\frac{5}{2}\right)$$
$$6x - 28z - 21 + 33z = -15$$
$$6x + 5z - 21 = -15$$
$$6x = -5z + 6$$
$$x = \frac{-5z + 6}{6}$$
$$x = -\frac{5}{6}z + 1$$

The solution to the system is

$$\left\{ (x, y, z) \,\middle|\, x = -\frac{5}{6}z + 1, \right.$$

$$\left. y = \frac{4}{3}z + 1, \; z \text{ is any real number} \right\}.$$

70. Write the augmented matrix of the system and then put it in row echelon form.

$$\begin{bmatrix} 5 & 3 & 7 & | & 9 \\ 3 & 5 & 4 & | & 8 \\ 1 & 3 & 3 & | & 9 \end{bmatrix} \quad (\text{Interchange } r_1 \text{ and } r_3)$$

$$= \begin{bmatrix} 1 & 3 & 3 & | & 9 \\ 3 & 5 & 4 & | & 8 \\ 5 & 3 & 7 & | & 9 \end{bmatrix} \quad \begin{pmatrix} R_2 = -3r_1 + r_2 \\ R_3 = -5r_1 + r_3 \end{pmatrix}$$

$$= \begin{bmatrix} 1 & 3 & 3 & | & 9 \\ 0 & -4 & -5 & | & -19 \\ 0 & -12 & -8 & | & -36 \end{bmatrix} \quad \left(R_2 = -\frac{1}{4}r_2 \right)$$

$$= \begin{bmatrix} 1 & 3 & 3 & | & 9 \\ 0 & 1 & \frac{5}{4} & | & \frac{19}{4} \\ 0 & -12 & -8 & | & -36 \end{bmatrix} \quad (R_3 = 12r_2 + r_3)$$

$$= \begin{bmatrix} 1 & 3 & 3 & | & 9 \\ 0 & 1 & \frac{5}{4} & | & \frac{19}{4} \\ 0 & 0 & 7 & | & 21 \end{bmatrix} \quad \left(R_3 = \frac{1}{7}r_3 \right)$$

$$= \begin{bmatrix} 1 & 3 & 3 & | & 9 \\ 0 & 1 & \frac{5}{4} & | & \frac{19}{4} \\ 0 & 0 & 1 & | & 3 \end{bmatrix}$$

Write the system of equations that corresponds to the row-echelon matrix

$$\begin{cases} x + 3y + 3z = 9 & (1) \\ \quad y + \dfrac{5}{4}z = \dfrac{19}{4} & (2) \\ \qquad z = 3 & (3) \end{cases}$$

This system is consistent and independent. Substituting 3 for z in equation (2), we obtain

$$y + \frac{5}{4}(3) = \frac{19}{4}$$
$$y + \frac{15}{4} = \frac{19}{4}$$
$$y = 1$$

Substituting 3 for z and 1 for y in equation (1), we obtain

$$x + 3(1) + 3(3) = 9$$
$$x + 3 + 9 = 9$$
$$x + 12 = 9$$
$$x = -3$$

The solution is the ordered triple $(-3, 1, 3)$.

71. $\begin{vmatrix} 3 & 4 \\ -1 & 2 \end{vmatrix} = 3(2) - (-1)(4) = 6 + 4 = 10$

72. $\begin{vmatrix} 2 & -3 \\ -6 & 9 \end{vmatrix} = 2(9) - (-6)(-3) = 18 - 18 = 0$

73. $\begin{vmatrix} -5 & 7 \\ -4 & 6 \end{vmatrix} = -5(6) - (-4)(7) = -30 + 28 = -2$

74. $\begin{vmatrix} -7 & 2 \\ 6 & -3 \end{vmatrix} = -7(-3) - 6(2) = 21 - 12 = 9$

75.
$$\begin{vmatrix} 5 & 0 & 1 \\ 2 & -3 & -1 \\ 3 & 6 & -4 \end{vmatrix}$$
$$= 5\begin{vmatrix} -3 & -1 \\ 6 & -4 \end{vmatrix} - 0\begin{vmatrix} 2 & -1 \\ 3 & -4 \end{vmatrix} + 1\begin{vmatrix} 2 & -3 \\ 3 & 6 \end{vmatrix}$$
$$= 5\big[-3(-4)-6(-1)\big] - 0\big[2(-4)-3(-1)\big]$$
$$\qquad\qquad + 1\big[2(6)-3(-3)\big]$$
$$= 5(18)-0(-5)+1(21)$$
$$= 90+0+21$$
$$= 111$$

76.
$$\begin{vmatrix} 1 & -4 & 5 \\ 0 & 1 & -3 \\ 2 & -6 & 4 \end{vmatrix}$$
$$= 1\begin{vmatrix} 1 & -3 \\ -6 & 4 \end{vmatrix} - (-4)\begin{vmatrix} 0 & -3 \\ 2 & 4 \end{vmatrix} + 5\begin{vmatrix} 0 & 1 \\ 2 & -6 \end{vmatrix}$$
$$= 1\big[1(4)-(-6)(-3)\big] + 4\big[0(4)-2(-3)\big]$$
$$\qquad\qquad + 5\big[0(-6)-2(1)\big]$$
$$= 1(-14)+4(6)+5(-2)$$
$$= -14+24-10$$
$$= 0$$

77.
$$\begin{vmatrix} 3 & 0 & -1 \\ 2 & 6 & 7 \\ 2 & 5 & 4 \end{vmatrix} = 3\begin{vmatrix} 6 & 7 \\ 5 & 4 \end{vmatrix} - 0\begin{vmatrix} 2 & 7 \\ 2 & 4 \end{vmatrix} + (-1)\begin{vmatrix} 2 & 6 \\ 2 & 5 \end{vmatrix}$$
$$= 3\big[6(4)-5(7)\big] - 0\big[2(4)-2(7)\big]$$
$$\qquad\qquad - 1\big[2(5)-2(6)\big]$$
$$= 3(-11)-0(-6)-1(-2)$$
$$= -33+0+2$$
$$= -31$$

78.
$$\begin{vmatrix} 2 & 3 & 1 \\ 1 & -3 & -7 \\ -5 & 4 & 8 \end{vmatrix}$$
$$= 2\begin{vmatrix} -3 & -7 \\ 4 & 8 \end{vmatrix} - 3\begin{vmatrix} 1 & -7 \\ -5 & 8 \end{vmatrix} + 1\begin{vmatrix} 1 & -3 \\ -5 & 4 \end{vmatrix}$$
$$= 2\big[-3(8)-4(-7)\big] - 3\big[1(8)-(-5)(-7)\big]$$
$$\qquad\qquad + 1\big[1(4)-(-5)(-3)\big]$$
$$= 2(4)-3(-27)+1(-11)$$
$$= 8+81-11$$
$$= 78$$

79. $D = \begin{vmatrix} 1 & 2 \\ 2 & 3 \end{vmatrix} = 1(3)-2(2) = 3-4 = -1$

$D_x = \begin{vmatrix} -1 & 2 \\ 1 & 3 \end{vmatrix} = -1(3)-1(2) = -3-2 = -5$

$D_y = \begin{vmatrix} 1 & -1 \\ 2 & 1 \end{vmatrix} = 1(1)-2(-1) = 1+2 = 3$

$x = \dfrac{D_x}{D} = \dfrac{-5}{-1} = 5$; $y = \dfrac{D_y}{D} = \dfrac{3}{-1} = -3$

Thus, the solution is the ordered pair (5, −3).

80. $D = \begin{vmatrix} 4 & 9 \\ -5 & 6 \end{vmatrix} = 4(6)-(-5)(9) = 24+45 = 69$

$D_x = \begin{vmatrix} -13 & 9 \\ 22 & 6 \end{vmatrix}$
$= -13(6)-22(9) = -78-198 = -276$

$D_y = \begin{vmatrix} 4 & -13 \\ -5 & 22 \end{vmatrix}$
$= 4(22)-(-5)(-13) = 88-65 = 23$

$x = \dfrac{D_x}{D} = \dfrac{-276}{69} = -4$; $y = \dfrac{D_y}{D} = \dfrac{23}{69} = \dfrac{1}{3}$

Thus, the solution is the ordered pair $\left(-4, \dfrac{1}{3}\right)$.

81. $D = \begin{vmatrix} 4 & -1 \\ 3 & 5 \end{vmatrix} = 4(5)-3(-1) = 20+3 = 23$

$D_x = \begin{vmatrix} -6 & -1 \\ 7 & 5 \end{vmatrix} = -6(5)-7(-1) = -30+7 = -23$

$D_y = \begin{vmatrix} 4 & -6 \\ 3 & 7 \end{vmatrix} = 4(7)-3(-6) = 28+18 = 46$

$x = \dfrac{D_x}{D} = \dfrac{-23}{23} = -1$; $y = \dfrac{D_y}{D} = \dfrac{46}{23} = 2$

Thus, the solution is the ordered pair (−1, 2).

82. $D = \begin{vmatrix} 1 & -1 \\ 2 & 1 \end{vmatrix} = 1(1)-2(-1) = 1+2 = 3$

$D_x = \begin{vmatrix} 2 & -1 \\ 5 & 1 \end{vmatrix} = 2(1)-5(-1) = 2+5 = 7$

$D_y = \begin{vmatrix} 1 & 2 \\ 2 & 5 \end{vmatrix} = 1(5)-2(2) = 5-4 = 1$

$x = \dfrac{D_x}{D} = \dfrac{7}{3}$; $y = \dfrac{D_y}{D} = \dfrac{1}{3}$

Thus, the solution is the ordered pair $\left(\dfrac{7}{3}, \dfrac{1}{3}\right)$.

83. $D = \begin{vmatrix} 6 & 2 \\ 15 & 5 \end{vmatrix} = 6(5) - 15(2) = 30 - 30 = 0$

Because $D = 0$, Cramer's Rule does not apply.

84. $D = \begin{vmatrix} 12 & 1 \\ -6 & 7 \end{vmatrix} = 12(7) - (-6)(1) = 84 + 6 = 90$

$D_x = \begin{vmatrix} 6 & 1 \\ 15 & 7 \end{vmatrix} = 6(7) - 15(1) = 42 - 15 = 27$

$D_y = \begin{vmatrix} 12 & 6 \\ -6 & 15 \end{vmatrix}$
$= 12(15) - (-6)(6) = 180 + 36 = 216$

$x = \dfrac{D_x}{D} = \dfrac{27}{90} = \dfrac{3}{10}$; $\ y = \dfrac{D_y}{D} = \dfrac{216}{90} = \dfrac{12}{5}$

Thus, the solution is the ordered pair $\left(\dfrac{3}{10}, \dfrac{12}{5}\right)$.

85. $D = \begin{vmatrix} 1 & -1 & 2 \\ 3 & 2 & -4 \\ 0 & 3 & 5 \end{vmatrix}$

$= 1\begin{vmatrix} 2 & -4 \\ 3 & 5 \end{vmatrix} - (-1)\begin{vmatrix} 3 & -4 \\ 0 & 5 \end{vmatrix} + 2\begin{vmatrix} 3 & 2 \\ 0 & 3 \end{vmatrix}$

$= 1[2(5) - 3(-4)] - (-1)[3(5) - 0(-4)]$
$\qquad\qquad\qquad + 2[3(3) - 0(2)]$

$= 1(22) - (-1)(15) + 2(9)$

$= 22 + 15 + 18$

$= 55$

$D_x = \begin{vmatrix} 9 & -1 & 2 \\ 7 & 2 & -4 \\ -1 & 3 & 5 \end{vmatrix}$

$= 9\begin{vmatrix} 2 & -4 \\ 3 & 5 \end{vmatrix} - (-1)\begin{vmatrix} 7 & -4 \\ -1 & 5 \end{vmatrix} + 2\begin{vmatrix} 7 & 2 \\ -1 & 3 \end{vmatrix}$

$= 9[2(5) - 3(-4)] - (-1)[7(5) - (-1)(-4)]$
$\qquad\qquad\qquad + 2[7(3) - (-1)(2)]$

$= 9(22) - (-1)(31) + 2(23)$

$= 198 + 31 + 46$

$= 275$

$D_y = \begin{vmatrix} 1 & 9 & 2 \\ 3 & 7 & -4 \\ 0 & -1 & 5 \end{vmatrix}$

$= 1\begin{vmatrix} 7 & -4 \\ -1 & 5 \end{vmatrix} - 9\begin{vmatrix} 3 & -4 \\ 0 & 5 \end{vmatrix} + 2\begin{vmatrix} 3 & 7 \\ 0 & -1 \end{vmatrix}$

$= 1[7(5) - (-1)(-4)] - 9[3(5) - 0(-4)]$
$\qquad\qquad\qquad + 2[3(-1) - 0(7)]$

$= 1(31) - 9(15) + 2(-3)$

$= 31 - 135 - 6$

$= -110$

$D_z = \begin{vmatrix} 1 & -1 & 9 \\ 3 & 2 & 7 \\ 0 & 3 & -1 \end{vmatrix}$

$= 1\begin{vmatrix} 2 & 7 \\ 3 & -1 \end{vmatrix} - (-1)\begin{vmatrix} 3 & 7 \\ 0 & -1 \end{vmatrix} + 9\begin{vmatrix} 3 & 2 \\ 0 & 3 \end{vmatrix}$

$= 1[2(-1) - 3(7)] - (-1)[3(-1) - 0(7)]$
$\qquad\qquad\qquad + 9[3(3) - 0(2)]$

$= 1(-23) - (-1)(-3) + 9(9)$

$= -23 - 3 + 81$

$= 55$

$x = \dfrac{D_x}{D} = \dfrac{275}{55} = 5$; $\ y = \dfrac{D_y}{D} = \dfrac{-110}{55} = -2$;

$z = \dfrac{D_z}{D} = \dfrac{55}{55} = 1$

Thus, the solution is the ordered triple $(5, -2, 1)$.

86. $D = \begin{vmatrix} 1 & -3 & -3 \\ 7 & 1 & -2 \\ 6 & -5 & -4 \end{vmatrix}$

$= 1\begin{vmatrix} 1 & -2 \\ -5 & -4 \end{vmatrix} - (-3)\begin{vmatrix} 7 & -2 \\ 6 & -4 \end{vmatrix} + (-3)\begin{vmatrix} 7 & 1 \\ 6 & -5 \end{vmatrix}$

$= 1[1(-4) - (-5)(-2)] - (-3)[7(-4) - 6(-2)]$
$\qquad\qquad\qquad + (-3)[7(-5) - 6(1)]$

$= 1(-14) - (-3)(-16) + (-3)(-41)$

$= -14 - 48 + 123$

$= 61$

$D_x = \begin{vmatrix} -5 & -3 & -3 \\ 24 & 1 & -2 \\ -9 & -5 & -4 \end{vmatrix}$

$= -5\begin{vmatrix} 1 & -2 \\ -5 & -4 \end{vmatrix} - (-3)\begin{vmatrix} 24 & -2 \\ -9 & -4 \end{vmatrix} + (-3)\begin{vmatrix} 24 & 1 \\ -9 & -5 \end{vmatrix}$

$= -5[1(-4) - (-5)(-2)] - (-3)[24(-4) - (-9)(-2)]$
$\qquad\qquad\qquad + (-3)[24(-5) - (-9)(1)]$

$= -5(-14) - (-3)(-114) + (-3)(-111)$

$= 70 - 342 + 333$

$= 61$

$D_y = \begin{vmatrix} 1 & -5 & -3 \\ 7 & 24 & -2 \\ 6 & -9 & -4 \end{vmatrix}$

$= 1\begin{vmatrix} 24 & -2 \\ -9 & -4 \end{vmatrix} - (-5)\begin{vmatrix} 7 & -2 \\ 6 & -4 \end{vmatrix} + (-3)\begin{vmatrix} 7 & 24 \\ 6 & -9 \end{vmatrix}$

$= 1[24(-4) - (-9)(-2)] - (-5)[7(-4) - 6(-2)]$
$\qquad\qquad\qquad + (-3)[7(-9) - 6(24)]$

$= 1(-114) - (-5)(-16) + (-3)(-207)$

$= -114 - 80 + 621$

$= 427$

$$D_z = \begin{vmatrix} 1 & -3 & -5 \\ 7 & 1 & 24 \\ 6 & -5 & -9 \end{vmatrix}$$

$$= 1\begin{vmatrix} 1 & 24 \\ -5 & -9 \end{vmatrix} - (-3)\begin{vmatrix} 7 & 24 \\ 6 & -9 \end{vmatrix} + (-5)\begin{vmatrix} 7 & 1 \\ 6 & -5 \end{vmatrix}$$

$$= 1[1(-9)-(-5)(24)] - (-3)[7(-9)-6(24)]$$
$$+ (-5)[7(-5)-6(1)]$$

$$= 1(111)-(-3)(-207)+(-5)(-41)$$

$$= 111 - 621 + 205$$

$$= -305$$

$$x = \frac{D_x}{D} = \frac{61}{61} = 1; \quad y = \frac{D_y}{D} = \frac{427}{61} = 7;$$

$$z = \frac{D_z}{D} = \frac{-305}{61} = -5$$

Thus, the solution is the ordered triple (1, 7, −5).

87. $D = \begin{vmatrix} 1 & 1 & 1 \\ 3 & 2 & 4 \\ 2 & 2 & 3 \end{vmatrix}$

$$= 1\begin{vmatrix} 2 & 4 \\ 2 & 3 \end{vmatrix} - 1\begin{vmatrix} 3 & 4 \\ 2 & 3 \end{vmatrix} + 1\begin{vmatrix} 3 & 2 \\ 2 & 2 \end{vmatrix}$$

$$= 1[2(3)-2(4)] - 1[3(3)-2(4)]$$
$$+ 1[3(2)-2(2)]$$

$$= -2 - 1 + 2$$

$$= -1$$

$$D_x = \begin{vmatrix} 1 & 1 & 1 \\ -1 & 2 & 4 \\ 0 & 2 & 3 \end{vmatrix}$$

$$= 1\begin{vmatrix} 2 & 4 \\ 2 & 3 \end{vmatrix} - 1\begin{vmatrix} -1 & 4 \\ 0 & 3 \end{vmatrix} + 1\begin{vmatrix} -1 & 2 \\ 0 & 2 \end{vmatrix}$$

$$= 1[2(3)-2(4)] - 1[-1(3)-0(4)]$$
$$+ 1[-1(2)-0(2)]$$

$$= -2 + 3 - 2$$

$$= -1$$

$$D_y = \begin{vmatrix} 1 & 1 & 1 \\ 3 & -1 & 4 \\ 2 & 0 & 3 \end{vmatrix}$$

$$= 1\begin{vmatrix} -1 & 4 \\ 0 & 3 \end{vmatrix} - 1\begin{vmatrix} 3 & 4 \\ 2 & 3 \end{vmatrix} + 1\begin{vmatrix} 3 & -1 \\ 2 & 0 \end{vmatrix}$$

$$= 1[-1(3)-0(4)] - 1[3(3)-2(4)]$$
$$+ 1[3(0)-2(-1)]$$

$$= -3 - 1 + 2$$

$$= -2$$

$$D_z = \begin{vmatrix} 1 & 1 & 1 \\ 3 & 2 & -1 \\ 2 & 2 & 0 \end{vmatrix}$$

$$= 1\begin{vmatrix} 2 & -1 \\ 2 & 0 \end{vmatrix} - 1\begin{vmatrix} 3 & -1 \\ 2 & 0 \end{vmatrix} + 1\begin{vmatrix} 3 & 2 \\ 2 & 2 \end{vmatrix}$$

$$= 1[2(0)-2(-1)] - 1[3(0)-2(-1)]$$
$$+ 1[3(2)-2(2)]$$

$$= 2 - 2 + 2$$

$$= 2$$

$$x = \frac{D_x}{D} = \frac{-1}{-1} = 1; \quad y = \frac{D_y}{D} = \frac{-2}{-1} = 2;$$

$$z = \frac{D_z}{D} = \frac{2}{-1} = -2$$

Thus, the solution is the ordered triple (1, 2, −2).

88. $D = \begin{vmatrix} 4 & -3 & 1 \\ 4 & -2 & 3 \\ 8 & -5 & -2 \end{vmatrix}$

$$= 4\begin{vmatrix} -2 & 3 \\ -5 & -2 \end{vmatrix} - (-3)\begin{vmatrix} 4 & 3 \\ 8 & -2 \end{vmatrix} + 1\begin{vmatrix} 4 & -2 \\ 8 & -5 \end{vmatrix}$$

$$= 4[-2(-2)-(-5)(3)] - (-3)[4(-2)-8(3)]$$
$$+ 1[4(-5)-8(-2)]$$

$$= 4(19)-(-3)(-32)+1(-4)$$

$$= 76 - 96 - 4$$

$$= -24$$

$$D_x = \begin{vmatrix} -6 & -3 & 1 \\ -3 & -2 & 3 \\ -12 & -5 & -2 \end{vmatrix}$$

$$= -6\begin{vmatrix} -2 & 3 \\ -5 & -2 \end{vmatrix} - (-3)\begin{vmatrix} -3 & 3 \\ -12 & -2 \end{vmatrix} + 1\begin{vmatrix} -3 & -2 \\ -12 & -5 \end{vmatrix}$$

$$= -6[-2(-2)-(-5)(3)] - (-3)[-3(-2)-(-12)(3)]$$
$$+ 1[-3(-5)-(-12)(-2)]$$

$$= -6(19)-(-3)(42)+1(-9)$$

$$= -114 + 126 - 9$$

$$= 3$$

$$D_y = \begin{vmatrix} 4 & -6 & 1 \\ 4 & -3 & 3 \\ 8 & -12 & -2 \end{vmatrix}$$

$$= 4\begin{vmatrix} -3 & 3 \\ -12 & -2 \end{vmatrix} - (-6)\begin{vmatrix} 4 & 3 \\ 8 & -2 \end{vmatrix} + 1\begin{vmatrix} 4 & -3 \\ 8 & -12 \end{vmatrix}$$

$$= 4[-3(-2)-(-12)(3)] - (-6)[4(-2)-8(3)]$$
$$+ 1[4(-12)-8(-3)]$$

$$= 4(42)-(-6)(-32)+1(-24)$$

$$= 168 - 192 - 24$$

$$= -48$$

$$D_z = \begin{vmatrix} 4 & -3 & -6 \\ 4 & -2 & -3 \\ 8 & -5 & -12 \end{vmatrix}$$

$$= 4\begin{vmatrix} -2 & -3 \\ -5 & -12 \end{vmatrix} - (-3)\begin{vmatrix} 4 & -3 \\ 8 & -12 \end{vmatrix} + (-6)\begin{vmatrix} 4 & -2 \\ 8 & -5 \end{vmatrix}$$

$$= 4[-2(-12)-(-5)(-3)] - (-3)[4(-12)-8(-3)]$$
$$\qquad\qquad + (-6)[4(-5)-8(-2)]$$

$$= 4(9)-(-3)(-24)+(-6)(-4)$$

$$= 36-72+24$$

$$= -12$$

$$x = \frac{D_x}{D} = \frac{3}{-24} = -\frac{1}{8}; \quad y = \frac{D_y}{D} = \frac{-48}{-24} = 2;$$

$$z = \frac{D_z}{D} = \frac{-12}{-24} = \frac{1}{2}$$

The solution is the ordered triple $\left(-\dfrac{1}{8},\ 2,\ \dfrac{1}{2}\right)$.

89. $D = \begin{vmatrix} 1 & -1 & -4 \\ 4 & -3 & -3 \\ 3 & -2 & 1 \end{vmatrix}$

$$= 1\begin{vmatrix} -3 & -3 \\ -2 & 1 \end{vmatrix} - (-1)\begin{vmatrix} 4 & -3 \\ 3 & 1 \end{vmatrix} + (-4)\begin{vmatrix} 4 & -3 \\ 3 & -2 \end{vmatrix}$$

$$= 1[-3(1)-(-2)(-3)] - (-1)[4(1)-3(-3)]$$
$$\qquad\qquad + (-4)[4(-2)-3(-3)]$$

$$= 1(-9)-(-1)(13)+(-4)(1)$$

$$= -9+13-4$$

$$= 0$$

Because $D = 0$, Cramer's Rule does not apply.

90. $D = \begin{vmatrix} 1 & 1 & 0 \\ 0 & 2 & -1 \\ 5 & 0 & 1 \end{vmatrix}$

$$= 1\begin{vmatrix} 2 & -1 \\ 0 & 1 \end{vmatrix} - 1\begin{vmatrix} 0 & -1 \\ 5 & 1 \end{vmatrix} + 0\begin{vmatrix} 0 & 2 \\ 5 & 0 \end{vmatrix}$$

$$= 1[2(1)-0(-1)] - 1[0(1)-5(-1)]$$
$$\qquad\qquad + 0[0(0)-5(2)]$$

$$= 1(2)-1(5)+0(-10)$$

$$= 2-5+0$$

$$= -3$$

$$D_x = \begin{vmatrix} -3 & 1 & 0 \\ -1 & 2 & -1 \\ 1 & 0 & 1 \end{vmatrix}$$

$$= -3\begin{vmatrix} 2 & -1 \\ 0 & 1 \end{vmatrix} - 1\begin{vmatrix} -1 & -1 \\ 1 & 1 \end{vmatrix} + 0\begin{vmatrix} -1 & 2 \\ 1 & 0 \end{vmatrix}$$

$$= -3[2(1)-0(-1)] - 1[-1(1)-1(-1)]$$
$$\qquad\qquad + 0[-1(0)-1(2)]$$

$$= -3(2)-1(0)+0(-2)$$

$$= -6-0+0$$

$$= -6$$

$$D_y = \begin{vmatrix} 1 & -3 & 0 \\ 0 & -1 & -1 \\ 5 & 1 & 1 \end{vmatrix}$$

$$= 1\begin{vmatrix} -1 & -1 \\ 1 & 1 \end{vmatrix} - (-3)\begin{vmatrix} 0 & -1 \\ 5 & 1 \end{vmatrix} + 0\begin{vmatrix} 0 & -1 \\ 5 & 1 \end{vmatrix}$$

$$= 1[-1(1)-1(-1)] - (-3)[0(1)-5(-1)]$$
$$\qquad\qquad + 0[0(1)-5(-1)]$$

$$= 1(0)-(-3)(5)+0(5)$$

$$= 0+15+0$$

$$= 15$$

$$D_z = \begin{vmatrix} 1 & 1 & -3 \\ 0 & 2 & -1 \\ 5 & 0 & 1 \end{vmatrix}$$

$$= 1\begin{vmatrix} 2 & -1 \\ 0 & 1 \end{vmatrix} - 1\begin{vmatrix} 0 & -1 \\ 5 & 1 \end{vmatrix} + (-3)\begin{vmatrix} 0 & 2 \\ 5 & 0 \end{vmatrix}$$

$$= 1[2(1)-0(-1)] - 1[0(1)-5(-1)]$$
$$\qquad\qquad + (-3)[0(0)-5(2)]$$

$$= 1(2)-1(5)+(-3)(-10)$$

$$= 2-5+30$$

$$= 27$$

$$x = \frac{D_x}{D} = \frac{-6}{-3} = 2; \quad y = \frac{D_y}{D} = \frac{15}{-3} = -5;$$

$$z = \frac{D_z}{D} = \frac{27}{-3} = -9$$

Thus, the solution is the ordered triple $(2, -5, -9)$.

91. a. Let $x = 5$ and $y = 2$ in each inequality.
$$\begin{cases} x-y = 5-2 = 3 > 2 \\ 2x+3y = 2(5)+3(2) = 16 > 8 \end{cases}$$

Both inequalities are true when $x = 5$ and $y = 2$, so $(5, 2)$ is a solution.

b. Let $x = -3$ and $y = 5$ in each inequality.
$$\begin{cases} x-y = -3-5 = -8 \not> 2 \\ 2x+3y = 2(-3)+3(5) = 9 > 8 \end{cases}$$

The inequality $x - y > 2$ is not true when $x = -3$ and $y = 5$, so $(-3, 5)$ is not a solution.

92. a. Let $x = -1$ and $y = 2$ in each inequality.
$$\begin{cases} 7x + 3y = 7(-1) + 3(2) = -1 \le 21 \\ -2x + y = -2(-1) + 2 = 4 \ngeq 5 \end{cases}$$
The inequality $-2x + y \ge 5$ is not true when $x = -1$ and $y = 2$, so $(-1, 2)$ is not a solution.

b. Let $x = -3$ and $y = 4$ in each inequality.
$$\begin{cases} 7x + 3y = 7(-3) + 3(4) = -9 \le 21 \\ -2x + y = -2(-3) + 4 = 10 \ge 5 \end{cases}$$
Both inequalities are true when $x = -3$ and $y = 4$, so $(-3, 4)$ is a solution.

93. a. Let $x = 2$ and $y = 1$ in each inequality.
$$\begin{cases} x + 2y = 2 + 2(1) = 4 \le 6 \\ 3x - y = 3(2) - 1 = 5 \ge 2 \\ x = 2 \ge 0 \\ y = 1 \ge 0 \end{cases}$$
All four inequalities are true when $x = 2$ and $y = 1$, so $(2, 1)$ is a solution.

b. Let $x = 1$ and $y = 2$ in each inequality.
$$\begin{cases} x + 2y = 1 + 2(2) = 5 \le 6 \\ 3x - y = 3(1) - 2 = 1 \ngeq 2 \\ x = 1 \ge 0 \\ y = 2 \ge 0 \end{cases}$$
The inequality $3x - y \ge 2$ is not true when $x = 1$ and $y = 2$, so $(1, 2)$ is not a solution.

94. a. Let $x = -2$ and $y = 3$ in each inequality.
$$\begin{cases} 3x - 2y = 3(-2) - 2(3) = -12 \le 12 \\ 2x + y = 2(-2) + 3 = -1 \le 15 \\ x = -2 \ngeq 0 \\ y = 3 \ge 0 \end{cases}$$
The inequality $x \ge 0$ is not true when $x = -2$ and $y = 3$, so $(-2, 3)$ is not a solution.

b. Let $x = 4$ and $y = 1$ in each inequality.
$$\begin{cases} 3x - 2y = 3(4) - 2(1) = 10 \le 12 \\ 2x + y = 2(4) + 1 = 9 \le 15 \\ x = 4 \ge 0 \\ y = 1 \ge 0 \end{cases}$$
All four inequalities are true when $x = 4$ and $y = 1$, so $(4, 1)$ is a solution.

95. $\begin{cases} x + y \ge 7 \\ 2x - y \ge 5 \end{cases}$

First, graph the inequality $x + y \ge 7$. To do so, replace the inequality symbol with an equal sign

to obtain $x + y = 7$. Because the inequality is non-strict, graph $x + y = 7$ $(y = -x + 7)$ using a solid line.

Test Point: $(0,0)$: $0 + 0 = 0 \ngeq 7$

Therefore, the half-plane not containing $(0, 0)$ is the solution set of $x + y \ge 7$.

Second, graph the inequality $2x - y \ge 5$. To do so, replace the inequality symbol with an equal sign to obtain $2x - y = 5$. Because the inequality is non-strict, graph $2x - y = 5$ $(y = 2x - 5)$ using a solid line.

Test Point: $(0,0)$: $2(0) - 0 = 0 \ngeq 5$

Therefore, the half-plane not containing $(0, 0)$ is the solution set of $2x - y \ge 5$.

The overlapping shaded region (that is, the shaded region in the graph below) is the solution to the system of linear inequalities.

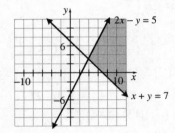

96. $\begin{cases} 2x + 3y > 9 \\ x - 3y > -18 \end{cases}$

First, graph the inequality $2x + 3y > 9$. To do so, replace the inequality symbol with an equal sign to obtain $2x + 3y = 9$. Because the inequality is strict, graph $2x + 3y = 9$ $\left(y = -\dfrac{2}{3}x + 3 \right)$ using a dashed line.

Test Point: $(0,0)$: $2(0) + 3(0) = 0 \ngtr 9$

Therefore, the half-plane not containing $(0, 0)$ is the solution set of $2x + 3y > 9$.

Second, graph the inequality $x - 3y > -18$. To do so, replace the inequality symbol with an equal sign to obtain $x - 3y = -18$. Because the inequality is strict, graph $x - 3y = -18$ $\left(y = \dfrac{1}{3}x + 6 \right)$ using a dashed line.

Test Point: $(0,0)$: $0 - 3(0) = 0 > -18$

Therefore, the half-plane containing $(0, 0)$ is the solution set of $x - 3y > -18$.

The overlapping shaded region (that is, the shaded region in the graph below) is the solution to the system of linear inequalities.

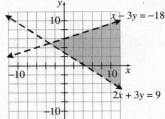

97. $\begin{cases} x - 4y > -4 \\ x + 2y \le 8 \end{cases}$

First, graph the inequality $x - 4y > -4$. To do so, replace the inequality symbol with an equal sign to obtain $x - 4y = -4$. Because the inequality is strict, graph $x - 4y = -4$ $\left(y = \dfrac{1}{4}x + 1 \right)$ using a dashed line.

Test Point: $(0,0)$: $0 - 4(0) = 0 > -4$

Therefore, the half-plane containing $(0, 0)$ is the solution set of $x - 4y > -4$.

Second, graph the inequality $x + 2y \le 8$. To do so, replace the inequality symbol with an equal sign to obtain $x + 2y = 8$. Because the inequality is non-strict, graph $x + 2y = 8$ $\left(y = -\dfrac{1}{2}x + 4 \right)$ using a solid line.

Test Point: $(0,0)$: $0 + 2(0) = 0 \le 8$

Therefore, the half-plane containing $(0, 0)$ is the solution set of $x + 2y \le 8$.

The overlapping shaded region (that is, the shaded region in the graph below) is the solution to the system of linear inequalities.

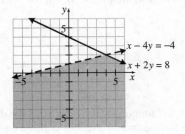

98. $\begin{cases} x - y < -2 \\ 2x - 2y > 6 \end{cases}$

First, graph the inequality $x - y < -2$. To do so, replace the inequality symbol with an equal sign

to obtain $x - y = -2$. Because the inequality is strict, graph $x - y = -2$ $(y = x + 2)$ using a dashed line.

Test Point: $(0,0)$: $0 - 0 = 0 \not< -2$

Therefore, the half-plane not containing $(0, 0)$ is the solution set of $x - y < -2$.

Second, graph the inequality $2x - 2y > 6$. To do so, replace the inequality symbol with an equal sign to obtain $2x - 2y = 6$. Because the inequality is strict, graph $2x - 2y = 6$ $(y = x - 3)$ using a dashed line.

Test Point: $(0,0)$: $2(0) - 2(0) = 0 \not> 6$

Therefore, the half-plane not containing $(0, 0)$ is the solution set of $2x - 2y > 6$.

Notice in the graph below that the two shaded regions do not overlap. Thus, there are no points in the Cartesian plane that satisfy both inequalities. The system has no solution, so the solution set is $\varnothing$ or { }.

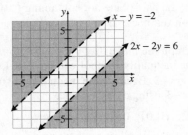

99. $\begin{cases} 2x - y \ge -2 \\ 2x - 3y \le 6 \end{cases}$

First, graph the inequality $2x - y \ge -2$. To do so, replace the inequality symbol with an equal sign to obtain $2x - y = -2$. Because the inequality is non-strict, graph $2x - y = -2$ $(y = 2x + 2)$ using a solid line.

Test Point: $(0,0)$: $2(0) - 0 = 0 \ge -2$

Therefore, the half-plane containing $(0, 0)$ is the solution set of $2x - y \ge -2$.

Second, graph the inequality $2x - 3y \le 6$. To do so, replace the inequality symbol with an equal sign to obtain $2x - 3y = 6$. Because the inequality is non-strict, graph $2x - 3y = 6$ $\left(y = \dfrac{2}{3}x - 2 \right)$ using a solid line.

Test Point: $(0,0)$: $2(0) - 3(0) = 0 \le 6$

Therefore, the half-plane containing (0, 0) is the solution set of $2x - 3y \leq 6$.

The overlapping shaded region (that is, the shaded region in the graph below) is the solution to the system of linear inequalities.

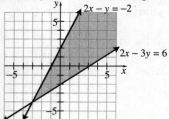

100. $\begin{cases} x \geq 2 \\ y < -3 \end{cases}$

First, graph the inequality $x \geq 2$. To do so, replace the inequality symbol with an equal sign to obtain $x = 2$. Because the inequality is non-strict, graph $x = 2$ using a solid line.

Test Point: $(0,0)$: $0 \not\geq 2$

Therefore, the half-plane not containing (0, 0) is the solution set of $x \geq 2$.

Second, graph the inequality $y < -3$. To do so, replace the inequality symbol with an equal sign to obtain $y = -3$. Because the inequality is strict, graph $y = -3$ using a dashed line.

Test Point: $(0,0)$: $0 \not< -3$

Therefore, the half-plane not containing (0, 0) is the solution set of $y < -3$.

The overlapping shaded region (that is, the shaded region in the graph below) is the solution to the system of linear inequalities.

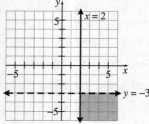

101. $\begin{cases} 3x + 5y \leq 30 \\ 4x - 5y \geq 5 \\ x \leq 8 \\ y \geq 0 \end{cases}$

First, graph the inequality $3x + 5y \leq 30$. To do so, replace the inequality symbol with an equal sign to obtain $3x + 5y = 30$. Because the

inequality is non-strict, graph $3x + 5y = 30$

$\left(y = -\dfrac{3}{5}x + 6 \right)$ using a solid line.

Test Point: $(0,0)$: $3(0) + 5(0) = 0 \leq 30$

Therefore, the half-plane containing (0, 0) is the solution set of $3x + 5y \leq 30$.

Next, graph the inequality $4x - 5y \geq 5$. To do so, replace the inequality symbol with an equal sign to obtain $4x - 5y = 5$. Because the inequality is

non-strict, graph $4x - 5y = 5$ $\left(y = \dfrac{4}{5}x - 1 \right)$

using a solid line.

Test Point: $(0,0)$: $4(0) - 5(0) = 0 \not\geq 5$

Therefore, the half-plane not containing (0, 0) is the solution set of $4x - 5y \geq 5$.

The inequality $x \leq 8$ requires that the graph be to the left of the solid vertical line $x = 8$.

The inequality $y \geq 0$ requires that the graph be above the solid x-axis.

The overlapping shaded region (that is, the shaded region in the graph below) is the solution to the system of linear inequalities.

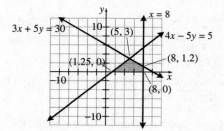

The graph is bounded. The four corner points of the bounded region are (1.25, 0), (8, 0), (8, 1.2), and (5, 3).

102. $\begin{cases} 3x + 2y \geq 10 \\ x + 2y \geq 6 \\ x \geq 0 \\ y \geq 0 \end{cases}$

First, graph the inequality $3x + 2y \geq 10$. To do so, replace the inequality symbol with an equal sign to obtain $3x + 2y = 10$. Because the inequality is non-strict, graph $3x + 2y = 10$

$\left(y = -\dfrac{3}{2}x + 5 \right)$ using a solid line.

Test Point: $(0,0)$: $3(0)+2(0)=0 \not\geq 10$

Therefore, the half-plane not containing (0, 0) is the solution set of $3x + 2y \geq 10$.

Second, graph the inequality $x + 2y \geq 6$. To do so, replace the inequality symbol with an equal sign to obtain $x + 2y = 6$. Because the inequality is non-strict, graph $x+2y=6$ $\left(y=-\dfrac{1}{2}x+3 \right)$ using a solid line.

Test Point: $(0,0)$: $0+2(0)=0 \not\geq 6$

Therefore, the half-plane not containing (0, 0) is the solution set of $x + 2y \geq 6$.

The two inequalities $x \geq 0$ and $y \geq 0$ require that the graph be in quadrant I.

The overlapping shaded region (that is, the shaded region in the graph below) is the solution to the system of linear inequalities.

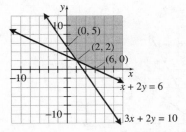

The graph is unbounded. The three corner points of the unbounded region are (0, 5), (2, 2), and (6, 0).

103. a. Let x represent the amount to be invested at 6% interest and let y represent the amount to be invested at 8% interest. Now, Anna has at most $4000 to invest, so $x+y \leq 4000$.

She wants to earn at least $275 per year in interest, so $0.06x+0.08y \geq 275$. The amount to be invested at 6% will be at least $500, so $x \geq 500$. The amount to be invested at 8% is at least $2500, so $y \geq 2500$. Thus, we have the following system:

$$\begin{cases} x+y \leq 4000 \\ 0.06x+0.08y \geq 275 \\ x \geq 500 \\ y \geq 2500 \end{cases}$$

b. First, graph the inequality $x+y \leq 4000$. To do so, replace the inequality symbol with an

equal sign to obtain $x+y=4000$. Because the inequality is non-strict, graph $x+y=4000$ $(y=-x+4000)$ using a solid line.

Test Point: $(0,0)$: $0+0=0 \leq 4000$

Thus, the solution set of $x+y \leq 4000$ is the half-plane containing (0, 0).

Second, graph the inequality $0.06x+0.08y \geq 275$. To do so, replace the inequality symbol with an equal sign to obtain $0.06x+0.08y = 275$. Because the inequality is non-strict, graph

$0.06x+0.08y=275$ $\left(y=-\dfrac{3}{4}x+3437.5 \right)$

using a solid line.

Test Point:
$(0,0)$: $0.06(0)+0.08(0)=0 \not\geq 275$

Thus, the solution set of $0.06x+0.08y \geq 275$ is the half-plane not containing (0, 0).

The inequality $x \geq 500$ requires that the graph be to the right of the solid vertical line $x = 500$.

The inequality $y \geq 2500$ requires that the graph be above the solid horizontal line $y = 2500$.

The overlapping shaded region (that is, the shaded region in the graph below) is the solution to the system of linear inequalities.

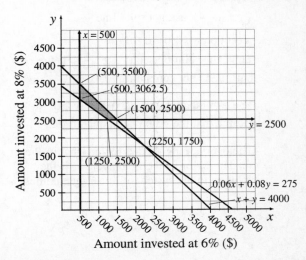

The four corner points are (500, 3500), (500, 3062.5), (1250, 2500), and (1500, 2500).

104. a. Let x represent the number of dozens of red tulip bulbs and let y represent the number of dozens of yellow tulip bulbs. Now, Jordan has a maximum of \$144 to spend, so $6x + 4y \leq 144$. She wants at least 4 more dozens of red tulip bulbs than yellow, so $x \geq y + 4$. Finally, Jordan cannot buy a negative number of dozens of bulbs, so we have the non-negativity constraints $x \geq 0$ and $y \geq 0$. Thus, we have the following system:

$$\begin{cases} 6x + 4y \leq 144 \\ x \geq y + 4 \\ x \geq 0 \\ y \geq 0 \end{cases}$$

b. First, graph the inequality $6x + 4y \leq 144$. To do so, replace the inequality symbol with an equal sign to obtain $6x + 4y = 144$. Because the inequality is non-strict, graph

$6x + 4y = 144 \left(y = -\dfrac{3}{2}x + 36 \right)$ using a

solid line.

Test Point: $(0,0)$: $6(0) + 4(0) = 0 \leq 144$

Thus, the solution set of $6x + 4y \leq 144$ is the half-plane containing (0, 0).

Second, graph the inequality $x \geq y + 4$. To do so, replace the inequality symbol with an equal sign to obtain $x = y + 4$. Because the inequality is non-strict, graph $x = y + 4$

$(y = x - 4)$ using a solid line.

Test Point: $(0,0)$: $0 \not\geq 0 + 4$

Thus, the solution set of $x \geq y + 4$ is the half-plane not containing (0, 0).

The two inequalities $x \geq 0$ and $y \geq 0$ require that the graph be in quadrant I.

The overlapping shaded region (that is, the shaded region in the graph below) is the solution to the system of linear inequalities.

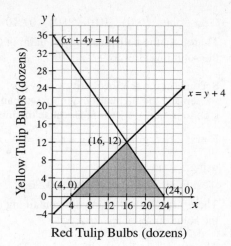

The three corner points are (4, 0), (24, 0), and (16, 12).

Chapter 3 Test

1. $\begin{cases} 2x - y = 0 \\ 4x - 5y = 12 \end{cases}$

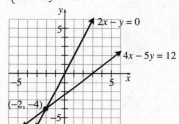

The graphs of the linear equations intersect at the point (−2, −4). Thus, the solution is (−2, −4).

2. $\begin{cases} 5x + 2y = -3 & (1) \\ y = 2x - 6 & (2) \end{cases}$

Because equation (2) is solved for y, we use substitution to solve the system.
Substituting $2x - 6$ for y in equation (1), we obtain

$5x + 2(2x - 6) = -3$
$5x + 4x - 12 = -3$
$9x = 9$
$x = 1$

Substituting 1 for x in equation (2), we obtain
$y = 2(1) - 6 = 2 - 6 = -4$.

The solution is the ordered pair (1, −4).

3. $\begin{cases} 9x + 3y = 1 & (1) \\ x - 2y = 4 & (2) \end{cases}$

Because the coefficient of x in equation (2) is 1, we use substitution to solve the system.

Equation (2) solved for x is $x = 2y + 4$.

Substituting $2y + 4$ for x in equation (1), we obtain

$$9(2y + 4) + 3y = 1$$
$$18y + 36 + 3y = 1$$
$$21y = -35$$
$$y = \frac{-35}{21} = -\frac{5}{3}$$

Substituting $-\dfrac{5}{3}$ for y in equation (1), we obtain

$$9x + 3\left(-\frac{5}{3}\right) = 1$$
$$9x - 5 = 1$$
$$9x = 6$$
$$x = \frac{6}{9} = \frac{2}{3}$$

The solution is the ordered pair $\left(\dfrac{2}{3}, -\dfrac{5}{3}\right)$.

4. $\begin{cases} 6x - 9y = 5 & (1) \\ 8x - 12y = 7 & (2) \end{cases}$

Because none of variables have a coefficient of 1, we use elimination to solve the system. Multiply both sides of equation (1) by 4, multiply both sides of equation (2) by −3, and add the results.
$$24x - 36y = 20$$
$$-24x + 36y = -21$$
$$\overline{\,0 = -1} \qquad \text{False}$$
The system has no solution. The solution set is $\varnothing$ or $\{\ \}$. The system is inconsistent.

5. $\begin{cases} 2x + y = -4 & (1) \\ \dfrac{1}{3}x + \dfrac{1}{2}y = 2 & (2) \end{cases}$

Because equation (2) contains fractions, we choose to use the elimination method to solve the system. Multiply both sides of equation (2) by -6 and add the result to equation (1).
$$-2x - 3y = -12$$
$$2x + y = -4$$
$$\overline{\,-2y = -16}$$
$$y = 8$$

Substituting 8 for y in equation (1), we obtain
$$2x + 8 = -4$$
$$2x = -12$$
$$x = -6$$
The solution is the ordered pair (−6, 8).

6. $\begin{cases} x - 2y + 3z = 1 & (1) \\ x + y - 3z = 7 & (2) \\ 3x - 4y + 5z = 7 & (3) \end{cases}$

Multiply both sides of equation (1) by −1 and add the result to equation (2).
$$-x + 2y - 3z = -1$$
$$x + y - 3z = 7$$
$$\overline{\,3y - 6z = 6} \qquad (4)$$

Multiply both sides of equation (1) by −3 and add the result to equation (3).
$$-3x + 6y - 9z = -3$$
$$3x - 4y + 5z = 7$$
$$\overline{\,2y - 4z = 4} \qquad (5)$$

Multiply both sides of equation (4) by 2, multiply both sides of equation (5) by −3, and add the results.
$$6y - 12z = 12$$
$$-6y + 12z = -12$$
$$\overline{\,0 = 0} \qquad \text{True}$$
Thus, the system is dependent and has an infinite number of solutions.

Solve equation (5) for y.
$$2y - 4z = 4$$
$$2y = 4z + 4$$
$$y = \frac{4z + 4}{2}$$
$$y = 2z + 2$$

Substituting $2z + 2$ for y in equation (1), we obtain

$$x - 2(2z + 2) + 3z = 1$$
$$x - 4z - 4 + 3z = 1$$
$$x - z - 4 = 1$$
$$x = z + 5$$

The solution to the system is $\{(x, y, z) | x = z + 5,$ $y = 2z + 2$, z is any real number$\}$.

7. $\begin{cases} 2x + 4y + 3z = 5 & (1) \\ 3x - y + 2z = 8 & (2) \\ x + y + 2z = 0 & (3) \end{cases}$

Multiply both sides of equation (3) by −2 and add the result to equation (1).
$$-2x - 2y - 4z = 0$$
$$2x + 4y + 3z = 5$$
$$\overline{\,2y - z = 5} \qquad (4)$$

     273

Multiply both sides of equation (3) by −3 and add the result to equation (2).

$$-3x - 3y - 6z = 0$$
$$3x - y + 2z = 8$$
$$-4y - 4z = 8 \qquad (5)$$

Multiply both sides of equation (4) by 2 and add the result to equation (5).

$$4y - 2z = 10$$
$$-4y - 4z = 8$$
$$-6z = 18$$
$$z = -3$$

Substituting −3 for z in equation (4), we obtain

$$2y - (-3) = 5$$
$$2y + 3 = 5$$
$$2y = 2$$
$$y = 1$$

Substituting −3 for z and 1 for y in equation (1), we obtain

$$2x + 4(1) + 3(-3) = 5$$
$$2x + 4 - 9 = 5$$
$$2x - 5 = 5$$
$$2x = 10$$
$$x = 5$$

The solution is the ordered triple (5, 1, −3).

8. a. $\begin{bmatrix} 1 & -3 & | & -2 \\ 2 & -4 & | & 8 \end{bmatrix}$ $\qquad (R_2 = -2r_1 + r_2)$

$$= \begin{bmatrix} 1 & -3 & | & -2 \\ -2(1)+2 & -2(-3)+(-4) & | & -2(-2)+8 \end{bmatrix}$$

$$= \begin{bmatrix} 1 & -3 & | & -2 \\ 0 & 2 & | & 12 \end{bmatrix}$$

b. $\begin{bmatrix} 1 & -3 & | & -2 \\ 0 & 2 & | & 12 \end{bmatrix}$ $\qquad \left(R_2 = \dfrac{1}{2}r_2\right)$

$$= \begin{bmatrix} 1 & -3 & | & -2 \\ \frac{1}{2}(0) & \frac{1}{2}(2) & | & \frac{1}{2}(12) \end{bmatrix}$$

$$= \begin{bmatrix} 1 & -3 & | & -2 \\ 0 & 1 & | & 6 \end{bmatrix}$$

9. a. $\begin{bmatrix} 1 & -2 & 1 & | & -2 \\ 3 & -5 & 2 & | & 1 \\ 0 & -4 & 5 & | & -32 \end{bmatrix}$ $\qquad (R_2 = -3r_1 + r_2)$

$$= \begin{bmatrix} 1 & -2 & 1 & | & -2 \\ -3(1)+3 & -3(-2)+(-5) & -3(1)+2 & | & -3(-2)+1 \\ 0 & -4 & 5 & | & -32 \end{bmatrix}$$

$$= \begin{bmatrix} 1 & -2 & 1 & | & -2 \\ 0 & 1 & -1 & | & 7 \\ 0 & -4 & 5 & | & -32 \end{bmatrix}$$

b. $\begin{bmatrix} 1 & -2 & 1 & | & -2 \\ 0 & 1 & -1 & | & 7 \\ 0 & -4 & 5 & | & -32 \end{bmatrix}$ $(R_3 = 4r_2 + r_3)$

$= \begin{bmatrix} 1 & -2 & 1 & | & -2 \\ 0 & 1 & -1 & | & 7 \\ 4(0)+0 & 4(1)+(-4) & 4(-1)+5 & | & 4(7)+(-32) \end{bmatrix}$

$= \begin{bmatrix} 1 & -2 & 1 & | & -2 \\ 0 & 1 & -1 & | & 7 \\ 0 & 0 & 1 & | & -4 \end{bmatrix}$

10. $\begin{cases} x - 5y = 2 & (1) \\ 2x + y = 4 & (2) \end{cases}$

Write the augmented matrix of the system and then put it in row echelon form.

$\begin{bmatrix} 1 & -5 & | & 2 \\ 2 & 1 & | & 4 \end{bmatrix}$ $(R_2 = -2r_1 + r_2)$

$= \begin{bmatrix} 1 & -5 & | & 2 \\ 0 & 11 & | & 0 \end{bmatrix}$ $\left(R_2 = \dfrac{1}{11}r_2\right)$

$= \begin{bmatrix} 1 & -5 & | & 2 \\ 0 & 1 & | & 0 \end{bmatrix}$

Write the system of equations that corresponds to the row-echelon matrix

$\begin{cases} x - 5y = 2 & (1) \\ y = 0 & (2) \end{cases}$

This system is consistent and independent.
Substituting 0 for y in equation (1), we obtain

$x - 5(0) = 2$

$\qquad x = 2$

The solution is the ordered pair (2, 0).

11. $\begin{cases} x + 2y + z = 3 & (1) \\ 4y + 3z = 5 & (2) \\ 2x + 3y = 1 & (3) \end{cases}$

Write the augmented matrix of the system and then put it in row echelon form.

$$\begin{bmatrix} 1 & 2 & 1 & | & 3 \\ 0 & 4 & 3 & | & 5 \\ 2 & 3 & 0 & | & 1 \end{bmatrix} \qquad (R_3 = -2r_1 + r_3)$$

$$= \begin{bmatrix} 1 & 2 & 1 & | & 3 \\ 0 & 4 & 3 & | & 5 \\ 0 & -1 & -2 & | & -5 \end{bmatrix} \qquad \left(R_2 = \frac{1}{4}r_2 \right)$$

$$= \begin{bmatrix} 1 & 2 & 1 & | & 3 \\ 0 & 1 & \frac{3}{4} & | & \frac{5}{4} \\ 0 & -1 & -2 & | & -5 \end{bmatrix} \qquad (R_3 = r_3 + r_2)$$

$$= \begin{bmatrix} 1 & 2 & 1 & | & 3 \\ 0 & 1 & \frac{3}{4} & | & \frac{5}{4} \\ 0 & 0 & -\frac{5}{4} & | & -\frac{15}{4} \end{bmatrix} \qquad \left(R_3 = -\frac{4}{5}r_3 \right)$$

$$= \begin{bmatrix} 1 & 2 & 1 & | & 3 \\ 0 & 1 & \frac{3}{4} & | & \frac{5}{4} \\ 0 & 0 & 1 & | & 3 \end{bmatrix}$$

Write the system of equations that corresponds to the row-echelon matrix

$$\begin{cases} x + 2y + z = 3 & (1) \\ y + \frac{3}{4}z = \frac{5}{4} & (2) \\ z = 3 & (3) \end{cases}$$

This system is consistent and independent.
Substituting 3 for z in equation (2), we obtain

$$y + \frac{3}{4}(3) = \frac{5}{4}$$
$$y + \frac{9}{4} = \frac{5}{4}$$
$$y = \frac{-4}{4} = -1$$

Substituting 3 for z and -1 for y in equation (1), we obtain

$$x + 2(-1) + 3 = 3$$
$$x - 2 + 3 = 3$$
$$x = 2$$

The solution is the ordered triple $(2, -1, 3)$.

12. $\begin{vmatrix} 3 & -5 \\ 4 & -8 \end{vmatrix} = 3(-8) - 4(-5)$
$$= -24 + 20$$
$$= -4$$

13. $\begin{vmatrix} 0 & 1 & 2 \\ 3 & 3 & -1 \\ -2 & 1 & 2 \end{vmatrix}$

$$= 0 \begin{vmatrix} 3 & -1 \\ 1 & 2 \end{vmatrix} - 1 \begin{vmatrix} 3 & -1 \\ -2 & 2 \end{vmatrix} + 2 \begin{vmatrix} 3 & 3 \\ -2 & 1 \end{vmatrix}$$

$$= 0 \left[3(2) - 1(-1) \right] - 1 \left[3(2) - (-2)(-1) \right]$$
$$\qquad + 2 \left[3(1) - (-2)(3) \right]$$

$$= 0(7) - 1(4) + 2(9)$$
$$= 0 - 4 + 18$$
$$= 14$$

14. $D = \begin{vmatrix} 1 & -1 \\ 5 & 3 \end{vmatrix} = 1(3) - 5(-1) = 3 + 5 = 8$

$$D_x = \begin{vmatrix} -2 & -1 \\ -8 & 3 \end{vmatrix}$$
$$= -2(3) - (-8)(-1) = -6 - 8 = -14$$

$$D_y = \begin{vmatrix} 1 & -2 \\ 5 & -8 \end{vmatrix} = 1(-8) - 5(-2) = -8 + 10 = 2$$

$$x = \frac{D_x}{D} = \frac{-14}{8} = -\frac{7}{4} \; ; \; y = \frac{D_y}{D} = \frac{2}{8} = \frac{1}{4}$$

Thus, the solution is the ordered pair $\left(-\frac{7}{4}, \frac{1}{4} \right)$.

15. $D = \begin{vmatrix} 1 & 1 & 1 \\ 1 & 1 & -2 \\ 4 & 2 & 3 \end{vmatrix}$

$$= 1 \begin{vmatrix} 1 & -2 \\ 2 & 3 \end{vmatrix} - 1 \begin{vmatrix} 1 & -2 \\ 4 & 3 \end{vmatrix} + 1 \begin{vmatrix} 1 & 1 \\ 4 & 2 \end{vmatrix}$$

$$= 1 \left[1(3) - 2(-2) \right] - 1 \left[1(3) - 4(-2) \right]$$
$$\qquad + 1 \left[1(2) - 4(1) \right]$$

$$= 1(7) - 1(11) + 1(-2)$$
$$= -6$$

$$D_x = \begin{vmatrix} -2 & 1 & 1 \\ 1 & 1 & -2 \\ -15 & 2 & 3 \end{vmatrix}$$

$$= -2 \begin{vmatrix} 1 & -2 \\ 2 & 3 \end{vmatrix} - 1 \begin{vmatrix} 1 & -2 \\ -15 & 3 \end{vmatrix} + 1 \begin{vmatrix} 1 & 1 \\ -15 & 2 \end{vmatrix}$$

$$= -2 \left[1(3) - 2(-2) \right] - 1 \left[1(3) - (-15)(-2) \right]$$
$$\qquad + 1 \left[1(2) - (-15)(1) \right]$$

$$= -2(7) - 1(-27) + 1(17)$$
$$= -14 + 27 + 17$$
$$= 30$$

$$D_y = \begin{vmatrix} 1 & -2 & 1 \\ 1 & 1 & -2 \\ 4 & -15 & 3 \end{vmatrix}$$

$$= 1\begin{vmatrix} 1 & -2 \\ -15 & 3 \end{vmatrix} - (-2)\begin{vmatrix} 1 & -2 \\ 4 & 3 \end{vmatrix} + 1\begin{vmatrix} 1 & 1 \\ 4 & -15 \end{vmatrix}$$

$$= 1\left[1(3)-(-15)(-2)\right] - (-2)\left[1(3)-4(-2)\right]$$
$$\qquad\qquad + 1\left[1(-15)-4(1)\right]$$

$$= 1(-27)-(-2)(11)+1(-19)$$

$$= -27 + 22 - 19$$

$$= -24$$

$$D_z = \begin{vmatrix} 1 & 1 & -2 \\ 1 & 1 & 1 \\ 4 & 2 & -15 \end{vmatrix}$$

$$= 1\begin{vmatrix} 1 & 1 \\ 2 & -15 \end{vmatrix} - 1\begin{vmatrix} 1 & 1 \\ 4 & -15 \end{vmatrix} + (-2)\begin{vmatrix} 1 & 1 \\ 4 & 2 \end{vmatrix}$$

$$= 1\left[1(-15)-2(1)\right] - 1\left[1(-15)-4(1)\right]$$
$$\qquad\qquad + (-2)\left[1(2)-4(1)\right]$$

$$= 1(-17)-1(-19)+(-2)(-2)$$

$$= -17 + 19 + 4$$

$$= 6$$

$$x = \frac{D_x}{D} = \frac{30}{-6} = -5; \quad y = \frac{D_y}{D} = \frac{-24}{-6} = 4;$$

$$z = \frac{D_z}{D} = \frac{6}{-6} = -1$$

Thus, the solution is the ordered triple $(-5, 4, -1)$.

16. $\begin{cases} 2x - y > -2 \\ x - 3y < 9 \end{cases}$

First, graph the inequality $2x - y > -2$. To do so, replace the inequality symbol with an equal sign to obtain $2x - y = -2$. Because the inequality is strict, graph $2x - y = -2$ $(y = 2x + 2)$ using a dashed line.

Test Point: $(0,0)$: $2(0) - 0 = 0 > -2$

Therefore, the half-plane containing (0, 0) is the solution set of $2x - y > -2$.

Second, graph the inequality $x - 3y < 9$. To do so, replace the inequality symbol with an equal sign to obtain $x - 3y = 9$. Because the inequality is strict, graph $x - 3y = 9$ $\left(y = \frac{1}{3}x - 3\right)$ using a dashed line.

Test Point: $(0,0)$: $0 - 3(0) = 0 < 9$

Therefore, the half-plane containing (0, 0) is the solution set of $x - 3y < 9$.

The overlapping shaded region (that is, the shaded region in the graph below) is the solution to the system of linear inequalities.

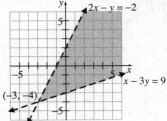

The corner point is $(-3, -4)$.

17. $\begin{cases} 3x + 2y \le 12 \\ x - 2y \ge -4 \\ x \ge 0 \\ y \ge 0 \end{cases}$

First, graph the inequality $3x + 2y \le 12$. To do so, replace the inequality symbol with an equal sign to obtain $3x + 2y = 12$. Because the inequality is non-strict, graph $3x + 2y = 12$ $\left(y = -\frac{3}{2}x + 6\right)$ using a solid line.

Test Point: $(0,0)$: $3(0) + 2(0) = 0 \le 12$

Therefore, the half-plane containing (0, 0) is the solution set of $3x + 2y \le 12$.

Next, graph the inequality $x - 2y \ge 4$. To do so, replace the inequality symbol with an equal sign to obtain $x - 2y = -4$. Because the inequality is non-strict, graph $x - 2y = -4$ $\left(y = \frac{1}{2}x + 2\right)$ using a solid line.

Test Point: $(0,0)$: $0 - 2(0) = 0 \ge -4$

Therefore, the half-plane containing (0, 0) is the solution set of $x - 2y \ge -4$.

The two inequalities $x \ge 0$ and $y \ge 0$ require that the graph be in quadrant I.

The overlapping shaded region (that is, the shaded region in the graph below) is the solution to the system of linear inequalities.

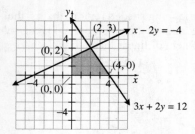

The graph is bounded. The four corner points of the bounded region are (0, 0), (4, 0), (2, 3), and (0, 2).

18. Let x represent the number of twenty-five-ton bins and y represent the number of twenty-ton bins.

$$\begin{cases} x+y=50 & (1) \\ 25x+20y=1160 & (2) \end{cases}$$

Equation (1) solved for y is $y=-x+50$.

Substituting $-x+50$ for y in equation (2), we obtain

$$25x+20(-x+50)=1160$$
$$25x-20x+1000=1160$$
$$5x+1000=1160$$
$$5x=160$$
$$x=32$$

Substituting 32 for x in equation (1), we obtain
$$32+y=50$$
$$y=18$$

The warehouse contains 32 twenty-five-ton bins and 18 twenty-ton bins.

19. Because the sum of the measures of the three angles in a triangle must equal 180°, we obtain the following system.

$$\begin{cases} x+y+z=180 & (1) \\ z=y+10 & (2) \\ 5x=y+z & (3) \end{cases}$$

Writing the equations in the system in standard form, we obtain

$$\begin{cases} x+y+z=180 & (1) \\ y-z=-10 & (2) \\ 5x-y-z=0 & (3) \end{cases}$$

Adding equations (1) and (2), we obtain
$$x+y+z=180$$
$$y-z=-10$$
$$x+2y\ \ \ =170 \qquad (4)$$

Adding equations (1) and (3), we obtain.
$$x+y+z=180$$
$$5x-y-z=0$$
$$6x\ \ \ \ \ \ \ \ =180$$
$$x=30$$

Substituting 30 for x in equation (4), we obtain
$$30+2y=170$$
$$2y=140$$
$$y=70$$

Substituting 30 for x and 70 for y and in equation (1), we obtain
$$30+70+z=180$$
$$100+z=180$$
$$z=80$$
Thus, angle $x=30°$, angle $y=70°$, and angle $z=80°$.

20. a. Margaret has a maximum of $180 to spend, so $12x+18y\le180$. She will buy no more than 13 items, so $x+y\le13$. Finally, Margaret cannot buy a negative number of blouses or sweaters, so we have the non-negativity constraints $x\ge0$ and $y\ge0$. Thus, we have the following system:

$$\begin{cases} 12x+18y\le180 \\ x+y\le13 \\ x\ge0 \\ y\ge0 \end{cases}$$

b. First, graph the inequality $12x+18y\le180$. To do so, replace the inequality symbol with an equal sign to obtain $12x+18y=180$. Because the inequality is non-strict, graph

$$12x+18y=180 \ \left(y=-\frac{2}{3}x+10\right) \text{ using a}$$

solid line.

Test Point: $(0,0)$: $12(0)+18(0)=0\le180$

Thus, the solution set of $12x+18y\le180$ is the half-plane containing (0, 0).

Second, graph the inequality $x+y\le13$. To do so, replace the inequality symbol with an equal sign to obtain $x+y=13$. Because the inequality is non-strict, graph $x+y=13$

$(y=-x+13)$ using a solid line.

Test Point: $(0,0)$: $0+0=0\le13$

Thus, the solution set of $x + y \le 13$ is the half-plane containing (0, 0).

The two inequalities $x \ge 0$ and $y \ge 0$ require that the graph be in quadrant I.

The overlapping shaded region (that is, the shaded region in the graph below) is the solution to the system of linear inequalities.

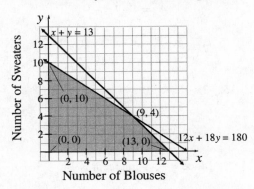

The four corner points are (0, 0), (13, 0), (9, 4), and (0, 10).

Cumulative Review Chapters R–3

1. **a.** $\{11\}$

   b. $\{-13, 0, 11\}$

   c. $\left\{-13, -\dfrac{7}{8}, 0, 2.7, 11\right\}$

   d. $\left\{\dfrac{\pi}{2}, 4\sqrt{2}\right\}$

   e. $\left\{-13, -\dfrac{7}{8}, 0, \dfrac{\pi}{2}, 2.7, 4\sqrt{2}, 11\right\}$

2. Let x = the number.
   $2(x + 4) = 3x - 6$

3. $\dfrac{3 - 7(5)}{4} = \dfrac{3 - 35}{4} = \dfrac{-32}{4} = -8$

4. $\dfrac{x^2 - 5x + 4}{x - 1}$ when $x = 7$

   $\dfrac{(7)^2 - 5(7) + 4}{(7) - 1} = \dfrac{49 - 35 + 4}{6}$

   $= \dfrac{14 + 4}{6}$

   $= \dfrac{18}{6} = 3$

5. $x(x - 4) - 7(x + 5) + 2(x^2 + 4)$
   $= x^2 - 4x - 7x - 35 + 2x^2 + 8$
   $= 3x^2 - 11x - 27$

6. $3|x + 7| - 4 = 20$
   $3|x + 7| = 24$
   $|x + 7| = 8$
   $x + 7 = 8$ or $x + 7 = -8$
   $x = 1$ $x = -15$
   The solution set is $\{-15, 1\}$.

7. $\dfrac{x + 1}{5} \ge \dfrac{5x + 29}{10}$

   $0 \ge \dfrac{5x + 29}{10} - \dfrac{x + 1}{5}$

   $10\left(\dfrac{5x + 29}{10} - \dfrac{x + 1}{5}\right) \le 10 \cdot 0$

   $5x + 29 - 2(x + 1) \le 0$
   $5x + 29 - 2x - 2 \le 0$
   $3x + 27 \le 0$
   $3x \le -27$
   $x \le -9$
   The solution set is $\{x | x \le -9\}$, or in interval notation $(-\infty, -9]$.

8. $|2x - 7| > 3$
   $2x - 7 > 3$ or $2x - 7 < -3$
   $2x > 10$ $2x < 4$
   $x > 5$ $x < 2$
   The solution set is $\{x | x < 2 \text{ or } x > 5\}$, or in interval notation $(-\infty, 2] \cup [5, \infty)$.

9. **a.** $A \cap B = \{0, 6, 12\}$

   b. $A \cup B = \{0, 2, 3, 4, 6, 8, 9, 10, 12, 15\}$

279

10.

x	$y = 2|x| - 5$	(x, y)
-3	$y = 2|-3| - 5 = 1$	$(-3, 1)$
-1	$y = 2|-1| - 5 = -3$	$(-1, -3)$
0	$y = 2|0| - 5 = -5$	$(0, -5)$
1	$y = 2|1| - 5 = -3$	$(1, -3)$
3	$y = 2|3| - 5 = 1$	$(3, 1)$

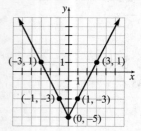

11. $f(x) = \dfrac{2}{3}x - 4$

$\dfrac{2}{3}x - 4 = 0$

$\dfrac{2}{3}x = 4$

$2x = 12$

$x = 6$

The zero of f is 6.

12. (a) and (b) are functions because they pass the vertical line test. (c) is not a function because it fails the vertical line test.

13. $f(x) = \dfrac{2x + 7}{2x - 1}$

The domain must exclude any values of x that cause us to divide by zero. Therefore, we must find the zero(s) of the denominator.

$2x - 1 = 0$

$2x = 1$

$x = \dfrac{1}{2}$

The domain of f must exclude $\dfrac{1}{2}$, so the domain is $\left\{ x \mid x \neq \dfrac{1}{2} \right\}$.

14. $g(x) = 3x^2 + 4x - 7$

a. $g(-4) = 3(-4)^2 + 4(-4) - 7$
$= 3 \cdot 16 - 16 - 7$
$= 48 - 16 - 7$
$= 25$

b. $g\left(\dfrac{2}{3}\right) = 3\left(\dfrac{2}{3}\right)^2 + 4\left(\dfrac{2}{3}\right) - 7$
$= 3 \cdot \dfrac{4}{9} + \dfrac{8}{3} - 7$
$= \dfrac{4}{3} + \dfrac{8}{3} - 7$
$= \dfrac{12}{3} - 7$
$= 4 - 7 = -3$

c. $g(-x) = 3(-x)^2 + 4(-x) - 7$
$= 3x^2 - 4x - 7$

15. $2x - 5y = 20$

$5y = 2x - 20$

$y = \dfrac{2}{5}x - 4$

Slope $= \dfrac{2}{5}$, y-intercept is -4.

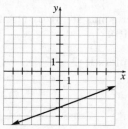

16. $\begin{cases} 3x - 9y = 5 \\ 5x - 15y = -27 \end{cases}$

$3x - 9y = 5$

$9y = 3x - 5$

$y = \dfrac{1}{3}x - \dfrac{5}{9}$

$m_1 = \dfrac{1}{3}$

$5x - 15y = -27$

$15y = 5x + 27$

$y = \dfrac{1}{3}x + \dfrac{9}{5}$

$m_2 = \dfrac{1}{3}$

The lines are parallel.

17. To begin, find the slope of the line using the two given points.

$m = \dfrac{-11 - 14}{9 - (-6)} = \dfrac{-25}{15} = -\dfrac{5}{3}$

Now use either point to find the equation of the line. We will use the point $(-6, 14)$.

$y - y_1 = m(x - x_1)$

$y - 14 = -\dfrac{5}{3}(x - (-6))$

$y - 14 = -\dfrac{5}{3}(x + 6)$

$y - 14 = -\dfrac{5}{3}x - 10$

$y = -\dfrac{5}{3}x + 4$

Or to write the line in standard form

$y = -\dfrac{5}{3}x + 4$

$3y = 3\left(-\dfrac{5}{3}x + 4\right)$

$3y = -5x + 12$

$5x + 3y = 12$

The equation of the line passing through the points (–6, 14) and (9, –11) is

$y = -\dfrac{5}{3}x + 4 \quad \text{or} \quad 5x + 3y = 12$

18. Let w = the weight of the quarters and q = the number of quarters. Since weight varies directly with quarters, we have

$w = k \cdot q$

Substituting 281 for w and 50 for q, we can solve for k.

$281 = k \cdot 50$

$k = \dfrac{281}{50} = 5.62$

Replace k with 5.62 in the model and find the weight of 35 quarters by letting $q = 35$.

$w = 5.62q$

$w = 5.62(35) = 196.7$

35 quarters weigh 196.7 grams.

19. **a.** The independent value is x, the number of years after 1930. The dependent variable is P, the population of New Orleans.

b. Since the model is only valid for the years 1930 ($x = 0$) and 2010 ($x = 80$), the domain is $\{x \mid 0 \le x \le 80\}$, or [0, 80].

c. $P(x) = -0.123x^2 + 8.543x + 447.873$

$P(60) = -0.123(60)^2 + 8.543(60) + 447.873$

$= 517.653$

This means the population of New Orleans in 1930 + 60 = 1990 was 517,653.

d. $x = 2010 - 1930 = 80$

$P(80) = -0.123(80)^2 + 8.543(80) + 447.873$

$= 344.113$

The population of New Orleans in 2010 was 344,113.

20. **a.** $C(x) = 4.95 + 0.08x$

b. The number of minutes cannot be negative. The implied domain is $\{0 \mid x \ge 0\}$.

c. $C(645) = 4.95 + 0.08(645)$

$= 4.95 + 51.6$

$= 56.55$

The cost for 645 minutes of long distance calls during one month is $56.55.

d. $130.95 = 4.95 + 0.08x$

$126 = 0.08x$

$x = 1575$

For $130.95, 1575 minutes of long distance calls can be made.

21. $\begin{cases} y = \dfrac{1}{2}x + 9 & (1) \\ 5x + 4y = -20 & (2) \end{cases}$

Begin by substituting $\dfrac{1}{2}x + 9$ for y in equation (2).

$5x + 4\left(\dfrac{1}{2}x + 9\right) = -20$

$5x + 2x + 36 = -20$

$7x = -56$

$x = -8$

Now substitute –8 for x in equation (1).

$y = \dfrac{1}{2}(-8) + 9$

$y = -4 + 9$

$y = 5$

The solution to the system of equations is the ordered pair (–8, 5).

22. Write the augmented matrix of the system and then put it in row echelon form.

$\begin{bmatrix} 6 & -4 & 3 & | & 3 \\ 3 & 10 & 1 & | & -4 \\ 9 & 2 & 4 & | & 1 \end{bmatrix} \quad \left(R_1 = \dfrac{1}{6}r_1\right)$

$$= \begin{bmatrix} 1 & -\dfrac{2}{3} & \dfrac{1}{2} & \bigg| & \dfrac{1}{2} \\ 3 & 10 & 1 & \bigg| & -4 \\ 9 & 2 & 4 & \bigg| & 1 \end{bmatrix} \quad \begin{pmatrix} R_2 = -3r_1 + r_2 \\ R_3 = -9r_1 + r_3 \end{pmatrix}$$

$$= \begin{bmatrix} 1 & -\dfrac{2}{3} & \dfrac{1}{2} & \bigg| & \dfrac{1}{2} \\ 0 & 12 & -\dfrac{1}{2} & \bigg| & -\dfrac{11}{12} \\ 0 & 8 & -\dfrac{1}{2} & \bigg| & -\dfrac{7}{2} \end{bmatrix} \quad \left(R_2 = \dfrac{1}{12}r_2 \right)$$

$$= \begin{bmatrix} 1 & -\dfrac{2}{3} & \dfrac{1}{2} & \bigg| & \dfrac{1}{2} \\ 0 & 1 & -\dfrac{1}{24} & \bigg| & -\dfrac{11}{24} \\ 0 & 8 & -\dfrac{1}{2} & \bigg| & -\dfrac{7}{2} \end{bmatrix} \quad (R_2 = -8r_2 + r_2)$$

$$= \begin{bmatrix} 1 & -\dfrac{2}{3} & \dfrac{1}{2} & \bigg| & \dfrac{1}{2} \\ 0 & 1 & -\dfrac{1}{24} & \bigg| & -\dfrac{11}{24} \\ 0 & 0 & -\dfrac{1}{6} & \bigg| & \dfrac{1}{6} \end{bmatrix} \quad (R_3 = -6r_3)$$

$$= \begin{bmatrix} 1 & -\dfrac{2}{3} & \dfrac{1}{2} & \bigg| & \dfrac{1}{2} \\ 0 & 1 & -\dfrac{1}{24} & \bigg| & -\dfrac{11}{24} \\ 0 & 0 & 1 & \bigg| & -1 \end{bmatrix}$$

Write the system of equations that corresponds to the row-echelon matrix

$$\begin{cases} x - \dfrac{2}{3}y + \dfrac{1}{2}z = \dfrac{1}{2} & (1) \\ y - \dfrac{1}{24}z = -\dfrac{11}{24} & (2) \\ z = -1 & (3) \end{cases}$$

Substituting −1 for z in equation (2), we obtain

$$y - \dfrac{1}{24}(-1) = -\dfrac{11}{24}$$

$$y + \dfrac{1}{24} = -\dfrac{11}{24}$$

$$y = -\dfrac{12}{24} = -\dfrac{1}{2}$$

Substituting $-\dfrac{1}{2}$ for y and −1 for z in equation (1), we obtain

$$x - \dfrac{2}{3}\left(-\dfrac{1}{2}\right) + \dfrac{1}{2}(-1) = \dfrac{1}{2}$$

$$x + \dfrac{1}{3} - \dfrac{1}{2} = \dfrac{1}{2}$$

$$x = \dfrac{2}{3}$$

The solution is the ordered triple $\left(\dfrac{2}{3}, -\dfrac{1}{2}, -1 \right)$.

23. $\begin{vmatrix} -4 & 2 \\ -5 & 3 \end{vmatrix} = (-4)(3) - (2)(-5)$

$$= -12 + 10$$

$$= -2$$

24. $\begin{vmatrix} 2 & 3 & 0 \\ -1 & 4 & 2 \\ 2 & -2 & -3 \end{vmatrix}$

$$= 2\begin{vmatrix} 4 & 2 \\ -2 & -3 \end{vmatrix} - 3\begin{vmatrix} -1 & 2 \\ 2 & -3 \end{vmatrix} + 0\begin{vmatrix} -1 & 4 \\ 2 & -2 \end{vmatrix}$$

$$= 2[4(-3) - 2(-2)] - 3[(-1)(-3) - 2(2)]$$

$$= 2[-12 + 4] - 3[3 - 4]$$

$$= 2[-8] - 3[-1]$$

$$= -16 + 3$$

$$= -13$$

25. $\begin{cases} x + 3y < 12 \\ x \geq 3 \end{cases}$

First, graph the inequality $x + 3y < 12$. To do so, replace the inequality symbol with an equal sign to obtain $x + 3y = 12$. Because the inequality is strict, graph $x + 3y = 12$ $\left(y = -\dfrac{1}{3}x + 4 \right)$ using a dashed line.

Test Point: $(0,0)$: $0 + 3 \cdot 0 = 0 < 12$

Therefore, the half-plane containing (0, 0) is the solution set of $x + 3y < 12$.

Next, graph the inequality $x \geq 3$. To do so, replace the inequality symbol with an equal sign to obtain $x = 3$. Because the inequality is non-strict, graph the horizontal line $x = 3$ using a solid line.

Test Point: $(0,0)$: $0 \not\geq 3$

Therefore, the half-plane not containing (0, 0) is the solution set of $x \geq 3$.

The overlapping shaded region (that is, the shaded region in the graph below) is the solution to the system of linear inequalities.

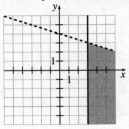

Chapter 4

1. In the notation a^n we call a the <u>base</u> and n the <u>power</u> or <u>exponent.</u>

2. If a is a real number, and m and n are positive integers, then $a^m \cdot a^n = \underline{a^{m+n}}$.

3. $5^2 \cdot 5 = 5^{2+1} = 5^3 = 125$

4. $(-3)^2 \cdot (-3)^3 = (-3)^{2+3} = (-3)^5 = -243$

5. $y^4 \cdot y^3 = y^{4+3} = y^7$

6. $\left(5x^2\right) \cdot \left(-2x^5\right) = 5 \cdot (-2) \cdot x^2 \cdot x^5$
$$= -10x^{2+5}$$
$$= -10x^7$$

7. $\left(6y^3\right) \cdot \left(-y^2\right) = 6 \cdot (-1) \cdot y^3 \cdot y^2$
$$= -6y^{3+2}$$
$$= -6y^5$$

8. True

9. $\dfrac{5^6}{5^4} = 5^{6-4} = 5^2 = 25$

10. $\dfrac{y^8}{y^6} = y^{8-6} = y^2$

11. $\dfrac{16a^6}{10a^5} = \dfrac{8 \cdot 2}{5 \cdot 2} \cdot \dfrac{a^6}{a^5} = \dfrac{8}{5} \cdot a^{6-5} = \dfrac{8}{5}a$

12. $\dfrac{-24b^5}{16b^3} = \dfrac{-3 \cdot 8}{2 \cdot 8} \cdot \dfrac{b^5}{b^3} = \dfrac{-3}{2} \cdot b^{5-3} = -\dfrac{3}{2}b^2$

13. $a^0 = \underline{1}$, provided $a \neq \underline{0}$.

14. $a^{-n} = \dfrac{1}{\underline{a^n}}$, provided $a \neq \underline{0}$.

15. $5^{-3} = \dfrac{1}{5^3} = \dfrac{1}{125}$

16. $5z^{-7} = 5 \cdot \dfrac{1}{z^7} = \dfrac{5}{z^7}$

17. $\dfrac{1}{x^{-4}} = x^4$

18. $\dfrac{5}{y^{-3}} = 5y^3$

19. $-4^0 = -1 \cdot 4^0 = -1 \cdot 1 = -1$

20. $(-10)^0 = 1$

21. $\left(\dfrac{4}{3}\right)^{-2} = \left(\dfrac{3}{4}\right)^2 = \dfrac{3}{4} \cdot \dfrac{3}{4} = \dfrac{9}{16}$

22. $\left(-\dfrac{1}{4}\right)^{-3} = (-4)^3 = (-4) \cdot (-4) \cdot (-4) = -64$

23. $\left(\dfrac{3}{x}\right)^{-2} = \left(\dfrac{x}{3}\right)^2 = \dfrac{x}{3} \cdot \dfrac{x}{3} = \dfrac{x^{1+1}}{9} = \dfrac{x^2}{9}$

24. $\dfrac{5}{2^{-2}} = 5 \cdot 2^2 = 5 \cdot 4 = 20$

25. $6^3 \cdot 6^{-5} = 6^{3+(-5)} = 6^{-2} = \dfrac{1}{6^2} = \dfrac{1}{36}$

26. $\dfrac{10^{-3}}{10^{-5}} = 10^{-3-(-5)} = 10^{-3+5} = 10^2 = 100$

27. $\left(4x^2y^3\right) \cdot \left(5xy^{-4}\right) = 4 \cdot 5 \cdot x^2 \cdot x \cdot y^3 \cdot y^{-4}$
$$= 20x^{2+1}y^{3+(-4)}$$
$$= 20x^3y^{-1}$$
$$= \dfrac{20x^3}{y}$$

28. $\left(\dfrac{3}{4}a^3b\right)\cdot\left(\dfrac{8}{9}a^{-2}b^3\right)=\dfrac{3}{4}\cdot\dfrac{8}{9}\cdot a^3\cdot a^{-2}\cdot b\cdot b^3$

$\qquad\qquad\qquad\qquad = \dfrac{3\cdot4\cdot2}{4\cdot3\cdot3}\cdot a^{3+(-2)}b^{1+3}$

$\qquad\qquad\qquad\qquad = \dfrac{2}{3}a^1b^4$

$\qquad\qquad\qquad\qquad = \dfrac{2}{3}ab^4$

29. $\dfrac{-24b^5}{16b^{-3}}=\dfrac{-24}{16}\cdot\dfrac{b^5}{b^{-3}}$

$\qquad\qquad = \dfrac{-3\cdot8}{2\cdot8}\cdot b^{5-(-3)}$

$\qquad\qquad = \dfrac{-3}{2}\cdot b^{5+3}$

$\qquad\qquad = -\dfrac{3}{2}b^8$

30. $\dfrac{50s^2t}{15s^5t^{-4}}=\dfrac{50}{15}\cdot\dfrac{s^2}{s^5}\cdot\dfrac{t}{t^{-4}}$

$\qquad\qquad = \dfrac{10\cdot5}{3\cdot5}\cdot s^{2-5}\cdot t^{1-(-4)}$

$\qquad\qquad = \dfrac{10}{3}s^{-3}t^{1+4}$

$\qquad\qquad = \dfrac{10t^5}{3s^3}$

31. $\left(2^2\right)^3=2^{2\cdot3}=2^6=64$

32. $\left(5^8\right)^0=5^{8\cdot0}=5^0=1$

33. $\left[(-4)^3\right]^2=(-4)^{3\cdot2}=(-4)^6=4096$

34. $\left(a^3\right)^5=a^{3\cdot5}=a^{15}$

35. $\left(z^3\right)^{-6}=z^{3\cdot(-6)}=z^{-18}=\dfrac{1}{z^{18}}$

36. $\left(s^{-3}\right)^{-7}=s^{(-3)\cdot(-7)}=s^{21}$

37. $(5y)^3=5^3\cdot y^3=125y^3$

38. $(6y)^0=6^0\cdot y^0=1\cdot1=1$

39. $\left(3x^2\right)^4=3^4\cdot\left(x^2\right)^4$

$\qquad\qquad = 81x^{2\cdot4}$

$\qquad\qquad = 81x^8$

40. $\left(4a^3\right)^{-2}=4^{-2}\cdot\left(a^3\right)^{-2}$

$\qquad\qquad = \dfrac{1}{4^2}\cdot a^{3\cdot(-2)}$

$\qquad\qquad = \dfrac{1}{16}a^{-6}$

$\qquad\qquad = \dfrac{1}{16a^6}$

41. $\left(\dfrac{z}{3}\right)^4=\dfrac{z^4}{3^4}=\dfrac{z^4}{81}$

42. $\left(\dfrac{x}{2}\right)^{-5}=\left(\dfrac{2}{x}\right)^5=\dfrac{2^5}{x^5}=\dfrac{32}{x^5}$

43. $\left(\dfrac{x^2}{y^3}\right)^4=\dfrac{\left(x^2\right)^4}{\left(y^3\right)^4}=\dfrac{x^{2\cdot4}}{y^{3\cdot4}}=\dfrac{x^8}{y^{12}}$

44. $\left(\dfrac{3a^{-2}}{b^4}\right)^3=\dfrac{3^3\cdot\left(a^{-2}\right)^3}{\left(b^4\right)^3}$

$\qquad\qquad = \dfrac{27a^{(-2)\cdot3}}{b^{4\cdot3}}$

$\qquad\qquad = \dfrac{27a^{-6}}{b^{12}}$

$\qquad\qquad = \dfrac{27}{a^6b^{12}}$

45. $\dfrac{\left(3x^2y\right)^2}{12xy^{-2}}=\dfrac{3^2\cdot\left(x^2\right)^2\cdot y^2}{12xy^{-2}}$

$\qquad\qquad = \dfrac{9x^4y^2}{12xy^{-2}}$

$\qquad\qquad = \dfrac{3\cdot3}{3\cdot4}\cdot x^{4-1}\cdot y^{2-(-2)}$

$\qquad\qquad = \dfrac{3x^3y^4}{4}$

46. $\left(3ab^3\right)^3 \cdot \left(6a^2b^2\right)^{-2}$

$= \left(3^3 \cdot a^3 \cdot \left(b^3\right)^3\right) \cdot \left(6^{-2} \cdot \left(a^2\right)^{-2} \cdot \left(b^2\right)^{-2}\right)$

$= \left(27a^3b^9\right) \cdot \left(\dfrac{1}{36}a^{-4}b^{-4}\right)$

$= \dfrac{27}{36}a^{3+(-4)}b^{9+(-4)}$

$= \dfrac{3 \cdot 9}{4 \cdot 9}a^{-1}b^5$

$= \dfrac{3b^5}{4a}$

47. $\left(\dfrac{2x^2y^{-1}}{x^{-2}y^2}\right)^2 \cdot \left(\dfrac{4x^3y^2}{xy^{-2}}\right)^{-1}$

$= \left(2x^{2-(-2)}y^{-1-2}\right)^2 \cdot \left(4x^{3-1}y^{2-(-2)}\right)^{-1}$

$= \left(2x^4y^{-3}\right)^2 \cdot \left(4x^2y^4\right)^{-1}$

$= \left(2^2 \cdot \left(x^4\right)^2 \cdot \left(y^{-3}\right)^2\right) \cdot \left(4^{-1} \cdot \left(x^2\right)^{-1} \cdot \left(y^4\right)^{-1}\right)$

$= \left(4x^8y^{-6}\right) \cdot \left(\dfrac{1}{4}x^{-2}y^{-4}\right)$

$= \dfrac{4}{4}x^{8+(-2)}y^{-6+(-4)}$

$= x^6y^{-10}$

$= \dfrac{x^6}{y^{10}}$

48. Because 532 is greater than 1, we move the decimal point to the left. We move it two places to the left until it is between the 5 and the 3.

$532 = 5.32 \times 10^2$

49. Because the absolute value of $-1,230,000$ is greater than 1, we move the decimal point to the left. We move it six places to the left until it is between the 1 and the 2.

$-1,230,000 = -1.23 \times 10^6$

50. Because the absolute value of 0.034 is less than 1, we move the decimal point to the right. We move it two places to the right until it is between the 3 and the 4.

$0.034 = 3.4 \times 10^{-2}$

51. Because the absolute value of -0.0000845 is less than 1, we move the decimal point to the right. We move it five places to the right until it is between the 8 and the 4.

$-0.0000845 = -8.45 \times 10^{-5}$

52. The exponent on 10 is 2 so we move the decimal point 2 places to the right.

$5 \times 10^2 = 5.00 \times 10^2 = 500$

53. The exponent on 10 is 5 so we move the decimal point 5 places to the right.

$9.1 \times 10^5 = 9.10000 \times 10^5 = 910,000$

54. The exponent on 10 is -4 so we move the decimal point 4 places to the left.

$1.8 \times 10^{-4} = 00001.8 \times 10^{-4} = 0.00018$

55. The exponent on 10 is -6 so we move the decimal point 6 places to the left.

$1 \times 10^{-6} = 0000001 \times 10^{-6} = 0.000001$

56. $\left(3 \times 10^3\right) \cdot \left(2 \times 10^5\right) = (3 \cdot 2) \times \left(10^3 \cdot 10^5\right)$

$\qquad\qquad\qquad\qquad = 6 \times 10^8$

57. $\left(2 \times 10^{-4}\right) \cdot \left(4 \times 10^{-7}\right) = (2 \cdot 4) \times \left(10^{-4} \cdot 10^{-7}\right)$

$\qquad\qquad\qquad\qquad\quad = 8 \times 10^{-11}$

58. $\left(6 \times 10^{-5}\right) \cdot \left(4 \times 10^8\right) = (6 \cdot 4) \times \left(10^{-5} \cdot 10^8\right)$

$\qquad\qquad\qquad\qquad = 24 \times 10^3$

$\qquad\qquad\qquad\qquad = \left(2.4 \times 10^1\right) \times 10^3$

$\qquad\qquad\qquad\qquad = 2.4 \times 10^4$

59. $\dfrac{6 \times 10^8}{3 \times 10^6} = \dfrac{6}{3} \times \dfrac{10^8}{10^6} = 2 \times 10^2$

60. $\dfrac{6.8 \times 10^{-8}}{3.4 \times 10^{-5}} = \dfrac{6.8}{3.4} \times \dfrac{10^{-8}}{10^{-5}}$

$\qquad\qquad\quad = 2 \times 10^{-8-(-5)}$

$\qquad\qquad\quad = 2 \times 10^{-3}$

61. $\dfrac{4.8 \times 10^7}{9.6 \times 10^3} = \dfrac{4.8}{9.6} \times \dfrac{10^7}{10^3}$

$\qquad = 0.5 \times 10^4$

$\qquad = \left(5 \times 10^{-1}\right) \times 10^4$

$\qquad = 5 \times 10^3$

62. $\dfrac{3 \times 10^{-5}}{8 \times 10^7} = \dfrac{3}{8} \times \dfrac{10^{-5}}{10^7}$

$\qquad = 0.375 \times 10^{-12}$

$\qquad = \left(3.75 \times 10^{-1}\right) \times 10^{-12}$

$\qquad = 3.75 \times 10^{-13}$

63. $(8,000,000) \cdot (30,000) = \left(8 \times 10^6\right) \cdot \left(3 \times 10^4\right)$

$\qquad = (8 \cdot 3) \times \left(10^6 \cdot 10^4\right)$

$\qquad = 24 \times 10^{10}$

$\qquad = \left(2.4 \times 10^1\right) \times 10^{10}$

$\qquad = 2.4 \times 10^{11}$

$\qquad = 240,000,000,000$

64. $\dfrac{0.000000012}{0.000004} = \dfrac{1.2 \times 10^{-8}}{4 \times 10^{-6}}$

$\qquad = \dfrac{1.2}{4} \times \dfrac{10^{-8}}{10^{-6}}$

$\qquad = 0.3 \times 10^{-2}$

$\qquad = \left(3 \times 10^{-1}\right) \times 10^{-2}$

$\qquad = 3 \times 10^{-3}$

$\qquad = 0.003$

65. $(25,000,000) \cdot (0.00003)$

$\qquad = \left(2.5 \times 10^7\right) \cdot \left(3 \times 10^{-5}\right)$

$\qquad = (2.5 \cdot 3) \times \left(10^7 \cdot 10^{-5}\right)$

$\qquad = 7.5 \times 10^2$

$\qquad = 750$

66. $\dfrac{0.000039}{13,000,000} = \dfrac{3.9 \times 10^{-5}}{1.3 \times 10^7}$

$\qquad = \dfrac{3.9}{1.3} \times \dfrac{10^{-5}}{10^7}$

$\qquad = 3 \times 10^{-12}$

$\qquad = 0.000000000003$

Getting Ready for Chapter 4 Exercises

67. $-5^2 = -1 \cdot 25 = -25$

69. $-5^{-2} = -\dfrac{1}{5^2} = -\dfrac{1}{25}$

71. $8^2 \cdot 8^{-2} = 8^{2+(-2)} = 8^0 = 1$

73. $\left(\dfrac{4}{9}\right)^{-2} = \left(\dfrac{9}{4}\right)^2 = \dfrac{9^2}{4^2} = \dfrac{81}{4}$

75. $(-3)^2 \cdot (-3)^{-5} = (-3)^{2+(-5)}$

$\qquad = (-3)^{-3}$

$\qquad = \dfrac{1}{(-3)^3}$

$\qquad = -\dfrac{1}{27}$

77. $\dfrac{(-4)^2}{(-4)^{-1}} = (-4)^{2-(-1)}$

$\qquad = (-4)^3$

$\qquad = -64$

79. $\dfrac{2^3 \cdot 3^{-2}}{2^{-2} \cdot 3^{-4}} = 2^{3-(-2)} \cdot 3^{-2-(-4)}$

$\qquad = 2^5 \cdot 3^2$

$\qquad = 32 \cdot 9$

$\qquad = 288$

81. $(6x)^3 (6x)^{-3} = (6x)^0 = 1$ (assuming $x \neq 0$)

83. $\left(2s^{-2}t^4\right)\left(-5s^2 t\right) = 2(-5)s^{-2+2}t^{4+1}$

$\qquad = -10s^0 t^5$

$\qquad = -10t^5$

85. $\left(\dfrac{1}{4}xy\right) \cdot \left(20xy^{-2}\right) = \dfrac{1}{4} \cdot 20 \cdot x^{1+1}y^{1+(-2)}$

$\qquad = 5x^2 y^{-1}$

$\qquad = 5x^2 \cdot \dfrac{1}{y}$

$\qquad = \dfrac{5x^2}{y}$

87. $\dfrac{36x^7y^3}{9x^5y^2} = \dfrac{36}{9} \cdot x^{7-5}y^{3-2}$

$\qquad = 4x^2y$

89. $\dfrac{21a^2b}{14a^3b^{-2}} = \dfrac{21}{14} \cdot a^{2-3}b^{1-(-2)}$

$\qquad = \dfrac{3}{2}a^{-1}b^3$

$\qquad = \dfrac{3b^3}{2a}$

91. $\left(x^{-2}\right)^4 = x^{-2\cdot 4}$

$\qquad = x^{-8}$

$\qquad = \dfrac{1}{x^8}$

93. $\left(3x^2y\right)^3 = 3^3\left(x^2\right)^3 y^3$

$\qquad = 27x^{2\cdot 3}y^3$

$\qquad = 27x^6y^3$

95. $\left(\dfrac{z}{4}\right)^{-3} = \dfrac{z^{-3}}{4^{-3}} = \dfrac{4^3}{z^3} = \dfrac{64}{z^3}$

97. $\left(3a^{-3}\right)^{-2} = 3^{-2}\left(a^{-3}\right)^{-2}$

$\qquad = \dfrac{1}{3^2} \cdot a^{-3\cdot -2}$

$\qquad = \dfrac{1}{9}a^6$

$\qquad = \dfrac{a^6}{9}$

99. $\left(-2a^2b^3\right)^{-4} = (-2)^{-4}\left(a^2\right)^{-4}\left(b^3\right)^{-4}$

$\qquad = \dfrac{1}{(-2)^4} \cdot a^{2\cdot -4} \cdot b^{3\cdot -4}$

$\qquad = \dfrac{1}{16} \cdot a^{-8} \cdot b^{-12}$

$\qquad = \dfrac{1}{16} \cdot \dfrac{1}{a^8} \cdot \dfrac{1}{b^{12}}$

$\qquad = \dfrac{1}{16a^8b^{12}}$

101. $\dfrac{2^3 \cdot xy^{-2}}{12\left(x^2\right)^{-2}y} = \dfrac{8xy^{-2}}{12x^{2\cdot -2}y}$

$\qquad = \dfrac{8xy^{-2}}{12x^{-4}y}$

$\qquad = \dfrac{8}{12}x^{1-(-4)}y^{-2-1}$

$\qquad = \dfrac{2}{3}x^5y^{-3}$

$\qquad = \dfrac{2x^5}{3y^3}$

103. $\left(\dfrac{15a^2b^3}{3a^{-4}b^5}\right)^{-2} = \left(5a^{2-(-4)}b^{3-5}\right)^{-2}$

$\qquad = \left(5a^6b^{-2}\right)^{-2}$

$\qquad = 5^{-2}\left(a^6\right)^{-2}\left(b^{-2}\right)^{-2}$

$\qquad = \dfrac{1}{5^2}a^{6\cdot -2}b^{-2\cdot -2}$

$\qquad = \dfrac{1}{25}a^{-12}b^4$

$\qquad = \dfrac{b^4}{25a^{12}}$

105. $\left(4x^4y^{-2}\right)^{-1} \cdot \left(2x^2y^{-1}\right)^2$

$\qquad = 4^{-1}\left(x^4\right)^{-1}\left(y^{-2}\right)^{-1} \cdot 2^2\left(x^2\right)^2\left(y^{-1}\right)^2$

$\qquad = \dfrac{1}{4}x^{-4}y^2 \cdot 4x^4y^{-2}$

$\qquad = \dfrac{4}{4}x^{-4+4}y^{2+(-2)}$

$\qquad = 1 \cdot x^0 \cdot y^0$

$\qquad = 1$

107. $\dfrac{(-2)^2x^3(yz)^2}{-4xy^{-2}z} = \dfrac{4x^3y^2z^2}{-4xy^{-2}z}$

$\qquad = \dfrac{4}{-4}x^{3-1}y^{2-(-2)}z^{2-1}$

$\qquad = -1 \cdot x^2y^4z$

$\qquad = -x^2y^4z$

109.
$$\frac{\left(3x^{-1}yz^2\right)^2}{\left(xy^{-2}z\right)^3} = \frac{3^2\left(x^{-1}\right)^2 y^2 \left(z^2\right)^2}{x^3\left(y^{-2}\right)^3 z^3}$$
$$= \frac{9x^{-2}y^2 z^4}{x^3 y^{-6} z^3}$$
$$= 9x^{-2-3} y^{2-(-6)} z^{4-3}$$
$$= 9x^{-5} y^8 z^1$$
$$= \frac{9y^8 z}{x^5}$$

111.
$$\frac{\left(6a^3 b^{-2}\right)^{-1}}{\left(2a^{-2}b\right)^{-2}} \cdot \left(\frac{3ab^3}{2a^2 b^{-3}}\right)^2$$
$$= \frac{6^{-1}\left(a^3\right)^{-1}\left(b^{-2}\right)^{-1}}{2^{-2}\left(a^{-2}\right)^{-2} b^{-2}} \cdot \left(\frac{3}{2}a^{1-2} b^{3-(-3)}\right)^2$$
$$= \frac{2^2 a^{-3} b^2}{6a^4 b^{-2}} \cdot \left(\frac{3}{2}a^{-1} b^6\right)^2$$
$$= \frac{2}{3}a^{-3-4} b^{2-(-2)} \cdot \left(\frac{3}{2}\right)^2 \left(a^{-1}\right)^2 \left(b^6\right)^2$$
$$= \frac{2}{3}a^{-7} b^4 \cdot \frac{9}{4}a^{-2} b^{12}$$
$$= \frac{2}{3} \cdot \frac{9}{4} a^{-7+(-2)} b^{4+12}$$
$$= \frac{3}{2}a^{-9} b^{16}$$
$$= \frac{3b^{16}}{2a^9}$$

113.
$$\left(-5.3\times10^{-4}\right)\cdot\left(2.8\times10^{-3}\right)$$
$$= (-5.3\cdot2.8)\times\left(10^{-4}\cdot10^{-3}\right)$$
$$= -14.84\times10^{-7}$$
$$= -\left(1.484\times10^1\right)\times10^{-7}$$
$$= -1.484\times10^{-6}$$

115.
$$\left(4\times10^6\right)^3 = 4^3\times\left(10^6\right)^3$$
$$= 64\times10^{18}$$
$$= \left(6.4\times10^1\right)\times10^{18}$$
$$= 6.4\times10^{19}$$

117.
$$\frac{5\times10^{-6}}{8\times10^{-4}} = \frac{5}{8}\times\frac{10^{-6}}{10^{-4}}$$
$$= 0.625\times10^{-6-(-4)}$$
$$= 6.25\times10^{-1}\times10^{-2}$$
$$= 6.25\times10^{-3}$$

119.
$$\frac{\left(4\times10^3\right)\left(6\times10^7\right)}{3\times10^4} = \frac{(4\cdot6)\times\left(10^3\cdot10^7\right)}{3\times10^4}$$
$$= \frac{24\times10^{10}}{3\times10^4}$$
$$= \frac{24}{3}\times\frac{10^{10}}{10^4}$$
$$= 8\times10^{10-4}$$
$$= 8\times10^6$$

121.
$$\frac{6.2\times10^{-3}}{\left(3.1\times10^4\right)\cdot\left(2\times10^{-7}\right)} = \frac{6.2\times10^{-3}}{(3.1\cdot2)\times\left(10^4\cdot10^{-7}\right)}$$
$$= \frac{6.2\times10^{-3}}{6.2\times10^{-3}}$$
$$= 1$$

123.
$$(4,000,000)\cdot(3,000,000) = \left(4\times10^6\right)\cdot\left(3\times10^6\right)$$
$$= (4\cdot3)\times\left(10^6\cdot10^6\right)$$
$$= 12\times10^{12}$$
$$= \left(1.2\times10^1\right)\times10^{12}$$
$$= 1.2\times10^{13}$$
$$= 12,000,000,000,000$$

125.
$$\frac{0.00008}{0.002} = \frac{8\times10^{-5}}{2\times10^{-3}}$$
$$= \frac{8}{2}\times\frac{10^{-5}}{10^{-3}}$$
$$= 4\times10^{-5-(-3)}$$
$$= 4\times10^{-2}$$
$$= 0.04$$

127.

$$\frac{(0.000004)\cdot 1,600,000}{(0.0008)\cdot(0.002)} = \frac{\left(4\times10^{-6}\right)\cdot\left(1.6\times10^6\right)}{\left(8\times10^{-4}\right)\cdot\left(2\times10^{-3}\right)}$$

$$= \frac{(4\cdot1.6)\times\left(10^{-6}\cdot10^6\right)}{(8\cdot2)\times\left(10^{-4}\cdot10^{-3}\right)}$$

$$= \frac{6.4\times10^0}{16\times10^{-7}}$$

$$= \frac{6.4}{16}\times10^{0-(-7)}$$

$$= 0.4\times10^7$$

$$= \left(4\times10^{-1}\right)\times10^7$$

$$= 4\times10^6$$

$$= 4,000,000$$

129. $V = s^3$

Since each side of the cube has a length of $s = x^2$, the volume is

$$V = \left(x^2\right)^3 = x^{2\cdot3} = x^6$$

The volume of the cube is x^6 cubic units.

131. $0.00001276 = 1.276\times10^{-5}$

The plant cell diameter is 1.276×10^{-5} meters.

133. $12,760,000 = 1.276\times10^7$

The diameter of Earth is 1.276×10^7 meters.

135. $1\times10^{-9} = 0.000000001$

One nanometer is 0.000000001 meters.

137. $4\times10^{19} = 40,000,000,000,000,000,000$

The Rubik's Cube has
$40,000,000,000,000,000,000$ possible states.

139.

$$\frac{1.53\times10^{13}}{3.1\times10^8} = \frac{1.53}{3.1}\times\frac{10^{13}}{10^8}$$

$$\approx 0.49355\times10^{13-8}$$

$$= \left(4.9355\times10^{-1}\right)\times10^5$$

$$= 4.9355\times10^4$$

$$= 49,355$$

The per capita GDP of the United States in 2011 was $49,355.

141. Multiply by 30, since April has 30 days.

$$30(1.75\times10^6) = (30\times1.75)\times10^6$$

$$= 52.5\times10^6$$

$$= (5.25\times10^1)\times10^6$$

$$= 5.25\times10^7$$

$$= 52,500,000$$

Facebook had 5.25×10^7 or 52,500,000 visitors in April 2012.

143. a.

$$\text{pop. density} = \frac{2.81\times10^8}{3.54\times10^6}$$

$$= \frac{2.81}{3.54}\times\frac{10^8}{10^6}$$

$$\approx 0.794\times10^{8-6}$$

$$= \left(7.94\times10^{-1}\right)\times10^2$$

$$= 7.94\times10^1$$

$$= 79.4$$

The population density of the U.S. in 2000 was roughly 79.4 people per square mile. This means that, on average, each square mile of land contained about 79 people.

b.

$$\text{pop. density} = \frac{3.11\times10^8}{3.54\times10^6}$$

$$= \frac{3.11}{3.54}\times\frac{10^8}{10^6}$$

$$\approx 0.879\times10^{8-6}$$

$$= \left(8.79\times10^{-1}\right)\times10^2$$

$$= 8.79\times10^1$$

$$= 87.9$$

The population density of the U.S. in 2011 was roughly 87.9 people per square mile. This means that, on average, each square mile of land contained about 88 people.

c. $87.9 - 79.4 = 8.5$

The population density increased by about 9 people per square mile. This means that, on average, the number of people living on each square mile of land increased by 9, thereby making living space more crowded.

145. $\dfrac{2^{x+3}}{4} = \dfrac{2^{x+3}}{2^2} = 2^{x+3-2} = 2^{x+1}$

147. $3^x \cdot 27^{3x+1} = 3^x \cdot \left(3^3\right)^{3x+1}$

$\qquad\qquad\quad = 3^x \cdot 3^{9x+3}$

$\qquad\qquad\quad = 3^{x+9x+3}$

$\qquad\qquad\quad = 3^{10x+3}$

149. $\qquad 3^x = 5$

$\qquad \left(3^x\right)^4 = (5)^4$

$\qquad\quad 3^{4x} = 5^4$

$\qquad\quad 3^{4x} = 625$

151. $\qquad 2^x = 7$

$\qquad \left(2^x\right)^{-4} = (7)^{-4}$

$\qquad\quad 2^{-4x} = 7^{-4}$

$\qquad\quad 2^{-4x} = \dfrac{1}{7^4}$

$\qquad\quad 2^{-4x} = \dfrac{1}{2401}$

153. No, your friend is incorrect. He added the exponents instead of multiplying.

$\left(x^4\right)^3 = x^{4 \cdot 3} = x^{12}$

155. The expression a^{m+n} is equivalent to the expression $a^m \cdot a^n$. If either m, n, or $m+n$ is 0, then a cannot be 0 or we would have the indeterminate form 0^0. If either m, n, or $m+n$ is negative, then a cannot be 0 or we would have division by zero which is undefined. For example, $0^{-2} = \dfrac{1}{0^2} = \dfrac{1}{0} =$ undefined .

157. If the exponent, n, is 0, then neither base can equal 0 or we would have the indeterminate form 0^0. Since the product of 0 and any number is still 0, the quantity $a \cdot b$ will equal 0 if either a or b is equal to 0. This would lead to the indeterminate form 0^0 if the exponent is 0.

If n is negative, then neither base can be 0 or we would have division by zero which is undefined. Since the product of 0 and any number is still 0, the quantity $a \cdot b$ will equal 0 if either a or b is equal to 0. This would lead to division by 0 since the negative exponent would put the quantity $a \cdot b = 0$ in the denominator.

159. Written response. Answers may vary. The idea can be illustrated as follows:

$$\dfrac{x^7}{x^3} = \dfrac{x \cdot x \cdot x \cdot x \cdot x \cdot x \cdot x}{x \cdot x \cdot x}$$

$$= \dfrac{x \cdot x \cdot x \cdot x \cdot \cancel{x \cdot x \cdot x}}{\cancel{x \cdot x \cdot x}}$$

$$= x \cdot x \cdot x \cdot x$$

$$= x^4$$

Notice that the bases are the same. When we divide the two expressions on the left, we will need to subtract the exponents to get the result on the right.

161. Written response. Answers may vary. The idea can be illustrated as follows:

Multiply $x \cdot y$ together 4 times to get $(x \cdot y)^4$.

$$(x \cdot y) \cdot (x \cdot y) \cdot (x \cdot y) \cdot (x \cdot y) = (x \cdot y)^4$$

The Commutative and Associative Laws for Multiplication says that we can group these multiplications in any order we wish. Group the x's together and group the y's together to get

$$(x \cdot x \cdot x \cdot x) \cdot (y \cdot y \cdot y \cdot y) = (x \cdot y)^4$$

Using the Product Rule for Exponents, rewrite the expressions inside the parentheses to get

$$x^4 \cdot y^4 = (x \cdot y)^4 .$$

163. Written response. Answers may vary. See Objective 8.

Section 4.1

Are You Ready for This Section?

R1. The coefficient is -4, the number multiplied in front of the variable expression.

R2. $\quad 5x^2 - 3x + 1 - 2x^2 - 6x + 3$

$\quad = 5x^2 - 2x^2 - 3x - 6x + 1 + 3$

$\quad = (5-2)x^2 + (-3-6)x + (1+3)$

$\quad = 3x^2 - 9x + 4$

R3. $\quad -4(x-3) = -4 \cdot x - 4 \cdot (-3)$

$\qquad\qquad\quad = -4x + 12$

R4. $\quad f(x) = -4x + 3$

$\qquad f(3) = -4(3) + 3$

$\qquad\qquad = -12 + 3$

$\qquad\qquad = -9$

Section 4.1 Quick Checks

1. A <u>monomial</u> in one variable is the product of a number and a variable raised to a nonnegative integer power.

2. $8x^5$ is a monomial because it is in the form ax^k where $k \geq 0$. The coefficient is $a = 8$ and the degree is $k = 5$.

3. $5x^{-2}$ is not a monomial because the exponent of the variable is -2 and -2 is not a nonnegative integer.

4. 12 is a monomial because it is in the form ax^k where $k \geq 0$. The coefficient is $a = 12$ and the degree is $k = 0$.

5. $x^{1/3}$ is not a monomial because the exponent of the variable is $\frac{1}{3}$ and $\frac{1}{3}$ is not a nonnegative integer.

6. The degree of a monomial in the form $ax^m y^n$ is $m + n$.

7. $3x^5 y^2$ is a monomial because the exponents of the variables are nonnegative integers. The coefficient is 3 and the degree is $n = 5 + 2 = 7$.

8. $-2m^3 n$ is a monomial because the exponents of the variables are nonnegative integers. The coefficient is -2 and the degree is $n = 3 + 1 = 4$.

9. $4ab^{1/2}$ is not a monomial because the exponent ½ is not an integer.

10. $-xy$ is a monomial because the exponents of the variables are nonnegative integers (both are 1 in this case). The coefficient is -1 and the degree is $n = 1 + 1 = 2$.

11. True

12. $-3x^3 + 7x^2 - x + 5$ is a polynomial with degree $n = 3$.

13. $5z^{-1} + 3$ is not a polynomial because the exponent on the first term, -1, is negative.

14. $\dfrac{x-1}{x+1}$ is not a polynomial because it is the quotient of two polynomials and the polynomial in the denominator has degree greater than 0.

15. $\dfrac{3x^2 - 9x + 27}{3} = x^2 - 3x + 9$ is a polynomial. The degree is $n = 2$.

16. $5p^3 q - 8pq^2 + pq$ is a polynomial with degree $n = 3 + 1 = 4$.

17. False, $5y^3 + 7y^3 = 12y^3$

18. $\left(2x^2 - 3x + 1\right) + \left(4x^2 + 5x - 3\right)$
    $= 2x^2 - 3x + 1 + 4x^2 + 5x - 3$
    $= 2x^2 + 4x^2 - 3x + 5x + 1 - 3$
    $= (2 + 4)x^2 + (-3 + 5)x + (1 - 3)$
    $= 6x^2 + 2x - 2$

19. $\left(5w^4 - 3w^3 + w - 8\right) + \left(-2w^4 + w^3 - 7w^2 + 3\right)$
    $= 5w^4 - 3w^3 + w - 8 - 2w^4 + w^3 - 7w^2 + 3$
    $= 5w^4 - 2w^4 - 3w^3 + w^3 - 7w^2 + w - 8 + 3$
    $= (5 - 2)w^4 + (-3 + 1)w^3 - 7w^2 + w + (-8 + 3)$
    $= 3w^4 - 2w^3 - 7w^2 + w - 5$

20. $\left(\dfrac{1}{2}x^2 - \dfrac{4}{3}x + 1\right) + \left(\dfrac{1}{4}x^2 + \dfrac{2}{3}x - 8\right)$
    $= \dfrac{1}{2}x^2 - \dfrac{4}{3}x + 1 + \dfrac{1}{4}x^2 + \dfrac{2}{3}x - 8$
    $= \dfrac{1}{2}x^2 + \dfrac{1}{4}x^2 - \dfrac{4}{3}x + \dfrac{2}{3}x + 1 - 8$
    $= \left(\dfrac{1}{2} + \dfrac{1}{4}\right)x^2 + \left(-\dfrac{4}{3} + \dfrac{2}{3}\right)x + (1 - 8)$
    $= \dfrac{3}{4}x^2 - \dfrac{2}{3}x - 7$

21. $\left(8x^2y + 2x^2y^2 - 7xy^2\right) + \left(-3x^2y + 5x^2y^2 + 3xy^2\right) = 8x^2y + 2x^2y^2 - 7xy^2 - 3x^2y + 5x^2y^2 + 3xy^2$

$$= 8x^2y - 3x^2y + 2x^2y^2 + 5x^2y^2 - 7xy^2 + 3xy^2$$
$$= (8-3)x^2y + (2+5)x^2y^2 + (-7+3)xy^2$$
$$= 5x^2y + 7x^2y^2 - 4xy^2$$

22. $\left(5x^3 - 6x^2 + x + 9\right) - \left(4x^3 + 10x^2 - 6x + 7\right) = 5x^3 - 6x^2 + x + 9 - 4x^3 - 10x^2 + 6x - 7$

$$= 5x^3 - 4x^3 - 6x^2 - 10x^2 + x + 6x + 9 - 7$$
$$= (5-4)x^3 + (-6-10)x^2 + (1+6)x + (9-7)$$
$$= x^3 - 16x^2 + 7x + 2$$

23. $\left(8y^3 - 5y^2 + 3y + 1\right) - \left(-3y^3 + 6y + 8\right) = 8y^3 - 5y^2 + 3y + 1 + 3y^3 - 6y - 8$

$$= 8y^3 + 3y^3 - 5y^2 + 3y - 6y + 1 - 8$$
$$= (8+3)y^3 - 5y^2 + (3-6)y + (1-8)$$
$$= 11y^3 - 5y^2 - 3y - 7$$

24. $\left(8x^2y + 2x^2y^2 - 7xy^2\right) - \left(-3x^2y + 5x^2y^2 + 3xy^2\right) = 8x^2y + 2x^2y^2 - 7xy^2 + 3x^2y - 5x^2y^2 - 3xy^2$

$$= 8x^2y + 3x^2y + 2x^2y^2 - 5x^2y^2 - 7xy^2 - 3xy^2$$
$$= (8+3)x^2y + (2-5)x^2y^2 + (-7-3)xy^2$$
$$= 11x^2y - 3x^2y^2 - 10xy^2$$

25. $g(x) = -2x^3 + 7x + 1$

     a. $g(0) = -2(0)^3 + 7(0) + 1$
                $= 0 + 0 + 1$
                $= 1$

     b. $g(2) = -2(2)^3 + 7(2) + 1$
                $= -2(8) + 7(2) + 1$
                $= -16 + 14 + 1$
                $= -1$

     c. $g(-3) = -2(-3)^3 + 7(-3) + 1$
                $= -2(-27) + 7(-3) + 1$
                $= 54 - 21 + 1$
                $= 34$

26. $B(a) = -1399a^2 + 70,573a - 495,702$

     a. $B(25) = -1399(25)^2 + 70,573(25) - 495,702$
                $= -874,375 + 1,764,325 - 495,702$
                $= 394,248$

     We estimate there will be 394,248 first births to 25-year-old women in 2006.

b. $B(40) = -1399(40)^2 + 70,573(40) - 495,702$
$= -2,238,400 + 2,822,920 - 495,702$
$= 88,818$

We estimate there will be 88,818 first births to 40-year-old women in 2006.

27. $f(x) = 3x^2 - x + 1$ and $g(x) = -x^2 + 5x - 6$

a. $(f + g)(x) = f(x) + g(x)$
$= (3x^2 - x + 1) + (-x^2 + 5x - 6)$
$= 3x^2 - x + 1 - x^2 + 5x - 6$
$= 3x^2 - x^2 - x + 5x + 1 - 6$
$= 2x^2 + 4x - 5$

b. $(f - g)(x) = f(x) - g(x)$
$= (3x^2 - x + 1) - (-x^2 + 5x - 6)$
$= 3x^2 - x + 1 + x^2 - 5x + 6$
$= 3x^2 + x^2 - x - 5x + 1 + 6$
$= 4x^2 - 6x + 7$

c. Because $(f + g)(x) = 2x^2 + 4x - 5$, we have
$(f + g)(1) = 2(1)^2 + 4(1) - 5$
$= 2(1) + 4(1) - 5$
$= 2 + 4 - 5$
$= 1$

d. Because $(f - g)(x) = 4x^2 - 6x + 7$, we have
$(f - g)(-2) = 4(-2)^2 - 6(-2) + 7$
$= 4(4) - 6(-2) + 7$
$= 16 + 12 + 7$
$= 35$

28. a. The revenue function will remain the same, so we have
$R(x) = 12x$

Both the variable and fixed costs will change. Therefore, our new cost function will be
$C(x) = 8x + 1250$

The profit function is given by
$P(x) = R(x) - C(x)$
$= (12x) - (8x + 1250)$
$= 12x - 8x - 1250$
$= 4x - 1250$

The new profit function will be $P(x) = 4x - 1250$.

b. $P(x) = 4x - 1250$
$$P(800) = 4(800) - 1250$$
$$= 3200 - 1250$$
$$= 1950$$
If the company manufactures and sells 800 calculators, its profit will be $1950.

4.1 Exercises

29. Coefficient: 3
Degree: 2

31. Coefficient: -8
Degree: $2 + 3 = 5$

33. Coefficient: $\dfrac{4}{3}$
Degree: 6

35. Coefficient: 2
Degree: 0

37. $2x^{-1} + 3x$ is not a polynomial because the exponent in the first term is not an integer that is greater than or equal to 0.

39. $\dfrac{4}{z-1}$ is not a polynomial because there is a variable in the denominator. The expression cannot be written in the standard form for a polynomial.

41. $5x^2 - 9x + 1$
Yes; trinomial; degree 2; already in standard form.

43. $\dfrac{-20}{n}$ is not a polynomial because there is a variable in the denominator.

45. $3y^{1/3} + 2$ is not a polynomial because there is a fractional exponent.

47. $\dfrac{5}{8}$
Yes; monomial; degree 0; already in standard form.

49. $5 - 8y + 2y^2$
Yes; trinomial; degree 2; standard form is $2y^2 - 8y + 5$

51. $7x^{-1} + 4$ is not a polynomial because of the negative exponent.

53. $3x^2y^2 + 2xy^4 + 4$
Yes; trinomial; degree 5; standard form is $2xy^4 + 3x^2y^2 + 4$

55. $4pqr + 2p^2q + 3pq^{1/4}$ is not a polynomial because there is a fractional exponent.

57. $5z^3 + 8z^3 = (5+8)z^3 = 13z^3$

59. $(x^2 + 5x + 1) + (3x^2 - 2x - 3)$
$$= x^2 + 5x + 1 + 3x^2 - 2x - 3$$
$$= x^2 + 3x^2 + 5x - 2x + 1 - 3$$
$$= (1+3)x^2 + (5-2)x + (1-3)$$
$$= 4x^2 + 3x - 2$$

61. $(6p^3 - p^2 + 3p - 4) + (2p^3 - 7p + 3)$
$$= 6p^3 - p^2 + 3p - 4 + 2p^3 - 7p + 3$$
$$= 6p^3 + 2p^3 - p^2 + 3p - 7p - 4 + 3$$
$$= (6+2)p^3 - p^2 + (3-7)p + (-4+3)$$
$$= 8p^3 - p^2 - 4p - 1$$

63.
$$\begin{array}{r} 8x^3 + 4x^2 - 3x + 1 \\ + \ -x^3 - 2x^2 - 3x + 7 \\ \hline 7x^3 + 2x^2 - 6x + 8 \end{array}$$

65. $(5x^2 + 9x + 4) - (3x^2 + 5x + 1)$
$$= 5x^2 + 9x + 4 - 3x^2 - 5x - 1$$
$$= 5x^2 - 3x^2 + 9x - 5x + 4 - 1$$
$$= (5-3)x^2 + (9-5)x + (4-1)$$
$$= 2x^2 + 4x + 3$$

67. $(7s^2t^3 + st^2 - 5t - 8) - (4s^2t^3 + 5st^2 - 7)$
$$= 7s^2t^3 + st^2 - 5t - 8 - 4s^2t^3 - 5st^2 + 7$$
$$= 7s^2t^3 - 4s^2t^3 + st^2 - 5st^2 - 5t - 8 + 7$$
$$= (7-4)s^2t^3 + (1-5)st^2 - 5t + (-8+7)$$
$$= 3s^2t^3 - 4st^2 - 5t - 1$$

69. $\left(3-5x+x^2\right)+\left(-2+3x-5x^2\right)$

$=3-5x+x^2-2+3x-5x^2$

$=x^2-5x^2-5x+3x+3-2$

$=(1-5)x^2+(-5+3)x+(3-2)$

$=-4x^2-2x+1$

71. $\left(6-2y+y^3\right)-\left(-2+y^2-2y^3\right)$

$=6-2y+y^3+2-y^2+2y^3$

$=y^3+2y^3-y^2-2y+6+2$

$=(1+2)y^3-y^2-2y+(6+2)$

$=3y^3-y^2-2y+8$

73. $\left(\dfrac{1}{4}x^2+\dfrac{3}{2}x+3\right)+\left(\dfrac{1}{2}x^2-\dfrac{1}{4}x-2\right)$

$=\dfrac{1}{4}x^2+\dfrac{3}{2}x+3+\dfrac{1}{2}x^2-\dfrac{1}{4}x-2$

$=\dfrac{1}{4}x^2+\dfrac{1}{2}x^2+\dfrac{3}{2}x-\dfrac{1}{4}x+3-2$

$=\left(\dfrac{1}{4}+\dfrac{1}{2}\right)x^2+\left(\dfrac{3}{2}-\dfrac{1}{4}\right)x+(3-2)$

$=\dfrac{3}{4}x^2+\dfrac{5}{4}x+1$

75. $\left(5x^2y^2-8x^2y+xy^2\right)+\left(3x^2y^2+x^2y-4xy^2\right)$

$=5x^2y^2-8x^2y+xy^2+3x^2y^2+x^2y-4xy^2$

$=5x^2y^2+3x^2y^2-8x^2y+x^2y+xy^2-4xy^2$

$=(5+3)x^2y^2+(-8+1)x^2y+(1-4)xy^2$

$=8x^2y^2-7x^2y-3xy^2$

77. $\left(3x^2y+7xy^2+xy\right)-\left(2x^2y-4xy^2-xy\right)$

$=3x^2y+7xy^2+xy-2x^2y+4xy^2+xy$

$=3x^2y-2x^2y+7xy^2+4xy^2+xy+xy$

$=(3-2)x^2y+(7+4)xy^2+(1+1)xy$

$=x^2y+11xy^2+2xy$

79. $\quad 9a^3+2a^2-5a-8$

$\underline{-\quad 5a^3-2a^2+a-6}$

$\quad 4a^3+4a^2-6a-2$

81. $\left(5x^3-5x+3\right)+\left(-4x^3+x^2-2x+1\right)$

$=5x^3-5x+3-4x^3+x^2-2x+1$

$=5x^3-4x^3+x^2-5x-2x+3+1$

$=(5-4)x^3+x^2+(-5-2)x+(3+1)$

$=x^3+x^2-7x+4$

83. $\left(2b^3+5b^2-b+3\right)-\left(4b^3-b^2+3b-1\right)$

$=2b^3+5b^2-b+3-4b^3+b^2-3b+1$

$=2b^3-4b^3+5b^2+b^2-b-3b+3+1$

$=(2-4)b^3+(5+1)b^2+(-1-3)b+(3+1)$

$=-2b^3+6b^2-4b+4$

85. $(2x^2-3x+1)+(x^2+9)-(4x^2-2x-5)$

$=2x^2-3x+1+x^2+9-4x^2+2x+5$

$=2x^2+x^2-4x^2-3x+2x+1+9+5$

$=(2+1-4)x^2+(-3+2)x+(1+9+5)$

$=-x^2-x+15$

87. $(4n-3)+(n^3-9)-(2n^2-7n+3)$

$=4n-3+n^3-9-2n^2+7n-3$

$=n^3-2n^2+4n+7n-3-9-3$

$=n^3-2n^2+(4+7)n+(-3-9-3)$

$=n^3-2n^2+11n-15$

89. a. $f(x)=x^2-4x+1$

$\quad f(0)=(0)^2-4(0)+1$

$\quad\quad =0-0+1$

$\quad\quad =1$

b. $f(x)=x^2-4x+1$

$\quad f(2)=(2)^2-4(2)+1$

$\quad\quad =4-8+1$

$\quad\quad =-3$

c. $f(x)=x^2-4x+1$

$\quad f(-3)=(-3)^2-4(-3)+1$

$\quad\quad =9+12+1$

$\quad\quad =22$

91. a. $f(x)=2x^3-7x+3$

$\quad f(0)=2(0)^3-7(0)+3$

$\quad\quad =0-0+3$

$\quad\quad =3$

b. $f(x) = 2x^3 - 7x + 3$

$$f(2) = 2(2)^3 - 7(2) + 3$$
$$= 2 \cdot 8 - 14 + 3$$
$$= 16 - 14 + 3$$
$$= 5$$

c. $f(x) = 2x^3 - 7x + 3$

$$f(-3) = 2(-3)^3 - 7(-3) + 3$$
$$= 2 \cdot -27 + 21 + 3$$
$$= -54 + 21 + 3$$
$$= -30$$

93. a. $f(x) = -x^3 + 3x^2 - 2x + 3$

$$f(0) = -(0)^3 + 3(0)^2 - 2(0) + 3$$
$$= 0 + 0 - 0 + 3$$
$$= 3$$

b. $f(x) = -x^3 + 3x^2 - 2x + 3$

$$f(2) = -(2)^3 + 3(2)^2 - 2(2) + 3$$
$$= -8 + 3 \cdot 4 - 4 + 3$$
$$= -8 + 12 - 4 + 3$$
$$= 3$$

c. $f(x) = -x^3 + 3x^2 - 2x + 3$

$$f(-3) = -(-3)^3 + 3(-3)^2 - 2(-3) + 3$$
$$= -(-27) + 3 \cdot 9 + 6 + 3$$
$$= 27 + 27 + 6 + 3$$
$$= 63$$

95. a. $(f+g)(x) = f(x) + g(x)$

$$= (2x+5) + (-5x+1)$$
$$= 2x + 5 - 5x + 1$$
$$= 2x - 5x + 5 + 1$$
$$= -3x + 6$$

b. $(f-g)(x) = f(x) - g(x)$

$$= (2x+5) - (-5x+1)$$
$$= 2x + 5 + 5x - 1$$
$$= 2x + 5x + 5 - 1$$
$$= 7x + 4$$

c.   Since $(f+g)(x) = -3x + 6$, we have

$$(f+g)(2) = -3(2) + 6$$
$$= -6 + 6$$
$$= 0$$

d.   Since $(f-g)(x) = 7x + 4$, we have

$$(f-g)(1) = 7(1) + 4$$
$$= 7 + 4$$
$$= 11$$

97. a. $(f+g)(x) = f(x) + g(x)$

$$= \left(x^2 - 5x + 3\right) + \left(2x^2 + 3\right)$$
$$= x^2 - 5x + 3 + 2x^2 + 3$$
$$= x^2 + 2x^2 - 5x + 3 + 3$$
$$= 3x^2 - 5x + 6$$

b. $(f-g)(x) = f(x) - g(x)$

$$= \left(x^2 - 5x + 3\right) - \left(2x^2 + 3\right)$$
$$= x^2 - 5x + 3 - 2x^2 - 3$$
$$= x^2 - 2x^2 - 5x + 3 - 3$$
$$= -x^2 - 5x$$

c.   Since $(f+g)(x) = 3x^2 - 5x + 6$, we have

$$(f+g)(2) = 3(2)^2 - 5(2) + 6$$
$$= 3 \cdot 4 - 10 + 6$$
$$= 12 - 10 + 6$$
$$= 8$$

d.   Since $(f-g)(x) = -x^2 - 5x$, we have

$$(f-g)(1) = -(1)^2 - 5(1)$$
$$= -1 - 5$$
$$= -6$$

99. a. $(f+g)(x) = f(x) + g(x)$

$$= \left(x^3 + 6x^2 + 12x + 2\right) + \left(x^3 - 8\right)$$
$$= x^3 + 6x^2 + 12x + 2 + x^3 - 8$$
$$= x^3 + x^3 + 6x^2 + 12x + 2 - 8$$
$$= 2x^3 + 6x^2 + 12x - 6$$

b. $(f-g)(x) = f(x) - g(x)$

$$= \left(x^3 + 6x^2 + 12x + 2\right) - \left(x^3 - 8\right)$$
$$= x^3 + 6x^2 + 12x + 2 - x^3 + 8$$
$$= x^3 - x^3 + 6x^2 + 12x + 2 + 8$$
$$= 6x^2 + 12x + 10$$

c.   Since $(f+g)(x)=2x^3+6x^2+12x-6$, we have

$$\begin{aligned}(f+g)(2)&=2(2)^3+6(2)^2+12(2)-6\\&=2\cdot8+6\cdot4+24-6\\&=16+24+24-6\\&=58\end{aligned}$$

d.   Since $(f-g)(x)=6x^2+12x+10$, we have

$$\begin{aligned}(f-g)(1)&=6(1)^2+12(1)+10\\&=6+12+10\\&=28\end{aligned}$$

101. a.   If the vertex is $(2,2)$, then $x=2$.

$$\begin{aligned}A(x)&=-2x^2+6x\\A(2)&=-2(2)^2+6(2)\\&=-2\cdot4+12\\&=-8+12\\&=4\end{aligned}$$

The area of the region is 4 square units.

b.   If the vertex is $(1,4)$, then $x=1$.

$$\begin{aligned}A(x)&=-2x^2+6x\\A(1)&=-2(1)^2+6(1)\\&=-2+6\\&=4\end{aligned}$$

The area of the region is 4 square units.

c.   Since the rectangular region is in quadrant I and has one corner at the origin, the coordinates of the vertex give us the length and width of the rectangle. Since the area of a rectangle can be found by multiplying the length and width, we can just multiply the coordinates together to get the area of the region.

$(2,2)\rightarrow 2\cdot2=4$ The area is 4 square units.

$(1,4)\rightarrow 1\cdot4=4$ The area is 4 square units.

103. $P(t)=-0.043t^3+0.879t^2-2.184t+65.100$

a.   In 1995, $t=0$.

$$\begin{aligned}P(0)&=-0.043(0)^3+0.879(0)^2-2.184(0)+65.100\\&=0-0+0+65.100\\&=65.10\end{aligned}$$

According to the model, the average price per square foot of a new single-family home in 1995 was roughly $65.10.

b.   In 2015, $t=20$.

$$\begin{aligned}P(20)&=-0.043(20)^3+0.879(20)^2-2.14(20)+65.100\\&=-344+351.6-42.8+65.100\\&=29.02\end{aligned}$$

According to the model, the average price per square foot of a new single-family home in 2015 will be roughly $29.02. No, this does not seem reasonable.

105. a. Profit is the difference between revenue and costs. Thus,

$$P(x) = R(x) - C(x)$$
$$= \left(-1.2x^2 + 220x\right) - \left(0.05x^3 - 2x^2 + 65x + 500\right)$$
$$= -1.2x^2 + 220x - 0.05x^3 + 2x^2 - 65x - 500$$
$$= -0.05x^3 - 1.2x^2 + 2x^2 + 220x - 65x - 500$$
$$= -0.05x^3 + 0.8x^2 + 155x - 500$$

The profit function is: $P(x) = -0.05x^3 + 0.8x^2 + 155x - 500$

b. $P(15) = -0.05(15)^3 + 0.8(15)^2 + 155(15) - 500$
$$= -168.75 + 180 + 2325 - 500$$
$$= 1836.25$$

If 15 cell phones are sold, there would be a profit of $1836.25.

c. $P(100) = -0.05(100)^3 + 0.8(100)^2 + 155(100) - 500$
$$= -0.05(1,000,000) + 0.8(10,000) + 15,500 - 500$$
$$= -50,000 + 8000 + 15,500 - 500$$
$$= -27,000$$

If 100 cell phones are sold, the profit function yields a negative value. Thus, there would be a loss of $27,000.

107. $f(x) = 3x + 2$; $g(x) = ax - 5$

$$(f + g)(x) = (3x + 2) + (ax - 5)$$
$$= 3x + 2 + ax - 5$$
$$= 3x + ax + 2 - 5$$
$$= (3 + a)x - 3$$

Since $(f + g)(3) = 12$, we have

$$(3 + a)(3) - 3 = 12$$
$$9 + 3a - 3 = 12$$
$$3a + 6 = 12$$
$$3a = 6$$
$$a = 2$$

109. a. $(f + g)(2) = f(2) + g(2)$
$$= 2 + 1$$
$$= 3$$

b. $(f + g)(4) = f(4) + g(4)$
$$= 1 + (-3)$$
$$= -2$$

c. $(f - g)(6) = f(6) - g(6)$
$$= 0 - (1)$$
$$= -1$$

d. $(g-f)(6) = g(6) - f(6)$
$$= 1 - 0$$
$$= 1$$

111. Taxes paid to the federal government are only part of the total taxes paid. To find the remaining amount paid in other taxes, subtract the federal amount from the total amount.
$$S(x) = T(x) - F(x)$$

113. A linear polynomial is a polynomial of degree 1.

115. Answers will vary. Examples:

Monomials: 3, $4x$, $-7x^2$

Binomials: $x + 2$, $-4x^3 + 3x$

Trinomials: $x^2 - 7x + 2$, $-2x^5 + 4x^2 - x$

Linear Polynomial: x, $3x + 5$, $-7x$.

117. Answers will vary. When adding polynomials we add like terms by adding the coefficients of variables with the same exponent. If the polynomials have the same degree, the sum will have that degree as well (assuming the leading terms do not cancel each other out). If one polynomial has a higher degree, there will be no like term to add to that leading term, so the leading term of the higher-degree polynomial will be the leading term of the sum.

119. a.–c. $f(x) = 4x^2 - 7x + 1$

$f(4) = 37$

$f(-2) = 31$

$f(6) = 103$

121. a.–c. $f(x) = 2x^3 - 5x^2 + x + 5$

$f(4) = 57$

$f(-2) = -33$

$f(6) = 263$

Section 4.2

Are Your Ready for This Section?

R1. $4x^2 \cdot 3x^3 = 4 \cdot 3 \cdot x^2 \cdot x^3$
$$= 12x^{2+3}$$
$$= 12x^5$$

R2. $(-3x)^2 = (-3)^2 \cdot x^2 = 9x^2$

R3. $4(x-5) = 4 \cdot x - 4 \cdot 5$
$$= 4x - 20$$

Section 4.2 Quick Checks

1. $(3x^5) \cdot (2x^2) = (3 \cdot 2)(x^5 \cdot x^2)$
$$= 6x^{5+2}$$
$$= 6x^7$$

2. $(-7a^3b^2) \cdot (3ab^4) = (-7 \cdot 3)(a^3 \cdot a)(b^2 \cdot b^4)$
$$= -21a^{3+1}b^{2+4}$$
$$= -21a^4b^6$$

3. $\left(\dfrac{2}{3}x^4\right) \cdot \left(\dfrac{15}{8}x\right) = \left(\dfrac{2}{3} \cdot \dfrac{15}{8}\right)(x^4 \cdot x)$
$$= \dfrac{5}{4}x^{4+1}$$
$$= \dfrac{5}{4}x^5$$

4. $-3(x+2) = -3 \cdot x + (-3) \cdot 2$
$$= -3x - 6$$

5. $5x\left(x^2 + 3x + 2\right) = 5x \cdot x^2 + 5x \cdot 3x + 5x \cdot 2$
$$= 5x^3 + 15x^2 + 10x$$

6. $2xy\left(3x^2 - 5xy + 2y^2\right)$
$$= 2xy \cdot 3x^2 - 2xy \cdot 5xy + 2xy \cdot 2y^2$$
$$= 6x^3y - 10x^2y^2 + 4xy^3$$

7. $\dfrac{3}{4}y^2\left(\dfrac{4}{3}y^2 + \dfrac{2}{9}y + \dfrac{16}{3}\right)$
$$= \dfrac{3}{4}y^2 \cdot \dfrac{4}{3}y^2 + \dfrac{3}{4}y^2 \cdot \dfrac{2}{9}y + \dfrac{3}{4}y^2 \cdot \dfrac{16}{3}$$
$$= \left(\dfrac{3}{4} \cdot \dfrac{4}{3}\right)y^2 \cdot y^2 + \left(\dfrac{3}{4} \cdot \dfrac{2}{9}\right)y^2 \cdot y + \left(\dfrac{3}{4} \cdot \dfrac{16}{3}\right)y^2$$
$$= y^4 + \dfrac{1}{6}y^3 + 4y^2$$

8. The acronym FOIL stands for <u>First</u>, <u>Outer</u>, <u>Inner</u>, <u>Last</u>.

9. $(x+4)(x+1) = x \cdot x + x \cdot 1 + 4 \cdot x + 4 \cdot 1$
$$= x^2 + x + 4x + 4$$
$$= x^2 + 5x + 4$$

10. $(3v+5)(2v-3) = 3v \cdot 2v - 3v \cdot 3 + 5 \cdot 2v - 5 \cdot 3$
$$= 6v^2 - 9v + 10v - 15$$
$$= 6v^2 + v - 15$$

11. $(2a-b)(a+5b) = 2a \cdot a + 2a \cdot 5b - b \cdot a - b \cdot 5b$
$$= 2a^2 + 10ab - ab - 5b^2$$
$$= 2a^2 + 9ab - 5b^2$$

12. $(2y-3)\left(y^2 + 4y + 5\right)$
$$= 2y \cdot y^2 + 2y \cdot 4y + 2y \cdot 5 - 3 \cdot y^2 - 3 \cdot 4y - 3 \cdot 5$$
$$= 2y^3 + 8y^2 + 10y - 3y^2 - 12y - 15$$
$$= 2y^3 + 5y^2 - 2y - 15$$

13. $\left(z^2 - 3z + 2\right)\left(2z^2 + z + 6\right)$
$$= z^2 \cdot 2z^2 + z^2 \cdot z + z^2 \cdot 6 - 3z \cdot 2z^2 - 3z \cdot z$$
$$\quad -3z \cdot 6 + 2 \cdot 2z^2 + 2 \cdot z + 2 \cdot 6$$
$$= 2z^4 + z^3 + 6z^2 - 6z^3 - 3z^2 - 18z$$
$$\quad + 4z^2 + 2z + 12$$
$$= 2z^4 - 5z^3 + 7z^2 - 16z + 12$$

14. False; consider $(x-2)(x+2) = x^2 - 4$.

15. $(A-B)(A+B) = \underline{A^2 - B^2}$.

16. $(5y+2)(5y-2) = (5y)^2 - (2)^2$
$$= 25y^2 - 4$$

17. $\left(7y + 2z^3\right)\left(7y - 2z^3\right) = (7y)^2 - \left(2z^3\right)^2$
$$= 49y^2 - 4z^6$$

18. $(A-B)^2 = \underline{A^2 - 2AB + B^2}$;
$$(A+B)^2 = \underline{A^2 + 2AB + B^2}$$

19. $x^2 + 2xy + y^2$ is referred to as a <u>perfect square</u> trinomial.

20. False; $(x+a)^2 = x^2 + 2ax + a^2$

21. $(z-8)^2 = z^2 - 2 \cdot z \cdot 8 + 8^2$
$$= z^2 - 16z + 64$$

22. $(6p+5)^2 = (6p)^2 + 2 \cdot 6p \cdot 5 + 5^2$
$$= 36p^2 + 60p + 25$$

23. $(4a-3b)^2 = (4a)^2 - 2 \cdot (4a) \cdot (3b) + (3b)^2$
$$= 16a^2 - 24ab + 9b^2$$

24. $(f \cdot g)(x) = \underline{f(x) \cdot g(x)}$.

25. $f(x) = 5x - 3$; $g(x) = x^2 + 3x + 1$

     a. $\quad f(2) = 5(2) - 3 \qquad g(2) = (2)^2 + 3(2) + 1$
$$\qquad\qquad = 10 - 3 \qquad\qquad = 4 + 6 + 1$$
$$\qquad\qquad = 7 \qquad\qquad\qquad = 11$$
$$\qquad f(2) \cdot g(2) = 7 \cdot 11 = 77$$

b. $(f \cdot g)(x)$
$= f(x) \cdot g(x)$
$= (5x-3)(x^2+3x+1)$
$= 5x \cdot x^2 + 5x \cdot 3x + 5x \cdot 1 - 3 \cdot x^2 - 3 \cdot 3x - 3 \cdot 1$
$= 5x^3 + 15x^2 + 5x - 3x^2 - 9x - 3$
$= 5x^3 + 12x^2 - 4x - 3$

c. $(f \cdot g)(2) = 5(2)^3 + 12(2)^2 - 4(2) - 3$
$= 5(8) + 12(4) - 4(2) - 3$
$= 40 + 48 - 8 - 3$
$= 77$

26. $f(x) = x^2 - 2x$

a. $f(x-3) = (x-3)^2 - 2(x-3)$
$= x^2 - 2 \cdot x \cdot 3 + 3^2 - 2 \cdot x + 2 \cdot 3$
$= x^2 - 6x + 9 - 2x + 6$
$= x^2 - 8x + 15$

b. $f(x+h) - f(x)$
$= \left[(x+h)^2 - 2(x+h)\right] - \left[x^2 - 2x\right]$
$= \left[x^2 + 2xh + h^2 - 2x - 2h\right] - \left[x^2 - 2x\right]$
$= x^2 + 2xh + h^2 - 2x - 2h - x^2 + 2x$
$= 2xh + h^2 - 2h$

4.2 Exercises

27. $(5xy^2)(-3x^2y^3) = (5 \cdot -3)(x \cdot x^2)(y^2 \cdot y^3)$
$= -15x^{1+2}y^{2+3}$
$= -15x^3y^5$

29. $\left(\dfrac{3}{4}yz^3\right)\left(\dfrac{20}{9}y^3z^2\right) = \left(\dfrac{3}{4} \cdot \dfrac{20}{9}\right)(y \cdot y^3)(z^3 \cdot z^2)$
$= \dfrac{5}{3}y^{1+3}z^{3+2}$
$= \dfrac{5}{3}y^4z^5$

31. $5x(x^2+4x+2) = 5x \cdot x^2 + 5x \cdot 4x + 5x \cdot 2$
$= 5x^3 + 20x^2 + 10x$

33. $-4a^2b(3a^2+2ab-b^2)$
$= -4a^2b \cdot 3a^2 + (-4a^2b) \cdot 2ab - (-4a^2b) \cdot b^2$
$= -12a^4b - 8a^3b^2 + 4a^2b^3$

35. $\dfrac{2}{3}ab\left(\dfrac{3}{4}a^2b - \dfrac{9}{8}ab^3 + 6ab\right)$
$= \dfrac{2}{3}ab \cdot \dfrac{3}{4}a^2b - \dfrac{2}{3}ab \cdot \dfrac{9}{8}ab^3 + \dfrac{2}{3}ab \cdot 6ab$
$= \dfrac{1}{2}a^3b^2 - \dfrac{3}{4}a^2b^4 + 4a^2b^2$

37. $0.4x^2(1.2x^2 - 0.8x + 1.5)$
$= 0.4x^2 \cdot 1.2x^2 - 0.4x^2 \cdot 0.8x + 0.4x^2 \cdot 1.5$
$= 0.48x^4 - 0.32x^3 + 0.6x^2$

39. $(x+3)(x+5) = x \cdot x + x \cdot 5 + 3 \cdot x + 3 \cdot 5$
$= x^2 + 5x + 3x + 15$
$= x^2 + 8x + 15$

41. $(a+5)(a-3) = a(a-3) + 5(a-3)$
$= a \cdot a - a \cdot 3 + 5 \cdot a - 5 \cdot 3$
$= a^2 - 3a + 5a - 15$
$= a^2 + 2a - 15$

43. $(4a+3)(3a-1) = 4a \cdot 3a - 4a \cdot 1 + 3 \cdot 3a - 3 \cdot 1$
$= 12a^2 - 4a + 9a - 3$
$= 12a^2 + 5a - 3$

45. $(4-5x)(3+2x) = 4 \cdot 3 + 4 \cdot 2x - 5x \cdot 3 - 5x \cdot 2x$
$= 12 + 8x - 15x - 10x^2$
$= 12 - 7x - 10x^2$
$= -10x^2 - 7x + 12$

47. $\left(\dfrac{2}{3}x+2\right)\left(\dfrac{1}{2}x-4\right)$
$= \dfrac{2}{3}x \cdot \dfrac{1}{2}x - \dfrac{2}{3}x \cdot 4 + 2 \cdot \dfrac{1}{2}x - 2 \cdot 4$
$= \dfrac{1}{3}x^2 - \dfrac{8}{3}x + x - 8$
$= \dfrac{1}{3}x^2 - \dfrac{5}{3}x - 8$

49. $(4a+3b)(a-5b)$
$$= 4a \cdot a - 4a \cdot 5b + 3b \cdot a - 3b \cdot 5b$$
$$= 4a^2 - 20ab + 3ab - 15b^2$$
$$= 4a^2 - 17ab - 15b^2$$

51. $(2x^2+1)(x^2-3)$
$$= 2x^2 \cdot x^2 + 2x^2(-3) + 1 \cdot x^2 + 1(-3)$$
$$= 2x^4 - 6x^2 + x^2 - 3$$
$$= 2x^4 - 5x^2 - 3$$

53. $(x+1)(x^2+4x+2)$
$$= x \cdot x^2 + x \cdot 4x + x \cdot 2 + 1 \cdot x^2 + 1 \cdot 4x + 1 \cdot 2$$
$$= x^3 + 4x^2 + 2x + x^2 + 4x + 2$$
$$= x^3 + 5x^2 + 6x + 2$$

55. $(3a-2)(2a^2+a-5)$
$$= 3a \cdot 2a^2 + 3a \cdot a - 3a \cdot 5 - 2 \cdot 2a^2 - 2 \cdot a - 2 \cdot (-5)$$
$$= 6a^3 + 3a^2 - 15a - 4a^2 - 2a + 10$$
$$= 6a^3 - a^2 - 17a + 10$$

57. $(5z^2+3z+2)(4z+3)$
$$= 5z^2 \cdot 4z + 5z^2 \cdot 3 + 3z \cdot 4z + 3z \cdot 3 + 2 \cdot 4z + 2 \cdot 3$$
$$= 20z^3 + 15z^2 + 12z^2 + 9z + 8z + 6$$
$$= 20z^3 + 27z^2 + 17z + 6$$

59. $-\dfrac{1}{2}x(2x+6)(x-3)$
$$= \left(-\frac{1}{2}x \cdot 2x + \left(-\frac{1}{2}x\right)(6)\right)(x-3)$$
$$= (-x^2 - 3x)(x-3)$$
$$= -x^2 \cdot x + (-x^2)(-3) + (-3x)x + (-3x)(-3)$$
$$= -x^3 + 3x^2 - 3x^2 + 9x$$
$$= -x^3 + 9x$$

61. $(4+y)(2y^2-3+5y)$
$$= 4 \cdot 2y^2 - 4 \cdot 3 + 4 \cdot 5y + y \cdot 2y^2 - y \cdot 3 + y \cdot 5y$$
$$= 8y^2 - 12 + 20y + 2y^3 - 3y + 5y^2$$
$$= 2y^3 + 13y^2 + 17y - 12$$

63. $(w^2+2w+1)(2w^2-3w+1)$
$$= w^2 \cdot 2w^2 - w^2 \cdot 3w + w^2 \cdot 1 + 2w \cdot 2w^2 - 2w \cdot 3w$$
$$\quad + 2w \cdot 1 + 1 \cdot 2w^2 - 1 \cdot 3w + 1 \cdot 1$$
$$= 2w^4 - 3w^3 + w^2 + 4w^3 - 6w^2 + 2w + 2w^2$$
$$\quad - 3w + 1$$
$$= 2w^4 + w^3 - 3w^2 - w + 1$$

65. $(b+1)(b-2)(b+3) = (b^2 - 2b + b - 2)(b+3)$
$$= (b^2 - b - 2)(b+3)$$
$$= b^3 + 3b^2 - b^2 - 3b - 2b - 6$$
$$= b^3 + 2b^2 - 5b - 6$$

67. $(2ab+5)(4a^2-2ab+b^2)$
$$= 2ab \cdot 4a^2 - 2ab \cdot 2ab + 2ab \cdot b^2 + 5 \cdot 4a^2$$
$$\quad - 5 \cdot 2ab + 5 \cdot b^2$$
$$= 8a^3b - 4a^2b^2 + 2ab^3 + 20a^2 - 10ab + 5b^2$$

69. $(x-6)(x+6) = x^2 - 6^2$
$$= x^2 - 36$$

71. $(a+8)^2 = a^2 + 2 \cdot a \cdot 8 + 8^2$
$$= a^2 + 16a + 64$$

73. $(3y-1)^2 = (3y)^2 - 2 \cdot 3y \cdot 1 + 1^2$
$$= 9y^2 - 6y + 1$$

75. $(5a+3b)(5a-3b) = (5a)^2 - (3b)^2$
$$= 25a^2 - 9b^2$$

77. $(8z+y)^2 = (8z)^2 + 2 \cdot 8z \cdot y + y^2$
$$= 64z^2 + 16zy + y^2$$
$$\text{or}$$
$$= y^2 + 16yz + 64z^2$$

79. $(10x-y)^2 = (10x)^2 - 2 \cdot 10x \cdot y + y^2$
$$= 100x^2 - 20xy + y^2$$

81. $(a^3+2b)(a^3-2b) = (a^3)^2 - (2b)^2$
$$= a^6 - 4b^2$$

83. a. $(f \cdot g)(x) = f(x) \cdot g(x)$
$$= (x+4)(x-1)$$
$$= x \cdot x - x \cdot 1 + 4 \cdot x - 4 \cdot 1$$
$$= x^2 - x + 4x - 4$$
$$= x^2 + 3x - 4$$

b. Since $(f \cdot g)(x) = x^2 + 3x - 4$, we have
$$(f \cdot g)(3) = (3)^2 + 3(3) - 4$$
$$= 9 + 9 - 4$$
$$= 14$$

85. a. $(f \cdot g)(x) = f(x) \cdot g(x)$
$$= (4x-3)(2x+5)$$
$$= 4x \cdot 2x + 4x \cdot 5 - 3 \cdot 2x - 3 \cdot 5$$
$$= 8x^2 + 20x - 6x - 15$$
$$= 8x^2 + 14x - 15$$

b. Since $(f \cdot g)(x) = 8x^2 + 14x - 15$, we have
$$(f \cdot g)(3) = 8(3)^2 + 14(3) - 15$$
$$= 8 \cdot 9 + 42 - 15$$
$$= 72 + 42 - 15$$
$$= 99$$

87. a. $(f \cdot g)(x) = f(x) \cdot g(x)$
$$= (x-2)\left(x^2 + 5x - 3\right)$$
$$= x \cdot x^2 + x \cdot 5x - x \cdot 3 - 2 \cdot x^2$$
$$\qquad - 2 \cdot 5x - 2 \cdot (-3)$$
$$= x^3 + 5x^2 - 3x - 2x^2 - 10x + 6$$
$$= x^3 + 3x^2 - 13x + 6$$

b. Since $(f \cdot g)(x) = x^3 + 3x^2 - 13x + 6$, we have
$$(f \cdot g)(3) = (3)^3 + 3(3)^2 - 13(3) + 6$$
$$= 27 + 3 \cdot 9 - 39 + 6$$
$$= 27 + 27 - 39 + 6$$
$$= 21$$

89. a. $f(x) = x^2 + 1$
$$f(x+2) = (x+2)^2 + 1$$
$$= x^2 + 2 \cdot x \cdot 2 + 2^2 + 1$$
$$= x^2 + 4x + 4 + 1$$
$$= x^2 + 4x + 5$$

b. $f(x+h) - f(x)$
$$= \left[(x+h)^2 + 1 \right] - \left[x^2 + 1 \right]$$
$$= \left[x^2 + 2xh + h^2 + 1 \right] - \left[x^2 + 1 \right]$$
$$= x^2 + 2xh + h^2 + 1 - x^2 - 1$$
$$= 2xh + h^2$$

91. a. $f(x) = x^2 + 5x - 2$
$$f(x+2) = (x+2)^2 + 5(x+2) - 2$$
$$= x^2 + 4x + 4 + 5x + 10 - 2$$
$$= x^2 + 9x + 12$$

b. $f(x+h) - f(x)$
$$= \left[(x+h)^2 + 5(x+h) - 2 \right] - \left[x^2 + 5x - 2 \right]$$
$$= \left[x^2 + 2xh + h^2 + 5x + 5h - 2 \right] - \left[x^2 + 5x - 2 \right]$$
$$= x^2 + 2xh + h^2 + 5x + 5h - 2 - x^2 - 5x + 2$$
$$= 2xh + 5h + h^2$$

93. a. $f(x) = 3x^2 - x + 1$
$$f(x+2) = 3(x+2)^2 - (x+2) + 1$$
$$= 3\left(x^2 + 4x + 4\right) - x - 2 + 1$$
$$= 3x^2 + 12x + 12 - x - 1$$
$$= 3x^2 + 11x + 11$$

b. $f(x+h) - f(x)$
$$= \left[3(x+h)^2 - (x+h) + 1 \right] - \left[3x^2 - x + 1 \right]$$
$$= \left[3\left(x^2 + 2xh + h^2\right) - x - h + 1 \right] - \left[3x^2 - x + 1 \right]$$
$$= 3x^2 + 6xh + 3h^2 - x - h + 1 - 3x^2 + x - 1$$
$$= 6xh - h + 3h^2$$

95. $5ab(a-b)^2 = 5ab\left(a^2 - 2 \cdot a \cdot b + b^2\right)$
$$= 5ab\left(a^2 - 2ab + b^2\right)$$
$$= 5ab \cdot a^2 - 5ab \cdot 2ab + 5ab \cdot b^2$$
$$= 5a^3b - 10a^2b^2 + 5ab^3$$

97. $(x^2 + 3)(x^2 - 3) = (x^2)^2 - (3)^2 = x^4 - 9$

99. $(z^2+9)(z-3)(z+3) = (z^2+9)(z^2-3^2)$
$$= (z^2+9)(z^2-9)$$
$$= (z^2)^2 - 9^2$$
$$= z^4 - 81$$

101. $(2x^3+3)^2 = (2x^3)^2 + 2(2x^3)(3) + 3^2$
$$= 4x^6 + 12x^3 + 9$$

103. $(2m-3n)(4m+n) - (m-2n)^2$
$$= 2m \cdot 4m + 2m \cdot n - 3n \cdot 4m - 3n \cdot n$$
$$\quad - \left(m^2 - 2 \cdot m \cdot 2n + (2n)^2\right)$$
$$= 8m^2 + 2mn - 12mn - 3n^2 - \left(m^2 - 4mn + 4n^2\right)$$
$$= 8m^2 - 10mn - 3n^2 - m^2 + 4mn - 4n^2$$
$$= 7m^2 - 6mn - 7n^2$$

105. $(x+3)\left(x^2-3x+9\right)$
$$= x \cdot x^2 - x \cdot 3x + x \cdot 9 + 3 \cdot x^2 - 3 \cdot 3x + 3 \cdot 9$$
$$= x^3 - 3x^2 + 9x + 3x^2 - 9x + 27$$
$$= x^3 + 27$$

107. $\left(2x - \dfrac{1}{2}\right)^2 = (2x)^2 - 2 \cdot 2x \cdot \dfrac{1}{2} + \left(\dfrac{1}{2}\right)^2$
$$= 4x^2 - 2x + \dfrac{1}{4}$$

109. $(p+2)^3$
$$= (p+2)(p+2)^2$$
$$= (p+2)\left(p^2 + 2 \cdot p \cdot 2 + 2^2\right)$$
$$= (p+2)\left(p^2 + 4p + 4\right)$$
$$= p \cdot p^2 + p \cdot 4p + p \cdot 4 + 2 \cdot p^2 + 2 \cdot 4p + 2 \cdot 4$$
$$= p^3 + 4p^2 + 4p + 2p^2 + 8p + 8$$
$$= p^3 + 6p^2 + 12p + 8$$

111. $(7x-5y+2)(3x-2y+1)$
$$= 7x \cdot 3x - 7x \cdot 2y + 7x \cdot 1 - 5y \cdot 3x + 5y \cdot 2y$$
$$\quad - 5y \cdot 1 + 2 \cdot 3x - 2 \cdot 2y + 2 \cdot 1$$
$$= 21x^2 - 14xy + 7x - 15xy + 10y^2 - 5y$$
$$\quad + 6x - 4y + 2$$
$$= 21x^2 - 29xy + 13x + 10y^2 - 9y + 2$$

113. $(2p-1)(p+3) + (p-3)(p+3)$
$$= 2p(p+3) - 1(p+3) + p^2 - 3^2$$
$$= 2p \cdot p + 2p \cdot 3 - 1 \cdot p - 1 \cdot 3 + p^2 - 9$$
$$= 2p^2 + 6p - p - 3 + p^2 - 9$$
$$= 3p^2 + 5p - 12$$

115. $(x+3)(x-3)\left(x^2-9\right) - (x+1)\left(x^2-3\right)$
$$= \left(x^2-3^2\right)\left(x^2-9\right) - \left[x \cdot x^2 - x \cdot 3 + 1 \cdot x^2 - 1 \cdot 3\right]$$
$$= \left(x^2-9\right)\left(x^2-9\right) - \left(x^3 - 3x + x^2 - 3\right)$$
$$= \left(x^2\right)^2 - 2\left(x^2\right)(9) + 9^2 - x^3 + 3x - x^2 + 3$$
$$= x^4 - 18x^2 + 81 - x^3 + 3x - x^2 + 3$$
$$= x^4 - x^3 - 19x^2 + 3x + 84$$

117.

$A = A_1 + A_2 + A_3$
$$= x^2 + 4x + 2x$$
$$= x^2 + 6x$$

An alternate method is to note that the length is $(x+4)$ and width is $(x+2)$. The area of the corner piece that is missing is $2(4) = 8$ square units. We can multiply the length and width, then subtract off the missing area:
$$A = (x+4)(x+2) - 2(4)$$
$$= x^2 + 2x + 4x + 8 - 8$$
$$= x^2 + 6x$$

119.

$A = A_1 + A_2 + A_3 + A_4$
$$= 4x(x) + 5(x) + 6(4x) + 6(5)$$
$$= 4x^2 + 5x + 24x + 30$$
$$= 4x^2 + 29x + 30$$

An alternate method is to note that the length is $(4x+5)$ and the width is $(x+6)$. Since the area of a rectangle is the product of the length and width, we would get:

$A = (4x+5)(x+6)$

$= 4x^2 + 24x + 5x + 30$

$= 4x^2 + 29x + 30$

Notice how the FOIL method gives each of the individual areas.

121. $A = (2x+4)(3x-1) - (x+2)(2x-5)$

$= (6x^2 - 2x + 12x - 4) - (2x^2 - 5x + 4x - 10)$

$= (6x^2 + 10x - 4) - (2x^2 - x - 10)$

$= 6x^2 + 10x - 4 - 2x^2 + x + 10$

$= 4x^2 + 11x + 6$

The area of the shaded region is $4x^2 + 11x + 6$ square units.

123. a. $A_1 = a^2$; $A_2 = ab$; $A_3 = ab$; $A_4 = b^2$

         a b

  a A_1 A_2

  b A_3 A_4

b. $A = A_1 + A_2 + A_3 + A_4$

$= a^2 + ab + ab + b^2$

$= a^2 + 2ab + b^2$

c. The length of the region is $a+b$ and the width is also $a+b$.

The area of the region would be

$A = (a+b)(a+b)$

$= (a+b)^2$

The result from part (b) is obtained by multiplying this expression out.

125. $[3x - (y+1)][3x + (y+1)]$

$= (3x)^2 - (y+1)^2$

$= 9x^2 - (y^2 + 2y + 1)$

$= 9x^2 - y^2 - 2y - 1$

127. $[2a + (b-3)]^2 = (2a)^2 + 2 \cdot 2a \cdot (b-3) + (b-3)^2$

$= 4a^2 + 4a(b-3) + (b^2 - 2 \cdot b \cdot 3 + 3^2)$

$= 4a^2 + 4ab - 12a + b^2 - 6b + 9$

129. $(2^x + 3)(2^x - 4) = 2^x \cdot 2^x - 2^x \cdot 4 + 3 \cdot 2^x - 3 \cdot 4$

$= 2^{x+x} - 2^x - 12$

$= 2^{2x} - 2^x - 12$

131. $(5^y - 1)^2 = (5^y)^2 - 2 \cdot 5^y \cdot 1 + 1^2$

$= 5^{2y} - 2(5^y) + 1$

Section 4.3

Are You Ready for This Section?

R1. $\dfrac{15x^5}{12x^3} = \dfrac{15}{12} \cdot \dfrac{x^5}{x^3} = \dfrac{5}{4} x^{5-3} = \dfrac{5}{4} x^2$

R2. $\dfrac{2}{7} + \dfrac{5}{7} = \dfrac{2+5}{7} = \dfrac{7}{7} = 1$

Section 4.3 Quick Checks

1. The first step in simplifying $\dfrac{4x^4 + 8x^2}{2x}$ is to

rewrite $\dfrac{4x^4 + 8x^2}{2x}$ as $\dfrac{4x^4}{2x} + \dfrac{8x^2}{2x}$.

2. $\dfrac{9p^4 - 12p^3 + 3p^2}{3p} = \dfrac{9p^4}{3p} - \dfrac{12p^3}{3p} + \dfrac{3p^2}{3p}$

$= \dfrac{9}{3} p^{4-1} - \dfrac{12}{3} p^{3-1} + \dfrac{3}{3} p^{2-1}$

$= 3p^3 - 4p^2 + p$

3. $\dfrac{20a^5 - 10a^2 + 1}{5a^3} = \dfrac{20a^5}{5a^3} - \dfrac{10a^2}{5a^3} + \dfrac{1}{5a^3}$

$= \dfrac{20}{5} a^{5-3} - \dfrac{10}{5} a^{2-3} + \dfrac{1}{5} a^{-3}$

$= 4a^2 - 2a^{-1} + \dfrac{1}{5} a^{-3}$

$= 4a^2 - \dfrac{2}{a} + \dfrac{1}{5a^3}$

4.
$$\frac{x^4 y^4 + 8x^2 y^2 - 4xy}{4x^3 y}$$

$$= \frac{x^4 y^4}{4x^3 y} + \frac{8x^2 y^2}{4x^3 y} - \frac{4xy}{4x^3 y}$$

$$= \frac{1}{4} x^{4-3} y^{4-1} + \frac{8}{4} x^{2-3} y^{2-1} - \frac{4}{4} x^{1-3} y^{1-1}$$

$$= \frac{1}{4} xy^3 + 2x^{-1} y - x^{-2}$$

$$= \frac{xy^3}{4} + \frac{2y}{x} - \frac{1}{x^2}$$

5. True

6. Because $\dfrac{3x^2 + 2x - 1}{x+1} = 3x - 1,$ the remainder

when dividing $3x^2 + 2x - 1$ by $x + 1$ is $\underline{0}$, and

$3x^2 + 2x - 1 = \underline{(3x-1)} \cdot \underline{(x+1)}$. We call $x + 1$ and

$3x - 1$ <u>factors</u> of $3x^2 + 2x - 1$.

7. Given that

$\dfrac{6x^3 - x^2 - 9x + 8}{x+1} = 6x^2 - 7x - 2 + \dfrac{10}{x+1},$ we call

$6x^3 - x^2 - 9x + 8$ the <u>dividend</u>, $x + 1$ the <u>divisor</u>,

and 10 the <u>remainder</u>.

8. To check a division problem, we verify
<u>Quotient</u> · <u>Divisor</u> + <u>Remainder</u> = <u>Dividend</u>.

9.
$$\begin{array}{r} x^2 + 7x - 3 \\ x-4\overline{\smash{\big)}\,x^3 + 3x^2 - 31x + 21} \\ \underline{-\left(x^3 - 4x^2\right)} \\ 7x^2 - 31x \\ \underline{-\left(7x^2 - 28x\right)} \\ -3x + 21 \\ \underline{-(-3x + 12)} \\ 9 \end{array}$$

$$\frac{x^3 + 3x^2 - 31x + 21}{x - 4} = x^2 + 7x - 3 + \frac{9}{x-4}$$

Check:

$$(x-4)\left(x^2 + 7x - 3\right) + 9$$

$$= x^3 + 7x^2 - 3x - 4x^2 - 28x + 12 + 9$$

$$= x^3 + 3x^2 - 31x + 21$$

so our answer checks.

10.
$$\begin{array}{r} x^2 + 5x + 4 \\ 2x-3\overline{\smash{\big)}\,2x^3 + 7x^2 - 7x - 12} \\ \underline{-\left(2x^3 - 3x^2\right)} \\ 10x^2 - 7x \\ \underline{-\left(10x^2 - 15x\right)} \\ 8x - 12 \\ \underline{-(8x - 12)} \\ 0 \end{array}$$

$$\frac{2x^3 + 7x^2 - 7x - 12}{2x - 3} = x^2 + 5x + 4$$

Check:

$$(2x-3)\left(x^2 + 5x + 4\right)$$

$$= 2x^3 + 10x^2 + 8x - 3x^2 - 15x - 12$$

$$= 2x^3 + 7x^2 - 7x - 12$$

so our answer checks.

11.
$$\begin{array}{r} x^3 - 5x^2 + 2 \\ x^2-2\overline{\smash{\big)}\,x^5 - 5x^4 - 2x^3 + 12x^2 + 2} \\ \underline{-\left(x^5 \qquad - 2x^3\right)} \\ -5x^4 \qquad + 12x^2 \\ \underline{-\left(-5x^4 \qquad + 10x^2\right)} \\ 2x^2 + 2 \\ \underline{-\left(2x^2 - 4\right)} \\ 6 \end{array}$$

$$\frac{2 + 12x^2 - 2x^3 - 5x^4 + x^5}{x^2 - 2} = x^3 - 5x^2 + 2 + \frac{6}{x^2 - 2}$$

Check:

$$\left(x^2 - 2\right)\left(x^3 - 5x^2 + 2\right) + 6$$

$$= x^5 - 5x^4 + 2x^2 - 2x^3 + 10x^2 - 4 + 6$$

$$= x^5 - 5x^4 - 2x^3 + 12x^2 + 2$$

so our answer checks.

12. False; to use synthetic division, the divisor must be linear and in the form $x - c$.

13. False; to use synthetic division, the divisor must be linear and in the form $x - c$.

14. To divide a polynomial by $x + 7$, we use $c = \underline{-7}$

15. The divisor is $x-2$ so $c=2$.

$$2\overline{)\begin{array}{rrr} 2 & 1 & -7 & -13 \\ & 4 & 10 & 6 \\ \hline 2 & 5 & 3 & -7 \end{array}}$$

$$\frac{2x^3+x^2-7x-13}{x-2}=2x^2+5x+3-\frac{7}{x-2}$$

16. $\dfrac{x+1-3x^2+4x^3}{x+2}=\dfrac{4x^3-3x^2+x+1}{x+2}$

The divisor is $x+2$ so $c=-2$.

$$-2\overline{)\begin{array}{rrrr} 4 & -3 & 1 & 1 \\ & -8 & -8 & 22 & -46 \\ \hline 4 & -11 & 23 & -45 \end{array}}$$

$$\frac{x+1-3x^2+4x^3}{x+2}=4x^2-11x+23-\frac{45}{x+2}$$

17. The divisor is $x+3$ so $c=-3$.

$$-3\overline{)\begin{array}{rrrrr} 1 & 8 & 15 & -2 & -6 \\ & -3 & -15 & 0 & 6 \\ \hline 1 & 5 & 0 & -2 & 0 \end{array}}$$

$$\frac{x^4+8x^3+15x^2-2x-6}{x+3}=x^3+5x^2-2$$

18. $f(x)=3x^4-4x^3-3x^2+10x-5$;

$\qquad g(x)=x^2-2$

a.
$$\begin{array}{r} 3x^2-4x+3 \\ x^2-2\overline{)3x^4-4x^3-3x^2+10x-5} \\ \underline{-\left(3x^4\qquad -6x^2\right)} \\ -4x^3+3x^2+10x \\ \underline{-\left(-4x^3\qquad +8x\right)} \\ 3x^2+2x-5 \\ \underline{-\left(3x^2\qquad -6\right)} \\ 2x+1 \end{array}$$

Thus, $\left(\dfrac{f}{g}\right)(x)=3x^2-4x+3+\dfrac{2x+1}{x^2-2}$.

b. $\left(\dfrac{f}{g}\right)(3)=3(3)^2-4(3)+3+\dfrac{2(3)+1}{(3)^2-2}$

$\qquad\qquad =3(9)-4(3)+3+\dfrac{6+1}{9-2}$

$\qquad\qquad =27-12+3+\dfrac{7}{7}$

$\qquad\qquad =19$

19. $f(x)=3x^3+10x^2-9x-4$

a. The divisor is $x-2$ so the Remainder Theorem says that the remainder is $f(2)$.

$\quad f(2)=3(2)^3+10(2)^2-9(2)-4$

$\qquad\quad =3(8)+10(4)-9(2)-4$

$\qquad\quad =24+40-18-4$

$\qquad\quad =42$

When $f(x)=3x^3+10x^2-9x-4$ is divided by $x-2$, the remainder is 42.

b. The divisor is $x+4=x-(-4)$, so the Remainder Theorem says that the remainder is $f(-4)$.

$\quad f(-4)=3(-4)^3+10(-4)^2-9(-4)-4$

$\qquad\quad =3(-64)+10(16)-9(-4)-4$

$\qquad\quad =-192+160+36-4$

$\qquad\quad =0$

When $f(x)=3x^3+10x^2-9x-4$ is divided by $x+4$, the remainder is 0.

20. $f(x)=2x^3-9x^2-6x+5$

a. $f(-2)=2(-2)^3-9(-2)^2-6(-2)+5$

$\qquad\quad =2(-8)-9(4)-6(-2)+5$

$\qquad\quad =-16-36+12+5$

$\qquad\quad =-35$

Since $f(c)=f(-2)\neq 0$, we know that $x-c=x+2$ is not a factor of $f(x)$.

b. $f(5)=2(5)^3-9(5)^2-6(5)+5$

$\qquad\quad =2(125)-9(25)-6(5)+5$

$\qquad\quad =250-225-30+5$

$\qquad\quad =0$

Since $f(c)=f(5)=0$, we know that $x-c=x-5$ is a factor of $f(x)$.

$$5\overline{)\;2\quad -9\quad -6\quad\;\;5\;}$$
$$\underline{\;\;\;10\quad\;5\quad -5\;}$$
$$2\quad\;\;1\quad -1\quad\;\;0$$

$$f(x) = (x-5)\left(2x^2 + x - 1\right)$$

4.3 Exercises

21. $\dfrac{8x^2 + 12x}{4x} = \dfrac{8x^2}{4x} + \dfrac{12x}{4x}$

$$= \dfrac{8}{4}x^{2-1} + \dfrac{12}{4}x^{1-1}$$

$$= 2x + 3$$

23. $\dfrac{2a^3 - 15a^2 + 10a}{5a} = \dfrac{2a^3}{5a} - \dfrac{15a^2}{5a} + \dfrac{10a}{5a}$

$$= \dfrac{2}{5}a^{3-1} - \dfrac{15}{5}a^{2-1} + \dfrac{10}{5}a^{1-1}$$

$$= \dfrac{2}{5}a^2 - 3a + 2$$

25. $\dfrac{2y^3 + 6y}{4y^2} = \dfrac{2y^3}{4y^2} + \dfrac{6y}{4y^2}$

$$= \dfrac{2}{4}y^{3-2} + \dfrac{6}{4}y^{1-2}$$

$$= \dfrac{1}{2}y + \dfrac{3}{2}y^{-1}$$

$$= \dfrac{1}{2}y + \dfrac{3}{2y}$$

27. $\dfrac{4m^2n^2 + 6m^2n - 18mn^2}{4m^2n^2}$

$$= \dfrac{4m^2n^2}{4m^2n^2} + \dfrac{6m^2n}{4m^2n^2} - \dfrac{18mn^2}{4m^2n^2}$$

$$= \dfrac{4}{4}m^{2-2}n^{2-2} + \dfrac{6}{4}m^{2-2}n^{1-2} - \dfrac{18}{4}m^{1-2}n^{2-2}$$

$$= m^0n^0 + \dfrac{3}{2}m^0n^{-1} - \dfrac{9}{2}m^{-1}n^0$$

$$= 1 + \dfrac{3}{2n} - \dfrac{9}{2m}$$

29.
$$x+2\overline{)\;x^2 + 5x + 6\;}$$
with quotient $x+3$
$$\underline{-\left(x^2 + 2x\right)}$$
$$3x + 6$$
$$\underline{-(3x+6)}$$
$$0$$

$$\dfrac{x^2 + 5x + 6}{x+2} = x+3$$

31.
$$x-2\overline{)\;2x^2 + x - 4\;}$$
with quotient $2x+5$
$$\underline{-\left(2x^2 - 4x\right)}$$
$$5x - 4$$
$$\underline{-(5x-10)}$$
$$6$$

$$\dfrac{2x^2 + x - 4}{x-2} = 2x + 5 + \dfrac{6}{x-2}$$

33.
$$2w-7\overline{)\;2w^2 + 5w - 49\;}$$
with quotient $w+6$
$$\underline{-\left(2w^2 - 7w\right)}$$
$$12w - 49$$
$$\underline{-(12w-42)}$$
$$-7$$

$$\dfrac{2w^2 + 5w - 49}{2w-7} = w + 6 - \dfrac{7}{2w-7}$$

35.
$$x+3\overline{)\;x^3 + 8x^2 + x - 42\;}$$
with quotient $x^2 + 5x - 14$
$$\underline{-\left(x^3 + 3x^2\right)}$$
$$5x^2 + x$$
$$\underline{-\left(5x^2 + 15x\right)}$$
$$-14x - 42$$
$$\underline{-(-14x - 42)}$$
$$0$$

$$\dfrac{x^3 + 8x^2 + x - 42}{x+3} = x^2 + 5x - 14$$

37.

$$
\begin{array}{r}
w^2-4w-5 \\
w+4\overline{\smash{\big)}\,w^3+0w^2-21w-20} \\
\underline{-\left(w^3+4w^2\right)} \\
-4w^2-21w \\
\underline{-\left(-4w^2-16w\right)} \\
-5w-20 \\
\underline{-(-5w-20)} \\
0
\end{array}
$$

$$\frac{w^3-21w-20}{w+4}=w^2-4w-5$$

39.

$$
\begin{array}{r}
3x^2-14x-5 \\
2x+5\overline{\smash{\big)}\,6x^3-13x^2-80x-25} \\
\underline{-\left(6x^3+15x^2\right)} \\
-28x^2-80x \\
\underline{-\left(-28x^2-70x\right)} \\
-10x-25 \\
\underline{-(-10x-25)} \\
0
\end{array}
$$

$$\frac{6x^3-13x^2-80x-25}{2x+5}=3x^2-14x-5$$

41.

$$
\begin{array}{r}
x-7 \\
x^2+3\overline{\smash{\big)}\,x^3-7x^2+3x-21} \\
\underline{-\left(x^3+3x\right)} \\
-7x^2-21 \\
\underline{-\left(-7x^2-21\right)} \\
0
\end{array}
$$

$$\frac{x^3-7x^2+3x-21}{x^2+3}=x-7$$

43.

$$3z^2+1\overline{)3z^3+21z^2+5z+2}\quad\overset{z+7}{}$$

$$\underline{-\left(3z^3\qquad+z\right)}$$

$$21z^2+4z+2$$

$$\underline{-\left(21z^2\qquad+7\right)}$$

$$4z-5$$

$$\frac{3z^3+21z^2+5z+2}{3z^2+1}=z+7+\frac{4z-5}{3z^2+1}$$

45.

$$x^2+2x+5\overline{)2x^4-11x^3+8x^2-22x+144}\quad\overset{2x^2-15x+28}{}$$

$$\underline{-\left(2x^4+4x^3+10x^2\right)}$$

$$-15x^3-2x^2-22x$$

$$\underline{-\left(-15x^3-30x^2-75x\right)}$$

$$28x^2+53x+144$$

$$\underline{-\left(28x^2+56x+140\right)}$$

$$-3x+4$$

$$\frac{2x^4-11x^3+8x^2-22x+144}{x^2+2x+5}=2x^2-15x+28+\frac{-3x+4}{x^2+2x+5}$$

47. The divisor is $x-5$ so $c=5$.

$$5\overline{)\begin{array}{rrr}1 & -3 & -10\\ & 5 & 10\end{array}}$$
$$\begin{array}{rrr}1 & 2 & 0\end{array}$$

$$\frac{x^2-3x-10}{x-5}=x+2$$

49. The divisor is $x+4$ so $c=-4$.

$$-4\overline{)\begin{array}{rrr}2 & 11 & 12\\ & -8 & -12\end{array}}$$
$$\begin{array}{rrr}2 & 3 & 0\end{array}$$

$$\frac{2x^2+11x+12}{x+4}=2x+3$$

51. The divisor is $x-6$ so $c=6$.

$$6\overline{)\begin{array}{rrr}1 & -3 & -14\\ & 6 & 18\end{array}}$$
$$\begin{array}{rrr}1 & 3 & 4\end{array}$$

$$\frac{x^2-3x-14}{x-6}=x+3+\frac{4}{x-6}$$

53. The divisor is $x-5$ so $c=5$.

$$\begin{array}{r|rrr} 5 & 1 & 0 & -19 & -15 \\ & & 5 & 25 & 30 \\ \hline & 1 & 5 & 6 & 15 \end{array}$$

$$\frac{x^3-19x-15}{x-5}=x^2+5x+6+\frac{15}{x-5}$$

55. The divisor is $x-3$ so $c=3$.

$$\begin{array}{r|rrrrr} 3 & 3 & -5 & -21 & 17 & 25 \\ & & 9 & 12 & -27 & -30 \\ \hline & 3 & 4 & -9 & -10 & -5 \end{array}$$

$$\frac{3x^4-5x^3-21x^2+17x+25}{x-3}$$
$$=3x^3+4x^2-9x-10-\frac{5}{x-3}$$

57. The divisor is $x+6$ so $c=-6$.

$$\begin{array}{r|rrrrr} -6 & 1 & 0 & -40 & 0 & 109 \\ & & -6 & 36 & 24 & -144 \\ \hline & 1 & -6 & -4 & 24 & -35 \end{array}$$

$$\frac{x^4-40x^2+109}{x+6}=x^3-6x^2-4x+24-\frac{35}{x+6}$$

59. The divisor is $x-\frac{5}{2}$ so $c=\frac{5}{2}$.

$$\begin{array}{r|rrrr} \frac{5}{2} & 2 & 3 & -14 & -15 \\ & & 5 & 20 & 15 \\ \hline & 2 & 8 & 6 & 0 \end{array}$$

$$\frac{2x^3+3x^2-14x-15}{x-\frac{5}{2}}=2x^2+8x+6$$

61. a. $\left(\frac{f}{g}\right)(x)=\frac{4x^3-8x^2+12x}{4x}$

$$=\frac{4x^3}{4x}-\frac{8x^2}{4x}+\frac{12x}{4x}$$
$$=\frac{4}{4}x^{3-1}-\frac{8}{4}x^{2-1}+\frac{12}{4}x^{1-1}$$
$$=x^2-2x+3$$

b. Using the result from part **a**, we have:

$$\left(\frac{f}{g}\right)(2)=(2)^2-2(2)+3$$
$$=4-4+3$$
$$=3$$

63. a. $\left(\frac{f}{g}\right)(x)=\frac{x^2-x-12}{x-4}$

The divisor is $x-4$ so $c=4$.

$$\begin{array}{r|rrr} 4 & 1 & -1 & -12 \\ & & 4 & 12 \\ \hline & 1 & 3 & 0 \end{array}$$

$$\left(\frac{f}{g}\right)(x)=x+3$$

b. $\left(\frac{f}{g}\right)(2)=2+3=5$

65. a. $\left(\frac{f}{g}\right)(x)=\frac{2x^2+5x-1}{x+3}$

The divisor is $x+3$ so $c=-3$.

$$\begin{array}{r|rrr} -3 & 2 & 5 & -1 \\ & & -6 & 3 \\ \hline & 2 & -1 & 2 \end{array}$$

$$\left(\frac{f}{g}\right)(x)=2x-1+\frac{2}{x+3}$$

b. $\left(\frac{f}{g}\right)(2)=2(2)-1+\frac{2}{2+3}$
$$=4-1+\frac{2}{5}$$
$$=\frac{17}{5}$$

67. a. $\left(\frac{f}{g}\right)(x)=\frac{2x^3+9x^2+x-12}{2x+3}$

$$\begin{array}{r} x^2+3x-4 \\ 2x+3{\overline{\smash{\big)}\,2x^3+9x^2+x-12}} \\ \underline{-(2x^3+3x^2)} \\ 6x^2+x \\ \underline{-(6x^2+9x)} \\ -8x-12 \\ \underline{-(-8x-12)} \\ 0 \end{array}$$

$$\frac{2x^3+9x^2+x-12}{2x+3}=x^2+3x-4$$

b. $\left(\dfrac{f}{g}\right)(2) = (2)^2 + 3(2) - 4$

$\qquad\qquad = 4 + 6 - 4$

$\qquad\qquad = 6$

69. a. $\left(\dfrac{f}{g}\right)(x) = \dfrac{x^3 - 13x - 12}{x^2 - 9}$

$$
\begin{array}{r}
x \\
x^2 - 9 \overline{) x^3 + 0x^2 - 13x - 12} \\
-\left(x^3 \qquad -9x \right) \\
\hline
-4x - 12
\end{array}
$$

$\dfrac{x^3 - 13x - 12}{x^2 - 9} = x - \dfrac{4x + 12}{x^2 - 9}$

$\qquad\qquad\qquad = x - \dfrac{4(x+3)}{(x-3)(x+3)}$

$\qquad\qquad\qquad = x - \dfrac{4}{x-3}$

b. $\left(\dfrac{f}{g}\right)(2) = 2 - \dfrac{4(2)+12}{(2)^2 - 9}$

$\qquad\qquad = 2 - \dfrac{8+12}{4-9}$

$\qquad\qquad = 2 - \dfrac{20}{-5}$

$\qquad\qquad = 2 + 4$

$\qquad\qquad = 6$

71. The divisor is $x - 2$ so $c = 2$.

$f(2) = (2)^2 - 5(2) + 1$

$\qquad = 4 - 10 + 1$

$\qquad = -5$

The remainder is -5.

73. The divisor is $x + 4$ so $c = -4$.

$f(-4) = (-4)^3 - 2(-4)^2 + 5(-4) - 3$

$\qquad = -64 - 2(16) - 20 - 3$

$\qquad = -64 - 32 - 20 - 3$

$\qquad = -119$

The remainder is -119.

75. The divisor is $x - 5$ so $c = 5$.

$f(5) = 2(5)^3 - 4(5) + 1$

$\qquad = 2(125) - 20 + 1$

$\qquad = 250 - 20 + 1$

$\qquad = 231$

The remainder is 231.

77. The divisor is $x - 1$ so $c = 1$.

$f(1) = (1)^4 + 1$

$\qquad = 1 + 1$

$\qquad = 2$

The remainder is 2.

79. $c = 2$

$f(2) = (2)^2 - 3(2) + 2$

$\qquad = 4 - 6 + 2$

$\qquad = 0$

Since the remainder is 0, $x - 2$ is a factor.

$$
\begin{array}{r}
2 {\overline{\smash{)}\,1 \quad -3 \quad 2}} \\
2 \quad -2 \\
\hline
1 \quad -1 \quad 0
\end{array}
$$

$f(x) = (x-2)(x-1)$

81. $c = -2$

$f(-2) = 2(-2)^2 + 5(-2) + 2$

$\qquad = 2(4) - 10 + 2$

$\qquad = 8 - 10 + 2$

$\qquad = 0$

Since the remainder is 0, $x + 2$ is a factor.

$$
\begin{array}{r}
-2 {\overline{\smash{)}\,2 \quad 5 \quad 2}} \\
-4 \quad -2 \\
\hline
2 \quad 1 \quad 0
\end{array}
$$

$f(x) = (x+2)(2x+1)$

83. $c = 3$

$f(3) = 4(3)^3 - 9(3)^2 - 49(3) - 30$

$\qquad = 4(27) - 9(9) - 147 - 30$

$\qquad = 108 - 81 - 147 - 30$

$\qquad = -150$

Since the remainder is not 0, $x - 3$ is not a factor of $f(x)$.

85. $c = -1$

$f(-1) = 4(-1)^3 - 7(-1)^2 - 5(-1) + 6$

$\qquad = -4 - 7 + 5 + 6$

$\qquad = 0$

Since the remainder is 0, $x + 1$ is a factor.

$$
\begin{array}{r}
-1 {\overline{\smash{)}\,4 \quad -7 \quad -5 \quad 6}} \\
-4 \quad 11 \quad -6 \\
\hline
4 \quad -11 \quad 6 \quad 0
\end{array}
$$

$f(x) = (x+1)\left(4x^2 - 11x + 6\right)$

87. $\dfrac{3a^3b^2 - 9a^2b + 18ab}{3ab}$

$= \dfrac{3a^3b^2}{3ab} - \dfrac{9a^2b}{3ab} + \dfrac{18ab}{3ab}$

$= \dfrac{3}{3}a^{3-1}b^{2-1} - \dfrac{9}{3}a^{2-1}b^{1-1} + \dfrac{18}{3}a^{1-1}b^{1-1}$

$= a^2b - 3ab^0 + 6a^0b^0$

$= a^2b - 3a + 6$

89.
$$\begin{array}{r} y+3 \\ 3y+2\overline{)3y^2+11y+6} \end{array}$$
$$\underline{-\left(3y^2+2y\right)}$$
$$9y+6$$
$$\underline{-(9y+6)}$$
$$0$$

$\dfrac{3y^2+11y+6}{3y+2} = y+3$

91.
$$\begin{array}{r} x+6 \\ x^2+5\overline{)x^3+6x^2+8x+32} \end{array}$$
$$\underline{-\left(x^3 \qquad +5x\right)}$$
$$6x^2+3x+32$$
$$\underline{-\left(6x^2 \qquad +30\right)}$$
$$3x+2$$

$\dfrac{x^3+6x^2+8x+32}{x^2+5} = x+6+\dfrac{3x+2}{x^2+5}$

93. $\dfrac{8x^3+6x}{12x^2} = \dfrac{8x^3}{12x^2} + \dfrac{6x}{12x^2} = \dfrac{2x^{3-2}}{3} + \dfrac{1}{2x^{2-1}}$

$= \dfrac{2}{3}x + \dfrac{1}{2x}$

95. The divisor is $x+4$ so $c=-4$.
$$\begin{array}{r} -4\overline{)\,1 \quad 7 \quad 2 \quad -46} \\ \underline{\quad\; -4 \;\; -12 \;\; 40} \\ 1 \quad 3 \;\; -10 \;\; -6 \end{array}$$

$\dfrac{x^3+7x^2+2x-46}{x+4} = x^2+3x-10-\dfrac{6}{x+4}$

97.
$$\begin{array}{r} x^2-8 \\ 3x+5\overline{)3x^3+5x^2-24x-40} \end{array}$$
$$\underline{-\left(3x^3+5x^2\right)}$$
$$0-24x-40$$
$$\underline{-(-24x-40)}$$
$$0$$

$\dfrac{3x^3+5x^2-24x-40}{3x+5} = x^2-8$

99.
$$\begin{array}{r} 4x^2+6x+9 \\ 2x-3\overline{)8x^3+\;0x^2+\;0x\;-27} \end{array}$$
$$\underline{-\left(8x^3-12x^2\right)}$$
$$12x^2 \qquad -27$$
$$\underline{-\left(12x^2-18x\right)}$$
$$18x-27$$
$$\underline{-(18x-27)}$$
$$0$$

$\dfrac{8x^3-27}{2x-3} = 4x^2+6x+9$

101. $\dfrac{f(x)}{x-5} = 3x+5$

$f(x) = (x-5)(3x+5)$

$= x\cdot 3x + x\cdot 5 - 5\cdot 3x - 5\cdot 5$

$= 3x^2+5x-15x-25$

$= 3x^2-10x-25$

103. $\dfrac{f(x)}{x-3} = x+8+\dfrac{4}{x-3}$

$f(x) = (x-3)\left(x+8+\dfrac{4}{x-3}\right)$

$= x(x-3)+8(x-3)+\dfrac{4}{x-3}(x-3)$

$= x^2-3x+8x-24+4$

$= x^2+5x-20$

105. Since the area of a rectangle is the product of the length and width, we can find the length by dividing the area by the width.

$$
\begin{array}{r}
5x+2 \\
3x-1{\overline{\smash{\big)}\,15x^2 + x - 2}} \\
\underline{-\left(15x^2 - 5x\right)} \\
6x - 2 \\
\underline{-(6x - 2)} \\
0
\end{array}
$$

The area of the rectangle is given by

$A = 15x^2 + x - 2 = (3x-1)(5x+2)$.

The length of the rectangle is $5x+2$ feet.

107. The volume of a rectangular box is the product of the length, width, and height. The product of the width and height is given by

$(x+5)(x-3) = x^2 - 3x + 5x - 15$

$\qquad\qquad\quad = x^2 + 2x - 15$

To find the length, we need to divide the volume by this quantity.

$$
\begin{array}{r}
2x+5 \\
x^2+2x-15{\overline{\smash{\big)}\,2x^3 + 9x^2 - 20x - 75}} \\
\underline{-\left(2x^3 + 4x^2 - 30x\right)} \\
5x^2 + 10x - 75 \\
\underline{-\left(5x^2 + 10x - 75\right)} \\
0
\end{array}
$$

The volume of the box is given by

$V = 2x^3 + 9x^2 - 20x - 75 = (2x+5)(x+5)(x-3)$

The length of the box is $2x+5$ centimeters.

109. a. $\overline{C}(x) = \dfrac{0.01x^3 - 0.4x^2 + 13x + 400}{x}$

$\qquad = \dfrac{0.01x^3}{x} - \dfrac{0.4x^2}{x} + \dfrac{13x}{x} + \dfrac{400}{x}$

$\qquad = 0.01x^2 - 0.4x + 13 + \dfrac{400}{x}$

b. $\overline{C}(50) = 0.01(50)^2 - 0.4(50) + 13 + \dfrac{400}{50}$

$\qquad\quad = 25 - 20 + 13 + 8$

$\qquad\quad = 26$

The average cost of manufacturing $x = 50$ wristwatches in a day is \$26.

111. $\dfrac{2x^3 - 3x^2 - 26x - 37}{x+2} = ax^2 + bx + c + \dfrac{d}{x+2}$

Since the divisor is linear, we can use synthetic division.

$$
\begin{array}{r|rrrr}
-2 & 2 & -3 & -26 & -37 \\
   &   & -4 & 14 & 24 \\
\hline
   & 2 & -7 & -12 & -13
\end{array}
$$

$\dfrac{2x^3 - 3x^2 - 26x - 37}{x+2} = 2x^2 - 7x - 12 + \dfrac{-13}{x+2}$.

This gives us $a = 2$, $b = -7$, $c = -12$, and $d = -13$.

Thus,

$a + b + c + d = 2 + (-7) + (-12) + (-13) = -30$

113. For polynomial division we have

$\text{dividend} = \text{divisor} \cdot \text{quotient} + \text{remainder}$

The dividend is the polynomial f and has degree n. Since the remainder must be 0 or a polynomial that has a lower degree than f, the degree n must be obtained from the product of the divisor and the quotient. The divisor is $x+4$ which is of degree 1 so the quotient must be of degree $n-1$.

115. Yes, $3x+4$ is a factor since the remainder is 0. We can write

$6x^3 - x^2 - 9x + 4 = (3x+4)(2x^2 - 3x + 1)$

Putting the Concepts Together (Sections 4.1–4.3)

1. $5m^4 - 2m^3 + 3m + 8$

Degree: 4

2. $\left(7a^2 - 4a^3 + 7a - 1\right) + \left(2a^2 - 6a - 7\right)$

$= 7a^2 - 4a^3 + 7a - 1 + 2a^2 - 6a - 7$

$= -4a^3 + 7a^2 + 2a^2 + 7a - 6a - 1 - 7$

$= -4a^3 + (7+2)a^2 + (7-6)a + (-1-7)$

$= -4a^3 + 9a^2 + a - 8$

3. $\left(\tfrac{1}{5}y^2 + 2y - 6\right) - \left(4y^2 - y + 2\right)$

$= \tfrac{1}{5}y^2 + 2y - 6 - 4y^2 + y - 2$

$= \tfrac{1}{5}y^2 - 4y^2 + 2y + y - 6 - 2$

$= -\tfrac{19}{5}y^2 + 3y - 8$

4. $f(x) = 2x^3 - x^2 + 4x + 9$

$f(2) = 2(2)^3 - (2)^2 + 4(2) + 9$

$\quad = 2(8) - 4 + 8 + 9$

$\quad = 16 - 4 + 8 + 9$

$\quad = 29$

5. $(f+g)(x) = (6x+5) + (-x^2 + 2x + 3)$

$\quad = 6x + 5 - x^2 + 2x + 3$

$\quad = -x^2 + 8x + 8$

$(f+g)(-3) = -(-3)^2 + 8(-3) + 8$

$\quad = -9 - 24 + 8$

$\quad = -25$

6. $(f-g)(x) = (2x^2 + 7) - (x^2 - 4x - 3)$

$\quad = 2x^2 + 7 - x^2 + 4x + 3$

$\quad = 2x^2 - x^2 + 4x + 7 + 3$

$\quad = x^2 + 4x + 10$

7. $2mn^3(m^2n - 4mn + 6)$

$= 2mn^3 \cdot m^2n - 2mn^3 \cdot 4mn + 2mn^3 \cdot 6$

$= 2m^{1+2}n^{3+1} - 8m^{1+1}n^{3+1} + 12mn^3$

$= 2m^3n^4 - 8m^2n^4 + 12mn^3$

8. $(3a - 5b)^2 = (3a)^2 - 2(3a)(5b) + (5b)^2$

$\quad = 9a^2 - 30ab + 25b^2$

9. $(7n^2 + 3)(7n^2 - 3) = (7n^2)^2 - (3)^2$

$\quad = 49n^4 - 9$

10. $(3a + 2b)(6a^2 - 2ab + b^2)$

$= 3a \cdot 6a^2 - 3a \cdot 2ab + 3a \cdot b^2 + 2b \cdot 6a^2$

$\quad - 2b \cdot 2ab + 2b \cdot b^2$

$= 18a^3 - 6a^2b + 3ab^2 + 12a^2b - 4ab^2 + 2b^3$

$= 18a^3 + 6a^2b - ab^2 + 2b^3$

11. $(f \cdot g)(x) = (x + 2)(x^2 - 4x + 11)$

$\quad = x \cdot x^2 - x \cdot 4x + x \cdot 11 + 2 \cdot x^2$

$\quad\quad - 2 \cdot 4x + 2 \cdot 11$

$\quad = x^3 - 4x^2 + 11x + 2x^2 - 8x + 22$

$\quad = x^3 - 2x^2 + 3x + 22$

12.

$$
\begin{array}{r}
5z^2 + 3z - 7 \\
2z+7\overline{\smash{\big)}\,10z^3 + 41z^2 + 7z - 49} \\
\underline{-(10z^3 + 35z^2)} \\
6z^2 + 7z \\
\underline{-(6z^2 + 21z)} \\
-14z - 49 \\
\underline{-(-14z - 49)} \\
0
\end{array}
$$

$\dfrac{10z^3 + 41z^2 + 7z - 49}{2z + 7} = 5z^2 + 3z - 7$

13. The divisor is $x + 9$ so $c = -9$.

$$
\begin{array}{r|rrrr}
-9 & 2 & 25 & 62 & -6 \\
   &   & -18 & -63 & 9 \\
\hline
   & 2 & 7 & -1 & 3
\end{array}
$$

$\dfrac{2x^3 + 25x^2 + 62x - 6}{x + 9} = 2x^2 + 7x - 1 + \dfrac{3}{x + 9}$

14. $\left(\dfrac{f}{g}\right)(x) = \dfrac{x^3 + 2x^2 - 4x + 5}{x - 1}$

The divisor is $x - 1$ so $c = 1$.

$$
\begin{array}{r|rrrr}
1 & 1 & 2 & -4 & 5 \\
  &   & 1 & 3 & -1 \\
\hline
  & 1 & 3 & -1 & 4
\end{array}
$$

$\left(\dfrac{f}{g}\right)(x) = x^2 + 3x - 1 + \dfrac{4}{x - 1}$

15. The divisor is $x + 5$ so $c = -5$.

$f(x) = 3x^3 + 8x^2 - 23x + 60$

$f(-5) = 3(-5)^3 + 8(-5)^2 - 23(-5) + 60$

$\quad = 3(-125) + 8(25) + 115 + 60$

$\quad = -375 + 200 + 115 + 60$

$\quad = 0$

Since $f(-5) = 0$, the remainder is 0 and $x + 5$ is a factor of $f(x)$.

$$
\begin{array}{r|rrrr}
-5 & 3 & 8 & -23 & 60 \\
   &   & -15 & 35 & -60 \\
\hline
   & 3 & -7 & 12 & 0
\end{array}
$$

$f(x) = (x + 5)(3x^2 - 7x + 12)$

Section 4.4

Are You Ready for This Section?

R1. $24 = 8 \cdot 3$
$$= 4 \cdot 2 \cdot 3$$
$$= 2 \cdot 2 \cdot 2 \cdot 3$$

R2. $4(3x - 5) = 4 \cdot 3x - 4 \cdot 5$
$$= 4 \cdot 3 \cdot x - 4 \cdot 5$$
$$= 12x - 20$$

Section 4.4 Quick Checks

1. In $(3x+1)(x-5) = 3x^2 - 14x - 5$, the polynomials on the left side are called <u>factors</u> of the polynomial on the right side.

2. If a polynomial cannot be written as the product of two other polynomials (excluding 1 and −1), then the polynomial is said to be <u>prime</u>.

3. The <u>greatest common factor</u> of a polynomial is the largest polynomial that is a factor of all the terms in the polynomial.

4. To <u>factor</u> a polynomial means to write the polynomial as the product of two of more polynomials.

5. False. The factor 4 factors as $2 \cdot 2$, so $8 = 2 \cdot 2 \cdot 2$.

6. Look at the coefficients first. The largest number that divides into 5 and 15 evenly is 5, so 5 is part of the GCF. Because 15 does not have a variable factor, the GCF is 5.

7. Look at the coefficients first. The largest number that divides into 4, 10, and 12 evenly is 2, so 2 is part of the GCF. Now look at the variable expressions, z^3, z^2, and z. We choose the variable expression with the smallest exponent. The GCF is $2z$.

8. Look at the coefficients first. The largest number that divides into 6, 9, and 12 evenly is 3, so 3 is part of the GCF. Now look at each variable in the expressions. The smallest exponent on x is 1, so x is part of the GCF. The smallest exponent on y is 3, so y^3 is part of the GCF. The GCF is $3xy^3$.

9. First we find that the GCF is $7z$. Next we rewrite each term as the product of the GCF and a remaining factor, then factor out the GCF.
$$7z^2 - 14z = 7z \cdot z - 7z \cdot 2 = 7z(z - 2)$$

10. First we find that the GCF is $2y$. Next we rewrite each term as the product of the GCF and a remaining factor, then factor out the GCF.
$$6y^3 - 14y^2 + 10y = 2y \cdot 3y^2 - 2y \cdot 7y + 2y \cdot 5$$
$$= 2y(3y^2 - 7y + 5)$$

11. First we find that the GCF is $2m^2n^2$. Next we rewrite each term as the product of the GCF and a remaining factor, then factor out the GCF.
$$2m^4n^2 + 8m^3n^4 - 6m^2n^5$$
$$= 2m^2n^2 \cdot m^2 + 2m^2n^2 \cdot 4mn^2 - 2m^2n^2 \cdot 3n^3$$
$$= 2m^2n^2(m^2 + 4mn^2 - 3n^3)$$

12. First we find that the GCF is $-5y$. Next we rewrite each term as the product of the GCF and a remaining factor, then factor out the GCF.
$$-5y^2 + 10y = -5y \cdot y + (-5y) \cdot (-2) = -5y(y - 2)$$

13. First we find that the GCF is $-3a$. Next we rewrite each term as the product of the GCF and a remaining factor, then factor out the GCF.
$$-3a^3 + 6a^2 - 12a$$
$$= -3a \cdot a^2 + (-3a) \cdot (-2a) + (-3a) \cdot (4)$$
$$= -3a(a^2 - 2a + 4)$$

14. First we find that the GCF is $a - 3$. Next we rewrite each term as the product of the GCF and a remaining factor, then factor out the GCF.
$$4a(a-3) + 3(a-3) = (a-3) \cdot 4a + (a-3) \cdot 3$$
$$= (a-3)(4a+3)$$

15. First we find that the GCF is $w - 5$. Next we rewrite each term as the product of the GCF and a remaining factor, then factor out the GCF.
$$(w+2)(w-5) + (2w+1)(w-5)$$
$$= (w-5) \cdot (w+2) + (w-5) \cdot (2w+1)$$
$$= (w-5)(w+2+2w+1)$$
$$= (w-5)(3w+3)$$
Both terms in the second factor have a common factor of 3. Thus,
$$(w+2)(w-5) + (2w+1)(w-5) = 3(w-5)(w+1)$$

16.
$$5x + 5y + bx + by = (5x + 5y) + (bx + by)$$
$$= 5(x + y) + b(x + y)$$
$$= (x + y)(5 + b)$$
$$= (x + y)(b + 5)$$

17.
$$w^3 - 3w^2 + 4w - 12 = (w^3 - 3w^2) + (4w - 12)$$
$$= w^2(w - 3) + 4(w - 3)$$
$$= (w - 3)(w^2 + 4)$$

18.
$$2x^2 + x - 10x - 5 = (2x^2 + x) + (-10x - 5)$$
$$= x(2x + 1) + (-5)(2x + 1)$$
$$= (2x + 1)(x - 5)$$

4.4 Exercises

19. $5a + 35$
GCF: 5
$$5a + 35 = 5 \cdot a + 5 \cdot 7$$
$$= 5(a + 7)$$

21. $-3y + 21$
GCF: -3
$$-3y + 21 = -3 \cdot y + (-3) \cdot (-7)$$
$$= -3(y - 7)$$

23. $14x^2 - 21x$
GCF: $7x$
$$14x^2 - 21x = 7x \cdot 2x - 7x \cdot 3$$
$$= 7x(2x - 3)$$

25. $3z^3 - 6z^2 + 18z$
GCF: $3z$
$$3z^3 - 6z^2 + 18z = 3z \cdot z^2 - (3z) \cdot 2z + 3z \cdot 6$$
$$= 3z(z^2 - 2z + 6)$$

27. $-5p^4 + 10p^3 - 25p^2$
GCF: $-5p^2$
$$-5p^4 + 10p^3 - 25p^2$$
$$= -5p^2 \cdot p^2 + (-5p^2) \cdot (-2p) + (-5p^2) \cdot 5$$
$$= -5p^2(p^2 - 2p + 5)$$

29. $49m^3n + 84mn^3 - 35m^4n^2$
GCF: $7mn$
$$49m^3n + 84mn^3 - 35m^4n^2$$
$$= 7mn \cdot 7m^2 + 7mn \cdot 12n^2 + 7mn \cdot (-5m^3n)$$
$$= 7mn(7m^2 + 12n^2 - 5m^3n)$$

31. $-18z^3 + 14z^2 + 4z$
GCF: $-2z$
$$-18z^3 + 14z^2 + 4z$$
$$= -2z \cdot 9z^2 + (-2z)(-7z) + (-2z)(-2)$$
$$= -2z(9z^2 - 7z - 2)$$

33. $5c(3c - 2) - 3(3c - 2)$
GCF: $3c - 2$
$$5c(3c - 2) - 3(3c - 2) = (3c - 2) \cdot 5c + (3c - 2) \cdot (-3)$$
$$= (3c - 2)(5c - 3)$$

35. $(4a + 3)(a - 3) + (2a - 7)(a - 3)$
GCF: $a - 3$
$$(4a + 3)(a - 3) + (2a - 7)(a - 3)$$
$$= (a - 3) \cdot (4a + 3) + (a - 3)(2a - 7)$$
$$= (a - 3)(4a + 3 + 2a - 7)$$
$$= (a - 3)(6a - 4)$$
The binomial $6a - 4$ has a greatest common factor of 2. We can write the result as
$$(a - 3)(2 \cdot 3a - 2 \cdot 2) = (a - 3) \cdot 2(3a - 2)$$
Thus,
$$(4a + 3)(a - 3) + (2a - 7)(a - 3) = 2(a - 3)(3a - 2)$$

37.
$$5x + 5y + ax + ay = (5x + 5y) + (ax + ay)$$
$$= 5(x + y) + a(x + y)$$
$$= (x + y)(5 + a)$$

39.
$$2z^3 + 10z^2 - 5z - 25 = (2z^3 + 10z^2) + (-5z - 25)$$
$$= 2z^2(z + 5) + (-5)(z + 5)$$
$$= (z + 5)(2z^2 - 5)$$

41.
$$w^2 - 5w + 3w - 15 = (w^2 - 5w) + (3w - 15)$$
$$= w(w - 5) + 3(w - 5)$$
$$= (w - 5)(w + 3)$$

43. Pull out the GCF first, then factor by grouping.

$$2x^2 - 8x - 4x + 16 = 2\left(x^2 - 4x - 2x + 8\right)$$
$$= 2\left[\left(x^2 - 4x\right) + \left(-2x + 8\right)\right]$$
$$= 2\left[x(x-4) + (-2)(x-4)\right]$$
$$= 2(x-4)(x-2)$$

45. Pull out the GCF first, then factor by grouping.

$$3x^3 + 15x^2 - 12x^2 - 60x$$
$$= 3x\left(x^2 + 5x - 4x - 20\right)$$
$$= 3x\left[\left(x^2 + 5x\right) + \left(-4x - 20\right)\right]$$
$$= 3x\left[x(x+5) + (-4)(x+5)\right]$$
$$= 3x(x+5)(x-4)$$

47.
$$2ax - 2ay - bx + by = (2ax - 2ay) + (-bx + by)$$
$$= 2a(x-y) + (-b)(x-y)$$
$$= (x-y)(2a-b)$$

49. $(w+3)(w-3) - (w-2)(w-3)$

GCF: $(w-3)$

$$(w+3)(w-3) - (w-2)(w-3)$$
$$= (w-3)\cdot(w+3) - (w-3)\cdot(w-2)$$
$$= (w-3)(w+3-w+2)$$
$$= (w-3)(5)$$
$$= 5(w-3)$$

51.
$$2y^2 + 5y - 4y - 10 = \left(2y^2 + 5y\right) + (-4y - 10)$$
$$= y(2y+5) + (-2)(2y+5)$$
$$= (2y+5)(y-2)$$

53. $6x^3y^3 + 9x^2y - 21x^3y^2$

GCF: $3x^2y$

$$6x^3y^3 + 9x^2y - 21x^3y^2$$
$$= 3x^2y\cdot 2xy^2 + 3x^2y\cdot 3 + 3x^2y\cdot(-7xy)$$
$$= 3x^2y\left(2xy^2 + 3 - 7xy\right)$$
$$= 3x^2y\left(2xy^2 - 7xy + 3\right)$$

55.
$$x^3 + x^2 + 3x + 3 = \left(x^3 + x^2\right) + (3x + 3)$$
$$= x^2(x+1) + 3(x+1)$$
$$= (x+1)\left(x^2 + 3\right)$$

57. $x(x-2) + 3(x-2)^2$

GCF: $(x-2)$

$$x(x-2) + 3(x-2)(x-2)$$
$$= (x-2)\cdot x + (x-2)(3)(x-2)$$
$$= (x-2)(x + 3(x-2))$$
$$= (x-2)(x + 3x - 6)$$
$$= (x-2)(4x - 6)$$
$$= (x-2)(2\cdot 2x - 2\cdot 3)$$
$$= 2(x-2)(2x-3)$$

59. $3x^2(4x+1)^2 + 8x^3(4x+1)$

GCF: $x^2(4x+1)$

$$3x^2(4x+1)^2 + 8x^3(4x+1)$$
$$= x^2(4x+1)\cdot 3(4x+1) + x^2(4x+1)\cdot 8x$$
$$= x^2(4x+1)\left[3(4x+1) + 8x\right]$$
$$= x^2(4x+1)(12x + 3 + 8x)$$
$$= x^2(4x+1)(20x + 3)$$

61. $3(x+9)^2(2x+5) + 2(x+9)^3$

GCF: $(x+9)^2$

$$3(x+9)^2(2x+5) + 2(x+9)^3$$
$$= (x+9)^2\cdot 3(2x+5) + (x+9)^2\cdot 2(x+9)$$
$$= (x+9)^2\left[3(2x+5) + 2(x+9)\right]$$
$$= (x+9)^2(6x + 15 + 2x + 18)$$
$$= (x+9)^2(8x + 33)$$

63. $4(2x-1)(x-5)^3 + 3(x-5)^2(2x-1)^2$

GCF: $(2x-1)(x-5)^2$

$$4(2x-1)(x-5)^3 + 3(x-5)^2(2x-1)^2$$
$$= (2x-1)(x-5)^2\cdot 4(x-5)$$
$$\qquad + (2x-1)(x-5)^2\cdot 3(2x-1)$$
$$= (2x-1)(x-5)^2\left[4(x-5) + 3(2x-1)\right]$$
$$= (2x-1)(x-5)^2(4x - 20 + 6x - 3)$$
$$= (2x-1)(x-5)^2(10x - 23)$$

65. $2(x^2+1)\cdot 2x(4x-3)^3 + 3(4x-3)^2\cdot 4(x^2+1)^2$

GCF: $4(x^2+1)(4x-3)^2$

$2(x^2+1)\cdot 2x(4x-3)^3 + 3(4x-3)^2\cdot 4(x^2+1)^2$

$= 4(x^2+1)(4x-3)^2\cdot x(4x-3)$

$\qquad +4(x^2+1)(4x-3)^2\cdot 3(x^2+1)$

$= 4(x^2+1)(4x-3)^2\left[x(4x-3)+3(x^2+1)\right]$

$= 4(x^2+1)(4x-3)^2\left(4x^2-3x+3x^2+3\right)$

$= 4(x^2+1)(4x-3)^2\left(7x^2-3x+3\right)$

67. The shaded area is the difference between the area of the square and the area of the circle. The radius of the circle is x, and this is also half the length of one side of the square. Thus, we have

$A = A_{square} - A_{circle}$

$\quad = (2x)^2 - \pi x^2$

$\quad = 4x^2 - \pi x^2$

$\quad = x^2(4-\pi)$

The area of the shaded region is

$x^2(4-\pi)$ square units.

69. $S = 2\pi r^2 + 8\pi r$

$\quad = 2\pi r\cdot r + 2\pi r\cdot 4$

$\quad = 2\pi r(r+4)$

The surface area is $S = 2\pi r(r+4)$ square inches.

71. a. The selling price of the shirt is the supplier's cost plus the markup. That is,

$x+0.4x = 1.4x$

b. The discount is 40% of the selling price. We obtain the sale price by subtracting this amount from the original price. That is,

$1.4x - 0.4(1.4x)$

c. $1.4x - 0.4(1.4x) = 1.4x(1-0.4)$

$\qquad\qquad\qquad = 0.6(1.4x)$

$\qquad\qquad\qquad = 0.84x$

d. No, the sale price is less than the amount that the store paid for the shirt. The 40% increase is on a smaller amount than the 40% decrease. Therefore, the decrease will be larger in magnitude than the increase.

73. a. The stock price is the original price plus the increase. That is,

$x+0.15x = 1.15x$

b. The second year increase is based on the ending price for the first year. That is,

$1.15x + 0.1(1.15x)$.

c. $(1.15x)+0.1(1.15x) = (1.15x)(1+0.1)$

$\qquad\qquad\qquad\qquad = 1.1(1.15x)$

$\qquad\qquad\qquad\qquad = 1.265x$

d. $x = 20$

$1.265(20) = 25.30$

After two years, the value of the stock will be $25.30.

75. $\dfrac{1}{4}x - \dfrac{7}{4} = \dfrac{1}{4}\cdot x - \dfrac{1}{4}\cdot 7$

$\qquad = \dfrac{1}{4}(x-7)$

77. $\dfrac{1}{5}b^3 + \dfrac{8}{25}b = \dfrac{5}{25}b^3 + \dfrac{8}{25}b$

$\qquad\qquad = \dfrac{1}{25}b\cdot 5b^2 + \dfrac{1}{25}b\cdot 8$

$\qquad\qquad = \dfrac{1}{25}b\left(5b^2+8\right)$

79. $x^n + 3x^{n+1} + 6x^{2n}$

GCF: x^n

$x^n + 3x^{n+1} + 6x^{2n}$

$= x^n + 3x^n\cdot x^1 + 6x^n\cdot x^n$

$= x^n\cdot 1 + x^n\cdot 3x + x^n\cdot 6x^n$

$= x^n\left(1+3x+6x^n\right)$

81. $4y^{n+3} - 8y^{n+2} + 6y^{n+5}$

GCF: $2y^{n+2}$

$4y^{n+3} - 8y^{n+2} + 6y^{n+5}$

$= 4y^{n+2}\cdot y^1 - 8y^{n+2} + 6y^{n+2}\cdot y^3$

$= 2y^{n+2}\cdot 2y - 2y^{n+2}\cdot 4 + 2y^{n+2}\cdot 3y^3$

$= 2y^{n+2}\left(2y-4+3y^3\right)$

Section 4.5

Are You Ready for This Section?

R1. $4x^2 - 9x + 2$
The coefficients are the numeric factors of the terms. The coefficients are 4, -9, and 2.

R2.

factors	1, 6	2, 3
sum	7	5

2 and 3 are two integers whose sum is 5 and product is 6.

R3.

factors	1, -32	2, -16	4, -8	8, -4	16, -2	32, -1
sum	-31	-14	-4	4	14	31

4 and -8 are two integers whose sum is -4 and product is -32.

R4.

factor	1, -40	2, -20	4, -10	5, -8	8, -5	10, -4	20, -2	40, -1
sum	-39	-18	-6	-3	3	6	18	39

10 and -4 are two integers whose sum is 6 and product is -40.

R5.

factors	1, 32	2, 16	4, 8	$-1, -32$	$-2, -16$	$-4, -8$
sum	33	18	12	-33	-18	-12

-4 and -8 are two integers whose sum is -12 and product is 32.

R6.

factors	1, 18	2, 9	3, 6
sum	19	11	9

9 and 2 are the factors of 18 that sum to 11.

R7.

factors	1, -24	2, -12	3, -8	4, -6
sum	-23	-10	-5	-2

-6 and 4 are the factors of -24 that sum to -2.
(note: since the sum is negative, we restrict our choice of factors so that the larger factor (in absolute value) is negative.

Section 4.5 Quick Checks

1. A <u>quadratic trinomial</u> is a polynomial of the form $ax^2 + bx + c$, where a, b, and c are integers.

2. When factoring a trinomial of the form $x^2 + bx + c$, we need to find pairs of integers m and n such that <u>$m \cdot n = c$</u> and <u>$m + n = b$</u>.

3. True

4. $y^2 + 9y + 18$
   We are looking for two factors of $c = 18$ whose sum is $b = 9$. Since both c and b are positive, the two factors must both be positive.

Factors	1,18	2,9	3,6
Sum	19	11	9

$$y^2 + 9y + 18 = (y + 6)(y + 3)$$

5. $p^2 + 14p + 24$

We are looking for two factors of $c = 24$ whose sum is $b = 14$. Since both c and b are positive, the two factors must both be positive.

Factors	1, 24	2, 12	3, 8	4, 6
Sum	25	14	11	10

$$p^2 + 14p + 24 = (p + 2)(p + 12)$$

6. $q^2 - 6q + 8$

We are looking for two factors of $c = 8$ whose sum is $b = -6$. Since c is positive the factors have the same sign, and since b is negative both factors are negative.

Factors	-1, -8	-2, -4
Sum	-9	-6

$$q^2 - 6q + 8 = (q - 4)(q - 2)$$

7. $x^2 - 8x + 12$

We are looking for two factors of $c = 12$ whose sum is $b = -8$. Since c is positive the factors have the same sign, and since b is negative the two factors are negative.

Factors	-1, -12	-2, -6	-3, -4
Sum	-13	-8	-7

$$x^2 - 8x + 12 = (x - 2)(x - 6)$$

8. $w^2 - 4w - 21$

We are looking for two factors of $c = -21$ whose sum is $b = -4$. Since c is negative, the two factors will have opposite signs. Since b is negative, the factor with the larger absolute value will be negative.

Factors	-21,1	-7,3
Sum	-20	-4

$$w^2 - 4w - 21 = (w - 7)(w + 3)$$

9. $q^2 - 9q - 36$

We are looking for two factors of $c = -36$ whose sum is $b = -9$. Since c is negative, the two factors will have opposite signs. Since b is negative, the factor with the larger absolute value will be negative.

Factors	-36,1	-18,2	-12,3	-9,4	-6,6
Sum	-35	-16	-9	-5	0

$$q^2 - 9q - 36 = (q - 12)(q + 3)$$

10. A polynomial that cannot be written as the product of two other polynomials (other than 1 or −1) is a <u>prime</u> polynomial.

11. $z^2 - 5z + 8$

We are looking for two factors of $c = 8$ whose sum is $b = -5$. Since c is positive the factors have the same sign, and since b is negative the two factors are negative.

Factors	-1, -8	-2, -4
Sum	-9	-6

None of the possible factors yield the desired sum, so $t^2 - 5t + 8$ is prime.

12. $q^2 + 4q - 45$

We are looking for two factors of $c = -45$ whose sum is $b = 4$. Since c is negative, the two factors will have opposite signs. Since b is positive, the factor with the larger absolute value will be positive.

Factors	1, 45	-3, 15	-5, 9
Sum	44	12	4

$$q^2 + 4q - 45 = (q + 9)(q - 5)$$

13. $x^2 + 8xy + 15y^2$

We are looking for two factors of $c = 15$ whose sum is $b = 8$. Since c and b are both positive, the two factors will be positive.

Factors	1,15	3,5
Sum	16	8

$$x^2 + 8xy + 15y^2 = (x + 3y)(x + 5y)$$

14. $m^2 + mn - 20n^2$

We are looking for two factors of $c = -20$ whose sum is $b = 1$. Since c is negative, the factors will have opposite signs. Since b is positive, the factor with the larger absolute value will be positive.

Factors	$-1, 20$	$-2, 10$	$-4, 5$
Sum	19	8	1

$m^2 + mn - 20n^2 = (m - 4n)(m + 5n)$

15. False; the second factor, $3x - 15$, has a common factor of 3.

16. $2x^3 - 12x^2 - 54x$

First we factor out the GCF.

$2x^3 - 12x^2 - 54x = 2x(x^2 - 6x - 27)$

We are looking for two factors of $c = -27$ whose sum is $b = -6$. Since c is negative, the factors will have opposite signs. Since b is negative, the factor with the larger absolute value will be negative.

Factors	$-27, 1$	$-9, 3$
Sum	-26	-6

$2x^3 - 12x^2 - 54x = 2x(x^2 - 6x - 27)$
$ = 2x(x - 9)(x + 3)$

17. $-3z^2 - 21z - 30$

Begin by factoring out the GCF.

$-3z^2 - 21z - 30 = -3(z^2 + 7z + 10)$

We are looking for two factors of $c = 10$ whose sum is $b = 7$. Since both c and b are positive, the two factors will be positive.

Factors	$1, 10$	$2, 5$
Sum	11	7

$-3z^2 - 21z - 30 = -3(z^2 + 7z + 10)$
$ = -3(z + 5)(z + 2)$

18. To factor $2x^2 - 13x + 6$ using grouping, begin by finding factors whose product is $\underline{12}$, and whose sum is $\underline{-13}$.

19. $2b^2 + 7b - 15$

There are no common factors, so $a = 2$, $b = 7$, and $c = -15$. Thus, $a \cdot c = 2 \cdot (-15) = -30$. We are looking for two factors of -30 whose sum is $b = 7$. Since the product is negative, the two factors will have opposite signs. Since the sum is positive, the factor with the larger absolute value will be positive.

factor 1	factor 2	sum	
-2	15	13	too big
-5	6	1	too small
-3	10	7	← okay

$2b^2 + 7b - 15 = 2b^2 - 3b + 10b - 15$
$ = (2b^2 - 3b) + (10b - 15)$
$ = b(2b - 3) + 5(2b - 3)$
$ = (2b - 3)(b + 5)$

20. $10x^2 + 27x + 18$

There are no common factors, so $a = 10$, $b = 27$, $c = 18$. We find that $a \cdot c = 10 \cdot 18 = 180$. Therefore, we are looking for two factors of 180 whose sum is $b = 27$. Since the product and the sum are both positive, both factors must be positive and neither can be greater than 27.

factor 1	factor 2	sum	
10	18	28	too big
9	20	29	too big
12	15	27	← okay

$10x^2 + 27x + 18 = 10x^2 + 12x + 15x + 18$
$ = (10x^2 + 12x) + (15x + 18)$
$ = 2x(5x + 6) + 3(5x + 6)$
$ = (5x + 6)(2x + 3)$

21. $8x^2 + 14x + 5$

First note that there are no common factors and that $a = 8$, $b = 14$, and $c = 5$.

Since c is positive the signs in our factors will be the same. Since b is positive, the two signs will be positive. We will consider factorizations with this form:

$(\underline{}x + \underline{})(\underline{}x + \underline{})$.

Since $a = 8$ can be factored as $1 \cdot 8$ and $2 \cdot 4$, we have the following forms:

$(x + \underline{\quad})(8x + \underline{\quad})$

$(2x + \underline{\quad})(4x + \underline{\quad})$

$|c| = |5| = 5$ can be factored as $1 \cdot 5$. This gives us the following possibilities:

$(x + 1)(8x + 5) \rightarrow 8x^2 + 13x + 5$

$(2x + 1)(4x + 5) \rightarrow 8x^2 + 14x + 5$

$(x + 5)(8x + 1) \rightarrow 8x^2 + 41x + 5$

$(2x + 5)(4x + 1) \rightarrow 8x^2 + 22x + 5$

The correct factorization is

$8x^2 + 14x + 5 = (2x + 1)(4x + 5)$.

22. $12y^2 + 32y - 35$

First note that there are no common factors and that $a = 12$, $b = 32$, and $c = -35$.
Since c is negative, the signs in our factors will be opposites. We will consider factorizations with this form:

$(\underline{\quad} y + \underline{\quad})(\underline{\quad} y - \underline{\quad})$.

If our choice results in a middle term with the wrong sign, we simply switch the signs of the factors.
Since $a = 12$ can be factored as $1 \cdot 12$, $2 \cdot 6$, and $3 \cdot 4$, we have the following forms:

$(y + \underline{\quad})(12y - \underline{\quad})$

$(2y + \underline{\quad})(6y - \underline{\quad})$

$(3y + \underline{\quad})(4y - \underline{\quad})$

$|c| = |-35| = 35$ can be factored as $1 \cdot 35$ and $5 \cdot 7$. This gives us the following possibilities:

$(y + 1)(12y - 35) \rightarrow 12y^2 - 23y - 35$

$(2y + 1)(6y - 35) \rightarrow 12y^2 - 64y - 35$

$(3y + 1)(4y - 35) \rightarrow 12y^2 - 101y - 35$

$(y + 5)(12y - 7) \rightarrow 12y^2 + 53y - 35$

$(2y + 5)(6y - 7) \rightarrow 12y^2 + 16y - 35$

$(3y + 5)(4y - 7) \rightarrow 12y^2 - y - 35$

$(y + 35)(12y - 1) \rightarrow 12y^2 + 419y - 35$

$(2y + 35)(6y - 1) \rightarrow 12y^2 + 208y - 35$

$(3y + 35)(4y - 1) \rightarrow 12y^2 + 137y - 35$

$(y + 7)(12y - 5) \rightarrow 12y^2 + 79y - 35$

$(2y + 7)(6y - 5) \rightarrow 12y^2 + 32y - 35$

$(3y + 7)(4y - 5) \rightarrow 12y^2 + 13y - 35$

The correct factorization is

$12y^2 + 32y - 35 = (2y + 7)(6y - 5)$

23. $30x^2 + 7xy - 2y^2$

First note that there are no common factors and that $a = 30$, $b = 7$, and $c = -2$.
Since c is negative, the signs in our factors will be opposites. We will consider factorizations of the form:

$(\underline{\quad} x + \underline{\quad} y)(\underline{\quad} x - \underline{\quad} y)$.

If our choice results in a middle term with the wrong sign, we simply switch the signs of the factors.
Since $a = 30$ can be factored as $1 \cdot 30$, $2 \cdot 15$, $3 \cdot 10$, and $5 \cdot 6$, we have the following forms:

$(x + \underline{\quad} y)(30x - \underline{\quad} y)$

$(2x + \underline{\quad} y)(15x - \underline{\quad} y)$

$(3x + \underline{\quad} y)(10x - \underline{\quad} y)$

$(5x + \underline{\quad} y)(6x - \underline{\quad} y)$

$|c| = |-2| = 2$ can be factored as $1 \cdot 2$. Since the original expression had no common factors, the binomials we select cannot have a common factor. This gives us the following possibilities:

$(x + 2y)(30x - y) \rightarrow 30x^2 + 59xy - 2y^2$

$(2x + y)(15x - 2y) \rightarrow 30x^2 + 11xy - 2y^2$

$(3x + 2y)(10x - y) \rightarrow 30x^2 + 17xy - 2y^2$

$(5x + 2y)(6x - y) \rightarrow 30x^2 + 7xy - 2y^2$

The correct factorization is

$30x^2 + 7xy - 2y^2 = (5x + 2y)(6x - y)$.

24. $8x^2 - 10xy - 42y^2$

Begin by factoring out 2.

$8x^2 - 10xy - 42y^2 = 2(4x^2 - 5xy - 21y^2)$

Focusing on the reduced trinomial, we have $a = 4$, $b = -5$, $c = -21$. Since c is negative, the signs in our factors will be opposites. We will consider factorizations of the form:

$(\underline{\quad} x + \underline{\quad} y)(\underline{\quad} x - \underline{\quad} y)$.

If our choice results in a middle term with the wrong sign, we simply switch the signs of the factors.
Since $a = 4$ can be factored as $1 \cdot 4$ and $2 \cdot 2$, we have the following forms:

$(x + \underline{\quad} y)(4x - \underline{\quad} y)$

$(2x + \underline{\quad} y)(2x - \underline{\quad} y)$

$(4x + \underline{\quad} y)(x - \underline{\quad} y)$

$|c| = |-21| = 21$ can be factored as $1 \cdot 21$ and $3 \cdot 7$. Since the reduced expression had no common factors, the binomials we select cannot have a common factor. This gives us the following possibilities:

$$(x+y)(4x-21y) \rightarrow 4x^2 - 17xy - 21y^2$$
$$(x+3y)(4x-7y) \rightarrow 4x^2 + 5xy - 21y^2$$
$$(x+7y)(4x-3y) \rightarrow 4x^2 + 25xy - 21y^2$$
$$(x+21y)(4x-y) \rightarrow 4x^2 + 83xy - 21y^2$$
$$(2x+y)(2x-21y) \rightarrow 4x^2 - 40xy - 21y^2$$
$$(2x+3y)(2x-7y) \rightarrow 4x^2 - 8xy - 21y^2$$
$$(2x+7y)(2x-3y) \rightarrow 4x^2 + 8xy - 21y^2$$
$$(2x+21y)(2x-y) \rightarrow 4x^2 + 40xy - 21y^2$$
$$(4x+y)(x-21y) \rightarrow 4x^2 - 83xy - 21y^2$$
$$(4x+3y)(x-7y) \rightarrow 4x^2 - 25xy - 21y^2$$
$$(4x+7y)(x-3y) \rightarrow 4x^2 - 5xy - 21y^2$$
$$(4x+21y)(x-y) \rightarrow 4x^2 + 17xy - 21y^2$$

Recalling the GCF of 2, the correct factorization is $8x^2 - 10xy - 42y^2 = 2(4x+7y)(x-3y)$.

25. $-6y^2 + 23y + 4$

Begin by factoring out -1.

$$-6y^2 + 23y + 4 = -(6y^2 - 23y - 4)$$

Focusing on the reduced trinomial, we have $a = 6$, $b = -23$, and $c = -4$. Thus, $a \cdot c = 6 \cdot (-4) = -24$. We are looking for two factors of -24 whose sum is $b = -23$. Since the product is negative, the two factors have opposite signs. Since the sum is negative, the factor with the larger absolute value will be negative.

factor 1	factor 2	sum
−8	3	−5 too big
−12	2	−10 too big
−24	1	−23 ← okay

$$-6y^2 + 23y + 4 = -(6y^2 - 23y - 4)$$
$$= -(6y^2 - 24y + y - 4)$$
$$= -[(6y^2 - 24y) + (y - 4)]$$
$$= -[6y(y - 4) + 1(y - 4)]$$
$$= -(y - 4)(6y + 1)$$

26. $-9x^2 - 21xy - 10y^2$

Begin by factoring out -1.

$$-9x^2 - 21xy - 10y^2 = -(9x^2 + 21xy + 10y^2)$$

Focusing on the reduced trinomial, we have $a = 9$, $b = 21$, and $c = 10$. Thus, $a \cdot c = 9 \cdot 10 = 90$. We are looking for two factors of 90 whose sum is $b = 21$. Since the product is positive, the two factors have the same sign. Since the sum is also positive, the factors will both be positive.

factor 1	factor 2	sum
3	30	33 too big
9	10	19 too small
6	15	21 ← okay

$$-9x^2 - 21xy - 10y^2$$
$$= -(9x^2 + 21xy + 10y^2)$$
$$= -(9x^2 + 6xy + 15xy + 10y^2)$$
$$= -[(9x^2 + 6xy) + (15xy + 10y^2)]$$
$$= -[3x(3x + 2y) + 5y(3x + 2y)]$$
$$= -(3x + 2y)(3x + 5y)$$

27. When factoring $6x^4 - x^2 - 2$ by substitution, let $u = \underline{x^2}$.

28. When factoring $3(2x-3)^2 + 5(2x-3) + 2$ by substitution, let $u = \underline{2x - 3}$.

29. $y^4 - 2y^2 - 24$

Let $u = y^2$. Then $u^2 = \left(y^2\right)^2 = y^4$ and we get

$$y^4 - 2y^2 - 24 = \left(y^2\right)^2 - 2\left(y^2\right) - 24$$
$$= u^2 - 2u - 24$$
$$= (u - 6)(u + 4)$$
$$= \left(y^2 - 6\right)\left(y^2 + 4\right)$$

30. $4(x-3)^2 + 5(x-3) - 6$

Let $u = x - 3$. Then $u^2 = (x-3)^2$ and we get

$$4(x-3)^2 + 5(x-3) - 6$$
$$= 4u^2 + 5u - 6$$
$$= (4u - 3)(u + 2)$$
$$= (4(x-3) - 3)((x-3) + 2)$$
$$= (4x - 12 - 3)(x - 3 + 2)$$
$$= (4x - 15)(x - 1)$$

4.5 Exercises

31. $x^2 + 8x + 15$

We are looking for two factors of $c = 15$ whose sum is $b = 8$. Since both c and b are positive, the two factors must be positive.

Factors	1, 15	3, 5
Sum	16	8

$x^2 + 8x + 15 = (x+3)(x+5)$

33. $p^2 + 3p - 18$

We are looking for two factors of $c = -18$ whose sum is $b = 3$. Since c is negative the two factors have opposite signs, and since b is positive the factor with the larger absolute value must be positive.

Factors	−1,18	−2,9	−3, 6
Sum	17	7	3

$p^2 + 3p - 18 = (p+6)(p-3)$

35. $r^2 + 10r + 25$

We are looking for two factors of $c = 25$ whose sum is $b = 10$. Since both c and b are positive, the two factors must be positive.

Factors	1, 25	5, 5
Sum	26	10

$r^2 + 10r + 25 = (r+5)(r+5) = (r+5)^2$

37. $s^2 + 7s - 60$

We are looking for two factors of $c = -60$ whose sum is $b = 7$. Since c is negative the two factors have opposite signs, and since b is positive the factor with the larger absolute value must be positive.

Factors	−1,60	−2,30	−3,20	−4,15	−5,12	−6,10
Sum	59	28	17	11	7	4

$s^2 + 7s - 60 = (s+12)(s-5)$

39. $2x^2 - 30x + 112$

Start by factoring out 2.

$2x^2 - 30x + 112 = 2\left(x^2 - 15x + 56\right)$

We are looking for two factors of $c = 56$ whose sum is $b = -15$. Since c is positive the two factors have the same sign, and since b is negative the factors are both negative.

Factors	−1,−56	−2,−28	−4,−14	−7,−8
Sum	−57	−30	−18	−15

$2x^2 - 30x + 112 = 2(x-7)(x-8)$

41. $-w^2 - 2w + 24$

Start by factoring out a -1.

$-1\left(w^2 + 2w - 24\right)$

We are looking for two factors of $c = -24$ whose sum is $b = 2$. Since c is negative the two factors have opposite signs, and since b is positive the factor with the larger absolute value must be positive.

Factors	−1,24	−2,12	−3,8	−4,6
Sum	23	10	5	2

$-w^2 - 2w + 24 = -(w-4)(w+6)$

43. $x^2 + 7xy + 12y^2$

We are looking for two factors of $c = 12$ that sum to $b = 7$. Since both c and b are positive, the two factors are both positive.

Factors	1,12	2,6	3,4
Sum	13	8	7

$x^2 + 7xy + 12y^2 = (x+3y)(x+4y)$

45. $p^3 + 2p^2q - 24pq^2$

Start by factoring out p.

$p^3 + 2p^2q - 24pq^2 = p(p^2 + 2pq - 24q^2)$

We are looking for two factors of $c = -24$ whose sum is $b = 2$. Since c is negative the two factors have opposite signs. Since b is positive the factor with the larger absolute value must be positive.

Factors	−1,24	−2,12	−3,8	−4,6
Sum	23	10	5	2

$p^3 + 2p^2q - 24pq^2 = p(p-4q)(p+6q)$

47. <u>*ac* Method:</u>

$a \cdot c = 2 \cdot -8 = -16$

We are looking for two factors of -16 whose sum is -15. Since the product is negative, the factors will have opposite signs. The sum is negative so the factor with the largest absolute value will be negative.

factor 1	factor 2	sum
1	-16	$-15 \leftarrow$ okay
2	-8	-6
4	-4	0

$$2p^2 - 15p - 8 = 2p^2 + p - 16p - 8$$
$$= p(2p+1) - 8(2p+1)$$
$$= (2p+1)(p-8)$$

<u>Trial and Error Method:</u>

First note that there are no common factors and that $a = 2$, $b = -15$, and $c = -8$. Since c is negative, the signs in our factors will be opposites. We will consider factorizations with this form:

$$(\underline{}\, p + \underline{})(\underline{}\, p - \underline{})$$

If our choice results in a middle term with the wrong sign, we simply switch the signs of the factors.

Since $a = 2$ can be factored as $2 \cdot 1$, we have the following form:

$$(2p + \underline{})(p - \underline{})$$

$|c| = |-8| = 8$ can be factored as $1 \cdot 8$ and $2 \cdot 4$.

Since the original expression had no common factors, the binomials we select cannot have a common factor.

$$(2p+1)(p-8) \rightarrow 2p^2 - 15p - 8$$

The correct factorization is

$$2p^2 - 15p - 8 = (2p+1)(p-8)$$

49. <u>*ac* Method:</u>

$a \cdot c = 4 \cdot 6 = 24$

We are looking for two factors of 24 whose sum is -11. Since the product is positive, the factors will have the same sign. The sum is negative so the factors will be negative.

factor 1	factor 2	sum
-1	-24	-25
-2	-12	-14
-3	-8	$-11 \leftarrow$ okay
-4	-6	-10

$$4y^2 - 11y + 6 = 4y^2 - 8y - 3y + 6$$
$$= 4y(y-2) - 3(y-2)$$
$$= (y-2)(4y-3)$$

<u>Trial and Error Method:</u>

First note that there are no common factors and that $a = 4$, $b = -11$, and $c = 6$. Since c is positive the signs of our factors will be the same. Since b is negative, the signs in our factors will be negative. We will consider factorizations with this form:

$$(\underline{}\, y - \underline{})(\underline{}\, y - \underline{})$$

Since $a = 4$ can be factored as $1 \cdot 4$ or $2 \cdot 2$, we have the following forms:

$$(y - \underline{})(4y - \underline{})$$
$$(2y - \underline{})(2y - \underline{})$$

$|c| = |6| = 6$ can be factored as $1 \cdot 6$ and $2 \cdot 3$.

Since the original expression had no common factors, the binomials we select cannot have a common factor.

$$(y-6)(4y-1) \rightarrow 4y^2 - 25y + 6$$
$$(y-2)(4y-3) \rightarrow 4y^2 - 11y + 6$$

The correct factorization is

$$4y^2 - 11y + 6 = (y-2)(4y-3)$$

51. <u>*ac* Method:</u>

$a \cdot c = 8 \cdot -3 = -24$

We are looking for two factors of -24 whose sum is 2. Since the product is negative, the factors will have opposite signs. The sum is positive so the factor with the larger absolute value will be positive.

factor 1	factor 2	sum
-1	24	23
-2	12	10
-3	8	5
-4	6	$2 \leftarrow$ okay

$$8s^2 + 2s - 3 = 8s^2 - 4s + 6s - 3$$
$$= 4s(2s-1) + 3(2s-1)$$
$$= (2s-1)(4s+3)$$

<u>Trial and Error Method:</u>

First note that there are no common factors and that $a = 8$, $b = 2$, and $c = -3$. Since c is negative, the signs in our factors will be opposites. We will consider factorizations with this form:

$$(\underline{}\, s + \underline{})(\underline{}\, s - \underline{})$$

If our choice results in a middle term with the wrong sign, we simply switch the signs of the factors.

Since $a = 8$ can be factored as $1 \cdot 8$ and $2 \cdot 4$, we have the following forms:

$$(s + \underline{\ \ })(8s - \underline{\ \ })$$
$$(2s + \underline{\ \ })(4s - \underline{\ \ })$$

$|c| = |-3| = 3$ can be factored as $1 \cdot 3$. Since the original expression had no common factors, the binomials we select cannot have a common factor.

$$(s+1)(8s-3) \rightarrow 8s^2 + 5s - 3$$
$$(s+3)(8s-1) \rightarrow 8s^2 + 23s - 3$$
$$(2s+1)(4s-3) \rightarrow 8s^2 - 2s - 3$$
$$(2s+3)(4s-1) \rightarrow 8s^2 + 10s - 3$$

Notice that our third choice is correct except that the sign of the middle term is wrong. Therefore, we simply need to switch the signs of the factors. The correct factorization is

$$8s^2 + 2s - 3 = (2s-1)(4s+3)$$

53. *ac* Method:
$$a \cdot c = 16 \cdot -15 = -240$$
We are looking for two factors of -240 whose sum is 8. Since the product is negative, the factors will have opposite signs. The sum is positive so the factor with the largest absolute value will be positive.

factor 1	factor 2	sum
-8	30	22 too large
-10	24	14 too large
-12	20	$8 \leftarrow$ okay

$$16z^2 + 8z - 15 = 16z^2 - 12z + 20z - 15$$
$$= 4z(4z-3) + 5(4z-3)$$
$$= (4z-3)(4z+5)$$

Trial and Error Method:
First note that there are no common factors and that $a = 16$, $b = 8$, and $c = -15$. Since c is negative, the signs in our factors will be opposites. We will consider factorizations with this form:

$$(\underline{\ \ }z + \underline{\ \ })(\underline{\ \ }z - \underline{\ \ })$$

If our choice results in a middle term with the wrong sign, we simply switch the signs of the factors.

Since $a = 16$ can be factored as $1 \cdot 16$, $2 \cdot 8$, and $4 \cdot 4$, we have the following forms:

$$(z + \underline{\ \ })(16z - \underline{\ \ })$$
$$(2z + \underline{\ \ })(8z - \underline{\ \ })$$
$$(4z + \underline{\ \ })(4z - \underline{\ \ })$$

$|c| = |-15| = 15$ can be factored as $1 \cdot 15$ and $3 \cdot 5$. Since the original expression had no common factors, the binomials we select cannot have a common factor.

$$(z+1)(16z-15) \rightarrow 16z^2 + z - 15$$
$$(z+15)(16z-1) \rightarrow 16z^2 + 239z - 15$$
$$(z+3)(16z-5) \rightarrow 16z^2 + 43z - 15$$
$$(z+5)(16z-3) \rightarrow 16z^2 + 77z - 15$$
$$(2z+1)(8z-15) \rightarrow 16z^2 - 22z - 15$$
$$(2z+15)(8z-1) \rightarrow 16z^2 + 118z - 15$$
$$(2z+3)(8z-5) \rightarrow 16z^2 + 14z - 15$$
$$(2z+5)(8z-3) \rightarrow 16z^2 + 34z - 15$$
$$(4z+1)(4z-15) \rightarrow 16z^2 - 56z - 15$$
$$(4z+3)(4z-5) \rightarrow 16z^2 - 8z - 15$$

Notice that our last choice is correct except for the sign on the middle term. Therefore, we simply need to switch the signs of our factors. The correct factorization is

$$16z^2 + 8z - 15 = (4z-3)(4z+5)$$

55. $-18y^2 - 17y - 4 = -(18y^2 + 17y + 4)$

ac Method:
$$a \cdot c = 18 \cdot 4 = 72$$
We are looking for two factors of 72 whose sum is 17. Since the product is positive, the factors will have the same sign. The sum is positive so the factors will both be positive. Since we are adding two positive numbers to get 17, neither factor can exceed 17.

factor 1	factor 2	sum
6	12	18
8	9	$17 \leftarrow$ okay

$$18y^2 + 17y + 4 = 18y^2 + 9y + 8y + 4$$
$$= 9y(2y+1) + 4(2y+1)$$
$$= (2y+1)(9y+4)$$

$$-18y^2 - 17y - 4 = -(2y+1)(9y+4)$$

Trial and Error Method:
First note that there are no common factors and that $a = 18$, $b = 17$, and $c = 4$. Since c is positive, the signs in our factors will be the same. Since b is also positive, the signs of our factors will be positive. We will consider

factorizations with this form:

$$(\underline{\quad}y+\underline{\quad})(\underline{\quad}y+\underline{\quad})$$

Since $a=18$ can be factored as $1 \cdot 18$, $2 \cdot 9$, and $3 \cdot 6$, we have the following forms:

$$(y+\underline{\quad})(18y+\underline{\quad})$$
$$(2y+\underline{\quad})(9y+\underline{\quad})$$
$$(3y+\underline{\quad})(6y+\underline{\quad})$$

$|c|=|4|=4$ can be factored as $1 \cdot 4$ and $2 \cdot 2$.

Since the original expression had no common factors, the binomials we select cannot have a common factor.

$$(y+4)(18y+1) \rightarrow 18y^2 +73y+4$$
$$(2y+1)(9y+4) \rightarrow 18y^2 +17y+4$$
$$(3y+4)(6y+1) \rightarrow 18y^2 +27y+4$$

The correct factorization is

$$18y^2 +17y+4 = (2y+1)(9y+4)$$
$$-18y^2 -17y-4 = -(2y+1)(9y+4)$$

57. *ac* Method:

$a \cdot c = 2 \cdot -21 = -42$

We are looking for two factors of -42 whose sum is 11. Since the product is negative, the factors will have opposite signs. The sum is positive so the factor with the larger absolute value will be positive. Since the sum is not too large, we want factors whose absolute values are near each other.

factor 1	factor 2	sum
21	-2	19 too large
7	-6	1 too small
14	-3	11 ← okay

$$2x^2 +11xy -21y^2 = 2x^2 +14xy -3xy -21y^2$$
$$= 2x(x+7y)-3y(x+7y)$$
$$= (x+7y)(2x-3y)$$

Trial and Error Method:

$a=2$, $b=11$, and $c=-21$. Since c is negative, the signs in our factors will be opposites. We will consider factorizations with this form:

$$(\underline{\quad}x+\underline{\quad}y)(\underline{\quad}x-\underline{\quad}y)$$

If our choice results in a middle term with the wrong sign, we simply switch the signs of the factors.

Since $a=2$ can be factored as $1 \cdot 2$, we have the following form:

$$(x+\underline{\quad}y)(2x-\underline{\quad}y)$$

$|c|=|-21|=21$ can be factored as $1 \cdot 21$ and

$3 \cdot 7$. Since the original expression had no common factors, the binomials we select cannot have a common factor.

$$(x+y)(2x-21y) \rightarrow 2x^2 -19xy -21y^2$$
$$(x+21y)(2x-y) \rightarrow 2x^2 +41xy -21y^2$$
$$(x+3y)(2x-7y) \rightarrow 2x^2 -xy -21y^2$$
$$(x+7y)(2x-3y) \rightarrow 2x^2 +11xy -21y^2$$

The correct factorization is

$$2x^2 +11xy -21y^2 = (x+7y)(2x-3y)$$

59. $12r^2 -69rs +45s^2 = 3(4r^2 -23rs +15s^2)$

ac Method:

$a \cdot c = 4 \cdot 15 = 60$

We are looking for two factors of 60 whose sum is -23. Since the product is positive, the factors will have the same sign. The sum is negative so the factors will be negative. Since we are adding two negatives to get -23, neither factor can be less than -23.

factor 1	factor 2	sum
-10	-6	-16 too large
-15	-4	-19 too large
-20	-3	-23 ← okay

$$4r^2 -23rs +15s^2 = 4r^2 -20rs -3rs +15s^2$$
$$= 4r(r-5s)-3s(r-5s)$$
$$= (r-5s)(4r-3s)$$

$$12r^2 -69rs +45s^2 = 3(r-5s)(4r-3s)$$

Trial and Error Method:

$a=4$, $b=-23$, and $c=15$. Since c is positive, the signs in our factors will be the same. Since b is negative, the two factors will be negative. We will consider factorizations with this form:

$$(\underline{\quad}r-\underline{\quad}s)(\underline{\quad}r-\underline{\quad}s)$$

Since $a=4$ can be factored as $1 \cdot 4$ and $2 \cdot 2$, we have the following forms:

$$(r-\underline{\quad}s)(4r-\underline{\quad}s)$$
$$(2r-\underline{\quad}s)(2r-\underline{\quad}s)$$

$|c|=|15|=15$ can be factored as $1 \cdot 15$ and $3 \cdot 5$.

Since the original expression had no common factors, the binomials we select cannot have a common factor.

$$(r-s)(4r-15s) \rightarrow 4r^2 - 19rs + 15s^2$$
$$(r-15s)(4r-s) \rightarrow 4r^2 - 61rs + 15s^2$$
$$(r-3s)(4r-5s) \rightarrow 4r^2 - 17rs + 15s^2$$
$$(r-5s)(4r-3s) \rightarrow 4r^2 - 23rs + 15s^2$$
$$(2r-s)(2r-15s) \rightarrow 4r^2 - 32rs + 15s^2$$
$$(2r-3s)(2r-5s) \rightarrow 4r^2 - 16rs + 15s^2$$

The correct factorization is

$$4r^2 - 23rs + 15s^2 = (r-5s)(4r-3s)$$
$$12r^2 - 69rs + 45s^2 = 3(r-5s)(4r-3s)$$

61. *ac* Method:
First note that there are no common factors and that $a = 24$, $b = 23$, and $c = -12$.
$a \cdot c = 24 \cdot -12 = -288$. We want to determine two factors whose product is -288 and whose sum is 23. Since the product is negative we know the two factors have opposite signs. Since the sum is positive we know that the factor with the larger absolute value must be positive. Since 23 is not very large, we want to keep the absolute value of the factors somewhat close to each other.

factor 1	factor 2	sum
−16	18	2 too small
−12	24	12 too small
−9	32	23 ← okay

$$24r^2 + 23rs - 12s^2 = 24r^2 + 32rs - 9rs - 12s^2$$
$$= 8r(3r+4s) - 3s(3r+4s)$$
$$= (3r+4s)(8r-3s)$$

Trial and Error Method:
First note that there are no common factors and that $a = 24$, $b = 23$, and $c = -12$.
Since c is negative the signs in our factors will be opposite. We will consider factorizations with this form:

$$(\_r + \_s)(\_r - \_s).$$

If our choice results in a middle term with the wrong sign, we simply switch the signs of the factors.
Since $a = 24$ can be factored as $1 \cdot 24$, $2 \cdot 12$, $3 \cdot 8$, and $4 \cdot 6$, we have the following forms:

$$(r + \_s)(24r - \_s)$$
$$(2r + \_s)(12r - \_s)$$
$$(3r + \_s)(8r - \_s)$$
$$(4r + \_s)(6r - \_s)$$

$|c| = |-12| = 12$ can be factored as $1 \cdot 12$, $2 \cdot 6$,

and $3 \cdot 4$. Since the original expression had no common factors, the binomials we select cannot have a common factor.

$$(r+12s)(24r-s) \rightarrow 24r^2 + 287rs - 12s^2$$
$$(3r+4s)(8r-3s) \rightarrow 24r^2 + 23rs - 12s^2$$

The correct factorization is

$$24r^2 + 23rs - 12s^2 = (3r+4s)(8r-3s)$$

63. $x^4 + 3x^2 + 2$

Let $u = x^2$. This gives us $u^2 + 3u + 2$. We need two factors of 2 whose sum is 3. Since the product is positive, the two factors will have the same sign. Since the sum is positive, the signs will both be positive. The only possibility is:

$$u^2 + 3u + 2 = (u+2)(u+1)$$

Now get back in terms of *x*.

$$x^4 + 3x^2 + 2 = (x^2 + 2)(x^2 + 1)$$

65. $m^2n^2 + 5mn - 14$

Let $u = mn$. This gives us $u^2 + 5u - 14$.
We need two factors of -14 whose sum is 5. Since the product is negative, the factors will have opposite signs. Since the sum is positive, the factor with the larger absolute value will be positive.

$$u^2 + 5u - 14 = (u+7)(u-2)$$

Now get back in terms of *m* and *n*.

$$m^2n^2 + 5mn - 14 = (mn+7)(mn-2)$$

67. $(x+1)^2 - 6(x+1) - 16$

Let $u = x+1$. This gives us $u^2 - 6u - 16$.
We need two factors of -16 whose sum is -6. Since the product is negative, the factors will have opposite signs. Since the sum is negative, the factor with the largest absolute value will be negative.

$$u^2 - 6u - 16 = (u-8)(u+2)$$

Now get back in terms of *x*.

$$(x+1)^2 - 6(x+1) - 16 = (x+1-8)(x+1+2)$$
$$= (x-7)(x+3)$$

69. $(3r-1)^2 - 9(3r-1) + 20$

Let $u = 3r-1$. This gives us $u^2 - 9u + 20$.
We need two factors of 20 whose sum is -9.
Since the product is positive, the factors will

have the same sign. Since the sum is negative, the two factors will be negative.

$$u^2 - 9u + 20 = (u-5)(u-4)$$

Now get back in terms of r.

$$(3r-1)^2 - 9(3r-1) + 20 = (3r-1-5)(3r-1-4)$$
$$= (3r-6)(3r-5)$$
$$= 3(r-2)(3r-5)$$

71. $2(y-3)^2 + 13(y-3) + 15$

Let $u = y - 3$. This gives us $2u^2 + 13u + 15$.

$a \cdot c = 2 \cdot 15 = 30$

We need two factors of 30 whose sum is 13. Since the product is positive and the sum is positive, both factors will be positive.

factor 1	factor 2	sum
1	30	31
2	15	17
3	10	13 ← okay
5	6	11

$$2u^2 + 13u + 15 = 2u^2 + 10u + 3u + 15$$
$$= 2u(u+5) + 3(u+5)$$
$$= (u+5)(2u+3)$$

Now get back in terms of y.

$$2(y-3)^2 + 13(y-3) + 15$$
$$= (y-3+5)(2(y-3)+3)$$
$$= (y+2)(2y-6+3)$$
$$= (y+2)(2y-3)$$

73. *ac* Method:

$a \cdot c = 10 \cdot 21 = 210$

We are looking for two factors of 210 whose sum is 41. Since the product is positive, the factors will have the same sign. The sum is positive so the factors will both be positive.

factor 1	factor 2	sum
1	210	211
2	105	107
3	70	73
5	42	47
6	35	41 ← okay

$$10w^2 + 41w + 21 = 10w^2 + 6w + 35w + 21$$
$$= 2w(5w+3) + 7(5w+3)$$
$$= (5w+3)(2w+7)$$

Trial and Error Method:
First note that there are no common factors and that $a = 10$, $b = 41$, and $c = 21$. Since c is

positive, the signs in our factors will be the same. Since b is also positive, the signs of our factors will be positive. We will consider factorizations with this form:

$$(\underline{}w + \underline{})(\underline{}w + \underline{})$$

Since $a = 10$ can be factored as $1 \cdot 10$ and $2 \cdot 5$, we have the following forms:

$$(w + \underline{})(10w + \underline{})$$
$$(2w + \underline{})(5w + \underline{})$$

$|c| = |21| = 21$ can be factored as $1 \cdot 21$ or $3 \cdot 7$. Since the original expression had no common factors, the binomials we select cannot have a common factor.

$$(w+1)(10w+21) \to 10w^2 + 31w + 21$$
$$(w+21)(10w+1) \to 10w^2 + 211w + 21$$
$$(2w+1)(5w+21) \to 10w^2 + 47w + 21$$
$$(2w+21)(5w+1) \to 10w^2 + 107w + 21$$

$$(w+3)(10w+7) \to 10w^2 + 37w + 21$$
$$(w+7)(10w+3) \to 10w^2 + 73w + 21$$
$$(2w+3)(5w+7) \to 10w^2 + 29w + 21$$
$$(2w+7)(5w+3) \to 10w^2 + 41w + 21$$

The correct factorization is

$$10w^2 + 41w + 21 = (5w+3)(2w+7)$$

75. $4(2y+1)^2 - 3(2y+1) - 1$

Let $u = 2y + 1$. This gives us $4u^2 - 3u - 1$.

$a \cdot c = 4 \cdot -1 = -4$

We need two factors of -4 whose sum is -3. Since the product is negative, the factors will have opposite signs. Since the sum is also negative, the factor with the larger absolute value will be negative.

factor 1	factor 2	sum
1	−4	−3 ← okay
2	−2	0

$$4u^2 - 3u - 1 = 4u^2 - 4u + u - 1$$
$$= 4u(u-1) + (u-1)$$
$$= (u-1)(4u+1)$$

Now get back in terms of y.

$$4(2y+1)^2 - 3(2y+1) - 1$$
$$= (2y+1-1)(4(2y+1)+1)$$
$$= (2y)(8y+4+1) = 2y(8y+5)$$

77. *ac* Method:
First note that there are no common factors and
that $a = 12$, $b = 23$, and $c = 10$.
$a \cdot c = 12 \cdot 10 = 120$. We want to determine two
factors whose product is 120 and whose sum is
23. Since the product is positive we know the
two integers have the same sign. Since the sum is
positive we know that they must both be positive
and neither factor can be larger than 23.

factor 1	factor 2	sum
10	12	22 too small
20	6	26 too large
15	8	23 ← okay

$$12x^2 + 23xy + 10y^2 = 12x^2 + 8xy + 15xy + 10y^2$$
$$= 4x(3x + 2y) + 5y(3x + 2y)$$
$$= (3x + 2y)(4x + 5y)$$

Trial and Error Method:
First note that there are no common factors and
that $a = 12$, $b = 23$, and $c = 10$.
Since c is positive the signs in our factors will be
the same, and since b is positive we know that
the signs will both be positive. We will consider
factorizations with this form:
$$(\_\,x + \_\,y)(\_\,x + \_\,y).$$
Since $a = 12$ can be factored as $1 \cdot 12$, $2 \cdot 6$, and
$3 \cdot 4$, we have the following forms:
$$(1x + \_\,y)(12x + \_\,y)$$
$$(2x + \_\,y)(6x + \_\,y)$$
$$(3x + \_\,y)(4x + \_\,y)$$
$|c| = |10| = 10$ can be factored as $1 \cdot 10$ or $2 \cdot 5$.
Since the original expression had no common
factors, the binomials we select cannot have a
common factor.
$$(x + 10y)(12x + y) \rightarrow 12x^2 + 121xy + 10y^2$$
$$(3x + 10y)(4x + y) \rightarrow 12x^2 + 43xy + 10y^2$$
$$(x + 2y)(12x + 5y) \rightarrow 12x^2 + 29xy + 10y^2$$
$$(3x + 2y)(4x + 5y) \rightarrow 12x^2 + 23xy + 10y^2$$
The correct factorization is
$$12x^2 + 23xy + 10y^2 = (3x + 2y)(4x + 5y)$$

79. $y^2 + 2y - 27$
We are looking for two factors of $c = -27$
whose sum is $b = 2$. Since c is negative the two
factors have opposite signs, and since b is
positive the factor with the larger absolute value
must be positive.

Factors	$-1, 27$	$-3, 9$
Sum	26	6

Since none of the possibilities work,
$y^2 + 2y - 27$ is prime.

81. $x^4 + 6x^2 + 8$
Let $u = x^2$. This gives us $u^2 + 6u + 8$.
We need two factors of $c = 8$ whose sum is
$b = 6$. Since the product is positive, the two
factors will have the same sign. Since the sum is
also positive, the factors will both be positive.
$$u^2 + 6u + 8 = (u + 4)(u + 2)$$
Now get back in terms of x.
$$x^4 + 6x^2 + 8 = (x^2 + 4)(x^2 + 2)$$

83. $z^6 + 9z^3 + 20$
Let $u = z^3$. This gives us $u^2 + 9u + 20$.
We need two factors of 20 whose sum is 9. Since
the product is positive, the two factors will have
the same sign. Since the sum is also positive, the
factors will both be positive.
$$u^2 + 9u + 20 = (u + 5)(u + 4)$$
Now get back in terms of z.
$$z^6 + 5z^3 + 4 = (z^3 + 5)(z^3 + 4)$$

85. $r^2 - 12rs + 32s^2$
We are looking for two factors of $c = 32$ whose
sum is $b = -12$. Since c is positive the two
factors must have the same sign, and since b is
negative the factors must both be negative.

Factors	$-1, -32$	$-2, -16$	$-4, -8$
Sum	-33	-18	-12

$$r^2 - 12rs + 32s^2 = (r - 4s)(r - 8s)$$

87. $8(z + 1)^2 + 2(z + 1) - 1$
Let $u = z + 1$. This gives us $8u^2 + 2u - 1$.
$a \cdot c = 8 \cdot -1 = -8$
We need two factors of -8 whose sum is 2.
Since the product is negative, the factors will
have opposite signs. Since the sum is positive,
the factor with the larger absolute value will be
positive.

factor 1	factor 2	sum
−1	8	7
−2	4	2 ← okay

$$8u^2 + 2u - 1 = 8u^2 + 4u - 2u - 1$$
$$= 4u(2u+1) - 1(2u+1)$$
$$= (2u+1)(4u-1)$$

Now get back in terms of z.

$$8(z+1)^2 + 2(z+1) - 1$$
$$= (2(z+1)+1)(4(z+1)-1)$$
$$= (2z+2+1)(4z+4-1)$$
$$= (2z+3)(4z+3)$$

89. $3x^2 - 7x - 12$

We need two factors of $c = -12$ whose sum is $b = -7$. Since c is negative, the two factors will have opposite signs. Since b is also negative, the factor with the larger absolute value will be negative.

Factors	1, −12	2, −6	3, −4
Sum	−11	−4	−1

Since none of the possibilities work,
$3x^2 - 7x - 12$ is prime.

91. $2x^2 + 12x - 54$

Start by factoring out the GCF.

$$2\left(x^2 + 6x - 27\right)$$

We are looking for two factors of $c = -27$ whose sum is $b = 6$. Since c is negative the two factors have opposite signs, and since b is positive the factor with the larger absolute value must be positive.

Factors	−1, 27	−3, 9
Sum	26	6

$$2x^2 + 12x - 54 = 2(x-3)(x+9)$$

93. $-3r^2 + 39r - 120$

Start by factoring out the GCF.

$$-3\left(r^2 - 13r + 40\right)$$

We are looking for two factors of $c = 40$ whose sum is $b = -13$. Since c is positive the two factors must have the same sign, and since b is negative the two factors must be negative.

Factors	−1,−40	−2,−20	−4,−10	−5,−8
Sum	−41	−22	−14	−13

$$-3r^2 + 39r - 120 = -3(r-5)(r-8)$$

95. <u>*ac* Method:</u>
Start by factoring out the GCF.

$$-16m^2 + 12m + 70 = -2\left(8m^2 - 6m - 35\right)$$

Now focus on the reduced trinomial.
$a \cdot c = 8 \cdot -35 = -280$
We are looking for two factors of -280 whose sum is -6. Since the product is negative, the factors will have opposite signs. The sum is negative so the factor with the larger absolute value will be negative. Since the sum is not very large, we want factors whose absolute values are near each other.

factor 1	factor 2	sum
8	−35	−27 too small
10	−28	−18 too small
14	−20	−6 ← okay

$$-16m^2 + 12m + 70 = -2\left(8m^2 - 6m - 35\right)$$
$$= -2\left(8m^2 - 20m + 14m - 35\right)$$
$$= -2\left(4m(2m-5) + 7(2m-5)\right)$$
$$= -2(2m-5)(4m+7)$$

<u>Trial and Error Method:</u>
Start by factoring out the GCF.

$$-16m^2 + 12m + 70 = -2\left(8m^2 - 6m - 35\right)$$

Now focus on the reduced trinomial.
$a = 8$, $b = -6$, and $c = -35$. Since c is negative, the signs in our factors will be opposites. We will consider factorizations with this form:

$$(\underline{\quad}m + \underline{\quad})(\underline{\quad}m - \underline{\quad})$$

If our choice results in a middle term with the wrong sign, we simply switch the signs of the factors.
Since $a = 8$ can be factored as $1 \cdot 8$ and $2 \cdot 4$, we have the following forms:

$$(m + \underline{\quad})(8m - \underline{\quad})$$
$$(2m + \underline{\quad})(4m - \underline{\quad})$$

$|c| = |-35| = 35$ can be factored as $1 \cdot 35$ and $5 \cdot 7$. Since the original expression had no common factors, the binomials we select cannot have a common factor.

$$(m+1)(8m-35) \rightarrow 8m^2 - 27m - 35$$
$$(m+35)(8m-1) \rightarrow 8m^2 + 279m - 35$$
$$(m+5)(8m-7) \rightarrow 8m^2 + 33m - 35$$
$$(m+7)(8m-5) \rightarrow 8m^2 + 51m - 35$$
$$(2m+1)(4m-35) \rightarrow 8m^2 - 66m - 35$$
$$(2m+35)(4m-1) \rightarrow 8m^2 + 138m - 35$$
$$(2m+5)(4m-7) \rightarrow 8m^2 + 6m - 35$$
$$(2m+7)(4m-5) \rightarrow 8m^2 + 18m - 35$$

Notice that our seventh choice is correct except for the sign on the middle term. Therefore, we simply need to switch the signs of the factors. The correct factorization is

$$-16m^2 + 12m + 70 = -2(2m-5)(4m+7)$$

97. *ac* Method:
Start by factoring out the GCF.

$$48z^2 + 124z + 28 = 4(12z^2 + 31z + 7)$$

Now focus on the reduced trinomial.
$$a \cdot c = 12 \cdot 7 = 84$$
We are looking for two factors of 84 whose sum is 31. Since the product is positive, the factors will have the same sign. The sum is also positive so the factors will both be positive. Since we are adding two positive numbers to get 31, both factors need to be less than 31.

factor 1	factor 2	sum
7	12	19 too small
4	21	25 too small
3	28	31 ← okay

$$48z^2 + 124z + 28 = 4(12z^2 + 31z + 7)$$
$$= 4(12z^2 + 3z + 28z + 7)$$
$$= 4(3z(4z+1) + 7(4z+1))$$
$$= 4(4z+1)(3z+7)$$

Trial and Error Method:
Start by factoring out the GCF.

$$48z^2 + 124z + 28 = 4(12z^2 + 31z + 7)$$

Now focus on the reduced trinomial.
$a = 12, b = 31$, and $c = 7$. Since c is positive, the signs in our factors will be the same. Since b is also positive, the two factors will be positive. We will consider factorizations with this form:

$$(\underline{}z + \underline{})(\underline{}z + \underline{})$$

Since $a = 12$ can be factored as $1 \cdot 12$, $2 \cdot 6$, and $3 \cdot 4$, we have the following forms:

$$(z + \underline{})(12z + \underline{})$$
$$(2z + \underline{})(6z + \underline{})$$
$$(3z + \underline{})(4z + \underline{})$$

$|c| = |7| = 7$ can be factored as $1 \cdot 7$. Since the original expression had no common factors, the binomials we select cannot have a common factor.

$$(z+1)(12z+7) \rightarrow 12z^2 + 19z + 7$$
$$(z+7)(12z+1) \rightarrow 12z^2 + 85z + 7$$
$$(2z+1)(6z+7) \rightarrow 12z^2 + 20z + 7$$
$$(2z+7)(6z+1) \rightarrow 12z^2 + 44z + 7$$
$$(3z+1)(4z+7) \rightarrow 12z^2 + 25z + 7$$
$$(3z+7)(4z+1) \rightarrow 12z^2 + 31z + 7$$

The correct factorization is

$$48z^2 + 124z + 28 = 4(12z^2 + 31z + 7)$$
$$= 4(3z+7)(4z+1)$$

99. Start by factoring out the GCF.

$$3x^3 - 6x^2 - 240x = 3x(x^2 - 2x - 80)$$

Now we focus on the reduced trinomial.
We need two factors of -80 whose sum is -2. Since the product is negative, the factors will have opposite signs. Since the sum is also negative, the factor with the larger absolute value will be negative.

Factors	1,−80	2,−40	4,−20	5,−16	8,−10
Sum	−79	−38	−16	−11	−2

$$3x^3 - 6x^2 - 240x = 3x(x^2 - 2x - 80)$$
$$= 3x(x-10)(x+8)$$

101. Start by factoring out the GCF.

$$8x^3 y^2 - 76x^2 y^2 + 140xy^2$$
$$= 4xy^2(2x^2 - 19x + 35)$$

Now we focus on the reduced trinomial.
$$a \cdot c = 2 \cdot 35 = 70$$
We need two factors of 70 whose sum is -19. Since the product is positive, the two factors will have the same sign. Since the sum is negative, the two factors will be negative.

factor 1	factor 2	sum
-1	-70	-71
-2	-35	-37
-5	-14	$-19 \leftarrow$ okay
-7	-10	-17

$$8x^3 y^2 - 76x^2 y^2 + 140xy^2$$
$$= 4xy^2 \left(2x^2 - 19x + 35\right)$$
$$= 4xy^2 \left(2x^2 - 14x - 5x + 35\right)$$
$$= 4xy^2 \left(2x(x-7) - 5(x-7)\right)$$
$$= 4xy^2 (x-7)(2x-5)$$

103. Start by factoring out the GCF.
$$70r^4 s - 36r^3 s - 16r^2 s = 2r^2 s \left(35r^2 - 18r - 8\right)$$

Now we focus on the reduced trinomial.
$$a \cdot c = 35 \cdot (-8) = -280$$

We need two factors of -280 whose sum is -18. Since the product is negative, the two factors will have opposite signs. Since the sum is negative, the factor with the larger absolute value will be negative.

factor 1	factor 2	sum
4	-70	-66
5	-56	-51
7	-40	-33
10	-28	$-18 \leftarrow$ okay

$$70r^4 s - 36r^3 s - 16r^2 s$$
$$= 2r^2 s \left(35r^2 - 18r - 8\right)$$
$$= 2r^2 s \left(35r^2 - 28r + 10r - 8\right)$$
$$= 2r^2 s \left(7r(5r-4) + 2(5r-4)\right)$$
$$= 2r^2 s (5r-4)(7r+2)$$

105. a.
$$V(3) = 4(3)^3 - 108(3)^2 + 720(3)$$
$$= 4(27) - 108(9) + 720(3)$$
$$= 108 - 972 + 2160$$
$$= 1296$$

When 3-inch square corners are cut from the piece of cardboard, the resulting box will have a volume of 1296 cubic inches.

b.
$$V(x) = 4x^3 - 108x^2 + 720x$$
$$= 4x\left(x^2 - 27x + 180\right)$$
$$= 4x(x-15)(x-12)$$

c.
$$V(3) = 4(3)(3-15)(3-12)$$
$$= 12(-12)(-9)$$
$$= -144(-9)$$
$$= 1296$$

The volume is 1296 cubic inches.

d. Answers may vary. Often it will be easier in factored form.

107. If we know one factor, we can find the other by dividing the polynomial by the known factor.

$$\begin{array}{r} 2x-5 \\ 3x+2 \overline{)6x^2 - 11x - 10} \\ \underline{-\left(6x^2 + 4x\right)} \\ -15x - 10 \\ \underline{-(-15x - 10)} \\ 0 \end{array}$$

The other factor is $2x - 5$.
$$6x^2 - 11x - 10 = (3x+2)(2x-5)$$

109.
$$\frac{1}{2}x^2 + 3x + 4 = \frac{1}{2}x^2 + \frac{6}{2}x + \frac{8}{2}$$
$$= \frac{1}{2}\left(x^2 + 6x + 8\right)$$
$$= \frac{1}{2}(x+4)(x+2)$$

111.
$$\frac{1}{3}p^2 - \frac{2}{3}p - 1 = \frac{1}{3}p^2 - \frac{2}{3}p - \frac{3}{3}$$
$$= \frac{1}{3}\left(p^2 - 2p - 3\right)$$
$$= \frac{1}{3}(p-3)(p+1)$$

113.
$$\frac{4}{3}a^2 - \frac{8}{3}a - 32 = \frac{4}{3}a^2 - \frac{8}{3}a - \frac{96}{3}$$
$$= \frac{4}{3}a^2 - \frac{4}{3} \cdot 2a - \frac{4}{3} \cdot 24$$
$$= \frac{4}{3}\left(a^2 - 2a - 24\right)$$
$$= \frac{4}{3}(a-6)(a+4)$$

115. $2^{2n} - 4 \cdot 2^n - 5$

Let $u = 2^n$. This gives us $u^2 - 4u - 5$. We need two factors of -5 whose sum is -4. Since the product is negative, the factors will have opposite signs. Since the sum is negative, the factor with the largest absolute value will be negative.

$$u^2 - 4u - 5 = (u - 5)(u + 1)$$

Now get back in terms of n.

$$2^{2n} - 4 \cdot 2^n - 5 = \left(2^n - 5\right)\left(2^n + 1\right)$$

117. $4^{2x} - 12 \cdot 4^x + 32$

Let $u = 4^x$. This gives us $u^2 - 12u + 32$. We need two factors of 32 whose sum is -12. Since the product is positive and the sum is negative, the factors will both be negative.

$$u^2 - 12u + 32 = (u - 8)(u - 4)$$

Now get back in terms of x.

$$4^{2x} - 12 \cdot 4^x + 32 = (4^x - 8)(4^x - 4)$$
$$= (4^x - 8)(2^x - 2)(2^x + 2)$$

119. The Trial and Error method would be better when $a \cdot c$ gets large and there are lots of factors of the product whose sums must be determined.

121. The problem was not completely factored. The first binomial has a common factor of 3 than can be pulled out.

$$3p^2 - 9p - 30 = 3\left(p^2 - 3p - 10\right)$$
$$= 3(p + 2)(p - 5)$$

Section 4.6

Are You Ready for This Section?

R1. $1^2 = 1 \cdot 1 = 1$; 　 $2^2 = 2 \cdot 2 = 4$;
$3^2 = 3 \cdot 3 = 9$; 　 $4^2 = 4 \cdot 4 = 16$;
$5^2 = 5 \cdot 5 = 25$

R2. $\left(-\dfrac{3}{2}\right)^2 = \left(-\dfrac{3}{2}\right)\left(-\dfrac{3}{2}\right) = \dfrac{9}{4}$

Section 4.6 Quick Checks

1. A trinomial of the form $A^2 + 2AB + B^2$ or $A^2 - 2AB + B^2$ is called a <u>perfect square trinomial</u>.

2. $A^2 - 2AB + B^2 = \underline{(A - B)^2}$.

3. False; $4x^2 + 12x + 9$ would be a perfect square trinomial since it could be written as $(2x)^2 + 2 \cdot 2x \cdot 3 + (3)^2$.

4. $x^2 - 18x + 81 = x^2 - 2 \cdot 9 \cdot x + 9^2$
$$= (x - 9)^2$$

5. $4x^2 + 20xy + 25y^2 = (2x)^2 + 2 \cdot 2x \cdot 5y + (5y)^2$
$$= (2x + 5y)^2$$

6. $18p^4 - 84p^2 + 98$
$$= 2\left(9p^4 - 42p^2 + 49\right)$$
$$= 2\left(\left(3p^2\right)^2 - 2 \cdot \left(3p^2\right) \cdot 7 + 7^2\right)$$
$$= 2\left(3p^2 - 7\right)^2$$

7. $16z^2 - 25$ is called a <u>difference</u> of <u>two squares</u> and factors into two binomials.

8. $P^2 - Q^2 = \underline{(P - Q)(P + Q)}$

9. $z^2 - 16 = z^2 - 4^2$
$$= (z - 4)(z + 4)$$

10. $16m^2 - 81n^2 = (4m)^2 - (9n)^2$
$$= (4m - 9n)(4m + 9n)$$

11. $4a^2 - 9b^4 = (2a)^2 - \left(3b^2\right)^2$
$$= \left(2a - 3b^2\right)\left(2a + 3b^2\right)$$

12. True

13. $3b^4 - 48 = 3\left(b^4 - 16\right)$
$$= 3\left(\left(b^2\right)^2 - 4^2\right)$$
$$= 3\left(b^2 - 4\right)\left(b^2 + 4\right)$$
$$= 3\left(b^2 - 2^2\right)\left(b^2 + 4\right)$$
$$= 3(b - 2)(b + 2)\left(b^2 + 4\right)$$

14. $p^2 - 8p + 16 - q^2 = \left(p^2 - 8p + 16\right) - q^2$
$$= \left(p^2 - 2 \cdot 4 \cdot p + 4^2\right) - q^2$$
$$= (p-4)^2 - q^2$$
$$= (p-4-q)(p-4+q)$$

15. $A^3 + B^3 = \underline{(A+B)(A^2 - AB + B^2)}$

16. $A^3 - B^3 = \underline{(A-B)(A^2 + AB + B^2)}$

17. $z^3 + 64 = z^3 + 4^4$
$$= (z+4)\left(z^2 - 4 \cdot z + 4^2\right)$$
$$= (z+4)\left(z^2 - 4z + 16\right)$$

18. $125p^3 - 216q^6$
$$= (5p)^3 - \left(6q^2\right)^3$$
$$= \left(5p - 6q^2\right)\left((5p)^2 + 5p \cdot 6q^2 + \left(6q^2\right)^2\right)$$
$$= \left(5p - 6q^2\right)\left(25p^2 + 30pq^2 + 36q^4\right)$$

19. $32m^3 + 500n^6$
$$= 4\left(8m^3 + 125n^6\right)$$
$$= 4\left((2m)^3 + \left(5n^2\right)^3\right)$$
$$= 4\left(2m + 5n^2\right)\left((2m)^2 - 2m \cdot 5n^2 + \left(5n^2\right)^2\right)$$
$$= 4\left(2m + 5n^2\right)\left(4m^2 - 10mn^2 + 25n^4\right)$$

20. $(x+1)^3 - 27x^3$
$$= (x+1)^3 - (3x)^3$$
$$= (x+1-3x)\left((x+1)^2 + (x+1) \cdot 3x + (3x)^2\right)$$
$$= (-2x+1)\left(x^2 + 2x + 1 + 3x^2 + 3x + 9x^2\right)$$
$$= (-2x+1)\left(13x^2 + 5x + 1\right)$$

4.6 Exercises

21. $x^2 + 4x + 4 = x^2 + 2 \cdot x \cdot 2 + 2^2$
$$= (x+2)^2$$

23. $36 + 12w + w^2 = 6^2 + 2 \cdot 6 \cdot w + w^2$
$$= (6+w)^2 \quad \text{or} \quad (w+6)^2$$

25. $4x^2 + 4x + 1 = (2x)^2 + 2 \cdot 2x \cdot 1 + 1^2$
$$= (2x+1)^2$$

27. $9p^2 - 30p + 25 = (3p)^2 - 2 \cdot 3p \cdot 5 + 5^2$
$$= (3p-5)^2$$

29. $25a^2 + 90a + 81 = (5a)^2 + 2 \cdot 5a \cdot 9 + 9^2$
$$= (5a+9)^2$$

31. $9x^2 + 24xy + 16y^2 = (3x)^2 + 2(3x)(4y) + (4y)^2$
$$= (3x+4y)^2$$

33. $3w^2 - 30w + 75 = 3\left(w^2 - 10w + 25\right)$
$$= 3\left(w^2 - 2 \cdot w \cdot 5 + 5^2\right)$$
$$= 3(w-5)^2$$

35. $-5t^2 - 70t - 245 = -5\left(t^2 + 14t + 49\right)$
$$= -5\left(t^2 + 2 \cdot t \cdot 7 + 7^2\right)$$
$$= -5(t+7)^2$$

37. $32a^2 - 80ab + 50b^2 = 2\left(16a^2 - 40ab + 25b^2\right)$
$$= 2\left((4a)^2 - 2 \cdot 4a \cdot 5b + (5b)^2\right)$$
$$= 2(4a-5b)^2$$

39. $z^4 - 6z^2 + 9 = \left(z^2\right)^2 - 2 \cdot z^2 \cdot 3 + 3^2$
$$= \left(z^2 - 3\right)^2$$

41. $x^2 - 9 = x^2 - 3^2$
$$= (x-3)(x+3)$$

43. $4 - y^2 = 2^2 - y^2$
$$= (2-y)(2+y)$$

45. $4z^2 - 9 = (2z)^2 - 3^2$
$$= (2z-3)(2z+3)$$

47. $100m^2 - 81n^2 = (10m)^2 - (9n)^2$
$$= (10m - 9n)(10m + 9n)$$

49. $m^4 - 36n^6 = \left(m^2\right)^2 - \left(6n^3\right)^2$
$$= \left(m^2 - 6n^3\right)\left(m^2 + 6n^3\right)$$

51. $8p^2 - 18q^2 = 2\left(4p^2 - 9q^2\right)$
$$= 2\left((2p)^2 - (3q)^2\right)$$
$$= 2(2p - 3q)(2p + 3q)$$

53. $80p^2r - 245b^2r = 5r\left(16p^2 - 49b^2\right)$
$$= 5r\left((4p)^2 - (7b)^2\right)$$
$$= 5r(4p - 7b)(4p + 7b)$$

55. $(x + y)^2 - 9 = (x + y)^2 - 3^2$
$$= (x + y - 3)(x + y + 3)$$

57. $x^3 - 8 = x^3 - 2^3$
$$= (x - 2)\left(x^2 + x(2) + 2^2\right)$$
$$= (x - 2)\left(x^2 + 2x + 4\right)$$

59. $125 + m^3 = 5^3 + m^3$
$$= (5 + m)\left(5^2 - 5m + m^2\right)$$
$$= (5 + m)\left(25 - 5m + m^2\right)$$
$$= (m + 5)\left(m^2 - 5m + 25\right)$$

61. $x^6 - 64y^3 = \left(x^2\right)^3 - (4y)^3$
$$= \left(x^2 - 4y\right)\left(\left(x^2\right)^2 + x^2 \cdot 4y + (4y)^2\right)$$
$$= \left(x^2 - 4y\right)\left(x^4 + 4x^2 y + 16y^2\right)$$

63. $24x^3 - 375y^3$
$$= 3\left(8x^3 - 125y^3\right)$$
$$= 3\left((2x)^3 - (5y)^3\right)$$
$$= 3(2x - 5y)\left((2x)^2 + 2x \cdot 5y + (5y)^2\right)$$
$$= 3(2x - 5y)\left(4x^2 + 10xy + 25y^2\right)$$

65. $(p + 1)^3 - 27$
$$= (p + 1)^3 - 3^3$$
$$= (p + 1 - 3)\left((p + 1)^2 + (p + 1)\cdot 3 + 3^2\right)$$
$$= (p - 2)\left(p^2 + 2p + 1 + 3p + 3 + 9\right)$$
$$= (p - 2)\left(p^2 + 5p + 13\right)$$

67. $(3y + 1)^3 + 8y^3$
$$= (3y + 1)^3 + (2y)^3$$
$$= (3y + 1 + 2y)\left((3y + 1)^2 - (3y + 1)\cdot 2y + (2y)^2\right)$$
$$= (5y + 1)\left(9y^2 + 6y + 1 - 6y^2 - 2y + 4y^2\right)$$
$$= (5y + 1)\left(7y^2 + 4y + 1\right)$$

69. $y^6 + z^9 = \left(y^2\right)^3 + \left(z^3\right)^3$
$$= \left(y^2 + z^3\right)\left(\left(y^2\right)^2 - y^2 z^3 + \left(z^3\right)^2\right)$$
$$= \left(y^2 + z^3\right)\left(y^4 - y^2 z^3 + z^6\right)$$

71. $y^9 - 1 = \left(y^3\right)^3 - 1^3$
$$= \left(y^3 - 1\right)\left(\left(y^3\right)^2 + y^3 \cdot 1 + 1^2\right)$$
$$= \left(y^3 - 1\right)\left(y^6 + y^3 + 1\right)$$
$$= \left(y^3 - 1^3\right)\left(y^6 + y^3 + 1\right)$$
$$= (y - 1)\left(y^2 + y + 1\right)\left(y^6 + y^3 + 1\right)$$

73. $25x^2 - y^2 = (5x)^2 - y^2$
$$= (5x - y)(5x + y)$$

75. $8x^3 + 27 = (2x)^3 + 3^3$
$$= (2x + 3)\left((2x)^2 - 2x \cdot 3 + 3^2\right)$$
$$= (2x + 3)\left(4x^2 - 6x + 9\right)$$

77. $z^2 - 8z + 16 = z^2 - 2 \cdot z \cdot 4 + 4^2$
$$= (z - 4)^2$$

79. $5x^4 - 40xy^3 = 5x\left(x^3 - 8y^3\right)$

$$= 5x\left(x^3 - (2y)^3\right)$$

$$= 5x(x - 2y)\left(x^2 + x \cdot 2y + (2y)^2\right)$$

$$= 5x(x - 2y)\left(x^2 + 2xy + 4y^2\right)$$

81. $49m^2 - 42mn + 9n^2 = (7m)^2 - 2(7m)(3n) + (3n)^2$

$$= (7m - 3n)^2$$

83. $4x^2 + 16 = 4\left(x^2 + 4\right)$

85. $y^4 - 8y^2 + 16 = \left(y^2\right)^2 - 2 \cdot y^2 \cdot 4 + 4^2$

$$= \left(y^2 - 4\right)^2$$

$$= \left[y^2 - 2^2\right]^2$$

$$= \left[(y - 2)(y + 2)\right]^2$$

$$= (y - 2)^2 (y + 2)^2$$

87. $4a^2b^2 + 12ab + 9 = (2ab)^2 + 2 \cdot 2ab \cdot 3 + 3^2$

$$= (2ab + 3)^2$$

89. $p^2 + 2p + 4 = p^2 + 2 \cdot p + (2)^2$

This polynomial is not a perfect square trinomial because it does not have the necessary middle term. Try to factor it using the $a \cdot c$ method. Here we need two numbers which multiply to 4 and add to 2. There are no such numbers, so the polynomial is prime.

91. $n^2 + 13n + 36 = (n + 9)(n + 4)$

93. $2n^2 - 2m^2 + 40m - 200$

$$= 2\left(n^2 - m^2 + 20m - 100\right)$$

$$= 2\left(n^2 - \left(m^2 - 20m + 100\right)\right)$$

$$= 2\left(n^2 - \left(m^2 - 2 \cdot m \cdot 10 + 10^2\right)\right)$$

$$= 2\left(n^2 - (m - 10)^2\right)$$

$$= 2\left[(n - (m - 10))(n + (m - 10))\right]$$

$$= 2(n - m + 10)(n + m - 10)$$

95. $3y^3 - 24 = 3\left(y^3 - 8\right)$

$$= 3\left(y^3 - 2^3\right)$$

$$= 3(y - 2)\left(y^2 + 2y + 4\right)$$

97. $16x^2 + 24xy + 9y^2 - 100$

$$= \left(16x^2 + 24xy + 9y^2\right) - 100$$

$$= \left((4x)^2 + 2 \cdot 4x \cdot 3y + (3y)^2\right) - 10^2$$

$$= (4x + 3y)^2 - 10^2$$

$$= (4x + 3y - 10)(4x + 3y + 10)$$

99. The area of the shaded region is the area of the larger square minus the area of the smaller square.

$$x^2 - 3^2 = (x - 3)(x + 3)$$

The area of the shaded region is $(x - 3)(x + 3)$ square units.

101. The area of the shaded region is the area of the large square minus the total area of the four square corners.

$$x^2 - 4\left(2^2\right) = x^2 - 16$$

$$= x^2 - 4^2$$

$$= (x - 4)(x + 4)$$

The area of the shaded region is $(x - 4)(x + 4)$ square units.

103. The area of the shaded region is the area of the larger circle minus the area of the smaller circle.

$$\pi R^2 - \pi r^2 = \pi\left(R^2 - r^2\right)$$

$$= \pi(R - r)(R + r)$$

The area of the shaded region is $\pi(R - r)(R + r)$ square units.

105. The volume of the shaded region is the volume of the whole solid minus the volume of the rectangular solid in the middle.

$$V = a \cdot 2a \cdot 5a - b \cdot 2b \cdot 5a$$

$$= 10a^3 - 10ab^2$$

$$= 10a\left(a^2 - b^2\right)$$

$$= 10a(a - b)(a + b)$$

The volume of the shaded region is $10a(a - b)(a + b)$ cubic units.

107. $A = a \cdot a + a \cdot b + a \cdot b + b \cdot b$
$$= a^2 + ab + ab + b^2$$
$$= a^2 + 2ab + b^2$$
$$= (a+b)^2$$

109. $4x^2 + bx + 81$
To be a perfect square trinomial, the middle term must equal $\pm$ twice the product of the quantities that are squared to get the first and last terms.
$$4x^2 = (2x)^2$$
$$81 = 9^2$$
Thus, the middle term must equal $\pm$ twice the product of $2x$ and 9.
$$bx = \pm 2 \cdot 2x \cdot 9$$
$$bx = \pm 36x$$
Therefore, $b = \pm 36$.

111. $x^2 + 18x$
To be a perfect square trinomial, we need to have the form $a^2 + 2ab + b^2$. Here we know that $a = x$ and we want $2ab = 18x$.
$$2ab = 18x$$
$$2(x)b = 18x$$
$$2bx = 18x$$
$$2b = 18$$
$$b = 9$$
Since $b = 9$, we need to add $9^2 = 81$ to make a perfect square trinomial.
$$x^2 + 18x + 81 = (x+9)^2$$

113. $b^2 - 0.4b + 0.04 = b^2 - 2 \cdot b \cdot (0.2) + (0.2)^2$
$$= (b-0.2)^2$$

115. $9b^2 - \dfrac{1}{25} = (3b)^2 - \left(\dfrac{1}{5}\right)^2$
$$= \left(3b - \dfrac{1}{5}\right)\left(3b + \dfrac{1}{5}\right)$$

117. $\dfrac{x^2}{9} - \dfrac{y^2}{25} = \left(\dfrac{x}{3}\right)^2 - \left(\dfrac{y}{5}\right)^2$
$$= \left(\dfrac{x}{3} - \dfrac{y}{5}\right)\left(\dfrac{x}{3} + \dfrac{y}{5}\right)$$

119. $\dfrac{x^3}{8} - \dfrac{y^3}{27} = \left(\dfrac{x}{2}\right)^3 - \left(\dfrac{y}{3}\right)^3$
$$= \left(\dfrac{x}{2} - \dfrac{y}{3}\right)\left(\left(\dfrac{x}{2}\right)^2 + \dfrac{x}{2} \cdot \dfrac{y}{3} + \left(\dfrac{y}{3}\right)^2\right)$$
$$= \left(\dfrac{x}{2} - \dfrac{y}{3}\right)\left(\dfrac{x^2}{4} + \dfrac{xy}{6} + \dfrac{y^2}{9}\right)$$

Section 4.7

Section 4.7 Quick Checks

1. $2p^2q - 8pq^2 - 90q^3 = 2q\left(p^2 - 4pq - 45q^2\right)$
$$= 2q(p-9q)(p+5q)$$

2. $-45x^2y + 66xy + 27y = -3y\left(15x^2 - 22x - 9\right)$
$$= -3y(3x+1)(5x-9)$$

3. $81x^2 - 100y^2 = (9x)^2 - (10y)^2$
$$= (9x-10y)(9x+10y)$$

4. $-3m^2n + 147n = -3n\left(m^2 - 49\right)$
$$= -3n\left(m^2 - 7^2\right)$$
$$= -3n(m-7)(m+7)$$

5. $p^2 - 16pq + 64q^2 = p^2 - 2 \cdot p \cdot 8q + (8q)^2$
$$= (p-8q)^2$$

6. $20x^2 + 60x + 45 = 5\left(4x^2 + 12x + 9\right)$
$$= 5\left((2x)^2 + 2 \cdot 2x \cdot 3 + 3^2\right)$$
$$= 5(2x+3)^2$$

7. $64y^3 - 125 = (4y)^3 - 5^3$
$$= (4y-5)\left((4y)^2 + 4y \cdot 5 + 5^2\right)$$
$$= (4y-5)\left(16y^2 + 20y + 25\right)$$

8.
$$-16m^3 - 2n^3 = -2\left(8m^3 + n^3\right)$$
$$= -2\left((2m)^3 + n^3\right)$$
$$= -2(2m+n)\left((2m)^2 - 2m \cdot n + n^2\right)$$
$$= -2(2m+n)\left(4m^2 - 2mn + n^2\right)$$

9. $10z^2 - 15z + 35 = 5\left(2z^2 - 3z + 7\right)$

Using $a = 2$, $b = -3$, and $c = 7$, we get $a \cdot c = 2 \cdot 7 = 14$. We are looking for two factors of 14 whose sum is $b = -3$. There are no such factors, so the above factorization is complete.

10. $6xy^2 + 81x^3 = 3x\left(2y^2 + 27x^2\right)$

The expression in parentheses cannot be factored any further.

11.
$$2x^3 + 5x^2 + 4x + 10 = \left(2x^3 + 5x^2\right) + (4x + 10)$$
$$= x^2(2x+5) + 2(2x+5)$$
$$= (2x+5)\left(x^2+2\right)$$
$$= \left(x^2+2\right)(2x+5)$$

12.
$$9x^3 + 3x^2 - 9x - 3 = 3\left(3x^3 + x^2 - 3x - 1\right)$$
$$= 3\left(\left(3x^3 + x^2\right) + (-3x - 1)\right)$$
$$= 3\left(x^2(3x+1) - 1(3x+1)\right)$$
$$= 3(3x+1)\left(x^2-1\right)$$
$$= 3(3x+1)(x-1)(x+1)$$

13.
$$4x^2 + 4xy + y^2 - 81$$
$$= \left(4x^2 + 4xy + y^2\right) - 81$$
$$= \left((2x)^2 + 2 \cdot 2x \cdot y + y^2\right) - 81$$
$$= (2x+y)^2 - 9^2$$
$$= (2x+y-9)(2x+y+9)$$

14.
$$16 - m^2 - 8mn - 16n^2$$
$$= 16 - \left(m^2 + 8mn + 16n^2\right)$$
$$= 16 - \left(m^2 + 2 \cdot m \cdot 4n + (4n)^2\right)$$
$$= 4^2 - (m+4n)^2$$
$$= (4 - (m+4n))(4 + (m+4n))$$
$$= (4 - m - 4n)(4 + m + 4n)$$

15. $G(x) = 2x^2 + 3x - 14$
$$= (2x+7)(x-2)$$

16. $F(p) = 27p^2 - 12$
$$= 3\left(9p^2 - 4\right)$$
$$= 3\left((3p)^2 - (2)^2\right)$$
$$= 3(3p+2)(3p-2)$$

4.7 Exercises

17. $2x^2 - 12x - 144 = 2\left(x^2 - 6x - 72\right)$
$$= 2(x-12)(x+6)$$

19. $-3y^2 + 27 = -3\left(y^2 - 9\right)$
$$= -3\left(y^2 - 3^2\right)$$
$$= -3(y-3)(y+3)$$

21. $4b^2 + 20b + 25 = (2b)^2 + 2 \cdot 2b \cdot 5 + 5^2$
$$= (2b+5)^2$$

23.
$$16w^3 + 2y^6$$
$$= 2\left(8w^3 + y^6\right)$$
$$= 2\left((2w)^3 + \left(y^2\right)^3\right)$$
$$= 2\left(2w + y^2\right)\left((2w)^2 - 2w \cdot y^2 + \left(y^2\right)^2\right)$$
$$= 2\left(2w + y^2\right)\left(4w^2 - 2wy^2 + y^4\right)$$

25. $-3z^2 + 12z - 18 = -3\left(z^2 - 4z + 6\right)$

27. $a \cdot c = 20 \cdot -18 = -360$

We need two factors of -360 that add to get -9. Since the product is negative, they must have different signs, and since the sum is negative, the factor with the larger absolute value must be negative. The sum is also not too far from 0 so we want the factors to be near each other in magnitude.

factor 1	factor 2	sum
12	−30	−18 too small
18	−20	−2 too large
15	−24	−9 okay

$20y^2 - 9y - 18 = 20y^2 + 15y - 24y - 18$
$= 5y(4y+3) - 6(4y+3)$
$= (4y+3)(5y-6)$

29. $x^3 - 4x^2 + 5x - 20 = (x^3 - 4x^2) + (5x - 20)$
$= x^2(x-4) + 5(x-4)$
$= (x-4)(x^2+5)$

31. $200x^2 + 18y^2 = 2(100x^2 + 9y^2)$

33. $x^4 - 81 = (x^2)^2 - (9)^2$
$= (x^2 - 9)(x^2 + 9)$
$= (x^2 - 3^2)(x^2 + 9)$
$= (x-3)(x+3)(x^2+9)$

35. $a \cdot c = 3 \cdot -16 = -48$

We need two factors of -48 that add to get -7. Since the product is negative, they must have different signs, and since the sum is negative, the factor with the larger absolute value must be negative. The sum is also not too far from 0 so we want the factors to be near each other in magnitude.

factor 1	factor 2	sum
1	−48	−47
2	−24	−22
3	−16	−13
4	−12	−8
6	−8	−2

None of the possibilities work. The expression is prime.

37. $36q^3 + 24q^2 + 4q = 4q(9q^2 + 6q + 1)$
$= 4q((3q)^2 + 2 \cdot 3q \cdot 1 + 1^2)$
$= 4q(3q+1)^2$

39. $24m^3n - 66m^2n - 63mn = 3mn(8m^2 - 22m - 21)$

Focusing on the reduced trinomial we get:
$a \cdot c = 8 \cdot -21 = -168$
We need two factors of -168 that add to get -22. Since the product is negative, they must have different signs, and since the sum is negative, the factor with the larger absolute value must be negative.

factor 1	factor 2	sum
8	−21	−13 too large
4	−42	−38 too small
6	−28	−22 okay

$3mn(8m^2 - 22m - 21)$
$= 3mn(8m^2 + 6m - 28m - 21)$
$= 3mn(2m(4m+3) - 7(4m+3))$
$= 3mn(4m+3)(2m-7)$

41. $3r^5 - 24r^2s^3 = 3r^2(r^3 - 8s^3)$
$= 3r^2(r^3 - (2s)^3)$
$= 3r^2(r-2s)(r^2 + r \cdot 2s + (2s)^2)$
$= 3r^2(r-2s)(r^2 + 2rs + 4s^2)$

43. $2x^3 + 8x^2 - 18x - 72$
$= 2(x^3 + 4x^2 - 9x - 36)$
$= 2((x^3 + 4x^2) + (-9x - 36))$
$= 2(x^2(x+4) - 9(x+4))$
$= 2(x+4)(x^2 - 9)$
$= 2(x+4)(x^2 - 3^2)$
$= 2(x+4)(x-3)(x+3)$

45. $9x^4 - 1 = (3x^2)^2 - 1^2$
$= (3x^2 - 1)(3x^2 + 1)$

47. Let $u = w^2$. This gives

$3u^2 + 4u - 15$

$a \cdot c = 3 \cdot -15 = -45$

We need factors of -45 that add to get 4. Since the product is negative, they must have different signs, and since the sum is positive, the factor with the larger absolute value must be positive.

factor 1	factor 2	sum	
15	-3	12	too large
9	-5	4	okay

$$3u^2 + 4u - 15 = 3u^2 + 9u - 5u - 15$$
$$= 3u(u+3) - 5(u+3)$$
$$= (u+3)(3u-5)$$

Now get back in terms of w.

$$3w^4 + 4w^2 - 15 = \left(w^2 + 3\right)\left(3w^2 - 5\right)$$

49. $(2y+3)^2 - 5(2y+3) + 6$

Let $u = 2y + 3$. This gives

$$u^2 - 5u + 6 = (u-3)(u-2)$$

Now get back in terms of y.

$$(2y+3)^2 - 5(2y+3) + 6 = (2y+3-3)(2y+3-2)$$
$$= 2y(2y+1)$$

51. $p^2 - 10p + 25 - 36q^2$

$$= \left(p^2 - 10p + 25\right) - 36q^2$$
$$= \left(p^2 - 2 \cdot p \cdot 5 + 5^2\right) - 36q^2$$
$$= (p-5)^2 - 36q^2$$
$$= (p-5)^2 - (6q)^2$$
$$= (p-5-6q)(p-5+6q)$$
$$= (p-6q-5)(p+6q-5)$$

53. $y^6 + 6y^3 - 16$

Let $u = y^3$. This gives

$$u^2 + 6u - 16 = (u+8)(u-2)$$

Now get back in terms of y.

$$y^6 + 6y^3 - 16 = \left(y^3 + 8\right)\left(y^3 - 2\right)$$
$$= \left(y^3 + 2^3\right)\left(y^3 - 2\right)$$
$$= (y+2)\left(y^2 - 2 \cdot y + 2^2\right)\left(y^3 - 2\right)$$
$$= (y+2)\left(y^2 - 2y + 4\right)\left(y^3 - 2\right)$$

55. $p^6 - 1 = \left(p^3\right)^2 - 1^2$

$$= \left(p^3 - 1\right)\left(p^3 + 1\right)$$
$$= (p-1)\left(p^2 + p + 1\right)(p+1)\left(p^2 - p + 1\right)$$
$$= (p-1)(p+1)\left(p^2 + p + 1\right)\left(p^2 - p + 1\right)$$

57. $-3x^3 - 15x^2 + 27x + 135$

$$= -3\left(x^3 + 5x^2 - 9x - 45\right)$$
$$= -3\left[\left(x^3 + 5x^2\right) + (-9x - 45)\right]$$
$$= -3\left[x^2(x+5) - 9(x+5)\right]$$
$$= -3(x+5)\left(x^2 - 9\right)$$
$$= -3(x+5)(x-3)(x+3)$$

59. $3a - 27a^3 = 3a\left(1 - 9a^2\right)$

$$= 3a\left(1^2 - (3a)^2\right)$$
$$= 3a(1-3a)(1+3a)$$

61. $8t^5 + 14t^3 - 72t = 2t\left(4t^4 + 7t^2 - 36\right)$

Focusing on the reduced trinomial we get:

$a \cdot c = 4 \cdot -36 = -144$

We need two factors of -144 that add to get 7. Since the product is negative, they must have different signs, and since the sum is positive, the factor with the larger absolute value must be positive.

factor 1	factor 2	sum	
-6	24	18	too large
-12	12	0	too small
-9	16	7	okay

$$2t\left(4t^4 + 7t^2 - 36\right)$$
$$= 2t\left(4t^4 + 16t^2 - 9t^2 - 36\right)$$
$$= 2t\left(4t^2\left(t^2 + 4\right) - 9\left(t^2 + 4\right)\right)$$
$$= 2t\left(t^2 + 4\right)\left(4t^2 - 9\right)$$
$$= 2t\left(t^2 + 4\right)\left((2t)^2 - 3^2\right)$$
$$= 2t\left(t^2 + 4\right)(2t-3)(2t+3)$$

63. $2x^4y + 10x^3y - 18x^2y - 90xy$
$$= 2xy\left(x^3 + 5x^2 - 9x - 45\right)$$
$$= 2xy\left(x^2(x+5) - 9(x+5)\right)$$
$$= 2xy(x+5)\left(x^2 - 9\right)$$
$$= 2xy(x+5)\left(x^2 - 3^2\right)$$
$$= 2xy(x+5)(x-3)(x+3)$$

65. $f(x) = x^2 - 2x - 63$
$$= (x-9)(x+7)$$

67. $P(m) = 7m^2 + 31m + 12$
$$= (7m+3)(m+4)$$

69. $G(x) = -12x^2 + 147$
$$= -3\left(4x^2 - 49\right)$$
$$= -3\left((2x)^2 - (7)^2\right)$$
$$= -3(2x+7)(2x-7)$$

71. $s(t) = -16t^2 + 96t + 256$
$$= -16\left(t^2 - 6t - 16\right)$$
$$= -16(t-8)(t+2)$$

73. $H(a) = 2a^3 + 5a^2 - 32a - 80$
$$= \left(2a^3 + 5a^2\right) + (-32a - 80)$$
$$= a^2(2a+5) + (-16)(2a+5)$$
$$= (2a+5)\left(a^2 - 16\right)$$
$$= (2a+5)\left((a)^2 - (4)^2\right)$$
$$= (2a+5)(a+4)(a-4)$$

75. The area of the shaded region is the area of the large square minus the area of the small square.
$$(2x+5)^2 - (x+2)^2$$
$$= \left[(2x+5) - (x+2)\right]\left[(2x+5) + (x+2)\right]$$
$$= (2x+5-x-2)(2x+5+x+2)$$
$$= (x+3)(3x+7)$$
The area of the shaded region is
$(x+3)(3x+7)$ square units.

77. The area of the shaded region is the area of the large rectangle minus the area of the unshaded rectangle.
$$x(x+y) - y(x+y) = (x+y)(x-y)$$
The area of the shaded region is
$(x+y)(x-y)$ square units.

79. $(3x)^3 - (4)^3 = (3x-4)\left((3x)^2 + 3x \cdot 4 + (4)^2\right)$
$$= (3x-4)\left(9x^2 + 12x + 16\right)$$
The volumes of the boxes differ by
$(3x-4)\left(9x^2 + 12x + 16\right)$ cubic units.

81. Answers will vary. The binomial $x^2 + 9$ is the sum of two squares. Since the sum of two squares does not factor over the integers (or the real numbers), $x^2 + 9$ is prime.

83. $x^{5/2} - 9x^{1/2} = x^{1/2}\left(x^2 - 9\right)$
$$= x^{1/2}\left(x^2 - 3^2\right)$$
$$= x^{1/2}(x-3)(x+3)$$

85. $1 + 6x^{-1} + 8x^{-2} = x^{-2}\left(x^2 + 6x + 8\right)$
$$= x^{-2}(x+4)(x+2)$$

87. Answers will vary.
1) Factor out the GCF
2) Count the number of terms
   a) Check for special two-term cases.
   b) Check for special three-term cases.
   c) Try to factor by grouping with four terms.
   d) Trial and error

Section 4.8

Are You Ready for This Section?

R1. $x + 4 = 0$
$$x + 4 - 4 = 0 - 4$$
$$x = -4$$
Solution set: $\{-4\}$

R2. $3(x-2)-12=0$
$$3x-6-12=0$$
$$3x-18=0$$
$$3x-18+18=0+18$$
$$3x=18$$
$$\frac{3x}{3}=\frac{18}{3}$$
$$x=6$$
Solution set: $\{6\}$

R3. $2x^2+3x+1$

  a. When $x=2$, we get
$$2x^2+3x+1=2(2)^2+3(2)+1$$
$$=2(4)+3(2)+1$$
$$=8+6+1$$
$$=15$$

  b. When $x=-1$, we get
$$2x^2+3x+1=2(-1)^2+3(-1)+1$$
$$=2(1)+3(-1)+1$$
$$=2-3+1$$
$$=0$$

R4. $f(x)=4x+3$
$$11=4x+3$$
$$11-3=4x+3-3$$
$$8=4x$$
$$\frac{8}{4}=\frac{4x}{4}$$
$$2=x$$
Solution set: $\{2\}$

The point $(2,11)$ is on the graph of f.

R5. $f(x)=-2x+8$
$$f(4)=-2(4)+8$$
$$=-8+8$$
$$=0$$
The point $(4,0)$ is on the graph of f.

R6. $f(x)=\frac{2}{3}x-6$
$$\frac{2}{3}x-6=0$$
$$\frac{2}{3}x=6$$
$$2x=18$$
$$x=9$$
9 is the zero of $f(x)$.

Section 4.8 Quick Checks

  1. A <u>polynomial equation</u> is an equation that can be written in the form polynomial expression equals zero.

  2. The Zero-Product Property states that if $a \cdot b = 0$, then <u>$a=0$</u> or <u>$b=0$</u>.

  3. $x(x+7)=0$
$$x=0 \quad \text{or} \quad x+7=0$$
$$x=-7$$
Check: $(-7)(-7+7)\overset{?}{=}0$
$$(-7)(0)\overset{?}{=}0$$
$$0=0 \ ✓$$
The solution set is $\{-7,0\}$.

  4. $(x-3)(4x+3)=0$
$$x-3=0 \quad \text{or} \quad 4x+3=0$$
$$x=3 \qquad\qquad 4x=-3$$
$$x=-\frac{3}{4}$$
Check:
$$(3-3)(4(3)+3)\overset{?}{=}0$$
$$0(15)\overset{?}{=}0$$
$$0=0 \ ✓$$
$$\left(-\frac{3}{4}-3\right)\left(4\left(-\frac{3}{4}\right)+3\right)\overset{?}{=}0$$
$$\left(-\frac{15}{4}\right)(0)\overset{?}{=}0$$
$$0=0 \ ✓$$
The solution set is $\left\{-\frac{3}{4},3\right\}$.

  5. A <u>quadratic equation</u> is an equation equivalent to one of the form $ax^2+bx+c=0$, where a, b, and c are real numbers and $a \neq 0$.

6. Quadratic equations are also known as <u>second</u> degree equations.

7. False; the standard form of this equation is
$x^2 - 7x + 4 = 0$.

8.
$$p^2 - 5p + 6 = 0$$
$$(p-3)(p-2) = 0$$
$$p - 3 = 0 \quad \text{or} \quad p - 2 = 0$$
$$p = 3 \qquad\qquad p = 2$$
Check:
$$(2)^2 - 5(2) + 6 \stackrel{?}{=} 0$$
$$4 - 10 + 6 \stackrel{?}{=} 0$$
$$0 = 0 \ \checkmark$$
$$(3)^2 - 5(3) + 6 \stackrel{?}{=} 0$$
$$9 - 15 + 6 \stackrel{?}{=} 0$$
$$0 = 0 \ \checkmark$$
The solution set is {2, 3}.

9.
$$3t^2 - 14t = 5$$
$$3t^2 - 14t - 5 = 0$$
$$(3t+1)(t-5) = 0$$
$$3t + 1 = 0 \quad \text{or} \quad t - 5 = 0$$
$$3t = -1 \qquad\qquad t = 5$$
$$t = -\frac{1}{3}$$
Check:
$$3\left(-\frac{1}{3}\right)^2 - 14\left(-\frac{1}{3}\right) \stackrel{?}{=} 5$$
$$\frac{1}{3} + \frac{14}{3} \stackrel{?}{=} 5$$
$$5 = 5 \ \checkmark$$
$$3(5)^2 - 14(5) \stackrel{?}{=} 5$$
$$75 - 70 \stackrel{?}{=} 5$$
$$5 = 5 \ \checkmark$$
The solution set is $\left\{-\frac{1}{3}, 5\right\}$.

10.
$$4y^2 + 8y + 3 = y^2 - 1$$
$$3y^2 + 8y + 4 = 0$$
$$(3y+2)(y+2) = 0$$
$$3y + 2 = 0 \quad \text{or} \quad y + 2 = 0$$
$$3y = -2 \qquad\qquad y = -2$$
$$y = -\frac{2}{3}$$

Check:
$$4(-2)^2 + 8(-2) + 3 \stackrel{?}{=} (-2)^2 - 1$$
$$16 - 16 + 3 \stackrel{?}{=} 4 - 1$$
$$3 = 3 \ \checkmark$$
$$4\left(-\frac{2}{3}\right)^2 + 8\left(-\frac{2}{3}\right) + 3 \stackrel{?}{=} \left(-\frac{2}{3}\right)^2 - 1$$
$$\frac{16}{9} - \frac{16}{3} + 3 \stackrel{?}{=} \frac{4}{9} - 1$$
$$-\frac{5}{9} = -\frac{5}{9} \ \checkmark$$
The solution set is $\left\{-2, -\frac{2}{3}\right\}$.

11. False; the Zero-Product Property can be applied only when the product is equal to zero.

12.
$$x(x+3) = -2$$
$$x^2 + 3x = -2$$
$$x^2 + 3x + 2 = 0$$
$$(x+1)(x+2) = 0$$
$$x + 1 = 0 \quad \text{or} \quad x + 2 = 0$$
$$x = -1 \qquad\qquad x = -2$$
Check:
$$(-2)(-2+3) \stackrel{?}{=} -2$$
$$-2(1) \stackrel{?}{=} -2$$
$$-2 = -2 \ \checkmark$$
$$(-1)(-1+3) \stackrel{?}{=} -2$$
$$(-1)(2) \stackrel{?}{=} -2$$
$$-2 = -2 \ \checkmark$$
The solution set is {−2, −1}.

13.
$$(x-3)(x+5) = 9$$
$$x^2 - 3x + 5x - 15 = 9$$
$$x^2 + 2x - 24 = 0$$
$$(x+6)(x-4) = 0$$
$$x + 6 = 0 \quad \text{or} \quad x - 4 = 0$$
$$x = -6 \qquad\qquad x = 4$$
Check:
$$(-6-3)(-6+5) \stackrel{?}{=} 9$$
$$(-9)(-1) \stackrel{?}{=} 9$$
$$9 = 9 \ \checkmark$$
$$(4-3)(4+5) \stackrel{?}{=} 9$$
$$(1)(9) \stackrel{?}{=} 9$$
$$9 = 9 \ \checkmark$$
The solution set is {−6, 4}.

14. True; the left side factors as $x(x-4)(x+2)$ and each factor can be set equal to zero.

15.
$$y^3 - y^2 - 9y + 9 = 0$$
$$\left(y^3 - y^2\right) + \left(-9y + 9\right) = 0$$
$$y^2(y-1) - 9(y-1) = 0$$
$$(y-1)\left(y^2 - 9\right) = 0$$
$$(y-1)(y-3)(y+3) = 0$$
$$y - 1 = 0 \quad \text{or} \quad y - 3 = 0 \quad \text{or} \quad y + 3 = 0$$
$$y = 1 \qquad\qquad y = 3 \qquad\qquad y = -3$$

Check:
$$(-3)^3 - (-3)^2 - 9(-3) + 9 \overset{?}{=} 0$$
$$-27 - 9 + 27 + 9 \overset{?}{=} 0$$
$$0 = 0 \checkmark$$

$$(1)^3 - (1)^2 - 9(1) + 9 \overset{?}{=} 0$$
$$1 - 1 - 9 + 9 \overset{?}{=} 0$$
$$0 = 0 \checkmark$$

$$(3)^3 - (3)^2 - 9(3) + 9 \overset{?}{=} 0$$
$$27 - 9 - 27 + 9 \overset{?}{=} 0$$
$$0 = 0 \checkmark$$

The solution set is $\{-3, 1, 3\}$.

16.
$$2n^3 - 3n^2 - 2n = 0$$
$$n(2n^2 - 3n - 2) = 0$$
$$n(2n+1)(n-2) = 0$$
$$n = 0 \quad \text{or} \quad 2n+1 = 0 \quad \text{or} \quad n-2 = 0$$
$$2n = -1 \qquad\qquad n = 2$$
$$n = -\frac{1}{2}$$

Check: $2(0)^3 - 3(0)^2 - 2(0) \overset{?}{=} 0$
$$0 = 0 \checkmark$$

$$2\left(-\frac{1}{2}\right)^3 - 3\left(-\frac{1}{2}\right)^2 - 2\left(-\frac{1}{2}\right) \overset{?}{=} 0$$
$$-\frac{1}{4} - \frac{3}{4} + 1 \overset{?}{=} 0$$
$$0 = 0 \checkmark$$

$$2(2)^3 - 3(2)^2 - 2(2) \overset{?}{=} 0$$
$$16 - 12 - 4 \overset{?}{=} 0$$
$$0 = 0 \checkmark$$

17. $g(x) = x^2 - 8x + 3$

a.
$$g(x) = 12$$
$$x^2 - 8x + 3 = 12$$
$$x^2 - 8x - 9 = 0$$
$$(x-9)(x+1) = 0$$
$$x - 9 = 0 \quad \text{or} \quad x + 1 = 0$$
$$x = 9 \qquad\qquad x = -1$$

The solution set is $\{-1, 9\}$. The points $(-1, 12)$ and $(9, 12)$ are on the graph of g.

b.
$$g(x) = -4$$
$$x^2 - 8x + 3 = -4$$
$$x^2 - 8x + 7 = 0$$
$$(x-7)(x-1) = 0$$
$$x - 7 = 0 \quad \text{or} \quad x - 1 = 0$$
$$x = 7 \qquad\qquad x = 1$$

The solution set is $\{1, 7\}$. The points $(1, -4)$ and $(7, -4)$ are on the graph of g.

18.
$$h(x) = 2x^2 + 3x - 20$$
$$h(x) = 0$$
$$2x^2 + 3x - 20 = 0$$
$$(2x-5)(x+4) = 0$$
$$2x - 5 = 0 \quad \text{or} \quad x + 4 = 0$$
$$2x = 5 \qquad\qquad x = -4$$
$$x = \frac{5}{2}$$

The zeros are -4 and $\frac{5}{2}$; the x-intercepts are $(-4, 0)$ and $\left(\frac{5}{2}, 0\right)$.

19. $F(x) = 5x^2 - 3x - 2$
$$F(x) = 0$$
$$5x^2 - 3x - 2 = 0$$
$$(5x+2)(x-1) = 0$$
$$5x + 2 = 0 \quad \text{or} \quad x - 1 = 0$$
$$5x = -2 \qquad\qquad x = 1$$
$$x = -\frac{2}{5}$$

The zeros are $-\frac{2}{5}$ and 1; the x-intercepts are $\left(-\frac{2}{5}, 0\right)$ and $(1, 0)$.

20. Let w = the width of the plot. Then the length is given by $w+6$.

$$A = \text{length} \cdot \text{width}$$
$$= (w+6) \cdot w$$
$$135 = w(w+6)$$
$$135 = w^2 + 6w$$
$$0 = w^2 + 6w - 135$$
$$0 = (w+15)(w-9)$$

$$w+15 = 0 \quad \text{or} \quad w-9 = 0$$
$$w = -15 \qquad\qquad w = 9$$

Since the width must be positive, we discard the negative solution. The width is 9 miles and the length is 9 + 6 = 15 miles.

21. Let x = the number of boxes ordered. The price per box is then given by $100 - (x-30) = 130 - x$.

$$\text{Cost} = \text{price} \cdot \text{quantity}$$
$$4200 = (130 - x)x$$
$$4200 = 130x - x^2$$
$$x^2 - 130x + 4200 = 0$$
$$(x-60)(x-70) = 0$$

$$x-60 = 0 \quad \text{or} \quad x-70 = 0$$
$$x = 60 \qquad\qquad x = 70$$

Since the number of boxes ordered must be between 30 and 65, we discard the second solution. The customer ordered 60 boxes of CDs.

22. $s(t) = -16t^2 + 160t$

a.
$$s(t) = 384$$
$$-16t^2 + 160t = 384$$
$$16t^2 - 160t + 384 = 0$$
$$t^2 - 10t + 24 = 0$$
$$(t-4)(t-6) = 0$$
$$t-4 = 0 \quad \text{or} \quad t-6 = 0$$
$$t = 4 \qquad\qquad t = 6$$

The rocket will be 384 feet above the ground after 4 seconds and after 6 seconds.

b.
$$s(t) = 0$$
$$-16t^2 + 160t = 0$$
$$t^2 - 10t = 0$$
$$t(t-10) = 0$$
$$t = 0 \quad \text{or} \quad t-10 = 0$$
$$t = 10$$

The rocket will strike the ground after 10 seconds.

4.8 Exercises

23. $(x-3)(x+1) = 0$
$$x-3 = 0 \quad \text{or} \quad x+1 = 0$$
$$x = 3 \quad \text{or} \quad x = -1$$
The solution set is $\{-1, 3\}$.

25. $2x(3x+4) = 0$
$$2x = 0 \quad \text{or} \quad 3x+4 = 0$$
$$x = 0 \qquad\qquad 3x = -4$$
$$x = -\frac{4}{3}$$
The solution set is $\left\{-\frac{4}{3}, 0\right\}$.

27. $y(y-5)(y+3) = 0$
$$y = 0 \quad \text{or} \quad y-5 = 0 \quad \text{or} \quad y+3 = 0$$
$$y = 5 \qquad\qquad y = -3$$
The solution set is $\{-3, 0, 5\}$.

29. $3p^2 - 12p = 0$
$$3p(p-4) = 0$$
$$3p = 0 \quad \text{or} \quad p-4 = 0$$
$$p = 0 \qquad\qquad p = 4$$
The solution set is $\{0, 4\}$.

31.
$$2w^2 = 16w$$
$$2w^2 - 16w = 0$$
$$2w(w-8) = 0$$
$$2w = 0 \quad \text{or} \quad w-8 = 0$$
$$w = 0 \qquad\qquad w = 8$$
The solution set is $\{0, 8\}$.

33. $m^2 + 2m - 15 = 0$
$$(m+5)(m-3) = 0$$
$$m+5 = 0 \quad \text{or} \quad m-3 = 0$$
$$m = -5 \qquad\qquad m = 3$$
The solution set is $\{-5, 3\}$.

35.
$$w^2 - 13w = -36$$
$$w^2 - 13w + 36 = 0$$
$$(w-9)(w-4) = 0$$
$$w-9 = 0 \quad \text{or} \quad w-4 = 0$$
$$w = 9 \qquad\qquad w = 4$$
The solution set is $\{4, 9\}$.

37. $p^2 - 6p + 9 = 0$
$(p-3)^2 = 0$
$p - 3 = 0$
$p = 3$
The solution set is {3}.

39. $5x^2 = 2x + 3$
$5x^2 - 2x - 3 = 0$
$(5x+3)(x-1) = 0$
$5x + 3 = 0$ or $x - 1 = 0$
$5x = -3$ $x = 1$
$x = -\dfrac{3}{5}$
The solution set is $\left\{-\dfrac{3}{5}, 1\right\}$.

41. $m^2 + 7m + 9 = m$
$m^2 + 6m + 9 = 0$
$(m+3)^2 = 0$
$m + 3 = 0$
$m = -3$
The solution set is {−3}.

43. $3p^2 + 9p - 120 = 0$
$3\left(p^2 + 3p - 40\right) = 0$
$3(p+8)(p-5) = 0$
$p + 8 = 0$ or $p - 5 = 0$
$p = -8$ $p = 5$
The solution set is {−8, 5}.

45. $-4b^2 - 14b + 60 = 0$
$-2\left(2b^2 + 7b - 30\right) = 0$
$-2(2b-5)(b+6) = 0$
$2b - 5 = 0$ or $b + 6 = 0$
$2b = 5$ $b = -6$
$b = \dfrac{5}{2}$
The solution set is $\left\{-6, \dfrac{5}{2}\right\}$.

47. $\dfrac{1}{2}x^2 + 2x - 6 = 0$
$2\left(\dfrac{1}{2}x^2 + 2x - 6\right) = 2 \cdot 0$
$x^2 + 4x - 12 = 0$
$(x+6)(x-2) = 0$
$x + 6 = 0$ or $x - 2 = 0$
$x = -6$ $x = 2$
The solution set is {−6, 2}.

49. $\dfrac{2}{3}x^2 + x = \dfrac{14}{3}$
$3\left(\dfrac{2}{3}x^2 + x\right) = 3 \cdot \dfrac{14}{3}$
$2x^2 + 3x = 14$
$2x^2 + 3x - 14 = 0$
$(2x+7)(x-2) = 0$
$2x + 7 = 0$ or $x - 2 = 0$
$2x = -7$ $x = 2$
$x = -\dfrac{7}{2}$
The solution set is $\left\{-\dfrac{7}{2}, 2\right\}$.

51. $x(x+8) = 33$
$x^2 + 8x = 33$
$x^2 + 8x - 33 = 0$
$(x+11)(x-3) = 0$
$x + 11 = 0$ or $x - 3 = 0$
$x = -11$ $x = 3$
The solution set is {−11, 3}.

53. $(x+6)(x-1) = 9 + 3x$
$x^2 + 5x - 6 = 9 + 3x$
$x^2 + 2x - 15 = 0$
$(x+5)(x-3) = 0$
$x + 5 = 0$ or $x - 3 = 0$
$x = -5$ $x = 3$
The solution set is {−5, 3}.

55.
$$2z^3 - 5z^2 = 3z$$
$$2z^3 - 5z^2 - 3z = 0$$
$$z(2z^2 - 5z - 3) = 0$$
$$z(2z+1)(z-3) = 0$$
$$z = 0 \quad \text{or} \quad 2z+1 = 0 \quad \text{or} \quad z-3 = 0$$
$$2z = -1 \qquad z = 3$$
$$z = -\frac{1}{2}$$

The solution set is $\left\{-\frac{1}{2}, 0, 3\right\}$.

57.
$$2p^3 + 5p^2 - 8p - 20 = 0$$
$$p^2(2p+5) - 4(2p+5) = 0$$
$$(2p+5)(p^2 - 4) = 0$$
$$(2p+5)(p-2)(p+2) = 0$$
$$2p+5 = 0 \quad \text{or} \quad p-2 = 0 \quad \text{or} \quad p+2 = 0$$
$$2p = -5 \qquad p = 2 \qquad p = -2$$
$$p = -\frac{5}{2}$$

The solution set is $\left\{-\frac{5}{2}, -2, 2\right\}$.

59.
$$-30b^3 - 38b^2 = 12b$$
$$-30b^3 - 38b^2 - 12b = 0$$
$$-2b(15b^2 + 19b + 6) = 0$$
$$-2b(5b+3)(3b+2) = 0$$

$$-2b = 0 \quad \text{or} \quad 5b+3 = 0 \quad \text{or} \quad 3b+2 = 0$$
$$b = 0 \qquad 5b = -3 \qquad 3b = -2$$
$$b = -\frac{3}{5} \qquad b = -\frac{2}{3}$$

The solution set is $\left\{-\frac{2}{3}, -\frac{3}{5}, 0\right\}$.

61.
$$(x-2)^3 = x^3 - 2x$$
$$(x-2)(x-2)(x-2) = x^3 - 2x$$
$$(x^2 - 4x + 4)(x-2) = x^3 - 2x$$
$$x^3 - 4x^2 + 4x - 2x^2 + 8x - 8 = x^3 - 2x$$
$$x^3 - 6x^2 + 12x - 8 = x^3 - 2x$$
$$-6x^2 + 14x - 8 = 0$$
$$-2(3x^2 - 7x + 4) = 0$$
$$-2(3x-4)(x-1) = 0$$

$$3x-4 = 0 \quad \text{or} \quad x-1 = 0$$
$$3x = 4 \qquad x = 1$$
$$x = \frac{4}{3}$$

The solution set is $\left\{1, \frac{4}{3}\right\}$.

63. a.
$$f(x) = 2$$
$$x^2 + 7x + 12 = 2$$
$$x^2 + 7x + 10 = 0$$
$$(x+5)(x+2) = 0$$
$$x+5 = 0 \quad \text{or} \quad x+2 = 0$$
$$x = -5 \qquad x = -2$$
The solution set is $\{-5, -2\}$.

b.
$$f(x) = 20$$
$$x^2 + 7x + 12 = 20$$
$$x^2 + 7x - 8 = 0$$
$$(x+8)(x-1) = 0$$
$$x+8 = 0 \quad \text{or} \quad x-1 = 0$$
$$x = -8 \qquad x = 1$$
The solution set is $\{-8, 1\}$.

The points $(-5, 2)$, $(-2, 2)$, $(-8, 20)$, and $(1, 20)$ are on the graph of f.

65. a.
$$g(x) = 3$$
$$2x^2 - 6x - 5 = 3$$
$$2x^2 - 6x - 8 = 0$$
$$2(x^2 - 3x - 4) = 0$$
$$2(x-4)(x+1) = 0$$
$$x-4 = 0 \quad \text{or} \quad x+1 = 0$$
$$x = 4 \qquad x = -1$$
The solution set is $\{-1, 4\}$.

b.
$$g(x) = 15$$
$$2x^2 - 6x - 5 = 15$$
$$2x^2 - 6x - 20 = 0$$
$$2(x^2 - 3x - 10) = 0$$
$$2(x-5)(x+2) = 0$$
$$x-5 = 0 \quad \text{or} \quad x+2 = 0$$
$$x = 5 \qquad x = -2$$
The solution set is $\{-2, 5\}$.

The points $(-1, 3)$, $(4, 3)$, $(-2, 15)$, and $(5, 15)$ are on the graph of g.

67. a.
$$F(x) = 5$$
$$-3x^2 + 12x + 5 = 5$$
$$-3x^2 + 12x = 0$$
$$-3x(x - 4) = 0$$
$$-3x = 0 \quad \text{or} \quad x - 4 = 0$$
$$x = 0 \qquad\qquad x = 4$$
The solution set is $\{0, 4\}$.

b.
$$F(x) = -10$$
$$-3x^2 + 12x + 5 = -10$$
$$-3x^2 + 12x + 15 = 0$$
$$-3(x^2 - 4x - 5) = 0$$
$$-3(x - 5)(x + 1) = 0$$
$$x - 5 = 0 \quad \text{or} \quad x + 1 = 0$$
$$x = 5 \qquad\qquad x = -1$$
The solution set is $\{-1, 5\}$.

The points $(0, 5)$, $(4, 5)$, $(-1, -10)$, and $(5, -10)$ are on the graph of F.

69. $f(x) = x^2 + 9x + 14$
$$0 = x^2 + 9x + 14$$
$$0 = (x + 7)(x + 2)$$
$$x + 7 = 0 \quad \text{or} \quad x + 2 = 0$$
$$x = -7 \qquad\qquad x = -2$$
The zeros of $f(x)$ are -7 and -2.
The x-intercepts are $(-7, 0)$ and $(-2, 0)$.

71. $g(t) = 6t^2 - 25t - 9$
$$0 = 6t^2 - 25t - 9$$
$$0 = (3t + 1)(2t - 9)$$
$$3t + 1 = 0 \quad \text{or} \quad 2t - 9 = 0$$
$$3t = -1 \qquad\qquad 2t = 9$$
$$t = -\frac{1}{3} \qquad\qquad t = \frac{9}{2}$$
The zeros of $g(t)$ are $-\frac{1}{3}$ and $\frac{9}{2}$.

The x-intercepts are $\left(-\frac{1}{3}, 0\right)$ and $\left(\frac{9}{2}, 0\right)$.

73. $s(d) = 2d^3 + 2d^2 - 40d$
$$0 = 2d^3 + 2d^2 - 40d$$
$$0 = 2d(d^2 + d - 20)$$
$$0 = 2d(d + 5)(d - 4)$$
$$2d = 0 \quad \text{or} \quad d + 5 = 0 \quad \text{or} \quad d - 4 = 0$$
$$d = 0 \qquad\quad d = -5 \qquad\quad d = 4$$
The zeros of $s(d)$ are -5, 0, and 4.
The x-intercepts are $(-5, 0)$, $(0, 0)$, and $(4, 0)$.

75.
$$(x + 3)(x - 5) = 9$$
$$x^2 + 3x - 5x - 15 = 9$$
$$x^2 - 2x - 15 = 9$$
$$x^2 - 2x - 24 = 0$$
$$(x - 6)(x + 4) = 0$$
$$x - 6 = 0 \quad \text{or} \quad x + 4 = 0$$
$$x = 6 \qquad\qquad x = -4$$
The solution set is $\{-4, 6\}$.

77.
$$2q^2 + 3q - 14 = 0$$
$$(2q + 7)(q - 2) = 0$$
$$2q + 7 = 0 \quad \text{or} \quad q - 2 = 0$$
$$2q = -7 \qquad\qquad q = 2$$
$$q = -\frac{7}{2}$$
The solution set is $\left\{-\frac{7}{2}, 2\right\}$.

79.
$$-3b^2 + 21b = 0$$
$$-3b(b - 7) = 0$$
$$-3b = 0 \quad \text{or} \quad b - 7 = 0$$
$$b = 0 \qquad\qquad b = 7$$
The solution set is $\{0, 7\}$.

81. $(x + 2)(x + 3) = x(x - 2)$
$$x^2 + 5x + 6 = x^2 - 2x$$
$$7x + 6 = 0$$
$$7x = -6$$
$$x = -\frac{6}{7}$$
The solution set is $\left\{-\frac{6}{7}\right\}$.

83. $x^3 + 5x^2 - 4x - 20 = 0$

$x^2(x+5) - 4(x+5) = 0$

$(x+5)(x^2 - 4) = 0$

$(x+5)(x-2)(x+2) = 0$

$x + 5 = 0$ or $x - 2 = 0$ or $x + 2 = 0$

   $x = -5$ $x = 2$ $x = -2$

The solution set is $\{-5, -2, 2\}$.

85. $(2x+1)(x-3) - x^2 = (x-2)(x-3)$

$2x^2 - 6x + x - 3 - x^2 = x^2 - 3x - 2x + 6$

$x^2 - 5x - 3 = x^2 - 5x + 6$

$-3 = -6$

Because we arrived at this contradiction, the solution set is $\varnothing$ or $\{\ \}$.

87. $x^3 + x^2 + x + 6 = 3x^2 + 6x$

$x^3 - 2x^2 - 5x + 6 = 0$

Use synthetic division to find a factor. Try $c = 1$.

$$
\begin{array}{r|rrrr}
1 & 1 & -2 & -5 & 6 \\
  &   & 1 & -1 & -6 \\
\hline
  & 1 & -1 & -6 & 0
\end{array}
$$

So $x - 1$ is a factor of $x^3 - 2x^2 - 5x + 6$, which factors as $(x-1)(x^2 - x - 6)$.

$(x-1)(x^2 - x - 6) = 0$

$(x-1)(x-3)(x+2) = 0$

$x - 1 = 0$ or $x - 3 = 0$ or $x + 2 = 0$

   $x = 1$ $x = 3$ $x = -2$

The solution set is $\{-2, 1, 3\}$.

89. $4x^4 - 17x^2 + 4 = 0$

$(4x^2 - 1)(x^2 - 4) = 0$

$((2x)^2 - 1^2)(x^2 - 2^2) = 0$

$(2x-1)(2x+1)(x-2)(x+2) = 0$

$2x - 1 = 0$ or $2x + 1 = 0$ or $x - 2 = 0$ or $x + 2 = 0$

$2x = 1$ $2x = -1$ $x = 2$ $x = -2$

$x = \dfrac{1}{2}$ $x = -\dfrac{1}{2}$

The solution set is $\left\{-2, -\dfrac{1}{2}, \dfrac{1}{2}, 2\right\}$.

91. $(a+3)^2 - 5(a+3) = -6$

Let $u = a + 3$.

$u^2 - 5u = -6$

$u^2 - 5u + 6 = 0$

$(u-3)(u-2) = 0$

$u - 3 = 0$ or $u - 2 = 0$

   $u = 3$ $u = 2$

Now get back in terms of a.

$a + 3 = 3$ or $a + 3 = 2$

   $a = 0$ $a = -1$

The solution set is $\{-1, 0\}$.

93. $f(x) = \dfrac{5}{x^2 - 4}$

The domain is all real numbers except where the denominator equals 0.

$x^2 - 4 = 0$

$(x-2)(x+2) = 0$

$x - 2 = 0$ or $x + 2 = 0$

   $x = 2$ $x = -2$

The domain is all real numbers except ± 2. That is, $\{x \mid x \neq \pm 2\}$.

95. $g(x) = \dfrac{4x+3}{2x^2 - 3x + 1}$

The domain is all real numbers except where the denominator equals 0.

$2x^2 - 3x + 1 = 0$

$(2x-1)(x-1) = 0$

$2x - 1 = 0$ or $x - 1 = 0$

$2x = 1$ $x = 1$

$x = \dfrac{1}{2}$

The domain is all real numbers except $\dfrac{1}{2}$ and 1.

That is $\left\{x \mid x \neq \dfrac{1}{2}, 1\right\}$.

97. Let $x = $ width , then $x - 8 = $ length .

$A = $ width $\cdot$ length

$A = x(x-8)$

$x(x-8) = 128$

$x^2 - 8x = 128$

$x^2 - 8x - 128 = 0$

$(x-16)(x+8) = 0$

$x - 16 = 0$ or $x + 8 = 0$

   $x = 16$ $\cancel{x = -8}$

The width of the rectangle is 16 centimeters and the length is 8 centimeters.

99. Let $x = \text{base}$, then $x + 12 = \text{height}$.

$$A = \frac{1}{2} \cdot \text{base} \cdot \text{height}$$
$$= \frac{1}{2} x (x + 12)$$

$$\frac{1}{2} x (x + 12) = 110$$
$$\frac{1}{2} x^2 + 6x = 110$$
$$x^2 + 12x = 220$$
$$x^2 + 12x - 220 = 0$$
$$(x + 22)(x - 10) = 0$$
$$x + 22 = 0 \quad \text{or} \quad x - 10 = 0$$
$$\cancel{x = -22} \qquad\qquad x = 10$$

The base of the triangle is 10 feet and the height is 22 feet.

101.
$$D = \frac{n(n-3)}{2}$$
$$20 = \frac{n(n-3)}{2}$$
$$40 = n(n-3)$$
$$40 = n^2 - 3n$$
$$n^2 - 3n - 40 = 0$$
$$(n - 8)(n + 5) = 0$$
$$n - 8 = 0 \quad \text{or} \quad n + 5 = 0$$
$$n = 8 \qquad\qquad \cancel{n = -5}$$

A convex polygon with 20 diagonals has 8 sides (octagon).

103. Let $x = \text{length of plot bordering river}$. The width must be $\dfrac{100 - x}{2}$.

$$A = \text{length} \cdot \text{width}$$
$$= x \cdot \frac{(100 - x)}{2}$$
$$= \frac{x(100 - x)}{2}$$

$$800 = \frac{x(100 - x)}{2}$$
$$1600 = x(100 - x)$$
$$1600 = 100x - x^2$$
$$x^2 - 100x + 1600 = 0$$
$$(x - 80)(x - 20) = 0$$
$$x - 80 = 0 \quad \text{or} \quad x - 20 = 0$$
$$x = 80 \qquad\qquad x = 20$$

$$\frac{100 - 80}{2} = 10 \qquad\qquad \frac{100 - 20}{2} = 40$$

The dimensions of the plot are 40 meters by 20 meters (along the river) or 10 meters by 80 meters (along the river).

105. Let $x = \text{width of walkway}$.

$$(8 + 2x)(12 + 2x) = 252$$
$$96 + 24x + 16x + 4x^2 = 252$$
$$4x^2 + 40x - 156 = 0$$
$$4(x^2 + 10x - 39) = 0$$
$$4(x + 13)(x - 3) = 0$$
$$x + 13 = 0 \quad \text{or} \quad x - 3 = 0$$
$$\cancel{x = -13} \qquad\qquad x = 3$$

The walkway is 3 feet wide.

107. Let $x = \text{width of the piece of cardboard}$, then $x + 5 = \text{length}$.

To obtain the length and width of the box, we subtract 4 from both the length and width of the cardboard because a corner is cut from each end.

$$V = \text{length} \cdot \text{width} \cdot \text{height}$$
$$168 = (x + 5 - 4)(x - 4)(2)$$
$$168 = 2(x + 1)(x - 4)$$
$$84 = (x + 1)(x - 4)$$
$$84 = x^2 + x - 4x - 4$$
$$0 = x^2 - 3x - 88$$
$$0 = (x - 11)(x + 8)$$
$$x - 11 = 0 \quad \text{or} \quad x + 8 = 0$$
$$x = 11 \qquad\qquad \cancel{x = -8}$$

The width of the piece of cardboard is 11 inches and the length is 16 inches.

109. a. $C(x) = x^2 - 40x + 600$

$$C(30) = (30)^2 - 40(30) + 600$$
$$= 900 - 1200 + 600$$
$$= 300$$

The marginal cost for producing the 30th bicycle is $300.

b.
$$x^2 - 40x + 600 = 200$$
$$x^2 - 40x + 400 = 0$$
$$(x - 20)^2 = 0$$
$$x - 20 = 0$$
$$x = 20$$

20 bicycles must be manufactured to have a marginal cost of $200.

c.
$$x^2 - 40x + 600 = 225$$
$$x^2 - 40x + 375 = 0$$
$$(x - 25)(x - 15) = 0$$
$$x - 25 = 0 \quad \text{or} \quad x - 15 = 0$$
$$x = 25 \qquad\qquad x = 15$$

Since production would continue until marginal revenue equals marginal cost, the company should manufacture 15 bicycles.

111. a.
$$s(t) = -16t^2 + 120t$$
$$200 = -16t^2 + 120t$$
$$16t^2 - 120t + 200 = 0$$
$$8(2t^2 - 15t + 25) = 0$$
$$8(t - 5)(2t - 5) = 0$$
$$t - 5 = 0 \quad \text{or} \quad 2t - 5 = 0$$
$$t = 5 \qquad\qquad 2t = 5$$
$$t = \frac{5}{2}$$

The ball will be at a height of 200 feet after 2.5 seconds (on the way up), and again after 5 seconds (on the way down).

b. The ball will hit the ground when $s(t) = 0$.
$$-16t^2 + 120t = 0$$
$$-8t(2t - 15) = 0$$
$$-8t = 0 \quad \text{or} \quad 2t - 15 = 0$$
$$t = 0 \qquad\qquad 2t = 15$$
$$t = \frac{15}{2}$$

The ball will hit the ground after 7.5 seconds.

113. $2x^2 - 7x - 5 = 0$

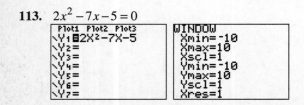

115. $0.2x^2 - 5.1x + 3 = 0$

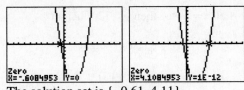

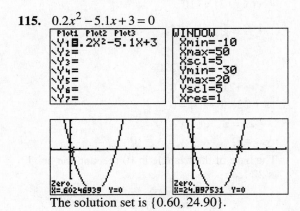

The solution set is $\{0.60, 24.90\}$.

117. $-x^2 + 0.6x = -2 \quad \text{or} \quad -x^2 + 0.6x + 2 = 0$

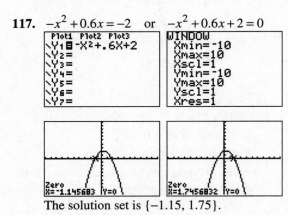

The solution set is $\{-1.15, 1.75\}$.

The solution set is $\{-0.61, 4.11\}$.

Chapter 4 Review

1. Coefficient: -7
Degree: 4

2. Coefficient: $\dfrac{1}{9}$

Degree: 3

3. $7x^3 - 2x^2 + x - 8$
Degree: 3

4. $y^4 - 3y^2 + 2y + 3$
Degree: 4

5. $\left(x^2+2x-7\right)+\left(3x^2-x-4\right)$

$\quad = x^2+2x-7+3x^2-x-4$

$\quad = x^2+3x^2+2x-x-7-4$

$\quad = (1+3)x^2+(2-1)x+(-7-4)$

$\quad = 4x^2+x-11$

6. $\left(4x^3-3x^2+x-5\right)-\left(x^4+2x^2-7x+1\right)$

$\quad = 4x^3-3x^2+x-5-x^4-2x^2+7x-1$

$\quad = -x^4+4x^3-3x^2-2x^2+x+7x-5-1$

$\quad = -x^4+4x^3+(-3-2)x^2+(1+7)x+(-5-1)$

$\quad = -x^4+4x^3-5x^2+8x-6$

7. $\left(\dfrac{1}{4}x^2-\dfrac{1}{2}x\right)-\left(4x-\dfrac{1}{6}\right)$

$\quad = \dfrac{1}{4}x^2-\dfrac{1}{2}x-4x+\dfrac{1}{6}$

$\quad = \dfrac{1}{4}x^2+\left(-\dfrac{1}{2}-4\right)x+\dfrac{1}{6}$

$\quad = \dfrac{1}{4}x^2-\dfrac{9}{2}x+\dfrac{1}{6}$

8. $\left(\dfrac{1}{2}x^2-x+\dfrac{1}{4}\right)+\left(\dfrac{1}{3}x^2+\dfrac{2}{5}\right)$

$\quad = \dfrac{1}{2}x^2-x+\dfrac{1}{4}+\dfrac{1}{3}x^2+\dfrac{2}{5}$

$\quad = \dfrac{1}{2}x^2+\dfrac{1}{3}x^2-x+\dfrac{1}{4}+\dfrac{2}{5}$

$\quad = \left(\dfrac{1}{2}+\dfrac{1}{3}\right)x^2-x+\left(\dfrac{1}{4}+\dfrac{2}{5}\right)$

$\quad = \dfrac{5}{6}x^2-x+\dfrac{13}{20}$

9. $\left(x^3y^2+6x^2y^2-xy\right)+\left(-x^3y^2+4x^2y^2+xy\right)$

$\quad = x^3y^2+6x^2y^2-xy-x^3y^2+4x^2y^2+xy$

$\quad = x^3y^2-x^3y^2+6x^2y^2+4x^2y^2-xy+xy$

$\quad = (1-1)x^3y^2+(6+4)x^2y^2+(-1+1)xy$

$\quad = 10x^2y^2$

10. $\left(a^2b-4ab^2+3\right)-\left(2a^2b+2ab^2+7\right)$

$\quad = a^2b-4ab^2+3-2a^2b-2ab^2-7$

$\quad = a^2b-2a^2b-4ab^2-2ab^2+3-7$

$\quad = (1-2)a^2b+(-4-2)ab^2+(3-7)$

$\quad = -a^2b-6ab^2-4$

11. a. $f(x)=-3x^2+2x-8$

$\quad f(-2)=-3(-2)^2+2(-2)-8$

$\quad\quad = -3(4)-4-8$

$\quad\quad = -12-4-8$

$\quad\quad = -24$

b. $f(x)=-3x^2+2x-8$

$\quad f(0)=-3(0)^2+2(0)-8$

$\quad\quad = 0+0-8$

$\quad\quad = -8$

c. $f(x)=-3x^2+2x-8$

$\quad f(3)=-3(3)^2+2(3)-8$

$\quad\quad = -3(9)+6-8$

$\quad\quad = -27+6-8$

$\quad\quad = -29$

12. a. $f(x)=x^3-5x^2+3x-1$

$\quad f(-3)=(-3)^3-5(-3)^2+3(-3)-1$

$\quad\quad = -27-5(9)-9-1$

$\quad\quad = -27-45-9-1$

$\quad\quad = -82$

b. $f(x)=x^3-5x^2+3x-1$

$\quad f(0)=(0)^3-5(0)^2+3(0)-1$

$\quad\quad = 0-0+0-1$

$\quad\quad = -1$

c. $f(x)=x^3-5x^2+3x-1$

$\quad f(2)=(2)^3-5(2)^2+3(2)-1$

$\quad\quad = 8-5(4)+6-1$

$\quad\quad = 8-20+6-1$

$\quad\quad = -7$

13. a. $(f+g)(x)=(4x-3)+\left(x^2+3x+2\right)$

$\quad\quad = 4x-3+x^2+3x+2$

$\quad\quad = x^2+4x+3x-3+2$

$\quad\quad = x^2+7x-1$

b. Since $(f+g)(x)=x^2+7x-1$, we have

$\quad (f+g)(3)=(3)^2+7(3)-1$

$\quad\quad = 9+21-1$

$\quad\quad = 29$

14. a. $(f-g)(x) = (2x^3 + x^2 - 7) - (3x^2 - x + 5)$
$$= 2x^3 + x^2 - 7 - 3x^2 + x - 5$$
$$= 2x^3 + x^2 - 3x^2 + x - 7 - 5$$
$$= 2x^3 - 2x^2 + x - 12$$

b. Since $(f-g)(x) = 2x^3 - 2x^2 + x - 12$, we have
$$(f-g)(2) = 2(2)^3 - 2(2)^2 + (2) - 12$$
$$= 2(8) - 2(4) + 2 - 12$$
$$= 16 - 8 + 2 - 12$$
$$= -2$$

15. a. Profit is the difference between revenue and costs. Thus,
$$P(x) = R(x) - C(x)$$
$$= (-1.5x^2 + 180x) - (x^2 - 100x + 3290)$$
$$= -1.5x^2 + 180x - x^2 + 100x - 3290$$
$$= -1.5x^2 - x^2 + 180x + 100x - 3290$$
$$= -2.5x^2 + 280x - 3290$$
The profit function is:
$$P(x) = -2.5x^2 + 280x - 3290$$

b. $P(25) = -2.5(25)^2 + 280(25) - 3290$
$$= -2.5(625) + 7000 - 3290$$
$$= -1562.5 + 7000 - 3290$$
$$= 2147.5$$
If 25 graphing calculators are sold, there would be a profit of $2147.50.

16. a. If the point is $(2,1)$, then $x = 2$.
$$A(x) = -x^2 + 5x$$
$$A(2) = -(2)^2 + 5(2)$$
$$= -4 + 10$$
$$= 6$$
The area of the region would be 6 square units.

b. If the point is $(1,3)$, then $x = 1$.
$$A(x) = -x^2 + 5x$$
$$A(1) = -(1)^2 + 5(1)$$
$$= -1 + 5$$
$$= 4$$
The area of the region would be 4 square units.

17. $(-3x^3 y)(4xy^2) = (-3 \cdot 4)(x^3 \cdot x)(y \cdot y^2)$
$$= -12x^{3+1} y^{1+2}$$
$$= -12x^4 y^3$$

18. $\left(\frac{1}{3} mn^4\right)(18m^3 n^3) = \left(\frac{1}{3} \cdot 18\right)(m \cdot m^3)(n^4 \cdot n^3)$
$$= \frac{18}{3} m^{1+3} n^{4+3}$$
$$= 6m^4 n^7$$

19. $5ab(-2a^2 b + ab^2 - 3ab)$
$$= -5ab \cdot 2a^2 b + 5ab \cdot ab^2 - 5ab \cdot 3ab$$
$$= -10a^3 b^2 + 5a^2 b^3 - 15a^2 b^2$$

20. $0.5c(1.7c^2 + 4.3c + 8.9)$
$$= 0.5c \cdot 1.7c^2 + 0.5c \cdot 4.3c + 0.5c \cdot 8.9$$
$$= 0.85c^3 + 2.15c^2 + 4.45c$$

21. $(x+2)(x-9) = x \cdot x - x \cdot 9 + 2 \cdot x - 2 \cdot 9$
$$= x^2 - 9x + 2x - 18$$
$$= x^2 - 7x - 18$$

22. $(-3x+1)(2x-8)$
$$= -3x \cdot 2x - 3x \cdot (-8) + 1 \cdot 2x - 1 \cdot 8$$
$$= -6x^2 + 24x + 2x - 8$$
$$= -6x^2 + 26x - 8$$

23. $(m-4n)(2m+n)$
$$= m \cdot 2m + m \cdot n - 4n \cdot 2m - 4n \cdot n$$
$$= 2m^2 + mn - 8mn - 4n^2$$
$$= 2m^2 - 7mn - 4n^2$$

24. $(2a+15)(-a+3)$
$$= 2a \cdot (-a) + 2a \cdot 3 + 15 \cdot (-a) + 15 \cdot 3$$
$$= -2a^2 + 6a - 15a + 45$$
$$= -2a^2 - 9a + 45$$

25. $(x+2)(3x^2 - 5x + 1)$
$$= x \cdot 3x^2 - x \cdot 5x + x \cdot 1 + 2 \cdot 3x^2 - 2 \cdot 5x + 2 \cdot 1$$
$$= 3x^3 - 5x^2 + x + 6x^2 - 10x + 2$$
$$= 3x^3 + x^2 - 9x + 2$$

26. $(w-4)(w^2+w-8)$

$= w \cdot w^2 + w \cdot w - w \cdot 8 - 4 \cdot w^2 - 4 \cdot w - 4 \cdot (-8)$

$= w^3 + w^2 - 8w - 4w^2 - 4w + 32$

$= w^3 - 3w^2 - 12w + 32$

27. $(m^2-2m+3)(2m^2+5m-7)$

$= m^2 \cdot 2m^2 + m^2 \cdot 5m - m^2 \cdot 7 - 2m \cdot 2m^2$

$\quad - 2m \cdot 5m - 2m \cdot (-7) + 3 \cdot 2m^2$

$\quad + 3 \cdot 5m - 3 \cdot 7$

$= 2m^4 + 5m^3 - 7m^2 - 4m^3 - 10m^2 + 14m$

$\quad + 6m^2 + 15m - 21$

$= 2m^4 + m^3 - 11m^2 + 29m - 21$

28. $(2p-3q)(p^2+7pq-4q^2)$

$= 2p \cdot p^2 + 2p \cdot 7pq - 2p \cdot 4q^2 - 3q \cdot p^2$

$\quad - 3q \cdot 7pq - 3q \cdot (-4q^2)$

$= 2p^3 + 14p^2q - 8pq^2 - 3p^2q - 21pq^2 + 12q^3$

$= 2p^3 + 11p^2q - 29pq^2 + 12q^3$

29. $(3w+1)(3w-1) = (3w)^2 - (1)^2$

$\qquad\qquad\qquad\quad = 9w^2 - 1$

30. $(2x-5y)(2x+5y) = (2x)^2 - (5y)^2$

$\qquad\qquad\qquad\qquad\quad = 4x^2 - 25y^2$

31. $(6k-5)^2 = (6k)^2 - 2(6k)(5) + (5)^2$

$\qquad\qquad\quad = 36k^2 - 60k + 25$

32. $(3a+2b)^2 = (3a)^2 + 2(3a)(2b) + (2b)^2$

$\qquad\qquad\quad = 9a^2 + 12ab + 4b^2$

33. $(x+2)(x^2-2x+4)$

$= x \cdot x^2 - x \cdot 2x + x \cdot 4 + 2 \cdot x^2 - 2 \cdot 2x + 2 \cdot 4$

$= x^3 - 2x^2 + 4x + 2x^2 - 4x + 8$

$= x^3 + 8$

34. $(2x-3)(4x^2+6x+9)$

$= 2x \cdot 4x^2 + 2x \cdot 6x + 2x \cdot 9 - 3 \cdot 4x^2 - 3 \cdot 6x - 3 \cdot 9$

$= 8x^3 + 12x^2 + 18x - 12x^2 - 18x - 27$

$= 8x^3 - 27$

35. a. $(f \cdot g)(x) = f(x) \cdot g(x)$

$\qquad\qquad = (3x-7)(6x+5)$

$\qquad\qquad = 3x \cdot 6x + 3x \cdot 5 - 7 \cdot 6x - 7 \cdot 5$

$\qquad\qquad = 18x^2 + 15x - 42x - 35$

$\qquad\qquad = 18x^2 - 27x - 35$

b. Since $(f \cdot g)(x) = 18x^2 - 27x - 35,$ we have

$(f \cdot g)(-2) = 18(-2)^2 - 27(-2) - 35$

$\qquad\qquad = 18(4) + 54 - 35$

$\qquad\qquad = 72 + 54 - 35$

$\qquad\qquad = 91$

36. a. $(f \cdot g)(x) = f(x) \cdot g(x)$

$\qquad\qquad = (x+2)(3x^2-x+1)$

$\qquad\qquad = x \cdot 3x^2 - x \cdot x + x \cdot 1 + 2 \cdot 3x^2$

$\qquad\qquad\quad - 2 \cdot x + 2 \cdot 1$

$\qquad\qquad = 3x^3 - x^2 + x + 6x^2 - 2x + 2$

$\qquad\qquad = 3x^3 + 5x^2 - x + 2$

b. Since $(f \cdot g)(x) = 3x^3 + 5x^2 - x + 2$, we have

$(f \cdot g)(4) = 3(4)^3 + 5(4)^2 - (4) + 2$

$\qquad\qquad = 3(64) + 5(16) - 4 + 2$

$\qquad\qquad = 192 + 80 - 4 + 2$

$\qquad\qquad = 270$

37. $\qquad f(x) = 5x^2 + 8$

$f(x-3) = 5(x-3)^2 + 8$

$\qquad\quad = 5(x^2 - 2 \cdot x \cdot 3 + 3^2) + 8$

$\qquad\quad = 5(x^2 - 6x + 9) + 8$

$\qquad\quad = 5x^2 - 30x + 45 + 8$

$\qquad\quad = 5x^2 - 30x + 53$

38. $f(x+h) - f(x)$

$= \left[-(x+h)^2 + 3(x+h) - 5 \right]$

$\quad - \left[-x^2 + 3x - 5 \right]$

$= \left[-x^2 - 2xh - h^2 + 3x + 3h - 5 \right]$

$\quad - \left[-x^2 + 3x - 5 \right]$

$= -x^2 - 2xh - h^2 + 3x + 3h - 5$

$\quad + x^2 - 3x + 5$

$= -2xh + 3h - h^2$

39. $\dfrac{12x^3 - 6x^2}{3x} = \dfrac{12x^3}{3x} - \dfrac{6x^2}{3x}$

$\qquad\qquad = \dfrac{12}{3}x^{3-1} - \dfrac{6}{3}x^{2-1}$

$\qquad\qquad = 4x^2 - 2x$

40. $\dfrac{15w^5 - 5w^3 + 25w^2 + 10w}{5w}$

$= \dfrac{15w^5}{5w} - \dfrac{5w^3}{5w} + \dfrac{25w^2}{5w} + \dfrac{10w}{5w}$

$= \dfrac{15}{5}w^{5-1} - \dfrac{5}{5}w^{3-1} + \dfrac{25}{5}w^{2-1} + \dfrac{10}{5}w^{1-1}$

$= 3w^4 - w^2 + 5w + 2$

41. $\dfrac{7y^3 + 12y^2 - 6y}{2y} = \dfrac{7y^3}{2y} + \dfrac{12y^2}{2y} - \dfrac{6y}{2y}$

$\qquad\qquad = \dfrac{7}{2}y^{3-1} + \dfrac{12}{2}y^{2-1} - \dfrac{6}{2}y^{1-1}$

$\qquad\qquad = \dfrac{7}{2}y^2 + 6y - 3$

42. $\dfrac{2m^3n^2 + 8m^2n^2 - 14mn^3}{4m^2n^3}$

$= \dfrac{2m^3n^2}{4m^2n^3} + \dfrac{8m^2n^2}{4m^2n^3} - \dfrac{14mn^3}{4m^2n^3}$

$= \dfrac{2}{4}m^{3-2}n^{2-3} + \dfrac{8}{4}m^{2-2}n^{2-3} - \dfrac{14}{4}m^{1-2}n^{3-3}$

$= \dfrac{1}{2}mn^{-1} + 2n^{-1} - \dfrac{7}{2}m^{-1}$

$= \dfrac{m}{2n} + \dfrac{2}{n} - \dfrac{7}{2m}$

43.
$$
\begin{array}{r}
3x+4 \\
x-2\overline{)3x^2 - 2x - 8} \\
\underline{-\left(3x^2 - 6x\right)} \\
4x - 8 \\
\underline{-(4x - 8)} \\
0
\end{array}
$$

$\dfrac{3x^2 - 2x - 8}{x - 2} = 3x + 4$

44.
$$
\begin{array}{r}
-2x + 7 \\
x+5\overline{)-2x^2 - 3x + 40} \\
\underline{-\left(-2x^2 - 10x\right)} \\
7x + 40 \\
\underline{-(7x + 35)} \\
5
\end{array}
$$

$\dfrac{-2x^2 - 3x + 40}{x + 5} = -2x + 7 + \dfrac{5}{x + 5}$

45.
$$
\begin{array}{r}
3z^2 + 2 \\
2z+3\overline{)6z^3 + 9z^2 + 4z - 6} \\
\underline{-\left(6z^3 + 9z^2\right)} \\
4z - 6 \\
\underline{-(4z + 6)} \\
-12
\end{array}
$$

$\dfrac{6z^3 + 9z^2 + 4z - 6}{2z + 3} = 3z^2 + 2 - \dfrac{12}{2z + 3}$

46.
$$
\begin{array}{r}
4k^2 + k - 2 \\
3k-8\overline{)12k^3 - 29k^2 - 14k + 16} \\
\underline{-\left(12k^3 - 32k^2\right)} \\
3k^2 - 14k \\
\underline{-\left(3k^2 - 8k\right)} \\
-6k + 16 \\
\underline{-(-6k + 16)} \\
0
\end{array}
$$

$\dfrac{12k^3 - 29k^2 - 14k + 16}{3k - 8} = 4k^2 + k - 2$

47.

$$
2x-3 \overline{\smash{\big)}\, 16x^4 + 0x^3 + 0x^2 + 0x - 81} \quad \overset{\displaystyle 8x^3 + 12x^2 + 18x + 27}{}
$$

$$
\underline{-\left(16x^4 - 24x^3\right)}
$$
$$
24x^3 + 0x^2
$$
$$
\underline{-\left(24x^3 - 36x^2\right)}
$$
$$
36x^2 + 0x
$$
$$
\underline{-\left(36x^2 - 54x\right)}
$$
$$
54x - 81
$$
$$
\underline{-\left(54x - 81\right)}
$$
$$
0
$$

$$\frac{16x^4 - 81}{2x - 3} = 8x^3 + 12x^2 + 18x + 27$$

48.

$$
x^2 - 3x + 4 \overline{\smash{\big)}\, 2x^4 - 11x^3 + 35x^2 - 54x + 55} \quad \overset{\displaystyle 2x^2 - 5x + 12}{}
$$

$$
\underline{-\left(2x^4 - 6x^3 + 8x^2\right)}
$$
$$
-5x^3 + 27x^2 - 54x
$$
$$
\underline{-\left(-5x^3 + 15x^2 - 20x\right)}
$$
$$
12x^2 - 34x + 55
$$
$$
\underline{-\left(12x^2 - 36x + 48\right)}
$$
$$
2x + 7
$$

$$\frac{2x^4 - 11x^3 + 35x^2 - 54x + 55}{x^2 - 3x + 4}$$
$$= 2x^2 - 5x + 12 + \frac{2x + 7}{x^2 - 3x + 4}$$

49. The divisor is $x + 2$ so $c = -2$.

$$
-2 \overline{\smash{\big)}\, 5 \quad\;\; 11 \quad\;\; 8}
$$
$$
\underline{\quad\;\; -10 \quad -2}
$$
$$
5 \quad\;\; 1 \quad\;\; 6
$$

$$\frac{5x^2 + 11x + 8}{x + 2} = 5x + 1 + \frac{6}{x + 2}$$

50. The divisor is $a - 2$ so $c = 2$.

$$
2 \overline{\smash{\big)}\, 9 \quad -14 \quad -8}
$$
$$
\underline{\quad\;\; 18 \quad\;\; 8}
$$
$$
9 \quad\;\; 4 \quad\;\; 0
$$

$$\frac{9a^2 - 14a - 8}{a - 2} = 9a + 4$$

51. The divisor is $m + 3$ so $c = -3$.

$$
-3 \overline{\smash{\big)}\, 3 \quad\;\; 11 \quad -5 \quad -33}
$$
$$
\underline{\quad\;\; -9 \quad -6 \quad\;\; 33}
$$
$$
3 \quad\;\; 2 \quad -11 \quad\;\; 0
$$

$$\frac{3m^3 + 11m^2 - 5m - 33}{m + 3} = 3m^2 + 2m - 11$$

52. The divisor is $n - 4$ so $c = 4$.

$$
4 \overline{\smash{\big)}\, 1 \quad 2 \quad -39 \quad\;\; 67}
$$
$$
\underline{\quad\;\; 4 \quad\;\; 24 \quad -60}
$$
$$
1 \quad 6 \quad -15 \quad\;\; 7
$$

$$\frac{n^3 + 2n^2 - 39n + 67}{n - 4} = n^2 + 6n - 15 + \frac{7}{n - 4}$$

53. The divisor is $x + 1$ so $c = -1$.

$$
-1 \overline{\smash{\big)}\, 1 \quad\;\; 0 \quad\;\; 6 \quad\;\; 0 \quad -7}
$$
$$
\underline{\quad\;\; -1 \quad\;\; 1 \quad -7 \quad\;\; 7}
$$
$$
1 \quad -1 \quad\;\; 7 \quad -7 \quad\;\; 0
$$

$$\frac{x^4 + 6x^2 - 7}{x + 1} = x^3 - x^2 + 7x - 7$$

54. The divisor is $x + 2$ so $c = -2$.

$$
-2 \overline{\smash{\big)}\, 2 \quad\;\; 0 \quad\;\; 5 \quad\;\; -8}
$$
$$
\underline{\quad\;\; -4 \quad\;\; 8 \quad -26}
$$
$$
2 \quad -4 \quad 13 \quad -34
$$

$$\frac{2x^3 + 5x - 8}{x + 2} = 2x^2 - 4x + 13 - \frac{34}{x + 2}$$

55. a. $\left(\dfrac{f}{g}\right)(x) = \dfrac{5x^3 + 25x^2 - 15x}{5x}$

$$= \frac{5x^3}{5x} + \frac{25x^2}{5x} - \frac{15x}{5x}$$
$$= \frac{5}{5}x^{3-1} + \frac{25}{5}x^{2-1} - \frac{15}{5}x^{1-1}$$
$$= x^2 + 5x - 3$$

$$\left(\frac{f}{g}\right)(x) = x^2 + 5x - 3$$

b. Since $\left(\dfrac{f}{g}\right)(x) = x^2 + 5x - 3$, we have

$$\left(\frac{f}{g}\right)(2) = (2)^2 + 5(2) - 3$$
$$= 4 + 10 - 3$$
$$= 11$$

56. a. $\left(\dfrac{f}{g}\right)(x) = \dfrac{9x^2 + 54x - 31}{3x - 2}$

$$\begin{array}{r} 3x + 20 \\ 3x-2\overline{\smash{\big)}9x^2 + 54x - 31} \\ \underline{-\left(9x^2 - 6x\right)} \\ 60x - 31 \\ \underline{-(60x - 40)} \\ 9 \end{array}$$

$\left(\dfrac{f}{g}\right)(x) = 3x + 20 + \dfrac{9}{3x - 2}$

b. Since $\left(\dfrac{f}{g}\right)(x) = 3x + 20 + \dfrac{9}{3x-2}$, we have

$\left(\dfrac{f}{g}\right)(-3) = 3(-3) + 20 + \dfrac{9}{3(-3)-2}$

$ = -9 + 20 + \dfrac{9}{-11}$

$ = 11 - \dfrac{9}{11}$

$ = \dfrac{112}{11}$

57. a. $\left(\dfrac{f}{g}\right)(x) = \dfrac{2x^3 + 12x^2 + 9x - 28}{x + 4}$

The divisor is $x + 4$ so $c = -4$.

$$\begin{array}{r|rrrr} -4 & 2 & 12 & 9 & -28 \\ & & -8 & -16 & 28 \\ \hline & 2 & 4 & -7 & 0 \end{array}$$

$\left(\dfrac{f}{g}\right)(x) = 2x^2 + 4x - 7$

b. Since $\left(\dfrac{f}{g}\right)(x) = 2x^2 + 4x - 7$, we have

$\left(\dfrac{f}{g}\right)(-2) = 2(-2)^2 + 4(-2) - 7$

$ = 2(4) - 8 - 7$

$ = 8 - 8 - 7$

$ = -7$

58. a. $\left(\dfrac{f}{g}\right)(x) = \dfrac{3x^4 - 14x^3 + 31x^2 - 58x + 22}{x^2 - x + 5}$

$$\begin{array}{r} 3x^2 - 11x + 5 \\ x^2-x+5\overline{\smash{\big)}3x^4 - 14x^3 + 31x^2 - 58x + 22} \\ \underline{-\left(3x^4 - 3x^3 + 15x^2\right)} \\ -11x^3 + 16x^2 - 58x \\ \underline{-\left(-11x^3 + 11x^2 - 55x\right)} \\ 5x^2 - 3x + 22 \\ \underline{-\left(5x^2 - 5x + 25\right)} \\ 2x - 3 \end{array}$$

$\left(\dfrac{f}{g}\right)(x) = 3x^2 - 11x + 5 + \dfrac{2x - 3}{x^2 - x + 5}$

b. Since $\left(\dfrac{f}{g}\right)(x) = 3x^2 - 11x + 5 + \dfrac{2x-3}{x^2-x+5}$, we have

$\left(\dfrac{f}{g}\right)(4) = 3(4)^2 - 11(4) + 5 + \dfrac{2(4)-3}{(4)^2 - (4) + 5}$

$ = 3(16) - 44 + 5 + \dfrac{8-3}{16 - 4 + 5}$

$ = 48 - 44 + 5 + \dfrac{5}{17}$

$ = 9 + \dfrac{5}{17}$

$ = \dfrac{158}{17}$

59. The divisor is $x - 4$ so $c = 4$

$ f(4) = 4(4)^2 - 7(4) + 23$

$ = 4(16) - 28 + 23$

$ = 64 - 28 + 23$

$ = 59$

The remainder is 59.

60. The divisor is $x + 2$ so $c = -2$

$ f(-2) = (-2)^3 - 2(-2)^2 + 12(-2) - 5$

$ = -8 - 2(4) - 24 - 5$

$ = -8 - 8 - 24 - 5 = -45$

The remainder is -45.

61. $c = 2$

$$f(2) = 3(2)^2 + (2) - 14$$
$$= 3(4) + 2 - 14$$
$$= 12 + 2 - 14$$
$$= 0$$

Since the remainder is 0, $x - 2$ is a factor.

$$2\overline{)3 \quad 1 \quad -14}$$
$$\underline{\quad\;\; 6 \quad 14\;\;}$$
$$3 \quad 7 \quad 0$$

$$f(x) = (x - 2)(3x + 7)$$

62. $c = -4$

$$f(-4) = 2(-4)^2 + 13(-4) + 22$$
$$= 2(16) - 52 + 22$$
$$= 32 - 52 + 22 = 2$$

Since the remainder is not 0, $x + 4$ is not a factor of $f(x)$.

63. Since $A = L \cdot W$ we get $L = \dfrac{A}{W}$

$$\begin{array}{r} 5x + 1 \\ 4x - 3 \overline{) 20x^2 - 11x - 3} \\ \underline{-(20x^2 - 15x)} \\ 4x - 3 \\ \underline{-(4x - 3)} \\ 0 \end{array}$$

The length of the rectangle is $(5x + 1)$ meters.

64. We have $V = L \cdot W \cdot H$ and the area of the top of the box is $L \cdot W$. We can rewrite to get

$$L \cdot W = \dfrac{V}{H}$$

$$2\overline{)2 \quad 1 \quad -7 \quad -6}$$
$$\underline{\quad\;\; 4 \quad 10 \quad 6\;\;}$$
$$2 \quad 5 \quad 3 \quad 0$$

The area of the top of the box is $(2x^2 + 5x + 3)$ square centimeters.

65. $4z + 24$

GCF: 4

$$4z + 24 = 4 \cdot z + 4 \cdot 6$$
$$= 4(z + 6)$$

66. $-7y^2 + 91y$

GCF: $-7y$

$$-7y^2 + 91y = -7y \cdot y - 7y \cdot (-13)$$
$$= -7y(y - 13)$$

67. $14x^3 y^2 + 2xy^2 - 8x^2 y$

GCF: $2xy$

$$14x^3 y^2 + 2xy^2 - 8x^2 y$$
$$= 2xy \cdot 7x^2 y + 2xy \cdot y + 2xy \cdot (-4x)$$
$$= 2xy(7x^2 y + y - 4x)$$

68. $30a^4 b^3 + 15a^3 b - 25a^2 b^2$

GCF: $5a^2 b$

$$30a^4 b^3 + 15a^3 b - 25a^2 b^2$$
$$= 5a^2 b \cdot 6a^2 b^2 + 5a^2 b \cdot 3a + 5a^2 b \cdot (-5b)$$
$$= 5a^2 b(6a^2 b^2 + 3a - 5b)$$

69. $3x(x + 5) - 4(x + 5)$

GCF: $x + 5$

$$3x(x + 5) - 4(x + 5) = 3x \cdot (x + 5) - 4 \cdot (x + 5)$$
$$= (x + 5)(3x - 4)$$

70. $-4c(2c + 9) + 3(2c + 9)$

GCF: $2c + 9$

$$-4c(2c + 9) + 3(2c + 9)$$
$$= -4c \cdot (2c + 9) + 3 \cdot (2c + 9)$$
$$= (2c + 9)(-4c + 3)$$
$$= (2c + 9)(3 - 4c)$$

71. $(5x + 3)(x - 5y) + (x + 2)(x - 5y)$

GCF: $x - 5y$

$$(5x + 3)(x - 5y) + (x + 2)(x - 5y)$$
$$= (x - 5y)\big((5x + 3) + (x + 2)\big)$$
$$= (x - 5y)(5x + 3 + x + 2)$$
$$= (x - 5y)(6x + 5)$$

72. $(3a - b)(a + 7) - (a + 1)(a + 7)$

GCF: $a + 7$

$$(3a - b)(a + 7) - (a + 1)(a + 7)$$
$$= (a + 7)\big((3a - b) - (a + 1)\big)$$
$$= (a + 7)(3a - b - a - 1)$$
$$= (a + 7)(2a - b - 1)$$

73. $x^2 + 6x - 3x - 18 = \left(x^2 + 6x\right) + (-3x - 18)$
$$= x(x+6) - 3(x+6)$$
$$= (x+6)(x-3)$$

74. $c^2 + 2c - 5c - 10 = \left(c^2 + 2c\right) + (-5c - 10)$
$$= c(c+2) - 5(c+2)$$
$$= (c+2)(c-5)$$

75. $14z^2 + 16z - 21z - 24$
$$= \left(14z^2 + 16z\right) + (-21z - 24)$$
$$= 2z(7z+8) - 3(7z+8)$$
$$= (7z+8)(2z-3)$$

76. $21w^2 - 28w + 6w - 8 = \left(21w^2 - 28w\right) + (6w - 8)$
$$= 7w(3w-4) + 2(3w-4)$$
$$= (3w-4)(7w+2)$$

77. Pull out the GCF first and then factor by grouping.
$$2x^3 + 2x^2 - 18x^2 - 18x$$
$$= 2x\left(x^2 + x - 9x - 9\right)$$
$$= 2x\left(\left(x^2 + x\right) + (-9x - 9)\right)$$
$$= 2x\left(x(x+1) - 9(x+1)\right)$$
$$= 2x(x+1)(x-9)$$

78. Pull out the GCF first and then factor by grouping.
$$10a^4 + 15a^3 + 70a^3 + 105a^2$$
$$= 5a^2\left(2a^2 + 3a + 14a + 21\right)$$
$$= 5a^2\left(\left(2a^2 + 3a\right) + (14a + 21)\right)$$
$$= 5a^2\left(a(2a+3) + 7(2a+3)\right)$$
$$= 5a^2(2a+3)(a+7)$$

79. a. $\dfrac{1}{2}n^2 + \dfrac{1}{2}n = \dfrac{1}{2}n \cdot n + \dfrac{1}{2}n \cdot 1$
$$= \dfrac{1}{2}n(n+1)$$

  b. $n = 32$
$$\dfrac{1}{2}(32)(32+1) = 16(33) = 528$$

80. $R(x) = 5200x - 2x^3$
$$= 2x \cdot 2600 - 2x \cdot x^2$$
$$= 2x\left(2600 - x^2\right)$$

The revenue function is $R(x) = 2x\left(2600 - x^2\right)$.

81. $w^2 - 11w - 26$
We are looking for two factors of $c = -26$ whose sum is $b = -11$. Since c is negative, the factors will have opposite signs, and since b is negative the factor with the larger absolute value must be negative.

Factors	$1, -26$	$2, -13$
Sum	-25	-11

$$w^2 - 11w - 26 = (w+2)(w-13)$$

82. $x^2 - 9x + 15$
We are looking for two factors of $c = 15$ whose sum is $b = -9$. Since c is positive the two factors have the same sign, and since b is negative the factors are both negative.

Factors	$-1, -15$	$-3, -5$
Sum	-16	-8

Since none of the possibilities work,
$x^2 - 9x + 15$ is prime.

83. $-t^2 + 6t + 72$
Start by factoring out -1.
$$-t^2 + 6t + 72 = -1\left(t^2 - 6t - 72\right)$$

We are looking for two factors of $c = -72$ whose sum is $b = -6$. Since c is negative, the two factors will have opposite signs. Since b is negative the factor with the larger absolute value will be negative.

Factors	$1, -72$	$2, -36$	$3, -24$	$4, -18$	$6, -12$	$8, -9$
Sum	-71	-34	-21	-14	-6	-1

$$-t^2 + 6t + 72 = -1\left(t^2 - 6t - 72\right)$$
$$= -1(t - 12)(t + 6)$$

84. $m^2 + 10m + 21$

We are looking for two factors of $c = 21$ whose sum is $b = 10$. Since both c and b are positive, the two factors must be positive.

Factors	1, 21	3, 7
Sum	22	10

$m^2 + 10m + 21 = (m+3)(m+7)$

85. $x^2 + 4xy - 320y^2$

We are looking for two factors of $c = -320$ whose sum is $b = 4$. Since c is negative, the two factors have opposite signs, and since b is positive the factor with the larger absolute value must be positive.

Factors	−1,320	−2,160	−4,80	−5,64
Sum	319	158	76	59

Factors	−8,40	−10,32	−16,20
Sum	32	22	4

$x^2 + 4xy - 320y^2 = (x+20y)(x-16y)$

86. $r^2 - 5rs + 6s^2$

We are looking for two factors of $c = 6$ whose sum is $b = -5$. Since c is positive, the factors will have the same sign, and since b is negative the factors will both be negative.

Factors	−1,−6	−2,−3
Sum	−7	−5

$r^2 - 5rs + 6s^2 = (r-2s)(r-3s)$

87. <u>ac Method:</u>

$a \cdot c = 5 \cdot -6 = -30$

We are looking for two factors of -30 whose sum is 13. Since the product is negative, the factors will have opposite signs. The sum is positive so the factor with the larger absolute value will be positive.

factor 1	factor 2	sum
−1	30	29
−2	15	13 ← okay
−3	10	7
−5	6	1

$5x^2 + 13x - 6 = 5x^2 - 2x + 15x - 6$
$$= x(5x-2) + 3(5x-2)$$
$$= (x+3)(5x-2)$$

<u>Trial and Error Method:</u>

First note that there are no common factors and that $a = 5$, $b = 13$, and $c = -6$. Since c is negative, the signs in our factors will be opposites. We will consider factorizations with this form:

$(\_\_x + \_\_)(\_\_x - \_\_)$

If our choice results in a middle term with the wrong sign, we simply switch the signs of the factors.

Since $a = 5$ can be factored as $1 \cdot 5$, we have the following form:

$(x + \_\_)(5x - \_\_)$

$|c| = |-6| = 6$ can be factored as $1 \cdot 6$ or $2 \cdot 3$.

$(x+1)(5x-6) \to 5x^2 - x - 6$

$(x+6)(5x-1) \to 5x^2 + 29x - 6$

$(x+2)(5x-3) \to 5x^2 + 7x - 6$

$(x+3)(5x-2) \to 5x^2 + 13x - 6$

The correct factorization is

$5x^2 + 13x - 6 = (x+3)(5x-2)$

88. <u>ac Method:</u>

$a \cdot c = 6 \cdot 44 = 264$

We are looking for two factors of 264 whose sum is 41. Since the product is positive, the factors will have the same sign. The sum is positive so the factors will both be positive, and both must be less than 41.

factor 1	factor 2	sum
8	33	41 ← okay
11	24	35
12	22	34

$6m^2 + 41m + 44 = 6m^2 + 8m + 33m + 44$
$$= 2m(3m+4) + 11(3m+4)$$
$$= (2m+11)(3m+4)$$

<u>Trial and Error Method:</u>

First note that there are no common factors and that $a = 6$, $b = 41$, and $c = 44$. Since c is positive, the signs in our factors will be the same. Since b is positive, both factors will be positive. We will consider factorizations with this form:

$$(\_\,m+\_\,)(\_\,m+\_\,)$$

Since $a=6$ can be factored as $1\cdot6$ or $2\cdot3$, we have the following forms:
$$(m+\_\,)(6m+\_\,)$$
$$(2m+\_\,)(3m+\_\,)$$

$c=44$ can be factored as $1\cdot44$, $2\cdot22$, or $4\cdot11$. Since the original expression had no common factors, the binomials we select cannot have a common factor.

$$(m+44)(6m+1)\to 6m^2+265m+44$$
$$(2m+11)(3m+4)\to 6m^2+41m+44$$
$$(2m+1)(3m+44)\to 6m^2+91m+44$$
$$(m+4)(6m+11)\to 6m^2+35m+44$$

The correct factorization is
$$6m^2+41m+44=(2m+11)(3m+4)$$

89. *ac* Method:
$$a\cdot c=4\cdot7=28$$
We are looking for two factors of 28 whose sum is -5. Since the product is positive, the factors will have the same sign. The sum is negative so the factors will both be negative.

factor 1	factor 2	sum
-1	-28	-29
-2	-14	-16
-4	-7	-11

Since none of the possibilities work, the expression is prime.

Trial and Error Method:
First note that there are no common factors and that $a=4$, $b=-5$, and $c=7$. Since c is positive, the signs in our factors will be the same. Since b is negative, the signs will both be negative. We will consider factorizations with this form:
$$(\_\,y-\_\,)(\_\,y-\_\,)$$

Since $a=4$ can be factored as $1\cdot4$ or $2\cdot2$, we have the following forms:
$$(y-\_\,)(4y-\_\,)$$
$$(2y-\_\,)(2y-\_\,)$$

$c=7$ can be factored as $1\cdot7$.
$$(y-1)(4y-7)\to 4y^2-11y+7$$
$$(y-7)(4y-1)\to 4y^2-29y+7$$
$$(2y-1)(2y-7)\to 4y^2-16y+7$$

None of the possibilities yield the given expression. Therefore, the expression is prime.

90. *ac* Method:
Start by factoring out the GCF.
$$8t^2+22t-6=2\left(4t^2+11t-3\right)$$

Now focus on the reduced trinomial.
$$a\cdot c=4\cdot-3=-12$$
We are looking for two factors of -12 whose sum is 11. Since the product is negative, the two factors will have opposite signs. The sum is positive so the factor with the larger absolute value will be positive.

factor 1	factor 2	sum
-1	12	$11\ \leftarrow$ okay
-2	6	4
-3	4	1

$$8t^2+22t-6=2\left(4t^2+11t-3\right)$$
$$=2\left(4t^2+12t-t-3\right)$$
$$=2\left(4t(t+3)-1(t+3)\right)$$
$$=2(t+3)(4t-1)$$

Trial and Error Method:
Start by factoring out the GCF.
$$8t^2+22t-6=2\left(4t^2+11t-3\right)$$

Now focus on the reduced trinomial with $a=4$, $b=11$, and $c=-3$. Since c is negative, our factors will have opposite signs. We will consider factorizations of this form:
$$(\_\,t+\_\,)(\_\,t-\_\,)$$

If our choice results in a middle term with the wrong sign, we simply switch the signs in our factors to obtain the correct result.
Since $a=4$ can be factored as $1\cdot4$ or $2\cdot2$, we have the following forms:
$$(t+\_\,)(4t-\_\,)$$
$$(2t+\_\,)(2t-\_\,)$$

$|c|=|-3|=3$ can be factored as $1\cdot3$.
$$(t+1)(4t-3)\to 4t^2+t-3$$
$$(t+3)(4t-1)\to 4t^2+11t-3$$
$$(2t+1)(2t-3)\to 4t^2-4t-3$$

The correct factorization is
$$8t^2+22t-6=2(t+3)(4t-1)$$

91. *ac* Method:
$$a\cdot c=6\cdot5=30$$
We are looking for two factors of 30 whose sum is -13. Since the product is positive, the factors will have the same sign. The sum is negative so

Copyright © 2014 Pearson Education, Inc.

the factors will both be negative.

factor 1	factor 2	sum
−1	−30	−31
−2	−15	−17
−3	−10	−13 ← okay
−5	−6	−11

$$6x^2 - 13x + 5 = 6x^2 - 3x - 10x + 5$$
$$= 3x(2x-1) - 5(2x-1)$$
$$= (2x-1)(3x-5)$$

Trial and Error Method:
First note that there are no common factors and that $a = 6$, $b = -13$, and $c = 5$. Since c is positive, the signs in our factors will be the same. Since b is negative, the signs will both be negative. We will consider factorizations with this form:
$$(\_\,x - \_)(\_\,x - \_)$$
Since $a = 6$ can be factored as $1 \cdot 6$ or $2 \cdot 3$, we have the following forms:
$$(x - \_)(6x - \_)$$
$$(2x - \_)(3x - \_)$$
$c = 5$ can be factored as $1 \cdot 5$.
$$(x-1)(6x-5) \rightarrow 6x^2 - 11x + 5$$
$$(x-5)(6x-1) \rightarrow 6x^2 - 31x + 5$$
$$(2x-1)(3x-5) \rightarrow 6x^2 - 13x + 5$$
$$(2x-5)(3x-1) \rightarrow 6x^2 - 17x + 5$$
The correct factorization is
$$6x^2 - 13x + 5 = (2x-1)(3x-5)$$

92. *ac* **Method:**
$a \cdot c = 21 \cdot -2 = -42$
We are looking for two factors of −42 whose sum is −1. Since the product is negative, the two factors will have opposite signs. Since the sum is negative, the factor with the larger absolute value will be negative.

factor 1	factor 2	sum
1	−42	−41
2	−21	−19
3	−14	−11
6	−7	−1 ← okay

$$21r^2 - rs - 2s^2 = 21r^2 + 6rs - 7rs - 2s^2$$
$$= 3r(7r+2s) - s(7r+2s)$$
$$= (7r+2s)(3r-s)$$

Trial and Error Method:
First note that there are no common factors and

that $a = 21$, $b = -1$, and $c = -2$. Since c is negative, the factors will have opposite signs. We will consider factorizations with this form:
$$(\_\,r + \_\,s)(\_\,r - \_\,s)$$
If our choice results in a middle term with the wrong sign, we simply switch the signs in our factors to obtain the correct result.
Since $a = 21$ can be factored as $1 \cdot 21$ or $3 \cdot 7$, we have the following forms:
$$(r + \_\,s)(21r - \_\,s)$$
$$(3r + \_\,s)(7r - \_\,s)$$
$|c| = |-2| = 2$ can be factored as $1 \cdot 2$.
$$(r+s)(21r-2s) \rightarrow 21r^2 + 19rs - 2s^2$$
$$(r+2s)(21r-s) \rightarrow 21r^2 + 41rs - 2s^2$$
$$(3r+s)(7r-2s) \rightarrow 21r^2 + rs - 2s^2$$
$$(3r+2s)(7r-s) \rightarrow 21r^2 + 11rs - 2s^2$$
The third result is almost what we need, except the sign on the middle term is wrong. We simply change the signs of the factors to get the correct result.
$$(3r-s)(7r+2s) \rightarrow 21r^2 - rs - 2s^2$$
The correct factorization is
$$21r^2 - rs - 2s^2 = (3r-s)(7r+2s)$$

93. *ac* **Method:**
$a \cdot c = 20 \cdot 27 = 540$
We are looking for two factors of 540 whose sum is −57. Since the product is positive, the two factors will have the same sign. Since the sum is negative, the factors will both be negative. We are adding two negative numbers to get −57 so neither can be less than −57.

factor 1	factor 2	sum
−10	−54	−64 too small
−15	−36	−51 too large
−12	−45	−57 ← okay

$$20x^2 - 57xy + 27y^2 = 20x^2 - 12xy - 45xy + 27y^2$$
$$= 4x(5x-3y) - 9y(5x-3y)$$
$$= (5x-3y)(4x-9y)$$

Trial and Error Method
First note that there are no common factors and that $a = 20$, $b = -57$, and $c = 27$. Since c is positive, the signs in our factors will be the same. Since b is negative, the signs will both be negative. We will consider factorizations with this form:
$$(\_\,x - \_\,y)(\_\,x - \_\,y)$$

Since $a = 20$ can be factored as $1 \cdot 20$, $2 \cdot 10$, or $4 \cdot 5$, we have the following forms:

$$(x - \underline{\quad}\, y)(20x - \underline{\quad}\, y)$$
$$(2x - \underline{\quad}\, y)(10x - \underline{\quad}\, y)$$
$$(4x - \underline{\quad}\, y)(5x - \underline{\quad}\, y)$$

$c = 27$ can be factored as $1 \cdot 27$ or $3 \cdot 9$.

$$(x - y)(20x - 27y) \rightarrow 20x^2 - 47xy + 27y^2$$
$$(x - 27y)(20x - y) \rightarrow 20x^2 - 541xy + 27y^2$$
$$(2x - y)(10x - 27y) \rightarrow 20x^2 - 64xy + 27y^2$$
$$(2x - 27y)(10x - y) \rightarrow 20x^2 - 272xy + 27y^2$$
$$(4x - y)(5x - 27y) \rightarrow 20x^2 - 113xy + 27y^2$$
$$(4x - 27y)(5x - y) \rightarrow 20x^2 - 139xy + 27y^2$$
$$(x - 3y)(20x - 9y) \rightarrow 20x^2 - 69xy + 27y^2$$
$$(x - 9y)(20x - 3y) \rightarrow 20x^2 - 183xy + 27y^2$$
$$(2x - 3y)(10x - 9y) \rightarrow 20x^2 - 48xy + 27y^2$$
$$(2x - 9y)(10x - 3y) \rightarrow 20x^2 - 96xy + 27y^2$$
$$(4x - 3y)(5x - 9y) \rightarrow 20x^2 - 51xy + 27y^2$$
$$(4x - 9y)(5x - 3y) \rightarrow 20x^2 - 57xy + 27y^2$$

The correct factorization is
$$20x^2 - 57xy + 27y^2 = (4x - 9y)(5x - 3y)$$

94. Begin by factoring out the GCF.

$$-2s^2 + 12s + 14 = -2\left(s^2 - 6s - 7\right)$$

Now focus on the reduced trinomial. We are looking for two factors of -7 whose sum is -6. Since the product is negative, the factors have opposite signs. Since the sum is negative, the factor with the larger absolute value must be negative.

$$-2s^2 + 12s + 14 = -2\left(s^2 - 6s - 7\right)$$
$$= -2(s - 7)(s + 1)$$

95. $x^4 - 10x^2 - 11$

Let $u = x^2$. This gives $u^2 - 10u - 11$.
We need two factors of -11 whose sum is -10. Since the product is negative, the two factors will have opposite signs. Since the sum is negative, the factor with the larger absolute value will be negative.

$$u^2 - 10u - 11 = (u + 1)(u - 11)$$

Now get back in terms of x.

$$x^4 - 10x^2 - 11 = \left(x^2 + 1\right)\left(x^2 - 11\right)$$

96. $10x^2 y^2 + 41xy + 4$

Let $u = xy$. This gives $10u^2 + 41u + 4$.
$a \cdot c = 10 \cdot 4 = 40$
We need two factors of 40 whose sum is 41. Since the product is positive, the two factors will have the same sign. Since the sum is also positive, the two factors are both positive.

factor 1	factor 2	sum
1	40	41 ← okay
2	20	22
4	10	14
5	8	13

$$10u^2 + 41u + 4 = 10u^2 + 40u + u + 4$$
$$= 10u(u + 4) + 1(u + 4)$$
$$= (u + 4)(10u + 1)$$

Now get back in terms of x and y.

$$10x^2 y^2 + 41xy + 4 = (xy + 4)(10xy + 1)$$

97. $(a + 4)^2 - 9(a + 4) - 36$

Let $u = a + 4$. This gives $u^2 - 9u - 36$.
We need two factors of -36 whose sum is -9. Since the product is negative, the two factors will have opposite signs. Since the sum is negative, the factor with the larger absolute value will be negative.

factor 1	factor 2	sum
1	−36	−35
2	−18	−16
3	−12	−9 ← okay
4	−9	−5
6	−6	0

$$u^2 - 9u - 36 = (u + 3)(u - 12)$$

Now get back in terms of a.

$$(a + 4)^2 - 9(a + 4) - 36 = (a + 4 + 3)(a + 4 - 12)$$
$$= (a + 7)(a - 8)$$

98. $2(w - 1)^2 + 11(w - 1) + 9$

Let $u = w - 1$. This gives $2u^2 + 11u + 9$.
$a \cdot c = 2 \cdot 9 = 18$
We are looking for two factors of 18 whose sum is 11. Since the product is positive, the factors will have the same sign. Since the sum is also positive, the two factors will both be positive.

factor 1	factor 2	sum
1	18	19
2	9	11 ← okay
3	6	9

$$2u^2 + 11u + 9 = 2u^2 + 2u + 9u + 9$$
$$= 2u(u+1) + 9(u+1)$$
$$= (u+1)(2u+9)$$

Now go back in terms of *w*.

$$2(w-1)^2 + 11(w-1) + 9$$
$$= (w-1+1)(2(w-1)+9)$$
$$= w(2w-2+9)$$
$$= w(2w+7)$$

99. $x^2 + 22x + 121 = x^2 + 2 \cdot 11 \cdot x + 11^2$
$$= (x+11)^2$$

100. $w^2 - 34w + 289 = w^2 - 2 \cdot 17 \cdot w + 17^2$
$$= (w-17)^2$$

101. $144 - 24c + c^2 = 12^2 - 2 \cdot 12 \cdot c + c^2$
$$= (12-c)^2$$

102. $x^2 - 8x + 16 = x^2 - 2 \cdot (4) \cdot x + (4)^2$
$$= (x-4)^2$$

103. $64y^2 + 80y + 25 = (8y)^2 + 2 \cdot (8y) \cdot 5 + 5^2$
$$= (8y+5)^2$$

104. $12z^2 + 48z + 48 = 12(z^2 + 4z + 4)$
$$= 12(z^2 + 2 \cdot 2 \cdot z + 2^2)$$
$$= 12(z+2)^2$$

105. $x^2 - 196 = x^2 - 14^2$
$$= (x-14)(x+14)$$

106. $49 - y^2 = 7^2 - y^2$
$$= (7-y)(7+y)$$

107. $t^2 - 225 = t^2 - (15)^2$
$$= (t-15)(t+15)$$

108. $4w^2 - 81 = (2w)^2 - 9^2$
$$= (2w-9)(2w+9)$$

109. $36x^4 - 25y^2 = (6x^2)^2 - (5y)^2$
$$= (6x^2 - 5y)(6x^2 + 5y)$$

110. $80mn^2 - 20m = 20m(4n^2 - 1)$
$$= 20m((2n)^2 - 1^2)$$
$$= 20m(2n-1)(2n+1)$$

111. $x^3 - 343 = x^3 - 7^3$
$$= (x-7)(x^2 + (7)x + 7^2)$$
$$= (x-7)(x^2 + 7x + 49)$$

112. $729 - y^3 = 9^3 - y^3$
$$= (9-y)(9^2 + 9(y) + y^2)$$
$$= (9-y)(81 + 9y + y^2)$$
$$= (9-y)(y^2 + 9y + 81)$$

113. $27x^3 - 125y^3$
$$= (3x)^3 - (5y)^3$$
$$= (3x-5y)((3x)^2 + (3x)(5y) + (5y)^2)$$
$$= (3x-5y)(9x^2 + 15xy + 25y^2)$$

114. $8m^6 + 27n^3$
$$= (2m^2)^3 + (3n)^3$$
$$= (2m^2 + 3n)((2m^2)^2 - (2m^2)(3n) + (3n)^2)$$
$$= (2m^2 + 3n)(4m^4 - 6m^2n + 9n^2)$$

115. $2a^6 - 2b^6$

$= 2\left(a^6 - b^6\right)$

$= 2\left(\left(a^3\right)^2 - \left(b^3\right)^2\right)$

$= 2\left(a^3 - b^3\right)\left(a^3 + b^3\right)$

$= 2(a-b)\left(a^2 + ab + b^2\right)(a+b)\left(a^2 - ab + b^2\right)$

$= 2(a-b)(a+b)\left(a^2 + ab + b^2\right)\left(a^2 - ab + b^2\right)$

116. $(y-1)^3 + 64$

$= (y-1)^3 + 4^3$

$= (y-1+4)\left((y-1)^2 - (y-1)(4) + 4^2\right)$

$= (y+3)\left(y^2 - 2y + 1 - 4y + 4 + 16\right)$

$= (y+3)\left(y^2 - 6y + 21\right)$

117. $x^2 + 7x + 6 = (x+1)(x+6)$

118. $-8x^2 y^3 + 12xy^3 = -4xy^3(2x-3)$

119. $7x^3 - 35x^2 + 28x = 7x\left(x^2 - 5x + 4\right)$

$= 7x(x-1)(x-4)$

120. $3x^2 - 3x - 18 = 3\left(x^2 - x - 6\right)$

$= 3(x-3)(x+2)$

121. $4z^2 - 60z + 225 = (2z)^2 - 2(2z)(15) + 15^2$

$= (2z-15)^2$

122. $12x^2 + 7x - 49$

$a \cdot c = 12 \cdot -49 = -588$

We are looking for two factors of -588 whose sum is 7. Since the product is negative, the factors will have opposite signs. Since the sum is positive, the factor with the larger absolute value will be positive. The difference is relatively small, so we should select factors whose absolute values are close to each other.

factor 1	factor 2	sum
-14	42	28
-21	28	$7 \leftarrow$ okay

$12x^2 + 7x - 49 = 12x^2 - 21x + 28x - 49$

$= 3x(4x-7) + 7(4x-7)$

$= (3x+7)(4x-7)$

123. $10n^2 - 33n - 7$

$a \cdot c = 10 \cdot -7 = -70$

We are looking for two factors of -70 whose sum is -33. Since the product is negative, the factors will have opposite signs. Since the sum is also negative, the factor with the larger absolute value will be negative.

factor 1	factor 2	sum
1	-70	-69
2	-35	$-33 \leftarrow$ okay

$10n^2 - 33n - 7 = 10n^2 + 2n - 35n - 7$

$= 2n(5n+1) - 7(5n+1)$

$= (2n-7)(5n+1)$

124. $8 - 2y - y^2 = -\left(y^2 + 2y - 8\right)$

$= -(y-2)(y+4)$

125. $2x^3 - 10x^2 + 6x - 30 = 2\left(x^3 - 5x^2 + 3x - 15\right)$

$= 2\left(x^2(x-5) + 3(x-5)\right)$

$= 2(x-5)\left(x^2 + 3\right)$

126. $(3h+2)^3 + 64$

$= (3h+2)^3 + 4^3$

$= (3h+2+4)\left((3h+2)^2 - (3h+2)(4) + 4^2\right)$

$= (3h+6)\left((3h)^2 + 2(3h)(2) + 2^2 - 12h - 8 + 16\right)$

$= 3(h+2)\left(9h^2 + 12h + 4 - 12h + 8\right)$

$= 3(h+2)\left(9h^2 + 12\right)$

$= 3(h+2) \cdot 3\left(3h^2 + 4\right)$

$= 9(h+2)\left(3h^2 + 4\right)$

127. $5p^3 q^2 - 80p = 5p\left(p^2 q^2 - 16\right)$

$= 5p\left((pq)^2 - 4^2\right)$

$= 5p(pq - 4)(pq + 4)$

128. $m^4 - 5m^2 + 4 = (m^2)^2 - 5(m^2) + 4$

Let $u = m^2$. This gives $u^2 - 5u + 4$.

$u^2 - 5u + 4 = (u - 1)(u - 4)$

Now go back in terms of m.

$m^4 - 5m^2 + 4 = (m^2 - 1)(m^2 - 4)$
$$= (m - 1)(m + 1)(m - 2)(m + 2)$$

129. $686 - 16m^6$
$$= -2(8m^6 - 343)$$
$$= -2\left((2m^2)^3 - 7^3\right)$$
$$= -2(2m^2 - 7)\left((2m^2)^2 + (2m^2)(7) + 7^2\right)$$
$$= -2(2m^2 - 7)(4m^4 + 14m^2 + 49)$$

130. $h^3 + 2h^2 - h - 2 = h^2(h + 2) - 1(h + 2)$
$$= (h^2 - 1)(h + 2)$$
$$= (h - 1)(h + 1)(h + 2)$$

131. $108x^3 + 4y^3 = 4(27x^3 + y^3)$
$$= 4\left((3x)^3 + y^3\right)$$
$$= 4(3x + y)\left((3x)^2 - (3x)(y) + y^2\right)$$
$$= 4(3x + y)(9x^2 - 3xy + y^2)$$

132. $F(c) = c^2 - 24c + 144$
$$= (c)^2 - 2 \cdot 12 \cdot c + (12)^2$$
$$= (c - 12)^2$$

133. $f(x) = -3x^2 + 6x + 45$
$$= -3(x^2 - 2x - 15)$$
$$= -3(x - 5)(x + 3)$$

134. $g(x) = 16x^2 - 100$
$$= 4(4x^2 - 25)$$
$$= 4\left((2x)^2 - 5^2\right)$$
$$= 4(2x + 5)(2x - 5)$$

135. $G(y) = 16y^3 + 250$
$$= 2(8y^3 + 125)$$
$$= 2\left((2y)^3 + 5^3\right)$$
$$= 2(2y + 5)(4y^2 - 10y + 25)$$

136. $f(x) = -4x^2 - 16$
$$= -4(x^2 + 4)$$

137. The area of the large square is x^2 square units, while the area of the small square is 5^2 square units. The area of the shaded region is
$x^2 - 5^2 = (x - 5)(x + 5)$ square units.

138. Each side of the square has length $2x$ units, so the area of the square is $(2x)^2 = 4x^2$ square units. The area of the circle is πx^2 square units. The area of the shaded region is
$4x^2 - \pi x^2 = (4 - \pi)x^2$ square units.

139. $(w + 5)(w - 13) = 0$
$w + 5 = 0 \quad$ or $\quad w - 13 = 0$
$\quad w = -5 \quad$ or $\quad\quad w = 13$
The solution set is $\{-5, 13\}$.

140. $x^2 + 21x + 54 = 0$
$(x + 18)(x + 3) = 0$
$x + 18 = 0 \quad$ or $\quad x + 3 = 0$
$x = -18 \quad$ or $\quad x = -3$
The solution set is $\{-18, -3\}$.

141. $\qquad y^2 + 2y = 15$
$y^2 + 2y - 15 = 0$
$(y + 5)(y - 3) = 0$
$y + 5 = 0 \quad$ or $\quad y - 3 = 0$
$\quad y = -5 \quad$ or $\quad\quad y = 3$
The solution set is $\{-5, 3\}$.

142. $\qquad 5a^2 = -20a$
$5a^2 + 20a = 0$
$5a(a + 4) = 0$
$5a = 0 \quad$ or $\quad a + 4 = 0$
$\quad a = 0 \quad$ or $\quad\quad a = -4$
The solution set is $\{-4, 0\}$.

143.
$$x(x+1)=110$$
$$x^2+x=110$$
$$x^2+x-110=0$$
$$(x+11)(x-10)=0$$
$$x+11=0 \quad \text{or} \quad x-10=0$$
$$x=-11 \quad \text{or} \qquad x=10$$
The solution set is {−11, 10}.

144. $15x^2+29x-14=0$
$$(5x-2)(3x+7)=0$$
$$5x-2=0 \quad \text{or} \quad 3x+7=0$$
$$5x=2 \quad \text{or} \qquad 3x=-7$$
$$x=\frac{2}{5} \quad \text{or} \qquad x=-\frac{7}{3}$$
The solution set is $\left\{-\frac{7}{3}, \frac{2}{5}\right\}$.

145. $\frac{1}{2}x^2+5x+12=0$
$$2\left(\frac{1}{2}x^2+5x+12\right)=2\cdot 0$$
$$x^2+10x+24=0$$
$$(x+6)(x+4)=0$$
$$x+6=0 \quad \text{or} \quad x+4=0$$
$$x=-6 \quad \text{or} \qquad x=-4$$
The solution set is {−6, −4}.

146. $(b+1)(b-3)=5$
$$b^2+b-3b-3=5$$
$$b^2-2b-3=5$$
$$b^2-2b-8=0$$
$$(b-4)(b+2)=0$$
$$b-4=0 \quad \text{or} \quad b+2=0$$
$$b=4 \quad \text{or} \qquad b=-2$$
The solution set is {−2, 4}.

147.
$$2x^3+5x^2-8x=20$$
$$2x^3+5x^2-8x-20=0$$
$$x^2(2x+5)-4(2x+5)=0$$
$$(2x+5)(x^2-4)=0$$
$$(2x+5)(x-2)(x+2)=0$$
$$2x+5=0 \quad \text{or} \quad x-2=0 \quad \text{or} \quad x+2=0$$
$$2x=-5 \qquad\qquad x=2 \qquad\qquad x=-2$$
$$x=-\frac{5}{2}$$
The solution set is $\left\{-\frac{5}{2}, -2, 2\right\}$.

148.
$$5x^3+x^2-45x-9=0$$
$$x^2(5x+1)-9(5x+1)=0$$
$$(5x+1)(x^2-9)=0$$
$$(5x+1)(x-3)(x+3)=0$$
$$5x+1=0 \quad \text{or} \quad x-3=0 \quad \text{or} \quad x+3=0$$
$$5x=-1 \qquad\qquad x=3 \qquad\qquad x=-3$$
$$x=-\frac{1}{5}$$
The solution set is $\left\{-3, -\frac{1}{5}, 3\right\}$.

149. a. $f(x)=x^2+5x-18$
$$6=x^2+5x-18$$
$$0=x^2+5x-24$$
$$0=(x+8)(x-3)$$
$$x+8=0 \quad \text{or} \quad x-3=0$$
$$x=-8 \quad \text{or} \qquad x=3$$
The solution set is {−8, 3}.

b. $f(x)=x^2+5x-18$
$$-4=x^2+5x-18$$
$$0=x^2+5x-14$$
$$0=(x+7)(x-2)$$
$$x+7=0 \quad \text{or} \quad x-2=0$$
$$x=-7 \quad \text{or} \qquad x=2$$
The solution set is {−7, 2}.

The points (−8, 6), (3, 6), (−7, −4), and (2, −4) are on the graph.

150. a. $f(x) = 5x^2 - 4x + 3$

$$3 = 5x^2 - 4x + 3$$

$$0 = 5x^2 - 4x$$

$$0 = x(5x - 4)$$

$$x = 0 \text{ or } 5x - 4 = 0$$

$$x = 0 \text{ or } \quad x = \frac{4}{5}$$

The solution set is $\left\{0, \dfrac{4}{5}\right\}$.

b. $f(x) = 5x^2 - 4x + 3$

$$4 = 5x^2 - 4x + 3$$

$$0 = 5x^2 - 4x - 1$$

$$0 = (5x + 1)(x - 1)$$

$$5x + 1 = 0 \quad \text{or } x - 1 = 0$$

$$x = -\frac{1}{5} \text{ or } \quad x = 1$$

The solution set is $\left\{-\dfrac{1}{5}, 1\right\}$.

The points $(0, 3)$, $\left(\dfrac{4}{5}, 3\right)$, $\left(-\dfrac{1}{5}, 4\right)$ and $(1, 4)$ are on the graph of f.

151. $f(x) = 3x^3 + 18x^2 + 24x$

$$0 = 3x\left(x^2 + 6x + 8\right)$$

$$0 = 3x(x + 4)(x + 2)$$

$$3x = 0 \text{ or } x + 4 = 0 \quad \text{or } x + 2 = 0$$

$$x = 0 \text{ or } \quad x = -4 \text{ or } \quad x = -2$$

The zeros of $f(x)$ are -4, -2, and 0. The x-intercepts are $(-4, 0)$, $(-2, 0)$, and $(0, 0)$.

152. $f(x) = -4x^2 + 22x + 42$

$$0 = -4x^2 + 22x + 42$$

$$0 = 2x^2 - 11x - 21$$

$$0 = (2x + 3)(x - 7)$$

$$2x + 3 = 0 \quad \text{or } x - 7 = 0$$

$$x = -\frac{3}{2} \text{ or } \quad x = 7$$

The zeros of $f(x)$ are $-\dfrac{3}{2}$ and 7. The x-intercepts are $\left(-\dfrac{3}{2}, 0\right)$ and $(7, 0)$.

153. $s(t) = -16t^2 + 2064$

$\qquad 1280 = -16t^2 + 2064$

$\qquad\quad 0 = -16t^2 + 784$

$\qquad\quad 0 = t^2 - 49$

$\qquad\quad 0 = (t-7)(t+7)$

$\qquad t - 7 = 0$ or $t + 7 = 0$

$\qquad\quad t = 7$ or $\quad \cancel{t = -7}$

The object will be 1280 feet above the ground after 7 seconds.

154. a. $R = 1 - (1-r)^2$

$\qquad 0.96 = 1 - (1-r)^2$

$\qquad\quad 0 = (1-r)^2 - 0.04$

$\qquad\quad 0 = ((1-r) - 0.2)((1-r) + 0.2)$

$\qquad\quad 0 = (0.8 - r)(1.2 - r)$

$\qquad 0.8 - r = 0$ or $1.2 - r = 0$

$\qquad\quad r = 0.8$ or $\quad \cancel{r = 1.2}$

The component reliability is 0.80. (Reliability cannot exceed 1)

b. $R = 1 - (1-r)^2$

$\qquad 0.99 = 1 - (1-r)^2$

$\qquad\quad 0 = (1-r)^2 - 0.01$

$\qquad\quad 0 = ((1-r) - 0.1)((1-r) + 0.1)$

$\qquad\quad 0 = (0.9 - r)(1.1 - r)$

$\qquad 0.9 - r = 0$ or $1.1 - r = 0$

$\qquad\quad r = 0.9$ or $\quad \cancel{r = 1.1}$

The component reliability is 0.90. (Reliability cannot exceed 1)

Chapter 4 Test

1. $-5x^7 + x^4 + 7x^2 - x + 1$
Degree: 7

2. $\left(-2a^3b^2 + 5a^2b + ab + 1\right) + \left(\frac{1}{3}a^3b^2 + 4a^2b - 6ab - 5\right) = -2a^3b^2 + 5a^2b + ab + 1 + \frac{1}{3}a^3b^2 + 4a^2b - 6ab - 5$

$\qquad\qquad\qquad\qquad\qquad\qquad\qquad\qquad = -2a^3b^2 + \frac{1}{3}a^3b^2 + 5a^2b + 4a^2b + ab - 6ab + 1 - 5$

$\qquad\qquad\qquad\qquad\qquad\qquad\qquad\qquad = -\frac{5}{3}a^3b^2 + 9a^2b - 5ab - 4$

3. $f(x) = x^3 + 3x^2 - x + 1$

$\quad f(-2) = (-2)^3 + 3(-2)^2 - (-2) + 1$

$\qquad\quad = -8 + 3(4) + 2 + 1$

$\qquad\quad = -8 + 12 + 2 + 1$

$\qquad\quad = 7$

4. $(f-g)(x) = (7x^3-1) - (4x^2+3x-2)$

$\qquad = 7x^3 - 1 - 4x^2 - 3x + 2$

$\qquad = 7x^3 - 4x^2 - 3x + 1$

5. $\frac{1}{2}a^2b(4ab^2 - 6ab + 8)$

$\qquad = \frac{1}{2}a^2b \cdot 4ab^2 - \frac{1}{2}a^2b \cdot 6ab + \frac{1}{2}a^2b \cdot 8$

$\qquad = 2a^{2+1}b^{1+2} - 3a^{2+1}b^{1+1} + 4a^2b$

$\qquad = 2a^3b^3 - 3a^3b^2 + 4a^2b$

6. $(3x-1)(4x+17) = 3x \cdot 4x + 3x \cdot 17 - 1 \cdot 4x - 1 \cdot 17$

$\qquad\qquad\qquad\qquad = 12x^2 + 51x - 4x - 17$

$\qquad\qquad\qquad\qquad = 12x^2 + 47x - 17$

7. $(2m-n)^2 = (2m)^2 - 2(2m)(n) + n^2$

$\qquad\qquad\quad = 4m^2 - 4mn + n^2$

8. $2z^2+1 \overline{)6z^3 - 14z^2 + z + 4}$ $3z-7$

$\qquad\quad \underline{-(6z^3 \qquad\quad +3z)}$

$\qquad\qquad\qquad -14z^2 - 2z + 4$

$\qquad\qquad\quad \underline{-(-14z^2 \qquad -7)}$

$\qquad\qquad\qquad\qquad -2z + 11$

$\dfrac{6z^3 - 14z^2 + z + 4}{2z^2 + 1} = 3z - 7 + \dfrac{-2z+11}{2z^2+1}$

9. $6\overline{)5 \quad -27 \quad -18}$

$\qquad \underline{\quad\;\; 30 \qquad 18}$

$\qquad\; 5 \quad\;\; 3 \qquad 0$

$\dfrac{5x^2 - 27x - 18}{x-6} = 5x + 3$

10. $2x+3\overline{)6x^2 + x - 12}$ $3x-4$

$\qquad\quad \underline{-(6x^2 + 9x)}$

$\qquad\qquad\qquad -8x - 12$

$\qquad\qquad\quad \underline{-(-8x-12)}$

$\qquad\qquad\qquad\qquad\quad 0$

$\left(\dfrac{f}{g}\right)(x) = 3x - 4$

$\left(\dfrac{f}{g}\right)(2) = 3(2) - 4 = 6 - 4 = 2$

11. The divisor is $x-3$ so $c = 3$.

$f(3) = 2(3)^3 - 3(3)^2 - 4(3) + 7$

$\qquad = 2(27) - 3(9) - 12 + 7$

$\qquad = 54 - 27 - 12 + 7$

$\qquad = 22$

12. GCF: $4ab^2$

$12a^3b^2 + 8a^2b^2 - 16ab^3$

$= 4ab^2 \cdot 3a^2 + 4ab^2 \cdot 2a - 4ab^2 \cdot 4b$

$= 4ab^2(3a^2 + 2a - 4b)$

13. $6c^2 + 21c - 4c - 14 = 3c(2c+7) - 2(2c+7)$

$\qquad\qquad\qquad\qquad\quad = (2c+7)(3c-2)$

14. $x^2 - 13x - 48$

We need two factors of -48 whose sum is -13. Since the product is negative, the factors will have opposite signs. Since the sum is also negative, the factor with the larger absolute value will be negative.

Factors	1,−48	2,−24	3,−16	4,−12	6,−8
Sum	−47	−22	−13	−8	−2

$x^2 - 13x - 48 = (x-16)(x+3)$

15. $-14p^2 - 17p + 6$

$a \cdot c = -14 \cdot 6 = -84$

We need two factors of -84 whose sum is -17. The product is negative, so the two factors will have opposite signs. Since the sum is also negative, the factor with the larger absolute value will be negative.

factor 1	factor 2	sum
3	−28	−25 ← too small
6	−14	−8 ← too large
4	−21	−17 ← okay

$$-14p^2 - 17p + 6 = -14p^2 - 21p + 4p + 6$$
$$= -7p(2p+3) + 2(2p+3)$$
$$= (2p+3)(-7p+2)$$
$$\text{or}$$
$$= (2p+3)(2-7p)$$

16. $5(z-1)^2 + 17(z-1) - 12$

Let $u = z - 1$. This gives $5u^2 + 17u - 12$.
$a \cdot c = 5 \cdot -12 = -60$
We need two factors of −60 whose sum is 17.
The product is negative, so the two factors will
have opposite signs. Since the sum is positive,
the factor with the larger absolute value will be
positive.

factor 1	factor 2	sum
15	−4	11 ← too small
30	−2	28 ← too large
20	−3	17 ← okay

$$5u^2 + 17u - 12 = 5u^2 + 20u - 3u - 12$$
$$= 5u(u+4) - 3(u+4)$$
$$= (u+4)(5u-3)$$

Now go back in terms of z.
$$5(z-1)^2 + 17(z-1) - 12$$
$$= (z-1+4)(5(z-1)-3)$$
$$= (z+3)(5z-5-3)$$
$$= (z+3)(5z-8)$$

17. $-98x^2 + 112x - 32 = -2(49x^2 - 56x + 16)$
$$= -2((7x)^2 - 2(7x)(4) + 4^2)$$
$$= -2(7x-4)^2$$

18. $16x^2 - 196 = 4(4x^2 - 49)$
$$= 4((2x)^2 - 7^2)$$
$$= 4(2x-7)(2x+7)$$

19.
$$3m^2 - 5m = 5m - 7$$
$$3m^2 - 5m - 5m + 7 = 0$$
$$3m^2 - 10m + 7 = 0$$
$$(3m-7)(m-1) = 0$$
$$3m - 7 = 0 \text{ or } m - 1 = 0$$
$$m = \frac{7}{3} \text{ or } m = 1$$

The solution set is $\left\{ 1, \frac{7}{3} \right\}$.

20. Let $x =$ the width (shorter side). Then $x+3$ is
the length (longer side).
From the formula for the area of a rectangle, we
have:
$$A = (\text{length})(\text{width})$$
$$= (x+3)(x)$$
$$= x^2 + 3x$$
We know the area is 108 square meters, so we
have:
$$108 = x^2 + 3x$$
$$0 = x^2 + 3x - 108$$
$$0 = (x-9)(x+12)$$
$$x - 9 = 0 \text{ or } x + 12 = 0$$
$$x = 9 \text{ or } \cancel{x = -12}$$
The patio is 9 meters wide and $9 + 3 = 12$ meters
long.

Chapter 5

Getting Ready for Chapter 5

Getting Ready for Chapter 5 Quick Checks

1. When a rational number is written so there are no common factors in the numerator and the denominator, we say that the rational number is in <u>lowest terms</u>.

2. $\dfrac{13 \cdot 5}{13 \cdot 6} = \dfrac{\cancel{13}^{1} \cdot 5}{\cancel{13}_{1} \cdot 6} = \dfrac{5}{6}$

3. $\dfrac{80}{12} = \dfrac{20 \cdot 4}{3 \cdot 4} = \dfrac{20 \cdot \cancel{4}^{1}}{3 \cdot \cancel{4}_{1}} = \dfrac{20}{3}$

4. $\dfrac{5}{7} \cdot \left(-\dfrac{21}{10} \right) = -\dfrac{5}{7} \cdot \dfrac{3 \cdot 7}{2 \cdot 5} = -\dfrac{\cancel{5}^{1}}{\cancel{7}_{1}} \cdot \dfrac{3 \cdot \cancel{7}^{1}}{2 \cdot \cancel{5}_{1}} = -\dfrac{3}{2}$

5. $\dfrac{35}{15} \cdot \dfrac{3}{14} = \dfrac{5 \cdot 7}{3 \cdot 5} \cdot \dfrac{3}{2 \cdot 7} = \dfrac{\cancel{5}^{1} \cdot \cancel{7}^{1}}{\cancel{3}_{1} \cdot \cancel{5}_{1}} \cdot \dfrac{\cancel{3}^{1}}{2 \cdot \cancel{7}_{1}} = \dfrac{1}{2}$

6. $\dfrac{\frac{4}{5}}{\frac{12}{25}} = \dfrac{4}{5} \cdot \dfrac{25}{12} = \dfrac{\cancel{4}^{1}}{\cancel{5}_{1}} \cdot \dfrac{\cancel{5}^{1} \cdot 5}{\cancel{4}_{1} \cdot 3} = \dfrac{5}{3}$

7. $\dfrac{24}{35} \div \left(-\dfrac{8}{7} \right) = -\dfrac{24}{35} \cdot \dfrac{7}{8}$

   $= -\dfrac{8 \cdot 3}{5 \cdot 7} \cdot \dfrac{7}{8}$

   $= -\dfrac{\cancel{8}^{1} \cdot 3}{5 \cdot \cancel{7}_{1}} \cdot \dfrac{\cancel{7}^{1}}{\cancel{8}_{1}}$

   $= -\dfrac{3}{5}$

8. $\dfrac{11}{12} + \dfrac{5}{12} = \dfrac{11 + 5}{12} = \dfrac{16}{12} = \dfrac{4 \cdot 4}{4 \cdot 3} = \dfrac{4}{3}$

9. $\dfrac{3}{18} - \dfrac{13}{18} = \dfrac{3 - 13}{18} = \dfrac{-10}{18} = -\dfrac{2 \cdot 5}{2 \cdot 9} = -\dfrac{5}{9}$

10. $25 = 5 \cdot 5$

    $15 = 3 \cdot 5$

    LCD $= 3 \cdot 5 \cdot 5 = 75$

    $\dfrac{3}{25} = \dfrac{3}{25} \cdot \dfrac{3}{3} = \dfrac{9}{75}$

    $\dfrac{2}{15} = \dfrac{2}{15} \cdot \dfrac{5}{5} = \dfrac{10}{75}$

11. $18 = 2 \cdot 3 \cdot 3$

    $63 = 3 \cdot 3 \cdot 7$

    LCD $= 2 \cdot 3 \cdot 3 \cdot 7 = 126$

    $\dfrac{5}{18} = \dfrac{5}{18} \cdot \dfrac{7}{7} = \dfrac{35}{126}$

    $-\dfrac{1}{63} = -\dfrac{1}{63} \cdot \dfrac{2}{2} = -\dfrac{2}{126}$

12. $4 = 2 \cdot 2$

    5 is prime

    LCD $= 2 \cdot 2 \cdot 5 = 20$

    $\dfrac{3}{4} + \dfrac{1}{5} = \dfrac{3}{4} \cdot \dfrac{5}{5} + \dfrac{1}{5} \cdot \dfrac{4}{4}$

    $= \dfrac{15}{20} + \dfrac{4}{20} = \dfrac{15 + 4}{20}$

    $= \dfrac{19}{20}$

13. $20 = 2 \cdot 2 \cdot 5$

    $15 = 3 \cdot 5$

    LCD $= 2 \cdot 2 \cdot 3 \cdot 5 = 60$

    $\dfrac{3}{20} + \dfrac{2}{15} = \dfrac{3}{20} \cdot \dfrac{3}{3} + \dfrac{2}{15} \cdot \dfrac{4}{4}$

    $= \dfrac{9}{60} + \dfrac{8}{60} = \dfrac{9 + 8}{60}$

    $= \dfrac{17}{60}$

14. $14 = 2 \cdot 7$

    $21 = 3 \cdot 7$

    LCD $= 2 \cdot 3 \cdot 7 = 42$

    $\dfrac{5}{14} - \dfrac{11}{21} = \dfrac{5}{14} \cdot \dfrac{3}{3} - \dfrac{11}{21} \cdot \dfrac{2}{2}$

    $= \dfrac{15}{42} - \dfrac{22}{42} = \dfrac{15 - 22}{42}$

    $= -\dfrac{7}{42} = -\dfrac{1 \cdot 7}{6 \cdot 7}$

    $= -\dfrac{1}{6}$

Getting Ready for Chapter 5 Exercises

15. $\dfrac{4}{12} = \dfrac{\cancel{2} \cdot \cancel{2}}{\cancel{2} \cdot \cancel{2} \cdot 3} = \dfrac{1}{3}$

17. $\dfrac{-15}{35} = \dfrac{-1 \cdot 3 \cdot \cancel{5}}{\cancel{5} \cdot 7} = \dfrac{-3}{7} = -\dfrac{3}{7}$

19. $\dfrac{-50}{-10} = \dfrac{\cancel{-1} \cdot 5 \cdot \cancel{5} \cdot \cancel{2}}{\cancel{-1} \cdot \cancel{5} \cdot \cancel{2}} = \dfrac{5}{1} = 5$

21. $\dfrac{3}{4} \cdot \dfrac{20}{9} = \dfrac{\cancel{3}}{\cancel{2} \cdot \cancel{2}} \cdot \dfrac{\cancel{2} \cdot \cancel{2} \cdot 5}{\cancel{3} \cdot 3} = \dfrac{5}{3}$

23. $-\dfrac{5}{6} \cdot \dfrac{18}{5} = \dfrac{-1 \cdot \cancel{5}}{\cancel{2} \cdot \cancel{3}} \cdot \dfrac{\cancel{2} \cdot \cancel{3} \cdot 3}{\cancel{5}} = \dfrac{-3}{1} = -3$

25. $\dfrac{5}{8} \cdot \dfrac{2}{15} = \dfrac{\cancel{5}}{\cancel{2} \cdot 2 \cdot 2} \cdot \dfrac{\cancel{2}}{3 \cdot \cancel{5}} = \dfrac{1}{12}$

27. $\dfrac{5}{2} \div \dfrac{25}{4} = \dfrac{5}{2} \cdot \dfrac{4}{25} = \dfrac{\cancel{5}}{\cancel{2}} \cdot \dfrac{\cancel{2} \cdot 2}{\cancel{5} \cdot 5} = \dfrac{2}{5}$

29. $\dfrac{\frac{-6}{5}}{\frac{8}{15}} = \dfrac{-6}{5} \cdot \dfrac{15}{8} = \dfrac{-1 \cdot \cancel{2} \cdot 3}{\cancel{5}} \cdot \dfrac{3 \cdot \cancel{5}}{\cancel{2} \cdot 2 \cdot 2} = \dfrac{-9}{4} = -\dfrac{9}{4}$

31. $\dfrac{\frac{-9}{2}}{\frac{-3}{4}} = -\dfrac{9}{2} \cdot \dfrac{4}{-3} = \dfrac{\cancel{-1} \cdot \cancel{3} \cdot 3}{\cancel{2}} \cdot \dfrac{\cancel{2} \cdot 2}{\cancel{-1} \cdot \cancel{3}} = \dfrac{6}{1} = 6$

33. $\dfrac{5}{3} + \dfrac{1}{3} = \dfrac{5+1}{3} = \dfrac{6}{3} = \dfrac{2 \cdot \cancel{3}}{\cancel{3}} = \dfrac{2}{1} = 2$

35. $\dfrac{11}{6} - \dfrac{1}{6} = \dfrac{11-1}{6} = \dfrac{10}{6} = \dfrac{\cancel{2} \cdot 5}{\cancel{2} \cdot 3} = \dfrac{5}{3}$

37. $4 = 2 \cdot 2$
3 is prime
The LCD $= 2 \cdot 2 \cdot 3 = 12$

$-\dfrac{3}{4} + \dfrac{1}{3} = -\dfrac{3}{4} \cdot \dfrac{3}{3} + \dfrac{1}{3} \cdot \dfrac{4}{4}$

$= -\dfrac{9}{12} + \dfrac{4}{12}$

$= \dfrac{-9+4}{12}$

$= -\dfrac{5}{12}$

39. $4 = 2 \cdot 2$
$6 = 2 \cdot 3$
LCD $= 2 \cdot 2 \cdot 3 = 12$

$\dfrac{1}{4} + \dfrac{5}{6} = \dfrac{1}{4} \cdot \dfrac{3}{3} + \dfrac{5}{6} \cdot \dfrac{2}{2} = \dfrac{3}{12} + \dfrac{10}{12} = \dfrac{3+10}{12} = \dfrac{13}{12}$

41. $10 = 2 \cdot 5$
$15 = 5 \cdot 3$
LCD $= 2 \cdot 5 \cdot 3 = 30$

$-\dfrac{3}{10} - \dfrac{7}{15} = -\dfrac{3}{10} \cdot \dfrac{3}{3} - \dfrac{7}{15} \cdot \dfrac{2}{2}$

$= -\dfrac{9}{30} - \dfrac{14}{30}$

$= \dfrac{-9-14}{30}$

$= -\dfrac{23}{30}$

43. $18 = 2 \cdot 3 \cdot 3$
$15 = 3 \cdot 5$
LCD $= 2 \cdot 3 \cdot 3 \cdot 5 = 90$

$\dfrac{5}{18} - \dfrac{11}{15} = \dfrac{5}{18} \cdot \dfrac{5}{5} - \dfrac{11}{15} \cdot \dfrac{6}{6}$

$= \dfrac{25}{90} - \dfrac{66}{90}$

$= \dfrac{25-66}{90}$

$= -\dfrac{41}{90}$

45. $8 = 2 \cdot 2 \cdot 2$
$10 = 2 \cdot 5$
LCD $= 2 \cdot 2 \cdot 2 \cdot 5 = 40$

$-\dfrac{7}{8} + \dfrac{3}{10} = -\dfrac{7}{8} \cdot \dfrac{5}{5} + \dfrac{3}{10} \cdot \dfrac{4}{4}$

$= -\dfrac{35}{40} + \dfrac{12}{40}$

$= \dfrac{-35+12}{40}$

$= -\dfrac{23}{40}$

47. $24 = 3 \cdot 2 \cdot 2 \cdot 2$

$20 = \quad\; 2 \cdot 2 \cdot 5$

$\text{LCD} = 3 \cdot 2 \cdot 2 \cdot 2 \cdot 5 = 120$

$$\frac{7}{24} - \frac{3}{20} = \frac{7}{24} \cdot \frac{5}{5} - \frac{3}{20} \cdot \frac{6}{6}$$

$$= \frac{35}{120} - \frac{18}{120}$$

$$= \frac{35 - 18}{120}$$

$$= \frac{17}{120}$$

49. $9 = 3 \cdot 3$

$15 = \quad 3 \cdot 5$

$\text{LCD} = 3 \cdot 3 \cdot 5 = 45$

$$-\frac{7}{9} - \frac{2}{15} = -\frac{7}{9} \cdot \frac{5}{5} - \frac{2}{15} \cdot \frac{3}{3}$$

$$= \frac{-35}{45} - \frac{6}{45}$$

$$= \frac{-35 - 6}{45}$$

$$= -\frac{41}{45}$$

51. $25 = 5 \cdot 5$

$30 = \quad 5 \cdot 2 \cdot 3$

$\text{LCD} = 5 \cdot 5 \cdot 2 \cdot 3 = 150$

$$\frac{-4}{25} - \frac{7}{30} = \frac{-4}{25} \cdot \frac{6}{6} - \frac{7}{30} \cdot \frac{5}{5}$$

$$= \frac{-24}{150} - \frac{35}{150}$$

$$= \frac{-24 - 35}{150}$$

$$= -\frac{59}{150}$$

53. Answers may vary.

Factor each denominator completely into their prime factors. List each factor that is common to the denominators, then list the factors that are not common. The LCD is the product of the listed factors.

Section 5.1

Are You Ready for This Section?

R1. $a \cdot c = 2(-21) = -42$

The factors of -42 that add to -11 (the linear coefficient) are -14 and 3. Thus,

$$2x^2 - 11x - 21 = 2x^2 - 14x + 3x - 21$$

$$= 2x(x - 7) + 3(x - 7)$$

$$= (2x + 3)(x - 7)$$

R2. $\qquad q^2 - 16 = 0$

$(q + 4)(q - 4) = 0$

$q + 4 = 0 \quad \text{or} \quad q - 4 = 0$

$\quad q = -4 \quad \text{or} \quad q = 4$

The solution set is $\{-4,\ 4\}$.

R3. The reciprocal of $\dfrac{5}{2}$ is $\dfrac{2}{5}$ because $\dfrac{5}{2} \cdot \dfrac{2}{5} = 1$.

R4. The domain is the set of all inputs for which the relation (or function) is defined.

Section 5.1 Quick Checks

1. The quotient of two polynomials is called a <u>rational expression</u>.

2. In the expression $\dfrac{x+3}{3x-5}$, we call $x + 3$ the <u>numerator</u>, and $3x - 5$ is called the <u>denominator</u>.

3. False. The domain of rational functions is all real numbers except those that cause the denominator to equal 0.

4. We need to find all values of x that cause the denominator $x + 6$ to equal 0.

$x + 6 = 0$

$\quad\; x = -6$

Thus, the domain of $\dfrac{x-4}{x+6}$ is $\{x | x \neq -6\}$.

5. We need to find all values of x that cause the denominator $z^2 + 3z - 28$ to equal 0.

$$z^2 + 3z - 28 = 0$$
$$(z+7)(x-4) = 0$$
$$z+7 = 0 \quad \text{or} \quad z-4 = 0$$
$$z = -7 \quad \text{or} \qquad z = 4$$

Thus, the domain of $\dfrac{z^2 - 9}{z^2 + 3z - 28}$ is

$\{z \mid z \neq 4, z \neq -7\}$.

6. False. Since x^2 is not a factor of either the numerator or denominator, you can not reduce this rational expression in this manner.

7. $\dfrac{x^2 - 7x + 12}{x^2 + 4x - 21} = \dfrac{(x-3)(x-4)}{(x-3)(x+7)}$

$= \dfrac{\cancel{(x-3)}(x-4)}{\cancel{(x-3)}(x+7)} = \dfrac{x-4}{x+7}$

8. $\dfrac{z^3 - 64}{2z^2 - 3z - 20} = \dfrac{(z-4)\left(z^2 + 4z + 16\right)}{(2z+5)(z-4)}$

$= \dfrac{\cancel{(z-4)}\left(z^2 + 4z + 16\right)}{(2z+5)\cancel{(z-4)}}$

$= \dfrac{z^2 + 4z + 16}{2z+5}$

9. $\dfrac{3w^2 + 13w - 10}{2 - 3w} = \dfrac{(3w-2)(w+5)}{-1(3w-2)}$

$= \dfrac{\cancel{(3w-2)}(w+5)}{-1\cancel{(3w-2)}}$

$= \dfrac{w+5}{-1} = -(w+5)$

10. $\dfrac{p^2 - 9}{p^2 + 5p + 6} \cdot \dfrac{3p^2 - p - 2}{2p - 6}$

$= \dfrac{(p-3)(p+3)}{(p+2)(p+3)} \cdot \dfrac{(3p+2)(p-1)}{2(p-3)}$

$= \dfrac{\cancel{(p-3)}\cancel{(p+3)}}{(p+2)\cancel{(p+3)}} \cdot \dfrac{(3p+2)(p-1)}{2\cancel{(p-3)}}$

$= \dfrac{(3p+2)(p-1)}{2(p+2)}$

11. $\dfrac{2x+8}{2x^2 + 11x + 12} \cdot \dfrac{2x^2 - 3x - 9}{6 - 2x}$

$= \dfrac{2(x+4)}{(2x+3)(x+4)} \cdot \dfrac{(2x+3)(x-3)}{-1 \cdot 2(x-3)}$

$= \dfrac{\cancel{2}\cancel{(x+4)}}{\cancel{(2x+3)}\cancel{(x+4)}} \cdot \dfrac{\cancel{(2x+3)}\cancel{(x-3)}}{-1 \cdot \cancel{2}\cancel{(x-3)}}$

$= \dfrac{1}{-1} = -1$

12. $\dfrac{m^2 + 2mn + n^2}{2m^2 + 3mn + n^2} \cdot \dfrac{2m^2 - 5mn - 3n^2}{3n - m}$

$= \dfrac{(m+n)(m+n)}{(2m+n)(m+n)} \cdot \dfrac{(2m+n)(m-3n)}{-1(m-3n)}$

$= \dfrac{\cancel{(m+n)}(m+n)}{\cancel{(2m+n)}\cancel{(m+n)}} \cdot \dfrac{\cancel{(2m+n)}\cancel{(m-3n)}}{-1\cancel{(m-3n)}}$

$= \dfrac{m+n}{-1} = -(m+n)$

13. $\dfrac{\dfrac{12a^4}{5b^2}}{\dfrac{4a^2}{15b^5}} = \dfrac{12a^4}{5b^2} \cdot \dfrac{15b^5}{4a^2}$

$= \dfrac{\overset{3}{\cancel{12}} \cdot \cancel{a^4} \cdot a^2}{\cancel{5} \cdot \cancel{b^2}} \cdot \dfrac{\overset{3}{\cancel{15}} \cdot \cancel{b^5} \cdot b^3}{4\cancel{a^2}} = 9a^2 b^3$

14. $\dfrac{\dfrac{m^2 - 5m}{m-7}}{\dfrac{2m}{m^2 - 6m - 7}} = \dfrac{m^2 - 5m}{m-7} \cdot \dfrac{m^2 - 6m - 7}{2m}$

$= \dfrac{m(m-5)}{m-7} \cdot \dfrac{(m-7)(m+1)}{2 \cdot m}$

$= \dfrac{\cancel{m}(m-5)}{\cancel{m-7}} \cdot \dfrac{\cancel{(m-7)}(m+1)}{2 \cdot \cancel{m}}$

$= \dfrac{(m-5)(m+1)}{2}$

15. We need to find all values of x that cause the denominator $x^2 + x - 30$ to equal 0.

$$x^2 + x - 30 = 0$$
$$(x+6)(x-5) = 0$$
$$x+6 = 0 \quad \text{or} \quad x-5 = 0$$
$$x = -6 \quad \text{or} \qquad x = 5$$

Thus, the domain of $R(x) = \dfrac{2x}{x^2 + x - 30}$ is

$\{x \mid x \neq -6, x \neq 5\}$.

16. **a.** $R(x) = f(x) \cdot g(x)$

$= \dfrac{x^2 - 4x - 5}{3x - 5} \cdot \dfrac{3x^2 + 4x - 15}{x^2 - 2x - 15}$

$= \dfrac{(x-5)(x+1)}{3x - 5} \cdot \dfrac{(3x-5)(x+3)}{(x-5)(x+3)}$

$= \dfrac{\cancel{(x-5)}(x+1)}{\cancel{3x-5}} \cdot \dfrac{\cancel{(3x-5)}\cancel{(x+3)}}{\cancel{(x-5)}\cancel{(x+3)}}$

$= x + 1$

The domain of $f(x)$ is $\left\{ x \middle| x \neq \dfrac{5}{3} \right\}$.

The domain of $g(x)$ is $\{x \mid x \neq -3, x \neq 5\}$.

Therefore, the domain of $R(x)$ is

$\left\{ x \middle| x \neq -3, x \neq \dfrac{5}{3}, x \neq 5 \right\}$.

b. $H(x) = \dfrac{f(x)}{h(x)} = \dfrac{\dfrac{x^2 - 4x - 5}{3x - 5}}{\dfrac{4x^2 + 7x + 3}{9x^2 - 15x}}$

$= \dfrac{x^2 - 4x - 5}{3x - 5} \cdot \dfrac{9x^2 - 15x}{4x^2 + 7x + 3}$

$= \dfrac{(x-5)(x+1)}{3x - 5} \cdot \dfrac{3x(3x-5)}{(4x+3)(x+1)}$

$= \dfrac{(x-5)\cancel{(x+1)}}{\cancel{3x-5}} \cdot \dfrac{3x\cancel{(3x-5)}}{(4x+3)\cancel{(x+1)}}$

$= \dfrac{3x(x-5)}{4x+3}$

The domain of $f(x)$ is $\left\{ x \middle| x \neq \dfrac{5}{3} \right\}$.

The domain of $h(x)$ is $\left\{ x \middle| x \neq 0, x \neq \dfrac{5}{3} \right\}$.

Because the denominator of $H(x)$ cannot equal 0, we must exclude those values of x such that $4x^2 + 7x + 3 = 0$. These values are $x = -\dfrac{3}{4}$ and $x = -1$. Therefore, the

domain of $H(x)$ is

$\left\{ x \middle| x \neq -1, x \neq -\dfrac{3}{4}, x \neq 0, x \neq \dfrac{5}{3} \right\}$.

5.1 Exercises

17. We need to find all values of x that cause the denominator $x + 5$ to equal 0.

$x + 5 = 0$

$x = -5$

Thus, the domain of $\dfrac{3}{x+5}$ is $\{x \mid x \neq -5\}$.

19. We need to find all values of x that cause the denominator $x^2 - 6x - 16$ to equal 0.

$x^2 - 6x - 16 = 0$

$(x+2)(x-8) = 0$

$x + 2 = 0$ or $x - 8 = 0$

$x = -2$ or $x = 8$

Thus, the domain of $\dfrac{x-1}{x^2 - 6x - 16}$ is

$\{x \mid x \neq -2, x \neq 8\}$.

21. We need to find all values of p that cause the denominator $2p^2 + p - 10$ to equal 0.

$2p^2 + p - 10 = 0$

$(2p+5)(p-2) = 0$

$2p + 5 = 0$ or $p - 2 = 0$

$p = -\dfrac{5}{2}$ or $p = 2$

Thus, the domain of $\dfrac{p^2 - 4}{2p^2 + p - 10}$ is

$\left\{ p \middle| p \neq -\dfrac{5}{2}, p \neq 2 \right\}$.

23. We need to find all values of x that cause the denominator $x^2 + 1$ to equal 0. However, if x is a real number, then $x^2 + 1$ can never equal 0.

Thus, the domain of $\dfrac{x+1}{x^2 + 1}$ is

$\{x \mid x \text{ is any real number}\}$.

25. We need to find all values of x that cause the denominator $(x-1)^2$ to equal 0.

$$(x-1)^2 = 0$$
$$x-1 = 0$$
$$x = 1$$

Thus, the domain of $\dfrac{3x-2}{(x-1)^2}$ is $\{x \mid x \neq 1\}$.

27. $\dfrac{2x+8}{x^2-16} = \dfrac{2\cancel{(x+4)}}{\cancel{(x+4)}(x-4)} = \dfrac{2}{x-4}$

29. $\dfrac{p^2+4p+3}{p+1} = \dfrac{\cancel{(p+1)}(p+3)}{\cancel{p+1}} = \dfrac{p+3}{1} = p+3$

31. $\dfrac{5x+25}{x^3+5x^2} = \dfrac{5\cancel{(x+5)}}{x^2\cancel{(x+5)}} = \dfrac{5}{x^2}$

33. $\dfrac{q^2-3q-18}{q^2-8q+12} = \dfrac{(q+3)\cancel{(q-6)}}{(q-2)\cancel{(q-6)}} = \dfrac{q+3}{q-2}$

35. $\dfrac{2y^2-3y-20}{2y^2+15y+25} = \dfrac{\cancel{(2y+5)}(y-4)}{\cancel{(2y+5)}(y+5)} = \dfrac{y-4}{y+5}$

37. $\dfrac{9-x^2}{x^2+2x-15} = \dfrac{-1\cancel{(x-3)}(x+3)}{\cancel{(x-3)}(x+5)}$

$$= \dfrac{-1(x+3)}{x+5} = -\dfrac{x+3}{x+5}$$

39. $\dfrac{x^3+2x^2-8x}{2x^4-32x^2} = \dfrac{\cancel{x}(x-2)\cancel{(x+4)}}{\underset{2x}{\cancel{2x^2}}(x-4)\cancel{(x+4)}}$

$$= \dfrac{x-2}{2x(x-4)}$$

41. $\dfrac{x^2-xy-6y^2}{x^2-4y^2} = \dfrac{(x-3y)\cancel{(x+2y)}}{(x-2y)\cancel{(x+2y)}} = \dfrac{x-3y}{x-2y}$

43. $\dfrac{x^3-5x^2+3x-15}{x^2-10x+25} = \dfrac{\cancel{(x-5)}(x^2+3)}{\cancel{(x-5)}(x-5)} = \dfrac{x^2+3}{x-5}$

45. $\dfrac{x^3+8}{x^2-5x-14} = \dfrac{\cancel{(x+2)}(x^2-2x+4)}{\cancel{(x+2)}(x-7)}$

$$= \dfrac{x^2-2x+4}{x-7}$$

47. $\dfrac{3x}{x^2-x-12} \cdot \dfrac{x-4}{12x^2} = \dfrac{3x}{(x-4)(x+3)} \cdot \dfrac{x-4}{3x \cdot 4x}$

$$= \dfrac{3x}{\cancel{(x-4)}(x+3)} \cdot \dfrac{\cancel{x-4}}{3x \cdot 4x}$$

$$= \dfrac{1}{4x(x+3)}$$

49. $\dfrac{2x^2-x-6}{x^2+3x-4} \cdot \dfrac{x^2-x-20}{2x^2-7x-15}$

$$= \dfrac{(2x+3)(x-2)}{(x-1)(x+4)} \cdot \dfrac{(x-5)(x+4)}{(2x+3)(x-5)}$$

$$= \dfrac{\cancel{(2x+3)}(x-2)}{(x-1)\cancel{(x+4)}} \cdot \dfrac{\cancel{(x-5)}\cancel{(x+4)}}{\cancel{(2x+3)}\cancel{(x-5)}} = \dfrac{x-2}{x-1}$$

51. $\dfrac{x^2-9}{x^2-25} \cdot \dfrac{x^2-2x-15}{x^2+4x-21}$

$$= \dfrac{(x-3)(x+3)}{(x+5)(x-5)} \cdot \dfrac{(x-5)(x+3)}{(x-3)(x+7)}$$

$$= \dfrac{\cancel{(x-3)}(x+3)}{(x+5)\cancel{(x-5)}} \cdot \dfrac{\cancel{(x-5)}(x+3)}{\cancel{(x-3)}(x+7)}$$

$$= \dfrac{(x+3)^2}{(x+5)(x+7)}$$

53. $\dfrac{2q^2-5q-3}{3q^2+19q+6} \cdot \dfrac{3q^2+7q+2}{3-q}$

$$= \dfrac{(2q+1)(q-3)}{(3q+1)(q+6)} \cdot \dfrac{(3q+1)(q+2)}{-1(q-3)}$$

$$= \dfrac{(2q+1)\cancel{(q-3)}}{\cancel{(3q+1)}(q+6)} \cdot \dfrac{\cancel{(3q+1)}(q+2)}{-1\cancel{(q-3)}}$$

$$= \dfrac{(2q+1)(q+2)}{-1(q+6)} \quad \text{or} \quad -\dfrac{(2q+1)(q+2)}{q+6}$$

55. $\dfrac{x^2-5x+6}{x^2+2x-8}\cdot(x+4)=\dfrac{(x-2)(x-3)}{(x-2)(x+4)}\cdot(x+4)$

$\qquad\qquad\qquad = \dfrac{\cancel{(x-2)}\,(x-3)}{\cancel{(x-2)}\,\cancel{(x+4)}}\cdot\cancel{(x+4)}$

$\qquad\qquad\qquad = x-3$

57. $\dfrac{m^2-n^2}{5m-5n}\cdot\dfrac{10m+5n}{2m^2+3mn+n^2}$

$\quad = \dfrac{(m-n)(m+n)}{5(m-n)}\cdot\dfrac{5(2m+n)}{(2m+n)(m+n)}$

$\quad = \dfrac{\cancel{(m-n)}\,\cancel{(m+n)}}{\cancel{5}\,\cancel{(m-n)}}\cdot\dfrac{\cancel{5}\,\cancel{(2m+n)}}{\cancel{(2m+n)}\,\cancel{(m+n)}}=1$

59. $\dfrac{\dfrac{x+3}{2x-8}}{\dfrac{4x}{9}}=\dfrac{x+3}{2x-8}\cdot\dfrac{9}{4x}=\dfrac{x+3}{2(x-4)}\cdot\dfrac{9}{4x}=\dfrac{9(x+3)}{8x(x-4)}$

61. $\dfrac{\dfrac{4a}{b^2}}{\dfrac{2a^2}{b}}=\dfrac{4a}{b^2}\cdot\dfrac{b}{2a^2}=\dfrac{\overset{2}{\cancel{4}}\,\cancel{a}}{\cancel{b}\cdot b}\cdot\dfrac{\cancel{b}}{\cancel{2}\,\cancel{a}\cdot a}=\dfrac{2}{ab}$

63. $\dfrac{\dfrac{p^2-4p-5}{2p^2-3p-2}}{\dfrac{p^2+p}{p^2+p-6}}$

$\quad = \dfrac{p^2-4p-5}{2p^2-3p-2}\cdot\dfrac{p^2+p-6}{p^2+p}$

$\quad = \dfrac{(p-5)(p+1)}{(2p+1)(p-2)}\cdot\dfrac{(p-2)(p+3)}{p(p+1)}$

$\quad = \dfrac{(p-5)\,\cancel{(p+1)}}{(2p+1)\,\cancel{(p-2)}}\cdot\dfrac{\cancel{(p-2)}\,(p+3)}{p\,\cancel{(p+1)}}$

$\quad = \dfrac{(p-5)(p+3)}{p(2p+1)}$

65. $\dfrac{\dfrac{x^3-1}{x^2-1}}{\dfrac{3x^2+3x+3}{x^2+3x+1}}$

$\quad = \dfrac{x^3-1}{x^2-1}\cdot\dfrac{x^2+3x+1}{3x^2+3x+3}$

$\quad = \dfrac{(x-1)\left(x^2+x+1\right)}{(x-1)(x+1)}\cdot\dfrac{x^2+3x+1}{3\left(x^2+x+1\right)}$

$\quad = \dfrac{\cancel{(x-1)}\,\cancel{\left(x^2+x+1\right)}}{\cancel{(x-1)}\,(x+1)}\cdot\dfrac{x^2+3x+1}{3\,\cancel{\left(x^2+x+1\right)}}$

$\quad = \dfrac{x^2+3x+1}{3(x+1)}$

67. We need to find all values of x that cause the denominator $x-1$ to equal 0.

$\qquad x-1=0$

$\qquad\quad x=1$

Thus, the domain of $R(x)=\dfrac{2}{x-1}$ is $\{x\,|\,x\neq 1\}$.

69. We need to find all values of x that cause the denominator $(2x+1)(x-4)$ to equal 0.

$\qquad (2x+1)(x-4)=0$

$\qquad 2x+1=0\quad\text{or}\quad x-4=0$

$\qquad\qquad x=-\dfrac{1}{2}\;\text{or}\qquad x=4$

Thus, the domain of $R(x)=\dfrac{x-2}{(2x+1)(x-4)}$ is

$\left\{x\,\middle|\,x\neq-\dfrac{1}{2},x\neq 4\right\}.$

71. We need to find all values of x that cause the denominator x^2+6x+5 to equal 0.

$\qquad x^2+6x+5=0$

$\qquad (x+1)(x+5)=0$

$\qquad x+1=0\quad\text{or}\quad x+5=0$

$\qquad\quad x=-1\;\text{or}\qquad x=-5$

Thus, the domain of $R(x)=\dfrac{x+9}{x^2+6x+5}$ is

$\{x\,|\,x\neq-5,x\neq-1\}.$

73. We need to find all values of x that cause the denominator $2x^2 - 9x + 10$ to equal 0.

$$2x^2 - 9x + 10 = 0$$
$$(2x-5)(x-2) = 0$$
$$2x-5 = 0 \text{ or } x-2 = 0$$
$$x = \frac{5}{2} \text{ or } \quad x = 2$$

Thus, the domain of $R(x) = \dfrac{x-2}{2x^2 - 9x + 10}$ is

$$\left\{ x \,\middle|\, x \neq 2, x \neq \frac{5}{2} \right\}.$$

75. We need to find all values of x that cause the denominator $x^2 + 1$ to equal 0. However, $x^2 + 1$ cannot equal 0 if x is a real number, so the domain of $R(x) = \dfrac{x-1}{x^2+1}$ is $\{x \mid x$ is any real number$\}$.

77. a. $R(x) = f(x) \cdot g(x)$

$$= \frac{x^2 - 2x - 15}{x+6} \cdot \frac{x^2 + 5x - 6}{2x^2 - 7x - 15}$$
$$= \frac{(x-5)(x+3)}{x+6} \cdot \frac{(x-1)(x+6)}{(2x+3)(x-5)}$$
$$= \frac{\cancel{(x-5)}(x+3)}{\cancel{x+6}} \cdot \frac{(x-1)\cancel{(x+6)}}{(2x+3)\cancel{(x-5)}}$$
$$= \frac{(x+3)(x-1)}{2x+3}$$

The domain of $f(x)$ is $\{x \mid x \neq -6\}$. The domain of $g(x)$ is $\left\{ x \,\middle|\, x \neq -\frac{3}{2}, x \neq 5 \right\}$.

Therefore, the domain of $R(x)$ is

$$\left\{ x \,\middle|\, x \neq -6, x \neq -\frac{3}{2}, x \neq 5 \right\}.$$

b. $R(x) = \dfrac{f(x)}{h(x)} = \dfrac{\dfrac{x^2 - 2x - 15}{x+6}}{\dfrac{x+3}{3x^2 + 17x - 6}}$

$$= \frac{x^2 - 2x - 15}{x+6} \cdot \frac{3x^2 + 17x - 6}{x+3}$$
$$= \frac{(x-5)(x+3)}{x+6} \cdot \frac{(3x-1)(x+6)}{x+3}$$
$$= \frac{(x-5)\cancel{(x+3)}}{\cancel{x+6}} \cdot \frac{(3x-1)\cancel{(x+6)}}{\cancel{x+3}}$$
$$= (x-5)(3x-1)$$

The domain of $f(x)$ is $\{x \mid x \neq -6\}$. The domain of $h(x)$ is $\left\{ x \,\middle|\, x \neq -6, x \neq \frac{1}{3} \right\}$.

Because the denominator of $R(x)$ cannot equal 0, we must exclude those values of x such that the numerator of $h(x)$ is 0. That is, we must exclude the values of x such that $x+3 = 0$. The only such value is $x = -3$. Therefore, the domain of $R(x)$ is

$$\left\{ x \,\middle|\, x \neq -6, x \neq -3, x \neq \frac{1}{3} \right\}.$$

79. a. $R(x) = (f \cdot g)(x)$

$$= \frac{3x^2 - x - 10}{x^3 - 1} \cdot \frac{x^2 + 5x - 6}{2x^2 + 3x - 14}$$
$$= \frac{(3x+5)(x-2)}{(x-1)(x^2+x+1)} \cdot \frac{(x-1)(x+6)}{(2x+7)(x-2)}$$
$$= \frac{(3x+5)\cancel{(x-2)}}{\cancel{(x-1)}(x^2+x+1)} \cdot \frac{\cancel{(x-1)}(x+6)}{(2x+7)\cancel{(x-2)}}$$
$$= \frac{(3x+5)(x+6)}{(x^2+x+1)(2x+7)}$$

The domain of $f(x)$ is $\{x \mid x \neq 1\}$. (Note that $x^2 + x + 1 = 0$ has no real solution.) The domain of $g(x)$ is $\left\{ x \,\middle|\, x \neq -\frac{7}{2}, x \neq 2 \right\}$.

Therefore, the domain of $R(x)$ is

$$\left\{ x \,\middle|\, x \neq -\frac{7}{2}, x \neq 1, x \neq 2 \right\}.$$

b. $R(x) = \left(\dfrac{f}{h}\right)(x) = \dfrac{\dfrac{3x^2 - x - 10}{x^3 - 1}}{\dfrac{3x^2 + 8x + 5}{x^2 - 1}}$

$= \dfrac{3x^2 - x - 10}{x^3 - 1} \cdot \dfrac{x^2 - 1}{3x^2 + 8x + 5}$

$= \dfrac{(3x+5)(x-2)}{(x-1)(x^2 + x + 1)} \cdot \dfrac{(x-1)(x+1)}{(3x+5)(x+1)}$

$= \dfrac{\cancel{(3x+5)}(x-2)}{\cancel{(x-1)}(x^2 + x + 1)} \cdot \dfrac{\cancel{(x-1)}\cancel{(x+1)}}{\cancel{(3x+5)}\cancel{(x+1)}}$

$= \dfrac{x-2}{x^2 + x + 1}$

The domain of $f(x)$ is $\{x \mid x \neq 1\}$. (Note that $x^2 + x + 1 = 0$ has no real solution.) The domain of $h(x)$ is $\{x \mid x \neq -1, x \neq 1\}$. Because the denominator of $R(x)$ cannot equal 0, we must exclude those values of x such that the numerator of $h(x)$ is 0. That is, we must exclude the values of x such that $3x^2 + 8x + 5 = 0$. These values are $x = -\dfrac{5}{3}$ and $x = -1$. Therefore, the domain of $R(x)$ is $\left\{x \mid x \neq -\dfrac{5}{3}, x \neq -1, x \neq 1\right\}$.

81. $\dfrac{z^3 + 8}{z^2 - 3z - 10} \cdot \dfrac{z^2 - 2z - 15}{2z^2 - 4z + 8}$

$= \dfrac{(z+2)(z^2 - 2z + 4)}{(z-5)(z+2)} \cdot \dfrac{(z-5)(z+3)}{2(z^2 - 2z + 4)}$

$= \dfrac{\cancel{(z+2)}\cancel{(z^2 - 2z + 4)}}{\cancel{(z-5)}\cancel{(z+2)}} \cdot \dfrac{\cancel{(z-5)}(z+3)}{2\cancel{(z^2 - 2z + 4)}}$

$= \dfrac{z+3}{2}$

83. $\dfrac{\dfrac{m^2 - 4n^2}{m^3 - n^3}}{\dfrac{2m + 4n}{m^2 + mn + n^2}}$

$= \dfrac{m^2 - 4n^2}{m^3 - n^3} \cdot \dfrac{m^2 + mn + n^2}{2m + 4n}$

$= \dfrac{(m-2n)(m+2n)}{(m-n)(m^2 + mn + n^2)} \cdot \dfrac{m^2 + mn + n^2}{2(m+2n)}$

$= \dfrac{(m-2n)\cancel{(m+2n)}}{(m-n)\cancel{(m^2 + mn + n^2)}} \cdot \dfrac{\cancel{m^2 + mn + n^2}}{2\cancel{(m+2n)}}$

$= \dfrac{m-2n}{2(m-n)}$

85. $\dfrac{4w + 8}{w^2 - 4w} \cdot \dfrac{w^2 - 3w - 4}{w^2 + 3w + 2}$

$= \dfrac{4(w+2)}{w(w-4)} \cdot \dfrac{(w-4)(w+1)}{(w+1)(w+2)}$

$= \dfrac{4\cancel{(w+2)}}{w\cancel{(w-4)}} \cdot \dfrac{\cancel{(w-4)}\cancel{(w+1)}}{\cancel{(w+1)}\cancel{(w+2)}}$

$= \dfrac{4}{w}$

87. $\dfrac{\dfrac{2x - 6}{x^2 + x}}{\dfrac{x^2 - 4x + 3}{x^2}} \cdot \dfrac{x^2 + 3x + 2}{x^2 + x}$

$= \dfrac{2x - 6}{x^2 + x} \cdot \dfrac{x^2}{x^2 - 4x + 3} \cdot \dfrac{x^2 + 3x + 2}{x^2 + x}$

$= \dfrac{2\cancel{(x-3)}}{\cancel{x}(x+1)} \cdot \dfrac{\cancel{x} \cdot x}{\cancel{(x-3)}(x-1)} \cdot \dfrac{(x+2)\cancel{(x+1)}}{\cancel{x}(x+1)}$

$= \dfrac{2(x+2)}{(x-1)(x+1)}$

89. Answers will vary. Any rational expression with $x - 3$ as a factor of the denominator will be undefined at $x = 3$. One possible expression is $\dfrac{x}{x - 3}$.

91. Answers will vary. Any rational expression with both $x+4$ and $x-5$ as factors of the denominator will be undefined at $x=-4$ and $x=5$. One possible rational expression

$$\frac{7}{(x+4)(x-5)} = \frac{7}{x^2-x-20}.$$

93. The rational function will have both $x+2$ and $x-1$ as factors of the denominator. Consider the

function $R(x) = \dfrac{k}{(x+2)(x-1)} = \dfrac{k}{x^2+x-2}$,

where k is a real number. To determine the value of k, we solve the following for k:

$$R(-3) = 1$$

$$\frac{k}{(-3)^2 + (-3) - 2} = 1$$

$$\frac{k}{4} = 1$$

$$k = 4$$

Thus, a rational function R that is undefined at $x=-2$ and $x=1$ such that $R(-3)=1$ will be

$$R(x) = \frac{4}{x^2+x-2}.$$

95. a. $g(0) = \dfrac{3.99\times10^{14}}{\left(6.374\times10^6 + 0\right)^2} \approx 9.8208$ m/sec^2

b. $g(1600) = \dfrac{3.99\times10^{14}}{\left(6.374\times10^6 + 1600\right)^2}$

≈ 9.8159 m/sec^2

c. $g(8848) = \dfrac{3.99\times10^{14}}{\left(6.374\times10^6 + 8848\right)^2}$

≈ 9.7936 m/sec^2

97. a. We need to find all values of x that cause the denominator $x-2$ to equal 0:

$$x - 2 = 0$$
$$x = 2$$

The domain of $f(x) = \dfrac{1}{x-2}$ is $\{x \mid x \neq 2\}$.

b. $f(x) = \dfrac{1}{x-2}$

x	3	2.5	2.1	2.01	2.001	2.0001
$f(x)$	1	2	10	100	1000	10,000

As x approaches 2, but remains greater than 2, f gets larger in the positive direction (i.e. f approaches ∞).

c.

x	1	1.5	1.9	1.99	1.999	1.9999
$f(x)$	-1	-2	-10	-100	-1000	$-10,000$

As x approaches 2, but remains less than 2, f gets larger in the negative direction (i.e. f approaches $-\infty$).

d. As x approaches 2, but remains greater than 2, f gets larger in the positive direction (i.e. f approaches ∞). As x approaches 2, but remains less than 2, f gets larger in the negative direction (i.e. f approaches $-\infty$). These results are the same as those in parts (b) and (c).

99. $\dfrac{x^{-1}}{x+1} = \dfrac{\frac{1}{x}}{x+1} = \dfrac{1}{x} \cdot \dfrac{1}{x+1} = \dfrac{1}{x(x+1)}$ or $\dfrac{1}{x^2+x}$

101. Answers will vary. One possibility follows: A rational expression is in lowest terms if the numerator and denominator contain no common factors.

103. The expression $\dfrac{\sqrt{x}}{x+1}$ is not a rational expression because the numerator is not a polynomial.

105. For $f(x) = 2x-3$, the slope is 2 and the y-intercept is -3. Begin at $(0,-3)$ and move to the right 1 unit and up 2 units to find the point $(1,-1)$.

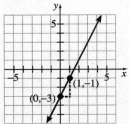

　　　　　Copyright © 2014 Pearson Education, Inc.

107. For $F(x) = -4x + 8$, the slope is -4 and the y-intercept is 8. Begin at $(0,8)$ and move to the right 1 unit and down 4 units to find the point $(1,4)$.

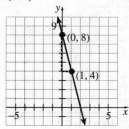

109. $h(x) = x^2 - 4$

Let $x = -3, -2, -1, 0, 1, 2,$ and 3.

$h(-3) = (-3)^2 - 4 = 9 - 4 = 5$

$h(-2) = (-2)^2 - 4 = 4 - 4 = 0$

$h(-1) = (-1)^2 - 4 = 1 - 4 = -3$

$h(0) = 0^2 - 4 = 0 - 4 = -4$

$h(1) = 1^2 - 4 = 1 - 4 = -3$

$h(2) = 2^2 - 4 = 4 - 4 = 0$

$h(3) = 3^2 - 4 = 9 - 4 = 5$

Thus, the points $(-3,5)$, $(-2,0)$, $(-1,-3)$, $(0,-4)$, $(1,-3)$, $(2,0)$, and $(3,5)$ are on the graph. We plot these points and connect them with a smooth curve.

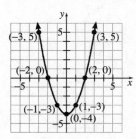

111. Answers will vary.

Section 5.2

Are You Ready for This Section?

R1. $\dfrac{1}{6} - \dfrac{5}{8} = \dfrac{4}{24} - \dfrac{15}{24} = \dfrac{4-15}{24} = -\dfrac{11}{24}$

R2. a. The additive inverse of 5 is -5 because $5 + (-5) = 0$.

    b. The additive inverse of $x - 2$ is $-(x-2) = -x + 2 = 2 - x$ because $(x-2) + (2-x) = 0$.

Section 5.2 Quick Checks

1. $\dfrac{x^2 - 3x - 1}{x - 2} + \dfrac{x^2 - 2x + 3}{x - 2}$

$= \dfrac{x^2 - 3x - 1 + \left(x^2 - 2x + 3\right)}{x - 2}$

$= \dfrac{2x^2 - 5x + 2}{x - 2}$

$= \dfrac{(2x-1)(x-2)}{x-2}$

$= \dfrac{(2x-1)\,\cancel{(x-2)}}{\cancel{x-2}}$

$= 2x - 1$

2. $\dfrac{4x+3}{x+5} - \dfrac{x-6}{x+5} = \dfrac{4x+3-(x-6)}{x+5}$

$= \dfrac{4x+3-x+6}{x+5}$

$= \dfrac{3x+9}{x+5}$ or $\dfrac{3(x+3)}{x+5}$

3. $\dfrac{4x}{x-5} + \dfrac{3}{5-x} = \dfrac{4x}{x-5} + \dfrac{3}{-1(x-5)}$

$= \dfrac{4x}{x-5} + \dfrac{-3}{x-5} = \dfrac{4x-3}{x-5}$

4. The <u>least common denominator</u> is the smallest polynomial that is a multiple of each denominator in the rational expressions to be added or subtracted.

5. $8x^2 y = 2^3 \cdot x^2 \cdot y$

$12xy^3 = 2^2 \cdot 3 \cdot x \cdot y^3$

$\text{LCD} = 2^3 \cdot 3 \cdot x^2 \cdot y^3 = 24x^2 y^3$

6. $x^2 - 5x - 14 = (x+2)(x-7)$

$x^2 + 4x + 4 = (x+2)^2$

$\text{LCD} = (x+2)^2 (x-7)$

7. $10a = 2 \cdot 5 \cdot a$

$15a^2 = 3 \cdot 5 \cdot a^2$

$\text{LCD} = 2 \cdot 3 \cdot 5 \cdot a^2 = 30a^2$

$$\frac{3}{10a} + \frac{4}{15a^2} = \frac{3}{10a} \cdot \frac{3a}{3a} + \frac{4}{15a^2} \cdot \frac{2}{2}$$

$$= \frac{9a}{30a^2} + \frac{8}{30a^2} = \frac{9a+8}{30a^2}$$

8. $8y = 2^3 \cdot y$

$24y = 2^3 \cdot 3 \cdot y$

$\text{LCD} = 2^3 \cdot 3 \cdot y = 24y$

$$\frac{3}{8y} - \frac{13}{24y} = \frac{3}{8y} \cdot \frac{3}{3} - \frac{13}{24y}$$

$$= \frac{9}{24y} - \frac{13}{24y}$$

$$= \frac{9-13}{24y}$$

$$= \frac{-4}{24y}$$

$$= \frac{-1 \cdot \cancel{4}}{\cancel{4} \cdot 6y} = -\frac{1}{6y}$$

9. $x-1$

$x+2$

$\text{LCD} = (x-1)(x+2)$

$$\frac{3x}{x-1} + \frac{x+5}{x+2} = \frac{3x}{x-1} \cdot \frac{x+2}{x+2} + \frac{x+5}{x+2} \cdot \frac{x-1}{x-1}$$

$$= \frac{3x^2+6x}{(x-1)(x+2)} + \frac{x^2+4x-5}{(x-1)(x+2)}$$

$$= \frac{3x^2+6x+\left(x^2+4x-5\right)}{(x-1)(x+2)}$$

$$= \frac{4x^2+10x-5}{(x-1)(x+2)}$$

10. $2x^2 + 7x + 6 = (2x+3)(x+2)$

$x^2 + 6x + 8 = (x+2)(x+4)$

$\text{LCD} = (2x+3)(x+2)(x+4)$

$$\frac{x-1}{2x^2+7x+6} + \frac{x-1}{x^2+6x+8}$$

$$= \frac{x-1}{(2x+3)(x+2)} + \frac{x-1}{(x+2)(x+4)}$$

$$= \frac{x-1}{(2x+3)(x+2)} \cdot \frac{x+4}{x+4}$$

$$\qquad\qquad + \frac{x-1}{(x+2)(x+4)} \cdot \frac{2x+3}{2x+3}$$

$$= \frac{x^2+3x-4}{(2x+3)(x+2)(x+4)}$$

$$\qquad\qquad + \frac{2x^2+x-3}{(2x+3)(x+2)(x+4)}$$

$$= \frac{x^2+3x-4+\left(2x^2+x-3\right)}{(2x+3)(x+2)(x+4)}$$

$$= \frac{3x^2+4x-7}{(2x+3)(x+2)(x+4)}$$

$$= \frac{(3x+7)(x-1)}{(2x+3)(x+2)(x+4)}$$

11. $2x^2 + x - 6 = (2x-3)(x+2)$

$x^2 + 4x + 4 = (x+2)^2$

$\text{LCD} = (2x-3)(x+2)^2$

$$\frac{3x+4}{2x^2+x-6} - \frac{x-1}{x^2+4x+4}$$

$$= \frac{3x+4}{(2x-3)(x+2)} - \frac{x-1}{(x+2)^2}$$

$$= \frac{3x+4}{(2x-3)(x+2)} \cdot \frac{x+2}{x+2} - \frac{x-1}{(x+2)^2} \cdot \frac{2x-3}{2x-3}$$

$$= \frac{3x^2+10x+8}{(2x-3)(x+2)^2} - \frac{2x^2-5x+3}{(2x-3)(x+2)^2}$$

$$= \frac{3x^2+10x+8-\left(2x^2-5x+3\right)}{(2x-3)(x+2)^2}$$

$$= \frac{3x^2+10x+8-2x^2+5x-3}{(2x-3)(x+2)^2}$$

$$= \frac{x^2+15x+5}{(2x-3)(x+2)^2}$$

12. $x^2 - 4 = (x-2)(x+2)$
$x-2$
$x+2$
$\text{LCD} = (x-2)(x+2)$

$$\frac{4}{x^2-4} - \frac{x+3}{x-2} + \frac{x+3}{x+2}$$

$$= \frac{4}{(x-2)(x+2)} - \frac{x+3}{x-2} + \frac{x+3}{x+2}$$

$$= \frac{4}{(x-2)(x+2)} - \frac{x+3}{x-2} \cdot \frac{x+2}{x+2} + \frac{x+3}{x+2} \cdot \frac{x-2}{x-2}$$

$$= \frac{4}{(x-2)(x+2)} - \frac{x^2+5x+6}{(x-2)(x+2)}$$
$$\qquad\qquad + \frac{x^2+x-6}{(x-2)(x+2)}$$

$$= \frac{4 - (x^2+5x+6) + x^2+x-6}{(x-2)(x+2)}$$

$$= \frac{4 - x^2 - 5x - 6 + x^2 + x - 6}{(x-2)(x+2)}$$

$$= \frac{-4x-8}{(x-2)(x+2)}$$

$$= \frac{-4(x+2)}{(x-2)(x+2)}$$

$$= \frac{-4}{x-2}$$

5.2 Exercises

13. $\dfrac{3x}{x+1} + \dfrac{5}{x+1} = \dfrac{3x+5}{x+1}$

15. $\dfrac{2x}{2x+5} - \dfrac{1}{2x+5} = \dfrac{2x-1}{2x+5}$

17. $\dfrac{2x}{2x^2-7x-15} + \dfrac{3}{2x^2-7x-15}$
$$= \frac{2x+3}{2x^2-7x-15} = \frac{2x+3}{(2x+3)(x-5)} = \frac{1}{x-5}$$

19. $\dfrac{2x^2-5x+7}{x^2-2x-15} - \dfrac{x^2+3x-8}{x^2-2x-15}$

$$= \frac{2x^2-5x+7 - (x^2+3x-8)}{x^2-2x-15}$$

$$= \frac{2x^2-5x+7 - x^2-3x+8}{x^2-2x-15}$$

$$= \frac{x^2-8x+15}{x^2-2x-15} = \frac{(x-5)(x-3)}{(x-5)(x+3)} = \frac{x-3}{x+3}$$

21. $\dfrac{3x}{x-5} + \dfrac{1}{5-x} = \dfrac{3x}{x-5} + \dfrac{1}{-1(x-5)}$

$$= \frac{3x}{x-5} + \frac{-1}{x-5} = \frac{3x-1}{x-5}$$

23. $\dfrac{2x^2-4x-1}{x-3} - \dfrac{x^2-6x+4}{3-x}$

$$= \frac{2x^2-4x-1}{x-3} - \frac{x^2-6x+4}{-1(x-3)}$$

$$= \frac{2x^2-4x-1}{x-3} + \frac{x^2-6x+4}{x-3}$$

$$= \frac{2x^2-4x-1+x^2-6x+4}{x-3}$$

$$= \frac{3x^2-10x+3}{x-3}$$

$$= \frac{(3x-1)(x-3)}{x-3}$$

$$= 3x-1$$

25. $4x^3 = 2^2 \cdot x^3$
$8x = 2^3 \cdot x$
$\text{LCD} = 2^3 \cdot x^3 = 8x^3$

27. $15xy^2 = 3 \cdot 5 \cdot x \cdot y^2$
$18x^3 y = 2 \cdot 3^2 \cdot x^3 \cdot y$
$\text{LCD} = 3^2 \cdot x^3 \cdot y^2 \cdot 2 \cdot 5 = 90x^3 y^2$

29. $x-4$
$x+2$
$\text{LCD} = (x-4)(x+2)$

31. $x^2 - x - 12 = (x-4)(x+3)$
$x^2 - 9x + 20 = (x-5)(x-4)$
$\text{LCD} = (x-4)(x+3)(x-5)$

33. $2p^2 + 3p - 2 = (2p - 1)(p + 2)$

$p^3 + 2p^2 = p^2(p + 2)$

$\text{LCD} = p^2(p + 2)(2p - 1)$

35. $4x^2 = 2^2 \cdot x^2$

$8x = 2^3 \cdot x$

$\text{LCD} = 2^3 \cdot x^2 = 8x^2$

$$\frac{3}{4x^2} + \frac{5}{8x} = \frac{3}{4x^2} \cdot \frac{2}{2} + \frac{5}{8x} \cdot \frac{x}{x}$$

$$= \frac{6}{8x^2} + \frac{5x}{8x^2} = \frac{6 + 5x}{8x^2} \text{ or } \frac{5x + 6}{8x^2}$$

37. $12a^2b = 2^2 \cdot 3 \cdot a^2 \cdot b$

$15ab^2 = 3 \cdot 5 \cdot a \cdot b^2$

$\text{LCD} = 3 \cdot a^2 \cdot b^2 \cdot 2^2 \cdot 5 = 60a^2b^2$

$$\frac{5}{12a^2b} - \frac{4}{15ab^2} = \frac{5}{12a^2b} \cdot \frac{5b}{5b} - \frac{4}{15ab^2} \cdot \frac{4a}{4a}$$

$$= \frac{25b}{60a^2b^2} - \frac{16a}{60a^2b^2} = \frac{25b - 16a}{60a^2b^2}$$

39. $y - 5$

$y + 3$

$\text{LCD} = (y - 5)(y + 3)$

$$\frac{y + 2}{y - 5} - \frac{y - 4}{y + 3} = \frac{y + 2}{y - 5} \cdot \frac{y + 3}{y + 3} - \frac{y - 4}{y + 3} \cdot \frac{y - 5}{y - 5}$$

$$= \frac{y^2 + 5y + 6}{(y - 5)(y + 3)} - \frac{y^2 - 9y + 20}{(y - 5)(y + 3)}$$

$$= \frac{y^2 + 5y + 6 - \left(y^2 - 9y + 20\right)}{(y - 5)(y + 3)}$$

$$= \frac{y^2 + 5y + 6 - y^2 + 9y - 20}{(y - 5)(y + 3)}$$

$$= \frac{14y - 14}{(y - 5)(y + 3)} \text{ or } \frac{14(y - 1)}{(y - 5)(y + 3)}$$

41. $a - 2$

$a^2 - 4 = (a - 2)(a + 2)$

$\text{LCD} = (a - 2)(a + 2)$

$$\frac{a + 5}{a - 2} - \frac{5a + 18}{a^2 - 4} = \frac{a + 5}{a - 2} \cdot \frac{a + 2}{a + 2} - \frac{5a + 18}{(a - 2)(a + 2)}$$

$$= \frac{a^2 + 7a + 10}{(a - 2)(a + 2)} - \frac{5a + 18}{(a - 2)(a + 2)}$$

$$= \frac{a^2 + 7a + 10 - (5a + 18)}{(a - 2)(a + 2)}$$

$$= \frac{a^2 + 7a + 10 - 5a - 18}{(a - 2)(a + 2)}$$

$$= \frac{a^2 + 2a - 8}{(a - 2)(a + 2)}$$

$$= \frac{(a - 2)(a + 4)}{(a - 2)(a + 2)} = \frac{a + 4}{a + 2}$$

43. $(x - 2)(x + 3)$

$(x + 3)(x + 4)$

$\text{LCD} = (x + 3)(x - 2)(x + 4)$

$$\frac{3}{(x - 2)(x + 3)} - \frac{5}{(x + 3)(x + 4)}$$

$$= \frac{3}{(x - 2)(x + 3)} \cdot \frac{x + 4}{x + 4} - \frac{5}{(x + 3)(x + 4)} \cdot \frac{x - 2}{x - 2}$$

$$= \frac{3x + 12}{(x + 3)(x - 2)(x + 4)} - \frac{5x - 10}{(x + 3)(x - 2)(x + 4)}$$

$$= \frac{3x + 12 - (5x - 10)}{(x + 3)(x - 2)(x + 4)}$$

$$= \frac{3x + 12 - 5x + 10}{(x + 3)(x - 2)(x + 4)}$$

$$= \frac{-2x + 22}{(x + 3)(x - 2)(x + 4)}$$

$$\text{or } \frac{-2(x - 11)}{(x + 3)(x - 2)(x + 4)}$$

45. $x^2 + 3x + 2 = (x+1)(x+2)$

$x^2 - 4 = (x-2)(x+2)$

$\text{LCD} = (x+2)(x+1)(x-2)$

$$\frac{x-3}{x^2+3x+2} + \frac{x-1}{x^2-4}$$

$$= \frac{x-3}{(x+1)(x+2)} \cdot \frac{x-2}{x-2} + \frac{x-1}{(x-2)(x+2)} \cdot \frac{x+1}{x+1}$$

$$= \frac{x^2-5x+6}{(x+2)(x+1)(x-2)} + \frac{x^2-1}{(x+2)(x+1)(x-2)}$$

$$= \frac{x^2-5x+6+x^2-1}{(x+2)(x+1)(x-2)} = \frac{2x^2-5x+5}{(x+2)(x+1)(x-2)}$$

47. $2w^2 + 3w + 1 = (2w+1)(w+1)$

$2w^2 - 5w - 3 = (2w+1)(w-3)$

$\text{LCD} = (2w+1)(w+1)(w-3)$

$$\frac{w-4}{2w^2+3w+1} - \frac{w+3}{2w^2-5w-3}$$

$$= \frac{w-4}{(2w+1)(w+1)} \cdot \frac{w-3}{w-3}$$

$$\qquad\qquad - \frac{w+3}{(2w+1)(w-3)} \cdot \frac{w+1}{w+1}$$

$$= \frac{w^2-7w+12}{(2w+1)(w+1)(w-3)}$$

$$\qquad\qquad - \frac{w^2+4w+3}{(2w+1)(w+1)(w-3)}$$

$$= \frac{w^2-7w+12-\left(w^2+4w+3\right)}{(2w+1)(w+1)(w-3)}$$

$$= \frac{w^2-7w+12-w^2-4w-3}{(2w+1)(w+1)(w-3)}$$

$$= \frac{-11w+9}{(2w+1)(w+1)(w-3)}$$

49. $x^2 - 6xy + 9y^2 = (x-3y)^2$

$x^2 - 2xy - 3y^2 = (x+y)(x-3y)$

$\text{LCD} = (x-3y)^2(x+y)$

$$\frac{x+y}{x^2-6xy+9y^2} + \frac{x+2y}{x^2-2xy-3y^2}$$

$$= \frac{x+y}{(x-3y)^2} \cdot \frac{x+y}{x+y} + \frac{x+2y}{(x+y)(x-3y)} \cdot \frac{x-3y}{x-3y}$$

$$= \frac{x^2+2xy+y^2}{(x-3y)^2(x+y)} + \frac{x^2-xy-6y^2}{(x-3y)^2(x+y)}$$

$$= \frac{x^2+2xy+y^2+x^2-xy-6y^2}{(x-3y)^2(x+y)}$$

$$= \frac{2x^2+xy-5y^2}{(x-3y)^2(x+y)}$$

51. $x^2 - 4x - 5 = (x-5)(x+1)$

$x^2 - 6x + 5 = (x-5)(x-1)$

$\text{LCD} = (x-5)(x+1)(x-1)$

$$\frac{3}{x^2-4x-5} - \frac{2}{x^2-6x+5}$$

$$= \frac{3}{(x-5)(x+1)} \cdot \frac{x-1}{x-1} - \frac{2}{(x-5)(x-1)} \cdot \frac{x+1}{x+1}$$

$$= \frac{3x-3}{(x-5)(x+1)(x-1)} - \frac{2x+2}{(x-5)(x+1)(x-1)}$$

$$= \frac{3x-3-(2x+2)}{(x-5)(x+1)(x-1)} = \frac{3x-3-2x-2}{(x-5)(x+1)(x-1)}$$

$$= \frac{x-5}{(x-5)(x+1)(x-1)} = \frac{1}{(x+1)(x-1)}$$

53. $p^2 - 16 = (p-4)(p+4)$

$16 - p^2 = -1(p-4)(p+4)$

$\text{LCD} = (p-4)(p+4)$

$\dfrac{p^2 - 3p - 10}{p^2 - 16} + \dfrac{p^2 - 3p - 10}{16 - p^2}$

$= \dfrac{p^2 - 3p - 10}{(p-4)(p+4)} + \dfrac{p^2 - 3p - 10}{-1(p-4)(p+4)} \cdot \dfrac{-1}{-1}$

$= \dfrac{p^2 - 3p - 10}{(p-4)(p+4)} + \dfrac{-p^2 + 3p + 10}{(p-4)(p+4)}$

$= \dfrac{p^2 - 3p - 10 - p^2 + 3p + 10}{(p-4)(p+4)}$

$= \dfrac{0}{(p-4)(p+4)} = 0$

55. $w + 2$

w

$w^2 - 4 = (w-2)(w+2)$

$\text{LCD} = w(w-2)(w+2)$

$\dfrac{2}{w+2} - \dfrac{3}{w} + \dfrac{w+10}{w^2 - 4}$

$= \dfrac{2}{w+2} \cdot \dfrac{w(w-2)}{w(w-2)} - \dfrac{3}{w} \cdot \dfrac{(w-2)(w+2)}{(w-2)(w+2)} +$

$\qquad \dfrac{w+10}{(w-2)(w+2)} \cdot \dfrac{w}{w}$

$= \dfrac{2w^2 - 4w}{w(w-2)(w+2)} - \dfrac{3w^2 - 12}{w(w-2)(w+2)} +$

$\qquad \dfrac{w^2 + 10w}{w(w-2)(w+2)}$

$= \dfrac{2w^2 - 4w - (3w^2 - 12) + w^2 + 10w}{w(w-2)(w+2)}$

$= \dfrac{2w^2 - 4w - 3w^2 + 12 + w^2 + 10w}{w(w-2)(w+2)}$

$= \dfrac{6w + 12}{w(w-2)(w+2)}$

$= \dfrac{6(w+2)}{w(w-2)(w+2)}$

$= \dfrac{6}{w(w-2)}$

57. $p^2 + 6p + 9 = (p+3)^2$

$p + 3$

$2p^2 + p - 15 = (2p-5)(p+3)$

$\text{LCD} = (p+3)^2 (2p-5)$

$\dfrac{p-2}{p^2 + 6p + 9} + \dfrac{1}{p+3} - \dfrac{2p+1}{2p^2 + p - 15}$

$= \dfrac{p-2}{(p+3)^2} \cdot \dfrac{2p-5}{2p-5} + \dfrac{1}{p+3} \cdot \dfrac{(p+3)(2p-5)}{(p+3)(2p-5)} -$

$\qquad \dfrac{2p+1}{(2p-5)(p+3)} \cdot \dfrac{p+3}{p+3}$

$= \dfrac{2p^2 - 9p + 10}{(p+3)^2 (2p-5)} + \dfrac{2p^2 + p - 15}{(p+3)^2 (2p-5)} -$

$\qquad \dfrac{2p^2 + 7p + 3}{(p+3)^2 (2p-5)}$

$= \dfrac{2p^2 - 9p + 10 + 2p^2 + p - 15 - \left(2p^2 + 7p + 3\right)}{(p+3)^2 (2p-5)}$

$= \dfrac{2p^2 - 9p + 10 + 2p^2 + p - 15 - 2p^2 - 7p - 3}{(p+3)^2 (2p-5)}$

$= \dfrac{2p^2 - 15p - 8}{(p+3)^2 (2p-5)}$ or $\dfrac{(2p+1)(p-8)}{(p+3)^2 (2p-5)}$

59. x

$x - 1$

$(x-1)^2$

$\text{LCD} = x(x-1)^2$

$\dfrac{2}{x} - \dfrac{2}{x-1} + \dfrac{3}{(x-1)^2}$

$= \dfrac{2}{x} \cdot \dfrac{(x-1)^2}{(x-1)^2} - \dfrac{2}{x-1} \cdot \dfrac{x(x-1)}{x(x-1)} + \dfrac{3}{(x-1)^2} \cdot \dfrac{x}{x}$

$= \dfrac{2x^2 - 4x + 2}{x(x-1)^2} - \dfrac{2x^2 - 2x}{x(x-1)^2} + \dfrac{3x}{x(x-1)^2}$

$= \dfrac{2x^2 - 4x + 2 - \left(2x^2 - 2x\right) + 3x}{x(x-1)^2}$

$= \dfrac{2x^2 - 4x + 2 - 2x^2 + 2x + 3x}{x(x-1)^2}$

$= \dfrac{x+2}{x(x-1)^2}$

61. $x - 3$

$x^3 - 27 = (x-3)(x^2 + 3x + 9)$

$\text{LCD} = (x-3)(x^2 + 3x + 9)$

$\dfrac{1}{x-3} - \dfrac{x^2 + 18}{x^3 - 27}$

$= \dfrac{1}{x-3} \cdot \dfrac{x^2 + 3x + 9}{x^2 + 3x + 9} - \dfrac{x^2 + 18}{(x-3)(x^2 + 3x + 9)}$

$= \dfrac{x^2 + 3x + 9}{(x-3)(x^2 + 3x + 9)} - \dfrac{x^2 + 18}{(x-3)(x^2 + 3x + 9)}$

$= \dfrac{x^2 + 3x + 9 - (x^2 + 18)}{(x-3)(x^2 + 3x + 9)}$

$= \dfrac{x^2 + 3x + 9 - x^2 - 18}{(x-3)(x^2 + 3x + 9)}$

$= \dfrac{3x - 9}{(x-3)(x^2 + 3x + 9)}$

$= \dfrac{3(x-3)}{(x-3)(x^2 + 3x + 9)} = \dfrac{3}{x^2 + 3x + 9}$

63. $\dfrac{2x^2 - x}{x+3} + \dfrac{3x}{x+3} - \dfrac{x^2 + 3}{x+3}$

$= \dfrac{2x^2 - x + 3x - (x^2 + 3)}{x+3}$

$= \dfrac{2x^2 - x + 3x - x^2 - 3}{x+3}$

$= \dfrac{x^2 + 2x - 3}{x+3} = \dfrac{(x-1)(x+3)}{x+3} = x - 1$

65. 1

$x + 3$

$\text{LCD} = x + 3$

$6 + \dfrac{x-3}{x+3} = \dfrac{6}{1} \cdot \dfrac{x+3}{x+3} + \dfrac{x-3}{x+3}$

$\phantom{6 + \dfrac{x-3}{x+3}} = \dfrac{6x+18}{x+3} + \dfrac{x-3}{x+3}$

$\phantom{6 + \dfrac{x-3}{x+3}} = \dfrac{6x + 18 + x - 3}{x+3} = \dfrac{7x + 15}{x+3}$

67. $b^2 + 2b - 8 = (b-2)(b+4)$

$b^2 - 4 = (b-2)(b+2)$

$\text{LCD} = (b-2)(b+2)(b+4)$

$\dfrac{b+3}{b^2 + 2b - 8} - \dfrac{b+2}{b^2 - 4}$

$= \dfrac{b+3}{(b-2)(b+4)} \cdot \dfrac{b+2}{b+2} - \dfrac{b+2}{(b-2)(b+2)} \cdot \dfrac{b+4}{b+4}$

$= \dfrac{b^2 + 5b + 6}{(b-2)(b+2)(b+4)} - \dfrac{b^2 + 6b + 8}{(b-2)(b+2)(b+4)}$

$= \dfrac{b^2 + 5b + 6 - (b^2 + 6b + 8)}{(b-2)(b+2)(b+4)}$

$= \dfrac{b^2 + 5b + 6 - b^2 - 6b - 8}{(b-2)(b+2)(b+4)}$

$= \dfrac{-b - 2}{(b-2)(b+2)(b+4)}$

$= \dfrac{-(b+2)}{(b-2)(b+2)(b+4)} = \dfrac{-1}{(b-2)(b+4)}$

69. $y + 4$

$y + 3$

$y^2 + 7y + 12 = (y+3)(y+4)$

$\text{LCD} = (y+3)(y+4)$

$\dfrac{y-1}{y+4} + \dfrac{y-2}{y+3} - \dfrac{y^2 + 3y + 1}{y^2 + 7y + 12}$

$= \dfrac{y-1}{y+4} \cdot \dfrac{y+3}{y+3} + \dfrac{y-2}{y+3} \cdot \dfrac{y+4}{y+4} - \dfrac{y^2 + 3y + 1}{(y+3)(y+4)}$

$= \dfrac{y^2 + 2y - 3}{(y+3)(y+4)} + \dfrac{y^2 + 2y - 8}{(y+3)(y+4)}$

$ - \dfrac{y^2 + 3y + 1}{(y+3)(y+4)}$

$= \dfrac{y^2 + 2y - 3 + (y^2 + 2y - 8) - (y^2 + 3y + 1)}{(y+3)(y+4)}$

$= \dfrac{y^2 + 2y - 3 + y^2 + 2y - 8 - y^2 - 3y - 1}{(y+3)(y+4)}$

$= \dfrac{y^2 + y - 12}{(y+3)(y+4)} = \dfrac{(y-3)(y+4)}{(y+3)(y+4)} = \dfrac{y-3}{y+3}$

71. $x^2 - 5x + 6 = (x-2)(x-3)$

$x^2 - 2x - 3 = (x-3)(x+1)$

$x^2 - x - 2 = (x+1)(x-2)$

LCD $= (x-2)(x-3)(x+1)$

$\dfrac{x+4}{x^2-5x+6} + \dfrac{x-1}{x^2-2x-3} - \dfrac{2x+1}{x^2-x-2}$

$= \dfrac{x+4}{(x-2)(x-3)} \cdot \dfrac{x+1}{x+1} + \dfrac{x-1}{(x-3)(x+1)} \cdot \dfrac{x-2}{x-2}$

$\quad - \dfrac{2x+1}{(x+1)(x-2)} \cdot \dfrac{x-3}{x-3}$

$= \dfrac{x^2+5x+4}{(x-2)(x-3)(x+1)} + \dfrac{x^2-3x+2}{(x-2)(x-3)(x+1)}$

$\quad - \dfrac{2x^2-5x-3}{(x-2)(x-3)(x+1)}$

$= \dfrac{x^2+5x+4+\left(x^2-3x+2\right)-\left(2x^2-5x-3\right)}{(x-2)(x-3)(x+1)}$

$= \dfrac{x^2+5x+4+x^2-3x+2-2x^2+5x+3}{(x-2)(x-3)(x+1)}$

$= \dfrac{7x+9}{(x-2)(x-3)(x+1)}$

73. a. $x-2$

$x+1$

LCD $= (x-2)(x+1)$

$R(x) = f(x) + g(x) = \dfrac{3}{x-2} + \dfrac{2}{x+1}$

$= \dfrac{3}{x-2} \cdot \dfrac{x+1}{x+1} + \dfrac{2}{x+1} \cdot \dfrac{x-2}{x-2}$

$= \dfrac{3x+3}{(x-2)(x+1)} + \dfrac{2x-4}{(x-2)(x+1)}$

$= \dfrac{3x+3+2x-4}{(x-2)(x+1)}$

$= \dfrac{5x-1}{(x-2)(x+1)}$

b. Since R is the sum of f and g, the restrictions for the domain of R will consist of all the restrictions for the domains of both f and g. Since 2 is restricted from the domain of f and since -1 is restricted from the domain of g, both 2 and -1 are restricted from the domain of R. That is, the domain of $R(x)$ is

$\{x \mid x \neq -1, x \neq 2\}$.

75. $x^2 - 3x - 4 = (x-4)(x+1)$

$x^2 - x - 12 = (x-4)(x+3)$

LCD $= (x-4)(x+1)(x+3)$

a. $R(x) = f(x) + g(x)$

$= \dfrac{x+1}{x^2-3x-4} + \dfrac{x+4}{x^2-x-12}$

$= \dfrac{x+1}{(x-4)(x+1)} \cdot \dfrac{x+3}{x+3} + \dfrac{x+4}{(x-4)(x+3)} \cdot \dfrac{x+1}{x+1}$

$= \dfrac{x^2+4x+3}{(x-4)(x+1)(x+3)}$

$\quad + \dfrac{x^2+5x+4}{(x-4)(x+1)(x+3)}$

$= \dfrac{x^2+4x+3+x^2+5x+4}{(x-4)(x+1)(x+3)}$

$= \dfrac{2x^2+9x+7}{(x-4)(x+1)(x+3)}$

$= \dfrac{(2x+7)(x+1)}{(x-4)(x+1)(x+3)} = \dfrac{2x+7}{(x-4)(x+3)}$

b. Since R is the sum of f and g, the restrictions for the domain of R will consist of all the restrictions for the domains of both f and g. Since 4 and -1 are restricted from the domain of f and since 4 and -3 are restricted from the domain of g, then 4, -1, and -3 are restricted from the domain of R. That is, the domain of $R(x)$ is

$\{x \mid x \neq -3, x \neq -1, x \neq 4\}$.

c. $H(x) = f(x) - g(x)$

$$= \frac{x+1}{x^2-3x-4} - \frac{x+4}{x^2-x-12}$$

$$= \frac{x+1}{(x-4)(x+1)} \cdot \frac{x+3}{x+3}$$

$$\qquad\qquad - \frac{x+4}{(x-4)(x+3)} \cdot \frac{x+1}{x+1}$$

$$= \frac{x^2+4x+3}{(x-4)(x+1)(x+3)}$$

$$\qquad\qquad - \frac{x^2+5x+4}{(x-4)(x+1)(x+3)}$$

$$= \frac{x^2+4x+3-\left(x^2+5x+4\right)}{(x-4)(x+1)(x+3)}$$

$$= \frac{x^2+4x+3-x^2-5x-4}{(x-4)(x+1)(x+3)}$$

$$= \frac{-x-1}{(x-4)(x+1)(x+3)}$$

$$= \frac{-1(x+1)}{(x-4)(x+1)(x+3)} = \frac{-1}{(x-4)(x+3)}$$

d. Since H is the difference of f and g, the restrictions for the domain of H will consist of all the restrictions for the domains of both f and g. Since 4 and −1 are restricted from the domain of f and since 4 and −3 are restricted from the domain of g, then 4, −1, and −3 are restricted from the domain of H. That is, the domain of $H(x)$ is

$$\{x \mid x \neq -3, x \neq -1, x \neq 4\}.$$

77. a. $S(x) = 2x^2 + \dfrac{8000}{x}$ (Note: LCD = x)

$$= \frac{2x^2}{1} \cdot \frac{x}{x} + \frac{8000}{x}$$

$$= \frac{2x^3}{x} + \frac{8000}{x}$$

$$= \frac{2x^3+8000}{x}$$

b. $S(10) = \dfrac{2(10)^3+8000}{10} = \dfrac{10,000}{10} = 1000$

For the square-based, closed box with volume 2000 cubic inches, if the length of the base is 10 inches, then the surface area of the box will be 1000 square inches.

79. a. s

$s+10$

LCD $= s(s+10)$

$$T(s) = \frac{50}{s} + \frac{150}{s+10}$$

$$= \frac{50}{s} \cdot \frac{s+10}{s+10} + \frac{150}{s+10} \cdot \frac{s}{s}$$

$$= \frac{50s+500}{s(s+10)} + \frac{150s}{s(s+10)}$$

$$= \frac{200s+500}{s(s+10)} \quad \text{or} \quad \frac{100(2s+5)}{s(s+10)}$$

b. $T(50) = \dfrac{200(50)+500}{50(50+10)} = \dfrac{10,500}{3000} = 3.5$

If the average speed for the first 50 miles of the trip is 50 miles per hour, then it will take a total time of 3.5 hours to get to the neighboring university.

81. $x^{-1} + y^{-1} = \dfrac{1}{x} + \dfrac{1}{y}$ (Note: LCD $= xy$)

$$= \frac{1}{x} \cdot \frac{y}{y} + \frac{1}{y} \cdot \frac{x}{x}$$

$$= \frac{y}{xy} + \frac{x}{xy}$$

$$= \frac{y+x}{xy} \quad \text{or} \quad \frac{x+y}{xy}$$

83. Answers may vary. One possibility follows: To find the least common denominator, first factor each denominator completely. Second, list the factors that are common to all denominators. Finally, list the factors that are not common.

85. $4a(a-3) = 4a \cdot a - 4a \cdot 3 = 4a^2 - 12a$

87. $(p-3)(p+3) = p^2 + 3p - 3p - 9 = p^2 - 9$

89. $(w-2)\left(w^2+2w+4\right)$

$$= w^3 + 2w^2 + 4w - 2w^2 - 4w - 8$$

$$= w^3 - 8$$

91. $\dfrac{x^3-8}{2x^2+5x+2}\cdot\dfrac{2x^2-x-1}{x^2-4}\cdot\dfrac{x^2+4x+4}{x-1}$

$=\dfrac{(x-2)\left(x^2+2x+4\right)}{(2x+1)(x+2)}\cdot\dfrac{(2x+1)(x-1)}{(x+2)(x-2)}\cdot\dfrac{(x+2)(x+2)}{(x-1)}$

$=\dfrac{(x-2)\left(x^2+2x+4\right)(2x+1)\,(x-1)\,(x+2)(x+2)}{(2x+1)\,(x+2)(x+2)\,(x-2)\,(x-1)}$

$=x^2+2x+4$

Section 5.3

Are You Ready for This Section?

R1. $a\cdot c=6(-6)=-36$

The factors of -36 that add to -5 (the linear coefficient) are -9 and 4. Thus,

$$6y^2-5y-6=6y^2-9y+4y-6$$
$$=3y(2y-3)+2(2y-3)$$
$$=(3y+2)(2y-3)$$

R2. $\left(\dfrac{3ab^2}{2a^{-1}b^5}\right)^{-2}=\left(\dfrac{2a^{-1}b^5}{3ab^2}\right)^2=\left(\dfrac{2b^{5-2}}{3a^{1+1}}\right)^2$

$=\left(\dfrac{2b^3}{3a^2}\right)^2=\dfrac{2^2b^{3\cdot2}}{3^2a^{2\cdot2}}=\dfrac{4b^6}{9a^4}$

Section 5.3 Quick Checks

1. An expression such as $\dfrac{\frac{x}{5}-\frac{y}{7}}{\frac{xy}{35}}$ is called a <u>complex rational expression</u>.

2. True

3. Write the numerator of the complex rational expression as a single rational expression. Note that the LCD of the rational expressions in the numerator is 6.

$$\dfrac{3}{2}-\dfrac{1}{3}=\dfrac{3}{2}\cdot\dfrac{3}{3}-\dfrac{1}{3}\cdot\dfrac{2}{2}=\dfrac{9}{6}-\dfrac{2}{6}=\dfrac{9-2}{6}=\dfrac{7}{6}$$

The denominator of the complex rational expression is already a single rational expression: $\dfrac{5}{6}$.

Rewrite the complex rational expression using the single rational expressions and simplify.

$$\dfrac{\frac{3}{2}-\frac{1}{3}}{\frac{5}{6}}=\dfrac{\frac{7}{6}}{\frac{5}{6}}=\dfrac{7}{6}\cdot\dfrac{6}{5}=\dfrac{7\cdot6}{6\cdot5}=\dfrac{7}{5}$$

4. Write the numerator of the complex rational expression as a single rational expression. Note that the LCD of the rational expressions in the numerator is $4z$.

$$\frac{z}{4} - \frac{4}{z} = \frac{z}{4} \cdot \frac{z}{z} - \frac{4}{z} \cdot \frac{4}{4} = \frac{z^2}{4z} - \frac{16}{4z} = \frac{z^2-16}{4z}$$

The denominator of the complex rational expression is already a single rational expression: $\dfrac{z+4}{16}$.

Rewrite the complex rational expression using the single rational expressions and simplify.

$$\frac{\dfrac{z}{4} - \dfrac{4}{z}}{\dfrac{z+4}{16}} = \frac{\dfrac{z^2-16}{4z}}{\dfrac{z+4}{16}} = \frac{z^2-16}{4z} \cdot \frac{16}{z+4}$$

$$= \frac{(z-4)(z+4)}{4z} \cdot \frac{16}{z+4}$$

$$= \frac{(z-4)\,\cancel{(z+4)}}{\cancel{4}z} \cdot \frac{\overset{4}{\cancel{16}}}{\cancel{z+4}} = \frac{4(z-4)}{z}$$

5. False; $\dfrac{x-y}{\dfrac{1}{x} + \dfrac{1}{y}} = \dfrac{x-y}{\dfrac{y}{xy} + \dfrac{x}{xy}}$

$$= \frac{x-y}{\dfrac{y+x}{xy}}$$

$$= (x-y) \cdot \frac{xy}{x+y}$$

$$= \frac{xy(x-y)}{x+y}$$

6. Write the numerator of the complex rational expression as a single rational expression. Note that the LCD of the rational expressions in the numerator is $(x+1)(x+2)$.

$$\frac{2x}{x+1} - \frac{x^2-3}{x^2+3x+2}$$

$$= \frac{2x}{x+1} - \frac{x^2-3}{(x+1)(x+2)}$$

$$= \frac{2x}{x+1} \cdot \frac{x+2}{x+2} - \frac{x^2-3}{(x+1)(x+2)}$$

$$= \frac{2x^2+4x}{(x+1)(x+2)} - \frac{x^2-3}{(x+1)(x+2)}$$

$$= \frac{2x^2+4x-\left(x^2-3\right)}{(x+1)(x+2)}$$

$$= \frac{2x^2+4x-x^2+3}{(x+1)(x+2)}$$

$$= \frac{x^2+4x+3}{(x+1)(x+2)}$$

Write the denominator of the complex rational expression as a single rational expression. Note that the LCD of the rational expressions in the denominator is $x+2$.

$$4 + \frac{4}{x+2} = 4 \cdot \frac{x+2}{x+2} + \frac{4}{x+2}$$

$$= \frac{4x+8}{x+2} + \frac{4}{x+2} = \frac{4x+12}{x+2} = \frac{4(x+3)}{x+2}$$

Rewrite the complex rational expression using the single rational expressions and simplify.

$$\frac{\dfrac{2x}{x+1} - \dfrac{x^2-3}{x^2+3x+2}}{4 + \dfrac{4}{x+2}} = \frac{\dfrac{x^2+4x+3}{(x+1)(x+2)}}{\dfrac{4(x+3)}{x+2}}$$

$$= \frac{x^2+4x+3}{(x+1)(x+2)} \cdot \frac{x+2}{4(x+3)}$$

$$= \frac{(x+1)(x+3)}{(x+1)(x+2)} \cdot \frac{x+2}{4(x+3)}$$

$$= \frac{\cancel{(x+1)}\,\cancel{(x+3)}}{\cancel{(x+1)}\,\cancel{(x+2)}} \cdot \frac{\cancel{x+2}}{4\cancel{(x+3)}}$$

$$= \frac{1}{4}$$

7. The LCD of all the denominators in the complex rational expression is 6.

$$\frac{\dfrac{3}{2}-\dfrac{1}{3}}{\dfrac{5}{6}} = \frac{\dfrac{3}{2}-\dfrac{1}{3}}{\dfrac{5}{6}}\cdot\frac{6}{6}$$

$$= \frac{\dfrac{3}{2}\cdot 6-\dfrac{1}{3}\cdot 6}{\dfrac{5}{6}\cdot 6}$$

$$= \frac{\dfrac{3}{2}\cdot\overset{3}{6}-\dfrac{1}{3}\cdot\overset{2}{6}}{\dfrac{5}{6}\cdot 6}$$

$$= \frac{9-2}{5} = \frac{7}{5}$$

8. The LCD of all the denominators in the complex rational expression is $16z$.

$$\frac{\dfrac{z}{4}-\dfrac{4}{z}}{\dfrac{z+4}{16}} = \frac{\dfrac{z}{4}-\dfrac{4}{z}}{\dfrac{z+4}{16}}\cdot\frac{16z}{16z} = \frac{\dfrac{z}{4}\cdot 16z-\dfrac{4}{z}\cdot 16z}{\dfrac{z+4}{16}\cdot 16z}$$

$$= \frac{\dfrac{z}{4}\cdot\overset{4}{16}z-\dfrac{4}{z}\cdot 16z}{\dfrac{z+4}{16}\cdot 16z}$$

$$= \frac{4z^2-64}{z(z+4)} = \frac{4(z-4)(z+4)}{z(z+4)}$$

$$= \frac{4(z-4)(z+4)}{z(z+4)} = \frac{4(z-4)}{z}$$

9. The LCD of all the denominators in the complex rational expression is $(x+5)(x+1)$.

$$\frac{\dfrac{x+2}{x+5}-\dfrac{x+2}{x+1}}{\dfrac{2x+1}{x+1}-1} = \frac{\dfrac{x+2}{x+5}-\dfrac{x+2}{x+1}}{\dfrac{2x+1}{x+1}-1}\cdot\frac{(x+5)(x+1)}{(x+5)(x+1)}$$

$$= \frac{\dfrac{x+2}{x+5}\cdot(x+5)(x+1)-\dfrac{x+2}{x+1}\cdot(x+5)(x+1)}{\dfrac{2x+1}{x+1}\cdot(x+5)(x+1)-1\cdot(x+5)(x+1)}$$

$$= \frac{\dfrac{x+2}{x+5}\cdot(x+5)(x+1)+\dfrac{x+2}{x+1}\cdot(x+5)(x+1)}{\dfrac{2x+1}{x+1}\cdot(x+5)(x+1)-1\cdot(x+5)(x+1)}$$

$$= \frac{(x+2)(x+1)-(x+2)(x+5)}{(2x+1)(x+5)-(x+5)(x+1)}$$

$$= \frac{(x+2)\big[x+1-(x+5)\big]}{(x+5)\big[2x+1-(x+1)\big]}$$

$$= \frac{(x+2)\big[x+1-x-5\big]}{(x+5)\big[2x+1-x-1\big]}$$

$$= \frac{(x+2)(-4)}{(x+5)x} = \frac{-4(x+2)}{x(x+5)}$$

10. Rewrite the expression so that it does not contain any negative exponents.

$$\frac{3a^{-1}+b^{-1}}{9a^{-2}-b^{-2}} = \frac{\dfrac{3}{a}+\dfrac{1}{b}}{\dfrac{9}{a^2}-\dfrac{1}{b^2}}$$

Using Method 1, begin by writing the numerator of the complex rational expression as a single rational expression. Note that the LCD of the rational expressions in the numerator is ab.

$$\frac{3}{a}+\frac{1}{b} = \frac{3}{a}\cdot\frac{b}{b}+\frac{1}{b}\cdot\frac{a}{a} = \frac{3b}{ab}+\frac{a}{ab} = \frac{3b+a}{ab}$$

Next, write the denominator of the complex rational expression as a single rational expression. Note that the LCD of the rational expressions in the denominator is a^2b^2.

$$\frac{9}{a^2}-\frac{1}{b^2} = \frac{9}{a^2}\cdot\frac{b^2}{b^2}-\frac{1}{b^2}\cdot\frac{a^2}{a^2}$$

$$= \frac{9b^2}{a^2b^2}-\frac{a^2}{a^2b^2} = \frac{9b^2-a^2}{a^2b^2}$$

Rewrite the complex rational expression using the single rational expressions and simplify.

$$\dfrac{\dfrac{3}{a}+\dfrac{1}{b}}{\dfrac{9}{a^2}-\dfrac{1}{b^2}}=\dfrac{\dfrac{3b+a}{ab}}{\dfrac{9b^2-a^2}{a^2b^2}}=\dfrac{3b+a}{ab}\cdot\dfrac{a^2b^2}{9b^2-a^2}$$

$$=\dfrac{3b+a}{ab}\cdot\dfrac{a^2b^2}{(3b-a)(3b+a)}$$

$$=\dfrac{\cancel{3b+a}}{\cancel{ab}}\cdot\dfrac{\cancel{a^2b^2}^{\,ab}}{(3b-a)\,\cancel{(3b+a)}}$$

$$=\dfrac{ab}{3b-a}$$

Using Method 2, the LCD of all the denominators in the complex rational expression is a^2b^2.

$$\dfrac{3a^{-1}+b^{-1}}{9a^{-2}-b^{-2}}=\dfrac{\dfrac{3}{a}+\dfrac{1}{b}}{\dfrac{9}{a^2}-\dfrac{1}{b^2}}=\dfrac{\dfrac{3}{a}+\dfrac{1}{b}}{\dfrac{9}{a^2}-\dfrac{1}{b^2}}\cdot\dfrac{a^2b^2}{a^2b^2}$$

$$=\dfrac{\dfrac{3}{a}\cdot a^2b^2+\dfrac{1}{b}\cdot a^2b^2}{\dfrac{9}{a^2}\cdot a^2b^2-\dfrac{1}{b^2}\cdot a^2b^2}$$

$$=\dfrac{3ab^2+a^2b}{9b^2-a^2}$$

$$=\dfrac{ab(3b+a)}{(3b-a)(3b+a)}$$

$$=\dfrac{ab\,\cancel{(3b+a)}}{(3b-a)\,\cancel{(3b+a)}}$$

$$=\dfrac{ab}{3b-a}$$

5.3 Exercises

11. Write the numerator of the complex rational expression as a single rational expression. Note that the LCD of the rational expressions in the numerator is 18.

$$\dfrac{5}{6}+\dfrac{1}{9}=\dfrac{5}{6}\cdot\dfrac{3}{3}+\dfrac{1}{9}\cdot\dfrac{2}{2}=\dfrac{15}{18}+\dfrac{2}{18}=\dfrac{17}{18}$$

Write the denominator of the complex rational expression as a single rational expression. Note that the LCD of the rational expressions in the denominator is 8.

$$\dfrac{5}{2}-\dfrac{3}{8}=\dfrac{5}{2}\cdot\dfrac{4}{4}-\dfrac{3}{8}=\dfrac{20}{8}-\dfrac{3}{8}=\dfrac{17}{8}$$

Rewrite the complex rational expression using the single rational expressions and simplify.

$$\dfrac{\dfrac{5}{6}+\dfrac{1}{9}}{\dfrac{5}{2}-\dfrac{3}{8}}=\dfrac{\dfrac{17}{18}}{\dfrac{17}{8}}=\dfrac{\cancel{17}}{18}\cdot\dfrac{8}{\cancel{17}}=\dfrac{8}{18}=\dfrac{4}{9}$$

13. Write the numerator of the complex rational expression as a single rational expression. Note that the LCD of the rational expressions in the numerator is x.

$$1+\dfrac{1}{x}=\dfrac{1}{1}\cdot\dfrac{x}{x}+\dfrac{1}{x}=\dfrac{x}{x}+\dfrac{1}{x}=\dfrac{x+1}{x}$$

Write the denominator of the complex rational expression as a single rational expression. Note that the LCD of the rational expressions in the denominator is x.

$$1-\dfrac{1}{x}=\dfrac{1}{1}\cdot\dfrac{x}{x}-\dfrac{1}{x}=\dfrac{x}{x}-\dfrac{1}{x}=\dfrac{x-1}{x}$$

Rewrite the complex rational expression using the single rational expressions and simplify.

$$\dfrac{1+\dfrac{1}{x}}{1-\dfrac{1}{x}}=\dfrac{\dfrac{x+1}{x}}{\dfrac{x-1}{x}}=\dfrac{x+1}{x}\cdot\dfrac{x}{x-1}=\dfrac{x+1}{\cancel{x}}\cdot\dfrac{\cancel{x}}{x-1}=\dfrac{x+1}{x-1}$$

15. Write the numerator of the complex rational expression as a single rational expression. Note that the LCD of the rational expressions in the numerator is $a-2$.

$$1-\dfrac{a}{a-2}=\dfrac{1}{1}\cdot\dfrac{a-2}{a-2}-\dfrac{a}{a-2}$$

$$=\dfrac{a-2}{a-2}-\dfrac{a}{a-2}=\dfrac{a-2-a}{a-2}=\dfrac{-2}{a-2}$$

Write the denominator of the complex rational expression as a single rational expression. Note that the LCD of the rational expressions in the denominator is a.

$$2-\dfrac{a+2}{a}=\dfrac{2}{1}\cdot\dfrac{a}{a}-\dfrac{a+2}{a}=\dfrac{2a}{a}-\dfrac{a+2}{a}$$

$$=\dfrac{2a-(a+2)}{a}=\dfrac{2a-a-2}{a}=\dfrac{a-2}{a}$$

Rewrite the complex rational expression using the single rational expressions and simplify.

$$\dfrac{1-\dfrac{a}{a-2}}{2-\dfrac{a+2}{a}}=\dfrac{\dfrac{-2}{a-2}}{\dfrac{a-2}{a}}=\dfrac{-2}{a-2}\cdot\dfrac{a}{a-2}=\dfrac{-2a}{(a-2)^2}$$

17. Write the numerator of the complex rational expression as a single rational expression. Note that the LCD of the rational expressions in the numerator is $(x-1)(x+3)$.

$$\frac{x+2}{x-1}-\frac{x+5}{x+3}=\frac{x+2}{x-1}\cdot\frac{x+3}{x+3}-\frac{x+5}{x+3}\cdot\frac{x-1}{x-1}$$

$$=\frac{x^2+5x+6}{(x-1)(x+3)}-\frac{x^2+4x-5}{(x-1)(x+3)}$$

$$=\frac{x^2+5x+6-\left(x^2+4x-5\right)}{(x-1)(x+3)}$$

$$=\frac{x^2+5x+6-x^2-4x+5}{(x-1)(x+3)}$$

$$=\frac{x+11}{(x-1)(x+3)}$$

The denominator is already a single rational expression: $\dfrac{x+11}{1}$.

Rewrite the complex rational expression using the single rational expressions and simplify.

$$\frac{\dfrac{x+2}{x-1}-\dfrac{x+5}{x+3}}{x+11}=\frac{x+11}{(x-1)(x+3)}\cdot\frac{1}{x+11}$$

$$=\frac{\cancel{x+11}}{(x-1)(x+3)}\cdot\frac{1}{\cancel{x+11}}$$

$$=\frac{1}{(x-1)(x+3)}$$

19. The LCD of all the denominators in the complex rational expression is 30.

$$\frac{\dfrac{5}{6}+\dfrac{7}{10}}{\dfrac{1}{5}-1}=\frac{\dfrac{5}{6}+\dfrac{7}{10}}{\dfrac{1}{5}-1}\cdot\frac{30}{30}$$

$$=\frac{\dfrac{5}{6}\cdot30+\dfrac{7}{10}\cdot30}{\dfrac{1}{5}\cdot30-1\cdot30}$$

$$=\frac{\dfrac{5}{\cancel{6}}\cdot\overset{5}{\cancel{30}}+\dfrac{7}{\cancel{10}}\cdot\overset{3}{\cancel{30}}}{\dfrac{1}{\cancel{5}}\cdot\overset{6}{\cancel{30}}-1\cdot30}$$

$$=\frac{25+21}{6-30}=\frac{46}{-24}=-\frac{23}{12}$$

21. The LCD of all the denominators in the complex rational expression is w.

$$\frac{w-\dfrac{1}{w}}{w+\dfrac{1}{w}}=\frac{w-\dfrac{1}{w}}{w+\dfrac{1}{w}}\cdot\frac{w}{w}=\frac{w\cdot w-\dfrac{1}{w}\cdot w}{w\cdot w+\dfrac{1}{w}\cdot w}$$

$$=\frac{w^2-1}{w^2+1}\quad\text{or}\quad\frac{(w-1)(w+1)}{w^2+1}$$

23. The LCD of all the denominators in the complex rational expression is $(x-4)(x-2)$.

$$\frac{\dfrac{x-1}{x-4}-\dfrac{x}{x-2}}{1-\dfrac{3}{x-4}}=\frac{\dfrac{x-1}{x-4}-\dfrac{x}{x-2}}{1-\dfrac{3}{x-4}}\cdot\frac{(x-4)(x-2)}{(x-4)(x-2)}$$

$$=\frac{\dfrac{x-1}{x-4}\cdot(x-4)(x-2)-\dfrac{x}{x-2}\cdot(x-4)(x-2)}{1\cdot(x-4)(x-2)-\dfrac{3}{x-4}\cdot(x-4)(x-2)}$$

$$=\frac{\dfrac{x-1}{\cancel{x-4}}\cdot\cancel{(x-4)}(x-2)-\dfrac{x}{\cancel{x-2}}\cdot(x-4)\cancel{(x-2)}}{1\cdot(x-4)(x-2)-\dfrac{3}{\cancel{x-4}}\cdot\cancel{(x-4)}(x-2)}$$

$$=\frac{(x-1)(x-2)-x(x-4)}{(x-4)(x-2)-3(x-2)}$$

$$=\frac{x^2-3x+2-x^2+4x}{(x-2)\left[(x-4)-3\right]}=\frac{x+2}{(x-2)(x-7)}$$

25. The LCD of all the denominators in the complex rational expression is $(z-2)(z+2)$.

$$\frac{\dfrac{z+2}{z-2}+\dfrac{z-2}{z+2}}{\dfrac{z+2}{z-2}-\dfrac{z-2}{z+2}} = \frac{\dfrac{z+2}{z-2}+\dfrac{z-2}{z+2}}{\dfrac{z+2}{z-2}-\dfrac{z-2}{z+2}} \cdot \frac{(z-2)(z+2)}{(z-2)(z+2)}$$

$$= \frac{\dfrac{z+2}{z-2}\cdot(z-2)(z+2)+\dfrac{z-2}{z+2}\cdot(z-2)(z+2)}{\dfrac{z+2}{z-2}\cdot(z-2)(z+2)-\dfrac{z-2}{z+2}\cdot(z-2)(z+2)}$$

$$= \frac{\dfrac{z+2}{z-2}\cdot(z-2)(z+2)+\dfrac{z-2}{z+2}\cdot(z-2)(z+2)}{\dfrac{z+2}{z-2}\cdot(z-2)(z+2)-\dfrac{z-2}{z+2}\cdot(z-2)(z+2)}$$

$$= \frac{(z+2)^2+(z-2)^2}{(z+2)^2-(z-2)^2}$$

$$= \frac{(z^2+4z+4)+(z^2-4z+4)}{(z^2+4z+4)-(z^2-4z+4)}$$

$$= \frac{2z^2+8}{8z} = \frac{2(z^2+4)}{2\cdot4z} = \frac{z^2+4}{4z}$$

27. Using Method 2, the LCD of all the denominators in the complex rational expression is y.

$$\frac{\dfrac{3}{y}-1}{\dfrac{9}{y}-y} = \frac{\dfrac{3}{y}-1}{\dfrac{9}{y}-y}\cdot\frac{y}{y} = \frac{\dfrac{3}{y}\cdot y-1\cdot y}{\dfrac{9}{y}\cdot y-y\cdot y} = \frac{3-y}{9-y^2}$$

$$= \frac{3-y}{(3-y)(3+y)} = \frac{1}{3+y} \quad\text{or}\quad \frac{1}{y+3}$$

29. Using Method 2, the LCD of all the denominators in the complex rational expression is mn.

$$\frac{\dfrac{4n}{m}-\dfrac{4m}{n}}{\dfrac{2}{m}+\dfrac{2}{n}} = \frac{\dfrac{4n}{m}-\dfrac{4m}{n}}{\dfrac{2}{m}+\dfrac{2}{n}}\cdot\frac{mn}{mn} = \frac{\dfrac{4n}{m}\cdot mn-\dfrac{4m}{n}\cdot mn}{\dfrac{2}{m}\cdot mn+\dfrac{2}{n}\cdot mn}$$

$$= \frac{\dfrac{4n}{m}\cdot mn-\dfrac{4m}{n}\cdot mn}{\dfrac{2}{m}\cdot mn+\dfrac{2}{n}\cdot mn}$$

$$= \frac{4n^2-4m^2}{2n+2m}$$

$$= \frac{4(n^2-m^2)}{2(n+m)}$$

$$= \frac{2\cdot2(n-m)(n+m)}{2(n+m)}$$

$$= \frac{2\cdot2(n-m)(n+m)}{2(n+m)}$$

$$= 2(n-m)$$

31. Using Method 1, write the numerator of the complex rational expression as a single rational expression. Note that the LCD of the rational expressions in the numerator is x.

$$2+\frac{3}{x} = \frac{2}{1}\cdot\frac{x}{x}+\frac{3}{x} = \frac{2x}{x}+\frac{3}{x} = \frac{2x+3}{x}$$

Write the denominator of the complex rational expression as a single rational expression. Note that the LCD of the rational expressions in the denominator is $x+3$.

$$\frac{2x^2}{x+3}-3 = \frac{2x^2}{x+3}-\frac{3}{1}\cdot\frac{x+3}{x+3}$$

$$= \frac{2x^2}{x+3}-\frac{3x+9}{x+3} = \frac{2x^2-(3x+9)}{x+3}$$

$$= \frac{2x^2-3x-9}{x+3} = \frac{(2x+3)(x-3)}{x+3}$$

Rewrite the complex rational expression using the single rational expressions and simplify.

$$\frac{2+\dfrac{3}{x}}{\dfrac{2x^2}{x+3}-3} = \frac{\dfrac{2x+3}{x}}{\dfrac{(2x+3)(x-3)}{x+3}}$$

$$= \frac{2x+3}{x}\cdot\frac{x+3}{(2x+3)(x-3)}$$

$$= \frac{2x+3}{x}\cdot\frac{x+3}{(2x+3)(x-3)} = \frac{x+3}{x(x-3)}$$

33. Using Method 2, the LCD of all the denominators in the complex rational expression is $3x^2$.

$$\frac{\dfrac{x}{3}-\dfrac{3}{x}}{\dfrac{3}{x^2}-\dfrac{1}{3}}=\frac{\dfrac{x}{3}-\dfrac{3}{x}}{\dfrac{3}{x^2}-\dfrac{1}{3}}\cdot\frac{3x^2}{3x^2}$$

$$=\frac{\dfrac{x}{3}\cdot 3x^2-\dfrac{3}{x}\cdot 3x^2}{\dfrac{3}{x^2}\cdot 3x^2-\dfrac{1}{3}\cdot 3x^2}$$

$$=\frac{\dfrac{x}{\cancel{3}}\cdot \cancel{3}x^2-\dfrac{3}{\cancel{x}}\cdot 3x^{\cancel{2}}}{\dfrac{3}{\cancel{x^2}}\cdot 3\cancel{x^2}-\dfrac{1}{\cancel{3}}\cdot \cancel{3}x^2}$$

$$=\frac{x^3-9x}{9-x^2}$$

$$=\frac{-x(3-x)(3+x)}{(3-x)(3+x)}$$

$$=\frac{-x\cancel{(3-x)}\cancel{(3+x)}}{\cancel{(3-x)}\cancel{(3+x)}}=-x$$

35. Using Method 1, write the numerator of the complex rational expression as a single rational expression. Note that the LCD of the rational expressions in the numerator is $(x-1)(x-3)$.

$$\frac{2x+1}{x-1}-\frac{x-1}{x-3}=\frac{2x+1}{x-1}\cdot\frac{x-3}{x-3}-\frac{x-1}{x-3}\cdot\frac{x-1}{x-1}$$

$$=\frac{2x^2-5x-3}{(x-1)(x-3)}-\frac{x^2-2x+1}{(x-1)(x-3)}$$

$$=\frac{2x^2-5x-3-\left(x^2-2x+1\right)}{(x-1)(x-3)}$$

$$=\frac{2x^2-5x-3-x^2+2x-1}{(x-1)(x-3)}$$

$$=\frac{x^2-3x-4}{(x-1)(x-3)}$$

$$=\frac{(x-4)(x+1)}{(x-1)(x-3)}$$

The denominator is already a single rational expression: $x^2-3x-4=(x-4)(x+1)$.

Rewrite the complex rational expression using

the single rational expressions and simplify.

$$\frac{\dfrac{2x+1}{x-1}-\dfrac{x-1}{x-3}}{x^2-3x-4}=\frac{\dfrac{(x-4)(x+1)}{(x-1)(x-3)}}{\dfrac{(x-4)(x+1)}{1}}$$

$$=\frac{(x-4)(x+1)}{(x-1)(x-3)}\cdot\frac{1}{(x-4)(x+1)}$$

$$=\frac{\cancel{(x-4)}\cancel{(x+1)}}{(x-1)(x-3)}\cdot\frac{1}{\cancel{(x-4)}\cancel{(x+1)}}$$

$$=\frac{1}{(x-1)(x-3)}$$

37. Using Method 2, the LCD of all the denominators in the complex rational expression is $(x-2)(x+4)$.

$$\frac{\dfrac{x-4}{x+4}+\dfrac{x-4}{x-2}}{1-\dfrac{2}{x-2}}=\frac{\dfrac{x-4}{x+4}+\dfrac{x-4}{x-2}}{1-\dfrac{2}{x-2}}\cdot\frac{(x-2)(x+4)}{(x-2)(x+4)}$$

$$=\frac{\dfrac{x-4}{x+4}\cdot(x-2)(x+4)+\dfrac{x-4}{x-2}\cdot(x-2)(x+4)}{1\cdot(x-2)(x+4)-\dfrac{2}{x-2}\cdot(x-2)(x+4)}$$

$$=\frac{\dfrac{x-4}{\cancel{x+4}}\cdot(x-2)\cancel{(x+4)}+\dfrac{x-4}{\cancel{x-2}}\cdot\cancel{(x-2)}(x+4)}{1\cdot(x-2)(x+4)-\dfrac{2}{\cancel{x-2}}\cdot\cancel{(x-2)}(x+4)}$$

$$=\frac{(x-4)(x-2)+(x-4)(x+4)}{(x-2)(x+4)-2(x+4)}$$

$$=\frac{(x-4)\left[(x-2)+(x+4)\right]}{(x+4)\left[(x-2)-2\right]}$$

$$=\frac{(x-4)[2x+2]}{(x+4)[x-4]}$$

$$=\frac{2(x-4)(x+1)}{(x+4)(x-4)}$$

$$=\frac{2(x+1)}{x+4}$$

39. Using Method 1, write the numerator of the complex rational expression as a single rational expression. Note that the LCD of the rational expressions in the numerator is $(b-5)(b+5)$.

$$\frac{b^2}{b^2-25}-\frac{b}{b+5}=\frac{b^2}{(b-5)(b+5)}-\frac{b}{b+5}\cdot\frac{b-5}{b-5}$$

$$=\frac{b^2}{(b-5)(b+5)}-\frac{b^2-5b}{(b-5)(b+5)}$$

$$=\frac{b^2-\left(b^2-5b\right)}{(b-5)(b+5)}$$

$$=\frac{b^2-b^2+5b}{(b-5)(b+5)}$$

$$=\frac{5b}{(b-5)(b+5)}$$

Write the denominator of the complex rational expression as a single rational expression. Note that the LCD of the rational expressions in the denominator is $(b-5)(b+5)$.

$$\frac{b}{b^2-25}-\frac{1}{b-5}=\frac{b}{(b-5)(b+5)}-\frac{1}{b-5}\cdot\frac{b+5}{b+5}$$

$$=\frac{b}{(b-5)(b+5)}-\frac{b+5}{(b-5)(b+5)}$$

$$=\frac{b-(b+5)}{(b-5)(b+5)}$$

$$=\frac{b-b-5}{(b-5)(b+5)}$$

$$=\frac{-5}{(b-5)(b+5)}$$

Rewrite the complex rational expression using the single rational expressions and simplify.

$$\frac{\dfrac{b^2}{b^2-25}-\dfrac{b}{b+5}}{\dfrac{b}{b^2-25}-\dfrac{1}{b-5}}$$

$$=\frac{\dfrac{5b}{(b-5)(b+5)}}{\dfrac{-5}{(b-5)(b+5)}}$$

$$=\frac{5b}{(b-5)(b+5)}\cdot\frac{(b-5)(b+5)}{-5}$$

$$=\frac{\cancel{5}b}{\cancel{(b-5)}\,\cancel{(b+5)}}\cdot\frac{\cancel{(b-5)}\,\cancel{(b+5)}}{-1\cdot\cancel{5}}$$

$$=\frac{b}{-1}=-b$$

41. Using Method 2, the LCD of all the denominators in the complex expression is x.

$$\frac{3}{\dfrac{2}{x}-1}=\frac{3}{\dfrac{2}{x}-1}\cdot\frac{x}{x}=\frac{3x}{2-x}$$

43. Using Method 2, the LCD of all the denominators in the complex expression is $x-1$.

$$\frac{1}{\dfrac{4}{x-1}+2}=\frac{1}{\dfrac{4}{x-1}+2}\cdot\frac{x-1}{x-1}$$

$$=\frac{x-1}{4+2(x-1)}$$

$$=\frac{x-1}{4+2x-2}$$

$$=\frac{x-1}{2x+2}$$

$$=\frac{x-1}{2(x+1)}$$

45. Using Method 2, the LCD of all the denominators in the complex expression is $x-1$.

$$\frac{\dfrac{4}{x-1}-1}{\dfrac{4}{x-1}+3}=\frac{\dfrac{4}{x-1}-1}{\dfrac{4}{x-1}+3}\cdot\frac{x-1}{x-1}$$

$$=\frac{4-1(x-1)}{4+3(x-1)}$$

$$=\frac{4-x+1}{4+3x-3}$$

$$=\frac{-x+5}{3x+1}$$

47. Rewrite the expression so that it does not contain any negative exponents.

$$\frac{3x^{-1}+3y^{-1}}{x^{-2}-y^{-2}}=\frac{\dfrac{3}{x}+\dfrac{3}{y}}{\dfrac{1}{x^2}-\dfrac{1}{y^2}}$$

Using Method 2, the LCD of all the denominators in the complex rational expression is x^2y^2.

$$\frac{3x^{-1}+3y^{-1}}{x^{-2}-y^{-2}} = \frac{\dfrac{3}{x}+\dfrac{3}{y}}{\dfrac{1}{x^2}-\dfrac{1}{y^2}} = \frac{\dfrac{3}{x}+\dfrac{3}{y}}{\dfrac{1}{x^2}-\dfrac{1}{y^2}} \cdot \frac{x^2 y^2}{x^2 y^2}$$

$$= \frac{\dfrac{3}{x}\cdot x^2 y^2 + \dfrac{3}{y}\cdot x^2 y^2}{\dfrac{1}{x^2}\cdot x^2 y^2 - \dfrac{1}{y^2}\cdot x^2 y^2}$$

$$= \frac{3xy^2 + 3x^2 y}{y^2 - x^2}$$

$$= \frac{3xy(y+x)}{(y-x)(y+x)} = \frac{3xy}{y-x}$$

49. Rewrite the expression so that it does not contain any negative exponents.

$$\frac{(m+n)^{-1}}{m^{-1}+n^{-1}} = \frac{\dfrac{1}{m+n}}{\dfrac{1}{m}+\dfrac{1}{n}}$$

Using Method 1, note the numerator of the complex rational expression is already a single rational expression.

Write the denominator of the complex rational expression as a single rational expression. Note that the LCD of the rational expressions in the denominator is mn.

$$\frac{1}{m}+\frac{1}{n} = \frac{1}{m}\cdot\frac{n}{n}+\frac{1}{n}\cdot\frac{m}{m}$$

$$= \frac{n}{mn}+\frac{m}{mn} = \frac{n+m}{mn} = \frac{m+n}{mn}$$

Rewrite the complex rational expression using the single rational expressions and simplify.

$$\frac{(m+n)^{-1}}{m^{-1}+n^{-1}} = \frac{\dfrac{1}{m+n}}{\dfrac{1}{m}+\dfrac{1}{n}} = \frac{\dfrac{1}{m+n}}{\dfrac{m+n}{mn}}$$

$$= \frac{1}{m+n}\cdot\frac{mn}{m+n} = \frac{mn}{(m+n)^2}$$

51. Rewrite the expression so that it does not contain any negative exponents.

$$\frac{a^{-2}b^{-1}-a^{-1}b^{-2}}{4a^{-2}-4b^{-2}} = \frac{\dfrac{1}{a^2 b}-\dfrac{1}{ab^2}}{\dfrac{4}{a^2}-\dfrac{4}{b^2}}$$

Using Method 2, the LCD of all the denominators in the complex rational expression

is $a^2 b^2$.

$$\frac{a^{-2}b^{-1}-a^{-1}b^{-2}}{4a^{-2}-4b^{-2}} = \frac{\dfrac{1}{a^2 b}-\dfrac{1}{ab^2}}{\dfrac{4}{a^2}-\dfrac{4}{b^2}}$$

$$= \frac{\dfrac{1}{a^2 b}-\dfrac{1}{ab^2}}{\dfrac{4}{a^2}-\dfrac{4}{b^2}}\cdot\frac{a^2 b^2}{a^2 b^2}$$

$$= \frac{\dfrac{1}{a^2 b}\cdot a^2 b^2 - \dfrac{1}{ab^2}\cdot a^2 b^2}{\dfrac{4}{a^2}\cdot a^2 b^2 - \dfrac{4}{b^2}\cdot a^2 b^2}$$

$$= \frac{b-a}{4b^2 - 4a^2}$$

$$= \frac{b-a}{4(b-a)(b+a)}$$

$$= \frac{1}{4(b+a)}$$

53. a. Using Method 2, the LCD of all the denominators in the complex rational expression is $R_1 R_2$.

$$R = \frac{1}{\dfrac{1}{R_1}+\dfrac{1}{R_2}} = \frac{1}{\dfrac{1}{R_1}+\dfrac{1}{R_2}}\cdot\frac{R_1 R_2}{R_1 R_2}$$

$$= \frac{1\cdot R_1 R_2}{\dfrac{1}{R_1}\cdot R_1 R_2 + \dfrac{1}{R_2}\cdot R_1 R_2}$$

$$= \frac{1\cdot R_1 R_2}{\dfrac{1}{\cancel{R_1}}\cdot\cancel{R_1} R_2 + \dfrac{1}{\cancel{R_2}}\cdot R_1 \cancel{R_2}}$$

$$= \frac{R_1 R_2}{R_2 + R_1} \quad\text{or}\quad \frac{R_1 R_2}{R_1 + R_2}$$

b. For $R_1 = 4$ ohms and $R_2 = 10$ ohms,

$$R = \frac{R_1 R_2}{R_1 + R_2} = \frac{4\cdot 10}{4+10} = \frac{40}{14} = \frac{20}{7} \approx 2.857$$

ohms.

55. a. Using Method 2, the LCD of all the denominators in the complex rational expression is $R_1 R_2$.

$$f = \cfrac{1}{(n-1)\left[\dfrac{1}{R_1} + \dfrac{1}{R_2}\right]}$$

$$= \cfrac{1}{(n-1)\left[\dfrac{1}{R_1} + \dfrac{1}{R_2}\right]} \cdot \dfrac{R_1 R_2}{R_1 R_2}$$

$$= \cfrac{1 \cdot R_1 R_2}{(n-1)\left[\dfrac{1}{R_1} \cdot R_1 R_2 + \dfrac{1}{R_2} \cdot R_1 R_2\right]}$$

$$= \dfrac{R_1 R_2}{(n-1)\left[R_2 + R_1\right]}$$

$$\text{or } \dfrac{R_1 R_2}{nR_2 + nR_1 - R_2 - R_1}$$

b. If $n = 1.5$, $R_1 = 0.5$ meter, and $R_2 = 0.3$ meter, then

$$f = \dfrac{R_1 R_2}{(n-1)\left[R_2 + R_1\right]}$$

$$= \dfrac{(0.5)(0.3)}{(1.5-1)(0.3+0.5)} = 0.375 \text{ meter}$$

57. a.

$$1 + \cfrac{1}{1 + \dfrac{1}{x}} = 1 + \cfrac{1}{\dfrac{x}{x} + \dfrac{1}{x}}$$

$$= 1 + \cfrac{1}{\dfrac{x+1}{x}}$$

$$= 1 + 1 \cdot \dfrac{x}{x+1}$$

$$= 1 + \dfrac{x}{x+1}$$

$$= \dfrac{x+1}{x+1} + \dfrac{x}{x+1}$$

$$= \dfrac{x+1+x}{x+1}$$

$$= \dfrac{2x+1}{x+1}$$

b.

$$1 + \cfrac{1}{1 + \cfrac{1}{1 + \dfrac{1}{x}}} = 1 + \cfrac{1}{\dfrac{2x+1}{x+1}} \quad (\text{from part (a)})$$

$$= 1 + 1 \cdot \dfrac{x+1}{2x+1}$$

$$= 1 + \dfrac{x+1}{2x+1}$$

$$= \dfrac{2x+1}{2x+1} + \dfrac{x+1}{2x+1}$$

$$= \dfrac{2x+1+x+1}{2x+1}$$

$$= \dfrac{3x+2}{2x+1}$$

c.

$$1 + \cfrac{1}{1 + \cfrac{1}{1 + \cfrac{1}{1 + \dfrac{1}{x}}}} = 1 + \cfrac{1}{\dfrac{3x+2}{2x+1}} \quad (\text{from part (b)})$$

$$= 1 + 1 \cdot \dfrac{2x+1}{3x+2}$$

$$= 1 + \dfrac{2x+1}{3x+2}$$

$$= \dfrac{3x+2}{3x+2} + \dfrac{2x+1}{3x+2}$$

$$= \dfrac{3x+2+2x+1}{3x+2}$$

$$= \dfrac{5x+3}{3x+2}$$

d. $1 + \cfrac{1}{1 + \cfrac{1}{1 + \cfrac{1}{1 + \cfrac{1}{1 + \cfrac{1}{x}}}}}$

$= 1 + \dfrac{1}{\dfrac{5x+3}{3x+2}}$ (from part (c))

$= 1 + 1 \cdot \dfrac{3x+2}{5x+3}$

$= 1 + \dfrac{3x+2}{5x+3}$

$= \dfrac{5x+3}{5x+3} + \dfrac{3x+2}{5x+3}$

$= \dfrac{5x+3+3x+2}{5x+3}$

$= \dfrac{8x+5}{5x+3}$

e. From part (a), 2, 1, 1;
From part (b), 3, 2, 1;
From part (c), 5, 3, 2;
From part (d), 8, 5, 3.
Pattern: Each subsequent term is the sum of the previous two.
The first 6 terms in the Fibonacci sequence are: 1, 1, 2, 3, 5, 8.

59. $f(x) = \dfrac{1}{x^2}$

$\dfrac{f(x+h) - f(x)}{h} = \dfrac{\dfrac{1}{(x+h)^2} - \dfrac{1}{x^2}}{h}$

Using method 2, the LCD of all the denominators in the complex rational expression is $x^2 (x+h)^2$.

$$\frac{\dfrac{1}{(x+h)^2}-\dfrac{1}{x^2}}{h}=\frac{\dfrac{1}{(x+h)^2}-\dfrac{1}{x^2}}{h}\cdot\frac{x^2(x+h)^2}{x^2(x+h)^2}$$

$$=\frac{\dfrac{1}{(x+h)^2}\cdot x^2(x+h)^2-\dfrac{1}{x^2}\cdot x^2(x+h)^2}{h\cdot x^2(x+h)^2}$$

$$=\frac{x^2-(x+h)^2}{hx^2(x+h)^2}$$

$$=\frac{x^2-\left(x^2+2xh+h^2\right)}{hx^2(x+h)^2}$$

$$=\frac{x^2-x^2-2xh-h^2}{hx^2(x+h)^2}$$

$$=\frac{h(-2x-h)}{hx^2(x+h)^2}$$

$$=-\frac{2x+h}{x^2(x+h)^2}$$

61. $f(x)=\dfrac{1}{x-1}$

$$\frac{f(x+h)-f(x)}{h}=\frac{\dfrac{1}{x+h-1}-\dfrac{1}{x-1}}{h}$$

Using method 2, the LCD of all the denominators in the complex rational expression is $(x+h-1)(x-1)$.

$$\frac{\dfrac{1}{x+h-1}-\dfrac{1}{x-1}}{h}=\frac{\dfrac{1}{x+h-1}-\dfrac{1}{x-1}}{h}\cdot\frac{(x+h-1)(x-1)}{(x+h-1)(x-1)}$$

$$=\frac{\dfrac{1}{x+h-1}\cdot(x+h-1)(x-1)-\dfrac{1}{x-1}\cdot(x+h-1)(x-1)}{h\cdot(x+h-1)(x-1)}$$

$$=\frac{(x-1)-(x+h-1)}{h\cdot(x+h-1)(x-1)}$$

$$=\frac{x-1-x-h+1}{h\cdot(x+h-1)(x-1)}$$

$$=\frac{-h}{h\cdot(x+h-1)(x-1)}$$

$$=-\frac{1}{(x+h-1)(x-1)}$$

63. Answers will vary.

65. $-5a + 2 = 22$
$$-5a = 20$$
$$a = -4$$
The solution set is $\{-4\}$.

67.
$$\frac{2}{3}x - \frac{5}{7}(x+21) = \frac{11}{21}x + \frac{4}{7}$$
$$21\left[\frac{2}{3}x - \frac{5}{7}(x+21)\right] = 21\left[\frac{11}{21}x + \frac{4}{7}\right]$$
$$21 \cdot \frac{2}{3}x - 21 \cdot \frac{5}{7}(x+21) = 21 \cdot \frac{11}{21}x + 21 \cdot \frac{4}{7}$$
$$14x - 15(x+21) = 11x + 12$$
$$14x - 15x - 315 = 11x + 12$$
$$-x - 315 = 11x + 12$$
$$-12x - 315 = 12$$
$$-12x = 327$$
$$\frac{-12x}{-12} = \frac{327}{-12}$$
$$x = -\frac{109}{4}$$
The solution set is $\left\{-\frac{109}{4}\right\}$.

69.
$$3p^2 + 19p = 14$$
$$3p^2 + 19p - 14 = 0$$
$$(3p-2)(p+7) = 0$$
$$3p - 2 = 0 \quad \text{or} \quad p + 7 = 0$$
$$p = \frac{2}{3} \quad \text{or} \quad p = -7$$
The solution set is $\left\{-7, \frac{2}{3}\right\}$.

Putting the Concepts Together (Sections 5.1–5.3)

1. We need to find all values of x that cause the denominator $3x^2 - 17x - 6$ to equal 0.
$$3x^2 - 17x - 6 = 0$$
$$(3x+1)(x-6) = 0$$
$$3x + 1 = 0 \quad \text{or} \quad x - 6 = 0$$
$$3x = -1 \quad \text{or} \quad x = 6$$
$$x = -\frac{1}{3}$$
Thus, the domain of $g(x) = \dfrac{3x+1}{3x^2 - 17x - 6}$ is
$$\left\{x \,\middle|\, x \neq -\frac{1}{3},\ x \neq 6\right\}.$$

2. $\dfrac{24n - 4n^2}{2n^2 - 9n - 18} = \dfrac{-4n(n-6)}{(2n+3)(n-6)}$
$$= \frac{-4n\,\cancel{(n-6)}}{(2n+3)\,\cancel{(n-6)}}$$
$$= \frac{-4n}{2n+3}$$

3. $\dfrac{2p^2 - pq - 10q^2}{3p^2 + 2pq - 8q^2} = \dfrac{(2p-5q)(p+2q)}{(3p-4q)(p+2q)}$
$$= \frac{(2p-5q)\,\cancel{(p+2q)}}{(3p-4q)\,\cancel{(p+2q)}}$$
$$= \frac{2p-5q}{3p-4q}$$

4. $\dfrac{a^2 - 16}{12a^2 + 48a} \cdot \dfrac{6a^3 - 30a^2}{a^2 + 2a - 24}$
$$= \frac{(a-4)(a+4)}{12a(a+4)} \cdot \frac{6a^2(a-5)}{(a-4)(a+6)}$$
$$= \frac{\cancel{(a-4)}\,\cancel{(a+4)}}{2 \cdot \cancel{6a}\,\cancel{(a+4)}} \cdot \frac{\cancel{6a} \cdot a(a-5)}{\cancel{(a-4)}(a+6)} = \frac{a(a-5)}{2(a+6)}$$

5. $\dfrac{\dfrac{x^2 + x - 2}{3x^2 - 5x - 2}}{\dfrac{3x^2 - 2x - 1}{x^2 - 9x + 14}} = \dfrac{x^2 + x - 2}{3x^2 - 5x - 2} \cdot \dfrac{x^2 - 9x + 14}{3x^2 - 2x - 1}$
$$= \frac{(x-1)(x+2)}{(3x+1)(x-2)} \cdot \frac{(x-2)(x-7)}{(3x+1)(x-1)}$$
$$= \frac{\cancel{(x-1)}(x+2)}{(3x+1)\,\cancel{(x-2)}} \cdot \frac{\cancel{(x-2)}(x-7)}{(3x+1)\,\cancel{(x-1)}}$$
$$= \frac{(x+2)(x-7)}{(3x+1)^2}$$

6. $\dfrac{x^2 - 10}{x^2 - 4} - \dfrac{3x}{x^2 - 4} = \dfrac{x^2 - 10 - 3x}{x^2 - 4} = \dfrac{x^2 - 3x - 10}{x^2 - 4}$
$$= \frac{(x-5)(x+2)}{(x-2)(x+2)} = \frac{(x-5)\,\cancel{(x+2)}}{(x-2)\,\cancel{(x+2)}} = \frac{x-5}{x-2}$$

7. $n^2 - 7n + 10 = (n-2)(n-5)$

$n^2 - 8n + 15 = (n-3)(n-5)$

$\text{LCD} = (n-2)(n-3)(n-5)$

$$\frac{3n}{n^2 - 7n + 10} - \frac{2n}{n^2 - 8n + 15}$$

$$= \frac{3n}{(n-2)(n-5)} - \frac{2n}{(n-3)(n-5)}$$

$$= \frac{3n}{(n-2)(n-5)} \cdot \frac{n-3}{n-3} - \frac{2n}{(n-3)(n-5)} \cdot \frac{n-2}{n-2}$$

$$= \frac{3n^2 - 9n}{(n-2)(n-3)(n-5)} - \frac{2n^2 - 4n}{(n-2)(n-3)(n-5)}$$

$$= \frac{3n^2 - 9n - \left(2n^2 - 4n\right)}{(n-2)(n-3)(n-5)} = \frac{3n^2 - 9n - 2n^2 + 4n}{(n-2)(n-3)(n-5)}$$

$$= \frac{n^2 - 5n}{(n-2)(n-3)(n-5)} = \frac{n(n-5)}{(n-2)(n-3)(n-5)}$$

$$= \frac{n}{(n-2)(n-3)}$$

8. $y^2 + 5y - 24 = (y-3)(y+8)$

$y^2 + 4y - 32 = (y-4)(y+8)$

$\text{LCD} = (y-3)(y-4)(y+8)$

$$\frac{3y+2}{y^2 + 5y - 24} + \frac{7}{y^2 + 4y - 32}$$

$$= \frac{3y+2}{(y-3)(y+8)} \cdot \frac{y-4}{y-4} + \frac{7}{(y-4)(y+8)} \cdot \frac{y-3}{y-3}$$

$$= \frac{3y^2 - 10y - 8}{(y-3)(y-4)(y+8)} + \frac{7y - 21}{(y-3)(y-4)(y+8)}$$

$$= \frac{3y^2 - 10y - 8 + 7y - 21}{(y-3)(y-4)(y+8)}$$

$$= \frac{3y^2 - 3y - 29}{(y-3)(y-4)(y+8)}$$

9. $P(x) = f(x) \cdot g(x)$

$$= \frac{2x+1}{x^2 - 11x + 28} \cdot \frac{3x - 12}{4x^2 + 4x + 1}$$

$$= \frac{2x+1}{(x-4)(x-7)} \cdot \frac{3(x-4)}{(2x+1)(2x+1)}$$

$$= \frac{\cancel{2x+1}}{\cancel{(x-4)}(x-7)} \cdot \frac{3\cancel{(x-4)}}{\cancel{(2x+1)}(2x+1)}$$

$$= \frac{3}{(x-7)(2x+1)}$$

10. $x - 7 = x - 7$

$x^2 - 11x + 28 = (x-4)(x-7)$

$\text{LCD} = (x-4)(x-7)$

$D(x) = h(x) - f(x)$

$$= \frac{3x}{x-7} - \frac{2x+1}{x^2 - 11x + 28}$$

$$= \frac{3x}{x-7} - \frac{2x+1}{(x-4)(x-7)}$$

$$= \frac{3x}{x-7} \cdot \frac{x-4}{x-4} - \frac{2x+1}{(x-4)(x-7)}$$

$$= \frac{3x^2 - 12x}{(x-4)(x-7)} - \frac{2x+1}{(x-4)(x-7)}$$

$$= \frac{3x^2 - 12x - (2x+1)}{(x-4)(x-7)}$$

$$= \frac{3x^2 - 12x - 2x - 1}{(x-4)(x-7)}$$

$$= \frac{3x^2 - 14x - 1}{(x-4)(x-7)}$$

11. Using Method 2, the LCD of all the denominators in the complex rational expression is $m^2 n^2$.

$$\frac{\dfrac{1}{m^2} - \dfrac{1}{n^2}}{\dfrac{1}{m} - \dfrac{1}{n}} = \frac{\dfrac{1}{m^2} - \dfrac{1}{n^2}}{\dfrac{1}{m} - \dfrac{1}{n}} \cdot \frac{m^2 n^2}{m^2 n^2}$$

$$= \frac{\dfrac{1}{m^2} \cdot m^2 n^2 - \dfrac{1}{n^2} \cdot m^2 n^2}{\dfrac{1}{m} \cdot m^2 n^2 - \dfrac{1}{n} \cdot m^2 n^2}$$

$$= \frac{n^2 - m^2}{mn^2 - m^2 n}$$

$$= \frac{(n-m)(n+m)}{mn(n-m)}$$

$$= \frac{n+m}{mn}$$

     407

12. Using Method 2, the LCD of all the denominators in the complex rational expression is $(z-2)(z+2)$.

$$\frac{\dfrac{z^2-2}{z^2-4}+\dfrac{7}{z-2}}{\dfrac{z^2+z-24}{z^2-4}-\dfrac{2}{z+2}}=\frac{\dfrac{z^2-2}{(z-2)(z+2)}+\dfrac{7}{z-2}}{\dfrac{z^2+z-24}{(z-2)(z+2)}-\dfrac{2}{z+2}}$$

$$=\frac{\dfrac{z^2-2}{(z-2)(z+2)}+\dfrac{7}{z-2}}{\dfrac{z^2+z-24}{(z-2)(z+2)}-\dfrac{2}{z+2}}\cdot\frac{(z-2)(z+2)}{(z-2)(z+2)}$$

$$=\frac{\dfrac{z^2-2}{(z-2)(z+2)}\cdot(z-2)(z+2)+\dfrac{7}{z-2}\cdot(z-2)(z+2)}{\dfrac{z^2+z-24}{(z-2)(z+2)}\cdot(z-2)(z+2)-\dfrac{2}{z+2}\cdot(z-2)(z+2)}$$

$$=\frac{z^2-2+7(z+2)}{z^2+z-24-2(z-2)}$$

$$=\frac{z^2-2+7z+14}{z^2+z-24-2z+4}$$

$$=\frac{z^2+7z+12}{z^2-z-20}$$

$$=\frac{(z+3)(z+4)}{(z+4)(z-5)}=\frac{z+3}{z-5}$$

Section 5.4

Are You Ready for This Section?

R1.
$$\frac{2}{3}x+\frac{1}{2}=\frac{3}{4}$$
$$12\left(\frac{2}{3}x+\frac{1}{2}\right)=12\left(\frac{3}{4}\right)$$
$$8x+6=9$$
$$8x=3$$
$$x=\frac{3}{8}$$

The solution set is $\left\{\dfrac{3}{8}\right\}$.

R2. $a\cdot c=3(-4)=-12$

The factors of -12 that add to 11 (the linear coefficient) are -1 and 12. Thus,
$$3z^2+11z-4=3z^2+12z-z-4$$
$$=3z(z+4)-1(z+4)$$
$$=(3z-1)(z+4)$$

R3. $6y^2 - y - 12 = 0$

$(3y + 4)(2y - 3) = 0$

$3y + 4 = 0$ or $2y - 3 = 0$

     $3y = -4$ or $2y = 3$

       $y = -\dfrac{4}{3}$ or $y = \dfrac{3}{2}$

The solution set is $\left\{ -\dfrac{4}{3},\ \dfrac{3}{2} \right\}$.

R4.   **a.**   If $x = -4$, then

$$\frac{x+4}{x^2 - 5x - 24} = \frac{-4+4}{(-4)^2 - 5(-4) - 24}$$

$$= \frac{-4+4}{16 + 20 - 24} = \frac{0}{12} = 0$$

Since $\dfrac{x+4}{x^2 - 5x - 24}$ is defined for $x = -4$, -4 is in the domain.

     b.   If $x = 8$, then

$$\frac{x+4}{x^2 - 5x - 24} = \frac{8+4}{8^2 - 5(8) - 24}$$

$$= \frac{8+4}{64 - 40 - 24}$$

$$= \frac{12}{0}$$

$$= \text{undefined}$$

Since $\dfrac{x+4}{x^2 - 5x - 24}$ is not defined for $x = 8$, 8 is not in the domain.

R5. $f(x) = 3$

$x^2 - 3x - 15 = 3$

$x^2 - 3x - 18 = 0$

$(x + 3)(x - 6) = 0$

$x + 3 = 0$ or $x - 6 = 0$

     $x = -3$ or $x = 6$

The solution set is $\{-3,\ 6\}$.

R6. If $g(4) = 3$, means that $y = 3$ when $x = 4$. Thus, the point $(4, 3)$ is on the graph of g.

Section 5.4 Quick Checks

   1. A <u>rational equation</u> is an equation that contains a rational expression.

2.
$$\frac{x-4}{x^2+4} = \frac{3}{3x+2}$$

$$\left(x^2+4\right)\left(3x+2\right)\left(\frac{x-4}{x^2+4}\right) = \left(x^2+4\right)\left(3x+2\right)\left(\frac{3}{3x+2}\right)$$

$$(3x+2)(x-4) = 3\left(x^2+4\right)$$

$$3x^2 - 10x - 8 = 3x^2 + 12$$

$$-10x - 8 = 12$$

$$-10x = 20$$

$$x = -2$$

Check:

$$\frac{-2-4}{(-2)^2+4} \stackrel{?}{=} \frac{3}{3(-2)+2}$$

$$\frac{-2-4}{4+4} \stackrel{?}{=} \frac{3}{-6+2}$$

$$\frac{-6}{8} \stackrel{?}{=} \frac{3}{-4}$$

$$-\frac{3}{4} = -\frac{3}{4} \leftarrow \text{True}$$

The solution checks, so the solution set is $\{-2\}$.

3.
$$\frac{4}{x+2} = \frac{7}{x+4}$$

$$(x+2)(x+4)\left(\frac{4}{x+2}\right) = (x+2)(x+4)\left(\frac{7}{x+4}\right)$$

$$(x+4)(4) = (x+2)(7)$$

$$4x + 16 = 7x + 14$$

$$-3x = -2$$

$$x = \frac{2}{3}$$

Check:

$$\frac{4}{\frac{2}{3}+2} \stackrel{?}{=} \frac{7}{\frac{2}{3}+4}$$

$$\frac{4}{\frac{2}{3}+2} \cdot \frac{3}{3} \stackrel{?}{=} \frac{7}{\frac{2}{3}+4} \cdot \frac{3}{3}$$

$$\frac{12}{2+6} \stackrel{?}{=} \frac{21}{2+12}$$

$$\frac{12}{8} \stackrel{?}{=} \frac{21}{14}$$

$$\frac{3}{2} = \frac{3}{2} \leftarrow \text{True}$$

The solution checks, so the solution set is $\left\{\frac{2}{3}\right\}$.

4.
$$\frac{5}{x}+\frac{1}{4}=\frac{3}{2x}-\frac{3}{2}$$

$$4x\left(\frac{5}{x}+\frac{1}{4}\right)=4x\left(\frac{3}{2x}-\frac{3}{2}\right)$$

$$4x\cdot\frac{5}{x}+4x\cdot\frac{1}{4}=4x\cdot\frac{3}{2x}-4x\cdot\frac{3}{2}$$

$$20+x=6-6x$$

$$20+7x=6$$

$$7x=-14$$

$$x=-2$$

Check:

$$\frac{5}{-2}+\frac{1}{4}\overset{?}{=}\frac{3}{2(-2)}-\frac{3}{2}$$

$$-\frac{5}{2}+\frac{1}{4}\overset{?}{=}\frac{3}{-4}-\frac{3}{2}$$

$$-\frac{10}{4}+\frac{1}{4}\overset{?}{=}-\frac{3}{4}-\frac{6}{4}$$

$$-\frac{9}{4}=-\frac{9}{4}\quad\leftarrow\text{True}$$

The solution checks, so the solution set is $\{-2\}$.

5.
$$\frac{5}{x}+2=\frac{10}{3x}+1$$

$$3x\left(\frac{5}{x}+2\right)=3x\left(\frac{10}{3x}+1\right)$$

$$3x\cdot\frac{5}{x}+3x\cdot2=3x\cdot\frac{10}{3x}+3x\cdot1$$

$$15+6x=10+3x$$

$$3x=-5$$

$$x=-\frac{5}{3}$$

Check:

$$\frac{5}{\left(-\dfrac{5}{3}\right)}+2\overset{?}{=}\frac{10}{3\left(-\dfrac{5}{3}\right)}+1$$

$$5\cdot\left(-\frac{3}{5}\right)+2\overset{?}{=}\frac{10}{\cancel{3}\left(-\dfrac{5}{\cancel{3}}\right)}+1$$

$$-3+2\overset{?}{=}\frac{10}{-5}+1$$

$$-1\overset{?}{=}-2+1$$

$$-1=-1\leftarrow\text{True}$$

The solution checks, so the solution set is $\left\{-\dfrac{5}{3}\right\}$.

6. First, we find the domain of the variable, x.

$x^2 + 5x + 4 = (x+1)(x+4)$, so $x \neq -1$, $x \neq -4$ in the first term.

$x^2 - 3x - 4 = (x+1)(x-4)$, so $x \neq -1$, $x \neq 4$ in the second term.

$x^2 - 16 = (x+4)(x-4)$, so $x \neq -4$, $x \neq 4$ in the third term.

$$\frac{3}{x^2+5x+4} + \frac{2}{x^2-3x-4} = \frac{4}{x^2-16}$$

$$\frac{3}{(x+1)(x+4)} + \frac{2}{(x+1)(x-4)} = \frac{4}{(x+4)(x-4)}$$

$$(x+1)(x+4)(x-4)\left(\frac{3}{(x+1)(x+4)} + \frac{2}{(x+1)(x-4)}\right) = (x+1)(x+4)(x-4)\left(\frac{4}{(x+4)(x-4)}\right)$$

$$3(x-4) + 2(x+4) = 4(x+1)$$
$$3x - 12 + 2x + 8 = 4x + 4$$
$$5x - 4 = 4x + 4$$
$$x - 4 = 4$$
$$x = 8$$

Check:

$$\frac{3}{8^2+5(8)+4} + \frac{2}{8^2-3(8)-4} \overset{?}{=} \frac{4}{8^2-16}$$

$$\frac{3}{108} + \frac{2}{36} \overset{?}{=} \frac{4}{48}$$

$$\frac{3}{108} + \frac{6}{108} \overset{?}{=} \frac{1}{12}$$

$$\frac{9}{108} \overset{?}{=} \frac{1}{12}$$

$$\frac{1}{12} = \frac{1}{12} \leftarrow \text{True}$$

The solution set is $\{8\}$.

7. <u>Extraneous solutions</u> are results that develop through the solution process but do not satisfy the original equation.

8. True

9. First, we find the domain of the variable, z.

$z^2 + 2z - 3 = (z-1)(z+3)$, so $z \neq -3$, $z \neq 1$ in the first term.

$z^2 + z - 2 = (z-1)(z+2)$, so $z \neq -2$, $z \neq 1$ in the second term.

$z^2 + 5z + 6 = (z+2)(z+3)$, so $z \neq -3$, $z \neq -2$ in the third term.

$$\frac{5}{z^2+2z-3} - \frac{3}{z^2+z-2} = \frac{1}{z^2+5z+6}$$

$$\frac{5}{(z-1)(z+3)} - \frac{3}{(z-1)(z+2)} = \frac{1}{(z+2)(z+3)}$$

$$(z-1)(z+2)(z+3)\left(\frac{5}{(z-1)(z+3)} - \frac{3}{(z-1)(z+2)}\right) = (z-1)(z+2)(z+3)\left(\frac{1}{(z+2)(z+3)}\right)$$

$$5(z+2) - 3(z+3) = z-1$$
$$5z+10-3z-9 = z-1$$
$$2z+1 = z-1$$
$$z+1 = -1$$
$$z = -2$$

Notice that $z = -2$ is not in the domain of the variable z. Therefore, there is no solution to the equation, and the solution set is $\varnothing$ or $\{\ \}$.

10. First, we find the domain of the variable, x.

$x - 4$, so $x \neq 4$ in the first or third term.

$x - 2$, so $x \neq 2$ in the second term.

$$\frac{5}{x-4} + \frac{3}{x-2} = \frac{11}{x-4}$$

$$(x-4)(x-2)\left(\frac{5}{x-4} + \frac{3}{x-2}\right)$$
$$= (x-4)(x-2)\left(\frac{11}{x-4}\right)$$

$$5(x-2) + 3(x-4) = 11(x-2)$$
$$5x-10+3x-12 = 11x-22$$
$$8x-22 = 11x-22$$
$$-3x = 0$$
$$x = 0$$

Check:

$$\frac{5}{0-4} + \frac{3}{0-2} \overset{?}{=} \frac{11}{0-4}$$

$$-\frac{5}{4} - \frac{3}{2} \overset{?}{=} -\frac{11}{4}$$

$$-\frac{5}{4} - \frac{6}{4} \overset{?}{=} -\frac{11}{4}$$

$$-\frac{11}{4} = -\frac{11}{4} \leftarrow \text{True}$$

The solution set is $\{0\}$.

11. First, we find the domain of the variable, p.

$p+2$, so $p \neq -2$ in the second term.

p, so $p \neq 0$ in the third term.

$$2 - \frac{3}{p+2} = \frac{6}{p}$$

$$p(p+2)\left(2 - \frac{3}{p+2}\right) = p(p+2)\left(\frac{6}{p}\right)$$

$$2p(p+2) - 3p = 6(p+2)$$

$$2p^2 + 4p - 3p = 6p + 12$$

$$2p^2 - 5p - 12 = 0$$

$$(2p+3)(p-4) = 0$$

$$2p+3 = 0 \quad \text{or} \quad p-4 = 0$$

$$p = -\frac{3}{2} \quad \text{or} \quad p = 4$$

Check:

$$2 - \frac{3}{\left(-\frac{3}{2}\right)+2} \overset{?}{=} \frac{6}{\left(-\frac{3}{2}\right)}$$

$$2 - \frac{3}{\left(-\frac{3}{2}\right)+\frac{4}{2}} \overset{?}{=} \frac{6}{\left(-\frac{3}{2}\right)}$$

$$2 - \frac{3}{\frac{1}{2}} \overset{?}{=} \frac{6}{\left(-\frac{3}{2}\right)}$$

$$2 - 3 \cdot \frac{2}{1} \overset{?}{=} 6 \cdot \left(-\frac{2}{3}\right)$$

$$2 - 6 \overset{?}{=} -4$$

$$-4 = -4 \leftarrow \text{True}$$

$$2 - \frac{3}{4+2} \overset{?}{=} \frac{6}{4}$$

$$2 - \frac{3}{6} \overset{?}{=} \frac{3}{2}$$

$$\frac{4}{2} - \frac{1}{2} \overset{?}{=} \frac{3}{2}$$

$$\frac{3}{2} = \frac{3}{2} \leftarrow \text{True}$$

Both check, so the solution set is $\left\{-\frac{3}{2}, 4\right\}$.

12. First, we find the domain of the variable, z.

$z+4$, so $z \neq -4$ in the first term.

$z-3$, so $z \neq 3$ in the second term.

$z^2 + z - 12 = (z+4)(z-3)$, so $z \neq -4, z \neq 3$ in the third term.

$$\frac{z+1}{z+4} + \frac{z+1}{z-3} = \frac{z^2+z+16}{z^2+z-12}$$

$$\frac{z+1}{z+4} + \frac{z+1}{z-3} = \frac{z^2+z+16}{(z+4)(z-3)}$$

$$(z+4)(z-3)\left(\frac{z+1}{z+4} + \frac{z+1}{z-3}\right)$$

$$= (z+4)(z-3)\left(\frac{z^2+z+16}{(z+4)(z-3)}\right)$$

$$(z-3)(z+1) + (z+4)(z+1) = z^2+z+16$$

$$z^2 - 2z - 3 + z^2 + 5z + 4 = z^2 + z + 16$$

$$2z^2 + 3z + 1 = z^2 + z + 16$$

$$z^2 + 2z - 15 = 0$$

$$(z+5)(z-3) = 0$$

$$z+5 = 0 \quad \text{or} \quad z-3 = 0$$

$$z = -5 \quad \text{or} \quad z = 3$$

Since $z = 3$ is not in the domain of the variable, it is an extraneous solution.

Check $z = -5$:

$$\frac{-5+1}{-5+4} + \frac{-5+1}{-5-3} \overset{?}{=} \frac{(-5)^2 + (-5) + 16}{(-5)^2 + (-5) - 12}$$

$$\frac{-4}{-1} + \frac{-4}{-8} \overset{?}{=} \frac{25 + (-5) + 16}{25 + (-5) - 12}$$

$$4 + \frac{1}{2} \overset{?}{=} \frac{36}{8}$$

$$\frac{9}{2} = \frac{9}{2} \leftarrow \text{True}$$

The solution checks, so the solution set is $\{-5\}$.

13. False; to solve $f(x) = 4$ when $f(x) = x + \frac{3}{x}$, substitute 4 for $f(x)$ and then solve for x.

14.
$$g(x) = 4$$

$$x - \frac{5}{x} = 4$$

$$x\left(x - \frac{5}{x}\right) = x \cdot 4$$

$$x^2 - 5 = 4x$$

$$x^2 - 4x - 5 = 0$$

$$(x+1)(x-5) = 0$$

$$x+1 = 0 \quad \text{or} \quad x-5 = 0$$

$$x = -1 \quad \text{or} \quad x = 5$$

The solution set is $\{-1, 5\}$. Thus, $g(-1) = 4$ and $g(5) = 4$, so the points $(-1, 4)$ and $(5, 4)$ are on the graph of g.

15.
$$f(x) = 1$$
$$2x - \frac{3}{x} = 1$$
$$x\left(2x - \frac{3}{x}\right) = x \cdot 1$$
$$2x^2 - 3 = x$$
$$2x^2 - x - 3 = 0$$
$$(2x - 3)(x + 1) = 0$$
$$2x - 3 = 0 \quad \text{or} \quad x + 1 = 0$$
$$x = \frac{3}{2} \quad \text{or} \quad x = -1$$

The solution set is $\left\{-1, \frac{3}{2}\right\}$. Thus, $f(-1) = 1$

and $f\left(\frac{3}{2}\right) = 1$, so the points $(-1, 1)$ and

$\left(\frac{3}{2}, 1\right)$ are on the graph of f.

16.
$$C(t) = 4$$
$$\frac{50t}{t^2 + 6} = 4$$
$$\left(t^2 + 6\right)\left(\frac{50t}{t^2 + 6}\right) = \left(t^2 + 6\right) \cdot 4$$
$$50t = 4t^2 + 24$$
$$-4t^2 + 50t - 24 = 0$$
$$-2(2t - 1)(t - 12) = 0$$
$$2t - 1 = 0 \quad \text{or} \quad t - 12 = 0$$
$$t = \frac{1}{2} \quad \text{or} \quad t = 12$$

The concentration of the drug will be

4 milligrams per liter after $\frac{1}{2}$ hour and after

12 hours.

5.4 Exercises

17.
$$\frac{3}{z} - \frac{1}{2z} = -\frac{5}{8} \qquad \text{Check:}$$
$$8z\left(\frac{3}{z} - \frac{1}{2z}\right) = 8z\left(-\frac{5}{8}\right) \quad \frac{3}{-4} - \frac{1}{2(-4)} \overset{?}{=} -\frac{5}{8}$$
$$8(3) - 4(1) = z(-5) \qquad -\frac{3}{4} - \left(-\frac{1}{8}\right) \overset{?}{=} -\frac{5}{8}$$
$$24 - 4 = -5z \qquad\qquad -\frac{6}{8} + \frac{1}{8} \overset{?}{=} -\frac{5}{8}$$
$$20 = -5z \qquad\qquad\qquad -\frac{5}{8} = -\frac{5}{8}$$
$$-4 = z \qquad\qquad\qquad\qquad \uparrow$$
$$\qquad\qquad\qquad\qquad\qquad\qquad \text{True}$$

The solution checks, so the solution set is $\{-4\}$.

19.
$$\frac{y+2}{y-5} = \frac{y+6}{y+1}$$
$$(y-5)(y+1)\left(\frac{y+2}{y-5}\right) = (y-5)(y+1)\left(\frac{y+6}{y+1}\right)$$
$$(y+1)(y+2) = (y-5)(y+6)$$
$$y^2 + 3y + 2 = y^2 + y - 30$$
$$2y = -32$$
$$y = -16$$

Check: $\dfrac{-16+2}{-16-5} \overset{?}{=} \dfrac{-16+6}{-16+1}$

$$\frac{-14}{-21} \overset{?}{=} \frac{-10}{-15}$$
$$\frac{2}{3} = \frac{2}{3} \quad \leftarrow \text{True}$$

The solution checks, so the solution set is
$\{-16\}$.

21.
$$\frac{x+8}{x+4} = \frac{x+2}{x-2}$$
$$(x+4)(x-2)\left(\frac{x+8}{x+4}\right) = (x+4)(x-2)\left(\frac{x+2}{x-2}\right)$$
$$(x-2)(x+8) = (x+4)(x+2)$$
$$x^2 + 6x - 16 = x^2 + 6x + 8$$
$$-16 = 8$$

The final statement is a contradiction, so the
solution set is $\{\ \}$ or $\varnothing$.

23.
$$a - \frac{5}{a} = 4$$
$$a\left(a - \frac{5}{a}\right) = a \cdot 4$$
$$a^2 - 5 = 4a$$
$$a^2 - 4a - 5 = 0$$
$$(a-5)(a+1) = 0$$
$$a-5 = 0 \text{ or } a+1 = 0$$
$$a = 5 \text{ or } a = -1$$

Check: Check:

$$5 - \frac{5}{5} \overset{?}{=} 4 \qquad -1 - \frac{5}{(-1)} \overset{?}{=} 4$$
$$5 - 1 \overset{?}{=} 4 \qquad\qquad -1 + 5 \overset{?}{=} 4$$
$$4 = 4 \leftarrow \text{True} \qquad 4 = 4 \leftarrow \text{True}$$

Both solutions check, so the solution set is $\{-1, 5\}$.

25.
$$6p - \frac{3}{p} = 7$$
$$p\left(6p - \frac{3}{p}\right) = p \cdot 7$$
$$6p^2 - 3 = 7p$$
$$6p^2 - 7p - 3 = 0$$
$$(3p+1)(2p-3) = 0$$
$$3p+1 = 0 \text{ or } 2p-3 = 0$$
$$3p = -1 \text{ or } 2p = 3$$
$$p = -\frac{1}{3} \text{ or } p = \frac{3}{2}$$

Check: Check:

$$6\left(-\frac{1}{3}\right) - \frac{3}{\left(-\frac{1}{3}\right)} \overset{?}{=} 7 \qquad 6\left(\frac{3}{2}\right) - \frac{3}{\left(\frac{3}{2}\right)} \overset{?}{=} 7$$
$$-2 + 9 \overset{?}{=} 7 \qquad\qquad 9 - 2 \overset{?}{=} 7$$
$$7 = 7 \qquad\qquad\qquad 7 = 7$$
$$\uparrow \qquad\qquad\qquad\qquad \uparrow$$
$$\text{True} \qquad\qquad\qquad \text{True}$$

Both solutions check, so the solution set is $\left\{-\frac{1}{3}, \frac{3}{2}\right\}$.

27.
$$\frac{5-p}{p-5} + 2 = \frac{1}{p}$$
$$p(p-5)\left(\frac{5-p}{p-5} + 2\right) = p(p-5)\left(\frac{1}{p}\right)$$
$$p(5-p) + 2p(p-5) = p-5$$
$$5p - p^2 + 2p^2 - 10p = p-5$$
$$p^2 - 6p + 5 = 0$$
$$(p-5)(p-1) = 0$$
$$p-5 = 0 \text{ or } p-1 = 0$$
$$p = 5 \text{ or } p = 1$$

Since $p = 5$ is not in the domain of the variable, it is an extraneous solution.

Check $p = 1$:
$$\frac{5-1}{1-5} + 2 \overset{?}{=} \frac{1}{1}$$
$$\frac{4}{-4} + 2 \overset{?}{=} 1$$
$$-1 + 2 \overset{?}{=} 1$$
$$1 = 1 \leftarrow \text{True}$$

The solution set is $\{1\}$.

29.
$$\frac{3}{2} + \frac{5}{x-3} = \frac{x+9}{2x-6}$$
$$\frac{3}{2} + \frac{5}{x-3} = \frac{x+9}{2(x-3)}$$
$$2(x-3)\left(\frac{3}{2} + \frac{5}{x-3}\right) = 2(x-3)\left(\frac{x+9}{2(x-3)}\right)$$
$$3(x-3) + 2(5) = x+9$$
$$3x - 9 + 10 = x+9$$
$$2x = 8$$
$$x = 4$$

Check: $\dfrac{3}{2} + \dfrac{5}{4-3} \overset{?}{=} \dfrac{4+9}{2 \cdot 4 - 6}$
$$\frac{3}{2} + \frac{5}{1} \overset{?}{=} \frac{13}{8-6}$$
$$\frac{3}{2} + \frac{10}{2} \overset{?}{=} \frac{13}{2}$$
$$\frac{13}{2} = \frac{13}{2} \leftarrow \text{True}$$

The solution set is $\{4\}$.

31.
$$1 + \frac{3}{x+3} = \frac{4}{x-3}$$

$$(x-3)(x+3)\left(1+\frac{3}{x+3}\right) = (x-3)(x+3)\left(\frac{4}{x-3}\right)$$

$$(x-3)(x+3) + 3(x-3) = 4(x+3)$$

$$x^2 + 3x - 3x - 9 + 3x - 9 = 4x + 12$$

$$x^2 + 3x - 18 = 4x + 12$$

$$x^2 - x - 30 = 0$$

$$(x-6)(x+5) = 0$$

$$x - 6 = 0 \text{ or } x + 5 = 0$$

$$x = 6 \text{ or } \quad x = -5$$

Check: Check:

$$1 + \frac{3}{6+3} \overset{?}{=} \frac{4}{6-3} \qquad 1 + \frac{3}{-5+3} \overset{?}{=} \frac{4}{-5-3}$$

$$1 + \frac{3}{9} \overset{?}{=} \frac{4}{3} \qquad\qquad 1 + \frac{3}{-2} \overset{?}{=} \frac{4}{-8}$$

$$1 + \frac{1}{3} \overset{?}{=} \frac{4}{3} \qquad\qquad 1 - \frac{3}{2} \overset{?}{=} -\frac{1}{2}$$

$$\frac{4}{3} = \frac{4}{3} \leftarrow \text{True} \qquad -\frac{1}{2} = -\frac{1}{2} \leftarrow \text{True}$$

Both solutions check, so the solution set is $\{-5, 6\}$.

33.
$$\frac{4}{x-5} + \frac{3}{x-2} = \frac{x+1}{x^2 - 7x + 10}$$

$$\frac{4}{x-5} + \frac{3}{x-2} = \frac{x+1}{(x-5)(x-2)}$$

$$(x-5)(x-2)\left(\frac{4}{x-5} + \frac{3}{x-2}\right) =$$

$$(x-5)(x-2)\left(\frac{x+1}{(x-5)(x-2)}\right)$$

$$4(x-2) + 3(x-5) = x+1$$

$$7x - 23 = x + 1$$

$$6x = 24$$

$$x = 4$$

Check: $\dfrac{4}{4-5} + \dfrac{3}{4-2} \overset{?}{=} \dfrac{4+1}{4^2 - 7(4) + 10}$

$$\frac{4}{-1} + \frac{3}{2} \overset{?}{=} \frac{5}{16 - 28 + 10}$$

$$-\frac{8}{2} + \frac{3}{2} \overset{?}{=} \frac{5}{-2}$$

$$-\frac{5}{2} = -\frac{5}{2} \leftarrow \text{True}$$

The solution checks, so the solution set is $\{4\}$.

35.

$$\frac{1}{x+5} = \frac{2x}{x^2-25} - \frac{3}{x-5}$$

$$\frac{1}{x+5} = \frac{2x}{(x-5)(x+5)} - \frac{3}{x-5}$$

$$(x-5)(x+5)\left(\frac{1}{x+5}\right) = (x-5)(x+5)\left(\frac{2x}{(x-5)(x+5)} - \frac{3}{x-5}\right)$$

$$(x-5) = 2x - 3(x+5)$$
$$x-5 = 2x - 3x - 15$$
$$x-5 = -x - 15$$
$$2x = -10$$
$$x = -5$$

Since $x = -5$ is not in the domain of the variable, it is an extraneous solution. The solution set is the empty set, $\varnothing$ or $\{\ \}$.

37.

$$\frac{3}{x-1} - \frac{2}{x+4} = \frac{x^2+8x+6}{x^2+3x-4}$$

$$\frac{3}{x-1} - \frac{2}{x+4} = \frac{x^2+8x+6}{(x-1)(x+4)}$$

$$(x-1)(x+4)\left(\frac{3}{x-1} - \frac{2}{x+4}\right) = (x-1)(x+4)\left(\frac{x^2+8x+6}{(x-1)(x+4)}\right)$$

$$3(x+4) - 2(x-1) = x^2 + 8x + 6$$
$$3x + 12 - 2x + 2 = x^2 + 8x + 6$$
$$x + 14 = x^2 + 8x + 6$$
$$0 = x^2 + 7x - 8$$
$$0 = (x-1)(x+8)$$

$x - 1 = 0$ or $x + 8 = 0$
$\quad\ x = 1$ or $\quad x = -8$

Since $x = 1$ is not in the domain of the variable, it is an extraneous solution.

Check $x = -8$:

$$\frac{3}{(-8)-1} - \frac{2}{(-8)+4} \overset{?}{=} \frac{(-8)^2 + 8(-8) + 6}{(-8)^2 + 3(-8) - 4}$$

$$\frac{3}{-9} - \frac{2}{-4} \overset{?}{=} \frac{64 - 64 + 6}{64 - 24 - 4}$$

$$-\frac{12}{36} + \frac{18}{36} \overset{?}{=} \frac{6}{36}$$

$$\frac{6}{36} = \frac{6}{36} \leftarrow \text{True}$$

The solution set is $\{-8\}$.

39.

$$\frac{7}{y^2+y-12}-\frac{4y}{y^2+7y+12}=\frac{6}{y^2-9}$$

$$\frac{7}{(y-3)(y+4)}-\frac{4y}{(y+3)(y+4)}=\frac{6}{(y-3)(y+3)}$$

$$(y-3)(y+3)(y+4)\left(\frac{7}{(y-3)(y+4)}-\frac{4y}{(y+3)(y+4)}\right)=(y-3)(y+3)(y+4)\left(\frac{6}{(y-3)(y+3)}\right)$$

$$7(y+3)-4y(y-3)=6(y+4)$$

$$7y+21-4y^2+12y=6y+24$$

$$-4y^2+19y+21=6y+24$$

$$0=4y^2-13y+3$$

$$0=(4y-1)(y-3)$$

$$4y-1=0 \quad \text{or} \quad y-3=0$$

$$y=\frac{1}{4} \quad \text{or} \quad y=3$$

Since $y=3$ is not in the domain of the variable, it is an extraneous solution.

Check $y=\frac{1}{4}$:

$$\frac{7}{\left(\frac{1}{4}\right)^2+\left(\frac{1}{4}\right)-12}-\frac{4\left(\frac{1}{4}\right)}{\left(\frac{1}{4}\right)^2+7\left(\frac{1}{4}\right)+12}\overset{?}{=}\frac{6}{\left(\frac{1}{4}\right)^2-9}$$

$$\frac{7}{\frac{1}{16}+\frac{1}{4}-12}-\frac{1}{\frac{1}{16}+\frac{7}{4}+12}\overset{?}{=}\frac{6}{\frac{1}{16}-9}$$

$$\frac{7}{\frac{1}{16}+\frac{1}{4}-12}\cdot\frac{16}{16}-\frac{1}{\frac{1}{16}+\frac{7}{4}+12}\cdot\frac{16}{16}\overset{?}{=}\frac{6}{\frac{1}{16}-9}\cdot\frac{16}{16}$$

$$\frac{112}{1+4-192}-\frac{16}{1+28+192}\overset{?}{=}\frac{96}{1-144}$$

$$\frac{112}{-187}-\frac{16}{221}\overset{?}{=}\frac{96}{-143}$$

$$-\frac{1456}{2431}-\frac{176}{2431}\overset{?}{=}-\frac{96}{143}$$

$$-\frac{1632}{2431}\overset{?}{=}-\frac{96}{143}$$

$$-\frac{96}{143}=-\frac{96}{143}\quad\leftarrow\text{True}$$

The solution set is $\left\{\frac{1}{4}\right\}$.

41.
$$\frac{2x+3}{x-3}+\frac{x+6}{x-4}=\frac{x+6}{x-3}$$

$$(x-3)(x-4)\left(\frac{2x+3}{x-3}+\frac{x+6}{x-4}\right)=(x-3)(x-4)\left(\frac{x+6}{x-3}\right)$$

$$(x-4)(2x+3)+(x-3)(x+6)=(x-4)(x+6)$$

$$2x^2-5x-12+x^2+3x-18=x^2+2x-24$$

$$3x^2-2x-30=x^2+2x-24$$

$$2x^2-4x-6=0$$

$$2(x-3)(x+1)=0$$

$$x-3=0 \text{ or } x+1=0$$

$$x=3 \text{ or } x=-1$$

Since $x=3$ is not in the domain of the variable, it is an extraneous solution.

Check $x=-1$:

$$\frac{2(-1)+3}{(-1)-3}+\frac{(-1)+6}{(-1)-4}\overset{?}{=}\frac{(-1)+6}{(-1)-3}$$

$$\frac{-2+3}{-4}+\frac{5}{-5}\overset{?}{=}\frac{5}{-4}$$

$$-\frac{1}{4}-1\overset{?}{=}-\frac{5}{4}$$

$$-\frac{5}{4}=-\frac{5}{4}\leftarrow \text{True}$$

The solution set is $\{-1\}$.

43.
$$f(x)=10$$

$$x+\frac{9}{x}=10$$

$$x\left(x+\frac{9}{x}\right)=x\cdot10$$

$$x^2+9=10x$$

$$x^2-10x+9=0$$

$$(x-9)(x-1)=0$$

$$x-9=0 \text{ or } x-1=0$$

$$x=9 \text{ or } x=1$$

Thus, $f(1)=10$ and $f(9)=10$, so the points $(1,10)$ and $(9,10)$ are on the graph of f.

45.
$$f(x) = -9$$
$$2x + \frac{4}{x} = -9$$
$$x\left(2x + \frac{4}{x}\right) = x \cdot (-9)$$
$$2x^2 + 4 = -9x$$
$$2x^2 + 9x + 4 = 0$$
$$(2x + 1)(x + 4) = 0$$
$$2x + 1 = 0 \text{ or } x + 4 = 0$$
$$x = -\frac{1}{2} \text{ or } x = -4$$

Thus, $f\left(-\frac{1}{2}\right) = -9$ and $f(-4) = -9$, so the points $\left(-\frac{1}{2}, -9\right)$ and $(-4, -9)$ are on the graph of f.

47.
$$f(x) = \frac{9}{2}$$
$$\frac{x + 3}{x - 4} = \frac{9}{2}$$
$$2(x - 4)\left(\frac{x + 3}{x - 4}\right) = 2(x - 4) \cdot \left(\frac{9}{2}\right)$$
$$2(x + 3) = 9(x - 4)$$
$$2x + 6 = 9x - 36$$
$$42 = 7x$$
$$6 = x$$

Thus, $f(6) = \frac{9}{2}$, so the point $\left(6, \frac{9}{2}\right)$ is on the graph of f.

49.
$$f(x) = g(x)$$
$$\frac{x + 2}{2x + 9} = \frac{x - 1}{x + 3}$$
$$(2x + 9)(x + 3)\left(\frac{x + 2}{2x + 9}\right) = (2x + 9)(x + 3)\left(\frac{x - 1}{x + 3}\right)$$
$$(x + 3)(x + 2) = (2x + 9)(x - 1)$$
$$x^2 + 5x + 6 = 2x^2 + 7x - 9$$
$$0 = x^2 + 2x - 15$$
$$0 = (x + 5)(x - 3)$$
$$x + 5 = 0 \quad \text{or} \quad x - 3 = 0$$
$$x = -5 \quad \text{or} \qquad x = 3$$

Now, $f(-5) = \frac{-5 + 2}{2(-5) + 9} = \frac{-3}{-10 + 9} = \frac{-3}{-1} = 3$, $g(-5) = \frac{-5 - 1}{-5 + 3} = \frac{-6}{-2} = 3$, $f(3) = \frac{3 + 2}{2(3) + 9} = \frac{5}{15} = \frac{1}{3}$, and

$g(3) = \frac{3 - 1}{3 + 3} = \frac{2}{6} = \frac{1}{3}$, so the intersection points of f and g are $(-5, 3)$ and $\left(3, \frac{1}{3}\right)$.

51.
$$\frac{4}{z+4} - \frac{3}{4} = \frac{5z+2}{4z+16}$$

$$\frac{4}{z+4} - \frac{3}{4} = \frac{5z+2}{4(z+4)}$$

$$4(z+4)\left(\frac{4}{z+4} - \frac{3}{4}\right) = 4(z+4)\left(\frac{5z+2}{4(z+4)}\right)$$

$$4(4) - 3(z+4) = 5z+2$$

$$16 - 3z - 12 = 5z + 2$$

$$-3z + 4 = 5z + 2$$

$$-8z = -2$$

$$z = \frac{-2}{-8} = \frac{1}{4}$$

Check:

$$\frac{4}{\frac{1}{4}+4} - \frac{3}{4} \stackrel{?}{=} \frac{5\left(\frac{1}{4}\right)+2}{4\left(\frac{1}{4}\right)+16}$$

$$\frac{4}{\frac{1}{4}+4} - \frac{3}{4} \stackrel{?}{=} \frac{\frac{5}{4}+2}{1+16}$$

$$\frac{4}{\frac{1}{4}+4} \cdot \frac{4}{4} - \frac{3}{4} \stackrel{?}{=} \frac{\frac{5}{4}+2}{1+16} \cdot \frac{4}{4}$$

$$\frac{16}{1+16} - \frac{3}{4} \stackrel{?}{=} \frac{5+8}{4+64}$$

$$\frac{16}{17} - \frac{3}{4} \stackrel{?}{=} \frac{13}{68}$$

$$\frac{64}{68} - \frac{51}{68} \stackrel{?}{=} \frac{13}{68}$$

$$\frac{13}{68} = \frac{13}{68} \leftarrow \text{True}$$

The solution checks, so the solution set is $\left\{\frac{1}{4}\right\}$.

53.
$$x + \frac{9}{x} = 6$$

$$x\left(x + \frac{9}{x}\right) = x \cdot 6$$

$$x^2 + 9 = 6x$$

$$x^2 - 6x + 9 = 0$$

$$(x-3)(x-3) = 0$$

$$x - 3 = 0$$

$$x = 3$$

Check:

$$3 + \frac{9}{3} \stackrel{?}{=} 6$$

$$3 + 3 \stackrel{?}{=} 6$$

$$6 = 6 \leftarrow \text{True}$$

The solution checks, so the solution set is $\{3\}$.

55.

$$\frac{2}{z^2+2z-3}+\frac{3}{z^2+4z+3}=\frac{6}{z^2-1}$$

$$\frac{2}{(z-1)(z+3)}+\frac{3}{(z+1)(z+3)}=\frac{6}{(z-1)(z+1)}$$

$$(z-1)(z+1)(z+3)\left(\frac{2}{(z-1)(z+3)}+\frac{3}{(z+1)(z+3)}\right)=(z-1)(z+1)(z+3)\left(\frac{6}{(z-1)(z+1)}\right)$$

$$2(z+1)+3(z-1)=6(z+3)$$

$$2z+2+3z-3=6z+18$$

$$5z-1=6z+18$$

$$-19=z$$

Check:

$$\frac{2}{(-19)^2+2(-19)-3}+\frac{3}{(-19)^2+4(-19)+3}\overset{?}{=}\frac{6}{(-19)^2-1}$$

$$\frac{2}{361-38-3}+\frac{3}{361-76+3}\overset{?}{=}\frac{6}{361-1}$$

$$\frac{2}{320}+\frac{3}{288}\overset{?}{=}\frac{6}{360}$$

$$\frac{18}{2880}+\frac{30}{2880}\overset{?}{=}\frac{48}{2880}$$

$$\frac{48}{2880}=\frac{48}{2880}\leftarrow\text{True}$$

The solution set is $\{-19\}$.

57.

$$\frac{4}{x}-\frac{5}{2x}=\frac{3}{4}$$

$$4x\left(\frac{4}{x}-\frac{5}{2x}\right)=4x\cdot\frac{3}{4}$$

$$16-10=3x$$

$$6=3x$$

$$2=x$$

Check:

$$\frac{4}{2}-\frac{5}{2(2)}\overset{?}{=}\frac{3}{4}$$

$$2-\frac{5}{4}\overset{?}{=}\frac{3}{4}$$

$$\frac{3}{4}=\frac{3}{4}\leftarrow\text{True}$$

The solution checks, so the solution set is $\{2\}$.

59.
$$\frac{3y+1}{y-1}+3=\frac{y+2}{y+1}$$

$$(y-1)(y+1)\left(\frac{3y+1}{y-1}+3\right)$$

$$=(y-1)(y+1)\left(\frac{y+2}{y+1}\right)$$

$$(y+1)(3y+1)+3(y-1)(y+1)=(y-1)(y+2)$$

$$3y^2+4y+1+3y^2-3=y^2+y-2$$

$$6y^2+4y-2=y^2+y-2$$

$$5y^2+3y=0$$

$$y(5y+3)=0$$

$$y=0 \quad \text{or} \quad 5y+3=0$$

$$5y=-3$$

$$y=-\frac{3}{5}$$

Check: $y=0$

$$\frac{3(0)+1}{0-1}+3 \stackrel{?}{=} \frac{0+2}{0+1}$$

$$\frac{1}{-1}+3 \stackrel{?}{=} \frac{2}{1}$$

$$-1+3 \stackrel{?}{=} 2$$

$$2=2 \leftarrow \text{True}$$

Check: $y=-\frac{3}{5}$

$$\frac{3\left(-\frac{3}{5}\right)+1}{\left(-\frac{3}{5}\right)-1}+3 \stackrel{?}{=} \frac{\left(-\frac{3}{5}\right)+2}{\left(-\frac{3}{5}\right)+1}$$

$$\frac{-\frac{9}{5}+1}{-\frac{8}{5}}+3 \stackrel{?}{=} \frac{\frac{7}{5}}{\frac{2}{5}}$$

$$\frac{-\frac{4}{5}}{-\frac{8}{5}}+3 \stackrel{?}{=} \frac{\frac{7}{5}}{\frac{2}{5}}$$

$$-\frac{4}{5}\left(-\frac{5}{8}\right)+3 \stackrel{?}{=} \frac{7}{5}\left(\frac{5}{2}\right)$$

$$\frac{1}{2}+3 \stackrel{?}{=} \frac{7}{2}$$

$$\frac{7}{2}=\frac{7}{2} \leftarrow \text{True}$$

Both check, so the solution set is $\left\{-\frac{3}{5},0\right\}$.

61.
$$\frac{3x+1}{x^2-4}=0$$

$$\frac{3x+1}{(x+2)(x-2)}=0$$

$$(x+2)(x-2)\frac{3x+1}{(x+2)(x-2)}=0(x+2)(x-2)$$

$$3x+1=0$$

$$x=-\frac{1}{3}$$

The zero is $-\frac{1}{3}$, and the x-intercept is $\left(-\frac{1}{3},0\right)$.

63.
$$\frac{2x^2+5x-12}{3x^2+5x+2}=0$$

$$\frac{(2x-3)(x+4)}{(3x+2)(x+1)}=0$$

$$(3x+2)(x+1)\frac{(2x-3)(x+4)}{(3x+2)(x+1)}=0(3x+2)(x+1)$$

$$(2x-3)(x+4)=0$$

$$2x-3=0 \quad \text{or} \quad x+4=0$$

$$x=\frac{3}{2} \quad \text{or} \qquad x=-4$$

The zeros are $\frac{3}{2}$ and -4, the x-intercepts are

$\left(\frac{3}{2},0\right)$ and $(-4, 0)$.

65.
$$\frac{x^3+3x^2-4x-12}{4x^3+12x^2+x+3}=0$$

$$\frac{\left(x^3+3x^2\right)+(-4x-12)}{(4x^3+12x^2)+(x+3)}=0$$

$$\frac{x^2(x+3)-4(x+3)}{4x^2(x+3)+1(x+3)}=0$$

$$\frac{\left(x^2-4\right)(x+3)}{(4x^2+1)(x+3)}=0$$

$$\frac{(x+2)(x-2)}{4x^2+1}=0$$

$$(4x^2+1)\frac{(x+2)(x-2)}{4x^2+1}=0(4x^2+1)$$

$$(x+2)(x-2)=0$$

$$x+2=0 \quad \text{or} \quad x-2=0$$

$$x=-2 \quad \text{or} \qquad x=2$$

The zeros are -2 and 2, the x-intercepts are $(-2, 0)$ and $(2, 0)$.

67.
$$\overline{C}(x) = 225$$
$$\frac{x^2 + 75x + 5000}{x} = 225$$
$$x\left(\frac{x^2 + 75x + 5000}{x}\right) = x \cdot 225$$
$$x^2 + 75x + 5000 = 225x$$
$$x^2 - 150x + 5000 = 0$$
$$(x - 100)(x - 50) = 0$$
$$x - 100 = 0 \text{ or } x - 50 = 0$$
$$x = 100 \text{ or } x = 50$$

The average daily cost will be $225 when either 50 or 100 bicycles are manufactured.

69. a.
$$C(x) = 100$$
$$\frac{25x}{100 - x} = 100$$
$$(100 - x)\left(\frac{25x}{100 - x}\right) = (100 - x) \cdot 100$$
$$25x = 10,000 - 100x$$
$$125x = 10,000$$
$$x = 80$$

Thus, 80% of the pollutants can be removed with a budget of $100 million.

b.
$$C(x) = 225$$
$$\frac{25x}{100 - x} = 225$$
$$(100 - x)\left(\frac{25x}{100 - x}\right) = (100 - x) \cdot 225$$
$$25x = 22,500 - 225x$$
$$250x = 22,500$$
$$x = 90$$

Thus, 90% of the pollutants can be removed with a budget of $225 million.

71. Substitute $R = 750$, $h = 1400$, $t = 2250$, and $b = 5500$ into the given formula and solve for w:
$$R = \frac{(h + w)t}{b + w}$$
$$750 = \frac{(1400 + w)2250}{5500 + w}$$
$$750 = \frac{3,150,000 + 2250w}{5500 + w}$$
$$(5500 + w)750$$
$$= (5500 + w)\left(\frac{3,150,000 + 2250w}{5500 + w}\right)$$
$$4,125,000 + 750w = 3,150,000 + 2250w$$
$$975,000 = 1500w$$
$$650 = w$$

We predict that the Athletics had 650 walks.

73. Answers will vary. One possibility follows:

The equation $\dfrac{2}{x - 1} = \dfrac{3}{x + 1}$ has one real solution, namely $\{5\}$.

75.
$$\left(\frac{4}{x + 3}\right)^2 - 5\left(\frac{4}{x + 3}\right) + 6 = 0$$
$$\frac{16}{(x + 3)^2} - \frac{20}{x + 3} + 6 = 0$$
$$(x + 3)^2\left(\frac{16}{(x + 3)^2} - \frac{20}{x + 3} + 6\right) = 0(x + 3)^2$$
$$16 - 20(x + 3) + 6(x + 3)^2 = 0$$
$$16 - 20(x + 3) + 6\left(x^2 + 6x + 9\right) = 0$$
$$16 - 20x - 60 + 6x^2 + 36x + 54 = 0$$
$$6x^2 + 16x + 10 = 0$$
$$3x^2 + 8x + 5 = 0$$
$$(3x + 5)(x + 1) = 0$$
$$3x + 5 = 0 \text{ or } x + 1 = 0$$
$$x = -\frac{5}{3} \text{ or } \quad x = -1$$

Both solutions check. The solution set is $\left\{-\dfrac{5}{3}, -1\right\}$.

77. Answers will vary. One possibility follows: The solution(s) of the rational equation must be within the domain of the variable of the equation. As we work through the algebraic process of solving the equation, if we find a "solution" that is not in the domain of the variable, then that "solution" is not really a solution at all. It is an extraneous solution.

79.
$$\frac{2a^6}{\left(a^3\right)^2} - \frac{5a^2}{a^3} = \frac{3a}{a^3}$$

$$\frac{2a^6}{a^6} - \frac{5a^2}{a^3} = \frac{3a}{a^3}$$

$$2 - \frac{5}{a} = \frac{3}{a^2}$$

$$a^2\left(2 - \frac{5}{a}\right) = a^2\left(\frac{3}{a^2}\right)$$

$$2a^2 - 5a = 3$$

$$2a^2 - 5a - 3 = 0$$

$$(2a+1)(a-3) = 0$$

$$2a+1 = 0 \quad \text{or} \quad a-3 = 0$$

$$a = -\frac{1}{2} \quad \text{or} \quad a = 3$$

Both solutions check. The solution set is $\left\{-\frac{1}{2}, 3\right\}$.

81.
$$\frac{3}{x-2} - \frac{2x+1}{x+1} = \frac{3}{x-2} \cdot \frac{x+1}{x+1} - \frac{2x+1}{x+1} \cdot \frac{x-2}{x-2}$$

$$= \frac{3(x+1)}{(x-2)(x+1)} - \frac{(2x+1)(x-2)}{(x-2)(x+1)}$$

$$= \frac{3(x+1) - (2x+1)(x-2)}{(x-2)(x+1)}$$

$$= \frac{3x+3 - \left(2x^2 - 4x + x - 2\right)}{(x-2)(x+1)}$$

$$= \frac{3x+3 - 2x^2 + 4x - x + 2}{(x-2)(x+1)}$$

$$= \frac{-2x^2 + 6x + 5}{(x-2)(x+1)}$$

83.
$$\frac{x+1}{2x+3} - \frac{3}{x-4} = \frac{-3}{2x^2 - 5x - 12}$$

$$\frac{x+1}{2x+3} - \frac{3}{x-4} = \frac{-3}{(2x+3)(x-4)}$$

$$(2x+3)(x-4)\left(\frac{x+1}{2x+3} - \frac{3}{x-4}\right) =$$

$$(2x+3)(x-4)\left(\frac{-3}{(2x+3)(x-4)}\right)$$

$$(x-4)(x+1) - 3(2x+3) = -3$$

$$x^2 + x - 4x - 4 - 6x - 9 = -3$$

$$x^2 - 9x - 13 = -3$$

$$x^2 - 9x - 10 = 0$$

$$(x-10)(x+1) = 0$$

$$x - 10 = 0 \quad \text{or} \quad x+1 = 0$$

$$x = 10 \quad \text{or} \quad x = -1$$

Both solutions check. The solution set is $\{-1, 10\}$.

85. $\dfrac{x-4}{x+4} = \dfrac{1}{2}$

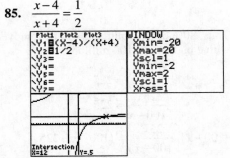

The solution set is $\{12\}$.

87. $\dfrac{3}{5} + \dfrac{4}{x+6} = \dfrac{x+12}{5x+30}$

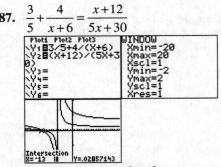

The solution set is $\{-13\}$.

89. $\dfrac{2x^2 + 11x + 12}{x+4} = -5$

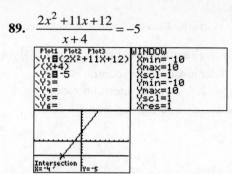

The apparent solution is $x = -4$. However, notice that -4 is not in the domain of the left side of the equation. Thus, the equation has no solution. The solution set is $\varnothing$ or { }.

Section 5.5

Are You Ready for This Section?

R1. The inequality $-1 < x \le 8$ in interval notation is $(-1,\ 8]$.

R2.
$$2x + 3 > 4x - 9$$
$$-4x + 2x + 3 > -4x + 4x - 9$$
$$-2x + 3 > -9$$
$$-2x + 3 - 3 > -9 - 3$$
$$-2x > -12$$
$$\frac{-2x}{-2} < \frac{-12}{-2}$$
$$x < 6$$

The solution set is $\{x \mid x < 6\}$ or, using interval notation, $(-\infty,\ 6)$.

Section 5.5 Quick Checks

1. The inequality $\dfrac{2x-3}{x+6} > 1$ is an example of a __rational__ inequality.

2. $\dfrac{x-7}{x+3} \ge 0$

The rational expression will equal 0 when $x = 7$. It is undefined when $x = -3$. Thus, we separate the real number line into the intervals $(-\infty, -3)$, $(-3,\ 7)$, and $(7,\ \infty)$. Determine where the numerator and denominator are positive and negative and where the quotient is positive and negative.

Interval	$(-\infty, -3)$	-3	$(-3, 7)$	7	$(7, \infty)$
$x - 7$	Neg	Neg	Neg	0	Pos
$x + 3$	Neg	0	Pos	Pos	Pos
$\dfrac{x-7}{x+3}$	Pos	Undef	Neg	0	Pos

The rational function is undefined at $x = -3$, so -3 is not part of the solution. The inequality is non-strict, so 7 is part of the solution. Now, $\dfrac{x-7}{x+3}$ is greater than zero where the quotient is positive. The solution is $\{x \mid x < -3 \text{ or } x \ge 7\}$ in set-builder notation; the solution is $(-\infty,\ -3) \cup [7,\ \infty)$ in interval notation.

3. $\dfrac{1-x}{x+5} > 0$

The rational expression will equal 0 when $x = 1$. It is undefined when $x = -5$. Thus we separate the real number line into intervals $(-\infty, -5)$, $(-5,1)$, and $(1, \infty)$. Determine where the numerator and denominator are positive and negative and where the quotient is positive and negative.

Interval	$(-\infty, -5)$	-5	$(-5, 1)$	1	$(1, \infty)$
$1 - x$	Pos	Pos	Pos	0	Neg
$x + 5$	Neg	0	Pos	Pos	Pos
$\dfrac{1-x}{x+5}$	Neg	Undef	Pos	0	Neg

The rational function is undefined at $x = -5$, so -5 is not part of the solution. The inequality is strict, so 1 is not part of the solution. Now, $\dfrac{1-x}{x+5}$ is greater than zero where the quotient is positive. The solution is $(-5,1)$ in interval notation.

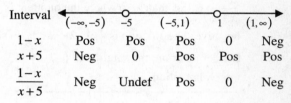

4.
$$\frac{4x+5}{x+2} < 3$$

$$\frac{4x+5}{x+2} - 3 < 0$$

$$\frac{4x+5}{x+2} - \frac{3(x+2)}{x+2} < 0$$

$$\frac{4x+5-3x-6}{x+2} < 0$$

$$\frac{x-1}{x+2} < 0$$

The rational expression will equal 0 when $x = 1$. It is undefined when $x = -2$. Thus, we separate the real number line into the intervals $(-\infty, -2)$, $(-2, 1)$, and $(1, \infty)$. Determine where the numerator and denominator are positive and negative and where the quotient is positive and negative.

Interval	$(-\infty, -2)$	-2	$(-2, 1)$	1	$(1, \infty)$
$x - 1$	Neg	Neg	Neg	0	Pos
$x + 2$	Neg	0	Pos	Pos	Pos
$\frac{x-1}{x+2}$	Pos	Undef	Neg	0	Pos

The rational function is undefined at $x = -2$, so -2 is not part of the solution. The inequality is strict, so 1 is not part of the solution. Now, $\frac{x-1}{x+2}$ is less than zero where the quotient is negative. The solution is $\{x \mid -2 < x < 1\}$ in set-builder notation; the solution is $(-2, 1)$ in interval notation.

(number line from -8 to 8 with open interval between -2 and 1)

5.5 Exercises

5. $\frac{x-4}{x+1} > 0$

The rational expression will equal 0 when $x = 4$. It is undefined when $x = -1$.
Determine where the numerator and denominator are positive and negative and where the quotient is positive and negative.

Interval	$(-\infty, -1)$	-1	$(-1, 4)$	4	$(4, \infty)$
$x - 4$	$----$	$-$	$----$	0	$++++$
$x + 1$	$----$	0	$++++$	$+$	$++++$
$\frac{x-4}{x+1}$	$++++$	$\varnothing$	$----$	0	$++++$

The rational function is undefined at $x = -1$, so -1 is not part of the solution. The inequality is strict, so 4 is not part of the solution. Now,

$\frac{x-4}{x+1}$ is greater than zero where the quotient is positive. The solution is $\{x \mid x < -1 \text{ or } x > 4\}$ in set-builder notation; the solution is $(-\infty, -1) \cup (4, \infty)$ in interval notation.

(number line from -8 to 6 with open rays left of -1 and right of 4)

7. $\frac{x+9}{x-3} < 0$

The rational expression will equal 0 when $x = -9$. It is undefined when $x = 3$.
Determine where the numerator and denominator are positive and negative and where the quotient is positive and negative.

Interval	$(-\infty, -9)$	-9	$(-9, 3)$	3	$(3, \infty)$
$x + 9$	$----$	0	$++++$	$+$	$++++$
$x - 3$	$----$	$-$	$----$	0	$++++$
$\frac{x+9}{x-3}$	$++++$	0	$----$	$\varnothing$	$++++$

The rational function is undefined at $x = 3$, so 3 is not part of the solution. The inequality is strict, so -9 is not part of the solution. Now, $\frac{x+9}{x-3}$ is less than zero where the quotient is negative. The solution is $\{x \mid -9 < x < 3\}$ in set-builder notation; the solution is $(-9, 3)$ in interval notation.

(number line from -11 to 5 with open interval between -9 and 3)

9. $\frac{x+10}{x-4} \geq 0$

The rational expression will equal 0 when $x = -10$. It is undefined when $x = 4$.
Determine where the numerator and denominator are positive and negative and where the quotient is positive and negative.

Interval	$(-\infty, -10)$	-10	$(-10, 4)$	4	$(4, \infty)$
$x + 10$	$----$	0	$++++$	$+$	$++++$
$x - 4$	$----$	$-$	$----$	0	$++++$
$\frac{x+10}{x-4}$	$++++$	0	$----$	$\varnothing$	$++++$

The rational function is undefined at $x = 4$, so 4 is not part of the solution. The inequality is non-strict, so -10 is part of the solution. Now,

$\frac{x+10}{x-4}$ is greater than zero where the quotient is positive. The solution is $\{x \mid x \leq -10 \text{ or } x > 4\}$ in set-builder notation; the solution is $(-\infty,\ -10] \cup (4,\ \infty)$ in interval notation.

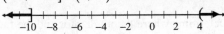

11. $\dfrac{(3x+5)(x+8)}{x-2} \le 0$

The rational expression will equal 0 when

$x = -\dfrac{5}{3}$ and when $x = -8$. It is undefined when

$x = 2$.

Determine where the factors of the numerator and denominator are positive and negative and where the quotient is positive and negative.

Interval	$(-\infty, -8)$	-8	$\left(-8, -\frac{5}{3}\right)$	$-\frac{5}{3}$	$\left(-\frac{5}{3}, 2\right)$	2	$(2, \infty)$
$3x+5$	$---$	$-$	$---$	0	$+++$	$+$	$+++$
$x+8$	$---$	0	$+++$	$+$	$+++$	$+$	$+++$
$x-2$	$---$	$-$	$---$	$-$	$---$	0	$+++$
$\frac{(3x+5)(x+8)}{x-2}$	$---$	0	$+++$	0	$---$	$\varnothing$	$+++$

The rational function is undefined at $x = 2$, so 2 is not part of the solution. The inequality is non-strict, so -8 and $-\dfrac{5}{3}$ are part of the solution.

Now, $\dfrac{(3x+5)(x+8)}{x-2}$ is less than zero where the quotient is negative. The solution is

$\left\{ x \mid x \le -8 \text{ or } -\dfrac{5}{3} \le x < 2 \right\}$ in set-builder

notation; the solution is $(-\infty,\ -8] \cup \left[-\dfrac{5}{3},\ 2\right)$ in interval notation.

13. $\dfrac{x-5}{x+1} < 1$

$\dfrac{x-5}{x+1} - 1 < 0$

$\dfrac{x-5}{x+1} - \dfrac{x+1}{x+1} < 0$

$\dfrac{x-5-x-1}{x+1} < 0$

$\dfrac{-6}{x+1} < 0$

The numerator is always negative, so the rational expression will never equal 0. However, it is undefined when $x = -1$.

Determine where the denominator is positive and negative and where the quotient is positive and negative.

Interval	$(-\infty, -1)$	-1	$(-1, \infty)$
-6	$----$	$-$	$----$
$x+1$	$----$	0	$++++$
$\frac{-6}{x+1}$	$++++$	$\varnothing$	$----$

The rational function is undefined at $x = -1$, so -1 is not part of the solution. Now, $\dfrac{-6}{x+1}$ is less than zero where the quotient is negative. The solution is $\{x \mid x > -1\}$ in set-builder notation; the solution is $(-1,\ \infty)$ in interval notation.

15. $\dfrac{2x-9}{x-3} > 4$

$\dfrac{2x-9}{x-3} - 4 > 0$

$\dfrac{2x-9}{x-3} - \dfrac{4(x-3)}{x-3} > 0$

$\dfrac{2x-9-4x+12}{x-3} > 0$

$\dfrac{-2x+3}{x-3} > 0$

The rational expression will equal 0 when

$x = \dfrac{3}{2}$. It is undefined when $x = 3$.

Determine where the numerator and denominator are positive and negative and where the quotient is positive and negative.

Interval	$\left(-\infty, \frac{3}{2}\right)$	$\frac{3}{2}$	$\left(\frac{3}{2}, 3\right)$	3	$(3, \infty)$
$-2x+3$	$++++$	0	$----$	$-$	$----$
$x-3$	$----$	$-$	$----$	0	$++++$
$\frac{-2x+3}{x-3}$	$----$	0	$++++$	$\varnothing$	$----$

The rational function is undefined at $x = 3$, so 3 is not part of the solution. The inequality is strict, so 3 is not part of the solution. Now, $\dfrac{-2x+3}{x-3}$ is greater than zero where the quotient is positive.

The solution is $\left\{ x \mid \dfrac{3}{2} < x < 3 \right\}$ in set-builder

notation; the solution is $\left(\dfrac{3}{2},\ 3\right)$ in interval notation.

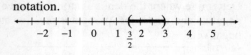

17.
$$\frac{3}{x-4}+\frac{1}{x}\ge 0$$

$$\frac{3x}{x(x-4)}+\frac{x-4}{x(x-4)}\ge 0$$

$$\frac{3x+x-4}{x(x-4)}\ge 0$$

$$\frac{4x-4}{x(x-4)}\ge 0$$

The rational expression will equal 0 when $x=1$. It is undefined when $x=0$ and when $x=4$. Determine where the numerator and the factors of the denominator are positive and negative and where the quotient is positive and negative.

Interval	$(-\infty,0)$	0	$(0,1)$	1	$(1,4)$	4	$(4,\infty)$
$4x-4$	$---$	$-$	$---$	0	$+++$	$+$	$+++$
x	$---$	0	$+++$	$+$	$+++$	$+$	$+++$
$x-4$	$---$	$-$	$---$	$-$	$---$	0	$+++$
$\frac{4x-4}{x(x-4)}$	$---$	$\varnothing$	$+++$	0	$---$	$\varnothing$	$+++$

The rational function is undefined at $x=0$ and $x=4$, so 0 and 4 are not part of the solution. The inequality is non-strict, so 1 is part of the solution. Now, $\dfrac{4x-4}{x(x-4)}$ is greater than zero where the quotient is positive. The solution is $\{x\,|\,0<x\le 1 \text{ or } x>4\}$ in set-builder notation; the solution is $(0,\,1]\cup(4,\,\infty)$ in interval notation.

19.
$$\frac{3}{x-2}\le\frac{4}{x+5}$$

$$\frac{3}{x-2}-\frac{4}{x+5}\le 0$$

$$\frac{3(x+5)}{(x+5)(x-2)}-\frac{4(x-2)}{(x+5)(x-2)}\le 0$$

$$\frac{3(x+5)-4(x-2)}{(x+5)(x-2)}\le 0$$

$$\frac{3x+15-4x+8}{(x+5)(x-2)}\le 0$$

$$\frac{23-x}{(x+5)(x-2)}\le 0$$

The rational expression will equal 0 when $x=23$. It is undefined when $x=2$ and when $x=-5$.

Determine where the numerator and the factors of the denominator are positive and negative and where the quotient is positive and negative.

Interval	$(-\infty,-5)$	-5	$(-5,2)$	2	$(2,23)$	23	$(23,\infty)$
$23-x$	$+++$	$+$	$+++$	$+$	$+++$	0	$---$
$x+5$	$---$	0	$+++$	$+$	$+++$	$+$	$+++$
$x-2$	$---$	$-$	$---$	0	$+++$	$+$	$+++$
$\frac{23-x}{(x+5)(x-2)}$	$+++$	$\varnothing$	$---$	$\varnothing$	$+++$	0	$---$

The rational function is undefined at $x=-5$ and $x=2$, so -5 and 2 are not part of the solution. The inequality is non-strict, so 23 is part of the solution. Now, $\dfrac{23-x}{(x+5)(x-2)}$ is less than zero where the quotient is negative. The solution is $\{x\,|\,-5<x<2 \text{ or } x\ge 23\}$ in set-builder notation; the solution is $(-5,2)\cup[23,\infty)$ in interval notation.

21.
$$\frac{(2x-1)(x+3)}{x-5}>0$$

The rational expression will equal 0 when $x=\dfrac{1}{2}$ and when $x=-3$. It is undefined when $x=5$. Determine where the factors of the numerator and denominator are positive and negative and where the quotient is positive and negative.

Interval	$(-\infty,-3)$	-3	$\left(-3,\frac{1}{2}\right)$	$\frac{1}{2}$	$\left(\frac{1}{2},5\right)$	5	$(5,\infty)$
$2x-1$	$---$	$-$	$---$	0	$+++$	$+$	$+++$
$x+3$	$---$	0	$+++$	$+$	$+++$	$+$	$+++$
$x-5$	$---$	$-$	$---$	$-$	$---$	0	$+++$
$\frac{(2x-1)(x+3)}{x-5}$	$---$	0	$+++$	0	$---$	$\varnothing$	$+++$

The rational function is undefined at $x=5$, so 5 is not part of the solution. The inequality is strict, so -3 and $\dfrac{1}{2}$ are not part of the solution. Now, $\dfrac{(2x-1)(x+3)}{x-5}$ is greater than zero where the quotient is positive. The solution is $\left\{x\,\middle|\,-3<x<\dfrac{1}{2} \text{ or } x>5\right\}$ in set-builder notation;

the solution is $\left(-3, \dfrac{1}{2}\right) \cup (5, \infty)$ in interval notation.

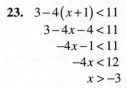

23. $3 - 4(x+1) < 11$

$\quad 3 - 4x - 4 < 11$

$\quad\quad -4x - 1 < 11$

$\quad\quad\quad -4x < 12$

$\quad\quad\quad\quad x > -3$

The solution is $\{x \mid x > -3\}$ in set-builder notation; the solution is $(-3, \infty)$ in interval notation.

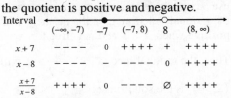

25. $\dfrac{x+7}{x-8} \le 0$

The rational expression will equal 0 when $x = -7$. It is undefined when $x = 8$.
Determine where the numerator and denominator are positive and negative and where the quotient is positive and negative.

Interval	$(-\infty, -7)$	-7	$(-7, 8)$	8	$(8, \infty)$
$x+7$	$----$	0	$++++$	$+$	$++++$
$x-8$	$----$	$-$	$----$	0	$++++$
$\dfrac{x+7}{x-8}$	$++++$	0	$----$	$\varnothing$	$++++$

The rational function is undefined at $x = 8$, so 8 is not part of the solution. The inequality is non-strict, so -7 is part of the solution. Now, $\dfrac{x+7}{x-8}$ is less than zero where the quotient is negative. The solution is $\{x \mid -7 \le x < 8\}$ in set-builder notation; the solution is $[-7, 8)$ in interval notation.

27. $\quad (x-2)(2x+1) \ge 2(x-1)^2$

$\quad\quad 2x^2 - 3x - 2 \ge 2\left(x^2 - 2x + 1\right)$

$\quad\quad 2x^2 - 3x - 2 \ge 2x^2 - 4x + 2$

$\quad 2x^2 - 3x - 2x^2 + 4x \ge 2 + 2$

$\quad\quad\quad\quad\quad\quad x \ge 4$

The solution is $\{x \mid x \ge 4\}$ in set-builder

notation; the solution is $[4, \infty)$ in interval notation.

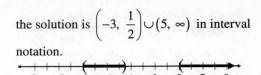

29. $\dfrac{3x-1}{x+4} \ge 2$

$\quad \dfrac{3x-1}{x+4} - 2 \ge 0$

$\quad \dfrac{3x-1}{x+4} - \dfrac{2(x+4)}{x+4} \ge 0$

$\quad\quad \dfrac{3x-1-2x-8}{x+4} \ge 0$

$\quad\quad\quad \dfrac{x-9}{x+4} \ge 0$

The rational expression will equal 0 when $x = 9$. It is undefined when $x = -4$.
Determine where the numerator and denominator are positive and negative and where the quotient is positive and negative.

Interval	$(-\infty, -4)$	-4	$(-4, 9)$	9	$(9, \infty)$
$x-9$	$----$	$-$	$----$	0	$++++$
$x+4$	$----$	0	$++++$	$+$	$++++$
$\dfrac{x-9}{x+4}$	$++++$	$\varnothing$	$----$	0	$++++$

The rational function is undefined at $x = -4$, so -4 is not part of the solution. The inequality is non-strict, so 9 is part of the solution. Now, $\dfrac{x-9}{x+4}$ is greater than zero where the quotient is positive. The solution is $\{x \mid x < -4 \text{ or } x \ge 9\}$ in set-builder notation; the solution is $(-\infty, -4) \cup [9, \infty)$ in interval notation.

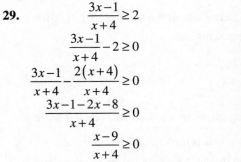

31. $R(x) \le 0$

$\quad \dfrac{x-6}{x+1} \le 0$

The rational expression will equal 0 when $x = 6$. It is undefined when $x = -1$.
Determine where the numerator and denominator are positive and negative and where the quotient is positive and negative.

Interval	$(-\infty, -1)$	-1	$(-1, 6)$	6	$(6, \infty)$
$x-6$	$----$	$-$	$----$	0	$++++$
$x+1$	$----$	0	$++++$	$+$	$++++$
$\dfrac{x-6}{x+1}$	$++++$	$\varnothing$	$----$	0	$++++$

The rational function is undefined at $x = -1$, so

-1 is not part of the solution. The inequality is non-strict, so 6 is part of the solution. Now,

$\dfrac{x-6}{x+1}$ is less than zero where the quotient is

negative. The solution is $\{x \mid -1 < x \le 6\}$ in set-builder notation; the solution is $(-1,\ 6]$ in interval notation.

33. $R(x) < 0$

$\dfrac{2x-5}{x+2} < 0$

The rational expression will equal 0 when

$x = \dfrac{5}{2}$. It is undefined when $x = -2$.

Determine where the numerator and denominator are positive and negative and where the quotient is positive and negative.

Interval	$(-\infty, -2)$	-2	$\left(-2, \frac{5}{2}\right)$	$\frac{5}{2}$	$\left(\frac{5}{2}, \infty\right)$
$2x-5$	$----$	$-$	$----$	0	$++++$
$x+2$	$----$	0	$++++$	$+$	$++++$
$\frac{2x-5}{x+2}$	$++++$	$\varnothing$	$----$	0	$++++$

The rational function is undefined at $x = -2$, so -2 is not part of the solution. The inequality is

strict, so $\dfrac{5}{2}$ is not part of the solution. Now,

$\dfrac{2x-5}{x+2}$ is less than zero where the quotient is

negative. The solution is $\left\{ x \,\middle|\, -2 < x < \dfrac{5}{2} \right\}$ in

set-builder notation; the solution is $\left(-2, \dfrac{5}{2}\right)$ in

interval notation.

35. The average cost will be no more than $130 when $\overline{C}(x) \le 130$.

$\dfrac{80x + 5000}{x} \le 130$

$\dfrac{80x + 5000}{x} - 130 \le 0$

$\dfrac{80x + 5000}{x} - \dfrac{130x}{x} \le 0$

$\dfrac{5000 - 50x}{x} \le 0$

The rational expression will equal 0 when $x = 100$. It is undefined when $x = 0$. Determine where the numerator and denominator are positive and negative and where

the quotient is positive and negative.

Interval	$(-\infty, 0)$	0	$(0, 100)$	100	$(100, \infty)$
$5000-50x$	$++++$	$+$	$++++$	0	$----$
x	$----$	0	$++++$	$+$	$++++$
$\frac{5000-50x}{x}$	$----$	$\varnothing$	$++++$	0	$----$

The rational function is undefined at $x = 0$, so 0 is not part of the solution. The inequality is non-strict, so 100 is part of the solution. Now,

$\dfrac{5000-50x}{x}$ is less than zero where the quotient

is negative. The solution is

$\{x \mid x < 0 \text{ or } x \ge 100\}$ in set-builder notation;

the solution is $(-\infty, 0) \cup [100, \infty)$ in interval

notation. However, for this problem, $x < 0$ is not meaningful. Thus, the average cost will be no more than $130 when 100 or more bicycles are produced each day.

37. Answers may vary. One possibility follows: The left endpoint of the interval is $x = 2$, so we can use $x - 2$ as a factor of the denominator of the rational function. Now, $x - 2$ will be positive for $x > 2$ and negative for $x < 2$. If we use any positive constant for the numerator of the rational function, then the quotient will be positive when $x > 2$ and negative for $x < 2$.

Thus, the rational inequality $\dfrac{10}{x-2} > 0$ will have

$(2, \infty)$ as the solution set.

NOTE: Because the solution set does not contain the endpoint 2, the inequality must be strict.

39. The statement $(-1, 4)$ indicates that neither endpoint is included in the solution set. The statement $\{x \mid -1 \le x \le 4\}$ indicates that both endpoints are included in the solution set. In

fact, the solution of the inequality $\dfrac{x-4}{x+1} \le 0$

should include the endpoint 4, but it should not include the endpoint -1. Therefore, the statement of the solution set should be

$\{x \mid -1 < x \le 4\}$ using set-builder notation or

$(-1, 4]$ using interval notation.

41. To find the x-intercept(s), we solve $F(x) = 0$:

$$F(x) = 0$$
$$6x - 12 = 0$$
$$6x = 12$$
$$x = 2$$

The x-intercept of F is $(2, 0)$.

43. To find the x-intercept(s), we solve $f(x) = 0$:

$$f(x) = 0$$
$$2x^2 + 3x - 14 = 0$$
$$(2x + 7)(x - 2) = 0$$
$$2x + 7 = 0 \quad \text{or} \quad x - 2 = 0$$
$$x = -\frac{7}{2} \quad \text{or} \quad x = 2$$

The x-intercepts of f are $\left(-\dfrac{7}{2}, 0\right)$ and $(2, 0)$.

45. To find the x-intercept(s), we solve $R(x) = 0$:

$$R(x) = 0$$
$$\frac{3x - 2}{x + 4} = 0$$
$$(x + 4)\left(\frac{3x - 2}{x + 4}\right) = (x + 4)0$$
$$3x - 2 = 0$$
$$3x = 2$$
$$x = \frac{2}{3}$$

The x-intercept of R is $\left(\dfrac{2}{3}, 0\right)$.

47. $\dfrac{x - 5}{x + 1} \le 3$

Let $Y_1 = \dfrac{x - 5}{x + 1}$ and $Y_2 = 3$. Graph the functions. Use the **INTERSECT** feature to find the x-coordinates of the point(s) of intersection.

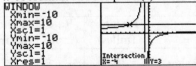

The rational function is undefined at $x = -1$, so -1 is not part of the solution. The inequality is non-strict, so -4 is part of the solution. From the graph, we can see that $\dfrac{x - 5}{x + 1} \le 3$ where $x \le -4$ and where $x > -1$. Thus, the solution set is $\{x \mid x \le -4 \text{ or } x > -1\}$ in set-builder notation;

the solution is $(-\infty, -4] \cup (-1, \infty)$ in interval notation.

49. $\dfrac{2x + 5}{x - 7} > 3$

Let $Y_1 = \dfrac{2x + 5}{x - 7}$ and $Y_2 = 3$. Graph the functions. Use the **INTERSECT** feature to find the x-coordinates of the point(s) of intersection.

The rational function is undefined at $x = 7$, so 7 is not part of the solution. The inequality is strict, so 26 is not part of the solution. From the graph, we can see that $\dfrac{2x + 5}{x - 7} > 3$ for $7 < x < 26$. Thus, the solution set is $\{x \mid 7 < x < 26\}$ in set-builder notation; the solution is $(7, 26)$ in interval notation.

Section 5.6

Are You Ready for This Section?

R1.
$$4x - 2y = 10$$
$$4x - 4x - 2y = 10 - 4x$$
$$-2y = -4x + 10$$
$$\frac{-2y}{-2} = \frac{-4x + 10}{-2}$$
$$y = 2x - 5$$

Section 5.6 Quick Checks

1. a.
$$Y = \frac{G}{1 - b}$$
$$(1 - b)Y = (1 - b)\left(\frac{G}{1 - b}\right)$$
$$Y - Yb = G$$
$$Y = G + Yb$$
$$Y - G = Yb$$
$$\frac{Y - G}{Y} = b \quad \text{or} \quad b = \frac{Y - G}{Y}$$

b. Substitute $G = \$100$ billion and
$Y = \$1000$ billion into $b = \dfrac{Y - G}{Y}$.

$$b = \frac{1000 \text{ billion} - 100 \text{ billion}}{1000 \text{ billion}}$$
$$= \frac{900 \text{ billion}}{1000 \text{ billion}} = 0.9$$

2. A <u>proportion</u> is a statement (equation) that two ratios are equal. Two figures are <u>similar</u> if their angles have the same measure and their corresponding sides are proportional.

3.
$$\frac{40}{h} = \frac{2}{6}$$
$$6h \cdot \frac{40}{h} = 6h \cdot \frac{2}{6}$$
$$240 = 2h$$
$$120 = h$$
The height of the building is 120 feet.

4.
$$\frac{AB}{AC} = \frac{DE}{DF}$$
$$\frac{3x}{8} = \frac{6}{x}$$
$$8x\left(\frac{3x}{8}\right) = 8x\left(\frac{6}{x}\right)$$
$$3x^2 = 48$$
$$3x^2 - 48 = 0$$
$$3(x+4)(x-4) = 0$$
$$x + 4 = 0 \quad \text{or} \quad x - 4 = 0$$
$$x = -4 \quad \text{or} \quad x = 4$$
Since the length of DF cannot be negative, we discard $x = -4$ and keep $x = 4$. The length of AB is $3x = 3 \cdot 4 = 12$. In summary, $AB = 12$ and $DF = 4$.

5. Let p represent the population of the United States in 2011.
$$\frac{170}{1000} = \frac{53,300,000}{p}$$
$$1000p\left(\frac{17}{1000}\right) = 1000p\left(\frac{53,300,000}{p}\right)$$
$$170p = 53,300,000,000$$
$$p = \frac{53,300,000,000}{170}$$
$$p = 313,529,411.8$$
In 2011, the population of the United States was approximately 313.5 million people.

6. Let t represent the time required to fill the pool using the two hoses together.

$$\begin{pmatrix} \text{Part done} \\ \text{by Juan and} \\ \text{Maria's hose} \\ \text{in 1 hour} \end{pmatrix} + \begin{pmatrix} \text{Part done by} \\ \text{the neighbor's} \\ \text{hose in 1 hour} \end{pmatrix} = \begin{pmatrix} \text{Part done} \\ \text{together} \\ \text{in 1 hour} \end{pmatrix}$$

$$\frac{1}{30} \quad + \quad \frac{1}{24} \quad = \quad \frac{1}{t}$$
$$120t\left(\frac{1}{30} + \frac{1}{24}\right) = 120t\left(\frac{1}{t}\right)$$
$$4t + 5t = 120$$
$$9t = 120$$
$$t = \frac{120}{9}$$
$$t = \frac{40}{3} = 13\frac{1}{3}$$

Using both hoses, it will take $\dfrac{40}{3}$ hours, or 13 hours and 20 minutes, to fill the pool.

7. Let t represent the time required to fill the pool.

$$\begin{pmatrix} \text{Portion of} \\ \text{pool filled in} \\ \text{1 minute by} \\ \text{the air pump} \end{pmatrix} - \begin{pmatrix} \text{Portion of} \\ \text{air released} \\ \text{from the pool} \\ \text{in 1 minute} \end{pmatrix} = \begin{pmatrix} \text{Portion of} \\ \text{pool filled in} \\ \text{1 minute with} \\ \text{valve open} \end{pmatrix}$$

$$\frac{1}{20} \quad - \quad \frac{1}{50} \quad = \quad \frac{1}{t}$$
$$100t\left(\frac{1}{20} - \frac{1}{50}\right) = 100t\left(\frac{1}{t}\right)$$
$$5t - 2t = 100$$
$$3t = 100$$
$$t = \frac{100}{3} \approx 33.3$$

To fill the pool with the valve open, it will take $\dfrac{100}{3} = 33\frac{1}{3}$ minutes or 33 minutes, 20 seconds.

8. Let x represent the speed of the canoe in still water.

	Distance (miles)	Rate (mph)	Time (hours)
Upstream	12	$x - 2$	$\dfrac{12}{x-2}$
Downstream	12	$x + 2$	$\dfrac{12}{x+2}$

Trip up + Trip down = 8

$$\frac{12}{x-2} + \frac{12}{x+2} = 8$$

$$(x-2)(x+2)\left(\frac{12}{x-2}+\frac{12}{x+2}\right) = (x-2)(x+2)\cdot 8$$

$$12(x+2)+12(x-2) = 8\left(x^2-4\right)$$

$$12x+24+12x-24 = 8\left(x^2-4\right)$$

$$24x = 8\left(x^2-4\right)$$

$$3x = x^2-4$$

$$0 = x^2-3x-4$$

$$0 = (x+1)(x-4)$$

$$x+1=0 \quad \text{or} \quad x-4=0$$
$$x=-1 \quad \text{or} \quad x=4$$

We disregard -1 because the speed of the canoe must be positive. Thus, the speed of the canoe in still water is 4 miles per hour.

5.6 Exercises

9.
$$\frac{V_1}{V_2} = \frac{P_2}{P_1}$$

$$V_2 P_1\left(\frac{V_1}{V_2}\right) = V_2 P_1\left(\frac{P_2}{P_1}\right)$$

$$P_1 V_1 = V_2 P_2$$

$$P_1 = \frac{V_2 P_2}{V_1}$$

11.
$$R = \frac{r}{1-t}$$

$$(1-t)\cdot R = (1-t)\left(\frac{r}{1-t}\right)$$

$$R - Rt = r$$

$$-Rt = r - R$$

$$t = \frac{r-R}{-R} \quad \text{or} \quad t = \frac{R-r}{R}$$

13.
$$m = \frac{y-y_1}{x-x_1}$$

$$(x-x_1)\cdot m = (x-x_1)\left(\frac{y-y_1}{x-x_1}\right)$$

$$mx - mx_1 = y - y_1$$

$$mx = y - y_1 + mx_1$$

$$x = \frac{y-y_1+mx_1}{m}$$

15.
$$\omega = \frac{rmv}{I+mr^2}$$

$$\left(I+mr^2\right)\cdot\omega = \left(I+mr^2\right)\left(\frac{rmv}{I+mr^2}\right)$$

$$\omega\left(I+mr^2\right) = rmv$$

$$\frac{\omega\left(I+mr^2\right)}{rm} = v$$

17.
$$V = \frac{mv}{M+m}$$

$$(M+m)\cdot V = (M+m)\left(\frac{mv}{M+m}\right)$$

$$MV + mV = mv$$

$$MV = mv - mV$$

$$MV = m(v-V)$$

$$\frac{MV}{v-V} = m$$

19.
$$\frac{AB}{AC} = \frac{DE}{DF}$$

$$\frac{x+11}{10} = \frac{8}{x}$$

$$10x\left(\frac{x+11}{10}\right) = 10x\left(\frac{8}{x}\right)$$

$$x(x+11) = 10\cdot 8$$

$$x^2 + 11x = 80$$

$$x^2 + 11x - 80 = 0$$

$$(x-5)(x+16) = 0$$

$$x-5=0 \quad \text{or} \quad x+16=0$$
$$x=5 \quad \text{or} \quad x=-16$$

Since the length of DF cannot be negative, we discard $x=-16$ and keep $x=5$. The length of AB is $x+11 = 5+11 = 16$.
In summary, $AB = 16$ and $DF = 5$.

21. Let x represent the population of the U.S. in 2009.

$$\frac{11.9}{100,000} = \frac{36,216}{x}$$

$$100,000x\left(\frac{11.9}{100,000}\right) = 100,000x\left(\frac{36,216}{x}\right)$$

$$11.9x = 3,621,600,000$$

$$x = \frac{3,621,600,000}{11.9}$$

$$x \approx 304,336,135$$

In 2009, the population of the United States was approximately 304,336,135 people.

23. Let x represent the amount of money each of the four people will have to contribute for the trip. Then $4x$ will be the total cost of the trip. Note that the trip is 105 miles each way. Thus, the entire trip consists of a distance of 210 miles.

$$\frac{8.2}{1} = \frac{210}{4x}$$
$$4x\left(\frac{8.2}{1}\right) = 4x\left(\frac{210}{4x}\right)$$
$$32.8x = 210$$
$$x = \frac{210}{32.8}$$
$$x \approx 6.4024$$

Each person must contribute approximately $6.40 for the trip.

25. Utilize Pascal's Principal.

$$\frac{F_1}{A_1} = \frac{F_2}{A_2}$$
$$\frac{30}{12} = \frac{F_2}{5}$$
$$60\left(\frac{30}{12}\right) = 60\left(\frac{F_2}{5}\right)$$
$$150 = 12F_2$$
$$12.5 = F_2$$

Thus, the force exerted by the right pipe is 12.5 pounds.

27. Let t represent the time required if Amiri and Horus deliver the papers together.

$$\begin{pmatrix} \text{Part done} \\ \text{by Amiri} \\ \text{in 1 minute} \end{pmatrix} + \begin{pmatrix} \text{Part done} \\ \text{by Horus} \\ \text{in 1 minute} \end{pmatrix} = \begin{pmatrix} \text{Part done} \\ \text{together} \\ \text{in 1 minute} \end{pmatrix}$$
$$\frac{1}{80} + \frac{1}{60} = \frac{1}{t}$$
$$240t\left(\frac{1}{80} + \frac{1}{60}\right) = 240t\left(\frac{1}{t}\right)$$
$$3t + 4t = 240$$
$$7t = 240$$
$$t = \frac{240}{7} \approx 34.3$$

Working together, it will take Amiri and Horus approximately 34.3 minutes to deliver the papers.

29. Let t represent the time required for Connor to cut the grass when working alone.

$$\begin{pmatrix} \text{Part done} \\ \text{by Avery} \\ \text{in 1 hour} \end{pmatrix} + \begin{pmatrix} \text{Part done} \\ \text{by Connor} \\ \text{in 1 hour} \end{pmatrix} = \begin{pmatrix} \text{Part done} \\ \text{together} \\ \text{in 1 hour} \end{pmatrix}$$
$$\frac{1}{3} + \frac{1}{t} = \frac{1}{2}$$
$$6t\left(\frac{1}{3} + \frac{1}{t}\right) = 6t\left(\frac{1}{2}\right)$$
$$2t + 6 = 3t$$
$$6 = t$$

Working alone, it will take Connor 6 hours to cut the grass.

31. Let t represent the time required for 10-horsepower pump to empty the pool when working alone.
Then $t + 5$ will represent the time required for the 6-horsepower pump to empty the pool when working alone.

$$\begin{pmatrix} \text{Part done by} \\ \text{10-horse pump} \\ \text{in 1 hour} \end{pmatrix} + \begin{pmatrix} \text{Part done by} \\ \text{6-horse pump} \\ \text{in 1 hour} \end{pmatrix} = \begin{pmatrix} \text{Part done} \\ \text{together} \\ \text{in 1 hour} \end{pmatrix}$$
$$\frac{1}{t} + \frac{1}{t+5} = \frac{1}{6}$$
$$6t(t+5)\left(\frac{1}{t} + \frac{1}{t+5}\right) = 6t(t+5)\left(\frac{1}{6}\right)$$
$$6(t+5) + 6t = t(t+5)$$
$$6t + 30 + 6t = t^2 + 5t$$
$$12t + 30 = t^2 + 5t$$
$$0 = t^2 - 7t - 30$$
$$0 = (t-10)(t+3)$$

$t - 10 = 0$ or $t + 3 = 0$
$t = 10$ or $t = -3$

Since time cannot be negative, we discard $t = -3$ and keep $t = 10$.
Working alone, it will take 10-horsepower pump 10 hours to empty the pool.

33. Let x represent the average speed going to work. Then $x - 7$ will represent the average speed on the way home.

	Distance (miles)	Rate (mph)	Time (hours)
Going to work	20	x	$\dfrac{20}{x}$
Coming home	16	$x - 7$	$\dfrac{16}{x-7}$

$$\frac{20}{x} = \frac{16}{x-7}$$

$$x(x-7)\left(\frac{20}{x}\right) = x(x-7)\left(\frac{16}{x-7}\right)$$

$$20(x-7) = 16x$$

$$20x - 140 = 16x$$

$$4x = 140$$

$$x = 35$$

The average speed going to work is 35 miles per hour.

35. Let x represent Hana's walking speed.

	Distance (feet)	Rate (ft/s)	Time (sec)
With moving walkway	152	$x+2$	$\frac{152}{x+2}$
Without moving walkway	72	x	$\frac{72}{x}$

$$\frac{152}{x+2} = \frac{72}{x}$$

$$x(x+2)\left(\frac{152}{x+2}\right) = x(x+2)\left(\frac{72}{x}\right)$$

$$152x = 72(x+2)$$

$$152x = 72x + 144$$

$$80x = 144$$

$$x = \frac{144}{80} = 1.8$$

Hana's walking speed is 1.8 feet per second.

37. Let x represent the distance Gronkowski has run when Urlacher catches him.
Then $x + 5$ will represent the distance Urlacher has run when he catches Shockey.

	Distance (yards)	Rate (yd/s)	Time (sec)
Shockey	x	$\frac{100}{12}$	$\frac{12x}{100}$
Urlacher	$x+5$	$\frac{100}{9}$	$\frac{9(x+5)}{100}$

$$\frac{12x}{100} = \frac{9(x+5)}{100}$$

$$100\left(\frac{12x}{100}\right) = 100\left(\frac{9(x+5)}{100}\right)$$

$$12x = 9x + 45$$

$$3x = 45$$

$$x = 15$$

Gronkowski will have run 15 yards when Urlacher catches up to him. Thus, Shockey will be on his own 35 yard line when Urlacher catches up to him.

39. Let x represent the speed of the bicyclist going uphill. Then $x + 8$ will represent the speed downhill.

	Distance (miles)	Rate (mph)	Time (hours)
Up	12	x	$\frac{12}{x}$
Down	12	$x+8$	$\frac{12}{x+8}$

$$\frac{12}{x} + \frac{12}{x+8} = \frac{9}{4}$$

$$4x(x+8)\left(\frac{12}{x} + \frac{12}{x+8}\right) = 4x(x+8)\left(\frac{9}{4}\right)$$

$$4\cdot12(x+8) + 4\cdot12x = 9x(x+8)$$

$$48x + 384 + 48x = 9x^2 + 72x$$

$$96x + 384 = 9x^2 + 72x$$

$$0 = 9x^2 - 24x - 384$$

$$0 = 3(3x+16)(x-8)$$

$$3x+16 = 0 \quad \text{or} \quad x-8 = 0$$

$$x = -\frac{16}{3} \quad \text{or} \quad x = 8$$

Since speed cannot be negative, we discard $x = -\frac{16}{3}$ and keep $x = 8$.

The speed of the bicyclist going uphill is 8 miles per hour.

41. Let x represent the speed driven for the last 68 miles of the trip.
Then $x + 9$ will represent the speed driven for the first 50 miles of the trip.

	Distance (miles)	Rate (mph)	Time (hours)
First part	50	$x + 9$	$\dfrac{50}{x+9}$
Last part	68	x	$\dfrac{68}{x}$

$$\binom{\text{Time of}}{\text{first part}} + \binom{\text{Time of}}{\text{second part}} = \binom{\text{Total}}{\text{time}}$$

$$\frac{50}{x+9} + \frac{68}{x} = 3$$

$$x(x+9)\left(\frac{50}{x+9} + \frac{68}{x}\right) = x(x+9)\cdot 3$$

$$50x + 68(x+9) = 3x(x+9)$$

$$50x + 68x + 612 = 3x^2 + 27x$$

$$118x + 612 = 3x^2 + 27x$$

$$0 = 3x^2 - 91x - 612$$

$$0 = (3x+17)(x-36)$$

$$3x+17 = 0 \text{ or } x-36 = 0$$

$$x = -\frac{17}{3} \text{ or } x = 36$$

Since speed cannot be negative, we discard $x = -\dfrac{17}{3}$ and keep $x = 36$. Thus, the speed for the last 68 miles of the trip was 36 miles per hour.
Now $x + 9 = 36 + 9 = 45$. Thus, the speed for the first 50 miles of the trip was 45 miles per hour.

43. Greene would require $9.79 - 9.58 = 0.2$ second more to finish the race than Bolt. Let x represent the distance that Greene can run in 0.21 second. Assuming Greene's speed will remain constant,

$$\frac{100}{9.79} = \frac{x}{0.21}$$

$$2.0559\left(\frac{100}{9.79}\right) = 2.0559\left(\frac{x}{0.21}\right)$$

$$21 = 9.79x$$

$$x \approx 2.145$$

Bolt would win by approximately 2.15 meters.

45. $\left(a^3\right)^5 = a^{3\cdot 5} = a^{15}$

47. $(ab)^{-2}\cdot\left(\dfrac{a^3}{b^2}\right)^2 = \dfrac{1}{(ab)^2}\cdot\dfrac{a^6}{b^4} = \dfrac{1}{a^2 b^2}\cdot\dfrac{a^6}{b^4} = \dfrac{a^4}{b^6}$

49. $\left(\dfrac{3m^3 n^{-1}}{mn^5}\right)^{-2} = \left(\dfrac{3m^2}{n^6}\right)^{-2}$

$$= \frac{3^{-2}m^{-4}}{n^{-12}} = \frac{n^{12}}{3^2 m^4} = \frac{n^{12}}{9m^4}$$

Section 5.7

Are You Ready for This Section?

R1. $30 = 5x$

$$\frac{30}{5} = \frac{5x}{5}$$

$$x = 6$$

The solution set is $\{6\}$.

R2. $4 = \dfrac{k}{3}$

$$4\cdot 3 = \frac{k}{\cancel{3}}\cdot\cancel{3}$$

$$k = 12$$

The solution set is $\{12\}$.

R3. $y = 3x$

$m = 3,\ b = 0$

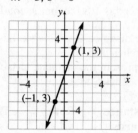

Section 5.7 Quick Checks

1. <u>Variation</u> refers to how one quantity varies in relation to some other quantity.

2. If x and y are two quantities, then y is directly proportional to x if there is a nonzero number k such that $\underline{y = k\cdot x}$.

3. a. Because C varies directly with g, we know that $C = kg$ for some constant k. Because $C = 33.60$ when $g = 8$ gallons, we obtain

$$C = kg$$
$$33.60 = k \cdot 8$$
$$k = \frac{33.60}{8} = 4.2$$

So, we have that $C = 4.2g$, or writing this as a linear function, $C(g) = 4.2g$.

b. $C(4.6) = 4.2(4.6) = 19.32$
So, the cost of 4.6 gallons would be \$19.32.

c. Label the horizontal axis g and the vertical axis C. From the problem we know the point $(8, 33.60)$ is on the graph, and from part (b), we know the point $(4.6, 19.32)$ is on the graph.

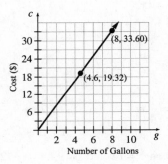

4. Suppose we let x and y represent two quantities. We say that y varies inversely with x, or y is inversely proportional to x, if there is a nonzero number k such that $\underline{y = \dfrac{k}{x}}$.

5. y varies inversely with x: $y = \dfrac{k}{x}$

a. $y = 2$ when $x = 3$
$$2 = \frac{k}{3}$$
$$6 = k$$
$$y = \frac{6}{x}$$

b. Find y when $x = 4$.
$$y = \frac{6}{4} = \frac{3}{2} = 1.5$$

6. a. $V = \dfrac{k}{l}$
$$500 = \frac{k}{30}$$
$$k = 500(30) = 15{,}000$$
Thus, the function that relates the rate of vibration to the length of the string is
$$V(l) = \frac{15{,}000}{l}.$$

b. $V(50) = \dfrac{15{,}000}{50} = 300$
The rate of vibration is 300 oscillations per second.

7. The equation $t = ksp$ is an example of <u>joint</u> variaton.

8. Let K represent the kinetic energy, m represent the mass, and v represent the velocity. Then,
$$K = kmv^2$$
$$4455 = k \cdot 110 \cdot 9^2$$
$$4455 = 8910k$$
$$\frac{1}{2} = k$$
Thus, $K = \dfrac{1}{2}mv^2$.
For $m = 140$ and $v = 5$,
$$K = \frac{1}{2}(140)(5)^2 = 1750.$$
The kinetic energy of a 140-kg lineman running at 5 meters per second is 1750 joules.

9. When direct and inverse variation occur at the same time, we have <u>combined variation</u>.

10. $R = \dfrac{kl}{d^2}$
$$1.24 = \frac{k \cdot 432}{4^2}$$
$$1.24 = \frac{432k}{16}$$
$$1.24 = 27k$$
$$k = \frac{1.24}{27} \approx 0.04593$$
Thus, $R = \dfrac{0.04593l}{d^2}$.

For $l = 282$ and $d = 3$, $R = \dfrac{0.04593(282)}{3^2} \approx 1.44$.

The resistance is approximately 1.44 ohms when the length of the wire is 282 feet and the diameter is 3 millimeters.

5.7 Exercises

11. a. $y = kx$
$30 = k \cdot 5$
$k = \dfrac{30}{5} = 6$

b. $y = 6x$

c. $y = 6(7) = 42$

13. a. $y = kx$
$3 = k \cdot 7$
$k = \dfrac{3}{7}$

b. $y = \dfrac{3}{7}x$

c. $y = \dfrac{3}{7}(28) = 12$

15. a. $y = kx$
$4 = k \cdot 8$
$k = \dfrac{4}{8} = \dfrac{1}{2}$

b. $y = \dfrac{1}{2}x$

c. $y = \dfrac{1}{2}(30) = 15$

17. a. $y = \dfrac{k}{x}$
$2 = \dfrac{k}{10}$
$10 \cdot 2 = 10\left(\dfrac{k}{10}\right)$
$20 = k$

b. $y = \dfrac{20}{x}$

c. When $x = 5$, $y = \dfrac{20}{5} = 4$.

19. a. $y = \dfrac{k}{x}$
$3 = \dfrac{k}{7}$
$7 \cdot 3 = 7\left(\dfrac{k}{7}\right)$
$21 = k$

b. $y = \dfrac{21}{x}$

c. When $x = 28$, $y = \dfrac{21}{28} = \dfrac{3}{4}$.

21. a. $y = kxz$
$10 = k \cdot 8 \cdot 5$
$10 = 40k$
$k = \dfrac{10}{40} = \dfrac{1}{4}$

b. $y = \dfrac{1}{4}xz$

c. When $x = 12$ and $z = 9$, $y = \dfrac{1}{4} \cdot 12 \cdot 9 = 27$.

23. a. $Q = \dfrac{kx}{y}$
$\dfrac{13}{12} = \dfrac{k \cdot 5}{6}$
$12\left(\dfrac{13}{12}\right) = 12\left(\dfrac{k \cdot 5}{6}\right)$
$13 = 10k$
$\dfrac{13}{10} = k$

b. $Q = \dfrac{(13/10)x}{y} = \dfrac{13x}{10y}$

c. When $x = 9$ and $y = 4$, $Q = \dfrac{13 \cdot 9}{10 \cdot 4} = \dfrac{117}{40}$.

25. a. Because p varies directly with b, we know that $p = kb$ for some constant k. We know that $p = \$700.29$ when $b = \$120,000$, so,

$$700.29 = k \cdot 120,000$$
$$k \approx 0.0058358$$

Therefore, we have $p = 0.0058358b$ or, using function notation,

$p(b) = 0.00538358b$.

b. $p(140,000) = 0.0058358(140,000)$
$$\approx 817.01$$

The monthly payment would be approximately \$817.01 if \$140,000 is borrowed.

c.

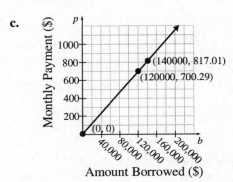

27. a. Because C varies directly with w, we know that $C = kw$ for some constant k. We know that $C = \$28$ when $w = 5$ lb, so,

$$28 = k \cdot 5$$
$$k = 5.6$$

Therefore, we have $C = 5.6w$ or, using function notation, $C(w) = 5.6w$.

b. $C(3.5) = 5.6(3.5) = \$19.60$

The cost of 3.5 pounds of chocolate covered almonds would be \$19.60.

c.

29. Because v varies directly with t, we know that $v = kt$ for some constant k. We know that $v = 64$ feet per second when $t = 2$ seconds, so

$$64 = k \cdot 2$$
$$k = 32$$

Thus, $v = 32t$ or, using function notation, $v(t) = 32t$. Finally, $v(3) = 32(3) = 96$.

Therefore, the velocity of a falling object will be 96 feet per second after 3 seconds.

31. a.
$$D = \frac{k}{p}$$
$$150 = \frac{k}{2.50}$$
$$2.50 \cdot 150 = 2.50\left(\frac{k}{2.50}\right)$$
$$375 = k$$

Thus, the demand of the candy as a function of its price is $D(p) = \dfrac{375}{p}$.

b. $D(3) = \dfrac{375}{3} = 125$

If the price of the candy is \$3.00, then 125 bags of candy will be sold.

33.
$$V = \frac{k}{P}$$
$$600 = \frac{k}{150}$$
$$150 \cdot 600 = 150\left(\frac{k}{150}\right)$$
$$90,000 = k$$
$$V(P) = \frac{90,000}{P}$$
$$V(200) = \frac{90,000}{200} = 450$$

When the pressure is 200 mm Hg, the volume of the gas will be 450 cc.

35. Let w represent the weight of an object that is a distance d from the center of the earth.

$$w = \frac{k}{d^2}$$

$$120 = \frac{k}{3960^2}$$

$$3960^2 \cdot 120 = 3960^2 \left(\frac{k}{3960^2} \right)$$

$$1,881,792,000 = k$$

$$w(d) = \frac{1,881,792,000}{d^2}$$

Now, Maria is $3960 + 3.8 = 3963.8$ miles from the center of Earth.

$$w(3963.8) = \frac{1,881,792,000}{3963.8^2} \approx 119.8$$

Maria's weight at the top of Mount McKinley is 119.8 pounds.

37. Let D represent drag force, A represent surface area, and v represent velocity.

$$D = kAv^2$$

$$1152 = k \cdot 2 \cdot 40^2$$

$$1152 = 3200k$$

$$0.36 = k$$

$$D = 0.36Av^2$$

For $A = 2.5$ and $v = 50$,

$$D = 0.36 \cdot 2.5 \cdot 50^2 = 2250.$$

The drag force on a parachutist whose surface area is 2.5 square meters falling at 50 meters per second is 2250 newtons.

39.

$$F = \frac{km_1 m_2}{r^2}$$

$$2.24112 \times 10^{-8} = \frac{k \cdot 105 \cdot 80}{5^2}$$

$$2.24112 \times 10^{-8} = 336k$$

$$\frac{2.24112 \times 10^{-8}}{336} = k$$

$$6.67 \times 10^{-11} = k$$

$$F = \frac{\left(6.67 \times 10^{-11} \right) m_1 m_2}{r^2}$$

For $m_1 = 105$, $m_2 = 80$, and $r = 2$,

$$F = \frac{\left(6.67 \times 10^{-11} \right) m_1 m_2}{r^2}$$

$$= \frac{\left(6.67 \times 10^{-11} \right)(105)(80)}{2^2}$$

$$= 1.4007 \times 10^{-7}.$$

When they are 2 meters apart, the force of gravity between the couple is 1.4007×10^{-7} newtons.

41. Let s represent stress, p represent internal pressure, d represent internal diameter, and x represent thickness.

$$s = \frac{kpd}{x}$$

$$100 = \frac{k \cdot 25 \cdot 5}{0.75}$$

$$0.75(100) = 0.75 \left(\frac{k \cdot 25 \cdot 5}{0.75} \right)$$

$$75 = 125k$$

$$0.6 = k$$

$$s = \frac{0.6pd}{x}$$

For $p = 50$, $d = 6$, and $x = 0.5$,

$$s = \frac{0.6 \cdot 50 \cdot 6}{0.5} = 360.$$

The stress is 360 pounds per square inch when the internal pressure is 50 pounds per square inch, the diameter is 6 inches, and the thickness is 0.5 inch.

43. a. $\omega = 2\pi \cdot 50 = 100\pi \approx 314.16$

     b. $v = \omega r = 314.16 \cdot 3 = 942.48$ meters per minute

$$v = \frac{942.48 \text{ m}}{1 \text{ min}} \cdot \frac{1 \text{ min}}{60 \text{ sec}} = 15.708 \text{ meters per second}$$

     c. $$F = \frac{kmv^2}{r}$$

$$2.3 = \frac{k \cdot 0.5 \cdot 15.708^2}{3}$$

$$2.3 = k \cdot 41.123544$$

$$\frac{2.3}{4.1123544} = k$$

$$0.056 \approx k$$

d. For spinning rate of 80 revolutions per minute, $\omega = 2\pi \cdot 80 = 160\pi \approx 502.65$.
$v = \omega r = 502.65 \cdot 4 = 2010.60$ meters per minute

$$v = \frac{2010.60 \text{ m}}{1 \text{ min}} \cdot \frac{1 \text{ min}}{60 \text{ sec}} = 33.51 \text{ meters per second}$$

$$F = \frac{0.056mv^2}{r} = \frac{0.056 \cdot 0.5 \cdot 33.51^2}{4} \approx 7.86$$

The force required to keep the stone in motion is approximately 7.86 newtons.

Chapter 5 Review

1. We need to find all values of x that cause the denominator $3x - 2$ to equal 0.
$$3x - 2 = 0$$
$$3x = 2$$
$$x = \frac{2}{3}$$
Thus, the domain of $\frac{x-5}{3x-2}$ is $\left\{ x \middle| x \neq \frac{2}{3} \right\}$

2. We need to find all values of a that cause the denominator $a^2 - 3a - 28$ to equal 0.
$$a^2 - 3a - 28 = 0$$
$$(a+4)(a-7) = 0$$
$$a + 4 = 0 \quad \text{or} \quad a - 7 = 0$$
$$a = -4 \quad \text{or} \quad a = 7$$
Thus, the domain of $\frac{a^2 - 16}{a^2 - 3a - 28}$ is $\{a \mid a \neq -4, \ a \neq 7\}$

3. We need to find all values of m that cause the denominator $m^2 + 9$ to equal 0. However, if x is a real number, then $m^2 + 9$ can never equal 0.
The domain of $\frac{m-3}{m^2 + 9}$ is $\{m \mid m$ is any real number$\}$.

4. We need to find all values of n that cause the denominator $n^2 - 2n - 8$ to equal 0.
$$n^2 - 2n - 8 = 0$$
$$(n+2)(n-4) = 0$$
$$n + 2 = 0 \quad \text{or} \quad n - 4 = 0$$
$$n = -2 \quad \text{or} \quad n = 4$$

Thus, the domain of $\frac{n^2 + 7n + 10}{n^2 - 2n - 8}$ is $\{n \mid n \neq -2, \ n \neq 4\}$.

5. $\dfrac{6x + 30}{x^2 - 25} = \dfrac{6(x+5)}{(x+5)(x-5)}$

$= \dfrac{6\cancel{(x+5)}}{\cancel{(x+5)}(x-5)}$

$= \dfrac{6}{x-5}$

6. $\dfrac{4y^2 - 28y}{2y^5 - 14y^4} = \dfrac{4y(y-7)}{2y^4(y-7)}$

$= \dfrac{2 \cdot 2y(y-7)}{2y \cdot y^3(y-7)}$

$= \dfrac{2 \cdot \cancel{2y}\cancel{(y-7)}}{\cancel{2y} \cdot y^3 \cancel{(y-7)}} = \dfrac{2}{y^3}$

7. $\dfrac{w^2 - 4w - 21}{w^2 + 7w + 12} = \dfrac{(w+3)(w-7)}{(w+3)(w+4)}$

$= \dfrac{\cancel{(w+3)}(w-7)}{\cancel{(w+3)}(w+4)} = \dfrac{w-7}{w+4}$

8. $\dfrac{6a^2 - 7ab - 3b^2}{10a^2 - 11ab - 6b^2} = \dfrac{(3a+b)(2a-3b)}{(5a+2b)(2a-3b)}$

$= \dfrac{(3a+b)\cancel{(2a-3b)}}{(5a+2b)\cancel{(2a-3b)}}$

$= \dfrac{3a+b}{5a+2b}$

9. $\dfrac{7-m}{3m^2 - 20m - 7} = \dfrac{-1(m-7)}{(3m+1)(m-7)}$

$= \dfrac{-1\cancel{(m-7)}}{(3m+1)\cancel{(m-7)}} = \dfrac{-1}{3m+1}$

10. $\dfrac{n^3 - 4n^2 + 3n - 12}{n^2 - 8n + 16} = \dfrac{(n-4)(n^2+3)}{(n-4)(n-4)}$

$= \dfrac{\cancel{(n-4)}(n^2+3)}{\cancel{(n-4)}(n-4)} = \dfrac{n^2+3}{n-4}$

443

11. $\dfrac{4p^2}{p^2-3p-18}\cdot\dfrac{p+3}{8p}=\dfrac{4p\cdot p}{(p+3)(p-6)}\cdot\dfrac{p+3}{2\cdot 4p}$

$\qquad\qquad = \dfrac{\cancel{4p}\cdot p}{\cancel{(p+3)}(p-6)}\cdot\dfrac{\cancel{p+3}}{2\cdot\cancel{4p}}$

$\qquad\qquad = \dfrac{p}{2(p-6)}$

12. $\dfrac{q^2+6q}{6q+12}\cdot\dfrac{4q+8}{q^2+q-30}$

$\qquad = \dfrac{q(q+6)}{6(q+2)}\cdot\dfrac{4(q+2)}{(q-5)(q+6)}$

$\qquad = \dfrac{q\cancel{(q+6)}}{\underset{3}{\cancel{6}}\cancel{(q+2)}}\cdot\dfrac{\overset{2}{\cancel{4}}\cancel{(q+2)}}{(q-5)\cancel{(q+6)}}$

$\qquad = \dfrac{2q}{3(q-5)}$

13. $\dfrac{x^3-4x^2}{x^2-4}\cdot\dfrac{x^2+4x-12}{x^3+2x^2}$

$\qquad = \dfrac{x^2(x-4)}{(x+2)(x-2)}\cdot\dfrac{(x-2)(x+6)}{x^2(x+2)}$

$\qquad = \dfrac{\cancel{x^2}(x-4)}{(x+2)\cancel{(x-2)}}\cdot\dfrac{\cancel{(x-2)}(x+6)}{\cancel{x^2}(x+2)}$

$\qquad = \dfrac{(x-4)(x+6)}{(x+2)^2}$

14. $\dfrac{y^2-3y-28}{y^3+4y^2}\cdot\dfrac{2y^2+10y}{y^2-12y+35}$

$\qquad = \dfrac{(y+4)(y-7)}{y^2(y+4)}\cdot\dfrac{2y(y+5)}{(y-5)(y-7)}$

$\qquad = \dfrac{(y+4)(y-7)}{y\cdot y\cdot(y+4)}\cdot\dfrac{2\cdot y\cdot(y+5)}{(y-5)(y-7)}$

$\qquad = \dfrac{\cancel{(y+4)}\cancel{(y-7)}}{\cancel{y}\cdot y\cdot\cancel{(y+4)}}\cdot\dfrac{2\cdot\cancel{y}\cdot(y+5)}{(y-5)\cancel{(y-7)}}=\dfrac{2(y+5)}{y(y-5)}$

15. $\dfrac{6a^2+ab-b^2}{3a^2+2ab-b^2}\cdot\dfrac{3a^2+4ab+b^2}{4a^2-b^2}$

$\qquad = \dfrac{(3a-b)(2a+b)}{(3a-b)(a+b)}\cdot\dfrac{(3a+b)(a+b)}{(2a+b)(2a-b)}$

$\qquad = \dfrac{\cancel{(3a-b)}\cancel{(2a+b)}}{\cancel{(3a-b)}\cancel{(a+b)}}\cdot\dfrac{(3a+b)\cancel{(a+b)}}{\cancel{(2a+b)}(2a-b)}$

$\qquad = \dfrac{3a+b}{2a-b}$

16. $\dfrac{m^2+m-20}{m^3-64}\cdot\dfrac{3m^2+12m+48}{m^2+3m-10}$

$\qquad = \dfrac{(m-4)(m+5)}{(m-4)(m^2+4m+16)}\cdot\dfrac{3(m^2+4m+16)}{(m-2)(m+5)}$

$\qquad = \dfrac{\cancel{(m-4)}\cancel{(m+5)}}{\cancel{(m-4)}\cancel{(m^2+4m+16)}}\cdot\dfrac{3\cancel{(m^2+4m+16)}}{(m-2)\cancel{(m+5)}}$

$\qquad = \dfrac{3}{m-2}$

17. $\dfrac{\dfrac{4c^2}{3d^4}}{\dfrac{8c}{27d}}=\dfrac{4c^2}{3d^4}\cdot\dfrac{27d}{8c}=\dfrac{\cancel{4c}\cdot c}{3d\cdot d^3}\cdot\dfrac{9\cdot 3d}{2\cdot\cancel{4c}}=\dfrac{9c}{2d^3}$

18. $\dfrac{\dfrac{6z-24}{7z+21}}{\dfrac{z-4}{z^2-9}}=\dfrac{6z-24}{7z+21}\cdot\dfrac{z^2-9}{z-4}$

$\qquad = \dfrac{6(z-4)}{7(z+3)}\cdot\dfrac{(z-3)(z+3)}{z-4}$

$\qquad = \dfrac{6\cancel{(z-4)}}{7\cancel{(z+3)}}\cdot\dfrac{(z-3)\cancel{(z+3)}}{\cancel{z-4}}=\dfrac{6(z-3)}{7}$

19. $\dfrac{\dfrac{x^2-11x+30}{x^2-8x+15}}{\dfrac{x^2-5x-6}{x^2+8x+7}}=\dfrac{x^2-11x+30}{x^2-8x+15}\cdot\dfrac{x^2+8x+7}{x^2-5x-6}$

$\qquad = \dfrac{(x-5)(x-6)}{(x-3)(x-5)}\cdot\dfrac{(x+1)(x+7)}{(x+1)(x-6)}$

$\qquad = \dfrac{\cancel{(x-5)}\cancel{(x-6)}}{(x-3)\cancel{(x-5)}}\cdot\dfrac{\cancel{(x+1)}(x+7)}{\cancel{(x+1)}\cancel{(x-6)}}$

$\qquad = \dfrac{x+7}{x-3}$

20.
$$\frac{\dfrac{m^2+mn-12n^2}{m^3-27n^3}}{\dfrac{m+5n}{m^2+3mn+9n^2}}$$

$$=\frac{m^2+mn-12n^2}{m^3-27n^3}\cdot\frac{m^2+3mn+9n^2}{m+5n}$$

$$=\frac{(m-3n)(m+4n)}{(m-3n)\left(m^2+3mn+9n^2\right)}\cdot\frac{m^2+3mn+9n^2}{m+5n}$$

$$=\frac{\cancel{(m-3n)}(m+4n)}{\cancel{(m-3n)}\cancel{\left(m^2+3mn+9n^2\right)}}\cdot\frac{\cancel{m^2+3mn+9n^2}}{m+5n}$$

$$=\frac{m+4n}{m+5n}$$

21.
$$\frac{\dfrac{4p^3-4pq^2}{p^2-5pq-24q^2}}{\dfrac{2p^3+4p^2q+2pq^2}{p^2-7pq-8q^2}}$$

$$=\frac{4p^3-4pq^2}{p^2-5pq-24q^2}\cdot\frac{p^2-7pq-8q^2}{2p^3+4p^2q+2pq^2}$$

$$=\frac{4p(p-q)(p+q)}{(p-8q)(p+3q)}\cdot\frac{(p+q)(p-8q)}{2p(p+q)(p+q)}$$

$$=\frac{2\cdot\cancel{2p}(p-q)\cancel{(p+q)}}{\cancel{(p-8q)}(p+3q)}\cdot\frac{\cancel{(p+q)}\cancel{(p-8q)}}{\cancel{2p}\cancel{(p+q)}\cancel{(p+q)}}$$

$$=\frac{2(p-q)}{p+3q}$$

22.
$$\frac{\dfrac{15a^2+11a-14}{25a^2-49}}{\dfrac{27a^3-8}{10a^2+11a-35}}$$

$$=\frac{15a^2+11a-14}{25a^2-49}\cdot\frac{10a^2+11a-35}{27a^3-8}$$

$$=\frac{(3a-2)(5a+7)}{(5a-7)(5a+7)}\cdot\frac{(2a+5)(5a-7)}{(3a-2)\left(9a^2+6a+4\right)}$$

$$=\frac{\cancel{(3a-2)}\cancel{(5a+7)}}{\cancel{(5a-7)}\cancel{(5a+7)}}\cdot\frac{(2a+5)\cancel{(5a-7)}}{\cancel{(3a-2)}\left(9a^2+6a+4\right)}$$

$$=\frac{2a+5}{9a^2+6a+4}$$

23. $P(x)=f(x)\cdot g(x)$

$$=\frac{2x^2+3x-2}{x-5}\cdot\frac{x^2-3x-10}{2x-1}$$

$$=\frac{(2x-1)(x+2)}{x-5}\cdot\frac{(x+2)(x-5)}{2x-1}$$

$$=\frac{\cancel{(2x-1)}(x+2)}{\cancel{x-5}}\cdot\frac{(x+2)\cancel{(x-5)}}{\cancel{2x-1}}$$

$$=(x+2)^2$$

The domain of $f(x)$ is $\left\{x\mid x\neq5\right\}$. The domain of $g(x)$ is $\left\{x\mid x\neq\dfrac{1}{2}\right\}$. Therefore, the domain of $P(x)$ is $\left\{x\mid x\neq\dfrac{1}{2},x\neq5\right\}$.

24. $R(x)=g(x)\cdot h(x)$

$$=\frac{x^2-3x-10}{2x-1}\cdot\frac{2x-1}{x^2+9x+14}$$

$$=\frac{(x+2)(x-5)}{2x-1}\cdot\frac{2x-1}{(x+2)(x+7)}$$

$$=\frac{\cancel{(x+2)}(x-5)}{\cancel{2x-1}}\cdot\frac{\cancel{2x-1}}{\cancel{(x+2)}(x+7)}$$

$$=\frac{x-5}{x+7}$$

The domain of $g(x)$ is $\left\{x\mid x\neq\dfrac{1}{2}\right\}$. The domain of $h(x)$ is $\left\{x\mid x\neq-7,x\neq-2\right\}$. Thus, the domain of $R(x)$ is $\left\{x\mid x\neq-7,x\neq-2,x\neq\dfrac{1}{2}\right\}$.

25. $Q(x)=\dfrac{g(x)}{f(x)}=\dfrac{\dfrac{x^2-3x-10}{2x-1}}{\dfrac{2x^2+3x-2}{x-5}}$

$$=\frac{x^2-3x-10}{2x-1}\cdot\frac{x-5}{2x^2+3x-2}$$

$$=\frac{(x+2)(x-5)}{2x-1}\cdot\frac{x-5}{(2x-1)(x+2)}$$

$$=\frac{\cancel{(x+2)}(x-5)}{2x-1}\cdot\frac{x-5}{(2x-1)\cancel{(x+2)}}$$

$$=\frac{(x-5)^2}{(2x-1)^2}$$

The domain of $g(x)$ is $\left\{x \mid x \neq \dfrac{1}{2}\right\}$. The domain

of $f(x)$ is $\{x \mid x \neq 5\}$. Because the denominator

of $Q(x)$ cannot equal 0, we must exclude those

values of x such that the numerator of $f(x)$ is 0.

That is, we must exclude the values of x such

that $2x^2 + 3x - 2 = 0$. These values are $x = -2$

and $x = \dfrac{1}{2}$. Therefore, the domain of $Q(x)$ is

$\left\{x \mid x \neq -2, x \neq \dfrac{1}{2}, x \neq 5\right\}$.

26. $T(x) = \dfrac{f(x)}{h(x)} = \dfrac{\dfrac{2x^2 + 3x - 2}{x - 5}}{\dfrac{2x - 1}{x^2 + 9x + 14}}$

$= \dfrac{2x^2 + 3x - 2}{x - 5} \cdot \dfrac{x^2 + 9x + 14}{2x - 1}$

$= \dfrac{(2x - 1)(x + 2)}{x - 5} \cdot \dfrac{(x + 2)(x + 7)}{2x - 1}$

$= \dfrac{\cancel{(2x - 1)}(x + 2)}{x - 5} \cdot \dfrac{(x + 2)(x + 7)}{\cancel{2x - 1}}$

$= \dfrac{(x + 2)^2 (x + 7)}{x - 5}$

The domain of $f(x)$ is $\{x \mid x \neq 5\}$. The domain

of $h(x)$ is $\{x \mid x \neq -7, x \neq -2\}$. Because the

denominator of $T(x)$ cannot equal 0, we must

exclude those values of x such that the numerator

of $h(x)$ is 0. That is, we must exclude the value

of x such that $2x - 1 = 0$. This value is $x = \dfrac{1}{2}$.

Therefore, the domain of $T(x)$ is

$\left\{x \mid x \neq -7, x \neq -2, x \neq \dfrac{1}{2}, x \neq 5\right\}$.

27. $\dfrac{4x}{x - 5} + \dfrac{3}{x - 5} = \dfrac{4x + 3}{x - 5}$

28. $\dfrac{4y}{y - 3} - \dfrac{12}{y - 3} = \dfrac{4y - 12}{y - 3} = \dfrac{4(y - 3)}{y - 3} = 4$

29. $\dfrac{a^2 - 2a - 4}{a^2 - 6a + 8} + \dfrac{4a - 20}{a^2 - 6a + 8} = \dfrac{a^2 - 2a - 4 + 4a - 20}{a^2 - 6a + 8}$

$= \dfrac{a^2 + 2a - 24}{a^2 - 6a + 8}$

$= \dfrac{(a - 4)(a + 6)}{(a - 4)(a - 2)}$

$= \dfrac{a + 6}{a - 2}$

30. $\dfrac{3b^2 + 8b - 5}{2b^2 - 5b - 12} - \dfrac{2b^2 + 7b + 15}{2b^2 - 5b - 12}$

$= \dfrac{3b^2 + 8b - 5 - \left(2b^2 + 7b + 15\right)}{2b^2 - 5b - 12}$

$= \dfrac{3b^2 + 8b - 5 - 2b^2 - 7b - 15}{2b^2 - 5b - 12}$

$= \dfrac{b^2 + b - 20}{2b^2 - 5b - 12} = \dfrac{(b + 5)(b - 4)}{(2b + 3)(b - 4)} = \dfrac{b + 5}{2b + 3}$

31. $\dfrac{5c^2 - 8c}{c - 8} + \dfrac{2c^2 + 16c}{8 - c} = \dfrac{5c^2 - 8c}{c - 8} + \dfrac{2c^2 + 16c}{-1(c - 8)}$

$= \dfrac{5c^2 - 8c}{c - 8} - \dfrac{2c^2 + 16c}{c - 8}$

$= \dfrac{5c^2 - 8c - \left(2c^2 + 16c\right)}{c - 8}$

$= \dfrac{5c^2 - 8c - 2c^2 - 16c}{c - 8}$

$= \dfrac{3c^2 - 24c}{c - 8}$

$= \dfrac{3c(c - 8)}{c - 8}$

$= 3c$

32. $\dfrac{2d^2+d}{d^2-1}-\dfrac{d^2+1}{d^2-1}+\dfrac{d-2}{d^2-1}$

$=\dfrac{2d^2+d-\left(d^2+1\right)+d-2}{d^2-1}$

$=\dfrac{2d^2+d-d^2-1+d-2}{d^2-1}$

$=\dfrac{d^2+2d-3}{d^2-1}$

$=\dfrac{(d-1)(d+3)}{(d+1)(d-1)}$

$=\dfrac{d+3}{d+1}$

33. $9x^4=3^2\cdot x^4$

$12x^2=2^2\cdot 3\cdot x$

$\text{LCD}=2^2\cdot 3^2\cdot x^4=36x^4$

34. $y-9$

$y+2$

$\text{LCD}=(y+2)(y-9)$

35. $2p^2-3p-20=(2p+5)(p-4)$

$2p^3+5p^2=p^2(2p+5)$

$\text{LCD}=p^2(2p+5)(p-4)$

36. $q^2+4q-5=(q-1)(q+5)$

$q^2+2q-15=(q-3)(q+5)$

$\text{LCD}=(q+5)(q-1)(q-3)$

37. mn^4

m^3n^2

$\text{LCD}=m^3n^4$

$\dfrac{1}{mn^4}+\dfrac{4}{m^3n^2}=\dfrac{1}{mn^4}\cdot\dfrac{m^2}{m^2}+\dfrac{4}{m^3n^2}\cdot\dfrac{n^2}{n^2}$

$=\dfrac{m^2}{m^3n^4}+\dfrac{4n^2}{m^3n^4}$

$=\dfrac{m^2+4n^2}{m^3n^4}$

38. $2xy^3=2\cdot x\cdot y^3$

$6x^2y=2\cdot 3\cdot x^2\cdot y$

$\text{LCD}=2\cdot 3\cdot x^2\cdot y^3=6x^2y^3$

$\dfrac{3}{2xy^3}-\dfrac{7}{6x^2y}=\dfrac{3}{2xy^3}\cdot\dfrac{3x}{3x}-\dfrac{7}{6x^2y}\cdot\dfrac{y^2}{y^2}$

$=\dfrac{9x}{6x^2y^3}-\dfrac{7y^2}{6x^2y^3}$

$=\dfrac{9x-7y^2}{6x^2y^3}$

39. $p-q$

$p+q$

$\text{LCD}=(p-q)(p+q)$

$\dfrac{p}{p-q}-\dfrac{q}{p+q}=\dfrac{p}{p-q}\cdot\dfrac{p+q}{p+q}-\dfrac{q}{p+q}\cdot\dfrac{p-q}{p-q}$

$=\dfrac{p^2+pq}{(p-q)(p+q)}-\dfrac{pq-q^2}{(p-q)(p+q)}$

$=\dfrac{p^2+pq-\left(pq-q^2\right)}{(p-q)(p+q)}$

$=\dfrac{p^2+pq-pq+q^2}{(p-q)(p+q)}=\dfrac{p^2+q^2}{(p-q)(p+q)}$

40. $x^2-10x+21=(x-3)(x-7)$

$x^2-3x-28=(x+4)(x-7)$

$\text{LCD}=(x-7)(x-3)(x+4)$

$\dfrac{x+8}{x^2-10x+21}-\dfrac{x-5}{x^2-3x-28}$

$=\dfrac{x+8}{(x-3)(x-7)}\cdot\dfrac{x+4}{x+4}-\dfrac{x-5}{(x+4)(x-7)}\cdot\dfrac{x-3}{x-3}$

$=\dfrac{x^2+12x+32}{(x-7)(x-3)(x+4)}-\dfrac{x^2-8x+15}{(x-7)(x-3)(x+4)}$

$=\dfrac{x^2+12x+32-\left(x^2-8x+15\right)}{(x-7)(x-3)(x+4)}$

$=\dfrac{x^2+12x+32-x^2+8x-15}{(x-7)(x-3)(x+4)}$

$=\dfrac{20x+17}{(x-7)(x-3)(x+4)}$

41. $y^2 - 2y + 1 = (y-1)^2$

$y^2 + y - 2 = (y-1)(y+2)$

$\text{LCD} = (y-1)^2(y+2)$

$\dfrac{3}{y^2-2y+1} - \dfrac{2}{y^2+y-2}$

$= \dfrac{3}{(y-1)^2} \cdot \dfrac{y+2}{y+2} - \dfrac{2}{(y-1)(y+2)} \cdot \dfrac{y-1}{y-1}$

$= \dfrac{3y+6}{(y-1)^2(y+2)} - \dfrac{2y-2}{(y-1)^2(y+2)}$

$= \dfrac{3y+6-(2y-2)}{(y-1)^2(y+2)}$

$= \dfrac{3y+6-2y+2}{(y-1)^2(y+2)} = \dfrac{y+8}{(y-1)^2(y+2)}$

42. $4a^2 - 9b^2 = (2a-3b)(2a+3b)$

$2a-3b$

$\text{LCD} = (2a-3b)(2a+3b)$

$\dfrac{3a-5b}{4a^2-9b^2} + \dfrac{4}{2a-3b}$

$= \dfrac{3a-5b}{(2a-3b)(2a+3b)} + \dfrac{4}{2a-3b} \cdot \dfrac{2a+3b}{2a+3b}$

$= \dfrac{3a-5b}{(2a-3b)(2a+3b)} + \dfrac{8a+12b}{(2a-3b)(2a+3b)}$

$= \dfrac{3a-5b+8a+12b}{(2a-3b)(2a+3b)} = \dfrac{11a+7b}{(2a-3b)(2a+3b)}$

43. $x^2 - 9 = (x-3)(x+3)$

$9 - x^2 = -1(x-3)(x+3)$

$\text{LCD} = (x-3)(x+3)$

$\dfrac{4x^2-10x}{x^2-9} + \dfrac{8x-2x^2}{9-x^2}$

$= \dfrac{4x^2-10x}{(x-3)(x+3)} + \dfrac{8x-2x^2}{-1(x-3)(x+3)} \cdot \dfrac{-1}{-1}$

$= \dfrac{4x^2-10x}{(x-3)(x+3)} + \dfrac{-8x+2x^2}{(x-3)(x+3)}$

$= \dfrac{4x^2-10x-8x+2x^2}{(x-3)(x+3)}$

$= \dfrac{6x^2-18x}{(x-3)(x+3)} = \dfrac{6x(x-3)}{(x-3)(x+3)} = \dfrac{6x}{x+3}$

44. $n+5$

$n^3 + 125 = (n+5)(n^2-5n+25)$

$\text{LCD} = (n+5)(n^2-5n+25)$

$\dfrac{1}{n+5} - \dfrac{n^2-10n}{n^3+125}$

$= \dfrac{1}{n+5} \cdot \dfrac{n^2-5n+25}{n^2-5n+25} - \dfrac{n^2-10n}{(n+5)(n^2-5n+25)}$

$= \dfrac{n^2-5n+25}{(n+5)(n^2-5n+25)} - \dfrac{n^2-10n}{(n+5)(n^2-5n+25)}$

$= \dfrac{n^2-5n+25-(n^2-10n)}{(n+5)(n^2-5n+25)}$

$= \dfrac{n^2-5n+25-n^2+10n}{(n+5)(n^2-5n+25)} = \dfrac{5n+25}{(n+5)(n^2-5n+25)}$

$= \dfrac{5(n+5)}{(n+5)(n^2-5n+25)} = \dfrac{5}{n^2-5n+25}$

45. $m+3n$

$m-7n$

$m^2 - 4mn - 21n^2 = (m+3n)(m-7n)$

$\text{LCD} = (m+3n)(m-7n)$

$\dfrac{m+n}{m+3n} - \dfrac{m-4n}{m-7n} + \dfrac{7mn+n^2}{m^2-4mn-21n^2}$

$= \dfrac{m+n}{m+3n} \cdot \dfrac{m-7n}{m-7n} - \dfrac{m-4n}{m-7n} \cdot \dfrac{m+3n}{m+3n} +$

$\dfrac{7mn+n^2}{(m+3n)(m-7n)}$

$= \dfrac{m^2-6mn-7n^2}{(m+3n)(m-7n)} - \dfrac{m^2-mn-12n^2}{(m+3n)(m-7n)} +$

$\dfrac{7mn+n^2}{(m+3n)(m-7n)}$

$= \dfrac{m^2-6mn-7n^2-(m^2-mn-12n^2)+7mn+n^2}{(m+3n)(m-7n)}$

$= \dfrac{m^2-6mn-7n^2-m^2+mn+12n^2+7mn+n^2}{(m+3n)(m-7n)}$

$= \dfrac{2mn+6n^2}{(m+3n)(m-7n)}$

$= \dfrac{2n(m+3n)}{(m+3n)(m-7n)} = \dfrac{2n}{m-7n}$

46. $z^2 - 9 = (z-3)(z+3)$

$z - 3$

$z + 3$

$\text{LCD} = (z-3)(z+3)$

$\dfrac{z^2+10z+3}{z^2-9} - \dfrac{2z}{z-3} + \dfrac{z}{z+3}$

$= \dfrac{z^2+10z+3}{(z-3)(z+3)} - \dfrac{2z}{z-3} \cdot \dfrac{z+3}{z+3} + \dfrac{z}{z+3} \cdot \dfrac{z-3}{z-3}$

$= \dfrac{z^2+10z+3}{(z-3)(z+3)} - \dfrac{2z^2+6z}{(z-3)(z+3)} + \dfrac{z^2-3z}{(z-3)(z+3)}$

$= \dfrac{z^2+10z+3 - \left(2z^2+6z\right) + z^2-3z}{(z-3)(z+3)}$

$= \dfrac{z^2+10z+3 - 2z^2 - 6z + z^2 - 3z}{(z-3)(z+3)}$

$= \dfrac{z+3}{(z-3)(z+3)}$

$= \dfrac{1}{z-3}$

47. $y - 2$

$y + 2$

$y^2 - 4 = (y-2)(y+2)$

$\text{LCD} = (y-2)(y+2)$

$\dfrac{y-1}{y-2} - \dfrac{y+1}{y+2} + \dfrac{y-6}{y^2-4}$

$= \dfrac{y-1}{y-2} \cdot \dfrac{y+2}{y+2} - \dfrac{y+1}{y+2} \cdot \dfrac{y-2}{y-2} + \dfrac{y-6}{(y-2)(y+2)}$

$= \dfrac{y^2+y-2}{(y-2)(y+2)} - \dfrac{y^2-y-2}{(y-2)(y+2)}$

$\qquad\qquad + \dfrac{y-6}{(y-2)(y+2)}$

$= \dfrac{y^2+y-2 - \left(y^2-y-2\right) + y-6}{(y-2)(y+2)}$

$= \dfrac{y^2+y-2 - y^2+y+2 + y-6}{(y-2)(y+2)}$

$= \dfrac{3y-6}{(y-2)(y+2)}$

$= \dfrac{3(y-2)}{(y-2)(y+2)} = \dfrac{3}{y+2}$

48. $a^2 - 16 = (a-4)(a+4)$

$a - 4$

$a^2 + 2a - 8 = (a-2)(a+4)$

$\text{LCD} = (a-4)(a+4)(a-2)$

$\dfrac{2a}{a^2-16} - \dfrac{1}{a-4} - \dfrac{1}{a^2+2a-8}$

$= \dfrac{2a}{(a-4)(a+4)} \cdot \dfrac{a-2}{a-2} - \dfrac{1}{a-4} \cdot \dfrac{(a+4)(a-2)}{(a+4)(a-2)} -$

$\quad \dfrac{1}{(a+4)(a-2)} \cdot \dfrac{a-4}{a-4}$

$= \dfrac{2a^2-4a}{(a-4)(a+4)(a-2)} - \dfrac{a^2+2a-8}{(a-4)(a+4)(a-2)} -$

$\quad \dfrac{a-4}{(a-4)(a+4)(a-2)}$

$= \dfrac{2a^2-4a - \left(a^2+2a-8\right) - (a-4)}{(a-4)(a+4)(a-2)}$

$= \dfrac{2a^2-4a - a^2 - 2a + 8 - a + 4}{(a-4)(a+4)(a-2)}$

$= \dfrac{a^2-7a+12}{(a-4)(a+4)(a-2)}$

$= \dfrac{(a-3)(a-4)}{(a-4)(a+4)(a-2)} = \dfrac{a-3}{(a+4)(a-2)}$

49. a. $x - 4$

$x + 2$

$\text{LCD} = (x-4)(x+2)$

$S(x) = f(x) + g(x) = \dfrac{5}{x-4} + \dfrac{x}{x+2}$

$= \dfrac{5}{x-4} \cdot \dfrac{x+2}{x+2} + \dfrac{x}{x+2} \cdot \dfrac{x-4}{x-4}$

$= \dfrac{5x+10}{(x-4)(x+2)} + \dfrac{x^2-4x}{(x-4)(x+2)}$

$= \dfrac{5x+10+x^2-4x}{(x-4)(x+2)}$

$= \dfrac{x^2+x+10}{(x-4)(x+2)}$

b. Since S is the sum of f and g, the restrictions for the domain of S will consist of all the restrictions for the domains of both f and g. Since 4 is restricted from the domain of f and since -2 is restricted from the domain of g, both 4 and -2 are restricted from the domain of S. That is, the domain of $S(x)$ is

$$\{x \mid x \neq -2, x \neq 4\}.$$

50. a. $2x^2 + x - 15 = (2x-5)(x+3)$

$4x^2 - 8x - 5 = (2x-5)(2x+1)$

$\text{LCD} = (2x-5)(2x+1)(x+3)$

$D(x) = f(x) - g(x)$

$= \dfrac{x+3}{2x^2+x-15} - \dfrac{x-7}{4x^2-8x-5}$

$= \dfrac{x+3}{(2x-5)(x+3)} \cdot \dfrac{2x+1}{2x+1}$

$\qquad - \dfrac{x-7}{(2x-5)(2x+1)} \cdot \dfrac{x+3}{x+3}$

$= \dfrac{2x^2+7x+3}{(2x-5)(2x+1)(x+3)}$

$\qquad - \dfrac{x^2-4x-21}{(2x-5)(2x+1)(x+3)}$

$= \dfrac{2x^2+7x+3-(x^2-4x-21)}{(2x-5)(2x+1)(x+3)}$

$= \dfrac{2x^2+7x+3-x^2+4x+21}{(2x-5)(2x+1)(x+3)}$

$= \dfrac{x^2+11x+24}{(2x-5)(2x+1)(x+3)}$

$= \dfrac{(x+3)(x+8)}{(2x-5)(2x+1)(x+3)}$

$= \dfrac{x+8}{(2x-5)(2x+1)}$

b. Since D is the difference of f and g, the restrictions for the domain of D will consist of all the restrictions for the domains of both f and g. Since -3 and $\dfrac{5}{2}$ are restricted from the domain of f and since $-\dfrac{1}{2}$ and $\dfrac{5}{2}$ are restricted from the domain of g, then -3, $-\dfrac{1}{2}$, and $\dfrac{5}{2}$ are restricted from the domain of D. That is, the domain of $D(x)$ is

$$\left\{ x \,\middle|\, x \neq -3, x \neq -\frac{1}{2}, x \neq \frac{5}{2} \right\}.$$

51. Write the numerator of the complex rational expression as a single rational expression. Note that the LCD of the rational expressions in the numerator is x.

$$x - \frac{1}{x} = \frac{x}{1} \cdot \frac{x}{x} - \frac{1}{x} = \frac{x^2}{x} - \frac{1}{x} = \frac{x^2-1}{x}$$

Write the denominator of the complex rational expression as a single rational expression. Note that the LCD of the rational expressions in the denominator is x.

$$1 - \frac{1}{x} = \frac{1}{1} \cdot \frac{x}{x} - \frac{1}{x} = \frac{x}{x} - \frac{1}{x} = \frac{x-1}{x}$$

Rewrite the complex rational expression using the single rational expressions and simplify.

$$\frac{x - \dfrac{1}{x}}{1 - \dfrac{1}{x}} = \frac{\dfrac{x^2-1}{x}}{\dfrac{x-1}{x}} = \frac{x^2-1}{x} \cdot \frac{x}{x-1}$$

$$= \frac{(x-1)(x+1)}{x} \cdot \frac{x}{x-1}$$

$$= \frac{(x-1)(x+1)}{x} \cdot \frac{x}{x-1}$$

$$= x+1$$

52. Write the numerator of the complex rational expression as a single rational expression. Note that the LCD of the rational expressions in the numerator is xy.

$$\frac{1}{x} - \frac{1}{y} = \frac{1}{x} \cdot \frac{y}{y} - \frac{1}{y} \cdot \frac{x}{x} = \frac{y}{xy} - \frac{x}{xy} = \frac{y-x}{xy}$$

Write the denominator of the complex rational expression as a single rational expression. Note that the LCD of the rational expressions in the denominator is $x^2 y^2$.

$$\frac{1}{x^2} - \frac{1}{y^2} = \frac{1}{x^2} \cdot \frac{y^2}{y^2} - \frac{1}{y^2} \cdot \frac{x^2}{x^2}$$

$$= \frac{y^2}{x^2 y^2} - \frac{x^2}{x^2 y^2} = \frac{y^2-x^2}{x^2 y^2}$$

Rewrite the complex rational expression using the single rational expressions and simplify.

$$\frac{\dfrac{1}{x}-\dfrac{1}{y}}{\dfrac{1}{x^2}-\dfrac{1}{y^2}}=\frac{\dfrac{y-x}{xy}}{\dfrac{y^2-x^2}{x^2y^2}}=\frac{y-x}{xy}\cdot\frac{x^2y^2}{y^2-x^2}$$

$$=\frac{y-x}{xy}\cdot\frac{x^2y^2}{(y-x)(y+x)}$$

$$=\frac{\cancel{y-x}}{\cancel{xy}}\cdot\frac{\cancel{x}\cdot x\cdot\cancel{y}\cdot y}{(\cancel{y-x})(y+x)}$$

$$=\frac{xy}{y+x}\quad\text{or}\quad\frac{xy}{x+y}$$

53. Write the numerator of the complex rational expression as a single rational expression. Note that the LCD of the rational expressions in the numerator is $b(a+b)$.

$$\frac{a}{b}-\frac{a-b}{a+b}=\frac{a}{b}\cdot\frac{a+b}{a+b}-\frac{a-b}{a+b}\cdot\frac{b}{b}$$

$$=\frac{a^2+ab}{b(a+b)}-\frac{ab-b^2}{b(a+b)}$$

$$=\frac{a^2+ab-\left(ab-b^2\right)}{b(a+b)}$$

$$=\frac{a^2+ab-ab+b^2}{b(a+b)}$$

$$=\frac{a^2+b^2}{b(a+b)}$$

Write the denominator of the complex rational expression as a single rational expression. Note that the LCD of the rational expressions in the denominator is $b(a-b)$.

$$\frac{a}{b}+\frac{a+b}{a-b}=\frac{a}{b}\cdot\frac{a-b}{a-b}+\frac{a+b}{a-b}\cdot\frac{b}{b}$$

$$=\frac{a^2-ab}{b(a-b)}+\frac{ab+b^2}{b(a-b)}$$

$$=\frac{a^2+b^2}{b(a-b)}$$

Rewrite the complex rational expression using the single rational expressions and simplify.

$$\frac{\dfrac{a}{b}-\dfrac{a-b}{a+b}}{\dfrac{a}{b}+\dfrac{a+b}{a-b}}=\frac{\dfrac{a^2+b^2}{b(a+b)}}{\dfrac{a^2+b^2}{b(a-b)}}=\frac{a^2+b^2}{b(a+b)}\cdot\frac{b(a-b)}{a^2+b^2}$$

$$=\frac{\cancel{a^2+b^2}}{\cancel{b}(a+b)}\cdot\frac{\cancel{b}(a-b)}{\cancel{a^2+b^2}}$$

$$=\frac{a-b}{a+b}$$

54. Write the numerator of the complex rational expression as a single rational expression. Note that the LCD of the rational expressions in the numerator is $a+2$.

$$\frac{2}{a+2}-1=\frac{2}{a+2}-\frac{1}{1}\cdot\frac{a+2}{a+2}=\frac{2}{a+2}-\frac{a+2}{a+2}$$

$$=\frac{2-(a+2)}{a+2}=\frac{2-a-2}{a+2}=\frac{-a}{a+2}$$

Write the denominator of the complex rational expression as a single rational expression. Note that the LCD of the rational expressions in the denominator is $a+2$.

$$\frac{1}{a+2}+1=\frac{1}{a+2}+\frac{1}{1}\cdot\frac{a+2}{a+2}$$

$$=\frac{1}{a+2}+\frac{a+2}{a+2}=\frac{1+a+2}{a+2}=\frac{a+3}{a+2}$$

Rewrite the complex rational expression using the single rational expressions and simplify.

$$\frac{\dfrac{2}{a+2}-1}{\dfrac{1}{a+2}+1}=\frac{\dfrac{-a}{a+2}}{\dfrac{a+3}{a+2}}=\frac{-a}{a+2}\cdot\frac{a+2}{a+3}$$

$$=\frac{-a}{\cancel{a+2}}\cdot\frac{\cancel{a+2}}{a+3}=\frac{-a}{a+3}$$

55. The LCD of all the denominators in the complex rational expression is t^2.

$$\frac{\dfrac{3}{t}+\dfrac{4}{t^2}}{5+\dfrac{1}{t^2}}=\frac{\dfrac{3}{t}+\dfrac{4}{t^2}}{5+\dfrac{1}{t^2}}\cdot\frac{t^2}{t^2}=\frac{\dfrac{3}{t}\cdot t^2+\dfrac{4}{t^2}\cdot t^2}{5\cdot t^2+\dfrac{1}{t^2}\cdot t^2}$$

$$=\frac{\dfrac{3}{\cancel{t}}\cdot t^{\cancel{2}}+\dfrac{4}{\cancel{t^2}}\cdot\cancel{t^2}}{5\cdot t^2+\dfrac{1}{\cancel{t^2}}\cdot\cancel{t^2}}=\frac{3t+4}{5t^2+1}$$

56. The LCD of all the denominators in the complex rational expression is ab.

$$\dfrac{\dfrac{1}{a}-\dfrac{1}{b}}{\dfrac{b}{a}-\dfrac{a}{b}}=\dfrac{\dfrac{1}{a}-\dfrac{1}{b}}{\dfrac{b}{a}-\dfrac{a}{b}}\cdot\dfrac{ab}{ab}=\dfrac{\dfrac{1}{a}\cdot ab-\dfrac{1}{b}\cdot ab}{\dfrac{b}{a}\cdot ab-\dfrac{a}{b}\cdot ab}$$

$$=\dfrac{\dfrac{1}{\cancel{a}}\cdot \cancel{a}b-\dfrac{1}{\cancel{b}}\cdot a\cancel{b}}{\dfrac{b}{\cancel{a}}\cdot \cancel{a}b-\dfrac{a}{\cancel{b}}\cdot a\cancel{b}}=\dfrac{b-a}{b^2-a^2}$$

$$=\dfrac{b-a}{(b-a)(b+a)}=\dfrac{\cancel{b-a}}{\cancel{(b-a)}(b+a)}$$

$$=\dfrac{1}{b+a}\quad\text{or}\quad\dfrac{1}{a+b}$$

57. The LCD of all the denominators in the complex rational expression is $z(z-1)(z+1)$.

$$\dfrac{\dfrac{1}{z-1}-\dfrac{1}{z}}{\dfrac{1}{z}-\dfrac{1}{z+1}}=\dfrac{\dfrac{1}{z-1}-\dfrac{1}{z}}{\dfrac{1}{z}-\dfrac{1}{z+1}}\cdot\dfrac{z(z-1)(z+1)}{z(z-1)(z+1)}$$

$$=\dfrac{\dfrac{1}{z-1}\cdot z(z-1)(z+1)-\dfrac{1}{z}\cdot z(z-1)(z+1)}{\dfrac{1}{z}\cdot z(z-1)(z+1)-\dfrac{1}{z+1}\cdot z(z-1)(z+1)}$$

$$=\dfrac{\dfrac{1}{\cancel{z-1}}\cdot z\cancel{(z-1)}(z+1)-\dfrac{1}{\cancel{z}}\cdot \cancel{z}(z-1)(z+1)}{\dfrac{1}{\cancel{z}}\cdot \cancel{z}(z-1)(z+1)-\dfrac{1}{\cancel{z+1}}\cdot z(z-1)\cancel{(z+1)}}$$

$$=\dfrac{z(z+1)-(z-1)(z+1)}{(z-1)(z+1)-z(z-1)}$$

$$=\dfrac{z^2+z-\left(z^2-1\right)}{z^2-1-z^2+z}$$

$$=\dfrac{z^2+z-z^2+1}{z-1}=\dfrac{z+1}{z-1}$$

58. The LCD of all the denominators in the complex rational expression is $(x-1)(x+1)$.

$$\dfrac{1+\dfrac{x}{x+1}}{\dfrac{2x+1}{x-1}}=\dfrac{1+\dfrac{x}{x+1}}{\dfrac{2x+1}{x-1}}\cdot\dfrac{(x-1)(x+1)}{(x-1)(x+1)}$$

$$=\dfrac{1\cdot(x-1)(x+1)+\dfrac{x}{x+1}\cdot(x-1)(x+1)}{\dfrac{2x+1}{x-1}\cdot(x-1)(x+1)}$$

$$=\dfrac{1\cdot(x-1)(x+1)+\dfrac{x}{\cancel{x+1}}\cdot(x-1)\cancel{(x+1)}}{\dfrac{2x+1}{\cancel{x-1}}\cdot\cancel{(x-1)}(x+1)}$$

$$=\dfrac{(x-1)(x+1)+x(x-1)}{(2x+1)(x+1)}$$

$$=\dfrac{x^2-1+x^2-x}{(2x+1)(x+1)}$$

$$=\dfrac{2x^2-x-1}{(2x+1)(x+1)}$$

$$=\dfrac{(2x+1)(x-1)}{(2x+1)(x+1)}$$

$$=\dfrac{\cancel{(2x+1)}(x-1)}{\cancel{(2x+1)}(x+1)}=\dfrac{x-1}{x+1}$$

59. Using Method 2, the LCD of all the denominators in the complex rational expression is y.

$$\dfrac{\dfrac{x}{y}+1}{\dfrac{x}{y}-1}=\dfrac{\dfrac{x}{y}+1}{\dfrac{x}{y}-1}\cdot\dfrac{y}{y}=\dfrac{\dfrac{x}{y}\cdot y+1\cdot y}{\dfrac{x}{y}\cdot y-1\cdot y}$$

$$=\dfrac{\dfrac{x}{\cancel{y}}\cdot\cancel{y}+1\cdot y}{\dfrac{x}{\cancel{y}}\cdot\cancel{y}-1\cdot y}=\dfrac{x+y}{x-y}$$

60. Using Method 2, the LCD of all the denominators in the complex rational expression is $(a-b)(a+b)$.

$$\dfrac{\dfrac{a}{a-b}-\dfrac{b}{a+b}}{\dfrac{b}{a-b}+\dfrac{a}{a+b}} = \dfrac{\dfrac{a}{a-b}-\dfrac{b}{a+b}}{\dfrac{b}{a-b}+\dfrac{a}{a+b}}\cdot\dfrac{(a-b)(a+b)}{(a-b)(a+b)}$$

$$= \dfrac{\dfrac{a}{a-b}\cdot(a-b)(a+b)-\dfrac{b}{a+b}\cdot(a-b)(a+b)}{\dfrac{b}{a-b}\cdot(a-b)(a+b)+\dfrac{a}{a+b}\cdot(a-b)(a+b)}$$

$$= \dfrac{\dfrac{a}{\cancel{a-b}}\cdot\cancel{(a-b)}(a+b)-\dfrac{b}{\cancel{a+b}}\cdot(a-b)\cancel{(a+b)}}{\dfrac{b}{\cancel{a-b}}\cdot\cancel{(a-b)}(a+b)+\dfrac{a}{\cancel{a+b}}\cdot(a-b)\cancel{(a+b)}}$$

$$= \dfrac{a(a+b)-b(a-b)}{b(a+b)+a(a-b)}$$

$$= \dfrac{a^2+ab-ab+b^2}{ab+b^2+a^2-ab}=\dfrac{a^2+b^2}{a^2+b^2}=1$$

61. Using Method 1, write the numerator of the complex rational expression as a single rational expression. Note that the LCD of the rational expressions in the numerator is $(x-2)(x+2)$.

$$\dfrac{1}{x-2}-\dfrac{x}{x^2-4} = \dfrac{1}{x-2}-\dfrac{x}{(x-2)(x+2)}$$

$$= \dfrac{1}{x-2}\cdot\dfrac{x+2}{x+2}-\dfrac{x}{(x-2)(x+2)}$$

$$= \dfrac{x+2}{(x-2)(x+2)}-\dfrac{x}{(x-2)(x+2)}$$

$$= \dfrac{x+2-x}{(x-2)(x+2)}=\dfrac{2}{(x-2)(x+2)}$$

Write the denominator of the complex rational expression as a single rational expression. Note that the LCD of the rational expressions in the denominator is $x+2$.

$$1-\dfrac{2}{x+2} = \dfrac{1}{1}\cdot\dfrac{x+2}{x+2}-\dfrac{2}{x+2}$$

$$= \dfrac{x+2}{x+2}-\dfrac{2}{x+2}=\dfrac{x+2-2}{x+2}=\dfrac{x}{x+2}$$

Rewrite the complex rational expression using the single rational expressions and simplify.

$$\dfrac{\dfrac{1}{x-2}-\dfrac{x}{x^2-4}}{1-\dfrac{2}{x+2}} = \dfrac{\dfrac{2}{(x-2)(x+2)}}{\dfrac{x}{x+2}}$$

$$= \dfrac{2}{(x-2)(x+2)}\cdot\dfrac{x+2}{x}$$

$$= \dfrac{2}{(x-2)\cancel{(x+2)}}\cdot\dfrac{\cancel{x+2}}{x}$$

$$= \dfrac{2}{x(x-2)}$$

62. Using Method 1, write the numerator of the complex rational expression as a single rational expression. Note that the LCD of the rational expressions in the numerator is $z+5$.

$$z-\dfrac{5z}{z+5} = \dfrac{z}{1}\cdot\dfrac{z+5}{z+5}-\dfrac{5z}{z+5}$$

$$= \dfrac{z^2+5z}{z+5}-\dfrac{5z}{z+5}$$

$$= \dfrac{z^2+5z-5z}{z+5}=\dfrac{z^2}{z+5}$$

Write the denominator of the complex rational expression as a single rational expression. Note that the LCD of the rational expressions in the denominator is $z-5$.

$$z+\dfrac{5z}{z-5} = \dfrac{z}{1}\cdot\dfrac{z-5}{z-5}+\dfrac{5z}{z-5}$$

$$= \dfrac{z^2-5z}{z-5}+\dfrac{5z}{z-5}$$

$$= \dfrac{z^2-5z+5z}{z-5}=\dfrac{z^2}{z-5}$$

Rewrite the complex rational expression using the single rational expressions and simplify.

$$\dfrac{z-\dfrac{5z}{z+5}}{z+\dfrac{5z}{z-5}} = \dfrac{\dfrac{z^2}{z+5}}{\dfrac{z^2}{z-5}}$$

$$= \dfrac{z^2}{z+5}\cdot\dfrac{z-5}{z^2}$$

$$= \dfrac{\cancel{z^2}}{z+5}\cdot\dfrac{z-5}{\cancel{z^2}}$$

$$= \dfrac{z-5}{z+5}$$

63. Using Method 2, the LCD of all the denominators in the complex rational expression is $mn(m+n)$.

$$\dfrac{\dfrac{m-n}{m+n}+\dfrac{n}{m}}{\dfrac{m}{n}-\dfrac{m-n}{m+n}}$$

$$=\dfrac{\dfrac{m-n}{m+n}+\dfrac{n}{m}}{\dfrac{m}{n}-\dfrac{m-n}{m+n}}\cdot\dfrac{mn(m+n)}{mn(m+n)}$$

$$=\dfrac{\dfrac{m-n}{m+n}\cdot mn(m+n)+\dfrac{n}{m}\cdot mn(m+n)}{\dfrac{m}{n}\cdot mn(m+n)-\dfrac{m-n}{m+n}\cdot mn(m+n)}$$

$$=\dfrac{\dfrac{m-n}{\cancel{m+n}}\cdot mn\,(\cancel{m+n})+\dfrac{n}{\cancel{m}}\cdot \cancel{m}n(m+n)}{\dfrac{m}{\cancel{n}}\cdot m\cancel{n}(m+n)-\dfrac{m-n}{\cancel{m+n}}\cdot mn\,(\cancel{m+n})}$$

$$=\dfrac{mn(m-n)+n^2(m+n)}{m^2(m+n)-mn(m-n)}$$

$$=\dfrac{m^2n-mn^2+mn^2+n^3}{m^3+m^2n-m^2n+mn^2}$$

$$=\dfrac{m^2n+n^3}{m^3+mn^2}=\dfrac{n\left(m^2+n^2\right)}{m\left(m^2+n^2\right)}$$

$$=\dfrac{n\left(\cancel{m^2+n^2}\right)}{m\left(\cancel{m^2+n^2}\right)}=\dfrac{n}{m}$$

64. Using Method 1, write the numerator of the complex rational expression as a single rational expression. Note that the LCD of the rational expressions in the numerator is $(x+1)(x-2)$.

$$\dfrac{x+4}{x-2}-\dfrac{x-3}{x+1}=\dfrac{x+4}{x-2}\cdot\dfrac{x+1}{x+1}-\dfrac{x-3}{x+1}\cdot\dfrac{x-2}{x-2}$$

$$=\dfrac{x^2+5x+4}{(x+1)(x-2)}-\dfrac{x^2-5x+6}{(x+1)(x-2)}$$

$$=\dfrac{x^2+5x+4-\left(x^2-5x+6\right)}{(x+1)(x-2)}$$

$$=\dfrac{x^2+5x+4-x^2+5x-6}{(x+1)(x-2)}$$

$$=\dfrac{10x-2}{(x+1)(x-2)}$$

$$=\dfrac{2(5x-1)}{(x+1)(x-2)}$$

The denominator is already a single rational

expression: $5x^2+4x-1=(5x-1)(x+1)$.

Rewrite the complex rational expression using the single rational expressions and simplify.

$$\dfrac{\dfrac{x+4}{x-2}-\dfrac{x-3}{x+1}}{5x^2+4x-1}=\dfrac{\dfrac{2(5x-1)}{(x+1)(x-2)}}{(5x-1)(x+1)}$$

$$=\dfrac{2(5x-1)}{(x+1)(x-2)}\cdot\dfrac{1}{(5x-1)(x+1)}$$

$$=\dfrac{2(5\cancel{x-1})}{(x+1)(x-2)}\cdot\dfrac{1}{(5\cancel{x-1})(x+1)}$$

$$=\dfrac{2}{(x+1)^2(x-2)}$$

65. Rewrite the expression so that it does not contain any negative exponents.

$$\dfrac{3x^{-1}-3y^{-1}}{(x+y)^{-1}}=\dfrac{\dfrac{3}{x}-\dfrac{3}{y}}{\dfrac{1}{x+y}}$$

Using Method 2, the LCD of all the denominators in the complex rational expression is $xy(x+y)$.

$$\dfrac{3x^{-1}-3y^{-1}}{(x+y)^{-1}}=\dfrac{\dfrac{3}{x}-\dfrac{3}{y}}{\dfrac{1}{x+y}}=\dfrac{\dfrac{3}{x}-\dfrac{3}{y}}{\dfrac{1}{x+y}}\cdot\dfrac{xy(x+y)}{xy(x+y)}$$

$$=\dfrac{\dfrac{3}{x}\cdot xy(x+y)-\dfrac{3}{y}\cdot xy(x+y)}{\dfrac{1}{x+y}\cdot xy(x+y)}$$

$$=\dfrac{\dfrac{3}{\cancel{x}}\cdot\cancel{x}y(x+y)-\dfrac{3}{\cancel{y}}\cdot x\cancel{y}(x+y)}{\dfrac{1}{\cancel{x+y}}\cdot xy\,(\cancel{x+y})}$$

$$=\dfrac{3y(x+y)-3x(x+y)}{xy}$$

$$=\dfrac{3xy+3y^2-3x^2-3xy}{xy}$$

$$=\dfrac{3y^2-3x^2}{xy}\quad\text{or}\quad\dfrac{3(y-x)(y+x)}{xy}$$

66. Rewrite the expression so that it does not contain any negative exponents.

$$\frac{2c^{-1}-(3d)^{-1}}{(6d)^{-1}}=\frac{\dfrac{2}{c}-\dfrac{1}{3d}}{\dfrac{1}{6d}}$$

Using Method 2, the LCD of all the denominators in the complex rational expression is $6cd$.

$$\frac{2c^{-1}-(3d)^{-1}}{(6d)^{-1}}=\frac{\dfrac{2}{c}-\dfrac{1}{3d}}{\dfrac{1}{6d}}=\frac{\dfrac{2}{c}-\dfrac{1}{3d}}{\dfrac{1}{6d}}\cdot\frac{6cd}{6cd}$$

$$=\frac{\dfrac{2}{c}\cdot 6cd-\dfrac{1}{3d}\cdot 6cd}{\dfrac{1}{6d}\cdot 6cd}=\frac{\dfrac{2}{\cancel{c}}\cdot 6\cancel{c}d-\dfrac{1}{\cancel{3}\cancel{d}}\cdot\overset{2}{\cancel{6}}c\cancel{d}}{\dfrac{1}{\cancel{6}\cancel{d}}\cdot\cancel{6}c\cancel{d}}$$

$$=\frac{12d-2c}{c}\quad\text{or}\quad\frac{2(6d-c)}{c}$$

67.
$$\frac{2}{z}-\frac{1}{3z}=\frac{1}{6}$$
$$6z\left(\frac{2}{z}-\frac{1}{3z}\right)=6z\left(\frac{1}{6}\right)$$
$$12-2=z$$
$$10=z$$

Check:
$$\frac{2}{10}-\frac{1}{3\cdot10}\overset{?}{=}\frac{1}{6}$$
$$\frac{6}{30}-\frac{1}{30}\overset{?}{=}\frac{1}{6}$$
$$\frac{5}{30}\overset{?}{=}\frac{1}{6}$$
$$\frac{1}{6}=\frac{1}{6}\ \leftarrow\text{True}$$

The solution checks, so the solution set is $\{10\}$.

68.
$$\frac{4}{m-4}=\frac{-5}{m+2}$$
$$(m-4)(m+2)\left(\frac{4}{m-4}\right)=(m-4)(m+2)\left(\frac{-5}{m+2}\right)$$
$$4(m+2)=-5(m-4)$$
$$4m+8=-5m+20$$
$$9m=12$$
$$m=\frac{12}{9}=\frac{4}{3}$$

Check:

$$\frac{4}{\frac{4}{3}-4} \overset{?}{=} \frac{-5}{\frac{4}{3}+2}$$

$$\frac{4}{\frac{4}{3}-\frac{12}{3}} \overset{?}{=} \frac{-5}{\frac{4}{3}+\frac{6}{3}}$$

$$\frac{4}{-\frac{8}{3}} \overset{?}{=} \frac{-5}{\frac{10}{3}}$$

$$4\cdot\left(-\frac{3}{8}\right) \overset{?}{=} -5\cdot\frac{3}{10}$$

$$-\frac{3}{2}=-\frac{3}{2} \quad \leftarrow \text{True}$$

The solution checks, so the solution set is $\left\{\dfrac{4}{3}\right\}$.

69.
$$m-\frac{14}{m}=5$$

$$m\left(m-\frac{14}{m}\right)=m(5)$$

$$m^2-14=5m$$

$$m^2-5m-14=0$$

$$(m+2)(m-7)=0$$

$$m+2=0 \quad \text{or} \quad m-7=0$$
$$m=-2 \quad \text{or} \qquad m=7$$

Check:

$$-2-\frac{14}{-2} \overset{?}{=} 5$$

$$-2+7 \overset{?}{=} 5$$

$$-5=5 \quad \leftarrow \text{True}$$

Check:

$$7-\frac{14}{7} \overset{?}{=} 5$$

$$7-2 \overset{?}{=} 5$$

$$5=5 \quad \leftarrow \text{True}$$

Both solutions check, so the solution set is $\{-2,\ 7\}$.

70.
$$\frac{2}{n+3}=\frac{1}{n-3}$$

$$(n-3)(n+3)\left(\frac{2}{n+3}\right)=(n-3)(n+3)\left(\frac{1}{n-3}\right)$$

$$2(n-3)=n+3$$

$$2n-6=n+3$$

$$n=9$$

Check: $\dfrac{2}{9+3} \overset{?}{=} \dfrac{1}{9-3}$

$$\frac{2}{12} \overset{?}{=} \frac{1}{6}$$

$$\frac{1}{6}=\frac{1}{6} \quad \leftarrow \text{True}$$

The solution checks, so the solution set is $\{9\}$.

71.
$$\frac{s}{s-1}=1+\frac{2}{s}$$

$$s(s-1)\left(\frac{s}{s-1}\right)=s(s-1)\left(1+\frac{2}{s}\right)$$

$$s^2=s(s-1)+2(s-1)$$

$$s^2=s^2-s+2s-2$$

$$s^2=s^2+s-2$$

$$0=s-2$$

$$2=s$$

Check: $\dfrac{2}{2-1} \overset{?}{=} 1+\dfrac{2}{2}$

$$\frac{2}{1} \overset{?}{=} 1+1$$

$$2=2 \quad \leftarrow \text{True}$$

The solution checks, so the solution set is $\{2\}$.

72.
$$\frac{3}{x^2-7x+10}+2=\frac{x-4}{x-5}$$

$$\frac{3}{(x-2)(x-5)}+2=\frac{x-4}{x-5}$$

$$(x-2)(x-5)\left(\frac{3}{(x-2)(x-5)}+2\right)=$$

$$(x-2)(x-5)\left(\frac{x-4}{x-5}\right)$$

$$3+2(x-2)(x-5)=(x-4)(x-2)$$

$$3+2x^2-14x+20=x^2-6x+8$$

$$2x^2-14x+23=x^2-6x+8$$

$$x^2-8x+15=0$$

$$(x-3)(x-5)=0$$

$$x-3=0 \quad \text{or} \quad x-5=0$$
$$x=3 \quad \text{or} \qquad x=5$$

Since $x=5$ is not in the domain of the variable, it is an extraneous solution.

Check $x = 3$: $\dfrac{3}{3^2 - 7 \cdot 3 + 10} + 2 \overset{?}{=} \dfrac{3-4}{3-5}$

$$\dfrac{3}{9 - 21 + 10} + 2 \overset{?}{=} \dfrac{-1}{-2}$$

$$\dfrac{3}{-2} + 2 \overset{?}{=} \dfrac{1}{2}$$

$$\dfrac{1}{2} = \dfrac{1}{2} \leftarrow \text{True}$$

The solution set is $\{3\}$.

73.
$$\dfrac{1}{k-1} + \dfrac{1}{k+2} = \dfrac{3}{k^2 + k - 2}$$

$$\dfrac{1}{k-1} + \dfrac{1}{k+2} = \dfrac{3}{(k-1)(k+2)}$$

$$(k-1)(k+2)\left(\dfrac{1}{k-1} + \dfrac{1}{k+2}\right) =$$

$$(k-1)(k+2)\left(\dfrac{3}{(k-1)(k+2)}\right)$$

$$(k+2) + (k-1) = 3$$

$$2k + 1 = 3$$

$$2k = 2$$

$$k = 1$$

Since $k = 1$ is not in the domain of the variable, it is an extraneous solution. The solution set is $\varnothing$ or { }.

74.
$$x + \dfrac{3x}{x-3} = \dfrac{9}{x-3}$$

$$(x-3)\left(x + \dfrac{3x}{x-3}\right) = (x-3)\left(\dfrac{9}{x-3}\right)$$

$$x(x-3) + 3x = 9$$

$$x^2 - 3x + 3x = 9$$

$$x^2 = 9$$

$$x^2 - 9 = 0$$

$$(x-3)(x+3) = 0$$

$$x - 3 = 0 \quad \text{or} \quad x + 3 = 0$$

$$x = 3 \quad \text{or} \quad x = -3$$

Since $x = 3$ is not in the domain of the variable, it is an extraneous solution.

Check $x = -3$: $-3 + \dfrac{3(-3)}{-3-3} \overset{?}{=} \dfrac{9}{-3-3}$

$$-3 + \dfrac{-9}{-6} \overset{?}{=} \dfrac{9}{-6}$$

$$-3 + \dfrac{3}{2} \overset{?}{=} -\dfrac{3}{2}$$

$$-\dfrac{3}{2} = -\dfrac{3}{2} \leftarrow \text{True}$$

The solution set is $\{-3\}$.

75.
$$\dfrac{2}{a+3} - \dfrac{4}{a^2 - 4} = \dfrac{a+1}{a^2 + 5a + 6}$$

$$\dfrac{2}{a+3} - \dfrac{4}{(a-2)(a+2)} = \dfrac{a+1}{(a+2)(a+3)}$$

$$(a-2)(a+2)(a+3)\left(\dfrac{2}{a+3} - \dfrac{4}{(a-2)(a+2)}\right) =$$

$$(a-2)(a+2)(a+3)\left(\dfrac{a+1}{(a+2)(a+3)}\right)$$

$$2(a-2)(a+2) - 4(a+3) = (a+1)(a-2)$$

$$2a^2 - 8 - 4a - 12 = a^2 - a - 2$$

$$2a^2 - 4a - 20 = a^2 - a - 2$$

$$a^2 - 3a - 18 = 0$$

$$(a+3)(a-6) = 0$$

$$a + 3 = 0 \quad \text{or} \quad a - 6 = 0$$

$$a = -3 \quad \text{or} \quad a = 6$$

Since $a = -3$ is not in the domain of the variable, it is an extraneous solution.

Check $a = 6$:

$$\dfrac{2}{6+3} - \dfrac{4}{6^2 - 4} \overset{?}{=} \dfrac{6+1}{6^2 + 5 \cdot 6 + 6}$$

$$\dfrac{2}{9} - \dfrac{4}{36 - 4} \overset{?}{=} \dfrac{7}{36 + 30 + 6}$$

$$\dfrac{2}{9} - \dfrac{4}{32} \overset{?}{=} \dfrac{7}{72}$$

$$\dfrac{64}{288} - \dfrac{36}{288} \overset{?}{=} \dfrac{7}{72}$$

$$\dfrac{28}{288} \overset{?}{=} \dfrac{7}{72}$$

$$\dfrac{7}{72} = \dfrac{7}{72} \leftarrow \text{True}$$

The solution set is $\{6\}$.

76. $\dfrac{2}{z^2 + 2z - 8} = \dfrac{1}{z^2 + 9z + 20} + \dfrac{4}{z^2 + 3z - 10}$

$\dfrac{2}{(z-2)(z+4)} = \dfrac{1}{(z+4)(z+5)} + \dfrac{4}{(z-2)(z+5)}$

$(z-2)(z+4)(z+5)\left(\dfrac{2}{(z-2)(z+4)}\right) =$

$(z-2)(z+4)(z+5)\left(\dfrac{1}{(z+4)(z+5)} + \dfrac{4}{(z-2)(z+5)}\right)$

$2(z+5) = (z-2) + 4(z+4)$

$2z + 10 = z - 2 + 4z + 16$

$2z + 10 = 5z + 14$

$-3z = 4$

$z = -\dfrac{4}{3}$

Check:

$\dfrac{2}{\left(-\dfrac{4}{3}\right)^2 + 2\left(-\dfrac{4}{3}\right) - 8} \overset{?}{=}$

$\dfrac{1}{\left(-\dfrac{4}{3}\right)^2 + 9\left(-\dfrac{4}{3}\right) + 20} + \dfrac{4}{\left(-\dfrac{4}{3}\right)^2 + 3\left(-\dfrac{4}{3}\right) - 10}$

$\dfrac{2}{\dfrac{16}{9} - \dfrac{8}{3} - 8} \overset{?}{=} \dfrac{1}{\dfrac{16}{9} - 12 + 20} + \dfrac{4}{\dfrac{16}{9} - 4 - 10}$

$\dfrac{2}{-\dfrac{80}{9}} \overset{?}{=} \dfrac{1}{\dfrac{88}{9}} + \dfrac{4}{-\dfrac{110}{9}}$

$-\dfrac{9}{40} \overset{?}{=} \dfrac{9}{88} - \dfrac{18}{55}$

$-\dfrac{9}{40} \overset{?}{=} \dfrac{45}{440} - \dfrac{144}{440}$

$-\dfrac{9}{40} \overset{?}{=} -\dfrac{99}{440}$

$-\dfrac{9}{40} = -\dfrac{9}{40} \leftarrow$ True

The solution set is $\left\{-\dfrac{4}{3}\right\}$.

77. $\dfrac{x-3}{x+4} = \dfrac{14}{x^2 + 6x + 8}$

$\dfrac{x-3}{x+4} = \dfrac{14}{(x+2)(x+4)}$

$(x+2)(x+4)\left(\dfrac{x-3}{x+4}\right) = (x+2)(x+4)\left(\dfrac{14}{(x+2)(x+4)}\right)$

$(x+2)(x-3) = 14$

$x^2 - x - 6 = 14$

$x^2 - x - 20 = 0$

$(x+4)(x-5) = 0$

$x + 4 = 0 \quad$ or $\quad x - 5 = 0$

$x = -4 \quad$ or $\qquad x = 5$

Since $x = -4$ is not in the domain of the variable, it is an extraneous solution.

Check $x = 5$:

$\dfrac{5-3}{5+4} \overset{?}{=} \dfrac{14}{5^2 + 6 \cdot 5 + 8}$

$\dfrac{2}{9} \overset{?}{=} \dfrac{14}{25 + 30 + 8}$

$\dfrac{2}{9} \overset{?}{=} \dfrac{14}{63}$

$\dfrac{2}{9} = \dfrac{2}{9} \leftarrow$ True

The solution set is $\{5\}$.

78. $\dfrac{5}{y-5} + 4 = \dfrac{3y - 10}{y - 5}$

$(y-5)\left(\dfrac{5}{y-5} + 4\right) = (y-5)\left(\dfrac{3y-10}{y-5}\right)$

$5 + 4(y-5) = 3y - 10$

$5 + 4y - 20 = 3y - 10$

$4y - 15 = 3y - 10$

$y = 5$

Since $y = 5$ is not in the domain of the variable, it is an extraneous solution. The solution set is $\varnothing$ or { }.

79.
$$f(x) = 2$$
$$\frac{6}{x-2} = 2$$
$$(x-2)\left(\frac{6}{x-2}\right) = (x-2) \cdot 2$$
$$6 = 2x - 4$$
$$10 = 2x$$
$$5 = x$$

Thus, $f(5) = 2$, so the point (5, 2) is on the graph of *f*.

80.
$$g(x) = 4$$
$$x - \frac{21}{x} = 4$$
$$x\left(x - \frac{21}{x}\right) = x \cdot 4$$
$$x^2 - 21 = 4x$$
$$x^2 - 4x - 21 = 0$$
$$(x+3)(x-7) = 0$$
$$x + 3 = 0 \quad \text{or} \quad x - 7 = 0$$
$$x = -3 \quad \text{or} \quad x = 7$$

Thus, $g(-3) = 4$ and $g(7) = 4$, so the points
(−3, 4) and (7, 4) are on the graph of *f*.

81. $\dfrac{x-4}{x+2} \geq 0$

The rational expression will equal 0 when $x = 4$.
It is undefined when $x = -2$.
Determine where the numerator and
denominator are positive and negative and where
the quotient is positive and negative.

Interval	$(-\infty, -2)$	−2	$(-2, 4)$	4	$(4, \infty)$
$x - 4$	− − − −	−	− − − −	0	+ + + +
$x + 2$	− − − −	0	+ + + +	+	+ + + +
$\frac{x-4}{x+2}$	+ + + +	$\varnothing$	− − − −	0	+ + + +

The rational function is undefined at $x = -2$, so
−2 is not part of the solution. The inequality is
non-strict, so 4 is part of the solution. Now,
$\dfrac{x-4}{x+2}$ is greater than zero where the quotient is

positive. The solution is $\{x \,|\, x < -2 \text{ or } x \geq 4\}$ in
set-builder notation; the solution is
$(-\infty, \ -2) \cup [4, \ \infty)$ in interval notation.

82. $\dfrac{y-5}{y+4} < 0$

The rational expression will equal 0 when
$y = 5$. It is undefined when $y = -4$.
Determine where the numerator and
denominator are positive and negative and where
the quotient is positive and negative.

Interval	$(-\infty, -4)$	−4	$(-4, 5)$	5	$(5, \infty)$
$y - 5$	− − − −	−	− − − −	0	+ + + +
$y + 4$	− − − −	0	+ + + +	+	+ + + +
$\frac{y-5}{y+4}$	+ + + +	$\varnothing$	− − − −	0	+ + + +

The rational function is undefined at $y = -4$, so
−4 is not part of the solution. The inequality is
strict, so 5 is not part of the solution. Now,
$\dfrac{y-5}{y+4}$ is less than zero where the quotient is

negative. The solution is $\{y \,|\, -4 < y < 5\}$ in set-
builder notation; the solution is $(-4, \ 5)$ in
interval notation.

83. $\dfrac{4}{z^2 - 9} \leq 0$
$$\frac{4}{(z-3)(z+3)} \leq 0$$

Because the numerator is a constant, this rational
expression cannot equal 0. However, it is
undefined when $z = 3$ and when $z = -3$.
Determine where the factors of the denominator
are positive and negative and where the quotient
is positive and negative.

Interval	$(-\infty, -3)$	−3	$(-3, 3)$	3	$(3, \infty)$
4	+ + + +	+	+ + + +	+	+ + + +
$z - 3$	− − − −	−	− − − −	0	+ + + +
$z + 3$	− − − −	0	+ + + +	+	+ + + +
$\frac{4}{(z-3)(z+3)}$	+ + + +	$\varnothing$	− − − −	$\varnothing$	+ + + +

The rational function is undefined at $z = -3$ and
$z = 3$, so −3 and 3 are not part of the solution.

Now, $\dfrac{4}{(z-3)(z+3)}$ is less than zero where the

quotient is negative. The solution is
$\{z \,|\, -3 < z < 3\}$ in set-builder notation; the

solution is $(-3,\ 3)$ in interval notation.

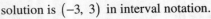
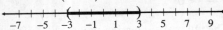

84. $\dfrac{w^2 + 5w - 14}{w - 4} < 0$

$\dfrac{(w-2)(w+7)}{w-4} < 0$

The rational expression will equal 0 when $w = 2$ and when $w = -7$. It is undefined when $w = 4$. Determine where the factors of the numerator and the denominator are positive and negative and where the quotient is positive and negative.

Interval	$(-\infty,-7)$	-7	$(-7,2)$	2	$(2,4)$	4	$(4,\infty)$
$w-2$	$---$	$-$	$---$	0	$+++$	$+$	$+++$
$w+7$	$---$	0	$+++$	$+$	$+++$	$+$	$+++$
$w-4$	$---$	$-$	$---$	$-$	$---$	0	$+++$
$\frac{(w-2)(w+7)}{w-4}$	$---$	0	$+++$	0	$---$	$\varnothing$	$+++$

The rational function is undefined at $w = 4$, so 4 is not part of the solution. The inequality is strict, so -7 and 2 are not part of the solution.

Now, $\dfrac{(w-2)(w+7)}{w-4}$ is less than zero where the quotient is negative. The solution is $\{w \mid w < -7 \text{ or } 2 < w < 4\}$ in set-builder notation; the solution is $(-\infty, -7) \cup (2,\ 4)$ in interval notation.

85. $\dfrac{m-5}{m^2 + 3m - 10} \geq 0$

$\dfrac{m-5}{(m-2)(m+5)} \geq 0$

The rational expression will equal 0 when $m = 5$. It is undefined when $m = 2$ and when $m = -5$.

Determine where the numerator and the factors of the denominator are positive and negative and where the quotient is positive and negative.

Interval	$(-\infty,-5)$	-5	$(-5,2)$	2	$(2,5)$	5	$(5,\infty)$
$m-5$	$---$	$-$	$---$	$-$	$---$	0	$+++$
$m-2$	$---$	$-$	$---$	0	$+++$	$+$	$+++$
$m+5$	$---$	0	$+++$	$+$	$+++$	$+$	$+++$
$\frac{m-5}{(m-2)(m+5)}$	$---$	$\varnothing$	$+++$	$\varnothing$	$---$	0	$+++$

The rational function is undefined at $m = -5$ and when $m = 2$, so -5 and 2 are not part of the solution. The inequality is non-strict, so 5 is part of the solution. Now, $\dfrac{m-5}{(m-2)(m+5)}$ is greater than zero where the quotient is positive. The solution is $\{m \mid -5 < m < 2 \text{ or } m \geq 5\}$ in set-builder notation; the solution is $(-5,\ 2) \cup [5,\ \infty)$ in interval notation.

86. $\dfrac{4}{n-2} \leq -2$

$\dfrac{4}{n-2} + 2 \leq 0$

$\dfrac{4}{n-2} + \dfrac{2(n-2)}{n-2} \leq 0$

$\dfrac{4 + 2n - 4}{n-2} \leq 0$

$\dfrac{2n}{n-2} \leq 0$

The rational expression will equal 0 when $n = 0$. It is undefined when $n = 2$.

Determine where the numerator and denominator are positive and negative and where the quotient is positive and negative.

Interval	$(-\infty,0)$	0	$(0,2)$	2	$(2,\infty)$
$2n$	$----$	0	$++++$	$+$	$++++$
$n-2$	$----$	$-$	$----$	0	$++++$
$\frac{2n}{n-2}$	$++++$	0	$----$	$\varnothing$	$++++$

The rational function is undefined at $n = 2$, so 2 is not part of the solution. The inequality is non-strict, so 0 is part of the solution. Now, $\dfrac{2n}{n-2}$ is less than zero where the quotient is negative. The solution is $\{n \mid 0 \leq n < 2\}$ in set-builder notation; the solution is $[0,\ 2)$ in interval notation.

87.
$$\frac{a+1}{a-2} > 3$$

$$\frac{a+1}{a-2} - 3 > 0$$

$$\frac{a+1}{a-2} - \frac{3(a-2)}{a-2} > 0$$

$$\frac{a+1-3a+6}{a-2} > 0$$

$$\frac{-2a+7}{a-2} > 0$$

The rational expression will equal 0 when
$a = \dfrac{7}{2}$. It is undefined when $a = 2$.

Determine where the numerator and denominator are positive and negative and where the quotient is positive and negative.

Interval	$(-\infty, 2)$	2	$\left(2, \frac{7}{2}\right)$	$\frac{7}{2}$	$\left(\frac{7}{2}, \infty\right)$
$-2a+7$	$++++$	$+$	$++++$	0	$----$
$a-2$	$----$	0	$++++$	$+$	$++++$
$\frac{-2a+7}{a-2}$	$----$	$\varnothing$	$++++$	0	$----$

The rational function is undefined at $a = 2$, so 2 is not part of the solution. The inequality is strict, so $\dfrac{7}{2}$ is not part of the solution. Now,

$\dfrac{-2a+7}{a-2}$ is greater than zero where the quotient

is positive. The solution is $\left\{ a \,\middle|\, 2 < a < \dfrac{7}{2} \right\}$ in

set-builder notation; the solution is $\left(2, \dfrac{7}{2}\right)$ in

interval notation.

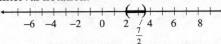

88.
$$\frac{4}{c-2} - \frac{3}{c} < 0$$

$$\frac{4c}{c(c-2)} - \frac{3(c-2)}{c(c-2)} < 0$$

$$\frac{4c-3c+6}{c(c-2)} < 0$$

$$\frac{c+6}{c(c-2)} < 0$$

The rational expression will equal 0 when
$c = -6$. It is undefined when $c = 0$ and when $c = 2$.
Determine where the numerator and the factors

of the denominator are positive and negative and where the quotient is positive and negative.

Interval	$(-\infty, -6)$	-6	$(-6, 0)$	0	$(0, 2)$	2	$(2, \infty)$
$c+6$	$---$	0	$+++$	$+$	$+++$	$+$	$+++$
c	$---$	$-$	$---$	0	$+++$	$+$	$+++$
$c-2$	$---$	$-$	$---$	$-$	$---$	0	$+++$
$\frac{c+6}{c(c-2)}$	$---$	0	$+++$	$\varnothing$	$---$	$\varnothing$	$+++$

The rational function is undefined at $c = 0$ and $c = 2$, so 0 and 2 are not part of the solution. The inequality is strict, so -6 is not part of the solution. Now, $\dfrac{c+6}{c(c-2)}$ is less than zero where

the quotient is negative. The solution is $\left\{ c \,\middle|\, c < -6 \text{ or } 0 < c < 2 \right\}$ in set-builder notation; the solution is $(-\infty, -6) \cup (0, 2)$ in interval notation.

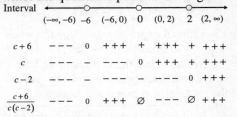

89. $Q(x) < 0$

$$\frac{2x+3}{x-4} < 0$$

The rational expression will equal 0 when
$x = -\dfrac{3}{2}$. It is undefined when $x = 4$.

Determine where the numerator and denominator are positive and negative and where the quotient is positive and negative.

Interval	$\left(-\infty, -\frac{3}{2}\right)$	$-\frac{3}{2}$	$\left(-\frac{3}{2}, 4\right)$	4	$(4, \infty)$
$2x+3$	$----$	0	$++++$	$+$	$++++$
$x-4$	$----$	$-$	$----$	0	$++++$
$\frac{2x+3}{x-4}$	$++++$	0	$----$	$\varnothing$	$++++$

The rational function is undefined at $x = 4$, so 4 is not part of the solution. The inequality is strict, so $-\dfrac{3}{2}$ is not part of the solution. Now, $\dfrac{2x+3}{x-4}$ is less than zero where the quotient is negative. The solution is $\left\{ x \,\middle|\, -\dfrac{3}{2} < x < 4 \right\}$ in set-builder notation; the solution is $\left(-\dfrac{3}{2}, 4\right)$ in interval notation.

90. $R(x) \geq 0$

$\dfrac{x+5}{x+1} \geq 0$

The rational expression will equal 0 when $x = -5$. It is undefined when $x = -1$. Determine where the numerator and the denominator are positive and negative and where the quotient is positive and negative.

Interval	$(-\infty, -5)$	-5	$(-5, -1)$	-1	$(-1, \infty)$
$x+5$	$----$	0	$++++$	$+$	$++++$
$x+1$	$----$	$-$	$----$	0	$++++$
$\dfrac{x+5}{x+1}$	$++++$	0	$----$	$\varnothing$	$++++$

The rational function is undefined at $x = -1$, so -1 is not part of the solution. The inequality is non-strict, so -5 is part of the solution. Now, $\dfrac{x+5}{x+1}$ is greater than zero where the quotient is positive. The solution is $\{x \mid x \leq -5 \text{ or } x > -1\}$ in set-builder notation; the solution is $(-\infty,\ -5] \cup (-1,\ \infty)$ in interval notation.

91.

$\dfrac{1}{C_1} + \dfrac{1}{C_2} = \dfrac{1}{C}$

$CC_1C_2\left(\dfrac{1}{C_1} + \dfrac{1}{C_2}\right) = CC_1C_2\left(\dfrac{1}{C}\right)$

$CC_2 + CC_1 = C_1C_2$

$C(C_2 + C_1) = C_1C_2$

$C = \dfrac{C_1C_2}{C_2 + C_1}$

92.

$\dfrac{P_1V_1}{T_1} = \dfrac{P_2V_2}{T_2}$

$T_1T_2\left(\dfrac{P_1V_1}{T_1}\right) = T_1T_2\left(\dfrac{P_2V_2}{T_2}\right)$

$T_2P_1V_1 = T_1P_2V_2$

$T_2 = \dfrac{T_1P_2V_2}{P_1V_1}$

93.

$T = \dfrac{4\pi^2 a^2}{MG}$

$MG \cdot T = MG\left(\dfrac{4\pi^2 a^2}{MG}\right)$

$MGT = 4\pi^2 a^2$

$G = \dfrac{4\pi^2 a^2}{MT}$

94.

$z = \dfrac{x - \mu}{\sigma}$

$z \cdot \sigma = \left(\dfrac{x - \mu}{\sigma}\right)\sigma$

$z \cdot \sigma = x - \mu$

$z \cdot \sigma + \mu = x$

95.

$\dfrac{AB}{AC} = \dfrac{DE}{DF}$

$\dfrac{3x-6}{16} = \dfrac{15}{x}$

$16x\left(\dfrac{3x-6}{16}\right) = 16x\left(\dfrac{15}{x}\right)$

$x(3x-6) = 16 \cdot 15$

$3x^2 - 6x = 240$

$3x^2 - 6x - 240 = 0$

$3(x^2 - 2x - 80) = 0$

$3(x+8)(x-10) = 0$

$x + 8 = 0 \quad \text{or} \quad x - 10 = 0$

$x = -8 \quad \text{or} \quad x = 10$

Since the length of DF cannot be negative, we discard $x = -8$ and keep $x = 10$. The length of AB is $3x - 6 = 3 \cdot 10 - 6 = 30 - 6 = 24$. In summary, $AB = 24$ and $DF = 10$.

96. Let x represent the height of the tree.

$\dfrac{\text{tree's height}}{\text{tree's shadow length}} = \dfrac{\text{post's height}}{\text{post's shadow length}}$

$\dfrac{x}{30} = \dfrac{5}{8}$

$120\left(\dfrac{x}{30}\right) = 120\left(\dfrac{5}{8}\right)$

$4x = 75$

$x = \dfrac{75}{4} = 18.75$

The height of the pine tree is 18.75 feet.

97. Let x represent the number of grams of total carbohydrates in the 3-cup bowl.

$$\frac{\frac{3}{4}}{26} = \frac{3}{x}$$

$$26x\left(\frac{\frac{3}{4}}{26}\right) = 26x\left(\frac{3}{x}\right)$$

$$\frac{3}{4}x = 78$$

$$\frac{4}{3}\left(\frac{3}{4}x\right) = \frac{4}{3}(78)$$

$$x = 104$$

Three bowls of Honey Nut Chex® will contain 104 grams of total carbohydrates.

98. Let x represent the amount Jeri will earn working an 8-hour day.

$$\frac{48.75}{5} = \frac{x}{8}$$

$$40\left(\frac{48.75}{5}\right) = 40\left(\frac{x}{8}\right)$$

$$8(48.75) = 5x$$

$$390 = 5x$$

$$78 = x$$

Jeri will earn $78.00 for working an 8-hour day.

99. Let t represent the time required to fill the tank if both pipes are used.
NOTE: 1 hour and 12 minutes = 72 minutes

$$\begin{pmatrix}\text{Part filled by}\\\text{first pipe}\\\text{in 1 minute}\end{pmatrix} + \begin{pmatrix}\text{Part filled by}\\\text{second pipe}\\\text{in 1 minute}\end{pmatrix} = \begin{pmatrix}\text{Part filled}\\\text{together}\\\text{in 1 minute}\end{pmatrix}$$

$$\frac{1}{48} \quad + \quad \frac{1}{72} \quad = \quad \frac{1}{t}$$

$$144t\left(\frac{1}{48} + \frac{1}{72}\right) = 144t\left(\frac{1}{t}\right)$$

$$3t + 2t = 144$$

$$5t = 144$$

$$t = \frac{144}{5} = 28.8$$

If both pipes are used, it will take 28.8 minutes (or 28 minutes and 48 seconds) to fill the tank.

100. Let t represent the time required for Craig to mow the lawn when working alone.

NOTE: 1 hour and 10 minutes = $\frac{7}{6}$ hours

$$\begin{pmatrix}\text{Part done}\\\text{by Diane}\\\text{in 1 hour}\end{pmatrix} + \begin{pmatrix}\text{Part done}\\\text{by Craig}\\\text{in 1 hour}\end{pmatrix} = \begin{pmatrix}\text{Part done}\\\text{together}\\\text{in 1 hour}\end{pmatrix}$$

$$\frac{1}{2} \quad + \quad \frac{1}{t} \quad = \quad \frac{1}{\frac{7}{6}}$$

$$\frac{1}{2} + \frac{1}{t} = \frac{6}{7}$$

$$14t\left(\frac{1}{2} + \frac{1}{t}\right) = 14t\left(\frac{6}{7}\right)$$

$$7t + 14 = 12t$$

$$14 = 5t$$

$$t = \frac{14}{5} = 2.8$$

Working alone, it will take Craig 2.8 hours (or 2 hours and 48 minutes) to mow the lawn.

101. Let t represent the time required for John to carpet the room when working alone.
Then $t - 7$ will represent the time required for Rick to carpet the room when working alone.

$$\begin{pmatrix}\text{Part done}\\\text{by John}\\\text{in 1 hour}\end{pmatrix} + \begin{pmatrix}\text{Part done}\\\text{by Rick}\\\text{in 1 hour}\end{pmatrix} = \begin{pmatrix}\text{Part done}\\\text{together}\\\text{in 1 hour}\end{pmatrix}$$

$$\frac{1}{t} \quad + \quad \frac{1}{t-7} \quad = \quad \frac{1}{12}$$

$$12t(t-7)\left(\frac{1}{t} + \frac{1}{t-7}\right) = 12t(t-7)\left(\frac{1}{12}\right)$$

$$12(t-7) + 12t = t(t-7)$$

$$12t - 84 + 12t = t^2 - 7t$$

$$24t - 84 = t^2 - 7t$$

$$0 = t^2 - 31t + 84$$

$$0 = (t-3)(t-28)$$

$t - 3 = 0$ or $t - 28 = 0$
$t = 3$ or $t = 28$

If John's time is $t = 3$, then Rick's time will be $t - 7 = 3 - 7 = -4$. Since time cannot be negative, we discard $t = 3$ and keep $t = 28$.
Working alone, it would take John 28 hours to carpet the room. It would take Rick $28 - 7 = 21$ hours to carpet the room alone.

102. Let t represent the time required for the drain to empty a full sink.
NOTE: 1 minute 30 seconds = 1.5 minutes.

$$\begin{pmatrix}\text{Part filled in}\\ \text{1 minute if}\\ \text{the drain is}\\ \text{plugged}\end{pmatrix} - \begin{pmatrix}\text{Part emptied}\\ \text{in 1 minute}\\ \text{if the drain}\\ \text{is unplugged}\end{pmatrix} = \begin{pmatrix}\text{Part filled in}\\ \text{1 minute if}\\ \text{the drain is}\\ \text{unplugged}\end{pmatrix}$$

$$\frac{1}{1} - \frac{1}{t} = \frac{1}{1.5}$$

$$3t\left(1 - \frac{1}{t}\right) = 3t\left(\frac{1}{1.5}\right)$$

$$3t - 3 = 2t$$

$$t = 3$$

It will take 3 minutes for the drain to empty a full sink.

103. Let x represent the speed of the wind.

NOTE: 1 hour and 15 minutes = $\dfrac{5}{4}$ hours.

	Distance (miles)	Rate (mph)	Time (hours)
With the wind	100	$180 + x$	$\dfrac{100}{180+x}$
Against the wind	100	$180 - x$	$\dfrac{100}{180-x}$
Total	200		$\dfrac{5}{4}$

$$\frac{100}{180+x} + \frac{100}{180-x} = \frac{5}{4}$$

$$4(180+x)(180-x)\left(\frac{100}{180+x} + \frac{100}{180-x}\right) =$$

$$4(180+x)(180-x)\left(\frac{5}{4}\right)$$

$$400(180-x) + 400(180+x) = 5(180+x)(180-x)$$

$$72,000 - 400x + 72,000 + 400x = 162,000 - 5x^2$$

$$144,000 = 162,000 - 5x^2$$

$$\frac{144,000}{5} = \frac{162,000 - 5x^2}{5}$$

$$28,800 = 32,400 - x^2$$

$$x^2 - 3600 = 0$$

$$(x-60)(x+60) = 0$$

$$x - 60 = 0 \quad \text{or} \quad x + 60 = 0$$

$$x = 60 \quad \text{or} \quad x = -60$$

Because the rate cannot be negative, we discard $x = -60$ and keep $x = 60$. The speed of the wind was 60 miles per hour.

104. Let x represent the speed of Jesse's boat in still water. Then $x + 5$ will represent the speed of his boat downstream and $x - 5$ will be the speed of his boat up stream.

	Distance (miles)	Rate (mph)	Time (hours)
Down	20	$x + 5$	$\dfrac{20}{x+5}$
Up	10	$x - 5$	$\dfrac{10}{x-5}$

$$\frac{20}{x+5} = \frac{10}{x-5}$$

$$(x+5)(x-5)\left(\frac{20}{x+5}\right) = (x+5)(x-5)\left(\frac{10}{x-5}\right)$$

$$20(x-5) = 10(x+5)$$

$$20x - 100 = 10x + 50$$

$$10x = 150$$

$$x = 15$$

Jesse's boat can travel 15 miles per hour in still water.

105. Let x represent the average speed at which Todd walks. Then $4x$ will represent the average speed at which he runs.

NOTE: 35 minutes = $\dfrac{7}{12}$ hours.

	Distance (miles)	Rate (mph)	Time (hours)
Walking	1	x	$\dfrac{1}{x}$
Running	3	$4x$	$\dfrac{3}{4x}$

$$\frac{1}{x} + \frac{3}{4x} = \frac{7}{12}$$

$$12x\left(\frac{1}{x} + \frac{3}{4x}\right) = 12x\left(\frac{7}{12}\right)$$

$$12 + 9 = 7x$$

$$21 = 7x$$

$$3 = x$$

Todd's average walking speed is 3 miles per hour. His average running speed is $4(3) = 12$ miles per hour.

106. Let x represent the average speed for the last 80 miles of the trip.

	Distance (miles)	Rate (mph)	Time (hours)
First 20 miles	20	30	$\dfrac{20}{30}=\dfrac{2}{3}$
Last 80 miles	80	x	$\dfrac{80}{x}$
Whole trip	100	50	$\dfrac{100}{50}=2$

$$\frac{2}{3}+\frac{80}{x}=2$$
$$3x\left(\frac{2}{3}+\frac{80}{x}\right)=3x\cdot 2$$
$$2x+240=6x$$
$$240=4x$$
$$60=x$$

Danielle's average speed for the last 80 miles of her trip was 60 miles per hour.

107. a. $y=kx$
$$30=k\cdot 6$$
$$k=5$$

b. $y=5x$

c. $y=5(10)$
$$y=50$$

108. a. $y=kx$
$$18=k\cdot(-3)$$
$$k=-6$$

b. $y=-6x$

c. $y=-6(8)$
$$y=-48$$

109. a. $y=\dfrac{k}{x}$
$$15=\frac{k}{4}$$
$$60=k$$

b. $y=\dfrac{60}{x}$

c. If $x=5$, then $y=\dfrac{60}{5}=12$.

110. a. $y=kxz$
$$45=k\cdot 6\cdot 10$$
$$45=60k$$
$$k=\frac{45}{60}=\frac{3}{4}$$

b. $y=\dfrac{3}{4}xz$

c. If $x=8$ and $z=7$, then $y=\dfrac{3}{4}\cdot 8\cdot 7=42$.

111. a. $s=\dfrac{k}{t^2}$
$$18=\frac{k}{2^2}$$
$$18=\frac{k}{4}$$
$$72=k$$

b. $s=\dfrac{72}{t^2}$

c. If $t=3$, then $s=\dfrac{72}{3^2}=\dfrac{72}{9}=8$.

112. a. $w=k\cdot\dfrac{x}{z}$
$$\frac{4}{3}=k\cdot\frac{10}{12}$$
$$k=\frac{4}{3}\cdot\frac{12}{10}=\frac{8}{5}$$

b. $w=\dfrac{8}{5}\cdot\dfrac{x}{z}$ or $w=\dfrac{8x}{5z}$

c. If $x=9$ and $z=16$, then $w=\dfrac{8\cdot 9}{5\cdot 16}=\dfrac{9}{10}$.

113. Let w represent the amount of water in d inches of snow.
$$w=kd$$
$$4.8=k\cdot 40$$
$$k=0.12$$
$$w=0.12d$$
$$w=0.12(50)$$
$$w=6$$
There are 6 inches of water contained in 50 inches of snow.

114.
$$p = kb$$
$$293.49 = k \cdot 15,000$$
$$k = 0.019566$$
$$p = 0.019566b$$
$$p = 0.019566(18,000)$$
$$p = 352.188$$

If Roberta borrows $18,000, her payment would be $352.19.

115. Let *f* represent the frequency of a radio signal with wavelength *l*.

$$f = \frac{k}{l}$$

$$800 = \frac{k}{375}$$

$$375 \cdot 800 = 375\left(\frac{k}{375}\right)$$

$$300,000 = k$$

$$f(l) = \frac{300,000}{l}$$

$$f(250) = \frac{300,000}{250} = 1200$$

When the wavelength is 250 meters, the frequency is 1200 kilohertz.

116. Let *C* represent the electrical current when the resistance is *R*.

$$C = \frac{k}{R}$$

$$8 = \frac{k}{15}$$

$$15 \cdot 8 = 15\left(\frac{k}{15}\right)$$

$$120 = k$$

$$C(R) = \frac{120}{R}$$

$$C(R) = 10$$

$$\frac{120}{R} = 10$$

$$120 = 10R$$

$$12 = R$$

If the electrical current is 10 amperes, then the resistance is 12 ohms.

117.
$$V = khd^2$$
$$231 = k \cdot 6 \cdot 7^2$$
$$231 = 294k$$
$$k = \frac{231}{294} = \frac{11}{14}$$
$$V = \frac{11}{14}hd^2$$

For $h = 14$ and $d = 8$, $V = \frac{11}{14}(14)(8)^2 = 704$.

If the diameter is 8 centimeters and the height is 14 centimeters, then the volume of the cylinder is 704 cubic centimeters.

118.
$$V = kBh$$
$$270 = k(81)(10)$$
$$270 = 810k$$
$$k = \frac{270}{810} = \frac{1}{3}$$
$$V = \frac{1}{3}Bh$$

For $B = 125$ and $h = 9$, $V = \frac{1}{3}(125)(9) = 375$.

When the area of the base is 125 square inches and the height is 9 inches, the volume of the pyramid is 375 cubic inches.

Chapter 5 Test

1. We need to find all values of *x* that cause the denominator $2x^2 - 13x - 7$ to equal 0.

$$2x^2 - 13x - 7 = 0$$
$$(2x+1)(x-7) = 0$$

$$2x + 1 = 0 \quad \text{or} \quad x - 7 = 0$$
$$2x = -1 \quad \text{or} \qquad x = 7$$
$$x = -\frac{1}{2}$$

Thus, the domain of $f(x) = \dfrac{2x+1}{2x^2 - 13x - 7}$ is

$$\left\{ x \,\middle|\, x \neq -\frac{1}{2},\ x \neq 7 \right\}.$$

2.
$$\frac{2m^2 + 5m - 12}{3m^2 + 11m - 4} = \frac{(2m-3)(m+4)}{(3m-1)(m+4)}$$

$$= \frac{(2m-3)\cancel{(m+4)}}{(3m-1)\cancel{(m+4)}}$$

$$= \frac{2m-3}{3m-1}$$

3. $\dfrac{2b-3a}{3a^2+10ab-8b^2} = \dfrac{-1(3a-2b)}{(3a-2b)(a+4b)}$

$$= \dfrac{-1\cancel{(3a-2b)}}{\cancel{(3a-2b)}(a+4b)}$$

$$= \dfrac{-1}{a+4b}$$

4. $\dfrac{4x^2-12x}{x^2-9} \cdot \dfrac{2x^2+11x+15}{8x^3-32x^2}$

$$= \dfrac{4x(x-3)}{(x+3)(x-3)} \cdot \dfrac{(2x+5)(x+3)}{8x^2(x-4)}$$

$$= \dfrac{\cancel{4x}\cancel{(x-3)}}{\cancel{(x+3)}\cancel{(x-3)}} \cdot \dfrac{(2x+5)\cancel{(x+3)}}{2x \cdot \cancel{4x}(x-4)}$$

$$= \dfrac{2x+5}{2x(x-4)}$$

5. $\dfrac{\dfrac{y^2+2y-8}{4y^2-5y-6}}{\dfrac{3y^2-14y-5}{4y^2-17y-15}}$

$$= \dfrac{y^2+2y-8}{4y^2-5y-6} \cdot \dfrac{4y^2-17y-15}{3y^2-14y-5}$$

$$= \dfrac{(y-2)(y+4)}{(4y+3)(y-2)} \cdot \dfrac{(4y+3)(y-5)}{(3y+1)(y-5)}$$

$$= \dfrac{\cancel{(y-2)}(y+4)}{\cancel{(4y+3)}\cancel{(y-2)}} \cdot \dfrac{\cancel{(4y+3)}\cancel{(y-5)}}{(3y+1)\cancel{(y-5)}}$$

$$= \dfrac{y+4}{3y+1}$$

6. $p^2-q^2 = (p-q)(p+q)$

$p-q$

$\text{LCD} = (p-q)(p+q)$

$$\dfrac{3p^2+3pq}{p^2-q^2} - \dfrac{3p-2q}{p-q}$$

$$= \dfrac{3p^2+3pq}{(p-q)(p+q)} - \dfrac{3p-2q}{p-q} \cdot \dfrac{p+q}{p+q}$$

$$= \dfrac{3p^2+3pq}{(p-q)(p+q)} - \dfrac{3p^2+pq-2q^2}{(p-q)(p+q)}$$

$$= \dfrac{3p^2+3pq-\left(3p^2+pq-2q^2\right)}{(p-q)(p+q)}$$

$$= \dfrac{3p^2+3pq-3p^2-pq+2q^2}{(p-q)(p+q)}$$

$$= \dfrac{2pq+2q^2}{(p-q)(p+q)}$$

$$= \dfrac{2q(p+q)}{(p-q)(p+q)}$$

$$= \dfrac{2q}{p-q}$$

7. $3c^2-2c-8 = (3c+4)(c-2)$

$3c^2+c-4 = (3c+4)(c-1)$

$\text{LCD} = (3c+4)(c-2)(c-1)$

$$\dfrac{9c+2}{3c^2-2c-8} + \dfrac{7}{3c^2+c-4}$$

$$= \dfrac{9c+2}{(3c+4)(c-2)} \cdot \dfrac{c-1}{c-1} + \dfrac{7}{(3c+4)(c-1)} \cdot \dfrac{c-2}{c-2}$$

$$= \dfrac{9c^2-7c-2}{(3c+4)(c-2)(c-1)} + \dfrac{7c-14}{(3c+4)(c-2)(c-1)}$$

$$= \dfrac{9c^2-7c-2+7c-14}{(3c+4)(c-2)(c-1)}$$

$$= \dfrac{9c^2-16}{(3c+4)(c-2)(c-1)}$$

$$= \dfrac{(3c+4)(3c-4)}{(3c+4)(c-2)(c-1)}$$

$$= \dfrac{3c-4}{(c-2)(c-1)}$$

8. $Q(x) = \dfrac{f(x)}{h(x)} = \dfrac{\dfrac{3x}{x^2-4}}{\dfrac{9x^2-45x}{x^2-2x-8}}$

$= \dfrac{3x}{x^2-4} \cdot \dfrac{x^2-2x-8}{9x^2-45x}$

$= \dfrac{3x}{(x+2)(x-2)} \cdot \dfrac{(x-4)(x+2)}{9x(x-5)}$

$= \dfrac{3\cancel{x}}{(\cancel{x+2})(x-2)} \cdot \dfrac{(x-4)\cancel{(x+2)}}{3 \cdot 3\cancel{x}(x-5)}$

$= \dfrac{x-4}{3(x-2)(x-5)}$

The domain of $f(x)$ is $\{x \mid x \neq -2, x \neq 2\}$. The domain of $h(x)$ is $\{x \mid x \neq -2, x \neq 4\}$. Because the denominator of $Q(x)$ cannot equal 0, we must exclude those values of x such that the numerator of $h(x)$ is 0. That is, we must exclude the values of x such that $9x^2 - 45x$. These values are $x = 0$ and $x = 5$. Therefore, the domain of $Q(x)$ is

$\{x \mid x \neq -2, x \neq 0, x \neq 2, x \neq 4, x \neq 5\}$.

9. $x^2 - 4 = (x+2)(x-2)$

$x^2 + 2x = x(x+2)$

$LCD = x(x+2)(x-2)$

$S(x) = f(x) + g(x)$

$= \dfrac{3x}{x^2-4} + \dfrac{6}{x^2+2x}$

$= \dfrac{3x}{(x+2)(x-2)} \cdot \dfrac{x}{x} + \dfrac{6}{x(x+2)} \cdot \dfrac{x-2}{x-2}$

$= \dfrac{3x^2}{x(x+2)(x-2)} + \dfrac{6x-12}{x(x+2)(x-2)}$

$= \dfrac{3x^2+6x-12}{x(x+2)(x-2)}$ or $\dfrac{3(x^2+2x-4)}{x(x+2)(x-2)}$

The domain of $f(x)$ is $\{x \mid x \neq -2, x \neq 2\}$. The domain of $g(x)$ is $\{x \mid x \neq -2, x \neq 0\}$. Therefore, the domain of $S(x)$ is

$\{x \mid x \neq -2, x \neq 0, x \neq 2\}$.

10. Using Method 2, the LCD of all the denominators in the complex rational expression is a^2.

$\dfrac{1-\dfrac{1}{a}}{1-\dfrac{1}{a^2}} = \dfrac{1-\dfrac{1}{a}}{1-\dfrac{1}{a^2}} \cdot \dfrac{a^2}{a^2} = \dfrac{1 \cdot a^2 - \dfrac{1}{a} \cdot a^2}{1 \cdot a^2 - \dfrac{1}{a^2} \cdot a^2}$

$= \dfrac{a^2-a}{a^2-1} = \dfrac{a(a-1)}{(a+1)(a-1)} = \dfrac{a}{a+1}$

11. Using Method 2, the LCD of all the denominators in the complex rational expression is $(d+2)(d-2)$.

$\dfrac{\dfrac{5}{d+2}-\dfrac{1}{d-2}}{\dfrac{3}{d+2}-\dfrac{6}{d-2}} = \dfrac{\dfrac{5}{d+2}-\dfrac{1}{d-2}}{\dfrac{3}{d+2}-\dfrac{6}{d-2}} \cdot \dfrac{(d+2)(d-2)}{(d+2)(d-2)}$

$= \dfrac{\dfrac{5}{d+2} \cdot (d+2)(d-2) - \dfrac{1}{d-2} \cdot (d+2)(d-2)}{\dfrac{3}{d+2} \cdot (d+2)(d-2) - \dfrac{6}{d-2} \cdot (d+2)(d-2)}$

$= \dfrac{5(d-2)-(d+2)}{3(d-2)-6(d+2)}$

$= \dfrac{5d-10-d-2}{3d-6-6d-12} = \dfrac{4d-12}{-3d-18}$ or $\dfrac{4(d-3)}{-3(d+6)}$

12.

$\dfrac{1}{6x} - \dfrac{1}{3} = \dfrac{5}{4x} + \dfrac{3}{4}$

$12x\left(\dfrac{1}{6x} - \dfrac{1}{3}\right) = 12x\left(\dfrac{5}{4x} + \dfrac{3}{4}\right)$

$12x\left(\dfrac{1}{6x}\right) - 12x\left(\dfrac{1}{3}\right) = 12x\left(\dfrac{5}{4x}\right) + 12x\left(\dfrac{3}{4}\right)$

$2 - 4x = 15 + 9x$

$-13x = 13$

$x = -1$

Check:

$\dfrac{1}{6(-1)} - \dfrac{1}{3} \overset{?}{=} \dfrac{5}{4(-1)} + \dfrac{3}{4}$

$\dfrac{1}{-6} - \dfrac{1}{3} \overset{?}{=} \dfrac{5}{-4} + \dfrac{3}{4}$

$-\dfrac{1}{6} - \dfrac{2}{6} \overset{?}{=} -\dfrac{5}{4} + \dfrac{3}{4}$

$\dfrac{-3}{6} \overset{?}{=} \dfrac{-2}{4}$

$-\dfrac{1}{2} = -\dfrac{1}{2}$ $\leftarrow$ True

The solution checks, so the solution set is $\{-1\}$.

13.
$$\frac{7n}{n+3}+\frac{21}{n-3}=\frac{126}{n^2-9}$$
$$\frac{7n}{n+3}+\frac{21}{n-3}=\frac{126}{(n+3)(n-3)}$$
$$(n+3)(n-3)\left(\frac{7n}{n+3}+\frac{21}{n-3}\right)=$$
$$(n+3)(n-3)\left(\frac{126}{(n+3)(n-3)}\right)$$
$$7n(n-3)+21(n+3)=126$$
$$7n^2-21n+21n+63=126$$
$$7n^2+63=126$$
$$7n^2-63=0$$
$$\frac{7n^2-63}{7}=\frac{0}{7}$$
$$n^2-9=0$$
$$(n+3)(n-3)=0$$
$$n+3=0 \quad \text{or} \quad n-3=0$$
$$n=-3 \quad \text{or} \quad n=3$$

Since $n=-3$ and $n=3$ are neither in the domain of the variable, they are both extraneous solutions. The solution set is $\varnothing$ or { }.

14.
$$\frac{x+5}{x-2}\ge 3$$
$$\frac{x+5}{x-2}-3\ge 0$$
$$\frac{x+5}{x-2}-\frac{3(x-2)}{x-2}\ge 0$$
$$\frac{x+5-3x+6}{x-2}\ge 0$$
$$\frac{-2x+11}{x-2}\ge 0$$

The rational expression will equal 0 when $x=\dfrac{11}{2}$. It is undefined when $x=2$.

Determine where the numerator and denominator are positive and negative and where the quotient is positive and negative.

Interval	$(-\infty, 2)$	2	$\left(2,\frac{11}{2}\right)$	$\frac{11}{2}$	$\left(\frac{11}{2},\infty\right)$
$-2x+11$	$++++$	$+$	$++++$	0	$----$
$x-2$	$----$	0	$++++$	$+$	$++++$
$\frac{-2x+11}{x-2}$	$----$	$\varnothing$	$++++$	0	$----$

The rational function is undefined at $x=2$, so 2 is not part of the solution. The inequality is

non-strict, so $\dfrac{11}{2}$ is part of the solution. Now, $\dfrac{-2x+11}{x-2}$ is greater than zero where the quotient is positive. The solution is $\left\{x \,\middle|\, 2<x\le\dfrac{11}{2}\right\}$ in set-builder notation; the solution is $\left(2,\dfrac{11}{2}\right]$ in interval notation.

15.
$$\frac{1}{F}=\frac{D^2}{kq_1q_2}$$
$$Fkq_1q_2\left(\frac{1}{F}\right)=Fkq_1q_2\left(\frac{D^2}{kq_1q_2}\right)$$
$$kq_1q_2=FD^2$$
$$k=\frac{FD^2}{q_1q_2}$$

16. Let t represent the time to print out the 48-page document.

$$\frac{\text{Number of pages}}{\text{time to print document}}=\frac{\text{Number of pages}}{\text{time to print document}}$$
$$\frac{10}{25}=\frac{48}{t}$$
$$25t\left(\frac{10}{25}\right)=25t\left(\frac{48}{t}\right)$$
$$10t=1200$$
$$t=120$$

The printer will take 120 seconds (or 2 minutes) to print out the 48-page document.

17. Let t represent the time required if Linnette and Darrell work together to clean their house.

$$\left(\begin{array}{c}\text{Part done}\\\text{by Linnette}\\\text{in 1 hour}\end{array}\right)+\left(\begin{array}{c}\text{Part done}\\\text{by Darrell}\\\text{in 1 hour}\end{array}\right)=\left(\begin{array}{c}\text{Part done}\\\text{together}\\\text{in 1 hour}\end{array}\right)$$
$$\frac{1}{4}\qquad+\qquad\frac{1}{6}\qquad=\qquad\frac{1}{t}$$
$$24t\left(\frac{1}{4}+\frac{1}{6}\right)=24t\left(\frac{1}{t}\right)$$
$$6t+4t=24$$
$$10t=24$$
$$t=\frac{24}{10}=2.4$$

Working together, it will take Linnette and Darrell 2.4 hours (or 2 hours and 24 minutes) to clean their house.

18. Let x represent the speed of the current.

	Distance (miles)	Rate (mph)	Time (hours)
Up	4	$7-x$	$\dfrac{4}{7-x}$
Down	10	$7+x$	$\dfrac{10}{7+x}$

$$\frac{4}{7-x}=\frac{10}{7+x}$$
$$(7-x)(7+x)\left(\frac{4}{7-x}\right)=(7-x)(7+x)\left(\frac{10}{7+x}\right)$$
$$4(7+x)=10(7-x)$$
$$28+4x=70-10x$$
$$14x=42$$
$$x=3$$

The rate of the current is 3 miles per hour.

19. $F=\dfrac{k}{l}$

$50=\dfrac{k}{4}$

$200=k$

$F=\dfrac{200}{l}$

If $l=10$, then $F=\dfrac{200}{10}=20$.

If the length of the force arm of the lever is 10 feet, the force required to lift the boulder will be 20 pounds.

20. $L=krh$

$528=k(7)(12)$

$528=84k$

$k=\dfrac{528}{84}=\dfrac{44}{7}$

$L=\dfrac{44}{7}rh$

For $r=9$ and $h=14$, $L=\dfrac{44}{7}(9)(14)=792$.

When the radius is 9 centimeters and the height is 14 centimeters, the lateral surface area of the right circular cylinder is 792 square centimeters.

Cumulative Review Chapters R–5

1. $\dfrac{4^3-6\cdot7+14}{4-1^2}=\dfrac{64-6\cdot7+14}{4-1}$
$$=\frac{64-42+14}{3}$$
$$=\frac{36}{3}=12$$

2. $2x(x-3)+4(x-2)+15=2x^2-6x+4x-8+15$
$$=2x^2-2x+7$$

3. $\dfrac{3(-2)^2-4(-2)-5}{(-2)-3}=\dfrac{3\cdot4+8-5}{-5}$
$$=\frac{12+8-5}{-5}$$
$$=\frac{15}{-5}=-3$$

4. $7x+9=3x-23$
$7x-3x=-23-9$
$4x=-32$
$x=-8$
The solution set is $\{-8\}$.

5. $|3x+7|\le8$
$-8\le3x+7\le8$
$-8-7\le3x\le8-7$
$-15\le3x\le1$
$-5\le x\le\dfrac{1}{3}$

The solution set is $\left\{x\mid-5\le x\le\dfrac{1}{3}\right\}$ in set builder notation; $\left[-5,\dfrac{1}{3}\right]$ in interval notation.

6. $h(x)=\dfrac{x-5}{2x^2-7x-15}$
$2x^2-7x-15=0$
$(2x+3)(x-5)=0$
$2x+3=0$　　or　　$x-5=0$
$x=-\dfrac{3}{2}$　　or　　　$x=5$

The domain of $h(x)$ is $\left\{x\mid x\ne-\dfrac{3}{2}, x\ne5\right\}$.

7. a.
$$f(x) = x^2 - 5x$$
$$f(-3) = (-3)^2 - 5(-3)$$
$$= 9 + 15$$
$$= 24$$

b.
$$f(x) = x^2 - 5x$$
$$f\left(\frac{1}{4}\right) = \left(\frac{1}{4}\right)^2 - 5\left(\frac{1}{4}\right)$$
$$= \frac{1}{16} - \frac{5}{4}$$
$$= \frac{1}{16} - \frac{20}{16}$$
$$= -\frac{19}{16}$$

c.
$$f(x) = x^2 - 5x$$
$$f(x+2) = (x+2)^2 - 5(x+2)$$
$$= x^2 + 4x + 4 - 5x - 10$$
$$= x^2 - x - 6$$

8. $4x + 3y = 15$
$$3y = -4x + 15$$
$$y = -\frac{4}{3}x + 5$$

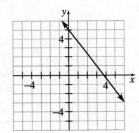

9. $m = \dfrac{y_2 - y_1}{x_2 - x_1} = \dfrac{-8 - 1}{10 - (-5)} = \dfrac{-9}{15} = -\dfrac{3}{5}$
$$y - y_1 = m(x - x_1)$$
$$y - 1 = -\frac{3}{5}(x + 5)$$
$$y - 1 = -\frac{3}{5}x - 3$$
$$y = -\frac{3}{5}x - 2$$
or
$$y = -\frac{3}{5}x - 2$$
$$5y = -3x - 10$$
$$3x + 5y = -10$$

10. $x + 4y = 20$
$$4y = -x + 20$$
$$y = -\frac{1}{4} + 5$$
$$m = -\frac{1}{4}$$
$$m_\perp = 4$$
$$y - y_1 = m(x - x_1)$$
$$y - 5 = 4(x - 3)$$
$$y - 5 = 4x - 12$$
$$y = 4x - 7$$
or
$$4x - y = 7$$

11. $D = kw$
$$1500 = 125k$$
$$k = \frac{1500}{125} = 12$$
$$D = 12w$$
$$D = 12 \cdot 180$$
$$D = 2160 \text{ milligrams}$$

12. $\begin{cases} 2x - 3y = 7 \\ 5x + 2y = 8 \end{cases}$

13. a. $\begin{bmatrix} 1 & 1 & 0 & | & -3 \\ 0 & 2 & -1 & | & -1 \\ 5 & 0 & 1 & | & 1 \end{bmatrix}$ $(R_3 = -5r_1 + r_3)$

$$= \begin{bmatrix} 1 & 1 & 0 & | & -3 \\ 0 & 2 & -1 & | & -1 \\ -5(1)+5 & -5(1)+0 & -5(0)+1 & | & -5(-3)+1 \end{bmatrix}$$

$$= \begin{bmatrix} 1 & 1 & 0 & | & -3 \\ 0 & 2 & -1 & | & -1 \\ 0 & -5 & 1 & | & 16 \end{bmatrix}$$

b. $\begin{bmatrix} 1 & 1 & 0 & | & -3 \\ 0 & 2 & -1 & | & -1 \\ 0 & -5 & 1 & | & 16 \end{bmatrix}$ $\left(R_2 = \dfrac{1}{2}r_2\right)$

$$= \begin{bmatrix} 1 & 1 & 0 & | & -3 \\ \dfrac{1}{2}(0) & \dfrac{1}{2}(2) & \dfrac{1}{2}(-1) & | & \dfrac{1}{2}(-1) \\ 0 & -5 & 1 & | & 16 \end{bmatrix}$$

$$= \begin{bmatrix} 1 & 1 & 0 & | & -3 \\ 0 & 1 & -\dfrac{1}{2} & | & -\dfrac{1}{2} \\ 0 & -5 & 1 & | & 16 \end{bmatrix}$$

14. $\begin{cases} x + 4y = -2 & (1) \\ 2x - 12y = -9 & (2) \end{cases}$

From equation (1), we have $x = -4y - 2$. Substitute into equation (2), and solve for y.

$$2(-4y - 2) - 12y = -9$$
$$-8y - 4 - 12y = -9$$
$$-20y - 4 = -9$$
$$-20y = -5$$
$$y = \frac{1}{4}$$

Substitute this value into $x = -4y - 2$:

$$x = -4\left(\frac{1}{4}\right) - 2$$
$$= -1 - 2 = -3$$

The solution to the system of equations is $\left(-3, \dfrac{1}{4}\right)$.

15. $\begin{vmatrix} 2 & 0 & 4 \\ 1 & -1 & -2 \\ 2 & -2 & 3 \end{vmatrix} = 2\begin{vmatrix} -1 & -2 \\ -2 & 3 \end{vmatrix} - 0\begin{vmatrix} 1 & -2 \\ 2 & 3 \end{vmatrix} + 4\begin{vmatrix} 1 & -1 \\ 2 & -2 \end{vmatrix}$

$$= 2((-1)(3) - (-2)(-2)) - 0 + 4((1)(-2) - (-1)(2))$$
$$= 2(-3 - 4) + 4(-2 + 2)$$
$$= 2(-7) + 4(0)$$
$$= -14$$

16. $\begin{cases} 3x+2y<8 & (1) \\ x-4y\geq12 & (2) \end{cases}$

Replace the inequality symbol in (1) with an equals sign to obtain:

$3x+2y=8$

$$y=-\frac{3}{2}x+4$$

Because the inequality is strict, graph

$y=-\frac{3}{2}x+4$ using a dashed line.

Test Point: $(0,0)$: $3(0)+2(0)\overset{?}{<}8$

$\overset{?}{0<8}$ True

Therefore, $(0,0)$ is a solution to $3x+2y<8$.

Shade the half-plane that contains $(0,0)$.

Replace the inequality symbol in (2) with an equals sign to obtain:

$x-4y=12$

$y=\frac{1}{4}x-3$

Because the inequality is non-strict, graph

$y=\frac{1}{4}x-3$ using a solid line.

Test Point: $(0,0)$: $(0)-4(0)\overset{?}{\geq}12$

$\overset{?}{0\geq12}$ False

Therefore, $(0,0)$ is not a solution to

$x-4y\geq12$. Shade the half-plane that does not

contain $(0,0)$.

The overlapping shaded region (that is, the shaded region in the graph below) is the solution to the system of linear inequalities.

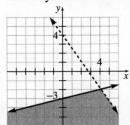

17. $\left(3x^2-4xy+7y^2\right)+\left(5x^2-9xy+2y^2\right)$

$=3x^2-4xy+7y^2+5x^2-9xy+2y^2$

$=8x^2-13xy+9y^2$

18. $\left(5x^2-3x+12\right)-\left(2x^2-4x-15\right)$

$=5x^2-3x+12-2x^2+4x+15$

$=3x^2+x+27$

19. $(2x-3)\left(x^2-4x+6\right)$

$=2x^3-8x^2+12x-3x^2+12x-18$

$=2x^3-11x^2+24x-18$

20. $\begin{array}{r} 2x^2-5x+9 \\ 2x+5\overline{\smash{\big)}4x^3+0x^2-7x+45} \end{array}$

$\underline{-\left(4x^3+10x^2\right)}$

$-10x^2-7x$

$\underline{-\left(-10x^2-25x\right)}$

$18x+45$

$\underline{-(18x+45)}$

0

$$\frac{4x^3-7x+45}{2x+5}=2x^2-5x+9$$

21. $x^4+5x^3-8x-40=\left(x^4+5x^3\right)+(-8x-40)$

$=x^3(x+5)-8(x+5)$

$=(x+5)\left(x^3-8\right)$

$=(x+5)(x-2)\left(x^2+2x+4\right)$

22. $6x^2+x-15=(2x-3)(3x+5)$

23. $\dfrac{3x^2-2x-1}{3x^2-5x-2}\cdot\dfrac{x^2-9x+14}{x^2+x-2}$

$=\dfrac{(3x+1)\,(x-1)}{(3x+1)\,(x-2)}\cdot\dfrac{(x-7)\,(x-2)}{(x+2)\,(x-1)}$

$=\dfrac{x-7}{x+2}$

24.
$$\frac{3}{x^2+x-6}-\frac{2}{x^2+2x-3}$$
$$=\frac{3}{(x+3)(x-2)}-\frac{2}{(x+3)(x-1)}$$
$$=\frac{3}{(x+3)(x-2)}\cdot\frac{(x-1)}{(x-1)}-\frac{2}{(x+3)(x-1)}\cdot\frac{(x-2)}{(x-2)}$$
$$=\frac{3(x-1)-2(x-2)}{(x+3)(x-2)(x-1)}$$
$$=\frac{3x-3-2x+4}{(x+3)(x-2)(x-1)}$$
$$=\frac{x+1}{(x+3)(x-2)(x-1)}$$

25. Let t represent the time required to grade the test papers if Bill and Karl work together.

$$\begin{pmatrix}\text{Part graded}\\\text{by Bill in}\\\text{1 minute}\end{pmatrix}+\begin{pmatrix}\text{Part graded}\\\text{by Karl in}\\\text{1 minute}\end{pmatrix}=\begin{pmatrix}\text{Part graded}\\\text{together}\\\text{in 1 minute}\end{pmatrix}$$

$$\frac{1}{80}\quad+\quad\frac{1}{120}\quad=\quad\frac{1}{t}$$

$$240t\left(\frac{1}{80}+\frac{1}{120}\right)=240t\left(\frac{1}{t}\right)$$
$$3t+2t=240$$
$$5t=240$$
$$t=48$$

If Bill and Karl work together, they will grade the test papers in 48 minutes.

Chapter 6

Getting Ready for Chapter 6

Getting Ready for Chapter 6 Quick Checks

1. The symbol $\sqrt{}$ is called a <u>radical sign</u>.

2. If a is a nonnegative real number, the nonnegative number b such that $b^2 = a$ is the <u>principal square root</u> of a and is denoted by $b = \sqrt{a}$.

3. The square roots of 16 are <u>-4</u> and <u>4</u>.

4. $\sqrt{81} = 9$ because $9^2 = 81$.

5. $\sqrt{900} = 30$ because $30^2 = 900$.

6. $\sqrt{\dfrac{9}{4}} = \dfrac{3}{2}$ because $\left(\dfrac{3}{2}\right)^2 = \dfrac{9}{4}$.

7. $\sqrt{0.16} = 0.4$ because $(0.4)^2 = 0.16$.

8. $\left(\sqrt{13}\right)^2 = 13$ because $\left(\sqrt{c}\right)^2 = c$ if $c \geq 0$.

9. $5\sqrt{9} = 5\sqrt{3^2} = 5 \cdot 3 = 15$

10. $\sqrt{36 + 64} = \sqrt{100} = \sqrt{10^2} = 10$

11. $\sqrt{36} + \sqrt{64} = \sqrt{6^2} + \sqrt{8^2} = 6 + 8 = 14$

12. $\sqrt{25 - 4 \cdot 3 \cdot (-2)} = \sqrt{25 + 24} = \sqrt{49} = \sqrt{7^2} = 7$

13. True

14. $\sqrt{400}$ is rational because $20^2 = 400$. Thus, $\sqrt{400} = 20$.

15. $\sqrt{40}$ is irrational because 40 is not a perfect square. There is no rational number whose square is 40. Using a calculator, we find $\sqrt{40} \approx 6.32$.

16. $\sqrt{-25}$ is not a real number. There is no real number whose square is -25.

17. $-\sqrt{196}$ is a rational number because $14^2 = 196$. Thus, $-\sqrt{196} = -14$.

18. $\sqrt{a^2} = |a|$

19. $\sqrt{(-14)^2} = |-14| = 14$

20. $\sqrt{z^2} = |z|$

21. $\sqrt{(2x+3)^2} = |2x+3|$

22. $\sqrt{p^2 - 12p + 36} = \sqrt{(p-6)^2} = |p-6|$

Getting Ready for Chapter 6 Exercises

23. $\sqrt{1} = \sqrt{1^2} = 1$

25. $-\sqrt{100} = -\sqrt{10^2} = -10$

27. $\sqrt{\dfrac{1}{4}} = \sqrt{\left(\dfrac{1}{2}\right)^2} = \dfrac{1}{2}$

29. $\sqrt{0.36} = \sqrt{(0.6)^2} = 0.6$

31. $\left(\sqrt{1.6}\right)^2 = 1.6$

33. $\sqrt{-14}$ is not a real number because the radicand of the square root is negative.

35. $\sqrt{64}$ is rational because 64 is a perfect square.
$\sqrt{64} = \sqrt{(8)^2} = 8$

37. $\sqrt{\dfrac{1}{16}}$ is rational because $\dfrac{1}{16}$ is a perfect square.
$\sqrt{\dfrac{1}{16}} = \sqrt{\left(\dfrac{1}{4}\right)^2} = \dfrac{1}{4}$

39. $\sqrt{44}$ is irrational because 44 is not a perfect square.
$\sqrt{44} \approx 6.63$

41. $\sqrt{50}$ is irrational because 50 is not a perfect square.

$\sqrt{50} \approx 7.07$

43. $\sqrt{-16}$ is not a real number because the radicand of the square root is negative.

45. $\sqrt{8^2} = 8$

47. $\sqrt{(-19)^2} = |-19| = 19$

49. $\sqrt{r^2} = |r|$

51. $\sqrt{(x+4)^2} = |x+4|$

53. $\sqrt{(4x-3)^2} = |4x-3|$

55. $\sqrt{4y^2 + 12y + 9} = \sqrt{(2y+3)^2} = |2y+3|$

57. $\sqrt{25+144} = \sqrt{169} = \sqrt{(13)^2} = 13$

59. $\sqrt{25} + \sqrt{144} = \sqrt{5^2} + \sqrt{12^2} = 5 + 12 = 17$

61. $\sqrt{-144}$ is not a real number.

63. $3\sqrt{25} = 3\sqrt{(5)^2} = 3 \cdot 5 = 15$

65. $5\sqrt{\dfrac{16}{25}} - \sqrt{144} = 5\sqrt{\left(\dfrac{4}{5}\right)^2} - \sqrt{12^2} = 5 \cdot \dfrac{4}{5} - 12$

$= 4 - 12 = -8$

67. $\sqrt{8^2 - 4 \cdot 1 \cdot 7} = \sqrt{64 - 28} = \sqrt{36} = \sqrt{6^2} = 6$

69. $\sqrt{(-5)^2 - 4 \cdot 2 \cdot 5} = \sqrt{25 - 40} = \sqrt{-15}$ is not a real number.

71. $\dfrac{-(-1) + \sqrt{(-1)^2 - 4 \cdot 6 \cdot (-2)}}{2 \cdot (-1)} = \dfrac{1 + \sqrt{1+48}}{-2}$

$= \dfrac{1 + \sqrt{49}}{-2} = \dfrac{1+7}{-2} = \dfrac{8}{-2} = -4$

73. $\sqrt{(6-1)^2 + (15-3)^2} = \sqrt{5^2 + 12^2}$

$= \sqrt{25 + 144} = \sqrt{169} = \sqrt{13^2} = 13$

75. The square roots of 36 are 6 and -6 because

$(6)^2 = 36$ and $(-6)^2 = 36$.

$\sqrt{36} = 6$ (the principal square root).

77. $Z = \dfrac{X - \mu}{\dfrac{\sigma}{\sqrt{n}}} = \dfrac{120 - 100}{\dfrac{15}{\sqrt{13}}} = \dfrac{20}{\dfrac{15}{\sqrt{13}}}$

$= 20 \cdot \dfrac{\sqrt{13}}{15} = \dfrac{4\sqrt{13}}{3} \approx 4.81$

79. Answers may vary. When taking a square root, we are looking for a number that we can multiply by itself to obtain the radicand. Since $a \cdot a = a^2$, it should follow that $\sqrt{a^2} = a$. However, since the product of two negative numbers is positive, it is impossible to determine from a^2 whether a was positive or negative. We define $\sqrt{a^2} = |a|$ to ensure that the principle square root is positive even if $a < 0$.
For example:

$2^2 = 4$ and $\sqrt{4} = 2$

$(-2)^2 = 4$ and $\sqrt{4} = 2$ not -2 since

$\sqrt{(-2)^2} = |-2| = 2$.

Section 6.1

Are You Ready for This Section?

R1. $\left(\dfrac{x^2 y}{xy^{-2}}\right)^{-3} = \left(x^{2-1} y^{1-(-2)}\right)^{-3}$

$= \left(xy^3\right)^{-3}$

$= \left(\dfrac{1}{xy^3}\right)^3$

$= \dfrac{1}{x^3 y^9}$

R2. $\left(\sqrt{7}\right)^2 = 7$

R3. $\sqrt{64} = 8$ since $8^2 = 64$.

R4. $\sqrt{(x+1)^2} = |x+1|$

R5. a. $3^{-2} = \dfrac{1}{3^2} = \dfrac{1}{9}$

    b. $x^{-4} = \dfrac{1}{x^4}$

Section 6.1 Quick Checks

1. In the notation $\sqrt[n]{a}$, the integer n, $n \geq 2$, is called the <u>index</u>.

2. $\sqrt[3]{64} = 4$ since $4^3 = 64$

3. $\sqrt[4]{81} = 3$ since $3^4 = 81$.

4. $\sqrt[3]{-216} = -6$ since $(-6)^3 = -216$.

5. $\sqrt[4]{-32}$ is not a real number.

6. $\sqrt[5]{\dfrac{1}{32}} = \dfrac{1}{2}$ since $\left(\dfrac{1}{2}\right)^5 = \dfrac{1}{32}$.

7. $\sqrt[3]{50} \approx 3.68$

```
³√50
        3.684031499
```

8. $\sqrt[4]{80} \approx 2.99$

```
⁴√80
        2.990697562
```

9. Because the index is even, we have
$\sqrt[4]{5^4} = |5| = 5$.

10. Because the index is even, we have $\sqrt[6]{z^6} = |z|$.

11. Because the index is odd, we have
$\sqrt[7]{(3x-2)^7} = 3x-2$.

12. Because the index is even, we have
$\sqrt[8]{(-2)^8} = |-2| = 2$.

13. Because the index is odd, we have
$$\sqrt[5]{-\dfrac{32}{243}} = \sqrt[5]{\dfrac{-32}{243}} = \sqrt[5]{\dfrac{(-2)^5}{3^5}} = \sqrt[5]{\left(-\dfrac{2}{3}\right)^5} = -\dfrac{2}{3}$$

14. If a is a nonnegative real number and $n \geq 2$ is an integer, then $a^{1/n} = \sqrt[n]{a}$.

15. $25^{1/2} = \sqrt{25} = \sqrt{5^2} = 5$

16. $(-27)^{1/3} = \sqrt[3]{-27} = \sqrt[3]{(-3)^3} = -3$

17. $-64^{1/2} = -\sqrt{64} = -\sqrt{8^2} = -8$

18. $(-64)^{1/2} = \sqrt{-64}$ is not a real number.

19. $b^{1/2} = \sqrt{b}$

20. $\sqrt[5]{8b} = (8b)^{1/5}$

21. $\sqrt[8]{\dfrac{mn^5}{3}} = \left(\dfrac{mn^5}{3}\right)^{1/8}$

22. $4\sqrt[3]{y} = 4y^{\frac{1}{3}}$

23. If a is a real number and $\dfrac{m}{n}$ is a rational number in lowest terms with $n \geq 2$, then $a^{m/n} = \sqrt[n]{a^m}$ or $\left(\sqrt[n]{a}\right)^m$, provided that $\sqrt[n]{a}$ exists.

24. $16^{3/2} = \left(\sqrt{16}\right)^3 = 4^3 = 64$

25. $27^{2/3} = \left(\sqrt[3]{27}\right)^2 = 3^2 = 9$

26. $-16^{3/4} = -\left(\sqrt[4]{16}\right)^3 = -2^3 = -8$

27. $(-64)^{2/3} = \left(\sqrt[3]{-64}\right)^2 = (-4)^2 = 16$

28. $(-25)^{5/2} = \left(\sqrt{-25}\right)^5$ is not a real number.

29. $50^{2/3} \approx 13.57$

```
50^2/3
            13.57208808
```

30. $40^{3/20} \approx 1.74$

```
40^3/20
            1.739037707
■
```

31. $\sqrt[8]{a^3} = \left(a^3\right)^{1/8} = a^{3/8}$

32. $\sqrt[4]{t^{12}} = (t^{12})^{1/4} = t^{12/4} = t^3$

33. $\left(\sqrt[4]{12ab^3}\right)^9 = \left(\left(12ab^3\right)^{1/4}\right)^9 = \left(12ab^3\right)^{9/4}$

34. $81^{-1/2} = \dfrac{1}{81^{1/2}} = \dfrac{1}{\sqrt{81}} = \dfrac{1}{9}$

35. $\dfrac{1}{8^{-2/3}} = 8^{2/3} = \left(\sqrt[3]{8}\right)^2 = 2^2 = 4$

36. $(13x)^{-3/2} = \dfrac{1}{(13x)^{3/2}}$

6.1 Exercises

37. $\sqrt[3]{125} = \sqrt[3]{5^3} = 5$

39. $\sqrt[3]{-27} = \sqrt[3]{(-3)^3} = -3$

41. $-\sqrt[4]{625} = -\sqrt[4]{5^4} = -5$

43. $\sqrt[3]{-\dfrac{1}{8}} = \sqrt[3]{\left(-\dfrac{1}{2}\right)^3} = -\dfrac{1}{2}$

45. $-\sqrt[5]{-243} = -\sqrt[5]{(-3)^5} = -(-3) = 3$

47. $\sqrt[3]{25} = 25^{\wedge}(1/3) \approx 2.92$

49. $\sqrt[4]{12} = 12^{\wedge}(1/4) \approx 1.86$

51. $\sqrt[3]{5^3} = 5$

53. $\sqrt[4]{m^4} = |m|$

55. $\sqrt[9]{(x-3)^9} = x-3$

57. $-\sqrt[4]{(3p+1)^4} = -|3p+1|$

59. $4^{1/2} = \sqrt{4} = \sqrt{2^2} = 2$

61. $-36^{1/2} = -\sqrt{36} = -\sqrt{6^2} = -6$

63. $8^{1/3} = \sqrt[3]{8} = \sqrt[3]{2^3} = 2$

65. $-16^{1/4} = -\sqrt[4]{16} = -\sqrt[4]{2^4} = -2$

67. $\left(\dfrac{4}{25}\right)^{1/2} = \sqrt{\dfrac{4}{25}} = \dfrac{\sqrt{4}}{\sqrt{25}} = \dfrac{\sqrt{2^2}}{\sqrt{5^2}} = \dfrac{2}{5}$

69. $(-125)^{1/3} = \sqrt[3]{-125} = \sqrt[3]{(-5)^3} = -5$

71. $(-4)^{1/2} = \sqrt{-4}$ is not a real number.

73. $\sqrt[3]{3x} = (3x)^{1/3}$

75. $\sqrt[4]{\dfrac{x}{3}} = \left(\dfrac{x}{3}\right)^{1/4}$

77. $4^{5/2} = \left(\sqrt{4}\right)^5 = 2^5 = 32$

79. $-16^{3/2} = -\left(\sqrt{16}\right)^3 = -4^3 = -64$

81. $8^{4/3} = \left(\sqrt[3]{8}\right)^4 = 2^4 = 16$

83. $(-64)^{2/3} = \left(\sqrt[3]{-64}\right)^2 = (-4)^2 = 16$

85. $-(-32)^{3/5} = -\left(\sqrt[5]{-32}\right)^3 = -(-2)^3 = -(-8) = 8$

87. $144^{-1/2} = \left(\sqrt{144}\right)^{-1} = (12)^{-1} = \dfrac{1}{12}$

89. $\dfrac{1}{25^{-3/2}} = 25^{3/2} = \left(\sqrt{25}\right)^3 = 5^3 = 125$

91. $\dfrac{1}{8^{-5/3}} = 8^{5/3} = \left(\sqrt[3]{8}\right)^5 = 2^5 = 32$

93. $\sqrt[4]{x^3} = x^{3/4}$

95. $\left(\sqrt[5]{3x}\right)^2 = (3x)^{2/5}$

97. $\sqrt{\left(\dfrac{5x}{y}\right)^3} = \left(\dfrac{5x}{y}\right)^{3/2}$

99. $\sqrt[3]{(9ab)^4} = (9ab)^{4/3}$

101. $20^{1/2} = \sqrt{20} \approx 4.47$

103. $4^{5/3} = 4\,\hat{}\,(5/3) \approx 10.08$

105. $10^{0.1} = 10\,\hat{}\,(0.1) \approx 1.26$

107. $\sqrt[3]{x^3} + 4\sqrt[5]{x^5} = x + 4x = 5x$

109. $-16^{3/4} - \sqrt[5]{32} = -\left(\sqrt[4]{16}\right)^3 - 2$
$$= -2^3 - 2$$
$$= -8 - 2$$
$$= -10$$

111. $\sqrt[3]{512} = \sqrt[3]{8^3} = 8$

113. $9^{2.5} = 9^{5/2} = \left(\sqrt{9}\right)^5 = 3^5 = 243$

115. $\sqrt[4]{-16}$ is not a real number because the index is even and the radicand is negative.

117. $144^{-1/2} \cdot 3^2 = \dfrac{1}{144^{1/2}} \cdot 9 = \dfrac{1}{\sqrt{144}} \cdot 9 = \dfrac{1}{12} \cdot 9 = \dfrac{3}{4}$

119. $\sqrt[3]{0.008} = \sqrt[3]{(0.2)^3} = 0.2$

121. $4^{0.5} + 25^{1.5} = 4^{1/2} + 25^{3/2}$
$$= \sqrt{4} + \left(\sqrt{25}\right)^3$$
$$= 2 + 5^3$$
$$= 2 + 125$$
$$= 127$$

123. $(-25)^{5/2} = \left(\sqrt{-25}\right)^5$ is not a real number.

125. $\sqrt[3]{(3p-5)^3} = 3p - 5$

127. $-9^{\frac{3}{2}} + \dfrac{1}{27^{-\frac{2}{3}}} = -\left(9^{1/2}\right)^3 + \dfrac{1}{\left(27^{1/3}\right)^{-2}}$
$$= -(3)^3 + \dfrac{1}{(3)^{-2}}$$
$$= -27 + 3^2$$
$$= -27 + 9$$
$$= -18$$

129. $\sqrt[6]{(-2)^6} = \sqrt[6]{64} = \sqrt[6]{2^6} = 2$

131. $f(x) = x^{3/2}$
$$f(4) = 4^{3/2} = \left(\sqrt{4}\right)^3 = 2^3 = 8$$

133. $F(z) = z^{4/3}$
$$F(-8) = (-8)^{4/3} = \left(\sqrt[3]{-8}\right)^4 = (-2)^4 = 16$$

135. 10 is the only cube root of 1000.
$$\sqrt[3]{1000} = \sqrt[3]{(10)^3} = 10$$

137. a. $W = 35.74 + 0.6215(30) - 35.75(10)^{0.16}$
$$+ 0.4275(30)(10)^{0.16}$$
$$\approx 21.25$$
The windchill would be about $21.25°F$.

b. $W = 35.74 + 0.6215(30) - 35.75(20)^{0.16}$
$$+ 0.4275(30)(20)^{0.16}$$
$$\approx 17.36$$
The windchill would be about $17.36°F$.

c. $W = 35.74 + 0.6215(0) - 35.75(10)^{0.16}$
$$+ 0.4275(0)(10)^{0.16}$$
$$\approx -15.93$$
The windchill would be about $-15.93°F$.

139. a. $v_t = \sqrt{\dfrac{2mg}{C\rho A}}$

$$= \sqrt{\dfrac{2 \cdot \frac{4}{3}\pi r^3 \rho_w \cdot g}{C\rho \cdot \pi r^2}}$$

$$= \sqrt{\dfrac{8r\rho_w g}{3C\rho}} \text{ m/s}$$

b. The radius is $1.5 \text{ mm} = 0.0015 \text{ m}$. We also
have $\rho_w = 1000 \text{ kg/m}^3$, $g \approx 9.81 \text{ m/s}^2$,
$C = 0.6$, and $\rho \approx 1.2 \text{ kg/m}^3$.

$$v_t = \sqrt{\dfrac{8(0.0015)(1000)(9.81)}{3(0.6)(1.2)}} \approx 7.38 \text{ m/s}$$

The terminal velocity is about 7.38 meters
per second.

141. Answers will vary.
$(-9)^{1/2}$ and $-9^{1/2}$ are different because of order
of operations. This is easier seen in a different
form:
$(-9)^{1/2} = \sqrt{-9}$ (radicand is negative, so the
result is not a real number)
$-9^{1/2} = -\left(9^{1/2}\right) = -\sqrt{9} = -3$ (radicand is
positive, so the result is a real number)

143. If $\dfrac{m}{n}$, in lowest terms, is positive, then $a^{\frac{m}{n}}$ is a
real number provided $\sqrt[n]{a}$ is a real number.

If $\dfrac{m}{n}$, in lowest terms, is negative, then $a^{\frac{m}{n}}$ is a
real number provided $a \neq 0$ and $\sqrt[n]{a}$ is a real
number.

145. $\dfrac{(x+2)^2(x-1)^4}{(x+2)(x-1)} = (x+2)^{2-1}(x-1)^{4-1}$
$$= (x+2)(x-1)^3$$

147. $\dfrac{\left(4z^2-7z+3\right)+\left(-3z^2-z+9\right)}{\left(4z^2-2z-7\right)+\left(-3z^2-z+9\right)}$

$$= \dfrac{4z^2-7z+3-3z^2-z+9}{4z^2-2z-7-3z^2-z+9}$$

$$= \dfrac{z^2-8z+12}{z^2-3z+2} = \dfrac{(z-6)(z-2)}{(z-1)(z-2)}$$

$$= \dfrac{z-6}{z-1}$$

Section 6.2

Are You Ready for This Section?

R1. $z^{-3} = \dfrac{1}{z^3}$

R2. $x^{-2} \cdot x^5 = x^{-2+5} = x^3$

R3. $\left(\dfrac{2a^2}{b^{-1}}\right)^3 = \left(2a^2 b\right)^3 = 2^3 \left(a^2\right)^3 b^3 = 8a^6 b^3$

R4. $\sqrt{64} = 8$ because $8^2 = 64$.

Section 6.2 Quick Checks

1. If a and b are real numbers and if r is a rational
number, then assuming the expression is defined,
$(ab)^r = \underline{a^r b^r}$.

2. If a and b are real numbers and if r and s are
rational numbers, then assuming the expression
is defined, $a^r \cdot a^s = \underline{a^{r-s}}$.

3. $5^{3/4} \cdot 5^{1/6} = 5^{\frac{3}{4}+\frac{1}{6}} = 5^{\frac{9}{12}+\frac{2}{12}} = 5^{\frac{11}{12}}$

4. $\dfrac{32^{6/5}}{32^{3/5}} = 32^{\frac{6}{5}-\frac{3}{5}} = 32^{3/5} = \left(\sqrt[5]{32}\right)^3 = 2^3 = 8$

5. $\left(100^{3/8}\right)^{4/3} = 100^{\frac{3}{8}\cdot\frac{4}{3}} = 100^{1/2} = \sqrt{100} = 10$

6. $\left(a^{3/2} \cdot b^{5/4}\right)^{2/3} = a^{\frac{3}{2}\cdot\frac{2}{3}} \cdot b^{\frac{5}{4}\cdot\frac{2}{3}} = ab^{5/6}$

7. $\dfrac{x^{1/2} \cdot x^{1/3}}{\left(x^{1/12}\right)^2} = \dfrac{x^{\frac{1}{2}+\frac{1}{3}}}{x^{\frac{1}{12}\cdot 2}}$

$= \dfrac{x^{\frac{3}{6}+\frac{2}{6}}}{x^{\frac{1}{6}}}$

$= \dfrac{x^{\frac{5}{6}}}{x^{\frac{1}{6}}}$

$= x^{\frac{5}{6}-\frac{1}{6}}$

$= x^{4/6}$

$= x^{2/3}$

8. $\left(8x^{3/4}y^{-1}\right)^{2/3} = 8^{2/3}\left(x^{3/4}\right)^{2/3}\left(y^{-1}\right)^{2/3}$

$= 4x^{\frac{3}{4}\cdot\frac{2}{3}}y^{-1\cdot\frac{2}{3}}$

$= 4x^{1/2}y^{-2/3}$

$= \dfrac{4x^{1/2}}{y^{2/3}}$

9. $\left(\dfrac{25x^{1/2}y^{3/4}}{x^{-3/4}y}\right)^{1/2} = \left(25x^{\frac{1}{2}-\left(-\frac{3}{4}\right)}y^{\frac{3}{4}-1}\right)^{1/2}$

$= \left(25x^{5/4}y^{-1/4}\right)^{1/2}$

$= 25^{1/2}\left(x^{5/4}\right)^{1/2}\left(y^{-1/4}\right)^{1/2}$

$= 5x^{\frac{5}{4}\cdot\frac{1}{2}}y^{-\frac{1}{4}\cdot\frac{1}{2}}$

$= 5x^{5/8}y^{-1/8}$

$= \dfrac{5x^{5/8}}{y^{1/8}}$

10. $8\left(125a^{3/4}b^{-1}\right)^{2/3} = 8\cdot125^{2/3}\left(a^{3/4}\right)^{2/3}\left(b^{-1}\right)^{2/3}$

$= 8\cdot25a^{\frac{3}{4}\cdot\frac{2}{3}}b^{-1\cdot\frac{2}{3}}$

$= 200a^{1/2}b^{-2/3}$

$= \dfrac{200a^{1/2}}{b^{2/3}}$

11. $\sqrt[10]{36^5} = \left(36^5\right)^{1/10} = 36^{5\cdot\frac{1}{10}} = 36^{1/2} = \sqrt{36} = 6$

12. $\sqrt[4]{16a^8b^{12}} = \left(16a^8b^{12}\right)^{1/4}$

$= 16^{1/4}\left(a^8\right)^{1/4}\left(b^{12}\right)^{1/4}$

$= 2a^{8\cdot\frac{1}{4}}b^{12\cdot\frac{1}{4}}$

$= 2a^2b^3$

13. $\dfrac{\sqrt[3]{x^2}}{\sqrt[4]{x}} = \dfrac{x^{2/3}}{x^{1/4}} = x^{\frac{2}{3}-\frac{1}{4}} = x^{5/12} = \sqrt[12]{x^5}$

14. $\sqrt[4]{\sqrt[3]{a^2}} = \sqrt[4]{a^{2/3}} = \left(a^{2/3}\right)^{1/4} = a^{\frac{2}{3}\cdot\frac{1}{4}} = a^{1/6} = \sqrt[6]{a}$

15. $8x^{3/2} + 3x^{1/2}(4x+3) = 8x^{\frac{2}{2}+\frac{1}{2}} + 3x^{1/2}(4x+3)$

$= 8x\cdot x^{1/2} + 3x^{1/2}(4x+3)$

$= x^{1/2}\left(8x+3(4x+3)\right)$

$= x^{1/2}\left(8x+12x+9\right)$

$= x^{1/2}(20x+9)$

16. $9x^{1/3} + x^{-2/3}(3x+1) = 9x^{\frac{3}{3}-\frac{2}{3}} + x^{-2/3}(3x+1)$

$= 9x\cdot x^{-2/3} + x^{-2/3}(3x+1)$

$= x^{-2/3}\left(9x+(3x+1)\right)$

$= x^{-2/3}(12x+1)$

$= \dfrac{12x+1}{x^{2/3}}$

6.2 Exercises

17. $5^{1/2}\cdot5^{3/2} = 5^{\frac{1}{2}+\frac{3}{2}} = 5^{\frac{4}{2}} = 5^2 = 25$

19. $\dfrac{8^{5/4}}{8^{1/4}} = 8^{\frac{5}{4}-\frac{1}{4}} = 8^{4/4} = 8^1 = 8$

21. $2^{1/3}\cdot2^{-3/2} = 2^{\frac{1}{3}+\left(-\frac{3}{2}\right)} = 2^{\frac{2}{6}-\frac{9}{6}} = 2^{-7/6} = \dfrac{1}{2^{7/6}}$

23. $\dfrac{x^{1/4}}{x^{5/6}} = x^{\frac{1}{4}-\frac{5}{6}} = x^{\frac{3}{12}-\frac{10}{12}} = x^{-7/12} = \dfrac{1}{x^{7/12}}$

25. $\left(4^{4/3}\right)^{3/8} = 4^{\frac{4}{3}\cdot\frac{3}{8}} = 4^{1/2} = \sqrt{4} = 2$

27. $\left(25^{3/4} \cdot 4^{-3/4}\right)^2 = \left[\left(25 \cdot 4^{-1}\right)^{3/4}\right]^2$

$\qquad = \left(\dfrac{25}{4}\right)^{\frac{3}{4} \cdot 2} = \left(\dfrac{25}{4}\right)^{3/2}$

$\qquad = \left(\sqrt{\dfrac{25}{4}}\right)^3 = \left(\dfrac{5}{2}\right)^3 = \dfrac{125}{8}$

29. $\left(x^{3/4} \cdot y^{1/3}\right)^{2/3} = \left(x^{3/4}\right)^{2/3} \cdot \left(y^{1/3}\right)^{2/3}$

$\qquad = x^{\frac{3}{4} \cdot \frac{2}{3}} \cdot y^{\frac{1}{3} \cdot \frac{2}{3}}$

$\qquad = x^{1/2} y^{2/9}$

31. $\left(x^{-1/3} \cdot y\right)\left(x^{1/2} \cdot y^{-4/3}\right) = x^{-\frac{1}{3}+\frac{1}{2}} \cdot y^{1+\left(-\frac{4}{3}\right)}$

$\qquad = x^{-\frac{2}{6}+\frac{3}{6}} \cdot y^{\frac{3}{3}-\frac{4}{3}}$

$\qquad = x^{1/6} \cdot y^{-1/3}$

$\qquad = \dfrac{x^{1/6}}{y^{1/3}}$

33. $\left(4a^2 b^{-3/2}\right)^{1/2} = 4^{1/2} \cdot \left(a^2\right)^{1/2} \cdot \left(b^{-3/2}\right)^{1/2}$

$\qquad = 2 \cdot a^{2 \cdot \frac{1}{2}} \cdot b^{-\frac{3}{2} \cdot \frac{1}{2}}$

$\qquad = 2a^1 b^{-3/4}$

$\qquad = \dfrac{2a}{b^{3/4}}$

35. $\left(\dfrac{x^{2/3} y^{-1/3}}{8x^{1/2} y}\right)^{1/3} = \left(\dfrac{1}{8} x^{(2/3)-(1/2)} y^{(-1/3)-1}\right)^{1/3}$

$\qquad = \left(\dfrac{1}{8} x^{1/6} y^{-4/3}\right)^{1/3}$

$\qquad = \left(\dfrac{1}{8}\right)^{1/3} \left(x^{1/6}\right)^{1/3} \left(y^{-4/3}\right)^{1/3}$

$\qquad = \dfrac{1}{2} x^{1/18} y^{-4/9}$

$\qquad = \dfrac{x^{1/18}}{2y^{4/9}}$

37. $\left(\dfrac{50x^{3/4} y}{2x^{1/2}}\right)^{1/2} + \left(\dfrac{x^{1/2} y^{1/2}}{9x^{3/4} y^{3/2}}\right)^{-1/2}$

$\qquad = \left(25x^{\frac{3}{4}-\frac{1}{2}} y\right)^{1/2} + \left(\dfrac{1}{9} x^{\frac{1}{2}-\frac{3}{4}} y^{\frac{1}{2}-\frac{3}{2}}\right)^{-1/2}$

$\qquad = \left(25x^{1/4} y\right)^{1/2} + \left(\dfrac{1}{9} x^{-1/4} y^{-1}\right)^{-1/2}$

$\qquad = \left(25x^{1/4} y\right)^{1/2} + \left(\dfrac{1}{9x^{1/4} y^1}\right)^{-1/2}$

$\qquad = \left(25x^{1/4} y\right)^{1/2} + \left(9x^{1/4} y\right)^{1/2}$

$\qquad = 25^{1/2} \left(x^{1/4}\right)^{1/2} y^{1/2} + 9^{1/2} \left(x^{1/4}\right)^{1/2} y^{1/2}$

$\qquad = 5x^{1/8} y^{1/2} + 3x^{1/8} y^{1/2}$

$\qquad = 8x^{1/8} y^{1/2}$

39. $\sqrt{x^8} = \left(x^8\right)^{1/2} = x^{8 \cdot \frac{1}{2}} = x^4$

41. $\sqrt[12]{8^4} = \left(8^4\right)^{1/12} = 8^{4 \cdot \frac{1}{12}} = 8^{1/3} = \sqrt[3]{8} = 2$

43. $\sqrt[3]{8a^3 b^{12}} = \left(8a^3 b^{12}\right)^{1/3}$

$\qquad = 8^{1/3} \left(a^3\right)^{1/3} \left(b^{12}\right)^{1/3}$

$\qquad = 2ab^4$

45. $\dfrac{\sqrt{x}}{\sqrt[4]{x}} = \dfrac{x^{1/2}}{x^{1/4}} = x^{\frac{1}{2}-\frac{1}{4}} = x^{\frac{2}{4}-\frac{1}{4}} = x^{1/4} = \sqrt[4]{x}$

47. $\sqrt{x} \cdot \sqrt[3]{x} = x^{1/2} \cdot x^{1/3} = x^{\frac{1}{2}+\frac{1}{3}}$

$\qquad = x^{\frac{3}{6}+\frac{2}{6}} = x^{5/6}$

$\qquad = \sqrt[6]{x^5}$

49. $\sqrt{\sqrt[4]{x^3}} = \left(\left(x^3\right)^{1/4}\right)^{1/2} = \left(x^{3/4}\right)^{1/2} = x^{3/8} = \sqrt[8]{x^3}$

51. $\sqrt{3} \cdot \sqrt[3]{9} = \sqrt{3} \cdot \sqrt[3]{3^2}$
$= 3^{1/2} \cdot 3^{2/3}$
$= 3^{\frac{1}{2} + \frac{2}{3}}$
$= 3^{7/6}$
$= \sqrt[6]{3^7}$

53. $\dfrac{\sqrt{6}}{\sqrt[4]{36}} = \dfrac{\sqrt{6}}{\sqrt[4]{6^2}} = \dfrac{6^{1/2}}{6^{2/4}} = 6^{\frac{1}{2} - \frac{2}{4}} = 6^{\frac{1}{2} - \frac{1}{2}} = 6^0 = 1$

55. $2x^{3/2} + 3x^{1/2}(x+5) = 2x^{1/2} \cdot x + 3x^{1/2}(x+5)$
$= x^{1/2}(2x + 3(x+5))$
$= x^{1/2}(2x + 3x + 15)$
$= x^{1/2}(5x + 15)$
$= 5x^{1/2}(x+3)$

57. $5(x+2)^{2/3}(3x-2) + 9(x+2)^{5/3}$
$= 5(x+2)^{2/3}(3x-2) + 9(x+2)^{2/3} \cdot (x+2)^{3/3}$
$= (x+2)^{2/3}(5(3x-2) + 9(x+2))$
$= (x+2)^{2/3}(15x - 10 + 9x + 18)$
$= (x+2)^{2/3}(24x + 8)$
$= 8(x+2)^{2/3}(3x+1)$

59. $x^{-1/2}(2x+5) + 4x^{1/2}$
$= x^{-1/2}(2x+5) + 4x^{-1/2} \cdot x^{2/2}$
$= x^{-1/2}(2x + 5 + 4x)$
$= x^{-1/2}(6x + 5)$
$= \dfrac{6x+5}{x^{1/2}}$

61. $2(x-4)^{-1/3}(4x-3) + 12(x-4)^{2/3}$
$= 2(x-4)^{-1/3}(4x-3) + 12(x-4)^{-1/3}(x-4)^{3/3}$
$= 2(x-4)^{-1/3}((4x-3) + 6(x-4))$
$= 2(x-4)^{-1/3}(4x - 3 + 6x - 24)$
$= 2(x-4)^{-1/3}(10x - 27)$
$= \dfrac{2(10x-27)}{(x-4)^{1/3}}$

63. $15x(x^2+4)^{1/2} + 5(x^2+4)^{3/2}$
$= 15x(x^2+4)^{1/2} + 5(x^2+4)^{1/2}(x^2+4)^{2/2}$
$= 5(x^2+4)^{1/2}(3x + x^2 + 4)$
$= 5(x^2+4)^{1/2}(x^2 + 3x + 4)$

65. $\sqrt[8]{4^4} = 4^{4/8} = 4^{1/2} = (2^2)^{1/2} = 2$

67. $2^{1/2} \cdot 2^{3/2} = 2^{4/2} = 2^2 = 4$

69. $\left(100^{1/3}\right)^{3/2} = 100^{\frac{1}{3} \cdot \frac{3}{2}} = 100^{1/2} = \sqrt{100} = 10$

71. $\left(\sqrt[4]{25}\right)^2 = 25^{2/4} = 25^{1/2} = \sqrt{25} = 5$

73. $\sqrt[4]{x^2} - \dfrac{\sqrt[4]{x^6}}{x} = x^{2/4} - \dfrac{x^{6/4}}{x}$
$= x^{1/2} - x^{\frac{3}{2}-1}$
$= x^{1/2} - x^{1/2}$
$= 0$

75. $\left(4 \cdot 9^{1/4}\right)^{-2} = 4^{-2} \cdot \left(9^{1/4}\right)^{-2}$
$= \dfrac{1}{4^2} \cdot 9^{-1/2} = \dfrac{1}{16 \cdot 9^{1/2}}$
$= \dfrac{1}{16\sqrt{9}} = \dfrac{1}{16(3)} = \dfrac{1}{48}$

77. $x^{1/2}\left(x^{3/2} - 2\right) = x^{1/2} \cdot x^{3/2} - x^{1/2} \cdot 2$
$= x^{4/2} - 2x^{1/2}$
$= x^2 - 2x^{1/2}$

79. $2y^{-1/3}(1+3y) = 2y^{-1/3} \cdot 1 + 2y^{-1/3} \cdot 3y$
$= 2y^{-1/3} + 6y^{-\frac{1}{3}+1}$
$= \dfrac{2}{y^{1/3}} + 6y^{2/3}$

81. $4z^{3/2}\left(z^{3/2}-8z^{-3/2}\right)$

$=4z^{3/2}\cdot z^{3/2}-4z^{3/2}\cdot 8z^{-3/2}$

$=4z^{\frac{3}{2}+\frac{3}{2}}-32z^{\frac{3}{2}+\left(-\frac{3}{2}\right)}$

$=4z^{6/2}-32z^0$

$=4z^3-32$

$=4\left(z^3-8\right)$

$=4(z-2)\left(z^2+2z+4\right)$

83. $3^x=25$

$3^{x/2}=3^{x\cdot\frac{1}{2}}=\left(3^x\right)^{1/2}=(25)^{1/2}=5$

Thus, $3^{x/2}=5$.

85. $7^x=9$

$\sqrt{7^x}=\sqrt{9}=3$

87. $\sqrt[4]{\sqrt[3]{\sqrt{x}}}=\left(\sqrt[3]{\sqrt{x}}\right)^{1/4}$

$=\left(\left(\sqrt{x}\right)^{1/3}\right)^{1/4}$

$=\left(\sqrt{x}\right)^{1/12}$

$=\left(x^{1/2}\right)^{1/12}$

$=x^{1/24}$

$=\sqrt[24]{x}$

89. $\left(6^{\sqrt{2}}\right)^{\sqrt{2}}=6^{\sqrt{2}\cdot\sqrt{2}}=6^2=36$

91. $f(x)=(x+3)^{1/2}(x+1)^{-1/2}=\dfrac{\sqrt{x+3}}{\sqrt{x+1}}$

We need $x+3\ge 0$ and $x+1>0$.

$x+3\ge 0$ and $x+1>0$

$x\ge -3$ $x>-1$

The domain is the set of values that satisfy both inequalities. Thus, the domain is $x>-1$.

Domain: $\{x\,|\,x>-1\}$ or $(-1,\infty)$

We can check this graphically:

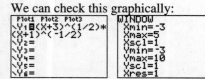

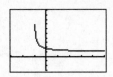

93. $3a(a-3)+(a+3)(a-2)$

$=3a^2-9a+a^2-2a+3a-6$

$=4a^2-8a-6$

95. $\dfrac{x^2-4}{x+2}\cdot(x+5)-(x+4)(x-1)$

$=\dfrac{(x+2)(x-2)(x+5)}{(x+2)}-(x+4)(x-1)$

$=(x-2)(x+5)-(x+4)(x-1)$

$=\left(x^2+3x-10\right)-\left(x^2+3x-4\right)$

$=x^2+3x-10-x^2-3x+4$

$=-6$

Section 6.3

Are You Ready for This Section?

R1. $1^2=1$, $2^2=4$, $3^2=9$, $4^2=16$, $5^2=25$,

$6^2=36$, $7^2=49$, $8^2=64$, $9^2=81$,

$10^2=100$, $11^2=121$, $12^2=144$, $13^2=169$,

and $14^2=196$.

R2. $1^3=1$, $2^3=8$, $3^3=27$, $4^3=64$, and

$5^3=125$.

R3. a. $\sqrt{16}=\sqrt{4^2}=4$

b. $\sqrt{p^2}=|p|$

Section 6.3 Quick Checks

1. If $\sqrt[n]{a}$ and $\sqrt[n]{b}$ are real numbers and $n\ge 2$ is an integer, then $\sqrt[n]{a}\cdot\sqrt[n]{b}=\sqrt[n]{ab}$.

2. $\sqrt{11}\cdot\sqrt{7}=\sqrt{11\cdot 7}=\sqrt{77}$

3. $\sqrt[4]{6}\cdot\sqrt[4]{7}=\sqrt[4]{6\cdot 7}=\sqrt[4]{42}$

4. $\sqrt{x-5}\cdot\sqrt{x+5}=\sqrt{(x-5)(x+5)}=\sqrt{x^2-25}$,

$x\ge 5$

5. $\sqrt[7]{5p} \cdot \sqrt[7]{4p^3} = \sqrt[7]{5p \cdot 4p^3} = \sqrt[7]{20p^4}$

6. $0 = 0^2$, $1 = 1^2$, $4 = 2^2$, $9 = 3^2$, $16 = 4^2$,

$25 = 5^2$

7. $1 = 1^3$, $8 = 2^3$, $27 = 3^3$, $64 = 4^3$, $125 = 5^3$,

$216 = 6^3$

8. $\sqrt{48} = \sqrt{16 \cdot 3} = \sqrt{16} \cdot \sqrt{3} = 4\sqrt{3}$

9. $4\sqrt[3]{54} = 4\sqrt[3]{27 \cdot 2} = 4\sqrt[3]{27} \cdot \sqrt[3]{2} = 4 \cdot 3\sqrt[3]{2} = 12\sqrt[3]{2}$

10. $\sqrt{200a^2} = \sqrt{200} \cdot \sqrt{a^2}$

$= \sqrt{100 \cdot 2} \cdot \sqrt{a^2}$

$= \sqrt{100} \cdot \sqrt{2} \cdot \sqrt{a^2}$

$= 10\sqrt{2}|a|$ or $10|a|\sqrt{2}$

11. $\sqrt[4]{40}$ cannot be simplified further.

12. $\dfrac{6 + \sqrt{45}}{3} = \dfrac{6 + \sqrt{9 \cdot 5}}{3} = \dfrac{6 + \sqrt{9} \cdot \sqrt{5}}{3}$

$= \dfrac{6 + 3\sqrt{5}}{3} = \dfrac{3(2 + \sqrt{5})}{3}$

$= 2 + \sqrt{5}$

13. $\dfrac{-2 + \sqrt{32}}{4} = \dfrac{-2 + \sqrt{16 \cdot 2}}{4} = \dfrac{-2 + \sqrt{16} \cdot \sqrt{2}}{4}$

$= \dfrac{-2 + 4\sqrt{2}}{4} = \dfrac{2(-1 + 2\sqrt{2})}{4}$

$= \dfrac{-1 + 2\sqrt{2}}{2}$

14. $\sqrt{75a^6} = \sqrt{25 \cdot 3 \cdot a^6}$

$= \sqrt{25} \cdot \sqrt{3} \cdot \sqrt{a^6}$

$= \sqrt{25} \cdot \sqrt{3} \cdot \sqrt{(a^3)^2}$

$= 5\sqrt{3} \cdot a^3$

$= 5a^3\sqrt{3}$

15. $\sqrt{18a^5} = \sqrt{9 \cdot 2 \cdot a^4 \cdot a}$

$= \sqrt{9} \cdot \sqrt{2} \cdot \sqrt{a^4} \cdot \sqrt{a}$

$= \sqrt{9} \cdot \sqrt{2} \cdot \sqrt{(a^2)^2} \cdot \sqrt{a}$

$= 3\sqrt{2} \cdot a^2 \sqrt{a}$

$= 3a^2\sqrt{2a}$

16. $\sqrt[3]{128x^6y^{10}} = \sqrt[3]{64 \cdot 2 \cdot x^6 \cdot y^9 \cdot y}$

$= \sqrt[3]{64} \cdot \sqrt[3]{2} \cdot \sqrt[3]{x^6} \cdot \sqrt[3]{y^9} \cdot \sqrt[3]{y}$

$= \sqrt[3]{64} \cdot \sqrt[3]{2} \cdot \sqrt[3]{(x^2)^3} \cdot \sqrt[3]{(y^3)^3} \cdot \sqrt[3]{y}$

$= 4x^2 y^3 \sqrt[3]{2y}$

17. $\sqrt[4]{16a^5b^{11}} = \sqrt[4]{16 \cdot a^4 \cdot a \cdot b^8 \cdot b^3}$

$= \sqrt[4]{16} \cdot \sqrt[4]{a^4} \cdot \sqrt[4]{a} \cdot \sqrt[4]{b^8} \cdot \sqrt[4]{b^3}$

$= \sqrt[4]{16} \cdot \sqrt[4]{a^4} \cdot \sqrt[4]{a} \cdot \sqrt[4]{(b^2)^4} \cdot \sqrt[4]{b^3}$

$= 2ab^2 \sqrt[4]{ab^3}$

18. $\sqrt{6} \cdot \sqrt{8} = \sqrt{6 \cdot 8}$

$= \sqrt{48}$

$= \sqrt{16 \cdot 3}$

$= \sqrt{16} \cdot \sqrt{3}$

$= 4\sqrt{3}$

19. $\sqrt[3]{12a^2} \cdot \sqrt[3]{10a^4} = \sqrt[3]{12a^2 \cdot 10a^4}$

$= \sqrt[3]{120a^6}$

$= \sqrt[3]{8 \cdot 15 \cdot a^6}$

$= \sqrt[3]{8} \cdot \sqrt[3]{a^6} \cdot \sqrt[3]{15}$

$= 2a^2 \sqrt[3]{15}$

20. $4\sqrt[3]{8a^2b^5} \cdot \sqrt[3]{6a^2b^4} = 4\sqrt[3]{8a^2b^5 \cdot 6a^2b^4}$

$= 4\sqrt[3]{48a^4b^9}$

$= 4\sqrt[3]{8 \cdot 6 \cdot a^3 \cdot a \cdot b^9}$

$= 4\sqrt[3]{8} \cdot \sqrt[3]{6} \cdot \sqrt[3]{a^3} \cdot \sqrt[3]{a} \cdot \sqrt[3]{b^9}$

$= 4 \cdot 2ab^3 \cdot \sqrt[3]{6a}$

$= 8ab^3 \sqrt[3]{6a}$

21. $\sqrt{\dfrac{13}{49}} = \dfrac{\sqrt{13}}{\sqrt{49}} = \dfrac{\sqrt{13}}{7}$

22. $\sqrt[3]{\dfrac{27p^3}{8}} = \dfrac{\sqrt[3]{27p^3}}{\sqrt[3]{8}} = \dfrac{3p}{2}$

23. $\sqrt[4]{\dfrac{3q^4}{16}} = \dfrac{\sqrt[4]{3q^4}}{\sqrt[4]{16}} = \dfrac{q\sqrt[4]{3}}{2}$

24. $\dfrac{\sqrt{12a^5}}{\sqrt{3a}} = \sqrt{\dfrac{12a^5}{3a}} = \sqrt{4a^4} = 2a^2$

25. $\dfrac{\sqrt[3]{-24x^2}}{\sqrt[3]{3x^{-1}}} = \sqrt[3]{\dfrac{-24x^2}{3x^{-1}}} = \sqrt[3]{-8x^3} = -2x$

26. $\dfrac{\sqrt[3]{250a^5b^{-2}}}{\sqrt[3]{2ab}} = \sqrt[3]{\dfrac{250a^5b^{-2}}{2ab}}$

$\qquad = \sqrt[3]{125a^4b^{-3}}$

$\qquad = \sqrt[3]{\dfrac{125a^4}{b^3}}$

$\qquad = \dfrac{5a\sqrt[3]{a}}{b}$

27. $\sqrt[4]{5} \cdot \sqrt[3]{3} = 5^{1/4} \cdot 3^{1/3}$

$\qquad = 5^{3/12} \cdot 3^{4/12}$

$\qquad = \left(5^3\right)^{1/12} \cdot \left(3^4\right)^{1/12}$

$\qquad = \left[\left(5^3\right)\left(3^4\right)\right]^{1/12}$

$\qquad = \left(10,125\right)^{1/12}$

$\qquad = \sqrt[12]{10,125}$

28. $\sqrt{10} \cdot \sqrt[3]{12} = 10^{1/2} \cdot 12^{1/3}$

$\qquad = 10^{3/6} \cdot 12^{2/6}$

$\qquad = \left(10^3\right)^{1/6} \cdot \left(12^2\right)^{1/6}$

$\qquad = \left[\left(10^3\right)\left(12^2\right)\right]^{1/6}$

$\qquad = \left(2^3 \cdot 5^3 \cdot 4^2 \cdot 3^2\right)^{1/6}$

$\qquad = \left(2^3 \cdot 5^3 \cdot 2^4 \cdot 3^2\right)^{1/6}$

$\qquad = \left(2^7 \cdot 5^3 \cdot 3^2\right)^{1/6}$

$\qquad = 2\left(2 \cdot 5^3 \cdot 3^2\right)^{1/6}$

$\qquad = 2\left(2250\right)^{1/6}$

$\qquad = 2\sqrt[6]{2250}$

6.3 Exercises

29. $\sqrt[3]{6} \cdot \sqrt[3]{10} = \sqrt[3]{6 \cdot 10} = \sqrt[3]{60}$

31. $\sqrt{3a} \cdot \sqrt{5b} = \sqrt{3a \cdot 5b} = \sqrt{15ab}$ if $a, b \ge 0$

33. $\sqrt{x-7} \cdot \sqrt{x+7} = \sqrt{(x-7)(x+7)} = \sqrt{x^2 - 49}$

35. $\sqrt{\dfrac{5x}{3}} \cdot \sqrt{\dfrac{3}{x}} = \sqrt{\dfrac{5x}{3} \cdot \dfrac{3}{x}} = \sqrt{5}$

37. $\sqrt{50} = \sqrt{25 \cdot 2} = \sqrt{25} \cdot \sqrt{2} = 5\sqrt{2}$

39. $\sqrt[3]{54} = \sqrt[3]{27 \cdot 2} = \sqrt[3]{27} \cdot \sqrt[3]{2} = 3\sqrt[3]{2}$

41. $\sqrt{48x^2} = \sqrt{16x^2 \cdot 3} = \sqrt{16x^2} \cdot \sqrt{3} = 4|x|\sqrt{3}$

43. $\sqrt[3]{-27x^3} = -3x$

45. $\sqrt[4]{32m^4} = \sqrt[4]{16m^4 \cdot 2} = \sqrt[4]{16m^4} \cdot \sqrt[4]{2} = 2|m|\sqrt[4]{2}$

47. $\sqrt{12p^2q} = \sqrt{4p^2 \cdot 3q} = \sqrt{4p^2} \cdot \sqrt{3q} = 2|p|\sqrt{3q}$

49. $\sqrt{162m^4} = \sqrt{81m^4 \cdot 2} = \sqrt{81m^4} \cdot \sqrt{2} = 9m^2\sqrt{2}$

51. $\sqrt{y^{13}} = \sqrt{y^{12} \cdot y} = \sqrt{y^{12}} \cdot \sqrt{y} = y^6\sqrt{y}$

53. $\sqrt[3]{c^8} = \sqrt[3]{c^6 \cdot c^2} = \sqrt[3]{c^6} \cdot \sqrt[3]{c^2} = c^2\sqrt[3]{c^2}$

55. $\sqrt{125p^3q^4} = \sqrt{25p^2q^4 \cdot 5p} = \sqrt{25p^2q^4} \cdot \sqrt{5p}$
$$= 5|p|q^2\sqrt{5p}$$

57. $\sqrt[3]{-16x^9} = \sqrt[3]{-8x^9 \cdot 2} = \sqrt[3]{-8x^9} \cdot \sqrt[3]{2} = -2x^3\sqrt[3]{2}$

59. $\sqrt[5]{-16m^8n^2} = \sqrt[5]{-m^5 \cdot 16m^3n^2}$
$$= \sqrt[5]{-m^5} \cdot \sqrt[5]{16m^3n^2}$$
$$= -m\sqrt[5]{16m^3n^2}$$

61. $\sqrt[4]{(x-y)^5} = \sqrt[4]{(x-y)^4 \cdot (x-y)}$
$$= \sqrt[4]{(x-y)^4} \cdot \sqrt[4]{x-y}$$
$$= (x-y)\sqrt[4]{x-y}$$

63. $\sqrt[3]{8x^3 - 8y^3} = \sqrt[3]{8(x^3 - y^3)}$
$$= \sqrt[3]{8} \cdot \sqrt[3]{x^3 - y^3}$$
$$= 2\sqrt[3]{x^3 - y^3}$$

65. $\dfrac{4 + \sqrt{36}}{2} = \dfrac{4 + 6}{2} = \dfrac{10}{2} = 5$

67. $\dfrac{9 + \sqrt{18}}{3} = \dfrac{9 + \sqrt{9 \cdot 2}}{3} = \dfrac{9 + \sqrt{9} \cdot \sqrt{2}}{3}$
$$= \dfrac{9 + 3\sqrt{2}}{3} = \dfrac{3(3 + \sqrt{2})}{3}$$
$$= 3 + \sqrt{2}$$

69. $\dfrac{7 - \sqrt{98}}{14} = \dfrac{7 - \sqrt{49 \cdot 2}}{14} = \dfrac{7 - \sqrt{49} \cdot \sqrt{2}}{14}$
$$= \dfrac{7 - 7\sqrt{2}}{14} = \dfrac{7(1 - \sqrt{2})}{7 \cdot 2}$$
$$= \dfrac{1 - \sqrt{2}}{2}$$

71. $\sqrt{5} \cdot \sqrt{5} = \sqrt{5 \cdot 5} = \sqrt{25} = 5$

73. $\sqrt{2} \cdot \sqrt{8} = \sqrt{2 \cdot 8} = \sqrt{16} = 4$

75. $\sqrt[3]{4} \cdot \sqrt[3]{2} = \sqrt[3]{4 \cdot 2} = \sqrt[3]{8} = 2$

77. $\sqrt{5x} \cdot \sqrt{15x} = \sqrt{5x \cdot 15x}$
$$= \sqrt{75x^2}$$
$$= \sqrt{25x^2 \cdot 3}$$
$$= \sqrt{25x^2} \cdot \sqrt{3}$$
$$= 5x\sqrt{3}$$

79. $\sqrt[3]{4b^2} \cdot \sqrt[3]{6b^2} = \sqrt[3]{4b^2 \cdot 6b^2}$
$$= \sqrt[3]{24b^4}$$
$$= \sqrt[3]{8b^3 \cdot 3b}$$
$$= \sqrt[3]{8b^3} \cdot \sqrt[3]{3b}$$
$$= 2b\sqrt[3]{3b}$$

81. $2\sqrt{6ab} \cdot 3\sqrt{15ab^3} = 6\sqrt{6ab \cdot 15ab^3}$
$$= 6\sqrt{90a^2b^4}$$
$$= 6\sqrt{9a^2b^4 \cdot 10}$$
$$= 6\sqrt{9a^2b^4} \cdot \sqrt{10}$$
$$= 18ab^2\sqrt{10}$$

83. $\sqrt[4]{27p^3q^2} \cdot \sqrt[4]{12p^2q^2} = \sqrt[4]{27p^3q^2 \cdot 12p^2q^2}$
$$= \sqrt[4]{324p^5q^4}$$
$$= \sqrt[4]{81p^4q^4 \cdot 4p}$$
$$= \sqrt[4]{81p^4q^4} \cdot \sqrt[4]{4p}$$
$$= 3pq\sqrt[4]{4p}$$

85. $\sqrt[5]{-8a^3b^4} \cdot \sqrt[5]{12a^3b} = \sqrt[5]{-8a^3b^4 \cdot 12a^3b}$
$$= \sqrt[5]{-96a^6b^5}$$
$$= \sqrt[5]{-32a^5b^5 \cdot 3a}$$
$$= \sqrt[5]{-32a^5b^5} \cdot \sqrt[5]{3a}$$
$$= -2ab\sqrt[5]{3a}$$

87. $\sqrt[4]{8(x-y)^2} \cdot \sqrt[4]{6(x-y)^3}$
$$= \sqrt[4]{8(x-y)^2 \cdot 6(x-y)^3}$$
$$= \sqrt[4]{48(x-y)^5}$$
$$= \sqrt[4]{16(x-y)^4 \cdot 3(x-y)}$$
$$= \sqrt[4]{16(x-y)^4} \cdot \sqrt[4]{3(x-y)}$$
$$= 2(x-y)\sqrt[4]{3(x-y)}$$

89. $\sqrt{\dfrac{3}{16}} = \dfrac{\sqrt{3}}{\sqrt{16}} = \dfrac{\sqrt{3}}{4}$

91. $\sqrt[4]{\dfrac{5x^4}{16}} = \dfrac{\sqrt[4]{5x^4}}{\sqrt[4]{16}} = \dfrac{x\sqrt[4]{5}}{2}$

93. $\sqrt{\dfrac{9y^2}{25x^2}} = \dfrac{\sqrt{9y^2}}{\sqrt{25x^2}} = \dfrac{3y}{5x}$

95. $\sqrt[3]{\dfrac{-27x^9}{64y^{12}}} = \dfrac{\sqrt[3]{-27x^9}}{\sqrt[3]{64y^{12}}} = \dfrac{-3x^3}{4y^4} = -\dfrac{3x^3}{4y^4}$

97. $\dfrac{\sqrt{8}}{\sqrt{2}} = \sqrt{\dfrac{8}{2}} = \sqrt{4} = 2$

99. $\dfrac{\sqrt[3]{128}}{\sqrt[3]{2}} = \sqrt[3]{\dfrac{128}{2}} = \sqrt[3]{64} = 4$

101. $\dfrac{\sqrt{48a^3}}{\sqrt{6a}} = \sqrt{\dfrac{48a^3}{6a}} = \sqrt{8a^2} = 2a\sqrt{2}$

103. $\dfrac{\sqrt{24a^5b}}{\sqrt{3ab^3}} = \sqrt{\dfrac{24a^5b}{3ab^3}} = \sqrt{\dfrac{8a^4}{b^2}} = \dfrac{\sqrt{8a^4}}{\sqrt{b^2}} = \dfrac{2a^2\sqrt{2}}{b}$

105. $\dfrac{\sqrt{512a^7b}}{3\sqrt{2ab^3}} = \dfrac{1}{3}\sqrt{\dfrac{512a^7b}{2ab^3}} = \dfrac{1}{3}\sqrt{\dfrac{256a^6}{b^2}} = \dfrac{16a^3}{3b}$

107. $\dfrac{\sqrt[3]{104a^5}}{\sqrt[3]{4a^{-1}}} = \sqrt[3]{\dfrac{104a^5}{4a^{-1}}} = \sqrt[3]{26a^6} = a^2\sqrt[3]{26}$

109. $\dfrac{\sqrt{90x^3y^{-1}}}{\sqrt{2x^{-3}y}} = \sqrt{\dfrac{90x^3y^{-1}}{2x^{-3}y}}$

$= \sqrt{\dfrac{45x^6}{y^2}}$

$= \dfrac{\sqrt{45x^6}}{\sqrt{y^2}}$

$= \dfrac{3x^3\sqrt{5}}{y}$

111. $\sqrt{3} \cdot \sqrt[3]{4} = 3^{1/2} \cdot 4^{1/3}$
$= 3^{3/6} \cdot 4^{2/6}$
$= \left(3^3\right)^{1/6} \cdot \left(4^2\right)^{1/6}$
$= \left(3^3 \cdot 4^2\right)^{1/6}$
$= (432)^{1/6}$
$= \sqrt[6]{432}$

113. $\sqrt[3]{2} \cdot \sqrt[6]{3} = 2^{1/3} \cdot 3^{1/6}$
$= 2^{2/6} \cdot 3^{1/6}$
$= \left(2^2\right)^{1/6} \cdot (3)^{1/6}$
$= \left(2^2 \cdot 3\right)^{1/6}$
$= (12)^{1/6}$
$= \sqrt[6]{12}$

115. $\sqrt{3} \cdot \sqrt[3]{18} = 3^{1/2} \cdot 18^{1/3}$
$= 3^{3/6} \cdot 18^{2/6}$
$= \left(3^3\right)^{1/6} \cdot \left(18^2\right)^{1/6}$
$= \left(3^3 \cdot 18^2\right)^{1/6}$
$= \left(3^3 \cdot (2 \cdot 3 \cdot 3)^2\right)^{1/6}$
$= \left(3^3 \cdot 2^2 \cdot 3^2 \cdot 3^2\right)^{1/6}$
$= \left(3^7 \cdot 2^2\right)^{1/6}$
$= 3\left(3 \cdot 2^2\right)^{1/6}$
$= 3(12)^{1/6}$
$= 3\sqrt[6]{12}$

117. $\sqrt[4]{9} \cdot \sqrt[6]{12} = 9^{1/4} \cdot 12^{1/6}$

$= \left(3^2\right)^{1/4} \cdot 12^{1/6}$

$= 3^{1/2} \cdot 12^{1/6}$

$= 3^{3/6} \cdot 12^{1/6}$

$= \left(3^3\right)^{1/6} \cdot (12)^{1/6}$

$= \left(3^3 \cdot 12\right)^{1/6}$

$= \left(3^3 \cdot 3 \cdot 2^2\right)^{1/6}$

$= \left(3^4 \cdot 2^2\right)^{1/6}$

$= \left(\left(3^2 \cdot 2\right)^2\right)^{1/6}$

$= \left(3^2 \cdot 2\right)^{1/3}$

$= \sqrt[3]{18}$

119. $\sqrt[3]{\dfrac{5x}{8}} = \dfrac{\sqrt[3]{5x}}{\sqrt[3]{8}} = \dfrac{\sqrt[3]{5x}}{2}$

121. $\sqrt[3]{5a} \cdot \sqrt[3]{9a} = \sqrt[3]{5a \cdot 9a} = \sqrt[3]{45a^2}$

123. $\sqrt{72a^4} = \sqrt{36a^4 \cdot 2} = \sqrt{36a^4} \cdot \sqrt{2} = 6a^2\sqrt{2}$

125. $\sqrt[3]{6a^2b} \cdot \sqrt[3]{9ab} = \sqrt[3]{6a^2b \cdot 9ab}$

$= \sqrt[3]{54a^3b^2}$

$= \sqrt[3]{27a^3 \cdot 2b^2}$

$= \sqrt[3]{27a^3} \cdot \sqrt[3]{2b^2}$

$= 3a\sqrt[3]{2b^2}$

127. $\dfrac{\sqrt[3]{-32a}}{\sqrt[3]{2a^4}} = \sqrt[3]{\dfrac{-32a}{2a^4}}$

$= \sqrt[3]{\dfrac{-16}{a^3}}$

$= \dfrac{\sqrt[3]{-16}}{\sqrt[3]{a^3}}$

$= \dfrac{-2\sqrt[3]{2}}{a}$

$= -\dfrac{2\sqrt[3]{2}}{a}$

129. $-5\sqrt[3]{32m^3} = -5\sqrt[3]{8m^3 \cdot 4}$

$= -5 \cdot \sqrt[3]{8m^3} \cdot \sqrt[3]{4}$

$= -5 \cdot 2m \cdot \sqrt[3]{4}$

$= -10m\sqrt[3]{4}$

131. $\sqrt[3]{81a^4b^7} = \sqrt[3]{27a^3b^6 \cdot 3ab}$

$= \sqrt[3]{27a^3b^6} \cdot \sqrt[3]{3ab}$

$= 3ab^2\sqrt[3]{3ab}$

133. $\sqrt[3]{12} \cdot \sqrt[3]{18} = \sqrt[3]{12 \cdot 18}$

$= \sqrt[3]{216}$

$= 6$

135. a.

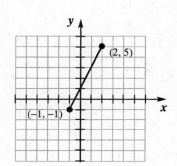

b. $\sqrt{\left(5-(-1)\right)^2 + \left(2-(-1)\right)^2}$

$= \sqrt{\left(5+1\right)^2 + \left(2+1\right)^2}$

$= \sqrt{6^2 + 3^2}$

$= \sqrt{36+9}$

$= \sqrt{45}$

$= 3\sqrt{5}$

The line segment has a length of $3\sqrt{5}$ units.

137. a. $R(8) = \sqrt[3]{\dfrac{8}{2}} = \sqrt[3]{4} \approx 1.587$

The company's annual revenue will be roughly $1,587,000 after 8 years.

b. $R(27) = \sqrt[3]{\dfrac{27}{2}} = \dfrac{\sqrt[3]{27}}{\sqrt[3]{2}} = \dfrac{3}{\sqrt[3]{2}} \approx 2.381$

The company's annual revenue will be roughly $2,381,000 after 27 years.

     489

139. a.
$$(f \cdot g)(x) = f(x) \cdot g(x)$$
$$= \sqrt{2x} \cdot \sqrt{8x^3}$$
$$= \sqrt{2x \cdot 8x^3}$$
$$= \sqrt{16x^4}$$
$$= 4x^2$$

b. Using the result from part (a), we get
$$(f \cdot g)(3) = 4(3)^2 = 36$$

141. $a = 1, b = 6, c = 3$

$$x = \frac{-6 \pm \sqrt{6^2 - 4(1)(3)}}{2(1)}$$
$$= \frac{-6 \pm \sqrt{36 - 12}}{2}$$
$$= \frac{-6 \pm \sqrt{24}}{2}$$
$$= \frac{-6 \pm 2\sqrt{6}}{2}$$
$$= -3 \pm \sqrt{6}$$
$$x = -3 - \sqrt{6} \quad \text{or} \quad x = -3 + \sqrt{6}$$

143. $a = 3, b = 4, c = -1$

$$x = \frac{-4 \pm \sqrt{4^2 - 4(3)(-1)}}{2(3)}$$
$$= \frac{-4 \pm \sqrt{16 + 12}}{6}$$
$$= \frac{-4 \pm \sqrt{28}}{6}$$
$$= \frac{-4 \pm 2\sqrt{7}}{6}$$
$$= \frac{-2 \pm \sqrt{7}}{3}$$
$$x = \frac{-2 - \sqrt{7}}{3} \quad \text{or} \quad x = \frac{-2 + \sqrt{7}}{3}$$

145. The indexes must be the same.

147. $4x + 3 = 13$
$$4x = 10$$
$$x = \frac{10}{4}$$
$$x = \frac{5}{2}$$

The solution set is $\left\{ \frac{5}{2} \right\}$.

149. $\frac{3}{5}x + 2 \le 28$
$$\frac{3}{5}x \le 26$$
$$x \le \frac{130}{3}$$

The solution set is $\left\{ x \mid x \le \frac{130}{3} \right\}$, or in interval notation we would write $\left(-\infty, \frac{130}{3} \right]$.

151. Answers will vary.

Section 6.4

Are You Ready for This Section?

R1. $4y^3 - 2y^2 + 8y - 1 + \left(-2y^3 + 7y^2 - 3y + 9 \right)$
$$= 4y^3 - 2y^3 - 2y^2 + 7y^2 + 8y - 3y - 1 + 9$$
$$= 2y^3 + 5y^2 + 5y + 8$$

R2. $5z^2 + 6 - \left(3z^2 - 8z - 3 \right)$
$$= 5z^2 + 6 - 3z^2 + 8z + 3$$
$$= 5z^2 - 3z^2 + 8z + 6 + 3$$
$$= 2z^2 + 8z + 9$$

R3. $(4x + 3)(x - 5) = 4x \cdot x - 4x \cdot 5 + 3 \cdot x - 3 \cdot 5$
$$= 4x^2 - 20x + 3x - 15$$
$$= 4x^2 - 17x - 15$$

R4. $(2y - 3)(2y + 3) = (2y)^2 - 3^2$
$$= 4y^2 - 9$$

Section 6.4 Quick Checks

1. Two radicals are <u>like radicals</u> if each radical has the same index and the same radicand.

2. False; $\sqrt{7} + \sqrt{5} \approx 4.9$ but $\sqrt{7 + 5} = \sqrt{12} \approx 3.5$.

3. $9\sqrt{13y} + 4\sqrt{13y} = (9 + 4)\sqrt{13y} = 13\sqrt{13y}$

4. $\sqrt[4]{5} + 9\sqrt[4]{5} - 3\sqrt[4]{5} = (1 + 9 - 3)\sqrt[4]{5} = 7\sqrt[4]{5}$

5. $4\sqrt{18} - 3\sqrt{8} = 4 \cdot \sqrt{9} \cdot \sqrt{2} - 3 \cdot \sqrt{4} \cdot \sqrt{2}$
$= 4 \cdot 3\sqrt{2} - 3 \cdot 2\sqrt{2}$
$= 12\sqrt{2} - 6\sqrt{2}$
$= (12-6)\sqrt{2}$
$= 6\sqrt{2}$

6. $-5x\sqrt[3]{54x} + 7\sqrt[3]{2x^4}$
$= -5x \cdot \sqrt[3]{27} \cdot \sqrt[3]{2x} + 7 \cdot \sqrt[3]{x^3} \cdot \sqrt[3]{2x}$
$= -5x \cdot 3\sqrt[3]{2x} + 7x \cdot \sqrt[3]{2x}$
$= -15x\sqrt[3]{2x} + 7x\sqrt[3]{2x}$
$= (-15+7)x\sqrt[3]{2x}$
$= -8x\sqrt[3]{2x}$

7. $7\sqrt{10} - 6\sqrt{3}$ cannot be simplified further.

8. $\sqrt[3]{8z^4} - 2z\sqrt[3]{-27z} + \sqrt[3]{125z}$
$= \sqrt[3]{8z^3} \cdot \sqrt[3]{z} - 2z \cdot \sqrt[3]{-27} \cdot \sqrt[3]{z} + \sqrt[3]{125} \cdot \sqrt[3]{z}$
$= 2z\sqrt[3]{z} - 2z(-3)\sqrt[3]{z} + 5\sqrt[3]{z}$
$= 2z\sqrt[3]{z} + 6z\sqrt[3]{z} + 5\sqrt[3]{z}$
$= (2z+6z+5)\sqrt[3]{z}$
$= (8z+5)\sqrt[3]{z}$

9. $\sqrt{25m} - 3\sqrt[4]{m^2} = \sqrt{25m} - 3m^{2/4}$
$= \sqrt{25m} - 3m^{1/2}$
$= \sqrt{25m} - 3\sqrt{m}$
$= \sqrt{25} \cdot \sqrt{m} - 3\sqrt{m}$
$= 5\sqrt{m} - 3\sqrt{m}$
$= (5-3)\sqrt{m}$
$= 2\sqrt{m}$

10. $\sqrt{6}(3 - 5\sqrt{6}) = \sqrt{6} \cdot 3 - 5\sqrt{6} \cdot \sqrt{6}$
$= 3\sqrt{6} - 5\sqrt{36}$
$= 3\sqrt{6} - 5 \cdot 6$
$= 3\sqrt{6} - 30$
$= 3(\sqrt{6} - 10)$

11. $\sqrt[3]{12}(3 - \sqrt[3]{2}) = \sqrt[3]{12} \cdot 3 - \sqrt[3]{12} \cdot \sqrt[3]{2}$
$= 3\sqrt[3]{12} - \sqrt[3]{24}$
$= 3\sqrt[3]{12} - 2\sqrt[3]{3}$

12. $(2 - 7\sqrt{3})(5 + 4\sqrt{3})$
$= 2 \cdot 5 + 2 \cdot 4\sqrt{3} - 7\sqrt{3} \cdot 5 - 7\sqrt{3} \cdot 4\sqrt{3}$
$= 10 + 8\sqrt{3} - 35\sqrt{3} - 28\sqrt{9}$
$= 10 - 27\sqrt{3} - 28 \cdot 3$
$= 10 - 27\sqrt{3} - 84$
$= -74 - 27\sqrt{3}$

13. False; $\left(\sqrt{a} + \sqrt{b}\right)^2 = \left(\sqrt{a}\right)^2 + 2\sqrt{a}\sqrt{b} + \left(\sqrt{b}\right)^2$
$= a + 2\sqrt{ab} + b$

14. The radical expressions $4 + \sqrt{5}$ and $4 - \sqrt{5}$ are examples of <u>conjugates</u>.

15. False. The conjugate of $-5 + \sqrt{2}$ is $-5 - \sqrt{2}$.

16. $(2 - \sqrt{5})^2 = (2)^2 - 2(2)\left(\sqrt{5}\right) + \left(\sqrt{5}\right)^2$
$= 4 - 4\sqrt{5} + 5$
$= 9 - 4\sqrt{5}$

17. $\left(\sqrt{7} - 3\sqrt{2}\right)^2 = \left(\sqrt{7} - 3\sqrt{2}\right)\left(\sqrt{7} - 3\sqrt{2}\right)$
$= \left(\sqrt{7}\right)^2 - 2\left(\sqrt{7}\right)\left(3\sqrt{2}\right) + \left(3\sqrt{2}\right)^2$
$= \sqrt{49} - 6\sqrt{14} + 9\sqrt{4}$
$= 7 - 6\sqrt{14} + 9 \cdot 2$
$= 7 - 6\sqrt{14} + 18$
$= 25 - 6\sqrt{14}$

18. $\left(\sqrt{3} + \sqrt{2}\right)\left(\sqrt{3} - \sqrt{2}\right) = \left(\sqrt{3}\right)^2 - \left(\sqrt{2}\right)^2$
$= \sqrt{9} - \sqrt{4}$
$= 3 - 2$
$= 1$

6.4 Exercises

19. $3\sqrt{2} + 7\sqrt{2} = (3+7)\sqrt{2} = 10\sqrt{2}$

21. $5\sqrt[3]{x} - 3\sqrt[3]{x} = (5-3)\sqrt[3]{x} = 2\sqrt[3]{x}$

23. $8\sqrt{5x} - 3\sqrt{5x} + 9\sqrt{5x} = (8-3+9)\sqrt{5x}$
$= 14\sqrt{5x}$

25. $4\sqrt[3]{5} - 3\sqrt{5} + 7\sqrt[3]{5} - 8\sqrt{5}$
$= (4+7)\sqrt[3]{5} + (-3-8)\sqrt{5}$
$= 11\sqrt[3]{5} - 11\sqrt{5}$

27.
$$\sqrt{8} + 6\sqrt{2} = \sqrt{4} \cdot \sqrt{2} + 6\sqrt{2}$$
$$= 2\sqrt{2} + 6\sqrt{2}$$
$$= (2+6)\sqrt{2}$$
$$= 8\sqrt{2}$$

29.
$$\sqrt[3]{24} - 4\sqrt[3]{3} = \sqrt[3]{8} \cdot \sqrt[3]{3} - 4\sqrt[3]{3}$$
$$= 2\sqrt[3]{3} - 4\sqrt[3]{3}$$
$$= (2-4)\sqrt[3]{3}$$
$$= -2\sqrt[3]{3}$$

31.
$$\sqrt[3]{54} - 7\sqrt[3]{128} = \sqrt[3]{27} \cdot \sqrt[3]{2} - 7 \cdot \sqrt[3]{64} \cdot \sqrt[3]{2}$$
$$= 3\sqrt[3]{2} - 7 \cdot 4 \cdot \sqrt[3]{2}$$
$$= 3\sqrt[3]{2} - 28\sqrt[3]{2}$$
$$= (3-28)\sqrt[3]{2}$$
$$= -25\sqrt[3]{2}$$

33.
$$5\sqrt{54x} - 3\sqrt{24x}$$
$$= 5 \cdot \sqrt{9} \cdot \sqrt{6x} - 3 \cdot \sqrt{4} \cdot \sqrt{6x}$$
$$= 5 \cdot 3 \cdot \sqrt{6x} - 3 \cdot 2 \cdot \sqrt{6x}$$
$$= 15\sqrt{6x} - 6\sqrt{6x}$$
$$= (15-6)\sqrt{6x}$$
$$= 9\sqrt{6x}$$

35.
$$2\sqrt{8} + 3\sqrt{10} = 2\sqrt{4} \cdot \sqrt{2} + 3\sqrt{10}$$
$$= 2 \cdot 2\sqrt{2} + 3\sqrt{10}$$
$$= 4\sqrt{2} + 3\sqrt{10}$$
Cannot be simplified further.

37.
$$\sqrt{12x^3} + 5x\sqrt{108x}$$
$$= \sqrt{4x^2} \cdot \sqrt{3x} + 5x \cdot \sqrt{36} \cdot \sqrt{3x}$$
$$= 2x\sqrt{3x} + 5x \cdot 6 \cdot \sqrt{3x}$$
$$= 2x\sqrt{3x} + 30x\sqrt{3x}$$
$$= (2x+30x)\sqrt{3x}$$
$$= 32x\sqrt{3x}$$

39.
$$\sqrt{12x^2} + 3x\sqrt{2} - 2\sqrt{98x^2}$$
$$= \sqrt{4x^2} \cdot \sqrt{3} + 3x\sqrt{2} - 2\sqrt{49x^2} \cdot \sqrt{2}$$
$$= 2x\sqrt{3} + 3x\sqrt{2} - 2 \cdot 7x\sqrt{2}$$
$$= 2x\sqrt{3} + 3x\sqrt{2} - 14x\sqrt{2}$$
$$= 2x\sqrt{3} + (3-14)x\sqrt{2}$$
$$= 2x\sqrt{3} - 11x\sqrt{2}$$

41.
$$\sqrt[3]{-54x^3} + 3x\sqrt[3]{16} - 2\sqrt[3]{128}$$
$$= \sqrt[3]{-27x^3} \cdot \sqrt[3]{2} + 3x\sqrt[3]{8} \cdot \sqrt[3]{2} - 2\sqrt[3]{64} \cdot \sqrt[3]{2}$$
$$= -3x\sqrt[3]{2} + 3x \cdot 2\sqrt[3]{2} - 2 \cdot 4\sqrt[3]{2}$$
$$= -3x\sqrt[3]{2} + 6x\sqrt[3]{2} - 8\sqrt[3]{2}$$
$$= (-3x+6x-8)\sqrt[3]{2}$$
$$= (3x-8)\sqrt[3]{2}$$

43.
$$\sqrt{9x-9} + \sqrt{4x-4}$$
$$= \sqrt{9(x-1)} + \sqrt{4(x-1)}$$
$$= \sqrt{9} \cdot \sqrt{x-1} + \sqrt{4} \cdot \sqrt{x-1}$$
$$= 3\sqrt{x-1} + 2\sqrt{x-1}$$
$$= (3+2)\sqrt{x-1}$$
$$= 5\sqrt{x-1}$$

45.
$$\sqrt{16x} - \sqrt[6]{x^3} = \sqrt{16x} - x^{3/6}$$
$$= \sqrt{16} \cdot \sqrt{x} - x^{1/2}$$
$$= 4\sqrt{x} - \sqrt{x}$$
$$= (4-1)\sqrt{x}$$
$$= 3\sqrt{x}$$

47.
$$\sqrt[3]{27x} + 2\sqrt[9]{x^3} = \sqrt[3]{27} \cdot \sqrt[3]{x} + 2 \cdot x^{3/9}$$
$$= 3\sqrt[3]{x} + 2x^{1/3}$$
$$= 3\sqrt[3]{x} + 2\sqrt[3]{x}$$
$$= (3+2)\sqrt[3]{x}$$
$$= 5\sqrt[3]{x}$$

49.
$$\sqrt{3}\left(2 - 3\sqrt{2}\right) = \sqrt{3} \cdot 2 - \sqrt{3} \cdot 3\sqrt{2}$$
$$= 2\sqrt{3} - 3\sqrt{2 \cdot 3}$$
$$= 2\sqrt{3} - 3\sqrt{6}$$

51.
$$\sqrt{3}\left(\sqrt{2} + \sqrt{6}\right) = \sqrt{3} \cdot \sqrt{2} + \sqrt{3} \cdot \sqrt{6}$$
$$= \sqrt{3 \cdot 2} + \sqrt{3 \cdot 6}$$
$$= \sqrt{6} + \sqrt{18}$$
$$= \sqrt{6} + \sqrt{9} \cdot \sqrt{2}$$
$$= \sqrt{6} + 3\sqrt{2}$$

53.
$$\sqrt[3]{4}\left(\sqrt[3]{3} - \sqrt[3]{6}\right) = \sqrt[3]{4} \cdot \sqrt[3]{3} - \sqrt[3]{4} \cdot \sqrt[3]{6}$$
$$= \sqrt[3]{4 \cdot 3} - \sqrt[3]{4 \cdot 6}$$
$$= \sqrt[3]{12} - \sqrt[3]{24}$$
$$= \sqrt[3]{12} - \sqrt[3]{8} \cdot \sqrt[3]{3}$$
$$= \sqrt[3]{12} - 2\sqrt[3]{3}$$

55. $\sqrt{2x}\left(3-\sqrt{10x}\right)=\sqrt{2x}\cdot 3-\sqrt{2x}\cdot\sqrt{10x}$
$$=3\sqrt{2x}-\sqrt{20x^2}$$
$$=3\sqrt{2x}-\sqrt{4x^2}\cdot\sqrt{5}$$
$$=3\sqrt{2x}-2x\sqrt{5}$$

57. $\left(3+\sqrt{2}\right)\left(4+\sqrt{3}\right)$
$$=3\cdot 4+3\cdot\sqrt{3}+\sqrt{2}\cdot 4+\sqrt{2}\cdot\sqrt{3}$$
$$=12+3\sqrt{3}+4\sqrt{2}+\sqrt{6}$$

59. $\left(6+\sqrt{3}\right)\left(2-\sqrt{7}\right)$
$$=6\cdot 2-6\cdot\sqrt{7}+\sqrt{3}\cdot 2-\sqrt{3}\cdot\sqrt{7}$$
$$=12-6\sqrt{7}+2\sqrt{3}-\sqrt{21}$$

61. $\left(4-2\sqrt{7}\right)\left(3+3\sqrt{7}\right)$
$$=4\cdot 3+4\cdot 3\sqrt{7}-2\sqrt{7}\cdot 3-2\sqrt{7}\cdot 3\sqrt{7}$$
$$=12+12\sqrt{7}-6\sqrt{7}-6\sqrt{49}$$
$$=12+(12-6)\sqrt{7}-6\cdot 7$$
$$=12+6\sqrt{7}-42$$
$$=6\sqrt{7}-30$$

63. $\left(\sqrt{2}+3\sqrt{6}\right)\left(\sqrt{3}-2\sqrt{2}\right)$
$$=\sqrt{2}\cdot\sqrt{3}-\sqrt{2}\cdot 2\sqrt{2}+3\sqrt{6}\cdot\sqrt{3}-3\sqrt{6}\cdot 2\sqrt{2}$$
$$=\sqrt{6}-2\sqrt{4}+3\sqrt{18}-6\sqrt{12}$$
$$=\sqrt{6}-2\cdot 2+3\cdot 3\sqrt{2}-6\cdot 2\sqrt{3}$$
$$=\sqrt{6}-4+9\sqrt{2}-12\sqrt{3}$$
$$=\sqrt{6}-12\sqrt{3}+9\sqrt{2}-4$$

65. $\left(2\sqrt{5}+\sqrt{3}\right)\left(4\sqrt{5}-3\sqrt{3}\right)$
$$=2\sqrt{5}\cdot 4\sqrt{5}-2\sqrt{5}\cdot 3\sqrt{3}+\sqrt{3}\cdot 4\sqrt{5}-\sqrt{3}\cdot 3\sqrt{3}$$
$$=8\sqrt{25}-6\sqrt{15}+4\sqrt{15}-3\sqrt{9}$$
$$=8\cdot 5+(-6+4)\sqrt{15}-3\cdot 3$$
$$=40-2\sqrt{15}-9$$
$$=31-2\sqrt{15}$$

67. $\left(1+\sqrt{3}\right)^2=(1)^2+2(1)\left(\sqrt{3}\right)+\left(\sqrt{3}\right)^2$
$$=1+2\sqrt{3}+\sqrt{9}$$
$$=1+2\sqrt{3}+3$$
$$=4+2\sqrt{3}$$

69. $\left(\sqrt{2}-\sqrt{5}\right)^2=\left(\sqrt{2}\right)^2-2\left(\sqrt{2}\right)\left(\sqrt{5}\right)+\left(\sqrt{5}\right)^2$
$$=\sqrt{4}-2\sqrt{10}+\sqrt{25}$$
$$=2-2\sqrt{10}+5$$
$$=7-2\sqrt{10}$$

71. $\left(\sqrt{x}-\sqrt{2}\right)^2=\left(\sqrt{x}\right)^2-2\left(\sqrt{x}\right)\left(\sqrt{2}\right)+\left(\sqrt{2}\right)^2$
$$=\sqrt{x^2}-2\sqrt{2x}+\sqrt{4}$$
$$=x-2\sqrt{2x}+2$$

73. $\left(\sqrt{2}-1\right)\left(\sqrt{2}+1\right)=\left(\sqrt{2}\right)^2-(1)^2$
$$=\sqrt{4}-1$$
$$=2-1$$
$$=1$$

75. $\left(3-2\sqrt{5}\right)\left(3+2\sqrt{5}\right)=(3)^2-\left(2\sqrt{5}\right)^2$
$$=9-4\sqrt{25}$$
$$=9-4\cdot 5$$
$$=9-20$$
$$=-11$$

77. $\left(\sqrt{2x}+\sqrt{3y}\right)\left(\sqrt{2x}-\sqrt{3y}\right)$
$$=\left(\sqrt{2x}\right)^2-\left(\sqrt{3y}\right)^2$$
$$=\sqrt{4x^2}-\sqrt{9y^2}$$
$$=2x-3y$$

79. $\left(\sqrt[3]{x}+4\right)\left(\sqrt[3]{x}-3\right)$
$$=\sqrt[3]{x}\cdot\sqrt[3]{x}-\sqrt[3]{x}\cdot 3+4\cdot\sqrt[3]{x}-4\cdot 3$$
$$=\sqrt[3]{x^2}-3\sqrt[3]{x}+4\sqrt[3]{x}-12$$
$$=\sqrt[3]{x^2}+(-3+4)\sqrt[3]{x}-12$$
$$=\sqrt[3]{x^2}+\sqrt[3]{x}-12$$

81. $\left(\sqrt[3]{2a}-5\right)\left(\sqrt[3]{2a}+5\right)=\left(\sqrt[3]{2a}\right)^2-(5)^2$
$$=\sqrt[3]{4a^2}-25$$

83. $\sqrt{5}\left(\sqrt{3}+\sqrt{10}\right)=\sqrt{5}\cdot\sqrt{3}+\sqrt{5}\cdot\sqrt{10}$
$$=\sqrt{15}+\sqrt{50}$$
$$=\sqrt{15}+\sqrt{25}\cdot\sqrt{2}$$
$$=\sqrt{15}+5\sqrt{2}$$

85. $\sqrt{28x^5} - x\sqrt{7x^3} + 5\sqrt{175x^5}$

$= \sqrt{4x^4} \cdot \sqrt{7x} - x\sqrt{x^2} \cdot \sqrt{7x} + 5\sqrt{25x^4} \cdot \sqrt{7x}$

$= 2x^2\sqrt{7x} - x^2\sqrt{7x} + 25x^2\sqrt{7x}$

$= \left(2x^2 - x^2 + 25x^2\right)\sqrt{7x}$

$= 26x^2\sqrt{7x}$

87. $\left(2\sqrt{3} + 5\right)\left(2\sqrt{3} - 5\right) = \left(2\sqrt{3}\right)^2 - 5^2$

$= 4\sqrt{9} - 25$

$= 4 \cdot 3 - 25$

$= 12 - 25$

$= -13$

89. $\sqrt[3]{7}\left(2 + \sqrt[3]{4}\right) = \sqrt[3]{7} \cdot 2 + \sqrt[3]{7} \cdot \sqrt[3]{4}$

$= 2\sqrt[3]{7} + \sqrt[3]{7 \cdot 4}$

$= 2\sqrt[3]{7} + \sqrt[3]{28}$

91. $\left(2\sqrt{2} + 5\right)\left(4\sqrt{2} - 4\right)$

$= 2\sqrt{2} \cdot 4\sqrt{2} - 2\sqrt{2} \cdot 4 + 5 \cdot 4\sqrt{2} - 5 \cdot 4$

$= 8\sqrt{4} - 8\sqrt{2} + 20\sqrt{2} - 20$

$= 8 \cdot 2 + (-8 + 20)\sqrt{2} - 20$

$= 16 + 12\sqrt{2} - 20$

$= -4 + 12\sqrt{2}$

93. $4\sqrt{18} + 2\sqrt{32} = 4 \cdot \sqrt{9} \cdot \sqrt{2} + 2 \cdot \sqrt{16} \cdot \sqrt{2}$

$= 4 \cdot 3 \cdot \sqrt{2} + 2 \cdot 4 \cdot \sqrt{2}$

$= 12\sqrt{2} + 8\sqrt{2}$

$= (12 + 8)\sqrt{2}$

$= 20\sqrt{2}$

95. $\left(\sqrt{5} - \sqrt{3}\right)^2 - \sqrt{60}$

$= \left(\sqrt{5}\right)^2 - 2\left(\sqrt{5}\right)\left(\sqrt{3}\right) + \left(\sqrt{3}\right)^2 - \sqrt{4 \cdot 15}$

$= \sqrt{25} - 2\sqrt{15} + \sqrt{9} - \sqrt{4}\sqrt{15}$

$= 5 - 2\sqrt{15} + 3 - 2\sqrt{15}$

$= 8 - 4\sqrt{15}$

97. $3\sqrt[3]{5x^3 y} + \sqrt[3]{40y}$

$= 3 \cdot \sqrt[3]{x^3} \cdot \sqrt[3]{5y} + \sqrt[3]{8} \cdot \sqrt[3]{5y}$

$= 3x \cdot \sqrt[3]{5y} + 2 \cdot \sqrt[3]{5y}$

$= (3x + 2)\sqrt[3]{5y}$

99. $\left(\sqrt{2x} - \sqrt{7y}\right)\left(\sqrt{2x} + \sqrt{7y}\right) - 2\sqrt{x^2}$

$= \left(\sqrt{2x}\right)^2 - \left(\sqrt{7y}\right)^2 - 2 \cdot x$

$= \sqrt{4x^2} - \sqrt{49y^2} - 2x$

$= 2x - 7y - 2x$

$= 7y$

101. $\left(2 + \sqrt{x+1}\right)^2 = 2^2 + 2(2)\left(\sqrt{x+1}\right) + \left(\sqrt{x+1}\right)^2$

$= 4 + 4\sqrt{x+1} + x + 1$

$= 5 + x + 4\sqrt{x+1}$

103. $\left(5 - \sqrt{x+2}\right)^2 = 5^2 - 2(5)\left(\sqrt{x+2}\right) + \left(\sqrt{x+2}\right)^2$

$= 25 - 10\sqrt{x+2} + x + 2$

$= 27 + x - 10\sqrt{x+2}$

105. $-\dfrac{3}{5} \cdot \left(-\dfrac{\sqrt{5}}{5}\right) - \dfrac{4}{5} \cdot \left(-\dfrac{2\sqrt{5}}{5}\right)$

$= \dfrac{3\sqrt{5}}{25} + \dfrac{8\sqrt{5}}{25}$

$= \dfrac{3\sqrt{5} + 8\sqrt{5}}{25}$

$= \dfrac{(3+8)\sqrt{5}}{25}$

$= \dfrac{11\sqrt{5}}{25}$

107. a. $(f+g)(x) = \sqrt{3x} + \sqrt{12x}$

$= \sqrt{3x} + \sqrt{4} \cdot \sqrt{3x}$

$= \sqrt{3x} + 2\sqrt{3x}$

$= (1+2)\sqrt{3x}$

$= 3\sqrt{3x}$

b. Use the result from part (a).

$(f+g)(4) = 3\sqrt{3(4)}$

$= 3\sqrt{12}$

$= 3\sqrt{4} \cdot \sqrt{3}$

$= 3 \cdot 2\sqrt{3}$

$= 6\sqrt{3}$

c. $(f \cdot g)(x) = \sqrt{3x} \cdot \sqrt{12x}$

$= \sqrt{36x^2}$

$= 6x$

109. Check $x = -2 + \sqrt{5}$:

$0 \stackrel{?}{=} x^2 + 4x - 1$

$0 \stackrel{?}{=} \left(-2+\sqrt{5}\right)^2 + 4\left(-2+\sqrt{5}\right) - 1$

$0 \stackrel{?}{=} (-2)^2 + 2(-2)\left(\sqrt{5}\right) + \left(\sqrt{5}\right)^2 - 8 + 4\sqrt{5} - 1$

$0 \stackrel{?}{=} 4 - 4\sqrt{5} + \sqrt{25} - 8 + 4\sqrt{5} - 1$

$0 \stackrel{?}{=} 9 - 9$

$0 = 0$ true

Substituting $-2 + \sqrt{5}$ for x yields a true statement, therefore the value is a solution.

Check $x = -2 - \sqrt{5}$:

$0 \stackrel{?}{=} x^2 + 4x - 1$

$0 \stackrel{?}{=} \left(-2-\sqrt{5}\right)^2 + 4\left(-2-\sqrt{5}\right) - 1$

$0 \stackrel{?}{=} (-2)^2 - 2(-2)\left(\sqrt{5}\right) + \left(\sqrt{5}\right)^2 - 8 - 4\sqrt{5} - 1$

$0 \stackrel{?}{=} 4 + 4\sqrt{5} + \sqrt{25} - 8 - 4\sqrt{5} - 1$

$0 \stackrel{?}{=} 9 - 9$

$0 = 0$ true

Substituting $-2 - \sqrt{5}$ for x yields a true statement, therefore the value is a solution.

111. $A = l \cdot w$

$= \sqrt{162} \cdot \sqrt{72}$

$= \sqrt{81} \cdot \sqrt{2} \cdot \sqrt{36} \cdot \sqrt{2}$

$= 9\sqrt{2} \cdot 6\sqrt{2}$

$= 54\sqrt{4}$

$= 54 \cdot 2$

$= 108$

The area is 108 square units.

$P = 2l + 2w$

$= 2\left(\sqrt{162}\right) + 2\left(\sqrt{72}\right)$

$= 2\sqrt{81} \cdot \sqrt{2} + 2\sqrt{36} \cdot \sqrt{2}$

$= 18\sqrt{2} + 12\sqrt{2}$

$= (18 + 12)\sqrt{2}$

$= 30\sqrt{2}$

The perimeter is $30\sqrt{2}$ units.

113. Area of larger triangle:

$s = \frac{1}{2}(14+10+8) = \frac{1}{2}(32) = 16$

$A = \sqrt{16(16-14)(16-10)(16-8)}$

$= \sqrt{16(2)(6)(8)} = \sqrt{16 \cdot 16 \cdot 6}$

$= \sqrt{16} \cdot \sqrt{16} \cdot \sqrt{6} = 4 \cdot 4 \cdot \sqrt{6}$

$= 16\sqrt{6}$

Area of smaller triangle:

$s = \frac{1}{2}(7+5+4) = \frac{1}{2}(16) = 8$

$A = \sqrt{8(8-7)(8-5)(8-4)}$

$= \sqrt{8(1)(3)(4)} = \sqrt{96} = \sqrt{16} \cdot \sqrt{6}$

$= 4\sqrt{6}$

Area of shaded region:

area of larger $-$ area of smaller

$= 16\sqrt{6} - 4\sqrt{6} = (16-4)\sqrt{6} = 12\sqrt{6}$

The area of the shaded region is $12\sqrt{6}$ square units.

115. Answers may vary.

117. $\left(3a^3b\right)\left(4a^2b^4\right) = 3 \cdot 4a^{3+2}b^{1+4} = 12a^5b^5$

119. $(3y+2)(2y-1) = 3y(2y-1) + 2(2y-1)$

$= 6y^2 - 3y + 4y - 2$

$= 6y^2 + y - 2$

121. $(5w+2)(5w-2) = (5w)^2 - 2^2 = 25w^2 - 4$

Section 6.5

Are You Ready for This Section?

R1. Start by finding the prime factors of 12.

$12 = 3 \cdot 4 = 3 \cdot 2 \cdot 2$

To make a perfect square, we need to have each unique prime factor occur an even number of times. To make the smallest perfect square, we want to use the smallest even number possible for each factor. In this case, the factor 2 occurs twice so no additional factors of 2 are needed. However, the factor 3 only occurs once. Therefore, we need to multiply by one more factor of 3.

The smallest perfect square that is a multiple of 12 is $12 \cdot 3 = 36$.

R2. $\sqrt{25x^2} = \sqrt{(5x)^2} = 5x$

(since $x > 0$, we do not need $|x|$)

Section 6.5 Quick Checks

1. Rewriting a quotient to remove radicals from the denominator is called <u>rationalizing the denominator</u>.

2. To rationalize the denominator of $\dfrac{\sqrt{5}}{\sqrt{11}}$, we would multiply the numerator and denominator by $\underline{\sqrt{11}}$.

3. $\dfrac{1}{\sqrt{3}}=\dfrac{1}{\sqrt{3}}\cdot\dfrac{\sqrt{3}}{\sqrt{3}}=\dfrac{\sqrt{3}}{\sqrt{9}}=\dfrac{\sqrt{3}}{3}$

4. $\dfrac{\sqrt{5}}{\sqrt{8}}=\dfrac{\sqrt{5}}{2\sqrt{2}}=\dfrac{\sqrt{5}}{2\sqrt{2}}\cdot\dfrac{\sqrt{2}}{\sqrt{2}}=\dfrac{\sqrt{10}}{2\sqrt{4}}=\dfrac{\sqrt{10}}{2\cdot2}=\dfrac{\sqrt{10}}{4}$

5. $\dfrac{5}{\sqrt{10x}}=\dfrac{5}{\sqrt{10x}}\cdot\dfrac{\sqrt{10x}}{\sqrt{10x}}$

$=\dfrac{5\sqrt{10x}}{\sqrt{100x^2}}$

$=\dfrac{5\sqrt{10x}}{10x}$

$=\dfrac{\sqrt{10x}}{2x}$

6. $\dfrac{4}{\sqrt[3]{3}}=\dfrac{4}{\sqrt[3]{3}}\cdot\dfrac{\sqrt[3]{3^2}}{\sqrt[3]{3^2}}=\dfrac{4\sqrt[3]{9}}{\sqrt[3]{27}}=\dfrac{4\sqrt[3]{9}}{3}$

7. $\sqrt[3]{\dfrac{3}{20}}=\dfrac{\sqrt[3]{3}}{\sqrt[3]{20}}=\dfrac{\sqrt[3]{3}}{\sqrt[3]{2^2\cdot5}}\cdot\dfrac{\sqrt[3]{2\cdot5^2}}{\sqrt[3]{2\cdot5^2}}$

$=\dfrac{\sqrt[3]{150}}{\sqrt[3]{1000}}$

$=\dfrac{\sqrt[3]{150}}{10}$

8. $\dfrac{3}{\sqrt[4]{p}}=\dfrac{3}{\sqrt[4]{p}}\cdot\dfrac{\sqrt[4]{p^3}}{\sqrt[4]{p^3}}=\dfrac{3\sqrt[4]{p^3}}{p}$

9. To rationalize the denominator of $\dfrac{4-\sqrt{3}}{-2+\sqrt{7}}$, multiply the numerator and denominator by $\underline{-2-\sqrt{7}}$.

10. $\dfrac{4}{\sqrt{3}+1}=\dfrac{4}{\sqrt{3}+1}\cdot\dfrac{\sqrt{3}-1}{\sqrt{3}-1}$

$=\dfrac{4(\sqrt{3}-1)}{(\sqrt{3}+1)(\sqrt{3}-1)}$

$=\dfrac{4(\sqrt{3}-1)}{(\sqrt{3})^2-1^2}$

$=\dfrac{4(\sqrt{3}-1)}{3-1}$

$=\dfrac{4(\sqrt{3}-1)}{2}$

$=2(\sqrt{3}-1)$

11. $\dfrac{\sqrt{2}}{\sqrt{6}-\sqrt{2}}=\dfrac{\sqrt{2}}{\sqrt{6}-\sqrt{2}}\cdot\dfrac{\sqrt{6}+\sqrt{2}}{\sqrt{6}+\sqrt{2}}$

$=\dfrac{\sqrt{2}(\sqrt{6}+\sqrt{2})}{(\sqrt{6}-\sqrt{2})(\sqrt{6}+\sqrt{2})}$

$=\dfrac{\sqrt{2}(\sqrt{6}+\sqrt{2})}{(\sqrt{6})^2-(\sqrt{2})^2}$

$=\dfrac{\sqrt{12}+\sqrt{4}}{6-2}$

$=\dfrac{2\sqrt{3}+2}{4}$

$=\dfrac{2(\sqrt{3}+1)}{4}$

$=\dfrac{\sqrt{3}+1}{2}$

12. $\dfrac{\sqrt{5}+4}{\sqrt{5}-\sqrt{2}}=\dfrac{\sqrt{5}+4}{\sqrt{5}-\sqrt{2}}\cdot\dfrac{\sqrt{5}+\sqrt{2}}{\sqrt{5}+\sqrt{2}}$

$=\dfrac{(\sqrt{5}+4)(\sqrt{5}+\sqrt{2})}{(\sqrt{5}-\sqrt{2})(\sqrt{5}+\sqrt{2})}$

$=\dfrac{\sqrt{25}+\sqrt{10}+4\sqrt{5}+4\sqrt{2}}{(\sqrt{5})^2-(\sqrt{2})^2}$

$=\dfrac{5+\sqrt{10}+4\sqrt{5}+4\sqrt{2}}{5-2}$

$=\dfrac{5+\sqrt{10}+4\sqrt{5}+4\sqrt{2}}{3}$

6.5 Exercises

13. $\dfrac{1}{\sqrt{2}} = \dfrac{1}{\sqrt{2}} \cdot \dfrac{\sqrt{2}}{\sqrt{2}} = \dfrac{\sqrt{2}}{\sqrt{4}} = \dfrac{\sqrt{2}}{2}$

15. $-\dfrac{6}{5\sqrt{3}} = -\dfrac{6}{5\sqrt{3}} \cdot \dfrac{\sqrt{3}}{\sqrt{3}}$

$\qquad = -\dfrac{6\sqrt{3}}{5\sqrt{9}}$

$\qquad = -\dfrac{6\sqrt{3}}{5 \cdot 3}$

$\qquad = -\dfrac{2\sqrt{3}}{5}$

17. $\dfrac{3}{\sqrt{12}} = \dfrac{3}{2\sqrt{3}} = \dfrac{3}{2\sqrt{3}} \cdot \dfrac{\sqrt{3}}{\sqrt{3}}$

$\qquad = \dfrac{3\sqrt{3}}{2\sqrt{9}}$

$\qquad = \dfrac{3\sqrt{3}}{2 \cdot 3}$

$\qquad = \dfrac{\sqrt{3}}{2}$

19. $\dfrac{\sqrt{2}}{\sqrt{6}} = \dfrac{\sqrt{2}}{\sqrt{6}} \cdot \dfrac{\sqrt{6}}{\sqrt{6}} = \dfrac{\sqrt{12}}{\sqrt{36}} = \dfrac{2\sqrt{3}}{6} = \dfrac{\sqrt{3}}{3}$

21. $\sqrt{\dfrac{2}{p}} = \dfrac{\sqrt{2}}{\sqrt{p}} = \dfrac{\sqrt{2}}{\sqrt{p}} \cdot \dfrac{\sqrt{p}}{\sqrt{p}} = \dfrac{\sqrt{2p}}{\sqrt{p^2}} = \dfrac{\sqrt{2p}}{p}$

23. $\dfrac{\sqrt{8}}{\sqrt{y^3}} = \dfrac{2\sqrt{2}}{y\sqrt{y}} = \dfrac{2\sqrt{2}}{y\sqrt{y}} \cdot \dfrac{\sqrt{y}}{\sqrt{y}}$

$\qquad = \dfrac{2\sqrt{2y}}{y\sqrt{y^2}} = \dfrac{2\sqrt{2y}}{y \cdot y}$

$\qquad = \dfrac{2\sqrt{2y}}{y^2}$

25. $\dfrac{2}{\sqrt[3]{2}} = \dfrac{2}{\sqrt[3]{2}} \cdot \dfrac{\sqrt[3]{4}}{\sqrt[3]{4}} = \dfrac{2\sqrt[3]{4}}{\sqrt[3]{8}} = \dfrac{2\sqrt[3]{4}}{2} = \sqrt[3]{4}$

27. $\sqrt[3]{\dfrac{7}{q}} = \dfrac{\sqrt[3]{7}}{\sqrt[3]{q}} = \dfrac{\sqrt[3]{7}}{\sqrt[3]{q}} \cdot \dfrac{\sqrt[3]{q^2}}{\sqrt[3]{q^2}}$

$\qquad = \dfrac{\sqrt[3]{7q^2}}{\sqrt[3]{q^3}}$

$\qquad = \dfrac{\sqrt[3]{7q^2}}{q}$

29. $\sqrt[3]{\dfrac{-3}{50}} = \dfrac{\sqrt[3]{-3}}{\sqrt[3]{50}} = \dfrac{\sqrt[3]{-3}}{\sqrt[3]{50}} \cdot \dfrac{\sqrt[3]{20}}{\sqrt[3]{20}}$

$\qquad = \dfrac{\sqrt[3]{-60}}{\sqrt[3]{1000}}$

$\qquad = -\dfrac{\sqrt[3]{60}}{10}$

31. $\dfrac{2}{\sqrt[3]{20y}} = \dfrac{2}{\sqrt[3]{20y}} \cdot \dfrac{\sqrt[3]{50y^2}}{\sqrt[3]{50y^2}}$

$\qquad = \dfrac{2\sqrt[3]{50y^2}}{\sqrt[3]{1000y^3}} = \dfrac{2\sqrt[3]{50y^2}}{10y}$

$\qquad = \dfrac{\sqrt[3]{50y^2}}{5y}$

33. $\dfrac{-4}{\sqrt[4]{3x^3}} = \dfrac{-4}{\sqrt[4]{3x^3}} \cdot \dfrac{\sqrt[4]{27x}}{\sqrt[4]{27x}}$

$\qquad = \dfrac{-4\sqrt[4]{27x}}{\sqrt[4]{81x^4}}$

$\qquad = -\dfrac{4\sqrt[4]{27x}}{3x}$

35. $\dfrac{12}{\sqrt[5]{m^3n^2}} = \dfrac{12}{\sqrt[5]{m^3n^2}} \cdot \dfrac{\sqrt[5]{m^2n^3}}{\sqrt[5]{m^2n^3}}$

$\qquad = \dfrac{12\sqrt[5]{m^2n^3}}{\sqrt[5]{m^5n^5}}$

$\qquad = \dfrac{12\sqrt[5]{m^2n^3}}{mn}$

37. $\dfrac{4}{\sqrt{6}-2} = \dfrac{4}{\sqrt{6}-2} \cdot \dfrac{\sqrt{6}+2}{\sqrt{6}+2} = \dfrac{4\left(\sqrt{6}+2\right)}{\left(\sqrt{6}\right)^2 - 2^2}$

$\qquad = \dfrac{4\left(\sqrt{6}+2\right)}{6-4} = \dfrac{4\left(\sqrt{6}+2\right)}{2}$

$\qquad = 2\left(\sqrt{6}+2\right)$

39. $\dfrac{5}{\sqrt{5}+2} = \dfrac{5}{\sqrt{5}+2} \cdot \dfrac{\sqrt{5}-2}{\sqrt{5}-2}$

$\qquad = \dfrac{5\left(\sqrt{5}-2\right)}{\left(\sqrt{5}\right)^2 - 2^2} = \dfrac{5\left(\sqrt{5}-2\right)}{5-4}$

$\qquad = 5\left(\sqrt{5}-2\right)$

41. $\dfrac{8}{\sqrt{7}-\sqrt{3}} = \dfrac{8}{\sqrt{7}-\sqrt{3}} \cdot \dfrac{\sqrt{7}+\sqrt{3}}{\sqrt{7}+\sqrt{3}}$

$\qquad = \dfrac{8\left(\sqrt{7}+\sqrt{3}\right)}{\left(\sqrt{7}\right)^2 - \left(\sqrt{3}\right)^2}$

$\qquad = \dfrac{8\left(\sqrt{7}+\sqrt{3}\right)}{7-3}$

$\qquad = \dfrac{8\left(\sqrt{7}+\sqrt{3}\right)}{4}$

$\qquad = 2\left(\sqrt{7}+\sqrt{3}\right)$

43. $\dfrac{\sqrt{2}}{\sqrt{10}-\sqrt{6}} = \dfrac{\sqrt{2}}{\sqrt{2}\sqrt{5}-\sqrt{2}\sqrt{3}}$

$\qquad = \dfrac{1}{\sqrt{5}-\sqrt{3}}$

$\qquad = \dfrac{1}{\sqrt{5}-\sqrt{3}} \cdot \dfrac{\sqrt{5}+\sqrt{3}}{\sqrt{5}+\sqrt{3}}$

$\qquad = \dfrac{\sqrt{5}+\sqrt{3}}{\left(\sqrt{5}\right)^2 - \left(\sqrt{3}\right)^2}$

$\qquad = \dfrac{\sqrt{5}+\sqrt{3}}{5-3}$

$\qquad = \dfrac{\sqrt{5}+\sqrt{3}}{2}$

45. $\dfrac{\sqrt{p}}{\sqrt{p}+\sqrt{q}} = \dfrac{\sqrt{p}}{\sqrt{p}+\sqrt{q}} \cdot \dfrac{\sqrt{p}-\sqrt{q}}{\sqrt{p}-\sqrt{q}}$

$\qquad = \dfrac{\sqrt{p}\left(\sqrt{p}-\sqrt{q}\right)}{\left(\sqrt{p}\right)^2 - \left(\sqrt{q}\right)^2}$

$\qquad = \dfrac{\sqrt{p^2}-\sqrt{pq}}{p-q}$

$\qquad = \dfrac{p-\sqrt{pq}}{p-q}$

47. $\dfrac{18}{2\sqrt{3}+3\sqrt{2}} = \dfrac{18}{2\sqrt{3}+3\sqrt{2}} \cdot \dfrac{2\sqrt{3}-3\sqrt{2}}{2\sqrt{3}-3\sqrt{2}}$

$\qquad = \dfrac{18\left(2\sqrt{3}-3\sqrt{2}\right)}{\left(2\sqrt{3}\right)^2 - \left(3\sqrt{2}\right)^2}$

$\qquad = \dfrac{18\left(2\sqrt{3}-3\sqrt{2}\right)}{12-18}$

$\qquad = \dfrac{18\left(2\sqrt{3}-3\sqrt{2}\right)}{-6}$

$\qquad = -3\left(2\sqrt{3}-3\sqrt{2}\right)$ or $3\left(3\sqrt{2}-2\sqrt{3}\right)$

49. $\dfrac{\sqrt{7}+3}{\sqrt{7}-3} = \dfrac{\sqrt{7}+3}{\sqrt{7}-3} \cdot \dfrac{\sqrt{7}+3}{\sqrt{7}+3}$

$\qquad = \dfrac{\left(\sqrt{7}+3\right)\left(\sqrt{7}+3\right)}{\left(\sqrt{7}-3\right)\left(\sqrt{7}+3\right)}$

$\qquad = \dfrac{\left(\sqrt{7}\right)^2 + 2\left(\sqrt{7}\right)(3) + (3)^2}{\left(\sqrt{7}\right)^2 - (3)^2}$

$\qquad = \dfrac{7+6\sqrt{7}+9}{7-9}$

$\qquad = \dfrac{16+6\sqrt{7}}{-2}$

$\qquad = -8-3\sqrt{7}$ or $-3\sqrt{7}-8$

51.
$$\frac{\sqrt{3}-4\sqrt{2}}{2\sqrt{3}+5\sqrt{2}} = \frac{\sqrt{3}-4\sqrt{2}}{2\sqrt{3}+5\sqrt{2}} \cdot \frac{2\sqrt{3}-5\sqrt{2}}{2\sqrt{3}-5\sqrt{2}}$$
$$= \frac{\left(\sqrt{3}-4\sqrt{2}\right)\left(2\sqrt{3}-5\sqrt{2}\right)}{\left(2\sqrt{3}+5\sqrt{2}\right)\left(2\sqrt{3}-5\sqrt{2}\right)}$$
$$= \frac{2\sqrt{9}-5\sqrt{6}-8\sqrt{6}+20\sqrt{4}}{\left(2\sqrt{3}\right)^2 - \left(5\sqrt{2}\right)^2}$$
$$= \frac{6-13\sqrt{6}+40}{12-50}$$
$$= \frac{46-13\sqrt{6}}{-38}$$
$$= \frac{13\sqrt{6}-46}{38}$$

53.
$$\frac{\sqrt{p}+2}{\sqrt{p}-2} = \frac{\sqrt{p}+2}{\sqrt{p}-2} \cdot \frac{\sqrt{p}+2}{\sqrt{p}+2}$$
$$= \frac{\left(\sqrt{p}+2\right)\left(\sqrt{p}+2\right)}{\left(\sqrt{p}-2\right)\left(\sqrt{p}+2\right)}$$
$$= \frac{\left(\sqrt{p}\right)^2 + 2\left(\sqrt{p}\right)(2) + 2^2}{\left(\sqrt{p}\right)^2 - 2^2}$$
$$= \frac{p+4\sqrt{p}+4}{p-4}$$

55.
$$\frac{\sqrt{2}-3}{\sqrt{8}-\sqrt{2}} = \frac{\sqrt{2}-3}{2\sqrt{2}-\sqrt{2}}$$
$$= \frac{\sqrt{2}-3}{\sqrt{2}}$$
$$= \frac{\sqrt{2}-3}{\sqrt{2}} \cdot \frac{\sqrt{2}}{\sqrt{2}}$$
$$= \frac{\sqrt{4}-3\sqrt{2}}{\sqrt{4}}$$
$$= \frac{2-3\sqrt{2}}{2}$$

57.
$$\sqrt{3} + \frac{1}{\sqrt{3}} = \sqrt{3} + \frac{1}{\sqrt{3}} \cdot \frac{\sqrt{3}}{\sqrt{3}} = \sqrt{3} + \frac{\sqrt{3}}{3}$$
$$= \frac{3\sqrt{3}}{3} + \frac{\sqrt{3}}{3} = \frac{3\sqrt{3}+\sqrt{3}}{3}$$
$$= \frac{4\sqrt{3}}{3}$$

59.
$$\frac{\sqrt{10}}{2} - \frac{1}{\sqrt{2}} = \frac{\sqrt{10}}{2} - \frac{1}{\sqrt{2}} \cdot \frac{\sqrt{2}}{\sqrt{2}}$$
$$= \frac{\sqrt{10}}{2} - \frac{\sqrt{2}}{2}$$
$$= \frac{\sqrt{10}-\sqrt{2}}{2}$$

61.
$$\sqrt{\frac{1}{3}} + \sqrt{12} + \sqrt{75} = \frac{\sqrt{1}}{\sqrt{3}} + 2\sqrt{3} + 5\sqrt{3}$$
$$= \frac{1}{\sqrt{3}} + 7\sqrt{3}$$
$$= \frac{1}{\sqrt{3}} \cdot \frac{\sqrt{3}}{\sqrt{3}} + 7\sqrt{3}$$
$$= \frac{\sqrt{3}}{3} + 7\sqrt{3}$$
$$= \frac{\sqrt{3}}{3} + \frac{21\sqrt{3}}{3}$$
$$= \frac{\sqrt{3}+21\sqrt{3}}{3}$$
$$= \frac{22\sqrt{3}}{3}$$

63.
$$\frac{3}{\sqrt{18}} - \sqrt{\frac{1}{2}} = \frac{3}{3\sqrt{2}} - \frac{\sqrt{1}}{\sqrt{2}}$$
$$= \frac{1}{\sqrt{2}} - \frac{1}{\sqrt{2}}$$
$$= 0$$

65.
$$\frac{\sqrt{3}}{\sqrt{12}} = \frac{\sqrt{3}}{2\sqrt{3}} = \frac{1}{2}$$

67.
$$\frac{3}{\sqrt{72}} = \frac{3}{6\sqrt{2}} = \frac{1}{2\sqrt{2}}$$
$$= \frac{1}{2\sqrt{2}} \cdot \frac{\sqrt{2}}{\sqrt{2}} = \frac{\sqrt{2}}{2\cdot 2}$$
$$= \frac{\sqrt{2}}{4}$$

69.
$$\sqrt{\frac{4}{3}} = \frac{\sqrt{4}}{\sqrt{3}} = \frac{2}{\sqrt{3}} = \frac{2}{\sqrt{3}} \cdot \frac{\sqrt{3}}{\sqrt{3}} = \frac{2\sqrt{3}}{3}$$

71. $\dfrac{\sqrt{3}-3}{\sqrt{3}+3} = \dfrac{\sqrt{3}-3}{\sqrt{3}+3} \cdot \dfrac{\sqrt{3}-3}{\sqrt{3}-3}$

$\qquad = \dfrac{\left(\sqrt{3}\right)^2 - 2\left(\sqrt{3}\right)(3) + 3^2}{\left(\sqrt{3}\right)^2 - 3^2}$

$\qquad = \dfrac{3 - 6\sqrt{3} + 9}{3 - 9} = \dfrac{12 - 6\sqrt{3}}{-6}$

$\qquad = \sqrt{3} - 2$

73. $\dfrac{2}{\sqrt{5}+2} = \dfrac{2}{\sqrt{5}+2} \cdot \dfrac{\sqrt{5}-2}{\sqrt{5}-2}$

$\qquad = \dfrac{2\left(\sqrt{5}-2\right)}{\left(\sqrt{5}\right)^2 - 2^2} = \dfrac{2\left(\sqrt{5}-2\right)}{5-4}$

$\qquad = 2\left(\sqrt{5}-2\right)$

75. $\dfrac{\sqrt{8}}{\sqrt{2}} = \dfrac{2\sqrt{2}}{\sqrt{2}} = 2$

77. $\dfrac{1}{\sqrt{3}} = \dfrac{1}{\sqrt{3}} \cdot \dfrac{\sqrt{3}}{\sqrt{3}} = \dfrac{\sqrt{3}}{3}$

79. $\dfrac{1}{\sqrt[3]{12}} = \dfrac{1}{\sqrt[3]{12}} \cdot \dfrac{\sqrt[3]{18}}{\sqrt[3]{18}} = \dfrac{\sqrt[3]{18}}{\sqrt[3]{216}} = \dfrac{\sqrt[3]{18}}{6}$

(Since $12 = 2^2 \cdot 3$, we need to multiply by $2 \cdot 3^2$ to get powers of 3 on the factors.)

81. $\dfrac{1}{\sqrt{3}+5} = \dfrac{1}{\sqrt{3}+5} \cdot \dfrac{\sqrt{3}-5}{\sqrt{3}-5}$

$\qquad = \dfrac{\sqrt{3}-5}{\left(\sqrt{3}+5\right)\left(\sqrt{3}-5\right)}$

$\qquad = \dfrac{\sqrt{3}-5}{\left(\sqrt{3}\right)^2 - 5^2}$

$\qquad = \dfrac{\sqrt{3}-5}{3 - 25}$

$\qquad = \dfrac{\sqrt{3}-5}{-22}$

$\qquad = \dfrac{5 - \sqrt{3}}{22}$

83. $\dfrac{1}{\sqrt{2}} \cdot \dfrac{\sqrt{3}}{2} - \dfrac{1}{\sqrt{2}} \cdot \dfrac{1}{2} = \dfrac{\sqrt{3}}{2\sqrt{2}} - \dfrac{1}{2\sqrt{2}}$

$\qquad = \dfrac{\sqrt{3}-1}{2\sqrt{2}}$

$\qquad = \dfrac{\sqrt{3}-1}{2\sqrt{2}} \cdot \dfrac{\sqrt{2}}{\sqrt{2}}$

$\qquad = \dfrac{\sqrt{6} - \sqrt{2}}{4}$

85. $\dfrac{\sqrt{2}+1}{3} = \dfrac{\sqrt{2}+1}{3} \cdot \dfrac{\sqrt{2}-1}{\sqrt{2}-1}$

$\qquad = \dfrac{\left(\sqrt{2}+1\right)\left(\sqrt{2}-1\right)}{3\left(\sqrt{2}-1\right)}$

$\qquad = \dfrac{\left(\sqrt{2}\right)^2 - 1^2}{3\sqrt{2}-3}$

$\qquad = \dfrac{2-1}{3\sqrt{2}-3}$

$\qquad = \dfrac{1}{3\sqrt{2}-3} = \dfrac{1}{3\left(\sqrt{2}-1\right)}$

87. $\dfrac{\sqrt{x}-\sqrt{h}}{\sqrt{x}} = \dfrac{\sqrt{x}-\sqrt{h}}{\sqrt{x}} \cdot \dfrac{\sqrt{x}+\sqrt{h}}{\sqrt{x}+\sqrt{h}}$

$\qquad = \dfrac{\left(\sqrt{x}-\sqrt{h}\right)\left(\sqrt{x}+\sqrt{h}\right)}{\sqrt{x}\left(\sqrt{x}+\sqrt{h}\right)}$

$\qquad = \dfrac{\left(\sqrt{x}\right)^2 - \left(\sqrt{h}\right)^2}{\sqrt{x^2} + \sqrt{xh}}$

$\qquad = \dfrac{x - h}{x + \sqrt{xh}}$

89.

$$\left(\frac{\sqrt{6}+\sqrt{2}}{4}\right)^2 \stackrel{?}{=} \left(\frac{\sqrt{2+\sqrt{3}}}{2}\right)^2$$

$$\frac{\left(\sqrt{6}\right)^2 + 2\cdot\sqrt{6}\cdot\sqrt{2} + \left(\sqrt{2}\right)^2}{4^2} \stackrel{?}{=} \frac{2+\sqrt{3}}{4}$$

$$\frac{6+2\sqrt{12}+2}{16} \stackrel{?}{=} \frac{2+\sqrt{3}}{4}$$

$$\frac{8+2\cdot2\sqrt{3}}{16} \stackrel{?}{=} \frac{2+\sqrt{3}}{4}$$

$$\frac{8+4\sqrt{3}}{16} \stackrel{?}{=} \frac{2+\sqrt{3}}{4}$$

$$\frac{2+\sqrt{3}}{4} = \frac{2+\sqrt{3}}{4}$$

91. a.

$$\frac{\sqrt{x+h}-\sqrt{x}}{h}$$

$$=\frac{\sqrt{x+h}-\sqrt{x}}{h}\cdot\frac{\sqrt{x+h}+\sqrt{x}}{\sqrt{x+h}+\sqrt{x}}$$

$$=\frac{\left(\sqrt{x+h}-\sqrt{x}\right)\left(\sqrt{x+h}+\sqrt{x}\right)}{h\left(\sqrt{x+h}+\sqrt{x}\right)}$$

$$=\frac{(x+h)+\sqrt{x}\sqrt{x+h}-\sqrt{x}\sqrt{x+h}-(x)}{h\left(\sqrt{x+h}+\sqrt{x}\right)}$$

$$=\frac{x+h-x}{h\left(\sqrt{x+h}+\sqrt{x}\right)}$$

$$=\frac{\cancel{h}}{\cancel{h}\left(\sqrt{x+h}+\sqrt{x}\right)}$$

$$=\frac{1}{\sqrt{x+h}+\sqrt{x}}$$

b. $\dfrac{1}{\sqrt{x+0}+\sqrt{x}} = \dfrac{1}{\sqrt{x}+\sqrt{x}} = \dfrac{1}{2\sqrt{x}}$

c. $m = \dfrac{1}{2\sqrt{4}} = \dfrac{1}{2\cdot2} = \dfrac{1}{4}$

d. $f(x) = \sqrt{x}$

$f(4) = \sqrt{4} = 2$

The point (4, 2) is on the graph.

e.

$$y - y_1 = m(x - x_1)$$

$$y - 2 = \frac{1}{4}(x - 4)$$

$$y - 2 = \frac{1}{4}x - 1$$

$$y = \frac{1}{4}x + 1$$

f.

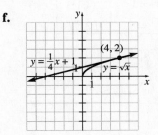

93. Answers will vary. The goal is to express a fraction containing irrational numbers in the denominator as a fraction containing only rational numbers in the denominator.

95. $g(x) = -3x + 9$

x	$y = -3x + 9$	(x, y)
-1	$y = -3(-1) + 9 = 12$	$(-1, 12)$
0	$y = -3(0) + 9 = 9$	$(0, 9)$
1	$y = -3(1) + 9 = 6$	$(1, 6)$
2	$y = -3(2) + 9 = 3$	$(2, 3)$
3	$y = -3(3) + 9 = 0$	$(3, 0)$

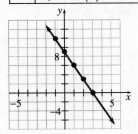

97. $F(x) = x^3$

x	$y = x^3$	(x, y)
-2	$y = (-2)^3 = -8$	$(-2, -8)$
-1	$y = (-1)^3 = -1$	$(-1, -1)$
0	$y = 0^3 = 0$	$(0, 0)$
1	$y = 1^3 = 1$	$(1, 1)$
2	$y = 2^3 = 8$	$(2, 8)$

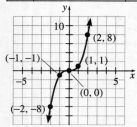

Putting the Concepts Together (Sections 6.1–6.5)

1. $-25^{1/2} = -\sqrt{25} = -5$

2. $(-64)^{-2/3} = \dfrac{1}{(-64)^{2/3}} = \dfrac{1}{\left(\sqrt[3]{-64}\right)^2} = \dfrac{1}{(-4)^2} = \dfrac{1}{16}$

3. $\sqrt[4]{3x^3} = \left(3x^3\right)^{1/4}$

4. $7z^{4/5} = 7\left(z^4\right)^{1/5} = 7\sqrt[5]{z^4}$

5. $\sqrt[3]{\sqrt{64x^3}} = \left(\left(64x^3\right)^{1/2}\right)^{1/3}$
$$= \left(64x^3\right)^{1/6}$$
$$= \left(2^6 x^3\right)^{1/6}$$
$$= 2^{6/6} x^{3/6}$$
$$= 2x^{1/2} \quad \text{or} \quad 2\sqrt{x}$$

6. $c^{1/2}\left(c^{3/2} + c^{5/2}\right) = c^{1/2} \cdot c^{3/2} + c^{1/2} \cdot c^{5/2}$
$$= c^{\frac{1}{2} + \frac{3}{2}} + c^{\frac{1}{2} + \frac{5}{2}}$$
$$= c^{\frac{4}{2}} + c^{\frac{6}{2}}$$
$$= c^2 + c^3$$

7. $\left(a^{2/3}b^{-1/3}\right)\left(a^{4/3}b^{-5/3}\right) = a^{\frac{2}{3}+\frac{4}{3}} b^{-\frac{1}{3}+\left(-\frac{5}{3}\right)}$
$$= a^{6/3} b^{-6/3}$$
$$= a^2 b^{-2}$$
$$= \dfrac{a^2}{b^2}$$

8. $\dfrac{x^{3/4}}{x^{1/8}} = x^{\frac{3}{4}-\frac{1}{8}} = x^{\frac{6}{8}-\frac{1}{8}} = x^{5/8} \text{ or } \sqrt[8]{x^5}$

9. $\left(x^{3/4} y^{-1/8}\right)^8 = \left(x^{3/4}\right)^8 \left(y^{-1/8}\right)^8$
$$= x^{\frac{3}{4} \cdot 8} y^{-\frac{1}{8} \cdot 8}$$
$$= x^6 y^{-1}$$
$$= \dfrac{x^6}{y}$$

10. $\sqrt{15a} \cdot \sqrt{2b} = \sqrt{15a \cdot 2b} = \sqrt{30ab}$

11. $\sqrt{10m^3n^2} \cdot \sqrt{20mn} = \sqrt{10m^3n^2 \cdot 20mn}$
$$= \sqrt{200m^4n^3}$$
$$= \sqrt{100m^4n^2 \cdot 2n}$$
$$= \sqrt{100m^4n^2} \cdot \sqrt{2n}$$
$$= 10m^2n\sqrt{2n}$$

12. $\sqrt[3]{\dfrac{-32xy^4}{4x^{-2}y}} = \sqrt[3]{-8x^3y^3} = \sqrt[3]{(-2)^3 x^3 y^3} = -2xy$

13. $2\sqrt{108} - 3\sqrt{75} + \sqrt{48}$
$$= 2 \cdot \sqrt{36 \cdot 3} - 3\sqrt{25 \cdot 3} + \sqrt{16 \cdot 3}$$
$$= 2\sqrt{36} \cdot \sqrt{3} - 3\sqrt{25} \cdot \sqrt{3} + \sqrt{16} \cdot \sqrt{3}$$
$$= 2 \cdot 6\sqrt{3} - 3 \cdot 5\sqrt{3} + 4\sqrt{3}$$
$$= 12\sqrt{3} - 15\sqrt{3} + 4\sqrt{3}$$
$$= (12 - 15 + 4)\sqrt{3}$$
$$= \sqrt{3}$$

14. $-5b\sqrt{8b} + 7\sqrt{18b^3} = -5b\sqrt{4 \cdot 2b} + 7\sqrt{9b^2 \cdot 2b}$

$$= -5b\sqrt{4} \cdot \sqrt{2b} + 7\sqrt{9b^2} \cdot \sqrt{2b}$$
$$= -5b \cdot 2\sqrt{2b} + 7 \cdot 3b\sqrt{2b}$$
$$= -10b\sqrt{2b} + 21b\sqrt{2b}$$
$$= (-10 + 21)b\sqrt{2b}$$
$$= 11b\sqrt{2b}$$

15. $\sqrt[3]{16y^4} - y\sqrt[3]{2y} = \sqrt[3]{8y^3 \cdot 2y} - y\sqrt[3]{2y}$

$$= \sqrt[3]{8y^3} \cdot \sqrt[3]{2y} - y\sqrt[3]{2y}$$
$$= 2y\sqrt[3]{2y} - y\sqrt[3]{2y}$$
$$= (2 - 1)y\sqrt[3]{2y}$$
$$= y\sqrt[3]{2y}$$

16. $\left(3\sqrt{x}\right)\left(4\sqrt{x}\right) = 3 \cdot 4 \cdot \sqrt{x \cdot x}$

$$= 12\sqrt{x^2}$$
$$= 12x$$

17. $3\sqrt{x} + 4\sqrt{x} = (3 + 4)\sqrt{x} = 7\sqrt{x}$

18. $\left(2 - 3\sqrt{2}\right)\left(10 + \sqrt{2}\right)$

$$= 2 \cdot 10 + 2 \cdot \sqrt{2} - 3\sqrt{2} \cdot 10 - 3\sqrt{2} \cdot \sqrt{2}$$
$$= 20 + 2\sqrt{2} - 30\sqrt{2} - 3 \cdot 2$$
$$= 20 - 28\sqrt{2} - 6$$
$$= 14 - 28\sqrt{2} = 14\left(1 - 2\sqrt{2}\right)$$

19. $\left(4\sqrt{2} - 3\right)^2 = \left(4\sqrt{2}\right)^2 - 2\left(4\sqrt{2}\right)(3) + (3)^2$

$$= 16 \cdot 2 - 24\sqrt{2} + 9$$
$$= 32 - 24\sqrt{2} + 9$$
$$= 41 - 24\sqrt{2}$$

20. $\dfrac{3}{2\sqrt{32}} = \dfrac{3}{2 \cdot 4\sqrt{2}} = \dfrac{3}{8\sqrt{2}}$

$$= \dfrac{3}{8\sqrt{2}} \cdot \dfrac{\sqrt{2}}{\sqrt{2}} = \dfrac{3\sqrt{2}}{8 \cdot 2}$$
$$= \dfrac{3\sqrt{2}}{16}$$

21. $\dfrac{4}{\sqrt{3} - 8} = \dfrac{4}{\sqrt{3} - 8} \cdot \dfrac{\sqrt{3} + 8}{\sqrt{3} + 8}$

$$= \dfrac{4\left(\sqrt{3} + 8\right)}{\left(\sqrt{3} - 8\right)\left(\sqrt{3} + 8\right)}$$
$$= \dfrac{4\left(\sqrt{3} + 8\right)}{\left(\sqrt{3}\right)^2 - 8^2}$$
$$= \dfrac{4\sqrt{3} + 32}{3 - 64}$$
$$= \dfrac{4\sqrt{3} + 32}{-61}$$
$$= -\dfrac{4\sqrt{3} + 32}{61} = -\dfrac{4\left(\sqrt{3} + 8\right)}{61}$$

Section 6.6

Are You Ready for This Section?

R1. $\sqrt{121} = \sqrt{11^2} = 11$

R2. $\sqrt{p^2} = |p|$

R3. $f(x) = x^2 - 4$

$$f(3) = (3)^2 - 4 = 9 - 4 = 5$$

R4.

$$-2x + 3 \geq 0$$
$$-2x + 3 - 3 \geq 0 - 3$$
$$-2x \geq -3$$
$$\dfrac{-2x}{-2} \leq \dfrac{-3}{-2}$$
$$x \leq \dfrac{3}{2}$$

The solution set is $\left\{x \mid x \leq \dfrac{3}{2}\right\}$, or $\left(-\infty, \dfrac{3}{2}\right]$ in interval notation.

R5. $f(x) = x^2 + 1$

x	$y = x^2 + 1$	(x, y)
-2	$y = (-2)^2 + 1 = 5$	$(-2, 5)$
-1	$y = (-1)^2 + 1 = 2$	$(-1, 2)$
0	$y = (0)^2 + 1 = 1$	$(0, 1)$
1	$y = (1)^2 + 1 = 2$	$(1, 2)$
2	$y = (2)^2 + 1 = 5$	$(2, 5)$

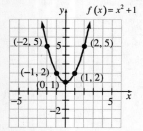

Section 6.6 Quick Checks

1. $f(x) = \sqrt{3x + 7}$

    a. $f(3) = \sqrt{3(3) + 7} = \sqrt{9 + 7} = \sqrt{16} = 4$

    b. $f(7) = \sqrt{3(7) + 7} = \sqrt{21 + 7} = \sqrt{28} = 2\sqrt{7}$

2. $g(x) = \sqrt[3]{2x + 7}$

    a. $g(-4) = \sqrt[3]{2(-4) + 7} = \sqrt[3]{-8 + 7} = \sqrt[3]{-1} = -1$

    b. $g(10) = \sqrt[3]{2(10) + 7} = \sqrt[3]{20 + 7} = \sqrt[3]{27} = 3$

3. If the index on a radical expression is <u>even</u>, then the radicand must be greater than or equal to zero. If the index on a radical expression is <u>odd</u>, then the radicand can be any real number.

4. $H(x) = \sqrt{x + 6}$

    The function tells us to take the square root of $x + 6$. We can only take the square root of numbers that are greater than or equal to 0. Thus, we need
    $x + 6 \geq 0$
    $\quad x \geq -6$
    The domain of H is $\{x \mid x \geq -6\}$ or the interval $[-6, \infty)$.

5. $g(t) = \sqrt[5]{3t - 1}$

    The function tells us to take the fifth root of $3t - 1$. Since we can take the fifth root of any real number, the domain of g is all real numbers $\{t \mid t \text{ is any real number}\}$ or the interval $(-\infty, \infty)$.

6. $F(m) = \sqrt[4]{6 - 3m}$

    The function tells us to take the fourth root of $6 - 3m$. We can only take the fourth root of numbers that are greater than or equal to 0. Thus, we need
    $6 - 3m \geq 0$
    $\quad -3m \geq -6$
    $\quad\quad m \leq 2$
    The domain of F is $\{m \mid m \leq 2\}$ or the interval $(-\infty, 2]$.

7. $f(x) = \sqrt{x + 3}$

    a. The function tells us to take the square root of $x + 3$. We can only take the square root of numbers that are greater than or equal to 0. Thus, we need
    $x + 3 \geq 0$
    $\quad x \geq -3$
    The domain of f is $\{x \mid x \geq -3\}$ or the interval $[-3, \infty)$.

    b.
    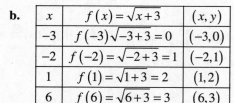

x	$f(x) = \sqrt{x + 3}$	(x, y)
-3	$f(-3)\sqrt{-3 + 3} = 0$	$(-3, 0)$
-2	$f(-2) = \sqrt{-2 + 3} = 1$	$(-2, 1)$
1	$f(1) = \sqrt{1 + 3} = 2$	$(1, 2)$
6	$f(6) = \sqrt{6 + 3} = 3$	$(6, 3)$

    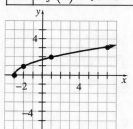

    c. Based on the graph, the range is $\{y \mid y \geq 0\}$ or the interval $[0, \infty)$.

8. $G(x) = \sqrt[3]{x} - 1$

  a. The function tells us to take the cube root of x and then subtract 1. We can take the cube root of any real number so the domain is all real numbers, $\{x \mid x \text{ is any real number}\}$ or the interval $(-\infty, \infty)$.

  b.

x	$G(x) = \sqrt[3]{x} - 1$	(x, y)
-8	$G(-8) = \sqrt[3]{-8} - 1 = -3$	$(-8, -3)$
-1	$G(-1) = \sqrt[3]{-1} - 1 = -2$	$(-1, -2)$
0	$G(0) = \sqrt[3]{0} - 1 = -1$	$(0, -1)$
1	$G(1) = \sqrt[3]{1} - 1 = 0$	$(1, 0)$
8	$G(8) = \sqrt[3]{8} - 1 = 1$	$(8, 1)$

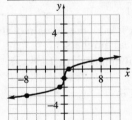

  c. Based on the graph, the range is all real numbers, $\{y \mid y \text{ is any real number}\}$ or the interval $(-\infty, \infty)$.

6.6 Exercises

9. $f(x) = \sqrt{x + 6}$

  a. $f(3) = \sqrt{3 + 6} = \sqrt{9} = 3$

  b. $f(8) = \sqrt{8 + 6} = \sqrt{14}$

  c. $f(-2) = \sqrt{-2 + 6} = \sqrt{4} = 2$

11. $g(x) = -\sqrt{2x + 3}$

  a. $g(11) = -\sqrt{2(11) + 3}$
       $= -\sqrt{25}$
       $= -5$

  b. $g(-1) = -\sqrt{2(-1) + 3}$
       $= -\sqrt{1}$
       $= -1$

  c. $g\left(\dfrac{1}{8}\right) = -\sqrt{2\left(\dfrac{1}{8}\right) + 3}$
       $= -\sqrt{\dfrac{13}{4}}$
       $= -\dfrac{\sqrt{13}}{\sqrt{4}}$
       $= -\dfrac{\sqrt{13}}{2}$

13. $G(m) = 2\sqrt{5m - 1}$

  a. $G(1) = 2\sqrt{5(1) - 1}$
       $= 2\sqrt{4}$
       $= 2 \cdot 2$
       $= 4$

  b. $G(5) = 2\sqrt{5(5) - 1}$
       $= 2\sqrt{24}$
       $= 4\sqrt{6}$

  c. $G\left(\dfrac{1}{2}\right) = 2\sqrt{5\left(\dfrac{1}{2}\right) - 1}$
       $= 2\sqrt{\dfrac{3}{2}}$
       $= \dfrac{2\sqrt{3}}{\sqrt{2}} \cdot \dfrac{\sqrt{2}}{\sqrt{2}}$
       $= \dfrac{2\sqrt{6}}{2}$
       $= \sqrt{6}$

15. $H(z) = \sqrt[3]{z + 4}$

  a. $H(4) = \sqrt[3]{4 + 4} = \sqrt[3]{8} = 2$

  b. $H(-12) = \sqrt[3]{-12 + 4} = \sqrt[3]{-8} = -2$

  c. $H(-20) = \sqrt[3]{-20 + 4}$
       $= \sqrt[3]{-16}$
       $= \sqrt[3]{-8} \cdot \sqrt[3]{2}$
       $= -2\sqrt[3]{2}$

17. $f(x) = \sqrt{\dfrac{x-2}{x+2}}$

   a. $f(7) = \sqrt{\dfrac{7-2}{7+2}} = \sqrt{\dfrac{5}{9}} = \dfrac{\sqrt{5}}{\sqrt{9}} = \dfrac{\sqrt{5}}{3}$

   b. $f(6) = \sqrt{\dfrac{6-2}{6+2}}$

          $= \sqrt{\dfrac{1}{2}}$

          $= \dfrac{1}{\sqrt{2}} \cdot \dfrac{\sqrt{2}}{\sqrt{2}}$

          $= \dfrac{\sqrt{2}}{2}$

   c. $f(10) = \sqrt{\dfrac{10-2}{10+2}} = \sqrt{\dfrac{2}{3}}$

             $= \dfrac{\sqrt{2}}{\sqrt{3}} \cdot \dfrac{\sqrt{3}}{\sqrt{3}}$

             $= \dfrac{\sqrt{6}}{3}$

19. $g(z) = \sqrt[3]{\dfrac{2z}{z-4}}$

   a. $g(-4) = \sqrt[3]{\dfrac{2(-4)}{-4-4}} = \sqrt[3]{\dfrac{-8}{-8}} = \sqrt[3]{1} = 1$

   b. $g(8) = \sqrt[3]{\dfrac{2(8)}{8-4}} = \sqrt[3]{\dfrac{16}{4}} = \sqrt[3]{4}$

   c. $g(12) = \sqrt[3]{\dfrac{2(12)}{12-4}} = \sqrt[3]{\dfrac{24}{8}} = \sqrt[3]{3}$

21. $f(x) = \sqrt{x-7}$

   $x - 7 \geq 0$

     $x \geq 7$

   The domain of the function is $\{x \mid x \geq 7\}$ or the interval $[7, \infty)$.

23. $g(x) = \sqrt{2x+7}$

   $2x + 7 \geq 0$

     $2x \geq -7$

       $x \geq -\dfrac{7}{2}$

The domain of the function is $\left\{x \mid x \geq -\dfrac{7}{2}\right\}$ or the interval $\left[-\dfrac{7}{2}, \infty\right)$.

25. $F(x) = \sqrt{4-3x}$

   $4 - 3x \geq 0$

     $-3x \geq -4$

       $x \leq \dfrac{4}{3}$

The domain of the function is $\left\{x \mid x \leq \dfrac{4}{3}\right\}$ or the interval $\left(-\infty, \dfrac{4}{3}\right]$.

27. $H(z) = \sqrt[3]{2z+1}$

Since the index is odd, the domain of the function is all real numbers.

$\{z \mid z \text{ is any real number}\}$ or $(-\infty, \infty)$.

29. $W(p) = \sqrt[4]{7p-2}$

   $7p - 2 \geq 0$

     $7p \geq 2$

       $p \geq \dfrac{2}{7}$

The domain of the function is $\left\{p \mid p \geq \dfrac{2}{7}\right\}$ or the interval $\left[\dfrac{2}{7}, \infty\right)$.

31. $g(x) = \sqrt[5]{x-3}$

Since the index is odd, the domain of the function is all real numbers.

$\{x \mid x \text{ is any real number}\}$ or $(-\infty, \infty)$.

33. $f(x) = \sqrt{\dfrac{3}{x+5}}$

The index is even so we need the radicand to be nonnegative. The numerator is positive which means we need the denominator to be positive as well.

   $x + 5 > 0$

     $x > -5$

The domain of the function is $\{x \mid x > -5\}$ or the interval $(-5, \infty)$.

35. $H(x) = \sqrt{\dfrac{x+3}{x-3}}$

The index is even so we need the radicand to be nonnegative. That is, we need to solve

$$\dfrac{x+3}{x-3} \geq 0$$

The radicand equals 0 when $x = -3$ and it is undefined when $x = 3$. We can use these two values to split up the real number line into subintervals.

	-3		3	
Interval	$(-\infty, -3)$	$(-3, 3)$		$(3, \infty)$
Num. chosen	-4	0		4
Value of radicand	$\frac{1}{7}$	-1		7
Conclusion	positive	negative		positive

Since we need the radicand to be positive or 0, the domain is $\{x \mid x \leq -3 \text{ or } x > 3\}$ or the interval $(-\infty, -3] \cup (3, \infty)$.

37. $f(x) = \sqrt{x-4}$

a. $x - 4 \geq 0$

$x \geq 4$

The domain is $\{x \mid x \geq 4\}$ or the interval $[4, \infty)$.

b.

x	$f(x) = \sqrt{x-4}$	(x, y)
4	$f(4) = \sqrt{4-4} = 0$	$(4, 0)$
5	$f(5) = 1$	$(5, 1)$
8	$f(8) = 2$	$(8, 2)$
13	$f(13) = 3$	$(13, 3)$
20	$f(20) = 4$	$(20, 4)$

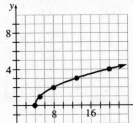

c. Based on the graph, the range is $[0, \infty)$.

39. $g(x) = \sqrt{x+2}$

a. $x + 2 \geq 0$

$x \geq -2$

The domain is $\{x \mid x \geq -2\}$ or the interval $[-2, \infty)$.

b.

x	$g(x) = \sqrt{x+2}$	(x, y)
-2	$g(-2) = \sqrt{-2+2} = 0$	$(-2, 0)$
-1	$g(-1) = 1$	$(-1, 1)$
2	$g(2) = 2$	$(2, 2)$
7	$g(7) = 3$	$(7, 3)$
14	$g(14) = 4$	$(14, 4)$

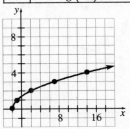

c. Based on the graph, the range is $[0, \infty)$.

41. $G(x) = \sqrt{2-x}$

a. $2 - x \geq 0$

$-x \geq -2$

$x \leq 2$

The domain is $\{x \mid x \leq 2\}$ or the interval $(-\infty, 2]$.

b.

x	$G(x) = \sqrt{2-x}$	(x, y)
-14	$G(-14) = \sqrt{2+14} = 4$	$(-14, 4)$
-7	$G(-7) = 3$	$(-7, 3)$
-2	$G(-2) = 2$	$(-2, 2)$
1	$G(1) = 1$	$(1, 1)$
2	$G(2) = 0$	$(2, 0)$

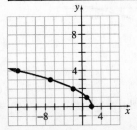

c. Based on the graph, the range is $[0, \infty)$.

43. $f(x) = \sqrt{x} + 3$

    a. The domain is $\{x \mid x \geq 0\}$ or the interval $[0, \infty)$.

    b.

x	$f(x) = \sqrt{x} + 3$	(x, y)
0	$f(0) = \sqrt{0} + 3 = 3$	$(0, 3)$
1	$f(1) = 4$	$(1, 4)$
4	$f(4) = 5$	$(4, 5)$
9	$f(9) = 6$	$(9, 6)$
16	$f(16) = 7$	$(16, 7)$

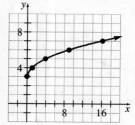

    c. Based on the graph, the range is $[3, \infty)$.

45. $g(x) = \sqrt{x} - 4$

    a. The domain is $\{x \mid x \geq 0\}$ or the interval $[0, \infty)$.

    b.

x	$g(x) = \sqrt{x} - 4$	(x, y)
0	$g(0) = \sqrt{0} - 4 = -4$	$(0, -4)$
1	$g(1) = -3$	$(1, -3)$
4	$g(4) = -2$	$(4, -2)$
9	$g(9) = -1$	$(9, -1)$
16	$g(16) = 0$	$(16, 0)$

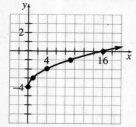

    c. Based on the graph, the range is $[-4, \infty)$.

47. $H(x) = 2\sqrt{x}$

    a. The domain is $\{x \mid x \geq 0\}$ or the interval $[0, \infty)$.

    b.

x	$H(x) = 2\sqrt{x}$	(x, y)
0	$H(0) = 2\sqrt{0} = 0$	$(0, 0)$
1	$H(1) = 2$	$(1, 2)$
4	$H(4) = 4$	$(4, 4)$
9	$H(9) = 6$	$(9, 6)$
16	$H(16) = 8$	$(16, 8)$

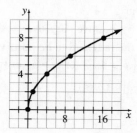

    c. Based on the graph, the range is $[0, \infty)$.

49. $f(x) = \dfrac{1}{2}\sqrt{x}$

    a. The domain is $\{x \mid x \geq 0\}$ or the interval $[0, \infty)$.

    b.

x	$f(x) = \dfrac{1}{2}\sqrt{x}$	(x, y)
0	$f(0) = \dfrac{1}{2}\sqrt{0} = 0$	$(0, 0)$
1	$f(1) = \dfrac{1}{2}$	$\left(1, \dfrac{1}{2}\right)$
4	$f(4) = 1$	$(4, 1)$
9	$f(9) = \dfrac{3}{2}$	$\left(9, \dfrac{3}{2}\right)$
16	$f(16) = 2$	$(16, 2)$

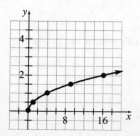

c. Based on the graph, the range is $[0,\infty)$.

51. $G(x) = -\sqrt{x}$

a. The domain is $\{x \mid x \geq 0\}$ or the interval $[0,\infty)$.

b.

x	$G(x) = -\sqrt{x}$	(x, y)
0	$G(0) = -\sqrt{0} = 0$	$(0,0)$
1	$G(1) = -1$	$(1,-1)$
4	$G(4) = -2$	$(4,-2)$
9	$G(9) = -3$	$(9,-3)$
16	$G(16) = -4$	$(16,-4)$

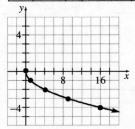

c. Based on the graph, the range is $(-\infty, 0]$.

53. $h(x) = \sqrt[3]{x+2}$

a. The index is odd so the domain is all real numbers, $\{x \mid x \text{ is any real number}\}$ or $(-\infty, \infty)$.

b.

x	$h(x) = \sqrt[3]{x+2}$	(x, y)
-10	$h(-10) = \sqrt[3]{-10+2} = -2$	$(-10,-2)$
-3	$h(-3) = -1$	$(-3,-1)$
-2	$h(-2) = 0$	$(-2,0)$
-1	$h(-1) = 1$	$(-1,1)$
6	$h(6) = 2$	$(6,2)$

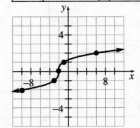

c. Based on the graph, the range is $(-\infty, \infty)$.

55. $f(x) = \sqrt[3]{x} - 3$

a. The index is odd so the domain is all real numbers, $\{x \mid x \text{ is any real number}\}$ or $(-\infty, \infty)$.

b.

x	$f(x) = \sqrt[3]{x} - 3$	(x, y)
-8	$f(-8) = \sqrt[3]{-8} - 3 = -5$	$(-8,-5)$
-1	$f(-1) = -4$	$(-1,-4)$
0	$f(0) = -3$	$(0,-3)$
1	$f(1) = -2$	$(1,-2)$
8	$f(8) = -1$	$(8,-1)$

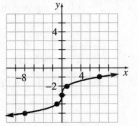

c. Based on the graph, the range is $(-\infty, \infty)$.

57. $G(x) = 2\sqrt[3]{x}$

a. The index is odd so the domain is all real numbers, $\{x \mid x \text{ is any real number}\}$ or $(-\infty, \infty)$.

b.

x	$G(x) = 2\sqrt[3]{x}$	(x, y)
-8	$G(-8) = 2\sqrt[3]{-8} = -4$	$(-8,-4)$
-1	$G(-1) = -2$	$(-1,-2)$
0	$G(0) = 0$	$(0,0)$
1	$G(1) = 2$	$(1,2)$
8	$G(8) = 4$	$(8,4)$

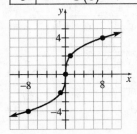

c. Based on the graph, the range is $(-\infty, \infty)$.

59. a. $d(0) = \sqrt{0^4 - 9(0)^2 + 25} = \sqrt{25} = 5$

The distance is 5 units.

b. $d(1) = \sqrt{1^4 - 9(1)^2 + 25} = \sqrt{17}$

The distance is $\sqrt{17} \approx 4.123$ units.

c. $d(5) = \sqrt{5^4 - 9(5)^2 + 25} = \sqrt{425} = 5\sqrt{17}$

The distance is $5\sqrt{17} \approx 20.616$ units.

61. a. $x = 1$

$A(1) = 2(1) \cdot \sqrt{9 - 1^2} = 2\sqrt{8} = 4\sqrt{2}$

The area is $4\sqrt{2} \approx 5.657$ square units.

b. $x = 2$

$A(2) = 2(2) \cdot \sqrt{9 - 2^2} = 4\sqrt{5}$

The area is $4\sqrt{5} \approx 8.944$ square units.

c. $x = \sqrt{2}$

$A(\sqrt{2}) = 2(\sqrt{2}) \cdot \sqrt{9 - (\sqrt{2})^2}$

$= 2\sqrt{2} \cdot \sqrt{7} = 2\sqrt{14}$

The area is $2\sqrt{14} \approx 7.483$ square units.

63. The graph of $g(x) = \sqrt{x + c}$ can be obtained from the graph of $f(x) = \sqrt{x}$ by shifting the graph of $f(x)$ c units to the right (if $c < 0$) or left (if $c > 0$).

65. $\dfrac{1}{3} + \dfrac{1}{2} = \dfrac{2}{6} + \dfrac{3}{6} = \dfrac{2+3}{6} = \dfrac{5}{6}$

67. $\dfrac{1}{x} + \dfrac{3}{x+1} = \dfrac{1}{x} \cdot \dfrac{(x+1)}{(x+1)} + \dfrac{3}{x+1} \cdot \dfrac{x}{x}$

$= \dfrac{x+1}{x(x+1)} + \dfrac{3x}{x(x+1)}$

$= \dfrac{x+1+3x}{x(x+1)}$

$= \dfrac{4x+1}{x(x+1)}$

69. $\dfrac{4}{x-1} + \dfrac{3}{x+1} = \dfrac{4}{x-1} \cdot \dfrac{x+1}{x+1} + \dfrac{3}{x+1} \cdot \dfrac{x-1}{x-1}$

$= \dfrac{4(x+1) + 3(x-1)}{(x-1)(x+1)}$

$= \dfrac{4x+4+3x-3}{(x-1)(x+1)}$

$= \dfrac{7x+1}{(x-1)(x+1)}$ or $\dfrac{7x+1}{x^2-1}$

71. $f(x) = \sqrt{x-4}$

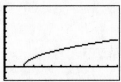

73. $g(x) = \sqrt{x+2}$

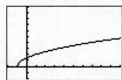

75. $G(x) = \sqrt{2-x}$

77. $f(x) = \sqrt{x} + 3$

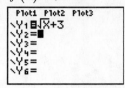

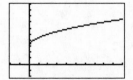

79. $g(x) = \sqrt{x} - 4$

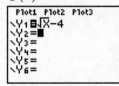

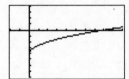

81. $H(x) = 2\sqrt{x}$

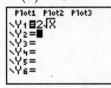

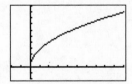

83. $f(x) = \frac{1}{2}\sqrt{x}$

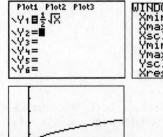

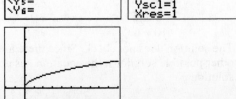

85. $G(x) = -\sqrt{x}$

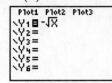

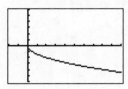

87. $h(x) = \sqrt[3]{x} + 2$

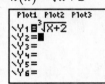

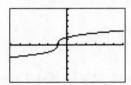

89. $f(x) = \sqrt[3]{x} - 3$

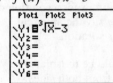

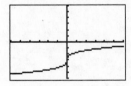

91. $G(x) = 2\sqrt[3]{x}$

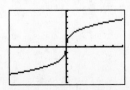

Section 6.7

Are You Ready for This Section?

R1. $3x - 5 = 0$

$3x = 5$

$x = \dfrac{5}{3}$

The solution set is $\left\{ \dfrac{5}{3} \right\}$.

R2. $2p^2 + 4p - 6 = 0$

$p^2 + 2p - 3 = 0$

$(p + 3)(p - 1) = 0$

$p + 3 = 0 \quad$ or $\quad p - 1 = 0$

$p = -3 \qquad\qquad p = 1$

The solution set is $\{-3, 1\}$.

R3. $\left(\sqrt[3]{x-5} \right)^3 = x - 5$

Section 6.7 Quick Checks

1. When the variable in an equation occurs in a radical, the equation is called a <u>radical equation</u>.

2. When an apparent solution is not a solution of the original equation, we say the apparent solution is an <u>extraneous</u> solution.

3. False. The first step would be to isolate the radical by subtracting x from both sides.

4. $\sqrt{3x+1} - 4 = 0$

$\sqrt{3x+1} = 4$

$\left(\sqrt{3x+1} \right)^2 = 4^2$

$3x + 1 = 16$

$3x + 1 - 1 = 16 - 1$

$3x = 15$

$\dfrac{3x}{3} = \dfrac{15}{3}$

$x = 5$

Check:

$\sqrt{3(5)+1} \overset{?}{=} 4$

$\sqrt{15+1} \overset{?}{=} 4$

$\sqrt{16} \overset{?}{=} 4$

$4 = 4\,\text{T}$

The solution set is $\{5\}$.

5. $\sqrt{2x+35} - 2 = 3$

$\sqrt{2x+35} - 2 + 2 = 3 + 2$

$\sqrt{2x+35} = 5$

$\left(\sqrt{2x+35} \right)^2 = 5^2$

$2x + 35 = 25$

$2x + 35 - 35 = 25 - 35$

$2x = -10$

$\dfrac{2x}{2} = \dfrac{-10}{2}$

$x = -5$

Check:

$\sqrt{2(-5)+35} - 2 \overset{?}{=} 3$

$\sqrt{25} - 2 \overset{?}{=} 3$

$5 - 2 \overset{?}{=} 3$

$3 = 3\,\text{T}$

The solution set is $\{-5\}$.

6. $\sqrt{2x+3} + 8 = 6$

$\sqrt{2x+3} + 8 - 8 = 6 - 8$

$\sqrt{2x+3} = -2$

$\left(\sqrt{2x+3} \right)^2 = (-2)^2$

$2x + 3 = 4$

$2x + 3 - 3 = 4 - 3$

$2x = 1$

$\dfrac{2x}{2} = \dfrac{1}{2}$

$x = \dfrac{1}{2}$

Check:

$\sqrt{2\left(\dfrac{1}{2}\right)+3} + 8 \overset{?}{=} 6$

$\sqrt{4} + 8 \overset{?}{=} 6$

$2 + 8 \overset{?}{=} 6$

$10 \neq 6$

The solution does not check. Since there are no other possible solutions, the equation has no real solution.

7.
$$\sqrt{2x+1} = x-1$$
$$\left(\sqrt{2x+1}\right)^2 = (x-1)^2$$
$$2x+1 = x^2 - 2x + 1$$
$$x^2 - 4x = 0$$
$$x(x-4) = 0$$
$$x = 0 \quad \text{or} \quad x-4 = 0$$
$$x = 4$$

Check:
$$\sqrt{2(0)+1} \stackrel{?}{=} 0-1 \qquad \sqrt{2(4)+1} \stackrel{?}{=} 4-1$$
$$\sqrt{1} \stackrel{?}{=} -1 \qquad\qquad \sqrt{9} \stackrel{?}{=} 3$$
$$1 \ne -1 \qquad\qquad\quad 3 = 3\,\text{T}$$

The solution set is $\{4\}$.

8.
$$\sqrt[3]{3x+1} - 4 = -6$$
$$\sqrt[3]{3x+1} - 4 + 4 = -6 + 4$$
$$\sqrt[3]{3x+1} = -2$$
$$\left(\sqrt[3]{3x+1}\right)^3 = (-2)^3$$
$$3x+1 = -8$$
$$3x+1-1 = -8-1$$
$$3x = -9$$
$$\frac{3x}{3} = \frac{-9}{3}$$
$$x = -3$$

Check:
$$\sqrt[3]{3(-3)+1} - 4 \stackrel{?}{=} -6$$
$$\sqrt[3]{-8} - 4 \stackrel{?}{=} -6$$
$$-2 - 4 \stackrel{?}{=} -6$$
$$-6 = -6\,\text{T}$$

The solution set is $\{-3\}$.

9.
$$(2x-3)^{1/3} - 7 = -4$$
$$(2x-3)^{1/3} - 7 + 7 = -4 + 7$$
$$(2x-3)^{1/3} = 3$$
$$\left[(2x-3)^{1/3}\right]^3 = 3^3$$
$$2x-3 = 27$$
$$2x-3+3 = 27+3$$
$$2x = 30$$
$$\frac{2x}{2} = \frac{30}{2}$$
$$x = 15$$

Check:
$$(2(15)-3)^{1/3} - 7 \stackrel{?}{=} -4$$
$$(27)^{1/3} - 7 \stackrel{?}{=} -4$$
$$3 - 7 \stackrel{?}{=} -4$$
$$-4 = -4\,\text{T}$$

The solution set is $\{15\}$.

10.
$$\sqrt[3]{m^2 + 4m + 4} = \sqrt[3]{2m+7}$$
$$\left(\sqrt[3]{m^2+4m+4}\right)^3 = \left(\sqrt[3]{2m+7}\right)^3$$
$$m^2 + 4m + 4 = 2m + 7$$
$$m^2 + 4m + 4 - 2m - 7 = 2m + 7 - 2m - 7$$
$$m^2 + 2m - 3 = 0$$
$$(m+3)(m-1) = 0$$
$$m+3 = 0 \quad \text{or} \quad m-1 = 0$$
$$m = -3 \qquad\qquad m = 1$$

Check:
$$\sqrt[3]{(-3)^2 + 4(-3) + 4} \stackrel{?}{=} \sqrt[3]{2(-3)+7}$$
$$\sqrt[3]{9 - 12 + 4} \stackrel{?}{=} \sqrt[3]{-6+7}$$
$$\sqrt[3]{1} \stackrel{?}{=} \sqrt[3]{1}$$
$$1 = 1\,\text{T}$$

$$\sqrt[3]{(1)^2 + 4(1) + 4} \stackrel{?}{=} \sqrt[3]{2(1)+7}$$
$$\sqrt[3]{1 + 4 + 4} \stackrel{?}{=} \sqrt[3]{2+7}$$
$$\sqrt[3]{9} = \sqrt[3]{9}\,\text{T}$$

The solution set is $\{-3, 1\}$.

11.
$$\sqrt{2x+1} - \sqrt{x+4} = 1$$
$$\sqrt{2x+1} = 1 + \sqrt{x+4}$$
$$\left(\sqrt{2x+1}\right)^2 = \left(1 + \sqrt{x+4}\right)^2$$
$$2x+1 = 1 + 2\sqrt{x+4} + \left(\sqrt{x+4}\right)^2$$
$$2x+1 = 1 + 2\sqrt{x+4} + x + 4$$
$$2x+1 = 5 + x + 2\sqrt{x+4}$$
$$x - 4 = 2\sqrt{x+4}$$
$$(x-4)^2 = \left(2\sqrt{x+4}\right)^2$$
$$x^2 - 8x + 16 = 4(x+4)$$
$$x^2 - 8x + 16 = 4x + 16$$
$$x^2 - 12x = 0$$
$$x(x-12) = 0$$
$$x = 0 \quad \text{or} \quad x-12 = 0$$
$$x = 12$$

Check:

$$\sqrt{2(0)+1} - \sqrt{0+4} \stackrel{?}{=} 1$$

$$\sqrt{1} - \sqrt{4} \stackrel{?}{=} 1$$

$$1 - 2 \stackrel{?}{=} 1$$

$$-1 \neq 1$$

$$\sqrt{2(12)+1} - \sqrt{(12)+4} \stackrel{?}{=} 1$$

$$\sqrt{25} - \sqrt{16} \stackrel{?}{=} 1$$

$$5 - 4 \stackrel{?}{=} 1$$

$$1 = 1 \text{ T}$$

The solution set is $\{12\}$.

12. a.
$$T = 2\pi\sqrt{\frac{L}{32}}$$

$$\frac{T}{2\pi} = \sqrt{\frac{L}{32}}$$

$$\left(\frac{T}{2\pi}\right)^2 = \left(\sqrt{\frac{L}{32}}\right)^2$$

$$\frac{T^2}{4\pi^2} = \frac{L}{32}$$

$$L = \frac{32T^2}{4\pi^2}$$

$$L = \frac{8T^2}{\pi^2}$$

b. $L = \dfrac{8T^2}{\pi^2} = \dfrac{8(2\pi)^2}{\pi^2} = \dfrac{8 \cdot 4\pi^2}{\pi^2} = 32$ feet

6.7 Exercises

13.
$$\sqrt{x} = 4$$

$$\left(\sqrt{x}\right)^2 = 4^2$$

$$x = 16$$

Check: $\sqrt{16} = 4$?

$$4 = 4 \text{ T}$$

The solution set is $\{16\}$.

15.
$$\sqrt{x-3} = 2$$

$$\left(\sqrt{x-3}\right)^2 = 2^2$$

$$x - 3 = 4$$

$$x = 7$$

Check: $\sqrt{7-3} = 2$?

$$\sqrt{4} = 2$$?

$$2 = 2 \text{ T}$$

The solution set is $\{7\}$.

17.
$$\sqrt{2t+3} = 5$$

$$\left(\sqrt{2t+3}\right)^2 = 5^2$$

$$2t + 3 = 25$$

$$2t = 22$$

$$t = 11$$

Check: $\sqrt{2(11)+3} = 5$?

$$\sqrt{25} = 5$$?

$$5 = 5 \text{ T}$$

The solution set is $\{11\}$.

19. $\sqrt{4x+3} = -2$

Since the principal square root is never negative, this equation has no real solution. Solving this equation by squaring both sides will yield an extraneous solution.

$$\sqrt{4x+3} = -2$$

$$\left(\sqrt{4x+3}\right)^2 = (-2)^2$$

$$4x + 3 = 4$$

$$4x = 1$$

$$x = \frac{1}{4}$$

Check: $\sqrt{4\left(\dfrac{1}{4}\right)+3} = -2$?

$$\sqrt{4} = -2$$?

$$2 = -2 \text{ False}$$

The solution does not check so the problem has no real solution.

21.
$$\sqrt[3]{4t} = 2$$

$$\left(\sqrt[3]{4t}\right)^3 = 2^3$$

$$4t = 8$$

$$t = 2$$

Check: $\sqrt[3]{4(2)} = 2$?

$$\sqrt[3]{8} = 2$$?

$$2 = 2 \text{ T}$$

The solution set is $\{2\}$.

23. $\sqrt[3]{5q+4} = 4$

$\left(\sqrt[3]{5q+4}\right)^3 = 4^3$

$5q+4 = 64$

$5q = 60$

$q = 12$

Check: $\sqrt[3]{5(12)+4} = 4$?

$\sqrt[3]{64} = 4$?

$4 = 4$ T

The solution set is $\{12\}$.

25. $\sqrt{y}+3 = 8$

$\sqrt{y} = 5$

$\left(\sqrt{y}\right)^2 = 5^2$

$y = 25$

Check: $\sqrt{25}+3 = 8$?

$5+3 = 8$?

$8 = 8$ T

The solution set is $\{25\}$.

27. $\sqrt{x+5}-3 = 1$

$\sqrt{x+5} = 4$

$\left(\sqrt{x+5}\right)^2 = 4^2$

$x+5 = 16$

$x = 11$

Check: $\sqrt{11+5}-3 = 1$?

$\sqrt{16}-3 = 1$?

$4-3 = 1$?

$1 = 1$ T

The solution set is $\{11\}$.

29. $\sqrt{2x+9}+5 = 6$

$\sqrt{2x+9} = 1$

$\left(\sqrt{2x+9}\right)^2 = (1)^2$

$2x+9 = 1$

$2x = -8$

$x = -4$

Check: $\sqrt{2(-4)+9}+5 = 6$?

$\sqrt{1}+5 = 6$?

$1+5 = 6$?

$6 = 6$ ✓

The solution set is $\{-4\}$.

31. $3\sqrt{x}+5 = 8$

$3\sqrt{x} = 3$

$\sqrt{x} = 1$

$\left(\sqrt{x}\right)^2 = 1^2$

$x = 1$

Check: $3\sqrt{1}+5 = 8$?

$3+5 = 8$?

$8 = 8$ T

The solution set is $\{1\}$.

33. $\sqrt{4-x}-3 = 0$

$\sqrt{4-x} = 3$

$\left(\sqrt{4-x}\right)^2 = 3^2$

$4-x = 9$

$-x = 5$

$x = -5$

Check: $\sqrt{4-(-5)}-3 = 0$?

$\sqrt{9}-3 = 0$?

$3-3 = 0$?

$0 = 0$ T

The solution set is $\{-5\}$.

35. $\sqrt{p} = 2p$

$\left(\sqrt{p}\right)^2 = (2p)^2$

$p = 4p^2$

$0 = 4p^2 - p$

$0 = p(4p-1)$

$p = 0$ or $4p-1 = 0$

$p = 0$ or $p = \dfrac{1}{4}$

Check:

$\sqrt{0} = 2(0)$?

$0 = 0$ T

$$\sqrt{\frac{1}{4}} = 2\left(\frac{1}{4}\right) \ ?$$

$$\frac{1}{2} = \frac{1}{2} \ \text{T}$$

The solution set is $\left\{0, \frac{1}{4}\right\}$.

37.
$$\sqrt{x+6} = x$$
$$\left(\sqrt{x+6}\right)^2 = x^2$$
$$x+6 = x^2$$
$$0 = x^2 - x - 6$$
$$0 = (x-3)(x+2)$$
$$x-3 = 0 \ \text{or} \ x+2 = 0$$
$$x = 3 \ \text{or} \ x = -2$$

Check:
$$\sqrt{3+6} = 3 \ ?$$
$$\sqrt{9} = 3 \ ?$$
$$3 = 3 \ \text{T}$$

$$\sqrt{-2+6} = -2 \ ?$$
$$\sqrt{4} = -2 \ ?$$
$$2 = -2 \ \text{False}$$

The second solution does not check. It is extraneous. The solution set is $\{3\}$.

39.
$$\sqrt{w} = 6 - w$$
$$\left(\sqrt{w}\right)^2 = (6-w)^2$$
$$w = 36 - 12w + w^2$$
$$0 = w^2 - 13w + 36$$
$$0 = (w-4)(w-9)$$
$$w-4 = 0 \ \text{or} \ w-9 = 0$$
$$w = 4 \ \text{or} \quad w = 9$$
Check:
$$\sqrt{4} = 6 - 4 \ ?$$
$$2 = 2 \ \checkmark$$

$$\sqrt{9} = 6 - 9 \ ?$$
$$3 \neq -3 \ \text{False}$$

The second solution does not check. It is extraneous. The solution set is $\{4\}$.

41.
$$\sqrt{17-2x} + 1 = x$$
$$\sqrt{17-2x} = x-1$$
$$\left(\sqrt{17-2x}\right)^2 = (x-1)^2$$
$$17 - 2x = x^2 - 2x + 1$$
$$0 = x^2 - 16$$
$$0 = (x-4)(x+4)$$
$$x-4 = 0 \ \text{or} \ x+4 = 0$$
$$x = 4 \ \text{or} \ x = -4$$

Check:
$$\sqrt{17-2(4)} + 1 = 4 \ ?$$
$$\sqrt{9} + 1 = 4 \ ?$$
$$3 + 1 = 4 \ ?$$
$$4 = 4 \ \text{T}$$

$$\sqrt{17-2(-4)} + 1 = -4 \ ?$$
$$\sqrt{17+8} + 1 = -4 \ ?$$
$$\sqrt{25} + 1 = -4 \ ?$$
$$5 + 1 = -4 \ ?$$
$$6 = -4 \ \text{False}$$

The second solution is extraneous, so the solution set is $\{4\}$.

43.
$$\sqrt{w^2 - 11} + 5 = w + 4$$
$$\sqrt{w^2 - 11} = w - 1$$
$$\left(\sqrt{w^2 - 11}\right)^2 = (w-1)^2$$
$$w^2 - 11 = w^2 - 2w + 1$$
$$0 = -2w + 12$$
$$2w = 12$$
$$w = 6$$

Check: $\sqrt{6^2 - 11} + 5 = 6 + 4 \ ?$
$$\sqrt{36-11} + 5 = 10 \ ?$$
$$\sqrt{25} + 5 = 10 \ ?$$
$$5 + 5 = 10 \ ?$$
$$10 = 10 \ \text{T}$$

The solution set is $\{6\}$.

45.
$$\sqrt{x+9} = \sqrt{2x+5}$$
$$\left(\sqrt{x+9}\right)^2 = \left(\sqrt{2x+5}\right)^2$$
$$x+9 = 2x+5$$
$$-x = -4$$
$$x = 4$$

Check: $\sqrt{4+9} = \sqrt{2(4)+5}$?

$$\sqrt{13} = \sqrt{8+5} \ ?$$

$$\sqrt{13} = \sqrt{13} \text{ T}$$

The solution set is $\{4\}$.

47. $\sqrt[3]{4x-3} = \sqrt[3]{2x-9}$

$$\left(\sqrt[3]{4x-3}\right)^3 = \left(\sqrt[3]{2x-9}\right)^3$$

$$4x-3 = 2x-9$$

$$2x = -6$$

$$x = -3$$

Check: $\sqrt[3]{4(-3)-3} = \sqrt[3]{2(-3)-9}$?

$$\sqrt[3]{-12-3} = \sqrt[3]{-6-9} \ ?$$

$$\sqrt[3]{-15} = \sqrt[3]{-15} \text{ T}$$

The solution set is $\{-3\}$.

49. $\sqrt{2w^2 - 3w - 4} = \sqrt{w^2 + 6w + 6}$

$$\left(\sqrt{2w^2 - 3w - 4}\right)^2 = \left(\sqrt{w^2 + 6w + 6}\right)^2$$

$$2w^2 - 3w - 4 = w^2 + 6w + 6$$

$$w^2 - 9w - 10 = 0$$

$$(w-10)(w+1) = 0$$

$$w - 10 = 0 \ \text{ or } \ w + 1 = 0$$

$$w = 10 \ \text{ or } \ w = -1$$

Check:

$$\sqrt{2(10)^2 - 3(10) - 4} = \sqrt{10^2 + 6(10) + 6} \ ?$$

$$\sqrt{200 - 30 - 4} = \sqrt{100 + 60 + 6} \ ?$$

$$\sqrt{166} = \sqrt{166} \text{ T}$$

$$\sqrt{2(-1)^2 - 3(-1) - 4} = \sqrt{(-1)^2 + 6(-1) + 6} \ ?$$

$$\sqrt{2 + 3 - 4} = \sqrt{1 - 6 + 6} \ ?$$

$$\sqrt{1} = \sqrt{1}$$

$$1 = 1 \text{ T}$$

The solution set is $\{-1, 10\}$.

51. $\sqrt{3w+4} = 2 + \sqrt{w}$

$$\left(\sqrt{3w+4}\right)^2 = \left(2 + \sqrt{w}\right)^2$$

$$3w + 4 = 2^2 + 2 \cdot 2\sqrt{w} + \left(\sqrt{w}\right)^2$$

$$3w + 4 = 4 + 4\sqrt{w} + w$$

$$2w = 4\sqrt{w}$$

$$w = 2\sqrt{w}$$

$$w^2 = \left(2\sqrt{w}\right)^2$$

$$w^2 = 4w$$

$$w^2 - 4w = 0$$

$$w(w-4) = 0$$

$$w = 0 \ \text{ or } \ w - 4 = 0$$

$$w = 0 \ \text{ or } \ w = 4$$

Check:

$$\sqrt{3(0)+4} = 2 + \sqrt{0} \ ?$$

$$\sqrt{4} = 2 \ ?$$

$$2 = 2 \text{ T}$$

$$\sqrt{3(4)+4} = 2 + \sqrt{4} \ ?$$

$$\sqrt{16} = 2 + 2 \ ?$$

$$4 = 4 \text{ T}$$

The solution set is $\{0, 4\}$.

53. $\sqrt{x+1} - \sqrt{x-2} = 1$

$$\sqrt{x+1} = \sqrt{x-2} + 1$$

$$\left(\sqrt{x+1}\right)^2 = \left(\sqrt{x-2} + 1\right)^2$$

$$x + 1 = (x-2) + 2\sqrt{x-2} + 1$$

$$x + 1 = x - 1 + 2\sqrt{x-2}$$

$$2 = 2\sqrt{x-2}$$

$$1 = \sqrt{x-2}$$

$$1^2 = \left(\sqrt{x-2}\right)^2$$

$$1 = x - 2$$

$$3 = x$$

Check: $\sqrt{3+1} - \sqrt{3-2} = 1$?

$$\sqrt{4} - \sqrt{1} = 1 \ ?$$

$$2 - 1 = 1 \ ?$$

$$1 = 1 \text{ T}$$

The solution set is $\{3\}$.

55. $\sqrt{2x+6}-\sqrt{x-1}=2$

$\sqrt{2x+6}=\sqrt{x-1}+2$

$\left(\sqrt{2x+6}\right)^2=\left(\sqrt{x-1}+2\right)^2$

$2x+6=(x-1)+2\cdot2\sqrt{x-1}+2^2$

$2x+6=x-1+4\sqrt{x-1}+4$

$x+3=4\sqrt{x-1}$

$(x+3)^2=\left(4\sqrt{x-1}\right)^2$

$x^2+6x+9=16(x-1)$

$x^2+6x+9=16x-16$

$x^2-10x+25=0$

$(x-5)^2=0$

$x-5=0$

$x=5$

Check: $\sqrt{2(5)+6}-\sqrt{5-1}=2$?

$\sqrt{16}-\sqrt{4}=2$?

$4-2=2$?

$2=2$ T

The solution set is $\{5\}$.

57. $\sqrt{2x+5}-\sqrt{x-1}=2$

$\sqrt{2x+5}=\sqrt{x-1}+2$

$\left(\sqrt{2x+5}\right)^2=\left(\sqrt{x-1}+2\right)^2$

$2x+5=(x-1)+2\cdot2\sqrt{x-1}+2^2$

$2x+5=x-1+4\sqrt{x-1}+4$

$x+2=4\sqrt{x-1}$

$(x+2)^2=\left(4\sqrt{x-1}\right)^2$

$x^2+4x+4=16(x-1)$

$x^2+4x+4=16x-16$

$x^2-12x+20=0$

$(x-10)(x-2)=0$

$x-10=0$ or $x-2=0$

$x=10$ or $x=2$

Check:

$\sqrt{2(10)+5}-\sqrt{10-1}=2$?

$\sqrt{25}-\sqrt{9}=2$?

$5-3=2$?

$2=2$ T

$\sqrt{2(2)+5}-\sqrt{2-1}=2$?

$\sqrt{9}-\sqrt{1}=2$?

$3-1=2$?

$2=2$ T

The solution set is $\{2,10\}$.

59. $(2x+3)^{1/2}=3$

$\left((2x+3)^{1/2}\right)^2=3^2$

$2x+3=9$

$2x=6$

$x=3$

Check: $(2(3)+3)^{1/2}=3$?

$(9)^{1/2}=3$?

$3=3$ T

The solution set is $\{3\}$.

61. $(6x-1)^{1/4}=(2x+15)^{1/4}$

$\left((6x-1)^{1/4}\right)^4=\left((2x+15)^{1/4}\right)^4$

$6x-1=2x+15$

$4x=16$

$x=4$

Check: $(6(4)-1)^{1/4}=(2(4)+15)^{1/4}$?

$(23)^{1/4}=(23)^{1/4}$ T

The solution set is $\{4\}$.

63. $(x+3)^{1/2}-(x-5)^{1/2}=2$

$(x+3)^{1/2}=(x-5)^{1/2}+2$

$\left((x+3)^{1/2}\right)^2=\left((x-5)^{1/2}+2\right)^2$

$x+3=(x-5)+2\cdot2(x-5)^{1/2}+2^2$

$x+3=x-5+4(x-5)^{1/2}+4$

$4=4(x-5)^{1/2}$

$1=(x-5)^{1/2}$

$1^2=\left((x-5)^{1/2}\right)^2$

$1=x-5$

$6=x$

Check: $(6+3)^{1/2} - (6-5)^{1/2} = 2$?

$\qquad 9^{1/2} - 1^{1/2} = 2$?

$\qquad\qquad 3 - 1 = 2$?

$\qquad\qquad\qquad 2 = 2$ T

The solution set is $\{6\}$.

65. $\quad (x-1)^{2/3} = 4$

$\qquad ((x-1)^{2/3})^3 = (4)^3$

$\qquad\quad (x-1)^2 = 64$

$\qquad x^2 - 2x + 1 = 64$

$\qquad x^2 - 2x - 63 = 0$

$\qquad (x-9)(x+7) = 0$

$\quad x - 9 = 0 \quad$ or $\quad x + 7 = 0$

$\qquad x = 9 \quad$ or $\qquad x = -7$

Check: $(9-1)^{2/3} = 4$?

$\qquad\quad 8^{2/3} = 4$?

$\qquad\qquad 4 = 4$ T

Check: $(-7-1)^{2/3} = 4$?

$\qquad\quad (-8)^{2/3} = 4$?

$\qquad\qquad 4 = 4$ T

The solution set is $\{-7, 9\}$.

67. $\quad (x-2)^{3/2} = 64$

$\qquad [(x-2)^{3/2}]^{2/3} = (64)^{2/3}$

$\qquad\qquad x - 2 = 16$

$\qquad\qquad x = 18$

Check: $(18-2)^{3/2} = 64$?

$\qquad\quad 16^{3/2} = 64$?

$\qquad\qquad 64 = 64$ T

The solution set is $\{18\}$.

69. $\quad (2x^4 - 81)^{1/4} + 3x = 4x$

$\qquad\quad (2x^4 - 81)^{1/4} = x$

$\qquad ((2x^4 - 81)^{1/4})^4 = (x)^4$

$\qquad\qquad 2x^4 - 81 = x^4$

$\qquad\qquad x^4 - 18 = 0$

$\qquad (x^2 + 9)(x^2 - 9) = 0$

$\quad (x^2 + 9)(x+3)(x-3) = 0$

$x^2 + 9 = 0 \quad$ or $\quad x + 3 = 0 \quad$ or $\quad x - 3 = 0$

$\quad x^2 = -9 \quad$ or $\qquad x = -3 \quad$ or $\qquad x = 3$

There are no real numbers x such that $x^2 = -9$.

Check: $(2(-3)^4 - 81)^{1/4} + 3(-3) = 4(-3)$?

$\qquad\qquad 81^{1/4} - 9 = -12$?

$\qquad\qquad\qquad 3 - 9 = -12$?

$\qquad\qquad\qquad\quad -6 = -12$ F

Check: $(2(3)^4 - 81)^{1/4} + 3(3) = 4(3)$?

$\qquad\qquad 81^{1/4} + 9 = 12$?

$\qquad\qquad\qquad 3 + 9 = 12$?

$\qquad\qquad\qquad\quad 12 = 12$ T

The solution set is $\{3\}$.

71. $\quad \dfrac{1}{(x-4)^{-2/3}} - 3 = 6$

$\qquad\quad \dfrac{1}{(x-4)^{-2/3}} = 9$

$\qquad\qquad (x-4)^{2/3} = 9$

$\qquad\qquad ((x-4)^{2/3})^3 = 9^3$

$\qquad\qquad\quad (x-4)^2 = 729$

$\qquad\qquad x^2 - 8x + 16 = 729$

$\qquad\qquad x^2 - 8x - 713 = 0$

$\qquad (x-31)(x+23) = 0$

$x - 31 = 0 \quad$ or $\quad x + 23 = 0$

$\quad x = 31 \quad$ or $\qquad x = -23$

Check: $\dfrac{1}{(31-4)^{-2/3}} - 3 = 6$?

$\qquad\qquad 9 - 3 = 6$?

$\qquad\qquad\quad 6 = 6$ T

Check: $\dfrac{1}{(-23-4)^{-2/3}} - 3 = 6$?

$\qquad\qquad 9 - 3 = 6$?

$\qquad\qquad\quad 6 = 6$ T

The solution set is $\{-23, 31\}$.

73. $\quad A = P\sqrt{1+r}$

$\qquad \dfrac{A}{P} = \sqrt{1+r}$

$\qquad \left(\dfrac{A}{P}\right)^2 = \left(\sqrt{1+r}\right)^2$

$\qquad \dfrac{A^2}{P^2} = 1 + r$

$\dfrac{A^2}{P^2} - 1 = r \quad$ or $\quad r = \dfrac{A^2}{P^2} - 1 = \dfrac{A^2 - P^2}{P^2}$

75.
$$r = \sqrt[3]{\frac{3V}{4\pi}}$$

$$r^3 = \left(\sqrt[3]{\frac{3V}{4\pi}}\right)^3$$

$$r^3 = \frac{3V}{4\pi}$$

$$4\pi r^3 = 3V$$

$$\frac{4\pi r^3}{3} = V \quad \text{or} \quad V = \frac{4}{3}\pi r^3$$

77.
$$r = \sqrt{\frac{4F\pi\varepsilon_0}{q_1 q_2}}$$

$$r^2 = \left(\sqrt{\frac{4F\pi\varepsilon_0}{q_1 q_2}}\right)^2$$

$$r^2 = \frac{4F\pi\varepsilon_0}{q_1 q_2}$$

$$q_1 q_2 r^2 = 4F\pi\varepsilon_0$$

$$\frac{q_1 q_2 r^2}{4\pi\varepsilon_0} = F \quad \text{or} \quad F = \frac{q_1 q_2 r^2}{4\pi\varepsilon_0}$$

79. $\sqrt{5p-3} + 7 = 3$
$$\sqrt{5p-3} = -4$$

At this point we see that there is no real solution. Since the principal square root is never negative, $\sqrt{5p-3} = -4$ has no real solution. Solving the usual way yields the same result.

$$\left(\sqrt{5p-3}\right)^2 = (-4)^2$$
$$5p-3 = 16$$
$$5p = 19$$
$$p = \frac{19}{5}$$

Check: $\sqrt{5\left(\frac{19}{5}\right)-3} + 7 = 3$?
$$\sqrt{19-3} + 7 = 3 \ ?$$
$$\sqrt{16} + 7 = 3 \ ?$$
$$4 + 7 = 3 \ ?$$
$$11 = 3 \quad \text{False}$$

The solution does not check, so the equation has no real solutions.
$\varnothing$ or { }

81.
$$\sqrt{x+12} = x$$
$$\left(\sqrt{x+12}\right)^2 = x^2$$
$$x+12 = x^2$$
$$0 = x^2 - x - 12$$
$$0 = (x-4)(x+3)$$
$$x-4 = 0 \quad \text{or} \quad x+3 = 0$$
$$x = 4 \qquad\qquad x = -3$$

Check:
$$\sqrt{4+12} = 4 \ ? \qquad \sqrt{-3+12} = -3 \ ?$$
$$\sqrt{16} = 4 \ ? \qquad\qquad \sqrt{9} = -3 \ ?$$
$$4 = 4 \ \text{T} \qquad\qquad 3 = -3 \ \text{False}$$

The second solution does not check. The solution set is $\{4\}$.

83.
$$\sqrt{2p+12} = 4$$
$$\left(\sqrt{2p+12}\right)^2 = 4^2$$
$$2p+12 = 16$$
$$2p = 4$$
$$p = 2$$

Check: $\sqrt{2(2)+12} = 4 \ ?$
$$\sqrt{16} = 4 \ ?$$
$$4 = 4 \ \text{T}$$
The solution set is $\{2\}$.

85.
$$\sqrt[4]{x+7} = 2$$
$$\left(\sqrt[4]{x+7}\right)^4 = 2^4$$
$$x+7 = 16$$
$$x = 9$$

Check: $\sqrt[4]{9+7} = 2 \ ?$
$$\sqrt[4]{16} = 2 \ ?$$
$$2 = 2 \ \text{T}$$
The solution set is $\{9\}$.

87. $(3x+1)^{1/3} + 2 = 0$
$$(3x+1)^{1/3} = -2$$
$$\left((3x+1)^{1/3}\right)^3 = (-2)^3$$
$$3x+1 = -8$$
$$3x = -9$$
$$x = -3$$

Check: $(3(-3)+1)^{1/3} + 2 = 0$?

$$(-8)^{1/3} + 2 = 0 \ ?$$
$$-2 + 2 = 0 \ ?$$
$$0 = 0 \ \text{T}$$

The solution set is $\{-3\}$.

89. $\sqrt{10-x} - x = 10$

$$\sqrt{10-x} = x + 10$$
$$\left(\sqrt{10-x}\right)^2 = (x+10)^2$$
$$10 - x = x^2 + 20x + 100$$
$$0 = x^2 + 21x + 90$$
$$0 = (x+6)(x+15)$$
$$x + 6 = 0 \quad \text{or} \quad x + 15 = 0$$
$$x = -6 \ \text{or} \quad\quad x = -15$$

Check:

$$\sqrt{10-(-6)} - (-6) = 10 \ ?$$
$$\sqrt{16} + 6 = 10 \ ?$$
$$4 + 6 = 10 \ ?$$
$$10 = 10 \ \checkmark$$

$$\sqrt{10-(-15)} - (-15) = 10 \ ?$$
$$\sqrt{25} + 15 = 10 \ ?$$
$$5 + 15 = 10 \ ?$$
$$20 \neq 10 \ \text{False}$$

The second solution does not check. It is extraneous. The solution set is $\{-6\}$.

91. $\sqrt{2x+5} = \sqrt{3x-4}$

$$\left(\sqrt{2x+5}\right)^2 = \left(\sqrt{3x-4}\right)^2$$
$$2x + 5 = 3x - 4$$
$$-x + 5 = -4$$
$$-x = -9$$
$$x = 9$$

Check: $\sqrt{2(9)+5} = \sqrt{3(9)-4}$?

$$\sqrt{23} = \sqrt{23} \ \text{T}$$

The solution set is $\{9\}$.

93. $\sqrt{x-1} + \sqrt{x+4} = 5$

$$\sqrt{x-1} = 5 - \sqrt{x+4}$$
$$\left(\sqrt{x-1}\right)^2 = \left(5 - \sqrt{x+4}\right)^2$$
$$x - 1 = 5^2 - 2 \cdot 5\sqrt{x+4} + (x+4)$$
$$x - 1 = 25 - 10\sqrt{x+4} + x + 4$$
$$-30 = -10\sqrt{x+4}$$
$$3 = \sqrt{x+4}$$
$$3^2 = \left(\sqrt{x+4}\right)^2$$
$$9 = x + 4$$
$$5 = x$$

Check: $\sqrt{5-1} + \sqrt{5+4} = 5$?

$$\sqrt{4} + \sqrt{9} = 5 \ ?$$
$$2 + 3 = 5 \ ?$$
$$5 = 5 \ \text{T}$$

The solution set is $\{5\}$.

95. $\sqrt{x^2} = x + 4$

$$\left(\sqrt{x^2}\right)^2 = (x+4)^2$$
$$x^2 = x^2 + 8x + 16$$
$$0 = 8x + 16$$
$$8x = -16$$
$$x = -2$$

Check:

$$\sqrt{(-2)^2} = -2 + 4 \ ?$$
$$\sqrt{4} = 2 \ ?$$
$$2 = 2 \ \checkmark$$

The solution set is $\{-2\}$.

$$\sqrt{x^2} = x + 4$$
$$|x| = x + 4$$
$$x = -(x+4) \quad \text{or} \quad x = x + 4$$
$$x = -x - 4 \quad \text{or} \quad 0 = 4 \quad \text{False}$$
$$2x = -4$$
$$x = -2$$

Preference will vary.

97. $f(x) = \sqrt{x-2}$

a.
$$f(x) = 0$$
$$\sqrt{x-2} = 0$$
$$\left(\sqrt{x-2}\right)^2 = 0^2$$
$$x - 2 = 0$$
$$x = 2$$
The point $(2,0)$ is on the graph of f.

b.
$$f(x) = 1$$
$$\sqrt{x-2} = 1$$
$$\left(\sqrt{x-2}\right)^2 = 1^2$$
$$x - 2 = 1$$
$$x = 3$$
The point $(3,1)$ is on the graph of f.

c.
$$f(x) = 2$$
$$\sqrt{x-2} = 2$$
$$\left(\sqrt{x-2}\right)^2 = 2^2$$
$$x - 2 = 4$$
$$x = 6$$
The point $(6,2)$ is on the graph of f.

d. The points $(2,0)$, $(3,1)$, and $(6,2)$ are on the graph.

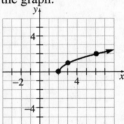

e. The equation $f(x) = -1$ has no solution because the graph of the function does not go below the x-axis. Therefore, the value of the function will never be negative.

99. a.
$$\sqrt{4^2 + (y-2)^2} = 5$$
$$\left(\sqrt{4^2 + (y-2)^2}\right)^2 = 5^2$$
$$4^2 + (y-2)^2 = 5^2$$
$$16 + (y-2)^2 = 25$$
$$(y-2)^2 = 9$$
$$\sqrt{(y-2)^2} = \sqrt{9}$$
$$|y-2| = 3$$
$$y - 2 = 3 \text{ or } y - 2 = -3$$
$$y = 5 \text{ or } y = -1$$

b. The points are $(3,2)$, $(-1,-1)$, and $(-1,5)$.

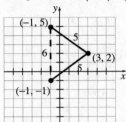

The figure is an isosceles triangle whose common side has a length of 5 units. The base has a length of 6 units. If the triangle were bisected from the vertex to the base, two 3-4-5 right triangles would be formed.

101. a.
$$\sqrt[3]{\frac{t}{2}} = 1$$
$$\left(\sqrt[3]{\frac{t}{2}}\right)^3 = 1^3$$
$$\frac{t}{2} = 1$$
$$t = 2$$
Annual revenue will be \$1 million after 2 years.

b.
$$\sqrt[3]{\frac{t}{2}} = 2$$
$$\left(\sqrt[3]{\frac{t}{2}}\right)^3 = 2^3$$
$$\frac{t}{2} = 8$$
$$t = 16$$
Annual revenue will be \$2 million after 16 years.

103. $R(t) = 26 \cdot \sqrt[10]{t}$

  a. $26 \cdot \sqrt[10]{t} = 39$

$$\sqrt[10]{t} = \frac{39}{26}$$

$$\left(\sqrt[10]{t}\right)^{10} = \left(\frac{39}{26}\right)^{10}$$

$$t \approx 57.67$$

The plural birth rate is predicted to be 39 in the year 2053.

  b. $26 \cdot \sqrt[10]{t} = 36$

$$\sqrt[10]{t} = \frac{18}{13}$$

$$\left(\sqrt[10]{t}\right)^{10} = \left(\frac{18}{13}\right)^{10}$$

$$t \approx 25.9$$

The plural birth rate is predicted to be 36 in the year 2021.

105. $\sqrt{3\sqrt{x+1}} = \sqrt{2x+3}$

$$\left(\sqrt{3\sqrt{x+1}}\right)^2 = \left(\sqrt{2x+3}\right)^2$$

$$3\sqrt{x+1} = 2x+3$$

$$\left(3\sqrt{x+1}\right)^2 = (2x+3)^2$$

$$9(x+1) = (2x)^2 + 2\cdot 3\cdot 2x + 3^2$$

$$9x+9 = 4x^2 + 12x + 9$$

$$0 = 4x^2 + 3x$$

$$0 = x(4x+3)$$

$x = 0$ or $4x+3 = 0$
$x = 0$ or $4x = -3$
$x = 0$ or $x = -\dfrac{3}{4}$

Check:

$$\sqrt{3\sqrt{0+1}} = \sqrt{2(0)+3} \ ?$$

$$\sqrt{3\sqrt{1}} = \sqrt{3} \ ?$$

$$\sqrt{3} = \sqrt{3} \ \text{T}$$

$$\sqrt{3\sqrt{-\frac{3}{4}+1}} = \sqrt{2\left(-\frac{3}{4}\right)+3} \ ?$$

$$\sqrt{3\cdot \frac{1}{2}} = \sqrt{-\frac{3}{2}+3} \ ?$$

$$\sqrt{\frac{3}{2}} = \sqrt{\frac{3}{2}} \ \text{T}$$

The solution set is $\left\{-\dfrac{3}{4}, 0\right\}$.

107. Answers will vary. It is necessary to check solutions to radical equations because manipulations of the equation may have introduced extraneous solutions.

109. Answers will vary. Radical equations with an even index may have extraneous solutions because often the domain is not all real numbers. When we clear the radical, we lose this restriction. Radical equations with an odd index do not have this problem, because they do not have domain restrictions.

111. 0, -4, and 12 are the integers.

113. $\sqrt{2^3}$, π, and $\sqrt[3]{-4}$ are irrational numbers.

115. Answers will vary.
A rational number is a real number that can be expressed as the ratio of two integers. An irrational number is a real number that cannot be written as the ratio of two integers. In decimal form, a rational number will terminate or repeat (e.g. 3.7 or $1.\overline{3}$) while an irrational number will be non-terminating and non-repeating (e.g. $\sqrt{2} = 1.414213562...$).

117.

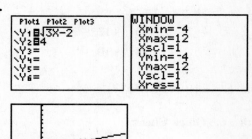

119.

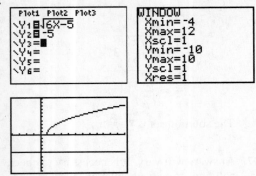

Section 6.8

Are You Ready for This Section?

R1. a. 8 is the only natural number in the set.

 b. 8 and 0 are the whole numbers in the set.

 c. -23, $-\dfrac{12}{3}$, 0, and 8 are the integers in the set.

 d. -23, $-\dfrac{12}{3}$, $-\dfrac{1}{3}$, 0, $1.\overline{26}$, and 8 are the rational numbers in the set.

 e. $\sqrt{2}$ is the only irrational number in the set.

 f. -23, $-\dfrac{12}{3}$, $-\dfrac{1}{3}$, 0, $1.\overline{26}$, $\sqrt{2}$, and 8 are the real numbers in the set.

R2. $3x(4x-3) = 3x\cdot 4x - 3x\cdot 3 = 12x^2 - 9x$

R3. $(z+4)(3z-2) = z\cdot 3z + 4\cdot 3z - 2\cdot z - 4\cdot 2$
$$= 3z^2 + 12z - 2z - 8$$
$$= 3z^2 + 10z - 8$$

R4. $(2y+5)(2y-5) = (2y)^2 - 5^2 = 4y^2 - 25$

Section 6.8 Quick Checks

 1. The <u>imaginary unit</u>, denoted by i, is the number whose square is -1.

 2. Any number of the form bi is called a <u>pure imaginary number</u>.

3. If N is a positive real number, we define the principal square root of $-N$, denoted by $\sqrt{-N}$, as $\sqrt{-N} = \sqrt{N}\,i$.

4. True

5. $\sqrt{-36} = \sqrt{36\cdot(-1)} = \sqrt{36}\cdot\sqrt{-1} = 6i$

6. $\sqrt{-5} = \sqrt{5\cdot(-1)} = \sqrt{5}\cdot\sqrt{-1} = \sqrt{5}\,i$

7. $\sqrt{-12} = \sqrt{12\cdot(-1)} = \sqrt{12}\cdot\sqrt{-1} = 2\sqrt{3}\,i$

8. $4 + \sqrt{-100} = 4 + \sqrt{100\cdot(-1)}$
$$= 4 + \sqrt{100}\cdot\sqrt{-1}$$
$$= 4 + 10i$$

9. $-2 - \sqrt{-8} = -2 - \sqrt{8\cdot(-1)}$
$$= -2 - \sqrt{8}\cdot\sqrt{-1}$$
$$= -2 - 2\sqrt{2}\,i$$

10. $\dfrac{6-\sqrt{-72}}{3} = \dfrac{6 - \sqrt{72\cdot(-1)}}{3}$
$$= \dfrac{6 - \sqrt{72}\cdot\sqrt{-1}}{3}$$
$$= \dfrac{6 - 6\sqrt{2}\,i}{3}$$
$$= \dfrac{3\left(2 - 2\sqrt{2}\,i\right)}{3}$$
$$= 2 - 2\sqrt{2}\,i$$

11. $(4+6i) + (-3+5i) = 4 + 6i - 3 + 5i$
$$= (4-3) + (6+5)i$$
$$= 1 + 11i$$

12. $(4-2i) - (-2+7i) = 4 - 2i + 2 - 7i$
$$= (4+2) + (-2-7)i$$
$$= 6 - 9i$$

13. $\left(4-\sqrt{-4}\right) + \left(-7+\sqrt{-9}\right) = (4-2i) + (-7+3i)$
$$= 4 - 2i - 7 + 3i$$
$$= (4-7) + (-2+3)i$$
$$= -3 + i$$

14. $3i(5-4i) = 3i \cdot 5 - 3i \cdot 4i$
    $= 15i - 12i^2$
    $= 15i - 12(-1)$
    $= 12 + 15i$

15. $(-2+5i)(4-2i) = -2 \cdot 4 - 2 \cdot (-2i) + 5i \cdot 4 - 5i \cdot 2i$
    $= -8 + 4i + 20i - 10i^2$
    $= -8 + 24i - 10(-1)$
    $= 2 + 24i$

16. $\sqrt{-9} \cdot \sqrt{-36} = 3i \cdot 6i = 18i^2 = -18$

17. $\left(2+\sqrt{-36}\right)\left(4-\sqrt{-25}\right)$
    $= (2+6i)(4-5i)$
    $= 2 \cdot 4 - 2 \cdot 5i + 6i \cdot 4 - 6i \cdot 5i$
    $= 8 - 10i + 24i - 30i^2$
    $= 8 + 14i - 30(-1)$
    $= 38 + 14i$

18. The complex conjugate of $-3 + 5i$ is $\underline{-3 - 5i}$.

19. $(3-8i)(3+8i) = 3^2 - (8i)^2$
    $= 9 - 64i^2$
    $= 9 - 64(-1)$
    $= 73$

20. $(-2+5i)(-2-5i) = (-2)^2 - (5i)^2$
    $= 4 - 25i^2$
    $= 4 - 25(-1)$
    $= 29$

21. $\dfrac{-4+i}{3i} = \dfrac{-4+i}{3i} \cdot \dfrac{3i}{3i} = \dfrac{-12i+3i^2}{9i^2}$
    $= \dfrac{-12i-3}{-9} = \dfrac{-3}{-9} + \dfrac{-12}{-9}i$
    $= \dfrac{1}{3} + \dfrac{4}{3}i$

22. $\dfrac{4+3i}{1-3i} = \dfrac{4+3i}{1-3i} \cdot \dfrac{1+3i}{1+3i}$
    $= \dfrac{(4+3i)(1+3i)}{(1-3i)(1+3i)}$
    $= \dfrac{4+12i+3i+9i^2}{1-9i^2}$
    $= \dfrac{4+15i-9}{1+9}$
    $= \dfrac{-5+15i}{10}$
    $= -\dfrac{5}{10} + \dfrac{15}{10}i$
    $= -\dfrac{1}{2} + \dfrac{3}{2}i$

23. $i^{43} = i^{40} \cdot i^3$
    $= \left(i^4\right)^{10} \cdot i^3$
    $= (1)^{10} \cdot (-i)$
    $= -i$

24. $i^{98} = i^{96} \cdot i^2$
    $= \left(i^4\right)^{24} \cdot i^2$
    $= (1)^{24} \cdot (-1)$
    $= -1$

6.8 Exercises

25. $\sqrt{-4} = \sqrt{-1} \cdot \sqrt{4} = i \cdot 2 = 2i$

27. $-\sqrt{-81} = -\sqrt{-1} \cdot \sqrt{81} = -i \cdot 9 = -9i$

29. $\sqrt{-45} = \sqrt{-1} \cdot \sqrt{9} \cdot \sqrt{5} = i \cdot 3 \cdot \sqrt{5} = 3\sqrt{5}\, i$

31. $\sqrt{-300} = \sqrt{-1} \cdot \sqrt{3} \cdot \sqrt{100} = i\sqrt{3} \cdot 10 = 10\sqrt{3}\, i$

33. $\sqrt{-7} = \sqrt{-1} \cdot \sqrt{7} = i\sqrt{7} = \sqrt{7}\, i$

35. $5+\sqrt{-49} = 5+\sqrt{-1} \cdot \sqrt{49}$
    $= 5 + i7$
    $= 5 + 7i$

37. $-2-\sqrt{-28} = -2-\sqrt{-1} \cdot \sqrt{4} \cdot \sqrt{7}$
    $= -2 - i \cdot 2\sqrt{7}$
    $= -2 - 2\sqrt{7}\, i$

39. $\dfrac{4+\sqrt{-4}}{2} = \dfrac{4+\sqrt{-1}\cdot\sqrt{4}}{2}$

$\qquad = \dfrac{4+i\cdot 2}{2}$

$\qquad = 2+i$

41. $\dfrac{4+\sqrt{-8}}{12} = \dfrac{4+\sqrt{-1}\cdot\sqrt{4}\cdot\sqrt{2}}{12}$

$\qquad = \dfrac{4+i\cdot 2\sqrt{2}}{12}$

$\qquad = \dfrac{2+i\sqrt{2}}{6}$

$\qquad = \dfrac{1}{3}+\dfrac{\sqrt{2}}{6}i$

43. $(4+5i)+(2-7i) = 4+5i+2-7i$

$\qquad = (4+2)+(5-7)i$

$\qquad = 6-2i$

45. $(4+i)-(8-5i) = 4+i-8+5i$

$\qquad = (4-8)+(1+5)i$

$\qquad = -4+6i$

47. $\left(4-\sqrt{-4}\right)-\left(2+\sqrt{-9}\right) = (4-2i)-(2+3i)$

$\qquad = 4-2i-2-3i$

$\qquad = (4-2)+(-2-3)i$

$\qquad = 2-5i$

49. $\left(-2+\sqrt{-18}\right)+\left(5-\sqrt{-50}\right)$

$\quad = \left(-2+3\sqrt{2}\,i\right)+\left(5-5\sqrt{2}\,i\right)$

$\quad = -2+3\sqrt{2}\,i+5-5\sqrt{2}\,i$

$\quad = (-2+5)+\left(3\sqrt{2}-5\sqrt{2}\right)i$

$\quad = 3-2\sqrt{2}\,i$

51. $6i(2-4i) = 6i\cdot 2-6i\cdot 4i$

$\qquad = 12i-24i^2$

$\qquad = 12i-24(-1)$

$\qquad = 24+12i$

53. $-\dfrac{1}{2}i(4-10i) = -\dfrac{1}{2}i\cdot 4+\dfrac{1}{2}i\cdot 10i$

$\qquad = -2i+5i^2$

$\qquad = -2i+5(-1)$

$\qquad = -5-2i$

55. $(2+i)(4+3i) = 2\cdot 4+2\cdot 3i+i\cdot 4+i\cdot 3i$

$\qquad = 8+6i+4i+3i^2$

$\qquad = 8+10i-3$

$\qquad = 5+10i$

57. $(-3-5i)(2+4i) = -3\cdot 2-3\cdot 4i-5i\cdot 2-5i\cdot 4i$

$\qquad = -6-12i-10i-20i^2$

$\qquad = -6-22i-20(-1)$

$\qquad = -6-22i+20$

$\qquad = 14-22i$

59. $(2-3i)(4+6i) = 2\cdot 4+2\cdot 6i-3i\cdot 4-3i\cdot 6i$

$\qquad = 8+12i-12i-18i^2$

$\qquad = 8-18(-1)$

$\qquad = 8+18$

$\qquad = 26$

61. $\left(3-\sqrt{2}i\right)\left(-2+\sqrt{2}i\right)$

$\quad = 3(-2)+3\left(\sqrt{2}\,i\right)-\sqrt{2}\,i(-2)-\left(\sqrt{2}\,i\right)^2$

$\quad = -6+3\sqrt{2}\,i+2\sqrt{2}\,i-2i^2$

$\quad = -6+5\sqrt{2}\,i-2(-1)$

$\quad = -4+5\sqrt{2}\,i$

63. $\left(\dfrac{1}{2}-\dfrac{1}{4}i\right)\left(\dfrac{2}{3}+\dfrac{3}{4}i\right)$

$\quad = \dfrac{1}{2}\cdot\dfrac{2}{3}+\dfrac{1}{2}\cdot\dfrac{3}{4}i-\dfrac{1}{4}i\cdot\dfrac{2}{3}-\dfrac{1}{4}i\cdot\dfrac{3}{4}i$

$\quad = \dfrac{1}{3}+\dfrac{3}{8}i-\dfrac{1}{6}i-\dfrac{3}{16}i^2$

$\quad = \dfrac{1}{3}+\dfrac{5}{24}i-\dfrac{3}{16}(-1)$

$\quad = \dfrac{25}{48}+\dfrac{5}{24}i$

65. $(3+2i)^2 = 3^2+2(3)(2i)+(2i)^2$

$\qquad = 9+12i+4i^2$

$\qquad = 9+12i+4(-1)$

$\qquad = 5+12i$

67. $(-4-5i)^2 = (-4)^2-2(-4)(5i)+(5i)^2$

$\qquad = 16+40i+25i^2$

$\qquad = 16+40i+25(-1)$

$\qquad = -9+40i$

69. $\sqrt{-9}\cdot\sqrt{-4} = 3i\cdot 2i = 6i^2 = -6$

71.
$$\sqrt{-8} \cdot \sqrt{-10} = 2\sqrt{2}\,i \cdot \sqrt{10}\,i$$
$$= 2\sqrt{20}\,i^2$$
$$= 2 \cdot 2\sqrt{5} \cdot (-1)$$
$$= -4\sqrt{5}$$

73.
$$\left(2 + \sqrt{-81}\right)\left(-3 - \sqrt{-100}\right)$$
$$= (2 + 9i)(-3 - 10i)$$
$$= 2(-3) - 2(10i) - 9i(3) - 9i(10i)$$
$$= -6 - 20i - 27i - 90i^2$$
$$= -6 - 47i - 90(-1)$$
$$= 84 - 47i$$

75. a. The conjugate of $3 + 5i$ is $3 - 5i$.

b.
$$(3 + 5i)(3 - 5i) = 3^2 - (5i)^2$$
$$= 9 - 25i^2$$
$$= 9 - 25(-1)$$
$$= 9 + 25$$
$$= 34$$

77. a. The conjugate of $2 - 7i$ is $2 + 7i$.

b.
$$(2 - 7i)(2 + 7i) = 2^2 - (7i)^2$$
$$= 4 - 49i^2$$
$$= 4 - 49(-1)$$
$$= 4 + 49$$
$$= 53$$

79. a. The conjugate of $-7 + 2i$ is $-7 - 2i$.

b.
$$(-7 + 2i)(-7 - 2i) = (-7)^2 - (2i)^2$$
$$= 49 - 4i^2$$
$$= 49 - 4(-1)$$
$$= 49 + 4$$
$$= 53$$

81.
$$\frac{1+i}{3i} = \frac{(1+i)}{3i} \cdot \frac{i}{i}$$
$$= \frac{i + i^2}{3i^2}$$
$$= \frac{i + (-1)}{3(-1)}$$
$$= \frac{i - 1}{-3}$$
$$= \frac{-1}{-3} + \frac{1}{-3}i$$
$$= \frac{1}{3} - \frac{1}{3}i$$

83.
$$\frac{-5 + 2i}{5i} = \frac{(-5 + 2i)}{5i} \cdot \frac{i}{i}$$
$$= \frac{-5i + 2i^2}{5i^2}$$
$$= \frac{-5i + 2(-1)}{5(-1)}$$
$$= \frac{-5i - 2}{-5}$$
$$= \frac{-2}{-5} + \frac{-5}{-5}i$$
$$= \frac{2}{5} + i$$

85.
$$\frac{3}{2+i} = \frac{3}{2+i} \cdot \frac{2-i}{2-i}$$
$$= \frac{6 - 3i}{2^2 - i^2}$$
$$= \frac{6 - 3i}{4 - (-1)}$$
$$= \frac{6 - 3i}{5}$$
$$= \frac{6}{5} - \frac{3}{5}i$$

87. $\dfrac{-2}{-3-7i} = \dfrac{-2}{-3-7i} \cdot \dfrac{-3+7i}{-3+7i}$

$= \dfrac{6-14i}{(-3)^2 - (7i)^2}$

$= \dfrac{6-14i}{9-49i^2}$

$= \dfrac{6-14i}{9-49(-1)}$

$= \dfrac{6-14i}{58}$

$= \dfrac{6}{58} - \dfrac{14}{58}i$

$= \dfrac{3}{29} - \dfrac{7}{29}i$

89. $\dfrac{2+3i}{3-2i} = \dfrac{2+3i}{3-2i} \cdot \dfrac{3+2i}{3+2i}$

$= \dfrac{2\cdot 3 + 2\cdot 2i + 3i\cdot 3 + 3i\cdot 2i}{3^2 - (2i)^2}$

$= \dfrac{6+4i+9i+6i^2}{9-4i^2}$

$= \dfrac{6+13i-6}{9+4}$

$= \dfrac{13i}{13}$

$= i$

91. $\dfrac{4+2i}{1-i} = \dfrac{4+2i}{1-i} \cdot \dfrac{1+i}{1+i}$

$= \dfrac{4\cdot 1 + 4\cdot i + 2i\cdot 1 + 2i\cdot i}{1^2 - i^2}$

$= \dfrac{4+4i+2i+2i^2}{1+1}$

$= \dfrac{4+6i-2}{2}$

$= \dfrac{2+6i}{2}$

$= 1+3i$

93. $\dfrac{4-2i}{1+3i} = \dfrac{4-2i}{1+3i} \cdot \dfrac{1-3i}{1-3i}$

$= \dfrac{4\cdot 1 - 4\cdot 3i - 2i\cdot 1 + 2i\cdot 3i}{1^2 - (3i)^2}$

$= \dfrac{4-12i-2i+6i^2}{1-9i^2}$

$= \dfrac{4-14i-6}{1+9}$

$= \dfrac{-2-14i}{10}$

$= \dfrac{-2}{10} + \dfrac{-14i}{10}$

$= -\dfrac{1}{5} - \dfrac{7}{5}i$

95. $i^{53} = i^{52} \cdot i = \left(i^4\right)^{13} \cdot i = 1 \cdot i = i$

97. $i^{43} = i^{40} \cdot i^3 = \left(i^4\right)^{10} \cdot i^3 = 1 \cdot (-i) = -i$

99. $i^{153} = i^{152} \cdot i = \left(i^4\right)^{38} \cdot i = 1 \cdot i = i$

101. $i^{-45} = i^{-48} \cdot i^3 = \left(i^4\right)^{-12} \cdot i^3 = 1 \cdot i^3 = i^3 = -i$

103. $(-4-i)(4+i) = -4\cdot 4 - 4\cdot i - i\cdot 4 - i\cdot i$

$= -16 - 4i - 4i - i^2$

$= -16 - 8i - (-1)$

$= -15 - 8i$

105. $(3+2i)^2 = 3^2 + 2(3)(2i) + (2i)^2$

$= 9 + 12i + 4i^2$

$= 9 + 12i - 4$

$= 5 + 12i$

107. $\dfrac{-3+2i}{3i} = \dfrac{-3+2i}{3i} \cdot \dfrac{i}{i}$

$= \dfrac{-3i + 2i^2}{3i^2}$

$= \dfrac{-3i - 2}{-3}$

$= \dfrac{2}{3} + i$

109. $\dfrac{-4+i}{-5-3i} = \dfrac{-4+i}{-5-3i} \cdot \dfrac{-5+3i}{-5+3i}$

$= \dfrac{-4(-5)-4\cdot 3i - i\cdot 5 + i\cdot 3i}{(-5)^2 - (3i)^2}$

$= \dfrac{20 - 12i - 5i + 3i^2}{25 - 9i^2}$

$= \dfrac{20 - 17i - 3}{25 + 9}$

$= \dfrac{17 - 17i}{34}$

$= \dfrac{17}{34} - \dfrac{17}{34}i$

$= \dfrac{1}{2} - \dfrac{1}{2}i$

111. $(10-3i)+(2+3i) = 10-3i+2+3i$

$= (10+2)+(-3+3)i$

$= 12 + 0i$

$= 12$

113. $5i^{37}(-4+3i) = 5\cdot i^{36}\cdot i(-4+3i)$

$= 5\left(i^4\right)^9 \cdot i(-4+3i)$

$= 5\cdot 1\cdot i(-4+3i)$

$= 5i(-4+3i)$

$= 5i\cdot(-4)+5i\cdot 3i$

$= -20i + 15i^2$

$= -20i - 15$

$= -15 - 20i$

115. $\sqrt{-10}\cdot\sqrt{-15} = \sqrt{-1}\cdot\sqrt{10}\cdot\sqrt{-1}\cdot\sqrt{15}$

$= i\cdot\sqrt{10}\cdot i\cdot\sqrt{15}$

$= i^2\cdot\sqrt{150}$

$= -1\cdot\sqrt{25}\cdot\sqrt{6}$

$= -5\sqrt{6}$

117. $\dfrac{1}{5i} = \dfrac{1}{5i}\cdot\dfrac{i}{i} = \dfrac{i}{5i^2} = \dfrac{i}{-5} = -\dfrac{1}{5}i$

119. $\dfrac{1}{2-i} = \dfrac{1}{2-i}\cdot\dfrac{2+i}{2+i}$

$= \dfrac{2+i}{2^2 - i^2}$

$= \dfrac{2+i}{4-(-1)}$

$= \dfrac{2+i}{5}$

$= \dfrac{2}{5} + \dfrac{1}{5}i$

121. $\dfrac{1}{-4+5i} = \dfrac{1}{-4+5i}\cdot\dfrac{-4-5i}{-4-5i}$

$= \dfrac{-4-5i}{(-4)^2 - (5i)^2}$

$= \dfrac{-4-5i}{16-25i^2}$

$= \dfrac{-4-5i}{16+25}$

$= \dfrac{-4-5i}{41}$

$= -\dfrac{4}{41} - \dfrac{5}{41}i$

123. $f(x) = x^2$

    a. $f(i) = i^2 = -1$

    b. $f(1+i) = (1+i)^2$

        $= 1^2 + 2i + i^2$

        $= 1 + 2i - 1$

        $= 2i$

125. $f(x) = x^2 + 2x + 2$

    a. $f(3i) = (3i)^2 + 2(3i) + 2$

        $= 9i^2 + 6i + 2$

        $= -9 + 6i + 2$

        $= -7 + 6i$

    b. $f(1-i) = (1-i)^2 + 2(1-i) + 2$

        $= 1^2 - 2i + i^2 + 2 - 2i + 2$

        $= 1 - 2i - 1 + 2 - 2i + 2$

        $= 4 - 4i$

127. a. To find the total impedance, we add the individual impedances.

$$(7+3i)+(3-4i) = 7+3i+3-4i$$
$$= (7+3)+(3-4)i$$
$$= 10-i$$

The total impedance is $10-i$ ohms.

b. The total resistance is the real part of the impedance. Thus, the total resistance is 10 ohms.

c. The total reactance is the imaginary part of the impedance. Thus, the total reactance is -1 ohm. (Since the reactance is negative, this would be called *capacitive reactance*.)

129. $f(x) = x^2 + 4x + 5$

a.
$$f(-2+i) = (-2+i)^2 + 4(-2+i) + 5$$
$$= (-2)^2 + 2(-2)i + i^2 - 8 + 4i + 5$$
$$= 4 - 4i - 1 - 8 + 4i + 5$$
$$= 0$$

b.
$$f(-2-i) = (-2-i)^2 + 4(-2-i) + 5$$
$$= (-2)^2 - 2(-2)i + i^2 - 8 - 4i + 5$$
$$= 4 + 4i - 1 - 8 - 4i + 5$$
$$= 0$$

131. $f(x) = x^3 + 1$

a. $f(-1) = (-1)^3 + 1 = -1 + 1 = 0$

b.
$$f\left(\frac{1}{2}+\frac{\sqrt{3}}{2}i\right) = \left(\frac{1}{2}+\frac{\sqrt{3}}{2}i\right)^3 + 1$$
$$= \left(\frac{1}{2}+\frac{\sqrt{3}}{2}i\right)\left(\frac{1}{2}+\frac{\sqrt{3}}{2}i\right)^2 + 1$$
$$= \left(\frac{1}{2}+\frac{\sqrt{3}}{2}i\right)\left(\frac{1}{4}+\frac{\sqrt{3}}{2}i+\frac{3}{4}i^2\right) + 1$$
$$= \left(\frac{1}{2}+\frac{\sqrt{3}}{2}i\right)\left(\frac{1}{4}+\frac{\sqrt{3}}{2}i-\frac{3}{4}\right) + 1$$
$$= \left(\frac{1}{2}+\frac{\sqrt{3}}{2}i\right)\left(-\frac{1}{2}+\frac{\sqrt{3}}{2}i\right) + 1$$
$$= \left(\frac{\sqrt{3}}{2}i\right)^2 - \left(\frac{1}{2}\right)^2 + 1$$
$$= \frac{3}{4}i^2 - \frac{1}{4} + 1$$
$$= -\frac{3}{4} + \frac{3}{4}$$
$$= 0$$

c.
$$f\left(\frac{1}{2}-\frac{\sqrt{3}}{2}i\right) = \left(\frac{1}{2}-\frac{\sqrt{3}}{2}i\right)^3 + 1$$
$$= \left(\frac{1}{2}-\frac{\sqrt{3}}{2}i\right)\left(\frac{1}{2}-\frac{\sqrt{3}}{2}i\right)^2 + 1$$
$$= \left(\frac{1}{2}-\frac{\sqrt{3}}{2}i\right)\left(\frac{1}{4}-\frac{\sqrt{3}}{2}i+\frac{3}{4}i^2\right) + 1$$
$$= \left(\frac{1}{2}-\frac{\sqrt{3}}{2}i\right)\left(\frac{1}{4}-\frac{\sqrt{3}}{2}i-\frac{3}{4}\right) + 1$$
$$= \left(\frac{1}{2}-\frac{\sqrt{3}}{2}i\right)\left(-\frac{1}{2}-\frac{\sqrt{3}}{2}i\right) + 1$$
$$= \left(-\frac{\sqrt{3}}{2}i\right)^2 - \left(\frac{1}{2}\right)^2 + 1$$
$$= \frac{3}{4}i^2 - \frac{1}{4} + 1$$
$$= -\frac{3}{4} + \frac{3}{4}$$
$$= 0$$

133. For a polynomial with real coefficients, the zeros will be real numbers or will occur in conjugate pairs. If the complex number $a+bi$ is a complex zero of the polynomial, then its conjugate $a-bi$ is also a complex conjugate.

135. Answers will vary.
The counting numbers are a subset of the whole numbers; the whole numbers are a subset of the integers; the integers are a subset of the rational numbers; rational numbers are a subset of the real numbers; and the real numbers are a subset of the complex numbers.

137. Answers will vary.
Complex numbers, in standard form, resemble binomials in that there are two terms separated by a $+$ or a $-$.

139. $(x+2)^3 = x^3 + 3x^2 \cdot 2 + 3x \cdot 2^2 + 2^3$
$= x^3 + 6x^2 + 12x + 8$

141. $(3+i)^3 = 3^3 + 3\cdot 3^2 \cdot i + 3\cdot 3 \cdot i^2 + i^3$
$= 27 + 27i + 9i^2 + i^3$
$= 27 + 27i - 9 - i$
$= 18 + 26i$

143. Answers will vary.

145.
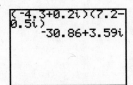

147.
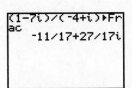

149.
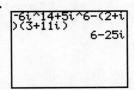

151.

Chapter 6 Review

1. $\sqrt[3]{343} = \sqrt[3]{7^3} = 7$

2. $\sqrt[3]{-125} = \sqrt[3]{(-5)^3} = -5$

3. $\sqrt[3]{\dfrac{8}{27}} = \sqrt[3]{\left(\dfrac{2}{3}\right)^3} = \dfrac{2}{3}$

4. $\sqrt[4]{81} = \sqrt[4]{3^4} = 3$

5. $-\sqrt[5]{-243} = -\sqrt[5]{(-3)^5} = -(-3) = 3$

6. $\sqrt[3]{10^3} = 10$

7. $\sqrt[5]{z^5} = z$

8. $\sqrt[4]{(5p-3)^4} = |5p-3|$

9. $81^{1/2} = \sqrt{81} = \sqrt{9^2} = 9$

10. $(-256)^{1/4} = \sqrt[4]{-256}$ is not a real number.

11. $-4^{1/2} = -\sqrt{4} = -\sqrt{2^2} = -2$

12. $729^{1/3} = \sqrt[3]{729} = \sqrt[3]{9^3} = 9$

13. $16^{7/4} = \left(\sqrt[4]{16}\right)^7 = \left(\sqrt[4]{2^4}\right)^7 = 2^7 = 128$

14. $-(-27)^{2/3} = -\left(\sqrt[3]{-27}\right)^2$
$= -\left(\sqrt[3]{(-3)^3}\right)^2$
$= -(-3)^2$
$= -9$

15. $-121^{3/2} = -\left(\sqrt{121}\right)^3$
$= -\left(\sqrt{11^2}\right)^3$
$= -(11)^3$
$= -1331$

16. $\dfrac{1}{36^{-1/2}} = 36^{1/2} = \sqrt{36} = \sqrt{6^2} = 6$

17. $(-65)^{1/3} \approx -4.02$

18. $4^{3/5} \approx 2.30$

19. $\sqrt[3]{100} \approx 4.64$

20. $\sqrt[4]{10} \approx 1.78$

21. $\sqrt[3]{5a} = (5a)^{1/3}$

22. $\sqrt[5]{p^7} = p^{7/5}$

23. $\left(\sqrt[4]{10z}\right)^3 = (10z)^{3/4}$

24. $\sqrt[6]{(2ab)^5} = (2ab)^{5/6}$

25. $4^{2/3} \cdot 4^{7/3} = 4^{\frac{2}{3}+\frac{7}{3}} = 4^{\frac{9}{3}} = 4^3 = 64$

26. $\dfrac{k^{1/2}}{k^{3/4}} = k^{\frac{1}{2}-\frac{3}{4}} = k^{\frac{2}{4}-\frac{3}{4}} = k^{-1/4} = \dfrac{1}{k^{1/4}}$

27. $\left(p^{4/3} \cdot q^4\right)^{3/2} = \left(p^{4/3}\right)^{3/2} \cdot \left(q^4\right)^{3/2}$

$\qquad = p^{\frac{4}{3}\cdot\frac{3}{2}} \cdot q^{4 \cdot \frac{3}{2}}$

$\qquad = p^2 \cdot q^6 \quad \text{or} \quad \left(p \cdot q^3\right)^2$

28. $\left(32a^{-3/2} \cdot b^{1/4}\right)^{1/5} = (32)^{1/5} \cdot \left(a^{-3/2}\right)^{1/5} \cdot \left(b^{1/4}\right)^{1/5}$

$\qquad = 2 \cdot a^{-3/10} \cdot b^{1/20}$

$\qquad = \dfrac{2b^{1/20}}{a^{3/10}} \quad \text{or} \quad 2\left(\dfrac{b}{a^6}\right)^{1/20}$

29. $5m^{-2/3}\left(2m + m^{-1/3}\right)$

$\qquad = 5m^{-2/3} \cdot 2m + 5m^{-2/3} \cdot m^{-1/3}$

$\qquad = 10m^{-2/3} \cdot m^{3/3} + 5m^{-3/3}$

$\qquad = 10m^{1/3} + 5m^{-1}$

$\qquad = 10m^{1/3} + \dfrac{5}{m}$

30. $\left(\dfrac{16x^{1/3}}{x^{-1/3}}\right)^{-1/2} + \left(\dfrac{x^{-3/2}}{64x^{-1/2}}\right)^{1/3}$

$\qquad = \left(16x^{\frac{1}{3}-\left(-\frac{1}{3}\right)}\right)^{-1/2} + \left(64^{-1}x^{-\frac{3}{2}-\left(-\frac{1}{2}\right)}\right)^{1/3}$

$\qquad = \left(16x^{2/3}\right)^{-1/2} + \left(\dfrac{1}{64}x^{-1}\right)^{1/3}$

$\qquad = \left(\dfrac{1}{16}x^{-2/3}\right)^{1/2} + \left(\dfrac{1}{64}x^{-1}\right)^{1/3}$

$\qquad = \left(\dfrac{1}{16}\right)^{1/2}\left(x^{-2/3}\right)^{1/2} + \left(\dfrac{1}{64}\right)^{1/3}\left(x^{-1}\right)^{1/3}$

$\qquad = \dfrac{1}{4}x^{-1/3} + \dfrac{1}{4}x^{-1/3}$

$\qquad = \dfrac{1}{2}x^{-1/3}$

$\qquad = \dfrac{1}{2x^{1/3}}$

31. $\sqrt[8]{x^6} = x^{6/8} = x^{3/4} = \sqrt[4]{x^3}$

32. $\sqrt{121x^4y^{10}} = \left(121x^4y^{10}\right)^{1/2}$

$\qquad = 121^{1/2}\left(x^4\right)^{1/2}\left(y^{10}\right)^{1/2}$

$\qquad = 11x^{4/2}y^{10/2}$

$\qquad = 11x^2y^5$

33. $\sqrt[3]{m^2} \cdot \sqrt{m^3} = m^{2/3} \cdot m^{3/2}$

$\qquad = m^{\frac{2}{3}+\frac{3}{2}}$

$\qquad = m^{\frac{4}{6}+\frac{9}{6}}$

$\qquad = m^{13/6}$

$\qquad = \sqrt[6]{m^{13}}$

$\qquad = m^2\sqrt[6]{m}$

34. $\dfrac{\sqrt[3]{c}}{\sqrt[6]{c^4}} = \dfrac{c^{1/3}}{c^{4/6}} = \dfrac{c^{1/3}}{c^{2/3}} = c^{\frac{1}{3}-\frac{2}{3}} = c^{-1/3} = \dfrac{1}{c^{1/3}} = \dfrac{1}{\sqrt[3]{c}}$

35. $2(3m-1)^{1/4}+(m-7)(3m-1)^{5/4}$

$=2\cdot(3m-1)^{1/4}+(m-7)(3m-1)^{1/4}\cdot(3m-1)$

$=(3m-1)^{1/4}\big(2+(m-7)(3m-1)\big)$

$=(3m-1)^{1/4}\big(2+3m^2-21m-m+7\big)$

$=(3m-1)^{1/4}\big(3m^2-22m+9\big)$

36. $3\big(x^2-5\big)^{1/3}-4x\big(x^2-5\big)^{-2/3}$

$=3\big(x^2-5\big)^{-2/3}\cdot\big(x^2-5\big)-4x\big(x^2-5\big)^{-2/3}$

$=\big(x^2-5\big)^{-2/3}\big(3\big(x^2-5\big)-4x\big)$

$=\big(x^2-5\big)^{-2/3}\big(3x^2-15-4x\big)$

$=\big(x^2-5\big)^{-2/3}\big(3x^2-4x-15\big)$

$=\dfrac{3x^2-4x-15}{\big(x^2-5\big)^{2/3}}=\dfrac{(3x+5)(x-3)}{\big(x^2-5\big)^{2/3}}$

37. $\sqrt{15}\cdot\sqrt{7}=\sqrt{15\cdot7}=\sqrt{105}$

38. $\sqrt[4]{2ab^2}\cdot\sqrt[4]{6a^2b}=\sqrt[4]{2ab^2\cdot6a^2b}=\sqrt[4]{12a^3b^3}$

39. $\sqrt{80}=\sqrt{16\cdot5}=\sqrt{16}\cdot\sqrt{5}=4\sqrt{5}$

40. $\sqrt[3]{-500}=\sqrt[3]{-125\cdot4}=\sqrt[3]{-125}\cdot\sqrt[3]{4}=-5\sqrt[3]{4}$

41. $\sqrt[3]{162m^6n^4}=\sqrt[3]{27m^6n^3\cdot6n}$

$=\sqrt[3]{27m^6n^3}\cdot\sqrt[3]{6n}$

$=3m^2n\sqrt[3]{6n}$

42. $\sqrt[4]{50p^8q^4}=\sqrt[4]{p^8q^4\cdot50}$

$=\sqrt[4]{p^8q^4}\cdot\sqrt[4]{50}$

$=p^2|q|\sqrt[4]{50}$

43. $2\sqrt{16x^6y}=2\sqrt{16x^6\cdot y}$

$=2\sqrt{16x^6}\cdot\sqrt{y}$

$=2\cdot4\big|x^3\big|\sqrt{y}$

$=8\big|x^3\big|\sqrt{y}$

as long as $y\ge0$. Otherwise the result is not a real number.

44. $\sqrt{(2x+1)^3}=\sqrt{(2x+1)^2\cdot(2x+1)}$

$=\sqrt{(2x+1)^2}\cdot\sqrt{2x+1}$

$=(2x+1)\sqrt{2x+1}$

as long as $2x+1\ge0$. Otherwise the result is not a real number.

45. $\sqrt{w^3z^2}=\sqrt{w^2z^2\cdot w}=\sqrt{w^2z^2}\cdot\sqrt{w}=wz\sqrt{w}$

46. $\sqrt{45x^4yz^3}=\sqrt{9x^4z^2\cdot5yz}$

$=\sqrt{9x^4z^2}\cdot\sqrt{5yz}$

$=3x^2z\sqrt{5yz}$

47. $\sqrt[3]{16a^{12}b^5}=\sqrt[3]{8a^{12}b^3\cdot2b^2}$

$=\sqrt[3]{8a^{12}b^3}\cdot\sqrt[3]{2b^2}$

$=2a^4b\sqrt[3]{2b^2}$

48. $\sqrt{4x^2+8x+4}=\sqrt{4\big(x^2+2x+1\big)}$

$=\sqrt{4\big(x+1\big)^2}$

$=2\big(x+1\big)$

49. $\sqrt{15}\cdot\sqrt{18}=\sqrt{15\cdot18}$

$=\sqrt{270}$

$=\sqrt{9\cdot30}$

$=\sqrt{9}\cdot\sqrt{30}$

$=3\sqrt{30}$

50. $\sqrt[3]{20}\cdot\sqrt[3]{30}=\sqrt[3]{20\cdot30}$

$=\sqrt[3]{600}$

$=\sqrt[3]{8\cdot75}$

$=\sqrt[3]{8}\cdot\sqrt[3]{75}$

$=2\sqrt[3]{75}$

51. $\sqrt[3]{-3x^4y^7}\cdot\sqrt[3]{24x^3y^2}=\sqrt[3]{-3x^4y^7\cdot24x^3y^2}$

$=\sqrt[3]{-72x^7y^9}$

$=\sqrt[3]{-8x^6y^9\cdot9x}$

$=\sqrt[3]{-8x^6y^9}\cdot\sqrt[3]{9x}$

$=-2x^2y^3\sqrt[3]{9x}$

52. $3\sqrt{4xy^2} \cdot 5\sqrt{3x^2 y} = 15\sqrt{4xy^2 \cdot 3x^2 y}$
$$= 15\sqrt{12x^3 y^3}$$
$$= 15\sqrt{4x^2 y^2 \cdot 3xy}$$
$$= 15\sqrt{4x^2 y^2} \cdot \sqrt{3xy}$$
$$= 15 \cdot 2xy \sqrt{3xy}$$
$$= 30xy \sqrt{3xy}$$

53. $\sqrt{\dfrac{121}{25}} = \dfrac{\sqrt{121}}{\sqrt{25}} = \dfrac{11}{5}$

54. $\sqrt{\dfrac{5a^4}{64b^2}} = \dfrac{\sqrt{5a^4}}{\sqrt{64b^2}} = \dfrac{a^2\sqrt{5}}{8b}$

55. $\sqrt[3]{\dfrac{54k^2}{9k^5}} = \sqrt[3]{\dfrac{6}{k^3}} = \dfrac{\sqrt[3]{6}}{\sqrt[3]{k^3}} = \dfrac{\sqrt[3]{6}}{k}$

56. $\sqrt[3]{\dfrac{-160w^{11}}{343w^{-4}}} = \sqrt[3]{\dfrac{-160w^{15}}{343}}$
$$= \dfrac{\sqrt[3]{-160w^{15}}}{\sqrt[3]{343}}$$
$$= \dfrac{\sqrt[3]{-8w^{15} \cdot 20}}{\sqrt[3]{7^3}}$$
$$= \dfrac{-2w^5 \sqrt[3]{20}}{7}$$

57. $\dfrac{\sqrt{12h^3}}{\sqrt{3h}} = \sqrt{\dfrac{12h^3}{3h}} = \sqrt{4h^2} = 2h$

58. $\dfrac{\sqrt{50a^3 b^3}}{\sqrt{8a^5 b^{-3}}} = \sqrt{\dfrac{50a^3 b^3}{8a^5 b^{-3}}} = \sqrt{\dfrac{25b^6}{4a^2}} = \dfrac{\sqrt{25b^6}}{\sqrt{4a^2}} = \dfrac{5b^3}{2a}$

59. $\dfrac{\sqrt[3]{-8x^7 y}}{\sqrt[3]{27xy^4}} = \sqrt[3]{\dfrac{-8x^7 y}{27xy^4}}$
$$= \sqrt[3]{\dfrac{-8x^6}{27y^3}}$$
$$= \dfrac{\sqrt[3]{-8x^6}}{\sqrt[3]{27y^3}}$$
$$= \dfrac{-2x^2}{3y}$$

60. $\dfrac{\sqrt[4]{48m^2 n^7}}{\sqrt[4]{3m^6 n}} = \sqrt[4]{\dfrac{48m^2 n^7}{3m^6 n}}$
$$= \sqrt[4]{\dfrac{16n^6}{m^4}}$$
$$= \dfrac{\sqrt[4]{16n^6}}{\sqrt[4]{m^4}}$$
$$= \dfrac{2n\sqrt[4]{n^2}}{m}$$
$$= \dfrac{2n\sqrt{n}}{m}$$

61. $\sqrt{5} \cdot \sqrt[3]{2} = 5^{1/2} \cdot 2^{1/3}$
$$= 5^{3/6} \cdot 2^{2/6}$$
$$= \left(5^3\right)^{1/6} \cdot \left(2^2\right)^{1/6}$$
$$= \left(5^3 \cdot 2^2\right)^{1/6}$$
$$= (125 \cdot 4)^{1/6}$$
$$= 500^{1/6}$$
$$= \sqrt[6]{500}$$

62. $\sqrt[4]{8} \cdot \sqrt[6]{4} = 8^{1/4} \cdot 4^{1/6}$
$$= \left(2^3\right)^{1/4} \cdot \left(2^2\right)^{1/6}$$
$$= 2^{3/4} \cdot 2^{1/3} = 2^{\frac{3}{4} + \frac{1}{3}}$$
$$= 2^{13/12} = 2^{\frac{12}{12}} \cdot 2^{\frac{1}{12}}$$
$$= 2\sqrt[12]{2}$$

63. $2\sqrt[4]{x} + 6\sqrt[4]{x} = (2+6)\sqrt[4]{x} = 8\sqrt[4]{x}$

64. $7\sqrt[3]{4y} + 2\sqrt[3]{4y} - 3\sqrt[3]{4y} = (7+2-3)\sqrt[3]{4y}$
$$= 6\sqrt[3]{4y}$$

65. $5\sqrt{2} - 2\sqrt{12} = 5\sqrt{2} - 2 \cdot \sqrt{4} \cdot \sqrt{3}$
$$= 5\sqrt{2} - 2 \cdot 2 \cdot \sqrt{3}$$
$$= 5\sqrt{2} - 4\sqrt{3}$$

Cannot be simplified further.

66. $\sqrt{18} + 2\sqrt{50} = \sqrt{9} \cdot \sqrt{2} + 2\sqrt{25} \cdot \sqrt{2}$
$$= 3 \cdot \sqrt{2} + 2 \cdot 5 \cdot \sqrt{2}$$
$$= 3\sqrt{2} + 10\sqrt{2}$$
$$= (3+10)\sqrt{2}$$
$$= 13\sqrt{2}$$

67.
$$\sqrt[3]{-16z} + \sqrt[3]{54z} = \sqrt[3]{-8} \cdot \sqrt[3]{2z} + \sqrt[3]{27} \cdot \sqrt[3]{2z}$$
$$= -2\sqrt[3]{2z} + 3\sqrt[3]{2z}$$
$$= (-2+3)\sqrt[3]{2z}$$
$$= \sqrt[3]{2z}$$

68.
$$7\sqrt[3]{8x^2} - \sqrt[3]{-27x^2} = 7 \cdot \sqrt[3]{8} \cdot \sqrt[3]{x^2} - \sqrt[3]{-27} \cdot \sqrt[3]{x^2}$$
$$= 7 \cdot 2 \cdot \sqrt[3]{x^2} - (-3) \cdot \sqrt[3]{x^2}$$
$$= 14\sqrt[3]{x^2} + 3\sqrt[3]{x^2}$$
$$= (14+3)\sqrt[3]{x^2}$$
$$= 17\sqrt[3]{x^2}$$

69.
$$\sqrt{16a} + \sqrt[6]{729a^3} = \sqrt{16} \cdot \sqrt{a} + \sqrt[6]{729} \cdot \sqrt[6]{a^3}$$
$$= \sqrt{16} \cdot \sqrt{a} + \sqrt[6]{729} \cdot a^{3/6}$$
$$= 4\sqrt{a} + 3 \cdot a^{1/2}$$
$$= 4\sqrt{a} + 3\sqrt{a}$$
$$= (4+3)\sqrt{a}$$
$$= 7\sqrt{a}$$

70.
$$\sqrt{27x^2} - x\sqrt{48} + 2\sqrt{75x^2}$$
$$= \sqrt{9x^2} \cdot \sqrt{3} - x \cdot \sqrt{16} \cdot \sqrt{3} + 2 \cdot \sqrt{25x^2} \cdot \sqrt{3}$$
$$= 3x\sqrt{3} - 4x\sqrt{3} + 10x\sqrt{3}$$
$$= (3 + 10 - 4)x\sqrt{3}$$
$$= 9x\sqrt{3}$$

71.
$$5\sqrt[3]{4m^5 y^2} - \sqrt[6]{16m^{10} y^4}$$
$$= 5 \cdot \sqrt[3]{m^3} \cdot \sqrt[3]{4m^2 y^2} - \sqrt[6]{m^6} \cdot \sqrt[6]{16m^4 y^4}$$
$$= 5m\sqrt[3]{4m^2 y^2} - m \cdot \sqrt[6]{\left(4m^2 y^2\right)^2}$$
$$= 5m\sqrt[3]{4m^2 y^2} - m \cdot \left(4m^2 y^2\right)^{2/6}$$
$$= 5m\sqrt[3]{4m^2 y^2} - m \cdot \left(4m^2 y^2\right)^{1/3}$$
$$= 5m\sqrt[3]{4m^2 y^2} - m\sqrt[3]{4m^2 y^2}$$
$$= (5-1)m\sqrt[3]{4m^2 y^2}$$
$$= 4m\sqrt[3]{4m^2 y^2}$$

72.
$$\sqrt{y^3 - 4y^2} - 2\sqrt{y-4} + \sqrt[4]{y^2 - 8y + 16}$$
$$= \sqrt{y^2} \cdot \sqrt{(y-4)} - 2\sqrt{y-4} + \sqrt[4]{(y-4)^2}$$
$$= y\sqrt{y-4} - 2\sqrt{y-4} + (y-4)^{2/4}$$
$$= y\sqrt{y-4} - 2\sqrt{y-4} + (y-4)^{1/2}$$
$$= y\sqrt{y-4} - 2\sqrt{y-4} + \sqrt{y-4}$$
$$= (y-2+1)\sqrt{y-4}$$
$$= (y-1)\sqrt{y-4}$$

73.
$$\sqrt{3}\left(\sqrt{5} - \sqrt{15}\right) = \sqrt{3} \cdot \sqrt{5} - \sqrt{3} \cdot \sqrt{15}$$
$$= \sqrt{3} \cdot \sqrt{5} - \sqrt{3 \cdot 15}$$
$$= \sqrt{3} \cdot \sqrt{5} - \sqrt{45}$$
$$= \sqrt{3} \cdot \sqrt{5} - \sqrt{9} \cdot \sqrt{5}$$
$$= \sqrt{3} \cdot \sqrt{5} - 3 \cdot \sqrt{5}$$
$$= \sqrt{15} - 3\sqrt{5}$$

74.
$$\sqrt[3]{5}\left(3 + \sqrt[3]{4}\right) = \sqrt[3]{5} \cdot 3 + \sqrt[3]{5} \cdot \sqrt[3]{4} = 3\sqrt[3]{5} + \sqrt[3]{20}$$

75.
$$(3 + \sqrt{5})(4 - \sqrt{5}) = 3 \cdot 4 - 3 \cdot \sqrt{5} + 4 \cdot \sqrt{5} - \sqrt{5} \cdot \sqrt{5}$$
$$= 12 - 3\sqrt{5} + 4\sqrt{5} - \sqrt{25}$$
$$= 12 + \sqrt{5} - 5$$
$$= 7 + \sqrt{5}$$

76.
$$(7 + \sqrt{3})(6 + \sqrt{2})$$
$$= 7 \cdot 6 + 7 \cdot \sqrt{2} + \sqrt{3} \cdot 6 + \sqrt{3} \cdot \sqrt{2}$$
$$= 42 + 7\sqrt{2} + 6\sqrt{3} + \sqrt{6}$$

77.
$$(1 - 3\sqrt{5})(1 + 3\sqrt{5}) = (1)^2 - (3\sqrt{5})^2$$
$$= 1 - 9\sqrt{25}$$
$$= 1 - 9 \cdot 5$$
$$= 1 - 45$$
$$= -44$$

78.
$$(\sqrt[3]{x} + 1)(9\sqrt[3]{x} - 4)$$
$$= \sqrt[3]{x} \cdot 9\sqrt[3]{x} + \sqrt[3]{x} \cdot (-4) + 1 \cdot 9\sqrt[3]{x} + 1 \cdot (-4)$$
$$= 9\sqrt[3]{x^2} - 4\sqrt[3]{x} + 9\sqrt[3]{x} - 4$$
$$= 9\sqrt[3]{x^2} + 5\sqrt[3]{x} - 4$$

79.
$$(\sqrt{x} - \sqrt{5})^2 = (\sqrt{x})^2 - 2(\sqrt{x})(\sqrt{5}) + (\sqrt{5})^2$$
$$= \sqrt{x^2} - 2\sqrt{5x} + \sqrt{25}$$
$$= x - 2\sqrt{5x} + 5$$

80. $\left(11\sqrt{2}+\sqrt{5}\right)^2$

$= \left(11\sqrt{2}\right)^2 + 2\left(11\sqrt{2}\right)\left(\sqrt{5}\right) + \left(\sqrt{5}\right)^2$

$= 121\sqrt{4} + 22\sqrt{10} + \sqrt{25}$

$= 121 \cdot 2 + 22\sqrt{10} + 5$

$= 242 + 22\sqrt{10} + 5$

$= 247 + 22\sqrt{10}$

81. $\left(\sqrt{2a}-b\right)\left(\sqrt{2a}+b\right) = \left(\sqrt{2a}\right)^2 - (b)^2$

$= \sqrt{4a^2} - b^2$

$= 2a - b^2$

82. $\left(\sqrt[3]{6s}+2\right)\left(\sqrt[3]{6s}-7\right)$

$= \sqrt[3]{6s} \cdot \sqrt[3]{6s} + \sqrt[3]{6s} \cdot (-7) + 2 \cdot \sqrt[3]{6s} + 2 \cdot (-7)$

$= \sqrt[3]{36s^2} - 7\sqrt[3]{6s} + 2\sqrt[3]{6s} - 14$

$= \sqrt[3]{36s^2} - 5\sqrt[3]{6s} - 14$

83. $\dfrac{2}{\sqrt{6}} = \dfrac{2}{\sqrt{6}} \cdot \dfrac{\sqrt{6}}{\sqrt{6}} = \dfrac{2\sqrt{6}}{\sqrt{36}} = \dfrac{2\sqrt{6}}{6} = \dfrac{\sqrt{6}}{3}$

84. $\dfrac{6}{\sqrt{3}} = \dfrac{6}{\sqrt{3}} \cdot \dfrac{\sqrt{3}}{\sqrt{3}} = \dfrac{6\sqrt{3}}{\sqrt{9}} = \dfrac{6\sqrt{3}}{3} = 2\sqrt{3}$

85. $\dfrac{\sqrt{48}}{\sqrt{p^3}} = \dfrac{4\sqrt{3}}{p\sqrt{p}} = \dfrac{4\sqrt{3}}{p\sqrt{p}} \cdot \dfrac{\sqrt{p}}{\sqrt{p}}$

$= \dfrac{4\sqrt{3p}}{p\sqrt{p^2}} = \dfrac{4\sqrt{3p}}{p \cdot p}$

$= \dfrac{4\sqrt{3p}}{p^2}$

86. $\dfrac{5}{\sqrt{2a}} = \dfrac{5}{\sqrt{2a}} \cdot \dfrac{\sqrt{2a}}{\sqrt{2a}} = \dfrac{5\sqrt{2a}}{\sqrt{4a^2}} = \dfrac{5\sqrt{2a}}{2a}$

87. $\dfrac{-2}{\sqrt{6y^3}} = \dfrac{-2}{y\sqrt{6y}} = \dfrac{-2}{y\sqrt{6y}} \cdot \dfrac{\sqrt{6y}}{\sqrt{6y}}$

$= \dfrac{-2\sqrt{6y}}{y\sqrt{36y^2}} = \dfrac{-2\sqrt{6y}}{y \cdot 6y}$

$= \dfrac{-2\sqrt{6y}}{6y^2}$

$= -\dfrac{\sqrt{6y}}{3y^2}$

88. $\dfrac{3}{\sqrt[3]{5}} = \dfrac{3}{\sqrt[3]{5}} \cdot \dfrac{\sqrt[3]{25}}{\sqrt[3]{25}} = \dfrac{3\sqrt[3]{25}}{\sqrt[3]{125}} = \dfrac{3\sqrt[3]{25}}{5}$

89. $\sqrt[3]{\dfrac{-4}{45}} = \dfrac{\sqrt[3]{-4}}{\sqrt[3]{45}} = \dfrac{\sqrt[3]{-4}}{\sqrt[3]{45}} \cdot \dfrac{\sqrt[3]{75}}{\sqrt[3]{75}} = \dfrac{\sqrt[3]{-300}}{\sqrt[3]{3375}} = -\dfrac{\sqrt[3]{300}}{15}$

90. $\dfrac{27}{\sqrt[5]{8p^3q^4}} = \dfrac{27}{\sqrt[5]{8p^3q^4}} \cdot \dfrac{\sqrt[5]{4p^2q}}{\sqrt[5]{4p^2q}}$

$= \dfrac{27\sqrt[5]{4p^2q}}{\sqrt[5]{32p^5q^5}}$

$= \dfrac{27\sqrt[5]{4p^2q}}{2pq}$

91. $\dfrac{6}{7-\sqrt{6}} = \dfrac{6}{7-\sqrt{6}} \cdot \dfrac{7+\sqrt{6}}{7+\sqrt{6}}$

$= \dfrac{6\left(7+\sqrt{6}\right)}{7^2 - \left(\sqrt{6}\right)^2}$

$= \dfrac{42+6\sqrt{6}}{49-6}$

$= \dfrac{42+6\sqrt{6}}{43}$

92. $\dfrac{3}{\sqrt{3}-9} = \dfrac{3}{\sqrt{3}-9} \cdot \dfrac{\sqrt{3}+9}{\sqrt{3}+9}$

$\qquad = \dfrac{3\left(\sqrt{3}+9\right)}{\left(\sqrt{3}\right)^2 - 9^2}$

$\qquad = \dfrac{3\sqrt{3}+27}{3-81}$

$\qquad = \dfrac{3\sqrt{3}+27}{-78}$

$\qquad = -\dfrac{3\sqrt{3}+27}{78}$

$\qquad = -\dfrac{\sqrt{3}+9}{26}$

93. $\dfrac{\sqrt{3}}{3+\sqrt{2}} = \dfrac{\sqrt{3}}{3+\sqrt{2}} \cdot \dfrac{3-\sqrt{2}}{3-\sqrt{2}}$

$\qquad = \dfrac{\sqrt{3}\left(3-\sqrt{2}\right)}{3^2 - \left(\sqrt{2}\right)^2}$

$\qquad = \dfrac{\sqrt{3}\left(3-\sqrt{2}\right)}{9-2}$

$\qquad = \dfrac{\sqrt{3}\left(3-\sqrt{2}\right)}{7}$ or $\dfrac{3\sqrt{3}-\sqrt{6}}{7}$

94. $\dfrac{\sqrt{k}}{\sqrt{k}-\sqrt{m}} = \dfrac{\sqrt{k}}{\sqrt{k}-\sqrt{m}} \cdot \dfrac{\sqrt{k}+\sqrt{m}}{\sqrt{k}+\sqrt{m}}$

$\qquad = \dfrac{\sqrt{k}\left(\sqrt{k}+\sqrt{m}\right)}{\left(\sqrt{k}\right)^2 - \left(\sqrt{m}\right)^2}$

$\qquad = \dfrac{k+\sqrt{km}}{k-m}$

95. $\dfrac{\sqrt{10}+2}{\sqrt{10}-2} = \dfrac{\sqrt{10}+2}{\sqrt{10}-2} \cdot \dfrac{\sqrt{10}+2}{\sqrt{10}+2}$

$\qquad = \dfrac{\left(\sqrt{10}\right)^2 + 2\cdot\sqrt{10}\cdot 2 + 2^2}{\left(\sqrt{10}\right)^2 - 2^2}$

$\qquad = \dfrac{10+4\sqrt{10}+4}{10-4} = \dfrac{14+4\sqrt{10}}{6}$

$\qquad = \dfrac{2\left(7+2\sqrt{10}\right)}{2(3)} = \dfrac{7+2\sqrt{10}}{3}$

96. $\dfrac{3-\sqrt{y}}{3+\sqrt{y}} = \dfrac{3-\sqrt{y}}{3+\sqrt{y}} \cdot \dfrac{3-\sqrt{y}}{3-\sqrt{y}}$

$\qquad = \dfrac{3^2 - 2\cdot 3\cdot\sqrt{y} + \left(\sqrt{y}\right)^2}{3^2 - \left(\sqrt{y}\right)^2}$

$\qquad = \dfrac{9-6\sqrt{y}+y}{9-y}$ or $\dfrac{y-6\sqrt{y}+9}{9-y}$

97. $\dfrac{4}{2\sqrt{3}+5\sqrt{2}} = \dfrac{4}{2\sqrt{3}+5\sqrt{2}} \cdot \dfrac{2\sqrt{3}-5\sqrt{2}}{2\sqrt{3}-5\sqrt{2}}$

$\qquad = \dfrac{4\left(2\sqrt{3}-5\sqrt{2}\right)}{\left(2\sqrt{3}\right)^2 - \left(5\sqrt{2}\right)^2}$

$\qquad = \dfrac{8\sqrt{3}-20\sqrt{2}}{4\cdot 3 - 25\cdot 2}$

$\qquad = \dfrac{8\sqrt{3}-20\sqrt{2}}{12-50}$

$\qquad = \dfrac{8\sqrt{3}-20\sqrt{2}}{-38}$

$\qquad = \dfrac{-2\left(-4\sqrt{3}+10\sqrt{2}\right)}{-2(19)}$

$\qquad = \dfrac{10\sqrt{2}-4\sqrt{3}}{19}$

98. $\dfrac{\sqrt{5}-\sqrt{6}}{\sqrt{10}+\sqrt{3}} = \dfrac{\sqrt{5}-\sqrt{6}}{\sqrt{10}+\sqrt{3}} \cdot \dfrac{\sqrt{10}-\sqrt{3}}{\sqrt{10}-\sqrt{3}}$

$\qquad = \dfrac{\left(\sqrt{5}-\sqrt{6}\right)\left(\sqrt{10}-\sqrt{3}\right)}{\left(\sqrt{10}+\sqrt{3}\right)\left(\sqrt{10}-\sqrt{3}\right)}$

$\qquad = \dfrac{\sqrt{50}-\sqrt{15}-\sqrt{60}+\sqrt{18}}{\left(\sqrt{10}\right)^2 - \left(\sqrt{3}\right)^2}$

$\qquad = \dfrac{5\sqrt{2}-\sqrt{15}-2\sqrt{15}+3\sqrt{2}}{10-3}$

$\qquad = \dfrac{8\sqrt{2}-3\sqrt{15}}{7}$

99.
$$\frac{\sqrt{7}}{3} + \frac{6}{\sqrt{7}} = \frac{\sqrt{7}}{3} \cdot \frac{\sqrt{7}}{\sqrt{7}} + \frac{6}{\sqrt{7}} \cdot \frac{3}{3}$$
$$= \frac{7}{3\sqrt{7}} + \frac{18}{3\sqrt{7}}$$
$$= \frac{25}{3\sqrt{7}}$$
$$= \frac{25}{3\sqrt{7}} \cdot \frac{\sqrt{7}}{\sqrt{7}}$$
$$= \frac{25\sqrt{7}}{3 \cdot 7}$$
$$= \frac{25\sqrt{7}}{21}$$

100.
$$\left(4 - \sqrt{7}\right)^{-1} = \frac{1}{4 - \sqrt{7}}$$
$$= \frac{1}{4 - \sqrt{7}} \cdot \frac{4 + \sqrt{7}}{4 + \sqrt{7}}$$
$$= \frac{4 + \sqrt{7}}{4^2 - \left(\sqrt{7}\right)^2}$$
$$= \frac{4 + \sqrt{7}}{16 - 7}$$
$$= \frac{4 + \sqrt{7}}{9}$$

101. $f(x) = \sqrt{x+4}$

    a. $f(-3) = \sqrt{-3+4} = \sqrt{1} = 1$

    b. $f(0) = \sqrt{0+4} = \sqrt{4} = 2$

    c. $f(5) = \sqrt{5+4} = \sqrt{9} = 3$

102. $g(x) = \sqrt{3x-2}$

    a. $g\left(\frac{2}{3}\right) = \sqrt{3\left(\frac{2}{3}\right) - 2} = \sqrt{2 - 2} = \sqrt{0} = 0$

    b. $g(2) = \sqrt{3(2) - 2} = \sqrt{6 - 2} = \sqrt{4} = 2$

    c. $g(6) = \sqrt{3(6) - 2} = \sqrt{18 - 2} = \sqrt{16} = 4$

103. $H(t) = \sqrt[3]{t+3}$

    a. $H(-2) = \sqrt[3]{-2+3} = \sqrt[3]{1} = 1$

    b. $H(-4) = \sqrt[3]{-4+3} = \sqrt[3]{-1} = -1$

    c. $H(5) = \sqrt[3]{5+3} = \sqrt[3]{8} = 2$

104. $G(z) = \sqrt{\dfrac{z-1}{z+2}}$

    a. $G(1) = \sqrt{\dfrac{1-1}{1+2}} = \sqrt{\dfrac{0}{3}} = \sqrt{0} = 0$

    b. $G(-3) = \sqrt{\dfrac{-3-1}{-3+2}} = \sqrt{\dfrac{-4}{-1}} = \sqrt{4} = 2$

    c. $G(2) = \sqrt{\dfrac{2-1}{2+2}} = \sqrt{\dfrac{1}{4}} = \dfrac{1}{2}$

105. $f(x) = \sqrt{3x-5}$
$$3x - 5 \geq 0$$
$$3x \geq 5$$
$$x \geq \frac{5}{3}$$

The domain of the function is $\left\{ x \mid x \geq \dfrac{5}{3} \right\}$ or the interval $\left[\dfrac{5}{3}, \infty\right)$.

106. $g(x) = \sqrt[3]{2x-7}$

Since the index is odd, the domain of the function is all real numbers, $\{ x \mid x \text{ is any real number} \}$ or the interval $(-\infty, \infty)$.

107. $h(x) = \sqrt[4]{6x+1}$
$$6x + 1 \geq 0$$
$$6x \geq -1$$
$$x \geq -\frac{1}{6}$$

The domain of the function is $\left\{ x \mid x \geq -\dfrac{1}{6} \right\}$ or the interval $\left[-\dfrac{1}{6}, \infty\right)$.

108. $F(x) = \sqrt[5]{2x-9}$

Since the index is odd, the domain of the function is all real numbers, $\{x \mid x \text{ is any real number}\}$, or the interval $(-\infty, \infty)$.

109. $G(x) = \sqrt{\dfrac{4}{x-2}}$

The index is even so we need the radicand to be nonnegative. The numerator is positive, so we need the denominator to be positive as well.
$$x - 2 > 0$$
$$x > 2$$
The domain of the function is $\{x \mid x > 2\}$ or the interval $(2, \infty)$.

110. $H(x) = \sqrt{\dfrac{x-3}{x}}$

The index is even so we need the radicand to be nonnegative. The radicand will equal zero when the numerator equals zero. The radicand will be positive when the numerator and denominator have the same sign.
$$x - 3 = 0$$
$$x = 3$$

$$x - 3 > 0 \text{ and } x > 0$$
$$x > 3 \text{ and } x > 0 \quad \rightarrow \quad x > 3$$

$$x - 3 < 0 \text{ and } x < 0$$
$$x < 3 \text{ and } x < 0 \quad \rightarrow \quad x < 0$$
The domain of the function is $\{x \mid x < 0 \text{ or } x \geq 3\}$ or the interval $(-\infty, 0) \cup [3, \infty)$.

111. $f(x) = \frac{1}{2}\sqrt{1-x}$

a. $1 - x \geq 0$
$$x \leq 1$$
The domain of the function is $\{x \mid x \leq 1\}$ or the interval $(-\infty, 1]$.

b.

x	$f(x) = \dfrac{1}{2}\sqrt{1-x}$	(x, y)
-15	$f(-15) = \dfrac{1}{2}\sqrt{1-(-15)} = 2$	$(-15, 2)$
-8	$f(-8) = \dfrac{3}{2}$	$\left(-8, \dfrac{3}{2}\right)$
-3	$f(-3) = 1$	$(-3, 1)$
0	$f(0) = \dfrac{1}{2}$	$\left(0, \dfrac{1}{2}\right)$
1	$f(1) = 0$	$(1, 0)$

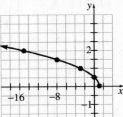

c. Based on the graph, the range is $[0, \infty)$.

112. $g(x) = \sqrt{x+1} - 2$

a. $x + 1 \geq 0 \quad \text{or} \quad x \geq -1$
The domain of the function is $\{x \mid x \geq -1\}$ or the interval $[-1, \infty)$.

b.

x	$g(x) = \sqrt{x+1} - 2$	(x, y)
-1	$g(-1) = \sqrt{-1+1} - 2 = -2$	$(-1, -2)$
0	$g(0) = -1$	$(0, -1)$
3	$g(3) = 0$	$(3, 0)$
8	$g(8) = 1$	$(8, 1)$
15	$g(15) = 2$	$(15, 2)$

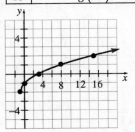

c. Based on the graph, the range is $[-2, \infty)$.

113. $h(x) = -\sqrt{x+3}$

a. $x + 3 \geq 0$

$x \geq -3$

The domain of the function is $\{x \mid x \geq -3\}$

or the interval $[-3, \infty)$.

b.

x	$h(x) = -\sqrt{x+3}$	(x, y)
-3	$h(-3) = -\sqrt{-3+3} = 0$	$(-3, 0)$
-2	$h(-2) = -1$	$(-2, -1)$
1	$h(1) = -2$	$(1, -2)$
6	$h(6) = -3$	$(6, -3)$
13	$h(13) = -4$	$(13, -4)$

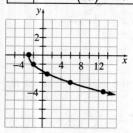

c. Based on the graph, the range is $(-\infty, 0]$.

114. $F(x) = \sqrt[3]{x+1}$

a. Since the index is odd, the domain of the function is all real numbers, $\{x \mid x \text{ is any real number}\}$, or the interval $(-\infty, \infty)$.

b.

x	$F(x) = \sqrt[3]{x+1}$	(x, y)
-9	$F(-9) = \sqrt[3]{-9+1} = -2$	$(-9, -2)$
-2	$F(-2) = -1$	$(-2, -1)$
-1	$F(-1) = 0$	$(-1, 0)$
0	$F(0) = 1$	$(0, 1)$
7	$F(7) = 2$	$(7, 2)$

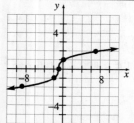

c. Based on the graph, the range is $(-\infty, \infty)$.

115. $\sqrt{m} = 13$

$\left(\sqrt{m}\right)^2 = 13^2$

$m = 169$

Check: $\sqrt{169} = 13$?

$13 = 13$ T

The solution set is $\{169\}$.

116. $\sqrt[3]{3t+1} = -2$

$\left(\sqrt[3]{3t+1}\right)^3 = (-2)^3$

$3t + 1 = -8$

$3t = -9$

$t = -3$

Check: $\sqrt[3]{3(-3)+1} = -2$?

$\sqrt[3]{-9+1} = -2$?

$\sqrt[3]{-8} = -2$?

$-2 = -2$ T

The solution set is $\{-3\}$.

117. $\sqrt[4]{3x-8} = 3$

$\left(\sqrt[4]{3x-8}\right)^4 = 3^4$

$3x - 8 = 81$

$3x = 89$

$x = \dfrac{89}{3}$

Check: $\sqrt[4]{3\left(\dfrac{89}{3}\right) - 8} = 3$?

$\sqrt[4]{89 - 8} = 3$?

$\sqrt[4]{81} = 3$?

$3 = 3$ T

The solution set is $\left\{\dfrac{89}{3}\right\}$.

118. $\sqrt{2x+5}+4=2$

$\sqrt{2x+5}=-2$

$\left(\sqrt{2x+5}\right)^2=(-2)^2$

$2x+5=4$

$2x=-1$

$x=-\dfrac{1}{2}$

Check: $\sqrt{2\left(-\dfrac{1}{2}\right)+5}+4=2$?

$\sqrt{-1+5}+4=2$?

$\sqrt{4}+4=2$?

$2+4=2$?

$6=2$ False

The solution does not check so the problem has no real solution.

119. $\sqrt{4-k}-3=0$

$\sqrt{4-k}=3$

$\left(\sqrt{4-k}\right)^2=3^2$

$4-k=9$

$-5=k$

Check: $\sqrt{4-(-5)}-3=0$?

$\sqrt{9}-3=0$?

$3-3=0$?

$0=0$ T

The solution set is $\{-5\}$.

120. $3\sqrt{t}-4=11$

$3\sqrt{t}=15$

$\sqrt{t}=5$

$\left(\sqrt{t}\right)^2=5^2$

$t=25$

Check: $3\sqrt{25}-4=11$?

$3\cdot5-4=11$?

$15-4=11$?

$11=11$ T

The solution set is $\{25\}$.

121. $2\sqrt[3]{m}+5=-11$

$2\sqrt[3]{m}=-16$

$\sqrt[3]{m}=-8$

$\left(\sqrt[3]{m}\right)^3=(-8)^3$

$m=-512$

Check: $2\sqrt[3]{-512}+5=-11$?

$2(-8)+5=-11$?

$-16+5=-11$?

$-11=-11$ T

The solution set is $\{-512\}$.

122. $\sqrt{q+2}=q$

$\left(\sqrt{q+2}\right)^2=q^2$

$q+2=q^2$

$q^2-q-2=0$

$(q-2)(q+1)=0$

$q-2=0$ or $q+1=0$

$q=2$ or $q=-1$

Check: $\sqrt{2+2}=2$?

$\sqrt{4}=2$?

$2=2$ T

$\sqrt{-1+2}=-1$?

$\sqrt{1}=-1$?

$1=-1$ False

The second solution is extraneous, so the solution set is $\{2\}$.

123. $\sqrt{w+11}+3=w+2$

$\sqrt{w+11}=w-1$

$\left(\sqrt{w+11}\right)^2=(w-1)^2$

$w+11=w^2-2w+1$

$w^2-3w-10=0$

$(w-5)(w+2)=0$

$w-5=0$ or $w+2=0$

$w=5$ or $w=-2$

Check: $\sqrt{5+11}+3=5+2$?

$\sqrt{16}+3=7$?

$4+3=7$?

$7=7$ T

$$\sqrt{-2+11}+3=-2+2 \ \ ?$$
$$\sqrt{9}+3=0 \ \ ?$$
$$3+3=0 \ \ ?$$
$$6=0 \ \ \text{False}$$

The second solution is extraneous, so the solution set is $\{5\}$.

124.
$$\sqrt{p^2-2p+9}=p+1$$
$$\left(\sqrt{p^2-2p+9}\right)^2=(p+1)^2$$
$$p^2-2p+9=p^2+2p+1$$
$$-4p+8=0$$
$$-4p=-8$$
$$p=2$$

Check: $\sqrt{2^2-2(2)+9}=2+1 \ \ ?$
$$\sqrt{4-4+9}=3 \ \ ?$$
$$\sqrt{9}=3 \ \ ?$$
$$3=3 \ \ \text{T}$$

The solution set is $\{2\}$.

125.
$$\sqrt{a+10}=\sqrt{2a-1}$$
$$\left(\sqrt{a+10}\right)^2=\left(\sqrt{2a-1}\right)^2$$
$$a+10=2a-1$$
$$11=a$$

Check: $\sqrt{11+10}=\sqrt{2(11)-1} \ \ ?$
$$\sqrt{21}=\sqrt{22-1} \ \ ?$$
$$\sqrt{21}=\sqrt{21} \ \ \text{T}$$

The solution set is $\{11\}$.

126.
$$\sqrt{5x+9}=\sqrt{7x-3}$$
$$\left(\sqrt{5x+9}\right)^2=\left(\sqrt{7x-3}\right)^2$$
$$5x+9=7x-3$$
$$12=2x$$
$$6=x$$

Check: $\sqrt{5(6)+9}=\sqrt{7(6)-3} \ \ ?$
$$\sqrt{30+9}=\sqrt{42-3} \ \ ?$$
$$\sqrt{39}=\sqrt{39} \ \ \text{T}$$

The solution set is $\{6\}$.

127.
$$\sqrt{c-8}+\sqrt{c}=4$$
$$\sqrt{c-8}=4-\sqrt{c}$$
$$\left(\sqrt{c-8}\right)^2=\left(4-\sqrt{c}\right)^2$$
$$c-8=4^2-2(4)\sqrt{c}+\left(\sqrt{c}\right)^2$$
$$c-8=16-8\sqrt{c}+c$$
$$8\sqrt{c}=24$$
$$\sqrt{c}=3$$
$$\left(\sqrt{c}\right)^2=3^2$$
$$c=9$$

Check: $\sqrt{9-8}+\sqrt{9}=4 \ \ ?$
$$\sqrt{1}+\sqrt{9}=4 \ \ ?$$
$$1+3=4 \ \ ?$$
$$4=4 \ \ \text{T}$$

The solution set is $\{9\}$.

128.
$$\sqrt{x+2}-\sqrt{x+9}=7$$
$$\sqrt{x+2}=\sqrt{x+9}+7$$
$$\left(\sqrt{x+2}\right)^2=\left(\sqrt{x+9}+7\right)^2$$
$$x+2=\left(\sqrt{x+9}\right)^2+2(7)\sqrt{x+9}+7^2$$
$$x+2=x+9+14\sqrt{x+9}+49$$
$$x+2=x+14\sqrt{x+9}+58$$
$$-56=14\sqrt{x+9}$$
$$-4=\sqrt{x+9}$$
$$(-4)^2=\left(\sqrt{x+9}\right)^2$$
$$16=x+9$$
$$7=x$$

Check: $\sqrt{7+2}-\sqrt{7+9}=7 \ \ ?$
$$\sqrt{9}-\sqrt{16}=7 \ \ ?$$
$$3-4=7 \ \ ?$$
$$-1=7 \ \ \text{False}$$

The solution does not check so the equation has no real solution.

129. $(4x-3)^{1/3} - 3 = 0$

$\quad\quad (4x-3)^{1/3} = 3$

$\quad\quad \left((4x-3)^{1/3}\right)^3 = 3^3$

$\quad\quad\quad 4x - 3 = 27$

$\quad\quad\quad\quad 4x = 30$

$\quad\quad\quad\quad x = \dfrac{30}{4} = \dfrac{15}{2}$

Check: $\left(4\left(\dfrac{15}{2}\right) - 3\right)^{1/3} - 3 = 0$?

$\quad\quad\quad (30-3)^{1/3} - 3 = 0$?

$\quad\quad\quad\quad 27^{1/3} - 3 = 0$?

$\quad\quad\quad\quad\quad 3 - 3 = 0$?

$\quad\quad\quad\quad\quad\quad 0 = 0$ T

The solution set is $\left\{\dfrac{15}{2}\right\}$.

130. $\left(x^2 - 9\right)^{1/4} = 2$

$\quad \left(\left(x^2 - 9\right)^{1/4}\right)^4 = 2^4$

$\quad\quad\quad x^2 - 9 = 16$

$\quad\quad\quad\quad x^2 = 25$

$\quad\quad\quad\quad x = \pm 5$

Check:

$\left(5^2 - 9\right)^{1/4} = 2$?

$(25-9)^{1/4} = 2$?

$\quad 16^{1/4} = 2$?

$\quad\quad 2 = 2$ T

$\left((-5)^2 - 9\right)^{1/4} = 2$?

$(25-9)^{1/4} = 2$?

$\quad 16^{1/4} = 2$?

$\quad\quad 2 = 2$ T

The solution set is $\{-5, 5\}$.

131. $r = \sqrt{\dfrac{3V}{\pi h}}$

$\quad (r)^2 = \left(\sqrt{\dfrac{3V}{\pi h}}\right)^2$

$\quad r^2 = \dfrac{3V}{\pi h}$

$\quad h \cdot r^2 = \dfrac{3V}{\pi}$

$\quad h = \dfrac{3V}{\pi r^2}$

132. $f_s = \sqrt[3]{\dfrac{30}{v}}$

$\quad (f_s)^3 = \left(\sqrt[3]{\dfrac{30}{v}}\right)^3$

$\quad f_s{}^3 = \dfrac{30}{v}$

$\quad v \cdot f_s{}^3 = 30$

$\quad v = \dfrac{30}{f_s{}^3}$

133. $\sqrt{-29} = \sqrt{-1} \cdot \sqrt{29} = i \cdot \sqrt{29} = \sqrt{29}\, i$

134. $\sqrt{-54} = \sqrt{-1} \cdot \sqrt{9} \cdot \sqrt{6} = i \cdot 3 \cdot \sqrt{6} = 3\sqrt{6}\, i$

135. $14 - \sqrt{-162} = 14 - \sqrt{-1} \cdot \sqrt{81} \cdot \sqrt{2}$

$\quad\quad\quad = 14 - i \cdot 9 \cdot \sqrt{2}$

$\quad\quad\quad = 14 - 9\sqrt{2}\, i$

136. $\dfrac{6 + \sqrt{-45}}{3} = \dfrac{6 + \sqrt{-1} \cdot \sqrt{9} \cdot \sqrt{5}}{3}$

$\quad\quad\quad = \dfrac{6 + i \cdot 3 \cdot \sqrt{5}}{3}$

$\quad\quad\quad = \dfrac{6 + 3\sqrt{5}\, i}{3}$

$\quad\quad\quad = \dfrac{3\left(2 + \sqrt{5}\, i\right)}{3}$

$\quad\quad\quad = 2 + \sqrt{5}\, i$

137. $(3 - 7i) + (-2 + 5i) = 3 - 2 - 7i + 5i$

$\quad\quad\quad = 1 - 2i$

138. $(4 + 2i) - (9 - 8i) = 4 + 2i - 9 + 8i$

$\quad\quad\quad = 4 - 9 + 2i + 8i$

$\quad\quad\quad = -5 + 10i$

139. $\left(8-\sqrt{-45}\right)-\left(3+\sqrt{-80}\right)$

$=\left(8-3\sqrt{5}\ i\right)-\left(3+4\sqrt{5}\ i\right)$

$=8-3\sqrt{5}\ i-3-4\sqrt{5}\ i$

$=8-3-3\sqrt{5}\ i-4\sqrt{5}\ i$

$=5-7\sqrt{5}\ i$

140. $\left(1+\sqrt{-9}\right)+\left(-6+\sqrt{-16}\right)=\left(1+3i\right)+\left(-6+4i\right)$

$=1+3i-6+4i$

$=1-6+3i+4i$

$=-5+7i$

141. $\left(4-5i\right)\left(3+7i\right)=4\cdot3+4\cdot7i-5i\cdot3-5i\cdot7i$

$=12+28i-15i-35i^2$

$=12+13i-35(-1)$

$=12+13i+35$

$=47+13i$

142. $\left(\dfrac{1}{2}+\dfrac{2}{3}i\right)\left(4-9i\right)$

$=\dfrac{1}{2}\cdot4+\dfrac{1}{2}(-9i)+\dfrac{2}{3}i\cdot4+\dfrac{2}{3}i(-9i)$

$=2-\dfrac{9}{2}i+\dfrac{8}{3}i-6i^2$

$=2-\dfrac{11}{6}i-6(-1)$

$=2-\dfrac{11}{6}i+6$

$=8-\dfrac{11}{6}i$

143. $\sqrt{-3}\cdot\sqrt{-27}=\sqrt{3}\ i\cdot3\sqrt{3}\ i$

$=3\left(\sqrt{3}\right)^2\cdot i^2$

$=3\cdot3\cdot(-1)$

$=-9$

Note: It is necessary to write each radical in terms of i prior to performing the operation. Otherwise the sign of the answer will be wrong.

$\sqrt{(-3)(-27)}=\sqrt{81}=9\neq-9=\sqrt{-3}\cdot\sqrt{-27}$

144. $\left(1+\sqrt{-36}\right)\left(-5-\sqrt{-144}\right)$

$=\left(1+6i\right)\left(-5-12i\right)$

$=1\cdot(-5)+1\cdot(-12i)+6i\cdot(-5)+6i\cdot(-12i)$

$=-5-12i-30i-72i^2$

$=-5-42i-72(-1)$

$=-5-42i+72$

$=67-42i$

145. $\left(1+12i\right)\left(1-12i\right)=1^2-\left(12i\right)^2$

$=1-144i^2$

$=1-144(-1)$

$=1+144$

$=145$

146. $\left(7+2i\right)\left(5+4i\right)=7\cdot5+7\cdot4i+2i\cdot5+2i\cdot4i$

$=35+28i+10i+8i^2$

$=35+38i+8(-1)$

$=35+38i-8$

$=27+38i$

147. $\dfrac{4}{3+5i}=\dfrac{4}{3+5i}\cdot\dfrac{3-5i}{3-5i}$

$=\dfrac{4(3-5i)}{3^2-(5i)^2}$

$=\dfrac{4(3-5i)}{9-25i^2}$

$=\dfrac{4(3-5i)}{9+25}$

$=\dfrac{4(3-5i)}{34}$

$=\dfrac{2\cdot2(3-5i)}{2\cdot17}$

$=\dfrac{2(3-5i)}{17}$

$=\dfrac{6-10i}{17}$

$=\dfrac{6}{17}-\dfrac{10}{17}i$

148.

$$\frac{-3}{7-2i} = \frac{-3}{7-2i} \cdot \frac{7+2i}{7+2i}$$

$$= \frac{-3(7+2i)}{7^2 - (2i)^2}$$

$$= \frac{-21-6i}{49-4i^2}$$

$$= \frac{-21-6i}{49-4(-1)}$$

$$= \frac{-21-6i}{49+4}$$

$$= \frac{-21-6i}{53}$$

$$= -\frac{21}{53} - \frac{6}{53}i$$

149.

$$\frac{2-3i}{5+2i} = \frac{2-3i}{5+2i} \cdot \frac{5-2i}{5-2i}$$

$$= \frac{(2-3i)(5-2i)}{(5+2i)(5-2i)}$$

$$= \frac{10-4i-15i+6i^2}{25-4i^2}$$

$$= \frac{10-19i+6(-1)}{25-4(-1)}$$

$$= \frac{10-19i-6}{25+4}$$

$$= \frac{4-19i}{29}$$

$$= \frac{4}{29} - \frac{19}{29}i$$

150.

$$\frac{4+3i}{1-i} = \frac{4+3i}{1-i} \cdot \frac{1+i}{1+i}$$

$$= \frac{(4+3i)(1+i)}{(1-i)(1+i)}$$

$$= \frac{4+4i+3i+3i^2}{1-i^2}$$

$$= \frac{4+7i+3(-1)}{1-(-1)}$$

$$= \frac{4+7i-3}{1+1}$$

$$= \frac{1+7i}{2}$$

$$= \frac{1}{2} + \frac{7}{2}i$$

151. $i^{59} = i^{56} \cdot i^3 = \left(i^4\right)^{14} \cdot i^3 = 1 \cdot i^3 = i^2 \cdot i = -1 \cdot i = -i$

152. $i^{173} = i^{172} \cdot i = \left(i^4\right)^{43} \cdot i = 1 \cdot i = i$

Chapter 6 Test

1. $49^{-1/2} = \frac{1}{49^{1/2}} = \frac{1}{\sqrt{49}} = \frac{1}{7}$

2.

$$\sqrt[3]{8x^{1/2}y^3} \cdot \sqrt{9xy^{1/2}}$$

$$= \left(8x^{1/2}y^3\right)^{1/3} \cdot \left(9xy^{1/2}\right)^{1/2}$$

$$= 8^{1/3}\left(x^{1/2}\right)^{1/3}\left(y^3\right)^{1/3} \cdot 9^{1/2}x^{1/2}\left(y^{1/2}\right)^{1/2}$$

$$= 2x^{1/6}y \cdot 3x^{1/2}y^{1/4}$$

$$= 6x^{\frac{1}{6}+\frac{1}{2}}y^{1+\frac{1}{4}}$$

$$= 6x^{2/3}y^{5/4}$$

$$= 6x^{8/12}y^{15/12}$$

$$= 6y\sqrt[12]{x^8y^3}$$

3. $\sqrt[5]{\left(2a^4b^3\right)^7} = \left(2a^4b^3\right)^{7/5} = 2^{7/5}a^{28/5}b^{21/5}$

$$= 2a^5b^4 \cdot 2^{2/5}a^{3/5}b^{1/5}$$

$$= 2a^5b^4\sqrt[5]{4a^3b}$$

4. $\sqrt{3m} \cdot \sqrt{13n} = \sqrt{3m \cdot 13n} = \sqrt{39mn}$

5.

$$\sqrt{32x^7y^4} = \sqrt{16x^6y^4 \cdot 2x}$$

$$= \sqrt{16x^6y^4} \cdot \sqrt{2x}$$

$$= 4x^3y^2\sqrt{2x}$$

6.

$$\frac{\sqrt{9a^3b^{-3}}}{\sqrt{4ab}} = \sqrt{\frac{9a^3b^{-3}}{4ab}}$$

$$= \sqrt{\frac{9a^2}{4b^4}}$$

$$= \frac{\sqrt{9a^2}}{\sqrt{4b^4}}$$

$$= \frac{3a}{2b^2}$$

7. $\sqrt{5x^3} + 2\sqrt{45x} = \sqrt{x^2 \cdot 5x} + 2\sqrt{9 \cdot 5x}$
$= \sqrt{x^2} \cdot \sqrt{5x} + 2\sqrt{9} \cdot \sqrt{5x}$
$= x\sqrt{5x} + 2 \cdot 3 \cdot \sqrt{5x}$
$= x\sqrt{5x} + 6\sqrt{5x}$
$= (x+6)\sqrt{5x}$

8. $\sqrt{9a^2b} - \sqrt[4]{16a^4b^2} = \sqrt{9a^2} \cdot \sqrt{b} - \sqrt[4]{16a^4} \cdot \sqrt[4]{b^2}$
$= 3a\sqrt{b} - 2a \cdot b^{2/4}$
$= 3a\sqrt{b} - 2a \cdot b^{1/2}$
$= 3a\sqrt{b} - 2a\sqrt{b}$
$= (3-2)a\sqrt{b}$
$= a\sqrt{b}$

9. $(11 + 2\sqrt{x})(3 - \sqrt{x})$
$= 11 \cdot 3 + 11(-\sqrt{x}) + 2\sqrt{x} \cdot 3 + 2\sqrt{x}(-\sqrt{x})$
$= 33 - 11\sqrt{x} + 6\sqrt{x} - 2(\sqrt{x})^2$
$= 33 + (-11+6)\sqrt{x} - 2x$
$= 33 - 5\sqrt{x} - 2x$

10. $\dfrac{-2}{3\sqrt{72}} = \dfrac{-2}{3 \cdot 6\sqrt{2}} = \dfrac{-2}{18\sqrt{2}}$
$= \dfrac{-1}{9\sqrt{2}} = \dfrac{-1}{9\sqrt{2}} \cdot \dfrac{\sqrt{2}}{\sqrt{2}}$
$= \dfrac{-\sqrt{2}}{9 \cdot 2}$
$= \dfrac{-\sqrt{2}}{18}$

11. $\dfrac{\sqrt{5}}{\sqrt{5}+2} = \dfrac{\sqrt{5}}{\sqrt{5}+2} \cdot \dfrac{\sqrt{5}-2}{\sqrt{5}-2}$
$= \dfrac{\sqrt{5}(\sqrt{5}-2)}{(\sqrt{5})^2 - 2^2}$
$= \dfrac{5 - 2\sqrt{5}}{5-4}$
$= 5 - 2\sqrt{5}$

12. $f(x) = \sqrt{-2x+3}$

a. $f(1) = \sqrt{-2(1)+3} = \sqrt{-2+3} = \sqrt{1} = 1$

b. $f(-3) = \sqrt{-2(-3)+3} = \sqrt{6+3} = \sqrt{9} = 3$

13. $g(x) = \sqrt{-3x+5}$
$-3x + 5 \geq 0$
$-3x \geq -5$
$x \leq \dfrac{5}{3}$

The domain of the function is $\left\{ x \mid x \leq \dfrac{5}{3} \right\}$ or the

interval $\left(-\infty, \dfrac{5}{3} \right]$.

14. $f(x) = \sqrt{x} - 3$

a. $x \geq 0$
The domain of the function is $\{x \mid x \geq 0\}$ or the interval $[0, \infty)$.

b.

x	$f(x) = \sqrt{x} - 3$	(x, y)
0	$f(0) = \sqrt{0} - 3 = -3$	$(0, -3)$
1	$f(1) = -2$	$(1, -2)$
4	$f(4) = -1$	$(4, -1)$
9	$f(9) = 0$	$(9, 0)$

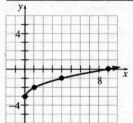

c. Based on the graph, the range is $[-3, \infty)$.

15. $\sqrt{x+3} = 4$
$(\sqrt{x+3})^2 = 4^2$
$x + 3 = 16$
$x = 13$

Check: $\sqrt{13+3} = 4$?
$\sqrt{16} = 4$?
$4 = 4$ T
The solution set is $\{13\}$.

16. $\sqrt{x+13} - 4 = x - 3$

$\sqrt{x+13} = x + 1$

$\left(\sqrt{x+13}\right)^2 = (x+1)^2$

$x + 13 = x^2 + 2x + 1$

$x^2 + x - 12 = 0$

$(x+4)(x-3) = 0$

$x + 4 = 0 \quad \text{or} \quad x - 3 = 0$

$x = -4 \quad \text{or} \quad x = 3$

Check: $\sqrt{-4+13} - 4 = -4 - 3$?

$\sqrt{9} - 4 = -7$?

$3 - 4 = -7$?

$-1 = -7$ False

$\sqrt{3+13} - 4 = 3 - 3$?

$\sqrt{16} - 4 = 0$?

$4 - 4 = 0$?

$0 = 0$ T

The first solution is extraneous, so the solution set is $\{3\}$.

17. $\sqrt{x-1} + \sqrt{x+2} = 3$

$\sqrt{x-1} = 3 - \sqrt{x+2}$

$\left(\sqrt{x-1}\right)^2 = \left(3 - \sqrt{x+2}\right)^2$

$x - 1 = 3^2 - 2(3)\sqrt{x+2} + \left(\sqrt{x+2}\right)^2$

$x - 1 = 9 - 6\sqrt{x+2} + x + 2$

$-12 = -6\sqrt{x+2}$

$2 = \sqrt{x+2}$

$2^2 = \left(\sqrt{x+2}\right)^2$

$4 = x + 2$

$2 = x$

Check: $\sqrt{2-1} + \sqrt{2+2} = 3$?

$\sqrt{1} + \sqrt{4} = 3$?

$1 + 2 = 3$?

$3 = 3$ T

The solution set is $\{2\}$.

18. $(13 + 2i) + (4 - 15i) = 13 + 2i + 4 - 15i$

$= 13 + 4 + 2i - 15i$

$= 17 - 13i$

19. $(4 - 7i)(2 + 3i) = 4 \cdot 2 + 4 \cdot 3i - 7i \cdot 2 - 7i \cdot 3i$

$= 8 + 12i - 14i - 21i^2$

$= 8 - 2i - 21(-1)$

$= 8 - 2i + 21$

$= 29 - 2i$

20. $\dfrac{7-i}{12+11i} = \dfrac{7-i}{12+11i} \cdot \dfrac{12-11i}{12-11i}$

$= \dfrac{(7-i)(12-11i)}{(12+11i)(12-11i)}$

$= \dfrac{84 - 77i - 12i + 11i^2}{144 - 121i^2}$

$= \dfrac{84 - 89i + 11(-1)}{144 - 121(-1)}$

$= \dfrac{84 - 89i - 11}{144 + 121}$

$= \dfrac{73 - 89i}{265}$

$= \dfrac{73}{265} - \dfrac{89}{265}i$

Chapter 7

Are You Ready for This Section?

R1. $(2p+3)^2 = (2p)^2 + 2(2p)(3) + 3^2$
$$= 4p^2 + 12p + 9$$

R2. $y^2 - 8y + 16 = y^2 - 2 \cdot y \cdot 4 + 4^2 = (y-4)^2$

R3. $x^2 + 5x - 14 = 0$
$(x+7)(x-2) = 0$
$x + 7 = 0$ or $x - 2 = 0$
$x = -7$ or $x = 2$
The solution set is $\{-7,\ 2\}$.

R4. $x^2 - 16 = 0$
$(x+4)(x-4) = 0$
$x + 4 = 0$ or $x - 4 = 0$
$x = -4$ or $x = 4$
The solution set is $\{-4,\ 4\}$.

R5. a. $\sqrt{36} = 6$ because $6^2 = 36$.

b. $\sqrt{45} = \sqrt{9 \cdot 5} = 3\sqrt{5}$

c. $\sqrt{-12} = \sqrt{-1 \cdot 4 \cdot 3} = 2\sqrt{3}\ i$

R6. The complex conjugate is $-3 - 2i$.

R7. $\sqrt{x^2} = |x|$ by definition.

R8. The absolute value of a number x, written $|x|$ is the distance from 0 to x on the real number line.
Alternatively, $|x| = \begin{cases} -x & \text{if} \quad x < 0 \\ x & \text{if} \quad x \geq 0 \end{cases}$.

Section 7.1 Quick Checks

1. If $x^2 = p$, then $x = \underline{\sqrt{p}}$ or $x = \underline{-\sqrt{p}}$.

2. $p^2 = 48$
$p = \pm\sqrt{48}$
$p = \pm 4\sqrt{3}$
The solution set is $\{-4\sqrt{3},\ 4\sqrt{3}\}$.

3. $3b^2 = 75$
$b^2 = 25$
$b = \pm\sqrt{25}$
$b = \pm 5$
The solution set is $\{-5,\ 5\}$.

4. $s^2 - 81 = 0$
$s^2 = 81$
$s = \pm\sqrt{81}$
$s = \pm 9$
The solution set is $\{-9,\ 9\}$.

5. $d^2 = -72$
$d = \pm\sqrt{-72}$
$d = \pm 6\sqrt{2}\ i$
The solution set is $\{-6\sqrt{2}\ i,\ 6\sqrt{2}\ i\}$.

6. $3q^2 + 27 = 0$
$3q^2 = -27$
$q^2 = -9$
$q = \pm\sqrt{-9}$
$q = \pm 3i$
The solution set is $\{-3i,\ 3i\}$.

7. $(y+3)^2 = 100$
$y + 3 = \pm\sqrt{100}$
$y + 3 = \pm 10$
$y = -3 \pm 10$
$y = -3 - 10$ or $y = -3 + 10$
$y = -13$ or $y = 7$
The solution set is $\{-13,\ 7\}$.

8. $(q-5)^2 + 20 = 4$
$(q-5)^2 = -16$
$q - 5 = \pm\sqrt{-16}$
$q - 5 = \pm 4i$
$q = 5 \pm 4i$
The solution set is $\{5 - 4i,\ 5 + 4i\}$.

9. Start: $p^2 + 14p$

Add: $\left(\dfrac{1}{2} \cdot 14\right)^2 = 49$

Result: $p^2 + 14p + 49$

Factored Form: $(p + 7)^2$

10. Start: $w^2 + 3w$

Add: $\left(\dfrac{1}{2} \cdot 3\right)^2 = \dfrac{9}{4}$

Result: $w^2 + 3w + \dfrac{9}{4}$

Factored Form: $\left(w + \dfrac{3}{2}\right)^2$

11.
$$b^2 + 2b - 8 = 0$$
$$b^2 + 2b \quad = 8$$
$$b^2 + 2b + \left(\dfrac{1}{2} \cdot 2\right)^2 = 8 + \left(\dfrac{1}{2} \cdot 2\right)^2$$
$$b^2 + 2b + 1 = 8 + 1$$
$$(b + 1)^2 = 9$$
$$b + 1 = \pm\sqrt{9}$$
$$b + 1 = \pm 3$$
$$b = -1 \pm 3$$
$$b = -4 \ \text{ or } \ b = 2$$

The solution set is $\{-4, \ 2\}$.

12.
$$z^2 - 8z + 9 = 0$$
$$z^2 - 8z \quad = -9$$
$$z^2 - 8z + \left(\dfrac{1}{2} \cdot (-8)\right)^2 = -9 + \left(\dfrac{1}{2} \cdot (-8)\right)^2$$
$$z^2 - 8z + 16 = -9 + 16$$
$$(z - 4)^2 = 7$$
$$z - 4 = \pm\sqrt{7}$$
$$z = 4 \pm \sqrt{7}$$

The solution set is $\left\{4 - \sqrt{7}, 4 + \sqrt{7}\right\}$.

13.
$$2q^2 + 6q - 1 = 0$$
$$\dfrac{2q^2 + 6q - 1}{2} = \dfrac{0}{2}$$
$$q^2 + 3q - \dfrac{1}{2} = 0$$
$$q^2 + 3q \quad = \dfrac{1}{2}$$
$$q^2 + 3q + \left[\dfrac{1}{2} \cdot 3\right]^2 = \dfrac{1}{2} + \left[\dfrac{1}{2} \cdot 3\right]^2$$
$$q^2 + 3q + \dfrac{9}{4} = \dfrac{1}{2} + \dfrac{9}{4}$$
$$\left(q + \dfrac{3}{2}\right)^2 = \dfrac{11}{4}$$
$$q + \dfrac{3}{2} = \pm\sqrt{\dfrac{11}{4}}$$
$$q + \dfrac{3}{2} = \pm\dfrac{\sqrt{11}}{2}$$
$$q = -\dfrac{3}{2} \pm \dfrac{\sqrt{11}}{2}$$

The solution set is $\left\{\dfrac{-3}{2} - \dfrac{\sqrt{11}}{2}, \ \dfrac{-3}{2} + \dfrac{\sqrt{11}}{2}\right\}$, or

$\left\{\dfrac{-3 - \sqrt{11}}{2}, \ \dfrac{-3 + \sqrt{11}}{2}\right\}$.

14.

$$3m^2 + 2m + 7 = 0$$

$$\frac{3m^2 + 2m + 7}{3} = \frac{0}{3}$$

$$m^2 + \frac{2}{3}m + \frac{7}{3} = 0$$

$$m^2 + \frac{2}{3}m = -\frac{7}{3}$$

$$m^2 + \frac{2}{3}m + \left[\frac{1}{2}\cdot\frac{2}{3}\right]^2 = -\frac{7}{3} + \left[\frac{1}{2}\cdot\frac{2}{3}\right]^2$$

$$m^2 + \frac{2}{3}m + \frac{1}{9} = -\frac{7}{3} + \frac{1}{9}$$

$$\left(m + \frac{1}{3}\right)^2 = -\frac{20}{9}$$

$$m + \frac{1}{3} = \pm\sqrt{-\frac{20}{9}}$$

$$m + \frac{1}{3} = \pm\frac{2\sqrt{5}}{3}i$$

$$m = -\frac{1}{3} \pm \frac{2\sqrt{5}}{3}i$$

The solution set is $\left\{-\frac{1}{3} - \frac{2\sqrt{5}}{3}i, -\frac{1}{3} + \frac{2\sqrt{5}}{3}i\right\}$.

15. The side of a right triangle opposite the 90° angle is called the <u>hypotenuse</u>; the remaining two sides are called <u>legs</u>.

16. False. The Pythagorean Theorem states that, in a *right* triangle, the *square* of the length of the hypotenuse is equal to the sum of the squares of the length of the legs.

17.

$$c^2 = a^2 + b^2$$
$$c^2 = 3^2 + 4^2$$
$$= 9 + 16$$
$$= 25$$
$$c = \sqrt{25}$$
$$= 5$$

The hypotenuse is 5 units long.

18. Let d represent the unknown distance.

Since the line of sight and the two lines drawn from the center of Earth form a right triangle, we

know $\text{Hypotenuse}^2 = \text{Leg}^2 + \text{Leg}^2$. We also know $200 \text{ feet} = \frac{200}{5280}$ mile, so the hypotenuse is $\left(3960 + \frac{200}{5280}\right)$ miles. Since one leg is the line of sight, d, and the other leg is 3960 miles, we have that

$$d^2 + 3960^2 = \left(3960 + \frac{200}{5280}\right)^2$$

$$d^2 = \left(3960 + \frac{200}{5280}\right)^2 - 3960^2$$

$$d^2 \approx 300.001435$$

$$d \approx \sqrt{300.001435}$$

$$\approx 17.32$$

The sailor can see approximately 17.32 miles.

7.1 Exercises

19.

$$y^2 = 100$$
$$y = \pm\sqrt{100}$$
$$y = \pm10$$

The solution set is $\{-10, 10\}$.

21.

$$p^2 = 50$$
$$p = \pm\sqrt{50}$$
$$p = \pm5\sqrt{2}$$

The solution set is $\left\{-5\sqrt{2}, 5\sqrt{2}\right\}$.

23.

$$m^2 = -25$$
$$m = \pm\sqrt{-25}$$
$$m = \pm5i$$

The solution set is $\{-5i, 5i\}$.

25.

$$w^2 = \frac{5}{4}$$
$$w = \pm\sqrt{\frac{5}{4}}$$
$$w = \pm\frac{\sqrt{5}}{2}$$

The solution set is $\left\{-\frac{\sqrt{5}}{2}, \frac{\sqrt{5}}{2}\right\}$.

27. $x^2 + 5 = 13$
$x^2 = 8$
$x = \pm\sqrt{8}$
$x = \pm 2\sqrt{2}$
The solution set is $\left\{-2\sqrt{2},\ 2\sqrt{2}\right\}$.

29. $3z^2 = 48$
$z^2 = 16$
$z = \pm\sqrt{16}$
$z = \pm 4$
The solution set is $\left\{-4,\ 4\right\}$.

31. $3x^2 = 8$
$x^2 = \dfrac{8}{3}$
$x = \pm\sqrt{\dfrac{8}{3}}$
$x = \pm\dfrac{\sqrt{8}}{\sqrt{3}}$
$x = \pm\dfrac{2\sqrt{2}}{\sqrt{3}} \cdot \dfrac{\sqrt{3}}{\sqrt{3}}$
$x = \pm\dfrac{2\sqrt{6}}{3}$
The solution set is $\left\{-\dfrac{2\sqrt{6}}{3},\ \dfrac{2\sqrt{6}}{3}\right\}$.

33. $2p^2 + 23 = 15$
$2p^2 = -8$
$p^2 = -4$
$p = \pm\sqrt{-4}$
$p = \pm 2i$
The solution set is $\left\{-2i,\ 2i\right\}$.

35. $(d-1)^2 = -18$
$d - 1 = \pm\sqrt{-18}$
$d - 1 = \pm 3\sqrt{2}i$
$d = 1 \pm 3\sqrt{2}i$
The solution set is $\left\{1 - 3\sqrt{2}i,\ 1 + 3\sqrt{2}i\right\}$.

37. $3(q+5)^2 - 1 = 8$
$3(q+5)^2 = 9$
$(q+5)^2 = 3$
$q + 5 = \pm\sqrt{3}$
$q = -5 \pm\sqrt{3}$
The solution set is $\left\{-5 - \sqrt{3},\ -5 + \sqrt{3}\right\}$.

39. $(3q+1)^2 = 9$
$3q + 1 = \pm\sqrt{9}$
$3q + 1 = \pm 3$
$3q = -1 \pm 3$
$3q = -1 - 3$ or $3q = -1 + 3$
$3q = -4$ or $3q = 2$
$q = -\dfrac{4}{3}$ or $q = \dfrac{2}{3}$
The solution set is $\left\{-\dfrac{4}{3},\ \dfrac{2}{3}\right\}$.

41. $\left(x - \dfrac{2}{3}\right)^2 = \dfrac{5}{9}$
$x - \dfrac{2}{3} = \pm\sqrt{\dfrac{5}{9}}$
$x - \dfrac{2}{3} = \pm\dfrac{\sqrt{5}}{3}$
$x = \dfrac{2}{3} \pm \dfrac{\sqrt{5}}{3}$
The solution set is $\left\{\dfrac{2}{3} - \dfrac{\sqrt{5}}{3},\ \dfrac{2}{3} + \dfrac{\sqrt{5}}{3}\right\}$.

43. $x^2 + 8x + 16 = 81$
$(x+4)^2 = 81$
$x + 4 = \pm\sqrt{81}$
$x + 4 = \pm 9$
$x = -4 \pm 9$
$x = -4 - 9$ or $x = -4 + 9$
$x = -13$ or $x = 5$
The solution set is $\left\{-13,\ 5\right\}$.

45. Start: $x^2 + 10x$
Add: $\left(\dfrac{1}{2} \cdot 10\right)^2 = 25$
Result: $x^2 + 10x + 25$
Factored Form: $(x+5)^2$

47. Start: $z^2 - 18z$

Add: $\left[\dfrac{1}{2} \cdot (-18) \right]^2 = 81$

Result: $z^2 - 18z + 81$

Factored Form: $(z - 9)^2$

49. Start: $y^2 + 7y$

Add: $\left(\dfrac{1}{2} \cdot 7 \right)^2 = \dfrac{49}{4}$

Result: $y^2 + 7y + \dfrac{49}{4}$

Factored Form: $\left(y + \dfrac{7}{2} \right)^2$

51. Start: $w^2 + \dfrac{1}{2} w$

Add: $\left(\dfrac{1}{2} \cdot \dfrac{1}{2} \right)^2 = \dfrac{1}{16}$

Result: $w^2 + \dfrac{1}{2} w + \dfrac{1}{16}$

Factored Form: $\left(w + \dfrac{1}{4} \right)^2$

53.
$$x^2 + 4x \qquad = 12$$
$$x^2 + 4x + \left(\dfrac{1}{2} \cdot 4 \right)^2 = 12 + \left(\dfrac{1}{2} \cdot 4 \right)^2$$
$$x^2 + 4x + 4 = 12 + 4$$
$$(x + 2)^2 = 16$$
$$x + 2 = \pm\sqrt{16}$$
$$x + 2 = \pm 4$$
$$x = -2 \pm 4$$
$$x = -6 \ \text{ or } \ x = 2$$
The solution set is $\{-6, \ 2\}$.

55.
$$x^2 - 4x + 1 = 0$$
$$x^2 - 4x \qquad = -1$$
$$x^2 - 4x + \left(\dfrac{1}{2} \cdot (-4) \right)^2 = -1 + \left(\dfrac{1}{2} \cdot (-4) \right)^2$$
$$x^2 - 4x + 4 = -1 + 4$$
$$(x - 2)^2 = 3$$
$$x - 2 = \pm\sqrt{3}$$
$$x = 2 \pm \sqrt{3}$$
The solution set is $\left\{ 2 - \sqrt{3}, \ 2 + \sqrt{3} \right\}$.

57.
$$a^2 - 4a + 5 = 0$$
$$a^2 - 4a \qquad = -5$$
$$a^2 - 4a + \left(\dfrac{1}{2} \cdot (-4) \right)^2 = -5 + \left(\dfrac{1}{2} \cdot (-4) \right)^2$$
$$a^2 - 4a + 4 = -5 + 4$$
$$(a - 2)^2 = -1$$
$$a - 2 = \pm\sqrt{-1}$$
$$a - 2 = \pm i$$
$$a = 2 \pm i$$
The solution set is $\{2 - i, \ 2 + i\}$.

59.
$$b^2 + 5b - 2 = 0$$
$$b^2 + 5b \qquad = 2$$
$$b^2 + 5b + \left(\dfrac{1}{2} \cdot 5 \right)^2 = 2 + \left(\dfrac{1}{2} \cdot 5 \right)^2$$
$$b^2 + 5b + \dfrac{25}{4} = 2 + \dfrac{25}{4}$$
$$\left(b + \dfrac{5}{2} \right)^2 = \dfrac{33}{4}$$
$$b + \dfrac{5}{2} = \pm\sqrt{\dfrac{33}{4}}$$
$$b + \dfrac{5}{2} = \pm\dfrac{\sqrt{33}}{2}$$
$$b = -\dfrac{5}{2} \pm \dfrac{\sqrt{33}}{2}$$

The solution set is $\left\{ -\dfrac{5}{2} - \dfrac{\sqrt{33}}{2}, \ -\dfrac{5}{2} + \dfrac{\sqrt{33}}{2} \right\}$.

61.
$$m^2 = 8m + 3$$
$$m^2 - 8m = 3$$
$$m^2 - 8m + \left(\frac{1}{2} \cdot (-8)\right)^2 = 3 + \left(\frac{1}{2} \cdot (-8)\right)^2$$
$$m^2 - 8m + 16 = 3 + 16$$
$$(m-4)^2 = 19$$
$$m - 4 = \pm\sqrt{19}$$
$$m = 4 \pm \sqrt{19}$$
The solution set is $\left\{4 - \sqrt{19}, \ 4 + \sqrt{19}\right\}$.

63.
$$p^2 - p + 3 = 0$$
$$p^2 - p = -3$$
$$p^2 - p + \left(\frac{1}{2} \cdot (-1)\right)^2 = -3 + \left(\frac{1}{2} \cdot (-1)\right)^2$$
$$p^2 - p + \frac{1}{4} = -3 + \frac{1}{4}$$
$$\left(p - \frac{1}{2}\right)^2 = -\frac{11}{4}$$
$$p - \frac{1}{2} = \pm\sqrt{-\frac{11}{4}}$$
$$p - \frac{1}{2} = \pm\frac{\sqrt{11}}{2}i$$
$$p = \frac{1}{2} \pm \frac{\sqrt{11}}{2}i$$
The solution set is $\left\{\frac{1}{2} - \frac{\sqrt{11}}{2}i, \ \frac{1}{2} + \frac{\sqrt{11}}{2}i\right\}$.

65.
$$2y^2 - 5y - 12 = 0$$
$$\frac{2y^2 - 5y - 12}{2} = \frac{0}{2}$$
$$y^2 - \frac{5}{2}y - 6 = 0$$
$$y^2 - \frac{5}{2}y = 6$$
$$y^2 - \frac{5}{2}y + \left[\frac{1}{2} \cdot \left(-\frac{5}{2}\right)\right]^2 = 6 + \left[\frac{1}{2} \cdot \left(-\frac{5}{2}\right)\right]^2$$
$$y^2 - \frac{5}{2}y + \frac{25}{16} = 6 + \frac{25}{16}$$
$$\left(y - \frac{5}{4}\right)^2 = \frac{121}{16}$$
$$y - \frac{5}{4} = \pm\sqrt{\frac{121}{16}}$$
$$y - \frac{5}{4} = \pm\frac{11}{4}$$
$$y = \frac{5}{4} \pm \frac{11}{4}$$
$$y = -\frac{3}{2} \ \text{ or } \ y = 4$$
The solution set is $\left\{-\frac{3}{2}, \ 4\right\}$.

67.
$$3y^2 - 6y + 2 = 0$$
$$\frac{3y^2 - 6y + 2}{3} = \frac{0}{3}$$
$$y^2 - 2y + \frac{2}{3} = 0$$
$$y^2 - 2y \quad = -\frac{2}{3}$$
$$y^2 - 2y + \left(\frac{1}{2} \cdot (-2)\right)^2 = -\frac{2}{3} + \left(\frac{1}{2} \cdot (-2)\right)^2$$
$$y^2 - 2y + 1 = -\frac{2}{3} + 1$$
$$(y-1)^2 = \frac{1}{3}$$
$$y - 1 = \pm\sqrt{\frac{1}{3}}$$
$$y - 1 = \pm\frac{1}{\sqrt{3}} \cdot \frac{\sqrt{3}}{\sqrt{3}}$$
$$y - 1 = \pm\frac{\sqrt{3}}{3}$$
$$y = 1 \pm \frac{\sqrt{3}}{3}$$

The solution set is $\left\{ 1 - \dfrac{\sqrt{3}}{3},\ 1 + \dfrac{\sqrt{3}}{3} \right\}$.

69.
$$2z^2 - 5z + 1 = 0$$
$$\frac{2z^2 - 5z + 1}{2} = \frac{0}{2}$$
$$z^2 - \frac{5}{2}z + \frac{1}{2} = 0$$
$$y^2 - \frac{5}{2}y \quad = -\frac{1}{2}$$
$$z^2 - \frac{5}{2}z + \left[\frac{1}{2} \cdot \left(-\frac{5}{2}\right)\right]^2 = -\frac{1}{2} + \left[\frac{1}{2} \cdot \left(-\frac{5}{2}\right)\right]^2$$
$$z^2 - \frac{5}{2}z + \frac{25}{16} = -\frac{1}{2} + \frac{25}{16}$$
$$\left(z - \frac{5}{4}\right)^2 = \frac{17}{16}$$
$$z - \frac{5}{4} = \pm\sqrt{\frac{17}{16}}$$
$$z - \frac{5}{4} = \pm\frac{\sqrt{17}}{4}$$
$$z = \frac{5}{4} \pm \frac{\sqrt{17}}{4}$$

The solution set is $\left\{ \dfrac{5}{4} - \dfrac{\sqrt{17}}{4},\ \dfrac{5}{4} + \dfrac{\sqrt{17}}{4} \right\}$.

71.
$$2x^2 + 4x + 5 = 0$$
$$\frac{2x^2 + 4x + 5}{2} = \frac{0}{2}$$
$$x^2 + 2x + \frac{5}{2} = 0$$
$$x^2 + 2x \quad = -\frac{5}{2}$$
$$x^2 + 2x + \left(\frac{1}{2} \cdot 2\right)^2 = -\frac{5}{2} + \left(\frac{1}{2} \cdot 2\right)^2$$
$$x^2 + 2x + 1 = -\frac{5}{2} + 1$$
$$(x+1)^2 = -\frac{3}{2}$$
$$x + 1 = \pm\sqrt{-\frac{3}{2}}$$
$$x + 1 = \pm\frac{\sqrt{3}}{\sqrt{2}}i$$
$$x + 1 = \pm\frac{\sqrt{6}}{2}i$$
$$x = -1 \pm \frac{\sqrt{6}}{2}i$$

The solution set is $\left\{ -1 - \dfrac{\sqrt{6}}{2}i,\ -1 + \dfrac{\sqrt{6}}{2}i \right\}$.

73.
$$c^2 = 6^2 + 8^2$$
$$= 36 + 64$$
$$= 100$$
$$c = \sqrt{100} = 10$$

75.
$$c^2 = 12^2 + 16^2$$
$$= 144 + 256$$
$$= 400$$
$$c = \sqrt{400} = 20$$

77.
$$c^2 = 5^2 + 5^2$$
$$= 25 + 25$$
$$= 50$$
$$c = \sqrt{50} = 5\sqrt{2} \approx 7.07$$

79.
$$c^2 = 1^2 + \left(\sqrt{3}\right)^2$$
$$= 1 + 3$$
$$= 4$$
$$c = \sqrt{4} = 2$$

81. $c^2 = 6^2 + 10^2$
      $= 36 + 100$
      $= 136$
    $c = \sqrt{136} = 2\sqrt{34} \approx 11.66$

83. $c^2 = a^2 + b^2$
    $8^2 = 4^2 + b^2$
   $64 = 16 + b^2$
   $48 = b^2$
     $b = \sqrt{48}$
     $b = 4\sqrt{3} \approx 6.93$

85. $c^2 = a^2 + b^2$
   $12^2 = a^2 + 8^2$
  $144 = a^2 + 64$
   $80 = a^2$
     $a = \sqrt{80}$
     $a = 4\sqrt{5} \approx 8.94$

87. $f(x) = 36$
    $(x-3)^2 = 36$
      $x - 3 = \pm\sqrt{36}$
      $x - 3 = \pm 6$
        $x = 3 \pm 6$
        $x = 3 - 6$ or $x = 3 + 6$
        $x = -3$ or $x = 9$
The solution set is $\{-3, \ 9\}$.
The points $(-3, 36)$ and $(9, 36)$ are on the graph of f.

89. $g(x) = 18$
    $(x+2)^2 = 18$
      $x + 2 = \pm\sqrt{18}$
      $x + 2 = \pm 3\sqrt{2}$
         $x = -2 \pm 3\sqrt{2}$
The solution set is $\left\{-2 - 3\sqrt{2}, \ -2 + 3\sqrt{2}\right\}$.
The points $\left(-2 - 3\sqrt{2}, 18\right)$ and $\left(-2 + 3\sqrt{2}, 18\right)$ are on the graph of g.

91. Let x represent the diagonal.
   $x^2 = 8^2 + 4^2$
   $x^2 = 64 + 16$
   $x^2 = 80$
     $x = \sqrt{80} = 4\sqrt{5} \approx 8.944$

93. Let x represent the distance from the ball to the center of the green.
   $x^2 = 30^2 + 100^2$
   $x^2 = 900 + 10{,}000$
   $x^2 = 10{,}900$
     $x = \sqrt{10{,}900} \approx 104.403$
The ball is approximately 104.403 yards from the center of the green.

95. Let x represent the length of the guy wire.
   $x^2 = 10^2 + 30^2$
   $x^2 = 100 + 900$
   $x^2 = 1000$
     $x = \sqrt{1000} \approx 31.623$
The guy wire is approximately 31.623 feet long.

97. a. Let x represent the height up the wall to which the ladder can reach.

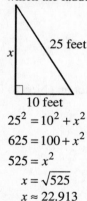

   $25^2 = 10^2 + x^2$
  $625 = 100 + x^2$
  $525 = x^2$
     $x = \sqrt{525}$
     $x \approx 22.913$
The ladder can reach approximately 22.913 feet up the wall.

    b. Let x represent the distance that the base of the ladder can be from the wall.

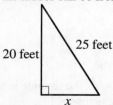

$$25^2 = x^2 + 20^2$$
$$625 = x^2 + 400$$
$$225 = x^2$$
$$x = \sqrt{225}$$
$$x = 15$$

The base of the ladder can be at most 15 feet from the wall.

99. a. $s = 16t^2$
$$16 = 16t^2$$
$$1 = t^2$$
$$t = \sqrt{1}$$
$$t = 1$$

It takes 1 second for an object to fall 16 feet.

b. $s = 16t^2$
$$48 = 16t^2$$
$$3 = t^2$$
$$t = \sqrt{3}$$
$$t \approx 1.732$$

It takes approximately 1.732 seconds for an object to fall 48 feet.

c. $s = 16t^2$
$$64 = 16t^2$$
$$4 = t^2$$
$$t = \sqrt{4}$$
$$t = 2$$

It takes 2 seconds for an object to fall 64 feet.

101. $1200 = 1000(1 + r)^2$
$$1.2 = (1 + r)^2$$
$$1 + r = \sqrt{1.2}$$
$$r = -1 + \sqrt{1.2}$$
$$r \approx 0.0954$$

The required rate of interest is approximately 9.54%.

103. $17^2 \overset{?}{=} 8^2 + 15^2$
$$289 \overset{?}{=} 64 + 225$$
$$289 = 289 \quad \leftarrow \text{True}$$

Because $c^2 = a^2 + b^2$, the triangle is a right triangle. The hypotenuse is 17.

105. $20^2 \overset{?}{=} 14^2 + 18^2$
$$400 \overset{?}{=} 196 + 324$$
$$400 \overset{?}{=} 520 \quad \leftarrow \text{False}$$

Because $c^2 \neq a^2 + b^2$, the triangle is not a right triangle.

107. $c^2 = \left(m^2 + n^2\right)^2 = m^4 + 2m^2n^2 + n^4$

$a^2 + b^2 = \left(m^2 - n^2\right)^2 + (2mn)^2$
$$= m^4 - 2m^2n^2 + n^4 + 4m^2n^2$$
$$= m^4 + 2m^2n^2 + n^4$$

Because c^2 and $a^2 + b^2$ result in the same expression, a, b, and c are the lengths of the sides of a right triangle.

109. $a^2 - 5a - 36 = 0$
$$(a - 9)(a + 4) = 0$$
$$a - 9 = 0 \quad \text{or} \quad a + 4 = 0$$
$$a = 9 \quad \text{or} \quad a = -4$$

The solution set is $\{-4, \ 9\}$.

111. $|4q + 1| = 3$
$$4q + 1 = 3 \quad \text{or} \quad 4q + 1 = -3$$
$$4q = 2 \quad \text{or} \quad 4q = -4$$
$$q = \frac{1}{2} \quad \text{or} \quad q = -1$$

The solution set is $\left\{-1, \ \frac{1}{2}\right\}$.

113. In both cases, the simpler equations are linear.

Section 7.2

Are You Ready for This Section?

R1. a. $\sqrt{121} = 11$ because $11^2 = 121$.

b. $\sqrt{54} = \sqrt{9 \cdot 6} = 3\sqrt{6}$

R2. a. $\sqrt{-9} = \sqrt{-1 \cdot 9} = 3i$

b. $\sqrt{-72} = \sqrt{-1 \cdot 36 \cdot 2} = 6\sqrt{2}\,i$

R3. $\dfrac{3+\sqrt{18}}{6} = \dfrac{3+3\sqrt{2}}{6}$

$\phantom{\dfrac{3+\sqrt{18}}{6}} = \dfrac{3\left(1+\sqrt{2}\right)}{6}$

$\phantom{\dfrac{3+\sqrt{18}}{6}} = \dfrac{1+\sqrt{2}}{2}$ or $\dfrac{1}{2}+\dfrac{\sqrt{2}}{2}$

Section 7.2 Quick Checks

1. The solution(s) to the quadratic equation $ax^2+bx+c=0$, $a \neq 0$, are given by the quadratic formula $x = \dfrac{-b \pm \sqrt{b^2-4ac}}{2a}$.

2. $2x^2-3x-9=0$

   For this equation, $a=2$, $b=-3$, and $c=-9$.

   $x = \dfrac{-(-3) \pm \sqrt{(-3)^2-4(2)(-9)}}{2(2)}$

   $ = \dfrac{3 \pm \sqrt{9+72}}{4}$

   $ = \dfrac{3 \pm \sqrt{81}}{4}$

   $ = \dfrac{3 \pm 9}{4}$

   $x = \dfrac{3-9}{4}$ or $x = \dfrac{3+9}{4}$

   $ = \dfrac{-6}{4}$ or $ = \dfrac{12}{4}$

   $ = -\dfrac{3}{2}$ or $ = 3$

   The solution set is $\left\{-\dfrac{3}{2},\, 3\right\}$.

3. $2x^2+7x=4$

   $2x^2+7x-4=0$

   For this equation, $a=2$, $b=7$, and $c=-4$.

   $x = \dfrac{-7 \pm \sqrt{7^2-4(2)(-4)}}{2(2)}$

   $ = \dfrac{-7 \pm \sqrt{49+32}}{4}$

   $ = \dfrac{-7 \pm \sqrt{81}}{4}$

   $ = \dfrac{-7 \pm 9}{4}$

   $x = \dfrac{-7-9}{4}$ or $x = \dfrac{-7+9}{4}$

   $ = \dfrac{-16}{4}$ or $ = \dfrac{2}{4}$

   $ = -4$ or $ = \dfrac{1}{2}$

   The solution set is $\left\{-4,\, \dfrac{1}{2}\right\}$.

4. False; $3x^2-2x=9$ or $3x^2-2x-9=0$, so $a=3$, $b=-2$, and $c=-9$.

5. $4z^2+1=8z$

   $4z^2-8z+1=0$

   For this equation, $a=4$, $b=-8$, and $c=1$.

   $z = \dfrac{-(-8) \pm \sqrt{(-8)^2-4(4)(1)}}{2(4)}$

   $ = \dfrac{8 \pm \sqrt{64-16}}{8}$

   $ = \dfrac{8 \pm \sqrt{48}}{8}$

   $ = \dfrac{8 \pm 4\sqrt{3}}{8}$

   $ = \dfrac{8}{8} \pm \dfrac{4\sqrt{3}}{8}$

   $ = 1 \pm \dfrac{\sqrt{3}}{2}$

   The solution set is $\left\{1-\dfrac{\sqrt{3}}{2},\, 1+\dfrac{\sqrt{3}}{2}\right\}$.

6.
$$4w + \frac{25}{w} = 20$$

$$w\left(4w + \frac{25}{w}\right) = w(20)$$

$$4w^2 + 25 = 20w$$

$$4w^2 - 20w + 25 = 0$$

For this equation, $a = 4$, $b = -20$, and $c = 25$.

$$w = \frac{-(-20) \pm \sqrt{(-20)^2 - 4(4)(25)}}{2(4)}$$

$$= \frac{20 \pm \sqrt{400 - 400}}{8}$$

$$= \frac{20 \pm \sqrt{0}}{8}$$

$$= \frac{20}{8}$$

$$= \frac{5}{2}$$

The solution set is $\left\{\dfrac{5}{2}\right\}$.

7.
$$2x = 8 - \frac{3}{x}$$

$$x(2x) = x\left(8 - \frac{3}{x}\right)$$

$$2x^2 = 8x - 3$$

$$2x^2 - 8x + 3 = 0$$

For this equation, $a = 2$, $b = -8$, and $c = 3$.

$$x = \frac{-(-8) \pm \sqrt{(-8)^2 - 4(2)(3)}}{2(2)}$$

$$= \frac{8 \pm \sqrt{64 - 24}}{4}$$

$$= \frac{8 \pm \sqrt{40}}{4}$$

$$= \frac{8 \pm 2\sqrt{10}}{4}$$

$$= \frac{8}{4} \pm \frac{2\sqrt{10}}{4}$$

$$= 2 \pm \frac{\sqrt{10}}{2}$$

The solution set is $\left\{2 - \dfrac{\sqrt{10}}{2},\ 2 + \dfrac{\sqrt{10}}{2}\right\}$.

8. $z^2 + 2z + 26 = 0$

For this equation, $a = 1$, $b = 2$, and $c = 26$.

$$z = \frac{-2 \pm \sqrt{2^2 - 4(1)(26)}}{2(1)}$$

$$= \frac{-2 \pm \sqrt{4 - 104}}{2}$$

$$= \frac{-2 \pm \sqrt{-100}}{2}$$

$$= \frac{-2 \pm 10i}{2}$$

$$= \frac{-2}{2} \pm \frac{10i}{2}$$

$$= -1 \pm 5i$$

The solution set is $\{-1 - 5i,\ -1 + 5i\}$.

9. In the quadratic formula, the quantity $b^2 - 4ac$ is called the <u>discriminant</u> of the quadratic equation.

10. False; the discriminant is $b^2 - 4ac$ or $25 - 4(4)(1) = 9$.

11. If the discriminant of a quadratic equation is <u>negative</u>, then the quadratic equation has two complex solutions that are not real.

12. False. The equation will have one repeated real solution.

13. True.

14. $2z^2 + 5z + 4 = 0$

For this equation, $a = 2$, $b = 5$, and $c = 4$.

$$b^2 - 4ac = 5^2 - 4(2)(4) = 25 - 32 = -7$$

Because $b^2 - 4ac = -7$ is negative, the quadratic equation will have two complex solutions that are not real. The solutions will be complex conjugates of each other.

15.
$$4y^2 + 12y = -9$$

$$4y^2 + 12y + 9 = 0$$

For this equation, $a = 4$, $b = 12$, and $c = 9$.

$$b^2 - 4ac = 12^2 - 4(4)(9) = 144 - 144 = 0$$

Because $b^2 - 4ac = 0$, the quadratic equation will have one repeated real solution.

16. $2x^2 - 4x + 1 = 0$

For this equation, $a = 2$, $b = -4$, and $c = 1$.

$b^2 - 4ac = (-4)^2 - 4(2)(1) = 16 - 8 = 8$

Because $b^2 - 4ac = 8$ is positive, but not a perfect square, the quadratic equation will have two irrational solutions.

17. True.

18. $5n^2 - 45 = 0$

Because this equation has no linear term, solve by using the square root method.

$5n^2 = 45$

$n^2 = 9$

$n = \pm\sqrt{9}$

$\quad = \pm 3$

The solution set is $\{-3, 3\}$.

19. $-2y^2 + 5y - 6 = 0$

$2y^2 - 5y + 6 = 0$

Because this equation does not easily factor, solve by using the quadratic formula. For this equation, $a = 2$, $b = -5$, and $c = 6$.

$y = \dfrac{-(-5) \pm \sqrt{(-5)^2 - 4(2)(6)}}{2(2)}$

$\quad = \dfrac{5 \pm \sqrt{25 - 48}}{4}$

$\quad = \dfrac{5 \pm \sqrt{-23}}{4}$

$\quad = \dfrac{5 \pm \sqrt{23}\, i}{4}$

The solution set is $\left\{ \dfrac{5 - \sqrt{23}\, i}{4},\ \dfrac{5 + \sqrt{23}\, i}{4} \right\}$ or

$\left\{ \dfrac{5}{4} - \dfrac{\sqrt{23}}{4} i,\ \dfrac{5}{4} + \dfrac{\sqrt{23}}{4} i \right\}$.

20. $3w^2 + 2w = 5$

$3w^2 + 2w - 5 = 0$

Because this equation factors easily, solve by factoring.

$(3w + 5)(w - 1) = 0$

$3w + 5 = 0 \quad$ or $\quad w - 1 = 0$

$\quad 3w = -5 \quad$ or $\quad\quad w = 1$

$\quad\quad w = -\dfrac{5}{3}$

The solution set is $\left\{ -\dfrac{5}{3}, 1 \right\}$.

21. a.
$$R(x) = 600$$
$$-0.005x^2 + 4x = 600$$
$$-0.005x^2 + 4x - 600 = 0$$

For this equation, $a = -0.005$, $b = 4$, and $c = -600$.

$x = \dfrac{-4 \pm \sqrt{4^2 - 4(-0.005)(-600)}}{2(-0.005)}$

$\quad = \dfrac{-4 \pm \sqrt{4}}{-0.01}$

$\quad = \dfrac{-4 \pm 2}{-0.01}$

$x = \dfrac{-6}{-0.01} \quad$ or $\quad x = \dfrac{-2}{-0.01}$

$\quad = 600 \quad\quad$ or $\quad\quad = 200$

If revenue is to be \$600 per day, then either 200 or 600 DVDs must be rented.

b.
$$R(x) = 800$$
$$-0.005x^2 + 4x = 800$$
$$-0.005x^2 + 4x - 800 = 0$$

For this equation, $a = -0.005$, $b = 4$, and $c = -800$.

$x = \dfrac{-4 \pm \sqrt{4^2 - 4(-0.005)(-800)}}{2(-0.005)}$

$\quad = \dfrac{-4 \pm \sqrt{0}}{-0.01}$

$\quad = \dfrac{-4 \pm 0}{-0.01}$

$x = \dfrac{-4}{-0.01} = 400$

If revenue is to be \$800 per day, then 400 DVDs must be rented.

22. Let w represent the width of the rectangle. Then $w + 14$ will represent the length.

$w^2 + (w + 14)^2 = 34^2$

$w^2 + w^2 + 28w + 196 = 1156$

$2w^2 + 28w - 960 = 0$

$w^2 + 14w - 480 = 0$

$(w - 16)(w + 30) = 0$

$w - 16 = 0 \quad$ or $\quad w + 30 = 0$

$\quad w = 16 \quad$ or $\quad\quad w = -30$

We disregard $w = -30$ because w represents the width of the rectangle, which must be positive. Thus, $w = 16$ is the only viable answer. Now, $w + 14 = 16 + 14 = 30$. Thus, the dimensions of the rectangle are 16 meters by 30 meters.

7.2 Exercises

23. $x^2 - 4x - 12 = 0$

For this equation, $a = 1$, $b = -4$, and $c = -12$.

$$x = \frac{-(-4) \pm \sqrt{(-4)^2 - 4(1)(-12)}}{2(1)}$$

$$= \frac{4 \pm \sqrt{16 + 48}}{2}$$

$$= \frac{4 \pm \sqrt{64}}{2}$$

$$= \frac{4 \pm 8}{2}$$

$$x = \frac{4 - 8}{2} \quad \text{or} \quad x = \frac{4 + 8}{2}$$

$$= \frac{-4}{2} \quad \text{or} \quad = \frac{12}{2}$$

$$= -2 \quad \text{or} \quad = 6$$

The solution set is $\{-2,\ 6\}$.

25. $6y^2 - y - 15 = 0$

For this equation, $a = 6$, $b = -1$, and $c = -15$.

$$y = \frac{-(-1) \pm \sqrt{(-1)^2 - 4(6)(-15)}}{2(6)}$$

$$= \frac{1 \pm \sqrt{1 + 360}}{12}$$

$$= \frac{1 \pm \sqrt{361}}{12}$$

$$= \frac{1 \pm 19}{12}$$

$$y = \frac{1 - 19}{12} \quad \text{or} \quad y = \frac{1 + 19}{12}$$

$$= \frac{-18}{12} \quad \text{or} \quad = \frac{20}{12}$$

$$= -\frac{3}{2} \quad \text{or} \quad = \frac{5}{3}$$

The solution set is $\left\{-\frac{3}{2},\ \frac{5}{3}\right\}$.

27. $4m^2 - 8m + 1 = 0$

For this equation, $a = 4$, $b = -8$, and $c = 1$.

$$m = \frac{-(-8) \pm \sqrt{(-8)^2 - 4(4)(1)}}{2(4)}$$

$$= \frac{8 \pm \sqrt{64 - 16}}{8}$$

$$= \frac{8 \pm \sqrt{48}}{8} = \frac{8}{8} \pm \frac{4\sqrt{3}}{8} = 1 \pm \frac{\sqrt{3}}{2}$$

The solution set is $\left\{1 - \frac{\sqrt{3}}{2},\ 1 + \frac{\sqrt{3}}{2}\right\}$.

29. $3w - 6 = \frac{1}{w}$

$$(3w - 6)w = \left(\frac{1}{w}\right)w$$

$$3w^2 - 6w = 1$$

$$3w^2 - 6w - 1 = 0$$

For this equation, $a = 3$, $b = -6$, and $c = -1$.

$$w = \frac{-(-6) \pm \sqrt{(-6)^2 - 4(3)(-1)}}{2(3)}$$

$$= \frac{6 \pm \sqrt{36 + 12}}{6}$$

$$= \frac{6 \pm \sqrt{48}}{6}$$

$$= \frac{6 \pm 4\sqrt{3}}{6} = \frac{6}{6} \pm \frac{4\sqrt{3}}{6} = 1 \pm \frac{2\sqrt{3}}{3}$$

The solution set is $\left\{1 - \frac{2\sqrt{3}}{3},\ 1 + \frac{2\sqrt{3}}{3}\right\}$.

31. $3p^2 = -2p + 4$

$$3p^2 + 2p - 4 = 0$$

For this equation, $a = 3$, $b = 2$, and $c = -4$.

$$p = \frac{-2 \pm \sqrt{2^2 - 4(3)(-4)}}{2(3)}$$

$$= \frac{-2 \pm \sqrt{4 + 48}}{6}$$

$$= \frac{-2 \pm \sqrt{52}}{6}$$

$$= \frac{-2 \pm 2\sqrt{13}}{6} = -\frac{2}{6} \pm \frac{2\sqrt{13}}{6} = -\frac{1}{3} \pm \frac{\sqrt{13}}{3}$$

The solution set is $\left\{-\frac{1}{3} - \frac{\sqrt{13}}{3},\ -\frac{1}{3} + \frac{\sqrt{13}}{3}\right\}$.

33. $x^2 - 2x + 7 = 0$

For this equation, $a = 1$, $b = -2$, and $c = 7$.

$$x = \frac{-(-2) \pm \sqrt{(-2)^2 - 4(1)(7)}}{2(1)}$$

$$= \frac{2 \pm \sqrt{4 - 28}}{2}$$

$$= \frac{2 \pm \sqrt{-24}}{2}$$

$$= \frac{2 \pm 2\sqrt{6}\ i}{2}$$

$$= \frac{2}{2} \pm \frac{2\sqrt{6}\ i}{2} = 1 \pm \sqrt{6}\ i$$

The solution set is $\left\{1 - \sqrt{6}\ i, 1 + \sqrt{6}\ i\right\}$.

35. $2z^2 + 7 = 2z$

$2z^2 - 2z + 7 = 0$

For this equation, $a = 2$, $b = -2$, and $c = 7$.

$$z = \frac{-(-2) \pm \sqrt{(-2)^2 - 4(2)(7)}}{2(2)}$$

$$= \frac{2 \pm \sqrt{4 - 56}}{4}$$

$$= \frac{2 \pm \sqrt{-52}}{4}$$

$$= \frac{2 \pm 2\sqrt{13}\ i}{4}$$

$$= \frac{2}{4} \pm \frac{2\sqrt{13}\ i}{4} = \frac{1}{2} \pm \frac{\sqrt{13}}{2}i$$

The solution set is $\left\{\frac{1}{2} - \frac{\sqrt{13}}{2}i, \frac{1}{2} + \frac{\sqrt{13}}{2}i\right\}$.

37. $4x^2 = 2x + 1$

$4x^2 - 2x - 1 = 0$

For this equation, $a = 4$, $b = -2$, and $c = -1$.

$$x = \frac{-(-2) \pm \sqrt{(-2)^2 - 4(4)(-1)}}{2(4)}$$

$$= \frac{2 \pm \sqrt{4 + 16}}{8}$$

$$= \frac{2 \pm \sqrt{20}}{8}$$

$$= \frac{2 \pm 2\sqrt{5}}{8} = \frac{2}{8} \pm \frac{2\sqrt{5}}{8} = \frac{1}{4} \pm \frac{\sqrt{5}}{4}$$

The solution set is $\left\{\frac{1}{4} - \frac{\sqrt{5}}{4}, \frac{1}{4} + \frac{\sqrt{5}}{4}\right\}$.

39. $1 = 3q^2 + 4q$

$0 = 3q^2 + 4q - 1$

For this equation, $a = 3$, $b = 4$, and $c = -1$.

$$q = \frac{-4 \pm \sqrt{4^2 - 4(3)(-1)}}{2(3)}$$

$$= \frac{-4 \pm \sqrt{16 + 12}}{6}$$

$$= \frac{-4 \pm \sqrt{28}}{6}$$

$$= \frac{-4 \pm 2\sqrt{7}}{6} = \frac{-4}{6} \pm \frac{2\sqrt{7}}{6} = -\frac{2}{3} \pm \frac{\sqrt{7}}{3}$$

The solution set is $\left\{-\frac{2}{3} - \frac{\sqrt{7}}{3}, -\frac{2}{3} + \frac{\sqrt{7}}{3}\right\}$.

41. $x^2 - 5x + 1 = 0$

For this equation, $a = 1$, $b = -5$, and $c = 1$.

$b^2 - 4ac = (-5)^2 - 4(1)(1) = 25 - 4 = 21$

Because $b^2 - 4ac = 21$ is positive, but not a perfect square, the quadratic equation will have two irrational solutions.

43. $3z^2 + 2z + 5 = 0$

For this equation, $a = 3$, $b = 2$, and $c = 5$.

$b^2 - 4ac = 2^2 - 4(3)(5) = 4 - 60 = -56$

Because $b^2 - 4ac = -56$ is negative, the quadratic equation will have two complex solutions that are not real. The solutions will be complex conjugates of each other.

45. $9q^2 - 6q + 1 = 0$

For this equation, $a = 9$, $b = -6$, and $c = 1$.

$b^2 - 4ac = (-6)^2 - 4(9)(1) = 36 - 36 = 0$

Because $b^2 - 4ac = 0$, the quadratic equation will have one repeated real solution.

47. $3w^2 = 4w - 2$

$3w^2 - 4w + 2 = 0$

For this equation, $a = 3$, $b = -4$, and $c = 2$.

$b^2 - 4ac = (-4)^2 - 4(3)(2) = 16 - 24 = -8$

Because $b^2 - 4ac = -8$ is negative, the quadratic equation will have two complex solutions that are not real. The solutions will be complex conjugates of each other.

49. $6x = 2x^2 - 1$

$0 = 2x^2 - 6x - 1$

For this equation, $a = 2$, $b = -6$, and $c = -1$.

$b^2 - 4ac = (-6)^2 - 4(2)(-1) = 36 + 8 = 44$

Because $b^2 - 4ac = 44$ is positive, but not a perfect square, the quadratic equation will have two irrational solutions.

51. $w^2 - 5w + 5 = 0$

Because this equation does not easily factor, solve by using the quadratic formula. For this equation, $a = 1$, $b = -5$, and $c = 5$.

$$w = \frac{-(-5) \pm \sqrt{(-5)^2 - 4(1)(5)}}{2(1)}$$

$$= \frac{5 \pm \sqrt{25 - 20}}{2} = \frac{5 \pm \sqrt{5}}{2} = \frac{5}{2} \pm \frac{\sqrt{5}}{2}$$

The solution set is $\left\{ \dfrac{5}{2} - \dfrac{\sqrt{5}}{2}, \dfrac{5}{2} + \dfrac{\sqrt{5}}{2} \right\}$.

53. $3x^2 + 5x = 8$

$3x^2 + 5x - 8 = 0$

Because this equation factors easily, solve by factoring.

$(3x + 8)(x - 1) = 0$

$3x + 8 = 0$ or $x - 1 = 0$

$3x = -8$ or $x = 1$

$x = -\dfrac{8}{3}$

The solution set is $\left\{ -\dfrac{8}{3}, 1 \right\}$.

55. $2x^2 = 3x + 35$

$2x^2 - 3x - 35 = 0$

Because this equation factors easily, solve by factoring.

$(2x + 7)(x - 5) = 0$

$2x + 7 = 0$ or $x - 5 = 0$

$2x = -7$ or $x = 5$

$x = -\dfrac{7}{2}$

The solution set is $\left\{ -\dfrac{7}{2}, 5 \right\}$.

57. $q^2 + 2q + 8 = 0$

Because this equation does not easily factor, solve by using the quadratic formula. For this equation, $a = 1$, $b = 2$, and $c = 8$.

$$q = \frac{-2 \pm \sqrt{2^2 - 4(1)(8)}}{2(1)}$$

$$= \frac{-2 \pm \sqrt{4 - 32}}{2}$$

$$= \frac{-2 \pm \sqrt{-28}}{2}$$

$$= \frac{-2 \pm 2\sqrt{7}i}{2} = \frac{-2}{2} \pm \frac{2\sqrt{7}i}{2} = -1 \pm \sqrt{7}i$$

The solution set is $\left\{ -1 - \sqrt{7}i, \ -1 + \sqrt{7}i \right\}$.

59. $2z^2 = 2(z + 3)^2$

$2z^2 = 2(z^2 + 6z + 9)$

$2z^2 = 2z^2 + 12z + 18$

$0 = 12z + 18$

$12z = -18$

$z = -\dfrac{18}{12} = -\dfrac{3}{2}$

The solution set is $\left\{ -\dfrac{3}{2} \right\}$.

61. $7q - 2 = \dfrac{4}{q}$

$(7q - 2)q = \left(\dfrac{4}{q} \right) q$

$7q^2 - 2q = 4$

$7q^2 - 2q - 4 = 0$

Because this equation does not easily factor, solve by using the quadratic formula. For this equation, $a = 7$, $b = -2$, and $c = -4$.

$$q = \frac{-(-2) \pm \sqrt{(-2)^2 - 4(7)(-4)}}{2(7)}$$

$$= \frac{2 \pm \sqrt{4 + 112}}{14} = \frac{2 \pm \sqrt{116}}{14}$$

$$= \frac{2 \pm 2\sqrt{29}}{14} = \frac{2}{14} \pm \frac{2\sqrt{29}}{14} = \frac{1}{7} \pm \frac{\sqrt{29}}{7}$$

The solution set is $\left\{ \dfrac{1}{7} - \dfrac{\sqrt{29}}{7}, \dfrac{1}{7} + \dfrac{\sqrt{29}}{7} \right\}$.

63. $5a^2 - 80 = 0$

Because this equation has no linear term, solve by using the square root method.

$5a^2 = 80$

$a^2 = 16$

$a = \pm\sqrt{16} = \pm 4$

The solution set is $\{-4,\ 4\}$.

65. $\qquad 8n^2 + 1 = 4n$

$8n^2 - 4n + 1 = 0$

Because this equation does not easily factor, solve by using the quadratic formula. For this equation, $a = 8$, $b = -4$, and $c = 1$.

$n = \dfrac{-(-4) \pm \sqrt{(-4)^2 - 4(8)(1)}}{2(8)}$

$= \dfrac{4 \pm \sqrt{16 - 32}}{16}$

$= \dfrac{4 \pm \sqrt{-16}}{16}$

$= \dfrac{4 \pm 4i}{16} = \dfrac{4}{16} \pm \dfrac{4i}{16} = \dfrac{1}{4} \pm \dfrac{1}{4}i$

The solution set is $\left\{\dfrac{1}{4} - \dfrac{1}{4}i,\ \dfrac{1}{4} + \dfrac{1}{4}i\right\}$.

67. $27x^2 + 36x + 12 = 0$

$\dfrac{27x^2 + 36x + 12}{3} = \dfrac{0}{3}$

$9x^2 + 12x + 4 = 0$

Because this equation factors easily, solve by factoring.

$(3x + 2)(3x + 2) = 0$

$3x + 2 = 0 \quad$ or $\quad 3x + 2 = 0$

$\quad 3x = -2 \quad$ or $\quad\quad 3x = -2$

$\quad\quad x = -\dfrac{2}{3} \quad$ or $\quad\quad x = -\dfrac{2}{3}$

The solution set is $\left\{-\dfrac{2}{3}\right\}$.

69. $\qquad \dfrac{1}{3}x^2 + \dfrac{2}{9}x - 1 = 0$

$9\left(\dfrac{1}{3}x^2 + \dfrac{2}{9}x - 1\right) = 9(0)$

$3x^2 + 2x - 9 = 0$

Because this equation does not easily factor, solve by using the quadratic formula. For this equation, $a = 3$, $b = 2$, and $c = -9$.

$x = \dfrac{-2 \pm \sqrt{2^2 - 4(3)(-9)}}{2(3)}$

$= \dfrac{-2 \pm \sqrt{4 + 108}}{6}$

$= \dfrac{-2 \pm \sqrt{112}}{6}$

$= \dfrac{-2 \pm 4\sqrt{7}}{6}$

$= \dfrac{-2}{6} \pm \dfrac{4\sqrt{7}}{6} = -\dfrac{1}{3} \pm \dfrac{2\sqrt{7}}{3}$

The solution set is $\left\{-\dfrac{1}{3} - \dfrac{2\sqrt{7}}{3},\ -\dfrac{1}{3} + \dfrac{2\sqrt{7}}{3}\right\}$.

71. $(x - 5)(x + 1) = 4$

$x^2 - 4x - 5 = 4$

$x^2 - 4x - 9 = 0$

Because this equation does not easily factor, solve by using the quadratic formula. For this equation, $a = 1$, $b = -4$, and $c = -9$.

$x = \dfrac{-(-4) \pm \sqrt{(-4)^2 - 4(1)(-9)}}{2(1)}$

$= \dfrac{4 \pm \sqrt{16 + 36}}{2}$

$= \dfrac{4 \pm \sqrt{52}}{2}$

$= \dfrac{4 \pm 2\sqrt{13}}{2}$

$= \dfrac{4}{2} \pm \dfrac{2\sqrt{13}}{2} = 2 \pm \sqrt{13}$

The solution set is $\left\{2 - \sqrt{13},\ 2 + \sqrt{13}\right\}$.

73. Note: $x \neq -2$.

$\dfrac{x - 2}{x + 2} = x - 3$

$(x + 2)\left(\dfrac{x - 2}{x + 2}\right) = (x + 2)(x - 3)$

$x - 2 = x^2 - x - 6$

$0 = x^2 - 2x - 4$

Because this equation does not easily factor, solve by using the quadratic formula. For this equation, $a = 1$, $b = -2$, and $c = -4$.

$$x = \frac{-(-2) \pm \sqrt{(-2)^2 - 4(1)(-4)}}{2(1)}$$

$$= \frac{2 \pm \sqrt{4 + 16}}{2}$$

$$= \frac{2 \pm \sqrt{20}}{2}$$

$$= \frac{2 \pm 2\sqrt{5}}{2}$$

$$= \frac{2}{2} \pm \frac{2\sqrt{5}}{2} = 1 \pm \sqrt{5}$$

The solution set is $\left\{1 - \sqrt{5},\ 1 + \sqrt{5}\right\}$.

75.
$$\frac{x-4}{x^2+2} = 2$$

$$\left(x^2 + 2\right)\left(\frac{x-4}{x^2+2}\right) = \left(x^2 + 2\right)(2)$$

$$x - 4 = 2x^2 + 4$$

$$0 = 2x^2 - x + 8$$

Because this equation does not easily factor, solve by using the quadratic formula. For this equation, $a = 2$, $b = -1$, and $c = 8$.

$$x = \frac{-(-1) \pm \sqrt{(-1)^2 - 4(2)(8)}}{2(2)}$$

$$= \frac{1 \pm \sqrt{1 - 64}}{4}$$

$$= \frac{1 \pm \sqrt{-63}}{4}$$

$$= \frac{1 \pm 3\sqrt{7}\,i}{4} = \frac{1}{4} \pm \frac{3\sqrt{7}}{4}i$$

The solution set is $\left\{\frac{1}{4} - \frac{3\sqrt{7}}{4}i,\ \frac{1}{4} + \frac{3\sqrt{7}}{4}i\right\}$.

77. a.
$$f(x) = 0$$

$$x^2 + 4x - 21 = 0$$

$$(x+7)(x-3) = 0$$

$$x + 7 = 0 \quad \text{or} \quad x - 3 = 0$$

$$x = -7 \quad \text{or} \quad x = 3$$

The solution set is $\{-7, 3\}$.

b.
$$f(x) = -21$$

$$x^2 + 4x - 21 = -21$$

$$x^2 + 4x = 0$$

$$x(x+4) = 0$$

$$x = 0 \quad \text{or} \quad x + 4 = 0$$

$$x = -4$$

The solution set is $\{-4, 0\}$.

The points $(-7, 0)$, $(3, 0)$, $(-4, 21)$ and $(0, -21)$ are on the graph of f.

79. a.
$$H(x) = 0$$

$$-2x^2 - 4x + 1 = 0$$

$$2x^2 + 4x - 1 = 0$$

For this equation, $a = 2$, $b = 4$, and $c = -1$.

$$x = \frac{-4 \pm \sqrt{4^2 - 4(2)(-1)}}{2(2)}$$

$$= \frac{-4 \pm \sqrt{16 + 8}}{4}$$

$$= \frac{-4 \pm \sqrt{24}}{4}$$

$$= \frac{-4 \pm 2\sqrt{6}}{4} = \frac{-4}{4} \pm \frac{2\sqrt{6}}{4} = -1 \pm \frac{\sqrt{6}}{2}$$

The solution set is $\left\{-1 - \frac{\sqrt{6}}{2},\ -1 + \frac{\sqrt{6}}{2}\right\}$.

b.
$$H(x) = 2$$

$$-2x^2 - 4x + 1 = 2$$

$$0 = 2x^2 + 4x + 1$$

For this equation, $a = 2$, $b = 4$, and $c = 1$.

$$x = \frac{-4 \pm \sqrt{4^2 - 4(2)(1)}}{2(2)}$$

$$= \frac{-4 \pm \sqrt{16 - 8}}{4}$$

$$= \frac{-4 \pm \sqrt{8}}{4}$$

$$= \frac{-4 \pm 2\sqrt{2}}{4} = \frac{-4}{4} \pm \frac{2\sqrt{2}}{4} = -1 \pm \frac{\sqrt{2}}{2}$$

The solution set is $\left\{-1 - \frac{\sqrt{2}}{2},\ -1 + \frac{\sqrt{2}}{2}\right\}$.

81. $G(x) = 3x^2 + 2x - 2$

$$3x^2 + 2x - 2 = 0$$

For this equation, $a = 3$, $b = 2$, and $c = -2$.

$$x = \frac{-2 \pm \sqrt{(2)^2 - 4(3)(-2)}}{2(3)}$$

$$= \frac{-2 \pm \sqrt{4 + 24}}{6}$$

$$= \frac{-2 \pm \sqrt{28}}{6}$$

$$= \frac{-2 \pm 2\sqrt{7}}{6}$$

$$= \frac{-2}{6} \pm \frac{2\sqrt{7}}{6}$$

$$= -\frac{1}{3} \pm \frac{\sqrt{7}}{3}$$

The solution set is $\left\{ -\frac{1}{3} - \frac{\sqrt{7}}{3}, \ -\frac{1}{3} + \frac{\sqrt{7}}{3} \right\}$.

The zeros of $G(x)$ are $x = \frac{-1 \pm \sqrt{7}}{3}$.

83.
$$x^2 + (x+1)^2 = (2x-1)^2$$
$$x^2 + x^2 + 2x + 1 = 4x^2 - 4x + 1$$
$$2x^2 + 2x + 1 = 4x^2 - 4x + 1$$
$$0 = 2x^2 - 6x$$
$$0 = 2x(x-3)$$
$$2x = 0 \ \text{ or } \ x - 3 = 0$$
$$x = 0 \ \text{ or } \quad x = 3$$

Disregard $x = 0$ because x represents the length of one leg of the triangle. Thus, $x = 3$ is the only viable answer. Now, $x + 1 = 3 + 1 = 4$ and $2x - 1 = 2(3) - 1 = 5$. The three measurements are 3, 4, and 5.

85.
$$(5x-1)^2 + (x+2)^2 = (5x)^2$$
$$25x^2 - 10x + 1 + x^2 + 4x + 4 = 25x^2$$
$$26x^2 - 6x + 5 = 25x^2$$
$$x^2 - 6x + 5 = 0$$
$$(x-1)(x-5) = 0$$
$$x - 1 = 0 \ \text{ or } \ x - 5 = 0$$
$$x = 1 \ \text{ or } \quad x = 5$$

For $x = 1$, $x + 2 = 1 + 2 = 3$, $5x - 1 = 5(1) - 1 = 4$, and $5x = 5(1) = 5$.

For $x = 5$, $x + 2 = 5 + 2 = 7$,
$5x - 1 = 5(5) - 1 = 24$, and $5x = 5(5) = 25$.

The three measurements can be either 3, 4, and 5, or 7, 24, and 25.

87. Let x represent the length of the rectangle. Then $x + 4$ will represent the width.
$$x(x+4) = 40$$
$$x^2 + 4x = 40$$
$$x^2 + 4x - 40 = 0$$
For this equation, $a = 1$, $b = 4$, and $c = -40$.

$$x = \frac{-4 \pm \sqrt{4^2 - 4(1)(-40)}}{2(1)}$$

$$= \frac{-4 \pm \sqrt{16 + 160}}{2} = \frac{-4 \pm \sqrt{176}}{2}$$

$$= \frac{-4 \pm 4\sqrt{11}}{2} = -\frac{4}{2} \pm \frac{4\sqrt{11}}{2} = -2 \pm 2\sqrt{11}$$

Disregard $x = -2 - 2\sqrt{11} \approx -8.633$ because x represents the length of the rectangle, which must be positive. Thus, $x = -2 + 2\sqrt{11} \approx 4.633$ is the only viable answer. Now,
$x + 4 = -2 + 2\sqrt{11} + 4 = 2 + 2\sqrt{11} \approx 8.633$. Thus, the dimensions of the rectangle are $-2 + 2\sqrt{11}$ inches by $2 + 2\sqrt{11}$ inches, which is approximately 4.633 inches by 8.633 inches.

89. Let x represent the base of the triangle. Then $x - 3$ will represent the height.
$$\frac{1}{2}x(x-3) = 25$$
$$\frac{1}{2}x^2 - \frac{3}{2}x - 25 = 0$$
$$2\left(\frac{1}{2}x^2 - \frac{3}{2}x - 25\right) = 2(0)$$
$$x^2 - 3x - 50 = 0$$
For this equation, $a = 1$, $b = -3$, and $c = -50$.

$$x = \frac{-(-3) \pm \sqrt{(-3)^2 - 4(1)(-50)}}{2(1)}$$

$$= \frac{3 \pm \sqrt{9 + 200}}{2} = \frac{3 \pm \sqrt{209}}{2} = \frac{3}{2} \pm \frac{\sqrt{209}}{2}$$

Disregard $x = \frac{3}{2} - \frac{\sqrt{209}}{2} \approx -5.728$ because x represents the base of the triangle, which must be positive. Thus, $x = \frac{3}{2} + \frac{\sqrt{209}}{2} \approx 8.728$ is the only viable answer. Now,

$$x - 3 = \frac{3}{2} + \frac{\sqrt{209}}{2} - 3 = -\frac{3}{2} + \frac{\sqrt{209}}{2} \approx 5.728.$$

Thus, the base of the triangle is

$\dfrac{3}{2} + \dfrac{\sqrt{209}}{2}$ inches, which is approximately

8.728 inches. The height of the triangle is

$-\dfrac{3}{2} + \dfrac{\sqrt{209}}{2}$ inches, which is approximately

5.728 inches.

91. a. $R(17) = -0.1(17)^2 + 70(17) = 1161.1$

If 17 pairs of sunglasses are sold per week, then the company's revenue will be $1161.10.

$R(25) = -0.1(25)^2 + 70(25) = 1687.5$

If 25 pairs of sunglasses are sold per week, then the company's revenue will be $1687.50.

b. $R(x) = 10,000$

$-0.1x^2 + 70x = 10,000$

$0 = 0.1x^2 - 70x + 10,000$

For this equation, $a = 0.1$, $b = -70$, and $c = 10,000$.

$x = \dfrac{-(-70) \pm \sqrt{(-70)^2 - 4(0.1)(10,000)}}{2(0.1)}$

$= \dfrac{70 \pm \sqrt{4900 - 4000}}{0.2}$

$= \dfrac{70 \pm \sqrt{900}}{0.2} = \dfrac{70 \pm 30}{0.2}$

$x = \dfrac{70 - 30}{0.2}$ or $x = \dfrac{70 + 30}{0.2}$

$= \dfrac{40}{0.2} = 200$ or $= \dfrac{100}{0.2} = 500$

The revenue will be $10,000 per week if either 200 or 500 pairs of sunglasses are sold per week.

c. $R(x) = 12,250$

$-0.1x^2 + 70x = 12,250$

$0 = 0.1x^2 - 70x + 12,250$

For this equation, $a = 0.1$, $b = -70$, and $c = 12,250$.

$x = \dfrac{-(-70) \pm \sqrt{(-70)^2 - 4(0.1)(12,250)}}{2(0.1)}$

$= \dfrac{70 \pm \sqrt{4900 - 4900}}{0.2}$

$= \dfrac{70 \pm \sqrt{0}}{0.2}$

$= \dfrac{70 \pm 0}{0.2} = \dfrac{70}{0.2} = 350$

The revenue will be $12,250 per week if 350 pairs of sunglasses are sold per week.

93. a. $s(t) = 40$

$40 = -16t^2 + 70t + 5$

$16t^2 - 70t + 35 = 0$

For this equation, $a = 16$, $b = -70$, and $c = 35$.

$x = \dfrac{-(-70) \pm \sqrt{(-70)^2 - 4(16)(35)}}{2(16)}$

$= \dfrac{70 \pm \sqrt{4900 - 2240}}{32}$

$= \dfrac{70 \pm \sqrt{2660}}{32}$

$= \dfrac{70 \pm 2\sqrt{665}}{32}$

$x = \dfrac{35 - \sqrt{665}}{16}$ or $x = \dfrac{35 + \sqrt{665}}{16}$

≈ 0.576 or ≈ 3.799

Rounding to the nearest tenth, the height of the ball will be 40 feet after approximately 0.6 second and after approximately 3.8 seconds.

b. $s(t) = 70$

$70 = -16t^2 + 70t + 5$

$16t^2 - 70t + 65 = 0$

For this equation, $a = 16$, $b = -70$, and $c = 65$.

$$x = \frac{-(-70) \pm \sqrt{(-70)^2 - 4(16)(65)}}{2(16)}$$

$$= \frac{70 \pm \sqrt{4900 - 4160}}{32}$$

$$= \frac{70 \pm \sqrt{740}}{32}$$

$$= \frac{70 \pm 2\sqrt{185}}{32}$$

$$= \frac{35 \pm \sqrt{185}}{16}$$

$$x = \frac{35 - \sqrt{185}}{16} \quad \text{or} \quad x = \frac{35 + \sqrt{185}}{16}$$

$$\approx 1.337 \quad \text{or} \quad \approx 3.038$$

Rounding to the nearest tenth, the height of the ball will be 70 feet after approximately 1.3 seconds and after approximately 3.0 seconds.

c.
$$s(t) = 150$$
$$150 = -16t^2 + 70t + 5$$
$$16t^2 - 70t + 145 = 0$$

For this equation, $a = 16$, $b = -70$, and $c = 145$.

$$x = \frac{-(-70) \pm \sqrt{(-70)^2 - 4(16)(145)}}{2(16)}$$

$$= \frac{70 \pm \sqrt{4900 - 9280}}{32}$$

$$= \frac{70 \pm \sqrt{-4380}}{32}$$

$$= \frac{70 \pm 2\sqrt{1095}\, i}{32}$$

$$= \frac{35 \pm \sqrt{1095}\, i}{16}$$

$$x = \frac{35}{16} - \frac{\sqrt{185}}{16}i \quad \text{or} \quad x = \frac{35}{16} + \frac{\sqrt{185}}{16}i$$

The ball will never reach a height of 150 feet. This is clear because the solutions to the equation above are complex solutions that are not real.

95. Because $\triangle ABC \sim \triangle DEC$, we know $\dfrac{\overline{AB}}{\overline{BC}} = \dfrac{\overline{DE}}{\overline{EC}}$.

$\overline{BC} = 24$, $\overline{DE} = 6$, $\overline{AB} = x$, and $\overline{EC} = x$,

$$\frac{x}{24} = \frac{6}{x}$$
$$x \cdot x = 24 \cdot 6$$
$$x^2 = 144$$
$$x = \pm\sqrt{144} = \pm 12$$

Disregard $x = -12$ because x represents a length, which must be positive. Thus, $x = 12$ is the only viable answer.

97. a. $I(a) = 40,000$

$$40,000 = -55a^2 + 5119a - 54,448$$
$$55a^2 - 5119a + 94,448 = 0$$

For this equation, $a = 55$, $b = -5119$, and $c = 94,448$.

$$x = \frac{-(-5119) \pm \sqrt{(-5119)^2 - 4(55)(94,448)}}{2(55)}$$

$$= \frac{5119 \pm \sqrt{26,204,161 - 20,778,560}}{110}$$

$$= \frac{5119 \pm \sqrt{5,425,601}}{110}$$

$$x = \frac{5119 - \sqrt{5,425,601}}{110} \approx 25.361$$

or

$$x = \frac{5119 + \sqrt{5,425,601}}{110} \approx 67.712$$

Rounding to the nearest year, the average income equals $40,000 at ages 25 and 68.

b. $I(a) = 50,000$

$$50,000 = -55a^2 + 5119a - 54,448$$
$$55a^2 - 5119a + 104,448 = 0$$

For this equation, $a = 55$, $b = -5119$, and $c = 104,448$.

$$x = \frac{-(-5119) \pm \sqrt{(-5119)^2 - 4(55)(104,448)}}{2(55)}$$

$$= \frac{5119 \pm \sqrt{26,204,161 - 22,978,560}}{110}$$

$$= \frac{5119 \pm \sqrt{3,225,601}}{110}$$

$$x = \frac{5119 - \sqrt{3,225,601}}{110} \approx 30.209$$

or

$$x = \frac{5119 + \sqrt{3,225,601}}{110} \approx 62.864$$

Rounding to the nearest year, the average income equals $50,000 at ages 30 and 63.

99. Let x represent speed of the current.

	Distance	**Rate**	**Time**
Up Stream	4	$5-x$	$\dfrac{4}{5-x}$
Down Stream	4	$5+x$	$\dfrac{4}{5+x}$

$$\frac{4}{5-x}+\frac{4}{5+x}=6$$

$$(5-x)(5+x)\left(\frac{4}{5-x}+\frac{4}{5+x}\right)=(5-x)(5+x)(6)$$

$$4(5+x)+4(5-x)=\left(25-x^2\right)(6)$$

$$20+4x+20-4x=150-6x^2$$

$$40=150-6x^2$$

$$6x^2=110$$

$$x^2=\frac{110}{6}$$

$$x=\pm\sqrt{\frac{110}{6}}$$

$$x\approx\pm4.282$$

The speed of the current is approximately 4.3 miles per hour.

101. Let t represent the time required for Susan to finish the route alone. Then $t-1$ will represent the time required for Robert to finish the route alone.

$$\begin{pmatrix}\text{Part done}\\ \text{by Susan}\\ \text{in 1 hour}\end{pmatrix}+\begin{pmatrix}\text{Part done}\\ \text{by Robert}\\ \text{in 1 hour}\end{pmatrix}=\begin{pmatrix}\text{Part done}\\ \text{together}\\ \text{in 1 hour}\end{pmatrix}$$

$$\frac{1}{t}\quad+\quad\frac{1}{t-1}\quad=\quad\frac{1}{2}$$

$$2t(t-1)\left(\frac{1}{t}+\frac{1}{t-1}\right)=2t(t-1)\left(\frac{1}{2}\right)$$

$$2(t-1)+2t=t(t-1)$$

$$2t-2+2t=t^2-t$$

$$4t-2=t^2-t$$

$$0=t^2-5t+2$$

For this equation, $a=1$, $b=-5$, and $c=2$.

$$t=\frac{-(-5)\pm\sqrt{(-5)^2-4(1)(2)}}{2(1)}$$

$$=\frac{5\pm\sqrt{25-8}}{2}=\frac{5\pm\sqrt{17}}{2}$$

$$t=\frac{5}{2}-\frac{\sqrt{17}}{2}\approx0.438 \text{ or } t=\frac{5}{2}+\frac{\sqrt{17}}{2}\approx4.562$$

Disregard $t\approx0.438$ because this value makes

Robert's time negative:
$t-1=0.438-1=-0.562$. The only viable answer is $t\approx4.562$ hours. Working alone, it will take Susan approximately 4.6 hours to finish the route.

103. By the quadratic formula, the solutions of the equation $ax^2+bx+c=0$ are

$$x=\frac{-b-\sqrt{b^2-4ac}}{2a} \text{ and } x=\frac{-b+\sqrt{b^2-4ac}}{2a}.$$

The sum of these two solutions is:

$$\frac{-b-\sqrt{b^2-4ac}}{2a}+\frac{-b+\sqrt{b^2-4ac}}{2a}=\frac{-2b}{2a}=-\frac{b}{a}.$$

105. Assume $b^2-4ac\geq0$. The solutions of

$$ax^2+bx+c=0 \text{ are } x=\frac{-b\pm\sqrt{b^2-4ac}}{2a}.$$

The solutions of $ax^2-bx+c=0$ are

$$x=\frac{-(-b)\pm\sqrt{(-b)^2-4ac}}{2a}=\frac{b\pm\sqrt{b^2-4ac}}{2a}.$$

Now, the negatives of the solutions to $ax^2-bx+c=0$ are

$$-\left(\frac{b\pm\sqrt{b^2-4ac}}{2a}\right)=\frac{-b\mp\sqrt{b^2-4ac}}{2a}.$$

$$=\frac{-b\pm\sqrt{b^2-4ac}}{2a}$$

which are the solutions to $ax^2+bx+c=0$.

Thus, the real solutions of $ax^2+bx+c=0$ are the negatives of the real solutions of $ax^2-bx+c=0$.

107. Answers may vary. One possibility follows: If the quadratic equation is in the form $ax^2+bx+c=0$ (or if it can easily be put in that form) and if the left side is easy to factor, then use factoring to solve the equation. Note that the discriminant must be a perfect square.

109. a. $f(x)=x^2+3x+2$

Let $x=-3,\ -2,\ -1.5,\ -1,$ and 0.

$$f(-3)=(-3)^2+3(-3)+2=9-9+2=2$$

$$f(-2)=(-2)^2+3(-2)+2=4-6+2=0$$

$$f(-1.5) = (-1.5)^2 + 3(-1.5) + 2$$
$$= 2.25 - 4.5 + 2$$
$$= -0.25$$

$$f(-1) = (-1)^2 + 3(-1) + 2 = 1 - 3 + 2 = 0$$

$$f(0) = 0^2 + 3(0) + 2 = 0 + 0 + 2 = 2$$

Thus, the points $(-3, 2)$, $(-2, 0)$,

$(-1.5, -0.25)$, $(-1, 0)$, and $(0, 2)$ are on

the graph of *f*. We plot the points and
connect them with a smooth curve.

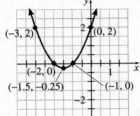

b. $x^2 + 3x + 2 = 0$
$(x+1)(x+2) = 0$
$x + 1 = 0$ or $x + 2 = 0$
$\quad x = -1$ or $\quad\quad x = -2$
The solution set is $\{-2, -1\}$.

c. From the graph in part (a), the *x*-intercepts
of the function $f(x) = x^2 + 3x + 2$ are -2
and -1, which are the same as the solutions
of the equation $x^2 + 3x + 2 = 0$. To find the
x-intercepts of a function, set the function
equal to zero and solve for *x*.

111. a. $g(x) = x^2 - 2x + 1$

Let $x = -1, 0, 1, 2,$ and 3.

$$f(-1) = (-1)^2 - 2(-1) + 1 = 1 + 2 + 1 = 4$$

$$f(0) = 0^2 - 2(0) + 1 = 0 - 0 + 1 = 1$$

$$f(1) = 1^2 - 2(1) + 1 = 1 - 2 + 1 = 0$$

$$f(2) = 2^2 - 2(2) + 1 = 4 - 4 + 1 = 1$$

$$f(3) = 3^2 - 2(3) + 1 = 9 - 6 + 1 = 4$$

Thus, the points $(-1, 4)$, $(0, 1)$, $(1, 0)$,

$(2, 1)$, and $(3, 4)$ are on the graph of *g*.

We plot the points and connect them with a
smooth curve.

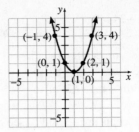

b. $x^2 - 2x + 1 = 0$
$(x-1)^2 = 0$
$x - 1 = 0$
$\quad x = 1$
The solution set is $\{1\}$.

c. From the graph in part (a), the *x*-intercept of
the function $g(x) = x^2 - 2x + 1$ is 1, which
is the same as the solution of the equation
$x^2 - 2x + 1 = 0$. To find the *x*-intercepts of
a function, set the function equal to zero and
solve for *x*.

113. $f(x) = 0$
$x^2 - 7x + 3 = 0$
For this equation, $a = 1$, $b = -7$, and $c = 3$.
$b^2 - 4ac = (-7)^2 - 4(1)(3) = 49 - 12 = 37$

Because $b^2 - 4ac = 37$ is positive, but not a
perfect square, the equation has two irrational
solutions. This conclusion based on the
discriminant is apparent in the graph because the
graph crosses the *x*-axis in two places. That is,
the graph has two *x*-intercepts.

115. $f(x) = 0$
$-x^2 - 3x - 4 = 0$
For this equation, $a = -1$, $b = -3$, and $c = -4$.
$b^2 - 4ac = (-3)^2 - 4(-1)(-4) = 9 - 16 = -7$

Because $b^2 - 4ac = -7$ is negative, the equation
has two complex solutions that are not real. This
conclusion based on the discriminant is apparent
in the graph because the graph does not cross the
x-axis. That is, the graph has no *x*-intercept.

117. a. $x^2 - 5x - 24 = 0$

$(x+3)(x-8) = 0$

$x+3 = 0$ or $x-8 = 0$

$x = -3$ or $x = 8$

The solution set is $\{-3, 8\}$.

b. The x-intercepts are -3 and 8 (see graph).

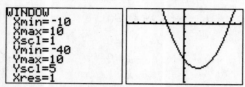

The x-intercepts of $y = x^2 - 5x - 24$ are the same as the solutions of $x^2 - 5x - 24 = 0$.

119. a. $x^2 - 6x + 9 = 0$

$(x-3)(x-3) = 0$

$x-3 = 0$ or $x-3 = 0$

$x = 3$ or $x = 3$

The solution set is $\{3\}$.

b. The x-intercept is 3 (see graph).

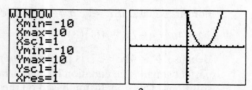

The x-intercept of $y = x^2 - 6x + 9$ is the same as the solutions of $x^2 - 6x + 9 = 0$.

121. a. $x^2 + 5x + 8 = 0$

For this equation, $a = 1$, $b = 5$, and $c = 8$.

$x = \dfrac{-5 \pm \sqrt{5^2 - 4(1)(8)}}{2(1)}$

$= \dfrac{-5 \pm \sqrt{25 - 32}}{2}$

$= \dfrac{-5 \pm \sqrt{-7}}{2}$

$= \dfrac{-5 \pm \sqrt{7}\, i}{2} = -\dfrac{5}{2} \pm \dfrac{\sqrt{7}}{2} i$

The solution set is

$\left\{ -\dfrac{5}{2} - \dfrac{\sqrt{7}}{2} i, \ -\dfrac{5}{2} + \dfrac{\sqrt{7}}{2} i \right\}$.

b. The graph has no x-intercepts (see graph).

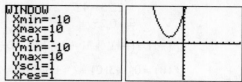

$y = x^2 + 5x + 8$ has no x-intercepts, and the solutions of $x^2 + 5x + 8 = 0$ are not real.

Section 7.3

Are You Ready for This Section?

R1. $x^4 - 5x^2 - 6 = \left(x^2 - 6\right)\left(x^2 + 1\right)$

R2. $2(p+3)^2 + 3(p+3) - 5$

$= \left[2(p+3)+5\right]\left[(p+3)-1\right]$

$= (2p+6+5)(p+3-1)$

$= (2p+11)(p+2)$

R3. a. $\left(x^2\right)^2 = x^{2\cdot 2} = x^4$

b. $\left(p^{-1}\right)^2 = p^{-1\cdot 2} = p^{-2} = \dfrac{1}{p^2}$

Section 7.3 Quick Checks

1. If a substitution u transforms an equation into one of the form $au^2 + bu + c = 0$, then the original equation is called an equation <u>quadratic in form</u>.

2. For the equation $2(3x+1)^2 - 5(3x+1) + 2 = 0$, an appropriate substitution would be $u = \underline{3x + 1}$.

3. True. Letting $u = \dfrac{x}{x-2}$ yields the quadratic equation $3u^2 - 5u + 3 = 0$.

4. $u = \dfrac{1}{x}$

5. $x^4 - 13x^2 + 36 = 0$

$\left(x^2\right)^2 - 13\left(x^2\right) + 36 = 0$

Let $u = x^2$.

$$u^2 - 13u + 36 = 0$$
$$(u-4)(u-9) = 0$$
$$u - 4 = 0 \quad \text{or} \quad u - 9 = 0$$
$$u = 4 \quad \text{or} \quad u = 9$$
$$x^2 = 4 \quad \text{or} \quad x^2 = 9$$
$$x = \pm\sqrt{4} \quad \text{or} \quad x = \pm\sqrt{9}$$
$$x = \pm 2 \quad \text{or} \quad x = \pm 3$$

Check:
$$x = -2: \quad (-2)^4 - 13(-2)^2 + 36 \overset{?}{=} 0$$
$$16 - 13 \cdot 4 + 36 \overset{?}{=} 0$$
$$16 - 52 + 36 \overset{?}{=} 0$$
$$0 = 0 \checkmark$$

$$x = 2: \quad 2^4 - 13(2)^2 + 36 \overset{?}{=} 0$$
$$16 - 13 \cdot 4 + 36 \overset{?}{=} 0$$
$$16 - 52 + 36 \overset{?}{=} 0$$
$$0 = 0 \checkmark$$

$$x = -3: \quad (-3)^4 - 13(-3)^2 + 36 \overset{?}{=} 0$$
$$81 - 13 \cdot 9 + 36 \overset{?}{=} 0$$
$$81 - 117 + 36 \overset{?}{=} 0$$
$$0 = 0 \checkmark$$

$$x = 3: \quad 3^4 - 13(3)^2 + 36 \overset{?}{=} 0$$
$$81 - 13 \cdot 9 + 36 \overset{?}{=} 0$$
$$81 - 117 + 36 \overset{?}{=} 0$$
$$0 = 0 \checkmark$$

All check; the solution set is $\{-3, -2, 2, 3\}$.

6.
$$p^4 - 7p^2 = 18$$
$$p^4 - 7p^2 - 18 = 0$$
$$\left(p^2\right)^2 - 7\left(p^2\right) - 18 = 0$$

Let $u = p^2$.
$$u^2 - 7u - 18 = 0$$
$$(u+2)(u-9) = 0$$
$$u + 2 = 0 \quad \text{or} \quad u - 9 = 0$$
$$u = -2 \quad \text{or} \quad u = 9$$

$$p^2 = -2 \quad \text{or} \quad p^2 = 9$$
$$p = \pm\sqrt{-2} \quad \text{or} \quad p = \pm\sqrt{9}$$
$$p = \pm\sqrt{2}\,i \quad \text{or} \quad p = \pm 3$$

Check:
$$p = -\sqrt{2}\,i: \quad \left(\sqrt{2}\,i\right)^4 - 7\left(\sqrt{2}\,i\right)^2 \overset{?}{=} 18$$
$$4i^4 - 7 \cdot 2i^2 \overset{?}{=} 18$$
$$4(1) - 7 \cdot 2(-1) \overset{?}{=} 18$$
$$4 - 7 \cdot 2(-1) \overset{?}{=} 18$$
$$4 + 14 \overset{?}{=} 18$$
$$18 = 18 \checkmark$$

$$p = \sqrt{2}\,i: \quad \left(-\sqrt{2}\,i\right)^4 - 7\left(-\sqrt{2}\,i\right)^2 \overset{?}{=} 18$$
$$4i^4 - 7 \cdot 2i^2 \overset{?}{=} 18$$
$$4(1) - 7 \cdot 2(-1) \overset{?}{=} 18$$
$$4 - 7 \cdot 2(-1) \overset{?}{=} 18$$
$$4 + 14 \overset{?}{=} 18$$
$$18 = 18 \checkmark$$

$$p = -3: \quad (-3)^4 - 7(-3)^2 \overset{?}{=} 18$$
$$81 - 7 \cdot 9 \overset{?}{=} 18$$
$$81 - 63 \overset{?}{=} 18$$
$$18 = 18 \checkmark$$

$$p = 3: \quad 3^4 - 7(3)^2 \overset{?}{=} 18$$
$$81 - 7 \cdot 9 \overset{?}{=} 18$$
$$81 - 63 \overset{?}{=} 18$$
$$18 = 18 \checkmark$$

All check; the solution set is
$$\left\{-\sqrt{2}\,i,\ \sqrt{2}\,i,\ -3,\ 3\right\}.$$

7. $\left(p^2 - 2\right)^2 - 9\left(p^2 - 2\right) + 14 = 0$

Let $u = p^2 - 2$.
$$u^2 - 9u + 14 = 0$$
$$(u-2)(u-7) = 0$$

$$u - 2 = 0 \quad \text{or} \quad u - 7 = 0$$
$$u = 2 \quad \text{or} \quad u = 7$$
$$p^2 - 2 = 2 \quad \text{or} \quad p^2 - 2 = 7$$
$$p^2 = 4 \quad \text{or} \quad p^2 = 9$$
$$p = \pm\sqrt{4} \quad \text{or} \quad p = \pm\sqrt{9}$$
$$p = \pm 2 \quad \text{or} \quad p = \pm 3$$

Check:

$p = -2$: $\left((-2)^2 - 2\right)^2 - 9\left((-2)^2 - 2\right) + 14 \stackrel{?}{=} 0$

$$(4-2)^2 - 9(4-2) + 14 \stackrel{?}{=} 0$$

$$2^2 - 9 \cdot 2 + 14 \stackrel{?}{=} 0$$

$$4 - 18 + 14 \stackrel{?}{=} 0$$

$$0 = 0 ✓$$

$p = 2$: $\left(2^2 - 2\right)^2 - 9\left(2^2 - 2\right) + 14 \stackrel{?}{=} 0$

$$(4-2)^2 - 9(4-2) + 14 \stackrel{?}{=} 0$$

$$2^2 - 9 \cdot 2 + 14 \stackrel{?}{=} 0$$

$$4 - 18 + 14 \stackrel{?}{=} 0$$

$$0 = 0 ✓$$

$p = -3$: $\left((-3)^2 - 2\right)^2 - 9\left((-3)^2 - 2\right) + 14 \stackrel{?}{=} 0$

$$(9-2)^2 - 9(9-2) + 14 \stackrel{?}{=} 0$$

$$7^2 - 9 \cdot 7 + 14 \stackrel{?}{=} 0$$

$$49 - 63 + 14 \stackrel{?}{=} 0$$

$$0 = 0 ✓$$

$p = 3$: $\left(3^2 - 2\right)^2 - 9\left(3^2 - 2\right) + 14 \stackrel{?}{=} 0$

$$(9-2)^2 - 9(9-2) + 14 \stackrel{?}{=} 0$$

$$7^2 - 9 \cdot 7 + 14 \stackrel{?}{=} 0$$

$$49 - 63 + 14 \stackrel{?}{=} 0$$

$$0 = 0 ✓$$

All check; the solution set is $\{-3, -2,\ 2,\ 3\}$.

8. $2\left(2z^2 - 1\right)^2 + 5\left(2z^2 - 1\right) - 3 = 0$

Let $u = 2z^2 - 1$.

$$2u^2 + 5u - 3 = 0$$

$$(2u - 1)(u + 3) = 0$$

$2u - 1 = 0 \qquad$ or $\qquad u + 3 = 0$

$u = \dfrac{1}{2} \qquad$ or $\qquad u = -3$

$2z^2 - 1 = \dfrac{1}{2} \qquad$ or $\quad 2z^2 - 1 = -3$

$2z^2 = \dfrac{3}{2} \qquad$ or $\qquad 2z^2 = -2$

$z^2 = \dfrac{3}{4} \qquad$ or $\qquad z^2 = -1$

$z = \pm\sqrt{\dfrac{3}{4}} \qquad$ or $\qquad z = \pm\sqrt{-1}$

$z = \pm\dfrac{\sqrt{3}}{2}$ or $\qquad z = \pm i$

Check:

$z = -\dfrac{\sqrt{3}}{2}$:

$$2\left(2\left(-\dfrac{\sqrt{3}}{2}\right)^2 - 1\right)^2 + 5\left(2\left(-\dfrac{\sqrt{3}}{2}\right)^2 - 1\right) - 3 \stackrel{?}{=} 0$$

$$2\left(2\left(\dfrac{3}{4}\right) - 1\right)^2 + 5\left(2\left(\dfrac{3}{4}\right) - 1\right) - 3 \stackrel{?}{=} 0$$

$$2\left(\dfrac{3}{2} - 1\right)^2 + 5\left(\dfrac{3}{2} - 1\right) - 3 \stackrel{?}{=} 0$$

$$2\left(\dfrac{1}{2}\right)^2 + 5\left(\dfrac{1}{2}\right) - 3 \stackrel{?}{=} 0$$

$$2\left(\dfrac{1}{4}\right) + 5\left(\dfrac{1}{2}\right) - 3 \stackrel{?}{=} 0$$

$$\dfrac{1}{2} + \dfrac{5}{2} - 3 \stackrel{?}{=} 0$$

$$0 = 0 ✓$$

$z = \dfrac{\sqrt{3}}{2}$:

$$2\left(2\left(\dfrac{\sqrt{3}}{2}\right)^2 - 1\right)^2 + 5\left(2\left(\dfrac{\sqrt{3}}{2}\right)^2 - 1\right) - 3 \overset{?}{=} 0$$

$$2\left(2\left(\dfrac{3}{4}\right) - 1\right)^2 + 5\left(2\left(\dfrac{3}{4}\right) - 1\right) - 3 \overset{?}{=} 0$$

$$2\left(\dfrac{3}{2} - 1\right)^2 + 5\left(\dfrac{3}{2} - 1\right) - 3 \overset{?}{=} 0$$

$$2\left(\dfrac{1}{2}\right)^2 + 5\left(\dfrac{1}{2}\right) - 3 \overset{?}{=} 0$$

$$2\left(\dfrac{1}{4}\right) + 5\left(\dfrac{1}{2}\right) - 3 \overset{?}{=} 0$$

$$\dfrac{1}{2} + \dfrac{5}{2} - 3 \overset{?}{=} 0$$

$$0 = 0 \checkmark$$

$z = -i$:

$$2\left(2(-i)^2 - 1\right)^2 + 5\left(2(-i)^2 - 1\right) - 3 \overset{?}{=} 0$$

$$2\left(2i^2 - 1\right)^2 + 5\left(2i^2 - 1\right) - 3 \overset{?}{=} 0$$

$$2\left(2(-1) - 1\right)^2 + 5\left(2(-1) - 1\right) - 3 \overset{?}{=} 0$$

$$2(-2 - 1)^2 + 5(-2 - 1) - 3 \overset{?}{=} 0$$

$$2(-3)^2 + 5(-3) - 3 \overset{?}{=} 0$$

$$2(9) + 5(-3) - 3 \overset{?}{=} 0$$

$$18 - 15 - 3 \overset{?}{=} 0$$

$$0 = 0 \checkmark$$

$z = i$:

$$2\left(2i^2 - 1\right)^2 + 5\left(2i^2 - 1\right) - 3 \overset{?}{=} 0$$

$$2\left(2(-1) - 1\right)^2 + 5\left(2(-1) - 1\right) - 3 \overset{?}{=} 0$$

$$2(-2 - 1)^2 + 5(-2 - 1) - 3 \overset{?}{=} 0$$

$$2(-3)^2 + 5(-3) - 3 \overset{?}{=} 0$$

$$2(9) + 5(-3) - 3 \overset{?}{=} 0$$

$$18 - 15 - 3 \overset{?}{=} 0$$

$$0 = 0 \checkmark$$

All check; the solution set is

$$\left\{ -\dfrac{\sqrt{3}}{2},\ \dfrac{\sqrt{3}}{2}, -i,\ i \right\}.$$

9.

$$3w - 14\sqrt{w} + 8 = 0$$

$$3\left(\sqrt{w}\right)^2 - 14\left(\sqrt{w}\right) + 8 = 0$$

Let $u = \sqrt{w}$.

$$3u^2 - 14u + 8 = 0$$

$$(3u - 2)(u - 4) = 0$$

$$3u - 2 = 0 \quad \text{or} \quad u - 4 = 0$$

$$u = \dfrac{2}{3} \quad \text{or} \quad u = 4$$

$$\sqrt{w} = \dfrac{2}{3} \quad \text{or} \quad \sqrt{w} = 4$$

$$w = \left(\dfrac{2}{3}\right)^2 \quad \text{or} \quad w = 4^2$$

$$w = \dfrac{4}{9} \quad \text{or} \quad w = 16$$

<u>Check</u>:

$$w = \dfrac{4}{9}: \quad 3\left(\dfrac{4}{9}\right) - 14\sqrt{\dfrac{4}{9}} + 8 \overset{?}{=} 0$$

$$\dfrac{4}{3} - 14 \cdot \dfrac{2}{3} + 8 \overset{?}{=} 0$$

$$\dfrac{4}{3} - \dfrac{28}{3} + 8 \overset{?}{=} 0$$

$$0 = 0 \checkmark$$

$$w = 16: \quad 3(16) - 14\sqrt{16} + 8 \overset{?}{=} 0$$

$$48 - 14 \cdot 4 + 8 \overset{?}{=} 0$$

$$48 - 56 + 8 \overset{?}{=} 0$$

$$0 = 0 \checkmark$$

Both check; the solution set is $\left\{ \dfrac{4}{9},\ 16 \right\}$.

10.

$$2q - 9\sqrt{q} - 5 = 0$$

$$2\left(\sqrt{q}\right)^2 - 9\left(\sqrt{q}\right) - 5 = 0$$

Let $u = \sqrt{q}$.

$$2u^2 - 9u - 5 = 0$$

$$(2u + 1)(u - 5) = 0$$

$$2u + 1 = 0 \quad \text{or} \quad u - 5 = 0$$

$$u = -\dfrac{1}{2} \quad \text{or} \quad u = 5$$

$$\sqrt{q} = -\frac{1}{2} \quad \text{or} \quad \sqrt{q} = 5$$

$$q = \left(-\frac{1}{2}\right)^2 \quad \text{or} \quad q = 5^2$$

$$q = \frac{1}{4} \quad \text{or} \quad q = 25$$

Check:

$$q = \frac{1}{4}: \quad 2\left(\frac{1}{4}\right) - 9\sqrt{\frac{1}{4}} - 5 \stackrel{?}{=} 0$$

$$\frac{1}{2} - 9 \cdot \frac{1}{2} - 5 \stackrel{?}{=} 0$$

$$\frac{1}{2} - \frac{9}{2} - 5 \stackrel{?}{=} 0$$

$$-9 \neq 0 \quad ✗$$

$$q = 25: \quad 2(25) - 9\sqrt{25} - 5 \stackrel{?}{=} 0$$

$$50 - 9 \cdot 5 - 5 \stackrel{?}{=} 0$$

$$50 - 45 - 5 \stackrel{?}{=} 0$$

$$0 = 0 \quad ✓$$

$q = \dfrac{1}{4}$ does not check; the solution set is $\{25\}$.

11.
$$5x^{-2} + 12x^{-1} + 4 = 0$$

$$5\left(x^{-1}\right)^2 + 12\left(x^{-1}\right) + 4 = 0$$

Let $u = x^{-1}$.

$$5u^2 + 12u + 4 = 0$$

$$(5u + 2)(u + 2) = 0$$

$$5u + 2 = 0 \quad \text{or} \quad u + 2 = 0$$

$$u = -\frac{2}{5} \quad \text{or} \quad u = -2$$

$$x^{-1} = -\frac{2}{5} \quad \text{or} \quad x^{-1} = -2$$

$$\frac{1}{x} = -\frac{2}{5} \quad \text{or} \quad \frac{1}{x} = -2$$

$$x = -\frac{5}{2} \quad \text{or} \quad x = -\frac{1}{2}$$

Check:

$$x = -\frac{5}{2}: \quad 5\left(-\frac{5}{2}\right)^{-2} + 12\left(-\frac{5}{2}\right)^{-1} + 4 \stackrel{?}{=} 0$$

$$5\left(-\frac{2}{5}\right)^2 + 12\left(-\frac{2}{5}\right) + 4 \stackrel{?}{=} 0$$

$$5\left(\frac{4}{25}\right) + 12\left(-\frac{2}{5}\right) + 4 \stackrel{?}{=} 0$$

$$\frac{4}{5} - \frac{24}{5} + 4 \stackrel{?}{=} 0$$

$$0 = 0 \quad ✓$$

$$x = -\frac{1}{2}: \quad 5\left(-\frac{1}{2}\right)^{-2} + 12\left(-\frac{1}{2}\right)^{-1} + 4 \stackrel{?}{=} 0$$

$$5(-2)^2 + 12(-2) + 4 \stackrel{?}{=} 0$$

$$5(4) + 12(-2) + 4 \stackrel{?}{=} 0$$

$$20 - 24 + 4 \stackrel{?}{=} 0$$

$$0 = 0 \quad ✓$$

Both check; the solution set is $\left\{-\dfrac{5}{2}, -\dfrac{1}{2}\right\}$.

12.
$$p^{2/3} - 4p^{1/3} - 5 = 0$$

$$\left(p^{1/3}\right)^2 - 4\left(p^{1/3}\right) - 5 = 0$$

Let $u = p^{1/3}$.

$$u^2 - 4u - 5 = 0$$

$$(u + 1)(u - 5) = 0$$

$$u + 1 = 0 \quad \text{or} \quad u - 5 = 0$$

$$u = -1 \quad \text{or} \quad u = 5$$

$$p^{1/3} = -1 \quad \text{or} \quad p^{1/3} = 5$$

$$\left(p^{1/3}\right)^3 = (-1)^3 \quad \text{or} \quad \left(p^{1/3}\right)^3 = 5^3$$

$$p = -1 \quad \text{or} \quad p = 125$$

Check:

$$p = -1: \quad (-1)^{2/3} - 4 \cdot (-1)^{1/3} - 5 \stackrel{?}{=} 0$$

$$\left(\sqrt[3]{-1}\right)^2 - 4\left(\sqrt[3]{-1}\right) - 5 \stackrel{?}{=} 0$$

$$(-1)^2 - 4(-1) - 5 \stackrel{?}{=} 0$$

$$1 + 4 - 5 \stackrel{?}{=} 0$$

$$0 = 0 \quad ✓$$

$p = 125:\ (125)^{2/3} - 4 \cdot (125)^{1/3} - 5 \overset{?}{=} 0$

$\left(\sqrt[3]{125}\right)^2 - 4\left(\sqrt[3]{125}\right) - 5 \overset{?}{=} 0$

$(5)^2 - 4(5) - 5 \overset{?}{=} 0$

$25 - 20 - 5 \overset{?}{=} 0$

$0 = 0\ \checkmark$

Both check; the solution set is $\{-1,\ 125\}$.

7.3 Exercises

13.　　$x^4 - 5x^2 + 4 = 0$

$\left(x^2\right)^2 - 5\left(x^2\right) + 4 = 0$

Let $u = x^2$.

$u^2 - 5u + 4 = 0$

$(u-1)(u-4) = 0$

$u - 1 = 0$　or　$u - 4 = 0$

$u = 1$　or　$u = 4$

$x^2 = 1$　or　$x^2 = 4$

$x = \pm\sqrt{1}$　or　$x = \pm\sqrt{4}$

$x = \pm 1$　or　$x = \pm 2$

Check:

$x = -1:\ (-1)^4 - 5(-1)^2 + 4 \overset{?}{=} 0$

$1 - 5 \cdot 1 + 4 \overset{?}{=} 0$

$1 - 5 + 4 \overset{?}{=} 0$

$0 = 0\ \checkmark$

$x = 1:\ 1^4 - 5(1)^2 + 4 \overset{?}{=} 0$

$1 - 5 \cdot 1 + 4 \overset{?}{=} 0$

$1 - 5 + 4 \overset{?}{=} 0$

$0 = 0\ \checkmark$

$x = -2:\ (-2)^4 - 5(-2)^2 + 4 \overset{?}{=} 0$

$16 - 5 \cdot 4 + 4 \overset{?}{=} 0$

$16 - 20 + 4 \overset{?}{=} 0$

$0 = 0\ \checkmark$

$x = 2:\ 2^4 - 5(2)^2 + 4 \overset{?}{=} 0$

$16 - 5 \cdot 4 + 4 \overset{?}{=} 0$

$16 - 20 + 4 \overset{?}{=} 0$

$0 = 0\ \checkmark$

All check; the solution set is $\{-2,\ -1,\ 1,\ 2\}$.

15.　　$q^4 + 13q^2 + 36 = 0$

$\left(q^2\right)^2 + 13\left(q^2\right) + 36 = 0$

Let $u = q^2$.

$u^2 + 13u + 36 = 0$

$(u+4)(u+9) = 0$

$q^2 = -4$　or　$q^2 = -9$

$q = \pm\sqrt{-4}$　or　$q = \pm\sqrt{-9}$

$q = \pm 2i$　or　$q = \pm 3i$

Check:

$q = -3i:\ (-3i)^4 + 13(-3i)^2 + 36 \overset{?}{=} 0$

$81i^4 + 13 \cdot 9i^2 + 36 \overset{?}{=} 0$

$81(1) + 13 \cdot 9 \cdot (-1) + 36 \overset{?}{=} 0$

$81 - 117 + 36 \overset{?}{=} 0$

$0 = 0\ \checkmark$

$q = 3i:\ (3i)^4 + 13(3i)^2 + 36 \overset{?}{=} 0$

$81i^4 + 13 \cdot 9i^2 + 36 \overset{?}{=} 0$

$81(1) + 13 \cdot 9 \cdot (-1) + 36 \overset{?}{=} 0$

$81 - 117 + 36 \overset{?}{=} 0$

$0 = 0\ \checkmark$

$q = -2i:\ (-2i)^4 + 13(-2i)^2 + 36 \overset{?}{=} 0$

$16i^4 + 13 \cdot 4i^2 + 36 \overset{?}{=} 0$

$16(1) + 13 \cdot 4 \cdot (-1) + 36 \overset{?}{=} 0$

$16 - 52 + 36 \overset{?}{=} 0$

$0 = 0\ \checkmark$

$q = 2i:\ (2i)^4 + 13(2i)^2 + 36 \overset{?}{=} 0$

$16i^4 + 13 \cdot 4i^2 + 36 \overset{?}{=} 0$

$16(1) + 13 \cdot 4 \cdot (-1) + 36 \overset{?}{=} 0$

$16 - 52 + 36 \overset{?}{=} 0$

$0 = 0\ \checkmark$

All check; the solution set is $\{-3i,\ 3i, -2i,\ 2i\}$.

17.　　$4a^4 - 17a^2 + 4 = 0$

$4\left(a^2\right)^2 - 17\left(a^2\right) + 4 = 0$

Let $u = a^2$.

$4u^2 - 17u + 4 = 0$

$(4u-1)(u-4) = 0$

$$4u - 1 = 0 \quad \text{or} \quad u - 4 = 0$$
$$4u = 1 \quad \text{or} \quad u = 4$$
$$u = \frac{1}{4}$$

$$a^2 = \frac{1}{4} \quad \text{or} \quad a^2 = 4$$
$$a = \pm\sqrt{\frac{1}{4}} \quad \text{or} \quad a = \pm\sqrt{4}$$
$$a = \pm\frac{1}{2} \quad \text{or} \quad a = \pm 2$$

Check:

$$a = -\frac{1}{2}: \quad 4\left(-\frac{1}{2}\right)^4 - 17\left(-\frac{1}{2}\right)^2 + 4 \stackrel{?}{=} 0$$
$$4\left(\frac{1}{16}\right) - 17\left(\frac{1}{4}\right) + 4 \stackrel{?}{=} 0$$
$$\frac{1}{4} - \frac{17}{4} + 4 \stackrel{?}{=} 0$$
$$0 = 0 \checkmark$$

$$a = \frac{1}{2}: \quad 4\left(\frac{1}{2}\right)^4 - 17\left(\frac{1}{2}\right)^2 + 4 \stackrel{?}{=} 0$$
$$4\left(\frac{1}{16}\right) - 17\left(\frac{1}{4}\right) + 4 \stackrel{?}{=} 0$$
$$\frac{1}{4} - \frac{17}{4} + 4 \stackrel{?}{=} 0$$
$$0 = 0 \checkmark$$

$$a = -2: \quad 4(-2)^4 - 17(-2)^2 + 4 \stackrel{?}{=} 0$$
$$4 \cdot 16 - 17 \cdot 4 + 4 \stackrel{?}{=} 0$$
$$64 - 68 + 4 \stackrel{?}{=} 0$$
$$0 = 0 \checkmark$$

$$a = 2: \quad 4(2)^4 - 17(2)^2 + 4 \stackrel{?}{=} 0$$
$$4 \cdot 16 - 17 \cdot 4 + 4 \stackrel{?}{=} 0$$
$$64 - 68 + 4 \stackrel{?}{=} 0$$
$$0 = 0 \checkmark$$

All check; the solution set is $\left\{-2, -\frac{1}{2}, \frac{1}{2}, 2\right\}$.

19.
$$p^4 + 6 = 5p^2$$
$$p^4 - 5p^2 + 6 = 0$$
$$\left(p^2\right)^2 - 5\left(p^2\right) + 6 = 0$$

Let $u = p^2$.

$$u^2 - 5u + 6 = 0$$
$$(u - 2)(u - 3) = 0$$
$$u - 2 = 0 \quad \text{or} \quad u - 3 = 0$$
$$u = 2 \quad \text{or} \quad u = 3$$

$$p^2 = 2 \quad \text{or} \quad p^2 = 3$$
$$p = \pm\sqrt{2} \quad \text{or} \quad p = \pm\sqrt{3}$$

Check:

$$p = -\sqrt{2}: \quad \left(-\sqrt{2}\right)^4 + 6 \stackrel{?}{=} 5\left(-\sqrt{2}\right)^2$$
$$4 + 6 \stackrel{?}{=} 5 \cdot 2$$
$$10 = 10 \checkmark$$

$$p = \sqrt{2}: \quad \left(\sqrt{2}\right)^4 + 6 \stackrel{?}{=} 5\left(\sqrt{2}\right)^2$$
$$4 + 6 \stackrel{?}{=} 5 \cdot 2$$
$$10 = 10 \checkmark$$

$$p = -\sqrt{3}: \quad \left(-\sqrt{3}\right)^4 + 6 \stackrel{?}{=} 5\left(-\sqrt{3}\right)^2$$
$$9 + 6 \stackrel{?}{=} 5 \cdot 3$$
$$15 = 15 \checkmark$$
$$15 = 15 \checkmark$$

$$p = \sqrt{3}: \quad \left(\sqrt{3}\right)^4 + 6 \stackrel{?}{=} 5\left(\sqrt{3}\right)^2$$
$$9 + 6 \stackrel{?}{=} 5 \cdot 3$$
$$15 = 15 \checkmark$$

All check; the solution set is
$$\left\{-\sqrt{3}, -\sqrt{2}, \sqrt{2}, \sqrt{3}\right\}.$$

21. $(x - 3)^2 - 6(x - 3) - 7 = 0$

Let $u = x - 3$.

$$u^2 - 6u - 7 = 0$$
$$(u - 7)(u + 1) = 0$$
$$u - 7 = 0 \quad \text{or} \quad u + 1 = 0$$
$$u = 7 \quad \text{or} \quad u = -1$$
$$x - 3 = 7 \quad \text{or} \quad x - 3 = -1$$
$$x = 10 \quad \text{or} \quad x = 2$$

Check:

$$x = 10: \quad (10 - 3)^2 - 6(10 - 3) - 7 \stackrel{?}{=} 0$$
$$7^2 - 6(7) - 7 \stackrel{?}{=} 0$$
$$49 - 42 - 7 \stackrel{?}{=} 0$$
$$0 = 0 \checkmark$$

$x=2: \ (2-3)^2-6(2-3)-7 \overset{?}{=} 0$

$\qquad (-1)^2-6(-1)-7 \overset{?}{=} 0$

$\qquad\qquad 1+6-7 \overset{?}{=} 0$

$\qquad\qquad\qquad 0=0 \ \checkmark$

Both check; the solution set is $\{2,\ 10\}$.

23. $\left(x^2-1\right)^2-11\left(x^2-1\right)+24=0$

Let $u=x^2-1$.

$u^2-11u+24=0$

$(u-3)(u-8)=0$

$u-3=0 \quad$ or $\quad u-8=0$

$\quad u=3 \quad$ or $\qquad u=8$

$x^2-1=3 \quad$ or $\quad x^2-1=8$

$\quad x^2=4 \quad$ or $\qquad x^2=9$

$\quad x=\pm\sqrt{4} \ $ or $\ x=\pm\sqrt{9}$

$\quad x=\pm 2 \quad$ or $\quad x=\pm 3$

Check:

$x=-2: \ \left((-2)^2-1\right)^2-11\left((-2)^2-1\right)+24 \overset{?}{=} 0$

$\qquad (4-1)^2-11(4-1)+24 \overset{?}{=} 0$

$\qquad\qquad 3^2-11\cdot 3+24 \overset{?}{=} 0$

$\qquad\qquad 9-33+24 \overset{?}{=} 0$

$\qquad\qquad\qquad 0=0 \ \checkmark$

$x=2: \ \left(2^2-1\right)^2-11\left(2^2-1\right)+24 \overset{?}{=} 0$

$\qquad (4-1)^2-11(4-1)+24 \overset{?}{=} 0$

$\qquad\qquad 3^2-11\cdot 3+24 \overset{?}{=} 0$

$\qquad\qquad 9-33+24 \overset{?}{=} 0$

$\qquad\qquad\qquad 0=0 \ \checkmark$

$x=-3: \ \left((-3)^2-1\right)^2-11\left((-3)^2-1\right)+24 \overset{?}{=} 0$

$\qquad (9-1)^2-11(9-1)+24 \overset{?}{=} 0$

$\qquad\qquad 8^2-11\cdot 8+24 \overset{?}{=} 0$

$\qquad\qquad 64-88+24 \overset{?}{=} 0$

$\qquad\qquad\qquad 0=0 \ \checkmark$

$x=3: \ \left(3^2-1\right)^2-11\left(3^2-1\right)+24 \overset{?}{=} 0$

$\qquad (9-1)^2-11(9-1)+24 \overset{?}{=} 0$

$\qquad\qquad 8^2-11\cdot 8+24 \overset{?}{=} 0$

$\qquad\qquad 64-88+24 \overset{?}{=} 0$

$\qquad\qquad\qquad 0=0 \ \checkmark$

All check; the solution set is $\{-3,-2,\ 2,\ 3\}$.

25. $\left(y^2+2\right)^2+7\left(y^2+2\right)+10=0$

Let $u=y^2+2$.

$u^2+7u+10=0$

$(u+2)(u+5)=0$

$u+2=0 \quad$ or $\quad u+5=0$

$\quad u=-2 \quad$ or $\qquad u=-5$

$y^2+2=-2 \quad$ or $\quad y^2+2=-5$

$\quad y^2=-4 \quad$ or $\qquad y^2=-7$

$\quad y=\pm\sqrt{-4} \ $ or $\ y=\pm\sqrt{-7}$

$\quad y=\pm 2i \quad$ or $\quad y=\pm\sqrt{7}\,i$

Check:

$y=-2i:$

$\left((-2i)^2+2\right)^2+7\left((-2i)^2+2\right)+10 \overset{?}{=} 0$

$\left(4i^2+2\right)^2+7\left(4i^2+2\right)+10 \overset{?}{=} 0$

$(-4+2)^2+7(-4+2)+10 \overset{?}{=} 0$

$(-2)^2+7(-2)+10 \overset{?}{=} 0$

$4-14+10 \overset{?}{=} 0$

$\qquad 0=0 \ \checkmark$

$y=2i: \ \left((2i)^2+2\right)^2+7\left((2i)^2+2\right)+10 \overset{?}{=} 0$

$\left(4i^2+2\right)^2+7\left(4i^2+2\right)+10 \overset{?}{=} 0$

$(-4+2)^2+7(-4+2)+10 \overset{?}{=} 0$

$(-2)^2+7(-2)+10 \overset{?}{=} 0$

$4-14+10 \overset{?}{=} 0$

$\qquad 0=0 \ \checkmark$

$y = -\sqrt{7}i:$

$$\left(\left(-\sqrt{7}i\right)^2 + 2\right)^2 + 7\left(\left(-\sqrt{7}i\right)^2 + 2\right) + 10 \stackrel{?}{=} 0$$

$$\left(7i^2 + 2\right)^2 + 7\left(7i^2 + 2\right) + 10 \stackrel{?}{=} 0$$

$$(-7+2)^2 + 7(-7+2) + 10 \stackrel{?}{=} 0$$

$$(-5)^2 + 7(-5) + 10 \stackrel{?}{=} 0$$

$$25 - 35 + 10 \stackrel{?}{=} 0$$

$$0 = 0 \checkmark$$

$y = \sqrt{7}i:$

$$\left(\left(\sqrt{7}i\right)^2 + 2\right)^2 + 7\left(\left(\sqrt{7}i\right)^2 + 2\right) + 10 \stackrel{?}{=} 0$$

$$\left(7i^2 + 2\right)^2 + 7\left(7i^2 + 2\right) + 10 \stackrel{?}{=} 0$$

$$(-7+2)^2 + 7(-7+2) + 10 \stackrel{?}{=} 0$$

$$(-5)^2 + 7(-5) + 10 \stackrel{?}{=} 0$$

$$25 - 35 + 10 \stackrel{?}{=} 0$$

$$0 = 0 \checkmark$$

All check; the solution set is
$\left\{-\sqrt{7}\ i,\ \sqrt{7}\ i, -2i,\ 2i\right\}$.

27. $x - 3\sqrt{x} - 4 = 0$

$$\left(\sqrt{x}\right)^2 - 3\left(\sqrt{x}\right) - 4 = 0$$

Let $u = \sqrt{x}$.

$$u^2 - 3u - 4 = 0$$
$$(u-4)(u+1) = 0$$
$$u - 4 = 0 \quad \text{or} \quad u + 1 = 0$$
$$u = 4 \quad \text{or} \quad u = -1$$

$$\sqrt{x} = 4 \quad \text{or} \quad \sqrt{x} = -1$$
$$x = 4^2 \quad \text{or} \quad x = (-1)^2$$
$$x = 16 \quad \text{or} \quad x = 1$$

Check:

$x = 16:\ 16 - 3\sqrt{16} - 4 \stackrel{?}{=} 0$
$$16 - 3 \cdot 4 - 4 \stackrel{?}{=} 0$$
$$16 - 12 - 4 \stackrel{?}{=} 0$$
$$0 = 0 \checkmark$$

$x = 1:\ 1 - 3\sqrt{1} - 4 \stackrel{?}{=} 0$
$$1 - 3 \cdot 1 - 4 \stackrel{?}{=} 0$$
$$1 - 3 - 4 \stackrel{?}{=} 0$$
$$-6 \neq 0 \ \text{✗}$$

$x = 1$ does not check; the solution set is $\{16\}$.

29. $w + 5\sqrt{w} + 6 = 0$

$$\left(\sqrt{w}\right)^2 + 5\left(\sqrt{w}\right) + 6 = 0$$

Let $u = \sqrt{w}$.

$$u^2 + 5u + 6 = 0$$
$$(u+2)(u+3) = 0$$
$$u + 2 = 0 \quad \text{or} \quad u + 3 = 0$$
$$u = -2 \quad \text{or} \quad u = -3$$

$$\sqrt{w} = -2 \quad \text{or} \quad \sqrt{w} = -3$$
$$w = (-2)^2 \quad \text{or} \quad w = (-3)^2$$
$$w = 4 \quad \text{or} \quad w = 9$$

Check:

$w = 4:\ 4 + 5\sqrt{4} + 6 \stackrel{?}{=} 0$
$$4 + 5 \cdot 2 + 6 \stackrel{?}{=} 0$$
$$4 + 10 + 6 \stackrel{?}{=} 0$$
$$20 \neq 0 \ \text{✗}$$

$w = 9:\ 9 + 5\sqrt{9} + 6 \stackrel{?}{=} 0$
$$9 + 5 \cdot 3 + 6 \stackrel{?}{=} 0$$
$$9 + 15 + 6 \stackrel{?}{=} 0$$
$$30 \neq 0 \ \text{✗}$$

Neither possibility checks; the equation has no solution. The solution set is $\{\ \}$ or $\varnothing$.

31. $2x + 5\sqrt{x} = 3$
$$2x + 5\sqrt{x} - 3 = 0$$
$$2\left(\sqrt{x}\right)^2 + 5\left(\sqrt{x}\right) - 3 = 0$$

Let $u = \sqrt{x}$.

$$2u^2 + 5u - 3 = 0$$
$$(2u-1)(u+3) = 0$$
$$2u - 1 = 0 \quad \text{or} \quad u + 3 = 0$$
$$2u = 1 \quad \text{or} \quad u = -3$$
$$u = \frac{1}{2}$$

$$\sqrt{x} = \frac{1}{2} \quad \text{or} \quad \sqrt{x} = -3$$

$$x = \left(\frac{1}{2}\right)^2 \quad \text{or} \quad x = (-3)^2$$

$$x = \frac{1}{4} \quad \text{or} \quad x = 9$$

Check:

$$x = \frac{1}{4}: \quad 2 \cdot \frac{1}{4} + 5\sqrt{\frac{1}{4}} \overset{?}{=} 3$$

$$\frac{1}{2} + 5 \cdot \frac{1}{2} \overset{?}{=} 3$$

$$\frac{1}{2} + \frac{5}{2} \overset{?}{=} 3$$

$$3 = 3 \checkmark$$

$$x = 9: \quad 2 \cdot 9 + 5\sqrt{9} \overset{?}{=} 3$$

$$2 \cdot 9 + 5 \cdot 3 \overset{?}{=} 3$$

$$18 + 15 \overset{?}{=} 3$$

$$33 \neq 3 \;\; \textbf{✗}$$

$x = 9$ does not check; the solution set is $\left\{\frac{1}{4}\right\}$.

33.
$$x^{-2} + 3x^{-1} = 28$$
$$x^{-2} + 3x^{-1} - 28 = 0$$
$$\left(x^{-1}\right)^2 + 3\left(x^{-1}\right) - 28 = 0$$

Let $u = x^{-1}$.

$$u^2 + 3u - 28 = 0$$
$$(u - 4)(u + 7) = 0$$
$$u - 4 = 0 \quad \text{or} \quad u + 7 = 0$$
$$u = 4 \quad \text{or} \quad u = -7$$
$$x^{-1} = 4 \quad \text{or} \quad x^{-1} = -7$$
$$\frac{1}{x} = 4 \quad \text{or} \quad \frac{1}{x} = -7$$
$$x = \frac{1}{4} \quad \text{or} \quad x = -\frac{1}{7}$$

Check:

$$x = \frac{1}{4}: \quad \left(\frac{1}{4}\right)^{-2} + 3\left(\frac{1}{4}\right)^{-1} \overset{?}{=} 28$$

$$(4)^2 + 3(4) \overset{?}{=} 28$$

$$16 + 12 \overset{?}{=} 28$$

$$28 = 28 \checkmark$$

$$x = -\frac{1}{7}: \quad \left(-\frac{1}{7}\right)^{-2} + 3\left(-\frac{1}{7}\right)^{-1} \overset{?}{=} 28$$

$$(-7)^2 + 3(-7) \overset{?}{=} 28$$

$$49 - 21 \overset{?}{=} 28$$

$$28 = 28 \checkmark$$

Both check; the solution set is $\left\{-\frac{1}{7}, \frac{1}{4}\right\}$.

35.
$$10z^{-2} + 11z^{-1} = 6$$
$$10z^{-2} + 11z^{-1} - 6 = 0$$
$$10\left(z^{-1}\right)^2 + 11\left(z^{-1}\right) - 6 = 0$$

Let $u = z^{-1}$.

$$10u^2 + 11u - 6 = 0$$
$$(5u - 2)(2u + 3) = 0$$
$$5u - 2 = 0 \quad \text{or} \quad 2u + 3 = 0$$
$$5u = 2 \quad \text{or} \quad 2u = -3$$
$$u = \frac{2}{5} \quad \text{or} \quad u = -\frac{3}{2}$$
$$z^{-1} = \frac{2}{5} \quad \text{or} \quad z^{-1} = -\frac{3}{2}$$
$$\frac{1}{z} = \frac{2}{5} \quad \text{or} \quad \frac{1}{z} = -\frac{3}{2}$$
$$z = \frac{5}{2} \quad \text{or} \quad z = -\frac{2}{3}$$

Check:

$$z = \frac{5}{2}:$$

$$10\left(\frac{5}{2}\right)^{-2} + 11\left(\frac{5}{2}\right)^{-1} \overset{?}{=} 6$$

$$10\left(\frac{2}{5}\right)^2 + 11\left(\frac{2}{5}\right) \overset{?}{=} 6$$

$$10\left(\frac{4}{25}\right) + 11\left(\frac{2}{5}\right) \overset{?}{=} 6$$

$$\frac{8}{5} + \frac{22}{5} \overset{?}{=} 6$$

$$\frac{30}{5} \overset{?}{=} 6$$

$$6 = 6 \checkmark$$

$z = -\dfrac{2}{3}:$

$$10\left(-\dfrac{2}{3}\right)^{-2} + 11\left(-\dfrac{2}{3}\right)^{-1} \overset{?}{=} 6$$

$$10\left(-\dfrac{3}{2}\right)^{2} + 11\left(-\dfrac{3}{2}\right) \overset{?}{=} 6$$

$$10\left(\dfrac{9}{4}\right) + 11\left(-\dfrac{3}{2}\right) \overset{?}{=} 6$$

$$\dfrac{45}{2} - \dfrac{33}{2} \overset{?}{=} 6$$

$$\dfrac{12}{2} \overset{?}{=} 6$$

$$6 = 6 \checkmark$$

Both check; the solution set is $\left\{-\dfrac{2}{3},\ \dfrac{5}{2}\right\}$.

37. $x^{2/3} + 3x^{1/3} - 4 = 0$

$$\left(x^{1/3}\right)^{2} + 3\left(x^{1/3}\right) - 4 = 0$$

Let $u = x^{1/3}$.

$$u^2 + 3u - 4 = 0$$
$$(u - 1)(u + 4) = 0$$

$$u - 1 = 0 \quad \text{or} \quad u + 4 = 0$$
$$u = 1 \quad \text{or} \quad u = -4$$
$$x^{1/3} = 1 \quad \text{or} \quad x^{1/3} = -4$$
$$\left(x^{1/3}\right)^3 = 1^3 \quad \text{or} \quad \left(x^{1/3}\right)^3 = (-4)^3$$
$$x = 1 \quad \text{or} \quad x = -64$$

Check:

$x = 1:$ $(1)^{2/3} + 3 \cdot (1)^{1/3} - 4 \overset{?}{=} 0$

$$\left(\sqrt[3]{1}\right)^2 + 3\left(\sqrt[3]{1}\right) - 4 \overset{?}{=} 0$$

$$(1)^2 + 3(1) - 4 \overset{?}{=} 0$$

$$1 + 3 - 4 \overset{?}{=} 0$$

$$0 = 0 \checkmark$$

$x = -64:$ $(-64)^{2/3} + 3 \cdot (-64)^{1/3} - 4 \overset{?}{=} 0$

$$\left(\sqrt[3]{-64}\right)^2 + 3\left(\sqrt[3]{-64}\right) - 4 \overset{?}{=} 0$$

$$(-4)^2 + 3(-4) - 4 \overset{?}{=} 0$$

$$16 - 12 - 4 \overset{?}{=} 0$$

$$0 = 0 \checkmark$$

Both check; the solution set is $\{-64,\ 1\}$.

39. $z^{2/3} - z^{1/3} = 2$

$$\left(z^{1/3}\right)^2 - \left(z^{1/3}\right) - 2 = 0$$

Let $u = z^{1/3}$.

$$u^2 - u - 2 = 0$$
$$(u - 2)(u + 1) = 0$$

$$u - 2 = 0 \quad \text{or} \quad u + 1 = 0$$
$$u = 2 \quad \text{or} \quad u = -1$$
$$z^{1/3} = 2 \quad \text{or} \quad z^{1/3} = -1$$
$$z = 2^3 = 8 \quad \text{or} \quad z = (-1)^3 = -1$$

Check:

$z = 8:$ $(8)^{2/3} - (8)^{1/3} \overset{?}{=} 2$

$$\left(\sqrt[3]{8}\right)^2 - \left(\sqrt[3]{8}\right) \overset{?}{=} 2$$

$$(2)^2 - 2 \overset{?}{=} 2$$

$$4 - 2 \overset{?}{=} 2$$

$$2 = 2 \checkmark$$

$z = -1:$ $(-1)^{2/3} - (-1)^{1/3} \overset{?}{=} 2$

$$\left(\sqrt[3]{-1}\right)^2 - \left(\sqrt[3]{-1}\right) \overset{?}{=} 2$$

$$(-1)^2 - (-1) \overset{?}{=} 2$$

$$1 + 1 \overset{?}{=} 2$$

$$2 = 2 \checkmark$$

Both check; the solution set is $\{-1,\ 8\}$.

41. $a + a^{1/2} = 30$

$$\left(a^{1/2}\right)^2 + \left(a^{1/2}\right) - 30 = 0$$

Let $u = a^{1/2}$.

$$u^2 + u - 30 = 0$$
$$(u - 5)(u + 6) = 0$$

$$u - 5 = 0 \quad \text{or} \quad u + 6 = 0$$
$$u = 5 \quad \text{or} \quad u = -6$$
$$a^{1/2} = 5 \quad \text{or} \quad a^{1/2} = -6$$
$$\left(a^{1/2}\right)^2 = 5^2 \quad \text{or} \quad \left(a^{1/2}\right)^2 = (-6)^2$$
$$a = 25 \quad \text{or} \quad a = 36$$

Check:

$a = 25$: $25 + (25)^{\frac{1}{2}} \overset{?}{=} 30$

$\qquad 25 + \sqrt{25} \overset{?}{=} 30$

$\qquad 25 + 5 \overset{?}{=} 30$

$\qquad 30 = 30 \checkmark$

$a = 36$: $36 + (36)^{\frac{1}{2}} \overset{?}{=} 30$

$\qquad 36 + \sqrt{36} \overset{?}{=} 30$

$\qquad 36 + 6 \overset{?}{=} 30$

$\qquad 42 \neq 30 \ \times$

$a = 36$ does not check; the solution set is $\{25\}$.

43. $\qquad \dfrac{1}{x^2} - \dfrac{5}{x} + 6 = 0$

$\left(\dfrac{1}{x}\right)^2 - 5\left(\dfrac{1}{x}\right) + 6 = 0$

Let $u = \dfrac{1}{x}$.

$u^2 - 5u + 6 = 0$

$(u - 2)(u - 3) = 0$

$u - 2 = 0 \quad$ or $\quad u - 3 = 0$

$\ \ u = 2 \quad$ or $\quad\ \ u = 3$

$\dfrac{1}{x} = 2 \quad$ or $\quad \dfrac{1}{x} = 3$

$\ \ x = \dfrac{1}{2} \quad$ or $\quad\ \ x = \dfrac{1}{3}$

Check:

$x = \dfrac{1}{2}$: $\dfrac{1}{\left(\frac{1}{2}\right)^2} - \dfrac{5}{\frac{1}{2}} + 6 \overset{?}{=} 0$

$\qquad \dfrac{1}{\frac{1}{4}} - \dfrac{5}{\frac{1}{2}} + 6 \overset{?}{=} 0$

$\qquad 4 - 10 + 6 \overset{?}{=} 0$

$\qquad 0 = 0 \checkmark$

$x = \dfrac{1}{3}$: $\dfrac{1}{\left(\frac{1}{3}\right)^2} - \dfrac{5}{\frac{1}{3}} + 6 \overset{?}{=} 0$

$\qquad \dfrac{1}{\frac{1}{9}} - \dfrac{5}{\frac{1}{3}} + 6 \overset{?}{=} 0$

$\qquad 9 - 15 + 6 \overset{?}{=} 0$

$\qquad 0 = 0 \checkmark$

Both check; the solution set is $\left\{\dfrac{1}{3}, \dfrac{1}{2}\right\}$.

45. $\qquad \left(\dfrac{1}{x+2}\right)^2 + \dfrac{4}{x+2} = 5$

$\left(\dfrac{1}{x+2}\right)^2 + 4\left(\dfrac{1}{x+2}\right) - 5 = 0$

Let $u = \dfrac{1}{x+2}$.

$u^2 + 4u - 5 = 0$

$(u - 1)(u + 5) = 0$

$u - 1 = 0 \quad$ or $\quad u + 5 = 0$

$\ \ u = 1 \quad$ or $\quad\ \ u = -5$

$\dfrac{1}{x+2} = 1 \quad$ or $\quad \dfrac{1}{x+2} = -5$

$x + 2 = 1 \quad$ or $\quad x + 2 = -\dfrac{1}{5}$

$\ x = -1 \quad$ or $\quad\ \ x = -\dfrac{11}{5}$

Check:

$x = -1$: $\left(\dfrac{1}{-1+2}\right)^2 + \dfrac{4}{-1+2} \overset{?}{=} 5$

$\qquad \left(\dfrac{1}{1}\right)^2 + \dfrac{4}{1} \overset{?}{=} 5$

$\qquad (1)^2 + 4 \overset{?}{=} 5$

$\qquad 1 + 4 \overset{?}{=} 5$

$\qquad 5 = 5 \checkmark$

$x = -\dfrac{11}{5}$: $\left(\dfrac{1}{-\frac{11}{5}+2}\right)^2 + \dfrac{4}{-\frac{11}{5}+2} \overset{?}{=} 5$

$\qquad \left(\dfrac{1}{-\frac{1}{5}}\right)^2 + \dfrac{4}{-\frac{1}{5}} \overset{?}{=} 5$

$\qquad (-5)^2 + 4(-5) \overset{?}{=} 5$

$\qquad 25 - 20 \overset{?}{=} 5$

$\qquad 5 = 5 \checkmark$

Both check; the solution set is $\left\{-\dfrac{11}{5}, -1\right\}$.

47. $\qquad p^6 - 28p^3 + 27 = 0$

$\left(p^3\right)^2 - 28\left(p^3\right) + 27 = 0$

Let $u = p^3$.

$$u^2 - 28u + 27 = 0$$
$$(u-1)(u-27) = 0$$
$$u-1 = 0 \quad \text{or} \quad u - 27 = 0$$
$$p^3 - 1 = 0 \quad \text{or} \quad p^3 - 27 = 0$$

First consider $p^3 - 1 = 0$. Note that the expression on the left is a difference of cubes.

$$p^3 - 1 = 0$$
$$(p-1)(p^2 + p + 1) = 0$$
$$p - 1 = 0 \quad \text{or} \quad p^2 + p + 1 = 0$$

$$p = 1 \quad \text{or} \quad p = \frac{-1 \pm \sqrt{1^2 - 4(1)(1)}}{2(1)}$$
$$= \frac{-1 \pm \sqrt{-3}}{2}$$
$$= \frac{-1 \pm \sqrt{3}\, i}{2}$$
$$= -\frac{1}{2} \pm \frac{\sqrt{3}}{2} i$$

Now consider $p^3 - 27 = 0$. Note that the expression on the left is a difference of cubes.

$$p^3 - 27 = 0$$
$$(p-3)(p^2 + 3p + 9) = 0$$
$$p - 3 = 0 \quad \text{or} \quad p^2 + 3p + 9 = 0$$

$$p = 3 \quad \text{or} \quad p = \frac{-3 \pm \sqrt{3^2 - 4(1)(9)}}{2(1)}$$
$$= \frac{-3 \pm \sqrt{-27}}{2}$$
$$= \frac{-3 \pm 3\sqrt{3}\, i}{2}$$
$$= -\frac{3}{2} \pm \frac{3\sqrt{3}}{2} i$$

All will check; the solution set is

$$\left\{ 1,\, 3,\, -\frac{1}{2} - \frac{\sqrt{3}}{2} i,\, -\frac{1}{2} + \frac{\sqrt{3}}{2} i,\, -\frac{3}{2} - \frac{3\sqrt{3}}{2} i, \right.$$
$$\left. -\frac{3}{2} + \frac{3\sqrt{3}}{2} i \right\}.$$

49.
$$8a^{-2} + 2a^{-1} = 1$$
$$8a^{-2} + 2a^{-1} - 1 = 0$$
$$8\left(a^{-1}\right)^2 + 2\left(a^{-1}\right) - 1 = 0$$

Let $u = a^{-1}$.
$$8u^2 + 2u - 1 = 0$$
$$(4u - 1)(2u + 1) = 0$$
$$4u - 1 = 0 \quad \text{or} \quad 2u + 1 = 0$$
$$4u = 1 \quad \text{or} \quad 2u = -1$$
$$u = \frac{1}{4} \quad \text{or} \quad u = -\frac{1}{2}$$
$$a^{-1} = \frac{1}{4} \quad \text{or} \quad a^{-1} = -\frac{1}{2}$$
$$\frac{1}{a} = \frac{1}{4} \quad \text{or} \quad \frac{1}{a} = -\frac{1}{2}$$
$$a = 4 \quad \text{or} \quad a = -2$$

Check:
$$a = 4: \quad 8(4)^{-2} + 2(4)^{-1} \overset{?}{=} 1$$
$$8\left(\frac{1}{4}\right)^2 + 2\left(\frac{1}{4}\right) \overset{?}{=} 1$$
$$8\left(\frac{1}{16}\right) + 2\left(\frac{1}{4}\right) \overset{?}{=} 1$$
$$\frac{1}{2} + \frac{1}{2} \overset{?}{=} 1$$
$$1 = 1 \checkmark$$

$$a = -2: \quad 8(-2)^{-2} + 2(-2)^{-1} \overset{?}{=} 1$$
$$8\left(-\frac{1}{2}\right)^2 + 2\left(-\frac{1}{2}\right) \overset{?}{=} 1$$
$$8\left(\frac{1}{4}\right) + 2\left(-\frac{1}{2}\right) \overset{?}{=} 1$$
$$2 - 1 \overset{?}{=} 1$$
$$1 = 1 \checkmark$$

Both check; the solution set is $\{-2,\, 4\}$.

51.
$$z^4 = 4z^2 + 32$$
$$z^4 - 4z^2 - 32 = 0$$
$$\left(z^2\right)^2 - 4\left(z^2\right) - 32 = 0$$

Let $u = z^2$.

$$u^2 - 4u - 32 = 0$$
$$(u-8)(u+4) = 0$$

$u - 8 = 0$	or	$u + 4 = 0$
$u = 8$	or	$u = -4$
$z^2 = 8$	or	$z^2 = -4$
$z = \pm\sqrt{8}$	or	$z = \pm\sqrt{-4}$
$z = \pm 2\sqrt{2}$	or	$z = \pm 2i$

Check:

$z = -2\sqrt{2}:\ \left(-2\sqrt{2}\right)^4 \overset{?}{=} 4\left(-2\sqrt{2}\right)^2 + 32$
$$64 \overset{?}{=} 4 \cdot 8 + 32$$
$$64 \overset{?}{=} 32 + 32$$
$$64 = 64 \ \checkmark$$

$z = 2\sqrt{2}:\ \left(2\sqrt{2}\right)^4 \overset{?}{=} 4\left(2\sqrt{2}\right)^2 + 32$
$$64 \overset{?}{=} 4 \cdot 8 + 32$$
$$64 \overset{?}{=} 32 + 32$$
$$64 = 64 \ \checkmark$$

$z = -2i:\ (-2i)^4 \overset{?}{=} 4(-2i)^2 + 32$
$$16i^4 \overset{?}{=} 4 \cdot 4i^2 + 32$$
$$16(1) \overset{?}{=} 4 \cdot (-4) + 32$$
$$16 \overset{?}{=} -16 + 32$$
$$16 = 16 \ \checkmark$$

$z = 2i:\ (2i)^4 \overset{?}{=} 4(2i)^2 + 32$
$$16i^4 \overset{?}{=} 4 \cdot 4i^2 + 32$$
$$16(1) \overset{?}{=} 4 \cdot (-4) + 32$$
$$16 \overset{?}{=} -16 + 32$$
$$16 = 16 \ \checkmark$$

All check; the solution set is
$$\left\{-2\sqrt{2},\ 2\sqrt{2}, -2i,\ 2i\right\}.$$

53. $\qquad x^{1/2} + x^{1/4} - 6 = 0$
$$\left(x^{1/4}\right)^2 + \left(x^{1/4}\right) - 6 = 0$$
Let $u = x^{1/4}$.

$$u^2 + u - 6 = 0$$
$$(u-2)(u+3) = 0$$

$u - 2 = 0$	or	$u + 3 = 0$
$u = 2$	or	$u = -3$
$x^{1/4} = 2$	or	$x^{1/4} = -3$
$\left(x^{1/4}\right)^4 = 2^4$	or	$\left(x^{1/4}\right)^4 = (-3)^4$
$x = 16$	or	$x = 81$

Check:

$x = 16:\ (16)^{1/2} + (16)^{1/4} - 6 \overset{?}{=} 0$
$$\sqrt{16} + \sqrt[4]{16} - 6 \overset{?}{=} 0$$
$$4 + 2 - 6 \overset{?}{=} 0$$
$$0 = 0 \ \checkmark$$

$x = 81:\ (81)^{1/2} + (81)^{1/4} - 6 \overset{?}{=} 0$
$$\sqrt{81} + \sqrt[4]{81} - 6 \overset{?}{=} 0$$
$$9 + 3 - 6 \overset{?}{=} 0$$
$$6 \ne 0 \ ✗$$

$x = 81$ does not check; the solution set is $\{16\}$.

55. $\qquad w^4 - 5w^2 - 36 = 0$
$$\left(w^2\right)^2 - 5\left(w^2\right) - 36 = 0$$

Let $u = w^2$.
$$u^2 - 5u - 36 = 0$$
$$(u-9)(u+4) = 0$$

$u - 9 = 0$	or	$u + 4 = 0$
$u = 9$	or	$u = -4$
$w^2 = 9$	or	$w^2 = -4$
$w = \pm\sqrt{9}$	or	$w = \pm\sqrt{-4}$
$w = \pm 3$	or	$w = \pm 2i$

Check:

$w = -3:\ (-3)^4 - 5(-3)^2 - 36 \overset{?}{=} 0$
$$81 - 5 \cdot 9 - 36 \overset{?}{=} 0$$
$$81 - 45 - 36 \overset{?}{=} 0$$
$$0 = 0 \ \checkmark$$

$w = 3:\ 3^4 - 5(3)^2 - 36 \overset{?}{=} 0$
$$81 - 5 \cdot 9 - 36 \overset{?}{=} 0$$
$$81 - 45 - 36 \overset{?}{=} 0$$
$$0 = 0 \ \checkmark$$

$$w = -2i : \quad (-2i)^4 - 5(-2i)^2 - 36 \overset{?}{=} 0$$
$$16i^4 - 5 \cdot 4i^2 - 36 \overset{?}{=} 0$$
$$16(1) - 5 \cdot 4(-1) - 36 \overset{?}{=} 0$$
$$16 + 20 - 36 \overset{?}{=} 0$$
$$0 = 0 \checkmark$$

$$w = 2i : \quad (2i)^4 - 5(2i)^2 - 36 \overset{?}{=} 0$$
$$16i^4 - 5 \cdot 4i^2 - 36 \overset{?}{=} 0$$
$$16(1) - 5 \cdot 4(-1) - 36 \overset{?}{=} 0$$
$$16 + 20 - 36 \overset{?}{=} 0$$
$$0 = 0 \checkmark$$

All check; the solution set is $\{-3, \ 3, -2i, \ 2i\}$.

57.
$$\left(\frac{1}{x+3}\right)^2 + \frac{2}{x+3} = 3$$
$$\left(\frac{1}{x+3}\right)^2 + 2\left(\frac{1}{x+3}\right) - 3 = 0$$

Let $u = \dfrac{1}{x+3}$.

$$u^2 + 2u - 3 = 0$$
$$(u-1)(u+3) = 0$$
$$u - 1 = 0 \quad \text{or} \quad u + 3 = 0$$
$$u = 1 \quad \text{or} \quad u = -3$$
$$\frac{1}{x+3} = 1 \quad \text{or} \quad \frac{1}{x+3} = -3$$
$$x + 3 = 1 \quad \text{or} \quad x + 3 = -\frac{1}{3}$$
$$x = -2 \text{ or} \quad\quad x = -\frac{10}{3}$$

Check:

$$x = -2 : \quad \left(\frac{1}{-2+3}\right)^2 + \frac{2}{-2+3} \overset{?}{=} 3$$
$$\left(\frac{1}{1}\right)^2 + \frac{2}{1} \overset{?}{=} 3$$
$$1 + 2 \overset{?}{=} 3$$
$$3 = 3 \checkmark$$

$$x = -\frac{10}{3} : \quad \left(\frac{1}{-\frac{10}{3}+3}\right)^2 + \frac{2}{-\frac{10}{3}+3} \overset{?}{=} 3$$
$$\left(\frac{1}{-\frac{1}{3}}\right)^2 + \frac{2}{-\frac{1}{3}} \overset{?}{=} 3$$
$$(-3)^2 + 2(-3) \overset{?}{=} 3$$
$$9 - 6 \overset{?}{=} 3$$
$$3 = 3 \checkmark$$

Both check; the solution set is $\left\{-\dfrac{10}{3}, -2\right\}$.

59.
$$x - 7\sqrt{x} + 12 = 0$$
$$\left(\sqrt{x}\right)^2 - 7\left(\sqrt{x}\right) + 12 = 0$$

Let $u = \sqrt{x}$.

$$u^2 - 7u + 12 = 0$$
$$(u-4)(u-3) = 0$$
$$u - 4 = 0 \quad \text{or} \quad u - 3 = 0$$
$$u = 4 \quad \text{or} \quad\quad u = 3$$

$$\sqrt{x} = 4 \quad \text{or} \quad \sqrt{x} = 3$$
$$x = 4^2 \quad \text{or} \quad\quad x = 3^2$$
$$x = 16 \quad \text{or} \quad\quad x = 9$$

Check:
$$x = 16 : \quad 16 - 7\sqrt{16} + 12 \overset{?}{=} 0$$
$$16 - 7 \cdot 4 + 12 \overset{?}{=} 0$$
$$16 - 28 + 12 \overset{?}{=} 0$$
$$0 = 0 \checkmark$$

$$x = 9 : \quad 9 - 7\sqrt{9} + 12 \overset{?}{=} 0$$
$$9 - 7 \cdot 3 + 12 \overset{?}{=} 0$$
$$9 - 21 + 12 \overset{?}{=} 0$$
$$0 = 0 \checkmark$$

Both check; the solution set is $\{9, \ 16\}$.

61.
$$2(x-1)^2 - 7(x-1) = 4$$
$$2(x-1)^2 - 7(x-1) - 4 = 0$$
Let $u = x - 1$.

$$2u^2 - 7u - 4 = 0$$
$$(2u+1)(u-4) = 0$$

$$2u+1 = 0 \quad \text{or} \quad u-4 = 0$$
$$u = -\frac{1}{2} \quad \text{or} \quad u = 4$$

$$x-1 = -\frac{1}{2} \quad \text{or} \quad x-1 = 4$$
$$x = \frac{1}{2} \quad \text{or} \quad x = 5$$

Check:

$$x = \frac{1}{2}: \quad 2\left(\frac{1}{2}-1\right)^2 - 7\left(\frac{1}{2}-1\right) \overset{?}{=} 4$$
$$2\left(-\frac{1}{2}\right)^2 - 7\left(-\frac{1}{2}\right) \overset{?}{=} 4$$
$$2\left(\frac{1}{4}\right) - 7\left(-\frac{1}{2}\right) \overset{?}{=} 4$$
$$\frac{1}{2} + \frac{7}{2} \overset{?}{=} 4$$
$$4 = 4 \ \checkmark$$

$$x = 5: \quad 2(5-1)^2 - 7(5-1) \overset{?}{=} 4$$
$$2(4)^2 - 7(4) \overset{?}{=} 4$$
$$2 \cdot 16 - 7 \cdot 4 \overset{?}{=} 4$$
$$32 - 28 \overset{?}{=} 4$$
$$4 = 4 \ \checkmark$$

Both check; the solution set is $\left\{\frac{1}{2}, 5\right\}$.

63. a.
$$f(x) = 12$$
$$x^4 + 7x^2 + 12 = 12$$
$$\left(x^2\right)^2 + 7\left(x^2\right) = 0$$

Let $u = x^2$.
$$u^2 + 7u = 0$$
$$u(u+7) = 0$$
$$u+7 = 0 \quad \text{or} \quad u = 0$$
$$u = -7$$
$$x^2 = -7 \quad \text{or} \quad x^2 = 0$$
$$x = \pm\sqrt{-7} \quad \text{or} \quad x = \pm\sqrt{0}$$
$$x = \pm\sqrt{7}\,i \quad \text{or} \quad x = 0$$

Check:

$$f\left(-\sqrt{7}\,i\right) = \left(-\sqrt{7}\,i\right)^4 + 7\left(-\sqrt{7}\,i\right)^2 + 12$$
$$= 49i^4 + 7 \cdot 7i^2 + 12$$
$$= 49(1) + 7 \cdot 7(-1) + 12$$
$$= 49 - 49 + 12$$
$$= 12 \ \checkmark$$

$$f\left(\sqrt{7}\,i\right) = \left(\sqrt{7}\,i\right)^4 + 7\left(\sqrt{7}\,i\right)^2 + 12$$
$$= 49i^4 + 7 \cdot 7i^2 + 12$$
$$= 49(1) + 7 \cdot 7(-1) + 12$$
$$= 49 - 49 + 12 = 12 \ \checkmark$$

$$f(0) = 0^4 + 7 \cdot 0^2 + 12$$
$$= 12 \ \checkmark$$

All check; the values that make $f(x) = 12$ are $\left\{-\sqrt{7}\,i,\ \sqrt{7}\,i,\ 0\right\}$.

b.
$$f(x) = 6$$
$$x^4 + 7x^2 + 12 = 6$$
$$\left(x^2\right)^2 + 7\left(x^2\right) + 6 = 0$$

Let $u = x^2$.
$$u^2 + 7u + 6 = 0$$
$$(u+6)(u+1) = 0$$
$$u+6 = 0 \quad \text{or} \quad u+1 = 0$$
$$u = -6 \quad \text{or} \quad u = -1$$
$$x^2 = -6 \quad \text{or} \quad x^2 = -1$$
$$x = \pm\sqrt{-6} \quad \text{or} \quad x = \pm\sqrt{-1}$$
$$x = \pm\sqrt{6}\,i \quad \text{or} \quad x = \pm i$$

Check:

$$f\left(-\sqrt{6}\,i\right) = \left(-\sqrt{6}\,i\right)^4 + 7\left(-\sqrt{6}\,i\right)^2 + 12$$
$$= 36i^4 + 7 \cdot 6i^2 + 12$$
$$= 36(1) + 7 \cdot 6(-1) + 12$$
$$= 36 - 42 + 12$$
$$= 6 \ \checkmark$$

$$f\left(\sqrt{6}\,i\right) = \left(\sqrt{6}\,i\right)^4 + 7\left(\sqrt{6}\,i\right)^2 + 12$$
$$= 36i^4 + 7 \cdot 6i^2 + 12$$
$$= 36(1) + 7 \cdot 6(-1) + 12$$
$$= 36 - 42 + 12$$
$$= 6 \ \checkmark$$

$$f(-i) = (-i)^4 + 7(-i)^2 + 12$$
$$= i^4 + 7i^2 + 12$$
$$= 1 + 7(-1) + 12$$
$$= 1 - 7 + 12$$
$$= 6 \checkmark$$

$$f(i) = i^4 + 7i^2 + 12$$
$$= 1 + 7(-1) + 12$$
$$= 1 - 7 + 12$$
$$= 6 \checkmark$$

All check; the values that make $f(x) = 6$ are $\left\{-\sqrt{6}\,i,\ \sqrt{6}\,i,\ -i,\ i\right\}$.

65. a.
$$g(x) = -5$$
$$2x^4 - 6x^2 - 5 = -5$$
$$2(x^2)^2 - 6(x^2) = 0$$

Let $u = x^2$.
$$2u^2 - 6u = 0$$
$$2u(u-3) = 0$$
$$2u = 0 \quad \text{or} \quad u - 3 = 0$$
$$u = 0 \quad \text{or} \quad u = 3$$
$$x^2 = 0 \quad \text{or} \quad x^2 = 3$$
$$x = \pm\sqrt{0} \quad \text{or} \quad x = \pm\sqrt{3}$$
$$x = 0$$

Check:
$$g(0) = 2 \cdot 0^4 - 6 \cdot 0^2 - 5$$
$$= -5 \checkmark$$

$$g(-\sqrt{3}) = 2(-\sqrt{3})^4 - 6(-\sqrt{3})^2 - 5$$
$$= 2 \cdot 9 - 6 \cdot 3 - 5$$
$$= 18 - 18 - 5$$
$$= -5 \checkmark$$

$$g(\sqrt{3}) = 2(\sqrt{3})^4 - 6(\sqrt{3})^2 - 5$$
$$= 2 \cdot 9 - 6 \cdot 3 - 5$$
$$= 18 - 18 - 5$$
$$= -5 \checkmark$$

All check; the values that make $g(x) = -5$ are $\left\{-\sqrt{3},\ 0,\ \sqrt{3}\right\}$.

b.
$$g(x) = 15$$
$$2x^4 - 6x^2 - 5 = 15$$
$$2x^4 - 6x^2 - 20 = 0$$
$$\frac{2x^4 - 6x^2 - 20}{2} = \frac{0}{2}$$
$$x^4 - 3x^2 - 10 = 0$$
$$(x^2)^2 - 3(x^2) - 10 = 0$$

Let $u = x^2$.
$$u^2 - 3u - 10 = 0$$
$$(u+2)(u-5) = 0$$

$$u + 2 = 0 \quad \text{or} \quad u - 5 = 0$$
$$u = -2 \quad \text{or} \quad u = 5$$
$$x^2 = -2 \quad \text{or} \quad x^2 = 5$$
$$x = \pm\sqrt{-2} \quad \text{or} \quad x = \pm\sqrt{5}$$
$$x = \pm\sqrt{2}\,i$$

Check:
$$g(-\sqrt{2}\,i) = 2(-\sqrt{2}\,i)^4 - 6(-\sqrt{2}\,i)^2 - 5$$
$$= 2 \cdot 4i^4 - 6 \cdot 2i^2 - 5$$
$$= 2 \cdot 4(1) - 6 \cdot 2(-1) - 5$$
$$= 8 + 12 - 5$$
$$= 15 \checkmark$$

$$g(\sqrt{2}\,i) = 2(\sqrt{2}\,i)^4 - 6(\sqrt{2}\,i)^2 - 5$$
$$= 2 \cdot 4i^4 - 6 \cdot 2i^2 - 5$$
$$= 2 \cdot 4(1) - 6 \cdot 2(-1) - 5$$
$$= 8 + 12 - 5$$
$$= 15 \checkmark$$

$$g(-\sqrt{5}) = 2(-\sqrt{5})^4 - 6(-\sqrt{5})^2 - 5$$
$$= 2 \cdot 25 - 6 \cdot 5 - 5$$
$$= 50 - 30 - 5$$
$$= 15 \checkmark$$

$$g(\sqrt{5}) = 2(\sqrt{5})^4 - 6(\sqrt{5})^2 - 5$$
$$= 2 \cdot 25 - 6 \cdot 5 - 5$$
$$= 50 - 30 - 5$$
$$= 15 \checkmark$$

All check; the values that make $g(x) = 15$ are $\left\{-\sqrt{2}\,i,\ \sqrt{2}\,i,\ -\sqrt{5},\ \sqrt{5}\right\}$.

67. a.
$$F(x) = 6$$
$$x^{-2} - 5x^{-1} = 6$$
$$\left(x^{-1}\right)^2 - 5\left(x^{-1}\right) - 6 = 0$$

Let $u = x^{-1}$.
$$u^2 - 5u - 6 = 0$$
$$(u - 6)(u + 1) = 0$$

$$u - 6 = 0 \quad \text{or} \quad u + 1 = 0$$
$$u = 6 \quad \text{or} \quad u = -1$$
$$x^{-1} = 6 \quad \text{or} \quad x^{-1} = -1$$
$$\frac{1}{x} = 6 \quad \text{or} \quad \frac{1}{x} = -1$$
$$x = \frac{1}{6} \quad \text{or} \quad x = -1$$

Check:
$$F\left(\frac{1}{6}\right) = \left(\frac{1}{6}\right)^{-2} - 5\left(\frac{1}{6}\right)^{-1}$$
$$= 6^2 - 5 \cdot 6^1$$
$$= 36 - 30$$
$$= 6 \checkmark$$

$$F(-1) = (-1)^{-2} - 5(-1)^{-1}$$
$$= (-1)^2 - 5(-1)^1$$
$$= 1 + 5$$
$$= 6 \checkmark$$

Both check; the values that make $F(x) = 6$ are $\left\{-1, \ \frac{1}{6}\right\}$.

b.
$$F(x) = 14$$
$$x^{-2} - 5x^{-1} = 14$$
$$\left(x^{-1}\right)^2 - 5\left(x^{-1}\right) - 14 = 0$$

Let $u = x^{-1}$.
$$u^2 - 5u - 14 = 0$$
$$(u - 7)(u + 2) = 0$$
$$u - 7 = 0 \quad \text{or} \quad u + 2 = 0$$
$$u = 7 \quad \text{or} \quad u = -2$$
$$x^{-1} = 7 \quad \text{or} \quad x^{-1} = -2$$
$$\frac{1}{x} = 7 \quad \text{or} \quad \frac{1}{x} = -2$$
$$x = \frac{1}{7} \quad \text{or} \quad x = -\frac{1}{2}$$

Check:
$$F\left(\frac{1}{7}\right) = \left(\frac{1}{7}\right)^{-2} - 5\left(\frac{1}{7}\right)^{-1}$$
$$= 7^2 - 5 \cdot 7^1$$
$$= 49 - 35$$
$$= 14 \checkmark$$

$$F\left(-\frac{1}{2}\right) = \left(-\frac{1}{2}\right)^{-2} - 5\left(-\frac{1}{2}\right)^{-1}$$
$$= (-2)^2 - 5 \cdot (-2)^1$$
$$= 4 + 10$$
$$= 14 \checkmark$$

Both check; the values that make $F(x) = 14$ are $\left\{-\frac{1}{2}, \ \frac{1}{7}\right\}$.

69.
$$x^4 + 9x^2 + 14 = 0$$
$$\left(x^2\right)^2 + 9\left(x^2\right) + 14 = 0$$

Let $u = x^2$.
$$u^2 + 9u + 14 = 0$$
$$(u + 7)(u + 2) = 0$$
$$u + 7 = 0 \quad \text{or} \quad u + 2 = 0$$
$$u = -7 \quad \text{or} \quad u = -2$$
$$x^2 = -7 \quad \text{or} \quad x^2 = -2$$
$$x = \pm\sqrt{-7} \quad \text{or} \quad x = \pm\sqrt{-2}$$
$$x = \pm\sqrt{7}\,i \quad \text{or} \quad x = \pm\sqrt{2}\,i$$

Check:
$$f\left(-\sqrt{7}\,i\right) = \left(-\sqrt{7}\,i\right)^4 + 9\left(-\sqrt{7}\,i\right)^2 + 14$$
$$= 49i^4 + 9 \cdot 7i^2 + 14$$
$$= 49(1) + 9 \cdot 7(-1) + 14$$
$$= 49 - 63 + 14$$
$$= 0 \checkmark$$

$$f\left(\sqrt{7}\,i\right) = \left(\sqrt{7}\,i\right)^4 + 9\left(\sqrt{7}\,i\right)^2 + 14$$
$$= 49i^4 + 9 \cdot 7i^2 + 14$$
$$= 49(1) + 9 \cdot 7(-1) + 14$$
$$= 49 - 63 + 14$$
$$= 0 \checkmark$$

$$f\left(-\sqrt{2}\,i\right) = \left(-\sqrt{2}\,i\right)^4 + 9\left(-\sqrt{2}\,i\right)^2 + 14$$
$$= 4i^4 + 9 \cdot 2i^2 + 14$$
$$= 4(1) + 9 \cdot 2(-1) + 14$$
$$= 4 - 18 + 14$$
$$= 0 \ \checkmark$$

$$f\left(\sqrt{2}\,i\right) = \left(\sqrt{2}\,i\right)^4 + 9\left(\sqrt{2}\,i\right)^2 + 14$$
$$= 4i^4 + 9 \cdot 2i^2 + 14$$
$$= 4(1) + 9 \cdot 2(-1) + 14$$
$$= 4 - 18 + 14$$
$$= 0 \ \checkmark$$

All check; the zeros of *f* are
$$\left\{-\sqrt{7}\,i,\ \sqrt{7}\,i, -\sqrt{2}\,i,\ \sqrt{2}\,i\right\}.$$

71.
$$6t - 25\sqrt{t} - 9 = 0$$
$$6\left(\sqrt{t}\right)^2 - 25\left(\sqrt{t}\right) - 9 = 0$$

Let $u = \sqrt{t}$.
$$6u^2 - 25u - 9 = 0$$
$$(2u - 9)(3u + 1) = 0$$

$$2u - 9 = 0 \quad \text{or} \quad 3u + 1 = 0$$
$$2u = 9 \quad \text{or} \quad 3u = -1$$
$$u = \frac{9}{2} \quad \text{or} \quad u = -\frac{1}{3}$$

$$\sqrt{t} = \frac{9}{2} \quad \text{or} \quad \sqrt{t} = -\frac{1}{3}$$
$$t = \left(\frac{9}{2}\right)^2 \quad \text{or} \quad t = \left(-\frac{1}{3}\right)^2$$
$$t = \frac{81}{4} \quad \text{or} \quad t = \frac{1}{9}$$

Check:
$$g\left(\frac{81}{4}\right) = 6 \cdot \frac{81}{4} - 25\sqrt{\frac{81}{4}} - 9$$
$$= 6 \cdot \frac{81}{4} - 25 \cdot \frac{9}{2} - 9$$
$$= \frac{243}{2} - \frac{225}{2} - 9$$
$$= 0 \ \checkmark$$

$$g\left(\frac{1}{9}\right) = 6 \cdot \frac{1}{9} - 25\sqrt{\frac{1}{9}} - 9$$
$$= 6 \cdot \frac{1}{9} - 25 \cdot \frac{1}{3} - 9$$
$$= \frac{2}{3} - \frac{25}{3} - 9$$
$$= -\frac{50}{3}$$
$$\neq 0 \ \textbf{✗}$$

$t = \dfrac{1}{9}$ does not check; the zero of *g* is $\left\{\dfrac{81}{4}\right\}$.

73.
$$\frac{1}{(d+3)^2} - \frac{4}{d+3} + 3 = 0$$
$$\left(\frac{1}{d+3}\right)^2 - 4\left(\frac{1}{d+3}\right) + 3 = 0$$

Let $u = \dfrac{1}{d+3}$.
$$u^2 - 4u + 3 = 0$$
$$(u-1)(u-3) = 0$$
$$u - 1 = 0 \quad \text{or} \quad u - 3 = 0$$
$$u = 1 \quad \text{or} \quad u = 3$$
$$\frac{1}{d+3} = 1 \quad \text{or} \quad \frac{1}{d+3} = 3$$
$$d + 3 = 1 \quad \text{or} \quad d + 3 = \frac{1}{3}$$
$$d = -2 \quad \text{or} \quad d = -\frac{8}{3}$$

Check:
$$s(-2) = \frac{1}{(-2+3)^2} - \frac{4}{-2+3} + 3$$
$$= \frac{1}{1^2} - \frac{4}{1} + 3$$
$$= 1 - 4 + 3$$
$$= 0 \ \checkmark$$

$$s\left(-\frac{8}{3}\right) = \frac{1}{\left(-\frac{8}{3}+3\right)^2} - \frac{4}{-\frac{8}{3}+3} + 3$$
$$= \frac{1}{\left(\frac{1}{3}\right)^2} - \frac{4}{\frac{1}{3}} + 3$$
$$= \frac{1}{\frac{1}{9}} - \frac{4}{\frac{1}{3}} + 3$$
$$= 9 - 12 + 3$$
$$= 0 \ \checkmark$$

Both check; the zeros of *s* are $\left\{-\dfrac{8}{3}, -2\right\}$.

75. a. $x^2 - 5x + 6 = 0$

$(x-2)(x-3) = 0$

$x - 2 = 0$ or $x - 3 = 0$

$\quad x = 2$ or $\quad\quad x = 3$

Both check; the solution set is $\{2,\ 3\}$.

b. $(x-3)^2 - 5(x-3) + 6 = 0$

Let $u = x - 3$.

$u^2 - 5u + 6 = 0$

$(u-2)(u-3) = 0$

$u - 2 = 0$ or $u - 3 = 0$

$\quad u = 2$ or $\quad\quad u = 3$

$x - 3 = 2$ or $x - 3 = 3$

$\quad x = 5$ or $\quad\quad x = 6$

Both check; the solution set is $\{5,\ 6\}$.

Comparing these solutions to those in part (a), we note that $5 = 2 + 3$ and $6 = 3 + 3$.

c. $(x+2)^2 - 5(x+2) + 6 = 0$

Let $u = x + 2$.

$u^2 - 5u + 6 = 0$

$(u-2)(u-3) = 0$

$u - 2 = 0$ or $u - 3 = 0$

$\quad u = 2$ or $\quad\quad u = 3$

$x + 2 = 2$ or $x + 2 = 3$

$\quad x = 0$ or $\quad\quad x = 1$

Both check; the solution set is $\{0,\ 1\}$.

Comparing these solutions to those in part (a), we note that $0 = 2 - 2$ and $1 = 3 - 2$.

d. $(x-5)^2 - 5(x-5) + 6 = 0$

Let $u = x - 5$.

$u^2 - 5u + 6 = 0$

$(u-2)(u-3) = 0$

$u - 2 = 0$ or $u - 3 = 0$

$\quad u = 2$ or $\quad\quad u = 3$

$x - 5 = 2$ or $x - 5 = 3$

$\quad x = 7$ or $\quad\quad x = 8$

Both check; the solution set is $\{7,\ 8\}$.

Comparing these solutions to those in part (a), we note that $7 = 2 + 5$ and $8 = 3 + 5$.

e. Conjecture: The solution set of the equation $(x-a)^2 - 5(x-a) + 6 = 0$ is $\{2+a,\ 3+a\}$.

NOTE: This conjecture can be shown to be true by using the techniques from parts (b) through (d).

77. a. $f(x) = 2x^2 - 3x + 1$

$2x^2 - 3x + 1 = 0$

$(2x-1)(x-1) = 0$

$2x - 1 = 0$ or $x - 1 = 0$

$\quad 2x = 1$ or $\quad\quad x = 1$

$\quad x = \dfrac{1}{2}$

Both check; the zeros of $f(x)$ are $\left\{\dfrac{1}{2},\ 1\right\}$.

b. $f(x-2) = 2(x-2)^2 - 3(x-2) + 1$

Let $u = x - 2$.

$2u^2 - 3u + 1 = 0$

$(2u-1)(u-1) = 0$

$2u - 1 = 0$ or $u - 1 = 0$

$\quad 2u = 1$ or $\quad\quad u = 1$

$\quad u = \dfrac{1}{2}$

$x - 2 = \dfrac{1}{2}$ or $x - 2 = 1$

$\quad x = \dfrac{5}{2}$ or $\quad\quad x = 3$

Both check; the zeros of $f(x-2)$ are $\left\{\dfrac{5}{2}, 3\right\}$. Comparing these solutions to those in part (a), we note that $\dfrac{5}{2} = \dfrac{1}{2} + 2$ and $3 = 1 + 2$.

c. $f(x-5) = 2(x-5)^2 - 3(x-5) + 1$

Let $u = x - 5$.

$2u^2 - 3u + 1 = 0$

$(2u-1)(u-1) = 0$

$2u - 1 = 0$ or $u - 1 = 0$

$\quad 2u = 1$ or $\quad\quad u = 1$

$\quad u = \dfrac{1}{2}$

$x - 5 = \dfrac{1}{2}$ or $x - 5 = 1$

$\quad x = \dfrac{11}{2}$ or $\quad\quad x = 6$

Both check; the zeros of $f(x-5)$ are $\left\{\dfrac{11}{2}, 6\right\}$. Comparing these solutions to those in part (a), we note that $\frac{11}{2} = \frac{1}{2} + 5$ and $6 = 1 + 5$.

d. Conjecture: For $f(x) = 2x^2 - 3x + 1$, the zeros of the $f(x-a)$ are $\left\{ \frac{1}{2} + a, \ 1 + a \right\}$.

NOTE: This conjecture can be shown to be true by using the techniques from parts (b) and (c).

79. a. $R(1990)$

$= \frac{(1990-1990)^2}{2} + \frac{3(1990-1990)}{2} + 3000$

$= \frac{0^2}{2} + \frac{3(0)}{2} + 3000$

$= 3000$

Interpretation: The revenue in 1990 was $3000 thousand (or $3,000,000).

b. $\frac{(x-1990)^2}{2} + \frac{3(x-1990)}{2} + 3000 = 3065$

$\frac{1}{2}(x-1990)^2 + \frac{3}{2}(x-1990) - 65 = 0$

Let $u = x - 1990$.

$\frac{1}{2}u^2 + \frac{3}{2}u - 65 = 0$

$2\left(\frac{1}{2}u^2 + \frac{3}{2}u - 65 \right) = 2(0)$

$u^2 + 3u - 130 = 0$

$(u-10)(u+13) = 0$

$u - 10 = 0$ or $u + 13 = 0$
$u = 10$ or $u = -13$
$x - 1990 = 10$ or $x - 1990 = -13$
$x = 2000$ or $x = 1977$

Since 1977 is before 1990, disregard it. The solution set is $\{2000\}$.

Interpretation: In the year 2000, revenue was $3065 thousand (or $3,065,000).

c. $\frac{(x-1990)^2}{2} + \frac{3(x-1990)}{2} + 3000 = 3350$

$\frac{1}{2}(x-1990)^2 + \frac{3}{2}(x-1990) - 350 = 0$

Let $u = x - 1990$.

$\frac{1}{2}u^2 + \frac{3}{2}u - 350 = 0$

$2\left(\frac{1}{2}u^2 + \frac{3}{2}u - 350 \right) = 2(0)$

$u^2 + 3u - 700 = 0$

$(u-25)(u+28) = 0$

$u - 25 = 0$ or $u + 28 = 0$
$u = 25$ or $u = -28$
$x - 1990 = 25$ or $x - 1990 = -28$
$x = 2015$ or $x = 1962$

Since 1962 is before 1990, disregard it. The solution set is $\{2015\}$. The model predicts that revenue will be $3350 thousand (or $3,350,000) in the year 2015.

81. $x^4 + 5x^2 + 2 = 0$

$\left(x^2 \right)^2 + 5\left(x^2 \right) + 2 = 0$

Let $u = x^2$.

$u^2 + 5u + 2 = 0$

For this equation, $a = 1$, $b = 5$, and $c = 2$.

$u = \frac{-5 \pm \sqrt{5^2 - 4(1)(2)}}{2(1)}$

$= \frac{-5 \pm \sqrt{25 - 8}}{2}$

$= \frac{-5 \pm \sqrt{17}}{2}$

$u = \frac{-5 - \sqrt{17}}{2}$ or $u = \frac{-5 + \sqrt{17}}{2}$

$x^2 = \frac{-5 - \sqrt{17}}{2}$ or $x^2 = \frac{-5 + \sqrt{17}}{2}$

$x = \pm\sqrt{\frac{-5 - \sqrt{17}}{2}}$ or $x = \pm\sqrt{\frac{-5 + \sqrt{17}}{2}}$

$x = \pm\sqrt{\frac{-1(5 + \sqrt{17})}{2}}$ $x = \pm\sqrt{\frac{-1(5 - \sqrt{17})}{2}}$

$x = \pm i\sqrt{\frac{5 + \sqrt{17}}{2}}$ $x = \pm i\sqrt{\frac{5 - \sqrt{17}}{2}}$

$x = \pm i\sqrt{\frac{5 + \sqrt{17}}{2} \cdot \frac{2}{2}}$ $x = \pm i\sqrt{\frac{5 - \sqrt{17}}{2} \cdot \frac{2}{2}}$

$x = \pm\frac{\sqrt{10 + 2\sqrt{17}}}{2}i$ $x = \pm\frac{\sqrt{10 - 2\sqrt{17}}}{2}i$

All check, the solution set is $\left\{ -\dfrac{\sqrt{10+2\sqrt{17}}}{2}\, i \,, \right.$

$\left. \dfrac{\sqrt{10+2\sqrt{17}}}{2}\, i,\ -\dfrac{\sqrt{10-2\sqrt{17}}}{2}\, i \,,\ \dfrac{\sqrt{10-2\sqrt{17}}}{2}\, i \right\}.$

83. $2(x-2)^2 + 8(x-2) - 1 = 0$

Let $u = x - 2$.

$2u^2 + 8u - 1 = 0$

For this equation, $a = 2$, $b = 8$, and $c = -1$.

$$u = \frac{-8 \pm \sqrt{8^2 - 4(2)(-1)}}{2(2)}$$

$$= \frac{-8 \pm \sqrt{64 + 8}}{4}$$

$$= \frac{-8 \pm \sqrt{72}}{4}$$

$$= \frac{-8 \pm 6\sqrt{2}}{4} = -2 \pm \frac{3\sqrt{2}}{2}$$

$u = -2 - \dfrac{3\sqrt{2}}{2}$ or $u = -2 + \dfrac{3\sqrt{2}}{2}$

$x - 2 = -2 - \dfrac{3\sqrt{2}}{2}$ or $x - 2 = -2 + \dfrac{3\sqrt{2}}{2}$

$x = -\dfrac{3\sqrt{2}}{2}$ or $x = \dfrac{3\sqrt{2}}{2}$

Both check, the solution set is $\left\{ -\dfrac{3\sqrt{2}}{2},\ \dfrac{3\sqrt{2}}{2} \right\}$.

85. Method 1:

$$x - 5\sqrt{x} - 6 = 0$$

$$\left(\sqrt{x}\right)^2 - 5\sqrt{x} - 6 = 0$$

Let $u = \sqrt{x}$

$$u^2 - 5u - 6 = 0$$

$$(u - 6)(u + 1) = 0$$

$u - 6 = 0$ or $u + 1 = 0$

$u = 6$ or $u = -1$

$\sqrt{x} = 6$ or $\sqrt{x} = -1$

$x = 36$ or $x = 1$

Check:

$x = 36:\ 36 - 5\sqrt{36} - 6 \overset{?}{=} 0$

$36 - 5\cdot 6 - 6 \overset{?}{=} 0$

$36 - 30 - 6 \overset{?}{=} 0$

$0 = 0\ \checkmark$

$x = 1:\ 1 - 5\sqrt{1} - 6 \overset{?}{=} 0$

$1 - 5\cdot 1 - 6 \overset{?}{=} 0$

$1 - 5 - 6 \overset{?}{=} 0$

$-10 \neq 0\ ✗$

The solution set is $\{36\}$.

Method 2:

$$x - 5\sqrt{x} - 6 = 0$$

$$x - 6 = 5\sqrt{x}$$

$$(x - 6)^2 = \left(5\sqrt{x}\right)^2$$

$$x^2 - 12x + 36 = 25x$$

$$x^2 - 37x + 36 = 0$$

$$(x - 36)(x - 1) = 0$$

$x - 36 = 0$ or $x - 1 = 0$

$x = 36$ or $x = 1$

Check:

$x = 36:\ 36 - 5\sqrt{36} - 6 \overset{?}{=} 0$

$36 - 5\cdot 6 - 6 \overset{?}{=} 0$

$36 - 30 - 6 \overset{?}{=} 0$

$0 = 0\ \checkmark$

$x = 1:\ 1 - 5\sqrt{1} - 6 \overset{?}{=} 0$

$1 - 5\cdot 1 - 6 \overset{?}{=} 0$

$1 - 5 - 6 \overset{?}{=} 0$

$-10 \neq 0\ ✗$

The solution set is $\{36\}$.

Preferences will vary.

87. Answers will vary. One possibility follows: Extraneous solutions may be introduced when we raise both sides of the equation to an even power. They may also occur when the equation involves rational expressions.

89. $\left(3p^{-2} - 4p^{-1} + 8\right) - \left(2p^{-2} - 8p^{-1} - 1\right)$

$= 3p^{-2} - 4p^{-1} + 8 - 2p^{-2} + 8p^{-1} + 1$

$= p^{-2} + 4p^{-1} + 9$

91. $\sqrt[3]{16a} + \sqrt[3]{54a} - \sqrt[3]{128a^4}$

$= \sqrt[3]{8\cdot 2a} + \sqrt[3]{27\cdot 2a} - \sqrt[3]{64a^3\cdot 2a}$

$= 2\sqrt[3]{2a} + 3\sqrt[3]{2a} - 4a\sqrt[3]{2a}$

$= 5\sqrt[3]{2a} - 4a\sqrt[3]{2a}$

93. Let $Y_1 = x^4 + 5x^2 - 14$.

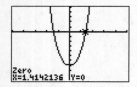

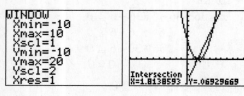

The solution set is approximately $\{-1.41,\ 1.41\}$.

95. Let $Y_1 = 2(x-2)^2$ and $Y_2 = 5(x-2)+1$.

The solution set is approximately $\{1.81,\ 4.69\}$.

97. Let $Y_1 = x - 5\sqrt{x}$ and $Y_2 = -3$.

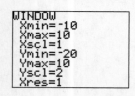

The solution set is approximately $\{0.49\}$.

99. For the graphs shown in parts (a) through (c) the WINDOW setting is the following:

a. $Y_1 = x^2 - 5x - 6$

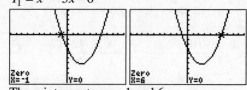

The *x*-intercepts are -1 and 6.

b. $Y_1 = (x+2)^2 - 5(x+2) - 6$

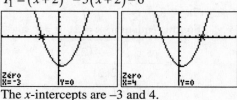

The *x*-intercepts are -3 and 4.

c. $Y_1 = (x+5)^2 - 5(x+5) - 6$

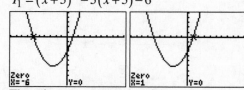

The *x*-intercepts are -6 and 1.

d. The *x*-intercepts of the graph of
$y = f(x) = x^2 - 5x - 6$ are -1 and 6.
The *x*-intercepts of the graph of
$y = f(x+a) = (x+a)^2 - 5(x+a) - 6$ are
$-1-a$ and $6-a$.

Putting the Concepts Together (Sections 7.1–7.3)

1. Start: $z^2 + 10z$

Add: $\left[\dfrac{1}{2} \cdot 10\right]^2 = 25$

Result: $z^2 + 10z + 25$

Factored Form: $(z+5)^2$

2. Start: $x^2 + 7x$

Add: $\left[\dfrac{1}{2} \cdot 7\right]^2 = \dfrac{49}{4}$

Result: $x^2 + 7x + \dfrac{49}{4}$

Factored Form: $\left(x + \dfrac{7}{2}\right)^2$

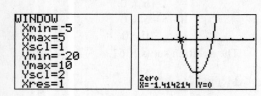

592　　　Copyright © 2014 Pearson Education, Inc.

3. Start: $n^2 - \dfrac{1}{4}n$

Add: $\left[\dfrac{1}{2} \cdot \left(-\dfrac{1}{4}\right)\right]^2 = \dfrac{1}{64}$

Result: $n^2 - \dfrac{1}{4}n + \dfrac{1}{64}$

Factored Form: $\left(n - \dfrac{1}{8}\right)^2$

4. $(2x-3)^2 - 5 = -1$

$\quad\quad (2x-3)^2 = 4$

$\quad\quad\quad 2x - 3 = \pm\sqrt{4}$

$\quad\quad\quad 2x - 3 = \pm 2$

$\quad\quad\quad\quad 2x = 3 \pm 2$

$\quad 2x = 1$ or $\quad 2x = 5$

$\quad x = \dfrac{1}{2}$ or $\quad x = \dfrac{5}{2}$

The solution set is $\left\{\dfrac{1}{2},\ \dfrac{5}{2}\right\}$.

5. $\quad\quad x^2 + 8x + 4 = 0$

$\quad\quad\quad x^2 + 8x \quad = -4$

$x^2 + 8x + \left(\dfrac{1}{2} \cdot 8\right)^2 = -4 + \left(\dfrac{1}{2} \cdot 8\right)^2$

$\quad\quad x^2 + 8x + 16 = -4 + 16$

$\quad\quad\quad (x+4)^2 = 12$

$\quad\quad\quad\quad x + 4 = \pm\sqrt{12}$

$\quad\quad\quad\quad x + 4 = \pm 2\sqrt{3}$

$\quad\quad\quad\quad\quad x = -4 \pm 2\sqrt{3}$

The solution set is $\left\{-4 - 2\sqrt{3},\ -4 + 2\sqrt{3}\right\}$.

6. $\quad x(x-6) = -7$

$x^2 - 6x + 7 = 0$

For this equation, $a = 1$, $b = -6$, and $c = 7$.

$x = \dfrac{-(-6) \pm \sqrt{(-6)^2 - 4(1)(7)}}{2(1)}$

$\quad = \dfrac{6 \pm \sqrt{36 - 28}}{2}$

$\quad = \dfrac{6 \pm \sqrt{8}}{2}$

$\quad = \dfrac{6 \pm 2\sqrt{2}}{2} = \dfrac{6}{2} \pm \dfrac{2\sqrt{2}}{2} = 3 \pm \sqrt{2}$

The solution set is $\left\{3 - \sqrt{2},\ 3 + \sqrt{2}\right\}$.

7. $49x^2 - 80 = 0$

$\quad 49x^2 = 80$

$\quad\quad x^2 = \dfrac{80}{49}$

$\quad\quad x = \pm\sqrt{\dfrac{80}{49}} = \pm\dfrac{\sqrt{80}}{\sqrt{49}} = \pm\dfrac{4\sqrt{5}}{7}$

The solution set is $\left\{-\dfrac{4\sqrt{5}}{7},\ \dfrac{4\sqrt{5}}{7}\right\}$.

8. $p^2 - 8p + 6 = 0$

Because this equation does not easily factor, solve by using the quadratic formula. For this equation, $a = 1$, $b = -8$, and $c = 6$.

$p = \dfrac{-(-8) \pm \sqrt{(-8)^2 - 4(1)(6)}}{2(1)}$

$\quad = \dfrac{8 \pm \sqrt{64 - 24}}{2}$

$\quad = \dfrac{8 \pm \sqrt{40}}{2}$

$\quad = \dfrac{8 \pm 2\sqrt{10}}{2}$

$\quad = \dfrac{8}{2} \pm \dfrac{2\sqrt{10}}{2} = 4 \pm \sqrt{10}$

The solution set is $\left\{4 - \sqrt{10},\ 4 + \sqrt{10}\right\}$.

9. $3y^2 + 6y + 4 = 0$

Because this equation does not easily factor, solve by using the quadratic formula. For this equation, $a = 3$, $b = 6$, and $c = 4$.

$$y = \frac{-6 \pm \sqrt{6^2 - 4(3)(4)}}{2(3)}$$

$$= \frac{-6 \pm \sqrt{36 - 48}}{6}$$

$$= \frac{-6 \pm \sqrt{-12}}{6}$$

$$= \frac{-6 \pm 2\sqrt{3}\,i}{6} = \frac{-6}{6} \pm \frac{2\sqrt{3}}{6}i = -1 \pm \frac{\sqrt{3}}{3}i$$

The solution set is $\left\{ -1 - \dfrac{\sqrt{3}}{3}i,\ -1 + \dfrac{\sqrt{3}}{3}i \right\}$.

10. $\dfrac{1}{4}n^2 + n = \dfrac{1}{6}$

$$12\left(\frac{1}{4}n^2 + n \right) = 12\left(\frac{1}{6} \right)$$

$$3n^2 + 12n = 2$$

$$3n^2 + 12n - 2 = 0$$

Because this equation does not easily factor, solve by using the quadratic formula. For this equation, $a = 3$, $b = 12$, and $c = -2$.

$$n = \frac{-12 \pm \sqrt{12^2 - 4(3)(-2)}}{2(3)}$$

$$= \frac{-12 \pm \sqrt{144 + 24}}{6}$$

$$= \frac{-12 \pm \sqrt{168}}{6}$$

$$= \frac{-12 \pm 2\sqrt{42}}{6}$$

$$= -\frac{12}{6} \pm \frac{2\sqrt{42}}{6} = -2 \pm \frac{\sqrt{42}}{3}$$

The solution set is $\left\{ -2 - \dfrac{\sqrt{42}}{3},\ -2 + \dfrac{\sqrt{42}}{3} \right\}$.

11. $9x^2 + 12x + 4 = 0$

For this equation, $a = 9$, $b = 12$, and $c = 4$.

$$b^2 - 4ac = 12^2 - 4(9)(4) = 144 - 144 = 0$$

Because $b^2 - 4ac = 0$, the quadratic equation will have one repeated real solution.

12. $3x^2 + 6x - 2 = 0$

For this equation, $a = 3$, $b = 6$, and $c = -2$.

$$b^2 - 4ac = 6^2 - 4(3)(-2) = 36 + 24 = 60$$

Because $b^2 - 4ac = 60$ is positive, but not a

perfect square, the quadratic equation will have two irrational solutions.

13. $2x^2 + 6x + 5 = 0$

For this equation, $a = 2$, $b = 6$, and $c = 5$.

$$b^2 - 4ac = 6^2 - 4(2)(5) = 36 - 40 = -4$$

Because $b^2 - 4ac = -4$ is negative, the quadratic equation will have two complex solutions that are not real. The solutions will be complex conjugates of each other.

14. $c^2 = 4^2 + 10^2 = 16 + 100 = 116$

$$c = \sqrt{116} = 2\sqrt{29}$$

15. $$2m + 7\sqrt{m} - 15 = 0$$

$$2\left(\sqrt{m}\right)^2 + 7\left(\sqrt{m}\right) - 15 = 0$$

Let $u = \sqrt{m}$.

$$2u^2 + 7u - 15 = 0$$

$$(2u - 3)(u + 5) = 0$$

$$2u - 3 = 0 \quad \text{or} \quad u + 5 = 0$$

$$2u = 3 \quad \text{or} \qquad u = -5$$

$$u = \frac{3}{2}$$

$$\sqrt{m} = \frac{3}{2} \quad \text{or} \quad \sqrt{m} = -5$$

$$m = \left(\frac{3}{2}\right)^2 \quad \text{or} \qquad m = (-5)^2$$

$$m = \frac{9}{4} \quad \text{or} \qquad m = 25$$

Check:

$$m = \frac{9}{4}:\quad 2 \cdot \frac{9}{4} + 7\sqrt{\frac{9}{4}} - 15 \overset{?}{=} 0$$

$$2 \cdot \frac{9}{4} + 7 \cdot \frac{3}{2} - 15 \overset{?}{=} 0$$

$$\frac{9}{2} + \frac{21}{2} - 15 \overset{?}{=} 0$$

$$0 = 0 \ \checkmark$$

$$m = 25:\quad 2 \cdot 25 + 7\sqrt{25} - 15 \overset{?}{=} 0$$

$$2 \cdot 25 + 7 \cdot 5 - 15 \overset{?}{=} 0$$

$$50 + 35 - 15 \overset{?}{=} 0$$

$$70 \ne 0 \ \times$$

$m = 25$ does not check; the solution set is $\left\{ \dfrac{9}{4} \right\}$.

16.
$$p^{-2} - 3p^{-1} - 18 = 0$$
$$\left(p^{-1}\right)^2 - 3\left(p^{-1}\right) - 18 = 0$$

Let $u = p^{-1}$.

$$u^2 - 3u - 18 = 0$$
$$(u+3)(u-6) = 0$$
$$u+3 = 0 \quad \text{or} \quad u-6 = 0$$
$$u = -3 \quad \text{or} \quad u = 6$$
$$p^{-1} = -3 \quad \text{or} \quad p^{-1} = 6$$
$$\frac{1}{p} = -3 \quad \text{or} \quad \frac{1}{p} = 6$$
$$p = -\frac{1}{3} \quad \text{or} \quad p = \frac{1}{6}$$

Check:

$$p = -\frac{1}{3}: \quad \left(-\frac{1}{3}\right)^{-2} - 3\left(-\frac{1}{3}\right)^{-1} - 18 \stackrel{?}{=} 0$$
$$(-3)^2 - 3(-3) - 18 \stackrel{?}{=} 0$$
$$9 + 9 - 18 \stackrel{?}{=} 0$$
$$0 = 0 \checkmark$$

$$p = \frac{1}{6}: \quad \left(\frac{1}{6}\right)^{-2} - 3\left(\frac{1}{6}\right)^{-1} - 18 \stackrel{?}{=} 0$$
$$(6)^2 - 3(6) - 18 \stackrel{?}{=} 0$$
$$36 - 18 - 18 \stackrel{?}{=} 0$$
$$0 = 0 \checkmark$$

Both check; the solution set is $\left\{\dfrac{1}{6}, \ -\dfrac{1}{3}\right\}$.

17.
$$12,000 = -0.4x^2 + 140x$$
$$0.4x^2 - 140x + 12,000 = 0$$

For this equation, $a = 0.4$, $b = -140$, and $c = 12,000$.

$$x = \frac{-(-140) \pm \sqrt{(-140)^2 - 4(0.4)(12,000)}}{2(0.4)}$$

$$= \frac{140 \pm \sqrt{400}}{0.8}$$

$$= \frac{140 \pm 20}{0.8}$$

$$x = \frac{140 + 20}{0.8} \quad \text{or} \quad x = \frac{140 - 20}{0.8}$$
$$= 200 \quad \text{or} \quad = 150$$

Revenue will be $12,000 when either 150 microwaves or 200 microwaves are sold.

18. Let x represent the speed of the wind.

	Distance	Rate	Time
Against Wind	300	$140 - x$	$\dfrac{300}{140 - x}$
With Wind	300	$140 + x$	$\dfrac{300}{140 + x}$

$$\frac{300}{140 - x} + \frac{300}{140 + x} = 5$$

$$(140 - x)(140 + x)\left(\frac{300}{140 - x} + \frac{300}{140 + x}\right)$$
$$= (140 - x)(140 + x)(5)$$

$$300(140 + x) + 300(140 - x) = \left(19,600 - x^2\right)(5)$$
$$42,000 + 300x + 42,000 - 300x = 98,000 - 5x^2$$
$$84,000 = 98,000 - 5x^2$$
$$-14,000 = -5x^2$$
$$2800 = x^2$$
$$x = \pm\sqrt{2800}$$
$$= \pm 20\sqrt{7}$$
$$\approx \pm 52.915$$

Because the speed should be positive, we disregard $x \approx -52.915$. Thus, the only viable answer is $x \approx 52.915$. Rounding to the nearest tenth, the speed of the wind was approximately 52.9 miles per hour.

Section 7.4

Are You Ready for This Section?

R1. Locate some points on the graph of $f(x) = x^2$.

x	$f(x) = x^2$	$(x, f(x))$
-3	$f(-3) = (-3)^2 = 9$	$(-3,\ 9)$
-2	$f(-2) = (-2)^2 = 4$	$(-2,\ 4)$
-1	$f(-1) = (-1)^2 = 1$	$(-1,\ 1)$
0	$f(0) = 0^2 = 0$	$(0,\ 0)$
1	$f(1) = 1^2 = 1$	$(1,\ 1)$
2	$f(2) = 2^2 = 4$	$(2,\ 4)$
3	$f(3) = 3^2 = 9$	$(3,\ 9)$

Plot the points and connect them with a smooth curve.

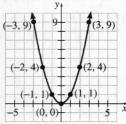

R2. Locate some points on the graph of
$f(x) = x^2 - 3$.

x	$f(x) = x^2 - 3$	$(x, f(x))$
-3	$f(-3) = (-3)^2 - 3 = 6$	$(-3,\ 6)$
-2	$f(-2) = (-2)^2 - 3 = 1$	$(-2,\ 1)$
-1	$f(-1) = (-1)^2 - 3 = -2$	$(-1, -2)$
0	$f(0) = 0^2 - 3 = -3$	$(0, -3)$
1	$f(1) = 1^2 - 3 = -2$	$(1, -2)$
2	$f(2) = 2^2 - 3 = 1$	$(2,\ 1)$
3	$f(3) = 3^2 - 3 = 6$	$(3,\ 6)$

Plot the points and connect them with a smooth curve.

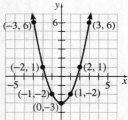

R3. For $f(x) = 2x^2 + 5x + 1$, the independent
variable x can take on any value on the real
number line. That is, f is defined for all real x
values. Thus, the domain is the set of all real
number, or $(-\infty, \infty)$.

Section 7.4 Quick Checks

1. A <u>quadratic function</u> is a function of the form
   $f(x) = ax^2 + bx + c$, where a, b, and c are real
   numbers and $a \ne 0$.

2. To graph $f(x) = x^2 + k$, $k > 0$, from the graph of
   $y = x^2$, shift the graph of $y = x^2$ <u>up</u> k units. To
   graph $f(x) = x^2 - k$, $k > 0$, from the graph of
   $y = x^2$ shift the graph of $y = x^2$ <u>down</u> k units.

3. Begin with the graph of $y = x^2$, then shift the
   graph up 5 unit to obtain the graph of
   $f(x) = x^2 + 5$.

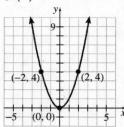

4. Begin with the graph of $y = x^2$, then shift the
   graph down 2 units to obtain the graph of
   $f(x) = x^2 - 2$.

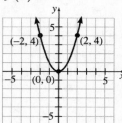

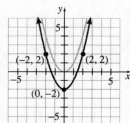

5. False. To obtain the graph of $f(x) = (x + 12)^2$
   we shift the graph of $y = x^2$ horizontally to the
   left 12 units.

6. Begin with the graph of $y = x^2$, then shift the
   graph to the left 5 units to obtain the graph of
   $f(x) = (x + 5)^2$.

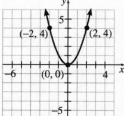

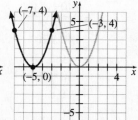

7. Begin with the graph of $y = x^2$, then shift the graph to the right 1 unit to obtain the graph of $f(x) = (x-1)^2$.

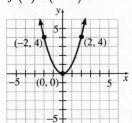 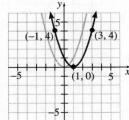

8. Begin with the graph of $y = x^2$, then shift the graph 3 units to the right to obtain the graph of $y = (x-3)^2$. Shift this graph up 2 units to obtain the graph of $f(x) = (x-3)^2 + 2$.

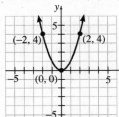 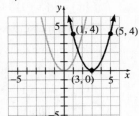

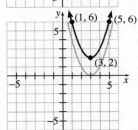

9. Begin with the graph of $y = x^2$, then shift the graph 1 unit to the left to obtain the graph of $y = (x+1)^2$. Shift this graph down 4 units to obtain the graph of $f(x) = (x+1)^2 - 4$.

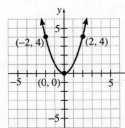 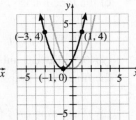

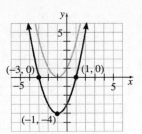

10. y-coordinate; a; vertically stretched; vertically compressed

11. Begin with the graph of $y = x^2$, then vertically stretch the graph by a factor of 3 (multiply each y-coordinate by 3) to obtain the graph of $f(x) = 3x^2$.

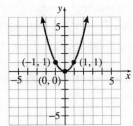 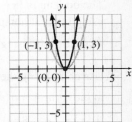

12. Begin with the graph of $y = x^2$, then multiply the y-coordinates by $-\dfrac{1}{4}$ to obtain the graph of $f(x) = -\dfrac{1}{4}x^2$.

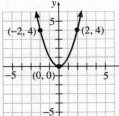 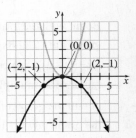

13. True. In this case, $a = -3$ which is less than zero, so the graph opens down.

14. Begin with the graph of $y = x^2$, then shift the graph 2 units to the left to obtain the graph of $y = (x+2)^2$. Multiply the y-coordinates by -3 to obtain the graph of $y = -3(x+2)^2$. Lastly, shift the graph up 1 unit to obtain the graph of $f(x) = -3(x+2)^2 + 1$.

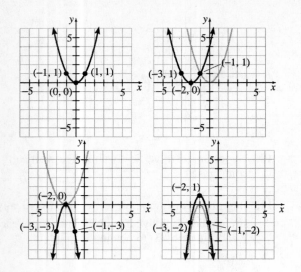

The domain is $\{x \mid x$ is and real number$\}$ or, using interval notation, $(-\infty, \infty)$. The range is $\{y \mid y \leq 1\}$ or using interval notation $(-\infty, 1]$.

15. Use completing the square to write the function in the form $y = a(x-h)^2 + k$.

$$f(x) = 2x^2 - 8x + 5$$
$$= \left(2x^2 - 8x\right) + 5$$
$$= 2\left(x^2 - 4x\right) + 5$$
$$= 2\left(x^2 - 4x + 4\right) + 5 - 8$$
$$= 2(x-2)^2 - 3$$

Begin with the graph of $y = x^2$, then shift the graph right 2 units to obtain the graph of $y = (x-2)^2$. Vertically stretch this graph by a factor of 2 (multiply the y-coordinates by 2) to obtain the graph of $y = 2(x-2)^2$. Lastly, shift the graph down 3 units to obtain the graph of $f(x) = 2(x-2)^2 - 3$.

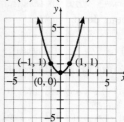
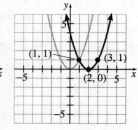

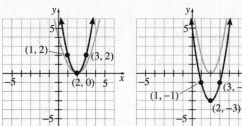

The domain is $\{x \mid x$ is and real number$\}$ or, using interval notation, $(-\infty, \infty)$. The range is $\{y \mid y \geq -3\}$ or, using interval notation, $[-3, \infty)$.

16. Consider the form $f(x) = a(x-h)^2 + k$. From the graph, we know that the vertex is $(-1, \ 2)$. So $h = -1$ and $k = 2$, and we have that

$$f(x) = a(x-(-1))^2 + 2$$
$$f(x) = a(x+1)^2 + 2$$

The graph also passes through the point $(0, \ 1)$ which means that $f(0) = 1$. Substituting these values into the function, we can solve for a:

$$f(x) = a(x+1)^2 + 2$$
$$1 = a(0+1)^2 + 2$$
$$1 = a(1)^2 + 2$$
$$1 = a + 2$$
$$-1 = a$$

The quadratic function is $f(x) = -(x+1)^2 + 2$.

7.4 Exercises

17. (I) The graph of $f(x) = x^2 + 3$ is the graph of $y = x^2$ shifted 3 units up. Thus, the graph of the function is graph (D).

(II) The graph of $f(x) = (x+3)^2$ is the graph of $y = x^2$ shifted 3 units to the left. Thus, the graph of the function is graph (A).

(III) The graph of $f(x) = x^2 - 3$ is the graph of $y = x^2$ shifted 3 units down. Thus, the graph of the function is graph (C).

(IV) The graph of $f(x) = (x-3)^2$ is the graph of $y = x^2$ shifted 3 units to the right. Thus, the graph of the function is graph (B).

19. To obtain the graph of $f(x) = (x+10)^2$, begin with the graph of $y = x^2$ and shift it 10 units to the left.

21. To obtain the graph of $F(x) = x^2 + 12$, begin with the graph of $y = x^2$ and shift it 12 units up.

23. To obtain the graph of $H(x) = 2(x-5)^2$, begin with the graph of $y = x^2$, shift it 5 units to the right, and vertically stretch it by a factor of 2 (multiply the *y*-coordinates by 2).

25. To obtain the graph of $f(x) = -3(x+5)^2 + 8$, begin with the graph of $y = x^2$, shift it 5 units to the left, multiply the *y*-coordinates by -3 (which means it opens down and is stretched vertically by a factor of 3), and shift the graph up 8 units.

27. Begin with the graph of $y = x^2$, then shift the graph up 1 unit to obtain the graph of $f(x) = x^2 + 1$.

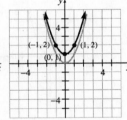

29. Begin with the graph of $y = x^2$, then shift the graph down 1 unit to obtain the graph of $f(x) = x^2 - 1$.

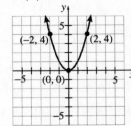

31. Begin with the graph of $y = x^2$, then shift the graph to the right 3 units to obtain the graph of $F(x) = (x-3)^2$.

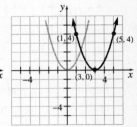

33. Begin with the graph of $y = x^2$, then shift the graph 2 units to the left to obtain the graph of $h(x) = (x+2)^2$.

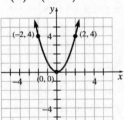

35. Begin with the graph of $y = x^2$, then vertically stretch the graph by a factor of 4 (multiply each *y*-coordinate by 4) to obtain the graph of $g(x) = 4x^2$.

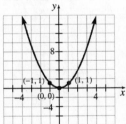

37. Begin with the graph of $y = x^2$, then vertically compress the graph by a factor of $\frac{1}{3}$ (multiply each *y*-coordinate by $\frac{1}{3}$) to obtain the graph of $H(x) = \frac{1}{3}x^2$.

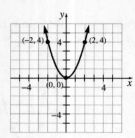

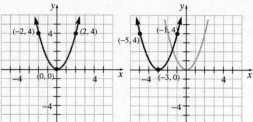

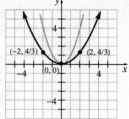

39. Begin with the graph of $y = x^2$, then multiply each y-coordinate by -1 to obtain the graph of $p(x) = -x^2$.

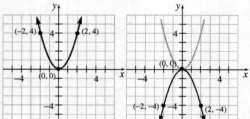

41. Begin with the graph of $y = x^2$, then shift the graph 1 unit to the right to obtain the graph of $y = (x-1)^2$. Shift this graph down 3 units to obtain the graph of $f(x) = (x-1)^2 - 3$.

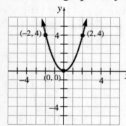

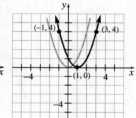

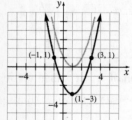

43. Begin with the graph of $y = x^2$, then shift the graph 3 units to the left to obtain the graph of $y = (x+3)^2$. Shift this graph up 1 unit to obtain the graph of $F(x) = (x+3)^2 + 1$.

45. Begin with the graph of $y = x^2$, then shift the graph 3 units to the left to obtain the graph of $y = (x+3)^2$. Multiply the y-coordinates by -1 to obtain the graph of $y = -(x+3)^2$. Lastly, shift the graph up 2 units to obtain the graph of $h(x) = -(x+3)^2 + 2$.

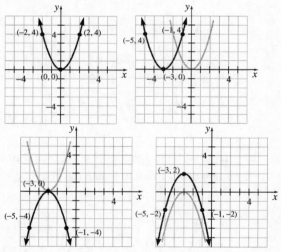

47. Begin with the graph of $y = x^2$, then shift the graph 1 unit to the left to obtain the graph of $y = (x+1)^2$. Vertically stretch this graph by a factor of 2 (multiply the y-coordinates by 2) to obtain the graph of $y = 2(x+1)^2$. Lastly, shift the graph down 2 units to obtain the graph of $G(x) = 2(x+1)^2 - 2$.

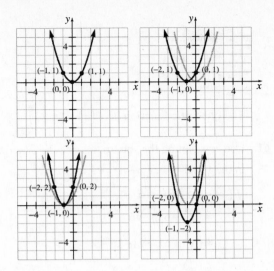

49. Begin with the graph of $y = x^2$, then shift the graph 5 units to the left to obtain the graph of $y = (x+5)^2$. Multiply the y-coordinates by $-\dfrac{1}{2}$ to obtain the graph of $y = -\dfrac{1}{2}(x+5)^2$. Lastly, shift the graph up 3 units to obtain the graph of $H(x) = -\dfrac{1}{2}(x+5)^2 + 3$.

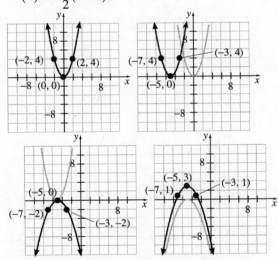

51. Use completing the square to write the function in the form $y = a(x-h)^2 + k$.

$$f(x) = x^2 + 2x - 4$$
$$= \left(x^2 + 2x + 1\right) - 4 - 1$$
$$= (x+1)^2 - 5$$

Begin with the graph of $y = x^2$, then shift the

graph 1 unit to the left to obtain the graph of $y = (x+1)^2$. Shift this graph down 5 units to obtain the graph of $f(x) = (x+1)^2 - 5$.

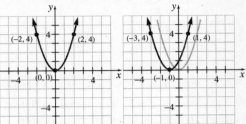

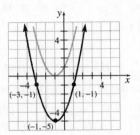

The vertex is $(h,k) = (-1,-5)$ and the axis of symmetry is $x = -1$. The domain is the set of all real numbers or, using interval notation, $(-\infty, \infty)$. The range is $\{y \mid y \geq -5\}$ or, using interval notation, $[-5, \infty)$.

53. Use completing the square to write the function in the form $y = a(x-h)^2 + k$.

$$g(x) = x^2 - 4x + 8 = \left(x^2 - 4x\right) + 8$$
$$= \left(x^2 - 4x + 4\right) + 8 - 4$$
$$= (x-2)^2 + 4$$

Begin with the graph of $y = x^2$, then shift the graph right 2 units to obtain the graph of $y = (x-2)^2$. Shift this graph up 4 units to obtain the graph of $g(x) = (x-2)^2 + 4$.

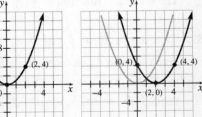

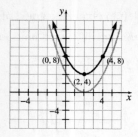

The vertex is $(h,k)=(2,4)$ and the axis of symmetry is $x=2$. The domain is the set of all real numbers or, using interval notation, $(-\infty,\infty)$. The range is $\{y\mid y\ge 4\}$ or, using interval notation, $[4,\infty)$.

55. Consider the form $y=a(x-h)^2+k$. From the graph we know that the vertex is $(-1,-3)$ so we have $h=-1$ and $k=-3$. The graph also passes through the point $(x,y)=(0,-2)$. Substituting these values for $x, y, h,$ and k, we can solve for a:

$$-2=a(0-(-1))^2+(-3)$$
$$-2=a(1)^2-3$$
$$-2=a-3$$
$$1=a$$

The quadratic function is $f(x)=(x+1)^2-3$.

57. Consider the form $y=a(x-h)^2+k$. From the graph we know that the vertex is $(3,7)$ so we have $h=3$ and $k=7$. The graph also passes through the point $(x,y)=(0,-11)$. Substituting these values for $x, y, h,$ and k, we can solve for a:

$$-11=a(0-3)^2+7$$
$$-11=a(-3)^2+7$$
$$-18=9a$$
$$-2=a$$

The quadratic function is $f(x)=-2(x-3)^2+7$.

59. Consider the form $y=a(x-h)^2+k$. From the graph we know that the vertex is $(-4,0)$ so we have $h=-4$ and $k=0$. The graph also passes through the point $(x,y)=(-3,1)$. Substituting these values for $x, y, h,$ and k, we can solve for a:

$$1=a(-3-(-4))^2+0$$
$$1=a(-3+4)^2$$
$$1=a(1)^2$$
$$1=a$$

The quadratic function is $f(x)=(x+4)^2$.

61. Use completing the square to write the function in the form $y=a(x-h)^2+k$.

$$\begin{aligned}f(x)&=x^2+6x-16\\&=\left(x^2+6x\right)-16\\&=\left(x^2+6x+9\right)-16-9\\&=(x+3)^2-25\end{aligned}$$

Begin with the graph of $y=x^2$, then shift the graph 3 units left to obtain the graph of $y=(x+3)^2$. Shift this result down 25 units to obtain the graph of $f(x)=(x+3)^2-25$.

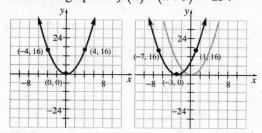

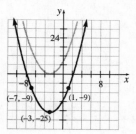

The vertex is $(h,k)=(-3,-25)$ and the axis of symmetry is $x=-3$. The domain is the set of all real numbers or, using interval notation, $(-\infty,\infty)$. The range is $\{y\mid y\ge -25\}$ or, using interval notation, $[-25,\infty)$.

63. Use completing the square to write the function in the form $y = a(x-h)^2 + k$.

$$F(x) = x^2 + x - 12$$
$$= \left(x^2 + x\right) - 12$$
$$= \left(x^2 + x + \frac{1}{4}\right) - 12 - \frac{1}{4}$$
$$= \left(x + \frac{1}{2}\right)^2 - \frac{49}{4}$$

Begin with the graph of $y = x^2$, then shift the graph left $\frac{1}{2}$ unit to obtain the graph of $y = \left(x + \frac{1}{2}\right)^2$. Shift this graph down $\frac{49}{4}$ units to obtain the graph of $F(x) = \left(x + \frac{1}{2}\right)^2 - \frac{49}{4}$.

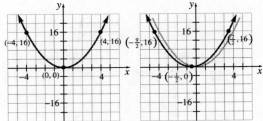

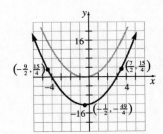

The vertex is $(h,k) = \left(-\frac{1}{2}, -\frac{49}{4}\right)$ and the axis of symmetry is $x = -\frac{1}{2}$. The domain is the set of all real numbers or, using interval notation, $(-\infty, \infty)$. The range is $\left\{ y \,\middle|\, y \geq -\frac{49}{4} \right\}$ or, using interval notation, $\left[-\frac{49}{4}, \infty \right)$.

65. Use completing the square to write the function in the form $y = a(x-h)^2 + k$.

$$H(x) = 2x^2 - 4x - 1$$
$$= \left(2x^2 - 4x\right) - 1$$
$$= 2\left(x^2 - 2x\right) - 1$$
$$= 2\left(x^2 - 2x + 1\right) - 1 - 2$$
$$= 2(x-1)^2 - 3$$

Begin with the graph of $y = x^2$, then shift the graph right 1 unit to obtain the graph of $y = (x-1)^2$. Vertically stretch this graph by a factor of 2 (multiply the *y*-coordinates by 2) to obtain the graph of $y = 2(x-1)^2$. Lastly, shift the graph down 3 units to obtain the graph of $H(x) = 2(x-1)^2 - 3$.

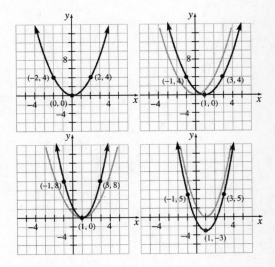

The vertex is $(h,k) = (1,-3)$ and the axis of symmetry is $x = 1$. The domain is the set of all real numbers or, using interval notation, $(-\infty, \infty)$. The range is $\{ y \,|\, y \geq -3 \}$ or, using interval notation, $[-3, \infty)$.

67. Use completing the square to write the function in the form $y = a(x-h)^2 + k$.

$$P(x) = 3x^2 + 12x + 13$$
$$= \left(3x^2 + 12x\right) + 13$$
$$= 3\left(x^2 + 4x\right) + 13$$
$$= 3\left(x^2 + 4x + 4\right) + 13 - 12$$
$$= 3(x+2)^2 + 1$$

Begin with the graph of $y = x^2$, then shift the graph left 2 units to obtain the graph of $y = (x+2)^2$. Vertically stretch this graph by a factor of 3 (multiply the y-coordinates by 3) to obtain the graph of $y = 3(x+2)^2$. Lastly, shift the graph up 1 unit to obtain the graph of $P(x) = 3(x+2)^2 + 1$.

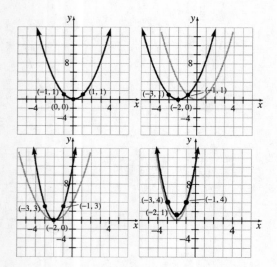

The vertex is $(h,k) = (-2,1)$ and the axis of symmetry is $x = -2$. The domain is the set of all real numbers or, using interval notation, $(-\infty, \infty)$. The range is $\{y \mid y \geq 1\}$ or, using interval notation, $[1, \infty)$.

69. Use completing the square to write the function in the form $y = a(x-h)^2 + k$.

$$F(x) = -x^2 - 10x - 21$$
$$= \left(-x^2 - 10x\right) - 21$$
$$= -\left(x^2 + 10x\right) - 21$$
$$= -\left(x^2 + 10x + 25\right) - 21 + 25$$
$$= -(x+5)^2 + 4$$

Begin with the graph of $y = x^2$, then shift the graph left 5 units to obtain the graph of $y = (x+5)^2$. Multiply the y-coordinates by -1 to obtain the graph of $y = -(x+5)^2$. Lastly, shift the graph up 4 units to obtain the graph of $F(x) = -(x+5)^2 + 4$.

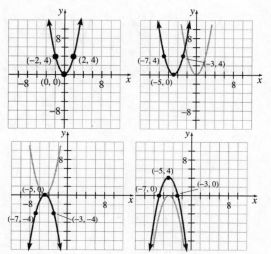

The vertex is $(h,k) = (-5,4)$ and the axis of symmetry is $x = -5$. The domain is the set of all real numbers or, using interval notation, $(-\infty, \infty)$. The range is $\{y \mid y \leq 4\}$ or, using interval notation, $(-\infty, 4]$.

71. Use completing the square to write the function in the form $y = a(x-h)^2 + k$.

$$g(x) = -x^2 + 6x - 1$$
$$= \left(-x^2 + 6x\right) - 1$$
$$= -\left(x^2 - 6x\right) - 1$$
$$= -\left(x^2 - 6x + 9\right) - 1 + 9$$
$$= -(x-3)^2 + 8$$

Begin with the graph of $y = x^2$, then shift the graph right 3 units to obtain the graph of $y = (x-3)^2$. Multiply the y-coordinates by -1 to obtain the graph of $y = -(x-3)^2$. Lastly, shift the graph up 8 units to obtain the graph of $g(x) = -(x-3)^2 + 8$.

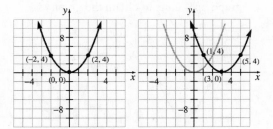

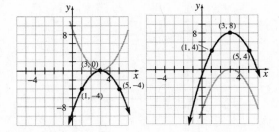

The vertex is $(h,k) = (3,8)$ and the axis of symmetry is $x = 3$. The domain is the set of all real numbers or, using interval notation, $(-\infty, \infty)$. The range is $\{y \mid y \le 8\}$ or, using interval notation, $(-\infty, 8]$.

73. Use completing the square to write the function in the form $y = a(x-h)^2 + k$.

$$H(x) = -2x^2 + 8x - 4$$
$$= \left(-2x^2 + 8x\right) - 4$$
$$= -2\left(x^2 - 4x\right) - 4$$
$$= -2\left(x^2 - 4x + 4\right) - 4 + 8$$
$$= -2(x-2)^2 + 4$$

Begin with the graph of $y = x^2$, then shift the graph right 2 units to obtain the graph of $y = (x-2)^2$. Multiply the y-coordinates by -2 to obtain the graph of $y = -2(x-2)^2$. Lastly, shift the graph up 4 units to obtain the graph of $H(x) = -2(x-2)^2 + 4$.

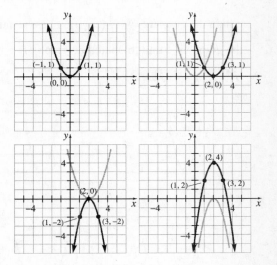

The vertex is $(h,k) = (2,4)$ and the axis of symmetry is $x = 2$. The domain is the set of all real numbers or, using interval notation, $(-\infty, \infty)$. The range is $\{y \mid y \le 4\}$ or, using interval notation, $(-\infty, 4]$.

75. Use completing the square to write the function in the form $y = a(x-h)^2 + k$.

$$f(x) = \frac{1}{3}x^2 - 2x + 4$$
$$= \left(\frac{1}{3}x^2 - 2x\right) + 4$$
$$= \frac{1}{3}\left(x^2 - 6x\right) + 4$$
$$= \frac{1}{3}\left(x^2 - 6x + 9\right) + 4 - 3$$
$$= \frac{1}{3}(x-3)^2 + 1$$

Begin with the graph of $y = x^2$, then shift the graph right 3 units to obtain the graph of $y = (x-3)^2$. Vertically compress the graph by a factor of $\frac{1}{3}$ (multiply the y-coordinates by $\frac{1}{3}$) to obtain the graph of $y = \frac{1}{3}(x-3)^2$. Lastly, shift the graph up 1 unit to obtain the graph of $f(x) = \frac{1}{3}(x-3)^2 + 1$.

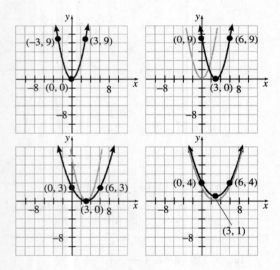

The vertex is $(h,k) = (3,1)$ and the axis of symmetry is $x = 1$. The domain is the set of all real numbers or, using interval notation, $(-\infty, \infty)$. The range is $\{y \mid y \geq 1\}$ or, using interval notation, $[1, \infty)$.

77. Use completing the square to write the function in the form $y = a(x-h)^2 + k$.

$$G(x) = -12x^2 - 12x + 1$$
$$= \left(-12x^2 - 12x\right) + 1$$
$$= -12\left(x^2 + x\right) + 1$$
$$= -12\left(x^2 + x + \frac{1}{4}\right) + 1 + 3$$
$$= -12\left(x + \frac{1}{2}\right)^2 + 4$$

Begin with the graph of $y = x^2$, then shift left $\frac{1}{2}$ unit to obtain the graph of $y = \left(x + \frac{1}{2}\right)^2$.

Multiply the y-coordinates by -12 to obtain the graph of $y = -12\left(x + \frac{1}{2}\right)^2$. Lastly, shift the graph up 4 units to obtain the graph of $G(x) = -12\left(x + \frac{1}{2}\right)^2 + 4$.

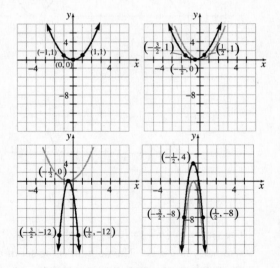

The vertex is $(h,k) = \left(-\frac{1}{2}, 4\right)$ and the axis of symmetry is $x = -\frac{1}{2}$. The domain is the set of all real numbers or, using interval notation, $(-\infty, \infty)$. The range is $\{y \mid y \leq 4\}$ or, using interval notation, $(-\infty, 4]$.

79. Answers may vary. Since the graph opens up, one possibility is to let $a = 1$. The vertex is $(3, 0)$ so we have $h = 3$ and $k = 0$. Substituting these values into the form $y = a(x - h)^2 + k$ gives $y = f(x) = (x - 3)^2$.

81. Answers may vary. Since the graph opens up, one possibility is to let $a = 1$. The vertex is $(-3, 1)$ so we have $h = -3$ and $k = 1$. Substituting these values into the form $y = a(x - h)^2 + k$ give $y = f(x) = (x + 3)^2 + 1$.

83. Answers may vary. Since the graph opens down, one possibility is to let $a = -1$. The vertex is $(5, -1)$ so we have $h = 5$ and $k = -1$. Substituting these values into the form $y = a(x - h)^2 + k$ gives $y = f(x) = -(x - 5)^2 - 1$.

85. Consider the form $y = a(x - h)^2 + k$. Since the graph opens up, we know that $a > 0$. The graph is vertically stretched by a factor of 4 so we know that $a = 4$. The vertex is $(9, -6)$ so we have $h = 9$ and $k = -6$. Substituting these values gives $y = f(x) = 4(x - 9)^2 - 6$.

87. Consider the form $y = a(x - h)^2 + k$. Since the graph opens down, we know that $a < 0$. The graph is vertically compressed by a factor of $\frac{1}{3}$ so we know that $a = -\frac{1}{3}$. The vertex is $(0, 6)$ so we have $h = 0$ and $k = 6$. Substituting these values gives $y = f(x) = -\frac{1}{3}x^2 + 6$.

89. The lowest or highest point on a parabola (the graph of $y = ax^2 + bx + c$) is called the *vertex*. If $a > 0$, the graph opens up and the vertex is the low point. If $a < 0$, the graph opens down and the vertex is the high point.

91. A quadratic function can never have a range of $(-\infty, \infty)$. The graph of a quadratic always has a turning point so it always has either a lowest point or a highest point. In either case, the range is limited by k, the y-coordinate of the vertex. The range will either be $(-\infty, k]$ if the graph opens down or $[k, \infty)$ if the graph opens up.

93.
$$12 \overline{)349}$$
$$\underline{24}$$
$$109$$
$$\underline{108}$$
$$1$$

So, $\dfrac{349}{12} = 29 + \dfrac{1}{12} = 29\dfrac{1}{12}$.

95.
$$2x - 1 \overline{)2x^4 - 11x^3 + 13x^2 - 8x} \quad\quad x^3 - 5x^2 + 4x - 2$$
$$\underline{2x^4 - \ x^3}$$
$$-10x^3 + 13x^2 - 8x$$
$$\underline{-10x^3 + \ 5x^2}$$
$$8x^2 - 8x$$
$$\underline{8x^2 - 4x}$$
$$-4x$$
$$\underline{-4x + 2}$$
$$-2$$

So,
$$\frac{2x^4 - 11x^3 + 13x^2 - 8x}{2x - 1} = x^3 - 5x^2 + 4x - 2 + \frac{-2}{2x - 1}$$

97. $f(x) = x^2 + 1.3$

Vertex: $(0, 1.3)$
Axis of symmetry: $x = 0$
Range: $\{y \mid y \geq 1.3\} = [1.3, \infty)$

99. $g(x) = (x - 2.5)^2$

Vertex: $(2.5, 0)$

Axis of symmetry: $x = 2.5$

Range: $\{y \mid y \geq 0\} = [0, \infty)$

101. $h(x) = 2.3(x - 1.4)^2 + 0.5$

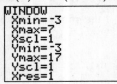 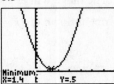

Vertex: $(1.4, 0.5)$

Axis of symmetry: $x = 1.4$

Range: $\{y \mid y \geq 0.5\} = [0.5, \infty)$

103. $F(x) = -3.4(x - 2.8)^2 + 5.9$

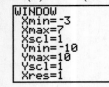 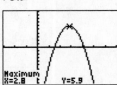

Vertex: $(2.8, 5.9)$

Axis of symmetry: $x = 2.8$

Range: $\{y \mid y \leq 5.9\} = (-\infty, 5.9]$

Section 7.5

Are You Ready for This Section?

R1. To find the y-intercept, let $x = 0$ and solve for y:

$2(0) + 5y = 20$

$5y = 20$

$y = 4$

To find the x-intercept, let $y = 0$ and solve for x:

$2x + 5(0) = 20$

$2x = 20$

$x = 10$

The intercepts are $(0, 4)$ and $(10, 0)$.

R2. $2x^2 - 3x - 20 = 0$

$(2x + 5)(x - 4) = 0$

$2x + 5 = 0 \quad$ or $\quad x - 4 = 0$

$x = -\dfrac{5}{2} \quad$ or $\quad x = 4$

The solution set is $\left\{ -\dfrac{5}{2}, 4 \right\}$.

R3. $f(x) = 0$

$x^2 - 3x - 4 = 0$

$(x + 1)(x - 4) = 0$

$x + 1 = 0 \quad$ or $\quad x - 4 = 0$

$x = -1 \quad$ or $\qquad x = 4$

The zeros are -1 and 4.

Section 7.5 Quick Checks

1. Any quadratic function $f(x) = ax^2 + bx + c$, $a \neq 0$, will have a vertex whose x-coordinate is

$x = -\dfrac{b}{2a}$.

2. The graph of $f(x) = ax^2 + bx + c$ will have two different x-intercepts if $b^2 - 4ac > 0$.

3. $f(x) = -2x^2 - 3x + 6$

$b^2 - 4ac = (-3)^2 - 4(-2)(6) = 9 + 48 = 57 > 0$

Since the discriminant is greater than zero, the graph will have two x-intercepts.

4. $f(x) = x^2 + 4x - 3$

$x = -\dfrac{b}{2a} = -\dfrac{4}{2} = -2$

$f(-2) = (-2)^2 + 4(-2) - 3$

$\qquad = 4 - 8 - 3 = -7$

The vertex is $(-2, -7)$.

5. For the function $f(x) = x^2 - 4x - 12$, we see that $a = 1$, $b = -4$, and $c = -12$. The parabola opens up because $a = 1 > 0$. The x-coordinate of the vertex is $x = -\dfrac{b}{2a} = -\dfrac{(-4)}{2(1)} = 2$. The y-coordinate of the vertex is

$f\left(-\dfrac{b}{2a} \right) = f(2)$

$\qquad = (2)^2 - 4(2) - 12$

$\qquad = 4 - 8 - 12$

$\qquad = -16$

Thus, the vertex is $(2, -16)$ and the axis of symmetry is the line $x = 2$.

The y-intercept is

$f(0) = (0)^2 - 4(0) - 12 = -12$.

Now, $b^2 - 4ac = (-4)^2 - 4(1)(-12) = 64 > 0$.
The parabola will have two distinct x-intercepts.
We find these by solving
$$f(x) = 0$$
$$x^2 - 4x - 12 = 0$$
$$(x-6)(x+2) = 0$$
$$x - 6 = 0 \text{ or } x + 2 = 0$$
$$x = 6 \text{ or } x = -2$$
Finally, the y-intercept point, $(0,-12)$, is two
units to the left of the axis of symmetry.
Therefore, if we move two units to the right of
the axis of symmetry, we obtain the point
$(4,-12)$ which must also be on the graph.

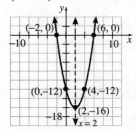

6. For the function $f(x) = -3x^2 + 12x - 7$, we see
that $a = -3$, $b = 12$, and $c = -7$. The parabola
opens down because $a = -3 < 0$. The
x-coordinate of the vertex is
$x = -\dfrac{b}{2a} = -\dfrac{12}{2(-3)} = 2$. The y-coordinate of
the vertex is
$$f\left(-\frac{b}{2a}\right) = f(2)$$
$$= -3(2)^2 + 12(2) - 7$$
$$= -12 + 24 - 7$$
$$= 5$$
Thus, the vertex is $(2,\ 5)$ and the axis of
symmetry is the line $x = 2$. The y-intercept is
$f(0) = -3(0)^2 + 12(0) - 7 = -7$.
Now, $b^2 - 4ac = 12^2 - 4(-3)(-7) = 60 > 0$. The
parabola will have two distinct x-intercepts. We
find these by solving
$$f(x) = 0$$
$$-3x^2 + 12x - 7 = 0$$

$$x = \frac{-12 \pm \sqrt{60}}{2(-3)}$$
$$= \frac{-12 \pm 2\sqrt{15}}{-6}$$
$$= \frac{6 \pm \sqrt{15}}{3}$$
$$x \approx 0.71 \text{ or } x \approx 3.29$$
Finally, the y-intercept point, $(0,-7)$, is two
units to the left of the axis of symmetry.
Therefore, if we move two units to the right of
the axis of symmetry, we obtain the point
$(4,-7)$ which must also be on the graph.

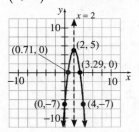

7. For the function $f(x) = x^2 + 6x + 9$, we see that
$a = 1$, $b = 6$, and $c = 9$. The parabola opens up
because $a = 1 > 0$. The x-coordinate of the
vertex is $x = -\dfrac{b}{2a} = -\dfrac{6}{2(1)} = -3$. The
y-coordinate of the vertex is
$$f\left(-\frac{b}{2a}\right) = f(-3)$$
$$= (-3)^2 + 6(-3) + 9$$
$$= 9 - 18 + 9$$
$$= 0$$
Thus, the vertex is $(-3,\ 0)$ and the axis of
symmetry is the line $x = -3$. The y-intercept is
$f(0) = 0^2 + 6(0) + 9 = 9$. Now,
$b^2 - 4ac = 6^2 - 4(1)(9) = 36 - 36 = 0$. Since the
discriminant is 0, the x-coordinate of the vertex
is the only x-intercept, $x = -3$. Finally, the y-
intercept point, $(0,\ 9)$, is three units to the right
of the axis of symmetry. Therefore, if we move
three units to the left of the axis of symmetry, we
obtain the point $(-6,\ 9)$ which must also be on
the graph.

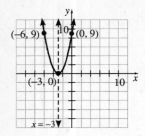

8. For the function $G(x) = -3x^2 + 9x - 8$, we see that $a = -3$, $b = 9$, and $c = -8$. The parabola opens down because $a = -3 < 0$. The x-coordinate of the vertex is

$$x = -\frac{b}{2a} = -\frac{9}{2(-3)} = \frac{9}{6} = \frac{3}{2}.$$

The y-coordinate of the vertex is

$$
\begin{aligned}
G\left(-\frac{b}{2a}\right) &= G\left(\frac{3}{2}\right) \\
&= -3\left(\frac{3}{2}\right)^2 + 9\left(\frac{3}{2}\right) - 8 \\
&= -\frac{27}{4} + \frac{27}{2} - 8 \\
&= -\frac{5}{4}
\end{aligned}
$$

Thus, the vertex is $\left(\frac{3}{2}, -\frac{5}{4}\right)$ and the axis of symmetry is the line $x = \frac{3}{2}$. The y-intercept is

$G(0) = -3(0)^2 + 9(0) - 8 = -8$. Now,

$b^2 - 4ac = 9^2 - 4(-3)(-8) = 81 - 96 = -15$.

Since the discriminant is negative, there are no x-intercepts. Finally, the y-intercept point, $(0, -8)$, is three-halves units to the left of the axis of symmetry. Therefore, if we move three-halves units to the right of the axis of symmetry, we obtain the point $(3, -8)$ which must also be on the graph.

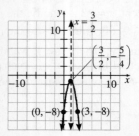

9. True. Since $a < 0$, the graph opens down. The vertex is the highest point on the graph, and therefore $f\left(-\dfrac{b}{2a}\right)$ is the maximum value of f.

10. If we compare $f(x) = 2x^2 - 8x + 1$ to
    $f(x) = ax^2 + bx + c$, we find that $a = 2$,
    $b = -8$, and $c = 1$. Because $a > 0$, we know the graph will open up, so the function will have a minimum value. The minimum value occurs at
    $$x = -\frac{b}{2a} = -\frac{(-8)}{2(2)} = 2.$$
    The minimum value of the function is
    $$f\left(-\frac{b}{2a}\right) = f(2) = 2(2)^2 - 8(2) + 1 = -7. \text{ So,}$$
    the minimum value is -7 and it occurs when $x = 2$.

11. If we compare $G(x) = -x^2 + 10x + 8$ to
    $G(x) = ax^2 + bx + c$, we find that $a = -1$,
    $b = 10$, and $c = 8$. Because $a < 0$, we know the graph will open down, so the function will have a maximum value. The maximum value occurs
    at $x = -\dfrac{b}{2a} = -\dfrac{10}{2(-1)} = 5$.
    The maximum value of the function is
    $$G\left(-\frac{b}{2a}\right) = G(5) = -5^2 + 10(5) + 8 = 33. \text{ So,}$$
    the maximum value is 33 and it occurs when $x = 5$.

12. **a.** We first recognize that the revenue function is a quadratic function whose graph opens down since $a = -0.5 < 0$. This means that the function indeed has a maximum value. The maximum value occurs when
    $$p = -\frac{b}{2a} = -\frac{75}{2(-0.5)} = 75.$$
    The revenue will be maximized when the calculators are sold at a price of \$75.

    b. The maximum revenue is obtained by evaluating the revenue function at the price found in part (a).
    $$R(75) = -0.5(75)^2 + 75(75) = 2812.5$$
    The maximum daily revenue is \$2812.50.

13. Let l = length and w = width.

The area of a rectangle is given by $A = l \cdot w$. Before we can work on maximizing the area, we need to get the function in terms of one independent variable.

The 1000 yards of fence will form the perimeter of the rectangle. That is, we have $2l + 2w = 1000$.

We can solve this equation for l and substitute the result in the area equation.

$2l + 2w = 1000$

$l + w = 500$

$l = 500 - w$

Thus, the area equation becomes

$A = l \cdot w$

$\quad = (500 - w) \cdot w$

$\quad = -w^2 + 500w$

The area function is a quadratic function whose graph opens down since $a = -1 < 0$. This means that the function indeed has a maximum. This

occurs when $w = -\dfrac{b}{2a} = -\dfrac{500}{2(-1)} = 250$.

The maximum area can be found by substituting this value for w in the area function.

$A = -250^2 + 500(250) = 62,500$ square yards

Since $l = 500 - w$ and $w = 250$, the length will be $l = 500 - 250 = 250$ yards.

The rectangular field will have a maximum area of 62,500 square yards when the field measures 250 yards by 250 yards.

14. Let x represent the number of boxes in excess of 30. Revenue is price times quantity. If 30 boxes of CDs are sold, the revenue will be $\$100(30)$.

If 31 boxes of CDs are sold, the revenue will be $\$99(31)$. If 32 boxes of CDs are sold, the

revenue will be $\$98(32)$. In general, if x boxes in excess of 30 are sold, then the number of boxes will be $30 + x$, and the price per box will be $100 - x$. Thus, revenue will be

$R(x) = (100 - x)(30 + x)$

$\quad = 3000 + 100x - 30x - x^2$

$\quad = -x^2 + 70x + 3000$

Now, this revenue function is a quadratic function whose graph opens down since $a = -1 < 0$. This means that the function indeed has a maximum value. The maximum value

occurs when $x = -\dfrac{b}{2a} = -\dfrac{70}{2(-1)} = 35$.

The maximum revenue will be

$R(35) = -35^2 + 70(35) + 3000 = 4225$.

Now recall that x is the number of boxes in excess of 30. Therefore, $30 + 30 = 65$ boxes of CDs should be sold in order to maximize revenue. The maximum revenue will be \$4225.

7.5 Exercises

15. a. $f(x) = x^2 - 6x - 16$

$x = -\dfrac{b}{2a} = -\dfrac{-6}{2} = 3$

$f(3) = (3)^2 - 6(3) - 16$

$\quad = 9 - 18 - 16$

$\quad = -25$

The vertex is $(3, -25)$.

b. $f(x) = x^2 - 6x - 16$;

$a = 1, b = -6, c = -16$;

$b^2 - 4ac = (-6)^2 - 4(1)(-16)$

$\quad = 36 + 64$

$\quad = 100$.

Because the discriminant is positive, the parabola will have two distinct x-intercepts.

The x-intercepts are:

$\qquad f(x) = 0$

$x^2 - 6x - 16 = 0$

$(x - 8)(x + 2) = 0$

$x - 8 = 0 \quad \text{or} \quad x + 2 = 0$

$\quad x = 8 \qquad\qquad x = -2$

$(8, 0)$ and $(-2, 0)$

17. a. $G(x) = -2x^2 + 4x - 5$

$x = -\dfrac{b}{2a} = -\dfrac{4}{-4} = 1$

$G(1) = -2(1)^2 + 4(1) - 5$

$\quad = -2 + 4 - 5$

$\quad = -3$

The vertex is $(1, -3)$.

b. $G(x) = -2x^2 + 4x - 5$

$a = -2,\ b = 4,\ c = -5;$

$$b^2 - 4ac = (4)^2 - 4(-2)(-5)$$
$$= 16 - 40$$
$$= -24$$

Because the discriminant is negative, the parabola will have no x-intercepts.

19. a. $h(x) = 4x^2 + 4x + 1$

$$x = -\frac{b}{2a} = -\frac{4}{8} = -\frac{1}{2}$$

$$h\left(-\frac{1}{2}\right) = 4\left(-\frac{1}{2}\right)^2 + 4\left(-\frac{1}{2}\right) + 1$$
$$= 1 - 2 + 1$$
$$= 0$$

The vertex is $\left(-\frac{1}{2}, 0\right)$.

b. $h(x) = 4x^2 + 4x + 1$;

$a = 4,\ b = 4,\ c = 1;$

$$b^2 - 4ac = 4^2 - 4(4)(1) = 16 - 16 = 0.$$

Because the discriminant is zero, the parabola will have one x-intercept.

The x-intercept is:
$$h(x) = 0$$
$$4x^2 + 4x + 1 = 0$$
$$(2x+1)^2 = 0$$
$$2x + 1 = 0$$
$$2x = -1$$
$$x = -\frac{1}{2}$$
$$\left(-\frac{1}{2}, 0\right)$$

21. a. $F(x) = 4x^2 - x - 1$

$$x = -\frac{b}{2a} = -\frac{-1}{8} = \frac{1}{8}$$

$$F\left(\frac{1}{8}\right) = 4\left(\frac{1}{8}\right)^2 - \left(\frac{1}{8}\right) - 1$$
$$= \frac{1}{16} - \frac{2}{16} - \frac{16}{16} = -\frac{17}{16}$$

The vertex is $\left(\frac{1}{8}, -\frac{17}{16}\right)$.

b. $F(x) = 4x^2 - x - 1$;

$a = 4,\ b = -1,\ c = -1;$

$$b^2 - 4ac = (-1)^2 - 4(4)(-1) = 1 + 16 = 17.$$

Because the discriminant is positive, the parabola will have two distinct x-intercepts.

The x-intercepts are:
$$F(x) = 0$$
$$4x^2 - x - 1 = 0$$

$$x = \frac{-(-1) \pm \sqrt{(-1)^2 - 4(4)(-1)}}{2(4)}$$

$$= \frac{1 \pm \sqrt{17}}{8}$$

$$x \approx -0.39 \quad \text{or} \quad x \approx 0.64$$
$(-0.39, 0)$ and $(0.64, 0)$

23. $f(x) = x^2 - 4x - 5$

$a = 1, b = -4, c = -5$

The graph opens up because $a > 0$.

<u>vertex:</u>

$$x = -\frac{b}{2a} = -\frac{(-4)}{2(1)} = 2$$

$$f(2) = (2)^2 - 4(2) - 5 = -9$$

The vertex is $(2, -9)$ and the axis of symmetry is $x = 2$.

<u>y-intercept:</u>

$$f(0) = (0)^2 - 4(0) - 5 = -5$$

<u>x-intercepts:</u>

$$b^2 - 4ac = (-4)^2 - 4(1)(-5) = 36 > 0$$

There are two distinct x-intercepts. We find these by solving

$$f(x) = 0$$
$$x^2 - 4x - 5 = 0$$
$$(x-5)(x+1) = 0$$
$$x - 5 = 0 \quad \text{or} \quad x + 1 = 0$$
$$x = 5 \quad \text{or} \quad x = -1$$

<u>Graph:</u>

The y-intercept point, $(0, -5)$, is two units to the left of the axis of symmetry. Therefore, if we move two units to the right of the axis of symmetry, we obtain the point $(4, -5)$ which must also be on the graph.

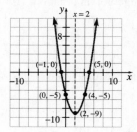

The domain is the set of all real numbers or, using interval notation, $(-\infty, \infty)$. The range is $\{y \mid y \geq -9\}$ or, using interval notation, $[-9, \infty)$.

25. $G(x) = x^2 + 12x + 32$

$a = 1, b = 12, c = 32$
The graph opens up because $a > 0$.

vertex:
$$x = -\frac{b}{2a} = -\frac{(12)}{2(1)} = -6$$
$$G(-6) = (-6)^2 + 12(-6) + 32 = -4$$
The vertex is $(-6, -4)$ and the axis of symmetry is $x = -6$.

y-intercept:
$$G(0) = (0)^2 + 12(0) + 32 = 32$$

x-intercepts:
$$b^2 - 4ac = (12)^2 - 4(1)(32) = 16 > 0$$
There are two distinct x-intercepts. We find these by solving
$$G(x) = 0$$
$$x^2 + 12x + 32 = 0$$
$$(x + 8)(x + 4) = 0$$
$$x + 8 = 0 \text{ or } x + 4 = 0$$
$$x = -8 \text{ or } x = -4$$

Graph:
The y-intercept point, $(0, 32)$, is six units to the right of the axis of symmetry. Therefore, if we move six units to the left of the axis of symmetry, we obtain the point $(-12, 32)$ which must also be on the graph.

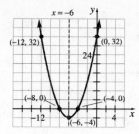

The domain is the set of all real numbers or, using interval notation, $(-\infty, \infty)$. The range is $\{y \mid y \geq -4\}$ or, using interval notation, $[-4, \infty)$.

27. $F(x) = -x^2 + 2x + 8$

$a = -1, b = 2, c = 8$
The graph opens down because $a < 0$.

vertex:
$$x = -\frac{b}{2a} = -\frac{(2)}{2(-1)} = 1$$
$$F(1) = -(1)^2 + 2(1) + 8 = 9$$
The vertex is $(1, 9)$ and the axis of symmetry is $x = 1$.

y-intercept:
$$F(0) = -(0)^2 + 2(0) + 8 = 8$$

x-intercepts:
$$b^2 - 4ac = (2)^2 - 4(-1)(8) = 36 > 0$$
There are two distinct x-intercepts. We find these by solving
$$F(x) = 0$$
$$-x^2 + 2x + 8 = 0$$
$$x^2 - 2x - 8 = 0$$
$$(x + 2)(x - 4) = 0$$
$$x + 2 = 0 \text{ or } x - 4 = 0$$
$$x = -2 \text{ or } x = 4$$

Graph:
The y-intercept point, $(0, 8)$, is one unit to the left of the axis of symmetry. Therefore, if we move one unit to the right of the axis of symmetry, we obtain the point $(2, 8)$ which must also be on the graph.

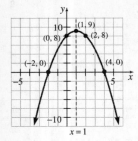

The domain is the set of all real numbers or, using interval notation, $(-\infty, \infty)$. The range is $\{y \mid y \le 9\}$ or, using interval notation, $(-\infty, 9]$.

29. $H(x) = x^2 - 4x + 4$

$a = 1, b = -4, c = 4$
The graph opens up because $a > 0$.

vertex:
$$x = -\frac{b}{2a} = -\frac{(-4)}{2(1)} = 2$$

$$H(2) = (2)^2 - 4(2) + 4 = 0$$

The vertex is $(2, 0)$ and the axis of symmetry is $x = 2$.

y-intercept:
$$H(0) = (0)^2 - 4(0) + 4 = 4$$

x-intercepts:
Since the discriminant is 0, the x-coordinate of the vertex is the only x-intercept, $x = 2$.

Graph:
The y-intercept point, $(0, 4)$, is two units to the left of the axis of symmetry. Therefore, if we move two units to the right of the axis of symmetry, we obtain the point $(4, 4)$ which must also be on the graph.

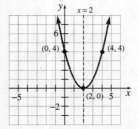

The domain is the set of all real numbers or, using interval notation, $(-\infty, \infty)$. The range is $\{y \mid y \ge 0\}$ or, using interval notation, $[0, \infty)$.

31. $g(x) = x^2 + 2x + 5$

$a = 1, b = 2, c = 5$
The graph opens up because $a > 0$.

vertex:
$$x = -\frac{b}{2a} = -\frac{(2)}{2(1)} = -1$$

$$g(-1) = (-1)^2 + 2(-1) + 5 = 4$$

The vertex is $(-1, 4)$ and the axis of symmetry is $x = -1$.

y-intercept:
$$g(0) = (0)^2 + 2(0) + 5 = 5$$

x-intercepts:
$$b^2 - 4ac = (2)^2 - 4(1)(5) = -16 < 0$$

There are no x-intercepts since the discriminant is negative.

Graph:
The y-intercept point, $(0, 5)$, is one unit to the right of the axis of symmetry. Therefore, if we move one unit to the left of the axis of symmetry, we obtain the point $(-2, 5)$ which must also be on the graph.

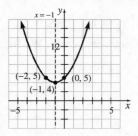

The domain is the set of all real numbers or, using interval notation, $(-\infty, \infty)$. The range is $\{y \mid y \ge 4\}$ or, using interval notation, $[4, \infty)$.

33. $h(x) = -x^2 - 10x - 25$

$a = -1, b = -10, c = -25$

The graph opens down because $a < 0$.

<u>vertex:</u>

$$x = -\frac{b}{2a} = -\frac{(-10)}{2(-1)} = -5$$

$$h(-5) = -(-5)^2 - 10(-5) - 25 = 0$$

The vertex is $(-5, 0)$ and the axis of symmetry is $x = -5$.

<u>y-intercept:</u>

$$h(0) = -(0)^2 - 10(0) - 25 = -25$$

<u>x-intercepts:</u>

$$b^2 - 4ac = (-10)^2 - 4(-1)(-25) = 100 - 100 = 0$$

Since the discriminant is 0, the x-coordinate of the vertex is the only x-intercept. $x = -5$.

<u>Graph:</u>

The y-intercept point, $(0, -25)$, is five units to the right of the axis of symmetry. Therefore, if we move five units to the left of the axis of symmetry, we obtain the point $(-10, -25)$ which must also be on the graph.

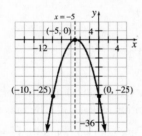

The domain is the set of all real numbers or, using interval notation, $(-\infty, \infty)$. The range is $\{y \mid y \le 0\}$ or, using interval notation, $(-\infty, 0]$.

35. $p(x) = -x^2 + 2x - 5$

$a = -1, b = 2, c = -5$

The graph opens down because $a < 0$.

<u>vertex:</u>

$$x = -\frac{b}{2a} = -\frac{(2)}{2(-1)} = 1$$

$$p(1) = -(1)^2 + 2(1) - 5 = -4$$

The vertex is $(1, -4)$ and the axis of symmetry is $x = 1$.

<u>y-intercept:</u>

$$p(0) = -(0)^2 + 2(0) - 5 = -5$$

<u>x-intercepts:</u>

$$b^2 - 4ac = (2)^2 - 4(-1)(-5) = -16 < 0$$

There are no x-intercepts since the discriminant is negative.

<u>Graph:</u>

The y-intercept point, $(0, -5)$, is one unit to the left of the axis of symmetry. Therefore, if we move one unit to the right of the axis of symmetry, we obtain the point $(2, -5)$ which must also be on the graph.

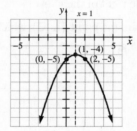

The domain is the set of all real numbers or, using interval notation, $(-\infty, \infty)$. The range is $\{y \mid y \le -4\}$ or, using interval notation, $(-\infty, -4]$.

37. $F(x) = 4x^2 - 4x - 3$

$a = 4, b = -4, c = -3$

The graph opens up because $a > 0$.

<u>vertex:</u>

$$x = -\frac{b}{2a} = -\frac{(-4)}{2(4)} = \frac{1}{2}$$

$$F\left(\frac{1}{2}\right) = 4\left(\frac{1}{2}\right)^2 - 4\left(\frac{1}{2}\right) - 3 = -4$$

The vertex is $\left(\frac{1}{2}, -4\right)$ and the axis of symmetry is $x = 2$.

y-intercept:

$$F(0) = 4(0)^2 - 4(0) - 3 = -3$$

x-intercepts:

$$b^2 - 4ac = (-4)^2 - 4(4)(-3) = 64 > 0$$

There are two distinct *x*-intercepts. We find these by solving

$$F(x) = 0$$
$$4x^2 - 4x - 3 = 0$$
$$(2x+1)(2x-3) = 0$$
$$2x+1 = 0 \quad \text{or} \quad 2x-3 = 0$$
$$2x = -1 \quad \text{or} \quad 2x = 3$$
$$x = -\frac{1}{2} \quad \text{or} \quad x = \frac{3}{2}$$

Graph:

The *y*-intercept point, $(0,-3)$, is $\frac{1}{2}$ unit to the left of the axis of symmetry. Therefore, if we move $\frac{1}{2}$ unit to the right of the axis of symmetry, we obtain the point $(1,-3)$ which must also be on the graph.

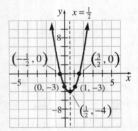

The domain is the set of all real numbers or, using interval notation, $(-\infty, \infty)$. The range is $\{y \mid y \geq -4\}$ or, using interval notation, $[-4, \infty)$.

39. $G(x) = -9x^2 + 18x + 7$

$a = -9, b = 18, c = 7$

The graph opens down because $a < 0$.

vertex:

$$x = -\frac{b}{2a} = -\frac{(18)}{2(-9)} = 1$$

$$G(1) = -9(1)^2 + 18(1) + 7 = 16$$

The vertex is $(1,16)$ and the axis of symmetry is $x = 1$.

y-intercept:

$$G(0) = -9(0)^2 + 18(0) + 7 = 7$$

x-intercepts:

$$b^2 - 4ac = (18)^2 - 4(-9)(7) = 576 > 0$$

There are two distinct *x*-intercepts. We find these by solving

$$G(x) = 0$$
$$-9x^2 + 18x + 7 = 0$$
$$9x^2 - 18x - 7 = 0$$
$$(3x+1)(3x-7) = 0$$
$$3x+1 = 0 \quad \text{or} \quad 3x-7 = 0$$
$$3x = -1 \quad \text{or} \quad 3x = 7$$
$$x = -\frac{1}{3} \quad \text{or} \quad x = \frac{7}{3}$$

Graph:

The *y*-intercept point, $(0,7)$, is one unit to the left of the axis of symmetry. Therefore, if we move one unit to the right of the axis of symmetry, we obtain the point $(2,7)$ which must also be on the graph.

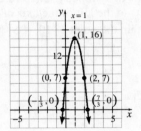

The domain is the set of all real numbers or, using interval notation, $(-\infty, \infty)$. The range is $\{y \mid y \leq 16\}$ or, using interval notation, $(-\infty, 16]$.

41. $H(x) = 4x^2 - 4x + 1$

$a = 4, b = -4, c = 1$

The graph opens up because $a > 0$.

vertex:

$$x = -\frac{b}{2a} = -\frac{(-4)}{2(4)} = \frac{1}{2}$$

$$H\left(\tfrac{1}{2}\right) = 4\left(\tfrac{1}{2}\right)^2 - 4\left(\tfrac{1}{2}\right) + 1 = 0$$

The vertex is $\left(\frac{1}{2},0\right)$ and the axis of symmetry is $x=\frac{1}{2}$.

y-intercept:
$$H(0)=4(0)^2-4(0)+1=1$$

x-intercepts:
$$b^2-4ac=(-4)^2-4(4)(1)=0$$
Since the discriminant is 0, the x-coordinate of the vertex is the only x-intercept, $x=\frac{1}{2}$.

Graph:
The y-intercept point, $(0,1)$, is $\frac{1}{2}$ unit to the left of the axis of symmetry. If we move $\frac{1}{2}$ unit to the right of the axis of symmetry, we obtain the point $(1,1)$ which must also be on the graph.

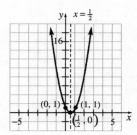

The domain is the set of all real numbers or, using interval notation, $(-\infty,\infty)$. The range is $\{y\mid y\ge 0\}$ or, using interval notation, $[0,\infty)$.

43. $f(x)=-16x^2-24x-9$
$a=-16, b=-24, c=-9$
The graph opens down because $a<0$.

vertex:
$$x=-\frac{b}{2a}=-\frac{(-24)}{2(-16)}=-\frac{3}{4}$$
$$f\left(-\frac{3}{4}\right)=-16\left(-\frac{3}{4}\right)^2-24\left(-\frac{3}{4}\right)-9=0$$
The vertex is $\left(-\frac{3}{4},0\right)$ and the axis of symmetry is $x=-\frac{3}{4}$.

y-intercept:
$$f(0)=-16(0)^2-24(0)-9=-9$$

x-intercepts:
$$b^2-4ac=(-24)^2-4(-16)(-9)=0$$
Since the discriminant is 0, the only x-intercept is the x-coordinate of the vertex, $x=-\frac{3}{4}$.

Graph:
The y-intercept point, $(0,-9)$, is $\frac{3}{4}$ unit to the right of the axis of symmetry. Therefore, if we move $\frac{3}{4}$ unit to the left of the axis of symmetry, we obtain the point $\left(-\frac{3}{2},-9\right)$ which must also be on the graph.

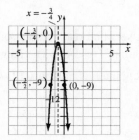

The domain is the set of all real numbers or, using interval notation, $(-\infty,\infty)$. The range is $\{y\mid y\le 0\}$ or, using interval notation, $(-\infty,0]$.

45. $f(x)=2x^2+8x+11$
$a=2, b=8, c=11$
The graph opens up because $a>0$.

vertex:
$$x=-\frac{b}{2a}=-\frac{(8)}{2(2)}=-2$$
$$f(-2)=2(-2)^2+8(-2)+11=3$$
The vertex is $(-2,3)$ and the axis of symmetry is $x=-2$.

y-intercept:
$$f(0)=2(0)^2+8(0)+11=11$$

x-intercepts:
$$b^2-4ac=(8)^2-4(2)(11)=-24<0$$
There are no x-intercepts since the discriminant is negative.

Graph:

The y-intercept point, $(0,11)$, is two units to the right of the axis of symmetry. Therefore, if we move two units to the left of the axis of symmetry, we obtain the point $(-4,11)$ which must also be on the graph.

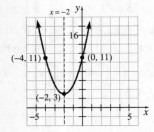

The domain is the set of all real numbers or, using interval notation, $(-\infty,\infty)$. The range is $\{y \mid y \geq 3\}$ or, using interval notation, $[3,\infty)$.

47. $P(x) = -4x^2 + 6x - 3$

$a = -4, b = 6, c = -3$

The graph opens down because $a < 0$.

vertex:

$$x = -\frac{b}{2a} = -\frac{(6)}{2(-4)} = \frac{3}{4}$$

$$P\left(\tfrac{3}{4}\right) = -4\left(\tfrac{3}{4}\right)^2 + 6\left(\tfrac{3}{4}\right) - 3 = -\frac{3}{4}$$

The vertex is $\left(\tfrac{3}{4}, -\tfrac{3}{4}\right)$ and the axis of symmetry is $x = \tfrac{3}{4}$.

y-intercept:

$$P(0) = -4(0)^2 + 6(0) - 3 = -3$$

x-intercepts:

$$b^2 - 4ac = (6)^2 - 4(-4)(-3) = -12 < 0$$

There are no x-intercepts since the discriminant is negative.

Graph:

The y-intercept point, $(0,-3)$, is $\tfrac{3}{4}$ unit to the left of the axis of symmetry. Therefore, if we move $\tfrac{3}{4}$ unit to the right of the axis of symmetry, we obtain the point $\left(\tfrac{3}{2}, -3\right)$ which must also be on the graph.

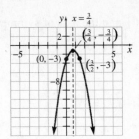

The domain is the set of all real numbers or, using interval notation, $(-\infty,\infty)$. The range is $\left\{y \mid y \leq -\dfrac{3}{4}\right\}$ or, using interval notation, $\left(-\infty, -\dfrac{3}{4}\right]$.

49. $h(x) = x^2 + 5x + 3$

$a = 1, b = 5, c = 3$

The graph opens up because $a > 0$.

vertex:

$$x = -\frac{b}{2a} = -\frac{(5)}{2(1)} = -\frac{5}{2}$$

$$h\left(-\tfrac{5}{2}\right) = \left(-\tfrac{5}{2}\right)^2 + 5\left(-\tfrac{5}{2}\right) + 3 = -\frac{13}{4}$$

The vertex is $\left(-\tfrac{5}{2}, -\tfrac{13}{4}\right)$ and the axis of symmetry is $x = -\tfrac{5}{2}$.

y-intercept:

$$h(0) = (0)^2 + 5(0) + 3 = 3$$

x-intercepts:

$$b^2 - 4ac = (5)^2 - 4(1)(3) = 13 > 0$$

There are two distinct x-intercepts. We find these by solving

$$h(x) = 0$$

$$x^2 + 5x + 3 = 0$$

$$x = \frac{-5 \pm \sqrt{13}}{2}$$

$$x \approx -0.70 \quad \text{or} \quad x \approx -4.30$$

Graph:

The y-intercept point, $(0,3)$, is $\tfrac{5}{2}$ units to the right of the axis of symmetry. Therefore, if we move $\tfrac{5}{2}$ units to the left of the axis of symmetry,

Copyright © 2014 Pearson Education, Inc.

we obtain the point $(-5, 3)$ which must also be on the graph.

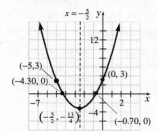

The domain is the set of all real numbers or, using interval notation, $(-\infty, \infty)$. The range is $\left\{ y \mid y \geq -\dfrac{13}{4} \right\}$ or, using interval notation, $\left[-\dfrac{13}{4}, \infty \right)$.

51. $G(x) = -3x^2 + 8x + 2$

$a = -3, b = 8, c = 2$

The graph opens down because $a < 0$.

vertex:

$$x = -\frac{b}{2a} = -\frac{(8)}{2(-3)} = \frac{4}{3}$$

$$G\left(\frac{4}{3}\right) = -3\left(\frac{4}{3}\right)^2 + 8\left(\frac{4}{3}\right) + 2 = \frac{22}{3}$$

The vertex is $\left(\frac{4}{3}, \frac{22}{3}\right)$ and the axis of symmetry is $x = \frac{4}{3}$.

y-intercept:

$$G(0) = -3(0)^2 + 8(0) + 2 = 2$$

x-intercepts:

$$b^2 - 4ac = (8)^2 - 4(-3)(2) = 88 > 0$$

There are two distinct x-intercepts. We find these by solving

$$G(x) = 0$$

$$-3x^2 + 8x + 2 = 0$$

$$x = \frac{-8 \pm \sqrt{88}}{2(-3)} = \frac{-8 \pm 2\sqrt{22}}{-6} = \frac{4 \pm \sqrt{22}}{3}$$

$$x \approx -0.23 \quad \text{or} \quad x \approx 2.90$$

Graph:

The y-intercept point, $(0, 2)$, is $\frac{4}{3}$ units to the left of the axis of symmetry. Therefore, if we move $\frac{4}{3}$ units to the right of the axis of symmetry, we obtain the point $\left(\frac{8}{3}, 2\right)$ which must also be on the graph.

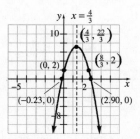

The domain is the set of all real numbers or, using interval notation, $(-\infty, \infty)$. The range is $\left\{ y \mid y \leq \dfrac{22}{3} \right\}$ or, using interval notation, $\left(-\infty, \dfrac{22}{3} \right]$.

53. $f(x) = 5x^2 - 5x + 2$

$a = 5, b = -5, c = 2$

The graph opens up because $a > 0$.

vertex:

$$x = -\frac{b}{2a} = -\frac{(-5)}{2(5)} = \frac{1}{2}$$

$$f\left(\frac{1}{2}\right) = 5\left(\frac{1}{2}\right)^2 - 5\left(\frac{1}{2}\right) + 2 = \frac{3}{4}$$

The vertex is $\left(\frac{1}{2}, \frac{3}{4}\right)$ and the axis of symmetry is $x = \frac{1}{2}$.

y-intercept:

$$f(0) = 5(0)^2 - 5(0) + 2 = 2$$

x-intercepts:

$$b^2 - 4ac = (-5)^2 - 4(5)(2) = -15 < 0$$

There are no x-intercepts since the discriminant is negative.

Graph:

The *y*-intercept point, $(0,2)$, is $\frac{1}{2}$ unit to the left of the axis of symmetry. Therefore, if we move one unit to the right of the axis of symmetry, we obtain the point $(1,2)$ which must also be on the graph.

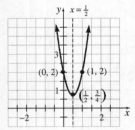

The domain is the set of all real numbers or, using interval notation, $(-\infty,\infty)$. The range is

$\left\{y \mid y \geq \dfrac{3}{4}\right\}$ or, using interval notation, $\left[\dfrac{3}{4},\infty\right)$.

55. $H(x)=-3x^2+6x$

$a=-3, b=6, c=0$

The graph opens down because $a<0$.

vertex:

$$x=-\frac{b}{2a}=-\frac{(6)}{2(-3)}=1$$

$$H(1)=-3(1)^2+6(1)=3$$

The vertex is $(1,3)$ and the axis of symmetry is $x=1$.

y-intercept:

$$H(0)=-3(0)^2+6(0)=0$$

x-intercepts:

$$b^2-4ac=(6)^2-4(-3)(0)=36>0$$

There are two distinct *x*-intercepts. We find these by solving

$$H(x)=0$$
$$-3x^2+6x=0$$
$$-3x(x-2)=0$$
$$-3x=0 \text{ or } x-2=0$$
$$\quad x=0 \text{ or } x=2$$

Graph:

The *y*-intercept point, $(0,0)$, is one unit to the left of the axis of symmetry. Therefore, if we move one unit to the right of the axis of symmetry, we obtain the point $(2,0)$ which must also be on the graph (these points are actually the *x*-intercept points).

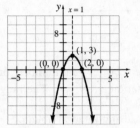

The domain is the set of all real numbers or, using interval notation, $(-\infty,\infty)$. The range is

$\{y \mid y \leq 3\}$ or, using interval notation, $(-\infty,3]$.

57. $f(x)=x^2-\dfrac{5}{2}x-\dfrac{3}{2}$

$a=1, b=-\frac{5}{2}, c=-\frac{3}{2}$

The graph opens up because $a>0$.

vertex:

$$x=-\frac{b}{2a}=-\frac{\left(-\frac{5}{2}\right)}{2(1)}=\frac{5}{4}$$

$$f\left(\tfrac{5}{4}\right)=\left(\tfrac{5}{4}\right)^2-\tfrac{5}{2}\left(\tfrac{5}{4}\right)-\tfrac{3}{2}=-\frac{49}{16}$$

The vertex is $\left(\tfrac{5}{4},-\tfrac{49}{16}\right)$ and the axis of symmetry is $x=\tfrac{5}{4}$.

y-intercept:

$$f(0)=(0)^2-\frac{5}{2}(0)-\frac{3}{2}=-\frac{3}{2}$$

x-intercepts:

$$b^2-4ac=\left(-\tfrac{5}{2}\right)^2-4(1)\left(-\tfrac{3}{2}\right)=\tfrac{49}{4}>0$$

There are two distinct *x*-intercepts. We find these by solving

$$f(x) = 0$$

$$x^2 - \frac{5}{2}x - \frac{3}{2} = 0$$

$$2x^2 - 5x - 3 = 0$$

$$(2x + 1)(x - 3) = 0$$

$$2x + 1 = 0 \text{ or } x - 3 = 0$$

$$2x = -1 \text{ or } x = 3$$

$$x = -\frac{1}{2} \text{ or } x = 3$$

Graph:

The *y*-intercept point, $\left(0, -\frac{3}{2}\right)$, is $\frac{5}{4}$ units to the left of the axis of symmetry. Therefore, if we move $\frac{5}{4}$ units to the right of the axis of symmetry, we obtain the point $\left(\frac{5}{2}, -\frac{3}{2}\right)$ which must also be on the graph.

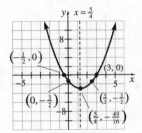

The domain is the set of all real numbers or, using interval notation, $(-\infty, \infty)$. The range is

$$\left\{ y \mid y \geq -\frac{49}{16} \right\} \text{ or, using interval notation,}$$

$$\left[-\frac{49}{16}, \infty \right).$$

59. $G(x) = \frac{1}{2}x^2 + 2x - 6$

$$a = \frac{1}{2}, b = 2, c = -6$$

The graph opens up because $a > 0$.

vertex:

$$x = -\frac{b}{2a} = -\frac{(2)}{2\left(\frac{1}{2}\right)} = -2$$

$$G(-2) = \frac{1}{2}(-2)^2 + 2(-2) - 6 = -8$$

The vertex is $(-2, -8)$ and the axis of symmetry is $x = -2$.

y-intercept:

$$G(0) = \frac{1}{2}(0)^2 + 2(0) - 6 = -6$$

x-intercepts:

$$b^2 - 4ac = (2)^2 - 4\left(\frac{1}{2}\right)(-6) = 16 > 0$$

There are two distinct *x*-intercepts. We find these by solving

$$G(x) = 0$$

$$\frac{1}{2}x + 2x - 6 = 0$$

$$x^2 + 4x - 12 = 0$$

$$(x + 6)(x - 2) = 0$$

$$x + 6 = 0 \text{ or } x - 2 = 0$$

$$x = -6 \text{ or } x = 2$$

Graph:

The *y*-intercept point, $(0, -6)$, is two units to the right of the axis of symmetry. Therefore, if we move two units to the left of the axis of symmetry, we obtain the point $(-4, -6)$ which must also be on the graph.

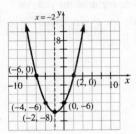

The domain is the set of all real numbers or, using interval notation, $(-\infty, \infty)$. The range is $\{ y \mid y \geq -8 \}$ or, using interval notation, $[-8, \infty)$.

61. $F(x) = -\frac{1}{4}x^2 + x + 15$

$$a = -\frac{1}{4}, b = 1, c = 15$$

The graph opens down because $a < 0$.

vertex:

$$x = -\frac{b}{2a} = -\frac{(1)}{2\left(-\frac{1}{4}\right)} = 2$$

$$F(2) = -\frac{1}{4}(2)^2 + (2) + 15 = 16$$

The vertex is $(2,16)$ and the axis of symmetry is $x = 2$.

y-intercept:

$$F(0) = -\frac{1}{4}(0)^2 + (0) + 15 = 15$$

x-intercepts:

$$b^2 - 4ac = (1)^2 - 4\left(-\frac{1}{4}\right)(15) = 16 > 0$$

There are two distinct *x*-intercepts. We find these by solving

$$F(x) = 0$$
$$-\frac{1}{4}x^2 + x + 15 = 0$$
$$x^2 - 4x - 60 = 0$$
$$(x-10)(x+6) = 0$$
$$x - 10 = 0 \text{ or } x + 6 = 0$$
$$x = 10 \text{ or } x = -6$$

Graph:

The *y*-intercept point, $(0,15)$, is two units to the left of the axis of symmetry. Therefore, if we move two units to the right of the axis of symmetry, we obtain the point $(4,15)$ which must also be on the graph.

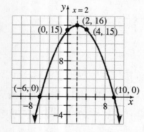

The domain is the set of all real numbers or, using interval notation, $(-\infty, \infty)$. The range is $\{y \mid y \le 16\}$ or, using interval notation, $(-\infty, 16]$.

63. If we compare $f(x) = x^2 + 8x + 13$ to

$f(x) = ax^2 + bx + c$, we find that $a = 1$, $b = 8$, and $c = 13$. Because $a > 0$, we know the graph will open up, so the function will have a minimum value.
The minimum value occurs at

$$x = -\frac{b}{2a} = -\frac{8}{2(1)} = -4.$$

The minimum value is

$$f(-4) = (-4)^2 + 8(-4) + 13 = -3.$$

So, the minimum value is -3 and it occurs when $x = -4$.

65. If we compare $G(x) = -x^2 - 10x + 3$ to

$G(x) = ax^2 + bx + c$, we find that $a = -1$, $b = -10$, and $c = 3$. Because $a < 0$, we know the graph will open down, so the function will have a maximum value.
The maximum value occurs at

$$x = -\frac{b}{2a} = -\frac{(-10)}{2(-1)} = -5.$$

The maximum value is

$$G(-5) = -(-5)^2 - 10(-5) + 3 = 28.$$

So, the maximum value is 28 and it occurs when $x = -5$.

67. If we compare $F(x) = -2x^2 + 12x + 5$ to

$F(x) = ax^2 + bx + c$, we find that $a = -2$, $b = 12$, and $c = 5$. Because $a < 0$, we know the graph will open down, so the function will have a maximum value.
The maximum value occurs at

$$x = -\frac{b}{2a} = -\frac{12}{2(-2)} = 3.$$

The maximum value is

$$F(3) = -2(3)^2 + 12(3) + 5 = 23.$$

So, the maximum value is 23 and it occurs when $x = 3$.

69. If we compare $h(x) = 4x^2 + 16x - 3$ to

$h(x) = ax^2 + bx + c$, we find that $a = 4$, $b = 16$, and $c = -3$. Because $a > 0$, we know the graph will open up, so the function will have a minimum value.
The minimum value occurs at

$$x = -\frac{b}{2a} = -\frac{16}{2(4)} = -2.$$

The minimum value is

$$h(-2) = 4(-2)^2 + 16(-2) - 3 = -19.$$

So, the minimum value is -19 and it occurs when $x = -2$.

71. If we compare $f(x) = 2x^2 - 5x + 1$ to

$f(x) = ax^2 + bx + c$, we find that $a = 2$,

$b = -5$, and $c = 1$. Because $a > 0$, we know the graph will open up, so the function will have a minimum value.

The minimum value occurs at

$x = -\dfrac{b}{2a} = -\dfrac{(-5)}{2(2)} = \dfrac{5}{4}$ or 1.25.

The minimum value is

$f\left(\dfrac{5}{4}\right) = 2\left(\dfrac{5}{4}\right)^2 - 5\left(\dfrac{5}{4}\right) + 1 = -\dfrac{17}{8}$ or -2.125.

So, the minimum value is $-\dfrac{17}{8}$ and it occurs

when $x = \dfrac{5}{4}$.

73. If we compare $H(x) = -3x^2 + 4x + 1$ to

$H(x) = ax^2 + bx + c$, we find that $a = -3$,

$b = 4$, and $c = 1$. Because $a < 0$, we know the graph will open down, so the function will have a maximum value.

The maximum value occurs at

$x = -\dfrac{b}{2a} = -\dfrac{4}{2(-3)} = \dfrac{2}{3}$.

The maximum value is

$H\left(\dfrac{2}{3}\right) = -3\left(\dfrac{2}{3}\right)^2 + 4\left(\dfrac{2}{3}\right) + 1 = \dfrac{7}{3}$.

So, the maximum value is $\dfrac{7}{3}$ and it occurs

when $x = \dfrac{2}{3}$.

75. a. We first recognize that the quadratic function has the leading coefficient $a = -2.5 < 0$. This means that the graph will open down and the function indeed has a maximum value. The maximum value

occurs when $p = -\dfrac{b}{2a} = -\dfrac{600}{2(-2.5)} = 120$.

The revenue will be maximized when the DVD players are sold at a price of $120.

b. The maximum revenue is obtained by evaluating the revenue function at the price found in part (a).

$R(120) = -2.5(120)^2 + 600(120) = 36{,}000$

The maximum revenue is $36,000.

77. First we recognize that the quadratic function has the leading coefficient $a = 0.05 > 0$. This means that the graph will open up and the function will indeed have a minimum value. The minimum

value occurs when $x = -\dfrac{b}{2a} = -\dfrac{(-6)}{2(0.05)} = 60$

The marginal cost will be minimized when 60 digital cameras are produced.

To find the minimum marginal cost, we evaluate the marginal cost function for $x = 60$.

$C(60) = 0.05(60)^2 - 6(60) + 215 = 35$

The minimum marginal cost is $35.

79. a. First we recognize that the quadratic function has the negative leading coefficient $a = -16 < 0$. This means the graph will open down and the function will indeed have a maximum.

The maximum height will occur when

$t = -\dfrac{b}{2a} = -\dfrac{240}{2(-16)} = 7.5$.

The pumpkin will reach a maximum height after 7.5 seconds.

b. The maximum height can be found by evaluating $s(t)$ for the value of t found in part (a).

$s(7.5) = -16(7.5)^2 + 240(7.5) + 10 = 910$

The pumpkin will reach a maximum height of 910 feet.

c. When the pumpkin is on the ground, it will have a height of 0 feet. Thus, we need to solve $s(t) = 0$.

$-16t^2 + 240t + 10 = 0$

$8t^2 - 120t - 5 = 0$

$a = 8, b = -120, c = -5$

$t = \dfrac{-(-120) \pm \sqrt{(-120)^2 - 4(8)(-5)}}{2(8)}$

$= \dfrac{120 \pm \sqrt{14560}}{16}$

$t \approx -0.042$ or $t \approx 15.042$

Since the time of flight cannot be negative, we discard the negative solution. The pumpkin will hit the ground in about 15.042 seconds.

81. a. We first recognize that the quadratic function has the leading coefficient $a = \dfrac{-32}{335^2} < 0$. This means the graph will open down and the function will indeed have a maximum.
The maximum height will occur when
$$x = -\frac{b}{2a} = -\frac{1}{2\left(-32/335^2\right)} \approx 1753.52 \,.$$
The pumpkin will reach a maximum height when it is about 1753.52 feet from the cannon.

b. The maximum height is obtained by evaluating $h(x)$ for the value of x found in part (a).
$$h(1753.52) = \frac{-32}{335^2}(1753.52)^2 + (1753.52) + 10$$
$$\approx 886.76$$
The pumpkin will reach a maximum height of about 886.76 feet.

c. When the pumpkin is on the ground, it will have a height of 0 feet. Thus, we need to solve $h(x) = 0$.
$$\frac{-32}{335^2}x^2 + x + 10 = 0$$
$$a = \frac{-32}{335^2}, \; b = 1, \; c = 10$$
$$x = \frac{-1 \pm \sqrt{1^2 - 4\left(\dfrac{-32}{335^2}\right)(10)}}{2\left(\dfrac{-32}{335^2}\right)}$$
$$x \approx -9.97 \quad \text{or} \quad x \approx 3517.00$$
Since the horizontal distance traveled cannot be negative, we discard the negative result. The pumpkin will hit the ground at a distance of about 3517 feet from the cannon.

d. The two answers are close. The difference is because the initial velocity component used in the formula for problem 73 was an approximation while the value used in the formula for this problem was exact. The exact formula for problem 73 would be
$$s(t) = -16t^2 + \frac{335}{\sqrt{2}}t + 10 \,.$$
If we had used this formula, the two results would be the same.

Note: The answer in part (b) of this problem is also an approximation because we approximated the time in part (a).

83. a. We first recognize that the quadratic function has the leading coefficient $a = -55 < 0$. This means the graph will open down and the function will indeed have a maximum.
The maximum value occurs when
$$x = -\frac{5119}{2(-55)} = \frac{5119}{110} \approx 46.5 \,.$$
Average income will be maximized at an age of about 46.5 years.

b. To determine the maximum average income, we evaluate $I(a)$ for the value of a found in part (a).
$$I(46.5)$$
$$= -55(46.5)^2 + 5119(46.5) - 54,448$$
$$= 64,661.75$$
The maximum average income is about $64,661.75.

85. Let x represent the first number. Then the second number must be $36 - x$. We can express the product of the two numbers as the function
$$p(x) = x(36 - x) = -x^2 + 36x$$
This is a quadratic function with $a = -1$, $b = 36$, and $c = 0$. The function is maximized when $x = -\dfrac{b}{2a} = -\dfrac{36}{2(-1)} = 18$.
The maximum product can be obtained by evaluating $p(x)$ when $x = 18$.
$$p(x) = 18(36 - 18) = 18(18) = 324$$
Two numbers that sum to 36 have a maximum product of 324 when both numbers are 18.

87. Let x represent the smaller number. Then the larger number must be $x + 18$. We can express the product of the two numbers as the function
$$p(x) = x(x + 18) = x^2 + 18x \,.$$
This is a quadratic function with $a = 1$, $b = 18$, and $c = 0$. The product will be a minimum when
$$x = -\frac{b}{2a} = -\frac{18}{2(1)} = -9 \,.$$
The minimum product can be found by evaluating $p(x)$ when $x = -9$.

$$p(-9) = -9(-9+18) = -9(9) = -81$$

Two numbers whose difference is 18 have a minimum product of −81 when the smaller number is −9 and the larger number is 9.

89. Let l = length and w = width.

The area of a rectangle is given by $A = l \cdot w$. Before we can work on maximizing area, we need to get the equation in terms of one independent variable.

The 500 yards of fencing will form the perimeter of the rectangle. That is, we have $2l + 2w = 500$.

We can solve this equation for l and substitute the result in the area equation.

$$2l + 2w = 500$$
$$l + w = 250$$
$$l = 250 - w$$

Thus, the area equation becomes

$$A = l \cdot w$$
$$= (250 - w) \cdot w$$
$$= -w^2 + 250w$$

Since $a = -1 < 0$, we know the graph opens down, so there will be a maximum area. This occurs when $w = -\dfrac{b}{2a} = -\dfrac{250}{2(-1)} = 125$.

The maximum area can be found by substituting this value for w in the area equation.

$$A = 125(250 - 125) = 125(125) = 15,625$$

The rectangular field will have a maximum area of 15,625 square yards when the field measures 125 yards × 125 yards.

91. The area of a rectangular region is the product of the length and width. From the figure, we can see that the area would be

$$A = (2000 - 2x) \cdot x = -2x^2 + 2000x$$

Since $a = -2 < 0$, we know the graph opens down and there will be a maximum area. This value occurs when

$$x = -\frac{b}{2a} = -\frac{2000}{2(-2)} = 500.$$

The maximum area can be found by substituting this value for x into the area equation.

$$A = 500(2000 - 2(500)) = 500,000$$

The rectangular field will have a maximum area of 500,000 square meters when the field measures 500 m × 1000 m and the long side is parallel to the river.

93. From the diagram, we can see that the cross-sectional area will be rectangular with a width of x inches and a length of $20 - 2x$ inches. The cross-sectional area is then

$$A = (20 - 2x) \cdot x = -2x^2 + 20x$$

Since $a = -2 < 0$, the graph will open down and the function will have a maximum value. This value occurs when

$$x = -\frac{b}{2a} = -\frac{20}{2(-2)} = 5$$

The maximum cross-sectional area can be found by substituting this value for x in the area equation.

$$A = (20 - 2(5)) \cdot 5 = 10 \cdot 5 = 50$$

The gutter will have a maximum cross-sectional area of 50 square inches if the gutter has a depth of 5 inches.

95. a. Since $R = x \cdot p$, we have

$$R = (-p + 110) \cdot p = -p^2 + 110p$$

b. The revenue function is quadratic with $a = -1 < 0$. This means the graph will open down and the function will have a maximum value. This value occurs when

$$p = -\frac{b}{2a} = -\frac{110}{2(-1)} = 55.$$

The maximum revenue can be found by substituting this value for p in the revenue equation.

$$R = (-(55) + 110) \cdot (55) = 55 \cdot 55 = 3025$$

There will be a maximum revenue of $3025 if the price is set at $55 for each pair of jeans.

c. To determine how many pairs of jeans will be sold, we substitute the maximizing price into the demand equation.

$$x = -p + 110$$
$$= -(55) + 110 = 55$$

When the price per pair is $55, the department store will sell 55 pairs.

97. a. $a = 1$:

$$f(x) = 1(x - 2)(x - 6) = x^2 - 8x + 12$$

$a = 2$:

$$f(x) = 2(x - 2)(x - 6) = 2x^2 - 16x + 24$$

$a = -2$:
$$f(x) = -2(x-2)(x-6) = -2x^2 + 16x - 24$$

b. The value of a has no effect on the x-intercepts. These depend only on the factors and are given in the problem. The value of a does have an effect on the y-intercept which can be expressed as $c = a \cdot r_1 \cdot r_2 = 12a$.

c. The value of a has no effect on the axis of symmetry. The axis of symmetry lies halfway between the two x-intercepts which are fixed. Note that in this case we have
$$f(x) = ax^2 - 8ax + 12a$$
The axis of symmetry would be
$$x = -\frac{(-8a)}{2(a)} = \frac{8a}{2a} = 4$$
which does not depend on a.

d. Consider the general function in this case written in the form $f(x) = a(x-h)^2 + k$.
$$\begin{aligned} f(x) &= ax^2 - 8ax + 12a \\ &= a\left(x^2 - 8x\right) + 12a \\ &= a\left(x^2 - 8x + 16\right) + 12a - 16a \\ &= a(x-4)^2 - 4a \end{aligned}$$

The x-coordinate of the vertex is 4, which does not depend on a. However, the y-coordinate is $-4a$ which does depend on a.

99. Answers may vary. One possibility follows:

If the discriminant $b^2 - 4ac > 0$, the graph of the quadratic function will have two different x-intercepts. If the discriminant $b^2 - 4ac = 0$, the graph will have one x-intercept. If the discriminant $b^2 - 4ac < 0$, the graph will not have any x-intercepts.

101. A revenue of $0 when the price charged is some positive number is in an indication that the price was too high to keep any consumer demand. Regardless of the price charged, if no items are sold, then no revenue can be generated.

103. $G(x) = \dfrac{1}{4}x - 2$

Let $x = -4,\ 0,$ and 4.

$$G(-4) = \frac{1}{4}(-4) - 2 = -1 - 2 = -3$$

$$G(0) = \frac{1}{4}(0) - 2 = 0 - 2 = -2$$

$$G(4) = \frac{1}{4}(4) - 2 = 1 - 2 = -1$$

Thus, the points $(-4,-3)$, $(0,-2)$, and $(4,-1)$ are on the graph.

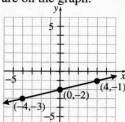

105. $f(x) = (x+2)^2 + 4$

Let $x = -4,\ -3,\ -2,\ -1,$ and 0.

$$f(-4) = (-4+2)^2 + 4 = (-2)^2 + 4 = 4 + 4 = 8$$

$$f(-3) = (-3+2)^2 + 4 = (-1)^2 + 4 = 1 + 4 = 5$$

$$f(-2) = (-2+2)^2 + 4 = (0)^2 + 4 = 0 + 4 = 4$$

$$f(-1) = (-1+2)^2 + 4 = (1)^2 + 4 = 1 + 4 = 5$$

$$f(0) = (0+2)^2 + 4 = (2)^2 + 4 = 4 + 4 = 8$$

Thus, the points $(-4,8)$, $(-3,5)$, $(-2,4)$, $(-1,5)$, and $(0,8)$ are on the graph.

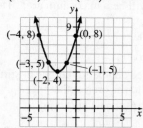

107. Vertex: $(3.5, -9.25)$

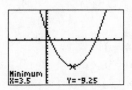

109. Vertex: $(3.5, 37.5)$

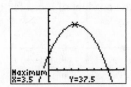

111. Vertex: $(-0.3, -20.45)$

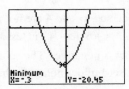

113. Vertex: $(0.67, 4.78)$

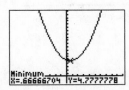

115. c is the y-intercept. In terms of transformations, c is also 1 larger than the vertical shift and thus, 1 larger than the y-coordinate of the vertex. This is because we can write:

$$f(x) = x^2 + 2x + c$$
$$= \left(x^2 + 2x\right) + c$$
$$= \left(x^2 + 2x + 1\right) + c - 1$$
$$= (x+1)^2 + (c-1)$$

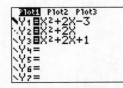

Section 7.6

Are You Ready for This Section?

R1. The inequality $-4 \le x < 5$ in interval notation is $[-4, 5)$.

R2.
$$3x + 5 > 5x - 3$$
$$3x + 5 - 5x > 5x - 3 - 5x$$
$$-2x + 5 > -3$$
$$-2x + 5 - 5 > -3 - 5$$
$$-2x > -8$$
$$\frac{-2x}{-2} < \frac{-8}{-2}$$
$$x < 4$$

The solution set is $\{x \mid x < 4\}$ or, using interval notation, $(-\infty, 4)$.

Section 7.6 Quick Checks

1. We graph $f(x) = x^2 + 3x - 10$. We see that $a = 1$, $b = 3$, and $c = -10$. The parabola opens up because $a = 1 > 0$. The x-coordinate of the vertex is $x = -\dfrac{b}{2a} = -\dfrac{3}{2(1)} = -\dfrac{3}{2}$. The y-coordinate of the vertex is

$$f\left(-\frac{b}{2a}\right) = f\left(-\frac{3}{2}\right)$$
$$= \left(-\frac{3}{2}\right)^2 + 3\left(-\frac{3}{2}\right) - 10$$
$$= \frac{9}{4} - \frac{9}{2} - 10$$
$$= -\frac{49}{4}$$

Thus, the vertex is $\left(-\dfrac{3}{2}, -\dfrac{49}{4}\right)$ and the axis of symmetry is the line $x = -\dfrac{3}{2}$.

The y-intercept is $f(0) = 0^2 + 3(0) - 10 = -10$.

Now, $b^2 - 4ac = 3^2 - 4(1)(-10) = 49 > 0$. The parabola will have two distinct x-intercepts. We find these by solving

$$f(x) = 0$$
$$x^2 + 3x - 10 = 0$$
$$(x+5)(x-2) = 0$$
$$x + 5 = 0 \quad \text{or} \quad x - 2 = 0$$
$$x = -5 \quad \text{or} \quad x = 2$$

Finally, the *y*-intercept point, $(0, -10)$, is three-haves units to the right of the axis of symmetry. Therefore, if we move three-halves units to the left of the axis of symmetry, we obtain the point $(-3, -10)$ which must also be on the graph.

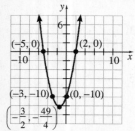

From the graph, we can see that $f(x) = x^2 + 3x - 10$ is greater than 0 for $x < -5$ or $x > 2$. Because the inequality is non-strict, we include the *x*-intercepts in the solution. So, the solution is $\{x \mid x \le -5 \text{ or } x \ge 2\}$ using set-builder notation; the solution is $(-\infty, -5] \cup [2, \infty)$ using interval notation.

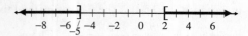

2. $x^2 + 3x - 10 \ge 0$

Solve: $x^2 + 3x - 10 = 0$
$$(x - 2)(x + 5) = 0$$
$$x - 2 = 0 \quad \text{or} \quad x + 5 = 0$$
$$x = 2 \quad \text{or} \quad x = -5$$

Determine where each factor is positive and negative and where the product of these factors is positive and negative.

Interval	$(-\infty, -5)$	-5	$(-5, 2)$	2	$(2, \infty)$
$x - 2$	Neg	Neg	Neg	0	Pos
$x + 5$	Neg	0	Pos	Pos	Pos
$(x-2)(x+5)$	Pos	0	Neg	0	Pos

The inequality is non-strict, so -5 and 2 are part of the solution. Now, $(x - 2)(x + 5)$ is greater than zero where the product is positive. The solution is $\{x \mid x \le -5 \text{ or } x \ge 2\}$ in set-builder notation; the solution is $(-\infty, -5] \cup [2, \infty)$ in interval notation.

3.
$$-x^2 > 2x - 24$$
$$0 > x^2 + 2x - 24$$
$$x^2 + 2x - 24 < 0$$

Solve: $x^2 + 2x - 24 = 0$
$$(x - 4)(x + 6) = 0$$
$$x - 4 = 0 \quad \text{or} \quad x + 6 = 0$$
$$x = 4 \quad \text{or} \quad x = -6$$

Determine where each factor is positive and negative and where the product of these factors is positive and negative.

Interval	$(-\infty, -6)$	-6	$(-6, 4)$	4	$(4, \infty)$
$x - 4$	Neg	Neg	Neg	0	Pos
$x + 6$	Neg	0	Pos	Pos	Pos
$(x-4)(x+6)$	Pos	0	Neg	0	Pos

The inequality is strict, so -6 and 4 are not part of the solution. Now, $(x + 6)(x - 4)$ is less than zero where the product is negative. The solution is $\{x \mid -6 < x < 4\}$ in set-builder notation; the solution is $(-6, 4)$ in interval notation.

4.
$$3x^2 > -x + 5$$
$$3x^2 + x - 5 > 0$$

Graphical Method:

To graph $f(x) = 3x^2 + x - 5$, we notice that $a = 3$, $b = 1$, and $c = -5$. The parabola opens up because $a = 3 > 0$. The *x*-coordinate of the vertex is $x = -\dfrac{b}{2a} = -\dfrac{1}{2(3)} = -\dfrac{1}{6}$. The *y*-coordinate of the vertex is

$$f\left(-\frac{b}{2a}\right) = f\left(-\frac{1}{6}\right)$$
$$= 3\left(-\frac{1}{6}\right)^2 + \left(-\frac{1}{6}\right) - 5$$
$$= -\frac{61}{12}$$

Thus, the vertex is $\left(-\dfrac{1}{6}, -\dfrac{61}{12}\right)$ and the axis of symmetry is the line $x = -\dfrac{1}{6}$. The *y*-intercept is

$f(0) = 3(0)^2 + 0 - 5 = -5$.

Now, $b^2 - 4ac = 1^2 - 4(3)(-5) = 61 > 0$. The parabola will have two distinct x-intercepts. We find these by solving

$$f(x) = 0$$
$$3x^2 + x - 5 = 0$$
$$x = \frac{-1 \pm \sqrt{61}}{2(3)}$$
$$= \frac{-1 \pm \sqrt{61}}{6}$$
$$x \approx -1.47 \quad \text{or} \quad x \approx 1.14$$

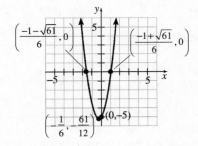

From the graph, we can see that $f(x) = 3x^2 + x - 5$ is greater than 0 for

$x < \dfrac{-1 - \sqrt{61}}{6}$ or $x > \dfrac{-1 + \sqrt{61}}{6}$. Because the

inequality is strict, we do not include the x-intercepts in the solution. So, the solution is

$\left\{ x \,\middle|\, x < \dfrac{-1 - \sqrt{61}}{6} \text{ or } x > \dfrac{-1 + \sqrt{61}}{6} \right\}$ using set-

builder notation; the solution is

$\left(-\infty, \dfrac{-1 - \sqrt{61}}{6} \right) \cup \left(\dfrac{-1 + \sqrt{61}}{6}, \infty \right)$ using

interval notation.

Algebraic Method:

Solve: $3x^2 + x - 5 = 0$

$$x = \frac{-1 \pm \sqrt{61}}{2(3)}$$
$$= \frac{-1 \pm \sqrt{61}}{6}$$
$$x \approx -1.47 \quad \text{or} \quad x \approx 1.14$$

Determine where each factor is positive and negative and where the product of these factors is positive and negative.

Interval	$(-\infty, -1.47)$	$(-1.47, 1.14)$	$(1.14, \infty)$
		$\dfrac{-1-\sqrt{61}}{6}$	$\dfrac{-1+\sqrt{61}}{6}$
$x - \left(\dfrac{-1-\sqrt{61}}{6} \right)$	Neg	0	Pos Pos Pos
$x - \left(\dfrac{-1+\sqrt{61}}{6} \right)$	Neg	Neg Neg	0 Pos
$\left[x - \left(\dfrac{-1-\sqrt{61}}{6} \right) \right]\left[x - \left(\dfrac{-1+\sqrt{61}}{6} \right) \right]$	Pos	0 Neg	0 Pos

The inequality is strict, so $\dfrac{-1-\sqrt{61}}{6}$ and

$\dfrac{-1+\sqrt{61}}{6}$ are not part of the solution. Now,

$\left[x - \left(\dfrac{-1-\sqrt{61}}{6} \right) \right]\left[x - \left(\dfrac{-1+\sqrt{61}}{6} \right) \right]$ is greater

than zero where the product is positive. So, the

solution is $\left\{ x \,\middle|\, x < \dfrac{-1-\sqrt{61}}{6} \text{ or } x > \dfrac{-1+\sqrt{61}}{6} \right\}$

using set-builder notation; the solution is

$\left(-\infty, \dfrac{-1-\sqrt{61}}{6} \right) \cup \left(\dfrac{-1+\sqrt{61}}{6}, \infty \right)$ using

interval notation.

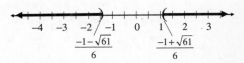

5. $(2x - 3)(x + 4)(x - 5) \le 0$

Solve.

$(2x - 3)(x + 4)(x - 5) = 0$

$2x - 3 = 0 \quad \text{or} \quad x + 4 = 0 \quad \text{or} \quad x - 5 = 0$

$\quad\quad x = \dfrac{3}{2} \quad \text{or} \quad\quad x = -4 \quad \text{or} \quad\quad x = 5$

Determine where each factor is positive and negative and where the product of these factors is positive and negative in the four intervals.

Interval	$(-\infty, -4)$	$\left(-4, \dfrac{3}{2}\right)$	$\left(\dfrac{3}{2}, 5\right)$	$(5, \infty)$
$(2x - 3)$	Neg Neg Neg	0	Pos Pos Pos	
$(x + 4)$	Neg	0 Pos	Pos Pos Pos	Pos
$(x - 5)$	Neg Neg Neg	Neg	Neg Neg	0 Pos
$(2x - 3)(x + 4)(x - 5)$	Neg	0 Pos	0 Neg	0 Pos

The inequality is non-strict, so -4, $\dfrac{3}{2}$, and 5 are

in the solution set. The product is less than zero

     629

for $(-\infty, -4)$ and $\left(\dfrac{3}{2}, 5\right)$. The solution

is $\left\{x \Big| x \le -4 \text{ or } \dfrac{3}{2} \le x \le 5\right\}$ in set-builder

notation, or $(-\infty, -4] \cup \left[\dfrac{3}{2}, 5\right]$ in interval

notation.

6. $x^3 + 7x^2 - 4x - 28 > 0$

$x^2(x+7) - 4(x+7) > 0$

$(x+7)(x^2 - 4) > 0$

$(x+7)(x+2)(x-2) > 0$

Solve $(x+7)(x+2)(x-2) = 0$.

$x + 7 = 0 \quad$ or $\quad x + 2 = 0 \quad$ or $\quad x - 2 = 0$

$x = -7 \quad$ or $\quad\quad x = -2 \quad$ or $\quad\quad x = 2$

Determine where each factor is positive and negative and where the product of these factors is positive and negative in the four intervals.

Interval	$(-\infty, -7)$	-7	$(-7, -2)$	-2	$(-2, 2)$	2	$(2, \infty)$
$(x+7)$	Neg	0	Pos	Pos	Pos	Pos	Pos
$(x+2)$	Neg	Neg	Neg	0	Pos	Pos	Pos
$(x-2)$	Neg	Neg	Neg	Neg	Neg	0	Pos
$(x+7)(x+2)(x-2)$	Neg	0	Pos	0	Neg	0	Pos

The inequality is strict, so -7, -2, and 2 are not in the solution set. The product is greater than zero for $(-7, -2)$ and $(2, \infty)$. The solution is $\{x|-7 < x < -2 \text{ or } x > 2\}$ in set-builder notation, or $(-7, -2) \cup (2, \infty)$ in interval notation.

7.6 Exercises

7. a. The graph is greater than 0 for $x < -6$ or $x > 5$. The solution is $\left\{x|x < -6 \text{ or } x > 5\right\}$ using set-builder notation; the solution is $(-\infty, -6) \cup (5, \infty)$ using interval notation.

b. The graph is 0 or less for $-6 \le x \le 5$. The solution is $\left\{x|-6 \le x \le 5\right\}$ using set-builder notation; the solution is $[-6, 5]$ using interval notation.

9. a. The graph is 0 or greater for $-6 \le x \le \dfrac{5}{2}$.

The solution is $\left\{x \Big| -6 \le x \le \dfrac{5}{2}\right\}$ using set-

builder notation; the solution is $\left[-6, \dfrac{5}{2}\right]$

using interval notation.

b. The graph is less than 0 for $x < -6$ or

$x > \dfrac{5}{2}$. The solution is $\left\{x \Big| x < -6 \text{ or } x > \dfrac{5}{2}\right\}$

using set-builder notation; the solution is

$(-\infty, -6) \cup \left(\dfrac{5}{2}, \infty\right)$ using interval

notation.

11. $(x-5)(x+2) \ge 0$

Solve: $(x-5)(x+2) = 0$

$x - 5 = 0 \quad$ or $\quad x + 2 = 0$

$x = 5 \quad$ or $\quad\quad x = -2$

Determine where each factor is positive and negative and where the product of these factors is positive and negative.

Interval	$(-\infty, -2)$	-2	$(-2, 5)$	5	$(5, \infty)$
$x - 5$	$----$	$-$	$----$	0	$++++$
$x + 2$	$----$	0	$++++$	$+$	$++++$
$(x-5)(x+2)$	$++++$	0	$----$	0	$++++$

The inequality is non-strict, so -2 and 5 are part of the solution. Now, $(x-5)(x+2)$ is greater than zero where the product is positive. The solution is $\{x \mid x \le -2 \text{ or } x \ge 5\}$ in set-builder notation; the solution is $(-\infty, -2] \cup [5, \infty)$ in interval notation.

13. $(x+3)(x+7) < 0$

Solve: $(x+3)(x+7) = 0$

$x + 3 = 0 \quad$ or $\quad x + 7 = 0$

$x = -3 \quad$ or $\quad\quad x = -7$

Determine where each factor is positive and negative and where the product of these factors is positive and negative.

Interval

	$(-\infty, -7)$	-7	$(-7, -3)$	-3	$(-3, \infty)$
$x + 7$	$----$	0	$++++$	$+$	$++++$
$x + 3$	$----$	$-$	$----$	0	$++++$
$(x+7)(x+3)$	$++++$	0	$----$	0	$++++$

The inequality is strict, so -7 and -3 are not part of the solution. Now, $(x+3)(x+7)$ is less than zero where the product is negative. The solution is $\{x \mid -7 < x < -3\}$ in set-builder notation; the solution is $(-7, \, -3)$ in interval notation.

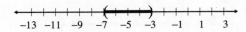

15. $x^2 - 2x - 35 > 0$

Solve: $x^2 - 2x - 35 = 0$

$(x+5)(x-7) = 0$

$x + 5 = 0$ or $x - 7 = 0$

     $x = -5$ or $x = 7$

Determine where each factor is positive and negative and where the product of these factors is positive and negative.

Interval

	$(-\infty, -5)$	-5	$(-5, 7)$	7	$(7, \infty)$
$x + 5$	$----$	0	$++++$	$+$	$++++$
$x - 7$	$----$	$-$	$----$	0	$++++$
$(x+5)(x-7)$	$++++$	0	$----$	0	$++++$

The inequality is strict, so -5 and 7 are not part of the solution. Now, $(x+5)(x-7)$ is greater than zero where the product is positive. The solution is $\{x \mid x < -5 \text{ or } x > 7\}$ in set-builder notation; the solution is $(-\infty, \, -5) \cup (7, \, \infty)$ in interval notation.

17. $n^2 - 6n - 8 \le 0$

Solve: $n^2 - 6n - 8 = 0$

$$n = \frac{-(-6) \pm \sqrt{(-6)^2 - 4(1)(-8)}}{2(1)}$$

$$= \frac{6 \pm \sqrt{68}}{2}$$

$$= \frac{6 \pm 2\sqrt{17}}{2}$$

$$= 3 \pm \sqrt{17}$$

$n = 3 - \sqrt{17}$ or $n = 3 + \sqrt{17}$

     ≈ -1.12 or ≈ 7.12

Determine where each factor is positive and negative and where the product of these factors is positive and negative.

Interval

	$(-\infty, -1.12)$	$3 - \sqrt{17}$ ≈ -1.12	$(-1.12, 7.12)$	$3 + \sqrt{17}$ ≈ 7.12	$(7.12, \infty)$
$n - (3 - \sqrt{17})$	$----$	0	$++++$	$+$	$++++$
$n - (3 + \sqrt{17})$	$----$	$-$	$----$	0	$++++$
$[n - (3 - \sqrt{17})][n - (3 + \sqrt{17})]$	$++++$	0	$----$	0	$++++$

The inequality is non-strict, so $3 - \sqrt{17}$ and $3 + \sqrt{17}$ are part of the solution. Now, $\left[n - \left(3 - \sqrt{17}\right)\right]\left[n - \left(3 + \sqrt{17}\right)\right]$ is less than zero where the product is negative. The solution is $\left\{n \mid 3 - \sqrt{17} \le n \le 3 + \sqrt{17}\right\}$ in set-builder notation; the solution is $\left[3 - \sqrt{17}, \, 3 + \sqrt{17}\right]$ in interval notation.

19. $m^2 + 5m \ge 14$

$m^2 + 5m - 14 \ge 0$

Solve: $m^2 + 5m - 14 = 0$

$(m+7)(m-2) = 0$

$m + 7 = 0$ or $m - 2 = 0$

     $m = -7$ or $m = 2$

Determine where each factor is positive and negative and where the product of these factors is positive and negative.

Interval

	$(-\infty, -7)$	-7	$(-7, 2)$	2	$(2, \infty)$
$m + 7$	$----$	0	$++++$	$+$	$++++$
$m - 2$	$----$	$-$	$----$	0	$++++$
$(m+7)(m-2)$	$++++$	0	$----$	0	$++++$

The inequality is non-strict, so -7 and 2 are part of the solution. Now, $(m+7)(m-2)$ is greater than zero where the product is positive. The solution is $\{m \mid m \le -7 \text{ or } m \ge 2\}$ in set-builder notation; the solution is $(-\infty, -7] \cup [2, \, \infty)$ in interval notation.

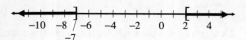

21. $2q^2 \geq q + 15$

$2q^2 - q - 15 \geq 0$

Solve: $2q^2 - q - 15 = 0$

$(2q + 5)(q - 3) = 0$

$2q + 5 = 0$ or $q - 3 = 0$

$q = -\dfrac{5}{2}$ or $q = 3$

Determine where each factor is positive and negative and where the product of these factors is positive and negative.

Interval	$\left(-\infty, -\frac{5}{2}\right)$	$-\frac{5}{2}$	$\left(-\frac{5}{2}, 3\right)$	3	$(3, \infty)$
$2q + 5$	$----$	0	$++++$	$+$	$++++$
$q - 3$	$----$	$-$	$----$	0	$++++$
$(2q + 5)(q - 3)$	$++++$	0	$----$	0	$++++$

The inequality is non-strict, so $-\dfrac{5}{2}$ and 3 are part of the solution. Now, $(2q + 5)(q - 3)$ is greater than zero where the product is positive. The solution is $\left\{ q \mid q \leq -\dfrac{5}{2} \text{ or } q \geq 3 \right\}$ in set-builder notation; the solution is $\left(-\infty, -\dfrac{5}{2}\right] \cup [3, \infty)$ in interval notation.

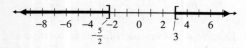

23. $3x + 4 \geq x^2$

$0 \geq x^2 - 3x - 4$

$x^2 - 3x - 4 \leq 0$

Solve: $x^2 - 3x - 4 = 0$

$(x + 1)(x - 4) = 0$

$x + 1 = 0$ or $x - 4 = 0$

$x = -1$ or $x = 4$

Determine where each factor is positive and negative and where the product of these factors is positive and negative.

Interval	$(-\infty, -1)$	-1		4	$(4, \infty)$
$x + 1$	$----$	0	$++++$	$+$	$++++$
$x - 4$	$----$	$-$	$----$	0	$++++$
$(x + 1)(x - 4)$	$++++$	0	$----$	0	$++++$

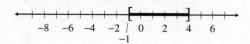

The inequality is non-strict, so -1 and 4 are part of the solution. Now, $(x + 1)(x - 4)$ is less than zero where the product is negative. The solution is $\{ x \mid -1 \leq x \leq 4 \}$ in set-builder notation; the solution is $[-1, \ 4]$ in interval notation.

25. $-x^2 + 3x < -10$

$0 < x^2 - 3x - 10$

$x^2 - 3x - 10 > 0$

Solve: $x^2 - 3x - 10 = 0$

$(x + 2)(x - 5) = 0$

$x + 2 = 0$ or $x - 5 = 0$

$x = -2$ or $x = 5$

Determine where each factor is positive and negative and where the product of these factors is positive and negative.

Interval	$(-\infty, -2)$	-2	$(-2, 5)$	5	$(5, \infty)$
$x + 2$	$----$	0	$++++$	$+$	$++++$
$x - 5$	$----$	$-$	$----$	0	$++++$
$(x + 2)(x - 5)$	$++++$	0	$----$	0	$++++$

The inequality is strict, so -2 and 5 are not part of the solution. Now, $(x + 2)(x - 5)$ is greater than zero where the product is positive. The solution is $\{ x \mid x < -2 \text{ or } x > 5 \}$ in set-builder notation; the solution is $(-\infty, -2) \cup (5, \ \infty)$ in interval notation.

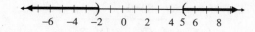

27. $-3x^2 \leq -10x - 8$

$0 \leq 3x^2 - 10x - 8$

$3x^2 - 10x - 8 \geq 0$

Solve: $3x^2 - 10x - 8 = 0$

$(3x + 2)(x - 4) = 0$

$$3x + 2 = 0 \quad \text{or} \quad x - 4 = 0$$
$$x = -\frac{2}{3} \quad \text{or} \quad x = 4$$

Determine where each factor is positive and negative and where the product of these factors is positive and negative.

The inequality is non-strict, so $-\dfrac{2}{3}$ and 4 are part of the solution. Now, $(3x+2)(x-4)$ is greater than zero where the product is positive. The solution is $\left\{ x \mid x \le -\dfrac{2}{3} \text{ or } x \ge 4 \right\}$ in set-builder notation; the solution is $\left(-\infty, \ -\dfrac{2}{3} \right] \cup [4, \ \infty)$ in interval notation.

29. $x^2 + 4x + 1 < 0$

Solve: $x^2 + 4x + 1 = 0$

$$x = \frac{-4 \pm \sqrt{4^2 - 4(1)(1)}}{2(1)}$$
$$= \frac{-4 \pm \sqrt{12}}{2}$$
$$= \frac{-4 \pm 2\sqrt{3}}{2}$$
$$= -2 \pm \sqrt{3}$$
$$x = -2 - \sqrt{3} \text{ or } x = -2 + \sqrt{3}$$
$$\approx -3.73 \quad \text{or} \quad \approx -0.27$$

Determine where each factor is positive and negative and where the product of these factors is positive and negative.

The inequality is strict, so $-2 - \sqrt{3}$ and $-2 + \sqrt{3}$ are not part of the solution. Now, $\left[x - \left(-2 - \sqrt{3} \right) \right]\left[x - \left(-2 + \sqrt{3} \right) \right]$ is less than zero where the product is negative. The solution is $\left\{ x \mid -2 - \sqrt{3} < x < -2 + \sqrt{3} \right\}$ in set-builder notation; the solution is $\left(-2 - \sqrt{3}, \ -2 + \sqrt{3} \right)$ in interval notation.

31. $-2a^2 + 7a \ge -4$
$$0 \ge 2a^2 - 7a - 4$$
$$2a^2 - 7a - 4 \le 0$$

Solve: $2a^2 - 7a - 4 = 0$
$$(2a + 1)(a - 4) = 0$$
$$2a + 1 = 0 \quad \text{or} \quad a - 4 = 0$$
$$a = -\frac{1}{2} \quad \text{or} \quad a = 4$$

Determine where each factor is positive and negative and where the product of these factors is positive and negative.

The inequality is non-strict, so $-\dfrac{1}{2}$ and 4 are part of the solution. Now, $(2a+1)(a-4)$ is less than zero where the product is negative. The solution is $\left\{ a \mid -\dfrac{1}{2} \le a \le 4 \right\}$ in set-builder notation; the solution is $\left[-\dfrac{1}{2}, \ 4 \right]$ in interval notation.

33. $z^2 + 2z + 3 > 0$

Solve: $z^2 + 2z + 3 = 0$

$$z = \frac{-2 \pm \sqrt{2^2 - 4(1)(3)}}{2(1)}$$

$$= \frac{-2 \pm \sqrt{-8}}{2}$$

$$= \frac{-2 \pm 2\sqrt{2}\, i}{2} = -1 \pm \sqrt{2}\, i$$

The solutions to the equation are non-real. This means that $z^2 + 2z + 3$ will not divide the number line into positive and negative intervals. Instead, $z^2 + 2z + 3$ will either be positive on the entire number line or be negative on the entire number line. The graph below shows that $f(z) = z^2 + 2z + 3$ is always positive.

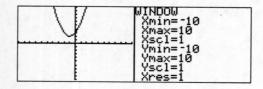

This means that $z^2 + 2z + 3$ is always greater than zero. The solution is $\{z \mid z$ is any real number$\}$; the solution is $(-\infty, \infty)$ in interval notation.

35. $2b^2 + 5b \leq -6$

$2b^2 + 5b + 6 \leq 0$

Solve: $2b^2 + 5b + 6 = 0$

$$b = \frac{-5 \pm \sqrt{5^2 - 4(2)(6)}}{2(1)}$$

$$= \frac{-5 \pm \sqrt{-23}}{2}$$

$$= \frac{-5 \pm \sqrt{23}\, i}{2}$$

$$= -\frac{5}{2} \pm \frac{\sqrt{23}}{2} i$$

The solutions to the equation are non-real. This means that $2b^2 + 5b + 6$ will not divide the number line into positive and negative intervals. Instead, $2b^2 + 5b + 6$ will either be positive on

the entire number line or be negative on the entire number line. The graph below shows that $f(b) = 2b^2 + 5b + 6$ is always positive.

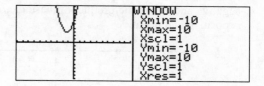

This means that $2b^2 + 5b + 6$ is never less than or equal to zero. The quadratic inequality has no solution: $\{\ \}$ or $\varnothing$.

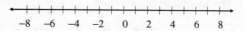

37. $x^2 - 6x + 9 > 0$

Solve: $x^2 - 6x + 9 = 0$

$(x - 3)(x - 3) = 0$

$x - 3 = 0$ or $x - 3 = 0$

$x = 3$ or $x = 3$

Determine where each factor is positive and negative and where the product of these factors is positive and negative.

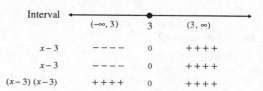

Interval	$(-\infty, 3)$	3	$(3, \infty)$
$x - 3$	$----$	0	$++++$
$x - 3$	$----$	0	$++++$
$(x-3)(x-3)$	$++++$	0	$++++$

The inequality is strict, so 3 is not part of the solution. Now, $(x - 3)(x - 3)$ is greater than zero where the product is positive. Thus, $x^2 - 6x + 9$ is always greater than zero when x is not equal to 3. The solution is $\{x \mid x \neq 3\}$ in set-builder notation; the solution is $(-\infty, 3) \cup (3, \infty)$ in interval notation.

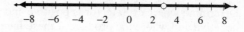

39. $(x+1)(x-2)(x-5) > 0$

Solve: $(x+1)(x-2)(x-5) = 0$

$x + 1 = 0$ or $x - 2 = 0$ or $x - 5 = 0$

$x = -1$ or $x = 2$ or $x = 5$

Determine where each factor is positive and negative and where the product of these factors is positive and negative.

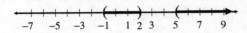

Interval $(-\infty, -1)$ -1 $(-1, 2)$ 2 $(2, 5)$ 5 $(5, \infty)$

$x+1$	$-\,-\,-$	0	$+++$	$+$	$+++$	$+$	$+++$					
$x-2$	$-\,-\,-$	$-$	$-\,-\,-$	0	$+++$	$+$	$+++$					
$x-5$	$-\,-\,-$	$-$	$-\,-\,-$	$-$	$-\,-\,-$	0	$+++$					
$(x+1)(x-2)(x-5)$	$-\,-\,-$	0	$+++$	0	$-\,-\,-$	0	$+++$					

The inequality is strict, so -1, 2, and 5 are not part of the solution. Now, $(x+1)(x-2)(x-5)$ is greater than zero where the product is positive. The solution is $\{x \mid -1 < x < 2 \text{ or } x > 5\}$ in set-builder notation; the solution is $(-1, 2) \cup (5, \infty)$ in interval notation.

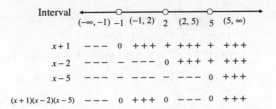

41. $(2x+1)(x-4)(x-9) \le 0$

Solve: $(2x+1)(x-4)(x-9) = 0$

$2x+1 = 0$ or $x-4 = 0$ or $x-9 = 0$

$x = -\dfrac{1}{2}$ or $x = 4$ or $x = 9$

Determine where each factor is positive and negative and where the product of these factors is positive and negative.

Interval $\left(-\infty, -\frac{1}{2}\right)$ $-\frac{1}{2}$ $\left(-\frac{1}{2}, 4\right)$ 4 $(4, 9)$ 9 $(9, \infty)$

$2x+1$	$-\,-\,-$	0	$+++$	$+$	$+++$	$+$	$+++$
$x-4$	$-\,-\,-$	$-$	$-\,-\,-$	0	$+++$	$+$	$+++$
$x-9$	$-\,-\,-$	$-$	$-\,-\,-$	$-$	$-\,-\,-$	0	$+++$
$(2x+1)(x-4)(x-9)$	$-\,-\,-$	0	$+++$	0	$-\,-\,-$	0	$+++$

The inequality is non-strict, so $-\dfrac{1}{2}$, 4, and 9 are part of the solution. Now, $(2x+1)(x-4)(x-9)$ is less than zero where the product is negative. The solution is $\left\{x \mid x \le -\dfrac{1}{2} \text{ or } 4 \le x \le 9\right\}$ in set-builder notation; the solution is $\left(-\infty, -\dfrac{1}{2}\right] \cup [4, 9]$ in interval notation.

43. $(x+3)(2x^2 - x - 1) \ge 0$

Solve $(x + 3)(2x + 1)(x - 1) = 0$.

$x+3 = 0$ or $2x+1 = 0$ or $x-1 = 0$

$x = -3$ or $x = -\dfrac{1}{2}$ or $x = 1$

Determine where each factor is positive and negative and where the product of these factors is positive and negative.

Interval -3 $-\frac{1}{2}$ 1

  $(-\infty, -3)$ $\left(-3, -\frac{1}{2}\right)$ $\left(-\frac{1}{2}, 1\right)$ $(1, \infty)$

$(x+3)$	$-\,-\,-$	0	$+++$	$+$	$+++$	$+$	$+++$
$(2x+1)$	$-\,-\,-$	$-$	$-\,-\,-$	0	$+++$	$+$	$+++$
$(x-1)$	$-\,-\,-$	$-$	$-\,-\,-$	$-$	$-\,-\,-$	0	$+++$
$(x+3)(2x+1)(x-1)$	$-\,-\,-$	0	$+++$	0	$-\,-\,-$	0	$+++$

The inequality is non-strict, so -3, $-\dfrac{1}{2}$, and 1 are part of the solution. Now, $(x + 3)(2x + 1)(x - 1)$ is greater than zero where the product is positive. The solution is $\left\{x \mid -3 \le x \le -\dfrac{1}{2} \text{ or } x \ge 1\right\}$ in set-builder notation; the solution is $\left[-\dfrac{3}{2}, -\dfrac{1}{2}\right] \cup [1, \infty)$ i interval notation.

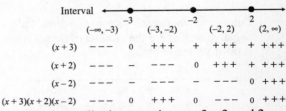

45. $x^3 + 3x^2 - 4x - 12 \le 0$

Solve $x^2(x+3) - 4(x+3) = 0$

$$(x+3)(x^2 - 4) = 0$$
$$(x+3)(x+2)(x-2) = 0$$

$x+3 = 0$ or $x+2 = 0$ or $x-2 = 0$

$x = -3$ or $x = -2$ or $x = 2$

Determine where each factor is positive and negative and where the product of these factors is positive and negative.

Interval -3 -2 2

  $(-\infty, -3)$ $(-3, -2)$ $(-2, 2)$ $(2, \infty)$

$(x+3)$	$-\,-\,-$	0	$+++$	$+$	$+++$	$+$	$+++$
$(x+2)$	$-\,-\,-$	$-$	$-\,-\,-$	0	$+++$	$+$	$+++$
$(x-2)$	$-\,-\,-$	$-$	$-\,-\,-$	$-$	$-\,-\,-$	0	$+++$
$(x+3)(x+2)(x-2)$	$-\,-\,-$	0	$+++$	0	$-\,-\,-$	0	$+++$

The inequality is non-strict, so -3, -2, and 2 are part of the solution. Now, $(x + 3)(x + 2)(x - 2)$ is less than zero where the product is negative. The solution is $\{x \mid x \le -3 \text{ or } -2 \le x \le 2\}$ in set-builder notation; the solution is $(-\infty, -3] \cup [-2, 2]$ in interval notation.

47. $4x^3 + 16x^2 - 9x - 36 > 0$

Solve $4x^2(x+4) - 9(x+4) = 0$

$$(x+4)(4x^2 - 9) = 0$$
$$(x+4)(2x+3)(2x-3) = 0$$
$$x+4 = 0 \quad \text{or} \quad 2x+3 = 0 \quad \text{or} \quad 2x-3 = 0$$
$$x = -4 \quad \text{or} \quad x = -\frac{3}{2} \quad \text{or} \quad x = \frac{3}{2}$$

Interval	$(-\infty, -4)$	-4	$\left(-4, -\frac{3}{2}\right)$	$-\frac{3}{2}$	$\left(-\frac{3}{2}, \frac{3}{2}\right)$	$\frac{3}{2}$	$\left(\frac{3}{2}, \infty\right)$
$(x+4)$	$-\,-\,-$	0	$+\,+\,+$	$+$	$+\,+\,+$	$+$	$+\,+\,+$
$(2x+3)$	$-\,-\,-$	$-$	$-\,-\,-$	0	$+\,+\,+$	$+$	$+\,+\,+$
$(2x-3)$	$-\,-\,-$	$-$	$-\,-\,-$	$-$	$-\,-\,-$	0	$+\,+\,+$
$(x+4)(2x+3)(2x-3)$	$-\,-\,-$	0	$+\,+\,+$	0	$-\,-\,-$	0	$+\,+\,+$

The inequality is strict, so -4, $-\frac{3}{2}$, and $\frac{3}{2}$ are not part of the solution. Now, $(x+4)(2x+3)(2x-3)$ is greater than zero when the product is positive. The solution is $\left\{x \mid -4 < x < -\frac{3}{2} \text{ or } x > \frac{3}{2}\right\}$ in set-builder notation; the solution is $\left(-4, -\frac{3}{2}\right) \cup \left(\frac{3}{2}, \infty\right)$ in interval notation.

49. $f(x) < 0$

$x^2 - 5x < 0$

Solve: $x^2 - 5x = 0$
$$x(x-5) = 0$$
$$x = 0 \quad \text{or} \quad x - 5 = 0$$
$$\text{or} \quad x = 5$$

Determine where each factor is positive and negative and where the product of these factors is positive and negative.

Interval	$(-\infty, 0)$	0	$(0, 5)$	5	$(5, \infty)$
x	$-\,-\,-\,-$	0	$+\,+\,+\,+$	$+$	$+\,+\,+\,+$
$x-5$	$-\,-\,-\,-$	$-$	$-\,-\,-\,-$	0	$+\,+\,+\,+$
$x(x-5)$	$+\,+\,+\,+$	0	$-\,-\,-\,-$	0	$+\,+\,+\,+$

The inequality is strict, so 0 and 5 are not part of

the solution. Now, $x(x-5)$ is less than zero where the product is negative. The solution is $\{x \mid 0 < x < 5\}$ in set-builder notation; the solution is $(0, 5)$ in interval notation.

51.
$$f(x) \geq 0$$
$$x^2 - 3x - 28 \geq 0$$

Solve: $x^2 - 3x - 28 = 0$
$$(x+4)(x-7) = 0$$
$$x+4 = 0 \quad \text{or} \quad x-7 = 0$$
$$x = -4 \quad \text{or} \quad x = 7$$

Determine where each factor is positive and negative and where the product of these factors is positive and negative.

Interval	$(-\infty, -4)$	-4	$(-4, 7)$	7	$(7, \infty)$
$x+4$	$-\,-\,-\,-$	0	$+\,+\,+\,+$	$+$	$+\,+\,+\,+$
$x-7$	$-\,-\,-\,-$	$-$	$-\,-\,-\,-$	0	$+\,+\,+\,+$
$(x+4)(x-7)$	$+\,+\,+\,+$	0	$-\,-\,-\,-$	0	$+\,+\,+\,+$

The inequality is non-strict, so -4 and 7 are part of the solution. Now, $(x+4)(x-7)$ is greater than zero where the product is positive. The solution is $\{x \mid x \leq -4 \text{ or } x \geq 7\}$ in set-builder notation; the solution is $(-\infty, -4] \cup [7, \infty)$ in interval notation.

53.
$$g(x) > 0$$
$$2x^2 + x - 10 > 0$$

Solve: $2x^2 + x - 10 = 0$
$$(2x+5)(x-2) = 0$$
$$2x+5 = 0 \quad \text{or} \quad x-2 = 0$$
$$2x = -5 \quad \text{or} \quad x = 2$$
$$x = -\frac{5}{2}$$

Determine where each factor is positive and negative and where the product of these factors is positive and negative.

Interval	$\left(-\infty, -\frac{5}{2}\right)$	$-\frac{5}{2}$	$\left(-\frac{5}{2}, 2\right)$	2	$(2, \infty)$
$2x+5$	$-\,-\,-\,-$	0	$+\,+\,+\,+$	$+$	$+\,+\,+\,+$
$x-2$	$-\,-\,-\,-$	$-$	$-\,-\,-\,-$	0	$+\,+\,+\,+$
$(2x+5)(x-2)$	$+\,+\,+\,+$	0	$-\,-\,-\,-$	0	$+\,+\,+\,+$

The inequality is strict, so $-\dfrac{5}{2}$ and 2 are not part

of the solution. Now, $(2x+5)(x-2)$ is greater

than zero where the product is positive. The

solution is $\left\{ x \middle| x < -\dfrac{5}{2} \text{ or } x > 2 \right\}$ in set-builder

notation; the solution is $\left(-\infty, \ -\dfrac{5}{2} \right) \cup (2, \ \infty)$ in

interval notation.

55. $f(x) < 0$

$4x^3 - x^2 - 14x < 0$

Solve: $4x^3 - x^2 - 14x = 0$

$x(4x^2 - x - 14) = 0$

$x(4x+7)(x-2) = 0$

$x = 0$ or $4x+7 = 0$ or $x - 2 = 0$

$x = -\dfrac{7}{4}$ or $x = 2$

Determine where each factor is positive and
negative and where the product of these factors
is positive and negative.

Interval	$\left(-\infty, -\frac{7}{4}\right)$	$-\frac{7}{4}$	$\left(-\frac{7}{4}, 0\right)$	0	$(0, 2)$	2	$(2, \infty)$
x	$---$	$-$	$---$	0	$+++$	$+$	$+++$
$(4x+7)$	$---$	0	$+++$	$+$	$+++$	$+$	$+++$
$(x-2)$	$---$	$-$	$---$	$-$	$---$	0	$+++$
$x(4x+7)(x-2)$	$---$	0	$+++$	0	$---$	0	$+++$

The inequality is strict, so $-\dfrac{7}{4}$, 0, and 2 are not

part of the solution. Now, $x(4x + 7)(x - 2)$ is less
than zero where the product is negative. The

solution is $\left\{ x \middle| x < -\dfrac{7}{4} \text{ or } 0 < x < 2 \right\}$ in set-

builder notation; the solution is

$\left(-\infty, \ -\dfrac{7}{4} \right) \cup (0, \ 2)$ in interval notation.

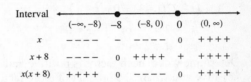

57. The domain of $f(x) = \sqrt{x^2 + 8x}$ will be the

solution set of $x^2 + 8x \geq 0$.

Solve: $x^2 + 8x = 0$

$x(x+8) = 0$

$x = 0$ or $x + 8 = 0$

or $x = -8$

Determine where each factor is positive and
negative and where the product of these factors
is positive and negative.

Interval	$(-\infty, -8)$	-8	$(-8, 0)$	0	$(0, \infty)$
x	$----$	$-$	$----$	0	$++++$
$x+8$	$----$	0	$++++$	$+$	$++++$
$x(x+8)$	$++++$	0	$----$	0	$++++$

The inequality is non-strict, so –8 and 0 are part
of the solution. Now, $x(x+8)$ is greater than
zero where the product is positive. Thus, the
domain of f is $\left\{ x \middle| x \leq -8 \text{ or } x \geq 0 \right\}$ in set-builder

notation; the domain is $(-\infty, -8] \cup [0, \ \infty)$ in
interval notation.

59. The domain of $g(x) = \sqrt{x^2 - x - 30}$ will be the

solution set of $x^2 - x - 30 \geq 0$.

Solve: $x^2 - x - 30 = 0$

$(x-6)(x+5) = 0$

$x - 6 = 0$ or $x + 5 = 0$

$x = 6$ or $x = -5$

Determine where each factor is positive and
negative and where the product of these factors
is positive and negative.

Interval	$(-\infty, -5)$	-5	$(-5, 6)$	6	$(6, \infty)$
$x-6$	$----$	$-$	$----$	0	$++++$
$x+5$	$----$	0	$++++$	$+$	$++++$
$(x-6)(x+5)$	$++++$	0	$----$	0	$++++$

The inequality is non-strict, so –6 and 5 are part
of the solution. Now, $(x-6)(x+5)$ is greater
than zero where the product is positive. Thus,
the domain of g is $\left\{ x \middle| x \leq -5 \text{ or } x \geq 6 \right\}$ in set-

builder notation; the domain is
$(-\infty, -5] \cup [6, \ \infty)$ in interval notation.

61. The ball will be more than 596 feet above sea level when $s(t) > 596$.

$$-16t^2 + 80t + 500 > 596$$
$$0 > 16t^2 - 80t + 96$$
$$16t^2 - 80t + 96 < 0$$
$$\frac{16t^2 - 80t + 96}{16} < \frac{0}{16}$$
$$t^2 - 5t + 6 < 0$$

Solve: $t^2 - 5t + 6 = 0$
$$(t-2)(t-3) = 0$$
$$t - 2 = 0 \quad \text{or} \quad t - 3 = 0$$
$$t = 2 \quad \text{or} \quad t = 3$$

Determine where each factor is positive and negative and where the product of these factors is positive and negative.

Interval	$(-\infty, 2)$	2	$(2, 3)$	3	$(3, \infty)$
$t - 2$	$----$	0	$++++$	+	$++++$
$t - 3$	$----$	-	$----$	0	$++++$
$(t-2)(t-3)$	$++++$	0	$----$	0	$++++$

The inequality is strict, so 2 and 3 are not part of the solution. Now, $(t-2)(t-3)$ is less than zero where the product is negative. The solution is $\{t \mid 2 < t < 3\}$ in set-builder notation; the solution is $(2, 3)$ in interval notation. Thus, the ball will be more than 596 feet above sea level when the time is between 2 and 3 seconds after the ball is thrown.

63. The revenue will exceed \$35,750 when $R(p) > 35,750$.

$$-2.5p^2 + 600p > 35,750$$
$$0 > 2.5p^2 - 600p + 35,750$$
$$2.5p^2 - 600p + 35,750 < 0$$
$$\frac{2.5p^2 - 600p + 35,750}{2.5} < \frac{0}{2.5}$$
$$p^2 - 240t + 14,300 < 0$$

Solve: $p^2 - 240t + 14,300 = 0$
$$(p - 110)(p - 130) = 0$$
$$p - 110 = 0 \quad \text{or} \quad p - 130 = 0$$
$$p = 110 \quad \text{or} \quad p = 130$$

Determine where each factor is positive and negative and where the product of these factors is positive and negative.

Interval	$(-\infty, 110)$	110	$(110, 130)$	130	$(130, \infty)$
$p - 110$	$----$	0	$++++$	+	$++++$
$p - 130$	$----$	-	$----$	0	$++++$
$(p-110)(p-130)$	$++++$	0	$----$	0	$++++$

The inequality is strict, so 110 and 130 are not part of the solution. Now, $(p-110)(p-130)$ is less than zero where the product is negative. The solution is $\{p \mid 110 < p < 130\}$ in set-builder notation; the solution is $(110, 130)$ in interval notation. Thus, the revenue will exceed \$35,750 when the DVD is sold for a price between \$110 and \$130.

65. By inspection, the only solution is $x = -3$. That is, the solution set is $\{-3\}$.

Explanation: The expression on the left side of the inequality is a perfect square. A perfect square cannot be negative (less than zero). Therefore, the only solution will be where the perfect square expression equals zero, which is -3.

67. By inspection, the solution set is the set of all real numbers.

Explanation: The expression on the left side of the inequality is a perfect square. A perfect square must always be zero or greater. Therefore, it must always be larger than -2. Thus, all values of x will make the inequality true. The solution is the set of all real numbers.

69. Answers may vary. One possibility follows: We want $x \geq -3$ and $x \leq 2$. Now, $x \geq -3$ means $x + 3 \geq 0$ (positive), and $x \leq 2$ means $x - 2 \leq 0$ (negative). If we multiply a positive by a negative, we get a negative result. Thus, $(x+3)(x-2) \leq 0$. Multiplying out the expression on the left, we get $x^2 + x - 6 \leq 0$. The solution set of $x^2 + x - 6 \leq 0$ is $[-3, 2]$.

NOTE: Because the solution contains the endpoints -3 and 2, the inequality must be non-strict.

71. $(x + 1)(x - 2)(x + 3)(x - 4) > 0$
Solve: $(x + 1)(x - 2)(x + 3)(x - 4) > 0$
$x + 1 = 0$ or $x - 2 = 0$ or $x + 3 = 0$ or $x - 4 = 0$
    $x = -1$ $x = 2$ $x = -3$ $x = 4$
Determine where each factor is positive and negative and where the product is positive and negative.

Interval	$(-\infty, -3)$	-3	$(-3, -1)$	-1	$(-1, 2)$	2	$(2, 4)$	4	$(4, \infty)$
$(x + 3)$	$---$	0	$+++$	$+$	$+++$	$+$	$+++$	$+$	$+++$
$(x + 1)$	$---$	$-$	$---$	0	$+++$	$+$	$+++$	$+$	$+++$
$(x - 2)$	$---$	$-$	$---$	$-$	$---$	0	$+++$	$+$	$+++$
$(x - 4)$	$---$	$-$	$---$	$-$	$---$	$-$	$---$	0	$+++$
product	$+++$	0	$---$	0	$+++$	0	$---$	0	$+++$

The inequality is strict, so -3, -1, 2, and 4 are not part of the solution. The solution set is
$\{x | x < -3 \text{ or } -1 < x < 2 \text{ or } x > 4\}$ in set-builder notation or $(-\infty, -3) \cup (-1, 2) \cup (4, \infty)$ in interval notation.

73. $$x^4 - 29x^2 + 100 \le 0$$
$$(x^2 - 25)(x^2 - 4) \le 0$$
$$(x - 5)(x + 5)(x - 2)(x + 2) \le 0$$
Solve: $(x - 5)(x + 5)(x - 2)(x + 2) \le 0$
$x - 5 = 0$ or $x + 5 = 0$ or $x - 2 = 0$ or $x + 2 = 0$
   $x = 5$ $x = -5$ $x = 2$ $x = -2$
Determine where each factor is positive and negative and where the product is positive and negative.

Interval	$(-\infty, -5)$	-5	$(-5, -2)$	-2	$(-2, 2)$	2	$(2, 5)$	5	$(5, \infty)$
$(x + 5)$	$---$	0	$+++$	$+$	$+++$	$+$	$+++$	$+$	$+++$
$(x + 2)$	$---$	$-$	$---$	0	$+++$	$+$	$+++$	$+$	$+++$
$(x - 2)$	$---$	$-$	$---$	$-$	$---$	0	$+++$	$+$	$+++$
$(x - 5)$	$---$	$-$	$---$	$-$	$---$	$-$	$---$	0	$+++$
product	$+++$	0	$---$	0	$+++$	0	$---$	0	$+++$

The inequality is non-strict, so -5, -2, 2, and 5 are included in the solution. The solution set is
$\{x | -5 \le x \le -2 \text{ or } 2 \le x \le 5\}$ in set-builder notation or $[-5, -2] \cup [2, 5]$ in interval notation.

75. $$\frac{x^2 + 5x + 6}{x - 2} > 0$$
$$\frac{(x + 3)(x + 2)}{x - 2} > 0$$
The rational expression will equal 0 when $x = -3$ and when $x = -2$. It is undefined when $x = 2$.
Determine where the factors of the numerator and the denominator are positive and negative and where the quotient is positive and negative.

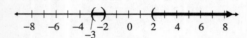

The inequality is strict, so -3, -2, and 2 are not part of the solution. Now, $\dfrac{(x+3)(x+2)}{x-2}$ is greater than zero where the quotient is positive. The solution is $\{x \mid -3 < x < -2 \text{ or } x > 2\}$ in set-builder notation; the solution is $(-3,-2) \cup (2, \infty)$ in interval notation.

77. Answer will vary. One possibility follows: The inequalities have the same solution set because they are equivalent:

$$(3x+2)^{-2} > \frac{1}{2}$$
$$\frac{1}{(3x+2)^2} > \frac{1}{2}$$
$$2(3x+2)^2 \cdot \frac{1}{(3x+2)^2} > 2(3x+2)^2 \cdot \frac{1}{2}$$
$$2 > (3x+2)^2$$
$$(3x+2)^2 < 2$$

Note: In the third step, we multiplied both sides by $2(3x+2)^2$ which must be positive, leaving the direction of the inequality symbol unchanged.

79. Answers may vary. One possibility follows: $x^2 \geq 0$ for all real values of x, so $x^2 - 1 \geq -1$ for all real values of x. That is, $x^2 - 1$ is always -1 or larger.

81. No. If $x = 0$, we obtain the following:
$$x^2 + 1 > 1$$
$$0^2 + 1 > 1$$
$$1 > 1 \leftarrow \text{False}$$
Thus, 0 is not a solution to the inequality. The inequality is true for all other real numbers.

Thus, the solution set to the inequality $x^2 + 1 > 1$ is all real numbers except 0. That is, the solution

set is $\{x \mid x \neq 0\}$ or, using interval notation, $(-\infty, 0) \cup (0, \infty)$.

83. To solve the quadratic inequality $f(x) > 0$ from the graph of $y = f(x)$, where f is a quadratic function, determine where the graph lies above the x-axis. This information will give you the solution set.

85. $\left(4mn^{-3}\right)\left(-2m^4n\right) = -8m^{4+1}n^{-3+1}$
$$= -8m^5n^{-2} = \frac{-8m^5}{n^2}$$

87. $\left(\dfrac{9a^{\frac{2}{3}}b^{\frac{1}{2}}}{a^{-\frac{1}{9}}b^{\frac{3}{4}}}\right)^{-1} = \left(\dfrac{a^{-\frac{1}{9}}b^{\frac{3}{4}}}{9a^{\frac{2}{3}}b^{\frac{1}{2}}}\right)^{1} = \dfrac{a^{-\frac{1}{9}}b^{\frac{3}{4}}}{9a^{\frac{2}{3}}b^{\frac{1}{2}}}$
$$= \frac{b^{\frac{3}{4}-\frac{1}{2}}}{9a^{\frac{2}{3}+\frac{1}{9}}} = \frac{b^{\frac{3}{4}-\frac{2}{4}}}{9a^{\frac{6}{9}+\frac{1}{9}}} = \frac{b^{\frac{1}{4}}}{9a^{\frac{7}{9}}}$$

89. $2x^2 + 7x - 49 > 0$
Let $Y_1 = 2x^2 + 7x - 49$. Graph the quadratic function. Use the ZERO feature to find the x-intercepts.

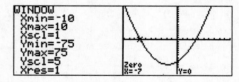

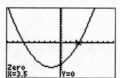

The inequality is strict, so -7 and 3.5 are not part of the solution. From the graph, we can see that $2x^2 + 7x - 49 > 0$ for $x < -7$ or $x > 3.5$. Thus, the solution set is $\{x \mid x < -7 \text{ or } x > 3.5\}$ in set-builder notation; the solution is $(-\infty, -7) \cup (3.5, \infty)$ in interval notation.

91. $\quad 6x^2 + x \leq 40$
$$6x^2 + x - 40 \leq 0$$
Let $Y_1 = 6x^2 + x - 40$. Graph the quadratic function. Use the ZERO feature to find the x-intercepts.

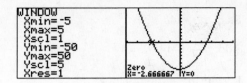

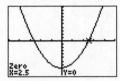

The inequality is non-strict, so $-2.\overline{6} = -\dfrac{8}{3}$ and

$2.5 = \dfrac{5}{2}$ are part of the solution. From the

graph, we can see that $6x^2 + x - 40 \le 0$ for

$-\dfrac{8}{3} \le x \le \dfrac{5}{2}$. Thus, the solution set is

$\left\{ x \,\middle|\, -\dfrac{8}{3} \le x \le \dfrac{5}{2} \right\}$ in set-builder notation; the

solution is $\left[-\dfrac{8}{3}, \dfrac{5}{2} \right]$ in interval notation.

Chapter 7 Review

1. $m^2 = 169$

$\quad m = \pm\sqrt{169}$

$\quad m = \pm 13$

The solution set is $\{-13,\ 13\}$.

2. $n^2 = 75$

$\quad n = \pm\sqrt{75}$

$\quad n = \pm 5\sqrt{3}$

The solution set is $\left\{ -5\sqrt{3},\ 5\sqrt{3} \right\}$.

3. $a^2 = -16$

$\quad a = \pm\sqrt{-16}$

$\quad a = \pm 4i$

The solution set is $\{-4i,\ 4i\}$.

4. $b^2 = \dfrac{8}{9}$

$\quad b = \pm\sqrt{\dfrac{8}{9}}$

$\quad b = \pm\dfrac{\sqrt{8}}{\sqrt{9}}$

$\quad b = \pm\dfrac{2\sqrt{2}}{3}$

The solution set is $\left\{ -\dfrac{2\sqrt{2}}{3},\ \dfrac{2\sqrt{2}}{3} \right\}$.

5. $(x-8)^2 = 81$

$\quad x - 8 = \pm\sqrt{81}$

$\quad x - 8 = \pm 9$

$\quad x = 8 \pm 9$

$\quad x = 8 - 9 \ \text{ or } \ x = 8 + 9$

$\quad x = -1 \quad \text{ or } \ x = 17$

The solution set is $\{-1,\ 17\}$.

6. $(y-2)^2 - 62 = 88$

$\quad (y-2)^2 = 150$

$\quad y - 2 = \pm\sqrt{150}$

$\quad y - 2 = \pm 5\sqrt{6}$

$\quad y = 2 \pm 5\sqrt{6}$

The solution set is $\left\{ 2 - 5\sqrt{6},\ 2 + 5\sqrt{6} \right\}$.

7. $(3z+5)^2 = 100$

$\quad 3z + 5 = \pm\sqrt{100}$

$\quad 3z + 5 = \pm 10$

$\quad 3z = -5 \pm 10$

$\quad 3z = -5 - 10 \ \text{ or } \ 3z = -5 + 10$

$\quad 3z = -15 \quad \text{ or } \ 3z = 5$

$\quad z = -5 \qquad \text{ or } \ z = \dfrac{5}{3}$

The solution set is $\left\{ -5,\ \dfrac{5}{3} \right\}$.

8. $7p^2 = 18$

$$p^2 = \frac{18}{7}$$

$$p = \pm\sqrt{\frac{18}{7}}$$

$$p = \pm\frac{\sqrt{18}}{\sqrt{7}} \cdot \frac{\sqrt{7}}{\sqrt{7}}$$

$$p = \pm\frac{\sqrt{126}}{\sqrt{49}}$$

$$p = \pm\frac{3\sqrt{14}}{7}$$

The solution set is $\left\{-\dfrac{3\sqrt{14}}{7}, \ \dfrac{3\sqrt{14}}{7}\right\}$.

9. $3q^2 + 251 = 11$

$$3q^2 = -240$$

$$q^2 = -80$$

$$q = \pm\sqrt{-80}$$

$$q = \pm 4\sqrt{5}\ i$$

The solution set is $\left\{-4\sqrt{5}\ i, \ 4\sqrt{5}\ i\right\}$.

10. $\left(x + \dfrac{3}{4}\right)^2 = \dfrac{13}{16}$

$$x + \frac{3}{4} = \pm\sqrt{\frac{13}{16}}$$

$$x + \frac{3}{4} = \pm\frac{\sqrt{13}}{4}$$

$$x = -\frac{3}{4} \pm \frac{\sqrt{13}}{4}$$

The solution set is $\left\{-\dfrac{3}{4} - \dfrac{\sqrt{13}}{4}, \ -\dfrac{3}{4} + \dfrac{\sqrt{13}}{4}\right\}$.

11. Start: $a^2 + 30a$

Add: $\left(\dfrac{1}{2} \cdot 30\right)^2 = 225$

Result: $a^2 + 30a + 225$

Factored Form: $(a + 15)^2$

12. Start: $b^2 - 14b$

Add: $\left[\dfrac{1}{2} \cdot (-14)\right]^2 = 49$

Result: $b^2 - 14b + 49$

Factored Form: $(b - 7)^2$

13. Start: $c^2 - 11c$

Add: $\left[\dfrac{1}{2} \cdot (-11)\right]^2 = \dfrac{121}{4}$

Result: $c^2 - 11c + \dfrac{121}{4}$

Factored Form: $\left(c - \dfrac{11}{2}\right)^2$

14. Start: $d^2 + 9d$

Add: $\left(\dfrac{1}{2} \cdot 9\right)^2 = \dfrac{81}{4}$

Result: $d^2 + 9d + \dfrac{81}{4}$

Factored Form: $\left(d + \dfrac{9}{2}\right)^2$

15. Start: $m^2 - \dfrac{1}{4}m$

Add: $\left[\dfrac{1}{2} \cdot \left(-\dfrac{1}{4}\right)\right]^2 = \dfrac{1}{64}$

Result: $m^2 - \dfrac{1}{4}m + \dfrac{1}{64}$

Factored Form: $\left(m - \dfrac{1}{8}\right)^2$

16. Start: $n^2 + \dfrac{6}{7}n$

Add: $\left(\dfrac{1}{2} \cdot \dfrac{6}{7}\right)^2 = \dfrac{9}{49}$

Result: $n^2 + \dfrac{6}{7}n + \dfrac{9}{49}$

Factored Form: $\left(n + \dfrac{3}{7}\right)^2$

17.
$$x^2 - 10x + 16 = 0$$
$$x^2 - 10x = -16$$
$$x^2 - 10x + \left(\frac{1}{2}\cdot(-10)\right)^2 = -16 + \left(\frac{1}{2}\cdot(-10)\right)^2$$
$$x^2 - 10x + 25 = -16 + 25$$
$$(x-5)^2 = 9$$
$$x - 5 = \pm\sqrt{9}$$
$$x - 5 = \pm 3$$
$$x = 5 \pm 3$$
$$x = 2 \text{ or } x = 8$$

The solution set is $\{2,\ 8\}$.

18.
$$y^2 - 3y - 28 = 0$$
$$y^2 - 3y = 28$$
$$y^2 - 3y + \left(\frac{1}{2}\cdot(-3)\right)^2 = 28 + \left(\frac{1}{2}\cdot(-3)\right)^2$$
$$y^2 - 3y + \frac{9}{4} = 28 + \frac{9}{4}$$
$$\left(y - \frac{3}{2}\right)^2 = \frac{121}{4}$$
$$y - \frac{3}{2} = \pm\sqrt{\frac{121}{4}}$$
$$y - \frac{3}{2} = \pm\frac{11}{2}$$
$$y = \frac{3}{2} \pm \frac{11}{2}$$
$$y = -4 \text{ or } y = 7$$

The solution set is $\{-4,\ 7\}$.

19.
$$z^2 - 6z - 3 = 0$$
$$z^2 - 6z = 3$$
$$z^2 - 6z + \left(\frac{1}{2}\cdot(-6)\right)^2 = 3 + \left(\frac{1}{2}\cdot(-6)\right)^2$$
$$z^2 - 6z + 9 = 3 + 9$$
$$(z-3)^2 = 12$$
$$z - 3 = \pm\sqrt{12}$$
$$z - 3 = \pm 2\sqrt{3}$$
$$z = 3 \pm 2\sqrt{3}$$

The solution set is $\left\{3 - 2\sqrt{3},\ 3 + 2\sqrt{3}\right\}$.

20.
$$a^2 - 5a - 7 = 0$$
$$a^2 - 5a = 7$$
$$a^2 - 5a + \left(\frac{1}{2}\cdot(-5)\right)^2 = 7 + \left(\frac{1}{2}\cdot(-5)\right)^2$$
$$a^2 - 5a + \frac{25}{4} = 7 + \frac{25}{4}$$
$$\left(a - \frac{5}{2}\right)^2 = \frac{53}{4}$$
$$a - \frac{5}{2} = \pm\sqrt{\frac{53}{4}}$$
$$a - \frac{5}{2} = \pm\frac{\sqrt{53}}{2}$$
$$a = \frac{5}{2} \pm \frac{\sqrt{53}}{2}$$

The solution set is $\left\{\dfrac{5}{2} - \dfrac{\sqrt{53}}{2},\ \dfrac{5}{2} + \dfrac{\sqrt{53}}{2}\right\}$.

21.
$$b^2 + b + 7 = 0$$
$$b^2 + b = -7$$
$$b^2 + b + \left(\frac{1}{2}\cdot 1\right)^2 = -7 + \left(\frac{1}{2}\cdot 1\right)^2$$
$$b^2 + b + \frac{1}{4} = -7 + \frac{1}{4}$$
$$\left(b + \frac{1}{2}\right)^2 = -\frac{27}{4}$$
$$b + \frac{1}{2} = \pm\sqrt{-\frac{27}{4}}$$
$$b + \frac{1}{2} = \pm\frac{3\sqrt{3}}{2}i$$
$$b = -\frac{1}{2} \pm \frac{3\sqrt{3}}{2}i$$

The solution set is $\left\{-\dfrac{1}{2} - \dfrac{3\sqrt{3}}{2}i,\ -\dfrac{1}{2} + \dfrac{3\sqrt{3}}{2}i\right\}$.

22.

$$c^2 - 6c + 17 = 0$$
$$c^2 - 6c = -17$$
$$c^2 - 6c + \left(\frac{1}{2} \cdot (-6)\right)^2 = -17 + \left(\frac{1}{2} \cdot (-6)\right)^2$$
$$c^2 - 6c + 9 = -17 + 9$$
$$(c - 3)^2 = -8$$
$$c - 3 = \pm\sqrt{-8}$$
$$c - 3 = \pm 2\sqrt{2}\, i$$
$$c = 3 \pm 2\sqrt{2}\, i$$

The solution set is $\left\{3 - 2\sqrt{2}\, i,\ 3 + 2\sqrt{2}\, i\right\}$.

23.

$$2d^2 - 7d + 3 = 0$$
$$\frac{2d^2 - 7d + 3}{2} = \frac{0}{2}$$
$$d^2 - \frac{7}{2}d + \frac{3}{2} = 0$$
$$d^2 - \frac{7}{2}d = -\frac{3}{2}$$
$$d^2 - \frac{7}{2}d + \left[\frac{1}{2} \cdot \left(-\frac{7}{2}\right)\right]^2 = -\frac{3}{2} + \left[\frac{1}{2} \cdot \left(-\frac{7}{2}\right)\right]^2$$
$$d^2 - \frac{7}{2}d + \frac{49}{16} = -\frac{3}{2} + \frac{49}{16}$$
$$\left(d - \frac{7}{4}\right)^2 = \frac{25}{16}$$
$$d - \frac{7}{4} = \pm\sqrt{\frac{25}{16}}$$
$$d - \frac{7}{4} = \pm\frac{5}{4}$$
$$d = \frac{7}{4} \pm \frac{5}{4}$$
$$d = \frac{1}{2} \ \text{ or } \ d = 3$$

The solution set is $\left\{\frac{1}{2},\ 3\right\}$.

24.

$$2w^2 + 2w + 5 = 0$$
$$\frac{2w^2 + 2w + 5}{2} = \frac{0}{2}$$
$$w^2 + w + \frac{5}{2} = 0$$
$$w^2 + w = -\frac{5}{2}$$
$$w^2 + w + \left(\frac{1}{2} \cdot 1\right)^2 = -\frac{5}{2} + \left(\frac{1}{2} \cdot 1\right)^2$$
$$w^2 + w + \frac{1}{4} = -\frac{5}{2} + \frac{1}{4}$$
$$\left(w + \frac{1}{2}\right)^2 = -\frac{9}{4}$$
$$w + \frac{1}{2} = \pm\sqrt{-\frac{9}{4}}$$
$$w + \frac{1}{2} = \pm\frac{3}{2}i$$
$$w = -\frac{1}{2} \pm \frac{3}{2}i$$

The solution set is $\left\{-\frac{1}{2} - \frac{3}{2}i,\ -\frac{1}{2} + \frac{3}{2}i\right\}$.

25.
$$3x^2 - 9x + 8 = 0$$

$$\frac{3x^2 - 9x + 8}{3} = \frac{0}{3}$$

$$x^2 - 3x + \frac{8}{3} = 0$$

$$x^2 - 3x = -\frac{8}{3}$$

$$x^2 - 3x + \left(\frac{1}{2} \cdot (-3)\right)^2 = -\frac{8}{3} + \left(\frac{1}{2} \cdot (-3)\right)^2$$

$$x^2 - 3x + \frac{9}{4} = -\frac{8}{3} + \frac{9}{4}$$

$$\left(x - \frac{3}{2}\right)^2 = -\frac{5}{12}$$

$$x - \frac{3}{2} = \pm \sqrt{-\frac{5}{12}}$$

$$x - \frac{3}{2} = \pm \frac{\sqrt{5}}{\sqrt{12}} i \cdot \frac{\sqrt{3}}{\sqrt{3}}$$

$$x - \frac{3}{2} = \pm \frac{\sqrt{15}}{\sqrt{36}} i$$

$$x - \frac{3}{2} = \pm \frac{\sqrt{15}}{6} i$$

$$x = \frac{3}{2} \pm \frac{\sqrt{15}}{6} i$$

The solution set is $\left\{ \frac{3}{2} - \frac{\sqrt{15}}{6} i, \ \frac{3}{2} + \frac{\sqrt{15}}{6} i \right\}$.

26.
$$3x^2 + 4x - 2 = 0$$

$$x^2 + \frac{4}{3}x - \frac{2}{3} = 0$$

$$x^2 + \frac{4}{3}x = \frac{2}{3}$$

$$x^2 + \frac{4}{3}x + \left(\frac{1}{2} \cdot \frac{4}{3}\right)^2 = \frac{2}{3} + \left(\frac{1}{2} \cdot \frac{4}{3}\right)^2$$

$$x^2 + \frac{4}{3}x + \frac{4}{9} = \frac{2}{3} + \frac{4}{9}$$

$$\left(x + \frac{2}{3}\right)^2 = \frac{10}{9}$$

$$x + \frac{2}{3} = \pm \sqrt{\frac{10}{9}}$$

$$x + \frac{2}{3} = \pm \frac{\sqrt{10}}{3}$$

$$x = -\frac{2}{3} \pm \frac{\sqrt{10}}{3}$$

The solution set is $\left\{ -\frac{2}{3} - \frac{\sqrt{10}}{3}, \ -\frac{2}{3} + \frac{\sqrt{10}}{3} \right\}$.

27. $c^2 = 9^2 + 12^2$
$$= 81 + 144$$
$$= 225$$
$$c = \sqrt{225}$$
$$= 15$$

28. $c^2 = 8^2 + 8^2$
$$= 64 + 64$$
$$= 128$$
$$c = \sqrt{128}$$
$$= 8\sqrt{2}$$

29. $c^2 = 3^2 + 6^2$
$$= 9 + 36$$
$$= 45$$
$$c = \sqrt{45}$$
$$= 3\sqrt{5}$$

30. $c^2 = 10^2 + 24^2$
$$= 100 + 576$$
$$= 676$$
$$c = \sqrt{676}$$
$$= 26$$

31. $c^2 = 5^2 + \left(\sqrt{11}\right)^2$
$$= 25 + 11$$
$$= 36$$
$$c = \sqrt{36}$$
$$= 6$$

32. $c^2 = 6^2 + \left(\sqrt{13}\right)^2$
$$= 36 + 13$$
$$= 49$$
$$c = \sqrt{49}$$
$$= 7$$

33. $c^2 = a^2 + b^2$
$$12^2 = 9^2 + b^2$$
$$144 = 81 + b^2$$
$$63 = b^2$$
$$b = \sqrt{63}$$
$$b = 3\sqrt{7}$$

34.
$$c^2 = a^2 + b^2$$
$$10^2 = a^2 + 5^2$$
$$100 = a^2 + 25$$
$$75 = a^2$$
$$a = \sqrt{75}$$
$$a = 5\sqrt{3}$$

35.
$$c^2 = a^2 + b^2$$
$$17^2 = a^2 + 6^2$$
$$289 = a^2 + 36$$
$$253 = a^2$$
$$a = \sqrt{253}$$

36. For this problem, we are actually looking for the hypotenuse of a right triangle that has legs 90 feet and 90 feet. We use the Pythagorean Theorem to find the desired distance:

$$c^2 = 90^2 + 90^2$$
$$= 8100 + 8100$$
$$= 16,200$$
$$c = \sqrt{16,200}$$
$$= 90\sqrt{2}$$
$$\approx 127.3$$

The distance from home plate to 2nd base is exactly $90\sqrt{2}$ feet or approximately 127.3 feet.

37. $x^2 - x - 20 = 0$

For this equation, $a = 1$, $b = -1$, and $c = -20$.

$$x = \frac{-(-1) \pm \sqrt{(-1)^2 - 4(1)(-20)}}{2(1)}$$
$$= \frac{1 \pm \sqrt{1 + 80}}{2}$$
$$= \frac{1 \pm \sqrt{81}}{2}$$
$$= \frac{1 \pm 9}{2}$$

$$x = \frac{1 - 9}{2} \quad \text{or} \quad x = \frac{1 + 9}{2}$$
$$= \frac{-8}{2} \quad \text{or} \quad = \frac{10}{2}$$
$$= -4 \quad \text{or} \quad = 5$$

The solution set is $\{-4,\ 5\}$.

38.
$$4y^2 = 8y + 21$$
$$4y^2 - 8y - 21 = 0$$

For this equation, $a = 4$, $b = -8$, and $c = -21$.

$$y = \frac{-(-8) \pm \sqrt{(-8)^2 - 4(4)(-21)}}{2(4)}$$
$$= \frac{8 \pm \sqrt{64 + 336}}{8}$$
$$= \frac{8 \pm \sqrt{400}}{8}$$
$$= \frac{8 \pm 20}{8}$$

$$y = \frac{8 - 20}{8} \quad \text{or} \quad y = \frac{8 + 20}{8}$$
$$= \frac{-12}{8} \quad \text{or} \quad = \frac{28}{8}$$
$$= -\frac{3}{2} \quad \text{or} \quad = \frac{7}{2}$$

The solution set is $\left\{-\dfrac{3}{2},\ \dfrac{7}{2}\right\}$.

39.
$$3p^2 + 8p = -3$$
$$3p^2 + 8p + 3 = 0$$

For this equation, $a = 3$, $b = 8$, and $c = 3$.

$$p = \frac{-8 \pm \sqrt{8^2 - 4(3)(3)}}{2(3)}$$
$$= \frac{-8 \pm \sqrt{64 - 36}}{6}$$
$$= \frac{-8 \pm \sqrt{28}}{6}$$
$$= \frac{-8 \pm 2\sqrt{7}}{6}$$
$$= \frac{-8}{6} \pm \frac{2\sqrt{7}}{6}$$
$$= -\frac{4}{3} \pm \frac{\sqrt{7}}{3}$$

The solution set is $\left\{-\dfrac{4}{3} - \dfrac{\sqrt{7}}{3},\ -\dfrac{4}{3} + \dfrac{\sqrt{7}}{3}\right\}$.

40.
$$2q^2 - 3 = 4q$$
$$2q^2 - 4q - 3 = 0$$
For this equation, $a = 2$, $b = -4$, and $c = -3$.
$$q = \frac{-(-4) \pm \sqrt{(-4)^2 - 4(2)(-3)}}{2(2)}$$
$$= \frac{4 \pm \sqrt{16 + 24}}{4}$$
$$= \frac{4 \pm \sqrt{40}}{4}$$
$$= \frac{4 \pm 2\sqrt{10}}{4}$$
$$= \frac{4}{4} \pm \frac{2\sqrt{10}}{4}$$
$$= 1 \pm \frac{\sqrt{10}}{2}$$
The solution set is $\left\{ 1 - \frac{\sqrt{10}}{2}, \ 1 + \frac{\sqrt{10}}{2} \right\}$.

41.
$$3w^2 + w = -3$$
$$3w^2 + w + 3 = 0$$
For this equation, $a = 3$, $b = 1$, and $c = 3$.
$$w = \frac{-1 \pm \sqrt{1^2 - 4(3)(3)}}{2(3)}$$
$$= \frac{-1 \pm \sqrt{1 - 36}}{6}$$
$$= \frac{-1 \pm \sqrt{-35}}{6}$$
$$= \frac{-1 \pm \sqrt{35}\, i}{6}$$
$$= -\frac{1}{6} \pm \frac{\sqrt{35}}{6} i$$
The solution set is $\left\{ -\frac{1}{6} - \frac{\sqrt{35}}{6} i, \ -\frac{1}{6} + \frac{\sqrt{35}}{6} i \right\}$.

42.
$$9z^2 + 16 = 24z$$
$$9z^2 - 24z + 16 = 0$$
For this equation, $a = 9$, $b = -24$, and $c = 16$.
$$z = \frac{-(-24) \pm \sqrt{(-24)^2 - 4(9)(16)}}{2(9)}$$
$$= \frac{24 \pm \sqrt{576 - 576}}{18}$$
$$= \frac{24 \pm \sqrt{0}}{18}$$
$$= \frac{24}{18}$$
$$= \frac{4}{3}$$
The solution set is $\left\{ \frac{4}{3} \right\}$. It is a double root.

43. $m^2 - 4m + 2 = 0$
For this equation, $a = 1$, $b = -4$, and $c = 2$.
$$m = \frac{-(-4) \pm \sqrt{(-4)^2 - 4(1)(2)}}{2(1)}$$
$$= \frac{4 \pm \sqrt{16 - 8}}{2}$$
$$= \frac{4 \pm \sqrt{8}}{2}$$
$$= \frac{4 \pm 2\sqrt{2}}{2}$$
$$= \frac{4}{2} \pm \frac{2\sqrt{2}}{2}$$
$$= 2 \pm \sqrt{2}$$
The solution set is $\left\{ 2 - \sqrt{2}, \ 2 + \sqrt{2} \right\}$.

44. $5n^2 + 4n + 1 = 0$
For this equation, $a = 5$, $b = 4$, and $c = 1$.
$$n = \frac{-4 \pm \sqrt{4^2 - 4(5)(1)}}{2(5)}$$
$$= \frac{-4 \pm \sqrt{16 - 20}}{10}$$
$$= \frac{-4 \pm \sqrt{-4}}{10}$$
$$= \frac{-4 \pm 2i}{10}$$
$$= -\frac{4}{10} \pm \frac{2i}{10} = -\frac{2}{5} \pm \frac{1}{5} i$$
The solution set is $\left\{ -\frac{2}{5} - \frac{1}{5} i, \ -\frac{2}{5} + \frac{1}{5} i \right\}$.

45. $5x + 13 = -x^2$

$x^2 + 5x + 13 = 0$

For this equation, $a = 1$, $b = 5$, and $c = 13$.

$$x = \frac{-5 \pm \sqrt{5^2 - 4(1)(13)}}{2(1)}$$

$$= \frac{-5 \pm \sqrt{25 - 52}}{2}$$

$$= \frac{-5 \pm \sqrt{-27}}{2}$$

$$= \frac{-5 \pm 3\sqrt{3}\, i}{2} = -\frac{5}{2} \pm \frac{3\sqrt{3}}{2} i$$

The solution set is $\left\{ -\dfrac{5}{2} - \dfrac{3\sqrt{3}}{2}i,\ -\dfrac{5}{2} + \dfrac{3\sqrt{3}}{2}i \right\}$.

46. $-2y^2 = 6y + 7$

$0 = 2y^2 + 6y + 7$

For this equation, $a = 2$, $b = 6$, and $c = 7$.

$$y = \frac{-6 \pm \sqrt{6^2 - 4(2)(7)}}{2(2)}$$

$$= \frac{-6 \pm \sqrt{36 - 56}}{4}$$

$$= \frac{-6 \pm \sqrt{-20}}{4}$$

$$= \frac{-6 \pm 2\sqrt{5}\, i}{4}$$

$$= -\frac{6}{4} \pm \frac{2\sqrt{5}}{4} i = -\frac{3}{2} \pm \frac{\sqrt{5}}{2} i$$

The solution set is $\left\{ -\dfrac{3}{2} - \dfrac{\sqrt{5}}{2}i,\ -\dfrac{3}{2} + \dfrac{\sqrt{5}}{2}i \right\}$.

47. $p^2 - 5p - 8 = 0$

For this equation, $a = 1$, $b = -5$, and $c = -8$.

$b^2 - 4ac = (-5)^2 - 4(1)(-8) = 25 + 32 = 57$

Because $b^2 - 4ac = 57$ is positive, but not a perfect square, the quadratic equation will have two irrational solutions.

48. $m^2 + 8m + 16 = 0$

For this equation, $a = 1$, $b = 8$, and $c = 16$.

$b^2 - 4ac = 8^2 - 4(1)(16) = 64 - 64 = 0$

Because $b^2 - 4ac = 0$, the quadratic equation will have one repeated real solution.

49. $3n^2 + n = -4$

$3n^2 + n + 4 = 0$

For this equation, $a = 3$, $b = 1$, and $c = 4$.

$b^2 - 4ac = 1^2 - 4(3)(4) = 1 - 48 = -47$

Because $b^2 - 4ac = -47$ is negative, the quadratic equation will have two complex solutions that are not real. The solutions will be complex conjugates of each other.

50. $7w^2 + 3 = 8w$

$7w^2 - 8w + 3 = 0$

For this equation, $a = 7$, $b = -8$, and $c = 3$.

$b^2 - 4ac = (-8)^2 - 4(7)(3) = 64 - 84 = -20$

Because $b^2 - 4ac = -20$ is negative, the quadratic equation will have two complex solutions that are not real. The solutions will be complex conjugates of each other.

51. $4x^2 + 49 = 28x$

$4x^2 - 28x + 49 = 0$

For this equation, $a = 4$, $b = -28$, and $c = 49$.

$b^2 - 4ac = (-28)^2 - 4(4)(49) = 784 - 784 = 0$

Because $b^2 - 4ac = 0$, the quadratic equation will have one repeated real solution.

52. $11z - 12 = 2z^2$

$0 = 2z^2 - 11z + 12$

For this equation, $a = 2$, $b = -11$, and $c = 12$.

$b^2 - 4ac = (-11)^2 - 4(2)(12) = 121 - 96 = 25$

Because $b^2 - 4ac = 25$ is positive and a perfect square, the quadratic equation will have two rational solutions.

53. $x^2 + 8x - 9 = 0$

Because this equation factors easily, solve by factoring.

$(x + 9)(x - 1) = 0$

$x + 9 = 0$ or $x - 1 = 0$

    $x = -9$ or $x = 1$

The solution set is $\{-9, 1\}$.

54. $6p^2 + 13p = 5$

$6p^2 + 13p - 5 = 0$

Because this equation factors easily, solve by factoring.

$(3p - 1)(2p + 5) = 0$

$3p - 1 = 0$ or $2p + 5 = 0$

$3p = 1$ or $2p = -5$

$p = \dfrac{1}{3}$ or $p = -\dfrac{5}{2}$

The solution set is $\left\{ -\dfrac{5}{2},\ \dfrac{1}{3} \right\}$.

55. $n^2 + 13 = -4n$

$n^2 + 4n + 13 = 0$

Because this equation does not easily factor, solve by using the quadratic formula. For this equation, $a = 1$, $b = 4$, and $c = 13$.

$$n = \frac{-4 \pm \sqrt{4^2 - 4(1)(13)}}{2(1)}$$

$$= \frac{-4 \pm \sqrt{16 - 52}}{2}$$

$$= \frac{-4 \pm \sqrt{-36}}{2}$$

$$= \frac{-4 \pm 6i}{2}$$

$$= -\frac{4}{2} \pm \frac{6i}{2}$$

$$= -2 \pm 3i$$

The solution set is $\{-2 - 3i,\ -2 + 3i\}$.

56. $5y^2 - 60 = 0$

Because this equation has no linear term, solve by using the square root method.

$5y^2 = 60$

$y^2 = 12$

$y = \pm\sqrt{12}$

$y = \pm 2\sqrt{3}$

The solution set is $\left\{ -2\sqrt{3},\ 2\sqrt{3} \right\}$.

57. $\dfrac{1}{4}q^2 - \dfrac{1}{2}q - \dfrac{3}{8} = 0$

$8\left(\dfrac{1}{4}q^2 - \dfrac{1}{2}q - \dfrac{3}{8} \right) = 8(0)$

$2q^2 - 4q - 3 = 0$

Because this equation does not easily factor, solve by using the quadratic formula. For this equation, $a = 2$, $b = -4$, and $c = -3$.

$$q = \frac{-(-4) \pm \sqrt{(-4)^2 - 4(2)(-3)}}{2(2)}$$

$$= \frac{4 \pm \sqrt{16 + 24}}{4}$$

$$= \frac{4 \pm \sqrt{40}}{4}$$

$$= \frac{4 \pm 2\sqrt{10}}{4}$$

$$= \frac{4}{4} \pm \frac{2\sqrt{10}}{4}$$

$$= 1 \pm \frac{\sqrt{10}}{2}$$

The solution set is $\left\{ 1 - \dfrac{\sqrt{10}}{2},\ 1 + \dfrac{\sqrt{10}}{2} \right\}$.

58. $\dfrac{1}{8}m^2 + m + \dfrac{5}{2} = 0$

$8\left(\dfrac{1}{8}m^2 + m + \dfrac{5}{2} \right) = 8(0)$

$m^2 + 8m + 20 = 0$

Because this equation does not easily factor, solve by using the quadratic formula. For this equation, $a = 1$, $b = 8$, and $c = 20$.

$$m = \frac{-8 \pm \sqrt{8^2 - 4(1)(20)}}{2(1)}$$

$$= \frac{-8 \pm \sqrt{64 - 80}}{2}$$

$$= \frac{-8 \pm \sqrt{-16}}{2}$$

$$= \frac{-8 \pm 4i}{2}$$

$$= -\frac{8}{2} \pm \frac{4i}{2}$$

$$= -4 \pm 2i$$

The solution set is $\{-4 - 2i,\ -4 + 2i\}$.

59. $(w - 8)(w + 6) = -33$

$w^2 - 2w - 48 = -33$

$w^2 - 2w - 15 = 0$

Because this equation factors easily, solve by factoring.

$$(w-5)(w+3)=0$$
$$w-5=0 \quad \text{or} \quad w+3=0$$
$$w=5 \quad \text{or} \quad w=-3$$
The solution set is $\{-3,\ 5\}$.

60. $(x-3)(x+1)=-2$
$$x^2-2x-3=-2$$
$$x^2-2x-1=0$$
Because this equation does not easily factor, solve by using the quadratic formula. For this equation, $a=1$, $b=-2$, and $c=-1$.

$$x=\frac{-(-2)\pm\sqrt{(-2)^2-4(1)(-1)}}{2(1)}$$
$$=\frac{2\pm\sqrt{4+4}}{2}$$
$$=\frac{2\pm\sqrt{8}}{2}$$
$$=\frac{2\pm 2\sqrt{2}}{2}$$
$$=\frac{2}{2}\pm\frac{2\sqrt{2}}{2}$$
$$=1\pm\sqrt{2}$$

The solution set is $\left\{1-\sqrt{2},\ 1+\sqrt{2}\right\}$.

61. $9z^2=16$
Because this equation has no linear term, solve by using the square root method.
$$z^2=\frac{16}{9}$$
$$z=\pm\sqrt{\frac{16}{9}}$$
$$z=\pm\frac{4}{3}$$

The solution set is $\left\{-\dfrac{4}{3},\ \dfrac{4}{3}\right\}$.

62.
$$\frac{1-2x}{x^2+5}=1$$
$$\left(x^2+5\right)\left(\frac{1-2x}{x^2+5}\right)=\left(x^2+5\right)(1)$$
$$1-2x=x^2+5$$
$$0=x^2+2x+4$$

Because this equation does not easily factor, solve by using the quadratic formula. For this equation, $a=1$, $b=2$, and $c=4$.

$$x=\frac{-2\pm\sqrt{2^2-4(1)(4)}}{2(1)}$$
$$=\frac{-2\pm\sqrt{4-16}}{2}$$
$$=\frac{-2\pm\sqrt{-12}}{2}$$
$$=\frac{-2\pm 2\sqrt{3}\ i}{2}$$
$$=-1\pm\sqrt{3}\ i$$

The solution set is $\left\{-1-\sqrt{3}\ i,\ -1+\sqrt{3}\ i\right\}$.

63.
$$(x+2)^2+(x-5)^2=(x+3)^2$$
$$x^2+4x+4+x^2-10x+25=x^2+6x+9$$
$$2x^2-6x+29=x^2+6x+9$$
$$x^2-12x+20=0$$
$$(x-2)(x-10)=0$$
$$x-2=0 \quad \text{or} \quad x-10=0$$
$$x=2 \quad \text{or} \quad x=10$$

Disregard $x=2$ because this value will cause the length of one of the legs to be negative: $x-5=2-5=-3$. Thus, $x=10$ is the only viable answer. Now, $x-5=10-5=5$, $x+2=10+2=12$, and $x+3=10+3=13$. The three measurements are 5, 12, and 13.

64. Let x represent the length of the rectangle. Then $x-3$ will represent the width.
$$x(x-3)=108$$
$$x^2-3x=108$$
$$x^2-3x-108=0$$
$$(x-12)(x+9)=0$$
$$x-12=0 \quad \text{or} \quad x+9=0$$
$$x=12 \quad \text{or} \quad x=-9$$

Disregard $x=-9$ because x represents the length of the rectangle, which must be positive. Thus, $x=12$ is the only viable answer. Now, $x-3=12-3=9$. Thus, the dimensions of the rectangle are 12 centimeters by 9 centimeters.

65. a. $-0.2x^2 + 180x = 36,000$

$$0 = 0.2x^2 - 180x + 36,000$$

For this equation, $a = 0.2$, $b = -180$, and $c = 36,000$.

$$x = \frac{-(-180) \pm \sqrt{(-180)^2 - 4(0.2)(36,000)}}{2(0.2)}$$

$$= \frac{180 \pm \sqrt{32,400 - 28,800}}{0.4}$$

$$= \frac{180 \pm \sqrt{3600}}{0.4}$$

$$= \frac{180 \pm 60}{0.4}$$

$$x = \frac{180 - 60}{0.4} \quad \text{or} \quad x = \frac{180 + 60}{0.4}$$
$$= 300 \quad \text{or} \quad = 600$$

The revenue will be \$36,000 per week if either 300 or 600 cellular phones are sold per week.

b. $-0.2x^2 + 180x = 40,500$

$$0 = 0.2x^2 - 180x + 40,500$$

For this equation, $a = 0.2$, $b = -180$, and $c = 40,500$.

$$x = \frac{-(-180) \pm \sqrt{(-180)^2 - 4(0.2)(40,500)}}{2(0.2)}$$

$$= \frac{180 \pm \sqrt{32,400 - 32,400}}{0.4}$$

$$= \frac{180 \pm \sqrt{0}}{0.4}$$

$$= \frac{180}{0.4}$$

$$= 450$$

The revenue will be \$40,500 per week if 450 cellular phones are sold per week.

66. a. $200 = -16t^2 + 50t + 180$

$$16t^2 - 50t + 20 = 0$$

For this equation, $a = 16$, $b = -50$, and $c = 20$.

$$t = \frac{-(-50) \pm \sqrt{(-50)^2 - 4(16)(20)}}{2(16)}$$

$$= \frac{50 \pm \sqrt{2500 - 1280}}{32}$$

$$= \frac{50 \pm \sqrt{1220}}{32}$$

$$= \frac{50 \pm 2\sqrt{305}}{32}$$

$$= \frac{50}{32} \pm \frac{2\sqrt{305}}{32}$$

$$= \frac{25}{16} \pm \frac{\sqrt{305}}{16}$$

$$t = \frac{25}{16} - \frac{\sqrt{305}}{16} \quad \text{or} \quad t = \frac{25}{16} + \frac{\sqrt{305}}{16}$$
$$\approx 0.471 \quad \text{or} \quad \approx 2.654$$

Rounding to the nearest tenth, the height of the ball will be 200 feet after approximately 0.5 seconds and after approximately 2.7 seconds.

b. $100 = -16t^2 + 50t + 180$

$$16t^2 - 50t - 80 = 0$$

For this equation, $a = 16$, $b = -50$, and $c = -80$.

$$t = \frac{-(-50) \pm \sqrt{(-50)^2 - 4(16)(-80)}}{2(16)}$$

$$= \frac{50 \pm \sqrt{2500 + 5120}}{32}$$

$$= \frac{50 \pm \sqrt{7620}}{32}$$

$$= \frac{50 \pm 2\sqrt{1905}}{32}$$

$$= \frac{50}{32} \pm \frac{2\sqrt{1905}}{32}$$

$$= \frac{25}{16} \pm \frac{\sqrt{1905}}{16}$$

$$t = \frac{25}{16} - \frac{\sqrt{1905}}{16} \quad \text{or} \quad t = \frac{25}{16} + \frac{\sqrt{1905}}{16}$$
$$\approx -1.165 \quad \text{or} \quad \approx 4.290$$

Because time cannot be negative, we disregard -1.165. Thus, $t \approx 4.290$ is the only viable answer. Rounding to the nearest tenth, the height of the ball will be 100 feet after approximately 4.3 seconds.

c.
$$300 = -16t^2 + 50t + 180$$
$$16t^2 - 50t + 120 = 0$$
For this equation, $a = 16$, $b = -50$, and $c = 120$.

$$t = \frac{-(-50) \pm \sqrt{(-50)^2 - 4(16)(120)}}{2(16)}$$

$$= \frac{50 \pm \sqrt{2500 - 7680}}{32}$$

$$= \frac{50 \pm \sqrt{-5180}}{32}$$

$$= \frac{50 \pm 2\sqrt{1295}\,i}{32}$$

$$= \frac{50}{32} \pm \frac{2\sqrt{1295}}{32}\,i$$

$$= \frac{25}{16} \pm \frac{\sqrt{1295}}{16}\,i$$

$$t = \frac{25}{16} - \frac{\sqrt{1295}}{16}\,i \quad \text{or} \quad t = \frac{25}{16} + \frac{\sqrt{1295}}{16}\,i$$

The ball will never reach a height of 300 feet. This is clear because the solutions to the equation above are complex solutions that are not real.

67. Let x represent the speed the boat would travel in still water.

	Distance	Rate	Time
Up Stream	10	$x-3$	$\dfrac{10}{x-3}$
Down Stream	10	$x+3$	$\dfrac{10}{x+3}$

$$\frac{10}{x-3} + \frac{10}{x+3} = 2$$

$$(x-3)(x+3)\left(\frac{10}{x-3} + \frac{10}{x+3}\right) = (x-3)(x+3)(2)$$

$$10(x+3) + 10(x-3) = \left(x^2 - 9\right)(2)$$

$$10x + 30 + 10x - 30 = 2x^2 - 18$$

$$20x = 2x^2 - 18$$

$$0 = 2x^2 - 20x - 18$$

For this equation, $a = 2$, $b = -20$, and $c = -18$.

$$x = \frac{-(-20) \pm \sqrt{(-20)^2 - 4(2)(-18)}}{2(2)}$$

$$= \frac{20 \pm \sqrt{400 + 144}}{4}$$

$$= \frac{20 \pm \sqrt{544}}{4}$$

$$= \frac{20 \pm 4\sqrt{34}}{4}$$

$$= 5 \pm \sqrt{34}$$

$$x = 5 - \sqrt{34} \quad \text{or} \quad x = 5 + \sqrt{34}$$

$$\approx -0.831 \quad \text{or} \quad \approx 10.831$$

Because the speed should be positive, we disregard $x \approx -0.831$. Thus, the only viable answer is $x \approx 10.831$. Rounding to the nearest tenth, the boat would travel approximately 10.8 miles per hour in still water.

68. Let t represent the time required for Beth to wash the car alone. Then $t - 14$ will represent the time required for Tom to wash the car alone.

$$\begin{pmatrix} \text{Part done} \\ \text{by Beth in} \\ 1 \text{ minute} \end{pmatrix} + \begin{pmatrix} \text{Part done} \\ \text{by Tom in} \\ 1 \text{ minute} \end{pmatrix} = \begin{pmatrix} \text{Part done} \\ \text{together in} \\ 1 \text{ minute} \end{pmatrix}$$

$$\frac{1}{t} + \frac{1}{t-14} = \frac{1}{30}$$

$$30t(t-14)\left(\frac{1}{t} + \frac{1}{t-14}\right) = 30t(t-14)\left(\frac{1}{30}\right)$$

$$30(t-14) + 30t = t(t-14)$$

$$30t - 420 + 30t = t^2 - 14t$$

$$60t - 420 = t^2 - 14t$$

$$0 = t^2 - 74t + 420$$

For this equation, $a = 1$, $b = -74$, and $c = 420$.

$$t = \frac{-(-74) \pm \sqrt{(-74)^2 - 4(1)(420)}}{2(1)}$$

$$= \frac{74 \pm \sqrt{5476 - 1680}}{2}$$

$$= \frac{74 \pm \sqrt{3796}}{2}$$

$$= \frac{74 \pm 2\sqrt{949}}{2}$$

$$= 37 \pm \sqrt{949}$$

$$t = 37 - \sqrt{949} \quad \text{or} \quad t = 37 + \sqrt{949}$$

$$\approx 6.194 \quad \text{or} \quad \approx 67.806$$

Disregard $t \approx 6.194$ because this value makes Tom's time negative:

$t - 14 = 6.194 - 14 = -7.806$. The only viable answer is $t \approx 67.806$ minutes. Working alone, it will take Beth approximately 67.8 minutes to wash the car.

69. $x^4 + 7x^2 - 144 = 0$

$$\left(x^2\right)^2 + 7\left(x^2\right) - 144 = 0$$

Let $u = x^2$.

$u^2 + 7u - 144 = 0$

$(u - 9)(u + 16) = 0$

$u - 9 = 0$ or $u + 16 = 0$

$u = 9$ or $u = -16$

$x^2 = 9$ or $x^2 = -16$

$x = \pm\sqrt{9}$ or $x = \pm\sqrt{-16}$

$x = \pm 3$ or $x = \pm 4i$

<u>Check:</u>

$x = -3:$ $(-3)^4 + 7(-3)^2 - 144 \stackrel{?}{=} 0$

$81 + 7 \cdot 9 - 144 \stackrel{?}{=} 0$

$81 + 63 - 144 \stackrel{?}{=} 0$

$144 = 144$ ✓

$x = 3:$ $3^4 + 7(3)^2 - 144 \stackrel{?}{=} 0$

$81 + 7 \cdot 9 - 144 \stackrel{?}{=} 0$

$81 + 63 - 144 \stackrel{?}{=} 0$

$0 = 0$ ✓

$x = -4i:$ $(-4i)^4 + 7(-4i)^2 - 144 \stackrel{?}{=} 0$

$256i^4 + 7 \cdot 16i^2 - 144 \stackrel{?}{=} 0$

$256(1) + 7 \cdot 16(-1) - 144 \stackrel{?}{=} 0$

$256 - 112 - 144 \stackrel{?}{=} 0$

$0 = 0$ ✓

$x = 4i:$ $(4i)^4 + 7(4i)^2 - 144 \stackrel{?}{=} 0$

$256i^4 + 7 \cdot 16i^2 - 144 \stackrel{?}{=} 0$

$256(1) + 7 \cdot 16(-1) - 144 \stackrel{?}{=} 0$

$256 - 112 - 144 \stackrel{?}{=} 0$

$0 = 0$ ✓

All check; the solution set is $\{-3,\ 3,\ -4i,\ 4i\}$.

70. $4w^4 + 5w^2 - 6 = 0$

$$4\left(w^2\right)^2 + 5\left(w^2\right) - 6 = 0$$

Let $u = w^2$.

$4u^2 + 5u - 6 = 0$

$(4u - 3)(u + 2) = 0$

$4u - 3 = 0$ or $u + 2 = 0$

$4u = 3$ or $u = -2$

$u = \dfrac{3}{4}$

$w^2 = \dfrac{3}{4}$ or $w^2 = -2$

$w = \pm\sqrt{\dfrac{3}{4}}$ or $w = \pm\sqrt{-2}$

$w = \pm\dfrac{\sqrt{3}}{2}$ or $w = \pm\sqrt{2}\,i$

<u>Check:</u>

$w = -\dfrac{\sqrt{3}}{2}:$ $4\left(-\dfrac{\sqrt{3}}{2}\right)^4 + 5\left(-\dfrac{\sqrt{3}}{2}\right)^2 - 6 \stackrel{?}{=} 0$

$4\left(\dfrac{9}{16}\right) + 5\left(\dfrac{3}{4}\right) - 6 \stackrel{?}{=} 0$

$\dfrac{9}{4} + \dfrac{15}{4} - 6 \stackrel{?}{=} 0$

$0 = 0$ ✓

$w = \dfrac{\sqrt{3}}{2}:$ $4\left(\dfrac{\sqrt{3}}{2}\right)^4 + 5\left(\dfrac{\sqrt{3}}{2}\right)^2 - 6 \stackrel{?}{=} 0$

$4\left(\dfrac{9}{16}\right) + 5\left(\dfrac{3}{4}\right) - 6 \stackrel{?}{=} 0$

$\dfrac{9}{4} + \dfrac{15}{4} - 6 \stackrel{?}{=} 0$

$0 = 0$ ✓

$w = -\sqrt{2}\,i:$ $4\left(-\sqrt{2}\,i\right)^4 + 5\left(-\sqrt{2}\,i\right)^2 - 6 \stackrel{?}{=} 0$

$4 \cdot 4i^4 + 5 \cdot 2i^2 - 6 \stackrel{?}{=} 0$

$4 \cdot 4(1) + 5 \cdot 2(-1) - 6 \stackrel{?}{=} 0$

$16 - 10 - 6 \stackrel{?}{=} 0$

$0 = 0$ ✓

$w = \sqrt{2}\,i:$ $4\left(\sqrt{2}\,i\right)^4 + 5\left(\sqrt{2}\,i\right)^2 - 6 \stackrel{?}{=} 0$

$4 \cdot 4i^4 + 5 \cdot 2i^2 - 6 \stackrel{?}{=} 0$

$4 \cdot 4(1) + 5 \cdot 2(-1) - 6 \stackrel{?}{=} 0$

$16 - 10 - 6 \stackrel{?}{=} 0$

$0 = 0$ ✓

All check; the solution set is

$$\left\{ -\dfrac{\sqrt{3}}{2},\ \dfrac{\sqrt{3}}{2},\ -\sqrt{2}\,i,\ \sqrt{2}\,i \right\} .$$

71. $3(a+4)^2 - 11(a+4) + 6 = 0$

Let $u = a + 4$.

$3u^2 - 11u + 6 = 0$

$(3u - 2)(u - 3) = 0$

$3u - 2 = 0$ or $u - 3 = 0$

$\quad 3u = 2$ or $u = 3$

$\qquad u = \dfrac{2}{3}$

$a + 4 = \dfrac{2}{3}$ or $a + 4 = 3$

$\quad\ a = -\dfrac{10}{3}$ or $a = -1$

Check:

$a = -\dfrac{10}{3}$:

$3\left(-\dfrac{10}{3} + 4\right)^2 - 11\left(-\dfrac{10}{3} + 4\right) + 6 \overset{?}{=} 0$

$\qquad 3\left(\dfrac{2}{3}\right)^2 - 11\left(\dfrac{2}{3}\right) + 6 \overset{?}{=} 0$

$\qquad\quad 3\left(\dfrac{4}{9}\right) - \dfrac{22}{3} + 6 \overset{?}{=} 0$

$\qquad\qquad \dfrac{4}{3} - \dfrac{22}{3} + 6 \overset{?}{=} 0$

$\qquad\qquad\qquad\qquad 0 = 0 \ \checkmark$

$a = -1:\ \ 3(-1+4)^2 - 11(-1+4) + 6 \overset{?}{=} 0$

$\qquad\qquad\quad 3(3)^2 - 11(3) + 6 \overset{?}{=} 0$

$\qquad\qquad\qquad 3(9) - 33 + 6 \overset{?}{=} 0$

$\qquad\qquad\qquad\quad 27 - 33 + 6 \overset{?}{=} 0$

$\qquad\qquad\qquad\qquad\qquad 0 = 0 \ \checkmark$

Both check; the solution set is $\left\{-\dfrac{10}{3},\ -1\right\}$.

72. $\left(q^2 - 11\right)^2 - 2\left(q^2 - 11\right) - 15 = 0$

Let $u = q^2 - 11$.

$u^2 - 2u - 15 = 0$

$(u + 3)(u - 5) = 0$

$u + 3 = 0$ or $u - 5 = 0$

$\quad u = -3$ or $u = 5$

$q^2 - 11 = -3$ or $q^2 - 11 = 5$

$\quad\ q^2 = 8$ or $q^2 = 16$

$\quad\ q = \pm\sqrt{8}$ or $q = \pm\sqrt{16}$

$\quad\ q = \pm 2\sqrt{2}$ or $q = \pm 4$

Check:

$q = -2\sqrt{2}$:

$\left(\left(-2\sqrt{2}\right)^2 - 11\right)^2 - 2\left(\left(-2\sqrt{2}\right)^2 - 11\right) - 15 \overset{?}{=} 0$

$\qquad\qquad\quad (8 - 11)^2 - 2(8 - 11) - 15 \overset{?}{=} 0$

$\qquad\qquad\quad\ (-3)^2 - 2(-3) - 15 \overset{?}{=} 0$

$\qquad\qquad\qquad\qquad 9 + 6 - 15 \overset{?}{=} 0$

$\qquad\qquad\qquad\qquad\qquad\quad 0 = 0 \ \checkmark$

$q = 2\sqrt{2}$:

$\left(\left(2\sqrt{2}\right)^2 - 11\right)^2 - 2\left(\left(2\sqrt{2}\right)^2 - 11\right) - 15 \overset{?}{=} 0$

$\qquad\qquad\quad (8 - 11)^2 - 2(8 - 11) - 15 \overset{?}{=} 0$

$\qquad\qquad\quad\ (-3)^2 - 2(-3) - 15 \overset{?}{=} 0$

$\qquad\qquad\qquad\qquad 9 + 6 - 15 \overset{?}{=} 0$

$\qquad\qquad\qquad\qquad\qquad\quad 0 = 0 \ \checkmark$

$q = -4$:

$\left((-4)^2 - 11\right)^2 - 2\left((-4)^2 - 11\right) - 15 \overset{?}{=} 0$

$\qquad\quad (16 - 11)^2 - 2(16 - 11) - 15 \overset{?}{=} 0$

$\qquad\qquad\quad\ (5)^2 - 2(5) - 15 \overset{?}{=} 0$

$\qquad\qquad\qquad 25 - 10 - 15 \overset{?}{=} 0$

$\qquad\qquad\qquad\qquad\qquad 0 = 0 \ \checkmark$

$q = 4:\ \ \left(4^2 - 11\right)^2 - 2\left(4^2 - 11\right) - 15 \overset{?}{=} 0$

$\qquad\qquad (16 - 11)^2 - 2(16 - 11) - 15 \overset{?}{=} 0$

$\qquad\qquad\qquad (5)^2 - 2(5) - 15 \overset{?}{=} 0$

$\qquad\qquad\qquad\ 25 - 10 - 15 \overset{?}{=} 0$

$\qquad\qquad\qquad\qquad\qquad\ 0 = 0 \ \checkmark$

All check; the solution set is
$\left\{-4,\ -2\sqrt{2},\ 2\sqrt{2},\ 4\right\}$.

73.
$$y - 13\sqrt{y} + 36 = 0$$
$$\left(\sqrt{y}\right)^2 - 13\left(\sqrt{y}\right) + 36 = 0$$

Let $u = \sqrt{y}$.
$$u^2 - 13u + 36 = 0$$
$$(u - 4)(u - 9) = 0$$
$$u - 4 = 0 \quad \text{or} \quad u - 9 = 0$$
$$u = 4 \quad \text{or} \quad u = 9$$

$$\sqrt{y} = 4 \quad \text{or} \quad \sqrt{y} = 9$$
$$y = 4^2 \quad \text{or} \quad y = 9^2$$
$$y = 16 \quad \text{or} \quad y = 81$$

Check:
$$y = 16: \quad 16 - 13\sqrt{16} + 36 \stackrel{?}{=} 0$$
$$16 - 13 \cdot 4 + 36 \stackrel{?}{=} 0$$
$$16 - 52 + 36 \stackrel{?}{=} 0$$
$$0 = 0 \ \checkmark$$

$$y = 16: \quad 81 - 13\sqrt{81} + 36 \stackrel{?}{=} 0$$
$$81 - 13 \cdot 9 + 36 \stackrel{?}{=} 0$$
$$81 - 117 + 36 \stackrel{?}{=} 0$$
$$0 = 0 \ \checkmark$$

Both check, the solution set is $\{16,\ 81\}$.

74.
$$5z + 2\sqrt{z} - 3 = 0$$
$$5\left(\sqrt{z}\right)^2 + 2\left(\sqrt{z}\right) - 3 = 0$$

Let $u = \sqrt{z}$.
$$5u^2 + 2u - 3 = 0$$
$$(5u - 3)(u + 1) = 0$$
$$5u - 3 = 0 \quad \text{or} \quad u + 1 = 0$$
$$5u = 3 \quad \text{or} \quad u = -1$$
$$u = \frac{3}{5}$$

$$\sqrt{z} = \frac{3}{5} \quad \text{or} \quad \sqrt{z} = -1$$
$$z = \left(\frac{3}{5}\right)^2 \quad \text{or} \quad z = (-1)^2$$
$$z = \frac{9}{25} \quad \text{or} \quad z = 1$$

Check:
$$z = \frac{9}{25}: \quad 5 \cdot \frac{9}{25} + 2\sqrt{\frac{9}{25}} - 3 \stackrel{?}{=} 0$$
$$\frac{9}{5} + 2 \cdot \frac{3}{5} - 3 \stackrel{?}{=} 0$$
$$\frac{9}{5} + \frac{6}{5} - 3 \stackrel{?}{=} 0$$
$$0 = 0 \ \checkmark$$

$$z = 1: \quad 5 \cdot 1 + 2\sqrt{1} - 3 \stackrel{?}{=} 0$$
$$5 + 2 \cdot 1 - 3 \stackrel{?}{=} 0$$
$$5 + 2 - 3 \stackrel{?}{=} 0$$
$$4 \neq 0 \ \text{✗}$$

$z = 1$ does not check; the solution set is $\left\{\dfrac{9}{25}\right\}$.

75.
$$p^{-2} - 4p^{-1} - 21 = 0$$
$$\left(p^{-1}\right)^2 - 4\left(p^{-1}\right) - 21 = 0$$

Let $u = p^{-1}$.
$$u^2 - 4u - 21 = 0$$
$$(u + 3)(u - 7) = 0$$
$$u + 3 = 0 \quad \text{or} \quad u - 7 = 0$$
$$u = -3 \quad \text{or} \quad u = 7$$
$$p^{-1} = -3 \quad \text{or} \quad p^{-1} = 7$$
$$\frac{1}{p} = -3 \quad \text{or} \quad \frac{1}{p} = 7$$
$$p = -\frac{1}{3} \quad \text{or} \quad p = \frac{1}{7}$$

Check:
$$p = -\frac{1}{3}: \quad \left(-\frac{1}{3}\right)^{-2} - 4\left(-\frac{1}{3}\right)^{-1} - 21 \stackrel{?}{=} 0$$
$$(-3)^2 - 4(-3) - 21 \stackrel{?}{=} 0$$
$$9 + 12 - 21 \stackrel{?}{=} 0$$
$$0 = 0 \ \checkmark$$

$$p = \frac{1}{7}: \quad \left(\frac{1}{7}\right)^{-2} - 4\left(\frac{1}{7}\right)^{-1} - 21 \stackrel{?}{=} 0$$
$$(7)^2 - 4(7) - 21 \stackrel{?}{=} 0$$
$$49 - 28 - 21 \stackrel{?}{=} 0$$
$$0 = 0 \ \checkmark$$

Both check; the solution set is $\left\{-\dfrac{1}{3},\ \dfrac{1}{7}\right\}$.

76.
$$2b^{2/3} + 13b^{1/3} - 7 = 0$$
$$2\left(b^{1/3}\right)^2 + 13\left(b^{1/3}\right) - 7 = 0$$

Let $u = b^{1/3}$.
$$2u^2 + 13u - 7 = 0$$
$$(2u - 1)(u + 7) = 0$$

$2u - 1 = 0$ or $u + 7 = 0$
$2u = 1$ or $u = -7$
$u = \dfrac{1}{2}$

$b^{1/3} = \dfrac{1}{2}$ or $b^{1/3} = -7$

$\left(b^{1/3}\right)^3 = \left(\dfrac{1}{2}\right)^3$ or $\left(b^{1/3}\right)^3 = (-7)^3$

$b = \dfrac{1}{8}$ or $b = -343$

Check:

$b = \dfrac{1}{8}$:

$$2\left(\frac{1}{8}\right)^{2/3} + 13\left(\frac{1}{8}\right)^{1/3} - 7 \overset{?}{=} 0$$
$$2\left(\sqrt[3]{\frac{1}{8}}\right)^2 + 13\left(\sqrt[3]{\frac{1}{8}}\right) - 7 \overset{?}{=} 0$$
$$2\left(\frac{1}{2}\right)^2 + 13\left(\frac{1}{2}\right) - 7 \overset{?}{=} 0$$
$$2\left(\frac{1}{4}\right) + 13\left(\frac{1}{2}\right) - 7 \overset{?}{=} 0$$
$$\frac{1}{2} + \frac{13}{2} - 7 \overset{?}{=} 0$$
$$0 = 0 \checkmark$$

$b = -343$: $2(-343)^{2/3} + 13(-343)^{1/3} - 7 \overset{?}{=} 0$
$$2\left(\sqrt[3]{-343}\right)^2 + 13\left(\sqrt[3]{-343}\right) - 7 \overset{?}{=} 0$$
$$2(-7)^2 + 13(-7) - 7 \overset{?}{=} 0$$
$$2(49) - 91 - 7 \overset{?}{=} 0$$
$$98 - 91 - 7 \overset{?}{=} 0$$
$$0 = 0 \checkmark$$

Both check; the solution set is $\left\{-343,\ \dfrac{1}{8}\right\}$.

77.
$$m^{1/2} + 2m^{1/4} - 8 = 0$$
$$\left(m^{1/4}\right)^2 + 2\left(m^{1/4}\right) - 8 = 0$$

Let $u = m^{1/4}$.
$$u^2 + 2u - 8 = 0$$
$$(u - 2)(u + 4) = 0$$

$u - 2 = 0$ or $u + 4 = 0$
$u = 2$ or $u = -4$
$m^{1/4} = 2$ or $m^{1/4} = -4$
$\left(m^{1/4}\right)^4 = 2^4$ or $\left(m^{1/4}\right)^4 = (-4)^4$
$m = 16$ or $m = 256$

Check:

$m = 16$: $(16)^{1/2} + 2(16)^{1/4} - 8 \overset{?}{=} 0$
$$\sqrt{16} + 2\sqrt[4]{16} - 8 \overset{?}{=} 0$$
$$4 + 2 \cdot 2 - 8 \overset{?}{=} 0$$
$$4 + 4 - 8 \overset{?}{=} 0$$
$$0 = 0 \checkmark$$

$m = 256$: $(256)^{1/2} + 2(256)^{1/4} - 8 \overset{?}{=} 0$
$$\sqrt{256} + 2\sqrt[4]{256} - 8 \overset{?}{=} 0$$
$$16 + 2 \cdot 4 - 8 \overset{?}{=} 0$$
$$16 + 8 - 8 \overset{?}{=} 0$$
$$16 \neq 0 \ \boldsymbol{\times}$$

$m = 256$ does not check; the solution set is $\{16\}$.

78.
$$\left(\frac{1}{x+5}\right)^2 + \frac{3}{x+5} = 28$$
$$\left(\frac{1}{x+5}\right)^2 + 3\left(\frac{1}{x+5}\right) - 28 = 0$$

Let $u = \dfrac{1}{x+5}$.
$$u^2 + 3u - 28 = 0$$
$$(u - 4)(u + 7) = 0$$

$u - 4 = 0$ or $u + 7 = 0$

$u = 4$ or $u = -7$

$\dfrac{1}{x+5} = 4$ or $\dfrac{1}{x+5} = -7$

$x + 5 = \dfrac{1}{4}$ or $x + 5 = -\dfrac{1}{7}$

$x = -\dfrac{19}{4}$ or $x = -\dfrac{36}{7}$

Check:

$x = -\dfrac{19}{4}:\quad \left(\dfrac{1}{-\frac{19}{4}+5}\right)^2 + \dfrac{3}{-\frac{19}{4}+5} \overset{?}{=} 28$

$\left(\dfrac{1}{\frac{1}{4}}\right)^2 + \dfrac{3}{\frac{1}{4}} \overset{?}{=} 28$

$(4)^2 + 12 \overset{?}{=} 28$

$16 + 12 \overset{?}{=} 28$

$28 = 28 \checkmark$

$x = -\dfrac{36}{7}:\quad \left(\dfrac{1}{-\frac{36}{7}+5}\right)^2 + \dfrac{3}{-\frac{36}{7}+5} \overset{?}{=} 28$

$\left(\dfrac{1}{-\frac{1}{7}}\right)^2 + \dfrac{3}{-\frac{1}{7}} \overset{?}{=} 28$

$(-7)^2 - 21 \overset{?}{=} 28$

$49 - 21 \overset{?}{=} 28$

$28 = 28 \checkmark$

Both check; the solution set is $\left\{-\dfrac{36}{7},\ -\dfrac{19}{4}\right\}$.

79. $\qquad 4x - 20\sqrt{x} + 21 = 0$

$4\left(\sqrt{x}\right)^2 - 20\left(\sqrt{x}\right) + 21 = 0$

Let $u = \sqrt{x}$.

$4u^2 - 20u + 21 = 0$

$(2u - 3)(2u - 7) = 0$

$2u - 3 = 0$ or $2u - 7 = 0$

$2u = 3$ or $2u = 7$

$u = \dfrac{3}{2}$ or $u = \dfrac{7}{2}$

$\sqrt{x} = \dfrac{3}{2}$ or $\sqrt{x} = \dfrac{7}{2}$

$x = \left(\dfrac{3}{2}\right)^2$ or $x = \left(\dfrac{7}{2}\right)^2$

$x = \dfrac{9}{4}$ or $x = \dfrac{49}{4}$

Check:

$f\left(\dfrac{9}{4}\right) = 4 \cdot \dfrac{9}{4} - 20\sqrt{\dfrac{9}{4}} + 21$

$= 4 \cdot \dfrac{9}{4} - 20 \cdot \dfrac{3}{2} + 21$

$= 9 - 30 + 21$

$= 0 \checkmark$

$f\left(\dfrac{49}{4}\right) = 4 \cdot \dfrac{49}{4} - 20\sqrt{\dfrac{49}{4}} + 21$

$= 4 \cdot \dfrac{49}{4} - 20 \cdot \dfrac{7}{2} + 21$

$= 49 - 70 + 21$

$= 0 \checkmark$

Both check; the zeros of f are $\left\{\dfrac{9}{4},\ \dfrac{49}{4}\right\}$.

80. $\qquad x^4 - 17x^2 + 60 = 0$

$\left(x^2\right)^2 - 17\left(x^2\right) + 60 = 0$

Let $u = x^2$.

$u^2 - 17u + 60 = 0$

$(u - 12)(u - 5) = 0$

$u - 12 = 0$ or $u - 5 = 0$

$u = 12$ or $u = 5$

$x^2 = 12$ or $x^2 = 5$

$x = \pm\sqrt{12}$ or $x = \pm\sqrt{5}$

$x = \pm 2\sqrt{3}$

Check:

$g\left(-2\sqrt{3}\right) = \left(-2\sqrt{3}\right)^4 - 17\left(-2\sqrt{3}\right)^2 + 60$

$= 144 - 17 \cdot 12 + 60$

$= 144 - 204 + 60$

$= 0 \checkmark$

$g\left(2\sqrt{3}\right) = \left(2\sqrt{3}\right)^4 - 17\left(2\sqrt{3}\right)^2 + 60$

$= 144 - 17 \cdot 12 + 60$

$= 144 - 204 + 60$

$= 0 \checkmark$

$$g\left(-\sqrt{5}\right) = \left(-\sqrt{5}\right)^4 - 17\left(-\sqrt{5}\right)^2 + 60$$
$$= 25 - 17 \cdot 5 + 60$$
$$= 25 - 85 + 60$$
$$= 0 \ \checkmark$$

$$g\left(\sqrt{5}\right) = \left(\sqrt{5}\right)^4 - 17\left(\sqrt{5}\right)^2 + 60$$
$$= 25 - 17 \cdot 5 + 60$$
$$= 25 - 85 + 60$$
$$= 0 \ \checkmark$$

All check; the zeros of g are
$\left\{-2\sqrt{3}, \ -\sqrt{5}, \ \sqrt{5}, \ 2\sqrt{3}\right\}$.

81. Begin with the graph of $y = x^2$, then shift the graph up 4 units to obtain the graph of $f(x) = x^2 + 4$.

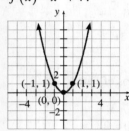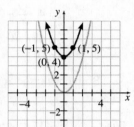

82. Begin with the graph of $y = x^2$, then shift the graph down 5 units to obtain the graph of $g(x) = x^2 - 5$.

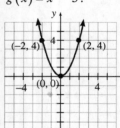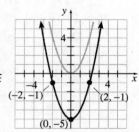

83. Begin with the graph of $y = x^2$, then shift the graph to the left 1 unit to obtain the graph of $h(x) = (x+1)^2$.

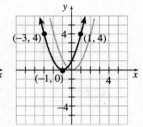

84. Begin with the graph of $y = x^2$, then shift the graph to the right 4 units to obtain the graph of $F(x) = (x-4)^2$.

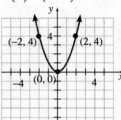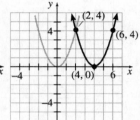

85. Begin with the graph of $y = x^2$, then multiply each y-coordinate by -4 to obtain the graph of $G(x) = -4x^2$.

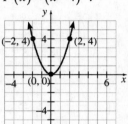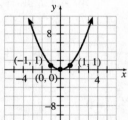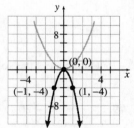

86. Begin with the graph of $y = x^2$, then vertically compress the graph by a factor of $\frac{1}{5}$ (multiply each y-coordinate by $\frac{1}{5}$) to obtain the graph of $H(x) = \frac{1}{5}x^2$.

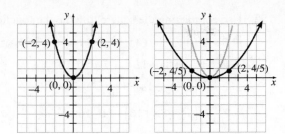

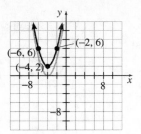

87. Begin with the graph of $y = x^2$, then shift the graph 4 units to the right to obtain the graph of $y = (x-4)^2$. Shift this graph down 3 units to obtain the graph of $p(x) = (x-4)^2 - 3.$

89. Begin with the graph of $y = x^2$, then shift the graph 1 unit to the right to obtain the graph of $y = (x-1)^2$. Multiply the y-coordinates by -1 to obtain the graph of $y = -(x-1)^2$. Shift this graph up 4 units to obtain the graph of $f(x) = -(x-1)^2 + 4.$

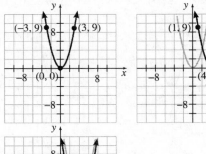

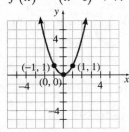 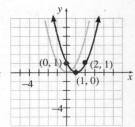

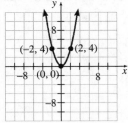 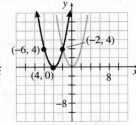

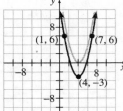 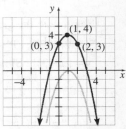

88. Begin with the graph of $y = x^2$, then shift the graph 4 units to the left to obtain the graph of $y = (x+4)^2$. Shift this graph up 2 units to obtain the graph of $P(x) = (x+4)^2 + 2.$

90. Begin with the graph of $y = x^2$, then shift the graph 2 unit to the left to obtain the graph of $y = (x+2)^2$. Multiply the y-coordinates by $\frac{1}{2}$ to obtain the graph of $y = \frac{1}{2}(x+2)^2$. Shift this graph down 1 unit to obtain the graph of $F(x) = \frac{1}{2}(x+2)^2 - 1$

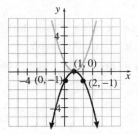

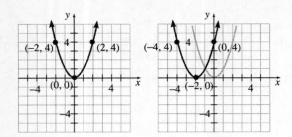

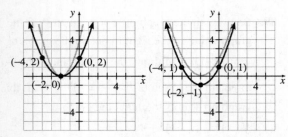

91. Use completing the square to write the function in the form $y = a(x-h)^2 + k$.

$$g(x) = x^2 - 6x + 10$$
$$= \left(x^2 - 6x\right) + 10$$
$$= \left(x^2 - 6x + 9\right) + 10 - 9$$
$$= (x-3)^2 + 1$$

Begin with the graph of $y = x^2$, then shift the graph right 3 units to obtain the graph of $y = (x-3)^2$. Shift this result up 1 unit to obtain the graph of $g(x) = (x-3)^2 + 1$.

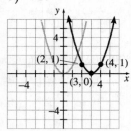

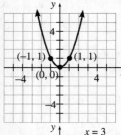

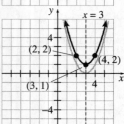

The vertex is $(3, 1)$; The axis of symmetry is $x = 3$.

92. Use completing the square to write the function in the form $y = a(x-h)^2 + k$.

$$G(x) = x^2 + 8x + 11$$
$$= \left(x^2 + 8x\right) + 11$$
$$= \left(x^2 + 8x + 16\right) + 11 - 16$$
$$= (x+4)^2 - 5$$

Begin with the graph of $y = x^2$, then shift the graph left 4 units to obtain the graph of $y = (x+4)^2$. Shift this result down 5 units to obtain the graph of $G(x) = (x+4)^2 - 5$.

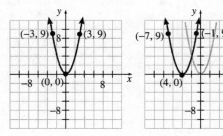

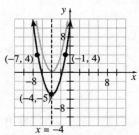

The vertex is $(-4, -5)$. The axis of symmetry is $x = -4$.

93. Use completing the square to write the function in the form $y = a(x-h)^2 + k$.

$$h(x) = 2x^2 - 4x - 3$$
$$= \left(2x^2 - 4x\right) - 3$$
$$= 2\left(x^2 - 2x\right) - 3$$
$$= 2\left(x^2 - 2x + 1\right) - 3 - 2$$
$$= 2(x-1)^2 - 5$$

Begin with the graph of $y = x^2$, then shift the graph right 1 unit to obtain the graph of $y = (x-1)^2$. Vertically stretch this graph by a factor of 2 (multiply the y-coordinates by 2) to obtain the graph of $y = 2(x-1)^2$. Lastly, shift the graph down 5 units to obtain the graph of $H(x) = 2(x-1)^2 - 5$.

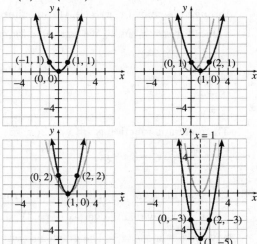

The vertex is $(1, \; -5)$. The axis of symmetry is $x = 1$.

94. Use completing the square to write the function in the form $y = a(x-h)^2 + k$.

$$H(x) = -x^2 - 6x - 10$$
$$= \left(-x^2 - 6x\right) - 10$$
$$= -\left(x^2 + 6x\right) - 10$$
$$= -\left(x^2 + 6x + 9\right) - 10 + 9$$
$$= -(x+3)^2 - 1$$

Begin with the graph of $y = x^2$, then shift the graph left 3 units to obtain the graph of $y = (x+3)^2$. Multiply the y-coordinates by -1 to obtain the graph of $y = -(x+3)^2$. Lastly, shift the graph down 1 unit to obtain the graph of $H(x) = -(x+3)^2 - 1$.

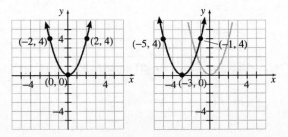

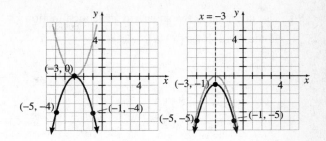

The vertex is $(-3, \; -1)$. The axis of symmetry is $x = -3$.

95. Use completing the square to write the function in the form $y = a(x-h)^2 + k$.

$$p(x) = -3x^2 + 12x - 8$$
$$= \left(-3x^2 + 12x\right) - 8$$
$$= -3\left(x^2 - 4x\right) - 8$$
$$= -3\left(x^2 - 4x + 4\right) - 8 + 12$$
$$= -3(x-2)^2 + 4$$

Begin with the graph of $y = x^2$, then shift the graph right 2 units to obtain the graph of $y = (x-2)^2$. Multiply the y-coordinates by -3 to obtain the graph of $y = -3(x-2)^2$. Lastly, shift the graph up 4 units to obtain the graph of $p(x) = -3(x-2)^2 + 4$.

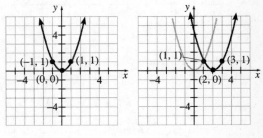

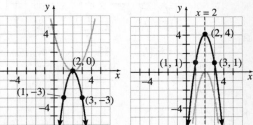

The vertex is $(2, \; 4)$. The axis of symmetry is $x = 2$.

96. Use completing the square to write the function in the form $y = a(x-h)^2 + k$.

$$P(x) = \frac{1}{2}x^2 - 2x + 5$$
$$= \left(\frac{1}{2}x^2 - 2x\right) + 5$$
$$= \frac{1}{2}\left(x^2 - 4x\right) + 5$$
$$= \frac{1}{2}\left(x^2 - 4x + 4\right) + 5 - 2$$
$$= \frac{1}{2}(x-2)^2 + 3$$

Begin with the graph of $y = x^2$, then shift the graph right 2 units to obtain the graph of $y = (x-2)^2$. Vertically compress this graph by a factor of $\frac{1}{2}$ (multiply the y-coordinates by $\frac{1}{2}$) to obtain the graph of $y = \frac{1}{2}(x-2)^2$. Lastly, shift the graph up 3 units to obtain the graph of $P(x) = \frac{1}{2}(x-2)^2 + 3$.

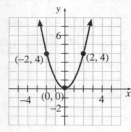
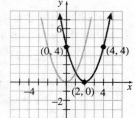
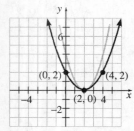
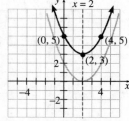

The vertex is $(2, 3)$. The axis of symmetry is $x = 2$.

97. Consider the form $y = a(x-h)^2 + k$. From the graph we know that the vertex is $(2,-4)$, so we have $h = 2$ and $k = -4$. The graph also passes through the point $(x, y) = (0, 4)$. Substituting values for x, y, h, and k, we can solve for a:

$$4 = a(0-2)^2 - 4$$
$$4 = a(-2)^2 - 4$$
$$8 = 4a$$
$$2 = a$$

The quadratic function is $f(x) = 2(x-2)^2 - 4$ or $f(x) = 2x^2 - 8x + 4$.

98. Consider the form $y = a(x-h)^2 + k$. From the graph we know that the vertex is $(4,3)$, so we have $h = 4$ and $k = 3$. The graph also passes through the point $(x, y) = (3, 2)$. Substituting values for x, y, h, and k, we can solve for a:

$$2 = a(3-4)^2 + 3$$
$$2 = a(-1)^2 + 3$$
$$2 = a + 3$$
$$-1 = a$$

The quadratic function is $f(x) = -(x-4)^2 + 3$ or $f(x) = -x^2 + 8x - 13$.

99. Consider the form $y = a(x-h)^2 + k$. From the graph we know that the vertex is $(-2,-1)$, so we have $h = -2$ and $k = -1$. The graph also passes through the point $(x, y) = (0, -3)$. Substituting values for x, y, h, and k, we can solve for a:

$$-3 = a(0-(-2))^2 - 1$$
$$-3 = a(0+2)^2 - 1$$
$$-3 = a(2)^2 - 1$$
$$-2 = 4a$$
$$-\frac{1}{2} = a$$

The quadratic function is $f(x) = -\frac{1}{2}(x+2)^2 - 1$ or $f(x) = -\frac{1}{2}x^2 - 2x - 3$.

100. Consider the form $y = a(x-h)^2 + k$. From the graph we know that the vertex is $(-2, 0)$, so we have $h = -2$ and $k = 0$. The graph also passes through the point $(x, y) = (-3, 3)$. Substituting values for $x, y, h,$ and k, we can solve for a:

$$3 = a(-3 - (-2))^2 - 0$$
$$3 = a(-3 + 2)^2$$
$$3 = a(-1)^2$$
$$3 = a$$

The quadratic function is $f(x) = 3(x+2)^2$ or $f(x) = 3x^2 + 12x + 12$.

101. $f(x) = x^2 + 2x - 8$

$a = 1, b = 2, c = -8$

The graph opens up because the coefficient on x^2 is positive.

vertex:

$$x = -\frac{b}{2a} = -\frac{2}{2(1)} = -1$$

$$f(-1) = (-1)^2 + 2(-1) - 8 = -9$$

The vertex is $(-1, -9)$ and the axis of symmetry is $x = -1$.

y-intercept:

$$f(0) = (0)^2 + 2(0) - 8 = -8$$

x-intercepts:

$$b^2 - 4ac = 2^2 - 4(1)(-8) = 36 > 0$$

There are two distinct x-intercepts. We find these by solving

$$f(x) = 0$$
$$x^2 + 2x - 8 = 0$$
$$(x - 2)(x + 4) = 0$$
$$x - 2 = 0 \text{ or } x + 4 = 0$$
$$x = 2 \text{ or } x = -4$$

Graph:

The y-intercept point, $(0, -8)$, is one unit to the right of the axis of symmetry. Therefore, if we move one unit to the left of the axis of symmetry, we obtain the point $(-2, -8)$ which must also be on the graph.

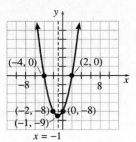

102. $F(x) = 2x^2 - 5x + 3$

$a = 2, b = -5, c = 3$

The graph opens up because the coefficient on x^2 is positive.

vertex:

$$x = -\frac{b}{2a} = -\frac{(-5)}{2(2)} = \frac{5}{4}$$

$$F\left(\frac{5}{4}\right) = 2\left(\frac{5}{4}\right)^2 - 5\left(\frac{5}{4}\right) + 3 = -\frac{1}{8}$$

The vertex is $\left(\frac{5}{4}, -\frac{1}{8}\right)$ and the axis of symmetry is $x = \frac{5}{4}$.

y-intercept:

$$F(0) = 2(0)^2 - 5(0) + 3 = 3$$

x-intercepts:

$$b^2 - 4ac = (-5)^2 - 4(2)(3) = 1 > 0$$

There are two distinct x-intercepts. We find these by solving

$$F(x) = 0$$
$$2x^2 - 5x + 3 = 0$$
$$(2x - 3)(x - 1) = 0$$
$$2x - 3 = 0 \text{ or } x - 1 = 0$$
$$2x = 3 \text{ or } x = 1$$
$$x = \frac{3}{2}$$

Graph:

The y-intercept point, $(0, 3)$, is five-fourths units to the left of the axis of symmetry. Therefore, if we move five-fourths units to the right of the axis of symmetry, we obtain the point $\left(\frac{5}{2}, 3\right)$ which must also be on the graph.

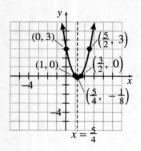

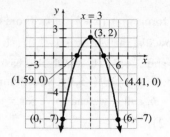

103. $g(x) = -x^2 + 6x - 7$

$a = -1, b = 6, c = -7$

The graph opens down because the coefficient on x^2 is negative.

vertex:

$$x = -\frac{b}{2a} = -\frac{6}{2(-1)} = 3$$

$$g(3) = -(3)^2 + 6(3) - 7 = 2$$

The vertex is $(3, 2)$ and the axis of symmetry is $x = 3$.

y-intercept:

$$g(0) = -(0)^2 + 6(0) - 7 = -7$$

x-intercepts:

$$b^2 - 4ac = 6^2 - 4(-1)(-7) = 8 > 0$$

There are two distinct x-intercepts. We find these by solving

$$g(x) = 0$$

$$-x^2 + 6x - 7 = 0$$

$$x = \frac{-6 \pm \sqrt{8}}{2(-1)}$$

$$= \frac{-6 \pm 2\sqrt{2}}{-2}$$

$$= 3 \pm \sqrt{2}$$

$$x \approx 1.59 \text{ or } x \approx 4.41$$

Graph:

The y-intercept point, $(0, -7)$, is three units to the left of the axis of symmetry. Therefore, if we move three units to the right of the axis of symmetry, we obtain the point $(6, -7)$ which must also be on the graph.

104. $G(x) = -2x^2 + 4x + 3$

$a = -2, b = 4, c = 3$

The graph opens down because the coefficient on x^2 is negative.

vertex:

$$x = -\frac{b}{2a} = -\frac{4}{2(-2)} = 1$$

$$G(1) = -2(1)^2 + 4(1) + 3 = 5$$

The vertex is $(1, 5)$ and the axis of symmetry is $x = 1$.

y-intercept:

$$G(0) = -2(0)^2 + 4(0) + 3 = 3$$

x-intercepts:

$$b^2 - 4ac = 4^2 - 4(-2)(3) = 40 > 0$$

There are two distinct x-intercepts. We find these by solving

$$G(x) = 0$$

$$-2x^2 + 4x + 3 = 0$$

$$x = \frac{-4 \pm \sqrt{40}}{2(-2)}$$

$$= \frac{-4 \pm 2\sqrt{10}}{-4}$$

$$= 1 \pm \frac{\sqrt{10}}{2}$$

$$x \approx -0.58 \text{ or } x \approx 2.58$$

Graph:

The y-intercept point, $(0, 3)$, is one unit to the left of the axis of symmetry. Therefore, if we move one unit to the right of the axis of symmetry, we obtain the point $(2, 3)$ which must also be on the graph.

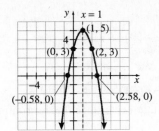

105. $h(x) = 4x^2 - 12x + 9$

$a = 4, b = -12, c = 9$

The graph opens up because the coefficient on x^2 is positive.

<u>vertex:</u>

$$x = -\frac{b}{2a} = -\frac{(-12)}{2(4)} = \frac{3}{2}$$

$$h\left(\frac{3}{2}\right) = 4\left(\frac{3}{2}\right)^2 - 12\left(\frac{3}{2}\right) + 9 = 0$$

The vertex is $\left(\frac{3}{2}, 0\right)$ and the axis of symmetry is

$x = \frac{3}{2}$.

<u>y-intercept:</u>

$$h(0) = 4(0)^2 - 12(0) + 9 = 9$$

<u>x-intercepts:</u>

$$b^2 - 4ac = (-12)^2 - 4(4)(9) = 0$$

Since the discriminant is 0, the x-coordinate of

the vertex is the only x-intercept, $x = \frac{3}{2}$.

<u>Graph:</u>

The y-intercept point, $(0,9)$, is three-halves units to the left of the axis of symmetry. Therefore, if we move three-halves units to the right of the axis of symmetry, we obtain the point $(3,9)$ which must also be on the graph.

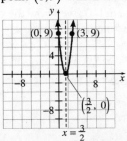

106. $H(x) = \frac{1}{3}x^2 + 2x + 3$

$a = \frac{1}{3}, b = 2, c = 3$

The graph opens up because the coefficient on x^2 is positive.

<u>vertex:</u>

$$x = -\frac{b}{2a} = -\frac{2}{2\left(\frac{1}{3}\right)} = -3$$

$$H(-3) = \frac{1}{3}(-3)^2 + 2(-3) + 3 = 0$$

The vertex is $(-3, 0)$ and the axis of symmetry is

$x = -3$.

<u>y-intercept:</u>

$$H(0) = \frac{1}{3}(0)^2 + 2(0) + 3 = 3$$

<u>x-intercepts:</u>

$$b^2 - 4ac = 2^2 - 4\left(\frac{1}{3}\right)(3) = 0$$

Since the discriminant is 0, the x-coordinate of the vertex is the only x-intercept, $x = -3$.

<u>Graph:</u>

The y-intercept point, $(0,3)$, is three units to the right of the axis of symmetry. Therefore, if we move three units to the left of the axis of symmetry, we obtain the point $(-6,3)$ which must also be on the graph.

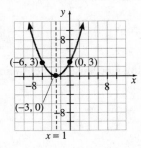

107. $p(x) = \frac{1}{4}x^2 + 3x + 10$

$a = \frac{1}{4}, b = 3, c = 10$

The graph opens up because the coefficient on x^2 is positive.

vertex:

$$x = -\frac{b}{2a} = -\frac{3}{2\left(\frac{1}{4}\right)} = -6$$

$$p(-6) = \frac{1}{4}(-6)^2 + 3(-6) + 10 = 1$$

The vertex is $(-6, 1)$ and the axis of symmetry is $x = -6$.

y-intercept:

$$p(0) = \frac{1}{4}(0)^2 + 3(0) + 10 = 10$$

x-intercepts:

$$b^2 - 4ac = 3^2 - 4\left(\frac{1}{4}\right)(10) = -1 < 0$$

There are no x-intercepts since the discriminant is negative.

Graph:

The y-intercept point, $(0, 10)$, is six units to the right of the axis of symmetry. Therefore, if we move six units to the left of the axis of symmetry, we obtain the point $(-12, 10)$ which must also be on the graph.

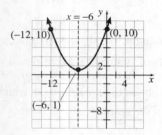

108. $P(x) = -x^2 + 4x - 9$

$a = -1, b = 4, c = -9$

The graph opens down because the coefficient on x^2 is negative.

vertex:

$$x = -\frac{b}{2a} = -\frac{4}{2(-1)} = 2$$

$$P(2) = -(2)^2 + 4(2) - 9 = -5$$

The vertex is $(2, -5)$ and the axis of symmetry is $x = 2$.

y-intercept:

$$P(0) = -(0)^2 + 4(0) - 9 = -9$$

x-intercepts:

$$b^2 - 4ac = 4^2 - 4(-1)(-9) = -20 < 0$$

There are no x-intercepts since the discriminant is negative.

Graph:

The y-intercept point, $(0, -9)$, is 2 units to the left of the axis of symmetry. Therefore, if we move 2 units to the right of the axis of symmetry, we obtain the point $(4, -9)$ which must also be on the graph.

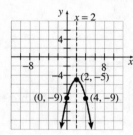

109. If we compare $f(x) = -2x^2 + 16x - 10$ to

$f(x) = ax^2 + bx + c$, we find that $a = -2$, $b = 16$, and $c = -10$. Because $a < 0$, we know the graph will open down, so the function will have a maximum value.
The maximum value occurs at

$$x = -\frac{b}{2a} = -\frac{16}{2(-2)} = 4.$$

The maximum value is

$$f(4) = -2(4)^2 + 16(4) - 10 = 22.$$

So, the maximum value is 22, and it occurs when $x = 4$.

110. If we compare $g(x) = 6x^2 - 3x - 1$ to

$g(x) = ax^2 + bx + c$, we find that $a = 6$, $b = -3$, and $c = -1$. Because $a > 0$, we know the graph will open up, so the function will have a minimum value.
The minimum value occurs at

$$x = -\frac{b}{2a} = -\frac{(-3)}{2(6)} = \frac{1}{4}.$$

The minimum value is

$g\left(\dfrac{1}{4}\right)=6\left(\dfrac{1}{4}\right)^{2}-3\left(\dfrac{1}{4}\right)-1=-\dfrac{11}{8}$.

So, the minimum value is $-\dfrac{11}{8}$, and it occurs

when $x=\dfrac{1}{4}$.

111. If we compare $h(x)=-4x^{2}+8x+3$ to

$h(x)=ax^{2}+bx+c$, we find that $a=-4$, $b=8$, and $c=3$. Because $a<0$, we know the graph will open down, so the function will have a maximum value.
The maximum value occurs at

$x=-\dfrac{b}{2a}=-\dfrac{8}{2(-4)}=1$.

The maximum value is

$h(4)=-4(1)^{2}+8(1)+3=7$.

So, the maximum value is 7, and it occurs when $x=1$.

112. If we compare $F(x)=-\dfrac{1}{3}x^{2}+4x-7$ to

$F(x)=ax^{2}+bx+c$, we find that $a=-\dfrac{1}{3}$,

$b=4$, and $c=-7$. Because $a<0$, we know the graph will open down, so the function will have a maximum value.
The maximum value occurs at

$x=-\dfrac{b}{2a}=-\dfrac{4}{2\left(-\frac{1}{3}\right)}=6$.

The maximum value is

$F(6)=-\dfrac{1}{3}(6)^{2}+4(6)-7=5$.

So, the maximum value is 5, and it occurs when $x=6$.

113. a. We first recognize that the quadratic function has a negative leading coefficient. This means that the graph will open down and the function indeed has a maximum value. The maximum value occurs when

$p=-\dfrac{b}{2a}=-\dfrac{150}{2\left(-\frac{1}{3}\right)}=225$.

The revenue will be maximized when the televisions are sold at a price of $225.

b. The maximum revenue is obtained by evaluating the revenue function at the price found in part (a).

$R(225)=-\dfrac{1}{3}(225)^{2}+150(225)=16,875$

The maximum revenue is $16,875.

114. a. We first recognize that the quadratic function has a negative leading coefficient. This means that the graph will open down and the function indeed has a maximum value. The maximum value occurs when

$I=-\dfrac{b}{2a}=-\dfrac{120}{2(-16)}=3.75$.

The power will be maximized when the current is 3.75 amperes.

b. The maximum power is obtained by evaluating the power function for the current found in part (a).

$P(3.75)=-16(3.75)^{2}+120(3.75)=225$

The maximum power is 225 watts.

115. Let x represent the first number. Then the second number must be $24-x$. We can express the product of the two numbers as the function

$p(x)=x(24-x)=-x^{2}+24x$

This is a quadratic function with $a=-1$, $b=24$, and $c=0$. The function is maximized

when $x=-\dfrac{b}{2a}=-\dfrac{24}{2(-1)}=12$.

The maximum product can be obtained by evaluating $p(x)$ when $x=12$.

$p(x)=12(24-12)=12(12)=144$

Two numbers that sum to 24 have a maximum product of 144 when both numbers are 12.

116. a. Let x represent the width of the rectangular kennel (the side that is not parallel to the garage). Then $15-2x$ is the length of the kennel (the side that is parallel to the garage). The area is the product of the length and width:

$A=(15-2x)\cdot x$

$\quad=-2x^{2}+15x$

The leading coefficient is negative so we know the graph opens down and there will be a maximum area. This value occurs when the width is

$$x = -\frac{b}{2a} = -\frac{15}{2(-2)} = 3.75.$$

Then the length is $15 - 2(3.75) = 7.5$.

The dimensions that maximize the area of the kennel are 3.75 yards by 7.5 yards.

b. The maximum area can be found by substituting 3.75 for x into the area function.

$$A = 3.75(15 - 2(3.75)) = 28.125$$

The maximum area of the kennel is 28.125 square yards.

117. a. First we recognize that the quadratic function has a negative leading coefficient. This means the graph will open down and the function will indeed have a maximum. The maximum height will occur when

$$x = -\frac{b}{2a} = -\frac{1}{2(-0.005)} = 100.$$

The ball will reach a maximum height when it is 100 feet from Ted.

b. The maximum height can be found by evaluating $h(x)$ for the value of x found in part (a).

$$h(100) = -0.005(100)^2 + 100 = 50$$

The ball will reach a maximum height of 50 feet.

c. When the ball is on the ground, it will have a height of 0 feet. Thus, we need to solve $h(x) = 0$.

$$-0.005x^2 + x = 0$$
$$200\left(-0.005x^2 + x\right) = 200(0)$$
$$-x^2 + 200x = 0$$
$$-x(x - 200) = 0$$
$$-x = 0 \quad \text{or} \quad x - 200 = 0$$
$$x = 0 \quad \text{or} \qquad x = 200$$

Zero (0) represents the distance from Ted before he kicks the ball. The ball will strike the ground again when it is 200 feet away from Ted.

118. a. Since $R = x \cdot p$, we have

$$R = x \cdot p$$
$$= (-0.002p + 60) \cdot p$$
$$= -0.002p^2 + 60p$$

b. The revenue function is quadratic with a negative leading coefficient. This means the graph will open down and the function will have a maximum value. This value occurs when

$$p = -\frac{b}{2a} = -\frac{60}{2(-0.002)} = 15,000.$$

The maximum revenue can be found by substituting this value for p in the revenue equation.

$$R = -0.002(15,000)^2 + 60(15,000)$$
$$= 450,000$$

The maximum revenue of \$450,000 will occur if the price of each automobile is set at \$15,000.

c. To determine how many automobiles will be sold, we substitute the maximizing price into the demand equation.

$$x = -0.002p + 60$$
$$= -0.002(15,000) + 60$$
$$= 30$$

When the price per automobile is \$15,000, the dealership will sell 30 automobiles per month.

119. a. The graph is greater than 0 for $x < -2$ or $x > 3$. The solution is $\{x | x < -2 \text{ or } x > 3\}$ in set-builder notation. The solution is $(-\infty, -2) \cup (3, \infty)$ using interval notation.

b. The graph is less than 0 for $-2 < x < 3$. The solution is $\{x | -2 < x < 3\}$ using set-builder notation. The solution is $(-2, 3)$ using interval notation.

120. a. The graph is 0 or greater for $-\frac{7}{2} \le x \le 1$.

The solution is $\left\{x \left| -\frac{7}{2} \le x \le 1\right.\right\}$ using set-builder notation. The solution is $\left[-\frac{7}{2}, 1\right]$ using interval notation.

b. The graph is 0 or less for $x \le -\dfrac{7}{2}$ or $x \ge 1$.

The solution is $\left\{ x \middle| x \le -\dfrac{7}{2} \text{ or } x \ge 1 \right\}$ using

set-builder notation. The solution is

$\left(-\infty, -\dfrac{7}{2} \right] \cup [1, \infty)$ using interval notation.

121. $x^2 - 2x - 24 \le 0$

Solve: $x^2 - 2x - 24 = 0$

$(x - 6)(x + 4) = 0$

$x - 6 = 0$ or $x + 4 = 0$

$x = 6$ or $x = -4$

Determine where each factor is positive and negative and where the product of these factors is positive and negative.

Interval	$(-\infty, -4)$	-4	$(-4, 6)$	6	$(6, \infty)$
$x - 6$	$----$	$-$	$----$	0	$++++$
$x + 4$	$----$	0	$++++$	$+$	$++++$
$(x-6)(x+4)$	$++++$	0	$----$	0	$++++$

The inequality is non-strict, so -4 and 6 are part of the solution. Now, $(x - 6)(x + 4)$ is less than zero where the product is negative. The solution is $\{ x \mid -4 \le x \le 6 \}$ or, using interval notation, $[-4, 6]$.

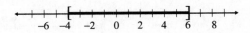

122. $y^2 + 7y - 8 \ge 0$

Solve: $y^2 + 7y - 8 = 0$

$(y - 1)(y + 8) = 0$

$y - 1 = 0$ or $y + 8 = 0$

$y = 1$ or $y = -8$

Determine where each factor is positive and negative and where the product of these factors is positive and negative.

Interval	$(-\infty, -8)$	-8	$(-8, 1)$	1	$(1, \infty)$
$y - 1$	$----$	$-$	$----$	0	$++++$
$y + 8$	$----$	0	$++++$	$+$	$++++$
$(y-1)(y+8)$	$++++$	0	$----$	0	$++++$

The inequality is non-strict, so -8 and 1 are part

of the solution. Now, $(y - 1)(y + 8)$ is greater than zero where the product is positive. The solution is $\{ y \mid y \le -8 \text{ or } y \ge 1 \}$ or, using interval notation; $(-\infty, -8] \cup [1, \infty)$.

123. $3z^2 - 19z + 20 > 0$

Solve: $3z^2 - 19z + 20 = 0$

$(3z - 4)(z - 5) = 0$

$3z - 4 = 0$ or $z - 5 = 0$

$3z = 4$ or $z = 5$

$z = \dfrac{4}{3}$

Determine where each factor is positive and negative and where the product of these factors is positive and negative.

Interval	$\left(-\infty, \dfrac{4}{3}\right)$	$\dfrac{4}{3}$	$\left(\dfrac{4}{3}, 5\right)$	5	$(5, \infty)$
$3z - 4$	$----$	0	$++++$	$+$	$++++$
$z - 5$	$----$	$-$	$----$	0	$++++$
$(3z-4)(z-5)$	$++++$	0	$----$	0	$++++$

The inequality is strict, so $\dfrac{4}{3}$ and 5 are not part

of the solution. Now, $(3z - 4)(z - 5)$ is greater than zero where the product is positive. The solution is $\left\{ z \middle| z < \dfrac{4}{3} \text{ or } z > 5 \right\}$ or, using interval

notation, $\left(-\infty, \dfrac{4}{3} \right) \cup (5, \infty)$.

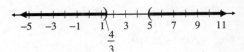

124. $p^2 + 4p - 2 < 0$

Solve: $p^2 + 4p - 2 = 0$

$$p = \frac{-4 \pm \sqrt{4^2 - 4(1)(-2)}}{2(1)}$$

$$= \frac{-4 \pm \sqrt{24}}{2}$$

$$= \frac{-4 \pm 2\sqrt{6}}{2}$$

$$= -2 \pm \sqrt{6}$$

$x = -2 - \sqrt{6}$ or $x = -2 + \sqrt{6}$

≈ -4.45 or ≈ 0.45

Determine where each factor is positive and negative and where the product of these factors is positive and negative.

Interval	$(-\infty, -2-\sqrt{6})$	$(-2-\sqrt{6}, -2+\sqrt{6})$	$(-2+\sqrt{6}, \infty)$
	$-2-\sqrt{6}$ ≈ -4.45	$-2+\sqrt{6}$ ≈ 0.45	
$p - (-2-\sqrt{6})$	$----$ 0	$++++$ $+$	$++++$
$p - (-2+\sqrt{6})$	$----$ $-$	$----$ 0	$++++$
$[p-(-2-\sqrt{6})][p-(-2+\sqrt{6})]$	$++++$ 0	$----$ 0	$++++$

The inequality is strict, so $-2-\sqrt{6}$ and $-2+\sqrt{6}$ are not part of the solution. Now, $\left[p - \left(-2-\sqrt{6}\right)\right]\left[p - \left(-2+\sqrt{6}\right)\right]$ is less than zero where the product is negative. The solution is $\left\{p \mid -2-\sqrt{6} < p < -2+\sqrt{6}\right\}$ or, using interval notation, $\left(-2-\sqrt{6},\ -2+\sqrt{6}\right)$.

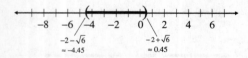

125. $4m^2 - 20m + 25 \geq 0$

Solve: $4m^2 - 20m + 25 = 0$

$(2m - 5)(2m - 5) = 0$

$2m - 5 = 0$ or $2m - 5 = 0$

$2m = 5$ or $2m = 5$

$m = \dfrac{5}{2}$ or $m = \dfrac{5}{2}$

Determine where each factor is positive and negative and where the product of these factors is positive and negative.

Interval	$\left(-\infty, \frac{5}{2}\right)$	$\frac{5}{2}$	$\left(\frac{5}{2}, \infty\right)$
$2m - 5$	$----$	0	$++++$
$2m - 5$	$----$	0	$++++$
$(2m-5)(2m-5)$	$++++$	0	$++++$

The inequality is non-strict, so $\dfrac{5}{2}$ is part of the solution. Now, $(2m-5)(2m-5)$ is greater than zero where the product is positive. Thus, $4m^2 - 20m + 25$ is always greater than or equal to zero. The solution is $\{m \mid m$ is any real number$\}$ or, using interval notation $(-\infty, \infty)$.

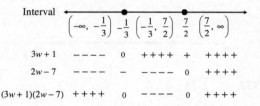

126. $6w^2 - 19w - 7 \leq 0$

Solve: $6w^2 - 19w - 7 = 0$

$(3w + 1)(2w - 7) = 0$

$3w + 1 = 0$ or $2w - 7 = 0$

$3w = -1$ or $2w = 7$

$w = -\dfrac{1}{3}$ or $w = \dfrac{7}{2}$

Determine where each factor is positive and negative and where the product of these factors is positive and negative.

Interval	$\left(-\infty, -\frac{1}{3}\right)$	$-\frac{1}{3}$	$\left(-\frac{1}{3}, \frac{7}{2}\right)$	$\frac{7}{2}$	$\left(\frac{7}{2}, \infty\right)$
$3w + 1$	$----$	0	$++++$ $+$		$++++$
$2w - 7$	$----$	$-$	$----$	0	$++++$
$(3w+1)(2w-7)$	$++++$	0	$----$	0	$++++$

The inequality is non-strict, so $-\dfrac{1}{3}$ and $\dfrac{7}{2}$ are part of the solution. Now, $(3w+1)(2w-7)$ is less than zero where the product is negative. The solution is $\left\{w \mid -\dfrac{1}{3} \leq w \leq \dfrac{7}{2}\right\}$ or, using interval notation, $\left[-\dfrac{1}{3},\ \dfrac{7}{2}\right]$.

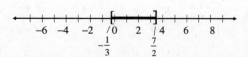

127. $(2x-3)(x+1)(x-2) < 0$

Solve $(2x-3)(x+1)(x-2) = 0$.

$2x-3=0$ or $x+1=0$ or $x-2=0$

$x=\dfrac{3}{2}$ or $x=-1$ or $x=2$

Determine where each factor is positive and negative and where the product of these factors is positive and negative.

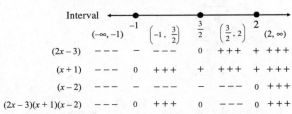

Interval	$(-\infty, -1)$	-1	$\left(-1, \frac{3}{2}\right)$	$\frac{3}{2}$	$\left(\frac{3}{2}, 2\right)$	2	$(2, \infty)$
$(2x-3)$	$---$	$-$	$---$	0	$+++$	$+$	$+++$
$(x+1)$	$---$	0	$+++$	$+$	$+++$	$+$	$+++$
$(x-2)$	$---$	$-$	$---$	$-$	$---$	0	$+++$
$(2x-3)(x+1)(x-2)$	$---$	0	$+++$	0	$---$	0	$+++$

The inequality is strict, so -1, $\dfrac{3}{2}$, and 2 are not part of the solution. Now, $(2x-3)(x+1)(x-2)$ is less than zero where the product is negative. The solution is $\left\{x \middle| x < -1 \text{ or } \dfrac{3}{2} < x < 2\right\}$ in set-builder notation and $(-\infty, -1) \cup \left(\dfrac{3}{2}, 2\right)$ in interval notation.

128. $x^3 + 5x^2 - 9x - 45 \geq 0$

Solve $x^3 + 5x^2 - 9x - 45 = 0$

$x^2(x+5) - 9(x+5) = 0$

$(x+5)(x^2-9) = 0$

$(x+5)(x+3)(x-3) = 0$

$x+5=0$ or $x+3=0$ or $x-3=0$

$x=-5$ or $x=-3$ or $x=3$

Determine where each factor is positive and negative and where the product of these factors is positive and negative.

Interval	$(-\infty, -5)$	-5	$(-5, -3)$	-3	$(-3, 3)$	3	$(3, \infty)$
$(x+5)$	$---$	0	$+++$	$+$	$+++$	$+$	$+++$
$(x+3)$	$---$	$-$	$---$	0	$+++$	$+$	$+++$
$(x-3)$	$---$	$-$	$---$	$-$	$---$	0	$+++$
$(x+5)(x+3)(x-3)$	$---$	0	$+++$	0	$---$	0	$+++$

The inequality is non-strict, so -5, -3, and 3 are part of the solution. Now, $(x+5)(x+3)(x-3)$ is greater than zero where the product is positive. The solution is $\{x | -5 \leq x \leq -3 \text{ or } x \geq 3\}$ in set-builder notation and $[-5, -3] \cup [3, \infty)$ in interval notation.

Chapter 7 Test

1. Start: $x^2 - 3x$

Add: $\left[\dfrac{1}{2} \cdot (-3)\right]^2 = \dfrac{9}{4}$

Result: $x^2 - 3x + \dfrac{9}{4}$

Factored Form: $\left(x - \dfrac{3}{2}\right)^2$

2. Start: $m^2 + \dfrac{2}{5}m$

Add: $\left[\dfrac{1}{2} \cdot \left(\dfrac{2}{5}\right)\right]^2 = \dfrac{1}{25}$

Result: $m^2 + \dfrac{2}{5}m + \dfrac{1}{25}$

Factored Form: $\left(m + \dfrac{1}{5}\right)^2$

3. $9\left(x + \dfrac{4}{3}\right)^2 = 1$

$\left(x + \dfrac{4}{3}\right)^2 = \dfrac{1}{9}$

$x + \dfrac{4}{3} = \pm\sqrt{\dfrac{1}{9}}$

$x + \dfrac{4}{3} = \pm\dfrac{1}{3}$

$x = -\dfrac{4}{3} \pm \dfrac{1}{3}$

$x = -\dfrac{4}{3} - \dfrac{1}{3}$ or $x = -\dfrac{4}{3} + \dfrac{1}{3}$

$x = -\dfrac{5}{3}$ or $x = -1$

The solution set is $\left\{-\dfrac{5}{3}, -1\right\}$.

4. $m^2 - 6m + 4 = 0$
Because this equation does not easily factor, solve by using the quadratic formula. For this equation, $a = 1$, $b = -6$, and $c = 4$.

$$m = \frac{-(-6) \pm \sqrt{(-6)^2 - 4(1)(4)}}{2(1)}$$
$$= \frac{6 \pm \sqrt{36 - 16}}{2}$$
$$= \frac{6 \pm \sqrt{20}}{2}$$
$$= \frac{6 \pm 2\sqrt{5}}{2}$$
$$= \frac{6}{2} \pm \frac{2\sqrt{5}}{2}$$
$$= 3 \pm \sqrt{5}$$

The solution set is. $\left\{3 - \sqrt{5},\ 3 + \sqrt{5}\right\}$.

5. $2w^2 - 4w + 3 = 0$
Because this equation does not easily factor, solve by using the quadratic formula. For this equation, $a = 2$, $b = -4$, and $c = 3$.

$$w = \frac{-(-4) \pm \sqrt{(-4)^2 - 4(2)(3)}}{2(2)}$$
$$= \frac{4 \pm \sqrt{16 - 24}}{4}$$
$$= \frac{4 \pm \sqrt{-8}}{4}$$
$$= \frac{4 \pm 2\sqrt{2}\,i}{4}$$
$$= \frac{4}{4} \pm \frac{2\sqrt{2}}{4}i$$
$$= 1 \pm \frac{\sqrt{2}}{2}i$$

The solution set is $\left\{1 - \frac{\sqrt{2}}{2}i,\ 1 + \frac{\sqrt{2}}{2}i\right\}$.

6. $\frac{1}{2}z^2 - \frac{3}{2}z = -\frac{7}{6}$
$$6\left(\frac{1}{2}z^2 - \frac{3}{2}z\right) = 6\left(-\frac{7}{6}\right)$$
$$3z^2 - 9z = -7$$
$$3z^2 - 9z + 7 = 0$$
Because this equation does not easily factor, solve by using the quadratic formula. For this equation, $a = 3$, $b = -9$, and $c = 7$.

$$z = \frac{-(-9) \pm \sqrt{(-9)^2 - 4(3)(7)}}{2(3)}$$
$$= \frac{9 \pm \sqrt{81 - 84}}{6}$$
$$= \frac{9 \pm \sqrt{-3}}{6}$$
$$= \frac{9 \pm \sqrt{3}\,i}{6}$$
$$= \frac{9}{6} \pm \frac{\sqrt{3}}{6}i$$
$$= \frac{3}{2} \pm \frac{\sqrt{3}}{6}i$$

The solution set is $\left\{\frac{3}{2} - \frac{\sqrt{3}}{6}i,\ \frac{3}{2} + \frac{\sqrt{3}}{6}i\right\}$.

7. $2x^2 + 5x = 4$
$2x^2 + 5x - 4 = 0$
For this equation, $a = 2$, $b = 5$, and $c = -4$.

$b^2 - 4ac = 5^2 - 4(2)(-4) = 25 + 32 = 57$

Because $b^2 - 4ac = 57$, but not a perfect square, the quadratic equation will have two irrational solutions.

8. $c^2 = a^2 + b^2$
$11^2 = a^2 + 7^2$
$121 = a^2 + 49$
$72 = a^2$
$a = \sqrt{72}$
$a = 6\sqrt{2}$

9. $x^4 - 5x^2 - 36 = 0$
$\left(x^2\right)^2 - 5\left(x^2\right) - 36 = 0$
Let $u = x^2$.
$u^2 - 5u - 36 = 0$
$(u - 9)(u + 4) = 0$
$u - 9 = 0$ or $u + 4 = 0$
$u = 9$ or $u = -4$
$x^2 = 9$ or $x^2 = -4$
$x = \pm\sqrt{9}$ or $x = \pm\sqrt{-4}$
$x = \pm 3$ or $x = \pm 2i$

Check:

$x = -3$: $(-3)^4 - 5(-3)^2 - 36 \overset{?}{=} 0$
$$81 - 5 \cdot 9 - 36 \overset{?}{=} 0$$
$$81 - 45 - 36 \overset{?}{=} 0$$
$$0 = 0 \checkmark$$

$x = 3$: $3^4 - 5(3)^2 - 36 \overset{?}{=} 0$
$$81 - 5 \cdot 9 - 36 \overset{?}{=} 0$$
$$81 - 45 - 36 \overset{?}{=} 0$$
$$0 = 0 \checkmark$$

$x = -2i$: $(-2i)^4 - 5(-2i)^2 - 36 \overset{?}{=} 0$
$$16i^4 - 5 \cdot 4i^2 - 36 \overset{?}{=} 0$$
$$16(1) - 5 \cdot 4(-1) - 36 \overset{?}{=} 0$$
$$16 + 20 - 36 \overset{?}{=} 0$$
$$0 = 0 \checkmark$$

$x = 2i$: $(2i)^4 - 5(2i)^2 - 36 \overset{?}{=} 0$
$$16i^4 - 5 \cdot 4i^2 - 36 \overset{?}{=} 0$$
$$16(1) - 5 \cdot 4(-1) - 36 \overset{?}{=} 0$$
$$16 + 20 - 36 \overset{?}{=} 0$$
$$0 = 0 \checkmark$$

All check; the solution set is $\{-3,\ 3,\ -2i,\ 2i\}$.

10. $\qquad 6y^{1/2} + 13y^{1/4} - 5 = 0$

$$6\left(y^{1/4}\right)^2 + 13\left(y^{1/4}\right) - 5 = 0$$

Let $u = y^{1/4}$.

$$6u^2 + 13u - 5 = 0$$
$$(3u - 1)(2u + 5) = 0$$

$3u - 1 = 0 \quad$ or $\quad 2u + 5 = 0$
$3u = 1 \quad$ or $\quad 2u = -5$
$u = \dfrac{1}{3} \quad$ or $\quad u = -\dfrac{5}{2}$
$y^{1/4} = \dfrac{1}{3} \quad$ or $\quad y^{1/4} = -\dfrac{5}{2}$
$\left(y^{1/4}\right)^4 = \left(\dfrac{1}{3}\right)^4 \quad$ or $\quad \left(y^{1/4}\right)^4 = \left(-\dfrac{5}{2}\right)^4$
$y = \dfrac{1}{81} \quad$ or $\quad y = \dfrac{625}{16}$

Check:

$y = \dfrac{1}{81}$: $6\left(\dfrac{1}{81}\right)^{1/2} + 13\left(\dfrac{1}{81}\right)^{1/4} - 5 \overset{?}{=} 0$

$$6\sqrt{\dfrac{1}{81}} + 13\sqrt[4]{\dfrac{1}{81}} - 5 \overset{?}{=} 0$$

$$6\left(\dfrac{1}{9}\right) + 13\left(\dfrac{1}{3}\right) - 5 \overset{?}{=} 0$$

$$\dfrac{2}{3} + \dfrac{13}{3} - 5 \overset{?}{=} 0$$

$$0 = 0 \checkmark$$

$y = \dfrac{625}{16}$: $6\left(\dfrac{625}{16}\right)^{1/2} + 13\left(\dfrac{625}{16}\right)^{1/4} - 5 \overset{?}{=} 0$

$$6\sqrt{\dfrac{625}{16}} + 13\sqrt[4]{\dfrac{625}{16}} - 5 \overset{?}{=} 0$$

$$6\left(\dfrac{25}{4}\right) + 13\left(\dfrac{5}{2}\right) - 5 \overset{?}{=} 0$$

$$\dfrac{75}{2} + \dfrac{65}{2} - 5 \overset{?}{=} 0$$

$$65 \neq 0 \ \textbf{✗}$$

$y = \dfrac{625}{16}$ does not check; the solution set is $\left\{\dfrac{1}{81}\right\}$.

11. Begin with the graph of $y = x^2$, then shift the graph 2 units to the left to obtain the graph of $y = (x+2)^2$. Shift this graph down 5 units to obtain the graph of $f(x) = (x+2)^2 - 5$.

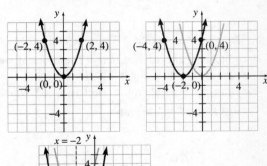

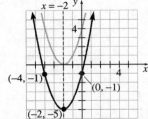

The vertex is $(-2, -5)$. The axis of symmetry is $x = -2$.

12. $g(x) = -2x^2 - 8x - 3$

$a = -2, b = -8, c = -3$

The graph opens down because the coefficient on x^2 is negative.

vertex:

$$x = -\frac{b}{2a} = -\frac{(-8)}{2(-2)} = -2$$

$$g(-2) = -2(-2)^2 - 8(-2) - 3 = 5$$

The vertex is $(-2, 5)$. The axis of symmetry is $x = -2$.

y-intercept:

$$g(0) = -2(0)^2 - 8(0) - 3 = -3$$

x-intercepts:

$$b^2 - 4ac = (-8)^2 - 4(-2)(-3) = 40 > 0$$

There are two distinct *x*-intercepts. We find these by solving

$$g(x) = 0$$

$$-2x^2 - 8x - 3 = 0$$

$$x = \frac{-(-8) \pm \sqrt{40}}{2(-2)}$$

$$= \frac{8 \pm 2\sqrt{10}}{-4}$$

$$= -2 \pm \frac{\sqrt{10}}{2}$$

$$x \approx -3.58 \quad \text{or} \quad x \approx -0.42$$

Graph:

The *y*-intercept point, $(0, -3)$, is two units to the right of the axis of symmetry. Therefore, if we move two units to the left of the axis of symmetry, we obtain the point $(-4, -3)$ which must also be on the graph.

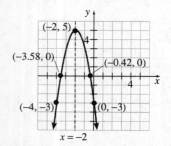

13. Consider the form $y = a(x - h)^2 + k$. From the graph we know that the vertex is $(-3, -5)$, so we have $h = -3$ and $k = -5$. The graph also passes through the point $(x, y) = (0, -2)$. Substituting values for *x, y, h,* and *k*, we can solve for *a*:

$$-2 = a(0 - (-3))^2 - 5$$

$$-2 = a(0 + 3)^2 - 5$$

$$-2 = a(3)^2 - 5$$

$$-2 = 9a - 5$$

$$3 = 9a$$

$$\frac{3}{9} = a$$

$$\frac{1}{3} = a$$

The quadratic function is $f(x) = \frac{1}{3}(x + 3)^2 - 5$

or $f(x) = \frac{1}{3}x^2 + 2x - 2$.

14. If we compare $h(x) = -\frac{1}{4}x^2 + x + 5$ to

$f(x) = ax^2 + bx + c$, we find that $a = -\frac{1}{4}$,

$b = 1$, and $c = 5$. Because $a < 0$, we know the graph will open down, so the function will have a maximum value.

The maximum value occurs at

$$x = -\frac{b}{2a} = -\frac{1}{2\left(-\frac{1}{4}\right)} = -\frac{1}{-\frac{1}{2}} = 2.$$

The maximum value is

$$h(2) = -\frac{1}{4}(2)^2 + 2 + 5 = 6.$$

So, the maximum value is 6, and it occurs when $x = 2$.

15. $2m^2 + m - 15 > 0$

Solve: $2m^2 + m - 15 = 0$

$(2m - 5)(m + 3) = 0$

$2m - 5 = 0 \quad \text{or} \quad m + 3 = 0$

$2m = 5 \quad \text{or} \quad m = -3$

$$m = \frac{5}{2}$$

Determine where each factor is positive and negative and where the product of these factors is positive and negative.

Interval $\xleftarrow{\hspace{0.5cm}}$

$(-\infty, -3) \quad -3 \quad \left(-3, \dfrac{5}{2}\right) \quad \dfrac{5}{2} \quad \left(\dfrac{5}{2}, \infty\right)$

$2m - 5 \quad ---- \quad - \quad ---- \quad 0 \quad ++++$

$m + 3 \quad ---- \quad 0 \quad ++++ \quad + \quad ++++$

$(2m - 5)(m + 3) \quad ++++ \quad 0 \quad ---- \quad 0 \quad ++++$

The inequality is strict, so -3 and $\dfrac{5}{2}$ are not part of the solution. Now, $(2m - 5)(m + 3)$ is greater than zero where the product is positive. The solution is $\left\{ m \;\middle|\; m < -3 \text{ or } m > \dfrac{5}{2} \right\}$ or, using interval notation, $(-\infty, -3) \cup \left(\dfrac{5}{2}, \infty\right)$.

$\xleftarrow{\hspace{0.3cm}}$ | | | |) | | | | | (| | | $\xrightarrow{\hspace{0.3cm}}$

  $-6 \quad -4 \; / \; -2 \quad 0 \quad 2 \; \backslash \; 4 \quad 6 \quad 8$
  $ -3 \dfrac{5}{2}$

16. $x^3 + 5x^2 - 4x - 20 \leq 0$

Solve $x^3 + 5x^2 - 4x - 20 = 0$

$x^2(x + 5) - 4(x + 5) = 0$

$(x + 5)(x^2 - 4) = 0$

$(x + 5)(x + 2)(x - 2) = 0$

$x + 5 = 0 \quad \text{or} \quad x + 2 = 0 \quad \text{or} \quad x - 2 = 0$

$x = -5 \quad \text{or} \quad\quad x = -2 \quad \text{or} \quad\quad x = 2$

Determine where each factor is positive and negative and where the product of these factors is positive and negative.

Interval $\xleftarrow{\hspace{0.5cm}}$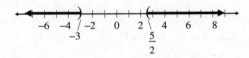

$ -5 -2 2$
$(-\infty, -5) (-5, -2) (-2, 2) (2, \infty)$

$(x + 5) \quad --- \quad 0 \quad +++ \quad + \quad +++ \quad + \quad +++$

$(x + 2) \quad --- \quad - \quad --- \quad 0 \quad +++ \quad + \quad +++$

$(x - 2) \quad --- \quad - \quad --- \quad - \quad --- \quad 0 \quad +++$

$(x + 5)(x + 2)(x - 2) \quad --- \quad 0 \quad +++ \quad 0 \quad --- \quad 0 \quad +++$

The inequality is non-strict, so -5, -2, and 2 are part of the solution. Now, $(x + 5)(x + 2)(x - 2)$ is less than zero where the product is negative. The solution is $\{x \mid x \leq -5 \text{ or } -2 \leq x \leq 2\}$ in set-builder notation and $(-\infty, -5] \cup [-2, 2]$ in interval notation.

$\xleftarrow{\hspace{0.3cm}}$ | | |] | | [| | |] | | | | | $\xrightarrow{\hspace{0.3cm}}$

  $-8 \quad -6 \quad -4 \quad -2 \quad 0 \quad 2 \quad 4 \quad 6 \quad 8$

17. $\qquad 50 = -16t^2 + 80t + 20$

$16t^2 - 80t + 30 = 0$

For this equation, $a = 16$, $b = -80$, and $c = 30$.

$t = \dfrac{-(-80) \pm \sqrt{(-80)^2 - 4(16)(30)}}{2(16)}$

$= \dfrac{80 \pm \sqrt{6400 - 1920}}{32}$

$= \dfrac{80 \pm \sqrt{4480}}{32}$

$= \dfrac{80 \pm 8\sqrt{70}}{32}$

$= \dfrac{80}{32} \pm \dfrac{8\sqrt{70}}{32}$

$= \dfrac{5}{2} \pm \dfrac{\sqrt{70}}{4}$

$t = \dfrac{5}{2} - \dfrac{\sqrt{70}}{4} \quad \text{or} \quad t = \dfrac{5}{2} - \dfrac{\sqrt{70}}{4}$

$\approx 0.408 \qquad \text{or} \qquad \approx 4.592$

Rounding to the nearest tenth, the height of the rock will be 50 feet at both 0.4 seconds and 4.6 seconds.

18. Let t represent the time required for Rupert to roof the house alone. Then $t - 4$ will represent the time required for Lex to roof the house alone.

$\begin{pmatrix} \text{Part done} \\ \text{by Rupert} \\ \text{in 1 hour} \end{pmatrix} + \begin{pmatrix} \text{Part done} \\ \text{by Lex} \\ \text{in 1 hour} \end{pmatrix} = \begin{pmatrix} \text{Part done} \\ \text{together} \\ \text{in 1 hour} \end{pmatrix}$

$\qquad \dfrac{1}{t} \quad + \quad \dfrac{1}{t - 4} \quad = \quad \dfrac{1}{16}$

$16t(t - 4)\left(\dfrac{1}{t} + \dfrac{1}{t - 4}\right) = 16t(t - 4)\left(\dfrac{1}{16}\right)$

$16(t - 4) + 16t = t(t - 4)$

$16t - 64 + 16t = t^2 - 4t$

$32t - 64 = t^2 - 4t$

$0 = t^2 - 36t + 64$

For this equation, $a = 1$, $b = -36$, and $c = 64$.

675

$$t = \frac{-(-36) \pm \sqrt{(-36)^2 - 4(1)(64)}}{2(1)}$$

$$= \frac{36 \pm \sqrt{1296 - 256}}{2}$$

$$= \frac{36 \pm \sqrt{1040}}{2}$$

$$= \frac{36 \pm 4\sqrt{65}}{2}$$

$$= 18 \pm 2\sqrt{65}$$

$t = 18 - 2\sqrt{65}$ or $t = 18 + 2\sqrt{65}$

≈ 1.875 or ≈ 34.125

Disregard $t \approx 1.875$ because this value makes Lex's time negative: $t - 4 = 1.875 - 4 = -2.125$. The only viable answer is $t \approx 34.125$ hours. Rounding to the nearest tenth, Rupert can roof the house in 34.1 hours when working alone.

19. a. We first recognize that the quadratic function has a negative leading coefficient. This means that the graph will open down and the function indeed has a maximum value. The maximum value occurs when

$$p = -\frac{b}{2a} = -\frac{170}{2(-0.25)} = 340.$$

The revenue will be maximized when the product is sold at a price of $340.

b. The maximum revenue is obtained by evaluating the revenue function at the price found in part (a).

$$R(340) = -0.25(340)^2 + 170(340)$$
$$= 28,900$$

The maximum weekly revenue is $28,900.

20. a. Let x represent the width of the base of the box. Then $\frac{50 - 2x}{2} = 25 - x$ is the length of the base. The volume is the product of the length, width, and height:

$$V = (25 - x) \cdot x \cdot 12$$
$$= -12x^2 + 300x$$

The leading coefficient is negative so we know the graph opens down and there will be a maximum volume. This value occurs when the width is

$$x = -\frac{b}{2a} = -\frac{300}{2(-12)} = 12.5.$$

Then the length is $25 - 12.5 = 12.5$. The dimensions that maximize the volume of the box are 12.5 in. by 12.5 in. by 12 in.

b. The maximum volume can be found by substituting 12.5 for x into the volume function.

$$V = (25 - 12.5) \cdot 12.5 \cdot 12 = 1875$$

The maximum volume of the box is 1875 cubic inches.

Cumulative Review Chapters R–7

1. $\dfrac{2^4 - 5 \cdot 7 + 3}{1 - 3^2} = \dfrac{16 - 35 + 3}{1 - 9} = \dfrac{-16}{-8} = 2$

2. $2(3c+1) - (c-9) - 2c = 6c + 2 - c + 9 - 2c$
$ = 3c + 11$

3. $p - 5 = 3(p-2) - 9$
$p - 5 = 3p - 6 - 9$
$p - 5 = 3p - 15$
$-5 + 15 = 3p - p$
$10 = 2p$
$5 = p$
The solution set is {5}.

4. $5 + 2|x+1| > 9$
$2|x+1| > 9 - 5$
$2|x+1| > 4$
$|x+1| > 2$
$x + 1 > 2$ or $x + 1 < -2$
$x > 1$ or $ x < -3$
The solution set is $(-\infty - 3) \cup (1, \infty)$.

5. $x^2 + x - 12 = 0$
$(x+4)(x-3) = 0$
$x + 4 = 0$ or $x - 3 = 0$
$x = -4$ or $ x = 3$
These are the values of x that make the denominator equal to zero, so they must be avoided. The domain of h is $\{x \mid x \neq -4, x \neq 3\}$.

6. $3x - 4y = 8$
$4y = 3x - 8$
$y = \dfrac{3}{4}x - 2$

From the slope-intercept form, we can see the

y-intercept is -2 and the slope is $\dfrac{3}{4}$.

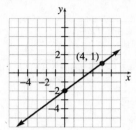

7. $2x + 5y = -15$

$5y = -2x - 15$

$y = -\dfrac{2}{5}x - 3$

Since $m = -\dfrac{2}{5}$, the slope of the parallel line will

be $m = -\dfrac{2}{5}$.

$y - y_1 = m(x - x_1)$

$y - (-1) = -\dfrac{2}{5}(x - 5)$

$y + 1 = -\dfrac{2}{5}x + 2$

$y = -\dfrac{2}{5}x + 1$

in slope-intercept form, or

$y = -\dfrac{2}{5}x + 1$

$5y = -2x + 5$

$2x + 5y = 5$

in general form.

8. $\begin{cases} 5x + 2y = 0 & (1) \\ 2x - y = -9 & (2) \end{cases}$

Multiply both sides of equation (2) by 2, and add the result to equation (1).

$5x + 2y = 0$

$\underline{4x - 2y = -18}$

$9x \quad\quad = -18$

$x = -2$

Substituting -2 for x in equation (1), we obtain

$5(-2) + 2y = 0$

$-10 + 2y = 0$

$2y = 10$

$y = 5$

The solution is the ordered pair $(-2, 5)$.

9. $\begin{vmatrix} 4 & 1 & 0 \\ 1 & 2 & -1 \\ 3 & -2 & 1 \end{vmatrix} = 4\begin{vmatrix} 2 & -1 \\ -2 & 1 \end{vmatrix} - 1\begin{vmatrix} 1 & -1 \\ 3 & 1 \end{vmatrix} + 0\begin{vmatrix} 1 & 2 \\ 3 & -2 \end{vmatrix}$

$= 4\big[2(1) - (-1)(-2)\big] - 1\big[1(1) - (-1)(3)\big]$
$\quad\quad + 0\big[1(-2) - 2(3)\big]$

$= 4(2 - 2) - 1(1 + 3) + 0$

$= 4(0) - 1(4)$

$= 0 - 4$

$= -4$

10. $\begin{cases} 2x + 3y < -3 \\ 2x + y > -5 \end{cases}$

First, graph the inequality $2x + 3y < -3$. To do so, replace the inequality symbol with an equal sign to obtain $2x + 3y = -3$. Because the inequality is strict, graph $2x + 3y = -3$

$\left(y = -\dfrac{2}{3}x - 1\right)$ using a dashed line.

Test Point: $(0,0): 2 \cdot 0 + 3 \cdot 0 = 0 \not< -3$

Therefore, the half-plane not containing $(0,0)$ is the solution set of $2x + 3y < -3$.

Next, graph the inequality $2x + y > -5$. To do so, replace the inequality symbol with an equal sign to obtain $2x + y = -5$. Because the inequality is strict, graph $2x + y = -5$

$(y = -2x - 5)$ using a dashed line.

Test Point: $(0,0): 2 \cdot 0 + 0 = 0 > -5$

Therefore, the half-plane containing $(0,0)$ is the solution set of $2x + y > -5$.

The overlapping shaded region (that is, the shaded region in the graph) is the solution to the system of linear inequalities.

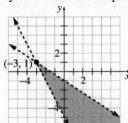

11. $\left(a^3 - 9a^2 + 11\right) + \left(7a^2 - 5a - 8\right)$

$= a^3 - 9a^2 + 11 + 7a^2 - 5a - 8$

$= a^3 - 2a^2 - 5a + 3$

12.

$$
2x^2 - x + 5 \overline{\smash{\big)}\, 8x^4 + 0x^3 + 12x^2 + 17x - 18} \quad \overset{\textstyle 4x^2 + 2x - 3}{}
$$

$$
\underline{-\left(8x^4 - 4x^3 + 20x^2\right)}
$$
$$
4x^3 - 8x^2 + 17x
$$
$$
\underline{-\left(4x^3 - 2x^2 + 10x\right)}
$$
$$
-6x^2 + 7x - 18
$$
$$
\underline{-\left(-6x^2 + 3x - 15\right)}
$$
$$
4x - 3
$$

$$
\frac{8x^4 + 12x^2 + 17x - 18}{2x^2 - x + 5}
$$
$$
= 4x^2 + 2x - 3 + \frac{4x - 3}{2x^2 - x + 5}
$$

13.
$$
8m^3 + 27n^3 = (2m)^3 + (3n)^3
$$
$$
= (2m + 3n)\left(4m^2 - 6mn + 9n^2\right)
$$

14. $7y^2 + 23y - 20 = (7y - 5)(y + 4)$

15.
$$
\frac{\dfrac{2w^2 - 11w + 12}{w^2 - 16}}{\dfrac{2w^2 - 7w + 6}{w^2 + 9w + 20}}
$$
$$
= \frac{2w^2 - 11w + 12}{w^2 - 16} \cdot \frac{w^2 + 9w + 20}{2w^2 - 7w + 6}
$$
$$
= \frac{(2w - 3)\,(w - 4)}{(w + 4)\,(w - 4)} \cdot \frac{(w + 4)\,(w + 5)}{(2w - 3)\,(w - 2)}
$$
$$
= \frac{w + 5}{w - 2}
$$

16.
$$
\frac{3}{k^2 - 5k + 4} - \frac{2}{k^2 + 4k - 5}
$$
$$
= \frac{3}{(k - 4)(k - 1)} - \frac{2}{(k + 5)(k - 1)}
$$
$$
= \frac{3(k + 5)}{(k - 4)(k - 1)(k + 5)} - \frac{2(k - 4)}{(k + 5)(k - 1)(k - 4)}
$$
$$
= \frac{3(k + 5) - 2(k - 4)}{(k + 5)(k - 1)(k - 4)}
$$
$$
= \frac{3k + 15 - 2k + 8}{(k + 5)(k - 1)(k - 4)}
$$
$$
= \frac{k + 23}{(k + 5)(k - 1)(k - 4)}
$$

17.
$$
\frac{1}{2n} - \frac{1}{6} = \frac{5}{4n} + \frac{1}{3}
$$
$$
12n\left(\frac{1}{2n} - \frac{1}{6}\right) = 12n\left(\frac{5}{4n} + \frac{1}{3}\right)
$$
$$
6 - 2n = 15 + 4n
$$
$$
6 - 15 = 4n + 2n
$$
$$
6n = -9
$$
$$
n = -\frac{3}{2}
$$

The solution set is $\left\{-\dfrac{3}{2}\right\}$.

18. $\dfrac{x - 4}{x + 5} \geq 0$

The rational expression will equal 0 when $x = 4$. It is undefined when $x = -5$. Thus we separate the real number line into intervals $(-\infty, -5)$, $(-5, 4)$, and $(4, \infty)$. Determine where the numerator and denominator are positive and negative and where the quotient is positive and negative.

Interval	$(-\infty, -5)$	-5	$(-5, 4)$	4	$(4, \infty)$
$x - 4$	Neg	Neg	Neg	0	Pos
$x + 5$	Neg	0	Pos	Pos	Pos
$\dfrac{x - 4}{x + 5}$	Pos	Undef	Neg	0	Pos

The rational function is undefined at $x = -5$, so -5 is not part of the solution. The inequality is non-strict, so 4 is part of the solution. Now,

$\dfrac{x-4}{x+5}$ is greater than zero where the quotient is positive. The solution is $(-\infty,-5)\cup[4,\infty)$ in interval notation.

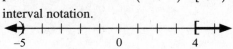

19. Recall $D=r\cdot t$.

Let r_s be the speed of the slower plane, then $D_s=r_s\cdot 2$ is the distance the slower plane traveled.

Let r_f be the speed of the faster plane, then $D_f=r_f\cdot 2$ is the distance the faster plane traveled. Since $r_f=r_s+16$, we can express D_f as $D_f=(r_s+16)\cdot 2$.

We are giving the total distance is 536 miles, so
$$D_s+D_f=536$$
$$r_s\cdot 2+(r_s+16)\cdot 2=536$$
$$2r_s+2r_s+32=536$$
$$4r_s+32=536$$
$$4r_s=504$$
$$r_s=126$$

The rate of the slower plane is 126 mph, and the rate of the faster plane is 142 mph.

20.
$$\sqrt{48}+\sqrt{75}-2\sqrt{3}=\sqrt{16\cdot 3}+\sqrt{25\cdot 3}-2\sqrt{3}$$
$$=4\sqrt{3}+5\sqrt{3}-2\sqrt{3}$$
$$=7\sqrt{3}$$

21. $\dfrac{5}{\sqrt[3]{16}}=\dfrac{5}{\sqrt[3]{16}}\cdot\dfrac{\sqrt[3]{4}}{\sqrt[3]{4}}=\dfrac{5\sqrt[3]{4}}{\sqrt[3]{64}}=\dfrac{5\sqrt[3]{4}}{4}$

22.
$$\sqrt{a-12}+\sqrt{a}=6$$
$$\sqrt{a-12}=6-\sqrt{a}$$
$$\left(\sqrt{a-12}\right)^2=\left(6-\sqrt{a}\right)^2$$
$$a-12=36-12\sqrt{a}+a$$
$$12\sqrt{a}=36+a-a+12$$
$$12\sqrt{a}=48$$
$$\sqrt{a}=4$$
$$\left(\sqrt{a}\right)^2=4^2$$
$$a=16$$

Check: $\sqrt{16-12}+\sqrt{16}\overset{?}{=}6$
$$\sqrt{4}+4\overset{?}{=}6$$
$$2+4\overset{?}{=}6$$
$$6=6\ \checkmark$$

The solution set is $\{16\}$.

23. $a^2+b^2=c^2$
$$(x-4)^2+(x+13)^2=(x+14)^2$$
$$x^2-8x+16+x^2+26x+169=x^2+28x+196$$
$$2x^2+18x+185=x^2+28x+196$$
$$2x^2+18x+185-x^2-28x-196=0$$
$$x^2-10x-11=0$$
$$(x-11)(x+1)=0$$

$$x-11=0\quad\text{or}\quad x+1=0$$
$$x=11\quad\text{or}\qquad x=-1$$

Since the value $x=-1$ would give a negative value for one of the measurements of the triangle, we ignore it. The value of x is 11.

24.
$$x^2-20=8x$$
$$x^2-8x-20=0$$
$$(x-10)(x+2)=0$$
$$x-10=0\quad\text{or}\quad x+2=0$$
$$x=10\quad\text{or}\qquad x=-2$$

The solution set is $\{-2,10\}$.

25. $f(x) = -2x^2 + 4x + 3$

$a = -2, b = 4, c = 3$

The graph opens up because $a > 0$.

<u>vertex</u>:

$$x = -\frac{b}{2a} = -\frac{4}{2(-2)} = 1$$

$$f(1) = -2(1)^2 + 4(1) + 3 = 5$$

<u>y-intercept</u>:

$$f(0) = -2(0)^2 + 4(0) + 3 = 3$$

<u>x-intercepts</u>:

$$b^2 - 4ac = (4)^2 - 4(-2)(3) = 40 > 0$$

There are two distinct x-intercepts. We find these by solving

$$f(x) = 0$$

$$-2x^2 + 4x + 3 = 0$$

$$x = \frac{-4 \pm \sqrt{4^2 - 4(-2)(3)}}{2(-2)}$$

$$= \frac{-4 \pm \sqrt{40}}{-4}$$

$$= \frac{-4 \pm 2\sqrt{10}}{-4}$$

$$= 1 \pm \frac{\sqrt{10}}{2}$$

$$x = 1 - \frac{\sqrt{10}}{2} \approx -0.58 \text{ or } x = 1 + \frac{\sqrt{10}}{2} \approx 2.58$$

<u>Graph</u>:

The y-intercept point, $(0,3)$, is one unit to the left of the axis of symmetry. Therefore, if we move one unit to the right of the axis of symmetry, we obtain the point $(2,3)$ which must also be on the graph.

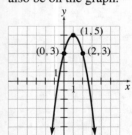

Chapter 8

Section 8.1

Are You Ready for This Section?

R1. We need to find all values of x that cause the denominator $x^2 + 3x - 28$ to equal 0.

$$x^2 + 3x - 28 = 0$$
$$(x+7)(x-4) = 0$$
$$x + 7 = 0 \quad \text{or} \quad x - 4 = 0$$
$$x = -7 \quad \text{or} \quad x = 4$$

Thus, the domain of $R(x) = \dfrac{x^2 - 9}{x^2 + 3x - 28}$ is

$\{x \mid x \neq -7, x \neq 4\}$.

R2. a. $f(-2) = 2(-2)^2 - (-2) + 1$
$$= 8 + 2 + 1$$
$$= 11$$

b. $f(a+1) = 2(a+1)^2 - (a+1) + 1$
$$= 2(a^2 + 2a + 1) - (a+1) + 1$$
$$= 2a^2 + 4a + 2 - a - 1 + 1$$
$$= 2a^2 + 3a + 2$$

R3. The graph shown is not a function because it fails the vertical line test.

Section 8.1 Quick Checks

1. Given two functions f and g, the <u>composite function</u>, denoted by $f \circ g$, is defined by $(f \circ g)(x) = f(g(x))$.

2. $f(x) = 4x - 3$; $g(x) = x^2 + 1$

a. $g(2) = (2)^2 + 1 = 4 + 1 = 5$
$$f(5) = 4(5) - 3 = 20 - 3 = 17$$
$$(f \circ g)(2) = f(g(2)) = f(5) = 17$$

b. $f(2) = 4(2) - 3 = 8 - 3 = 5$
$$g(5) = (5)^2 + 1 = 25 + 1 = 26$$
$$(g \circ f)(2) = g(f(2)) = g(5) = 26$$

c. $f(-3) = 4(-3) - 3 = -12 - 3 = -15$
$$f(-15) = 4(-15) - 3 = -60 - 3 = -63$$
$$(f \circ f)(-3) = f(f(-3)) = f(-15) = -63$$

3. False. $(f \circ g)(x) = f(g(x))$

4. $g(x) = 4x - 3$

5. $f(x) = x^2 - 3x + 1$; $g(x) = 3x + 2$

a. $(f \circ g)(x) = f(g(x))$
$$= (3x+2)^2 - 3(3x+2) + 1$$
$$= 9x^2 + 12x + 4 - 9x - 6 + 1$$
$$= 9x^2 + 3x - 1$$

b. $(g \circ f)(x) = g(f(x))$
$$= 3(x^2 - 3x + 1) + 2$$
$$= 3x^2 - 9x + 3 + 2$$
$$= 3x^2 - 9x + 5$$

c. $(f \circ g)(-2) = 9(-2)^2 + 3(-2) - 1$
$$= 9(4) + 3(-2) - 1$$
$$= 36 - 6 - 1$$
$$= 29$$

6. A function is <u>one-to-one</u> if any two different inputs in the domain correspond to two different outputs in the range. That is, if x_1 and x_2 are two different inputs to a function f, then $f(x_1) \neq f(x_2)$.

7. Since the two friends (inputs) Max and Yolanda share the same birthday (output) of November 8, this function is not one-to-one.

8. The function is one-to-one because there are no two distinct inputs that correspond to the same output.

9. a. The graph fails the horizontal line test. For example, the line $y = -1$ will intersect the graph of f in four places. Therefore, the function is not one-to-one.

b. The graph passes the horizontal line test because every horizontal line will intersect the graph of *f* exactly once. Thus, the function is one-to-one.

10. The inverse of the one-to-one function is:

Right Tibia → Right Humerus
36.05 → 24.80
35.57 → 24.59
34.58 → 24.29
34.20 → 23.81
34.73 → 24.87

The domain of the inverse function is {36.05, 35.57, 34.58, 34.20, 34.73}. The range of the inverse function is {24.80, 24.59, 24.29, 23.81, 24.87}.

11. To obtain the inverse, we switch the *x*- and *y*-coordinates:
$\{(3,-3),(2,-2),(1,-1),(0,0),(-1,1)\}$
The domain of the inverse function is $\{3,2,1,0,-1\}$. The range of the inverse function is $\{-3,-2,-1,0,1\}$.

12. To plot the inverse, switch the *x*- and *y*-coordinates of each point and connect the corresponding points. The graph of the function (shaded) and the line $y=x$ (dashed) are included for reference.

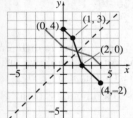

13. True.

14. False. The notation $f^{-1}(x)$ represents the inverse function for the function $f(x)$.

15.
$g(x)=5x-1$
$y=5x-1$
$x=5y-1$
$x+1=5y$
$\dfrac{x+1}{5}=y$
$g^{-1}(x)=\dfrac{x+1}{5}$

Check: $g\left(g^{-1}(x)\right)=5\left(\dfrac{x+1}{5}\right)-1=x+1-1=x$

$g^{-1}(g(x))=\dfrac{(5x-1)+1}{5}=\dfrac{5x}{5}=x$

16.
$f(x)=x^5+3$
$y=x^5+3$
$x=y^5+3$
$x-3=y^5$
$\sqrt[5]{x-3}=y$
$f^{-1}(x)=\sqrt[5]{x-3}$

Check:
$f\left(f^{-1}(x)\right)=\left(\sqrt[5]{x-3}\right)^5+3=x-3+3=x$

$f^{-1}\left(f(x)\right)=\sqrt[5]{x^5+3-3}=\sqrt[5]{x^5}=x$

8.1 Exercises

17. $f(x)=2x+5$; $g(x)=x-4$

a. $g(3)=3-4=-1$
$f(-1)=2(-1)+5=3$
$(f\circ g)(3)=f(g(3))=f(-1)=3$

b. $f(-2)=2(-2)+5=1$
$g(1)=1-4=-3$
$(g\circ f)(-2)=g(f(-2))=g(1)=-3$

c. $f(1)=2(1)+5=7$
$f(7)=2(7)+5=19$
$(f\circ f)(1)=f(f(1))=f(7)=19$

d. $g(-4) = -4 - 4 = -8$
$g(-8) = -8 - 4 = -12$
$(g \circ g)(-4) = g(g(-4)) = g(-8) = -12$

19. $f(x) = x^2 + 4$; $g(x) = 2x + 3$

a. $g(3) = 2(3) + 3 = 9$
$f(9) = (9)^2 + 4 = 85$
$(f \circ g)(3) = f(g(3)) = f(9) = 85$

b. $f(-2) = (-2)^2 + 4 = 8$
$g(8) = 2(8) + 3 = 19$
$(g \circ f)(-2) = g(f(-2)) = g(8) = 19$

c. $f(1) = (1)^2 + 4 = 5$
$f(5) = (5)^2 + 4 = 29$
$(f \circ f)(1) = f(f(1)) = f(5) = 29$

d. $g(-4) = 2(-4) + 3 = -5$
$g(-5) = 2(-5) + 3 = -7$
$(g \circ g)(-4) = g(g(-4)) = g(-5) = -7$

21. $f(x) = 2x^3$; $g(x) = -2x^2 + 5$

a. $g(3) = -2(3)^2 + 5 = -2(9) + 5 = -13$
$f(-13) = 2(-13)^3 = -4394$
$(f \circ g)(3) = f(g(3)) = f(-13) = -4394$

b. $f(-2) = 2(-2)^3 = 2(-8) = -16$
$g(-16) = -2(-16)^2 + 5$
$\qquad = -2(256) + 5$
$\qquad = -507$
$(g \circ f)(-2) = g(f(-2)) = g(-16) = -507$

c. $f(1) = 2(1)^3 = 2$
$f(2) = 2(2)^3 = 16$
$(f \circ f)(1) = f(f(1)) = f(2) = 16$

d. $g(-4) = -2(-4)^2 + 5 = -2(16) + 5 = -27$
$g(-27) = -2(-27)^2 + 5$
$\qquad = -2(729) + 5$
$\qquad = -1453$
$(g \circ g)(-4) = g(g(-4)) = g(-27) = -1453$

23. $f(x) = |x - 10|$; $g(x) = \dfrac{12}{x + 3}$

a. $g(3) = \dfrac{12}{3 + 3} = \dfrac{12}{6} = 2$
$f(2) = |2 - 10| = |-8| = 8$
$(f \circ g)(3) = f(g(3)) = f(2) = 8$

b. $f(-2) = |-2 - 10| = |-12| = 12$
$g(12) = \dfrac{12}{12 + 3} = \dfrac{12}{15} = \dfrac{4}{5}$
$(g \circ f)(-2) = g(f(-2)) = g(12) = \dfrac{4}{5}$

c. $f(1) = |1 - 10| = |-9| = 9$
$f(9) = |9 - 10| = |-1| = 1$
$(f \circ f)(1) = f(f(1)) = f(9) = 1$

d. $g(-4) = \dfrac{12}{-4 + 3} = \dfrac{12}{-1} = -12$
$g(-12) = \dfrac{12}{-12 + 3} = \dfrac{12}{-9} = -\dfrac{4}{3}$
$(g \circ g)(-4) = g(g(-4)) = g(-12) = -\dfrac{4}{3}$

25. $f(x) = x + 1$; $g(x) = 2x$

a. $(f \circ g)(x) = f(g(x)) = (2x) + 1 = 2x + 1$

b. $(g \circ f)(x) = g(f(x)) = 2(x + 1) = 2x + 2$

c. $(f \circ f)(x) = f(f(x)) = (x + 1) + 1 = x + 2$

d. $(g \circ g)(x) = g(g(x)) = 2(2x) = 4x$

27. $f(x)=2x+7$; $g(x)=-4x+5$

  a. $(f \circ g)(x)=f(g(x))$
$$=2(-4x+5)+7$$
$$=-8x+10+7$$
$$=-8x+17$$

  b. $(g \circ f)(x)=g(f(x))$
$$=-4(2x+7)+5$$
$$=-8x-28+5$$
$$=-8x-23$$

  c. $(f \circ f)(x)=f(f(x))$
$$=2(2x+7)+7$$
$$=4x+14+7$$
$$=4x+21$$

  d. $(g \circ g)(x)=g(g(x))$
$$=-4(-4x+5)+5$$
$$=16x-20+5$$
$$=16x-15$$

29. $f(x)=x^2$; $g(x)=x-3$

  a. $(f \circ g)(x)=f(g(x))$
$$=(x-3)^2=x^2-6x+9$$

  b. $(g \circ f)(x)=g(f(x))=(x^2)-3=x^2-3$

  c. $(f \circ f)(x)=f(f(x))=(x^2)^2=x^4$

  d. $(g \circ g)(x)=g(g(x))=(x-3)-3=x-6$

31. $f(x)=\sqrt{x}$; $g(x)=x+4$

  a. $(f \circ g)(x)=f(g(x))=\sqrt{x+4}$

  b. $(g \circ f)(x)=g(f(x))=\sqrt{x}+4$

  c. $(f \circ f)(x)=f(f(x))=\sqrt{\sqrt{x}}=\sqrt[4]{x}$

  d. $(g \circ g)(x)=g(g(x))=(x+4)+4=x+8$

33. $f(x)=|x+4|$; $g(x)=x^2-4$

  a. $(f \circ g)(x)=f(g(x))$
$$=|(x^2-4)+4|=|x^2|=x^2$$

  b. $(g \circ f)(x)=g(f(x))$
$$=(|x+4|)^2-4=(x+4)^2-4$$
$$=x^2+8x+16-4$$
$$=x^2+8x+12$$

  c. $(f \circ f)(x)=f(f(x))=||x+4|+4|$

  d. $(g \circ g)(x)=g(g(x))$
$$=(x^2-4)^2-4$$
$$=x^4-8x^2+16-4$$
$$=x^4-8x^2+12$$

35. $f(x)=\dfrac{2}{x+1}$; $g(x)=\dfrac{1}{x}$

  a. $(f \circ g)(x)=f(g(x))$
$$=\frac{2}{\frac{1}{x}+1}=\frac{2}{\frac{1+x}{x}}=\frac{2x}{x+1}$$
    where $x \ne -1,0$.

  b. $(g \circ f)(x)=g(f(x))$
$$=\frac{1}{\frac{2}{x+1}}=1 \cdot \frac{x+1}{2}=\frac{x+1}{2}$$
    where $x \ne -1$.

  c. $(f \circ f)(x)=f(f(x))$
$$=\frac{2}{\frac{2}{x+1}+1}=\frac{2}{\frac{2+x+1}{x+1}}$$
$$=2 \cdot \frac{x+1}{x+3}=\frac{2(x+1)}{x+3}$$
    where $x \ne -1,-3$.

  d. $(g \circ g)(x)=g(g(x))=\dfrac{1}{\frac{1}{x}}=1 \cdot \dfrac{x}{1}=x$
    where $x \ne 0$.

37. The function is one-to-one. Each element in the range corresponds to exactly one element in the domain.

39. The function is not one-to-one. There is an element in the range (57.3) that corresponds to more than one element in the domain (Spain and China).

41. The function is one-to-one. Each element in the range corresponds to exactly one element in the domain.

43. The function is not one-to-one. There are elements in the range (2 and 4) that correspond to more than one element in the domain.

45. The function is one-to-one. Each element in the range corresponds to exactly one element in the domain.

47. The graph passes the horizontal line test, so the graph is that of a one-to-one function.

49. The graph fails the horizontal line test. Therefore, the function is not one-to-one.

51. The graph passes the horizontal line test, so the graph is that of a one-to-one function.

53. Inverse:

Weight (g)	U. S. Coin
2.500 ⟶	Cent
5.000 ⟶	Nickel
2.268 ⟶	Dime
5.670 ⟶	Quarter
11.340 ⟶	Half Dollar
8.100 ⟶	Dollar

55. To obtain the inverse, we switch the *x*- and *y*-coordinates.

Inverse: $\{(3,0),(4,1),(5,2),(6,3)\}$

57. To obtain the inverse, we switch the *x*- and *y*-coordinates.

Inverse: $\{(3,-2),(1,-1),(-3,0),(9,1)\}$

59. To plot the inverse, switch the *x*- and *y*-coordinates in each point and connect the corresponding points. The graph of the function (shaded) and the line $y = x$ (dashed) are included for reference.

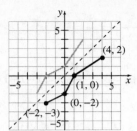

61. To plot the inverse, switch the *x*- and *y*-coordinates in each point and connect the corresponding points. The graph of the function (shaded) and the line $y = x$ (dashed) are included for reference.

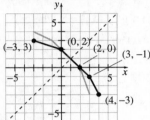

63. To plot the inverse, switch the *x*- and *y*-coordinates in each point and connect the corresponding points. The graph of the function (shaded) and the line $y = x$ (dashed) are included for reference.

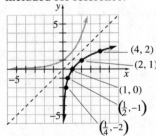

65. $f(g(x)) = (x-5)+5 = x$

$g(f(x)) = (x+5)-5 = x$

Since $f(g(x)) = g(f(x)) = x$, the two functions are inverses of each other.

67. $f(g(x)) = 5\left(\dfrac{x-7}{5}\right)+7 = x-7+7 = x$

$g(f(x)) = \dfrac{(5x+7)-7}{5} = \dfrac{5x}{5} = x$

Since $f(g(x)) = g(f(x)) = x$, the two functions are inverses of each other.

69. $f\left(g\left(x\right)\right)=\dfrac{3}{\left(\dfrac{3}{x}+1\right)-1}=\dfrac{3}{\dfrac{3}{x}}=3\cdot\dfrac{x}{3}=x$

$g\left(f\left(x\right)\right)=\dfrac{3}{\dfrac{3}{x-1}}+1=3\cdot\dfrac{x-1}{3}+1=x-1+1=x$

Since $f\left(g\left(x\right)\right)=g\left(f\left(x\right)\right)=x$, the two functions are inverses of each other.

71. $f\left(g\left(x\right)\right)=\sqrt[3]{\left(x^3-4\right)+4}=\sqrt[3]{x^3}=x$

$g\left(f\left(x\right)\right)=\left(\sqrt[3]{x+4}\right)^3-4=x+4-4=x$

Since $f\left(g\left(x\right)\right)=g\left(f\left(x\right)\right)=x$, the two functions are inverses of each other.

73. $f\left(x\right)=6x$

$y=6x$

$x=6y$

$\dfrac{x}{6}=y$

$f^{-1}\left(x\right)=\dfrac{x}{6}$

Check:

$f\left(f^{-1}\left(x\right)\right)=6\left(\dfrac{x}{6}\right)=x$

$f^{-1}\left(f\left(x\right)\right)=\dfrac{6x}{6}=x$

75. $f\left(x\right)=x+4$

$y=x+4$

$x=y+4$

$x-4=y$

$f^{-1}\left(x\right)=x-4$

Check:

$f\left(f^{-1}\left(x\right)\right)=\left(x-4\right)+4=x$

$f^{-1}\left(f\left(x\right)\right)=\left(x+4\right)-4=x$

77. $h\left(x\right)=2x-7$

$y=2x-7$

$x=2y-7$

$x+7=2y$

$\dfrac{x+7}{2}=y$

$h^{-1}\left(x\right)=\dfrac{x+7}{2}$

Check:

$h\left(h^{-1}\left(x\right)\right)=2\left(\dfrac{x+7}{2}\right)-7=x+7-7=x$

$h^{-1}\left(h\left(x\right)\right)=\dfrac{\left(2x-7\right)+7}{2}=\dfrac{2x}{2}=x$

79. $G\left(x\right)=2-5x$

$y=2-5x$

$x=2-5y$

$x-2=-5y$

$\dfrac{x-2}{-5}=y$

$\dfrac{2-x}{5}=y$

$G^{-1}\left(x\right)=\dfrac{2-x}{5}$

Check:

$G\left(G^{-1}\left(x\right)\right)=2-5\left(\dfrac{2-x}{5}\right)=2-\left(2-x\right)=x$

$G^{-1}\left(G\left(x\right)\right)=\dfrac{2-\left(2-5x\right)}{5}=\dfrac{5x}{5}=x$

81. $g\left(x\right)=x^3+3$

$y=x^3+3$

$x=y^3+3$

$x-3=y^3$

$\sqrt[3]{x-3}=y$

$g^{-1}\left(x\right)=\sqrt[3]{x-3}$

Check:

$g\left(g^{-1}\left(x\right)\right)=\left(\sqrt[3]{x-3}\right)^3+3=x-3+3=x$

$g^{-1}\left(g\left(x\right)\right)=\sqrt[3]{x^3+3-3}=\sqrt[3]{x^3}=x$

83. $p(x) = \dfrac{1}{x+3}$

$y = \dfrac{1}{x+3}$

$x = \dfrac{1}{y+3}$

$y+3 = \dfrac{1}{x}$

$y = \dfrac{1}{x} - 3$

$p^{-1}(x) = \dfrac{1}{x} - 3$

Check:

$p\left(p^{-1}(x)\right) = \dfrac{1}{\left(\dfrac{1}{x} - 3\right) + 3} = \dfrac{1}{\dfrac{1}{x}} = 1 \cdot \dfrac{x}{1} = x$

$p^{-1}\left(p(x)\right) = \dfrac{1}{\dfrac{1}{x+3}} - 3 = x + 3 - 3 = x$

85. $F(x) = \dfrac{5}{2-x}$

$y = \dfrac{5}{2-x}$

$x = \dfrac{5}{2-y}$

$2 - y = \dfrac{5}{x}$

$-y = -2 + \dfrac{5}{x}$

$y = 2 - \dfrac{5}{x}$

$F^{-1}(x) = 2 - \dfrac{5}{x}$

Check:

$F\left(F^{-1}(x)\right) = \dfrac{5}{2 - \left(2 - \dfrac{5}{x}\right)} = \dfrac{5}{\dfrac{5}{x}} = 5 \cdot \dfrac{x}{5} = x$

$F^{-1}\left(F(x)\right) = 2 - \dfrac{5}{\dfrac{5}{2-x}} = 2 - (2 - x) = x$

87. $f(x) = \sqrt[3]{x-2}$

$y = \sqrt[3]{x-2}$

$x = \sqrt[3]{y-2}$

$x^3 = y - 2$

$x^3 + 2 = y$

$f^{-1}(x) = x^3 + 2$

Check: $f\left(f^{-1}(x)\right) = \sqrt[3]{x^3 + 2 - 2} = \sqrt[3]{x^3} = x$

$f^{-1}\left(f(x)\right) = \left(\sqrt[3]{x-2}\right)^3 + 2 = x - 2 + 2 = x$

89. $R(x) = \dfrac{x}{x+2}$

$y = \dfrac{x}{x+2}$

$x = \dfrac{y}{y+2}$

$x(y+2) = y$

$xy + 2x = y$

$2x = y - xy$

$2x = y(1-x)$

$\dfrac{2x}{1-x} = y$

$R^{-1}(x) = \dfrac{2x}{1-x}$

Check: $R\left(R^{-1}(x)\right) = \dfrac{\dfrac{2x}{1-x}}{\dfrac{2x}{1-x} + 2} = \dfrac{\dfrac{2x}{1-x}}{\dfrac{2x+2-2x}{1-x}}$

$= \dfrac{2x}{1-x} \cdot \dfrac{1-x}{2} = x$

$R^{-1}\left(R(x)\right) = \dfrac{2\left(\dfrac{x}{x+2}\right)}{1 - \dfrac{x}{x+2}} = \dfrac{\dfrac{2x}{x+2}}{\dfrac{x+2-x}{x+2}}$

$= \dfrac{2x}{x+2} \cdot \dfrac{x+2}{2} = x$

91.
$$f(x) = \sqrt[3]{x-1} + 4$$
$$y = \sqrt[3]{x-1} + 4$$
$$x = \sqrt[3]{y-1} + 4$$
$$x - 4 = \sqrt[3]{y-1}$$
$$(x-4)^3 = y - 1$$
$$(x-4)^3 + 1 = y$$
$$f^{-1}(x) = (x-4)^3 + 1$$

Check:
$$f\left(f^{-1}(x)\right) = \sqrt[3]{(x-4)^3 + 1 - 1} + 4$$
$$= \sqrt[3]{(x-4)^3} + 4$$
$$= x - 4 + 4$$
$$= x$$

$$f^{-1}(f(x)) = \left[\left(\sqrt[3]{x-1}+4\right)-4\right]^3 + 1$$
$$= \left(\sqrt[3]{x-1}\right)^3 + 1$$
$$= x - 1 + 1$$
$$= x$$

93. $A(r) = \pi r^2$; $r(t) = 20t$
$$A(t) = A(r(t)) = \pi(20t)^2 = 400\pi t^2$$
$$A(3) = 400\pi(3)^2 = 3600\pi \approx 11,309.73 \text{ sq ft}$$

95. $A(x) = \dfrac{x}{9}$; $C(A) = 18A$

a. $C(x) = C(A(x)) = 18\left(\dfrac{x}{9}\right) = 2x$

b. The area of the room is $15 \cdot 21 = 315$ sq ft.
$$C(315) = 2(315) = 630$$
It will cost $630 to install the carpet.

97. $f^{-1}(12) = f^{-1}(f(4)) = 4$

99. To find the domain and range of the inverse, we simply switch the values for the domain and range of the function.
Domain of f^{-1}: $[-5, \infty)$
Range of f^{-1}: $[0, \infty)$

101. To find the domain and range of the inverse, we simply switch the values for the domain and range of the function.
Domain of g^{-1}: $(-6, 12)$
Range of g^{-1}: $[-4, 10]$

103.
$$T(x) = 0.15(x-8700) + 870$$
$$T = 0.15(x-8700) + 870$$
$$T = 0.15x - 1305 + 870$$
$$T = 0.15x - 435$$
$$T + 435 = 0.15x$$
$$\frac{T+435}{0.15} = x$$
$$x(T) = \frac{T+435}{0.15}$$
To determine the restrictions, we find
$T(8700) = 0.15(8700-8700) + 870 = 870$ and
$T(35,350) = 0.15(35,350-8700) + 870 = 4867.50$.

Thus, $x(T) = \dfrac{T+435}{0.15}$ for $870 \le T \le 4867.50$.

105. $f(x) = 2x^2 - x + 5$; $g(x) = x + a$
$$(f \circ g)(x) = f(g(x))$$
$$= 2(x+a)^2 - (x+a) + 5$$
$$= 2(x^2 + 2ax + a^2) - x - a + 5$$
$$= 2x^2 + 4ax + 2a^2 - x - a + 5$$

The y-intercept is found by letting $x = 0$ and solving for y. We know the y-intercept of $(f \circ g)(x)$ is $(0, 20)$. Therefore, we get
$$(f \circ g)(0) = 2(0)^2 + 4a(0) + 2a^2 - 0 - a + 5$$
$$20 = 2a^2 - a + 5$$
$$0 = 2a^2 - a - 15$$
$$0 = (2a+5)(a-3)$$
$$2a+5 = 0 \quad \text{or} \quad a - 3 = 0$$
$$2a = -5 \qquad\qquad a = 3$$
$$a = -\frac{5}{2}$$

The solution set for a is $\left\{-\dfrac{5}{2}, 3\right\}$.

107. Answers may vary. One possibility follows:
A function is one-to-one if each element of the range corresponds to no more than one element in the domain. If a function is not one-to-one, then there is at least one element in the range that corresponds to more than one element in the domain. For the inverse, we would switch the domain and range. Therefore, if a function is not one-to-one, the domain of its inverse will have at least one element that corresponds to more than one element in its range. Hence, the inverse would not be a function.

109. Answers may vary. One possibility follows:
By definition, the inverse of a function with ordered pairs of the form (a, b) is the set of ordered pairs of the form (b, a). This means that the set of y-coordinates (i.e., the range) of the original function will become the set of x-coordinates (i.e., the domain) of the inverse function. Likewise, the set of x-coordinates (i.e., the domain) of the original function will become the set of y-coordinates (i.e., the range) of the inverse function.

111. $f(x) = 2x + 5$; $g(x) = x - 4$

```
Plot1 Plot2 Plot3
\Y1 2X+5
\Y2 X-4
\Y3=
\Y4=
\Y5=
\Y6=
\Y7=
```

a. $(f \circ g)(3) = 3$
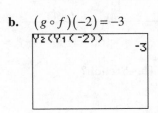
```
Y1(Y2(3))
                3
```

b. $(g \circ f)(-2) = -3$
```
Y2(Y1(-2))
               -3
```

c. $(f \circ f)(1) = 19$
```
Y1(Y1(1))
               19
```

d. $(g \circ g)(-4) = -12$
```
Y2(Y2(-4))
               -12
```

113. $f(x) = x^2 + 4$; $g(x) = 2x + 3$
```
Plot1 Plot2 Plot3
\Y1 X²+4
\Y2 2X+3
\Y3=
\Y4=
\Y5=
\Y6=
\Y7=
```

a. $(f \circ g)(3) = 85$
```
Y1(Y2(3))
               85
```

b. $(g \circ f)(-2) = 19$
```
Y2(Y1(-2))
               19
```

c. $(f \circ f)(1) = 29$
```
Y1(Y1(1))
               29
```

d. $(g \circ g)(-4) = -7$
```
Y2(Y2(-4))
               -7
```


115. $f(x) = 2x^3$; $g(x) = -2x^2 + 5$

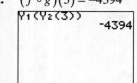

  a. $(f \circ g)(3) = -4394$

```
Y₁(Y₂(3))
              -4394
```

  b. $(g \circ f)(-2) = -507$

```
Y₂(Y₁(-2))
               -507
```

  c. $(f \circ f)(1) = 16$

```
Y₁(Y₁(1))
                 16
```

  d. $(g \circ g)(-4) = -1453$

```
Y₂(Y₂(-4))
              -1453
```

117. $f(x) = |x - 10|$; $g(x) = \dfrac{12}{x+3}$

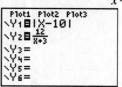

  a. $(f \circ g)(3) = 8$

```
Y₁(Y₂(3))
                  8
```

  b. $(g \circ f)(-2) = \dfrac{4}{5}$

```
Y₂(Y₁(-2))
                 4/5
```

  c. $(f \circ f)(1) = 1$

```
Y₁(Y₁(1))
                  1
```

  d. $(g \circ g)(-4) = -\dfrac{4}{3}$

```
Y₂(Y₂(-4))
                -4/3
■
```

119. $f(x) = x + 5$; $g(x) = x - 5$

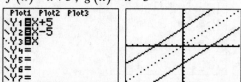

121. $f(x) = 5x + 7$; $g(x) = \dfrac{x-7}{5}$

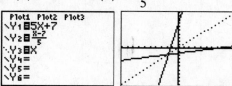

Section 8.2

Are You Ready for This Section?

  R1. **a.** $2^3 = 2 \cdot 2 \cdot 2 = 8$

        b. $2^{-1} = \dfrac{1}{2^1} = \dfrac{1}{2}$

        c. $3^4 = 3 \cdot 3 \cdot 3 \cdot 3 = 81$

R2. Locate some points on the graph of $f(x) = x^2$.

x	$f(x) = x^2$	$(x, f(x))$
-3	$f(-3) = (-3)^2 = 9$	$(-3, 9)$
-2	$f(-2) = (-2)^2 = 4$	$(-2, 4)$
-1	$f(-1) = (-1)^2 = 1$	$(-1, 1)$
0	$f(0) = 0^2 = 0$	$(0, 0)$
1	$f(1) = 1^2 = 1$	$(1, 1)$
2	$f(2) = 2^2 = 4$	$(2, 4)$
3	$f(3) = 3^2 = 9$	$(3, 9)$

Plot the points and connect them with a smooth curve.

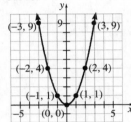

R3. A **rational number** is a number that can be expressed as a quotient $\dfrac{p}{q}$ of two integers. The integer p is called the *numerator*, and the integer q, which cannot be 0, is called the **denominator**. The set of rational numbers is the numbers

$$\mathbb{Q} = \left\{ x \mid x = \frac{p}{q}, \text{ where } p, q \text{ are integers and } q \neq 0 \right\}.$$

R4. An irrational number has a decimal representation that neither repeats nor terminates.

R5. a. Rounding to 4 decimal places, we obtain $3.20349193 \approx 3.2035$.

   b. Truncating to 4 decimal places, we obtain $3.20349193 \approx 3.2034$.

R6. a. $m^3 \cdot m^5 = m^{3+5} = m^8$

   b. $\dfrac{a^7}{a^2} = a^{7-2} = a^5$

c. $\left(z^3\right)^4 = z^{3 \cdot 4} = z^{12}$

R7. $\qquad x^2 - 5x = 14$

$x^2 - 5x - 14 = 0$

$(x-7)(x+2) = 0$

$x - 7 = 0 \ \text{ or } \ x + 2 = 0$

$\quad x = 7 \ \text{ or } \qquad x = -2$

The solution set is $\{-2, \ 7\}$.

Section 8.2 Quick Checks

1. An exponential function is a function of the form $f(x) = a^x$ where $a > 0$ and $a \neq 1$.

2. **a.** $2^{1.7} \approx 3.249009585$

   b. $2^{1.73} \approx 3.317278183$

   c. $2^{1.732} \approx 3.321880096$

   d. $2^{1.7321} \approx 3.32211036$

   e. $2^{\sqrt{3}} \approx 3.321997085$

3. Locate some points on the graph of $f(x) = 4^x$.

x	$f(x) = 4^x$	$(x, f(x))$
-2	$f(-2) = 4^{-2} = \dfrac{1}{4^2} = \dfrac{1}{16}$	$\left(-2, \dfrac{1}{16}\right)$
-1	$f(-1) = 4^{-1} = \dfrac{1}{4^1} = \dfrac{1}{4}$	$\left(-1, \dfrac{1}{4}\right)$
0	$f(0) = 4^0 = 1$	$(0, 1)$
1	$f(1) = 4^1 = 4$	$(1, 4)$

Plot the points and connect them with a smooth curve.

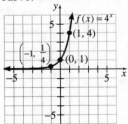

The domain of *f* is all real numbers or, using interval notation, $(-\infty, \infty)$. The range of *f* is $\{y \mid y > 0\}$ or , using interval notation, $(0, \infty)$.

4. The graph of every exponential function

$f(x) = a^x$ passes through three points: $\underline{\left(-1, \dfrac{1}{a}\right)}$,

$\underline{(0, 1)}$, and $\underline{(1, a)}$.

5. True.

6. False. The range of the exponential function $f(x) = a^x$, $a > 0$, $a \ne 1$ is $(0, \infty)$.

7. Locate some points on the graph of

$f(x) = \left(\dfrac{1}{4}\right)^x$.

x	$f(x) = \left(\dfrac{1}{4}\right)^x$	$(x, f(x))$
-1	$f(-1) = \left(\dfrac{1}{4}\right)^{-1} = 4^1 = 4$	$(-1, 4)$
0	$f(0) = \left(\dfrac{1}{4}\right)^0 = 1$	$(0, 1)$
1	$f(1) = \left(\dfrac{1}{4}\right)^1 = \dfrac{1}{4}$	$\left(1, \dfrac{1}{4}\right)$
2	$f(2) = \left(\dfrac{1}{4}\right)^2 = \dfrac{1}{16}$	$\left(2, \dfrac{1}{16}\right)$

Plot the points and connect them with a smooth curve.

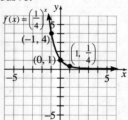

The domain of *f* is all real numbers or, using interval notation, $(-\infty, \infty)$. The range of *f* is $\{y \mid y > 0\}$ or , using interval notation, $(0, \infty)$.

8. Locate some points on the graph of $f(x) = 2^{x-1}$.

x	$f(x) = 2^{x-1}$	$(x, f(x))$
-1	$f(-1) = 2^{-1-1} = 2^{-2} = \dfrac{1}{2^2} = \dfrac{1}{4}$	$\left(-1, \dfrac{1}{4}\right)$
0	$f(0) = 2^{0-1} = 2^{-1} = \dfrac{1}{2^1} = \dfrac{1}{2}$	$\left(0, \dfrac{1}{2}\right)$
1	$f(1) = 2^{1-1} = 2^0 = 1$	$(1, 1)$
2	$f(2) = 2^{2-1} = 2^1 = 2$	$(2, 2)$
3	$f(3) = 2^{3-1} = 2^2 = 4$	$(3, 4)$
4	$f(4) = 2^{4-1} = 2^3 = 8$	$(4, 8)$

Plot the points and connect them with a smooth curve.

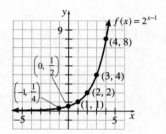

The domain of *f* is all real numbers or, using interval notation, $(-\infty, \infty)$. The range of *f* is $\{y \mid y > 0\}$ or , using interval notation, $(0, \infty)$.

9. Locate some points on the graph of

$f(x) = 3^x + 1$.

x	$f(x) = 3^x + 1$	$(x, f(x))$
-2	$f(-2) = 3^{-2} + 1 = \dfrac{1}{9} + 1 = \dfrac{10}{9}$	$\left(-2, \dfrac{10}{9}\right)$
-1	$f(-1) = 3^{-1} + 1 = \dfrac{1}{3} + 1 = \dfrac{4}{3}$	$\left(-1, \dfrac{4}{3}\right)$
0	$f(0) = 3^0 + 1 = 1 + 1 = 2$	$(0, 2)$
1	$f(1) = 3^1 + 1 = 3 + 1 = 4$	$(1, 4)$
2	$f(2) = 3^2 + 1 = 9 + 1 = 10$	$(2, 10)$

Plot the points and connect them with a smooth curve.

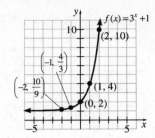

The domain of *f* is all real numbers or, using interval notation, $(-\infty, \infty)$. The range of *f* is $\{y \mid y > 1\}$ or, using interval notation, $(1, \infty)$.

10. $e \approx 2.71828$

11. a. $e^4 \approx 54.598$

    b. $e^{-4} \approx 0.018$

12. $5^{x-4} = 5^{-1}$
$x - 4 = -1$
$x = 3$
The solution set is $\{3\}$.

13. $3^{x+2} = 81$
$3^{x+2} = 3^4$
$x + 2 = 4$
$x = 2$
The solution set is $\{2\}$.

14. $e^{x^2} = e^x \cdot e^{4x}$
$e^{x^2} = e^{5x}$
$x^2 = 5x$
$x^2 - 5x = 0$
$x(x - 5) = 0$
$x = 0$ or $x - 5 = 0$
$x = 5$
The solution set is $\{0, \, 5\}$.

15. $\dfrac{2^{x^2}}{8} = 2^{2x}$

$\dfrac{2^{x^2}}{2^3} = 2^{2x}$

$2^{x^2 - 3} = 2^{2x}$

$x^2 - 3 = 2x$

$x^2 - 2x - 3 = 0$

$(x + 1)(x - 3) = 0$

$x + 1 = 0$ or $x - 3 = 0$

    $x = -1$ or $x = 3$

The solution set is $\{-1, \, 3\}$.

16. a. $F(10) = 1 - e^{-0.25(10)} \approx 0.918$.
The likelihood that a person will arrive within 10 minutes of 3:00 P.M. is 0.918, or 91.8%.

    b. $F(25) = 1 - e^{-0.25(25)} \approx 0.998$.
The likelihood that a person will arrive within 25 minutes of 3:00 P.M. is 0.998, or 99.8%.

17. a. $A(10) = 10\left(\dfrac{1}{2}\right)^{10/18.72} \approx 6.91$.
After 10 days, approximately 6.91 grams of thorium-227 will be left in the sample.

    b. $A(18.72) = 10\left(\dfrac{1}{2}\right)^{18.72/18.72} = 5$.
After 18.72 days, 5 grams of thorium-227 will be left in the sample.

    c. $A(74.88) = 10\left(\dfrac{1}{2}\right)^{74.88/18.72} = 0.625$.
After 74.88 days, 0.625 gram of thorium-227 will be left in the sample.

    d. $A(100) = 10\left(\dfrac{1}{2}\right)^{100/18.72} \approx 0.247$.
After 100 days, approximately 0.247 gram of thorium-227 will be left in the sample.

18. We use the compound interest formula with $P = \$2000$, $r = 0.05$, and $n = 12$, so that

$$A = 2000\left(1 + \frac{0.05}{12}\right)^{12t}$$

a. The value of the account after $t = 1$ years is

$$A = 2000\left(1 + \frac{0.05}{12}\right)^{12(1)}$$

$$= 2000\left(1 + \frac{0.05}{12}\right)^{12} \approx \$2102.32$$

b. The value of the account after $t = 15$ years

is $A = 2000\left(1 + \frac{0.05}{12}\right)^{12(15)}$

$$= 2000\left(1 + \frac{0.05}{12}\right)^{180} \approx \$4227.41$$

c. The value of the account after $t = 30$ years

is $A = 2000\left(1 + \frac{0.05}{12}\right)^{12(30)}$

$$= 2000\left(1 + \frac{0.05}{12}\right)^{360} \approx \$8935.49$$

8.2 Exercises

19. a. $3^{2.2} \approx 11.212$

b. $3^{2.23} \approx 11.587$

c. $3^{2.236} \approx 11.664$

d. $3^{2.2361} \approx 11.665$

e. $3^{\sqrt{5}} \approx 11.665$

21. a. $4^{3.1} \approx 73.517$

b. $4^{3.14} \approx 77.708$

c. $4^{3.142} \approx 77.924$

d. $4^{3.1416} \approx 77.881$

e. $4^{\pi} \approx 77.880$

23. g. $f(x) = 2^x - 1$ because the following points are on the graph:

x	$f(x) = 2^x - 1$	$(x, f(x))$
-1	$f(-1) = 2^{-1} - 1 = \frac{1}{2} - 1 = -\frac{1}{2}$	$\left(-1, -\frac{1}{2}\right)$
0	$f(0) = 2^0 - 1 = 1 - 1 = 0$	$(0, 0)$
1	$f(1) = 2^1 - 1 = 2 - 1 = 1$	$(1, 1)$

25. e. $f(x) = -2^x$ because the following points are on the graph:

x	$f(x) = -2^x$	$(x, f(x))$
-1	$f(-1) = -2^{-1} = -\frac{1}{2}$	$\left(-1, -\frac{1}{2}\right)$
0	$f(0) = -2^0 = -1$	$(0, -1)$
1	$f(1) = -2^1 = -2$	$(1, -2)$

27. f. $f(x) = 2^x + 1$ because the following points are on the graph:

x	$f(x) = 2^x + 1$	$(x, f(x))$
-1	$f(-1) = 2^{-1} + 1 = \frac{1}{2} + 1 = \frac{3}{2}$	$\left(-1, \frac{3}{2}\right)$
0	$f(0) = 2^0 + 1 = 1 + 1 = 2$	$(0, 2)$
1	$f(1) = 2^1 + 1 = 2 + 1 = 3$	$(1, 3)$

29. h. $f(x) = -2^{-x}$ because the following points are on the graph:

x	$f(x) = -2^{-x}$	$(x, f(x))$
-1	$f(-1) = -2^{-(-1)} = -2^1 = -2$	$(-1, -2)$
0	$f(0) = -2^{-0} = -2^0 = -1$	$(0, -1)$
1	$f(1) = -2^{-1} = -\frac{1}{2}$	$\left(1, -\frac{1}{2}\right)$

31. Locate some points on the graph of $f(x) = 5^x$.

x	$f(x) = 5^x$	$(x, f(x))$
-2	$f(-2) = 5^{-2} = \frac{1}{5^2} = \frac{1}{25}$	$\left(-2, \frac{1}{25}\right)$
-1	$f(-1) = 5^{-1} = \frac{1}{5^1} = \frac{1}{5}$	$\left(-1, \frac{1}{5}\right)$
0	$f(0) = 5^0 = 1$	$(0, 1)$
1	$f(1) = 5^1 = 5$	$(1, 5)$

Plot the points and connect them with a smooth

curve.

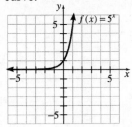

The domain of f is all real numbers or, using interval notation, $(-\infty, \infty)$. The range of f is $\{y \mid y > 0\}$ or, using interval notation, $(0, \infty)$.

33. Locate some points on the graph of

$$F(x) = \left(\frac{1}{5}\right)^x.$$

x	$F(x) = \left(\frac{1}{5}\right)^x$	$(x, F(x))$
-1	$F(-1) = \left(\frac{1}{5}\right)^{-1} = 5^1 = 5$	$(-1, 5)$
0	$F(0) = \left(\frac{1}{5}\right)^0 = 1$	$(0, 1)$
1	$F(1) = \left(\frac{1}{5}\right)^1 = \frac{1}{5}$	$\left(1, \frac{1}{5}\right)$
2	$F(2) = \left(\frac{1}{5}\right)^2 = \frac{1}{25}$	$\left(2, \frac{1}{25}\right)$

Plot the points and connect them with a smooth curve.

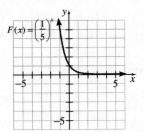

The domain of F is all real numbers or, using interval notation, $(-\infty, \infty)$. The range of F is $\{y \mid y > 0\}$ or, using interval notation, $(0, \infty)$.

35. Locate some points on the graph of $h(x) = 2^{x+2}$.

x	$h(x) = 2^{x+2}$	$(x, h(x))$
-5	$h(-5) = 2^{-5+2} = 2^{-3} = \dfrac{1}{2^3} = \dfrac{1}{8}$	$\left(-5, \dfrac{1}{8}\right)$
-4	$h(-4) = 2^{-4+2} = 2^{-2} = \dfrac{1}{2^2} = \dfrac{1}{4}$	$\left(-4, \dfrac{1}{4}\right)$
-3	$h(-3) = 2^{-3+2} = 2^{-1} = \dfrac{1}{2^1} = \dfrac{1}{2}$	$\left(-3, \dfrac{1}{2}\right)$
-2	$h(-2) = 2^{-2+2} = 2^0 = 1$	$(-2, 1)$
-1	$h(-1) = 2^{-1+2} = 2^1 = 2$	$(-1, 2)$
0	$h(0) = 2^{0+2} = 2^2 = 4$	$(0, 4)$
1	$h(1) = 2^{1+2} = 2^3 = 8$	$(1, 8)$

Plot the points and connect them with a smooth curve.

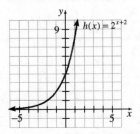

The domain of h is all real numbers or, using interval notation, $(-\infty, \infty)$. The range of h is $\{y \mid y > 0\}$ or , using interval notation, $(0, \infty)$.

37. Locate some points on the graph of

$$f(x) = 2^x + 3.$$

x	$f(x) = 2^x + 3$	$(x, f(x))$
-2	$f(-2) = 2^{-2} + 3 = \dfrac{1}{4} + 3 = \dfrac{13}{4}$	$\left(-2, \dfrac{13}{4}\right)$
-1	$f(-1) = 2^{-1} + 3 = \dfrac{1}{2} + 3 = \dfrac{7}{2}$	$\left(-1, \dfrac{7}{2}\right)$
0	$f(0) = 2^0 + 3 = 1 + 3 = 4$	$(0, 4)$
1	$f(1) = 2^1 + 3 = 2 + 3 = 5$	$(1, 5)$
2	$f(2) = 2^2 + 3 = 4 + 3 = 7$	$(2, 7)$

Plot the points and connect them with a smooth curve.

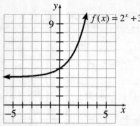

The domain of *f* is all real numbers or, using interval notation, $(-\infty, \infty)$. The range of *f* is $\{y \mid y > 3\}$ or, using interval notation, $(3, \infty)$.

39. Locate some points on the graph of
$F(x) = \left(\dfrac{1}{2}\right)^{x} - 1$.

x	$F(x) = \left(\dfrac{1}{2}\right)^{x} - 1$	$(x, F(x))$
−3	$F(-3) = \left(\dfrac{1}{2}\right)^{-3} - 1 = 8 - 1 = 7$	$(-3, 7)$
−2	$F(-2) = \left(\dfrac{1}{2}\right)^{-2} - 1 = 4 - 1 = 3$	$(-2, 3)$
−1	$F(-1) = \left(\dfrac{1}{2}\right)^{-1} - 1 = 2 - 1 = 1$	$(-1, 1)$
0	$F(0) = \left(\dfrac{1}{2}\right)^{0} - 1 = 1 - 1 = 0$	$(0, 0)$
1	$F(1) = \left(\dfrac{1}{2}\right)^{1} - 1 = \dfrac{1}{2} - 1 = -\dfrac{1}{2}$	$\left(1, -\dfrac{1}{2}\right)$
2	$F(2) = \left(\dfrac{1}{2}\right)^{2} - 1 = \dfrac{1}{4} - 1 = -\dfrac{3}{4}$	$\left(2, -\dfrac{3}{4}\right)$
3	$F(3) = \left(\dfrac{1}{2}\right)^{3} - 1 = \dfrac{1}{8} - 1 = -\dfrac{7}{8}$	$\left(3, -\dfrac{7}{8}\right)$

Plot the points and connect them with a smooth curve.

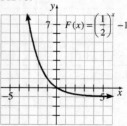

The domain of *F* is all real numbers or, using interval notation, $(-\infty, \infty)$. The range of *F* is $\{y \mid y > -1\}$ or, using interval notation, $(-1, \infty)$.

41. Locate some points on the graph of
$P(x) = \left(\dfrac{1}{3}\right)^{x-2}$.

x	$P(x) = \left(\dfrac{1}{3}\right)^{x-2}$	$(x, P(x))$
0	$P(0) = \left(\dfrac{1}{3}\right)^{0-2} = \left(\dfrac{1}{3}\right)^{-2} = 3^2 = 9$	$(0, 9)$
1	$P(1) = \left(\dfrac{1}{3}\right)^{1-2} = \left(\dfrac{1}{3}\right)^{-1} = 3^1 = 3$	$(1, 3)$
2	$P(2) = \left(\dfrac{1}{3}\right)^{2-2} = \left(\dfrac{1}{3}\right)^{0} = 1$	$(2, 1)$
3	$P(3) = \left(\dfrac{1}{3}\right)^{3-2} = \left(\dfrac{1}{3}\right)^{1} = \dfrac{1}{3}$	$\left(3, \dfrac{1}{3}\right)$
4	$P(4) = \left(\dfrac{1}{3}\right)^{4-2} = \left(\dfrac{1}{3}\right)^{2} = \dfrac{1}{9}$	$\left(4, \dfrac{1}{9}\right)$

Plot the points and connect them with a smooth curve.

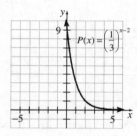

The domain of *P* is all real numbers or, using interval notation, $(-\infty, \infty)$. The range of *P* is $\{y \mid y > 0\}$ or, using interval notation, $(0, \infty)$.

43. a. $3.1^{2.7} \approx 21.217$

b. $3.14^{2.72} \approx 22.472$

c. $3.142^{2.718} \approx 22.460$

d. $3.1416^{2.7183} \approx 22.460$

e. $\pi^{e} \approx 22.459$

45. $e^{2} \approx 7.389$

47. $e^{-2} \approx 0.135$

49. $e^{2.3} \approx 9.974$

51. Locate some points on the graph of $g(x) = e^{x-1}$.

x	$g(x) = e^{x-1}$	$(x,\, g(x))$
-1	$g(-1) = e^{-1-1} = e^{-2} = \dfrac{1}{e^2} \approx 0.135$	$\left(-1,\, \dfrac{1}{e^2}\right)$
0	$g(0) = e^{0-1} = e^{-1} = \dfrac{1}{e} \approx 0.368$	$\left(0,\, \dfrac{1}{e}\right)$
1	$g(1) = e^{1-1} = e^0 = 1$	$(1,\, 1)$
2	$g(2) = e^{2-1} = e^1 = e \approx 2.718$	$(2,\, e)$
3	$g(3) = e^{3-1} = e^2 \approx 7.389$	$\left(3,\, e^2\right)$

Plot the points and connect them with a smooth curve.

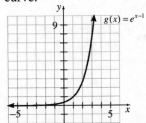

The domain of g is all real numbers or, using interval notation, $(-\infty, \infty)$. The range of g is $\{y \mid y > 0\}$ or, using interval notation, $(0, \infty)$.

53. Locate some points on the graph of $f(x) = -2e^x$.

x	$f(x) = -2e^x$	$(x, f(x))$
-1	$f(-1) = -2e^{-1} = -\dfrac{2}{e} \approx -0.736$	$\left(-1, -\dfrac{2}{e}\right)$
0	$f(0) = -2e^0 = -2$	$(0, -2)$
1	$f(1) = -2e^1 = -2e \approx -5.437$	$(1, -2e)$
2	$f(2) = -2e^2 \approx -14.778$	$\left(2, -2e^2\right)$
3	$f(3) = -2e^3 \approx -40.171$	$\left(3, -2e^3\right)$

Plot the points and connect them with a smooth curve.

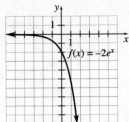

The domain of f is all real numbers, or using interval notation, $(-\infty, \infty)$. The range of f is $\{y \mid y < 0\}$ or, using interval notation, $(-\infty, 0)$.

55. $2^x = 2^5$
$x = 5$
The solution set is $\{5\}$.

57. $3^{-x} = 81$
$3^{-x} = 3^4$
$-x = 4$
$x = -4$
The solution set is $\{-4\}$.

59. $\left(\dfrac{1}{2}\right)^x = \dfrac{1}{32}$
$\left(\dfrac{1}{2}\right)^x = \left(\dfrac{1}{2}\right)^5$
$x = 5$
The solution set is $\{5\}$.

61. $5^{x-2} = 125$
$5^{x-2} = 5^3$
$x - 2 = 3$
$x = 5$
The solution set is $\{5\}$.

63. $\quad 4^x = 8$
$\left(2^2\right)^x = 2^3$
$2^{2x} = 2^3$
$2x = 3$
$x = \dfrac{3}{2}$
The solution set is $\left\{\dfrac{3}{2}\right\}$.

65. $2^{-x+5} = 16^x$

$2^{-x+5} = \left(2^4\right)^x$

$2^{-x+5} = 2^{4x}$

$-x+5 = 4x$

$5 = 5x$

$1 = x$

The solution set is $\{1\}$.

67. $3^{x^2-4} = 27^x$

$3^{x^2-4} = \left(3^3\right)^x$

$3^{x^2-4} = 3^{3x}$

$x^2 - 4 = 3x$

$x^2 - 3x - 4 = 0$

$(x-4)(x+1) = 0$

$x-4 = 0$ or $x+1 = 0$

$x = 4$ or $x = -1$

The solution set is $\{-1,\ 4\}$.

69. $4^x \cdot 2^{x^2} = 16^2$

$\left(2^2\right)^x \cdot 2^{x^2} = \left(2^4\right)^2$

$2^{2x} \cdot 2^{x^2} = 2^8$

$2^{x^2+2x} = 2^8$

$x^2 + 2x = 8$

$x^2 + 2x - 8 = 0$

$(x+4)(x-2) = 0$

$x+4 = 0$ or $x-2 = 0$

$x = -4$ or $x = 2$

The solution set is $\{-4,\ 2\}$.

71. $2^x \cdot 8 = 4^{x-3}$

$2^x \cdot 2^3 = \left(2^2\right)^{x-3}$

$2^{x+3} = 2^{2x-6}$

$x+3 = 2x-6$

$-x = -9$

$x = 9$

The solution set is $\{9\}$.

73. $\left(\dfrac{1}{5}\right)^x - 25 = 0$

$\left(\dfrac{1}{5}\right)^x = 25$

$\left(5^{-1}\right)^x = 5^2$

$5^{-x} = 5^2$

$-x = 2$

$x = -2$

The solution set is $\{-2\}$.

75. $\left(2^x\right)^x = 16$

$2^{x^2} = 2^4$

$x^2 = 4$

$x = \pm\sqrt{4}$

$x = \pm 2$

The solution set is $\{-2,\ 2\}$.

77. $e^x = e^{3x+4}$

$x = 3x+4$

$-2x = 4$

$x = -2$

The solution set is $\{-2\}$.

79. $\left(e^x\right)^2 = e^{3x-2}$

$e^{2x} = e^{3x-2}$

$2x = 3x-2$

$-x = -2$

$x = 2$

The solution set is $\{2\}$.

81. a. $f(3) = 2^3 = 8$.

The point (3, 8) is on the graph of f.

b. $f(x) = \dfrac{1}{8}$

$2^x = \dfrac{1}{8}$

$2^x = \dfrac{1}{2^3} = 2^{-3}$

$x = -3$

The point $\left(-3, \dfrac{1}{8}\right)$ is on the graph of f.

83. a. $g(-1) = 4^{-1} - 1 = \dfrac{1}{4} - 1 = -\dfrac{3}{4}$.

     The point $\left(-1, -\dfrac{3}{4}\right)$ is on the graph of g.

b. $g(x) = 15$

     $4^x - 1 = 15$

       $4^x = 16$

       $4^x = 4^2$

        $x = 2$

     The point $(2, 15)$ is on the graph of g.

85. a. $H(-3) = 3 \cdot \left(\dfrac{1}{2}\right)^{-3} = 3 \cdot 2^3 = 3 \cdot 8 = 24$.

     The point $(-3, \ 24)$ is on the graph of H.

b. $H(x) = \dfrac{3}{4}$

     $3 \cdot \left(\dfrac{1}{2}\right)^x = \dfrac{3}{4}$

       $\left(\dfrac{1}{2}\right)^x = \dfrac{1}{4}$

       $\left(\dfrac{1}{2}\right)^x = \left(\dfrac{1}{2}\right)^2$

        $x = 2$

     The point $\left(2, \dfrac{3}{4}\right)$ is on the graph of H.

87. a. $P(2015) = 313(1.011)^{2015-2012} \approx 323.443$

     According to the model, the population of the U.S. in 2015 will be approximately 323.4 million people.

b. $P(2050) = 313(1.011)^{2050-2012} \approx 474.338$

     According to the model, the population of the U.S. in 2042 will be approximately 474.3 million people.

c. $474.3 - 439 = 35.3$. The U.S. Census Bureau's prediction for the population in 2050 is 35.3 million people fewer than that of the model. Reasons given for the differences may vary. One possibility is that perhaps the U.S. Census Bureau expects the rate of population growth to decline below 1.1% per year over time.

89. We use the compound interest formula with $P = \$5000$, $r = 0.06$, and $n = 12$, so that

$$A = 5000\left(1 + \frac{0.06}{12}\right)^{12t} = 5000(1.005)^{12t}$$

a.   The value of the account after $t = 1$ year is

     $A = 5000(1.005)^{12(1)} \approx \$5308.39$.

b.   The value of the account after $t = 3$ years is

     $A = 5000(1.005)^{12(3)} \approx \$5983.40$.

c.   The value of the account after $t = 5$ years is

     $A = 5000(1.005)^{12(5)} \approx \$6744.25$.

91. a.   We use the compound interest formula with $P = \$2000$, $r = 0.03$, $t = 5$, and $n = 1$, so that $A = 2000\left(1 + \dfrac{0.03}{1}\right)^{1(5)} \approx \$2318.55$.

b.   We use the compound interest formula with $P = \$2000$, $r = 0.03$, $t = 5$, and $n = 4$, so that $A = 2000\left(1 + \dfrac{0.03}{4}\right)^{4(5)} \approx \$2322.37$.

c.   We use the compound interest formula with $P = \$2000$, $r = 0.03$, $t = 5$, and $n = 12$, so that $A = 2000\left(1 + \dfrac{0.03}{12}\right)^{12(5)} \approx \$2323.23$.

d.   We use the compound interest formula with $P = \$2000$, $r = 0.03$, $t = 5$, and $n = 365$, so that

$$A = 2000\left(1 + \frac{0.03}{365}\right)^{365(5)} \approx \$2323.65.$$

e.   Answers may vary. One possibility follows: All other things equal, the number of compounding periods does not have a very significant impact on the future value. In this case, for example, the difference between compounding annually $(n = 1)$ and daily $(n = 365)$ is only $5.10 over 5 years.

93. a. $V(0) = 19,841(0.88)^0 = 19,841$.

     According to the model, the value of a brand-new Focus is $19,841.

b. $V(2) = 19,841(0.88)^2 \approx 15,364.87$.

According to the model, the value of a 2-year-old Focus is \$15,365.

c. $V(5) = 19,841(0.88)^5 \approx 10,470.73$.

According to the model, the value of a 5-year-old Focus is \$10,471.

95. a. $A(1) = 100\left(\dfrac{1}{2}\right)^{1/13.81} \approx 95.105$.

After 1 second, approximately 95.105 grams of beryllium-11 will be left in the sample.

b. $A(13.81) = 100\left(\dfrac{1}{2}\right)^{13.81/13.81} = 50$.

After 13.81 seconds, 50 grams of beryllium-11 will be left in the sample.

c. $A(27.62) = 100\left(\dfrac{1}{2}\right)^{27.62/13.81} = 25$.

After 27.62 seconds, 25 grams of beryllium-11 will be left in the sample.

d. $A(100) = 100\left(\dfrac{1}{2}\right)^{100/13.81} \approx 0.661$.

After 100 seconds, approximately 0.661 gram of beryllium-11 will be left in the sample.

97. a. $u(5) = 70 + 330e^{-0.072(5)} \approx 300.233$.

According to the model, the temperature of the pizza after 5 minutes will be approximately $300.233°F$.

b. $u(10) = 70 + 330e^{-0.072(10)} \approx 230.628$.

According to the model, the temperature of the pizza after 10 minutes will be approximately $230.628°F$.

c. $u(13) = 70 + 330e^{-0.072(13)} \approx 199.424$.

According to the model, the temperature of the pizza after 13 minutes will be approximately $199.424°F$. Since this is below $200°F$, the pizza will be ready to eat after cooling for 13 minutes.

99. a. $L(45) = 200\left(1 - e^{-0.0035(45)}\right) \approx 29.14$.

According to the model, the student will learn approximately 29 words after 45 minutes.

b. $L(60) = 200\left(1 - e^{-0.0035(60)}\right) \approx 37.88$.

According to the model, the student will learn approximately 38 words after 60 minutes.

101. a. We use the equation with $E = 120$, $R = 10$, $L = 25$, and $t = 0.05$, so that

$$I = \frac{120}{10}\left[1 - e^{-(10/25)0.05}\right] \approx 0.238 \text{ ampere.}$$

b. We use the equation with $E = 240$, $R = 10$, $L = 25$, and $t = 0.05$, so that

$$I = \frac{240}{10}\left[1 - e^{-(10/25)0.05}\right] \approx 0.475 \text{ ampere.}$$

103. The exponential function will be of the form $y = a^x$. Now the function contains the points $\left(-1, \dfrac{1}{3}\right)$, $(0,1)$, and $(1,3)$, so $a = 3$. Thus, the equation of the function is $y = 3^x$.

105. As the base a increases, the steeper the graph of $f(x) = a^x$ $(a > 1)$ is for $x > 0$ and the closer the graph is to the x-axis for $x < 0$.

107. Answers may vary. One possibility follows: For exponential functions, the base is a constant and the exponent is the independent variable. For polynomial functions, the base is the independent variable and the exponents are constants (more specifically, whole numbers).

109. a. $x = -2$: $x^2 - 5x + 1 = (-2)^2 - 5(-2) + 1$
$$= 4 + 10 + 1$$
$$= 15$$

b. $x = 3$: $x^2 - 5x + 1 = 3^2 - 5(3) + 1$
$$= 9 - 15 + 1$$
$$= -5$$

111. a. $x = -2$: $\dfrac{4}{x+2} = \dfrac{4}{-2+2} = \dfrac{4}{0} =$ undefined

b. $x = 3$: $\dfrac{4}{x+2} = \dfrac{4}{3+2} = \dfrac{4}{5}$

113. a. $x = 2$: $\sqrt{2x+5} = \sqrt{2(2)+5}$
$$= \sqrt{4+5} = \sqrt{9} = 3$$

b. $x = 11$: $\sqrt{2x+5} = \sqrt{2(11)+5}$
$$= \sqrt{22+5}$$
$$= \sqrt{27} = \sqrt{9\cdot 3} = 3\sqrt{3}$$

115. Let $Y_1 = 1.5^x$.

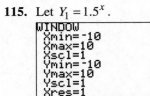

The domain of $f(x) = 1.5^x$ is all real numbers, or using interval notation $(-\infty, \infty)$. The range is $\{y \mid y > 0\}$, or using interval notation $(0, \infty)$.

117. Let $Y_1 = 0.9^x$.

The domain of $H(x) = 0.9^x$ is all real numbers, or using interval notation $(-\infty, \infty)$. The range is $\{y \mid y > 0\}$, or using interval notation $(0, \infty)$.

119. Let $Y_1 = 2.5^x + 3$.

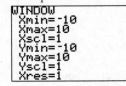

The domain of $g(x) = 2.5^x + 3$ is all real numbers, or using interval notation $(-\infty, \infty)$. The range is $\{y \mid y > 3\}$, or using interval notation $(3, \infty)$.

121. Let $Y_1 = 1.6^{x-3}$.

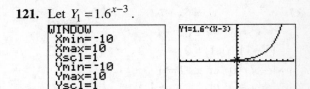

The domain of $F(x) = 1.6^{x-3}$ is all real numbers, or using interval notation $(-\infty, \infty)$. The range is $\{y \mid y > 0\}$, or using interval notation $(0, \infty)$.

Section 8.3

Are You Ready for This Section?

R1.
$$3x + 2 > 0$$
$$3x + 2 - 2 > 0 - 2$$
$$3x > -2$$
$$\frac{3x}{3} > \frac{-2}{3}$$
$$x > -\frac{2}{3}$$

The solution set is $\left\{ x \mid x > -\dfrac{2}{3} \right\}$ or, using interval notation, $\left(-\dfrac{2}{3}, \infty \right)$.

R2.
$$\sqrt{x+2} = x$$
$$\left(\sqrt{x+2}\right)^2 = (x)^2$$
$$x + 2 = x^2$$
$$0 = x^2 - x - 2$$
$$0 = (x-2)(x+1)$$
$$x - 2 = 0 \quad \text{or} \quad x + 1 = 0$$
$$x = 2 \quad \text{or} \quad x = -1$$
Check:
$$x = 2: \quad \sqrt{2+2} \overset{?}{=} 2$$
$$\sqrt{4} \overset{?}{=} 2$$
$$2 = 2 \checkmark$$

$$x = -1: \quad \sqrt{-1+2} \overset{?}{=} -1$$
$$\sqrt{1} \overset{?}{=} -1$$
$$1 \neq -1 \text{ ✗}$$

The potential solution $x = -1$ does not check; the solution set is $\{2\}$.

R3.
$$x^2 = 6x + 7$$
$$x^2 - 6x - 7 = 0$$
$$(x - 7)(x + 1) = 0$$
$$x - 7 = 0 \quad \text{or} \quad x + 1 = 0$$
$$x = 7 \quad \text{or} \quad x = -1$$
The solution set is $\{-1, 7\}$.

Section 8.3 Quick Checks

1. The logarithm to the base a of x, denoted $y = \log_a x$, can be expressed in exponential form as $\underline{x = a^y}$, where $a \geq 1$ and $a \neq 1$.

2. If $4^3 = 64$, then $3 = \log_4 64$.

3. If $p^{-2} = 8$, then $-2 = \log_p 8$.

4. If $4 = \log_2 16$, then $2^4 = 16$.

5. If $5 = \log_a 20$, then $a^5 = 20$.

6. If $-3 = \log_5 z$, then $5^{-3} = z$.

7. Let $y = \log_5 25$. Then,
$$5^y = 25$$
$$5^y = 5^2$$
$$y = 2$$
Thus, $\log_5 25 = 2$.

8. Let $y = \log_2 \dfrac{1}{8}$. Then,
$$2^y = \frac{1}{8}$$
$$2^y = \frac{1}{2^3} = 2^{-3}$$
$$y = -3$$
Thus, $\log_2 \dfrac{1}{8} = -3$.

9. $g(25)$ means to evaluate $\log_5 x$ at $x = 25$. So, we want to know the value of $\log_5 25$. Let $y = \log_5 25$. Then,
$$5^y = 25$$
$$5^y = 5^2$$
$$y = 2$$
Thus, $g(25) = 2$.

10. $g\left(\dfrac{1}{5}\right)$ means to evaluate $\log_5 x$ at $x = \dfrac{1}{5}$. So, we want to know the value of $\log_5 \left(\dfrac{1}{5}\right)$. Let $y = \log_5 \left(\dfrac{1}{5}\right)$. Then,
$$5^y = \frac{1}{5}$$
$$5^y = 5^{-1}$$
$$y = -1$$
Thus, $g\left(\dfrac{1}{5}\right) = -1$.

11. The domain of $g(x) = \log_8(x + 3)$ is the set of all real numbers x such that
$$x + 3 > 0$$
$$x > -3$$
Thus, the domain of $g(x) = \log_8(x + 3)$ is $\{x \mid x > -3\}$ or, using interval notation, $(-3, \infty)$.

12. The domain of $F(x) = \log_2(5 - 2x)$ is the set of all real numbers x such that
$$5 - 2x > 0$$
$$-2x > -5$$
$$x < \frac{-5}{-2}$$
$$x < \frac{5}{2}$$
Thus, the domain of $F(x) = \log_2(5 - 2x)$ is $\left\{x \mid x < \dfrac{5}{2}\right\}$ or, using interval notation, $\left(-\infty, \dfrac{5}{2}\right)$.

13. Rewrite $y = f(x) = \log_4 x$ as $x = 4^y$. Locate some points on the graph of $x = 4^y$.

y	$x = 4^y$	(x, y)
-2	$x = 4^{-2} = \dfrac{1}{4^2} = \dfrac{1}{16}$	$\left(\dfrac{1}{16}, -2\right)$
-1	$x = 4^{-1} = \dfrac{1}{4^1} = \dfrac{1}{4}$	$\left(\dfrac{1}{4}, -1\right)$
0	$x = 4^0 = 1$	$(1, 0)$
1	$x = 4^1 = 4$	$(4, 1)$

Plot the points and connect them with a smooth curve.

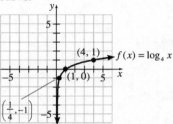

The domain of f is $\{x \mid x > 0\}$ or, using interval notation, $(0, \infty)$. The range of f is all real numbers or, using interval notation, $(-\infty, \infty)$.

14. Rewrite $y = f(x) = \log_{1/4} x$ as $x = \left(\dfrac{1}{4}\right)^y$.

Locate some points on the graph of $x = \left(\dfrac{1}{4}\right)^y$.

y	$x = \left(\dfrac{1}{4}\right)^y$	(x, y)
-1	$x = \left(\dfrac{1}{4}\right)^{-1} = 4^1 = 4$	$(4, -1)$
0	$x = \left(\dfrac{1}{4}\right)^0 = 1$	$(1, 0)$
1	$x = \left(\dfrac{1}{4}\right)^1 = \dfrac{1}{4}$	$\left(\dfrac{1}{4}, 1\right)$
2	$x = \left(\dfrac{1}{4}\right)^2 = \dfrac{1}{16}$	$\left(\dfrac{1}{16}, 2\right)$

Plot the points and connect them with a smooth curve.

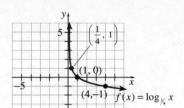

The domain of f is $\{x \mid x > 0\}$ or, using interval notation, $(0, \infty)$. The range of f is all real numbers or, using interval notation, $(-\infty, \infty)$.

15. $\log 1400 \approx 3.146$

16. $\ln 4.8 \approx 1.569$

17. $\log 0.3 \approx -0.523$

18. $\log_3(5x+1) = 4$
$$5x+1 = 3^4$$
$$5x+1 = 81$$
$$5x = 80$$
$$x = 16$$

Check: $\log_3(5 \cdot 16 + 1) \overset{?}{=} 4$
$$\log_3(80+1) \overset{?}{=} 4$$
$$\log_3(81) \overset{?}{=} 4$$
$$4 = 4 \checkmark$$

The solution set is $\{16\}$.

19. $\log_x 16 = 2$
$$x^2 = 16$$
$$x = \pm\sqrt{16}$$
$$x = \pm 4$$
Since the base of a logarithm must always be positive, we know that $a = -4$ is extraneous. We check the potential solution $x = 4$.

Check: $\log_4 16 \overset{?}{=} 2$
$$2 = 2 \checkmark$$

The solution set is $\{4\}$.

20. $\ln x = -2$

   $x = e^{-2}$

   Check: $\ln e^{-2} \overset{?}{=} -2$

   $\qquad\qquad -2 = -2$ ✓

   The solution set is $\left\{ e^{-2} \right\}$.

21. $\log(x - 20) = 4$

   $x - 20 = 10^4$

   $x - 20 = 10,000$

   $x = 10,020$

   Check: $\log(10,020 - 20) \overset{?}{=} 4$

   $\qquad\qquad \log(10,000) \overset{?}{=} 4$

   $\qquad\qquad\qquad\qquad 4 = 4$ ✓

   The solution set is $\{10,020\}$.

22. We evaluate $L(x) = 10\log\dfrac{x}{10^{-12}}$ at $x = 10^{-2}$.

   $$L\left(10^{-2}\right) = 10\log\dfrac{10^{-2}}{10^{-12}}$$

   $$= 10\log 10^{-2-(-12)}$$

   $$= 10\log 10^{10}$$

   $$= 10(10)$$

   $$= 100$$

   The loudness of an MP3 player on "full blast" is 100 decibels.

8.3 Exercises

23. If $64 = 4^3$, then $3 = \log_4 64$.

25. If $\dfrac{1}{8} = 2^{-3}$, then $-3 = \log_2\left(\dfrac{1}{8}\right)$.

27. If $a^3 = 19$, then $\log_a 19 = 3$.

29. If $5^{-6} = c$, then $\log_5 c = -6$.

31. If $\log_2 16 = 4$, then $2^4 = 16$.

33. If $\log_3\dfrac{1}{9} = -2$, then $3^{-2} = \dfrac{1}{9}$.

35. If $\log_5 a = -3$, then $5^{-3} = a$.

37. If $\log_a 4 = 2$, then $a^2 = 4$.

39. If $\log_{1/2} 12 = y$, then $\left(\dfrac{1}{2}\right)^y = 12$.

41. Let $y = \log_3 1$. Then,

   $3^y = 1$

   $3^y = 3^0$

   $y = 0$

   Thus, $\log_3 1 = 0$.

43. Let $y = \log_2 8$. Then,

   $2^y = 8$

   $2^y = 2^3$

   $y = 3$

   Thus, $\log_2 8 = 3$.

45. Let $y = \log_4\left(\dfrac{1}{16}\right)$. Then,

   $4^y = \dfrac{1}{16}$

   $4^y = \dfrac{1}{4^2} = 4^{-2}$

   $y = -2$

   Thus, $\log_4\dfrac{1}{16} = -2$.

47. Let $y = \log_{\sqrt{2}} 4$. Then,

   $\left(\sqrt{2}\right)^y = 4$

   $\left(2^{\frac{1}{2}}\right)^y = 2^2$

   $2^{\frac{1}{2}y} = 2^2$

   $\dfrac{1}{2}y = 2$

   $y = 4$

   Thus, $\log_{\sqrt{2}} 4 = 4$.

49. $f(81) = \log_3 81$. To determine the value, let $y = \log_3 81$. Then,

   $3^y = 81$

   $3^y = 3^4$

   $y = 4$

   Therefore, $f(81) = 4$.

51. $g\left(\sqrt{5}\right) = \log_5 \sqrt{5}$. To determine the value, let

$y = \log_5 \sqrt{5}$. Then,

$5^y = \sqrt{5}$

$5^y = 5^{\frac{1}{2}}$

$y = \frac{1}{2}$

Therefore, $g\left(\sqrt{5}\right) = \frac{1}{2}$.

53. The domain of $f(x) = \log_2(x-4)$ is the set of all real numbers x such that

$x - 4 > 0$

$\quad x > 4$

Thus, the domain of $f(x) = \log_2(x-4)$ is $\{x \mid x > 4\}$, or using interval notation, $(4, \infty)$.

55. The domain of $F(x) = \log_3(2x)$ is the set of all real numbers x such that

$2x > 0$

$\quad x > 0$

Thus, the domain of $F(x) = \log_3(2x)$ is $\{x \mid x > 0\}$, or using interval notation, $(0, \infty)$.

57. The domain of $f(x) = \log_8(3x-2)$ is the set of all real numbers x such that

$3x - 2 > 0$

$\quad 3x > 2$

$\quad x > \frac{2}{3}$

Thus, the domain of $f(x) = \log_8(3x-2)$ is $\left\{x \mid x > \frac{2}{3}\right\}$, or using interval notation, $\left(\frac{2}{3}, \infty\right)$.

59. The domain of $H(x) = \log_7(2x+1)$ is the set of all real numbers x such that

$2x + 1 > 0$

$\quad 2x > -1$

$\quad x > -\frac{1}{2}$

Thus, the domain of $H(x) = \log_7(2x+1)$ is $\left\{x \mid x > -\frac{1}{2}\right\}$, or using interval notation,

$\left(-\frac{1}{2}, \infty\right)$.

61. The domain of $H(x) = \log_2(1-4x)$ is the set of all real numbers x such that

$1 - 4x > 0$

$\quad -4x > -1$

$\quad x < \frac{-1}{-4}$

$\quad x < \frac{1}{4}$

Thus, the domain of $H(x) = \log_2(1-4x)$ is

$\left\{x \mid x < \frac{1}{4}\right\}$, or using interval notation, $\left(-\infty, \frac{1}{4}\right)$.

63. Rewrite $y = f(x) = \log_5 x$ as $x = 5^y$. Locate some points on the graph of $x = 5^y$.

y	$x = 5^y$	(x, y)
-2	$x = 5^{-2} = \dfrac{1}{5^2} = \dfrac{1}{25}$	$\left(\dfrac{1}{25}, -2\right)$
-1	$x = 5^{-1} = \dfrac{1}{5^1} = \dfrac{1}{5}$	$\left(\dfrac{1}{5}, -1\right)$
0	$x = 5^0 = 1$	$(1, 0)$
1	$x = 5^1 = 5$	$(5, 1)$

Plot the points and connect them with a smooth curve.

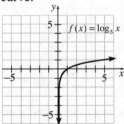

The domain of f is $\{x \mid x > 0\}$ or, using interval notation, $(0, \infty)$. The range of f is all real numbers or, using interval notation, $(-\infty, \infty)$.

65. Rewrite $y = g(x) = \log_6 x$ as $x = 6^y$. Locate some points on the graph of $x = 6^y$.

y	$x = 6^y$	(x, y)
-2	$x = 6^{-2} = \dfrac{1}{6^2} = \dfrac{1}{36}$	$\left(\dfrac{1}{36}, -2\right)$
-1	$x = 6^{-1} = \dfrac{1}{6^1} = \dfrac{1}{6}$	$\left(\dfrac{1}{6}, -1\right)$
0	$x = 6^0 = 1$	$(1, 0)$
1	$x = 6^1 = 6$	$(6, 1)$

Plot the points and connect them with a smooth

curve.

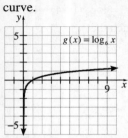

The domain of g is $\{x \mid x > 0\}$ or, using interval notation, $(0, \infty)$. The range of g is all real numbers or, using interval notation, $(-\infty, \infty)$.

67. Rewrite $y = F(x) = \log_{1/5} x$ as $x = \left(\dfrac{1}{5}\right)^y$.

    Locate some points on the graph of $x = \left(\dfrac{1}{5}\right)^y$.

y	$x = \left(\dfrac{1}{5}\right)^y$	(x, y)
-1	$x = \left(\dfrac{1}{5}\right)^{-1} = 5^1 = 5$	$(5, -1)$
0	$x = \left(\dfrac{1}{5}\right)^0 = 1$	$(1, 0)$
1	$x = \left(\dfrac{1}{5}\right)^1 = \dfrac{1}{5}$	$\left(\dfrac{1}{5}, 1\right)$
2	$x = \left(\dfrac{1}{5}\right)^2 = \dfrac{1}{25}$	$\left(\dfrac{1}{25}, 2\right)$

Plot the points and connect them with a smooth curve.

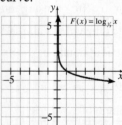

The domain of F is $\{x \mid x > 0\}$ or, using interval notation, $(0, \infty)$. The range of F is all real numbers or, using interval notation, $(-\infty, \infty)$.

69. If $e^x = 12$, then $\ln 12 = x$.

71. If $\ln x = 4$, then $e^4 = x$.

73. $H(0.1) = \log 0.1$. To determine the value, let $y = \log 0.1$. Then,

$$10^y = 0.1$$
$$10^y = \frac{1}{10}$$
$$10^y = 10^{-1}$$
$$y = -1$$

Therefore, $H(0.1) = -1$.

75. $P\left(e^3\right) = \ln e^3$. To determine the value, let $y = \ln e^3$. Then,

$$e^y = e^3$$
$$y = 3$$

Therefore, $P\left(e^3\right) = 3$.

77. $\log 67 \approx 1.826$

79. $\ln 5.4 \approx 1.686$

81. $\log 0.35 \approx -0.456$

83. $\ln 0.2 \approx -1.609$

85. $\log \dfrac{5}{4} \approx 0.097$

87. $\ln \dfrac{3}{8} \approx -0.981$

89. $\log_3(2x + 1) = 2$
$$2x + 1 = 3^2$$
$$2x + 1 = 9$$
$$2x = 8$$
$$x = 4$$
The solution set is $\{4\}$.

91. $\log_5(20x - 5) = 3$
$$20x - 5 = 5^3$$
$$20x - 5 = 125$$
$$20x = 130$$
$$x = \frac{130}{20} = \frac{13}{2}$$
The solution set is $\left\{\dfrac{13}{2}\right\}$.

93. $\log_a 36 = 2$

$\qquad a^2 = 36$

$\qquad a = \pm\sqrt{36}$

$\qquad a = \pm 6$

Since the base of a logarithm must always be positive, we know that $a = -6$ is extraneous. The solution set is $\{6\}$.

95. $\log_a 18 = 2$

$\qquad a^2 = 18$

$\qquad a = \pm\sqrt{18}$

$\qquad a = \pm 3\sqrt{2}$

Since the base of a logarithm must always be positive, we know that $a = -3\sqrt{2}$ is extraneous. The solution set is $\left\{3\sqrt{2}\right\}$.

97. $\log_a 1000 = 3$

$\qquad a^3 = 1000$

$\qquad a = \sqrt[3]{1000}$

$\qquad a = 10$

The solution set is $\{10\}$.

99. $\ln x = 5$

$\qquad x = e^5 \approx 148.413$

The solution set is $\left\{e^5\right\}$.

101. $\log(2x-1) = -1$

$\qquad 2x - 1 = 10^{-1}$

$\qquad 2x - 1 = \dfrac{1}{10}$

$\qquad 2x = \dfrac{11}{10}$

$\qquad x = \dfrac{1}{2} \cdot \dfrac{11}{10} = \dfrac{11}{20}$

The solution set is $\left\{\dfrac{11}{20}\right\}$.

103. $\ln e^x = -3$

$\qquad e^x = e^{-3}$

$\qquad x = -3$

The solution set is $\{-3\}$.

105. $\log_3(81) = x$

$\qquad 3^x = 81$

$\qquad 3^x = 3^4$

$\qquad x = 4$

The solution set is $\{4\}$.

107. $\log_2\left(x^2 - 1\right) = 3$

$\qquad x^2 - 1 = 2^3$

$\qquad x^2 - 1 = 8$

$\qquad x^2 = 9$

$\qquad x = \pm\sqrt{9}$

$\qquad x = \pm 3$

The solution set is $\{-3,\ 3\}$.

109. a. $f(16) = \log_2 16$. To determine the value, let $y = \log_2 16$. Then,

$\qquad 2^y = 16$

$\qquad 2^y = 2^4$

$\qquad y = 4$

Therefore, $f(16) = 4$, and the point $(16, 4)$ is on the graph of f.

b. $f(x) = -3$

$\qquad \log_2 x = -3$

$\qquad x = 2^{-3}$

$\qquad x = \dfrac{1}{2^3} = \dfrac{1}{8}$

The point $\left(\dfrac{1}{8}, -3\right)$ is on the graph of f.

111. a. $G(7) = \log_4(7+1) = \log_4 8$. To determine the value, let $y = \log_4 8$. Then,

$\qquad 4^y = 8$

$\qquad \left(2^2\right)^y = 2^3$

$\qquad 2^{2y} = 2^3$

$\qquad 2y = 3$

$\qquad y = \dfrac{3}{2}$

Therefore, $G(7) = \dfrac{3}{2}$, and the point $\left(7, \dfrac{3}{2}\right)$ is on the graph of G.

b.
$$G(x) = 2$$
$$\log_4(x+1) = 2$$
$$x+1 = 4^2$$
$$x+1 = 16$$
$$x = 15$$
The point $(15, 2)$ is on the graph of G.

113. If the graph of $f(x) = \log_a x$ contains the point $(16,\ 2)$, then
$$\log_a 16 = 2$$
$$a^2 = 16$$
$$a = \pm\sqrt{16} = \pm 4$$
Since the base of a logarithm must always be positive, we know that $a = -4$ is extraneous. Thus, $a = 4$.

115. The domain of $f(x) = \log_3\left(x^2 - 4x - 5\right)$ is the set of all real numbers x such that $x^2 - 4x - 5 > 0$.

Solve:
$$x^2 - 4x - 5 = 0$$
$$(x+1)(x-5) = 0$$
$$x+1 = 0 \quad \text{or} \quad x-5 = 0$$
$$x = -1 \quad \text{or} \quad x = 5$$
Determine where each factor is positive and negative and where the product of these factors is positive and negative.

Interval	$(-\infty,-1)$	-1	$(-1, 5)$	5	$(5, \infty)$
$x+1$	$----$	0	$++++$	$+$	$++++$
$x-5$	$----$	$-$	$----$	0	$++++$
$(x+1)(x-5)$	$++++$	0	$----$	0	$++++$

The inequality is strict, so –1 and 5 are not part of the solution. Now, $(x + 1)(x - 5)$ is greater than zero where the product is positive. Thus, the solution of the inequality and the domain of f is $\{x \mid x < -1 \text{ or } x > 5\}$, or using interval notation, $(-\infty, -1) \cup (5, \infty)$.

117. The domain of $f(x) = \log\left(\dfrac{x-4}{x+1}\right)$ is the set of all real numbers x such that $\dfrac{x-4}{x+1} > 0$.
The rational expression will equal 0 when $x = 3$. It is undefined when $x = -1$.
Determine where the numerator and the

denominator are positive and negative and where the quotient is positive and negative:

Interval	$(-\infty,-1)$	-1	$(-1, 4)$	4	$(4, \infty)$
$x-4$	$----$	$-$	$----$	0	$++++$
$x+1$	$----$	0	$++++$	$+$	$++++$
$\dfrac{x-4}{x+1}$	$++++$	$\varnothing$	$----$	0	$++++$

The rational function is undefined at $x = -1$, so -1 is not part of the solution. The inequality is strict, so 4 is not part of the solution. Now, $\dfrac{x-4}{x+1}$ is greater than zero where the quotient is positive. Thus, the solution of the inequality and the domain of f is $\{x \mid x < -1 \text{ or } x > 4\}$, or using interval notation, $(-\infty, -1) \cup (4, \infty)$.

119. The domain of $f(x) = \ln|x - 3|$ is the set of all real numbers x such that $|x - 3| > 0$.
Solve: $|x - 3| > 0$
$$x - 3 < 0 \quad \text{or} \quad x - 3 > 0$$
$$x < 3 \quad \text{or} \quad x > 3$$
The domain of f is $\{x \mid x \neq 3\}$ or $(-\infty, 3) \cup (3, \infty)$.

121. We evaluate $L(x) = 10\log\dfrac{x}{10^{-12}}$ at $x = 10^{-10}$.
$$L\left(10^{-10}\right) = 10\log\frac{10^{-10}}{10^{-12}}$$
$$= 10\log 10^{-10-(-12)}$$
$$= 10\log 10^2 = 10(2) = 20$$
The loudness of a whisper is 20 decibels.

123. We evaluate $L(x) = 10\log\dfrac{x}{10^{-12}}$ at $x = 10^1$.
$$L\left(10^1\right) = 10\log\frac{10^1}{10^{-12}}$$
$$= 10\log 10^{1-(-12)}$$
$$= 10\log 10^{13} = 10(13) = 130$$
The threshold of pain is 130 decibels.

125. We evaluate $M(x) = \log\left(\dfrac{x}{10^{-3}}\right)$ at $x = 63,096$.
$$M(63,096) = \log\left(\frac{63,096}{10^{-3}}\right)$$
$$= \log(63,096,000) \approx 7.8$$
The magnitude of the 1906 San Francisco earthquake was approximately 7.8 on the Richter scale.

127. We solve $M(x) = \log\left(\dfrac{x}{10^{-3}}\right)$ for $M(x) = 8.9$.

$$M(x) = 8.9$$

$$\log\left(\dfrac{x}{10^{-3}}\right) = 8.9$$

$$\dfrac{x}{10^{-3}} = 10^{8.9}$$

$$x = 10^{8.9} \cdot 10^{-3}$$

$$= 10^{8.9 + (-3)}$$

$$= 10^{5.9}$$

$$\approx 794,328$$

The seismographic reading of the 1911 Japan earthquake 100 kilometers from the epicenter was approximately 794,328 millimeters.

129. a. We evaluate $pH = -\log\left[H^+\right]$ for

$$\left[H^+\right] = 10^{-12}.$$

$$pH = -\log 10^{-12} = -(-12) = 12$$

The pH of household ammonia is 12, so household ammonia is basic.

b. We evaluate $pH = -\log\left[H^+\right]$ for

$$\left[H^+\right] = 10^{-5}.$$

$$pH = -\log 10^{-5} = -(-5) = 5$$

The pH of black coffee is 5, so black coffee is acidic.

c. We evaluate $pH = -\log\left[H^+\right]$ for

$$\left[H^+\right] = 10^{-2}.$$

$$pH = -\log 10^{-2} = -(-2) = 2$$

The pH of lemon juice is 2, so lemon juice is acidic.

d. We solve $pH = -\log\left[H^+\right]$ for $pH = 7.4$.

$$7.4 = -\log\left[H^+\right]$$

$$-7.4 = \log\left[H^+\right]$$

$$\left[H^+\right] = 10^{-7.4}$$

The concentration of hydrogen ions in human blood is $10^{-7.4}$ moles per liter.

131. The base of $f(x) = \log_a x$ cannot equal 1 because $y = \log_a x$ is equivalent to $x = a^y$ and a does not equal 1 in the exponential function because its graph is a vertical line $x = 1$.

133. The domain of $f(x) = \log_a\left(x^2 + 1\right)$ is the set of all real numbers because $x^2 + 1 > 0$ for all x.

135. $\left(2x^2 - 6x + 1\right) - \left(5x^2 + x - 9\right)$

$$= 2x^2 - 6x + 1 - 5x^2 - x + 9$$

$$= -3x^2 - 7x + 10$$

137. $x^2 - 1 = (x+1)(x-1)$

$x^2 + 3x + 2 = (x+1)(x+2)$

$LCD = (x+1)(x-1)(x+2)$

$$\dfrac{3x}{x^2 - 1} + \dfrac{x-3}{x^2 + 3x + 2}$$

$$= \dfrac{3x}{(x+1)(x-1)} \cdot \dfrac{x+2}{x+2} + \dfrac{x-3}{(x+1)(x+2)} \cdot \dfrac{x-1}{x-1}$$

$$= \dfrac{3x^2 + 6x}{(x+1)(x-1)(x+2)} + \dfrac{x^2 - x - 3x + 3}{(x+1)(x-1)(x+2)}$$

$$= \dfrac{3x^2 + 6x + x^2 - x - 3x + 3}{(x+1)(x-1)(x+2)}$$

$$= \dfrac{4x^2 + 2x + 3}{(x+1)(x-1)(x+2)}$$

139. $\sqrt{8x^3} + x\sqrt{18x} = \sqrt{4x^2 \cdot 2x} + x\sqrt{9 \cdot 2x}$

$$= 2x\sqrt{2x} + 3x\sqrt{2x} = 5x\sqrt{2x}$$

141. Let $Y_1 = \log(x+1)$.

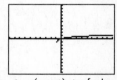

```
WINDOW
Xmin=-10
Xmax=10
Xscl=1
Ymin=-10
Ymax=10
Yscl=1
Xres=1
```

The domain of $f(x) = \log(x+1)$ is $\{x \mid x > -1\}$ or, using interval notation, $(-1, \infty)$. The range is all real numbers or, using interval notation, $(-\infty, \infty)$.

Note: The graph shown above is a bit misleading. The curve does not terminate at $x = -1$. Instead, the curve has an asymptote at $x = -1$. More specifically, as x approaches -1 (but stays larger than -1), f goes to $-\infty$.

143. Let $Y_1 = \ln(x)+1$.

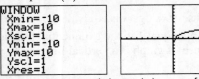

The domain of $G(x) = \ln(x)+1$ is $\{x\,|\,x>0\}$

or, using interval notation, $(0,\infty)$. The range is

all real numbers or, using interval notation,

$(-\infty,\infty)$.

Note: The graph shown above is a bit
misleading. The curve does not terminate at
$x=0$. Instead, the curve has an asymptote at
$x=0$. More specifically, as x approaches 0 (but
stays larger than 0), G goes to $-\infty$.

145. Let $Y_1 = 2\log(x-3)+1$.

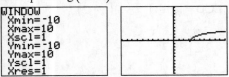

The domain of $f(x) = 2\log(x-3)+1$ is

$\{x\,|\,x>3\}$ or, using interval notation, $(3,\infty)$.

The range is all real numbers or, using interval

notation, $(-\infty,\infty)$.

Note: The graph shown above is a bit
misleading. The curve does not terminate at
$x=3$. Instead, the curve has an asymptote at
$x=3$. More specifically, as x approaches 3 (but
stays larger than 3), f goes to $-\infty$.

Putting the Concepts Together (Sections 8.1–8.3)

1. $f(x)=2x+3$; $g(x)=2x^2-4x$

a. $(f\circ g)(x) = f(g(x))$
$= 2(2x^2-4x)+3$
$= 4x^2-8x+3$

b. $(g\circ f)(x) = g(f(x))$
$= 2(2x+3)^2 - 4(2x+3)$
$= 2(4x^2+12x+9) - 4(2x+3)$
$= 8x^2+24x+18-8x-12$
$= 8x^2+16x+6$

c. Using the result from part (a):
$(f\circ g)(3) = 4(3)^2 - 8(3)+3$
$= 4(9) - 8(3)+3$
$= 36 - 24+3$
$= 15$

d. Using the result from part (b):
$(g\circ f)(-2) = 8(-2)^2 + 16(-2)+6$
$= 8(4)+16(-2)+6$
$= 32-32+6$
$= 6$

e. $f(1) = 2(1)+3 = 2+3 = 5$
$f(5) = 2(5)+3 = 10+3 = 13$
$(f\circ f)(1) = f(f(1)) = f(5) = 13$

2. a. $f(x) = 3x+4$
$y = 3x+4$
$x = 3y+4$
$x-4 = 3y$
$\dfrac{x-4}{3} = y$
$f^{-1}(x) = \dfrac{x-4}{3}$

Check:
$f\left(f^{-1}(x)\right) = 3\left(\dfrac{x-4}{3}\right)+4 = x-4+4 = x$
$f^{-1}(f(x)) = \dfrac{(3x+4)-4}{3} = \dfrac{3x}{3} = x$

b. $g(x) = x^3-4$
$y = x^3-4$
$x = y^3-4$
$x+4 = y^3$
$\sqrt[3]{x+4} = y$
$g^{-1}(x) = \sqrt[3]{x+4}$

Check:
$g\left(g^{-1}(x)\right) = \left(\sqrt[3]{x+4}\right)^3 - 4 = x+4-4 = x$
$g^{-1}(g(x)) = \sqrt[3]{(x^3-4)+4} = \sqrt[3]{x^3} = x$

3. To plot the inverse, switch the x- and
y-coordinates in each point and connect the
corresponding points. The graphs of the function

(shaded) and the line $y = x$ (dashed) are included for reference.

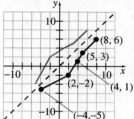

4. **a.** $2.7^{2.7} \approx 14.611$

    b. $2.72^{2.72} \approx 15.206$

    c. $2.718^{2.718} \approx 15.146$

    d. $2.7183^{2.7183} \approx 15.155$

    e. $e^e \approx 15.154$

5. **a.** If $a^4 = 6.4$, then $\log_a 6.4 = 4$.

    b. If $10^x = 278$, then $\log 278 = x$.

6. **a.** If $\log_2 x = 7$, then $2^7 = x$.

    b. If $\ln 16 = M$, then $e^M = 16$.

7. **a.** Let $y = \log_5 625$. Then,
    $$5^y = 625$$
    $$5^y = 5^4$$
    $$y = 4$$
    Thus, $\log_5 625 = 4$.

    b. Let $y = \log_{\frac{2}{3}}\left(\dfrac{9}{4}\right)$. Then,
    $$\left(\frac{2}{3}\right)^y = \frac{9}{4}$$
    $$\left(\frac{2}{3}\right)^y = \left(\frac{3}{2}\right)^2$$
    $$\left(\frac{2}{3}\right)^y = \left(\frac{2}{3}\right)^{-2}$$
    $$y = -2$$
    Thus, $\log_{\frac{2}{3}}\left(\dfrac{9}{4}\right) = -2$.

8. The domain of $f(x) = \log_{13}(2x+12)$ is the set of all real numbers x such that
    $$2x + 12 > 0$$
    $$2x > -12$$
    $$x > -6$$
    Thus, the domain of $f(x) = \log_{13}(2x+12)$ is $\{x \mid x > -6\}$, or using interval notation, $(-6, \infty)$.

9. Locate some points on the graph of
    $$f(x) = \left(\frac{1}{6}\right)^x.$$

x	$f(x) = \left(\dfrac{1}{6}\right)^x$	$(x, f(x))$
-1	$f(-1) = \left(\dfrac{1}{6}\right)^{-1} = \left(\dfrac{6}{1}\right)^1 = 6$	$(-1,\ 6)$
0	$f(0) = \left(\dfrac{1}{6}\right)^0 = 1$	$(0,\ 1)$
1	$f(1) = \left(\dfrac{1}{6}\right)^1 = \dfrac{1}{6}$	$\left(1,\ \dfrac{1}{6}\right)$
2	$f(2) = \left(\dfrac{1}{6}\right)^2 = \dfrac{1}{36}$	$\left(2,\ \dfrac{1}{36}\right)$

    Plot the points and connect them with a smooth curve.

    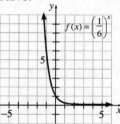

    The domain of f is all real numbers or, using interval notation, $(-\infty, \infty)$. The range of f is $\{y \mid y > 0\}$ or , using interval notation, $(0, \infty)$.

10. Rewrite $y = g(x) = \log_{\frac{3}{2}} x$ as $x = \left(\dfrac{3}{2}\right)^y$. Locate

some points on the graph of $x = \left(\dfrac{3}{2}\right)^y$.

y	$x = \left(\dfrac{3}{2}\right)^y$	(x, y)
-2	$x = \left(\dfrac{3}{2}\right)^{-2} = \left(\dfrac{2}{3}\right)^2 = \dfrac{4}{9}$	$\left(\dfrac{4}{9}, -2\right)$
-1	$x = \left(\dfrac{3}{2}\right)^{-1} = \left(\dfrac{2}{3}\right)^1 = \dfrac{2}{3}$	$\left(\dfrac{2}{3}, -1\right)$
0	$x = \left(\dfrac{3}{2}\right)^0 = 1$	$(1, 0)$
1	$x = \left(\dfrac{3}{2}\right)^1 = \dfrac{3}{2}$	$\left(\dfrac{3}{2}, 1\right)$
2	$x = \left(\dfrac{3}{2}\right)^2 = \dfrac{9}{4}$	$\left(\dfrac{9}{4}, 2\right)$

Plot the points and connect them with a smooth curve.

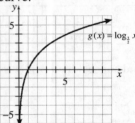

$g(x) = \log_{\frac{3}{2}} x$

The domain of g is $\{x \mid x > 0\}$ or, using interval notation, $(0, \infty)$. The range of g is all real numbers or, using interval notation, $(-\infty, \infty)$.

11. $3^{-x+2} = 27$
$3^{-x+2} = 3^3$
$-x + 2 = 3$
$-x = 1$
$x = -1$
The solution set is $\{-1\}$.

12. $e^x = e^{2x+5}$
$x = 2x + 5$
$-x = 5$
$x = -5$
The solution set is $\{-5\}$.

13. $\log_2 (2x + 5) = 4$
$2x + 5 = 2^4$
$2x + 5 = 16$
$2x = 11$
$x = \dfrac{11}{2}$
The solution set is $\left\{\dfrac{11}{2}\right\}$.

14. $\ln x = 7$
$x = e^7 \approx 1096.633$
The solution set is $\{e^7\}$.

15. $L(90) = 150\left(1 - e^{-0.0052(90)}\right) \approx 56.06$.

According to the model, the student will learn approximately 56 terms after 90 minutes.

Section 8.4

Are You Ready for This Section?

R1. $3.03468 \approx 3.035$

R2. If $a \neq 0$, then $a^0 = 1$.

Section 8.4 Quick Checks

1. $\log_5 1 = 0$

2. $\ln 1 = 0$

3. $\log_4 4 = 1$

4. $\log 10 = 1$

5. $12^{\log_{12} \sqrt{2}} = \sqrt{2}$

6. $10^{\log 0.2} = 0.2$

7. $\log_8 8^{1.2} = 1.2$

8. $\log 10^{-4} = -4$

9. False. $\log(x \cdot 4) = \log x + \log 4$, but there is no such rule for $\log(x + 4)$.

10. $\log_4 (9 \cdot 5) = \log_4 9 + \log_4 5$

11. $\log(5w) = \log 5 + \log w$

12. $\log_7\left(\dfrac{9}{5}\right) = \log_7 9 - \log_7 5$

13. $\ln\left(\dfrac{p}{3}\right) = \ln p - \ln 3$

14. $\log_2\left(\dfrac{3m}{n}\right) = \log_2(3m) - \log_2 n$
$$= \log_2 3 + \log_2 m - \log_2 n$$

15. $\ln\left(\dfrac{q}{3p}\right) = \ln q - \ln(3p)$
$$= \ln q - (\ln 3 + \ln p)$$
$$= \ln q - \ln 3 - \ln p$$

16. $\log_2 5^{1.6} = 1.6\log_2 5$

17. $\log b^5 = 5\log b$

18. $\log_4\left(a^2 b\right) = \log_4 a^2 + \log_4 b$
$$= 2\log_4 a + \log_4 b$$

19. $\log_3\left(\dfrac{9m^4}{\sqrt[3]{n}}\right) = \log_3\left(9m^4\right) - \log_3 \sqrt[3]{n}$
$$= \log_3\left(3^2 m^4\right) - \log_3 n^{1/3}$$
$$= \log_3 3^2 + \log_3 m^4 - \log_3 n^{1/3}$$
$$= 2 + 4\log_3 m - \frac{1}{3}\log_3 n$$

20. $\log_8 4 + \log_8 16 = \log_8(4 \cdot 16)$
$$= \log_8 64$$
$$= 2$$

21. $\log_3(x+4) - \log_3(x-1) = \log_3\left(\dfrac{x+4}{x-1}\right)$

22. $\log_5 x - 3\log_5 2 = \log_5 x - \log_5 2^3$
$$= \log_5 x - \log_5 8$$
$$= \log_5 \dfrac{x}{8}$$

23. $\log_2(x+1) + \log_2(x+2) - 2\log_2 x$
$$= \log_2\left[(x+1)(x+2)\right] - \log_2 x^2$$
$$= \log_2\left(x^2 + 3x + 2\right) - \log_2 x^2$$
$$= \log_2\left(\dfrac{x^2 + 3x + 2}{x^2}\right)$$

24. $\log_3 10 = \dfrac{\log 10}{\log 3} = \dfrac{\ln 10}{\ln 3}$

25. Using common logarithms:
$$\log_3 32 = \dfrac{\log 32}{\log 3} \approx 3.155$$

26. Using natural logarithms:
$$\log_{\sqrt{2}} \sqrt{7} = \dfrac{\ln \sqrt{7}}{\ln \sqrt{2}} \approx 2.807$$

8.4 Exercises

27. $\log_2 2^3 = 3$

29. $\ln e^{-7} = -7$

31. $3^{\log_3 5} = 5$

33. $e^{\ln 2} = 2$

35. $\log_7 7 = 1$

37. $\log 1 = 0$

39. $\ln 6 = \ln(2 \cdot 3) = \ln 2 + \ln 3 = a + b$

41. $\ln 9 = \ln 3^2 = 2\ln 3 = 2b$

43. $\ln 12 = \ln\left(2^2 \cdot 3\right)$
$$= \ln 2^2 + \ln 3$$
$$= 2\ln 2 + \ln 3$$
$$= 2a + b$$

45. $\ln \sqrt{2} = \ln 2^{1/2} = \dfrac{1}{2}\ln 2 = \dfrac{1}{2}a$

47. $\log(ab) = \log a + \log b$

49. $\log_5 x^4 = 4\log_5 x$

51. $\log_2\left(xy^2\right) = \log_2 x + \log_2 y^2 = \log_2 x + 2\log_2 y$

53. $\log_5\left(25x\right) = \log_5 25 + \log_5 x$

$\qquad = \log_5 5^2 + \log_5 x$

$\qquad = 2 + \log_5 x$

55. $\log_7\left(\dfrac{49}{y}\right) = \log_7 49 - \log_7 y$

$\qquad\qquad = \log_7 7^2 - \log_7 y$

$\qquad\qquad = 2 - \log_7 y$

57. $\ln\left(e^2 x\right) = \ln e^2 + \ln x = 2 + \ln x$

59. $\log_3\left(27\sqrt{x}\right) = \log_3 27 + \log_3 \sqrt{x}$

$\qquad\qquad = \log_3 3^3 + \log_3 x^{\frac{1}{2}}$

$\qquad\qquad = 3 + \dfrac{1}{2}\log_3 x$

61. $\log_5\left(x^2\sqrt{x^2+1}\right) = \log_5 x^2 + \log_5 \sqrt{x^2+1}$

$\qquad\qquad\qquad = \log_5 x^2 + \log_5\left(x^2+1\right)^{\frac{1}{2}}$

$\qquad\qquad\qquad = 2\log_5 x + \dfrac{1}{2}\log_5\left(x^2+1\right)$

63. $\log\left(\dfrac{x^4}{\sqrt[3]{x-1}}\right) = \log x^4 - \log\sqrt[3]{x-1}$

$\qquad\qquad\qquad = \log x^4 - \log(x-1)^{\frac{1}{3}}$

$\qquad\qquad\qquad = 4\log x - \dfrac{1}{3}\log(x-1)$

65. $\log_7\sqrt{\dfrac{x+1}{x}} = \log_7\left(\dfrac{x+1}{x}\right)^{\frac{1}{2}}$

$\qquad\qquad = \dfrac{1}{2}\log_7\left(\dfrac{x+1}{x}\right)$

$\qquad\qquad = \dfrac{1}{2}\left[\log_7(x+1) - \log_7 x\right]$

$\qquad\qquad = \dfrac{1}{2}\log_7(x+1) - \dfrac{1}{2}\log_7 x$

67. $\log_2\left[\dfrac{x(x-1)^2}{\sqrt{x+1}}\right]$

$= \log_2\left[x(x-1)^2\right] - \log_2\sqrt{x+1}$

$= \log_2 x + \log_2(x-1)^2 - \log_2\sqrt{x+1}$

$= \log_2 x + \log_2(x-1)^2 - \log_2(x+1)^{\frac{1}{2}}$

$= \log_2 x + 2\log_2(x-1) - \dfrac{1}{2}\log_2(x+1)$

69. $\log 25 + \log 4 = \log(25\cdot 4)$

$\qquad\qquad\qquad = \log 100$

$\qquad\qquad\qquad = \log 10^2$

$\qquad\qquad\qquad = 2$

71. $\log x + \log 3 = \log(3x)$

73. $\log_3 36 - \log_3 4 = \log_3\left(\dfrac{36}{4}\right)$

$\qquad\qquad\qquad = \log_3 9$

$\qquad\qquad\qquad = \log_3 3^2$

$\qquad\qquad\qquad = 2$

75. $10^{\log 8 - \log 2} = 10^{\log\left(\frac{8}{2}\right)} = 10^{\log 4} = 4$

77. $3\log_3 x = \log_3 x^3$

79. $\log_4(x+1) - \log_4 x = \log_4\left(\dfrac{x+1}{x}\right)$

81. $2\ln x + 3\ln y = \ln x^2 + \ln y^3 = \ln\left(x^2 y^3\right)$

83. $\dfrac{1}{2}\log_3 x + 3\log_3(x-1) = \log_3 x^{\frac{1}{2}} + \log_3(x-1)^3$

$\qquad\qquad\qquad\qquad = \log_3\sqrt{x} + \log_3(x-1)^3$

$\qquad\qquad\qquad\qquad = \log_3\left[\sqrt{x}(x-1)^3\right]$

85. $\log x^5 - 3\log x = \log x^5 - \log x^3$

$\qquad\qquad\qquad = \log\left(\dfrac{x^5}{x^3}\right)$

$\qquad\qquad\qquad = \log\left(x^2\right)$

87. $\dfrac{1}{2}\left[3\log x+\log y\right]=\dfrac{1}{2}\left[\log x^3+\log y\right]$

$=\dfrac{1}{2}\log\left(x^3y\right)$

$=\log\left(x^3y\right)^{1/2}$

$=\log\sqrt{x^3y}$

$=\log\left(x\sqrt{xy}\right)$

89. $\log_8\left(x^2-1\right)-\log_8\left(x+1\right)=\log_8\left(\dfrac{x^2-1}{x+1}\right)$

$=\log_8\left[\dfrac{(x-1)(x+1)}{x+1}\right]$

$=\log_8(x-1)$

91. $18\log\sqrt{x}+9\log\sqrt[3]{x}-\log 10$

$=18\log x^{1/2}+9\log x^{1/3}-\log 10$

$=\log\left(x^{1/2}\right)^{18}+\log\left(x^{1/3}\right)^{9}-\log 10$

$=\log x^9+\log x^3-\log 10$

$=\log\left(\dfrac{x^9\cdot x^3}{10}\right)$

$=\log\left(\dfrac{x^{12}}{10}\right)$

93. Using common logarithms:

$\log_2 10=\dfrac{\log 10}{\log 2}\approx 3.322$

95. Using common logarithms:

$\log_8 3=\dfrac{\log 3}{\log 8}\approx 0.528$

97. Using natural logarithms:

$\log_{1/3}19=\dfrac{\ln 19}{\ln\left(\dfrac{1}{3}\right)}\approx -2.680$

99. Using natural logarithms:

$\log_{\sqrt{2}}5=\dfrac{\ln 5}{\ln\sqrt{2}}\approx 4.644$

101. $\log_2 3\cdot\log_3 4\cdot\log_4 5\cdot\log_5 6\cdot\log_6 7\cdot\log_7 8$

$=\dfrac{\log 3}{\log 2}\cdot\dfrac{\log 4}{\log 3}\cdot\dfrac{\log 5}{\log 4}\cdot\dfrac{\log 6}{\log 5}\cdot\dfrac{\log 7}{\log 6}\cdot\dfrac{\log 8}{\log 7}$

$=\dfrac{\cancel{\log 3}}{\log 2}\cdot\dfrac{\cancel{\log 4}}{\cancel{\log 3}}\cdot\dfrac{\cancel{\log 5}}{\cancel{\log 4}}\cdot\dfrac{\cancel{\log 6}}{\cancel{\log 5}}\cdot\dfrac{\cancel{\log 7}}{\cancel{\log 6}}\cdot\dfrac{\log 8}{\cancel{\log 7}}$

$=\dfrac{\log 8}{\log 2}$

$=\log_2 8$

$=\log_2 2^3$

$=3$

103. $\log_2 3\cdot\log_3 4\cdot\ldots\cdot\log_n(n+1)\cdot\log_{n+1}2$

$=\dfrac{\log 3}{\log 2}\cdot\dfrac{\log 4}{\log 3}\cdot\ldots\cdot\dfrac{\log(n+1)}{\log n}\cdot\dfrac{\log 2}{\log(n+1)}$

$=\dfrac{\cancel{\log 3}}{\cancel{\log 2}}\cdot\dfrac{\log 4}{\cancel{\log 3}}\cdot\ldots\cdot\dfrac{\cancel{\log(n+1)}}{\log n}\cdot\dfrac{\cancel{\log 2}}{\cancel{\log(n+1)}}$

$=1$

Note: The expression "log 4" in the numerator of the second fraction will cancel with the expression "log 4" in the denominator of the third fraction (if the third fraction were written out). Similarly, the expression "log n" in the denominator of the next-to-the-last fraction will cancel with the expression "log n" in the numerator of the third-from-the-last fraction (if it were written out). Likewise, everything in between will cancel leaving an overall value of 1.

105. $\log_a\left(x+\sqrt{x^2-1}\right)+\log_a\left(x-\sqrt{x^2-1}\right)$

$=\log_a\left[\left(x+\sqrt{x^2-1}\right)\left(x-\sqrt{x^2-1}\right)\right]$

$=\log_a\left[x^2-x\sqrt{x^2-1}+x\sqrt{x^2-1}-\left(x^2-1\right)\right]$

$=\log_a\left(x^2-x^2+1\right)$

$=\log_a 1$

$=0$

107. If $f(x)=\log_a x$, then

$f(AB)=\log_a(AB)$

$=\log_a A+\log_a B$

$=f(A)+f(B)$

109. Answers may vary. One possibility follows: The log of a product is equal to the sum of the logs.

111. Answers may vary. One example follows:
Let $x = 2$ and $y = 1$. Then

$\log_2(x+y) = \log_2(2+1) = \log_2 3$, and

$\log_2 x + \log_2 y = \log_2 2 + \log_2 1 = 1 + 0 = 1$.

However, $\log_2 3 \neq 1$ because $2^1 \neq 3$. Thus,

$\log_2(x+y) \neq \log_2 x + \log_2 y$.

113. $4x + 3 = 13$

$\quad\quad 4x = 10$

$\quad\quad x = \dfrac{10}{4} = \dfrac{5}{2}$

The solution set is $\left\{\dfrac{5}{2}\right\}$.

115. $x^2 + 4x + 2 = 0$
Because this equation does not easily factor, solve by using the quadratic formula. For this equation, $a = 1$, $b = 4$, and $c = 2$.

$x = \dfrac{-4 \pm \sqrt{4^2 - 4(1)(2)}}{2(1)}$

$= \dfrac{-4 \pm \sqrt{16 - 8}}{2}$

$= \dfrac{-4 \pm \sqrt{8}}{2}$

$= \dfrac{-4 \pm 2\sqrt{2}}{2}$

$= -2 \pm \sqrt{2}$

The solution set is $\left\{-2 - \sqrt{2},\ -2 + \sqrt{2}\right\}$.

117. $\sqrt{x+2} - 3 = 4$

$\quad\quad \sqrt{x+2} = 7$

$\quad\quad \left(\sqrt{x+2}\right)^2 = 7^2$

$\quad\quad x + 2 = 49$

$\quad\quad x = 47$

Check:

$\sqrt{47 + 2} - 3 \overset{?}{=} 4$

$\sqrt{49} - 3 \overset{?}{=} 4$

$7 - 3 \overset{?}{=} 4$

$4 = 4$ ✓

The solution set is $\{47\}$.

119. Since $\log_3 x = \dfrac{\log x}{\log 3}$, let $Y_1 = \dfrac{\log x}{\log 3}$.

The domain of $f(x) = \log_3 x$ is $\{x \mid x > 0\}$, or using interval notation, $(0, \infty)$; the range is all real numbers or, using interval notation, $(-\infty, \infty)$.

Note: The graph shown above is a bit misleading. The curve does not terminate at $x = 0$. Instead, the curve has an asymptote at $x = 0$. More specifically, as x approaches 0 (but stays larger than 0), f goes to $-\infty$.

121. Since $\log_{1/2} x = \dfrac{\log x}{\log\left(\frac{1}{2}\right)}$, let $Y_1 = \dfrac{\log x}{\log\left(\frac{1}{2}\right)}$.

The domain of $F(x) = \log_{1/2} x$ is $\{x \mid x > 0\}$, or using interval notation, $(0, \infty)$; the range is all real numbers or, using interval notation, $(-\infty, \infty)$.

Note: The graph shown above is a bit misleading. The curve does not terminate at $x = 0$. Instead, the curve has an asymptote at $x = 0$. More specifically, as x approaches 0 (but stays larger than 0), F goes to ∞.

Section 8.5

Are You Ready for This Section?

R1. $\quad 2x + 5 = 13$

$\quad\quad 2x + 5 - 5 = 13 - 5$

$\quad\quad 2x = 8$

$\quad\quad \dfrac{2x}{2} = \dfrac{8}{2}$

$\quad\quad x = 4$

The solution set is $\{4\}$.

R2. $\quad x^2 - 4x = -3$

$\quad\quad x^2 - 4x + 3 = 0$

$\quad\quad (x-1)(x-3) = 0$

$\quad\quad x - 1 = 0 \quad \text{or} \quad x - 3 = 0$

$\quad\quad x = 1 \quad \text{or} \quad\quad x = 3$

The solution set is $\{1, 3\}$.

R3.　　　　$3a^2 = a + 5$

$3a^2 - a - 5 = 0$

For this equation, $a = 3$, $b = -1$, and $c = -5$.

$$a = \frac{-(-1) \pm \sqrt{(-1)^2 - 4(3)(-5)}}{2(3)}$$

$$= \frac{1 \pm \sqrt{1 + 60}}{6}$$

$$= \frac{1 \pm \sqrt{61}}{6}$$

The solution set is $\left\{ \dfrac{1 - \sqrt{61}}{6}, \dfrac{1 + \sqrt{61}}{6} \right\}$.

R4.　$(x+3)^2 + 2(x+3) - 8 = 0$

Let $u = x + 3$.

$$u^2 + 2u - 8 = 0$$

$$(u+4)(u-2) = 0$$

$u + 4 = 0$　or　$u - 2 = 0$

　$u = -4$　or　　$u = 2$

$x + 3 = -4$　or　$x + 3 = 2$

　$x = -7$　or　　$x = -1$

Check:

$x = -7$:　$(-7+3)^2 + 2(-7+3) - 8 \overset{?}{=} 0$

　　　　　　$(-4)^2 + 2(-4) - 8 \overset{?}{=} 0$

　　　　　　　　$16 - 8 - 8 \overset{?}{=} 0$

　　　　　　　　　　　$0 = 0$ ✓

$x = -1$:　$(-1+3)^2 + 2(-1+3) - \overset{?}{=} 0$

　　　　　　$(2)^2 + 2(2) - 8 \overset{?}{=} 0$

　　　　　　　　$4 + 4 - 8 \overset{?}{=} 0$

　　　　　　　　　　$0 = 0$ ✓

Both check; the solution set is $\{-7, \ -1\}$.

Section 8.5 Quick Checks

1. If $\log_a M = \log_a N$, then $\underline{M = N}$.

2. $2\log_4 x = \log_4 9$

$$\log_4 x^2 = \log_4 9$$

$$x^2 = 9$$

$$x = \pm\sqrt{9}$$

$$x = \pm 3$$

The apparent solution $x = -3$ is extraneous because the argument of a logarithm must be positive. The solution set is $\{3\}$.

3.　$\log_4 (x-6) + \log_4 x = 2$

$$\log_4 [x(x-6)] = 2$$

$$\log_4 (x^2 - 6x) = 2$$

$$x^2 - 6x = 4^2$$

$$x^2 - 6x = 16$$

$$x^2 - 6x - 16 = 0$$

$$(x+2)(x-8) = 0$$

$x + 2 = 0$　or　$x - 8 = 0$

　$x = -2$　or　　　$x = 8$

The apparent solution $x = -2$ is extraneous because it causes the argument of a logarithm to be negative. The solution set is $\{8\}$.

4.　$\log_3 (x+3) + \log_3 (x+5) = 1$

$$\log_3 [(x+3)(x+5)] = 1$$

$$\log_3 (x^2 + 8x + 15) = 1$$

$$x^2 + 8x + 15 = 3^1$$

$$x^2 + 8x + 12 = 0$$

$$(x+6)(x+2) = 0$$

$x + 6 = 0$　or　$x + 2 = 0$

　$x = -6$　or　　$x = -2$

The apparent solution $x = -6$ is extraneous because it causes the arguments of both logarithms to be negative. The solution set is $\{-2\}$.

5.　　　$2^x = 11$

$$\log 2^x = \log 11$$

$$x \log 2 = \log 11$$

$$x = \frac{\log 11}{\log 2} \approx 3.459$$

The solution set is $\left\{ \dfrac{\log 11}{\log 2} \right\} \approx \{3.459\}$. If we had taken the natural logarithm of both sides, the solution set would be $\left\{ \dfrac{\ln 11}{\ln 2} \right\} \approx \{3.459\}$.

6.　　　$5^{2x} = 3$

$$\log 5^{2x} = \log 3$$

$$2x \log 5 = \log 3$$

$$x = \frac{\log 3}{2\log 5} \approx 0.341$$

The solution set is $\left\{\dfrac{\log 3}{2\log 5}\right\} \approx \{0.341\}$. If we

had taken the natural logarithm of both sides, the

solution set would be $\left\{\dfrac{\ln 3}{2\ln 5}\right\} \approx \{0.341\}$.

7. $e^{2x} = 5$

$\ln e^{2x} = \ln 5$

$2x = \ln 5$

$x = \dfrac{\ln 5}{2} \approx 0.805$

The solution set is $\left\{\dfrac{\ln 5}{2}\right\} \approx \{0.805\}$.

8. $3e^{-4x} = 20$

$e^{-4x} = \dfrac{20}{3}$

$\ln e^{-4x} = \ln\left(\dfrac{20}{3}\right)$

$-4x = \ln\left(\dfrac{20}{3}\right)$

$x = \dfrac{\ln\left(\dfrac{20}{3}\right)}{-4} \approx -0.474$

The solution set is $\left\{\dfrac{\ln\left(\frac{20}{3}\right)}{-4}\right\} \approx \{-0.474\}$.

9. a. We need to determine the time until $A = 9$
grams. So we solve the equation

$9 = 10\left(\dfrac{1}{2}\right)^{t/18.72}$

$0.9 = \left(\dfrac{1}{2}\right)^{t/18.72}$

$\log 0.9 = \log\left(\dfrac{1}{2}\right)^{t/18.72}$

$\log 0.9 = \dfrac{t}{18.72}\log\left(\dfrac{1}{2}\right)$

$\dfrac{18.72 \cdot \log 0.9}{\log\left(\dfrac{1}{2}\right)} = t$

So, $t = \dfrac{18.72 \cdot \log 0.9}{\log\left(\dfrac{1}{2}\right)} \approx 2.85$ days.

Thus, 9 grams of thorium-227 will be left
after approximately 2.85 days.

b. We need to determine the time until $A = 3$
grams. So we solve the equation

$3 = 10\left(\dfrac{1}{2}\right)^{t/18.72}$

$0.3 = \left(\dfrac{1}{2}\right)^{t/18.72}$

$\log 0.3 = \log\left(\dfrac{1}{2}\right)^{t/18.72}$

$\log 0.3 = \dfrac{t}{18.72}\log\left(\dfrac{1}{2}\right)$

$\dfrac{18.72 \cdot \log 0.3}{\log\left(\dfrac{1}{2}\right)} = t$

So, $t = \dfrac{18.72 \cdot \log 0.3}{\log\left(\dfrac{1}{2}\right)} \approx 32.52$ days.

Thus, 3 grams of thorium-227 will be left
after approximately 32.52 days.

10. We first write the model with the parameters
$P = 2000$, $r = 0.06$, and $n = 12$ to obtain

$A = 2000\left(1 + \dfrac{0.06}{12}\right)^{12t}$ or $A = 2000(1.005)^{12t}$.

a. We need to determine the time until
$A = \$3000$, so we solve the equation

$3000 = 2000(1.005)^{12t}$

$1.5 = (1.005)^{12t}$

$\log 1.5 = \log(1.005)^{12t}$

$\log 1.5 = 12t\log 1.005$

$\dfrac{\log 1.5}{12\log 1.005} = t$

So, $t = \dfrac{\log 1.5}{12\log 1.005} \approx 6.77$. Thus, after

approximately 6.77 years (6 years, 9
months), the account will be worth \$3000.

b. We need to determine the time until
$A = \$4000$, so we solve the equation

$4000 = 2000(1.005)^{12t}$

$2 = (1.005)^{12t}$

$\log 2 = \log(1.005)^{12t}$

$\log 2 = 12t\log 1.005$

$\dfrac{\log 2}{12\log 1.005} = t$

So, $t = \dfrac{\log 2}{12\log 1.005} \approx 11.58$. Thus, after approximately 11.58 years (11 years, 7 months), the account will be worth $4000.

8.5 Exercises

11. $\log_2 x = \log_2 7$
$\qquad x = 7$
The solution set is $\{7\}$.

13. $2\log_3 x = \log_3 81$
$\qquad \log_3 x^2 = \log_3 81$
$\qquad\quad x^2 = 81$
$\qquad\quad x = \pm\sqrt{81} = \pm 9$
The apparent solution $x = -9$ is extraneous because the argument of a logarithm must be positive. The solution set is $\{9\}$.

15. $\log_6 (3x+1) = \log_6 10$
$\qquad\quad 3x+1 = 10$
$\qquad\qquad 3x = 9$
$\qquad\qquad x = 3$
The solution set is $\{3\}$.

17. $\dfrac{1}{2}\ln x = 2\ln 3$
$\qquad \ln x = 4\ln 3$
$\qquad \ln x = \ln 3^4$
$\qquad \ln x = \ln 81$
$\qquad\quad x = 81$
The solution set is $\{81\}$.

19. $\log_2 (x+3) + \log_2 x = 2$
$\qquad \log_2 [x(x+3)] = 2$
$\qquad \log_2 (x^2 + 3x) = 2$
$\qquad\qquad x^2 + 3x = 2^2$
$\qquad\qquad x^2 + 3x = 4$
$\qquad\qquad x^2 + 3x - 4 = 0$
$\qquad\quad (x+4)(x-1) = 0$
$\quad x+4 = 0 \quad$ or $\quad x-1 = 0$
$\qquad x = -4 \quad$ or $\qquad x = 1$
The apparent solution $x = -4$ is extraneous because it causes the argument of a logarithm to be negative. The solution set is $\{1\}$.

21. $\log_2 (x+2) + \log_2 (x+5) = \log_2 4$
$\qquad \log_2 [(x+2)(x+5)] = \log_2 4$
$\qquad\qquad x^2 + 7x + 10 = 4$
$\qquad\qquad x^2 + 7x + 6 = 0$
$\qquad\quad (x+6)(x+1) = 0$
$\quad x+6 = 0 \quad$ or $\quad x+1 = 0$
$\qquad x = -6 \quad$ or $\qquad x = -1$
The apparent solution $x = -6$ is extraneous because it causes the arguments of both logarithms to be negative. The solution set is $\{-1\}$.

23. $\log(x+3) - \log x = 1$
$\qquad \log\left(\dfrac{x+3}{x}\right) = 1$
$\qquad\qquad \dfrac{x+3}{x} = 10^1$
$\qquad x\left(\dfrac{x+3}{x}\right) = x(10)$
$\qquad\qquad x+3 = 10x$
$\qquad\qquad 3 = 9x$
$\qquad\qquad x = \dfrac{3}{9} = \dfrac{1}{3}$
The solution set is $\left\{\dfrac{1}{3}\right\}$.

25. $\log_4 (x+5) - \log_4 (x-1) = 2$
$\qquad \log_4\left(\dfrac{x+5}{x-1}\right) = 2$
$\qquad\qquad \dfrac{x+5}{x-1} = 4^2$
$\qquad (x-1)\left(\dfrac{x+5}{x-1}\right) = (x-1)16$
$\qquad\qquad x+5 = 16x - 16$
$\qquad\qquad -15x = -21$
$\qquad\qquad x = \dfrac{-21}{-15} = \dfrac{7}{5}$
The solution set is $\left\{\dfrac{7}{5}\right\}$.

27. $\log_4 (x+8) + \log_4 (x+6) = \log_4 3$

$\log_4 [(x+8)(x+6)] = \log_4 3$

$x^2 + 14x + 48 = 3$

$x^2 + 14x + 45 = 0$

$(x+9)(x+5) = 0$

$x+9 = 0 \quad \text{or} \quad x+5 = 0$

$x = -9 \quad \text{or} \quad x = -5$

The apparent solution $x = -9$ is extraneous because it causes the arguments of both logarithms to be negative. The solution set is $\{-5\}$.

29. $2^x = 10$

$\log 2^x = \log 10$

$x \log 2 = 1$

$x = \dfrac{1}{\log 2} \approx 3.322$

The solution set is $\left\{\dfrac{1}{\log 2}\right\} \approx \{3.322\}$. If we had taken the natural logarithm of both sides, the solution set would be $\left\{\dfrac{\ln 10}{\ln 2}\right\} \approx \{3.322\}$.

31. $5^x = 20$

$\log 5^x = \log 20$

$x \log 5 = \log 20$

$x = \dfrac{\log 20}{\log 5} \approx 1.861$

The solution set is $\left\{\dfrac{\log 20}{\log 5}\right\} \approx \{1.861\}$. If we had taken the natural logarithm of both sides, the solution set would be $\left\{\dfrac{\ln 20}{\ln 5}\right\} \approx \{1.861\}$.

33. $\left(\dfrac{1}{2}\right)^x = 7$

$\log\left(\dfrac{1}{2}\right)^x = \log 7$

$x \log\left(\dfrac{1}{2}\right) = \log 7$

$x = \dfrac{\log 7}{\log\left(\frac{1}{2}\right)} \approx -2.807$

The solution set is $\left\{\dfrac{\log 7}{\log\left(\frac{1}{2}\right)}\right\} \approx \{-2.807\}$. If we had taken the natural logarithm of both sides, the solution set would be $\left\{\dfrac{\ln 7}{\ln\left(\frac{1}{2}\right)}\right\} \approx \{-2.807\}$.

35. $e^x = 5$

$\ln e^x = \ln 5$

$x = \ln 5 \approx 1.609$

The solution set is $\{\ln 5\} \approx \{1.609\}$.

37. $10^x = 5$

$\log 10^x = \log 5$

$x = \log 5 \approx 0.699$

The solution set is $\{\log 5\} \approx \{0.699\}$.

39. $3^{2x} = 13$

$\log 3^{2x} = \log 13$

$2x \log 3 = \log 13$

$x = \dfrac{\log 13}{2 \log 3} \approx 1.167$

The solution set is $\left\{\dfrac{\log 13}{2 \log 3}\right\} \approx \{1.167\}$. If we had taken the natural logarithm of both sides, the solution set would be $\left\{\dfrac{\ln 13}{2 \ln 3}\right\} \approx \{1.167\}$.

41. $\left(\dfrac{1}{2}\right)^{4x} = 3$

$\log\left(\dfrac{1}{2}\right)^{4x} = \log 3$

$4x \log\left(\dfrac{1}{2}\right) = \log 3$

$x = \dfrac{\log 3}{4 \log\left(\frac{1}{2}\right)} \approx -0.396$

The solution set is $\left\{\dfrac{\log 3}{4 \log\left(\frac{1}{2}\right)}\right\} \approx \{-0.396\}$. If we had taken the natural logarithm of both sides, the solution set would be $\left\{\dfrac{\ln 3}{4 \ln\left(\frac{1}{2}\right)}\right\} \approx \{-0.396\}$.

43. $4 \cdot 2^x + 3 = 8$

$$4 \cdot 2^x = 5$$

$$2^x = \frac{5}{4}$$

$$\log 2^x = \log\left(\frac{5}{4}\right)$$

$$x \log 2 = \log\left(\frac{5}{4}\right)$$

$$x = \frac{\log\left(\frac{5}{4}\right)}{\log 2} \approx 0.322$$

The solution set is $\left\{\dfrac{\log\left(\frac{5}{4}\right)}{\log 2}\right\} \approx \{0.322\}$. If we

had taken the natural logarithm of both sides, the

solution set would be $\left\{\dfrac{\ln\left(\frac{5}{4}\right)}{\ln 2}\right\} \approx \{0.322\}$.

45. $-3e^x = -18$

$$e^x = 6$$

$$\ln e^x = \ln 6$$

$$x = \ln 6 \approx 1.792$$

The solution set is $\{\ln 6\} \approx \{1.792\}$.

47. $\qquad 0.2^{x+1} = 3^x$

$$\log 0.2^{x+1} = \log 3^x$$

$$(x+1)\log 0.2 = x \log 3$$

$$x \log 0.2 + \log 0.2 = x \log 3$$

$$\log 0.2 = x \log 3 - x \log 0.2$$

$$\log 0.2 = x(\log 3 - \log 0.2)$$

$$\frac{\log 0.2}{\log 3 - \log 0.2} = x$$

$$-0.594 \approx x$$

The solution set is $\left\{\dfrac{\log 0.2}{\log 3 - \log 0.2}\right\} \approx \{-0.594\}$.

If we had taken the natural logarithm of both

sides, the solution set would be

$\left\{\dfrac{\ln 0.2}{\ln 3 - \ln 0.2}\right\} \approx \{-0.594\}$.

49. $\log_4 x + \log_4 (x-6) = 2$

$$\log_4 [x(x-6)] = 2$$

$$\log_4 (x^2 - 6x) = 2$$

$$x^2 - 6x = 4^2$$

$$x^2 - 6x = 16$$

$$x^2 - 6x - 16 = 0$$

$$(x-8)(x+2) = 0$$

$$x - 8 = 0 \quad \text{or} \quad x + 2 = 0$$

$$x = 8 \quad \text{or} \quad x = -2$$

The apparent solution $x = -2$ is extraneous
because it causes the argument of a logarithm to
be negative. The solution set is $\{8\}$.

51. $\qquad 5^{3x} = 7$

$$\log 5^{3x} = \log 7$$

$$3x \log 5 = \log 7$$

$$x = \frac{\log 7}{3 \log 5} \approx 0.403$$

The solution set is $\left\{\dfrac{\log 7}{3 \log 5}\right\} \approx \{0.403\}$. If we

had taken the natural logarithm of both sides, the

solution set would be $\left\{\dfrac{\ln 7}{3 \ln 5}\right\} \approx \{0.403\}$.

53. $3 \log_2 x = \log_2 8$

$$\log_2 x^3 = \log_2 8$$

$$x^3 = 8$$

$$x = \sqrt[3]{8}$$

$$x = 2$$

The solution set is $\{2\}$.

55. $\dfrac{1}{3} e^x = 5$

$$e^x = 15$$

$$\ln e^x = \ln 15$$

$$x = \ln 15 \approx 2.708$$

The solution set is $\{\ln 15\} \approx \{2.708\}$.

57.

$$\left(\frac{1}{4}\right)^{x+1} = 8^x$$

$$\left(\frac{1}{2^2}\right)^{x+1} = \left(2^3\right)^x$$

$$\left(2^{-2}\right)^{x+1} = \left(2^3\right)^x$$

$$2^{-2(x+1)} = 2^{3x}$$

$$-2(x+1) = 3x$$

$$-2x - 2 = 3x$$

$$-2 = 5x$$

$$-\frac{2}{5} = x$$

The solution set is $\left\{-\dfrac{2}{5}\right\}$.

59.
$$\log_3 x^2 = \log_3 16$$
$$x^2 = 16$$
$$x = \pm 4$$
The solution set is $\{-4, 4\}$.

61.
$$\log_2 (x+4) + \log_2 (x+6) = \log_2 8$$
$$\log_2 \left[(x+4)(x+6)\right] = \log_2 8$$
$$x^2 + 10x + 24 = 8$$
$$x^2 + 10x + 16 = 0$$
$$(x+8)(x+2) = 0$$
$$x + 8 = 0 \quad \text{or} \quad x + 2 = 0$$
$$x = -8 \quad \text{or} \quad x = -2$$

The apparent solution $x = -8$ is extraneous because it causes the arguments of both logarithms to be negative. The solution set is $\{-2\}$.

63.
$$\log_4 x + \log_4 (x-4) = \log_4 3$$
$$\log_4 \left[x(x-4)\right] = \log_4 3$$
$$\log_4 \left(x^2 - 4x\right) = \log_4 3$$
$$x^2 - 4x = 3$$
$$x^2 - 4x - 3 = 0$$
$$a = 1,\ b = -4,\ c = -3$$
$$x = \frac{4 \pm \sqrt{16+12}}{2} = \frac{4 \pm \sqrt{28}}{2} = \frac{4 \pm 2\sqrt{7}}{2} = 2 \pm \sqrt{7}$$

The apparent solution $x = 2 - \sqrt{7}$ is extraneous because it causes the arguments of both logarithms to be negative. The solution set is $\left\{2 + \sqrt{7}\right\} = \{4.646\}$.

65. a. We need to determine the year when $P = 321$ million. So we solve the equation

$$321 = 313(1.011)^{t-2012}$$

$$\frac{321}{313} = (1.011)^{t-2012}$$

$$\log \frac{321}{313} = \log (1.011)^{t-2012}$$

$$\log \frac{321}{313} = (t - 2012) \log (1.011)$$

$$\frac{\log \dfrac{321}{313}}{\log 1.011} = t - 2012$$

$$\frac{\log \dfrac{321}{313}}{\log 1.011} + 2012 = t$$

$$2014.307 \approx t$$

Thus, according to the model, the population of the United States will reach 321 million people in about the year 2014.

b. We need to determine the year when $P = 471$ million. So we solve the equation

$$471 = 313(1.011)^{t-2012}$$

$$\frac{471}{313} = (1.011)^{t-2012}$$

$$\log \frac{471}{313} = \log (1.011)^{t-2012}$$

$$\log \frac{471}{313} = (t - 2012) \log (1.011)$$

$$\frac{\log \dfrac{471}{313}}{\log 1.011} = t - 2012$$

$$\frac{\log \dfrac{471}{313}}{\log 1.011} + 2012 = t$$

$$2049.354 \approx t$$

Thus, according to the model, the population of the United States will reach 471 million people in about the year 2048.

67. We first write the model with the parameters $P = 5000$, $r = 0.06$, and $n = 12$ to obtain

$$A = 5000\left(1 + \frac{0.06}{12}\right)^{12t} \text{ or } A = 5000(1.005)^{12t}.$$

a. We need to determine the time until $A = \$7000$, so we solve the equation

$$7000 = 5000(1.005)^{12t}$$
$$1.4 = (1.005)^{12t}$$
$$\log 1.4 = \log(1.005)^{12t}$$
$$\log 1.4 = 12t \log 1.005$$
$$\frac{\log 1.4}{12 \log 1.005} = t$$
$$5.622 \approx t$$

Thus, after approximately 5.6 years (5 years, 7 months), the account will be worth $7000.

b. We need to determine the time until $A = \$10,000$, so we solve the equation

$$10,000 = 5000(1.005)^{12t}$$
$$2 = (1.005)^{12t}$$
$$\log 2 = \log(1.005)^{12t}$$
$$\log 2 = 12t \log 1.005$$
$$\frac{\log 2}{12 \log 1.005} = t$$

So, $t = \dfrac{\log 2}{12 \log 1.005} \approx 11.581$. After about 11.6 years (11 years, 7 months), the account will be worth $7000.

69. a. We need to determine the time until $V = \$15,000$. So we solve the equation

$$15,000 = 19,841(0.88)^t$$
$$\frac{15,000}{19,841} = (0.88)^t$$
$$\log \frac{15,000}{19,841} = \log(0.88)^t$$
$$\log \frac{15,000}{19,841} = t \log 0.88$$
$$\frac{\log \frac{15,000}{19,841}}{\log 0.88} = t$$
$$2.188 \approx t$$

According to the model, the car will be worth $15,000 after about 2.188 years.

b. We need to determine the time until $V = \$5000$. So we solve the equation

$$5000 = 19,841(0.88)^t$$
$$\frac{5000}{19,841} = (0.88)^t$$
$$\log \frac{5000}{19,841} = \log(0.88)^t$$
$$\log \frac{5000}{19,841} = t \log 0.88$$
$$\frac{\log \frac{5000}{19,841}}{\log 0.88} = t$$
$$10.782 \approx t$$

According to the model, the car will be worth $5000 in about 10.782 years.

c. We need to determine the time until $V = \$1000$. So we solve the equation

$$1000 = 19,841(0.88)^t$$
$$\frac{1000}{19,841} = (0.88)^t$$
$$\log \frac{1000}{19,841} = \log(0.88)^t$$
$$\log \frac{1000}{19,841} = t \log 0.88$$
$$\frac{\log \frac{1000}{19,841}}{\log 0.88} = t$$
$$23.372 \approx t$$

According to the model, the car will be worth $1000 in about 23.372 years.

71. a. We need to determine the time until $A = 90$ grams. So we solve the equation

$$90 = 100\left(\frac{1}{2}\right)^{t/13.81}$$
$$0.9 = \left(\frac{1}{2}\right)^{t/13.81}$$
$$\log 0.9 = \log\left(\frac{1}{2}\right)^{t/13.81}$$
$$\log 0.9 = \frac{t}{13.81} \log\left(\frac{1}{2}\right)$$
$$\frac{13.81 \cdot \log 0.9}{\log(1/2)} = t$$
$$2.099 \approx t$$

Thus, 90 grams of beryllium-11 will be left after approximately 2.099 seconds.

b. We need to determine the time until $A = 25$ grams. So we solve the equation

$$25 = 100\left(\frac{1}{2}\right)^{t/13.81}$$

$$0.25 = \left(\frac{1}{2}\right)^{t/13.81}$$

$$\log 0.25 = \log\left(\frac{1}{2}\right)^{t/13.81}$$

$$\log 0.25 = \frac{t}{13.81}\log\left(\frac{1}{2}\right)$$

$$\frac{13.81 \cdot \log 0.25}{\log(1/2)} = t$$

$$27.62 = t$$

Thus, 25 grams of beryllium-11 will be left after 27.62 seconds.

c. We need to determine the time until $A = 10$ grams. So we solve the equation

$$10 = 100\left(\frac{1}{2}\right)^{t/13.81}$$

$$0.1 = \left(\frac{1}{2}\right)^{t/13.81}$$

$$\log 0.1 = \log\left(\frac{1}{2}\right)^{t/13.81}$$

$$\log 0.1 = \frac{t}{13.81}\log\left(\frac{1}{2}\right)$$

$$\frac{13.81 \cdot \log 0.1}{\log(1/2)} = t$$

$$45.876 \approx t$$

Thus, 10 grams of beryllium-11 will be left after approximately 45.876 seconds.

73. a. $u = 300°F$. So we solve the equation

$$300 = 70 + 330e^{-0.072t}$$

$$230 = 330e^{-0.072t}$$

$$\frac{230}{330} = e^{-0.072t}$$

$$\ln\left(\frac{230}{330}\right) = \ln e^{-0.072t}$$

$$\ln\left(\frac{230}{330}\right) = -0.072t$$

$$\frac{\ln\left(\frac{230}{330}\right)}{-0.072} = t$$

$$5.014 \approx t$$

According to the model, the temperature of the pizza will be $300°F$ after about 5 minutes.

b. We need to determine the time until $u = 220°F$. So we solve the equation

$$220 = 70 + 330e^{-0.072t}$$

$$150 = 330e^{-0.072t}$$

$$\frac{150}{330} = e^{-0.072t}$$

$$\ln\left(\frac{150}{330}\right) = \ln e^{-0.072t}$$

$$\ln\left(\frac{150}{330}\right) = -0.072t$$

$$\frac{\ln\left(\frac{150}{330}\right)}{-0.072} = t$$

$$10.951 \approx t$$

According to the model, the temperature of the pizza will be 220°F after about 11 minutes.

75. a. We need to determine the time at which $L = 50$ words. So we solve the equation

$$50 = 200\left(1 - e^{-0.0035t}\right)$$

$$0.25 = 1 - e^{-0.0035t}$$

$$e^{-0.0035t} = 0.75$$

$$\ln e^{-0.0035t} = \ln 0.75$$

$$-0.0035t = \ln 0.75$$

$$t = \frac{\ln 0.75}{-0.0035}$$

$$t \approx 82.195$$

According to the model, the student must study about 82 minutes in order to learn 50 words.

b. We need to determine the time at which $L = 150$ words. So we solve the equation

$$150 = 200\left(1 - e^{-0.0035t}\right)$$

$$0.75 = 1 - e^{-0.0035t}$$

$$e^{-0.0035t} = 0.25$$

$$\ln e^{-0.0035t} = \ln 0.25$$

$$-0.0035t = \ln 0.25$$

$$t = \frac{\ln 0.25}{-0.0035}$$

$$t \approx 396.084$$

According to the model, the student must study about 396 minutes (or 6.6 hours) in order to learn 150 words.

77. a. $t = \dfrac{72}{8} = 9$

According to the Rule of 72, an investment earning 8% annual interest will take about 9 years to double.

b. Let $A = 2P$ in the formula $A = P\left(1 + \dfrac{r}{n}\right)^{nt}$:

$$2P = P\left(1 + \frac{r}{n}\right)^{nt}$$

$$2 = \left(1 + \frac{r}{n}\right)^{nt}$$

$$\log 2 = \log\left(1 + \frac{r}{n}\right)^{nt}$$

$$\log 2 = nt \log\left(1 + \frac{r}{n}\right)$$

$$\frac{\log 2}{n \log\left(1 + \dfrac{r}{n}\right)} = t$$

c. Letting $r = 0.08$ and $n = 12$, we obtain

$$t = \frac{\log 2}{12 \cdot \log\left(1 + \dfrac{0.08}{12}\right)} \approx 8.693 .$$

According to our formula from part (b), an investment earning 8% annual interest will take about 8.693 years to double, which is about the same as the result from the Rule of 72.

79. Since the bacteria doubled every minute, the container would have been half full one minute before it was completely full. Now since it was full after 30 minutes, it would have been half full after $30 - 1 = 29$ minutes.

81. $f(x) = -2x + 7$

    a. $f(3) = -2(3) + 7 = -6 + 7 = 1$

    b. $f(-2) = -2(-2) + 7 = 4 + 7 = 11$

    c. $f(0) = -2(0) + 7 = 0 + 7 = 7$

83. $f(x) = \dfrac{x}{x - 5}$

    a. $f(3) = \dfrac{3}{3 - 5} = \dfrac{3}{-2} = -\dfrac{3}{2}$

    b. $f(-2) = \dfrac{-2}{-2 - 5} = \dfrac{-2}{-7} = \dfrac{2}{7}$

    c. $f(0) = \dfrac{0}{0 - 5} = \dfrac{0}{-5} = 0$

85. $f(x) = 2^x$

    a. $f(3) = 2^3 = 8$

    b. $f(-2) = 2^{-2} = \dfrac{1}{2^2} = \dfrac{1}{4}$

    c. $f(0) = 2^0 = 1$

87. Let $Y_1 = e^x$ and $Y_2 = -2x + 5$.

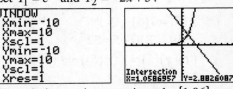

The solution set is approximately $\{1.06\}$.

89. Let $Y_1 = e^x$ and $Y_2 = x^2$.

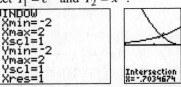

The solution set is approximately $\{-0.70\}$.

91. Let $Y_1 = e^x - \ln x$ and $Y_2 = 4$.

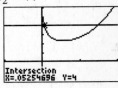

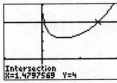

The solution set is approximately $\{0.05, 1.48\}$.

93. Let $Y_1 = \ln x$ and $Y_2 = x^2 - 1$.

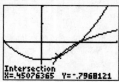

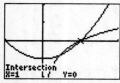

The solution set is approximately $\{0.45,\ 1\}$.

Chapter 8 Review

1. $f(x) = 3x + 5$; $g(x) = 2x - 1$

  a. $g(5) = 2(5) - 1 = 9$
$f(9) = 3(9) + 5 = 32$
$(f \circ g)(5) = f(g(5)) = f(9) = 32$

  b. $f(-3) = 3(-3) + 5 = -4$
$g(-4) = 2(-4) - 1 = -9$
$(g \circ f)(-3) = g(f(-3)) = g(-4) = -9$

  c. $f(-2) = 3(-2) + 5 = -1$
$f(-1) = 3(-1) + 5 = 2$
$(f \circ f)(-2) = f(f(-2)) = f(-1) = 2$

  d. $g(4) = 2(4) - 1 = 7$
$g(7) = 2(7) - 1 = 13$
$(g \circ g)(4) = g(g(4)) = g(7) = 13$

2. $f(x) = x - 3$; $g(x) = 5x + 2$

  a. $g(5) = 5(5) + 2 = 27$
$f(27) = 27 - 3 = 24$
$(f \circ g)(5) = f(g(5)) = f(27) = 24$

  b. $f(-3) = -3 - 3 = -6$
$g(-6) = 5(-6) + 2 = -28$
$(g \circ f)(-3) = g(f(-3)) = g(-6) = -28$

  c. $f(-2) = -2 - 3 = -5$
$f(-5) = -5 - 3 = -8$
$(f \circ f)(-2) = f(f(-2)) = f(-5) = -8$

  d. $g(4) = 5(4) + 2 = 22$
$g(22) = 5(22) + 2 = 112$
$(g \circ g)(4) = g(g(4)) = g(22) = 112$

3. $f(x) = 2x^2 + 1$; $g(x) = x + 5$

  a. $g(5) = 5 + 5 = 10$
$f(10) = 2(10)^2 + 1 = 201$
$(f \circ g)(5) = f(g(5)) = f(10) = 201$

  b. $f(-3) = 2(-3)^2 + 1 = 19$
$g(19) = 19 + 5 = 24$
$(g \circ f)(-3) = g(f(-3)) = g(19) = 24$

  c. $f(-2) = 2(-2)^2 + 1 = 9$
$f(9) = 2(9)^2 + 1 = 163$
$(f \circ f)(-2) = f(f(-2)) = f(9) = 163$

  d. $g(4) = 4 + 5 = 9$
$g(9) = 9 + 5 = 14$
$(g \circ g)(4) = g(g(4)) = g(9) = 14$

4. $f(x) = x - 3$; $g(x) = x^2 + 1$

  a. $g(5) = (5)^2 + 1 = 26$
$f(26) = 26 - 3 = 23$
$(f \circ g)(5) = f(g(5)) = f(26) = 23$

  b. $f(-3) = -3 - 3 = -6$
$g(-6) = (-6)^2 + 1 = 37$
$(g \circ f)(-3) = g(f(-3)) = g(-6) = 37$

  c. $f(-2) = -2 - 3 = -5$
$f(-5) = -5 - 3 = -8$
$(f \circ f)(-2) = f(f(-2)) = f(-5) = -8$

  d. $g(4) = (4)^2 + 1 = 17$

        $g(17) = (17)^2 + 1 = 290$

        $(g \circ g)(4) = g(g(4)) = g(17) = 290$

5. $f(x) = x + 1$; $g(x) = 5x$

  a. $(f \circ g)(x) = f(g(x)) = f(5x) = 5x + 1$

  b. $(g \circ f)(x) = g(f(x))$

          $= g(x+1)$

          $= 5(x+1)$

          $= 5x + 5$

  c. $(f \circ f)(x) = f(f(x))$

          $= f(x+1)$

          $= (x+1) + 1$

          $= x + 2$

  d. $(g \circ g)(x) = g(g(x))$

          $= g(5x)$

          $= 5(5x)$

          $= 25x$

6. $f(x) = 2x - 3$; $g(x) = x + 6$

  a. $(f \circ g)(x) = f(g(x))$

          $= f(x+6)$

          $= 2(x+6) - 3$

          $= 2x + 12 - 3$

          $= 2x + 9$

  b. $(g \circ f)(x) = g(f(x))$

          $= g(2x - 3)$

          $= (2x - 3) + 6$

          $= 2x + 3$

  c. $(f \circ f)(x) = f(f(x))$

          $= f(2x - 3)$

          $= 2(2x - 3) - 3$

          $= 4x - 6 - 3$

          $= 4x - 9$

  d. $(g \circ g)(x) = g(g(x))$

          $= g(x + 6)$

          $= (x + 6) + 6$

          $= x + 12$

7. $f(x) = x^2 + 1$; $g(x) = 2x + 1$

  a. $(f \circ g)(x) = f(g(x))$

          $= f(2x + 1)$

          $= (2x + 1)^2 + 1$

          $= 4x^2 + 4x + 1 + 1$

          $= 4x^2 + 4x + 2$

  b. $(g \circ f)(x) = g(f(x))$

          $= g(x^2 + 1)$

          $= 2(x^2 + 1) + 1$

          $= 2x^2 + 2 + 1$

          $= 2x^2 + 3$

  c. $(f \circ f)(x) = f(f(x))$

          $= f(x^2 + 1)$

          $= (x^2 + 1)^2 + 1$

          $= x^4 + 2x^2 + 1 + 1$

          $= x^4 + 2x^2 + 2$

  d. $(g \circ g)(x) = g(g(x))$

          $= g(2x + 1)$

          $= 2(2x + 1) + 1$

          $= 4x + 2 + 1$

          $= 4x + 3$

8. $f(x) = \dfrac{2}{x+1}$; $g(x) = \dfrac{1}{x}$

a. $(f \circ g)(x) = f(g(x))$

$$= f\left(\frac{1}{x}\right)$$

$$= \frac{2}{\frac{1}{x}+1}$$

$$= \frac{2}{\frac{1+x}{x}}$$

$$= \frac{2x}{x+1} \quad \text{where } x \neq -1, 0.$$

b. $(g \circ f)(x) = g(f(x))$

$$= g\left(\frac{2}{x+1}\right)$$

$$= \frac{1}{\frac{2}{x+1}}$$

$$= 1 \cdot \frac{x+1}{2}$$

$$= \frac{x+1}{2} \quad \text{where } x \neq -1.$$

c. $(f \circ f)(x) = f(f(x))$

$$= f\left(\frac{2}{x+1}\right)$$

$$= \frac{2}{\frac{2}{x+1}+1}$$

$$= \frac{2}{\frac{2+x+1}{x+1}}$$

$$= 2 \cdot \frac{x+1}{x+3}$$

$$= \frac{2(x+1)}{x+3} \quad \text{where } x \neq -1, -3.$$

d. $(g \circ g)(x) = g(g(x))$

$$= g\left(\frac{1}{x}\right)$$

$$= \frac{1}{\frac{1}{x}}$$

$$= 1 \cdot \frac{x}{1}$$

$$= x \quad \text{where } x \neq 0.$$

9. The function is not one-to-one. There is an element in the range (8) that corresponds to more than one element in the domain (-5 and -1).

10. The function is one-to-one. Each element in the range corresponds to exactly one element in the domain.

11. The graph passes the horizontal line test, so the graph is that of a one-to-one function.

12. The graph fails the horizontal line test. Therefore, the function is not one-to-one.

13. Inverse:

Height (inches)	Age
69	24
71	59
72	29
73	81
74	37

14. Inverse:

Quantity Demanded	Price ($)
112	300
129	200
144	170
161	150
176	130

15. To obtain the inverse, we switch the *x*- and *y*-coordinates.

Inverse: $\{(3,-5),(1,-3),(-3,1),(9,2)\}$

16. To obtain the inverse, we switch the *x*- and *y*-coordinates.

Inverse: $\{(1,-20),(4,-15),(3,5),(2,25)\}$

17. To plot the inverse, switch the *x*- and *y*-coordinates in each point and connect the corresponding points. The graphs of the function (shaded) and the line $y = x$ (dashed) are included for reference.

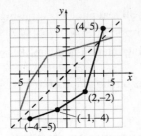

18. To plot the inverse, switch the x- and y-coordinates in each point and connect the corresponding points. The graphs of the function (shaded) and the line $y = x$ (dashed) are included for reference.

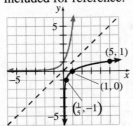

19. $f(x) = 5x$

$y = 5x$

$x = 5y$

$\dfrac{x}{5} = y$

$f^{-1}(x) = \dfrac{x}{5}$

Check: $f\left(f^{-1}(x)\right) = 5\left(\dfrac{x}{5}\right) = x$

$f^{-1}\left(f(x)\right) = \dfrac{5x}{5} = x$

20. $H(x) = 2x + 7$

$y = 2x + 7$

$x = 2y + 7$

$x - 7 = 2y$

$\dfrac{x - 7}{2} = y$

$H^{-1}(x) = \dfrac{x - 7}{2}$

Check:

$H\left(H^{-1}(x)\right) = 2\left(\dfrac{x-7}{2}\right) + 7 = x - 7 + 7 = x$

$H^{-1}\left(H(x)\right) = \dfrac{(2x+7) - 7}{2} = \dfrac{2x}{2} = x$

21. $P(x) = \dfrac{4}{x+2}$

$y = \dfrac{4}{x+2}$

$x = \dfrac{4}{y+2}$

$x(y+2) = 4$

$y + 2 = \dfrac{4}{x}$

$y = \dfrac{4}{x} - 2$

$P^{-1}(x) = \dfrac{4}{x} - 2$

Check:

$P\left(P^{-1}(x)\right) = \dfrac{4}{\left(\dfrac{4}{x} - 2\right) + 2} = \dfrac{4}{\dfrac{4}{x}} = 4 \cdot \dfrac{x}{4} = x$

$P^{-1}\left(P(x)\right) = \dfrac{4}{\dfrac{4}{x+2}} - 2$

$= 4 \cdot \dfrac{x+2}{4} - 2 = (x+2) - 2 = x$

22. $g(x) = 2x^3 - 1$

$y = 2x^3 - 1$

$x = 2y^3 - 1$

$x + 1 = 2y^3$

$\dfrac{x+1}{2} = y^3$

$\sqrt[3]{\dfrac{x+1}{2}} = y$

$g^{-1}(x) = \sqrt[3]{\dfrac{x+1}{2}}$

Check:

$g\left(g^{-1}(x)\right) = 2\left(\sqrt[3]{\dfrac{x+1}{2}}\right)^3 - 1$

$= 2\left(\dfrac{x+1}{2}\right) - 1 = (x+1) - 1 = x$

$g^{-1}\left(g(x)\right) = \sqrt[3]{\dfrac{(2x^3 - 1) + 1}{2}} = \sqrt[3]{\dfrac{2x^3}{2}} = \sqrt[3]{x^3} = x$

23. a. $7^{1.7} \approx 27.332$

729

    b. $7^{1.73} \approx 28.975$

    c. $7^{1.732} \approx 29.088$

    d. $7^{1.7321} \approx 29.093$

    e. $7^{\sqrt{3}} \approx 29.091$

24. a. $10^{3.1} \approx 1258.925$

    b. $10^{3.14} \approx 1380.384$

    c. $10^{3.142} \approx 1386.756$

    d. $10^{3.1416} \approx 1385.479$

    e. $10^{\pi} \approx 1385.456$

25. a. $e^{0.5} \approx 1.649$

    b. $e^{-1} \approx 0.368$

    c. $e^{1.5} \approx 4.482$

    d. $e^{-0.8} \approx 0.449$

    e. $e^{\sqrt{\pi}} \approx 5.885$

26. Locate some points on the graph of $f(x) = 9^x$.

x	$f(x) = 9^x$	$(x, f(x))$
-2	$f(-2) = 9^{-2} = \dfrac{1}{9^2} = \dfrac{1}{81}$	$\left(-2, \dfrac{1}{81}\right)$
-1	$f(-1) = 9^{-1} = \dfrac{1}{9^1} = \dfrac{1}{9}$	$\left(-1, \dfrac{1}{9}\right)$
0	$f(0) = 9^0 = 1$	$(0, 1)$
1	$f(1) = 9^1 = 9$	$(1, 9)$

Plot the points and connect them with a smooth curve.

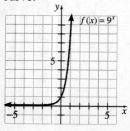

The domain of f is all real numbers or, using interval notation, $(-\infty, \infty)$. The range of f is $\{y \mid y > 0\}$ or , using interval notation, $(0, \infty)$.

27. Locate some points on the graph of $g(x) = \left(\dfrac{1}{9}\right)^x$.

x	$g(x) = \left(\dfrac{1}{9}\right)^x$	$(x, g(x))$
-1	$g(-1) = \left(\dfrac{1}{9}\right)^{-1} = 9^1 = 9$	$(-1, 9)$
0	$g(0) = \left(\dfrac{1}{9}\right)^0 = 1$	$(0, 1)$
1	$g(1) = \left(\dfrac{1}{9}\right)^1 = \dfrac{1}{9}$	$\left(1, \dfrac{1}{9}\right)$
2	$g(1) = \left(\dfrac{1}{9}\right)^2 = \dfrac{1}{81}$	$\left(2, \dfrac{1}{81}\right)$

Plot the points and connect them with a smooth curve.

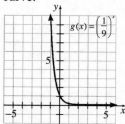

The domain of g is all real numbers or, using interval notation, $(-\infty, \infty)$. The range of g is $\{y \mid y > 0\}$ or , using interval notation, $(0, \infty)$.

28. Locate some points on the graph of $H(x) = 4^{x-2}$.

x	$H(x) = 4^{x-2}$	$(x, H(x))$
0	$H(0) = 4^{0-2} = 4^{-2} = \dfrac{1}{4^2} = \dfrac{1}{16}$	$\left(0, \dfrac{1}{16}\right)$
1	$H(1) = 4^{1-2} = 4^{-1} = \dfrac{1}{4^1} = \dfrac{1}{4}$	$\left(1, \dfrac{1}{4}\right)$
2	$H(2) = 4^{2-2} = 4^0 = 1$	$(2, 1)$
3	$H(3) = 4^{3-2} = 4^1 = 4$	$(3, 4)$

Plot the points and connect them with a smooth curve.

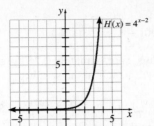

The domain of H is all real numbers or, using interval notation, $(-\infty, \infty)$. The range of H is $\{y \mid y > 0\}$ or , using interval notation, $(0, \infty)$.

29. Locate some points on the graph of $h(x) = 4^x - 2$.

x	$h(x) = 4^x - 2$	$(x,\ h(x))$
-2	$h(-2) = 4^{-2} - 2 = \dfrac{1}{16} - 2 = -\dfrac{31}{16}$	$\left(-2, -\dfrac{31}{16}\right)$
-1	$h(-1) = 4^{-1} - 2 = \dfrac{1}{4} - 2 = -\dfrac{7}{4}$	$\left(-1, -\dfrac{7}{4}\right)$
0	$h(0) = 4^0 - 2 = 1 - 2 = -1$	$(0, -1)$
1	$h(1) = 4^1 - 2 = 4 - 2 = 2$	$(1, 2)$

Plot the points and connect them with a smooth curve.

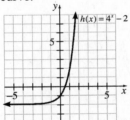

The domain of h is all real numbers or, using interval notation, $(-\infty, \infty)$. The range of h is $\{y \mid y > -2\}$ or , using interval notation, $(-2, \infty)$.

30. The number e is defined as the number that the expression $\left(1 + \dfrac{1}{n}\right)^n$ approaches as n becomes unbounded in the positive direction.

31. $2^x = 64$

$2^x = 2^6$

$x = 6$

The solution set is $\{6\}$.

32. $25^{x-2} = 125$

$\left(5^2\right)^{x-2} = 5^3$

$5^{2(x-2)} = 5^3$

$5^{2x-4} = 5^3$

$2x - 4 = 3$

$2x = 7$

$x = \dfrac{7}{2}$

The solution set is $\left\{\dfrac{7}{2}\right\}$.

33. $27^x \cdot 3^{x^2} = 9^2$

$\left(3^3\right)^x \cdot 3^{x^2} = \left(3^2\right)^2$

$3^{3x} \cdot 3^{x^2} = 3^4$

$3^{x^2+3x} = 3^4$

$x^2 + 3x = 4$

$x^2 + 3x - 4 = 0$

$(x+4)(x-1) = 0$

$x + 4 = 0 \quad \text{or} \quad x - 1 = 0$

$x = -4 \quad \text{or} \quad x = 1$

The solution set is $\{-4,\ 1\}$.

34. $\left(\dfrac{1}{4}\right)^x = 16$

$\left(4^{-1}\right)^x = 4^2$

$4^{-x} = 4^2$

$-x = 2$

$x = -2$

The solution set is $\{-2\}$.

35. $\left(e^2\right)^{x-1} = e^x \cdot e^7$

$e^{2(x-1)} = e^{x+7}$

$e^{2x-2} = e^{x+7}$

$2x - 2 = x + 7$

$x - 2 = 7$

$x = 9$

The solution set is $\{9\}$.

36. $\left(2^x\right)^x = 512$

$2^{x^2} = 2^9$

$x^2 = 9$

$x = \pm\sqrt{9}$

$x = \pm 3$

The solution set is $\{-3,\ 3\}$.

37. a. We use the compound interest formula with $P = \$2500$, $r = 0.045$, $t = 25$, and $n = 1$, so that

$$A = 2500\left(1 + \frac{0.045}{1}\right)^{1(25)} \approx \$7513.59.$$

b. We use the compound interest formula with $P = \$2500$, $r = 0.045$, $t = 25$, and $n = 4$, so that

$$A = 2500\left(1 + \frac{0.045}{4}\right)^{4(25)} \approx \$7652.33.$$

c. We use the compound interest formula with $P = \$2500$, $r = 0.045$, $t = 25$, and $n = 12$, so that

$$A = 2500\left(1 + \frac{0.045}{12}\right)^{12(25)} \approx \$7684.36.$$

d. We use the compound interest formula with $P = \$2500$, $r = 0.045$, $t = 25$, and $n = 365$, so that

$$A = 2500\left(1 + \frac{0.045}{365}\right)^{365(25)} \approx \$7700.01.$$

38. a. $A(1) = 100\left(\dfrac{1}{2}\right)^{\frac{1}{3.5}} \approx 82.034.$

After 1 day, approximately 82.034 grams of radon gas will be left in the sample.

b. $A(3.5) = 100\left(\dfrac{1}{2}\right)^{\frac{3.5}{3.5}} = 50.$

After 3.5 days, 50 grams of radon gas will be left in the sample.

c. $A(7) = 100\left(\dfrac{1}{2}\right)^{\frac{7}{3.5}} = 25.$

After 7 days, 25 grams of radon gas will be left in the sample.

d. $A(30) = 100\left(\dfrac{1}{2}\right)^{\frac{30}{3.5}} \approx 0.263.$

After 30 days, approximately 0.263 gram of radon gas will be left in the sample.

39. a. $P(2018) = 2.723(1.052)^{2018-2011} \approx 3.883$

According to the model, the population of Nevada in 2018 will be approximately 3.883 million people.

b. $P(2025) = 2.723(1.052)^{2025-2011} \approx 5.537$

According to the model, the population of Nevada in 2025 will be approximately 5.537 million people.

40. a. $u(15) = 72 + 278e^{-0.0835(15)} \approx 151.449.$

According to the model, the temperature of the cake after 15 minutes will be approximately $151.449°F$.

b. $u(30) = 72 + 278e^{-0.0835(30)} \approx 94.706.$

According to the model, the temperature of the cake after 30 minutes will be approximately $94.706°F$.

41. If $3^4 = 81$, then $\log_3 81 = 4$.

42. If $4^{-3} = \dfrac{1}{64}$, then $\log_4\left(\dfrac{1}{64}\right) = -3$.

43. If $b^3 = 5$, then $\log_b 5 = 3$.

44. If $10^{3.74} = x$, then $\log x = 3.74$.

45. If $\log_8 2 = \dfrac{1}{3}$, then $8^{1/3} = 2$.

46. If $\log_5 18 = r$, then $5^r = 18$.

47. If $\ln(x+3) = 2$, then $e^2 = x+3$.

48. If $\log x = -4$, then $10^{-4} = x$.

49. Let $y = \log_8 128$. Then,

$$8^y = 128$$
$$\left(2^3\right)^y = 2^7$$
$$2^{3y} = 2^7$$
$$3y = 7$$
$$y = \frac{7}{3}$$

Thus, $\log_8 128 = \frac{7}{3}$.

50. Let $y = \log_6 1$. Then,

$$6^y = 1$$
$$6^y = 6^0$$
$$y = 0$$

Thus, $\log_6 1 = 0$.

51. Let $y = \log \dfrac{1}{100}$. Then,

$$10^y = \frac{1}{100}$$
$$10^y = \frac{1}{10^2}$$
$$10^y = 10^{-2}$$
$$y = -2$$

Thus, $\log \dfrac{1}{100} = -2$.

52. Let $y = \log_9 27$. Then

$$9^y = 27$$
$$\left(3^2\right)^y = 3^3$$
$$3^{2y} = 3^3$$
$$2y = 3$$
$$y = \frac{3}{2}$$

Thus, $\log_9 27 = \frac{3}{2}$.

53. The domain of $f(x) = \log_2(x+5)$ is the set of all real numbers x such that
$$x + 5 > 0$$
$$x > -5$$
Thus, the domain of $f(x) = \log_2(x+5)$ is $\{x \mid x > -5\}$, or using interval notation, $(-5, \infty)$.

54. The domain of $g(x) = \log_8(7 - 3x)$ is the set of all real numbers x such that
$$7 - 3x > 0$$
$$-3x > -7$$
$$x < \frac{-7}{-3}$$
$$x < \frac{7}{3}$$
Thus, the domain of $g(x) = \log_8(7 - 3x)$ is $\left\{x \mid x < \dfrac{7}{3}\right\}$, or using interval notation, $\left(-\infty, \dfrac{7}{3}\right)$.

55. The domain of $h(x) = \ln(3x)$ is the set of all real numbers x such that
$$3x > 0$$
$$x > 0$$
Thus, the domain of $h(x) = \ln(3x)$ is $\{x \mid x > 0\}$, or using interval notation, $(0, \infty)$.

56. The domain of $F(x) = \log_{1/3}(4x + 10)$ is the set of all real numbers x such that
$$4x + 10 > 0$$
$$4x > -10$$
$$x > \frac{-10}{4}$$
$$x > -\frac{5}{2}$$
Thus, the domain of $F(x) = \log_{1/3}(4x + 10)$ is $\left\{x \mid x > -\dfrac{5}{2}\right\}$, or using interval notation,
$\left(-\dfrac{5}{2}, \infty\right)$.

57. Rewrite $y = f(x) = \log_{5/2} x$ as $x = \left(\dfrac{5}{2}\right)^y$.

Locate some points on the graph of $x = \left(\dfrac{5}{2}\right)^y$.

y	$x = \left(\dfrac{5}{2}\right)^{y}$	(x, y)
-2	$x = \left(\dfrac{5}{2}\right)^{-2} = \left(\dfrac{2}{5}\right)^{2} = \dfrac{4}{25}$	$\left(\dfrac{4}{25}, -2\right)$
-1	$x = \left(\dfrac{5}{2}\right)^{-1} = \left(\dfrac{2}{5}\right)^{1} = \dfrac{2}{5}$	$\left(\dfrac{2}{5}, -1\right)$
0	$x = \left(\dfrac{5}{2}\right)^{0} = 1$	$(1, 0)$
1	$x = \left(\dfrac{5}{2}\right)^{1} = \dfrac{5}{2}$	$\left(\dfrac{5}{2}, 1\right)$
2	$x = \left(\dfrac{5}{2}\right)^{2} = \dfrac{25}{4}$	$\left(\dfrac{25}{4}, 2\right)$

Plot the points and connect them with a smooth curve.

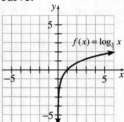

58. Rewrite $y = g(x) = \log_{2/5} x$ as $x = \left(\dfrac{2}{5}\right)^{y}$.

Locate some points on the graph of $x = \left(\dfrac{2}{5}\right)^{y}$.

y	$x = \left(\dfrac{2}{5}\right)^{y}$	(x, y)
-2	$x = \left(\dfrac{2}{5}\right)^{-2} = \left(\dfrac{5}{2}\right)^{2} = \dfrac{25}{4}$	$\left(\dfrac{25}{4}, -2\right)$
-1	$x = \left(\dfrac{2}{5}\right)^{-1} = \left(\dfrac{5}{2}\right)^{1} = \dfrac{5}{2}$	$\left(\dfrac{5}{2}, -1\right)$
0	$x = \left(\dfrac{2}{5}\right)^{0} = 1$	$(1, 0)$
1	$x = \left(\dfrac{2}{5}\right)^{1} = \dfrac{2}{5}$	$\left(\dfrac{2}{5}, 1\right)$
2	$x = \left(\dfrac{2}{5}\right)^{2} = \dfrac{4}{25}$	$\left(\dfrac{4}{25}, 2\right)$

Plot the points and connect them with a smooth curve.

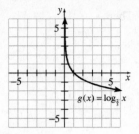

59. $\ln 24 \approx 3.178$

60. $\ln \dfrac{5}{6} \approx -0.182$

61. $\log 257 \approx 2.410$

62. $\log 0.124 \approx -0.907$

63. $\log_7 (4x - 19) = 2$
$$4x - 19 = 7^2$$
$$4x - 19 = 49$$
$$4x = 68$$
$$x = 17$$
The solution set is $\{17\}$.

64. $\log_{1/3} \left(x^2 + 8x\right) = -2$
$$x^2 + 8x = \left(\dfrac{1}{3}\right)^{-2}$$
$$x^2 + 8x = 3^2$$
$$x^2 + 8x = 9$$
$$x^2 + 8x - 9 = 0$$
$$(x + 9)(x - 1) = 0$$
$$x + 9 = 0 \quad \text{or} \quad x - 1 = 0$$
$$x = -9 \quad \text{or} \quad x = 1$$
The solution set is $\{-9, 1\}$.

65. $\log_a \dfrac{4}{9} = -2$
$$a^{-2} = \dfrac{4}{9}$$
$$a^2 = \dfrac{9}{4}$$
$$a = \pm\sqrt{\dfrac{9}{4}} = \pm\dfrac{3}{2}$$

Since the base of a logarithm must always be positive, we know that $a = -\dfrac{3}{2}$ is extraneous.

The solution set is $\left\{\dfrac{3}{2}\right\}$.

66. $\ln e^{5x} = 30$

$\qquad 5x = 30$

$\qquad x = 6$

The solution set is $\{6\}$.

67. $\log(6 - 7x) = 3$

$\qquad 6 - 7x = 10^3$

$\qquad 6 - 7x = 1000$

$\qquad -7x = 994$

$\qquad x = -142$

The solution set is $\{-142\}$.

68. $\log_b 75 = 2$

$\qquad b^2 = 75$

$\qquad b = \pm\sqrt{75}$

$\qquad b = \pm 5\sqrt{3}$

Since the base of a logarithm must always be positive, we know that $b = -5\sqrt{3}$ is extraneous. The solution set is $\left\{5\sqrt{3}\right\}$.

69. We evaluate $L(x) = 10\log\dfrac{x}{10^{-12}}$ at $x = 10^{-4}$.

$$L\left(10^{-4}\right) = 10\log\dfrac{10^{-4}}{10^{-12}}$$
$$= 10\log 10^{-4-(-12)}$$
$$= 10\log 10^{8}$$
$$= 10(8)$$
$$= 80$$

The loudness of the vacuum cleaner is 80 decibels.

70. We solve $M(x) = \log\left(\dfrac{x}{10^{-3}}\right)$ for $M(x) = 8$.

$$M(x) = 8$$
$$\log\left(\dfrac{x}{10^{-3}}\right) = 8$$
$$\dfrac{x}{10^{-3}} = 10^8$$
$$x = 10^8 \cdot 10^{-3}$$
$$= 10^{8+(-3)}$$
$$= 10^5$$
$$= 100,000$$

The seismographic reading of the Great New Madrid Earthquake 100 kilometers from the epicenter would have been 100,000 millimeters.

71. $\log_4 4^{21} = 21$

72. $7^{\log_7 9.34} = 9.34$

73. $\log_5 5 = 1$

74. $\log_9 1 = 0$

75. $\log_4 12 - \log_4 3 = \log_4 \dfrac{12}{3} = \log_4 4 = 1$

76. $12^{\log_{12} 2 + \log_{12} 8} = 12^{\log_{12}(2\cdot 8)} = 12^{\log_{12} 16} = 16$

77. $\log_7\left(\dfrac{xy}{z}\right) = \log_7(xy) - \log_7 z$

$\qquad\qquad\qquad = \log_7 x + \log_7 y - \log_7 z$

78. $\log_3\left(\dfrac{81}{x^2}\right) = \log_3 81 - \log_3 x^2$

$\qquad\qquad\qquad = \log_3 3^4 - \log_3 x^2$

$\qquad\qquad\qquad = 4 - 2\log_3 x$

79. $\log 1000 r^4 = \log 1000 + \log r^4$

$\qquad\qquad\quad = \log 10^3 + \log r^4$

$\qquad\qquad\quad = 3 + 4\log r$

80. $\ln\sqrt{\dfrac{x-1}{x}} = \ln\left(\dfrac{x-1}{x}\right)^{\frac{1}{2}}$

$\qquad\qquad\quad = \dfrac{1}{2}\ln\left(\dfrac{x-1}{x}\right)$

$\qquad\qquad\quad = \dfrac{1}{2}\left[\ln(x-1) - \ln x\right]$

$\qquad\qquad\quad = \dfrac{1}{2}\ln(x-1) - \dfrac{1}{2}\ln x$

81. $4\log_3 x + 2\log_3 y = \log_3 x^4 + \log_3 y^2$

$\qquad\qquad\qquad\qquad = \log_3\left(x^4 y^2\right)$

82. $\dfrac{1}{4}\ln x + \ln 7 - 2\ln 3 = \ln x^{\frac{1}{4}} + \ln 7 - \ln 3^2$

$\qquad\qquad\qquad\qquad = \ln\sqrt[4]{x} + \ln 7 - \ln 9$

$\qquad\qquad\qquad\qquad = \ln\left(7\sqrt[4]{x}\right) - \ln 9$

$\qquad\qquad\qquad\qquad = \ln\left(\dfrac{7\sqrt[4]{x}}{9}\right)$

83. $\log_2 3 - \log_2 6 = \log_2 \frac{3}{6} = \log_2 \frac{1}{2} = \log_2 2^{-1} = -1$

84. $\log_6 \left(x^2 - 7x + 12 \right) - \log_6 (x-3)$
$= \log_6 \left[(x-4)(x-3) \right] - \log_6 (x-3)$
$= \log_6 (x-4) + \log_6 (x-3) - \log_6 (x-3)$
$= \log_6 (x-4)$

85. Using common logarithms:
$$\log_6 50 = \frac{\log 50}{\log 6} \approx 2.183$$

86. Using common logarithms:
$$\log_\pi 2 = \frac{\log 2}{\log \pi} \approx 0.606$$

87. Using natural logarithms:
$$\log_{2/3} 6 = \frac{\ln 6}{\ln \left(\frac{2}{3} \right)} \approx -4.419$$

88. Using natural logarithms:
$$\log_{\sqrt{5}} 20 = \frac{\ln 20}{\ln \sqrt{5}} \approx 3.723$$

89. $3 \log_4 x = \log_4 1000$
$\log_4 x^3 = \log_4 1000$
$x^3 = 1000$
$x = \sqrt[3]{1000}$
$x = 10$
The solution set is $\{10\}$.

90. $\log_3 (x+7) + \log_3 (x+6) = \log_3 2$
$\log_3 \left[(x+7)(x+6) \right] = \log_3 2$
$x^2 + 13x + 42 = 2$
$x^2 + 13x + 40 = 0$
$(x+8)(x+5) = 0$
$x+8 = 0 \quad$ or $\quad x+5 = 0$
$x = -8 \quad$ or $\quad x = -5$
The apparent solution $x = -8$ is extraneous because it causes the arguments of both logarithms to be negative. The solution set is $\{-5\}$.

91. $\ln(x+2) - \ln x = \ln(x+1)$
$\ln \left(\frac{x+2}{x} \right) = \ln(x+1)$
$\frac{x+2}{x} = x+1$
$x(x+1) = x+2$
$x^2 + x = x+2$
$x^2 = 2$
$x = \pm \sqrt{2} \approx \pm 1.414$
The apparent solution $x = -\sqrt{2} \approx -1.414$ is extraneous because it causes the argument of a logarithm to be negative. The solution set is $\left\{ \sqrt{2} \right\} \approx \{1.414\}$.

92. $\frac{1}{3} \log_{12} x = 2 \log_{12} 2$
$\log_{12} x^{\frac{1}{3}} = \log_{12} 2^2$
$\log_{12} \sqrt[3]{x} = \log_{12} 4$
$\sqrt[3]{x} = 4$
$x = 4^3 = 64$
The solution set is $\{64\}$.

93. $\qquad 2^x = 15$
$\log 2^x = \log 15$
$x \log 2 = \log 15$
$x = \frac{\log 15}{\log 2} \approx 3.907$
The solution set is $\left\{ \frac{\log 15}{\log 2} \right\} \approx \{3.907\}$. If we had taken the natural logarithm of both sides, the solution set would be $\left\{ \frac{\ln 15}{\ln 2} \right\} \approx \{3.907\}$.

94. $\qquad 10^{3x} = 27$
$\log 10^{3x} = \log 27$
$3x = \log 27$
$x = \frac{\log 27}{3} \approx 0.477$
The solution set is $\left\{ \frac{\log 27}{3} \right\} \approx \{0.477\}$.

95. $\dfrac{1}{3}e^{7x} = 13$

$e^{7x} = 39$

$\ln e^{7x} = \ln 39$

$7x = \ln 39$

$x = \dfrac{\ln 39}{7} \approx 0.523$

The solution set is $\left\{\dfrac{\ln 39}{7}\right\} \approx \{0.523\}$.

96. $\qquad\qquad 3^x = 2^{x+1}$

$\log 3^x = \log 2^{x+1}$

$x \log 3 = (x+1)\log 2$

$x \log 3 = x \log 2 + \log 2$

$x \log 3 - x \log 2 = \log 2$

$x(\log 3 - \log 2) = \log 2$

$x = \dfrac{\log 2}{\log 3 - \log 2} \approx 1.710$

The solution set is $\left\{\dfrac{\log 2}{\log 3 - \log 2}\right\} \approx \{1.710\}$. If

we had taken the natural logarithm of both sides, the solution set would be

$\left\{\dfrac{\ln 2}{\ln 3 - \ln 2}\right\} \approx \{1.710\}$.

97. a. We need to determine the time until $A = 75$ grams. So we solve the equation

$$75 = 100\left(\frac{1}{2}\right)^{t/3.5}$$

$$0.75 = \left(\frac{1}{2}\right)^{t/3.5}$$

$$\log 0.75 = \log\left(\frac{1}{2}\right)^{t/3.5}$$

$$\log 0.75 = \frac{t}{3.5}\log\left(\frac{1}{2}\right)$$

$$\frac{3.5\log 0.75}{\log\left(\frac{1}{2}\right)} = t$$

So, $t = \dfrac{3.5\log 0.75}{\log\left(\frac{1}{2}\right)} \approx 1.453$ days.

Thus, 75 grams of radon gas will be left after approximately 1.453 days.

b. We need to determine the time until $A = 1$ gram. So we solve the equation

$$1 = 100\left(\frac{1}{2}\right)^{t/3.5}$$

$$0.01 = \left(\frac{1}{2}\right)^{t/3.5}$$

$$\log 0.01 = \log\left(\frac{1}{2}\right)^{t/3.5}$$

$$\log 0.01 = \frac{t}{3.5}\log\left(\frac{1}{2}\right)$$

$$\frac{3.5\log 0.01}{\log\left(\frac{1}{2}\right)} = t$$

So, $t = \dfrac{3.5\log 0.01}{\log\left(\frac{1}{2}\right)} \approx 23.253$ days.

Thus, 1 gram of radon gas will be left after approximately 23.253 days.

98. a. We need to determine the year when $P = 3.939$ million. So we solve the equation

$$3.939 = 2.723(1.052)^{t-2011}$$

$$\frac{3.939}{2.723} = (1.052)^{t-2011}$$

$$\log\frac{3.939}{2.723} = \log(1.052)^{t-2011}$$

$$\log\frac{3.939}{2.723} = (t-2011)\log(1.052)$$

$$\frac{\log\dfrac{3.939}{2.723}}{\log 1.052} = t - 2011$$

$$\frac{\log\dfrac{3.939}{2.496}}{\log 1.052} + 2011 = t$$

$$2018.283 \approx t$$

Thus, according to the model, the population of Nevada will reach 3.939 million people in about the year 2018.

b. We need to determine the year when $P = 8.426$ million. So we solve the equation

$$8.426 = 2.723(1.052)^{t-2011}$$

$$\frac{8.426}{2.723} = (1.052)^{t-2011}$$

$$\log\frac{8.426}{2.723} = \log(1.052)^{t-2011}$$

$$\log\frac{8.426}{2.723} = (t-2011)\log(1.052)$$

$$\frac{\log\frac{8.426}{2.723}}{\log 1.052} = t - 2011$$

$$\frac{\log\frac{8.426}{2.723}}{\log 1.052} + 2011 = t$$

$$2033.283 \approx t$$

Thus, according to the model, the population of Nevada will reach 8.426 million people in about the year 2033.

Chapter 8 Test

1. The function is not one-to-one. There is an element in the range (4) that corresponds to more than one element in the domain (1 and -1).

2.
$$f(x) = 4x - 3$$
$$y = 4x - 3$$
$$x = 4y - 3$$
$$x + 3 = 4y$$
$$\frac{x+3}{4} = y$$
$$f^{-1}(x) = \frac{x+3}{4}$$

Check:

$$f\left(f^{-1}(x)\right) = 4\left(\frac{x+3}{4}\right) - 3 = x + 3 - 3 = x$$

$$f^{-1}\left(f(x)\right) = \frac{(4x-3)+3}{4} = \frac{4x}{4} = x$$

3. a. $3.1^{3.1} \approx 33.360$

b. $3.14^{3.14} \approx 36.338$

c. $3.142^{3.142} \approx 36.494$

d. $3.1416^{3.1416} \approx 36.463$

e. $\pi^{\pi} \approx 36.462$

4. If $4^x = 19$, then $\log_4 19 = x$.

5. If $\log_b x = y$, then $b^y = x$.

6. a. Let $y = \log_3\left(\frac{1}{27}\right)$. Then,

$$3^y = \frac{1}{27}$$
$$3^y = \frac{1}{3^3}$$
$$3^y = 3^{-3}$$
$$y = -3$$

Thus, $\log_3\left(\frac{1}{27}\right) = -3$.

b. Let $y = \log 10,000$. Then,

$$10^y = 10,000$$
$$10^y = 10^4$$
$$y = 4$$

Thus, $\log 10,000 = 4$.

7. The domain of $f(x) = \log_5(7 - 4x)$ is the set of all real numbers x such that

$$7 - 4x > 0$$
$$-4x > -7$$
$$x < \frac{-7}{-4}$$
$$x < \frac{7}{4}$$

Thus, the domain of $f(x) = \log_5(7-4x)$ is $\left\{x \mid x < \frac{7}{4}\right\}$ or, using interval notation, $\left(-\infty, \frac{7}{4}\right)$.

8. Locate some points on the graph of $f(x) = 6^x$.

x	$f(x) = 6^x$	$(x, f(x))$
-2	$f(-2) = 6^{-2} = \frac{1}{6^2} = \frac{1}{36}$	$\left(-2, \frac{1}{36}\right)$
-1	$f(-1) = 6^{-1} = \frac{1}{6^1} = \frac{1}{6}$	$\left(-1, \frac{1}{6}\right)$
0	$f(0) = 6^0 = 1$	$(0, 1)$
1	$f(1) = 6^1 = 6$	$(1, 6)$

Plot the points and connect them with a smooth curve.

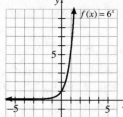

The domain of *f* is all real numbers or, using interval notation, $(-\infty, \infty)$. The range of *f* is $\{y \mid y > 0\}$ or, using interval notation, $(0, \infty)$.

9. Rewrite $y = g(x) = \log_{1/9} x$ as $x = \left(\dfrac{1}{9}\right)^y$.

Locate some points on the graph of $x = \left(\dfrac{1}{9}\right)^y$.

y	$x = \left(\dfrac{1}{9}\right)^y$	(x, y)
-1	$x = \left(\dfrac{1}{9}\right)^{-1} = \left(\dfrac{9}{1}\right)^1 = 9$	$(9, -1)$
0	$x = \left(\dfrac{1}{9}\right)^0 = 1$	$(1, 0)$
1	$x = \left(\dfrac{1}{9}\right)^1 = \dfrac{1}{9}$	$\left(\dfrac{1}{9}, 1\right)$
2	$x = \left(\dfrac{1}{9}\right)^2 = \dfrac{1}{81}$	$\left(\dfrac{1}{81}, 2\right)$

Plot the points and connect them with a smooth curve.

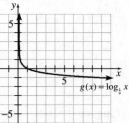

The domain of *g* is $\{x \mid x > 0\}$ or, using interval notation, $(0, \infty)$. The range of *g* is all real numbers or, using interval notation, $(-\infty, \infty)$.

10. a. $\log_7 7^{10} = 10$

    b. $3^{\log_3 15} = 15$

11. $\log_4 \dfrac{\sqrt{x}}{y^3} = \log_4 \sqrt{x} - \log_4 y^3$

$$= \log_4 x^{\frac{1}{2}} - \log_4 y^3$$

$$= \frac{1}{2}\log_4 x - 3\log_4 y$$

12. $4\log M + 3\log N = \log M^4 + \log N^3$

$$= \log\left(M^4 N^3\right)$$

13. Using common logarithms:

$$\log_{3/4} 10 = \frac{\log 10}{\log \dfrac{3}{4}} \approx -8.004$$

14.
$$4^{x+1} = 2^{3x+1}$$
$$\left(2^2\right)^{x+1} = 2^{3x+1}$$
$$2^{2x+2} = 2^{3x+1}$$
$$2x + 2 = 3x + 1$$
$$-x + 2 = 1$$
$$-x = -1$$
$$x = 1$$

The solution set is $\{1\}$.

15.
$$5^{x^2} \cdot 125 = 25^{2x}$$
$$5^{x^2} \cdot 5^3 = \left(5^2\right)^{2x}$$
$$5^{x^2+3} = 5^{4x}$$
$$x^2 + 3 = 4x$$
$$x^2 - 4x + 3 = 0$$
$$(x-1)(x-3) = 0$$
$$x - 1 = 0 \quad \text{or} \quad x - 3 = 0$$
$$x = 1 \quad \text{or} \quad x = 3$$

The solution set is $\{1, 3\}$.

16. $\log_a 64 = 3$
$$a^3 = 64$$
$$a = \sqrt[3]{64} = 4$$

The solution set is $\{4\}$.

17. $\log_2\left(x^2 - 33\right) = 8$

$$2^8 = x^2 - 33$$
$$256 = x^2 - 33$$
$$289 = x^2$$
$$x = \pm\sqrt{289} = \pm 17$$

The solution set is $\{-17,\ 17\}$.

18. $2\log_7(x-3) = \log_7 3 + \log_7 12$

$$\log_7(x-3)^2 = \log_7(3 \cdot 12)$$
$$\log_7(x-3)^2 = \log_7 36$$
$$(x-3)^2 = 36$$
$$x - 3 = \pm 6$$
$$x = 3 \pm 6$$
$$x = 9 \quad \text{or} \quad x = -3$$

The apparent solution $x = -3$ is extraneous because it causes the argument of a logarithm to be negative. The solution set is $\{9\}$.

19. $\qquad 3^{x-1} = 17$

$$\log 3^{x-1} = \log 17$$
$$(x-1)\log 3 = \log 17$$
$$x\log 3 - \log 3 = \log 17$$
$$x\log 3 = \log 17 + \log 3$$
$$x = \frac{\log 17 + \log 3}{\log 3} \approx 3.579$$

The solution set is $\left\{\dfrac{\log 17 + \log 3}{\log 3}\right\} \approx \{3.579\}$. If we had taken the natural logarithm of both sides, the solution set would be

$$\left\{\frac{\ln 17 + \ln 3}{\ln 3}\right\} \approx \{3.579\}.$$

20. $\log(x-2) + \log(x+2) = 2$

$$\log\left[(x-2)(x+2)\right] = 2$$
$$\log\left(x^2 - 4\right) = 2$$
$$10^2 = x^2 - 4$$
$$x^2 - 4 = 100$$
$$x^2 = 104$$
$$x = \pm\sqrt{104}$$
$$x = \pm 2\sqrt{26} \approx \pm 10.198$$

The apparent solution $x = -2\sqrt{26} \approx -10.198$ is extraneous because it causes the argument of a logarithm to be negative. The solution set is $\left\{2\sqrt{26}\right\} \approx \{10.198\}$.

21. a. Evaluate $P(t) = 34.1(1.008)^{t-2010}$ at $t = 2015$.

$$P(2015) = 34.1(1.008)^{2015-2010} \approx 35.486.$$

According to the model, the population of Canada in 2010 will be about 35.5 million people.

b. We need to determine the year when $P = 50$ million people. So we solve the equation

$$50 = 34.1(1.008)^{t-2010}$$
$$\frac{50}{34.1} = (1.008)^{t-2010}$$
$$\log\frac{50}{34.1} = \log(1.008)^{t-2010}$$
$$\log\frac{50}{34.1} = (t-2010)\log 1.008$$
$$\frac{\log\dfrac{50}{34.1}}{\log 1.008} = t - 2010$$
$$\frac{\log\dfrac{50}{34.1}}{\log 1.008} + 2010 = t$$
$$2058.032 \approx t$$

According to the model, the population of Canada will be 50 million people in about 2058.

22. We evaluate $L(x) = 10\log\dfrac{x}{10^{-12}}$ at $x = 10^{-11}$.

$$L\left(10^{-11}\right) = 10\log\frac{10^{-11}}{10^{-12}}$$
$$= 10\log 10^{-11-(-12)}$$
$$= 10\log 10^1$$
$$= 10(1)$$
$$= 10$$

The loudness of rustling leaves is 10 decibels.

Chapter 9

Are You Ready for This Section?

R1. **a.** $\sqrt{64} = 8$

  b. $\sqrt{24} = \sqrt{4 \cdot 6} = 2\sqrt{6}$

  c. $\sqrt{(x-2)^2} = |x-2|$

R2. $c^2 = a^2 + b^2$

$c^2 = 6^2 + 8^2$

$c^2 = 36 + 64$

$c^2 = 100$

$c = \sqrt{100} = 10$

The length of the hypotenuse is 10.

Section 9.1 Quick Checks

1. The distance between two points $P_1 = (x_1,\ y_1)$ and $P_2 = (x_2,\ y_2)$, denoted $d(P_1, P_2)$, is

$$d = \sqrt{(x_2 - x_1)^2 + (y_2 - y_1)^2}.$$

2. False. Distance must be positive.

3. Using the distance formula with $P_1 = (3,\ 8)$ and $P_2 = (0,\ 4)$, the length is

$$d(P_1, P_2) = \sqrt{(x_2 - x_1)^2 + (y_2 - y_1)^2}$$
$$= \sqrt{(0-3)^2 + (4-8)^2}$$
$$= \sqrt{(-3)^2 + (-4)^2}$$
$$= \sqrt{9+16}$$
$$= \sqrt{25}$$
$$= 5$$

4. Using the distance formula with $P_1 = (-2, -5)$ and $P_2 = (4,\ 7)$, the length is

$$d(P_1, P_2) = \sqrt{(x_2 - x_1)^2 + (y_2 - y_1)^2}$$
$$= \sqrt{(4-(-2))^2 + (7-(-5))^2}$$
$$= \sqrt{6^2 + 12^2}$$
$$= \sqrt{36 + 144}$$
$$= \sqrt{180}$$
$$= 6\sqrt{5} \approx 13.42$$

5. **a.**

  b. $d(A,B) = \sqrt{(4-(-2))^2 + (2-(-1))^2}$
$$= \sqrt{6^2 + 3^2}$$
$$= \sqrt{36 + 9}$$
$$= \sqrt{45}$$
$$= 3\sqrt{5}$$

$$d(B,C) = \sqrt{(0-4)^2 + (10-2)^2}$$
$$= \sqrt{(-4)^2 + 8^2}$$
$$= \sqrt{16 + 64}$$
$$= \sqrt{80}$$
$$= 4\sqrt{5}$$

$$d(A,C) = \sqrt{(0-(-2))^2 + (10-(-1))^2}$$
$$= \sqrt{2^2 + 11^2}$$
$$= \sqrt{4 + 121}$$
$$= \sqrt{125}$$
$$= 5\sqrt{5}$$

c. To determine if triangle ABC is a right triangle, we check to see if

$$[d(A,B)]^2 + [d(B,C)]^2 \overset{?}{=} [d(A,C)]^2$$

$$\left(3\sqrt{5}\right)^2 + \left(4\sqrt{5}\right)^2 \overset{?}{=} \left(5\sqrt{5}\right)^2$$

$$9 \cdot 5 + 16 \cdot 5 \overset{?}{=} 25 \cdot 5$$

$$45 + 80 \overset{?}{=} 125$$

$$125 = 125 \leftarrow \text{True}$$

Therefore, triangle ABC is a right triangle.

d. The length of the "base" of the triangle is $d(A,B) = 3\sqrt{5}$ and the length of the "height" of the triangle is $d(B,C) = 4\sqrt{5}$. Thus, the area of triangle ABC is

$$\text{Area} = \frac{1}{2} \cdot \text{base} \cdot \text{height} = \frac{1}{2} \cdot 3\sqrt{5} \cdot 4\sqrt{5} = 30$$

square units.

6. $M = \left(\dfrac{x_1 + x_2}{2}, \dfrac{y_1 + y_2}{2}\right)$

7. $M = \left(\dfrac{x_1 + x_2}{2}, \dfrac{y_1 + y_2}{2}\right)$

$= \left(\dfrac{3+0}{2}, \dfrac{8+4}{2}\right)$

$= \left(\dfrac{3}{2}, \dfrac{12}{2}\right)$

$= \left(\dfrac{3}{2}, 6\right)$

8. $M = \left(\dfrac{x_1 + x_2}{2}, \dfrac{y_1 + y_2}{2}\right)$

$= \left(\dfrac{-2+4}{2}, \dfrac{-5+10}{2}\right)$

$= \left(\dfrac{2}{2}, \dfrac{5}{2}\right)$

$= \left(1, \dfrac{5}{2}\right)$

9.1 Exercises

9. $d(P_1, P_2) = \sqrt{(x_2 - x_1)^2 + (y_2 - y_1)^2}$

$= \sqrt{(3-0)^2 + (4-0)^2}$

$= \sqrt{3^2 + 4^2}$

$= \sqrt{9+16}$

$= \sqrt{25}$

$= 5$

11. $d(P_1, P_2) = \sqrt{(x_2 - x_1)^2 + (y_2 - y_1)^2}$

$= \sqrt{(4-(-4))^2 + (1-5)^2}$

$= \sqrt{8^2 + (-4)^2}$

$= \sqrt{64+16}$

$= \sqrt{80}$

$= 4\sqrt{5} \approx 8.94$

13. $d(P_1, P_2) = \sqrt{(x_2 - x_1)^2 + (y_2 - y_1)^2}$

$= \sqrt{(6-2)^2 + (4-1)^2}$

$= \sqrt{4^2 + 3^2}$

$= \sqrt{16+9}$

$= \sqrt{25}$

$= 5$

15. $d(P_1, P_2) = \sqrt{(x_2 - x_1)^2 + (y_2 - y_1)^2}$

$= \sqrt{(9-(-3))^2 + (-3-2)^2}$

$= \sqrt{12^2 + (-5)^2}$

$= \sqrt{144+25}$

$= \sqrt{169}$

$= 13$

17. $d(P_1, P_2) = \sqrt{(x_2 - x_1)^2 + (y_2 - y_1)^2}$

$= \sqrt{(2-(-4))^2 + (2-2)^2}$

$= \sqrt{6^2 + 0^2}$

$= \sqrt{36+0}$

$= \sqrt{36}$

$= 6$

19. $d(P_1, P_2) = \sqrt{(x_2 - x_1)^2 + (y_2 - y_1)^2}$

$= \sqrt{(-3-0)^2 + (3-(-3))^2}$

$= \sqrt{(-3)^2 + 6^2}$

$= \sqrt{9+36}$

$= \sqrt{45}$

$= 3\sqrt{5} \approx 6.71$

21. $d(P_1, P_2) = \sqrt{(x_2 - x_1)^2 + (y_2 - y_1)^2}$

$= \sqrt{\left(5\sqrt{2} - 2\sqrt{2}\right)^2 + \left(4\sqrt{5} - \sqrt{5}\right)^2}$

$= \sqrt{\left(3\sqrt{2}\right)^2 + \left(3\sqrt{5}\right)^2}$

$= \sqrt{9(2) + 9(5)}$

$= \sqrt{18 + 45}$

$= \sqrt{63}$

$= 3\sqrt{7} \approx 7.94$

23. $d(P_1, P_2) = \sqrt{(x_2 - x_1)^2 + (y_2 - y_1)^2}$

$= \sqrt{(1.3 - 0.3)^2 + (0.1 - (-3.3))^2}$

$= \sqrt{1^2 + 3.4^2}$

$= \sqrt{1 + 11.56}$

$= \sqrt{12.56} \approx 3.54$

25. $M = \left(\dfrac{x_1 + x_2}{2}, \dfrac{y_1 + y_2}{2}\right)$

$= \left(\dfrac{2+6}{2}, \dfrac{2+4}{2}\right) = \left(\dfrac{8}{2}, \dfrac{6}{2}\right) = (4, 3)$

27. $M = \left(\dfrac{x_1 + x_2}{2}, \dfrac{y_1 + y_2}{2}\right)$

$= \left(\dfrac{-3+9}{2}, \dfrac{2+(-4)}{2}\right) = \left(\dfrac{6}{2}, \dfrac{-2}{2}\right) = (3, -1)$

29. $M = \left(\dfrac{x_1 + x_2}{2}, \dfrac{y_1 + y_2}{2}\right)$

$= \left(\dfrac{-4+2}{2}, \dfrac{3+4}{2}\right) = \left(\dfrac{-2}{2}, \dfrac{7}{2}\right) = \left(-1, \dfrac{7}{2}\right)$

31. $M = \left(\dfrac{x_1 + x_2}{2}, \dfrac{y_1 + y_2}{2}\right)$

$= \left(\dfrac{0+(-3)}{2}, \dfrac{-3+3}{2}\right) = \left(\dfrac{-3}{2}, \dfrac{0}{2}\right) = \left(-\dfrac{3}{2}, 0\right)$

33. $M = \left(\dfrac{x_1 + x_2}{2}, \dfrac{y_1 + y_2}{2}\right)$

$= \left(\dfrac{2\sqrt{2} + 5\sqrt{2}}{2}, \dfrac{\sqrt{5} + 4\sqrt{5}}{2}\right) = \left(\dfrac{7\sqrt{2}}{2}, \dfrac{5\sqrt{5}}{2}\right)$

35. $M = \left(\dfrac{x_1 + x_2}{2}, \dfrac{y_1 + y_2}{2}\right)$

$= \left(\dfrac{0.3 + 1.3}{2}, \dfrac{-3.3 + 0.1}{2}\right)$

$= \left(\dfrac{1.6}{2}, \dfrac{-3.2}{2}\right)$

$= (0.8, -1.6)$

37. a.

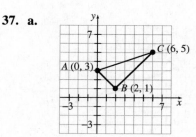

b. $d(A, B) = \sqrt{(2-0)^2 + (1-3)^2}$

$= \sqrt{2^2 + (-2)^2}$

$= \sqrt{4 + 4}$

$= \sqrt{8} = 2\sqrt{2} \approx 2.83$

$d(B, C) = \sqrt{(6-2)^2 + (5-1)^2}$

$= \sqrt{4^2 + 4^2}$

$= \sqrt{16 + 16}$

$= \sqrt{32} = 4\sqrt{2} \approx 5.66$

$d(A, C) = \sqrt{(6-0)^2 + (5-3)^2}$

$= \sqrt{6^2 + 2^2}$

$= \sqrt{36 + 4}$

$= \sqrt{40} = 2\sqrt{10} \approx 6.32$

c. To verify that triangle ABC is a right triangle, we show that

$$[d(A, B)]^2 + [d(B, C)]^2 \overset{?}{=} [d(A, C)]^2$$

$$\left(2\sqrt{2}\right)^2 + \left(4\sqrt{2}\right)^2 \overset{?}{=} \left(2\sqrt{10}\right)^2$$

$$4 \cdot 2 + 16 \cdot 2 \overset{?}{=} 4 \cdot 10$$

$$8 + 32 \overset{?}{=} 40$$

$$40 = 40 \leftarrow \text{True}$$

Therefore, triangle ABC is a right triangle.

d. The length of the "base" of the triangle is $d(B,C) = 4\sqrt{2}$ and the length of the "height" of the triangle is $d(A,B) = 2\sqrt{2}$. Thus, the area of triangle ABC is

$$\text{Area} = \frac{1}{2} \cdot \text{base} \cdot \text{height} = \frac{1}{2} \cdot 4\sqrt{2} \cdot 2\sqrt{2} = 8$$

square units.

39. a.

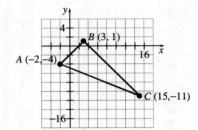

b.
$$d(A,B) = \sqrt{(3-(-2))^2 + (1-(-4))^2}$$
$$= \sqrt{5^2 + 5^2}$$
$$= \sqrt{25 + 25}$$
$$= \sqrt{50}$$
$$= 5\sqrt{2} \approx 7.07$$

$$d(B,C) = \sqrt{(15-3)^2 + (-11-1)^2}$$
$$= \sqrt{12^2 + (-12)^2}$$
$$= \sqrt{144 + 144}$$
$$= \sqrt{288}$$
$$= 12\sqrt{2} \approx 16.97$$

$$d(A,C) = \sqrt{(15-(-2))^2 + (-11-(-4))^2}$$
$$= \sqrt{17^2 + (-7)^2}$$
$$= \sqrt{289 + 49}$$
$$= \sqrt{338}$$
$$= 13\sqrt{2} \approx 18.38$$

c. To verify that triangle ABC is a right triangle, we show that

$$[d(A,B)]^2 + [d(B,C)]^2 \overset{?}{=} [d(A,C)]^2$$
$$\left(5\sqrt{2}\right)^2 + \left(12\sqrt{2}\right)^2 \overset{?}{=} \left(13\sqrt{2}\right)^2$$
$$25 \cdot 2 + 144 \cdot 2 \overset{?}{=} 169 \cdot 2$$
$$50 + 288 \overset{?}{=} 338$$
$$338 = 338 \leftarrow \text{True}$$

Therefore, triangle ABC is a right triangle.

d. The length of the "base" of the triangle is $d(B,C) = 12\sqrt{2}$ and the length of the "height" of the triangle is $d(A,B) = 5\sqrt{2}$. Thus, the area of triangle ABC is

$$\text{Area} = \frac{1}{2} \cdot \text{base} \cdot \text{height} = \frac{1}{2} \cdot 12\sqrt{2} \cdot 5\sqrt{2} = 60$$

square units.

41. We want to find y such that the distance between $P_1 = (5,1)$ and $P_2 = (2,y)$ is 5.

$$d(P_1, P_2) = \sqrt{(x_2 - x_1)^2 + (y_2 - y_1)^2}$$
$$5 = \sqrt{(2-5)^2 + (y-1)^2}$$
$$5 = \sqrt{(-3)^2 + (y-1)^2}$$
$$5 = \sqrt{9 + (y-1)^2}$$
$$25 = 9 + (y-1)^2$$
$$16 = (y-1)^2$$
$$y - 1 = \pm\sqrt{16}$$
$$y - 1 = \pm 4$$
$$y = 1 \pm 4$$
$$y = -3 \ \text{ or } \ y = 5$$

Thus, the distance between the points $(5,1)$ and $(2,-3)$ is 5, and the distance between $(5,1)$ and $(2,5)$ is 5.

43. We want to find x such that the distance between $P_1 = (2,3)$ and $P_2 = (x,-3)$ is 10.

$$d(P_1, P_2) = \sqrt{(x_2 - x_1)^2 + (y_2 - y_1)^2}$$
$$10 = \sqrt{(x-2)^2 + (-3-3)^2}$$
$$10 = \sqrt{(x-2)^2 + (-6)^2}$$
$$10 = \sqrt{(x-2)^2 + 36}$$
$$100 = (x-2)^2 + 36$$
$$64 = (x-2)^2$$
$$x - 2 = \pm\sqrt{64}$$
$$x - 2 = \pm 8$$
$$x = 2 \pm 8$$
$$x = -6 \ \text{ or } \ x = 10$$

Thus, the distance between the points $(2,3)$ and $(-6,-3)$ is 10, and the distance between $(2,3)$ and $(10,-3)$ is 10.

45. a. Treating the intersection of Madison and State Streets as the origin, $(0,0)$, of a Cartesian plane, Wrigley Field is located at the coordinates $(-10,36)$.

$$d = \sqrt{(-10-0)^2 + (36-0)^2}$$
$$= \sqrt{(-10)^2 + 36^2}$$
$$= \sqrt{100+1296}$$
$$= \sqrt{1396}$$
$$= 2\sqrt{349} \approx 37.36$$

The distance "as the crow flies" from Madison and State Street to Wrigley Field is approximately 37.36 blocks.

b. U.S. Cellular Field is located at the coordinates $(-3,-35)$.

$$d = \sqrt{(-3-0)^2 + (-35-0)^2}$$
$$= \sqrt{(-3)^2 + (-35)^2}$$
$$= \sqrt{9+1225}$$
$$= \sqrt{1234} \approx 35.13$$

The distance "as the crow flies" from Madison and State Street to U.S. Cellular Field is approximately 35.13 blocks.

c. $d = \sqrt{(-10-(-3))^2 + (36-(-35))^2}$
$$= \sqrt{(-7)^2 + 71^2}$$
$$= \sqrt{49+5041}$$
$$= \sqrt{5090} \approx 71.34$$

The distance "as the crow flies" from Wrigley Field to U.S. Cellular Field is approximately 71.34 blocks.

47. a. The distance from the right fielder to second base is

$$d = \sqrt{(320-90)^2 + (20-90)^2}$$
$$= \sqrt{230^2 + (-70)^2}$$
$$= \sqrt{52,900+4900}$$
$$= \sqrt{57,800}$$
$$= 170\sqrt{2} \approx 240.42 \text{ feet}$$

Now, $\text{time} = \dfrac{\text{distance}}{\text{speed}} \approx \dfrac{240.42}{130} \approx 1.85$.

It will take about 1.85 seconds for the ball to reach the second baseman from the right fielder.

b. The time for the runner to reach second base will be

$$\text{time} = \frac{\text{distance}}{\text{speed}} = \frac{90}{27} \approx 3.33 \text{ seconds.}$$

The total time for the ball to reach second base from the time the right fielder catches it will be $1.85 + 0.8 = 2.65$ seconds. Since the ball will reach second base before the runner, the first base coach should not send the runner.

49. Answers may vary. One possibility follows: The distance formula is derived from the Pythagorean Theorem. The absolute value of the difference between the x-coordinates, $|x_2 - x_1|$, is the length of one leg of a right triangle. The absolute value of the difference between the y-coordinates, $|y_2 - y_1|$, is the length of the other leg of the right triangle. The length of the hypotenuse is $d(P_1, P_2)$. By the Pythagorean Theorem,

$$\left[d(P_1, P_2)\right]^2 = |x_2 - x_1|^2 + |y_2 - y_1|^2$$
$$= (x_2 - x_1)^2 + (y_2 - y_1)^2$$
$$d(P_1, P_2) = \sqrt{(x_2 - x_1)^2 + (y_2 - y_1)^2}$$

51. $3^2 = 3 \cdot 3 = 9$; $\sqrt{9} = 3$

53. $(-3)^4 = (-3)(-3)(-3)(-3) = 81$; $\sqrt[4]{81} = |-3| = 3$

55. If n is a positive integer and $a^n = b$, then
$$\sqrt[n]{b} = \begin{cases} |a|, & \text{if } n \text{ is even} \\ a, & \text{if } n \text{ is odd} \end{cases}$$

Section 9.2

Are You Ready for This Section?

R1. Start: $x^2 - 8x$

Add: $\left(\dfrac{1}{2} \cdot (-8)\right)^2 = 16$

Result: $x^2 - 8x + 16$

Factored Form: $(x-4)^2$

Section 9.2 Quick Checks

1. A <u>circle</u> is the set of all points in the Cartesian plane that are a fixed distance r from a fixed point (h, k).

2. For a circle, the <u>radius</u> is the distance from the center to any point on the circle.

3. We are given that $h = 2$, $k = 4$, and $r = 5$. Thus, the equation of the circle is
$$(x-h)^2 + (y-k)^2 = r^2$$
$$(x-2)^2 + (y-4)^2 = 5^2$$
$$(x-2)^2 + (y-4)^2 = 25$$

4. We are given that $h = -2$, $k = 0$, and $r = \sqrt{2}$. Thus, the equation of the circle is
$$(x-h)^2 + (y-k)^2 = r^2$$
$$(x-(-2))^2 + (y-0)^2 = (\sqrt{2})^2$$
$$(x+2)^2 + y^2 = 2$$

5. False. The center is $(-1, 3)$.

6. True.

7. $(x-3)^2 + (y-1)^2 = 4$
$(x-3)^2 + (y-1)^2 = 2^2$
The center is $(h,k) = (3, 1)$; the radius is $r = 2$.

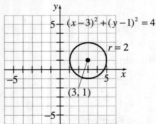

8. $(x+5)^2 + y^2 = 16$
$(x-(-5))^2 + (y-0)^2 = 4^2$
The center is $(h, k) = (-5, 0)$; the radius is $r = 4$.

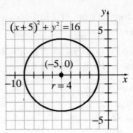

9. $x^2 + y^2 - 6x - 4y + 4 = 0$
$(x^2 - 6x) + (y^2 - 4y) = -4$
$(x^2 - 6x + 9) + (y^2 - 4y + 4) = -4 + 9 + 4$
$(x-3)^2 + (y-2)^2 = 9$
$(x-3)^2 + (y-2)^2 = 3^2$
The center is $(h, k) = (3, 2)$; the radius is $r = 3$.

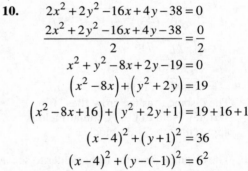

10. $2x^2 + 2y^2 - 16x + 4y - 38 = 0$
$$\frac{2x^2 + 2y^2 - 16x + 4y - 38}{2} = \frac{0}{2}$$
$x^2 + y^2 - 8x + 2y - 19 = 0$
$(x^2 - 8x) + (y^2 + 2y) = 19$
$(x^2 - 8x + 16) + (y^2 + 2y + 1) = 19 + 16 + 1$
$(x-4)^2 + (y+1)^2 = 36$
$(x-4)^2 + (y-(-1))^2 = 6^2$
The center is $(h, k) = (4, -1)$; the radius is $r = 6$.

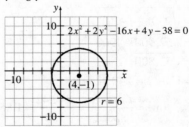

9.2 Exercises

11. The center of the circle is $(1, 2)$. The radius of the circle is the distance from the center point to the point $(-1, 2)$ on the circle. Thus,

$$r = \sqrt{(1-(-1))^2 + (2-2)^2} = 2.$$

The equation of the circle is

$$(x-h)^2 + (y-k)^2 = r^2$$
$$(x-1)^2 + (y-2)^2 = 2^2$$
$$(x-1)^2 + (y-2)^2 = 4$$

13. The center of the circle will be the midpoint of the line segment with endpoints $(-2, -1)$ and $(6, -1)$.

Thus, $(h, k) = \left(\dfrac{-2+6}{2}, \dfrac{-1+(-1)}{2} \right) = (2, -1)$. The

radius of the circle will be the distance from the center point $(2, -1)$ to a point on the circle, say

$(6, -1)$. Thus, $r = \sqrt{(6-2)^2 + (-1-(-1))^2} = 4$.

The equation of the circle is

$$(x-h)^2 + (y-k)^2 = r^2$$
$$(x-2)^2 + (y-(-1))^2 = 4^2$$
$$(x-2)^2 + (y+1)^2 = 16$$

15. $(x-h)^2 + (y-k)^2 = r^2$
$$(x-0)^2 + (y-0)^2 = 3^2$$
$$x^2 + y^2 = 9$$

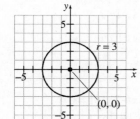

17. $(x-h)^2 + (y-k)^2 = r^2$
$$(x-1)^2 + (y-4)^2 = 2^2$$
$$(x-1)^2 + (y-4)^2 = 4$$

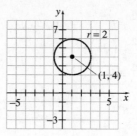

19. $(x-h)^2 + (y-k)^2 = r^2$
$$(x-(-2))^2 + (y-4)^2 = 6^2$$
$$(x+2)^2 + (y-4)^2 = 36$$

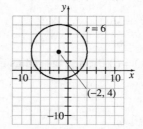

21. $(x-h)^2 + (y-k)^2 = r^2$
$$(x-0)^2 + (y-3)^2 = 4^2$$
$$x^2 + (y-3)^2 = 16$$

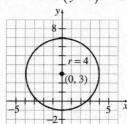

23. $(x-h)^2 + (y-k)^2 = r^2$
$$(x-5)^2 + (y-(-5))^2 = 5^2$$
$$(x-5)^2 + (y+5)^2 = 25$$

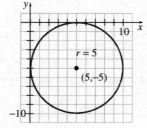

25. $(x-h)^2 + (y-k)^2 = r^2$

$(x-1)^2 + (y-2)^2 = \left(\sqrt{5}\right)^2$

$(x-1)^2 + (y-2)^2 = 5$

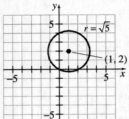

27. $x^2 + y^2 = 36$

$(x-0)^2 + (y-0)^2 = 6^2$

The center is $(h,k) = (0,0)$, and the radius is

$r = 6$.

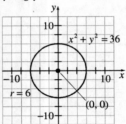

29. $(x-4)^2 + (y-1)^2 = 25$

$(x-4)^2 + (y-1)^2 = 5^2$

The center is $(h,k) = (4,1)$, and the radius is

$r = 5$.

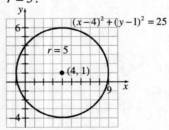

31. $(x+3)^2 + (y-2)^2 = 81$

$(x-(-3))^2 + (y-2)^2 = 9^2$

The center is $(h,k) = (-3,2)$, and the radius is

$r = 9$.

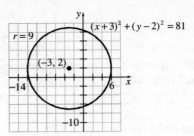

33. $x^2 + (y-3)^2 = 64$

$(x-0)^2 + (y-3)^2 = 8^2$

The center is $(h,k) = (0,3)$, and the radius is

$r = 8$.

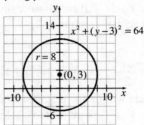

35. $(x-1)^2 + (y+1)^2 = \dfrac{1}{4}$

$(x-1)^2 + (y-(-1))^2 = \left(\dfrac{1}{2}\right)^2$

The center is $(h,k) = (1,-1)$, and the radius is

$r = \dfrac{1}{2}$.

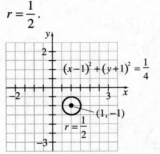

37. $x^2 + y^2 - 6x + 2y + 1 = 0$

$\left(x^2 - 6x\right) + \left(y^2 + 2y\right) = -1$

$\left(x^2 - 6x + 9\right) + \left(y^2 + 2y + 1\right) = -1 + 9 + 1$

$(x-3)^2 + (y+1)^2 = 9$

$(x-3)^2 + (y-(-1))^2 = 3^2$

The center is $(h,k) = (3,-1)$, and the radius is

$r = 3$.

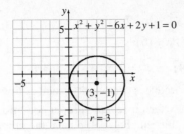

39.
$$x^2 + y^2 + 10x + 4y + 4 = 0$$
$$\left(x^2 + 10x\right) + \left(y^2 + 4y\right) = -4$$
$$\left(x^2 + 10x + 25\right) + \left(y^2 + 4y + 4\right) = -4 + 25 + 4$$
$$(x+5)^2 + (y+2)^2 = 25$$
$$\left(x - (-5)\right)^2 + \left(y - (-2)\right)^2 = 5^2$$

The center is $(h,k) = (-5,-2)$, and the radius is $r = 5$.

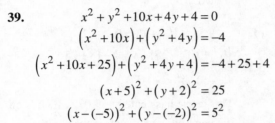

41.
$$2x^2 + 2y^2 - 12x + 24y - 72 = 0$$
$$\frac{2x^2 + 2y^2 - 12x + 24y - 72}{2} = \frac{0}{2}$$
$$x^2 + y^2 - 6x + 12y - 36 = 0$$
$$\left(x^2 - 6x\right) + \left(y^2 + 12y\right) = 36$$
$$\left(x^2 - 6x + 9\right) + \left(y^2 + 12y + 36\right) = 36 + 9 + 36$$
$$(x-3)^2 + (y+6)^2 = 81$$
$$(x-3)^2 + \left(y - (-6)\right)^2 = 9^2$$

The center is $(h,k) = (3,-6)$, and the radius is $r = 9$.

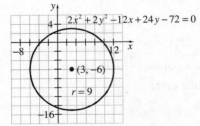

43. The radius of the circle will be the distance from the center point $(0,0)$ to the point on the circle $(4,-2)$. Thus,
$$r = \sqrt{(4-0)^2 + (-2-0)^2} = \sqrt{20} = 2\sqrt{5}.$$
The equation of the circle is
$$(x-h)^2 + (y-k)^2 = r^2$$
$$(x-0)^2 + (y-0)^2 = \left(2\sqrt{5}\right)^2$$
$$x^2 + y^2 = 20$$

45. Since the center of the circle is $(-3,2)$ and since the circle is tangent to the *y*-axis, the circle must contain the point $(0,2)$. The radius of the circle will be the distance from the center point $(-3,2)$ to the point on the circle $(0,2)$. Thus,
$$r = \sqrt{(-3-0)^2 + (2-2)^2} = 3.$$
The equation of the circle is
$$(x-h)^2 + (y-k)^2 = r^2$$
$$\left(x - (-3)\right)^2 + (y-2)^2 = 3^2$$
$$(x+3)^2 + (y-2)^2 = 9$$

47. The center of the circle will be the midpoint of the diameter with endpoints $(2,3)$ and $(-4,-5)$.

Thus, $(h,k) = \left(\dfrac{2 + (-4)}{2}, \dfrac{3 + (-5)}{2}\right) = (-1,-1)$.

The radius of the circle will be the distance from the center point $(-1,-1)$ to one of the endpoints of the diameter, say $(2,3)$. Thus,
$$r = \sqrt{\left(2 - (-1)\right)^2 + \left(3 - (-1)\right)^2} = 5.$$
The equation of the circle is
$$(x-h)^2 + (y-k)^2 = r^2$$
$$\left(x - (-1)\right)^2 + \left(y - (-1)\right)^2 = 5^2$$
$$(x+1)^2 + (y+1)^2 = 25$$

49. The radius of the circle $(x-3)^2 + (y-8)^2 = 64$ is $r = \sqrt{64} = 8$. Thus, the area of the circle is
$$A = \pi r^2 = \pi(8)^2 = 64\pi \text{ square units.}$$
The circumference of the circle is
$$C = 2\pi r = 2\pi(8) = 16\pi \text{ units.}$$

51. To find the area of the square, we must first find the length of the square's sides. To do so, we recognize that the x- and y-coordinates of the corner point in the first quadrant where the square touches the circle must be equal. That is, $x = y$.

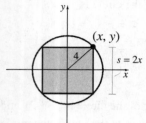

Thus, $x^2 + y^2 = 16$

$x^2 + x^2 = 16$

$2x^2 = 16$

$x^2 = 8$

$x = \sqrt{8} = 2\sqrt{2}$

This means the length of each side of the square is $s = 2x = 2\left(2\sqrt{2}\right) = 4\sqrt{2}$, and the area of the square is $A = s^2 = \left(4\sqrt{2}\right)^2 = 32$ square units.

53. The circle shown has its center point on the positive x-axis. Thus, the center point will be of the form $(h, 0)$ with $h > 0$. Also, the entire circle stays to the right of the y-axis, so $h > r$.

a. $(x-2)^2 + y^2 = 1$ has center $(2, 0)$ and radius $r = 1$, which meets the proper conditions. Thus, this equation <u>could</u> have the graph shown.

b. $x^2 + (y-2)^2 = 1$ has center $(0, 2)$, which is not on the x-axis. Thus, this equation could <u>not</u> have the graph shown.

c. $(x+4)^2 + y^2 = 9$ has center $(-4, 0)$, which is on the negative x-axis, not on the positive x-axis. Thus, this equation could <u>not</u> have the graph shown.

d. $(x-5)^2 + y^2 = 25$ has center $(5, 0)$ and radius $r = 5$. This means the graph of this circle would pass through the origin. Thus, this equation could <u>not</u> have the graph shown.

e. $x^2 + y^2 - 8x + 7 = 0$

$\left(x^2 - 8x + 16\right) + y^2 = -7 + 16$

$(x-4)^2 + y^2 = 9$

This circle has center $(4, 0)$ and radius $r = 3$, which meets the proper conditions. Thus, this equation <u>could</u> have the graph shown.

f. $x^2 + y^2 + 10x + 18 = 0$

$\left(x^2 + 10x + 25\right) + y^2 = -18 + 25$

$(x+5)^2 + y^2 = 7$

This circle has center $(-5, 0)$, which is on the negative x-axis, not on the positive x-axis. Thus, this equation could <u>not</u> have the graph shown.

55. Answers may vary. One possibility follows: If a circle has center (h, k) and radius r, then the distance from the center to any point (x, y) on that circle must always be r. By the distance formula, we have $r = \sqrt{(x-h)^2 + (y-k)^2}$.

Now, if we square both sides of this formula, we obtain the standard equation of the circle: $(x-h)^2 + (y-k)^2 = r^2$.

57. Yes, $x^2 = 36 - y^2$ is the equation of a circle. In standard form, the equation is

$$x^2 + y^2 = 36$$

$$(x-0)^2 + (y-0)^2 = 6^2$$

The center is $(0, 0)$, and the radius is 6.

59. $f(x) = 4x - 3$

To use point plotting, let $x = -1$, 0, and 1.

$f(-1) = 4(-1) - 3 = -4 - 3 = -7$

$f(0) = 4(0) - 3 = 0 - 3 = -3$

$f(1) = 4(1) - 3 = 4 - 3 = 1$

Thus, the points $(-1, -7)$, $(0, -3)$, and $(1, 1)$ are on the graph.

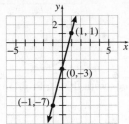

Using properties of linear functions, for $f(x) = 4x - 3$, the slope is 4 and the y-intercept is -3. Begin at $(0, -3)$ and move to the right 1 unit and up 4 units to find point $(1, 1)$.

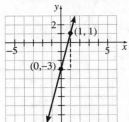

61. $g(x) = x^2 - 4x - 5$

To use point plotting, let $x = 0, 1, 2, 3,$ and 4.

$g(0) = 0^2 - 4(0) - 5 = 0 - 0 - 5 = -5$

$g(1) = 1^2 - 4(1) - 5 = 1 - 4 - 5 = -8$

$g(2) = 2^2 - 4(2) - 5 = 4 - 8 - 5 = -9$

$g(3) = 3^2 - 4(3) - 5 = 9 - 12 - 5 = -8$

$g(4) = 4^2 - 4(4) - 5 = 16 - 16 - 5 = -5$

Thus, the points $(0, -5)$, $(1, -8)$, $(2, -9)$, $(3, -8)$, and $(4, -5)$ are on the graph.

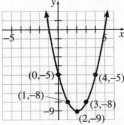

Using properties of quadratic functions, for $g(x) = x^2 - 4x - 5$, $a = 1$, $b = -4$, and $c = -5$. The graph opens up because the coefficient on x^2 is positive.

vertex:

$$x = -\frac{b}{2a} = -\frac{(-4)}{2(1)} = 2$$

$g(2) = (2)^2 - 4(2) - 5 = -9$

The vertex is $(2, -9)$ and the axis of symmetry is $x = 2$.

y-intercept:

$$g(0) = (0)^2 - 4(0) - 5 = -5$$

x-intercepts:

$$b^2 - 4ac = (-4)^2 - 4(1)(-5) = 36 > 0$$

There are two distinct x-intercepts. We find these by solving

$$g(x) = 0$$
$$x^2 - 4x - 5 = 0$$
$$(x - 5)(x + 1) = 0$$
$$x - 5 = 0 \text{ or } x + 1 = 0$$
$$x = 5 \text{ or } x = -1$$

Graph:

The y-intercept point, $(0, -5)$, is two units to the left of the axis of symmetry. Therefore, if we move two units to the right of the axis of symmetry, we obtain the point $(4, -5)$ which must also be on the graph.

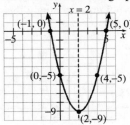

63. $G(x) = -2(x + 3)^2 - 5$

To use point plotting, let $x = -4, -3,$ and -2.

$$G(-4) = -2(-4 + 3)^2 - 5$$
$$= -2(-1)^2 - 5$$
$$= -2(1) - 5$$
$$= -2 - 5$$
$$= -7$$

$$G(-3) = -2(-3 + 3)^2 - 5$$
$$= -2(0)^2 - 5$$
$$= -2(0) - 5$$
$$= 0 - 5$$
$$= -5$$

$$G(-2) = -2(-2+3)^2 - 5$$
$$= -2(1)^2 - 5$$
$$= -2(1) - 5$$
$$= -2 - 5$$
$$= -7$$

Thus, the points $(-4,-7)$, $(-3,-5)$, and $(-2,-7)$ are on the graph.

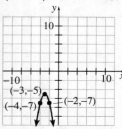

Using properties of quadratic functions, begin with the graph of $y = x^2$, then shift the graph 3 units to the left to obtain the graph of $y = (x+3)^2$. Multiply the y-coordinates by -2 to obtain the graph of $y = -2(x+3)^2$. Lastly, shift the graph down 5 units to obtain the graph of $G(x) = -2(x+3)^2 - 5$.

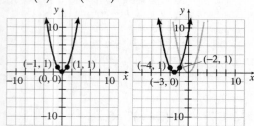

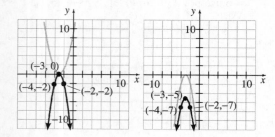

65. $x^2 + y^2 = 36$
$$y^2 = 36 - x^2$$
$$y = \pm\sqrt{36 - x^2}$$

Let $Y_1 = \sqrt{36 - x^2}$ and $Y_2 = -\sqrt{36 - x^2}$.

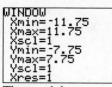
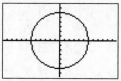

The graph here agrees with that in Problem 27.

67. $(x-4)^2 + (y-1)^2 = 25$
$$(y-1)^2 = 25 - (x-4)^2$$
$$y - 1 = \pm\sqrt{25 - (x-4)^2}$$
$$y = 1 \pm \sqrt{25 - (x-4)^2}$$

Let $Y_1 = 1 + \sqrt{25 - (x-4)^2}$ and $Y_2 = 1 - \sqrt{25 - (x-4)^2}$.

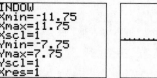

The graph here agrees with that in Problem 29.

69. $(x+3)^2 + (y-2)^2 = 81$
$$(y-2)^2 = 81 - (x+3)^2$$
$$y - 2 = \pm\sqrt{81 - (x+3)^2}$$
$$y = 2 \pm \sqrt{81 - (x+3)^2}$$

Let $Y_1 = 2 + \sqrt{81 - (x+3)^2}$ and $Y_2 = 2 - \sqrt{81 - (x+3)^2}$.

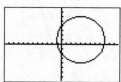

The graph here agrees with that in Problem 31.

71. $x^2 + (y-3)^2 = 64$
$$(y-3)^2 = 64 - x^2$$
$$y - 3 = \pm\sqrt{64 - x^2}$$
$$y = 3 \pm \sqrt{64 - x^2}$$

Let $Y_1 = 3 + \sqrt{64 - x^2}$ and $Y_2 = 3 - \sqrt{64 - x^2}$.

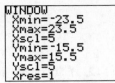

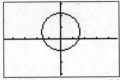

The graph here agrees with that in Problem 33.

73. $(x-1)^2 + (y+1)^2 = \dfrac{1}{4}$

$$(y+1)^2 = \dfrac{1}{4} - (x-1)^2$$

$$y+1 = \pm\sqrt{\dfrac{1}{4} - (x-1)^2}$$

$$y = -1 \pm \sqrt{\dfrac{1}{4} - (x-1)^2}$$

Let $Y_1 = -1 + \sqrt{\dfrac{1}{4} - (x-1)^2}$ and

$Y_2 = -1 - \sqrt{\dfrac{1}{4} - (x-1)^2}$.

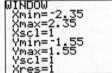

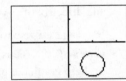

The graph here agrees with that in Problem 35.

Section 9.3

Are You Ready for This Section?

R1. For the function $f(x) = -3(x+4)^2 - 5$, we see that $a = -3$, $h = -4$, and $k = -5$. Thus, the vertex is $(h,\ k) = (-4, -5)$, and the axis of symmetry is $x = -4$. The parabola opens down since $a = -3 < 0$.

R2. For the function $f(x) = 2x^2 - 8x + 1$, we see that $a = 2$, $b = -8$, and $c = 1$. The x-coordinate of the vertex is $x = -\dfrac{b}{2a} = -\dfrac{(-8)}{2(2)} = 2$. The y-coordinate of the vertex is

$$f\left(-\dfrac{b}{2a}\right) = f(2)$$
$$= 2(2)^2 - 8(2) + 1$$
$$= 8 - 16 + 1$$
$$= -7$$

Thus, the vertex is $(2, -7)$ and the axis of symmetry is the line $x = 2$. The parabola opens up because $a = 2 > 0$.

R3. Start: $x^2 - 12x$

Add: $\left(\dfrac{1}{2} \cdot (-12)\right)^2 = 36$

Result: $x^2 - 12x + 36$

Factored Form: $(x-6)^2$

R4. $(x-3)^2 = 25$

$$x - 3 = \pm\sqrt{25}$$
$$x - 3 = \pm 5$$
$$x = 3 \pm 5$$
$$x = 3 - 5 \quad \text{or} \quad x = 3 + 5$$
$$x = -2 \quad \text{or} \quad x = 8$$

The solution set is $\{-2,\ 8\}$.

Section 9.3 Quick Checks

1. A <u>parabola</u> is the collection of all points P in the plane that are the same distance from a fixed point F as they are from a fixed line D.

2. The point of intersection of the parabola with its axis of symmetry is called the <u>vertex</u>.

3. The line through the focus and perpendicular to the directrix is called the <u>axis of symmetry</u>.

4. Notice that $y^2 = 8x$ is of the form $y^2 = 4ax$, where $4a = 8$, so that $a = 2$. Now, the graph of an equation of the form $y^2 = 4ax$ will be a parabola that opens to the right with the vertex at the origin, focus at $(a,\ 0)$, and directrix of $x = -a$. Thus, the graph of $y^2 = 8x$ is a parabola that opens to the right with vertex $(0,0)$, focus $(2,\ 0)$, and directrix $x = -2$. To help graph the parabola, we plot the two points on the graph above and below the focus. Let $x = 2$:

$$y^2 = 8(2)$$
$$y^2 = 16$$
$$y = \pm 4$$

The points $(2, -4)$ and $(2,\ 4)$ are on the graph.

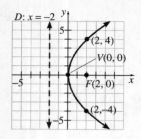

5. Notice that $y^2 = -20x$ is of the form

$y^2 = -4ax$, where $-4a = -20$, so that $a = 5$. Now, the graph of an equation of the form

$y^2 = -4ax$ will be a parabola that opens to the left with the vertex at the origin, focus at $(-a,\ 0)$, and directrix of $x = a$. Thus, the graph of $y^2 = -20x$ is a parabola that opens to the left with vertex $(0,0)$, focus $(-5,\ 0)$, and directrix $x = 5$. To help graph the parabola, we plot the two points on the graph above and below the focus. Let $x = 5$:

$$y^2 = -20(-5)$$
$$y^2 = 100$$
$$y = \pm 10$$

The points $(-5,-10)$ and $(-5,\ 10)$ are on the graph.

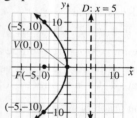

6. Notice that $x^2 = 4y$ is of the form $x^2 = 4ay$, where $4a = 4$, so that $a = 1$. Now, the graph of an equation of the form $x^2 = 4ay$ will be a parabola that opens up with the vertex at the origin, focus at $(0,\ a)$, and directrix of $y = -a$.

Thus, the graph of $x^2 = 4y$ is a parabola that opens up with vertex $(0,0)$, focus $(0,\ 1)$, and directrix $y = -1$. To help graph the parabola, we plot the two points on the graph to the left and to the right of the focus. Let $y = 1$:

$$x^2 = 4(1)$$
$$x^2 = 4$$
$$x = \pm 2$$

The points $(-2,\ 1)$ and $(2,\ 1)$ are on the graph.

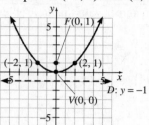

7. Notice that $x^2 = -12y$ is of the form

$x^2 = -4ay$, where $-4a = -12$, so that $a = 3$. Now, the graph of an equation of the form $x^2 = -4ay$ will be a parabola that opens down with the vertex at the origin, focus at $(0,-a)$, and directrix of $y = a$. Thus, the graph of

$x^2 = -12y$ is a parabola that opens down with vertex $(0,0)$, focus $(0,-3)$, and directrix $y = 3$. To help graph the parabola, we plot the two points on the graph to the left and to the right of the focus. Let $y = -3$:

$$x^2 = -12(-3)$$
$$x^2 = 36$$
$$x = \pm 6$$

The points $(-6,-3)$ and $(6,-3)$ are on the graph.

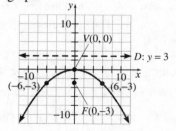

8. The distance from the vertex $(0,0)$ to the focus $(0,-8)$ is $a = 8$. Because the focus lies on the negative y-axis, we know that the parabola will open down and the axis of symmetry is the y-axis. This means the equation of the parabola is of the form $x^2 = -4ay$ with $a = 8$:

$$x^2 = -4(8)y$$
$$x^2 = -32y$$

The directrix is the line $y = 8$. To help graph the parabola, we plot the two points on the graph to the left and right of the focus. Let $y = -8$:

$$x^2 = -32(-8) = 256$$
$$x = \pm 16$$

The points $(-16, -8)$ and $(16, -8)$ are on the graph.

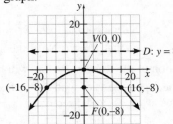

9. The vertex is at the origin and the axis of symmetry is the *x*-axis, so the parabola either opens left or right. Because the graph contains the point $(3,\ 2)$, which is in quadrant I, the parabola must open right. Therefore, the equation of the parabola is of the form $y^2 = 4ax$. Now $y = 2$ when $x = 3$, so

$$y^2 = 4ax$$
$$2^2 = 4a(3)$$
$$4 = 12a$$
$$a = \frac{4}{12} = \frac{1}{3}$$

The equation of the parabola is

$$y^2 = 4\left(\frac{1}{3}\right)x$$
$$y^2 = \frac{4}{3}x$$

With $a = \frac{1}{3}$, we know that the focus is $\left(\frac{1}{3},\ 0\right)$ and the directrix is the line $x = -\frac{1}{3}$. To help graph the parabola, we plot the two points on the graph to the left and right of the focus. Let $x = \frac{1}{3}$:

$$y^2 = \frac{4}{3}\left(\frac{1}{3}\right)$$
$$y^2 = \frac{4}{9}$$
$$y = \pm\frac{2}{3}$$

The points $\left(\frac{1}{3}, -\frac{2}{3}\right)$ and $\left(\frac{1}{3},\ \frac{2}{3}\right)$ are on the graph.

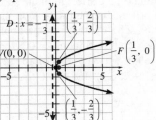

10. The vertex of the parabola $(x+3)^2 = -14(y-2)$ is (–3, 2).

11. We complete the square in *y* to write the equation in standard form:

$$y^2 - 4y - 12x - 32 = 0$$
$$y^2 - 4y = 12x + 32$$
$$y^2 - 4y + 4 = 12x + 32 + 4$$
$$y^2 - 4y + 4 = 12x + 36$$
$$(y-2)^2 = 12(x+3)$$

Notice that the equation is of the form $(y-k)^2 = 4a(x-h)$. The graph is a parabola that opens right with vertex $(h,\ k) = (-3,\ 2)$. Since $4a = 12$, we have that $a = 3$. The focus is $(h+a,\ k) = (-3+3,\ 2) = (0,\ 2)$, and the directrix is $x = h - a = -3 - 3 = -6$. To help graph the parabola, we plot the two points on the graph above and below the focus. Let $x = 0$:

$$(y-2)^2 = 12(0+3)$$
$$(y-2)^2 = 12(3)$$
$$(y-2)^2 = 36$$
$$y - 2 = \pm 6$$
$$y = 2 \pm 6$$
$$y = -4 \text{ or } y = 8$$

The points $(0, -4)$ and $(0,\ 8)$ are on the graph.

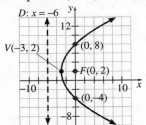

12. The receiver should be located at the focus of the satellite dish, so we need to find where the focus of the satellite dish is. To solve this problem, we draw a parabola on a Cartesian plane so that the vertex is the origin and the focus is on the positive y-axis. The width of the parabola is 4 feet, and the depth is 6 inches $= 0.5$ feet. Therefore, we know two points on the graph of the parabola: $(-2,\ 0.5)$ and $(2,\ 0.5)$.

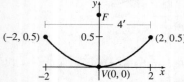

The equation of the parabola has the form $x^2 = 4ay$. Since $(2,\ 0.5)$ is a point on the graph, we have

$$2^2 = 4a(0.5)$$
$$4 = 2a$$
$$2 = a$$

The receiver should be located 2 feet from the base of the dish along its axis of symmetry.

9.3 Exercises

13. c. The parabola opens upward and has vertex $(0,0)$ and focus $(0,2)$, so the equation is of the form $x^2 = 4ay$ with $a=2$:

$$x^2 = 4(2)y$$
$$x^2 = 8y$$

15. a. The parabola opens to the right and has vertex $(0,0)$ and focus $(2,0)$, so the equation is of the form $y^2 = 4ax$ with $a=2$:

$$y^2 = 4(2)x$$
$$y^2 = 8x$$

17. b. The parabola opens to the left and has vertex $(0,0)$ and focus $(-2,0)$, so the equation is of the form $y^2 = -4ax$ with $a=2$:

$$y^2 = -4(2)x$$
$$y^2 = -8x$$

19. e. The parabola opens to the right and has vertex $(-1,2)$ and focus $(1,2)$, so the equation is of the form $(y-k)^2 = 4a(x-h)$, with $a = 1-(-1) = 2$, $h = -1$, and $k = 2$:

$$(y-2)^2 = 4\cdot 2\big(x-(-1)\big)$$
$$(y-2)^2 = 8(x+1)$$

21. Notice that $x^2 = 24y$ is of the form $x^2 = 4ay$, where $4a = 24$, so that $a = 6$. Now, the graph of an equation of the form $x^2 = 4ay$ will be a parabola that opens upward with the vertex at the origin, focus at $(0,a)$, and directrix of $y = -a$.

Thus, the graph of $x^2 = 24y$ is a parabola that opens upward with vertex $(0,0)$, focus $(0,6)$, and directrix $y = -6$. To help graph the parabola, we plot the two points on the graph to the left and to the right of the focus. Let $y = 6$:

$$x^2 = 24(6)$$
$$x^2 = 144$$
$$x = \pm 12$$

The points $(-12,\ 6)$ and $(12,\ 6)$ are on the graph.

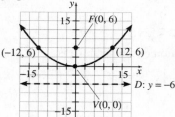

23. Notice that $y^2 = -6x$ is of the form $y^2 = -4ax$, where $-4a = -6$, so that $a = \dfrac{-6}{-4} = \dfrac{3}{2}$. Now, the graph of an equation of the form $y^2 = -4ax$ will be a parabola that opens to the left with the vertex at the origin, focus at $(-a,0)$, and directrix of $x = a$. Thus, the graph of $y^2 = -6x$ is a parabola that opens to the left with vertex $(0,0)$, focus $\left(-\dfrac{3}{2},\ 0\right)$, and directrix $x = \dfrac{3}{2}$. To help graph the parabola, we plot the two points on the graph above and below the focus. Let $x = -\dfrac{3}{2}$:

$$y^2 = -6\left(-\frac{3}{2}\right)$$

$$y^2 = 9$$

$$y = \pm 3$$

The points $\left(-\frac{3}{2}, -3\right)$ and $\left(-\frac{3}{2}, 3\right)$ are on the

graph.

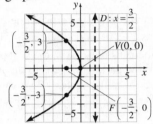

25. Notice that $x^2 = -8y$ is of the form $x^2 = -4ay$,
where $-4a = -8$, so that $a = 2$. Now, the graph
of an equation of the form $x^2 = -4ay$ will be a
parabola that opens downward with the vertex at
the origin, focus at $(0, -a)$, and directrix of
$y = a$. Thus, the graph of $x^2 = -8y$ is a
parabola that opens downward with vertex
$(0,0)$, focus $(0, -2)$, and directrix $y = 2$. To
help graph the parabola, we plot the two points
on the graph to the left and to the right of the
focus. Let $y = -2$:

$$x^2 = -8(-2)$$

$$x^2 = 16$$

$$x = \pm 4$$

The points $(-4, -2)$ and $(4, -2)$ are on the
graph.

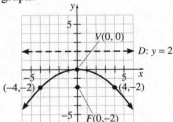

27. The distance from the vertex $(0,0)$ to the focus
$(5,0)$ is $a = 5$. Because the focus lies on the
positive *x*-axis, we know that the parabola will
open to the right and the axis of symmetry is the
x-axis. This means the equation of the parabola
is of the form $y^2 = 4ax$ with $a = 5$:

$$y^2 = 4(5)x$$

$$y^2 = 20x$$

The directrix is the line $x = -5$. To help graph
the parabola, we plot the two points on the graph
above and below the focus. Let $x = 5$:

$$y^2 = 20(5) = 100$$

$$y = \pm 10$$

The points $(5, -10)$ and $(5, 10)$ are on the graph.

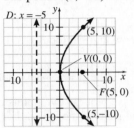

29. The distance from the vertex $(0,0)$ to the focus
$(0, -6)$ is $a = 6$. Because the focus lies on the
negative *y*-axis, we know that the parabola will
open downward and the axis of symmetry is the
y-axis. This means the equation of the parabola
is of the form $x^2 = -4ay$ with $a = 6$:

$$x^2 = -4(6)y$$

$$x^2 = -24y$$

The directrix is the line $y = 6$. To help graph
the parabola, we plot the two points on the graph
to the left and right of the focus. Let $y = -6$:

$$x^2 = -24(-6) = 144$$

$$x = \pm 12$$

The points $(-12, -6)$ and $(12, -6)$ are on the
graph.

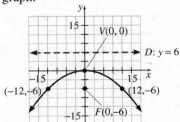

31. The vertex is at the origin and the axis of
symmetry is the *y*-axis, so the parabola either
opens upward or downward. Because the graph
contains the point $(6, 6)$, which is in quadrant I,
the parabola must open upward. Therefore, the
equation of the parabola is of the form
$x^2 = 4ay$. Now $y = 6$ when $x = 6$, so

$x^2 = 4ay$

$6^2 = 4a(6)$

$36 = 24a$

$a = \dfrac{36}{24} = \dfrac{3}{2}$

The equation of the parabola is

$x^2 = 4\left(\dfrac{3}{2}\right)y$

$x^2 = 6y$

With $a = \dfrac{3}{2}$, we know that the focus is $\left(0, \ \dfrac{3}{2}\right)$

and the directrix is the line $y = -\dfrac{3}{2}$. To help

graph the parabola, we plot the two points on the graph to the left and right of the focus. Let

$y = \dfrac{3}{2}$:

$x^2 = 6\left(\dfrac{3}{2}\right)$

$x^2 = 9$

$x = \pm 3$

The points $\left(-3, \ \dfrac{3}{2}\right)$ and $\left(3, \ \dfrac{3}{2}\right)$ are on the

graph.

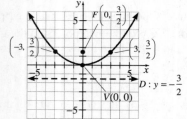

33. The vertex is at the origin and the directrix is the line $y = 3$, so the focus must be $(0, -3)$.

Accordingly, the parabola opens downward with $a = 3$. This means the equation of the parabola is of the form $x^2 = -4ay$ with $a = 3$:

$x^2 = -4(3)y$

$x^2 = -12y$

To help graph the parabola, we plot the two points on the graph to the left and right of the focus. Let $y = -3$:

$x^2 = -12(-3) = 36$

$x = \pm 6$

The points $(-6, -3)$ and $(6, -3)$ are on the graph.

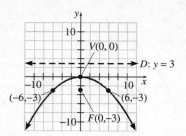

35. Notice that the directrix is the vertical line $x = 3$ and that the focus $(-3, 0)$ is on the x-axis. Thus, the axis of symmetry is the x-axis, the vertex is at the origin, and the parabola opens to the left. Now, the distance from the vertex $(0, 0)$ to the focus $(-3, 0)$ is $a = 3$, so the equation of the parabola is of the form $y^2 = -4ax$ with $a = 3$:

$y^2 = -4(3)x$

$y^2 = -12x$

To help graph the parabola, we plot the two points on the graph above and below the focus. Let $x = -3$:

$y^2 = -12(-3) = 36$

$y = \pm 6$

The points $(-3, -6)$ and $(-3, 6)$ are on the graph.

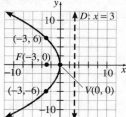

37. The parabola opens to the right and has vertex $(0, \ 0)$, so the equation must have the form

$y^2 = 4ax$. Because the point $(4, \ 2)$ is on the parabola, we let $x = 4$ and $y = 2$ to determine a:

$2^2 = 4a(4)$

$4 = 16a$

$a = \dfrac{4}{16} = \dfrac{1}{4}$

Thus, the equation of the parabola is

$y^2 = 4\left(\dfrac{1}{4}\right)x$

$y^2 = x$

39. Notice that $(x-2)^2 = 4(y-4)$ is of the form

$(x-h)^2 = 4a(y-k)$, where $4a = 4$, so that

$a = 1$ and $(h,k) = (2,\ 4)$. Now, the graph of an

equation of the form $(x-h)^2 = 4a(y-k)$ will be

a parabola that opens upward with vertex at

(h,k), focus at $(h, k+a)$, and directrix of

$y = k-a$. Note that $k+a = 4+1 = 5$ and

$k-a = 4-1 = 3$. Thus, the graph of

$(x-2)^2 = 4(y-4)$ is a parabola that opens

upward with vertex $(2,\ 4)$, focus $(2,\ 5)$, and

directrix $y = 3$. To help graph the parabola, we

plot the two points to the left and to the right of

the focus. Let $y = 5$:

$(x-2)^2 = 4(5-4)$

$(x-2)^2 = 4(1)$

$(x-2)^2 = 4$

$x - 2 = \pm 2$

$x = 2 \pm 2$

$x = 0$ or $x = 4$

The points $(0,\ 5)$ and $(4,\ 5)$ are on the graph.

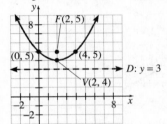

41. Notice that $(y+3)^2 = -8(x+2)$ is of the form

$(y-k)^2 = -4a(x-h)$, where $(h,k) = (-2,-3)$

and $-4a = -8$, so that $a = 2$. Now, the graph of

an equation of the form $(y-k)^2 = -4a(x-h)$

will be a parabola that opens to the left with the

vertex at (h,k), focus at $(h-a,k)$, and directrix

of $x = h+a$. Note that $h-a = -2-2 = -4$ and

$h+a = -2+2 = 0$. Thus, the graph of

$(y+3)^2 = -8(x+2)$ is a parabola that opens to

the left with vertex $(-2,-3)$, focus $(-4,-3)$,

and directrix $x = 0$. To help graph the parabola,

we plot the two points on the graph above and

below the focus. Let $x = -4$:

$(y+3)^2 = -8(-4+2)$

$(y+3)^2 = -8(-2)$

$(y+3)^2 = 16$

$y + 3 = \pm 4$

$y = -3 \pm 4$

$y = -7$ or $y = 1$

The points $(-4,-7)$ and $(-4,\ 1)$ are on the

graph.

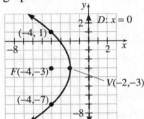

43. Notice that $(x+5)^2 = -20(y-1)$ is of the form

$(x-h)^2 = -4a(y-k)$, where $(h,k) = (-5,1)$ and

$-4a = -20$, so that $a = 5$. Now, the graph of an

equation of the form $(x-h)^2 = -4a(y-k)$ will

be a parabola that opens downward with the

vertex at (h,k), focus at $(h,k-a)$, and directrix

of $y = k+a$. Note that $k-a = 1-5 = -4$ and

$k+a = 1+5 = 6$. Thus, the graph of

$(x+5)^2 = -20(y-1)$ is a parabola that opens

downward with vertex $(-5,\ 1)$, focus $(-5,-4)$,

and directrix $y = 6$. To help graph the parabola,

we plot the two points on the graph to the left

and right of the focus. Let $y = -4$:

$(x+5)^2 = -20(-4-1)$

$(x+5)^2 = -20(-5)$

$(x+5)^2 = 100$

$x + 5 = \pm 10$

$x = -5 \pm 10$

$x = -15$ or $x = 5$

The points $(-15,-4)$ and $(5,-4)$ are on the

graph.

45. We complete the square in x to write the equation in standard form:

$$x^2 + 4x + 12y + 16 = 0$$
$$x^2 + 4x = -12y - 16$$
$$x^2 + 4x + 4 = -12y - 16 + 4$$
$$x^2 + 4x + 4 = -12y - 12$$
$$(x+2)^2 = -12(y+1)$$

Notice that $(x+2)^2 = -12(y+1)$ is of the form $(x-h)^2 = -4a(y-k)$, where $(h,k) = (-2,-1)$ and $-4a = -12$, so that $a = 3$. Now, the graph of an equation of the form $(x-h)^2 = -4a(y-k)$ will be a parabola that opens downward with the vertex at (h,k), focus at $(h, k-a)$, and directrix of $y = k+a$. Note that $k-a = -1-3 = -4$ and $k+a = -1+3 = 2$. Thus, the graph of $(x+2)^2 = -12(y+1)$ is a parabola that opens downward with vertex $(-2,-1)$, focus $(-2,-4)$, and directrix $y = 2$. To help graph the parabola, we plot the two points on the graph to the left and right of the focus. Let $y = -4$:

$$(x+2)^2 = -12(-4+1)$$
$$(x+2)^2 = -12(-3)$$
$$(x+2)^2 = 36$$
$$x+2 = \pm 6$$
$$x = -2 \pm 6$$
$$x = -8 \text{ or } x = 4$$

The points $(-8,-4)$ and $(4,-4)$ are on the graph.

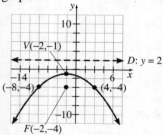

47. We complete the square in y to write the equation in standard form:

$$y^2 - 8y - 4x + 20 = 0$$
$$y^2 - 8y = 4x - 20$$
$$y^2 - 8y + 16 = 4x - 20 + 16$$
$$y^2 - 8y + 16 = 4x - 4$$
$$(y-4)^2 = 4(x-1)$$

Notice that $(y-4)^2 = 4(x-1)$ is of the form $(y-k)^2 = 4a(x-h)$, where $(h,k) = (1,4)$ and $4a = 4$, so that $a = 1$. Now, the graph of an equation of the form $(y-k)^2 = 4a(x-h)$ will be a parabola that opens to the right with the vertex at (h,k), focus at $(h+a,k)$, and directrix of $x = h-a$. Note that $h+a = 1+1 = 2$ and $h-a = 1-1 = 0$. Thus, the graph of $(y-4)^2 = 4(x-1)$ is a parabola that opens to the right with vertex $(1, 4)$, focus $(2, 4)$, and directrix $x = 0$. To help graph the parabola, we plot the two points on the graph above and below the focus. Let $x = 2$:

$$(y-4)^2 = 4(2-1)$$
$$(y-4)^2 = 4$$
$$y - 4 = \pm 2$$
$$y = 4 \pm 2$$
$$y = 2 \text{ or } y = 6$$

The points $(2, 2)$ and $(2, 6)$ are on the graph.

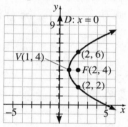

49. We complete the square in x to write the equation in standard form:

$$x^2 + 10x + 6y + 13 = 0$$
$$x^2 + 10x = -6y - 13$$
$$x^2 + 10x + 25 = -6y - 13 + 25$$
$$x^2 + 10x + 25 = -6y + 12$$
$$(x+5)^2 = -6(y-2)$$

Notice that $(x+5)^2 = -6(y-2)$ is of the form $(x-h)^2 = -4a(y-k)$, where $(h,k) = (-5,2)$ and $-4a = -6$, so that $a = \dfrac{-6}{-4} = \dfrac{3}{2}$. Now, the graph of an equation of the form $(x-h)^2 = -4a(y-k)$ will be a parabola that opens downward with the vertex at (h,k), focus at $(h, k-a)$, and directrix of $y = k+a$. Note

Copyright © 2014 Pearson Education, Inc.

that $k - a = 2 - \dfrac{3}{2} = \dfrac{1}{2}$ and $k + a = 2 + \dfrac{3}{2} = \dfrac{7}{2}$.

Thus, the graph of $(x+5)^2 = -6(y-2)$ is a parabola that opens downward with vertex $(-5, 2)$, focus $\left(-5, \dfrac{1}{2}\right)$, and directrix $y = \dfrac{7}{2}$.

To help graph the parabola, we plot the two points on the graph to the left and right of the focus. Let $y = \dfrac{1}{2}$:

$$(x+5)^2 = -6\left(\dfrac{1}{2} - 2\right)$$

$$(x+5)^2 = -6\left(-\dfrac{3}{2}\right)$$

$$(x+5)^2 = 9$$

$$x + 5 = \pm 3$$

$$x = -5 \pm 3$$

$$x = -8 \text{ or } x = -2$$

The points $\left(-8, \dfrac{1}{2}\right)$ and $\left(-2, \dfrac{1}{2}\right)$ are on the graph.

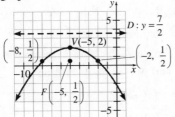

51. The light bulb should be located at the focus of the headlight, so we need to find where the focus of the headlight is. To solve this problem, we draw a parabola on a Cartesian plane so that the vertex is the origin and the focus is on the positive y-axis. The width of the parabola is 4 inches, and the depth is 1 inch. Therefore, we know two points on the graph of the parabola: $(-2, 1)$ and $(2, 1)$.

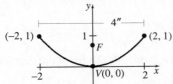

The equation of the parabola has the form $x^2 = 4ay$. Since $(2, 1)$ is a point on the graph, we have

$$2^2 = 4a(1)$$

$$4 = 4a$$

$$1 = a$$

The light bulb should be located 1 inch above the vertex, along its axis of symmetry.

53. To solve this problem, we draw a parabola on a Cartesian plane so that the vertex is the origin and the focus is on the positive y-axis. The distance between the towers is 500 feet, and the height of the towers is 60 feet. Therefore, we know two points on the graph of the parabola: $(-250, 60)$ and $(250, 60)$.

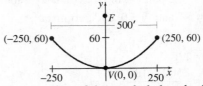

The equation of the parabola has the form $x^2 = 4ay$. Since $(250, 60)$ is a point on the graph, we have

$$250^2 = 4a(60)$$

$$62,500 = 240a$$

$$a = \dfrac{62,500}{240} = \dfrac{3125}{12}$$

Then the equation of the parabola is

$$x^2 = 4\left(\dfrac{3125}{12}\right)y$$

$$x^2 = \dfrac{3125}{3}y$$

To find the height of the cable at a point 150 feet from the center of the bridge, let $x = 150$:

$$150^2 = \dfrac{3125}{3}y$$

$$22,500 = \dfrac{3125}{3}y$$

$$y = \dfrac{3}{3125} \cdot 22,500 = 21.6$$

At a point 150 feet from the center of the bridge, the height of the cable is 21.6 feet.

55. To solve this problem, we draw a parabola on a Cartesian plane so that the vertex is on the positive y-axis and the base of the bridge is along the x-axis. The maximum height of the bridge is 30 feet, so we know the vertex of the parabola is $(0, 30)$. The span of the bridge is 100 feet, so we know two other points on the graph of the parabola: $(-50, 0)$ and $(50, 0)$.

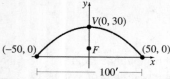

Because the parabola opens downward, its equation has the form $(x-h)^2 = -4a(y-k)$. Substituting the vertex $(h,k) = (0, \ 30)$ into this equation, we obtain

$$(x-0)^2 = -4a(y-30)$$
$$x^2 = -4a(y-30)$$

Because the point $(50, \ 0)$ is on the parabola, we let $x = 50$ and $y = 0$ to determine a:

$$50^2 = -4a(0-30)$$
$$2500 = 120a$$
$$a = \frac{2500}{120} = \frac{125}{6}$$

Thus, the equation of the parabola is

$$x^2 = -4 \cdot \frac{125}{6}(y-30)$$
$$x^2 = -\frac{250}{3}(y-30)$$

To find the height of the bridge at points 10, 30, and 50 feet from the center of the bridge, let $x = 10$, 30, and 50 respectively:

$x = 10$: $\quad 10^2 = -\dfrac{250}{3}(y-30)$

$$100 = -\frac{250}{3}(y-30)$$
$$300 = -250(y-30)$$
$$-1.2 = y-30$$
$$28.8 = y$$

$x = 30$: $\quad 30^2 = -\dfrac{250}{3}(y-30)$

$$900 = -\frac{250}{3}(y-30)$$
$$2700 = -250(y-30)$$
$$-10.8 = y-30$$
$$19.2 = y$$

$x = 50$: $\quad 50^2 = -\dfrac{250}{3}(y-30)$

$$2500 = -\frac{250}{3}(y-30)$$
$$7500 = -250(y-30)$$
$$-30 = y-30$$
$$0 = y$$

The height of the bridge is 28.8 feet at a distance of 10 feet from the center, 19.2 feet at a distance of 30 feet from the center, and 0 feet (i.e. ground level) at a distance of 50 feet from the center.

57. The parabola opens upward, so the equation must have the form $(x-h)^2 = 4a(y-k)$. Substituting the vertex $(h,k) = (3,-2)$ into this equation, we obtain $(x-3)^2 = 4a(y+2)$. Because the point $(5,-1)$ is on the parabola, we let $x = 5$ and $y = -1$ to determine a:

$$(5-3)^2 = 4a(-1+2)$$
$$2^2 = 4a(1)$$
$$4 = 4a$$
$$1 = a$$

Thus, the equation of the parabola is

$$(x-3)^2 = 4 \cdot 1(y+2)$$
$$(x-3)^2 = 4(y+2)$$

59. The parabola opens to the left, so the equation must have the form $(y-k)^2 = -4a(x-h)$. Substituting the vertex $(h,k) = (2, \ 3)$ into this equation, we obtain $(y-3)^2 = -4a(x-2)$. Because the point $(-2,-1)$ is on the parabola, we let $x = -2$ and $y = -1$ to determine a:

$$(-1-3)^2 = -4a(-2-2)$$
$$(-4)^2 = -4a(-4)$$
$$16 = 16a$$
$$1 = a$$

Thus, the equation of the parabola is

$$(y-3)^2 = -4 \cdot 1(x-2)$$
$$(y-3)^2 = -4(x-2)$$

61. $x^2 = 8y$

   a. Let $x = 4$ and $y = 2$:

   $$4^2 \overset{?}{=} 8 \cdot 2$$
   $$16 = 16 \ \leftarrow \text{True}$$

   Thus, the point $(4, \ 2)$ in on the parabola.

   b. Notice that $x^2 = 8y$ is of the form $x^2 = 4ay$, where $4a = 8$, so that $a = 2$. Thus, the focus of the parabola is $F(0, \ 2)$, and the directrix is $D: y = -2$. Now, the distance from the point $P(4,\ 2)$ to the focus

is $d(F,P) = \sqrt{(0-4)^2 + (2-2)^2} = \sqrt{16} = 4$,

and the distance from the point $P(4,\ 2)$ to

the directrix $D: y = -2$ is

$d(P,D) = 2 - (-2) = 4$.

Thus, $d(F,P) = d(P,D) = 4$.

63. If the distance from the point on the parabola to the focus is 8 units, then by the definition of a parabola, the distance from the same point on the parabola to the directrix must also be 8 units.

65. Answers may vary. One possibility follows:

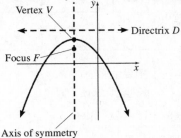

67. Begin with the graph of $y = x^2$, then shift the graph 3 units to the left to obtain the graph of $y = (x+3)^2$.

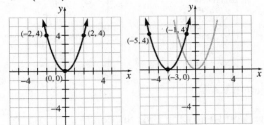

69. Notice that $y = (x+3)^2$ is of the form

$(x-h)^2 = 4a(y-k)$, where $4a = 1$, so that

$a = \dfrac{1}{4}$ and $(h,k) = (-3,\ 0)$. Now, the graph of

an equation of the form $(x-h)^2 = 4a(y-k)$ will be a parabola that opens upward with vertex at (h,k), focus at $(h,k+a)$, and directrix of

$y = k - a$. Note that $k + a = 0 + \dfrac{1}{4} = \dfrac{1}{4}$ and

$k - a = 0 - \dfrac{1}{4} = -\dfrac{1}{4}$. Thus, the graph of

$y = (x+3)^2$ is a parabola that opens upward

with vertex $(-3,\ 0)$, focus $\left(-3,\ \dfrac{1}{4}\right)$, and

directrix $y = -\dfrac{1}{4}$. To help graph the parabola, we plot the two points to the left and to the right

of the focus. Let $y = \dfrac{1}{4}$:

$\dfrac{1}{4} = (x+3)^2$

$\pm\dfrac{1}{2} = x + 3$

$-3 \pm \dfrac{1}{2} = x$

$x = -\dfrac{7}{2}$ or $x = -\dfrac{5}{2}$

The points $\left(-\dfrac{7}{2}, \dfrac{1}{4}\right)$ and $\left(-\dfrac{5}{2}, \dfrac{1}{4}\right)$ are on the

graph.

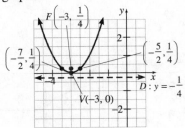

71. Note that $4(y+2) = (x-2)^2$ is of the form

$(x-h)^2 = 4a(y-k)$, where $4a = 4$, so that

$a = 1$ and $(h,k) = (2,-2)$. Now, the graph of an

equation of the form $(x-h)^2 = 4a(y-k)$ will be a parabola that opens upward with vertex at (h,k), focus at $(h,k+a)$, and directrix of

$y = k - a$. Note that $k + a = -2 + 1 = -1$ and

$k - a = -2 - 1 = -3$. Thus, the graph of

$4(y+2) = (x-2)^2$ is a parabola that opens

upward with vertex $(2,-2)$, focus $(2,-1)$, and

directrix $y = -3$. To help graph the parabola, we plot the two points to the left and to the right of the focus. Let $y = -1$:

$4(-1+2) = (x-2)^2$

$\qquad\quad 4 = (x-2)^2$

$\qquad \pm 2 = x - 2$

$\qquad 2 \pm 2 = x$

$\qquad x = 0$ or $x = 4$

The points $(0,-1)$ and $(4,-1)$ are on the graph.

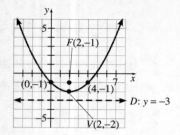

73. $x^2 = 24y$

$\dfrac{1}{24}x^2 = y$

Let $Y_1 = \dfrac{1}{24}x^2$.

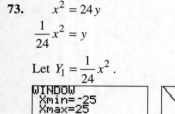

75. $y^2 = -6x$

$y = \pm\sqrt{-6x}$

Let $Y_1 = -\sqrt{-6x}$ and $Y_2 = \sqrt{-6x}$.

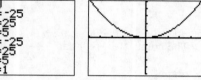

77. $(x-2)^2 = 4(y-4)$

$\dfrac{1}{4}(x-2)^2 = y-4$

$\dfrac{1}{4}(x-2)^2 + 4 = y$

Let $Y_1 = \dfrac{1}{4}(x-2)^2 + 4$.

79. $(y+3)^2 = -8(x+2)$

$y+3 = \pm\sqrt{-8(x+2)}$

$y = -3 \pm \sqrt{-8(x+2)}$

Let $Y_1 = -3 - \sqrt{-8(x+2)}$ and

$Y_2 = -3 + \sqrt{-8(x+2)}$.

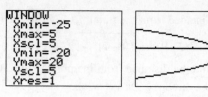

81. $x^2 + 4x + 12y + 16 = 0$

$x^2 + 4x + 16 = -12y$

$-\dfrac{1}{12}\left(x^2 + 4x + 16\right) = y$

Let $Y_1 = -\dfrac{1}{12}\left(x^2 + 4x + 16\right)$.

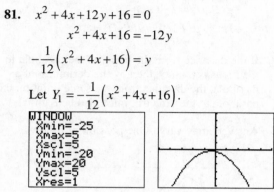

83. $y^2 - 8y - 4x + 20 = 0$

$y^2 - 8y = 4x - 20$

$y^2 - 8y + 16 = 4x - 20 + 16$

$(y-4)^2 = 4x - 4$

$y - 4 = \pm\sqrt{4x-4}$

$y = 4 \pm \sqrt{4x-4}$

Let $Y_1 = 4 - \sqrt{4x-4}$ and $Y_2 = 4 + \sqrt{4x-4}$.

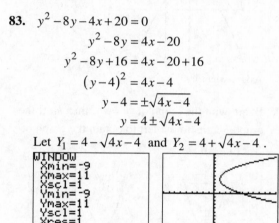

Section 9.4

Are You Ready for This Section?

R1. Start: $x^2 + 10x$

Add: $\left(\dfrac{1}{2}\cdot 10\right)^2 = 25$

Result: $x^2 + 10x + 25$

Factored Form: $(x+5)^2$

R2. Begin with the graph of $y = x^2$, then shift the graph 2 units to the left to obtain the graph of $y = (x+2)^2$. Shift this graph down 1 unit to obtain the graph of $f(x) = (x+2)^2 - 1$.

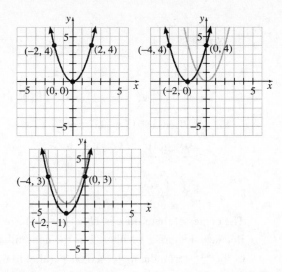

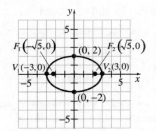

Section 9.4 Quick Checks

1. An <u>ellipse</u> is the collection of points in the plane such that the sum of the distances from two fixed points, called the <u>foci</u>, is a constant.

2. For an ellipse, the line containing the foci is called the <u>major axis</u>.

3. The two points of intersection of the ellipse and the major axis are the <u>vertices</u>, V_1 and V_2, of the ellipse.

4. False. The equation of an ellipse centered at the origin with vertex $(a,0)$ and focus $(c,0)$ is

$$\frac{x^2}{a^2}+\frac{y^2}{b^2}=1, \text{ where } b^2=a^2-c^2.$$

5. $\dfrac{x^2}{9}+\dfrac{y^2}{4}=1$

The larger number, 9, is in the denominator of the x^2-term. This means that the major axis is the *x*-axis and that the equation of the ellipse is

of the form $\dfrac{x^2}{a^2}+\dfrac{y^2}{b^2}=1$, so that $a^2=9$ and

$b^2=4$. The center of the ellipse is $(0,0)$.

Because $b^2=a^2-c^2$, or $c^2=a^2-b^2$, we have

that $c^2=9-4=5$, so that $c=\pm\sqrt{5}$. Since the

major axis is the *x*-axis, the foci are $\left(-\sqrt{5},0\right)$

and $\left(\sqrt{5},0\right)$. To find the *x*-intercepts (vertices),

let $y=0$; to find the *y*-intercepts, let $x=0$:

x-intercepts:	*y*-intercepts:
$\dfrac{x^2}{9}+\dfrac{0^2}{4}=1$	$\dfrac{0^2}{9}+\dfrac{y^2}{4}=1$
$\dfrac{x^2}{9}=1$	$\dfrac{y^2}{4}=1$
$x^2=9$	$y^2=4$
$x=\pm3$	$y=\pm2$

The intercepts are $(-3,0)$, $(3,0)$, $(0,-2)$, and $(0,2)$.

6. $\dfrac{x^2}{16}+\dfrac{y^2}{36}=1$

The larger number, 36, is in the denominator of the y^2-term. This means that the major axis is the *y*-axis and that the equation of the ellipse is

of the form $\dfrac{x^2}{b^2}+\dfrac{y^2}{a^2}=1$, so that $a^2=36$ and

$b^2=16$. The center of the ellipse is $(0,0)$.

Because $b^2=a^2-c^2$, or $c^2=a^2-b^2$, we have

that $c^2=36-16=20$, so that

$c=\pm\sqrt{20}=\pm2\sqrt{5}$. Since the major axis is the

y-axis, the foci are $\left(0,-2\sqrt{5}\right)$ and $\left(0,2\sqrt{5}\right)$. To

find the *x*-intercepts, let $y=0$; to find the

y-intercepts (vertices), let $x=0$:

x-intercepts:	*y*-intercepts:
$\dfrac{x^2}{16}+\dfrac{0^2}{36}=1$	$\dfrac{0^2}{16}+\dfrac{y^2}{36}=1$
$\dfrac{x^2}{16}=1$	$\dfrac{y^2}{36}=1$
$x^2=16$	$y^2=36$
$x=\pm4$	$y=\pm6$

The intercepts are $(-4,0)$, $(4,0)$, $(0,-6)$, and $(0,6)$.

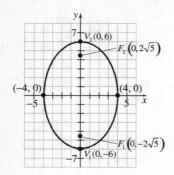

7. The given focus $(0, 3)$ and the given vertex $(0, 7)$ lie on the y-axis. Thus, the major axis is the y-axis, and the equation of the ellipse is of the form $\dfrac{x^2}{b^2} + \dfrac{y^2}{a^2} = 1$. The distance from the center of the ellipse to the vertex is $a = 7$ units. The distance from the center of the ellipse to the focus is $c = 3$ units. Because $b^2 = a^2 - c^2$, we have that $b^2 = 7^2 - 3^2 = 49 - 9 = 40$. So, the equation of the ellipse is $\dfrac{x^2}{40} + \dfrac{y^2}{49} = 1$.

To help graph the ellipse, find the x-intercepts:

Let $y = 0$: $\dfrac{x^2}{40} + \dfrac{0^2}{49} = 1$

$$\dfrac{x^2}{40} = 1$$

$$x^2 = 40$$

$$x = \pm\sqrt{40} = \pm 2\sqrt{10}$$

The x-intercepts are $\left(-2\sqrt{10}, 0\right)$ and $\left(2\sqrt{10}, 0\right)$.

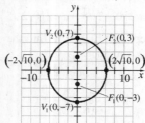

8. The center of $\dfrac{(x-3)^2}{25} + \dfrac{(y+1)^2}{16} = 1$ is $\underline{(3, -1)}$.

9.
$$9x^2 + y^2 + 54x - 2y + 73 = 0$$
$$9x^2 + 54x + y^2 - 2y = -73$$
$$9\left(x^2 + 6x\right) + \left(y^2 - 2y\right) = -73$$
$$9\left(x^2 + 6x + 9\right) + \left(y^2 - 2y + 1\right) = -73 + 9(9) + 1$$
$$9(x+3)^2 + (y-1)^2 = 9$$
$$\dfrac{9(x+3)^2 + (y-1)^2}{9} = \dfrac{9}{9}$$
$$\dfrac{(x+3)^2}{1} + \dfrac{(y-1)^2}{9} = 1$$

The center of the ellipse is $(h, k) = (-3, 1)$. Because the larger number, 9, is the denominator of the y^2-term, the major axis is parallel to the y-axis. Because $a^2 = 9$ and $b^2 = 1$, we have that $c^2 = a^2 - b^2 = 9 - 1 = 8$. The vertices are $a = 3$ units below and above the center at $V_1(-3, -2)$ and $V_2(-3, 4)$. The foci are $c = \sqrt{8} = 2\sqrt{2}$ units below and above the center at $F_1\left(-3, 1 - 2\sqrt{2}\right)$ and $F_2\left(-3, 1 + 2\sqrt{2}\right)$. We plot the points $b = 1$ unit to the left and right of the center point at $(-4, 1)$ and $(-2, 1)$.

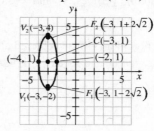

10. To solve the problem, we draw the ellipse on a Cartesian plane so that the center of the ellipse is at the origin and the major axis is along the x-axis. The equation of the ellipse is of the form $\dfrac{x^2}{a^2} + \dfrac{y^2}{b^2} = 1$. Since the length of the hall is 100 feet, the distance from the center of the room to each vertex is $a = \dfrac{100}{2} = 50$ feet. The distance from the center of the room to each focus is $c = 30$ feet.

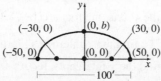

Now, because $b^2 = a^2 - c^2$, we have that $b^2 = 50^2 - 30^2 = 2500 - 900 = 1600$. Thus, the equation that describes the room is

$$\frac{x^2}{2500} + \frac{y^2}{1600} = 1.$$

The height of the room at its center is $b = \sqrt{1600} = 40$ feet.

9.4 Exercises

11. **c.** The center of the ellipse is the origin, the major axis is the x-axis, and the vertices are $(\pm 4, 0)$, and the y-intercepts are $(0, \pm 3)$. Thus,

the equation is of the form $\frac{x^2}{a^2} + \frac{y^2}{b^2} = 1$, with

$a = 4$ and $b = 3$. Thus, the equation is:

$$\frac{x^2}{4^2} + \frac{y^2}{3^2} = 1$$

$$\frac{x^2}{16} + \frac{y^2}{9} = 1$$

13. **d.** The center of the ellipse is the origin, the major axis is the y-axis, and the vertices are $(0, \pm 4)$, and the x-intercepts are $(\pm 3, 0)$. Thus,

the equation is of the form $\frac{x^2}{b^2} + \frac{y^2}{a^2} = 1$, with

$a = 3$ and $b = 4$. Thus, the equation is:

$$\frac{x^2}{3^2} + \frac{y^2}{4^2} = 1$$

$$\frac{x^2}{9} + \frac{y^2}{16} = 1$$

15. $\dfrac{x^2}{25} + \dfrac{y^2}{16} = 1$

The larger number, 25, is in the denominator of the x^2-term. This means that the major axis is the x-axis and that the equation of the ellipse is

of the form $\frac{x^2}{a^2} + \frac{y^2}{b^2} = 1$, so that $a^2 = 25$ and

$b^2 = 16$. The center of the ellipse is $(0, 0)$.

Because $b^2 = a^2 - c^2$, or $c^2 = a^2 - b^2$, we have

that $c^2 = 25 - 16 = 9$, so that $c = \pm\sqrt{9} = \pm 3$.

Since the major axis is the x-axis, the foci are $(-3, 0)$ and $(3, 0)$. To find the x-intercepts (vertices), let $y = 0$; to find the y-intercepts, let

$x = 0$:

x-intercepts:	y-intercepts:
$\dfrac{x^2}{25} + \dfrac{0^2}{16} = 1$	$\dfrac{0^2}{25} + \dfrac{y^2}{16} = 1$
$\dfrac{x^2}{25} = 1$	$\dfrac{y^2}{16} = 1$
$x^2 = 25$	$y^2 = 16$
$x = \pm 5$	$y = \pm 4$

The intercepts are $(-5, 0)$, $(5, 0)$, $(0, -4)$, and $(0, 4)$.

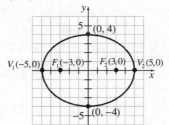

17. $\dfrac{x^2}{36} + \dfrac{y^2}{100} = 1$

The larger number, 100, is in the denominator of the y^2-term. This means that the major axis is the y-axis and that the equation of the ellipse is

of the form $\frac{x^2}{b^2} + \frac{y^2}{a^2} = 1$, so that $a^2 = 100$ and

$b^2 = 36$. The center of the ellipse is $(0, 0)$.

Because $b^2 = a^2 - c^2$, or $c^2 = a^2 - b^2$, we have

that $c^2 = 100 - 36 = 64$, so that $c = \pm\sqrt{64} = \pm 8$.

Since the major axis is the y-axis, the foci are $(0, -8)$ and $(0, 8)$. To find the x-intercepts, let $y = 0$; to find the y-intercepts (vertices), let

$x = 0$:

x-intercepts:	y-intercepts:
$\dfrac{x^2}{36} + \dfrac{0^2}{100} = 1$	$\dfrac{0^2}{36} + \dfrac{y^2}{100} = 1$
$\dfrac{x^2}{36} = 1$	$\dfrac{y^2}{100} = 1$
$x^2 = 36$	$y^2 = 100$
$x = \pm 6$	$y = \pm 10$

The intercepts are $(-6, 0)$, $(6, 0)$, $(0, -10)$, and $(0, 10)$.

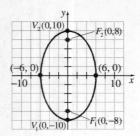

19. $\dfrac{x^2}{49} + \dfrac{y^2}{4} = 1$

The larger number, 49, is in the denominator of the x^2-term. This means that the major axis is the x-axis and that the equation of the ellipse is of the form $\dfrac{x^2}{a^2} + \dfrac{y^2}{b^2} = 1$, so that $a^2 = 49$ and $b^2 = 4$. The center of the ellipse is $(0,0)$. Because $b^2 = a^2 - c^2$, or $c^2 = a^2 - b^2$, we have that $c^2 = 49 - 4 = 45$, so that $c = \pm\sqrt{45} = \pm 3\sqrt{5}$. Since the major axis is the x-axis, the foci are $\left(-3\sqrt{5},0\right)$ and $\left(3\sqrt{5},0\right)$. To find the x-intercepts (vertices), let $y = 0$; to find the y-intercepts, let $x = 0$:

x-intercepts:

$\dfrac{x^2}{49} + \dfrac{0^2}{4} = 1$

$\dfrac{x^2}{49} = 1$

$x^2 = 49$

$x = \pm 7$

y-intercepts:

$\dfrac{0^2}{49} + \dfrac{y^2}{4} = 1$

$\dfrac{y^2}{4} = 1$

$y^2 = 4$

$y = \pm 2$

The intercepts are $(-7,0)$, $(7,0)$, $(0,-2)$, and $(0,2)$.

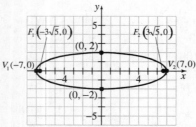

21. $x^2 + \dfrac{y^2}{49} = 1$

$\dfrac{x^2}{1} + \dfrac{y^2}{49} = 1$

The larger number, 49, is in the denominator of

the y^2-term. This means that the major axis is the y-axis and that the equation of the ellipse is of the form $\dfrac{x^2}{b^2} + \dfrac{y^2}{a^2} = 1$, so that $a^2 = 49$ and $b^2 = 1$. The center of the ellipse is $(0,0)$. Because $b^2 = a^2 - c^2$, or $c^2 = a^2 - b^2$, we have that $c^2 = 49 - 1 = 48$, so that $c = \pm\sqrt{48} = \pm 4\sqrt{3}$. Since the major axis is the y-axis, the foci are $\left(0,-4\sqrt{3}\right)$ and $\left(0,4\sqrt{3}\right)$. To find the x-intercepts, let $y = 0$; to find the y-intercepts (vertices), let $x = 0$:

x-intercepts:

$x^2 + \dfrac{0^2}{49} = 1$

$x^2 = 1$

$x = \pm 1$

y-intercepts:

$0^2 + \dfrac{y^2}{49} = 1$

$y^2 = 49$

$y = \pm 7$

The intercepts are $(-1,0)$, $(1,0)$, $(0,-7)$, and $(0,7)$.

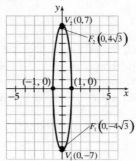

23. $4x^2 + y^2 = 16$

$\dfrac{4x^2 + y^2}{16} = \dfrac{16}{16}$

$\dfrac{x^2}{4} + \dfrac{y^2}{16} = 1$

The larger number, 16, is in the denominator of

the y^2-term. This means that the major axis is the y-axis and that the equation of the ellipse is of the form $\dfrac{x^2}{b^2} + \dfrac{y^2}{a^2} = 1$, so that $a^2 = 16$ and $b^2 = 4$. The center of the ellipse is $(0,0)$. Because $b^2 = a^2 - c^2$, or $c^2 = a^2 - b^2$, we have that $c^2 = 16 - 4 = 12$, so that $c = \pm\sqrt{12} = \pm 2\sqrt{3}$. Since the major axis is the y-axis, the foci are $\left(0,-2\sqrt{3}\right)$ and $\left(0,2\sqrt{3}\right)$. To find the

x-intercepts, let $y = 0$; to find the y-intercepts (vertices), let $x = 0$:

x-intercepts: y-intercepts:

$$4x^2 + 0^2 = 16 \qquad\qquad 4(0)^2 + y^2 = 16$$
$$4x^2 = 16 \qquad\qquad\qquad y^2 = 16$$
$$x^2 = 4 \qquad\qquad\qquad\qquad y = \pm 4$$
$$x = \pm 2$$

The intercepts are $(-2, 0)$, $(2, 0)$, $(0, -4)$, and $(0, 4)$.

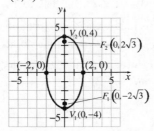

25. The given focus $(4, 0)$ and the given vertex $(6, 0)$ lie on the x-axis. Thus, the equation of the ellipse is of the form $\dfrac{x^2}{a^2} + \dfrac{y^2}{b^2} = 1$. The distance from the center of the ellipse to the vertex is $a = 6$ units. The distance from the center of the ellipse to the focus is $c = 4$ units. Because $b^2 = a^2 - c^2$, we have that $b^2 = 6^2 - 4^2 = 36 - 16 = 20$. Thus, the equation of the ellipse is $\dfrac{x^2}{36} + \dfrac{y^2}{20} = 1$.

To help graph the ellipse, find the y-intercepts:

Let $x = 0$: $\dfrac{0^2}{36} + \dfrac{y^2}{20} = 1$
$$\dfrac{y^2}{20} = 1$$
$$y^2 = 20$$
$$y = \pm\sqrt{20} = \pm 2\sqrt{5}$$

The y-intercepts are $\left(0, -2\sqrt{5}\right)$ and $\left(0, 2\sqrt{5}\right)$.

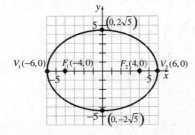

27. The given focus $(0, -4)$ and the given vertex $(0, 7)$ lie on the y-axis. Thus, the equation of the ellipse is of the form $\dfrac{x^2}{b^2} + \dfrac{y^2}{a^2} = 1$. The distance from the center of the ellipse to the vertex is $a = 7$ units. The distance from the center of the ellipse to the focus is $c = 4$ units. Because $b^2 = a^2 - c^2$, we have that $b^2 = 7^2 - 4^2 = 49 - 16 = 33$. Thus, the equation of the ellipse is $\dfrac{x^2}{33} + \dfrac{y^2}{49} = 1$.

To help graph the ellipse, find the x-intercepts:

Let $y = 0$: $\dfrac{x^2}{33} + \dfrac{0^2}{49} = 1$
$$\dfrac{x^2}{33} = 1$$
$$x^2 = 33$$
$$x = \pm\sqrt{33}$$

The x-intercepts are $\left(-\sqrt{33}, 0\right)$ and $\left(\sqrt{33}, 0\right)$.

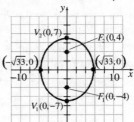

29. The given foci $(\pm 6, 0)$ and the given vertices $(\pm 10, 0)$ lie on the x-axis. The center of the ellipse is the midpoint between the two vertices (or foci). Thus, the center of the ellipse is $(0, 0)$, and its equation is of the form $\dfrac{x^2}{a^2} + \dfrac{y^2}{b^2} = 1$. The distance from the center of the ellipse to the vertex is $a = 10$ units. The distance from the center of the ellipse to the focus is $c = 6$ units. Because $b^2 = a^2 - c^2$, we have that $b^2 = 10^2 - 6^2 = 100 - 36 = 64$. Thus, the equation of the ellipse is $\dfrac{x^2}{100} + \dfrac{y^2}{64} = 1$.

To help graph the ellipse, find the y-intercepts:

Let $x = 0$: $\dfrac{0^2}{100} + \dfrac{y^2}{64} = 1$

$$\dfrac{y^2}{64} = 1$$

$$y^2 = 64$$

$$y = \pm 8$$

The y-intercepts are $(0, -8)$ and $(0, 8)$.

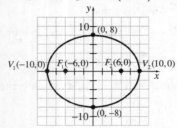

31. The given foci $(0, \pm 5)$ lie on the y-axis. The center of the ellipse is the midpoint between the two foci. Thus, center of the ellipse is $(0, 0)$ and its equation is of the form $\dfrac{x^2}{b^2} + \dfrac{y^2}{a^2} = 1$. Now, the length of the major axis is 16, so the vertices must be $(0, \pm 8)$, and the distance from the center of the ellipse to each vertex is $a = 8$ units. The distance from the center of the ellipse to each focus is $c = 5$ units. Because $b^2 = a^2 - c^2$, we have that $b^2 = 8^2 - 5^2 = 64 - 25 = 39$. Thus, the equation of the ellipse is $\dfrac{x^2}{39} + \dfrac{y^2}{64} = 1$.

To help graph the ellipse, we find the x-intercepts:

Let $y = 0$: $\dfrac{x^2}{39} + \dfrac{0^2}{64} = 1$

$$\dfrac{x^2}{39} = 1$$

$$x^2 = 39$$

$$x = \pm\sqrt{39}$$

The x-intercepts are $\left(-\sqrt{39}, 0\right)$ and $\left(\sqrt{39}, 0\right)$.

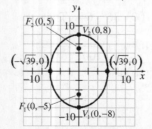

33. $\dfrac{(x-3)^2}{9} + \dfrac{(y+2)^2}{25} = 1$

The center of the ellipse is $(h, k) = (3, -2)$. Because the larger number, 25, is in the denominator of the y^2-term, the major axis is parallel to the y-axis. Because $a^2 = 25$ and $b^2 = 9$, we have that $c^2 = a^2 - b^2 = 25 - 9 = 16$. The vertices are $a = 5$ units below and above the center at $V_1(3, -7)$ and $V_2(3, 3)$. The foci are $c = \sqrt{16} = 4$ units below and above the center at $F_1(3, -6)$ and $F_2(3, 2)$. We plot the points $b = 3$ units to the left and right of the center point at $(0, -2)$ and $(6, -2)$.

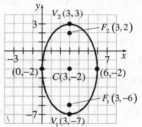

35. $\dfrac{(x+2)^2}{16} + \dfrac{(y-5)^2}{4} = 1$

The center of the ellipse is $(h, k) = (-2, 5)$. Because the larger number, 16, is in the denominator of the x^2-term, the major axis is parallel to the x-axis. Because $a^2 = 16$ and $b^2 = 4$, we have that $c^2 = a^2 - b^2 = 16 - 4 = 12$. The vertices are $a = 4$ units to the left and right of the center at $V_1(-6, 5)$ and $V_2(2, 5)$. The foci are $c = \sqrt{12} = 2\sqrt{3}$ units to the left and right of the center at $F_1\left(-2 - 2\sqrt{3},\ 5\right)$ and $F_2\left(-2 + 2\sqrt{3},\ 5\right)$. We plot the points $b = 2$ units above and below the center point at $(-2, 7)$ and $(-2, 3)$.

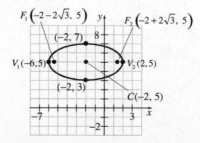

37. $(x-5)^2 + \dfrac{(y+1)^2}{49} = 1$

$\dfrac{(x-5)^2}{1} + \dfrac{(y+1)^2}{49} = 1$

The center of the ellipse is $(h,k) = (5,-1)$.
Because the larger number, 49, is in the
denominator of the y^2-term, the major axis is
parallel to the y-axis. Because $a^2 = 49$ and
$b^2 = 1$, we have that $c^2 = a^2 - b^2 = 49 - 1 = 48$.
The vertices are $a = 7$ units below and above
the center at $V_1(5,-8)$ and $V_2(5,6)$. The foci
are $c = \sqrt{48} = 4\sqrt{3}$ units below and above the
center at $F_1\left(5, -1 - 4\sqrt{3}\right)$ and $F_2\left(5, -1 + 4\sqrt{3}\right)$.
We plot the points $b = 1$ unit to the left and right
of the center point at $(4,-1)$ and $(6,-1)$.

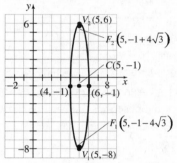

39. $4(x+2)^2 + 16(y-1)^2 = 64$

$\dfrac{4(x+2)^2 + 16(y-1)^2}{64} = \dfrac{64}{64}$

$\dfrac{(x+2)^2}{16} + \dfrac{(y-1)^2}{4} = 1$

The center of the ellipse is $(h,k) = (-2,1)$.
Because the larger number, 16, is in the
denominator of the x^2-term, the major axis is
parallel to the x-axis. Because $a^2 = 16$ and
$b^2 = 4$, we have that $c^2 = a^2 - b^2 = 16 - 4 = 12$.
The vertices are $a = 4$ units to the left and right
of the center at $V_1(-6,1)$ and $V_2(2,1)$. The foci
are $c = \sqrt{12} = 2\sqrt{3}$ units to the left and right of
the center at $F_1\left(-2 - 2\sqrt{3},\ 1\right)$ and

$F_2\left(-2 + 2\sqrt{3},\ 1\right)$. We plot the points $b = 2$

units above and below the center point at $(-2,3)$
and $(-2,-1)$.

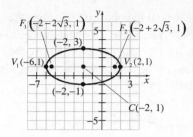

41. $4x^2 + y^2 - 24x + 2y - 63 = 0$

$4x^2 - 24x + y^2 + 2y = 63$

$4\left(x^2 - 6x\right) + \left(y^2 + 2y\right) = 63$

$4\left(x^2 - 6x + 9\right) + \left(y^2 + 2y + 1\right) = 63 + 4(9) + 1$

$4(x-3)^2 + (y+1)^2 = 100$

$\dfrac{4(x-3)^2 + (y+1)^2}{100} = \dfrac{100}{100}$

$\dfrac{(x-3)^2}{25} + \dfrac{(y+1)^2}{100} = 1$

The center of the ellipse is $(h,k) = (3,-1)$.
Because the larger number, 100, is in the
denominator of the y^2-term, the major axis is
parallel to the y-axis. Because $a^2 = 100$ and
$b^2 = 25$, we have that
$c^2 = a^2 - b^2 = 100 - 25 = 75$. The vertices are
$a = 10$ units below and above the center at
$V_1(3,-11)$ and $V_2(3,9)$. The foci are
$c = \sqrt{75} = 5\sqrt{3}$ units below and above the center
at $F_1\left(3, -1 - 5\sqrt{3}\right)$ and $F_2\left(3, -1 + 5\sqrt{3}\right)$. We
plot the points $b = 5$ units to the left and right of
the center point at $(-2,-1)$ and $(8,-1)$.

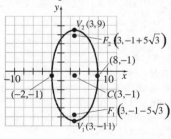

43. a. To solve this problem, we draw the ellipse
on a Cartesian plane so that the x-axis
coincides with the water and the y-axis
passes through the center of the arch. Thus,
the origin is the center of the ellipse. Now,
the "center" of the arch is 10 meters above

the water, so the point $(0,10)$ is on the ellipse. The river is 30 meters wide, so the two points $(-15,0)$ and $(15,0)$ are the two vertices of the ellipse and the major axis is along the x-axis.

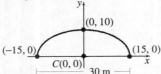

The equation of the ellipse must have the form $\dfrac{x^2}{a^2}+\dfrac{y^2}{b^2}=1$. The distance from the center of the ellipse to each vertex is $a=15$. Also, the height of the semi-ellipse is 10 feet, so $b=10$. Thus, the equation of the ellipse is

$$\frac{x^2}{15^2}+\frac{y^2}{10^2}=1$$
$$\frac{x^2}{225}+\frac{y^2}{100}=1$$

b. To determine if the barge can fit through the opening of the bridge, we center it beneath the arch so that the points $(-9,7)$ and $(9,7)$ represent the top corners of the barge. Now, we determine the height of the arch above the water at points 9 meters to the left and right of the center by substituting $x=9$ into the equation of the ellipse and solving for y:

$$\frac{9^2}{225}+\frac{y^2}{100}=1$$
$$\frac{9}{25}+\frac{y^2}{100}=1$$
$$\frac{y^2}{100}=\frac{16}{25}$$
$$y^2=64$$
$$y=8$$

At points 9 meters from the center of the river, the arch is 8 meters above the surface of the water. Thus, the barge can fit through the opening with about 1 meter of clearance above the top corners.

c. No. From part (b), the barge only has 1 meter of clearance when the water level is at it normal stage. Thus, if the water level increases by 1.1 meters, then the barge will not be able to pass under the bridge.

45. To find the perihelion of Earth, we recognize that $\text{Perihilion}=2(\text{Mean distance})-\text{Aphelion}$.
Thus, the perihelion of Earth is
$2\cdot93-94.5=91.5$ million miles.
To find the equation of the elliptical orbit of Earth, we draw the ellipse on a Cartesian plane so that the center of the ellipse is at the origin and the major axis is along the x-axis. The equation of the ellipse is of the form $\dfrac{x^2}{a^2}+\dfrac{y^2}{b^2}=1$. Now, the mean distance of Earth from the Sun is 93 million miles, so the distance from the center of the orbit to each vertex is $a=93$ million miles. Since the aphelion of Earth is 94.5 million miles, the distance from the center of the orbit to each focus is $c=94.5-93=1.5$ million miles.

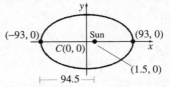

Now, because $b^2=a^2-c^2$, we have that $b^2=93^2-1.5^2=8649-2.25=8646.75$. Thus, the equation that describes the orbit of Earth is:
$$\frac{x^2}{8649}+\frac{y^2}{8646.75}=1.$$

47. The mean distance of Jupiter from the Sun is $507-23.2=483.8$ million miles.
To find the perihelion of Jupiter, we recognize that $\text{Perihilion}=2(\text{Mean distance})-\text{Aphelion}$.
Thus, the perihelion of Jupiter is
$2\cdot483.8-507=460.6$ million miles.
To find the equation of the elliptical orbit of Jupiter, we draw the ellipse on a Cartesian plane so that the center of the ellipse is at the origin and the major axis is along the x-axis. The equation of the ellipse is of the form $\dfrac{x^2}{a^2}+\dfrac{y^2}{b^2}=1$. Now, we found above that the mean distance of Jupiter from the Sun is 483.8 million miles, so the distance from the center of the orbit to each vertex is $a=483.8$ million miles. We are given the distance from the center of the orbit to the Sun (that is, to each focus) is $c=23.2$ million miles.

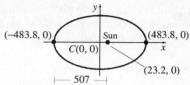

Now, because $b^2 = a^2 - c^2$, we have that

$$b^2 = 483.8^2 - 23.2^2$$
$$= 234,062.44 - 538.24$$
$$= 233,524.2$$

Thus, the equation that describes the orbit of

Jupiter is: $\dfrac{x^2}{234,062.44} + \dfrac{y^2}{233,524.2} = 1$.

49. The major axis of the ellipse is the line $y = 2$, which is parallel to the x-axis, so the equation of

the ellipse is of the form $\dfrac{(x-h)^2}{a^2} + \dfrac{(y-k)^2}{b^2} = 1$.

The center is $(h, k) = (1, 2)$. The vertices are $(-3,\ 2)$ and $(5,\ 2)$, so $a = 5 - 1 = 4$. The points $(1, -1)$ and $(1,\ 5)$ are the points on the ellipse that are below and above the center, so $b = 5 - 2 = 3$. Thus, the equation is

$$\frac{(x-1)^2}{4^2} + \frac{(y-2)^2}{3^2} = 1$$
$$\frac{(x-1)^2}{16} + \frac{(y-2)^2}{9} = 1$$

51. The major axis of the ellipse is the line $x = 2$, which is parallel to the y-axis, so the equation of

the ellipse is of the form $\dfrac{(x-h)^2}{b^2} + \dfrac{(y-k)^2}{a^2} = 1$.

The center is $(h, k) = (2, 0)$. The vertices are $(2, -4)$ and $(2,\ 4)$, so $a = 4 - 0 = 4$.

The points $(0,\ 0)$ and $(4,\ 0)$ are the points on the ellipse that are left and right of the center, so $b = 4 - 2 = 2$. Thus, the equation is

$$\frac{(x-2)^2}{2^2} + \frac{(y-0)^2}{4^2} = 1$$
$$\frac{(x-2)^2}{4} + \frac{y^2}{16} = 1$$

53. Let $a = b$, then

$$\frac{x^2}{a^2} + \frac{y^2}{b^2} = 1$$
$$\frac{x^2}{a^2} + \frac{y^2}{a^2} = 1$$
$$a^2\left(\frac{x^2}{a^2} + \frac{y^2}{a^2}\right) = a^2(1)$$
$$x^2 + y^2 = a^2$$

which is the equation of a circle with center $(0, 0)$ and radius a.

Because $c^2 = a^2 - b^2$, we have that

$c^2 = a^2 - a^2 = 0$, so that $c = 0$. This means that the foci are a distance of 0 units from the center point. In other words, the foci are located at the center point.

55. Since the center of the ellipse is the origin O, the major axis is along the x-axis, and point B is the y-intercept, then by definition, $F = (c, 0)$ and $B = (0, b)$. Thus, $d(O, F) = c$ and $d(O, B) = b$.

Now, the sum of the distances from the two foci to a point on the ellipse is defined to be $2a$. Since B is a point on the ellipse that is midway between the two foci, then the distance from F to B must be half this total distance:

$$d(F, B) = \frac{1}{2}(2a) = a.$$

57. $f(x) = \dfrac{5}{x+2}$

$$f(5) = \frac{5}{5+2} = \frac{5}{7} \approx 0.71429$$
$$f(10) = \frac{5}{10+2} = \frac{5}{12} \approx 0.41667$$
$$f(100) = \frac{5}{100+2} = \frac{5}{102} \approx 0.04902$$
$$f(1000) = \frac{5}{1000+2} = \frac{5}{1002} \approx 0.00499$$

x	5	10	100	100
$f(x)$	0.71429	0.41667	0.04902	0.00499

59. $f(x) = \dfrac{2x+1}{x-3}$

$f(5) = \dfrac{2(5)+1}{5-3} = \dfrac{10+1}{5-3} = \dfrac{11}{2} = 5.5$

$f(10) = \dfrac{2(10)+1}{10-3} = \dfrac{20+1}{10-3} = \dfrac{21}{7} = 3$

$f(100) = \dfrac{2(100)+1}{100-3} = \dfrac{200+1}{100-3} = \dfrac{201}{97} \approx 2.07216$

$f(1000) = \dfrac{2(1000)+1}{1000-3} = \dfrac{2000+1}{1000-3} = \dfrac{2001}{997} \approx 2.00702$

x	5	10	100	100
$f(x)$	5.5	3	2.07216	2.00702

61. $f(x) = \dfrac{x^2+3x+1}{x+1}$; $g(x) = x+2$

$f(5) = \dfrac{5^2+3(5)+1}{5+1} = \dfrac{41}{6} \approx 6.83333$

$f(10) = \dfrac{10^2+3(10)+1}{10+1} = \dfrac{131}{11} \approx 11.90909$

$f(100) = \dfrac{100^2+3(100)+1}{100+1} = \dfrac{10,301}{101} \approx 101.99010$

$f(1000) = \dfrac{1000^2+3(1000)+1}{1000+1}$

$\qquad = \dfrac{1,003,001}{1001} \approx 1001.99900$

$g(5) = 5+2 = 7$

$g(10) = 10+2 = 12$

$g(100) = 100+2 = 102$

$g(1000) = 1000+2 = 1002$

x	5	10	100	100
$f(x)$	6.83333	11.90909	101.99010	1001.99900
$g(x)$	7	12	102	1002

63. In Problems 57 and 58, the degrees of the numerators are less than the degrees of the denominators. In Problems 59 and 60, the degrees of the numerators and denominators are the same.

Conjecture 1: If the degree of the numerator of a rational function is less than the degree of the denominator, then as x increases, the value of the function will approach zero (0).

Conjecture 2: If the degree of the numerator of a rational function equals the degree of the

denominator, then as x increases, the value of the function will approach the ratio of the leading coefficients of the numerator and denominator.

65. $\dfrac{x^2}{25} + \dfrac{y^2}{16} = 1$

$\dfrac{y^2}{16} = 1 - \dfrac{x^2}{25}$

$y^2 = 16\left(1 - \dfrac{x^2}{25}\right)$

$y = \pm\sqrt{16\left(1 - \dfrac{x^2}{25}\right)} = \pm 4\sqrt{1 - \dfrac{x^2}{25}}$

Let $Y_1 = 4\sqrt{1 - \dfrac{x^2}{25}}$ and $Y_2 = -4\sqrt{1 - \dfrac{x^2}{25}}$.

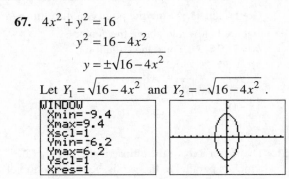

67. $4x^2 + y^2 = 16$

$y^2 = 16 - 4x^2$

$y = \pm\sqrt{16 - 4x^2}$

Let $Y_1 = \sqrt{16 - 4x^2}$ and $Y_2 = -\sqrt{16 - 4x^2}$.

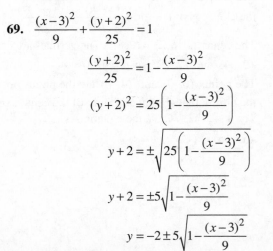

69. $\dfrac{(x-3)^2}{9} + \dfrac{(y+2)^2}{25} = 1$

$\dfrac{(y+2)^2}{25} = 1 - \dfrac{(x-3)^2}{9}$

$(y+2)^2 = 25\left(1 - \dfrac{(x-3)^2}{9}\right)$

$y+2 = \pm\sqrt{25\left(1 - \dfrac{(x-3)^2}{9}\right)}$

$y+2 = \pm 5\sqrt{1 - \dfrac{(x-3)^2}{9}}$

$y = -2 \pm 5\sqrt{1 - \dfrac{(x-3)^2}{9}}$

Let $Y_1 = -2 + 5\sqrt{1 - \dfrac{(x-3)^2}{9}}$ and

$Y_2 = -2 - 5\sqrt{1 - \dfrac{(x-3)^2}{9}}$.

```
WINDOW
 Xmin=-9.4
 Xmax=9.4
 Xscl=1
 Ymin=-8.2
 Ymax=4.2
 Yscl=1
 Xres=1
```

71. $\dfrac{(x+2)^2}{16} + \dfrac{(y-5)^2}{4} = 1$

$$\dfrac{(y-5)^2}{4} = 1 - \dfrac{(x+2)^2}{16}$$

$$(y-5)^2 = 4\left(1 - \dfrac{(x+2)^2}{16}\right)$$

$$y - 5 = \pm\sqrt{4\left(1 - \dfrac{(x+2)^2}{16}\right)}$$

$$y - 5 = \pm 2\sqrt{1 - \dfrac{(x+2)^2}{16}}$$

$$y = 5 \pm 2\sqrt{1 - \dfrac{(x+2)^2}{16}}$$

Let $Y_1 = 5 + 2\sqrt{1 - \dfrac{(x+2)^2}{16}}$ and

$Y_2 = 5 - 2\sqrt{1 - \dfrac{(x+2)^2}{16}}$.

```
WINDOW
 Xmin=-9.4
 Xmax=9.4
 Xscl=1
 Ymin=-4.2
 Ymax=8.2
 Yscl=1
 Xres=1
```

Section 9.5

Are You Ready for This Section?

R1. Start: $x^2 - 5x$

Add: $\left(\dfrac{1}{2} \cdot (-5)\right)^2 = \dfrac{25}{4}$

Result: $x^2 - 5x + \dfrac{25}{4}$

Factored Form: $\left(x - \dfrac{5}{2}\right)^2$

R2. $y^2 = 64$

$y = \pm\sqrt{64}$

$y = \pm 8$

The solution set is $\{-8,\ 8\}$.

Section 9.5 Quick Checks

1. A <u>hyperbola</u> is the collection of points in the plane the difference of whose distances from two fixed points is a positive constant.

2. For a hyperbola, the foci lie on a line called the <u>transverse axis.</u>

3. The line through the center of a hyperbola that is perpendicular to the transverse axis is called the <u>conjugate axis.</u>

4. $\dfrac{x^2}{36} - \dfrac{y^2}{64} = 1$

Notice the equation is of the form $\dfrac{x^2}{a^2} - \dfrac{y^2}{b^2} = 1$.

Because the x^2-term is first, the transverse axis is the *x*-axis and the hyperbola opens left and right. The center of the hyperbola is the origin. We have that $a^2 = 36$ and $b^2 = 64$. Because $c^2 = a^2 + b^2$, we have that $c^2 = 36 + 64 = 100$, so that $c = \sqrt{100} = 10$. The vertices are $(\pm a, 0) = (\pm 6, 0)$, and the foci are $(\pm c, 0) = (\pm 10, 0)$. To help graph the hyperbola, we plot the points on the graph above and below the foci. Let $x = \pm 10$:

$$\dfrac{(\pm 10)^2}{36} - \dfrac{y^2}{64} = 1$$

$$\dfrac{100}{36} - \dfrac{y^2}{64} = 1$$

$$\dfrac{25}{9} - \dfrac{y^2}{64} = 1$$

$$-\dfrac{y^2}{64} = -\dfrac{16}{9}$$

$$y^2 = \dfrac{1024}{9}$$

$$y = \pm\dfrac{32}{3}$$

The points above and below the foci are $\left(-10, -\dfrac{32}{3}\right)$, $\left(-10,\ \dfrac{32}{3}\right)$, $\left(10, -\dfrac{32}{3}\right)$, and

$\left(10, \dfrac{32}{3}\right).$

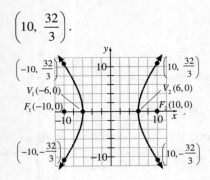

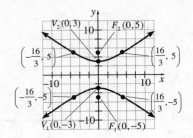

5. $\dfrac{y^2}{9} - \dfrac{x^2}{16} = 1$

Notice the equation is of the form $\dfrac{y^2}{a^2} - \dfrac{x^2}{b^2} = 1$.

Because the y^2-term is first, the transverse axis is the y-axis and the hyperbola opens up and down. The center of the hyperbola is the origin. We have that $a^2 = 9$ and $b^2 = 16$. Because $c^2 = a^2 + b^2$, we have that $c^2 = 9 + 16 = 25$, so that $c = \sqrt{25} = 5$. The vertices are $(0, \pm a) = (0, \pm 3)$, and the foci are $(0, \pm c) = (0, \pm 5)$. To help graph the hyperbola, we plot the points on the graph to the left and right of the foci. Let $y = \pm 5$:

$$\dfrac{(\pm 5)^2}{9} - \dfrac{x^2}{16} = 1$$

$$\dfrac{25}{9} - \dfrac{x^2}{16} = 1$$

$$-\dfrac{x^2}{16} = -\dfrac{16}{9}$$

$$x^2 = \dfrac{256}{9}$$

$$x = \pm \dfrac{16}{3}$$

The points to the left and right of the foci are $\left(-\dfrac{16}{3}, -5\right)$, $\left(-\dfrac{16}{3}, 5\right)$, $\left(\dfrac{16}{3}, -5\right)$, and $\left(\dfrac{16}{3}, 5\right)$.

6. False.

7. The given vertices $(\pm 4, 0)$ and focus $(6, 0)$ all lie on the x-axis. Thus, the transverse axis is the x-axis and the hyperbola opens left and right. The center of the hyperbola is the midpoint between the two vertices. Therefore, the center is $(0, 0)$ and the equation of the hyperbola is of the form $\dfrac{x^2}{a^2} - \dfrac{y^2}{b^2} = 1$. Now, the distance from the center $(0, 0)$ to each vertex is $a = 4$ units. Likewise, the distance from the center to the given focus $(6, 0)$ is $c = 6$ units. Also, $b^2 = c^2 - a^2$, so $b^2 = 6^2 - 4^2 = 36 - 16 = 20$. Thus, the equation of the hyperbola is $\dfrac{x^2}{16} - \dfrac{y^2}{20} = 1$.

To help graph the hyperbola, we first find the focus that was not given. Since the center is at $(0, 0)$ and since one focus is at $(6, 0)$, the other focus must be at $(-6, 0)$. Next, plot the points on the graph above and below the foci. Let $x = \pm 6$:

$$\dfrac{(\pm 6)^2}{16} - \dfrac{y^2}{20} = 1$$

$$\dfrac{36}{16} - \dfrac{y^2}{20} = 1$$

$$\dfrac{9}{4} - \dfrac{y^2}{20} = 1$$

$$-\dfrac{y^2}{20} = -\dfrac{5}{4}$$

$$y^2 = 25$$

$$y = \pm 5$$

The points above and below of the foci are $(-6, -5)$, $(-6, 5)$, $(6, -5)$, and $(6, 5)$.

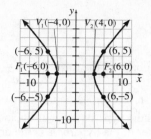

8. The asymptotes of the hyperbola $\dfrac{x^2}{a^2} - \dfrac{y^2}{b^2} = 1$

are $y = -\dfrac{b}{a}x$ and $y = \dfrac{b}{a}x$.

9. $x^2 - 9y^2 = 9$

$\dfrac{x^2 - 9y^2}{9} = \dfrac{9}{9}$

$\dfrac{x^2}{9} - \dfrac{y^2}{1} = 1$

Notice that the equation is of the form

$\dfrac{x^2}{a^2} - \dfrac{y^2}{b^2} = 1$. The center of the hyperbola is

$(0,0)$. Because the x^2-term is first, the
hyperbola opens left and right. The transverse
axis is along the x-axis. We have that $a^2 = 9$

and $b^2 = 1$. Because $c^2 = a^2 + b^2$, we have that

$c^2 = 9 + 1 = 10$, so that $c = \sqrt{10}$. The vertices
are $(\pm a, 0) = (\pm 3, 0)$, and the foci are

$(\pm c, 0) = \left(\pm\sqrt{10}, 0\right)$. The asymptotes are of the

form $y = -\dfrac{b}{a}x$ and $y = \dfrac{b}{a}x$. Since $a = 3$ and

$b = 1$, the equations of the asymptotes are

$y = -\dfrac{1}{3}x$ and $y = \dfrac{1}{3}x$. To help graph the

hyperbola, we form the rectangle using the
points $(\pm a, 0) = (\pm 3, 0)$ and $(0, \pm b) = (0, \pm 1)$.
The diagonals are the asymptotes.

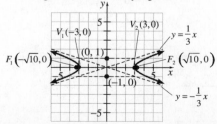

10. $\dfrac{y^2}{16} - \dfrac{x^2}{9} = 1$

Notice the equation is of the form $\dfrac{y^2}{a^2} - \dfrac{x^2}{b^2} = 1$.

Because the y^2-term is first, the transverse is
along the y-axis and the hyperbola opens up and
down. The center of the hyperbola is $(0,0)$. We

have that $a^2 = 16$ and $b^2 = 9$. Because

$c^2 = a^2 + b^2$, we have that $c^2 = 16 + 9 = 25$, so

that $c = \sqrt{25} = 5$. The vertices are
$(0, \pm a) = (0, \pm 4)$, and the foci are

$(0, \pm c) = (0, \pm 5)$. The asymptotes are of the

form $y = -\dfrac{a}{b}x$ and $y = \dfrac{a}{b}x$. Since $a = 4$ and

$b = 3$, the equations of the asymptotes are

$y = -\dfrac{4}{3}x$ and $y = \dfrac{4}{3}x$. To help graph the

hyperbola, we form the rectangle using the
points $(0, \pm a) = (0, \pm 4)$ and $(\pm b, 0) = (\pm 3, 0)$.
The diagonals are the asymptotes.

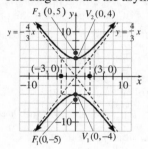

9.5 Exercises

11. b. The transverse axis is the x-axis and the
hyperbola opens left and right. Also, the
distance between the vertices $(\pm 1, 0)$ and the
center $(0,0)$ is $a = 1$ unit. This means that the
equation of the hyperbola is of the form

$\dfrac{x^2}{a^2} - \dfrac{y^2}{b^2} = 1$

$\dfrac{x^2}{1^2} - \dfrac{y^2}{b^2} = 1$

$x^2 - \dfrac{y^2}{b^2} = 1$

Of the list of provided equations, only equation

(b): $x^2 - \dfrac{y^2}{4} = 1$ is of this form.

13. a. The transverse axis is the x-axis and the hyperbola opens left and right. Also, the distance between the vertices $(\pm 2, 0)$ and the center $(0,0)$ is $a = 2$ units. This means that the equation of the hyperbola is of the form

$$\frac{x^2}{a^2} - \frac{y^2}{b^2} = 1$$

$$\frac{x^2}{2^2} - \frac{y^2}{b^2} = 1$$

$$\frac{x^2}{4} - \frac{y^2}{b^2} = 1$$

Of the list of provided equations, only equation

(a): $\frac{x^2}{4} - y^2 = 1$ is of this form.

15. $\frac{x^2}{4} - \frac{y^2}{16} = 1$

Notice the equation is of the form $\frac{x^2}{a^2} - \frac{y^2}{b^2} = 1$.

Because the x^2-term is first, the transverse axis is the x-axis and the hyperbola opens left and right. The center of the hyperbola is the origin. We have that $a^2 = 4$ and $b^2 = 16$. Because $c^2 = a^2 + b^2$, we have that $c^2 = 4 + 16 = 20$, so that $c = \sqrt{20} = 2\sqrt{5}$. The vertices are $(\pm a, 0) = (\pm 2, 0)$, and the foci are $(\pm c, 0) = \left(\pm 2\sqrt{5}, 0\right)$. Since $a = 2$ and $b = 4$, the equations of the asymptotes are

$y = \frac{4}{2}x = 2x$ and $y = -\frac{4}{2}x = -2x$. To help graph the hyperbola, we form the rectangle using the points $(\pm a, 0) = (\pm 2, 0)$ and $(0, \pm b) = (0, \pm 4)$. The diagonals are the asymptotes.

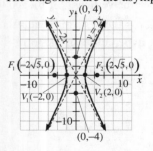

17. $\frac{y^2}{25} - \frac{x^2}{36} = 1$

Notice the equation is of the form $\frac{y^2}{a^2} - \frac{x^2}{b^2} = 1$.

Because the y^2-term is first, the transverse axis is the y-axis and the hyperbola opens up and down. The center of the hyperbola is the origin. We have that $a^2 = 25$ and $b^2 = 36$. Because $c^2 = a^2 + b^2$, we have that $c^2 = 25 + 36 = 61$, so that $c = \sqrt{61}$. The vertices are $(0, \pm a) = (0, \pm 5)$, and the foci are $(0, \pm c) = \left(0, \pm\sqrt{61}\right)$. To help graph the hyperbola, we plot the points on the graph to the left and right of the foci. Let $y = \pm\sqrt{61}$:

$$\frac{\left(\pm\sqrt{61}\right)^2}{25} - \frac{x^2}{36} = 1$$

$$\frac{61}{25} - \frac{x^2}{36} = 1$$

$$-\frac{x^2}{36} = -\frac{36}{25}$$

$$x^2 = \frac{1296}{25}$$

$$x = \pm\frac{36}{5}$$

The points to the left and right of the foci are $\left(\frac{36}{5}, \sqrt{61}\right)$, $\left(-\frac{36}{5}, \sqrt{61}\right)$, $\left(\frac{36}{5}, -\sqrt{61}\right)$, and $\left(-\frac{36}{5}, -\sqrt{61}\right)$.

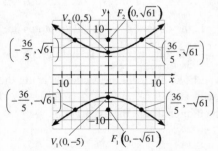

19. $4x^2 - y^2 = 36$

$$\frac{4x^2 - y^2}{36} = \frac{36}{36}$$

$$\frac{x^2}{9} - \frac{y^2}{36} = 1$$

Notice the equation is of the form $\dfrac{x^2}{a^2} - \dfrac{y^2}{b^2} = 1$.

Because the x^2-term is first, the transverse axis is the x-axis and the hyperbola opens left and right. The center of the hyperbola is the origin. We have that $a^2 = 9$ and $b^2 = 36$. Because $c^2 = a^2 + b^2$, we have that $c^2 = 9 + 36 = 45$, so that $c = \sqrt{45} = 3\sqrt{5}$. The vertices are $(\pm a, 0) = (\pm 3, 0)$, and the foci are $(\pm c, 0) = \left(\pm 3\sqrt{5}, 0\right)$. To help graph the hyperbola, we plot the points on the graph above and below the foci. Let $x = \pm 3\sqrt{5}$:

$$\frac{\left(\pm 3\sqrt{5}\right)^2}{9} - \frac{y^2}{36} = 1$$

$$\frac{45}{9} - \frac{y^2}{36} = 1$$

$$5 - \frac{y^2}{36} = 1$$

$$-\frac{y^2}{36} = -4$$

$$y^2 = 144$$

$$y = \pm 12$$

The points above and below the foci are $\left(3\sqrt{5}, 12\right)$, $\left(3\sqrt{5}, -12\right)$, $\left(-3\sqrt{5}, 12\right)$, and $\left(-3\sqrt{5}, -12\right)$.

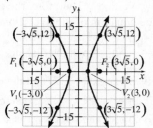

21. $25y^2 - x^2 = 100$

$$\frac{25y^2 - x^2}{100} = \frac{100}{100}$$

$$\frac{y^2}{4} - \frac{x^2}{100} = 1$$

Notice the equation is of the form $\dfrac{y^2}{a^2} - \dfrac{x^2}{b^2} = 1$.

Because the y^2-term is first, the transverse is along the y-axis and the hyperbola opens up and

down. The center of the hyperbola is $(0,0)$. We have that $a^2 = 4$ and $b^2 = 100$. Because $c^2 = a^2 + b^2$, we have that $c^2 = 100 + 4 = 104$, so that $c = \sqrt{104} = 2\sqrt{26}$. The vertices are $(0, \pm a) = (0, \pm 2)$, and the foci are $(0, \pm c) = \left(0, \pm 2\sqrt{26}\right)$. Since $a = 2$ and $b = 10$, the equations of the asymptotes are $y = \dfrac{2}{10}x = \dfrac{1}{5}x$ and $y = -\dfrac{2}{10}x = -\dfrac{1}{5}x$. To help graph the hyperbola, we form the rectangle using the points $(0, \pm a) = (0, \pm 2)$ and $(\pm b, 0) = (\pm 10, 0)$. The diagonals are the asymptotes.

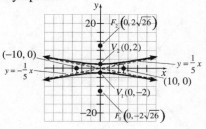

23. The given center $(0,0)$, focus $(3,0)$, and vertex $(2,0)$ all lie on the x-axis. Thus, the transverse axis is the x-axis and the hyperbola opens left and right. This means that the equation of the hyperbola is of the form $\dfrac{x^2}{a^2} - \dfrac{y^2}{b^2} = 1$. Now, the distance between the given vertex $(2,0)$ and the center $(0,0)$ is $a = 2$ units. Likewise, the distance between the given focus $(3,0)$ and the center is $c = 3$ units. Because $b^2 = c^2 - a^2$, we have that $b^2 = 3^2 - 2^2 = 9 - 4 = 5$. Thus, the equation of the hyperbola is $\dfrac{x^2}{4} - \dfrac{y^2}{5} = 1$.

To help graph the hyperbola, we first find the vertex and focus that were not given. Since the center is at $(0,0)$ and since one vertex is at $(2,0)$, the other vertex must be at $(-2,0)$. Similarly, one focus is at $(3,0)$, so the other focus is at $(-3,0)$. Next, plot the points on the graph above and below the foci. Let $x = \pm 3$:

$$\frac{(\pm 3)^2}{4} - \frac{y^2}{5} = 1$$

$$\frac{9}{4} - \frac{y^2}{5} = 1$$

$$-\frac{y^2}{5} = -\frac{5}{4}$$

$$y^2 = \frac{25}{4}$$

$$y = \pm\frac{5}{2}$$

The points above and below the foci are $\left(3, \frac{5}{2}\right)$,

$\left(3, -\frac{5}{2}\right)$, $\left(-3, \frac{5}{2}\right)$, and $\left(-3, -\frac{5}{2}\right)$.

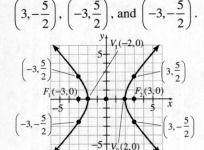

25. The given vertices $(0, \pm 5)$ and focus $(0, 7)$ all lie on the y-axis. Thus, the transverse axis is the y-axis and the hyperbola opens up and down. The center of the hyperbola is the midpoint between the two vertices. Therefore, the center is $(0, 0)$ and the equation of the hyperbola is of the form $\frac{y^2}{a^2} - \frac{x^2}{b^2} = 1$. Now, the distance from the center $(0, 0)$ to each vertex is $a = 5$ units. Likewise, the distance from the center to the given focus $(0, 7)$ is $c = 7$ units. Also, $b^2 = c^2 - a^2$, so $b^2 = 7^2 - 5^2 = 49 - 25 = 24$. Thus, the equation of the hyperbola is

$$\frac{y^2}{25} - \frac{x^2}{24} = 1.$$

To help graph the hyperbola, we first find the focus that was not given. Since the center is at $(0, 0)$ and since one focus is at $(0, 7)$, the other focus must be at $(0, -7)$. Next, plot the points on the graph to the left and right of the foci. Let $y = \pm 7$:

$$\frac{(\pm 7)^2}{25} - \frac{x^2}{24} = 1$$

$$\frac{49}{25} - \frac{x^2}{24} = 1$$

$$-\frac{x^2}{24} = -\frac{24}{25}$$

$$x^2 = \frac{576}{25}$$

$$x = \pm\frac{24}{5}$$

The points to the left and right of the foci are $\left(\frac{24}{5}, 7\right)$, $\left(-\frac{24}{5}, 7\right)$, $\left(\frac{24}{5}, -7\right)$, and $\left(-\frac{24}{5}, -7\right)$.

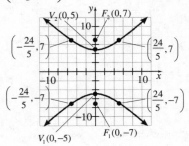

27. The given foci $(\pm 10, 0)$ and vertex $(-7, 0)$ all lie on the x-axis. Thus, the transverse axis is the x-axis and the hyperbola opens left and right. The center of the hyperbola is the midpoint between the two vertices. Therefore, the center is $(0, 0)$ and the equation of the hyperbola is of the form $\frac{x^2}{a^2} - \frac{y^2}{b^2} = 1$. Now, the distance between the given vertex $(-7, 0)$ and the center $(0, 0)$ is $a = 7$ units. Likewise, the distance between the given foci and the center is $c = 10$ units. Also, $b^2 = c^2 - a^2$, so $b^2 = 10^2 - 7^2 = 100 - 49 = 51$. Thus, the equation of the hyperbola is $\frac{x^2}{49} - \frac{y^2}{51} = 1$.

To help graph the hyperbola, we first find the vertex that was not given. Since the center is at $(0, 0)$ and since one vertex is at $(-7, 0)$, the other vertex must be at $(7, 0)$. Next, plot the points on the graph above and below the foci. Let $x = \pm 10$:

$$\frac{(\pm 10)^2}{49} - \frac{y^2}{51} = 1$$

$$\frac{100}{49} - \frac{y^2}{51} = 1$$

$$-\frac{y^2}{51} = -\frac{51}{49}$$

$$y^2 = \frac{2601}{49}$$

$$y = \pm \frac{51}{7}$$

The points above and below the foci are

$\left(10, \frac{51}{7}\right)$, $\left(10, -\frac{51}{7}\right)$, $\left(-10, \frac{51}{7}\right)$, and

$\left(-10, -\frac{51}{7}\right)$.

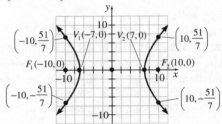

29. $\dfrac{x^2}{25} - \dfrac{y^2}{9} = 1$

Notice the equation is of the form $\dfrac{x^2}{a^2} - \dfrac{y^2}{b^2} = 1$.

Because the x^2-term is first, the transverse axis is the *x*-axis and the hyperbola opens left and right. The center of the hyperbola is the origin. We have that $a^2 = 25$ and $b^2 = 9$. Because $c^2 = a^2 + b^2$, we have that $c^2 = 25 + 9 = 34$, so that $c = \sqrt{34}$. The vertices are $(\pm a, 0) = (\pm 5, 0)$, and the foci are $(\pm c, 0) = \left(\pm \sqrt{34}, 0\right)$. Since $b^2 = 9$, we have that $b = 3$. The equations of the asymptotes are

$$y = \frac{b}{a}x = \frac{3}{5}x \text{ and } y = -\frac{b}{a}x = -\frac{3}{5}x.$$

To graph the hyperbola, we form a rectangle using the points $(\pm a, 0) = (\pm 5, 0)$ and $(0, \pm b) = (0, \pm 3)$. The diagonals help us draw the asymptotes.

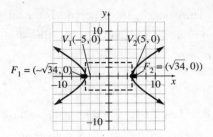

31. $\dfrac{y^2}{4} - \dfrac{x^2}{100} = 1$

Notice the equation is of the form $\dfrac{y^2}{a^2} - \dfrac{x^2}{b^2} = 1$.

Because the y^2-term is first, the transverse axis is the *y*-axis and the hyperbola opens up and down. The center of the hyperbola is the origin. We have that $a^2 = 4$ and $b^2 = 100$. Because $c^2 = a^2 + b^2$, we have that $c^2 = 4 + 100 = 104$, so that $c = \sqrt{104} = 2\sqrt{26}$. The vertices are $(0, \pm a) = (0, \pm 2)$, and the foci are $(0, \pm c) = \left(0, \pm 2\sqrt{26}\right)$. Since $b^2 = 100$, we have that $b = 10$. The equations of the asymptotes are

$$y = \frac{a}{b}x = \frac{2}{10}x = \frac{1}{5}x \text{ and}$$

$$y = -\frac{a}{b}x = -\frac{2}{10}x = -\frac{1}{5}x.$$

To graph the hyperbola, we form a rectangle using the points $(0, \pm a) = (0, \pm 2)$ and $(\pm b, 0) = (\pm 10, 0)$. The diagonals help us draw the asymptotes.

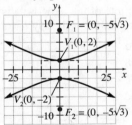

33. $x^2 - y^2 = 4$

$$\frac{x^2}{4} - \frac{y^2}{4} = 1$$

Notice the equation is of the form $\dfrac{x^2}{a^2} - \dfrac{y^2}{b^2} = 1$.

Because the x^2-term is first, the transverse axis is the *x*-axis and the hyperbola opens left and right. The center of the hyperbola is the origin. We have that $a^2 = 4$ and $b^2 = 4$. Because $c^2 = a^2 + b^2$, we have that $c^2 = 4 + 4 = 8$, so that $c = \sqrt{8} = 2\sqrt{2}$. The vertices are $(\pm a, 0) = (\pm 2, 0)$, and the foci are $(\pm c, 0) = (\pm 2\sqrt{2}, 0)$. Since $b^2 = 4$, we have that $b = 2$. The equations of the asymptotes are

$$y = \frac{b}{a}x = \frac{2}{2}x = x \text{ and } y = -\frac{b}{a}x = -\frac{2}{2}x = -x.$$

To graph the hyperbola, we form a rectangle using the points $(\pm a, 0) = (\pm 2, 0)$ and $(0, \pm b) = (0, \pm 2)$. The diagonals help us draw the asymptotes.

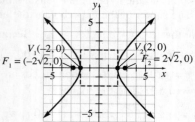

35. The given vertices $(0, \pm 8)$ both lie on the y-axis. Thus, the transverse axis is the y-axis and the hyperbola opens up and down. The center of the hyperbola is the midpoint between the two vertices. Therefore, the center is $(0, 0)$ and the equation of the hyperbola is of the form $\frac{y^2}{a^2} - \frac{x^2}{b^2} = 1$. Now, the distance from the center $(0, 0)$ to each vertex is $a = 8$ units. We are given that an asymptote of the hyperbola is $y = 2x$. Now the equation of the asymptote is of the form is $y = \frac{a}{b}x$ and $a = 8$. So,

$$\frac{a}{b} = 2$$
$$\frac{8}{b} = 2$$
$$8 = 2b$$
$$4 = b$$

Thus, the equation of the hyperbola is

$$\frac{y^2}{8^2} - \frac{x^2}{4^2} = 1$$
$$\frac{y^2}{64} - \frac{x^2}{16} = 1$$

To help graph the hyperbola, we first find the

foci. Since $c^2 = a^2 + b^2$, we have that $c^2 = 64 + 16 = 80$, so $c = 4\sqrt{5}$. Since the center is $(0, 0)$, the foci are $(0, \pm 4\sqrt{5})$. We form the rectangle using the points $(0, \pm a) = (0, \pm 8)$ and $(\pm b, 0) = (\pm 4, 0)$. The diagonals are the two asymptotes $y = 2x$ and $y = -2x$.

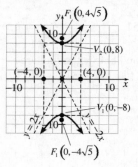

37. The given foci $(\pm 3, 0)$ both lie on the x-axis. Thus, the transverse axis is the x-axis and the hyperbola opens left and right. The center of the hyperbola is the midpoint between the two vertices. Therefore, the center is $(0, 0)$ and the equation of the hyperbola is of the form $\frac{x^2}{a^2} - \frac{y^2}{b^2} = 1$. We are given that an asymptote of the hyperbola is $y = x$. Now the equation of the asymptote is of the form $y = \frac{b}{a}x$. So, $\frac{b}{a} = 1$ which means $a = b$. Because each focus is 3 units from the center, we have $c = 3$ and $c^2 = 9$. So,

$$a^2 + b^2 = c^2$$
$$a^2 + a^2 = 9$$
$$2a^2 = 9$$
$$a^2 = \frac{9}{2}$$
$$a = \sqrt{\frac{9}{2}} = \frac{3}{\sqrt{2}} = \frac{3\sqrt{2}}{2}$$

Since $a = b$, we have that $a^2 = b^2 = \frac{9}{2} = 4.5$.

Thus, the equation of the hyperbola is

$$\frac{x^2}{4.5} - \frac{y^2}{4.5} = 1.$$

To help graph the hyperbola, we find the

vertices. Since $a = \dfrac{3\sqrt{2}}{2}$ and the center is

$(0,0)$, the vertices are $\left(\pm\dfrac{3\sqrt{2}}{2}, 0\right)$. We form

the rectangle using the points

$(\pm a, 0) = \left(\pm\dfrac{3\sqrt{2}}{2}, 0\right)$ and $(0, \pm b) = \left(0, \pm\dfrac{3\sqrt{2}}{2}\right)$.

The diagonals are the two asymptotes $y = x$ and
$y = -x$.

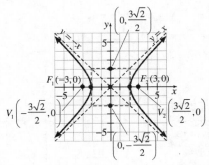

39. The graph of the hyperbola opens left and right
and the center is $(0,0)$, so the equation is of the

form $\dfrac{x^2}{a^2} - \dfrac{y^2}{b^2} = 1$. From the graph, we observe

that the asymptotes are $y = -x$ and $y = x$. Now
the equations of the asymptotes are of the form

$y = -\dfrac{b}{a}x$ and $y = \dfrac{b}{a}x$. So, $\dfrac{b}{a} = 1$ which means

$b = a$. We also observe that the vertices are
$(\pm1, 0)$, so the distance from the center to each
vertex is $a = 1$. Thus, $a = b = 1$ and
$a^2 = b^2 = 1$. Thus, the equation of the

hyperbola is $\dfrac{x^2}{1} - \dfrac{y^2}{1} = 1$ or $x^2 - y^2 = 1$.

41. The graph of the hyperbola opens up and down
and the center is $(0,0)$, so the equation is of the

form $\dfrac{y^2}{a^2} - \dfrac{x^2}{b^2} = 1$. From the graph, we observe

that the asymptotes are $y = -2x$ and $y = 2x$.
Now the equations of the asymptotes are of the

form $y = -\dfrac{a}{b}x$ and $y = \dfrac{a}{b}x$. So, $\dfrac{a}{b} = 2$ which

means $a = 2b$ or $b = \dfrac{1}{2}a$. We also observe that

the vertices are $(0, \pm 6)$, so the distance from the

center to each vertex is $a = 6$ and $b = \dfrac{1}{2}(6) = 3$.

Thus, $a^2 = 36$ and $b^2 = 9$, and the equation of

the hyperbola is $\dfrac{y^2}{36} - \dfrac{x^2}{9} = 1$.

43. We observe that the equation $\dfrac{x^2}{4} - y^2 = 1$ is of

the form $\dfrac{x^2}{a^2} - \dfrac{y^2}{b^2} = 1$ with $a = 2$ and $b = 1$.

The asymptotes of such a hyperbola are

$y = -\dfrac{b}{a}x$ and $y = \dfrac{b}{a}x$. Thus, the asymptotes

are $y = -\dfrac{1}{2}x$ and $y = \dfrac{1}{2}x$.

Similarly, we observe that the equation

$y^2 - \dfrac{x^2}{4} = 1$ is of the form $\dfrac{y^2}{a^2} - \dfrac{x^2}{b^2} = 1$ with

$a = 1$ and $b = 2$. The asymptotes of such a

hyperbola are $y = -\dfrac{a}{b}x$ and $y = \dfrac{a}{b}x$. Thus, the

asymptotes are $y = -\dfrac{1}{2}x$ and $y = \dfrac{1}{2}x$.

Since the two hyperbolas have the same
asymptotes, they are conjugate.

Now, the vertices of $y^2 - \dfrac{x^2}{4} = 1$ are $(0, \pm 1)$,

each of which is $a = 1$ unit above and below the
center. Utilizing the vertices and the asymptotes,

we draw the graph of $y^2 - \dfrac{x^2}{4} = 1$. Similarly,

the vertices of $\dfrac{x^2}{4} - y^2 = 1$ are $(\pm 2, 0)$, each of

which is $a = 2$ units to the left and right of the

center. We draw the graph of $\dfrac{x^2}{4} - y^2 = 1$ by

utilizing the vertices and the asymptotes.

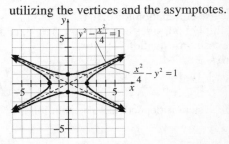

45. Answers may vary. One possibility follows: The asymptotes provide a means for determining the opening of each branch of the hyperbola without having to find and plot additional points.

47. $\begin{cases} 2x - 3y = -9 & (1) \\ -x + 5y = 8 & (2) \end{cases}$

Because equation (2) is easily solved for x, we use substitution to solve the system.
Equation (2) solved for x is $x = 5y - 8$.
Substituting $5y - 8$ for x in equation (1), we obtain
$$2(5y - 8) - 3y = -9$$
$$10y - 16 - 3y = -9$$
$$7y - 16 = -9$$
$$7y = 7$$
$$y = 1$$
Substituting 1 for y into equation (2), we obtain
$$-x + 5(1) = 8$$
$$-x + 5 = 8$$
$$-x = 3$$
$$x = -3$$
The solution is the ordered pair $(-3, 1)$.

49. $\begin{cases} 2x - 3y = 6 & (1) \\ -6x + 9y = -18 & (2) \end{cases}$

Because none of the variables have a coefficient of 1, we use elimination to solve the system.
Multiply both sides of equation (1) by 3, and add the result to equation (2).
$$6x - 9y = 18$$
$$-6x + 9y = -18$$
$$0 = 0$$
The system is dependent. The solution is
$\{(x, y) \mid 2x - 3y = 6\}$.

51. $\begin{cases} 6x + 3y = 4 & (1) \\ -2x - y = -\dfrac{4}{3} & (2) \end{cases}$

We use elimination to solve the system.
Multiply both sides of equation (2) by 3, and add the result to equation (1).
$$-6x - 3y = -4$$
$$6x + 3y = 4$$
$$0 = 0$$
The system is dependent. The solution is
$\{(x, y) \mid 6x + 3y = 4\}$.

53. $\dfrac{x^2}{4} - \dfrac{y^2}{16} = 1$

$$-\dfrac{y^2}{16} = 1 - \dfrac{x^2}{4}$$
$$-16\left(-\dfrac{y^2}{16}\right) = -16\left(1 - \dfrac{x^2}{4}\right)$$
$$y^2 = 16\left(\dfrac{x^2}{4} - 1\right)$$
$$y = \pm\sqrt{16\left(\dfrac{x^2}{4} - 1\right)}$$
$$y = \pm 4\sqrt{\dfrac{x^2}{4} - 1}$$

Let $Y_1 = 4\sqrt{\dfrac{x^2}{4} - 1}$ and $Y_2 = -4\sqrt{\dfrac{x^2}{4} - 1}$.

55. $\dfrac{y^2}{25} - \dfrac{x^2}{36} = 1$

$$\dfrac{y^2}{25} = \dfrac{x^2}{36} + 1$$
$$y^2 = 25\left(\dfrac{x^2}{36} + 1\right)$$
$$y = \pm\sqrt{25\left(\dfrac{x^2}{36} + 1\right)}$$
$$y = \pm 5\sqrt{\dfrac{x^2}{36} + 1}$$

Let $Y_1 = 5\sqrt{\dfrac{x^2}{36} + 1}$ and $Y_2 = -5\sqrt{\dfrac{x^2}{36} + 1}$.

57. $4x^2 - y^2 = 36$

$\qquad -y^2 = -4x^2 + 36$

$\qquad y^2 = 4x^2 - 36$

$\qquad y = \pm\sqrt{4x^2 - 36}$

Let $Y_1 = \sqrt{4x^2 - 36}$ and $Y_2 = -\sqrt{4x^2 - 36}$.

```
WINDOW
 Xmin=-9.4
 Xmax=9.4
 Xscl=1
 Ymin=-6.2
 Ymax=6.2
 Yscl=1
 Xres=1
```
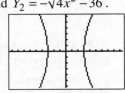

59. $25y^2 - x^2 = 100$

$\qquad 25y^2 = x^2 + 100$

$\qquad y^2 = \dfrac{x^2 + 100}{25}$

$\qquad y = \pm\sqrt{\dfrac{x^2 + 100}{25}}$

$\qquad y = \pm\dfrac{\sqrt{x^2 + 100}}{5}$

Let $Y_1 = \dfrac{\sqrt{x^2 + 100}}{5}$ and $Y_2 = -\dfrac{\sqrt{x^2 + 100}}{5}$.

```
WINDOW
 Xmin=-23.5
 Xmax=23.5
 Xscl=5
 Ymin=-15.5
 Ymax=15.5
 Yscl=5
 Xres=1
```

Putting the Concepts Together (Sections 9.1–9.5)

1. $d(P_1, P_2) = \sqrt{(x_2 - x_1)^2 + (y_2 - y_1)^2}$

$\qquad = \sqrt{(3 - (-6))^2 + (-2 - 4)^2}$

$\qquad = \sqrt{(9)^2 + (-6)^2}$

$\qquad = \sqrt{81 + 36}$

$\qquad = \sqrt{117}$

$\qquad = 3\sqrt{13}$

2. $M = \left(\dfrac{x_1 + x_2}{2}, \dfrac{y_1 + y_2}{2} \right)$

$\qquad = \left(\dfrac{-3 + 5}{2}, \dfrac{1 + (-7)}{2} \right) = \left(\dfrac{2}{2}, \dfrac{-6}{2} \right) = (1, -3)$

3. $(x + 2)^2 + (y - 8)^2 = 36$

$\qquad (x - (-2))^2 + (y - 8)^2 = 6^2$

The center is $(h, k) = (-2, 8)$, and the radius is $r = 6$.

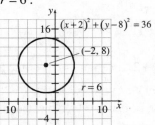

4. $x^2 + y^2 + 6x - 4y - 3 = 0$

$\qquad (x^2 + 6x) + (y^2 - 4y) = 3$

$\qquad (x^2 + 6x + 9) + (y^2 - 4y + 4) = 3 + 9 + 4$

$\qquad (x + 3)^2 + (y - 2)^2 = 16$

$\qquad (x - (-3))^2 + (y - 2)^2 = 4^2$

The center is $(h, k) = (-3, 2)$, and the radius is $r = 4$.

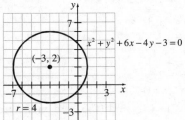

5. The radius of the circle will be the distance from the center point $(0, 0)$ to the point on the circle $(-5, 12)$. Thus, $r = \sqrt{(-5 - 0)^2 + (12 - 0)^2} = 13$. The equation of the circle is

$\qquad (x - h)^2 + (y - k)^2 = r^2$

$\qquad (x - 0)^2 + (y - 0)^2 = 13^2$

$\qquad x^2 + y^2 = 169$

6. The center of the circle will be the midpoint of the diameter with endpoints $(-1, 5)$ and $(5, -3)$.

Thus, $(h, k) = \left(\dfrac{-1 + 5}{2}, \dfrac{5 + (-3)}{2} \right) = (2, 1)$. The radius of the circle will be the distance from the center point $(2, 1)$ to one of the endpoints of the diameter, say $(-1, 5)$. Thus,

785

$r = \sqrt{(-1-2)^2 + (5-1)^2} = \sqrt{25} = 5$.

The equation of the circle is

$(x-h)^2 + (y-k)^2 = r^2$

$(x-2)^2 + (y-1)^2 = 5^2$

$(x-2)^2 + (y-1)^2 = 25$

7. Notice that $(x+2)^2 = -4(y-4)$ is of the form

$(x-h)^2 = -4a(y-k)$, where $(h,k) = (-2,4)$

and $-4a = -4$, so that $a = 1$. Now, the graph of

an equation of the form $(x-h)^2 = -4a(y-k)$

will be a parabola that opens downward with the

vertex at (h,k), focus at $(h,k-a)$, and directrix

of $y = k+a$. Note that $k-a = 4-1 = 3$ and

$k+a = 4+1 = 5$. Thus, the graph of

$(x+2)^2 = -4(y-4)$ is a parabola that opens

downward with vertex $(-2,4)$, focus $(-2,3)$,

and directrix $y = 5$. To help graph the parabola,

we plot the two points on the graph to the left

and right of the focus. Let $y = 3$:

$(x+2)^2 = -4(3-4)$

$(x+2)^2 = 4$

$x+2 = \pm 2$

$x = -2 \pm 2$

$x = -4$ or $x = 0$

The points $(-4,3)$ and $(0,3)$ are on the graph.

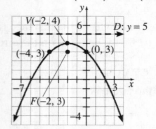

8. We complete the square in y to write the
equation in standard form:

$y^2 + 2y - 8x + 25 = 0$

$y^2 + 2y = 8x - 25$

$y^2 + 2y + 1 = 8x - 25 + 1$

$y^2 + 2y + 1 = 8x - 24$

$(y+1)^2 = 8(x-3)$

Notice that $(y+1)^2 = 8(x-3)$ is of the form

$(y-k)^2 = 4a(x-h)$, where $(h,k) = (3,-1)$ and

$4a = 8$, so that $a = 2$. Now, the graph of an

equation of the form $(y-k)^2 = 4a(x-h)$ will be

a parabola that opens to the right with the vertex

at (h,k), focus at $(h+a,k)$, and directrix of

$x = h-a$. Note that $h+a = 3+2 = 5$ and

$h-a = 3-2 = 1$. Thus, the graph of

$(y+1)^2 = 8(x-3)$ is a parabola that opens to the

right with vertex $(3,-1)$, focus $(5,-1)$, and

directrix $x = 1$. To help graph the parabola, we

plot the two points on the graph below and

above the focus. Let $x = 5$:

$(y+1)^2 = 8(5-3)$

$(y+1)^2 = 16$

$y+1 = \pm 4$

$y = -1 \pm 4$

$y = -5$ or $y = 3$

The points $(5,-5)$ and $(5,3)$ are on the graph.

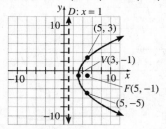

9. The distance from the vertex $(h,k) = (-1,-2)$ to

the focus $(-1,-5)$ is $a = 3$. The vertex and

focus both lie on the vertical line $x = -1$, which

is the axis of symmetry (parallel to the y-axis).

Because the focus is below the vertex, we know

that the parabola will open downward. This

means the equation of the parabola is of the form

$(x-h)^2 = -4a(y-k)$ with $a = 3$, $h = -1$, and

$k = -2$:

$(x-(-1))^2 = -4\cdot 3(y-(-2))$

$(x+1)^2 = -12(y+2)$

10. The vertex is $(-3,3)$ and the axis of symmetry is

parallel to the x-axis, so the axis of symmetry is

the line $y = 3$ and the parabola either opens to

the left or right. Because the graph contains the

point $(-1,7)$, which is to the right of the vertex,

the parabola must open to the right. Therefore,

the equation of the parabola is of the form

$(y-k)^2 = 4a(x-h)$, with $h = -3$ and $k = 3$.

Now $y = 7$ when $x = -1$, so

$$(7-3)^2 = 4a(-1-(-3))$$
$$(4)^2 = 4a(2)$$
$$16 = 8a$$
$$a = 2$$

The equation of the parabola is

$$(y-3)^2 = 4 \cdot 2(x-(-3))$$
$$(y-3)^2 = 8(x+3)$$

11. $x^2 + 9y^2 = 81$

$$\frac{x^2 + 9y^2}{81} = \frac{81}{81}$$

$$\frac{x^2}{81} + \frac{y^2}{9} = 1$$

The larger number, 81, is in the denominator of the x^2-term. This means that the major axis is the x-axis and that the equation of the ellipse is of the form $\dfrac{x^2}{a^2} + \dfrac{y^2}{b^2} = 1$, so that $a^2 = 81$ and

$b^2 = 9$. The center of the ellipse is $(0,0)$.

Because $c^2 = a^2 - b^2$, we have that

$c^2 = 81 - 9 = 72$, so that $c = \pm\sqrt{72} = \pm 6\sqrt{2}$.
Since the major axis is the x-axis, the foci are $\left(-6\sqrt{2}, 0\right)$ and $\left(6\sqrt{2}, 0\right)$. To find the x-intercepts (vertices), let $y = 0$; to find the y-intercepts, let $x = 0$:

x-intercepts:	y-intercepts:
$x^2 + 9(0)^2 = 81$	$0^2 + 9y^2 = 81$
$x^2 = 81$	$9y^2 = 81$
$x = \pm 9$	$y^2 = 9$
	$y = \pm 3$

The intercepts are $(-9, 0)$, $(9, 0)$, $(0, -3)$, and $(0, 3)$.

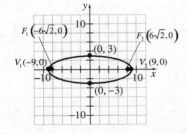

12. $\dfrac{(x+1)^2}{36} + \dfrac{(y-2)^2}{49} = 1$

The center of the ellipse is $(h, k) = (-1, 2)$. Because the larger number, 49, is the denominator of the y^2-term, the major axis is parallel to the y-axis. Because $a^2 = 49$ and $b^2 = 36$, we have that $c^2 = a^2 - b^2 = 49 - 36 = 13$. The vertices are $a = 7$ units below and above the center at $V_1(-1, -5)$ and $V_2(-1, 9)$. The foci are $c = \sqrt{13}$ units below and above the center at $F_1\left(-1, 2-\sqrt{13}\right)$ and $F_2\left(-1, 2+\sqrt{13}\right)$. We plot the points $b = 6$ units to the left and right of the center point at $(-7, 2)$ and $(5, 2)$.

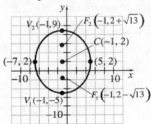

13. The given foci $(0, \pm 6)$ and the given vertices $(0, \pm 9)$ lie on the y-axis. The center of the ellipse is the midpoint between the two vertices (or foci). Thus, the center of the ellipse is $(0, 0)$ and its equation is of the form $\dfrac{x^2}{b^2} + \dfrac{y^2}{a^2} = 1$. The distance from the center of the ellipse to the vertex is $a = 9$ units. The distance from the center of the ellipse to the focus is $c = 6$ units. Because $b^2 = a^2 - c^2$, we have that $b^2 = 9^2 - 6^2 = 81 - 36 = 45$. Thus, the equation of the ellipse is $\dfrac{x^2}{45} + \dfrac{y^2}{81} = 1$.

14. The center is $(h, k) = (3, -4)$. The center, focus, and vertex given all lie on the horizontal line $y = -4$. Therefore, the major axis is parallel to the x-axis, and the equation of the ellipse is of the form $\dfrac{(x-h)^2}{a^2} + \dfrac{(y-k)^2}{b^2} = 1$. Now, $a = 4$ is the distance from the center $(3, -4)$ to a vertex $(7, -4)$, and $c = 3$ is the distance from the center to a focus $(6,-4)$. Also, $b^2 = a^2 - c^2$, so $b^2 = 4^2 - 3^2 = 16 - 9 = 7$. Thus, the equation of the ellipse is

$$\frac{(x-3)^2}{16} + \frac{(y-(-4))^2}{7} = 1$$

$$\frac{(x-3)^2}{16} + \frac{(y+4)^2}{7} = 1$$

15. $\dfrac{y^2}{81} - \dfrac{x^2}{9} = 1$

Notice the equation is of the form $\dfrac{y^2}{a^2} - \dfrac{x^2}{b^2} = 1$.

Because the y^2-term is first, the transverse is along the y-axis and the hyperbola opens up and down. The center of the hyperbola is $(0,0)$. We have that $a^2 = 81$ and $b^2 = 9$. Because $c^2 = a^2 + b^2$, we have that $c^2 = 81 + 9 = 90$, so that $c = \sqrt{90} = 3\sqrt{10}$. The vertices are $(0,\pm a) = (0,\pm 9)$, and the foci are $(0,\pm c) = \left(0,\pm 3\sqrt{10}\right)$. Since $a = 9$ and $b = 3$, the equations of the asymptotes are $y = \dfrac{9}{3}x = 3x$

and $y = -\dfrac{9}{3}x = -3x$. To help graph the hyperbola, we form the rectangle using the points $(0,\pm a) = (0,\pm 9)$ and $(\pm b, 0) = (\pm 3, 0)$. The diagonals are the asymptotes.

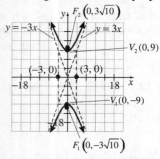

16. $25x^2 - y^2 = 25$

$$\frac{25x^2 - y^2}{25} = \frac{25}{25}$$

$$\frac{x^2}{1} - \frac{y^2}{25} = 1$$

Notice the equation is of the form $\dfrac{x^2}{a^2} - \dfrac{y^2}{b^2} = 1$.

Because the x^2-term is first, the transverse is along the x-axis and the hyperbola opens left and right. The center of the hyperbola is $(0,0)$. We have that $a^2 = 1$ and $b^2 = 25$. Because $c^2 = a^2 + b^2$, we have that $c^2 = 1 + 25 = 26$, so that $c = \sqrt{26}$. The vertices are $(\pm a, 0) = (\pm 1, 0)$, and the foci are $(\pm c, 0) = \left(\pm\sqrt{26}, 0\right)$. Since $a = 1$ and $b = 5$, the equations of the asymptotes are $y = \dfrac{5}{1}x = 5x$ and $y = -\dfrac{5}{1}x = -5x$. To help graph the hyperbola, we form the rectangle using the points $(\pm a, 0) = (\pm 1, 0)$ and $(0,\pm b) = (0,\pm 5)$. The diagonals are the asymptotes.

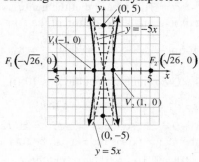

17. The given center $(0,0)$, focus $(0,-5)$, and vertex $(0,-2)$ all lie on the y-axis. Thus, the transverse axis is the y-axis and the hyperbola opens up and down. This means that the equation of the hyperbola is $\dfrac{y^2}{a^2} - \dfrac{x^2}{b^2} = 1$. Now, the distance between the given vertex $(0,-2)$ and the center $(0,0)$ is $a = 2$ units. Likewise, the distance between the given focus $(0,-5)$ and the center is $c = 5$ units. Because $b^2 = c^2 - a^2$, we have that $b^2 = 5^2 - 2^2 = 25 - 4 = 21$. Thus, the equation of the hyperbola is $\dfrac{y^2}{4} - \dfrac{x^2}{21} = 1$.

18. The light bulb should be located at the focus of the flood light, so we need to find where the focus of the flood light is. To solve this problem, we draw a parabola on a Cartesian plane so that the vertex is the origin and the focus is on the positive y-axis. The width of the parabola is 36 inches, and the depth is 12 inches.

Therefore, we know two points on the graph of the parabola: $(-18,12)$ and $(18,12)$.

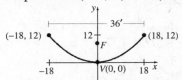

The equation of the parabola has the form $x^2 = 4ay$. Since $(18,12)$ is a point on the graph, we have: $18^2 = 4a(12)$

$$324 = 48a$$

$$a = \frac{324}{48} = 6.75$$

The light bulb should be located 6.75 inches above the vertex, along its axis of symmetry.

Section 9.6

Are You Ready for This Section?

R1. $\begin{cases} y = 2x - 5 & (1) \\ 2x - 3y = 7 & (2) \end{cases}$

Substituting $2x - 5$ for y in equation (2), we obtain

$$2x - 3(2x - 5) = 7$$

$$2x - 6x + 15 = 7$$

$$-4x + 15 = 7$$

$$-4x = -8$$

$$x = 2$$

Substituting 2 for x in equation (1), we obtain $y = 2(2) - 5 = 4 - 5 = -1$.

The solution is the ordered pair $(2, -1)$.

R2. $\begin{cases} 2x - 4y = -11 & (1) \\ -x + 5y = 13 & (2) \end{cases}$

Multiply both sides of equation (2) by 2, and add the result to equation (1).

$$2x - 4y = -11$$

$$-2x + 10y = 26$$

$$6y = 15$$

$$y = \frac{15}{6} = \frac{5}{2}$$

Substituting $\frac{5}{2}$ for y in equation (1), we obtain

$$2x - 4\left(\frac{5}{2}\right) = -11$$

$$2x - 10 = -11$$

$$2x = -1$$

$$x = -\frac{1}{2}$$

The solution is the ordered pair $\left(-\frac{1}{2}, \frac{5}{2}\right)$.

R3. $\begin{cases} 3x - 5y = 4 & (1) \\ -6x + 10y = -8 & (2) \end{cases}$

Multiply both sides of equation (1) by 2, and add the result to equation (2).

$$6x - 10y = 8$$

$$-6x + 10y = -8$$

$$0 = 0$$

The system is dependent. The solution is $\{(x, y) \mid 3x - 5y = 4\}$.

Section 9.6 Quick Checks

1. $\begin{cases} 2x + y = -1 \\ x^2 - y = 4 \end{cases}$

First, graph each equation in the system.

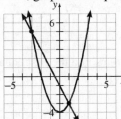

The system apparently has two solutions. Now solve the first equation for y: $y = -2x - 1$.

Substitute the result for y into the second equation:

$$x^2 - (-2x - 1) = 4$$

$$x^2 + 2x + 1 = 4$$

$$x^2 + 2x - 3 = 0$$

$$(x - 1)(x + 3) = 0$$

$$x = 1 \text{ or } x = -3$$

Substitute these x-values into the first equation to find the corresponding y-values:

$$x = 1: \ 2(1) + y = -1$$

$$2 + y = -1$$

$$y = -3$$

$x = -3:\ 2(-3) + y = -1$

$\qquad\qquad -6 + y = -1$

$\qquad\qquad\qquad\ y = 5$

Both pairs check, so the solutions are $(-3,\ 5)$ and $(1, -3)$.

2. $\begin{cases} 2x + y = 0 \\ (x-4)^2 + (y+2)^2 = 9 \end{cases}$

First, graph each equation in the system.

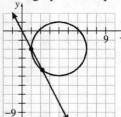

The system apparently has two solutions. Now solve the first equation for y: $y = -2x$.

Substitute the result for y into the second equation:

$$(x-4)^2 + (-2x+2)^2 = 9$$
$$x^2 - 8x + 16 + 4x^2 - 8x + 4 = 9$$
$$5x^2 - 16x + 11 = 0$$
$$(5x - 11)(x - 1) = 0$$
$$x = \frac{11}{5} \quad \text{or} \quad x = 1$$

Substitute these x-values into the first equation to find the corresponding y-values:

$x = \dfrac{11}{5}:\ 2\left(\dfrac{11}{5}\right) + y = 0$

$\qquad\qquad \dfrac{22}{5} + y = 0$

$\qquad\qquad\qquad y = -\dfrac{22}{5}$

$x = 1:\ 2(1) + y = 0$

$\qquad\quad\ 2 + y = 0$

$\qquad\qquad\ y = -2$

Both pairs check, so the solutions are $(1, -2)$ and $\left(\dfrac{11}{5}, -\dfrac{22}{5}\right)$.

3. $\begin{cases} x^2 + y^2 = 16 \\ x^2 - 2y = 8 \end{cases}$

First, graph each equation in the system.

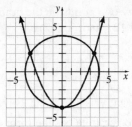

The system apparently has three solutions. Now multiply the second equation by -1 and add the result to the first equation:

$$\begin{array}{r} x^2 + y^2 = 16 \\ \underline{-x^2 + 2y = -8} \\ y^2 + 2y = 8 \end{array}$$

$$y^2 + 2y - 8 = 0$$
$$(y + 4)(y - 2) = 0$$
$$y = -4 \ \text{ or } \ y = 2$$

Substitute these y-values into the first equation to find the corresponding x-values:

$y = -4:\ x^2 + (-4)^2 = 16$

$\qquad\qquad\ x^2 + 16 = 16$

$\qquad\qquad\qquad\ x^2 = 0$

$\qquad\qquad\qquad\ x = 0$

$y = 2:\ x^2 + 2^2 = 16$

$\qquad\qquad x^2 + 4 = 16$

$\qquad\qquad\quad x^2 = 12$

$\qquad\qquad\quad\ x = \pm\sqrt{12} = \pm 2\sqrt{3}$

All three pairs check, so the solutions are $(0, -4)$, $\left(-2\sqrt{3},\ 2\right)$ and $\left(2\sqrt{3},\ 2\right)$.

4. $\begin{cases} x^2 - y = -4 \\ x^2 + y^2 = 9 \end{cases}$

First, graph each equation in the system.

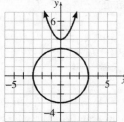

The system apparently has no solution. Now multiply the first equation by -1 and add the result to the second equation:

$$-x^2 \qquad + y = 4$$
$$\underline{x^2 + y^2 \qquad = 9}$$
$$y^2 + y = 13$$
$$y^2 + y - 13 = 0$$

$$y = \frac{-1 \pm \sqrt{1^2 - 4(1)(-13)}}{2(1)}$$

$$= \frac{-1 \pm \sqrt{53}}{2}$$

Substitute these y-values into the first equation to find the corresponding x-values:

$$y = \frac{-1 - \sqrt{53}}{2}: \quad x^2 - \left(\frac{-1 - \sqrt{53}}{2}\right) = -4$$

$$x^2 = -4 + \left(\frac{-1 - \sqrt{53}}{2}\right)$$

$$x^2 = \frac{-9 - \sqrt{53}}{2}$$

$$x = \pm \sqrt{\frac{-9 - \sqrt{53}}{2}} \quad \text{(not real)}$$

$$y = \frac{-1 + \sqrt{53}}{2}: \quad x^2 - \left(\frac{-1 + \sqrt{53}}{2}\right) = -4$$

$$x^2 = -4 + \left(\frac{-1 + \sqrt{53}}{2}\right)$$

$$x^2 = \frac{-9 + \sqrt{53}}{2}$$

$$x = \pm \sqrt{\frac{-9 + \sqrt{53}}{2}} \quad \text{(not real)}$$

Since both y-values result in non-real x-values (because the values under each of the radicals are negative), the system of equations is inconsistent. The solution set is $\varnothing$.

9.6 Exercises

5. $\begin{cases} y = x^2 + 4 \\ y = x + 4 \end{cases}$

First, graph each equation in the system.

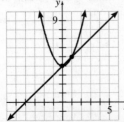

The system apparently has two solutions. Now

substitute $x^2 + 4$ for y into the second equation:

$$x^2 + 4 = x + 4$$
$$x^2 - x = 0$$
$$x(x - 1) = 0$$
$$x = 0 \quad \text{or} \quad x = 1$$

Substitute these x-values into the first equation to find the corresponding y-values:

$$x = 0: \quad y = 0^2 + 4 = 4$$

$$x = 1: \quad y = 1^2 + 4 = 1 + 4 = 5$$

Both pairs check, so the solutions are $(0, \ 4)$ and $(1, \ 5)$.

7. $\begin{cases} y = \sqrt{25 - x^2} \\ x + y = 7 \end{cases}$

First, graph each equation in the system.

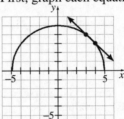

The system apparently has two solutions. Now substitute $\sqrt{25 - x^2}$ for y into the second equation:

$$x + \sqrt{25 - x^2} = 7$$
$$\sqrt{25 - x^2} = 7 - x$$
$$\left(\sqrt{25 - x^2}\right)^2 = (7 - x)^2$$
$$25 - x^2 = 49 - 14x + x^2$$
$$-2x^2 + 14x - 24 = 0$$
$$x^2 - 7x + 12 = 0$$
$$(x - 3)(x - 4) = 0$$
$$x = 3 \quad \text{or} \quad x = 4$$

Substitute these x-values into the first equation to find the corresponding y-values:

$$x = 3: \quad y = \sqrt{25 - 3^2} = \sqrt{25 - 9} = \sqrt{16} = 4$$

$$x = 4: \quad y = \sqrt{25 - 4^2} = \sqrt{25 - 16} = \sqrt{9} = 3$$

Both pairs check, so the solutions are $(3, 4)$ and $(4, 3)$.

9. $\begin{cases} x^2 + y^2 = 4 \\ y = x^2 - 2 \end{cases}$

First, graph each equation in the system.

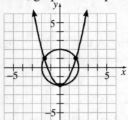

The system apparently has three solutions. Now solve the first equation for x^2: $x^2 = 4 - y^2$.

Substitute the result for x^2 into the second equation:

$$y = \left(4 - y^2\right) - 2$$
$$y = 2 - y^2$$
$$y^2 + y - 2 = 0$$
$$(y + 2)(y - 1) = 0$$
$$y = -2 \quad \text{or} \quad y = 1$$

Substitute these y-values into the first equation to find the corresponding x-values:

$$y = -2: \quad x^2 + (-2)^2 = 4$$
$$x^2 + 4 = 4$$
$$x^2 = 0$$
$$x = 0$$

$$y = 1: \quad x^2 + 1^2 = 4$$
$$x^2 + 1 = 4$$
$$x^2 = 3$$
$$x = \pm\sqrt{3}$$

All three pairs check, so the solutions are $(0, -2)$, $\left(-\sqrt{3}, 1\right)$, and $\left(\sqrt{3}, 1\right)$.

11. $\begin{cases} xy = 4 \\ x^2 + y^2 = 8 \end{cases}$

First, graph each equation in the system.

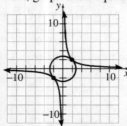

The system apparently has two solutions. Now

solve the first equation for y: $y = \dfrac{4}{x}$.

Substitute the result into the second equation.

$$x^2 + \left(\frac{4}{x}\right)^2 = 8$$
$$x^2 + \frac{16}{x^2} = 8$$
$$x^2\left(x^2 + \frac{16}{x^2}\right) = x^2(8)$$
$$x^4 + 16 = 8x^2$$
$$x^4 - 8x^2 + 16 = 0$$
$$\left(x^2 - 4\right)^2 = 0$$
$$x^2 - 4 = 0$$
$$(x + 2)(x - 2) = 0$$
$$x = -2 \quad \text{or} \quad x = 2$$

Substitute these x-values into the equation

$y = \dfrac{4}{x}$ to find the corresponding y-values.

$$x = -2: \quad y = \frac{4}{-2} = -2$$

$$x = 2: \quad y = \frac{4}{2} = 2$$

Both pairs check, so the solutions are $(-2, -2)$ and $(2, 2)$.

13. $\begin{cases} x^2 + y^2 = 4 \\ y^2 - x = 4 \end{cases}$

First, graph each equation in the system.

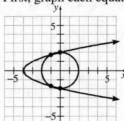

The system apparently has four solutions. Now multiply the second equation by -1 and add the result to the first equation:

$$\begin{array}{r} x^2 + y^2 = 4 \\ \underline{x - y^2 = -4} \\ x^2 + x \phantom{{}= y^2} = 0 \\ x(x + 1) = 0 \end{array}$$

$$x = 0 \quad \text{or} \quad x = -1$$

Substitute these x-values into the second equation to find the corresponding y-values:

$x = 0:\ y^2 - 0 = 4$
$\qquad\qquad y = \pm\sqrt{4} = \pm 2$
$x = -1:\ y^2 - (-1) = 4$
$\qquad\qquad y^2 + 1 = 4$
$\qquad\qquad y^2 = 3$
$\qquad\qquad y = \pm\sqrt{3}$

All four pairs check, so the solutions are $(0, -2)$, $(0,\ 2)$, $\left(-1, -\sqrt{3}\right)$ and $\left(-1,\ \sqrt{3}\right)$.

15. $\begin{cases} x^2 + y^2 = 7 \\ x^2 - y^2 = 25 \end{cases}$

First, graph each equation in the system.

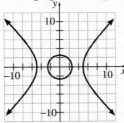

The system apparently has no solution. Now add the two equations:

$\quad x^2 + y^2 = 7$
$\quad \underline{x^2 - y^2 = 25}$
$\quad 2x^2 \qquad = 32$
$\qquad x^2 = 16$
$\qquad x = \pm 4$

Substitute 16 for x^2 into the first equation to find the y values.

$16 + y^2 = 7$
$\qquad y^2 = -9$
$\qquad y = \pm\sqrt{-9}$

which yields no real solutions. The system is inconsistent. The solution set is $\varnothing$.

17. $\begin{cases} x^2 + y^2 = 6y \\ x^2 = 3y \end{cases}$

First, graph each equation in the system.

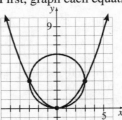

The system apparently has three solutions. Now multiply the second equation by -1 and add the result to the first equation:

$\quad x^2 + y^2 = 6y$
$\quad \underline{-x^2 \qquad = -3y}$
$\qquad\quad y^2 = 3y$
$\qquad y^2 - 3y = 0$
$\qquad y(y - 3) = 0$
$\qquad y = 0\ \text{ or }\ y = 3$

Substitute these y-values into the second equation to find the corresponding x-values:

$y = 0:\ x^2 = 3(0)$
$\qquad\qquad x^2 = 0$
$\qquad\qquad x = 0$
$y = 3:\ x^2 = 3(3)$
$\qquad\qquad x^2 = 9$
$\qquad\qquad x = \pm 3$

All three pairs check, so the solutions are $(0,\ 0)$, $(-3,\ 3)$, and $(3,\ 3)$.

19. $\begin{cases} x^2 - 2x - y = 8 \\ 6x + 2y = -4 \end{cases}$

First, graph each equation in the system.

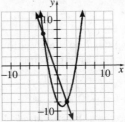

The system apparently has two solutions. Now divide the second equation by 2 and add the result to the first equation:

$$x^2 - 2x - y = 8$$
$$\underline{\qquad 3x + y = -2}$$
$$x^2 + x \qquad = 6$$
$$x^2 + x - 6 = 0$$
$$(x+3)(x-2) = 0$$
$$x = -3 \text{ or } x = 2$$

Substitute these results into the second equation to find the corresponding y-values.
$$x = -3: \quad 6(-3) + 2y = -4$$
$$-18 + 2y = -4$$
$$2y = 14$$
$$y = 7$$
$$x = 2: \quad 6(2) + 2y = -4$$
$$12 + 2y = -4$$
$$2y = -16$$
$$y = -8$$

Both pairs check, so the solutions are $(-3, 7)$ and $(2, -8)$.

21. $\begin{cases} y = x^2 - 6x + 4 \\ 5x + y = 6 \end{cases}$

First, graph each equation in the system.

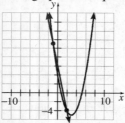

The system apparently has two solutions. Now solve the second equation for y: $y = -5x + 6$.

Substitute the result into the first equation:
$$-5x + 6 = x^2 - 6x + 4$$
$$0 = x^2 - x - 2$$
$$0 = (x+1)(x-2)$$
$$x = -1 \text{ or } x = 2$$

Substitute these x-values into the equation $y = -5x + 6$ to find the corresponding y-values:
$$x = -1: \quad y = -5(-1) + 6 = 5 + 6 = 11$$
$$x = 2: \quad y = -5(2) + 6 = -10 + 6 = -4$$

Both pairs check, so the solutions are $(-1, 11)$ and $(2, -4)$.

23. $\begin{cases} x^2 + y^2 = 16 \\ x^2 - y^2 = 16 \end{cases}$

First, graph each equation in the system.

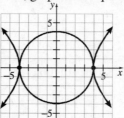

The system apparently has two solutions. Now add the two equations:
$$x^2 + y^2 = 16$$
$$\underline{x^2 - y^2 = 16}$$
$$2x^2 \qquad = 32$$
$$x^2 = 16$$
$$x = \pm 4$$

Substitute 16 for x^2 into the first equation to find the y-values.
$$16 + y^2 = 16$$
$$y^2 = 0$$
$$y = 0$$

Both pairs check, so the solutions are $(-4, 0)$ and $(4, 0)$.

25. $\begin{cases} (x-4)^2 + y^2 = 25 \\ x - y = -3 \end{cases}$

First, graph each equation in the system.

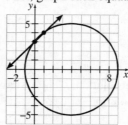

The system apparently has two solutions. Now solve the second equation for y: $y = x + 3$.

Substitute the result into the first equation:
$$(x-4)^2 + (x+3)^2 = 25$$
$$x^2 - 8x + 16 + x^2 + 6x + 9 = 25$$
$$2x^2 - 2x = 0$$
$$2x(x-1) = 0$$
$$x = 0 \text{ or } x = 1$$

Substitute these x-values into the equation

$y = x + 3$ to find the corresponding y-values:

$x = 0$: $y = 0 + 3 = 3$

$x = 1$: $y = 1 + 3 = 4$

Both pairs check, so the solutions are $(0,\ 3)$ and $(1,\ 4)$.

27. $\begin{cases} (x-1)^2 + (y+2)^2 = 4 \\ y^2 + 4y - x = -1 \end{cases}$

First, graph each equation in the system.

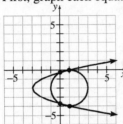

The system apparently has four solutions. Now expand the first equation and simplify:

$$(x-1)^2 + (y+2)^2 = 4$$

$$x^2 - 2x + 1 + y^2 + 4y + 4 = 4$$

$$x^2 - 2x + y^2 + 4y = -1$$

Multiply the second equation by -1 and add the result to the expanded form of the first equation:

$$\begin{array}{r} x^2 - 2x + y^2 + 4y = -1 \\ x - y^2 - 4y = 1 \\ \hline x^2 - x \qquad\qquad = 0 \end{array}$$

$$x(x-1) = 0$$

$$x = 0 \text{ or } x = 1$$

Substitute these x-values into the first equation to find the corresponding y-values.

$x = 0$: $(0-1)^2 + (y+2)^2 = 4$

$$1 + (y+2)^2 = 4$$

$$(y+2)^2 = 3$$

$$y + 2 = \pm\sqrt{3}$$

$$y = -2 \pm \sqrt{3}$$

$x = 1$: $(1-1)^2 + (y+2)^2 = 4$

$$(y+2)^2 = 4$$

$$y + 2 = \pm 2$$

$$y = -2 \pm 2$$

$$y = -4 \text{ or } y = 0$$

All four pairs check, so the solutions are $\left(0, -2-\sqrt{3}\right)$, $\left(0, -2+\sqrt{3}\right)$, $(1, -4)$, and $(1,\ 0)$.

29. $\begin{cases} (x+3)^2 + 4y^2 = 4 \\ x^2 + 6x - y = 13 \end{cases}$

First, graph each equation in the system.

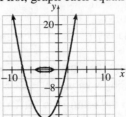

The system apparently has no solution. Now expand the first equation and simplify:

$$(x+3)^2 + 4y^2 = 4$$

$$x^2 + 6x + 9 + 4y^2 = 4$$

$$x^2 + 6x + 4y^2 = -5$$

Multiply the second equation by -1 and add the result to the expanded form of the first equation:

$$\begin{array}{r} x^2 + 6x + 4y^2 \quad = -5 \\ -x^2 - 6x \qquad + y = -13 \\ \hline 4y^2 + y = -18 \end{array}$$

$$4y^2 + y + 18 = 0$$

$$y = \frac{-1 \pm \sqrt{1^2 - 4(4)(18)}}{2(4)} = \frac{-1 \pm \sqrt{-287}}{8}$$

which yields no real solutions. The system is inconsistent. The solution set is $\varnothing$.

31. $\begin{cases} x^2 - y^2 = 21 \\ x + y = 7 \end{cases}$

First, graph each equation in the system.

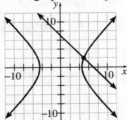

The system apparently has one solution. Now solve the second equation for y: $y = 7 - x$.

Substitute the result into the first equation:

$$x^2 - (7-x)^2 = 21$$

$$x^2 - \left(49 - 14x + x^2\right) = 21$$

$$x^2 - 49 + 14x - x^2 = 21$$

$$14x - 49 = 21$$

$$14x = 70$$

$$x = 5$$

Substitute 5 for x into the equation $y = 7 - x$ to find the corresponding y value: $y = 7 - 5 = 2$.

The pair checks, so the solution is $(5, 2)$.

33. $\begin{cases} x^2 + 2y^2 = 16 \\ 4x^2 - y^2 = 24 \end{cases}$

First, graph each equation in the system.

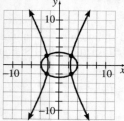

The system apparently has four solutions. Now multiply the second equation by 2 and add the result to the first equation:

$$x^2 + 2y^2 = 16$$

$$\underline{8x^2 - 2y^2 = 48}$$

$$9x^2 \qquad = 64$$

$$x^2 = \frac{64}{9}$$

$$x = \pm\frac{8}{3}$$

Substitute $\frac{64}{9}$ for x^2 into the first equation to find the y-values:

$$\frac{64}{9} + 2y^2 = 16$$

$$9\left(\frac{64}{9} + 2y^2\right) = 9(16)$$

$$64 + 18y^2 = 144$$

$$18y^2 = 80$$

$$y^2 = \frac{80}{18} = \frac{40}{9}$$

$$y = \pm\sqrt{\frac{40}{9}} = \pm\frac{\sqrt{40}}{\sqrt{9}} = \pm\frac{2\sqrt{10}}{3}$$

All four pairs check, so the solutions are

$$\left(-\frac{8}{3}, -\frac{2\sqrt{10}}{3}\right), \left(-\frac{8}{3}, \frac{2\sqrt{10}}{3}\right), \left(\frac{8}{3}, -\frac{2\sqrt{10}}{3}\right),$$

and $\left(\frac{8}{3}, \frac{2\sqrt{10}}{3}\right)$.

35. $\begin{cases} x^2 + y^2 = 25 \\ y = -x^2 + 6x - 5 \end{cases}$

First, graph each equation in the system.

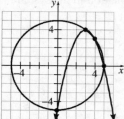

The system apparently has four solutions. Now substitute $-x^2 + 6x - 5$ for y into the first equation:

$$x^2 + \left(-x^2 + 6x - 5\right)^2 = 25$$

$$\left(x^2 - 25\right) + \left(-x^2 + 6x - 5\right)^2 = 0$$

$$(x - 5)(x + 5) + \left[-1(x - 5)(x - 1)\right]^2 = 0$$

$$(x - 5)(x + 5) + (x - 5)^2 (x - 1)^2 = 0$$

$$(x - 5)\left[(x + 5) + (x - 5)(x - 1)^2\right] = 0$$

$$(x - 5)\left[(x + 5) + \left(x^3 - 7x^2 + 11x - 5\right)\right] = 0$$

$$(x - 5)\left(x^3 - 7x^2 + 12x\right) = 0$$

$$x(x - 5)\left(x^2 - 7x + 12\right) = 0$$

$$x(x - 5)(x - 3)(x - 4) = 0$$

$$x = 0, \ x = 5, \ x = 3, \ \text{or} \ x = 4$$

Substitute these x-values into the second equation to find the corresponding y-values:

$$x = 0: \ y = -(0)^2 + 6(0) - 5 = 0 + 0 - 5 = -5$$

$$x = 3: \ y = -(3)^2 + 6(3) - 5 = -9 + 18 - 5 = 4$$

$$x = 4: \ y = -(4)^2 + 6(4) - 5 = -16 + 24 - 5 = 3$$

$$x = 5: \ y = -(5)^2 + 6(5) - 5 = -25 + 30 - 5 = 0$$

All four pairs check, so the solutions are $(0, -5)$, $(3, 4)$, $(4, 3)$, and $(5, 0)$.

37. Let x represent the larger of the two numbers and y represent the smaller of the two numbers.

$$\begin{cases} x - y = 2 \\ x^2 + y^2 = 34 \end{cases}$$

Solve the first equation for x: $x = y + 2$.

Substitute the result into the second equation:

$$(y+2)^2 + y^2 = 34$$
$$\left(y^2 + 4y + 4\right) + y^2 = 34$$
$$2y^2 + 4y - 30 = 0$$
$$y^2 + 2y - 15 = 0$$
$$(y+5)(y-3) = 0$$
$$y = -5 \text{ or } y = 3$$

Substitute these y-values into the equation $x = y + 2$ to find the corresponding x-values.

$$y = -5: \ x = -5 + 2 = -3$$
$$y = 3: \ x = 3 + 2 = 5$$

The numbers are either -5 and -3, or 3 and 5.

39. Let x represent the length and y represent the width of the rectangle.

$$\begin{cases} 2x + 2y = 48 \\ xy = 140 \end{cases}$$

Solve the first equation for y: $y = 24 - x$.

Substitute the result into the second equation:

$$x(24 - x) = 140$$
$$24x - x^2 = 140$$
$$x^2 - 24x + 140 = 0$$
$$(x-14)(x-10) = 0$$
$$x = 14 \text{ or } x = 10$$

Substitute these x-values into the second equation to find the corresponding y-values:

$$x = 14: \ 14y = 140$$
$$y = 10$$
$$x = 10: \ 10y = 140$$
$$y = 14$$

Note that the two outcomes result in the same overall dimensions. Assuming the length is the longer of the two sides, the length is 14 feet and the width is 10 feet.

41. Let x represent the length and y represent the width of the rectangular piece of cardboard.

The area of the cardboard is $A = xy = 190 \cdot \text{cm}^2$.

If 2-cm squares are cut from each corner of the rectangle in order to from the box, then the box will be $x - 4$ cm long, $y - 4$ cm wide, 2 cm high. Thus, the volume of the box will be

$V = 2(x-4)(y-4) = 180 \text{ cm}^3$.

The system of equations follow:

$$\begin{cases} xy = 190 \\ 2(x-4)(y-4) = 180 \end{cases}$$

Solve the first equation for y: $y = \dfrac{190}{x}$.

Substitute the result into the second equation:

$$2(x-4)(y-4) = 180$$
$$(x-4)(y-4) = 90$$
$$xy - 4x - 4y + 16 = 90$$
$$xy - 4x - 4y = 74$$
$$x\left(\frac{190}{x}\right) - 4x - 4\left(\frac{190}{x}\right) = 74$$
$$190 - 4x - \frac{760}{x} = 74$$
$$-4x + 116 - \frac{760}{x} = 0$$
$$x\left(-4x + 116 - \frac{760}{x}\right) = 0$$
$$-4x^2 + 116x - 760 = 0$$
$$-4\left(x^2 - 29x + 190\right) = 0$$
$$x^2 - 29x + 190 = 0$$
$$(x-19)(x-10) = 0$$
$$x = 19 \text{ or } x = 10$$

Substitute these x-values into the equation

$y = \dfrac{190}{x}$ to find the corresponding y-values:

$$x = 19: \ y = \frac{190}{19} = 10$$
$$x = 10: \ y = \frac{190}{10} = 19$$

Note that the two outcomes result in the same overall dimensions. The rectangular piece of cardboard must be 19 cm by 10 cm.

43. $$\begin{cases} y^2 + y + x^2 - x - 2 = 0 \\ y + 1 + \dfrac{x-2}{y} = 0 \end{cases}$$

Multiply the second equation by $-y$:

$$-y\left(y+1+\frac{x-2}{y}\right)=-y(0)$$

$$-y^2-y-x+2=0$$

Add the result to the first equation:

$$y^2+y+x^2-x-2=0$$

$$\underline{-y^2-y\quad\ -x+2=0}$$

$$x^2-2x=0$$

$$x(x-2)=0$$

$$x=0 \ \text{ or } \ x=2$$

Substitute these x-values into the first equation to find the corresponding y-values:

$$x=0: \ y^2+y+0^2-0-2=0$$

$$y^2+y-2=0$$

$$(y+2)(y-1)=0$$

$$y=-2 \ \text{ or } \ y=1$$

$$x=2: \ y^2+y+2^2-2-2=0$$

$$y^2+y+4-2-2=0$$

$$y^2+y=0$$

$$y(y+1)=0$$

$$y=0 \ \text{ or } \ y=-1$$

The apparent solution $y=0$ is extraneous because the denominator of a fraction cannot be 0.

The solutions are $(0,-2)$, $(0,\ 1)$, and $(2,-1)$.

45. $\begin{cases}\ln x=4\ln y\\ \log_3 x=2+2\log_3 y\end{cases}$

Write the first equation in exponential form:

$$\ln x=4\ln y$$

$$\ln x=\ln y^4$$

$$x=y^4$$

Write the second equation in exponential form:

$$\log_3 x=2+2\log_3 y$$

$$\log_3 x=\log_3 9+\log_3 y^2$$

$$\log_3 x=\log_3\left(9y^2\right)$$

$$x=9y^2$$

Thus, we have that

$$y^4=9y^2$$

$$y^4-9y^2=0$$

$$y^2\left(y^2-9\right)=0$$

$$y^2(y+3)(y-3)=0$$

$$y^2=0 \ \text{ or } \ y+3=0 \ \text{ or } \ y-3=0$$

$$y=0 \qquad\quad y=-3 \qquad\quad y=3$$

The apparent y-values 0 and -3 are extraneous because the argument of a logarithm must be positive. Thus, the only possible y-value is 3.

Substitute 3 for y into the equation $x=y^4$ to find the corresponding x-value: $x=3^4=81$.
The solution is $(81,\ 3)$.

47. $\begin{cases}r_1+r_2=-\dfrac{b}{a}\\[2mm] r_1r_2=\dfrac{c}{a}\end{cases}$

Solve the second equation for r_1: $r_1=\dfrac{c}{ar_2}$

Substitute the result into the first equation:

$$\frac{c}{ar_2}+r_2=-\frac{b}{a}$$

$$ar_2\left(\frac{c}{ar_2}+r_2\right)=ar_2\left(-\frac{b}{a}\right)$$

$$c+a\left(r_2\right)^2=-br_2$$

$$a\left(r_2\right)^2+br_2+c=0$$

$$r_2=\frac{-b\pm\sqrt{b^2-4ac}}{2a}$$

Substitute this result into the first equation to find the corresponding value of r_1:

$$r_1+\frac{-b\pm\sqrt{b^2-4ac}}{2a}=-\frac{b}{a}$$

$$r_1-\frac{b}{2a}\pm\frac{\sqrt{b^2-4ac}}{2a}=-\frac{b}{a}$$

$$r_1=-\frac{b}{a}+\frac{b}{2a}\mp\frac{\sqrt{b^2-4ac}}{2a}$$

$$r_1=-\frac{2b}{2a}+\frac{b}{2a}\mp\frac{\sqrt{b^2-4ac}}{2a}$$

$$r_1=-\frac{b}{2a}\pm\frac{\sqrt{b^2-4ac}}{2a}$$

$$r_1=\frac{-b\mp\sqrt{b^2-4ac}}{2a}$$

If $r_1=\dfrac{-b+\sqrt{b^2-4ac}}{2a}$, then

$$r_2 = \frac{-b - \sqrt{b^2 - 4ac}}{2a} \text{ ; if } r_1 = \frac{-b - \sqrt{b^2 - 4ac}}{2a},$$

then $r_2 = \dfrac{-b + \sqrt{b^2 - 4ac}}{2a}$.

49. a. $f(1) = 3(1) + 4 = 3 + 4 = 7$

 b. $g(1) = 2^1 = 2$

51. a. $f(3) = 3(3) + 4 = 9 + 4 = 13$

 b. $g(3) = 2^3 = 8$

53. a. $f(5) = 3(5) + 4 = 15 + 4 = 19$

 b. $g(5) = 2^5 = 32$

55. $\begin{cases} y = x^2 - 6x + 4 \\ 5x + y = 6 \quad (y = -5x + 6) \end{cases}$

Let $Y_1 = x^2 - 6x + 4$ and $Y_2 = -5x + 6$.

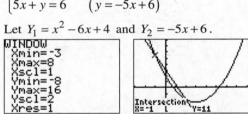

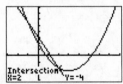

The solutions are $(-1, 11)$ and $(2, -4)$.

57. $\begin{cases} x^2 + y^2 = 16 \quad \left(y = \pm\sqrt{16 - x^2}\right) \\ x^2 - y^2 = 16 \quad \left(y = \pm\sqrt{x^2 - 16}\right) \end{cases}$

Let $Y_1 = \sqrt{16 - x^2}$, $Y_2 = -\sqrt{16 - x^2}$,

$Y_3 = \sqrt{x^2 - 16}$, and $Y_4 = -\sqrt{x^2 - 16}$.

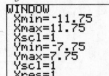
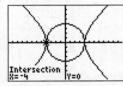

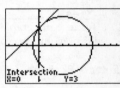

The solutions are $(-4, 0)$ and $(4, 0)$.

59. $\begin{cases} (x-4)^2 + y^2 = 25 \quad \left(y = \pm\sqrt{25 - (x-4)^2}\right) \\ x - y = -3 \qquad\qquad (y = x + 3) \end{cases}$

Let $Y_1 = \sqrt{25 - (x-4)^2}$, $Y_2 = -\sqrt{25 - (x-4)^2}$,

and $Y_3 = x + 3$.

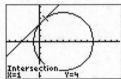

The solutions are $(0, 3)$ and $(1, 4)$.

61. $\begin{cases} (x-1)^2 + (y+2)^2 = 4 \quad \left(y = -2 \pm\sqrt{4 - (x-1)^2}\right) \\ x^2 + 4y - x = -1 \qquad\qquad \left(y = \dfrac{-x^2 + x - 1}{4}\right) \end{cases}$

Let $Y_1 = -2 + \sqrt{4 - (x-1)^2}$,

$Y_2 = -2 + \sqrt{4 - (x-1)^2}$, and $Y_3 = \dfrac{-x^2 + x - 1}{4}$.

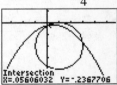

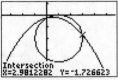

The solutions are approximately $(0.056, -0.237)$

and $(2.981, -1.727)$.

63. $\begin{cases} x^2 + 4y^2 = 4 & \left(y = \pm\sqrt{1 - \dfrac{x^2}{4}} \right) \\ x^2 + 6x - y = -13 & \left(y = x^2 + 6x + 13 \right) \end{cases}$

Let $Y_1 = \sqrt{1 - \dfrac{x^2}{4}}$, $Y_2 = -\sqrt{1 - \dfrac{x^2}{4}}$, and

$Y_3 = x^2 + 6x + 13$.

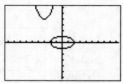

```
WINDOW
 Xmin=-9.4
 Xmax=9.4
 Xscl=1
 Ymin=-6.2
 Ymax=6.2
 Yscl=1
 Xres=1
```

Because the graphs do not intersect, the system has no solution, $\varnothing$.

Chapter 9 Review

1. $d(P_1, P_2) = \sqrt{(x_2 - x_1)^2 + (y_2 - y_1)^2}$
$= \sqrt{(-4 - 0)^2 + (-3 - 0)^2}$
$= \sqrt{(-4)^2 + (-3)^2}$
$= \sqrt{16 + 9} = \sqrt{25} = 5$

2. $d(P_1, P_2) = \sqrt{(x_2 - x_1)^2 + (y_2 - y_1)^2}$
$= \sqrt{(5 - (-3))^2 + (-4 - 2)^2}$
$= \sqrt{8^2 + (-6)^2}$
$= \sqrt{64 + 36} = \sqrt{100} = 10$

3. $d(P_1, P_2) = \sqrt{(x_2 - x_1)^2 + (y_2 - y_1)^2}$
$= \sqrt{(5 - (-1))^2 + (3 - 1)^2}$
$= \sqrt{6^2 + 2^2}$
$= \sqrt{36 + 4} = \sqrt{40} = 2\sqrt{10} \approx 6.32$

4. $d(P_1, P_2) = \sqrt{(x_2 - x_1)^2 + (y_2 - y_1)^2}$
$= \sqrt{(6 - 6)^2 + (-1 - (-7))^2}$
$= \sqrt{0^2 + 6^2} = \sqrt{0 + 36} = \sqrt{36} = 6$

5. $d(P_1, P_2) = \sqrt{(x_2 - x_1)^2 + (y_2 - y_1)^2}$
$= \sqrt{\left(4\sqrt{7} - \sqrt{7}\right)^2 + \left(5\sqrt{3} - \left(-\sqrt{3}\right)\right)^2}$
$= \sqrt{\left(3\sqrt{7}\right)^2 + \left(6\sqrt{3}\right)^2}$
$= \sqrt{9(7) + 36(3)}$
$= \sqrt{63 + 108} = \sqrt{171} = 3\sqrt{19} \approx 13.08$

6. $d(P_1, P_2) = \sqrt{(x_2 - x_1)^2 + (y_2 - y_1)^2}$
$= \sqrt{(1.3 - (-0.2))^2 + (3.7 - 1.7)^2}$
$= \sqrt{1.5^2 + 2^2}$
$= \sqrt{2.25 + 4} = \sqrt{6.25} = 2.5$

7. $M = \left(\dfrac{x_1 + x_2}{2}, \dfrac{y_1 + y_2}{2} \right)$
$= \left(\dfrac{-1 + (-3)}{2}, \dfrac{6 + 4}{2} \right) = \left(\dfrac{-4}{2}, \dfrac{10}{2} \right) = (-2, 5)$

8. $M = \left(\dfrac{x_1 + x_2}{2}, \dfrac{y_1 + y_2}{2} \right)$
$= \left(\dfrac{7 + 5}{2}, \dfrac{0 + (-4)}{2} \right) = \left(\dfrac{12}{2}, \dfrac{-4}{2} \right) = (6, -2)$

9. $M = \left(\dfrac{x_1 + x_2}{2}, \dfrac{y_1 + y_2}{2} \right)$
$= \left(\dfrac{-\sqrt{3} + \left(-7\sqrt{3}\right)}{2}, \dfrac{2\sqrt{6} + \left(-8\sqrt{6}\right)}{2} \right)$
$= \left(\dfrac{-8\sqrt{3}}{2}, \dfrac{-6\sqrt{6}}{2} \right) = \left(-4\sqrt{3}, -3\sqrt{6} \right)$

10. $M = \left(\dfrac{x_1 + x_2}{2}, \dfrac{y_1 + y_2}{2} \right)$
$= \left(\dfrac{5 + 0}{2}, \dfrac{-2 + 3}{2} \right) = \left(\dfrac{5}{2}, \dfrac{1}{2} \right)$

11. $M = \left(\dfrac{x_1 + x_2}{2}, \dfrac{y_1 + y_2}{2} \right)$
$= \left(\dfrac{\frac{1}{4} + \frac{5}{4}}{2}, \dfrac{\frac{2}{3} + \frac{1}{3}}{2} \right) = \left(\dfrac{\frac{3}{2}}{2}, \dfrac{1}{2} \right) = \left(\dfrac{3}{4}, \dfrac{1}{2} \right)$

12. a.

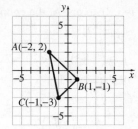

b. $d(A,B) = \sqrt{(1-(-2))^2 + (-1-2)^2}$

$\qquad = \sqrt{3^2 + (-3)^2}$

$\qquad = \sqrt{9+9}$

$\qquad = \sqrt{18}$

$\qquad = 3\sqrt{2} \approx 4.24$

$d(B,C) = \sqrt{(-1-1)^2 + (-3-(-1))^2}$

$\qquad = \sqrt{(-2)^2 + (-2)^2}$

$\qquad = \sqrt{4+4}$

$\qquad = \sqrt{8}$

$\qquad = 2\sqrt{2} \approx 2.83$

$d(A,C) = \sqrt{(-1-(-2))^2 + (-3-2)^2}$

$\qquad = \sqrt{1^2 + (-5)^2}$

$\qquad = \sqrt{1+25}$

$\qquad = \sqrt{26} \approx 5.10$

c. To determine if triangle ABC is a right triangle, we check to see if

$$[d(A,B)]^2 + [d(B,C)]^2 \overset{?}{=} [d(A,C)]^2$$

$$(3\sqrt{2})^2 + (2\sqrt{2})^2 \overset{?}{=} (\sqrt{26})^2$$

$$9 \cdot 2 + 4 \cdot 2 \overset{?}{=} 26$$

$$18 + 8 \overset{?}{=} 26$$

$$26 = 26 \leftarrow \text{True}$$

Therefore, triangle ABC is a right triangle.

d. The length of the "base" of the triangle is $d(B,C) = 2\sqrt{2}$ and the length of the "height" of the triangle is $d(A,B) = 3\sqrt{2}$. Thus, the area of triangle ABC is

$$\text{Area} = \frac{1}{2} \cdot \text{base} \cdot \text{height} = \frac{1}{2} \cdot 2\sqrt{2} \cdot 3\sqrt{2} = 6$$

square units.

13. The center of the circle will be the midpoint of the line segment with endpoints $(-6,1)$ and $(2,1)$. Thus, $(h,k) = \left(\dfrac{-6+2}{2}, \dfrac{1+1}{2} \right) = (-2,1)$.

The radius of the circle will be the distance from the center point $(-2,1)$ to a point on the circle, say $(2,1)$. Thus, $r = \sqrt{(2-(-2))^2 + (1-1)^2} = 4$.

The equation of the circle is

$$(x-h)^2 + (y-k)^2 = r^2$$

$$(x-(-2))^2 + (y-1)^2 = 4^2$$

$$(x+2)^2 + (y-1)^2 = 16$$

14. The center of the circle will be the midpoint of the line segment with endpoints $(2,3)$ and $(8,3)$. Thus, $(h,k) = \left(\dfrac{2+8}{2}, \dfrac{3+3}{2} \right) = (5,3)$.

The radius of the circle will be the distance from the center point $(5,3)$ to a point on the circle, say $(8,3)$. Thus, $r = \sqrt{(8-5)^2 + (3-3)^2} = 3$.

The equation of the circle is

$$(x-h)^2 + (y-k)^2 = r^2$$

$$(x-5)^2 + (y-3)^2 = 3^2$$

$$(x-5)^2 + (y-3)^2 = 9$$

15. $(x-h)^2 + (y-k)^2 = r^2$

$(x-0)^2 + (y-0)^2 = 4^2$

$x^2 + y^2 = 16$

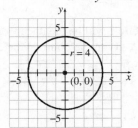

16. $(x-h)^2 + (y-k)^2 = r^2$

$(x-(-3))^2 + (y-1)^2 = 3^2$

$(x+3)^2 + (y-1)^2 = 9$

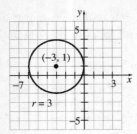

$r = 3$

17. $(x-h)^2 + (y-k)^2 = r^2$

$(x-5)^2 + (y-(-2))^2 = 1^2$

$(x-5)^2 + (y+2)^2 = 1$

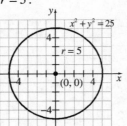

18. $(x-h)^2 + (y-k)^2 = r^2$

$(x-4)^2 + (y-0)^2 = (\sqrt{7})^2$

$(x-4)^2 + y^2 = 7$

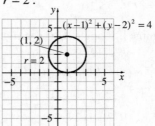

19. The radius of the circle will be the distance from the center point $(2,-1)$ to the point on the circle

$(5,3)$. Thus, $r = \sqrt{(5-2)^2 + (3-(-1))^2} = 5$.

The equation of the circle is

$(x-h)^2 + (y-k)^2 = r^2$

$(x-2)^2 + (y-(-1))^2 = 5^2$

$(x-2)^2 + (y+1)^2 = 25$

20. The center of the circle will be the midpoint of the diameter with endpoints $(-3,-1)$ and $(1,7)$.

Thus, $(h,k) = \left(\dfrac{-3+1}{2}, \dfrac{-1+7}{2}\right) = (-1,3)$. The

radius of the circle will be the distance from the

center point $(-1,3)$ to one of the endpoints of the diameter, say $(1,7)$. Thus,

$r = \sqrt{(1-(-1))^2 + (7-3)^2} = \sqrt{20} = 2\sqrt{5}$.

The equation of the circle is

$(x-h)^2 + (y-k)^2 = r^2$

$(x-(-1))^2 + (y-3)^2 = (2\sqrt{5})^2$

$(x+1)^2 + (y-3)^2 = 20$

21. $\qquad x^2 + y^2 = 25$

$(x-0)^2 + (y-0)^2 = 5^2$

The center is $(h,k) = (0,0)$, and the radius is $r = 5$.

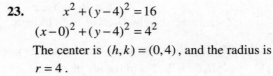

22. $(x-1)^2 + (y-2)^2 = 4$

$(x-1)^2 + (y-2)^2 = 2^2$

The center is $(h,k) = (1,2)$, and the radius is $r = 2$.

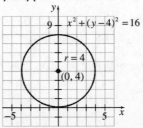

23. $\qquad x^2 + (y-4)^2 = 16$

$(x-0)^2 + (y-4)^2 = 4^2$

The center is $(h,k) = (0,4)$, and the radius is $r = 4$.

24.
$$(x+1)^2 + (y+6)^2 = 49$$
$$(x-(-1))^2 + (y-(-6))^2 = 7^2$$

The center is $(h,k)=(-1,-6)$, and the radius is $r=7$.

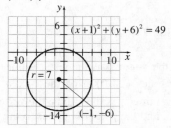

25.
$$(x+2)^2 + \left(y-\frac{3}{2}\right)^2 = \frac{1}{4}$$
$$(x-(-2))^2 + \left(y-\frac{3}{2}\right)^2 = \left(\frac{1}{2}\right)^2$$

The center is $(h,k)=\left(-2,\frac{3}{2}\right)$, and the radius is

$$r=\frac{1}{2}.$$

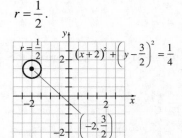

26.
$$(x+3)^2 + (y+3)^2 = 4$$
$$(x-(-3))^2 + (y-(-3))^2 = 2^2$$

The center is $(h,k)=(-3,-3)$, and the radius is $r=2$.

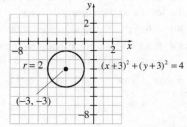

27.
$$x^2 + y^2 + 6x + 10y - 2 = 0$$
$$(x^2+6x)+(y^2+10y)=2$$
$$(x^2+6x+9)+(y^2+10y+25)=2+9+25$$
$$(x+3)^2+(y+5)^2=36$$
$$(x-(-3))^2+(y-(-5))^2=6^2$$

The center is $(h,k)=(-3,-5)$, and the radius is $r=6$.

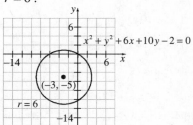

28.
$$x^2 + y^2 - 8x + 4y + 16 = 0$$
$$(x^2-8x)+(y^2+4y)=-16$$
$$(x^2-8x+16)+(y^2+4y+4)=-16+16+4$$
$$(x-4)^2+(y+2)^2=4$$
$$(x-4)^2+(y-(-2))^2=2^2$$

The center is $(h,k)=(4,-2)$, and the radius is $r=2$.

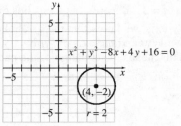

29.
$$x^2 + y^2 + 2x - 4y - 4 = 0$$
$$(x^2+2x)+(y^2-4y)=4$$
$$(x^2+2x+1)+(y^2-4y+4)=4+1+4$$
$$(x+1)^2+(y-2)^2=9$$
$$(x-(-1))^2+(y-2)^2=3^2$$

The center is $(h,k)=(-1,2)$, and the radius is $r=3$.

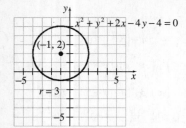

$$x^2 + y^2 + 2x - 4y - 4 = 0$$

$(-1, 2)$

$r = 3$

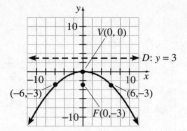

$V(0, 0)$

$D: y = 3$

$(-6, -3)$ $(6, -3)$

$F(0, -3)$

30.
$$x^2 + y^2 - 10x - 2y + 17 = 0$$
$$\left(x^2 - 10x\right) + \left(y^2 - 2y\right) = -17$$
$$\left(x^2 - 10x + 25\right) + \left(y^2 - 2y + 1\right) = -17 + 25 + 1$$
$$(x - 5)^2 + (y - 1)^2 = 9$$
$$(x - 5)^2 + (y - 1)^2 = 3^2$$
The center is $(h, k) = (5, 1)$, and the radius is
$r = 3$.

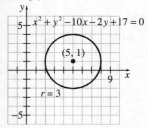

$x^2 + y^2 - 10x - 2y + 17 = 0$

$(5, 1)$

$r = 3$

31. The distance from the vertex $(0, 0)$ to the focus
$(0, -3)$ is $a = 3$. Because the focus lies on the
negative y-axis, we know that the parabola will
open downward and the axis of symmetry is the
y-axis. This means the equation of the parabola
is of the form $x^2 = -4ay$ with $a = 3$:
$$x^2 = -4(3)y$$
$$x^2 = -12y$$
The directrix is the line $y = 3$. To help graph
the parabola, we plot the two points on the graph
to the left and right of the focus. Let $y = -3$:
$$x^2 = -12(-3) = 36$$
$$x = \pm 6$$
The points $(-6, -3)$ and $(6, -3)$ are on the
graph.

32. Notice that the directrix is the vertical line $x = 4$
and that the focus $(-4, 0)$ is on the x-axis. Thus,
the axis of symmetry is the x-axis, the vertex is
at the origin, and the parabola opens to the left.
Now, the distance from the vertex $(0, 0)$ to the
focus $(-4, 0)$ is $a = 4$, so the equation of the
parabola is of the form $y^2 = -4ax$ with $a = 4$:
$$y^2 = -4(4)x$$
$$y^2 = -16x$$
To help graph the parabola, we plot the two
points on the graph above and below the focus.
Let $x = -4$:
$$y^2 = -16(-4) = 64$$
$$y = \pm 8$$
The points $(-4, -8)$ and $(-4, 8)$ are on the
graph.

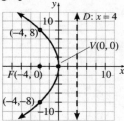

$D: x = 4$

$(-4, 8)$

$V(0, 0)$

$F(-4, 0)$

$(-4, -8)$

33. The vertex is at the origin and the axis of
symmetry is the x-axis, so the parabola either
opens to the left or to the right. Because the
graph contains the point $(8, -2)$, which is in
quadrant IV, the parabola must open to the right.
Therefore, the equation of the parabola is of the
form $y^2 = 4ax$. Now $y = -2$ when $x = 8$, so
$$y^2 = 4ax$$
$$(-2)^2 = 4a(8)$$
$$4 = 32a$$
$$a = \frac{4}{32} = \frac{1}{8}$$
The equation of the parabola is

$$y^2 = 4\left(\frac{1}{8}\right)x$$

$$y^2 = \frac{1}{2}x$$

With $a = \frac{1}{8}$, we know that the focus is $\left(\frac{1}{8}, 0\right)$

and the directrix is the line $x = -\frac{1}{8}$. To help

graph the parabola, we plot the two points on the

graph above and below the focus. Let $x = \frac{1}{8}$:

$$y^2 = \frac{1}{2}\left(\frac{1}{8}\right) = \frac{1}{16}$$

$$y = \pm\frac{1}{4}$$

The points $\left(\frac{1}{8}, -\frac{1}{4}\right)$ and $\left(\frac{1}{8}, \frac{1}{4}\right)$ are on the

graph.

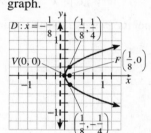

34. The vertex is at the origin and the directrix is the
line $y = -2$, so the focus must be $(0, 2)$.
Accordingly, the parabola opens upward with
$a = 2$. This means the equation of the parabola
is of the form $x^2 = 4ay$ with $a = 2$:

$$x^2 = 4(2)y$$

$$x^2 = 8y$$

To help graph the parabola, we plot the two
points on the graph to the left and right of the
focus. Let $y = 2$:

$$x^2 = 8(2) = 16$$

$$x = \pm 4$$

The points $(-4, 2)$ and $(4, 2)$ are on the graph.

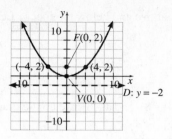

35. Notice that $x^2 = 2y$ is of the form $x^2 = 4ay$,

where $4a = 2$, so that $a = \frac{2}{4} = \frac{1}{2}$. Now, the

graph of an equation of the form $x^2 = 4ay$ will
be a parabola that opens upward with the vertex
at the origin, focus at $(0, a)$, and directrix of

$y = -a$. Thus, the graph of $x^2 = 2y$ is a

parabola that opens upward with vertex $(0, 0)$,

focus $\left(0, \frac{1}{2}\right)$, and directrix $y = -\frac{1}{2}$. To help

graph the parabola, we plot the two points on the
graph to the left and to the right of the focus.

Let $y = \frac{1}{2}$:

$$x^2 = 2\left(\frac{1}{2}\right)$$

$$x^2 = 1$$

$$x = \pm 1$$

The points $\left(-1, \frac{1}{2}\right)$ and $\left(1, \frac{1}{2}\right)$ are on the graph.

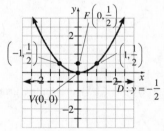

36. Notice that $y^2 = 16x$ is of the form $y^2 = 4ax$,
where $4a = 16$, so that $a = 4$. Now, the graph
of an equation of the form $y^2 = 4ax$ will be a
parabola that opens to the right with the vertex at
the origin, focus at $(a, 0)$, and directrix of

$x = -a$. Thus, the graph of $y^2 = 16x$ is a

parabola that opens to the right with vertex
$(0, 0)$, focus $(4, 0)$, and directrix $x = -4$. To
help graph the parabola, we plot the two points

on the graph above and below the focus. Let
$x = 4$:

$$y^2 = 16(4)$$
$$y^2 = 64$$
$$y = \pm 8$$

The points $(4, -8)$ and $(4, 8)$ are on the graph.

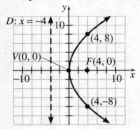

37. Notice that $(x+1)^2 = 8(y-3)$ is of the form
$(x-h)^2 = 4a(y-k)$, where $4a = 8$, so that
$a = 2$ and $(h, k) = (-1, 3)$. Now, the graph of an
equation of the form $(x-h)^2 = 4a(y-k)$ will be
a parabola that opens upward with vertex at
(h, k), focus at $(h, k+a)$, and directrix of
$y = k - a$. Note that $k + a = 3 + 2 = 5$ and
$k - a = 3 - 2 = 1$. Thus, the graph of
$(x+1)^2 = 8(y-3)$ is a parabola that opens
upward with vertex $(-1, 3)$, focus $(-1, 5)$, and
directrix $y = 1$. To help graph the parabola, we
plot the two points on the left and on the right of
the focus. Let $y = 5$:

$$(x+1)^2 = 8(5-3)$$
$$(x+1)^2 = 16$$
$$x+1 = \pm 4$$
$$x = -1 \pm 4$$
$$x = -5 \text{ or } x = 3$$

The points $(-5, 5)$ and $(3, 5)$ are on the graph.

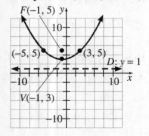

38. Notice that $(y-4)^2 = -2(x+3)$ is of the form
$(y-k)^2 = -4a(x-h)$, where $(h, k) = (-3, 4)$
and $-4a = -2$, so that $a = \dfrac{-2}{-4} = \dfrac{1}{2}$. Now, the
graph of an equation of the form
$(y-k)^2 = -4a(x-h)$ will be a parabola that
opens to the left with the vertex at (h, k), focus
at $(h-a, k)$, and directrix of $x = h + a$. Note
that $h - a = -3 - \dfrac{1}{2} = -\dfrac{7}{2}$ and
$h + a = -3 + \dfrac{1}{2} = -\dfrac{5}{2}$. Thus, the graph of
$(y-4)^2 = -2(x+3)$ is a parabola that opens to
the left with vertex $(-3, 4)$, focus $\left(-\dfrac{7}{2}, 4\right)$, and
directrix $x = -\dfrac{5}{2}$. To help graph the parabola,
we plot the two points on the graph above and
below the focus. Let $x = -\dfrac{7}{2}$:

$$(y-4)^2 = -2\left(-\dfrac{7}{2}+3\right)$$
$$(y-4)^2 = -2\left(-\dfrac{1}{2}\right)$$
$$(y-4)^2 = 1$$
$$y-4 = \pm 1$$
$$y = 4 \pm 1$$
$$y = 3 \text{ or } y = 5$$

The points $\left(-\dfrac{7}{2}, 3\right)$ and $\left(-\dfrac{7}{2}, 5\right)$ are on the
graph.

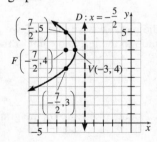

39. We complete the square in x to write the equation in standard form:

$$x^2 - 10x + 3y + 19 = 0$$
$$x^2 - 10x = -3y - 19$$
$$x^2 - 10x + 25 = -3y - 19 + 25$$
$$x^2 - 10x + 25 = -3y + 6$$
$$(x-5)^2 = -3(y-2)$$

Notice that $(x-5)^2 = -3(y-2)$ is of the form $(x-h)^2 = -4a(y-k)$, where $(h,k) = (5,2)$ and $-4a = -3$, so that $a = \dfrac{3}{4}$. Now, the graph of an equation of the form $(x-h)^2 = -4a(y-k)$ will be a parabola that opens downward with the vertex at (h,k), focus at $(h, k-a)$, and directrix of $y = k+a$. Note that $k - a = 2 - \dfrac{3}{4} = \dfrac{5}{4}$ and $k + a = 2 + \dfrac{3}{4} = \dfrac{11}{4}$. Thus, the graph of $(x-5)^2 = -3(y-2)$ is a parabola that opens downward with vertex $(5,2)$, focus $\left(5, \dfrac{5}{4}\right)$, and directrix $y = \dfrac{11}{4}$. To help graph the parabola, we plot the two points on the graph to the left and right of the focus. Let $y = \dfrac{5}{4}$:

$$(x-5)^2 = -3\left(\dfrac{5}{4} - 2\right)$$
$$(x-5)^2 = -3\left(-\dfrac{3}{4}\right)$$
$$(x-5)^2 = \dfrac{9}{4}$$
$$x - 5 = \pm\dfrac{3}{2}$$
$$x = 5 \pm \dfrac{3}{2}$$
$$x = \dfrac{7}{2} \text{ or } x = \dfrac{13}{2}$$

The points $\left(\dfrac{7}{2}, \dfrac{5}{4}\right)$ and $\left(\dfrac{13}{2}, \dfrac{5}{4}\right)$ are on the graph.

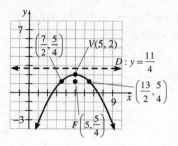

40. The receiver should be located at the focus of the dish, so we need to find where the focus of the dish is. To solve this problem, we draw a parabola on a Cartesian plane so that the vertex is the origin and the focus is on the positive y-axis. The width of the dish is 300 feet, and the depth is 44 feet. Therefore, we know two points on the graph of the parabola: $(-150, 44)$ and $(150, 44)$.

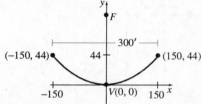

The equation of the parabola has the form $x^2 = 4ay$. Since $(150, 44)$ is a point on the graph, we have

$$150^2 = 4a(44)$$
$$22,500 = 176a$$
$$a = \dfrac{22,500}{176} \approx 127.84$$

The receiver should be located approximately 127.84 feet above the center of the dish, along its axis of symmetry.

41. $\dfrac{x^2}{9} + y^2 = 1$

$$\dfrac{x^2}{9} + \dfrac{y^2}{1} = 1$$

The larger number, 9, is in the denominator of the x^2-term. This means that the major axis is the x-axis and that the equation of the ellipse is of the form $\dfrac{x^2}{a^2} + \dfrac{y^2}{b^2} = 1$, so that $a^2 = 9$ and $b^2 = 1$. The center of the ellipse is $(0,0)$. Because $b^2 = a^2 - c^2$, or $c^2 = a^2 - b^2$, we have that $c^2 = 9 - 1 = 8$, so that $c = \pm\sqrt{8} = \pm 2\sqrt{2}$.

Since the major axis is the *x*-axis, the foci are $\left(-2\sqrt{2},0\right)$ and $\left(2\sqrt{2},0\right)$. To find the *x*-intercepts (vertices), let $y = 0$; to find the *y*-intercepts, let $x = 0$:

x-intercepts:

$$\frac{x^2}{9} + 0^2 = 1$$

$$\frac{x^2}{9} = 1$$

$$x^2 = 9$$

$$x = \pm 3$$

y-intercepts:

$$\frac{0^2}{9} + y^2 = 1$$

$$y^2 = 1$$

$$y = \pm 1$$

The intercepts are $(-3,0)$, $(3,0)$, $(0,-1)$, and $(0,1)$.

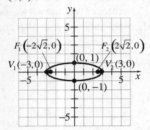

42. $9x^2 + 4y^2 = 36$

$$\frac{9x^2 + 4y^2}{36} = \frac{36}{36}$$

$$\frac{x^2}{4} + \frac{y^2}{9} = 1$$

The larger number, 9, is in the denominator of the y^2-term. This means that the major axis is the *y*-axis and that the equation of the ellipse is of the form $\frac{x^2}{b^2} + \frac{y^2}{a^2} = 1$, so that $a^2 = 9$ and $b^2 = 4$. The center of the ellipse is $(0,0)$. Because $b^2 = a^2 - c^2$, or $c^2 = a^2 - b^2$, we have that $c^2 = 9 - 4 = 5$, so that $c = \pm\sqrt{5}$. Since the major axis is the *y*-axis, the foci are $\left(0,-\sqrt{5}\right)$ and $\left(0,\sqrt{5}\right)$. To find the *x*-intercepts, let $y = 0$; to find the *y*-intercepts (vertices), let $x = 0$:

x-intercepts:

$$9x^2 + 4(0)^2 = 36$$

$$9x^2 = 36$$

$$x^2 = 4$$

$$x = \pm 2$$

y-intercepts:

$$9(0)^2 + 4y^2 = 36$$

$$4y^2 = 36$$

$$y^2 = 9$$

$$y = \pm 3$$

The intercepts are $(-2,0)$, $(2,0)$, $(0,-3)$, and

$(0,3)$.

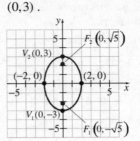

43. The given focus $(0,3)$ and the given vertex $(0,5)$ lie on the *y*-axis. Thus, the equation of the ellipse is of the form $\frac{x^2}{b^2} + \frac{y^2}{a^2} = 1$. The distance from the center of the ellipse to the vertex is $a = 5$ units. The distance from the center of the ellipse to the focus is $c = 3$ units. Because $b^2 = a^2 - c^2$, we have that $b^2 = 5^2 - 3^2 = 25 - 9 = 16$. Thus, the equation of the ellipse is $\frac{x^2}{16} + \frac{y^2}{25} = 1$.

To help graph the ellipse, find the *x*-intercepts:

Let $y = 0$:

$$\frac{x^2}{16} + \frac{0^2}{25} = 1$$

$$\frac{x^2}{16} = 1$$

$$x^2 = 16$$

$$x = \pm 4$$

The *x*-intercepts are $(-4,0)$ and $(4,0)$.

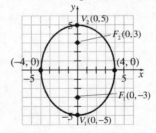

44. The given focus $(-2,0)$ and the given vertex $(-6,0)$ lie on the *x*-axis. Thus, the equation of the ellipse is of the form $\frac{x^2}{a^2} + \frac{y^2}{b^2} = 1$. The distance from the center of the ellipse to the vertex is $a = 6$ units. The distance from the center of the ellipse to the focus is $c = 2$ units. Because $b^2 = a^2 - c^2$, we have that $b^2 = 6^2 - 2^2 = 36 - 4 = 32$. Thus, the equation of the ellipse is $\dfrac{x^2}{36}+\dfrac{y^2}{32}=1$.

To help graph the ellipse, find the y-intercepts:

Let $x=0$: $\dfrac{0^2}{36}+\dfrac{y^2}{32}=1$

$$\dfrac{y^2}{32}=1$$

$$y^2=32$$

$$y=\pm\sqrt{32}=\pm 4\sqrt{2}$$

The y-intercepts are $\left(0,-4\sqrt{2}\right)$ and $\left(0,4\sqrt{2}\right)$.

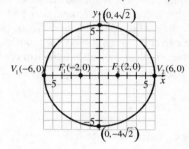

45. The given foci $(\pm 8,0)$ and the given vertices $(\pm 10,0)$ lie on the x-axis. The center of the ellipse is the midpoint between the two vertices (or foci). Thus, the center of the ellipse is $(0,0)$ and its equation is of the form $\dfrac{x^2}{a^2}+\dfrac{y^2}{b^2}=1$. The distance from the center of the ellipse to the vertex is $a=10$ units. The distance from the center of the ellipse to the focus is $c=8$ units. Because $b^2=a^2-c^2$, we have that $b^2=10^2-8^2=100-64=36$. Thus, the equation of the ellipse is $\dfrac{x^2}{100}+\dfrac{y^2}{36}=1$.

To help graph the ellipse, find the y-intercepts:

Let $x=0$: $\dfrac{0^2}{100}+\dfrac{y^2}{36}=1$

$$\dfrac{y^2}{36}=1$$

$$y^2=36$$

$$y=\pm 6$$

The y-intercepts are $(0,-6)$ and $(0,6)$.

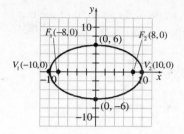

46. $\dfrac{(x-1)^2}{49}+\dfrac{(y+2)^2}{25}=1$

The center of the ellipse is $(h,k)=(1,-2)$. Because the larger number, 49, is the denominator of the x^2-term, the major axis is parallel to the x-axis. Because $a^2=49$ and $b^2=25$, we have that $c^2=a^2-b^2=49-25=24$. The vertices are $a=7$ units to the left and right of the center at $V_1(-6,-2)$ and $V_2(8,-2)$. The foci are $c=\sqrt{24}=2\sqrt{6}$ units to the left and right of the center at $F_1\left(1-2\sqrt{6},-2\right)$ and $F_2\left(1+2\sqrt{6},-2\right)$. We plot the points $b=5$ units above and below the center point at $(1,3)$ and $(1,-7)$.

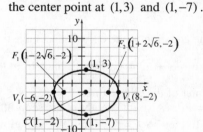

47. $25(x+3)^2+9(y-4)^2=225$

$$\dfrac{25(x+3)^2+9(y-4)^2}{225}=\dfrac{225}{225}$$

$$\dfrac{(x+3)^2}{9}+\dfrac{(y-4)^2}{25}=1$$

The center of the ellipse is $(h,k)=(-3,4)$. Because the larger number, 25, is the denominator of the y^2-term, the major axis is parallel to the y-axis. Because $a^2=25$ and $b^2=9$, we have that $c^2=a^2-b^2=25-9=16$. The vertices are $a=5$ units below and above the center at $V_1(-3,-1)$ and $V_2(-3,9)$. The foci are $c=\sqrt{16}=4$ units below and above the center at $F_1(-3,0)$ and $F_2(-3,8)$. We plot the points

$b = 3$ units to the left and right of the center point at $(-6, 4)$ and $(0, 4)$.

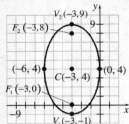

48. a. To solve this problem, we draw the ellipse on a Cartesian plane so that the x-axis coincides with the water and the y-axis passes through the center of the arch. Thus, the origin is the center of the ellipse. Now, the "center" of the arch is 16 feet above the water, so the point $(0, 16)$ is on the ellipse. The river is 60 feet wide, so the two points $(-30, 0)$ and $(30, 0)$ are the two vertices of the ellipse and the major axis is along the x-axis.

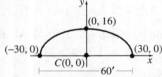

The equation of the ellipse must have the form $\dfrac{x^2}{a^2} + \dfrac{y^2}{b^2} = 1$. The distance from the center of the ellipse to each vertex is $a = 30$. Also, the height of the semi-ellipse is 16 feet, so $b = 16$. Thus, the equation of the ellipse is

$$\frac{x^2}{30^2} + \frac{y^2}{16^2} = 1$$

$$\frac{x^2}{900} + \frac{y^2}{256} = 1$$

b. To determine if the barge can fit through the opening of the bridge, we center it beneath the arch so that the points $(-12.5, 12)$ and $(12.5, 12)$ represent the top corners of the barge. Now, we determine the height of the arch above the water at points 12.5 feet to the left and right of the center by substituting $x = 12.5$ into the equation of the ellipse and solving for y:

$$\frac{12.5^2}{900} + \frac{y^2}{256} = 1$$

$$\frac{156.25}{900} + \frac{y^2}{256} = 1$$

$$\frac{y^2}{256} = \frac{119}{144}$$

$$y^2 = \frac{1904}{9}$$

$$y = \sqrt{\frac{1904}{9}} \approx 14.54$$

At points 12.5 feet from the center of the river, the arch is approximately 14.54 feet above the surface of the water. Thus, the barge can fit through the opening with about 2.54 feet of clearance above the top corners.

49. $\dfrac{x^2}{4} - \dfrac{y^2}{9} = 1$

Notice the equation is of the form $\dfrac{x^2}{a^2} - \dfrac{y^2}{b^2} = 1$.

Because the x^2-term is first, the transverse axis is the x-axis and the hyperbola opens left and right. The center of the hyperbola is the origin. We have that $a^2 = 4$ and $b^2 = 9$. Because $c^2 = a^2 + b^2$, we have that $c^2 = 4 + 9 = 13$, so that $c = \sqrt{13}$. The vertices are $(\pm a, 0) = (\pm 2, 0)$, and the foci are $(\pm c, 0) = \left(\pm\sqrt{13}, 0\right)$. To help graph the hyperbola, we plot the points on the graph above and below the foci. Let $x = \pm\sqrt{13}$:

$$\frac{\left(\pm\sqrt{13}\right)^2}{4} - \frac{y^2}{9} = 1$$

$$\frac{13}{4} - \frac{y^2}{9} = 1$$

$$-\frac{y^2}{9} = -\frac{9}{4}$$

$$y^2 = \frac{81}{4}$$

$$y = \pm\frac{9}{2}$$

The points above and below the foci are $\left(\sqrt{13}, \dfrac{9}{2}\right)$, $\left(\sqrt{13}, -\dfrac{9}{2}\right)$, $\left(-\sqrt{13}, \dfrac{9}{2}\right)$, and $\left(-\sqrt{13}, -\dfrac{9}{2}\right)$.

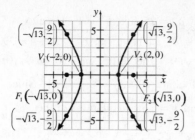

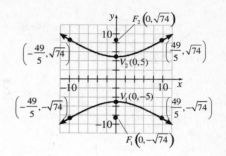

50. $\dfrac{y^2}{25} - \dfrac{x^2}{49} = 1$

Notice the equation is of the form $\dfrac{y^2}{a^2} - \dfrac{x^2}{b^2} = 1$.

Because the y^2-term is first, the transverse axis is the y-axis and the hyperbola opens up and down. The center of the hyperbola is the origin. We have that $a^2 = 25$ and $b^2 = 49$. Because $c^2 = a^2 + b^2$, we have that $c^2 = 25 + 49 = 74$, so that $c = \sqrt{74}$. The vertices are $(0, \pm a) = (0, \pm 5)$, and the foci are $(0, \pm c) = \left(0, \pm\sqrt{74}\right)$. To help graph the hyperbola, we plot the points on the graph to the left and right of the foci. Let $y = \pm\sqrt{74}$:

$$\dfrac{\left(\pm\sqrt{74}\right)^2}{25} - \dfrac{x^2}{49} = 1$$

$$\dfrac{74}{25} - \dfrac{x^2}{49} = 1$$

$$-\dfrac{x^2}{49} = -\dfrac{49}{25}$$

$$x^2 = \dfrac{2401}{25}$$

$$x = \pm\dfrac{49}{5}$$

The points to the left and right of the foci are $\left(\dfrac{49}{5}, \sqrt{74}\right)$, $\left(-\dfrac{49}{5}, \sqrt{74}\right)$, $\left(\dfrac{49}{5}, -\sqrt{74}\right)$, and $\left(-\dfrac{49}{5}, -\sqrt{74}\right)$.

51. $16y^2 - 25x^2 = 400$

$$\dfrac{16y^2 - 25x^2}{400} = \dfrac{400}{400}$$

$$\dfrac{y^2}{25} - \dfrac{x^2}{16} = 1$$

Notice the equation is of the form $\dfrac{y^2}{a^2} - \dfrac{x^2}{b^2} = 1$.

Because the y^2-term is first, the transverse axis is the y-axis and the hyperbola opens up and down. The center of the hyperbola is the origin. We have that $a^2 = 25$ and $b^2 = 16$. Because $c^2 = a^2 + b^2$, we have that $c^2 = 25 + 16 = 41$, so that $c = \sqrt{41}$. The vertices are $(0, \pm a) = (0, \pm 5)$, and the foci are $(0, \pm c) = \left(0, \pm\sqrt{41}\right)$. To help graph the hyperbola, we plot the points on the graph to the left and right of the foci. Let $y = \pm\sqrt{41}$:

$$\dfrac{\left(\pm\sqrt{41}\right)^2}{25} - \dfrac{x^2}{16} = 1$$

$$\dfrac{41}{25} - \dfrac{x^2}{16} = 1$$

$$-\dfrac{x^2}{16} = -\dfrac{16}{25}$$

$$x^2 = \dfrac{256}{25}$$

$$x = \pm\dfrac{16}{5}$$

The points to the left and right of the foci are $\left(\dfrac{16}{5}, \sqrt{41}\right)$, $\left(\dfrac{16}{5}, -\sqrt{41}\right)$, $\left(-\dfrac{16}{5}, \sqrt{41}\right)$, and $\left(-\dfrac{16}{5}, -\sqrt{41}\right)$.

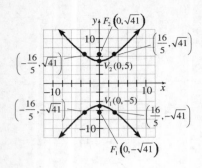

52. $\dfrac{x^2}{36} - \dfrac{y^2}{36} = 1$

Notice the equation is of the form $\dfrac{x^2}{a^2} - \dfrac{y^2}{b^2} = 1$.

Because the x^2-term is first, the transverse is along the x-axis and the hyperbola opens left and right. The center of the hyperbola is $(0,0)$. We have that $a^2 = 36$ and $b^2 = 36$. Because $c^2 = a^2 + b^2$, we have that $c^2 = 36 + 36 = 72$, so that $c = \sqrt{72} = 6\sqrt{2}$. The vertices are $(\pm a, 0) = (\pm 6, 0)$, and the foci are $(\pm c, 0) = \left(\pm 6\sqrt{2}, 0\right)$. Since $a = 6$ and $b = 6$, the equations of the asymptotes are $y = \dfrac{6}{6}x = x$ and $y = -\dfrac{6}{6}x = -x$. To help graph the hyperbola, we form the rectangle using the points $(\pm a, 0) = (\pm 6, 0)$ and $(0, \pm b) = (0, \pm 6)$. The diagonals are the asymptotes.

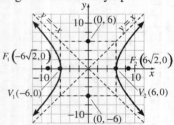

53. $\dfrac{y^2}{25} - \dfrac{x^2}{4} = 1$

Notice the equation is of the form $\dfrac{y^2}{a^2} - \dfrac{x^2}{b^2} = 1$.

Because the y^2-term is first, the transverse is along the y-axis and the hyperbola opens up and down. The center of the hyperbola is $(0,0)$. We

have that $a^2 = 25$ and $b^2 = 4$. Because $c^2 = a^2 + b^2$, we have that $c^2 = 25 + 4 = 29$, so that $c = \sqrt{29}$. The vertices are $(0, \pm a) = (0, \pm 5)$, and the foci are $(0, \pm c) = \left(0, \pm\sqrt{29}\right)$. Since $a = 5$ and $b = 2$, the equations of the asymptotes are $y = \dfrac{5}{2}x$ and $y = -\dfrac{5}{2}x$. To help graph the hyperbola, we form the rectangle using the points $(0, \pm a) = (0, \pm 5)$ and $(\pm b, 0) = (\pm 2, 0)$. The diagonals are the asymptotes.

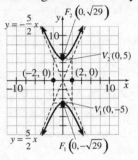

54. The given center $(0,0)$, focus $(-4,0)$, and vertex $(-3,0)$ all lie on the x-axis. Thus, the transverse axis is the x-axis and the hyperbola opens left and right. This means that the equation of the hyperbola is of the form $\dfrac{x^2}{a^2} - \dfrac{y^2}{b^2} = 1$. Now, the distance between the given vertex $(-3,0)$ and the center $(0,0)$ is $a = 3$ units. Likewise, the distance between the given focus $(-4,0)$ and the center is $c = 4$ units. Because $b^2 = c^2 - a^2$, we have that $b^2 = 4^2 - 3^2 = 16 - 9 = 7$. Thus, the equation of the hyperbola is $\dfrac{x^2}{9} - \dfrac{y^2}{7} = 1$.

To help graph the hyperbola, we first find the vertex and focus that were not given. Since the center is at $(0,0)$ and since one vertex is at $(-3,0)$, the other vertex must be at $(3,0)$. Similarly, one the focus is at $(-4,0)$, so the other focus is at $(4,0)$. Next, plot the points on the graph above and below the foci. Let $x = \pm 4$:

$$\frac{(\pm 4)^2}{9} - \frac{y^2}{7} = 1$$

$$\frac{16}{9} - \frac{y^2}{7} = 1$$

$$-\frac{y^2}{7} = -\frac{7}{9}$$

$$y^2 = \frac{49}{9}$$

$$y = \pm\frac{7}{3}$$

The points above and below the foci are $\left(4, \frac{7}{3}\right)$,

$\left(4, -\frac{7}{3}\right)$, $\left(-4, \frac{7}{3}\right)$, and $\left(-4, -\frac{7}{3}\right)$.

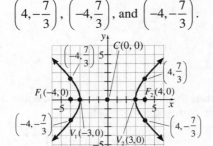

55. The given vertices $(0, \pm 3)$ and focus $(0, 5)$ all lie on the *y*-axis. Thus, the transverse axis is the *y*-axis and the hyperbola opens up and down. The center of the hyperbola is the midpoint between the two vertices. Therefore, the center is $(0, 0)$ and the equation of the hyperbola is of the form $\frac{y^2}{a^2} - \frac{x^2}{b^2} = 1$. Now, the distance from the center $(0, 0)$ to each vertex is $a = 3$ units. Likewise, the distance from the center to the given focus $(0, 5)$ is $c = 5$ units. Also, $b^2 = c^2 - a^2$, so $b^2 = 5^2 - 3^2 = 25 - 9 = 16$. Thus, the equation of the hyperbola is $\frac{y^2}{9} - \frac{x^2}{16} = 1$.

To help graph the hyperbola, we first find the focus that was not given. Since the center is at $(0, 0)$ and since one focus is at $(0, 5)$, the other focus must be at $(0, -5)$. Next, plot the points on the graph to the left and right of the foci. Let $y = \pm 5$:

$$\frac{(\pm 5)^2}{9} - \frac{x^2}{16} = 1$$

$$\frac{25}{9} - \frac{x^2}{16} = 1$$

$$-\frac{x^2}{16} = -\frac{16}{9}$$

$$x^2 = \frac{256}{9}$$

$$x = \pm\frac{16}{3}$$

The points to the left and right of the foci are $\left(\frac{16}{3}, 5\right)$, $\left(-\frac{16}{3}, 5\right)$, $\left(\frac{16}{3}, -5\right)$, and $\left(-\frac{16}{3}, -5\right)$.

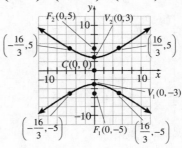

56. The given vertices $(0, \pm 4)$ both lie on the *y*-axis. Thus, the transverse axis is the *y*-axis and the hyperbola opens up and down. The center of the hyperbola is the midpoint between the two vertices. Therefore, the center is $(0, 0)$ and the equation of the hyperbola is of the form $\frac{y^2}{a^2} - \frac{x^2}{b^2} = 1$. Now, the distance from the center $(0, 0)$ to each vertex is $a = 4$ units. We are given that an asymptote of the hyperbola is $y = \frac{4}{3}x$. Now the equation of the asymptote is of the form is $y = \frac{a}{b}x$ and $a = 4$. So,

$$\frac{a}{b} = \frac{4}{3}$$

$$\frac{4}{b} = \frac{4}{3}$$

$$4b = 12$$

$$b = 3$$

Thus, the equation of the hyperbola is

$$\frac{y^2}{4^2} - \frac{x^2}{3^2} = 1.$$

$$\frac{y^2}{16} - \frac{x^2}{9} = 1$$

To help graph the hyperbola, we first find the foci. Since $c^2 = a^2 + b^2$, we have that $c^2 = 16 + 9 = 25$, so $c = 5$. Since the center is $(0,0)$, the foci are $(0, \pm 5)$. We form the rectangle using the points $(0, \pm a) = (0, \pm 4)$ and $(\pm b, 0) = (\pm 3, 0)$. The diagonals are the two asymptotes $y = \frac{4}{3}x$ and $y = -\frac{4}{3}x$.

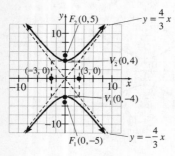

57. $\begin{cases} 4x^2 + y^2 = 10 \\ \quad\quad y = x \end{cases}$

First, graph each equation in the system.

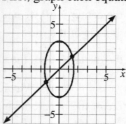

The system apparently has two solutions. Now substitute x for y into the first equation:

$$4x^2 + x^2 = 10$$
$$5x^2 = 10$$
$$x^2 = 2$$
$$x = \pm\sqrt{2}$$

Substitute these x-values into the second equation to find the corresponding y-values:

$x = \sqrt{2}: \quad y = x = \sqrt{2}$

$x = -\sqrt{2}: \quad y = x = -\sqrt{2}$

Both pairs check, so the solutions are $\left(\sqrt{2}, \sqrt{2}\right)$ and $\left(-\sqrt{2}, -\sqrt{2}\right)$.

58. $\begin{cases} y = 2x^2 + 1 \\ y = x + 2 \end{cases}$

First, graph each equation in the system.

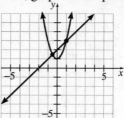

The system apparently has two solutions. Now substitute $2x^2 + 1$ for y into the second equation:

$$2x^2 + 1 = x + 2$$
$$2x^2 - x - 1 = 0$$
$$(2x + 1)(x - 1) = 0$$
$$x = -\frac{1}{2} \quad \text{or} \quad x = 1$$

Substitute these x-values into the first equation to find the corresponding y-values:

$x = -\frac{1}{2}: \quad y = 2\left(-\frac{1}{2}\right)^2 + 1 = 2\left(\frac{1}{4}\right) + 1 = \frac{1}{2} + 1 = \frac{3}{2}$

$x = 1: \quad y = 2(1)^2 + 1 = 2(1) + 1 = 2 + 1 = 3$

Both pairs check, so the solutions are $\left(-\frac{1}{2}, \frac{3}{2}\right)$ and $(1, 3)$.

59. $\begin{cases} 6x - y = 5 \\ \quad\quad xy = 1 \end{cases}$

First, graph each equation in the system.

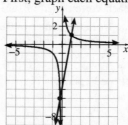

The system apparently has two solutions. Now solve the first equation for y: $y = 6x - 5$.

Substitute the result into the second equation.

$$x(6x - 5) = 1$$
$$6x^2 - 5x - 1 = 0$$
$$(6x + 1)(x - 1) = 0$$
$$x = -\frac{1}{6} \quad \text{or} \quad x = 1$$

Substitute these x-values into the equation

$y = 6x - 5$ to find the corresponding y-values.

$x = -\dfrac{1}{6}$: $\;y = 6\left(-\dfrac{1}{6}\right) - 5 = -1 - 5 = -6$

$x = 1$: $\;y = 6(1) - 5 = 6 - 5 = 1$

Both pairs check, so the solutions are $\left(-\dfrac{1}{6}, -6\right)$

and $(1, 1)$.

60. $\begin{cases} x^2 + y^2 = 26 \\ x^2 - 2y^2 = 23 \end{cases}$

First, graph each equation in the system.

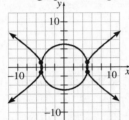

The system apparently has four solutions. Now
solve the first equation for x^2: $x^2 = 26 - y^2$.
Substitute the result into the second equation.

$\left(26 - y^2\right) - 2y^2 = 23$

$\qquad\qquad -3y^2 = -3$

$\qquad\qquad\quad y^2 = 1$

$\qquad\qquad\quad\; y = \pm 1$

Substitute 1 for y^2 into the first equation.

$x^2 + 1 = 26$

$\quad x^2 = 25$

$\quad\; x = \pm 5$

All four pairs check, so the solutions are
$(-5, -1)$, $(-5, 1)$, $(5, -1)$, and $(5, 1)$.

61. $\begin{cases} 4x - y^2 = 0 \\ 2x^2 + y^2 = 16 \end{cases}$

First, graph each equation in the system.

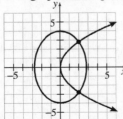

The system apparently has two solutions. Now
add the two equations:

$\begin{array}{r} 4x - y^2 = 0 \\ \underline{2x^2 + y^2 = 16} \\ 2x^2 + 4x = 16 \end{array}$

$2x^2 + 4x - 16 = 0$

$\quad x^2 + 2x - 8 = 0$

$\;(x + 4)(x - 2) = 0$

$\quad x = -4 \;$ or $\; x = 2$

Substitute these x-values into the first equation to
find the corresponding y-values:

$x = -4$: $\; 4(-4) - y^2 = 0$

$\qquad\qquad -16 - y^2 = 0$

$\qquad\qquad\quad -16 = y^2$

$\qquad\qquad\qquad\; y = \pm\sqrt{-16} \;$ (not real)

$x = 2$: $\; 4(2) - y^2 = 0$

$\qquad\qquad 8 - y^2 = 0$

$\qquad\qquad\quad 8 = y^2$

$\qquad\qquad\quad\; y = \pm\sqrt{8} = \pm 2\sqrt{2}$

Both real-number pairs check, so the solutions
are $\left(2, -2\sqrt{2}\right)$ and $\left(2, 2\sqrt{2}\right)$.

62. $\begin{cases} x^2 - y = -2 \\ x^2 + y = 4 \end{cases}$

First, graph each equation in the system.

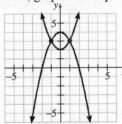

The system apparently has two solutions. Now
multiply the first equation by -1 and add the
result to the second equation:

$\begin{array}{r} -x^2 + y = 2 \\ \underline{x^2 + y = 4} \\ 2y = 6 \\ y = 3 \end{array}$

Substitute this y-value into the second equation
to find the corresponding x-values.

$x^2 + 3 = 4$

$\quad x^2 = 1$

$\quad\; x = \pm 1$

Both pairs check, so the solutions are $(-1, 3)$
and $(1, 3)$.

       815

63. $\begin{cases} 4x^2 - 2y^2 = 2 \\ -x^2 + y^2 = 2 \end{cases}$

First, graph each equation in the system.

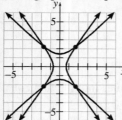

The system apparently has four solutions. Now multiply the second equation by 2 and add the result to the first equation:

$$4x^2 - 2y^2 = 2$$
$$-2x^2 + 2y^2 = 4$$
$$\overline{}$$
$$2x^2 = 6$$
$$x^2 = 3$$
$$x = \pm\sqrt{3}$$

Substitute 3 for x^2 into the second equation:

$$-3 + y^2 = 2$$
$$y^2 = 5$$
$$y = \pm\sqrt{5}$$

All four check, so the solutions are

$\left(-\sqrt{3}, \ -\sqrt{5}\right)$, $\left(-\sqrt{3}, \sqrt{5}\right)$, $\left(\sqrt{3}, \ -\sqrt{5}\right)$, and

$\left(\sqrt{3}, \sqrt{5}\right)$.

64. $\begin{cases} x^2 + y^2 = 8x \\ y^2 = 3x \end{cases}$

First, graph each equation in the system.

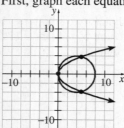

The system apparently has three solutions. Now multiply the second equation by -1 and add the result to the first equation:

$$x^2 + y^2 = 8x$$
$$\underline{ -y^2 = -3x}$$
$$x^2 = 5x$$
$$x^2 - 5x = 0$$
$$x(x-5) = 0$$
$$x = 0 \quad \text{or} \quad x = 5$$

Substitute these x-values into the second equation to find the corresponding y-values:

$$x = 0: \ y^2 = 3(0)$$
$$y^2 = 0$$
$$y = 0$$

$$x = 5: \ y^2 = 3(5)$$
$$y^2 = 15$$
$$y = \pm\sqrt{15}$$

All three pairs check, so the solutions are $(0,0)$,

$\left(5, -\sqrt{15}\right)$, and $\left(5, \sqrt{15}\right)$.

65. $\begin{cases} y = x + 2 \\ y = x^2 \end{cases}$

First, graph each equation in the system.

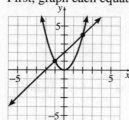

The system apparently has two solutions. Now Substitute x^2 for y into the first equation:

$$x^2 = x + 2$$
$$x^2 - x - 2 = 0$$
$$(x-2)(x+1) = 0$$
$$x = 2 \quad \text{or} \quad x = -1$$

Substitute these results into the first equation to find the corresponding y values.

$$x = 2: \ y = 2 + 2 = 4$$
$$x = -1: y = -1 + 2 = 1$$

Both pairs check, so the solutions are $(2, \ 4)$ and $(-1, \ 1)$.

66. $\begin{cases} x^2 + 2y = 9 \\ 5x - 2y = 5 \end{cases}$

First, graph each equation in the system.

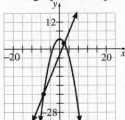

The system apparently has two solutions. Now add the two equations:

$$x^2 \quad + 2y = 9$$
$$\underline{\quad 5x - 2y = 5}$$
$$x^2 \quad + 5x = 14$$
$$x^2 + 5x - 14 = 0$$
$$(x + 7)(x - 2) = 0$$
$$x = -7 \text{ or } x = 2$$

Substitute these results into the first equation to find the corresponding y values.

$$x = -7: \quad (-7)^2 + 2y = 9$$
$$49 + 2y = 9$$
$$2y = -40$$
$$y = -20$$

$$x = 2: \quad (2)^2 + 2y = 9$$
$$4 + 2y = 9$$
$$2y = 5$$
$$y = \frac{5}{2}$$

Both pairs check, so the solutions are $(-7, -20)$ and $\left(2, \dfrac{5}{2}\right)$.

67. $\begin{cases} x^2 + y^2 = 36 \\ x - y = -6 \end{cases}$

First, graph each equation in the system.

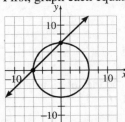

The system apparently has two solutions. Now solve the second equation for y: $y = x + 6$.

Substitute the result into the first equation:

$$x^2 + (x + 6)^2 = 36$$
$$x^2 + (x^2 + 12x + 36) = 36$$
$$2x^2 + 12x = 0$$
$$2x(x + 6) = 0$$
$$2x = 0 \text{ or } x + 6 = 0$$
$$x = 0 \text{ or } \quad x = -6$$

Substitute these x values into the equation $y = x + 6$ to find the corresponding y values:

$$x = 0: \quad y = 0 + 6 = 6$$
$$x = -6: \quad y = -6 + 6 = 0$$

Both pairs check, so the solutions are $(0,\ 6)$ and $(-6,\ 0)$.

68. $\begin{cases} y = 2x - 4 \\ y^2 = 4x \end{cases}$

First, graph each equation in the system.

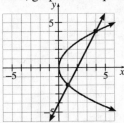

The system apparently has two solutions. Now substitute $2x - 4$ for y into the second equation:

$$(2x - 4)^2 = 4x$$
$$4x^2 - 16x + 16 = 4x$$
$$4x^2 - 20x + 16 = 0$$
$$x^2 - 5x + 4 = 0$$
$$(x - 4)(x - 1) = 0$$
$$x = 4 \text{ or } x = 1$$

Substitute these results into the first equation to find the corresponding y values.

$$x = 4: y = 2(4) - 4 = 4$$
$$x = 1: y = 2(1) - 4 = -2$$

Both pairs check, so the solutions are $(4,\ 4)$ and $(1, -2)$.

69. $\begin{cases} x^2 + y^2 = 9 \\ x + y = 7 \end{cases}$

First, graph each equation in the system.

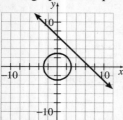

The system apparently has no solution. Now solve the second equation for y: $y = 7 - x$.

Substitute the result into the first equation:

$$x^2 + (7-x)^2 = 9$$
$$x^2 + (49 - 14x + x^2) = 9$$
$$2x^2 - 14x + 49 = 9$$
$$2x^2 - 14x + 40 = 0$$
$$x = \frac{-(-14) \pm \sqrt{(-14)^2 - 4(2)(40)}}{2(2)}$$
$$= \frac{14 \pm \sqrt{-124}}{4}$$
$$= \frac{14 \pm 2\sqrt{31}\,i}{4}$$
$$= \frac{7}{2} \pm \frac{\sqrt{31}}{2} i$$

Since the solutions are not real, the system is inconsistent. The solution set is $\varnothing$.

70. $\begin{cases} 2x^2 + 3y^2 = 14 \\ x^2 - y^2 = -3 \end{cases}$

First, graph each equation in the system.

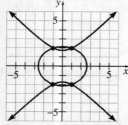

The system apparently has four solutions. Now multiply the second equation by 3 and add the result to the first equation.

$$3x^2 - 3y^2 = -9$$
$$2x^2 + 3y^2 = 14$$

$$5x^2 \quad\quad = 5$$
$$x^2 = 1$$
$$x = \pm 1$$

Substitute 1 for x^2 into the second equation to find the y values.

$$1 - y^2 = -3$$
$$-y^2 = -4$$
$$y^2 = 4$$
$$y = \pm 2$$

All four pairs check, so the solutions are $(-1, -2)$, $(-1, 2)$, $(1, -2)$, and $(1, 2)$.

71. $\begin{cases} x^2 + y^2 = 16 \\ x^2 + 4y = 16 \end{cases}$

First, graph each equation in the system.

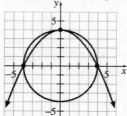

The system apparently has three solutions. Now multiply the second equation by -1 and add the result to the first equation:

$$x^2 + y^2 \quad\quad = 16$$
$$-x^2 \quad\; -4y = -16$$

$$y^2 - 4y = 0$$
$$y(y-4) = 0$$
$$y = 0 \quad \text{or} \quad y = 4$$

Substitute these y-values into the first equation to find the corresponding x-values:

$$y = 0: \quad x^2 + 0^2 = 16$$
$$x^2 = 16$$
$$x = \pm 4$$
$$y = 4: \quad x^2 + 4^2 = 16$$
$$x^2 + 16 = 16$$
$$x^2 = 0$$
$$x = 0$$

All three pairs check, so the solutions are $(-4, 0)$, $(4, 0)$, and $(0, 4)$.

72. $\begin{cases} x = 4 - y^2 \\ x = 2y + 4 \end{cases}$

First, graph each equation in the system.

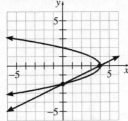

The system apparently has two solutions. Now substitute $2y + 4$ for x into the first equation:

$$2y + 4 = 4 - y^2$$
$$y^2 + 2y = 0$$
$$y(y + 2) = 0$$
$$y = 0 \text{ or } y = -2$$

Substitute these y-values into the second equation to find the corresponding x-values:

$y = 0$: $x = 2(0) + 4 = 4$

$y = -2$: $x = 2(-2) + 4 = 0$

Both pairs check, so the solutions are $(4, \ 0)$ and $(0, -2)$.

73. Let x represent the larger of the two numbers and y represent the smaller of the two numbers.

$$\begin{cases} x + y = 12 \\ x^2 - y^2 = 24 \end{cases}$$

Solve the first equation for y: $y = 12 - x$.

Substitute the result into the second equation:

$$x^2 - (12 - x)^2 = 24$$
$$x^2 - \left(144 - 24x + x^2\right) = 24$$
$$x^2 - 144 + 24x - x^2 = 24$$
$$24x = 168$$
$$x = 7$$

Substitute 7 for x into the first equation to find the corresponding y-value:

$$7 + y = 12$$
$$y = 5$$

The two numbers are 7 and 5.

74. Let x represent the length and y represent the width of the rectangle.

$$\begin{cases} 2x + 2y = 34 \\ xy = 60 \end{cases}$$

Solve the first equation for y: $y = 17 - x$.

Substitute the result into the second equation:

$$x(17 - x) = 60$$
$$17x - x^2 = 60$$
$$x^2 - 17x + 60 = 0$$
$$(x - 12)(x - 5) = 0$$
$$x = 12 \text{ or } x = 5$$

Substitute these x-values into the second equation to find the corresponding y-values:

$x = 12$: $12y = 60$

$\qquad\qquad y = 5$

$x = 5$: $5y = 60$

$\qquad\qquad y = 12$

Note that the two outcomes result in the same overall dimensions. Assuming the length is the longer of the two sides, the length is 12 centimeters and the width is 5 centimeters.

75. Let x represent the length and y represent the width of the rectangle.

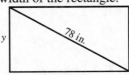

$$\begin{cases} xy = 2160 \\ x^2 + y^2 = 78^2 \end{cases}$$

Solve the first equation for y: $y = \dfrac{2160}{x}$.

Substitute the result into the second equation:

$$x^2 + \left(\frac{2160}{x}\right)^2 = 78^2$$
$$x^2 + \frac{4,665,600}{x^2} = 6084$$
$$x^4 + 4,665,600 = 6084x^2$$
$$x^4 - 6084x^2 + 4,665,600 = 0$$
$$\left(x^2 - 5184\right)\left(x^2 - 900\right) = 0$$
$$(x - 72)(x + 72)(x - 30)(x + 30) = 0$$
$$x = 72 \text{ or } x = -72 \text{ or } x = 30 \text{ or } x = -30$$

Since the length cannot be negative, we discard $x = -72$ and $x = -30$, so we are left with the possible answers $x = 72$ and $x = 30$. Substitute these x-values into the first equation to find the corresponding y-values:

$x = 72$: $72y = 2160$

$\qquad\qquad y = 30$

$x = 30$: $30y = 2160$

$\qquad\qquad y = 72$

Note that the two outcomes result in the same overall dimensions. Assuming the length is the longer of the two sides, the length is 72 inches and the width is 30 inches.

76. Let x and y represent the lengths of the two legs of the right triangle.

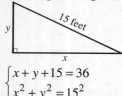

$$\begin{cases} x + y + 15 = 36 \\ x^2 + y^2 = 15^2 \end{cases}$$

Solve the first equation for y: $y = 21 - x$.

Substitute the result into the second equation:

$$x^2 + (21 - x)^2 = 15^2$$
$$x^2 + 441 - 42x + x^2 = 225$$
$$2x^2 - 42x + 216 = 0$$
$$x^2 - 21x + 108 = 0$$
$$(x - 12)(x - 9) = 0$$
$$x = 12 \text{ or } x = 9$$

Substitute these x-values into the first equation to find the corresponding y-values:

$x = 12: \quad 12 + y + 15 = 36$
$$y + 27 = 36$$
$$y = 9$$

$x = 9: \quad 9 + y + 15 = 36$
$$y + 24 = 36$$
$$y = 12$$

Note that the two outcomes result in the same overall dimensions. The lengths of the two legs are 12 inches and 9 inches.

Chapter 9 Test

1. $d(P_1, P_2) = \sqrt{(x_2 - x_1)^2 + (y_2 - y_1)^2}$

$\qquad = \sqrt{(3 - (-1))^2 + (-5 - 3)^2}$

$\qquad = \sqrt{4^2 + (-8)^2}$

$\qquad = \sqrt{16 + 64}$

$\qquad = \sqrt{80}$

$\qquad = 4\sqrt{5}$

2. $M = \left(\dfrac{x_1 + x_2}{2}, \dfrac{y_1 + y_2}{2} \right)$

$\qquad = \left(\dfrac{-7 + 5}{2}, \dfrac{6 + (-2)}{2} \right) = \left(\dfrac{-2}{2}, \dfrac{4}{2} \right) = (-1, 2)$

3. $\qquad (x - 4)^2 + (y + 1)^2 = 9$

$\qquad (x - 4)^2 + (y - (-1))^2 = 3^2$

The center is $(h, k) = (4, -1)$, and the radius is $r = 3$.

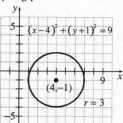

4. $\qquad x^2 + y^2 + 10x - 4y + 13 = 0$

$\qquad \left(x^2 + 10x + 25 \right) + \left(y^2 - 4y + 4 \right) = -13 + 25 + 4$

$\qquad\qquad (x + 5)^2 + (y - 2)^2 = 16$

$\qquad\qquad (x - (-5))^2 + (y - 2)^2 = 4^2$

The center is $(h, k) = (-5, 2)$, and the radius is $r = 4$.

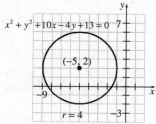

5. $\qquad (x - h)^2 + (y - k)^2 = r^2$

$\qquad (x - (-3))^2 + (y - 7)^2 = 6^2$

$\qquad\qquad (x + 3)^2 + (y - 7)^2 = 36$

6. The radius of the circle will be the distance from the center point $(-5, 8)$ to the point on the circle $(3, 2)$. Thus, $r = \sqrt{(3 - (-5))^2 + (2 - 8)^2} = 10$.

The equation of the circle is

$$(x - h)^2 + (y - k)^2 = r^2$$
$$(x - (-5))^2 + (y - 8)^2 = 10^2$$
$$(x + 5)^2 + (y - 8)^2 = 100$$

7. Notice that $(y + 2)^2 = 4(x - 1)$ is of the form $(y - k)^2 = 4a(x - h)$, where $(h, k) = (1, -2)$ and $4a = 4$, so that $a = 1$. Now, the graph of an equation of the form $(y - k)^2 = 4a(x - h)$ will be

Copyright © 2014 Pearson Education, Inc.

a parabola that opens to the right with the vertex at (h,k), focus at $(h+a,k)$, and directrix of $x=h-a$. Note that $h+a=1+1=2$ and $h-a=1-1=0$. Thus, the graph of $(y+2)^2=4(x-1)$ is a parabola that opens to the right with vertex $(1,-2)$, focus $(2,-2)$, and directrix $x=0$. To help graph the parabola, we plot the two points on the graph above and below the focus. Let $x=2$:

$$(y+2)^2=4(2-1)$$
$$(y+2)^2=4$$
$$y+2=\pm2$$
$$y=-2\pm2$$
$$y=-4 \ \text{ or } \ y=0$$

The points $(2,-4)$ and $(2,0)$ are on the graph.

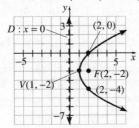

8. We complete the square in x to write the equation in standard form:

$$x^2-4x+3y-8=0$$
$$x^2-4x=-3y+8$$
$$x^2-4x+4=-3y+8+4$$
$$x^2-4x+4=-3y+12$$
$$(x-2)^2=-3(y-4)$$

Notice that $(x-2)^2=-3(y-4)$ is of the form $(x-h)^2=-4a(y-k)$, where $(h,k)=(2,4)$ and $-4a=-3$, so that $a=\dfrac{3}{4}$. Now, the graph of an equation of the form $(x-h)^2=-4a(y-k)$ will be a parabola that opens downward with the vertex at (h,k), focus at $(h,k-a)$, and directrix of $y=k+a$. Note that $k-a=4-\dfrac{3}{4}=\dfrac{13}{4}$ and $k+a=4+\dfrac{3}{4}=\dfrac{19}{4}$. Thus, the graph of $(x-2)^2=-3(y-4)$ is a parabola that opens downward with vertex $(2,4)$, focus $\left(2,\dfrac{13}{4}\right)$, and

directrix $y=\dfrac{19}{4}$. To help graph the parabola, we plot the two points on the graph to the left and right of the focus. Let $y=\dfrac{13}{4}$:

$$(x-2)^2=-3\left(\dfrac{13}{4}-4\right)$$
$$(x-2)^2=-3\left(-\dfrac{3}{4}\right)$$
$$(x-2)^2=\dfrac{9}{4}$$
$$x-2=\pm\dfrac{3}{2}$$
$$x=2\pm\dfrac{3}{2}$$
$$x=\dfrac{1}{2} \ \text{ or } \ x=\dfrac{7}{2}$$

The points $\left(\dfrac{1}{2},\dfrac{13}{4}\right)$ and $\left(\dfrac{7}{2},\dfrac{13}{4}\right)$ are on the graph.

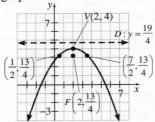

9. The distance from the vertex $(0,0)$ to the focus $(0,-4)$ is $a=4$. Because the focus lies on the negative y-axis, we know that the parabola will open downward and the axis of symmetry is the y-axis. This means the equation of the parabola is of the form $x^2=-4ay$ with $a=4$:

$$x^2=-4(4)y$$
$$x^2=-16y$$

10. Notice that the directrix $x=-1$ is a vertical line and that the focus is $(3,4)$. Therefore, the axis of symmetry must be the line $y=4$. Because the focus is to the right of the directrix, the parabola must open to the right. Since the vertex must be the point on the axis of symmetry that is midway between the focus and the directrix, the vertex is $(1,4)$. Now, the distance from the vertex $(1,4)$ to the focus $(3,4)$ is $a=2$, so the equation of the parabola is of the form

$(y-k)^2 = 4a(x-h)$ with $a=2$, $h=1$, and
$k=4$:
$$(y-4)^2 = 4\cdot 2(x-1)$$
$$(y-4)^2 = 8(x-1)$$

11. $9x^2 + 25y^2 = 225$
$$\frac{9x^2+25y^2}{225} = \frac{225}{225}$$
$$\frac{x^2}{25} + \frac{y^2}{9} = 1$$
The larger number, 25, is in the denominator of
the x^2-term. This means that the major axis is
the x-axis and that the equation of the ellipse is
of the form $\frac{x^2}{a^2} + \frac{y^2}{b^2} = 1$, so that $a^2 = 25$ and
$b^2 = 9$. The center of the ellipse is $(0,0)$.
Because $c^2 = a^2 - b^2$, we have that
$c^2 = 25-9 = 16$, so that $c = \pm\sqrt{16} = \pm 4$. Since
the major axis is the x-axis, the foci are $(-4,0)$
and $(4,0)$. To find the x-intercepts (vertices),
let $y=0$; to find the y-intercepts, let $x=0$:

x-intercepts: y-intercepts:
$9x^2 + 25(0)^2 = 225$ $9(0)^2 + 25y^2 = 225$
$\quad\quad 9x^2 = 225$ $\quad\quad 25y^2 = 225$
$\quad\quad\quad x^2 = 25$ $\quad\quad\quad y^2 = 9$
$\quad\quad\quad x = \pm 5$ $\quad\quad\quad y = \pm 3$

The intercepts are $(-5,0)$, $(5,0)$, $(0,-3)$, and
$(0,3)$.

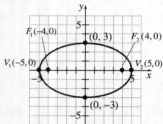

12. $\frac{(x-2)^2}{9} + \frac{(y+4)^2}{16} = 1$
The center of the ellipse is $(h,k) = (2,-4)$.
Because the larger number, 16, is the
denominator of the y^2-term, the major axis is
parallel to the y-axis. Because $a^2 = 16$ and
$b^2 = 9$, we have that $c^2 = a^2 - b^2 = 16-9 = 7$.

The vertices are $a=4$ units below and above
the center at $V_1(2,-8)$ and $V_2(2,0)$. The foci
are $c = \sqrt{7}$ units below and above the center at
$F_1\left(2,-4-\sqrt{7}\right)$ and $F_2\left(2,-4+\sqrt{7}\right)$. We plot
the points $b=3$ units to the left and right of the
center point at $(-1,-4)$ and $(5,-4)$.

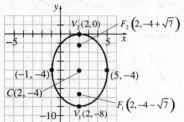

13. The given focus $(0,-4)$ and the given vertex
$(0,-5)$ lie on the y-axis. Thus, the equation of
the ellipse is of the form $\frac{x^2}{b^2} + \frac{y^2}{a^2} = 1$. The
distance from the center of the ellipse to the
vertex is $a=5$ units. The distance from the
center of the ellipse to the focus is $c=4$ units.
Because $b^2 = a^2 - c^2$, we have that
$b^2 = 5^2 - 4^2 = 25-16 = 9$. Thus, the equation
of the ellipse is $\frac{x^2}{9} + \frac{y^2}{25} = 1$.

14. The given vertices and focus all lie on the
vertical line $x=-1$. Therefore, the major axis is
parallel to the y-axis, and the equation of the
ellipse is of the form $\frac{(x-h)^2}{b^2} + \frac{(y-k)^2}{a^2} = 1$.
Now, the center will be the midpoint between the
two vertices $(-1,7)$ and $(-1,-3)$, which is
$(h,k) = (-1,2)$. Now, $a=5$ is the distance from
the center $(-1,2)$ to a vertex $(-1,7)$, and $c=3$
is the distance from the center to a focus
$(-1,-1)$.
Also, $b^2 = a^2 - c^2 = 5^2 - 3^2 = 25-9 = 16$.
Thus, the equation of the ellipse is
$$\frac{\left(x-(-1)\right)^2}{16} + \frac{\left(y-2\right)^2}{25} = 1$$
$$\frac{(x+1)^2}{16} + \frac{(y-2)^2}{25} = 1$$

Copyright © 2014 Pearson Education, Inc.

15. $x^2 - \dfrac{y^2}{4} = 1$

$\dfrac{x^2}{1} - \dfrac{y^2}{4} = 1$

Notice the equation is of the form $\dfrac{x^2}{a^2} - \dfrac{y^2}{b^2} = 1$.

Because the x^2-term is first, the transverse is along the x-axis and the hyperbola opens left and right. The center of the hyperbola is $(0,0)$. We have that $a^2 = 1$ and $b^2 = 4$. Because $c^2 = a^2 + b^2$, we have that $c^2 = 1 + 4 = 5$, so that $c = \sqrt{5}$. The vertices are $(\pm a, 0) = (\pm 1, 0)$, and the foci are $(\pm c, 0) = (\pm\sqrt{5}, 0)$. Since $a = 1$ and $b = 2$, the equations of the asymptotes are

$y = \dfrac{2}{1}x = 2x$ and $y = -\dfrac{2}{1}x = -2x$. To help graph the hyperbola, we form the rectangle using the points $(\pm a, 0) = (\pm 1, 0)$ and $(0, \pm b) = (0, \pm 2)$. The diagonals are the asymptotes.

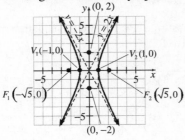

16. $16y^2 - 25x^2 = 1600$

$\dfrac{16y^2 - 25x^2}{1600} = \dfrac{1600}{1600}$

$\dfrac{y^2}{100} - \dfrac{x^2}{64} = 1$

Notice the equation is of the form $\dfrac{y^2}{a^2} - \dfrac{x^2}{b^2} = 1$.

Because the y^2-term is first, the transverse is along the y-axis and the hyperbola opens up and down. The center of the hyperbola is $(0,0)$. We have that $a^2 = 100$ and $b^2 = 64$. Because $c^2 = a^2 + b^2$, we have that $c^2 = 100 + 64 = 164$, so that $c = \sqrt{164} = 2\sqrt{41}$. The vertices are $(0, \pm a) = (0, \pm 10)$, and the foci are $(0, \pm c) = (0, \pm 2\sqrt{41})$. Since $a = 10$ and $b = 8$,

the equations of the asymptotes are

$y = \dfrac{10}{8}x = \dfrac{5}{4}x$ and $y = -\dfrac{10}{8}x = -\dfrac{5}{4}x$. To help graph the hyperbola, we form the rectangle using the points $(0, \pm a) = (0, \pm 10)$ and $(\pm b, 0) = (\pm 8, 0)$. The diagonals are the asymptotes.

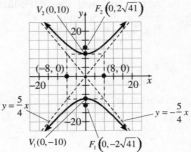

17. The given foci $(\pm 8, 0)$ and vertex $(-3, 0)$ all lie on the x-axis. Thus, the transverse axis is the x-axis and the hyperbola opens left and right. The center of the hyperbola is the midpoint between the two foci. Therefore, the center is $(0,0)$ and the equation of the hyperbola is of the form $\dfrac{x^2}{a^2} - \dfrac{y^2}{b^2} = 1$. Now, the distance between the given vertex $(-3, 0)$ and the center $(0,0)$ is $a = 3$ units. Likewise, the distance between the given foci and the center is $c = 8$ units. Also, $b^2 = c^2 - a^2$, so $b^2 = 8^2 - 3^2 = 64 - 9 = 55$. Thus, the equation of the hyperbola is

$\dfrac{x^2}{9} - \dfrac{y^2}{55} = 1$.

18. $\begin{cases} x^2 + y^2 = 17 \\ x + y = -3 \end{cases}$

First, graph each equation in the system.

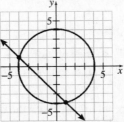

The system apparently has two solutions. Now solve the second equation for y: $y = -x - 3$.
Substitute the result into the first equation:

$$x^2+(-x-3)^2=17$$
$$x^2+(x^2+6x+9)=17$$
$$2x^2+6x-8=0$$
$$x^2+3x-4=0$$
$$(x+4)(x-1)=0$$
$$x=-4 \text{ or } x=1$$

Substitute these results into the first equation to find the corresponding y-values.
$$x=-4:\ -4+y=-3$$
$$y=1$$
$$x=1:\ 1+y=-3$$
$$y=-4$$

Both pairs check, so the solutions are $(-4, 1)$ and $(1,-4)$.

19. $\begin{cases} x^2+y^2=9 \\ 4x^2-y^2=16 \end{cases}$

First, graph each equation in the system.

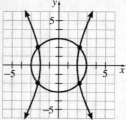

The system apparently has four solutions. Now add the two equations:
$$x^2+y^2=9$$
$$4x^2-y^2=16$$
$$5x^2=25$$
$$x^2=5$$
$$x=\pm\sqrt{5}$$

Substitute 5 for x^2 into the first equation:
$$5+y^2=9$$
$$y^2=4$$
$$y=\pm\sqrt{4}=\pm2$$

All four pairs check, so the solutions are $\left(-\sqrt{5},-2\right)$, $\left(-\sqrt{5},2\right)$, $\left(\sqrt{5},-2\right)$, and $\left(\sqrt{5},2\right)$.

20. **a.** To solve this problem, we draw the ellipse on a Cartesian plane so that the x-axis coincides with the water and the y-axis passes through the center of the arch. Thus, the origin is the center of the ellipse. Now, the "center" of the arch is 10 feet above the

water, so the point $(0,10)$ is on the ellipse. The creek is 30 feet wide, so the two points $(-15,0)$ and $(15,0)$ are the two vertices of the ellipse and the major axis is along the x-axis.

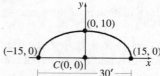

The equation of the ellipse must have the form $\frac{x^2}{a^2}+\frac{y^2}{b^2}=1$. The distance from the center of the ellipse to each vertex is $a=15$. Also, the height of the semi-ellipse is 10 feet, so $b=10$. Thus, the equation of the ellipse is
$$\frac{x^2}{15^2}+\frac{y^2}{10^2}=1$$
$$\frac{x^2}{225}+\frac{y^2}{100}=1$$

b. Substitute $x=12$ into the equation, and solve for y:
$$\frac{12^2}{225}+\frac{y^2}{100}=1$$
$$\frac{144}{225}+\frac{y^2}{100}=1$$
$$\frac{y^2}{100}=\frac{9}{25}$$
$$y^2=36$$
$$y=6$$

Thus, the height of the arch at a distance 12 feet from the center of the creek is 6 feet.

Cumulative Review Chapters R–9

1. To find the domain, we must first determine the values of x that make the denominator of the rational expression zero. Then we exclude those values from the domain of the function.
$$h(x)=\frac{4-x}{7-x}$$
$$7-x=0$$
$$7=x$$
Domain is all real numbers except 7, or $\{x\,|\,x\neq7\}$

2. $5x - 2y = 6$

$5x - 6 = 2y$

$\dfrac{2y}{2} = \dfrac{5x-6}{2}$

$y = \dfrac{5}{2}x - 3$

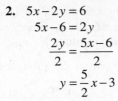

3. $4x - 3y = 15$

$4x - 15 = 3y$

$\dfrac{3y}{3} = \dfrac{4x-15}{3}$

$y = \dfrac{4}{3}x - 5$

The slope of this line is $\dfrac{4}{3}$, so a parallel line will

have the same slope.

$y - y_1 = m(x - x_1)$

$y - (-2) = \dfrac{4}{3}\left[x - (-3)\right]$

$y + 2 = \dfrac{4}{3}(x + 3)$

$y + 2 = \dfrac{4}{3}x + 4$

$y = \dfrac{4}{3}x + 2$

or

$y = \dfrac{4}{3}x + 2$

$3(y) = 3\left(\dfrac{4}{3}x + 2\right)$

$3y = 4x + 6$

$4x - 3y = -6$

4. $\begin{cases} 2x + y = -2 & (1) \\ 5x + 2y = -1 & (2) \end{cases}$

Multiply both sides of equation (1) by -2, and
add the result to equation (2).

$-4x - 2y = 4$

$\underline{5x + 2y = -1}$

$x \qquad = 3$

Substituting 3 for x in equation (1), we obtain

$2(3) + y = -2$

$6 + y = -2$

$y = -8$

The solution is the ordered pair $(3, -8)$.

5. $\begin{vmatrix} 1 & 2 & 0 \\ 7 & -1 & -2 \\ 4 & -2 & 1 \end{vmatrix} = 1\begin{vmatrix} -1 & -2 \\ -2 & 1 \end{vmatrix} - 2\begin{vmatrix} 7 & -2 \\ 4 & 1 \end{vmatrix} + 0\begin{vmatrix} 7 & -1 \\ 4 & -2 \end{vmatrix}$

$= 1\left[-1(1) - (-2)(-2)\right] - 2\left[7(1) - 4(-2)\right]$

$= \left[-1 - 4\right] - 2\left[7 + 8\right]$

$= -5 - 2\left[15\right]$

$= --5 - 30$

$= -35$

6. $8m^3 + n^3 = (2m)^3 + (n)^3$

$= (2m + n)\left(4m^2 - 2mn + n^2\right)$

7. There are no common factors, so
$a = 9$, $b = -18$, $c = 8$. We find that
$a \cdot c = 9 \cdot 8 = 72$. Therefore, we are looking for
two factors of 72 whose sum is $b = -18$. Since
the product is positive and the sum is negative,
both factors must be negative and neither can be
less than -18.

factor 1	factor 2	sum	
-4	-18	-22	too big
-6	-12	-18	← okay

$9p^2 - 18p + 8 = 9p^2 - 6p - 12p + 8$

$= \left(9p^2 - 6p\right) + \left(-12p + 8\right)$

$= 3p(3p - 2) - 4(3p - 2)$

$= (3p - 4)(3p - 2)$

8. $\dfrac{2a^2 - 7a - 4}{3a^2 - 13a + 4} \div \dfrac{a^2 - 2a - 35}{3a^2 + 14a - 5}$

$= \dfrac{(2a+1)\,\cancel{(a-4)}}{\cancel{(3a-1)}\,\cancel{(a-4)}} \cdot \dfrac{\cancel{(3a-1)}\,\cancel{(a+5)}}{(a-7)\,\cancel{(a+5)}}$

$= \dfrac{2a+1}{a-7}$

9.
$$\frac{5}{3t^2-7t+2}-\frac{2}{3t^2-4t+1}$$
$$=\frac{5}{(3t-1)(t-2)}-\frac{2}{(3t-1)(t-1)}$$
$$=\frac{5(t-1)}{(3t-1)(t-2)(t-1)}-\frac{2(t-2)}{(3t-1)(t-2)(t-1)}$$
$$=\frac{(5t-5)-(2t-4)}{(3t-1)(t-2)(t-1)}$$
$$=\frac{\cancel{3t-1}}{\cancel{(3t-1)}(t-2)(t-1)}$$
$$=\frac{1}{(t-2)(t-1)}$$

10. $1+\dfrac{1}{x-1}=\dfrac{x+2}{x}$

$$(x)(x-1)\left(1+\frac{1}{x-1}\right)=(x)(x-1)\left(\frac{x+2}{x}\right)$$
$$(x)(x-1)+x=(x-1)(x+2)$$
$$x^2-x+x=x^2+x-2$$
$$x^2=x^2+x-2$$
$$0=x-2$$
$$2=x$$

Check: $x=2$
$$1+\frac{1}{2-1}\overset{?}{=}\frac{2+2}{2}$$
$$1+1\overset{?}{=}\frac{4}{2}$$
$$2=2\ \checkmark$$

11. $\dfrac{x-5}{x+4}\ge 0$

The rational expression will equal 0 when $x=5$. It is undefined when $x=-4$. Thus we separate the real number line into intervals $(-\infty,-4)$, $(-4,5)$, and $(5,\infty)$. Determine where the numerator and denominator are positive and negative and where the quotient is positive and negative.

Interval	$(-\infty,-4)$	-4	$(-4,5)$	5	$(5,\infty)$
$x-5$	Neg	Neg	Neg	0	Pos
$x+4$	Neg	0	Pos	Pos	Pos
$\dfrac{x-5}{x+4}$	Pos	Undef	Neg	0	Pos

The rational function is undefined at $x=-4$, so -4 is not part of the solution. The inequality is not strict, so 5 is part of the solution. Now, $\dfrac{x-5}{x+4}$ is greater than zero where the quotient is positive. The solution is $(-\infty,-4)\cup[5,\infty)$ in interval notation.

12. $\sqrt{75}+7\sqrt{3}-\sqrt{27}=\sqrt{25\cdot3}+7\sqrt{3}-\sqrt{9\cdot3}$
$$=5\sqrt{3}+7\sqrt{3}-3\sqrt{3}$$
$$=9\sqrt{3}$$

13. $\dfrac{9}{2+\sqrt{7}}=\dfrac{9}{\left(2+\sqrt{7}\right)}\cdot\dfrac{\left(2-\sqrt{7}\right)}{\left(2-\sqrt{7}\right)}$
$$=\frac{9\left(2-\sqrt{7}\right)}{4-7}$$
$$=\frac{9\left(2-\sqrt{7}\right)}{-3}$$
$$=-3\left(2-\sqrt{7}\right)$$

14. $5-\sqrt{x+4}=2$
$$5-2=\sqrt{x+4}$$
$$\sqrt{x+4}=3$$
$$x+4=3^2$$
$$x+4=9$$
$$x=5$$
The solution set is $\{5\}$.

15. $(4-i)(7+3i)=28+12i-7i-3i^2$
$$=28+5i+3$$
$$=31+5i$$

16. $2x^2 - 4x - 3 = 0$

For this equation, $a = 2$, $b = -4$, and $c = -3$.

$$x = \frac{-(-4) \pm \sqrt{(-4)^2 - 4(2)(-3)}}{2(2)}$$

$$= \frac{4 \pm \sqrt{16 + 24}}{4}$$

$$= \frac{4 \pm \sqrt{40}}{4}$$

$$= \frac{4 \pm 2\sqrt{10}}{4}$$

$$= \frac{2 \pm \sqrt{10}}{2}$$

The solution set is $\left\{ \dfrac{2 - \sqrt{10}}{2}, \dfrac{2 + \sqrt{10}}{2} \right\}$.

17. $f(x) = -x^2 + 2x + 8$

$$x = -\frac{b}{2a} = -\frac{-2}{-2} = 1$$

$$f(1) = -(1)^2 + 2(1) + 8$$
$$= -1 + 2 + 8 = 9$$

The vertex is $(1, 9)$.

Because $a = -1 < 0$, the parabola opens downward.

$$-x^2 + 2x + 8 = 0$$
$$x^2 - 2x - 8 = 0$$
$$(x - 4)(x + 2) = 0$$

$x = 4$ or $x = -2$ are the *x*-intercepts of the graph.

$$f(0) = -0^2 + 2(0) + 8 = 8$$

$y = 8$ is the *y*-intercept.

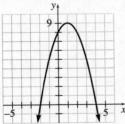

18. $x^2 - 3x - 4 \le 0$

$(x - 4)(x + 1) \le 0$

Solve: $(x - 4)(x + 1) = 0$

$$x - 4 = 0 \quad \text{or} \quad x + 1 = 0$$
$$x = 4 \quad \text{or} \quad x = -1$$

Determine where each factor is positive and negative and where the product of these factors is positive and negative.

Interval	$(-\infty, -1)$	-1	$(-1, 4)$	4	$(4, \infty)$
$x - 4$	$----$	0	$++++$	0	$++++$
$x + 1$	$----$	0	$----$	0	$++++$
$(x-4)(x+1)$	$++++$	0	$----$	0	$++++$

The inequality is not strict, so -1 and 4 are part of the solution. Now, $(x - 4)(x + 1)$ is less than zero where the product is negative. The solution is $\{x \mid -1 \le x \le 4\}$ in set-builder notation; the solution is $[-1, 4]$ in interval notation.

19. $f(x) = 2^{x+3}$

x	$f(x) = 2^{x+3}$
-4	$f(-4) = 2^{-4+3} = 2^{-1} = 0.5$
-3	$f(-3) = 2^{-3+3} = 2^0 = 1$
-2	$f(-2) = 2^{-2+3} = 2^1 = 2$
-1	$f(-1) = 2^{-1+3} = 2^2 = 4$

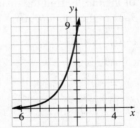

20. $g(x) = \log_4 x$

x	$g(x) = \log_4 x$
$\frac{1}{4}$	$g\left(\frac{1}{4}\right) = \log_4 \frac{1}{4} = -1$
1	$g(1) = \log_4 1 = 0$
4	$g(4) = \log_4 4 = 1$
16	$g(16) = \log_4 16 = 2$

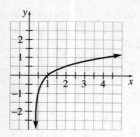

21. $\log_4 32 = \dfrac{\log_2 32}{\log_2 4} = \dfrac{5}{2}$

22. $6^{x+1} = 216$
$6^{x+1} = 6^3$
$x+1 = 3$
$\quad\ x = 2$
The solution set is $\{2\}$.

23. $\log_3\left(x^2 - 16\right) = 2$
$x^2 - 16 = 3^2$
$x^2 - 16 = 9$
$\quad\quad x^2 = 25$
$\quad\quad\ x = \pm 5$
The solution set is $\{-5,\ 5\}$

24. To find the radius of the circle, find the distance
from the center $(-3,\ 1)$ to the point $(0,\ 0)$.
$$r = \sqrt{(-3-0)^2 + (1-0)^2} = \sqrt{10}$$
The equation of the circle is
$$(x-h)^2 + (y-k)^2 = r^2$$
$$\left(x-(-3)\right)^2 + (y-1)^2 = \left(\sqrt{10}\right)^2$$
$$(x+3)^2 + (y-1)^2 = 10$$

25. Notice that $(y-3)^2 = 8(x+3)$ is of the form
$(y-k)^2 = 4a(x-h)$, where $(h,k) = (-3,3)$ and
$4a = 8$, so that $a = 2$. Now, the graph of an
equation of the form $(y-k)^2 = 4a(x-h)$ will be
a parabola that opens to the right with the vertex
at (h,k), focus at $(h+a,k)$, and directrix of
$x = h-a$. Note that $h+a = -3+2 = -1$ and
$h-a = -3-2 = -5$. Thus, the graph of
$(y-3)^2 = 8(x+3)$ is a parabola that opens to the
right with vertex $(-3,3)$, focus $(-1,-3)$, and
directrix $x = -5$. To help graph the parabola,
we plot the two points on the graph above and
below the focus. Let $x = -1$:
$$(y-3)^2 = 8(-1+3)$$
$$(y-3)^2 = 8(2)$$
$$(y-3)^2 = 16$$
$$y-3 = \pm 4$$
$$y = 3 \pm 4$$
$$y = -1 \text{ or } y = 7$$
The points $(-1,-1)$ and $(-1,\ 7)$ are on the
graph.

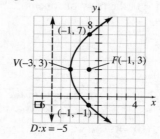

Chapter 10

Section 10.1

Are You Ready for This Section?

R1. $f(x) = x^2 - 4$

    a. $f(3) = (3)^2 - 4 = 9 - 4 = 5$

    b. $f(-7) = (-7)^2 - 4 = 49 - 4 = 45$

R2. $g(x) = 2x - 3$

$$g(1) = 2(1) - 3 = 2 - 3 = -1$$
$$g(2) = 2(2) - 3 = 4 - 3 = 1$$
$$g(3) = 2(3) - 3 = 6 - 3 = 3$$

Therefore,

$$g(1) + g(2) + g(3) = -1 + 1 + 3 = 3$$

R3. In the function $f(n) = n^2 - 4$, the independent variable is n (the input variable).

Section 10.1 Quick Checks

1. A <u>sequence</u> is a function whose domain is the set of positive integers.

2. A sequence that does not end is said to be an <u>infinite</u> sequence. A <u>finite</u> sequence has a domain that is the first n positive integers.

3. True

4. $a_n = 2n - 3$

$$a_1 = 2(1) - 3 = 2 - 3 = -1$$
$$a_2 = 2(2) - 3 = 4 - 3 = 1$$
$$a_3 = 2(3) - 3 = 6 - 3 = 3$$
$$a_4 = 2(4) - 3 = 8 - 3 = 5$$
$$a_5 = 2(5) - 3 = 10 - 3 = 7$$

The first five terms of the sequence are -1, 1, 3, 5, and 7.

5. $b_n = (-1)^n \cdot 4n$

$$b_1 = (-1)^1 \cdot 4(1) = -4$$
$$b_2 = (-1)^2 \cdot 4(2) = 8$$
$$b_3 = (-1)^3 \cdot 4(3) = -12$$
$$b_4 = (-1)^4 \cdot 4(4) = 16$$
$$b_5 = (-1)^5 \cdot 4(5) = -20$$

The first five terms of the sequence are -4, 8, -12, 16, and -20.

6. $5, 7, 9, 11, \ldots$

The terms are consecutive odd numbers with the first term being 5. We can write the terms as follows:

$$5 = 2(1) + 3$$
$$7 = 2(2) + 3$$
$$9 = 2(3) + 3$$
$$11 = 2(4) + 3$$

Notice that each term is 3 more than twice the term number. A formula for the nth term is given by $a_n = 2n + 3$.

7. $\dfrac{1}{2}, -\dfrac{1}{3}, \dfrac{1}{4}, -\dfrac{1}{5}, \ldots$

The terms alternate sign with the first term being positive. So, $(-1)^{n+1}$ must be part of the formula. Each term is a fraction with a numerator of 1 and a denominator that is 1 more than the term number. A formula for the nth term is given by

$$b_n = (-1)^{n+1} \cdot \frac{1}{n+1}.$$

8. When there is a finite number of terms to be added, the sum is called a <u>partial sum</u>.

9. $\displaystyle\sum_{i=1}^{3} (4i - 1) = (4 \cdot 1 - 1) + (4 \cdot 2 - 1) + (4 \cdot 3 - 1)$

$$= 3 + 7 + 11$$
$$= 21$$

10. $\displaystyle\sum_{i=1}^{5} (i^3 + 1) = (1^3 + 1) + (2^3 + 1) + (3^3 + 1)$

$$+ (4^3 + 1) + (5^3 + 1)$$
$$= (1 + 1) + (8 + 1) + (27 + 1)$$
$$+ (64 + 1) + (125 + 1)$$
$$= 2 + 9 + 28 + 65 + 126$$
$$= 230$$

11. $1 + 4 + 9 + \ldots + 144$

Notice that each term is a perfect square. We can rewrite the sum as $1^2 + 2^2 + 3^2 + \ldots + 12^2$. Thus, the sum has 12 terms, each of the form i^2.

$$1 + 4 + 9 + \ldots + 144 = \sum_{i=1}^{12} i^2$$

12. Begin by writing the first term as $\dfrac{1}{1}$. We notice that each term is a fraction with a numerator of 1 and a denominator that is a power of 2. We can write the sum as

$$\frac{1}{2^0} + \frac{1}{2^1} + \frac{1}{2^2} + \ldots + \frac{1}{2^5}$$

The exponent on 2 in the denominator is always 1 less than the term number. Thus, the sum has 6 terms, each of the form $\dfrac{1}{2^{i-1}}$.

$$1 + \frac{1}{2} + \frac{1}{4} + \ldots + \frac{1}{32} = \sum_{i=1}^{6} \left(\frac{1}{2^{i-1}} \right)$$

10.1 Exercises

13. $\{3n + 5\}$

$$a_1 = 3 \cdot 1 + 5 = 3 + 5 = 8$$
$$a_2 = 3 \cdot 2 + 5 = 6 + 5 = 11$$
$$a_3 = 3 \cdot 3 + 5 = 9 + 5 = 14$$
$$a_4 = 3 \cdot 4 + 5 = 12 + 5 = 17$$
$$a_5 = 3 \cdot 5 + 5 = 15 + 5 = 20$$

The first five terms of the sequence are 8, 11, 14, 17, and 20.

15. $\left\{ \dfrac{n}{n+2} \right\}$

$$a_1 = \frac{1}{1+2} = \frac{1}{3}$$
$$a_2 = \frac{2}{2+2} = \frac{2}{4} = \frac{1}{2}$$
$$a_3 = \frac{3}{3+2} = \frac{3}{5}$$
$$a_4 = \frac{4}{4+2} = \frac{4}{6} = \frac{2}{3}$$
$$a_5 = \frac{5}{5+2} = \frac{5}{7}$$

The first five terms of the sequence are $\dfrac{1}{3}$, $\dfrac{1}{2}$, $\dfrac{3}{5}$, $\dfrac{2}{3}$, and $\dfrac{5}{7}$.

17. $\left\{ (-1)^n n \right\}$

$$a_1 = (-1)^1 \cdot 1 = -1$$
$$a_2 = (-1)^2 \cdot 2 = 2$$
$$a_3 = (-1)^3 \cdot 3 = -3$$
$$a_4 = (-1)^4 \cdot 4 = 4$$
$$a_5 = (-1)^5 \cdot 5 = -5$$

The first five terms of the sequence are -1, 2, -3, 4, and -5.

19. $\left\{ 2^n + 1 \right\}$

$$a_1 = 2^1 + 1 = 2 + 1 = 3$$
$$a_2 = 2^2 + 1 = 4 + 1 = 5$$
$$a_3 = 2^3 + 1 = 8 + 1 = 9$$
$$a_4 = 2^4 + 1 = 16 + 1 = 17$$
$$a_5 = 2^5 + 1 = 32 + 1 = 33$$

The first five terms of the sequence are 3, 5, 9, 17, and 33.

21. $\left\{ \dfrac{2n}{2^n} \right\}$

$$a_1 = \frac{2 \cdot 1}{2^1} = \frac{2}{2} = 1$$
$$a_2 = \frac{2 \cdot 2}{2^2} = \frac{4}{4} = 1$$
$$a_3 = \frac{2 \cdot 3}{2^3} = \frac{6}{8} = \frac{3}{4}$$
$$a_4 = \frac{2 \cdot 4}{2^4} = \frac{8}{16} = \frac{1}{2}$$
$$a_5 = \frac{2 \cdot 5}{2^5} = \frac{10}{32} = \frac{5}{16}$$

The first five terms of the sequence are 1, 1, $\dfrac{3}{4}$, $\dfrac{1}{2}$, and $\dfrac{5}{16}$.

23. $\left\{ \dfrac{n}{e^n} \right\}$

$a_1 = \dfrac{1}{e^1} = \dfrac{1}{e}, \quad a_2 = \dfrac{2}{e^2}, \quad a_3 = \dfrac{3}{e^3}$

$a_4 = \dfrac{4}{e^4}, \quad a_5 = \dfrac{5}{e^5}$

The first five terms of the sequence are $\dfrac{1}{e}$, $\dfrac{2}{e^2}$,

$\dfrac{3}{e^3}$, $\dfrac{4}{e^4}$, and $\dfrac{5}{e^5}$.

25. The terms are all multiples of 2 with the first term equaling $2 \cdot 1$, the second term equaling $2 \cdot 2$, and so on. A formula for the nth term is given by $a_n = 2n$.

27. Each term is a fraction with the denominator equaling 1 more than the numerator. When $n = 1$, the numerator equals 1. Each subsequent numerator is one more than previous. A formula for the nth term is given by $a_n = \dfrac{n}{n+1}$.

29. When $n = 1$, we have that $a_1 = 3 = 1^2 + 2$; when $n = 2$, we have that $a_2 = 6 = 2^2 + 2$; when $n = 3$, we have that $a_3 = 11 = 3^2 + 2$. Notice that each term is equal to 2 more than the square of the term number. Therefore, a formula for the nth term is given by $a_n = n^2 + 2$.

31. Notice that the terms alternate signs with the first term being negative. Ignoring the signs, also notice that the terms are all perfect squares. Therefore, a formula for the nth term is given by $a_n = (-1)^n n^2$.

33. $\displaystyle\sum_{i=1}^{4} (5i+1)$

$= (5 \cdot 1 + 1) + (5 \cdot 2 + 1) + (5 \cdot 3 + 1) + (5 \cdot 4 + 1)$

$= 6 + 11 + 16 + 21$

$= 54$

35. $\displaystyle\sum_{i=1}^{5} \dfrac{i^2}{2} = \dfrac{1^2}{2} + \dfrac{2^2}{2} + \dfrac{3^2}{2} + \dfrac{4^2}{2} + \dfrac{5^2}{2}$

$= \dfrac{1}{2} + \dfrac{4}{2} + \dfrac{9}{2} + \dfrac{16}{2} + \dfrac{25}{2}$

$= \dfrac{55}{2}$

37. $\displaystyle\sum_{k=1}^{3} 2^k = 2^1 + 2^2 + 2^3 = 2 + 4 + 8 = 14$

39. $\displaystyle\sum_{k=1}^{5} \left[(-1)^{k+1} \cdot 2k \right]$

$= (-1)^{1+1} \cdot 2(1) + (-1)^{2+1} \cdot 2(2) + (-1)^{3+1} \cdot 2(3)$

$\quad + (-1)^{4+1} \cdot 2(4) + (-1)^{5+1} \cdot 2(5)$

$= (-1)^2 \cdot 2 + (-1)^3 \cdot 4 + (-1)^4 \cdot 6$

$\quad + (-1)^5 \cdot 8 + (-1)^6 \cdot 10$

$= 2 - 4 + 6 - 8 + 10$

$= 6$

41. $\displaystyle\sum_{j=1}^{10} 5 = 5+5+5+5+5+5+5+5+5+5 = 50$

43. $\displaystyle\sum_{k=3}^{7} (2k-1) = (2 \cdot 3 - 1) + (2 \cdot 4 - 1) + (2 \cdot 5 - 1)$

$\quad\quad\quad + (2 \cdot 6 - 1) + (2 \cdot 7 - 1)$

$\quad\quad\quad = 5 + 7 + 9 + 11 + 13$

$\quad\quad\quad = 45$

45. The sum $1 + 2 + 3 + \ldots + 15$ has 15 terms, each term has the form k, starts at $k = 1$, and ends at $k = 15$.

$1 + 2 + 3 + \ldots + 15 = \displaystyle\sum_{k=1}^{15} k$

47. The sum $1 + \dfrac{1}{2} + \dfrac{1}{3} + \ldots + \dfrac{1}{12}$ has 12 terms, each term has the form $\dfrac{1}{i}$, starts at $i = 1$, and ends at $i = 12$.

$1 + \dfrac{1}{2} + \dfrac{1}{3} + \ldots + \dfrac{1}{12} = \displaystyle\sum_{i=1}^{12} \dfrac{1}{i}$

49. To see the pattern, note that the first term can be written as $1 = \dfrac{1}{1}$. We see that the denominators are all powers of 3 and that the terms alternate signs with the first term being positive. When $n = 1$, we have the denominator $1 = 3^0$; when $n = 2$, we have the denominator $3 = 3^1$; when $n = 3$, we have the denominator $9 = 3^2$; and so on. Thus, the power on 3 in the denominator is always one less than the term number. Looking at the form of the last term, we see that there will be 9 terms and that the power on -1 will be one more than the term number.

$$1 - \frac{1}{3} + \frac{1}{9} - \frac{1}{27} + \ldots + (-1)^{9+1}\left(\frac{1}{3^{9-1}}\right)$$

$$= \sum_{i=1}^{9} (-1)^{i+1}\left(\frac{1}{3^{i-1}}\right)$$

51. To see the pattern, note that the first term can be written as $5 = (5 + 2 \cdot 0)$. Each term is 5 more than a multiple of two. The multiple is one less than the term number. Therefore, there are 11 terms being added together.

$$5 + (5 + 2 \cdot 1) + (5 + 2 \cdot 2) + (5 + 2 \cdot 3) + \ldots + (5 + 2 \cdot 10)$$

$$= \sum_{k=1}^{11} [5 + 2(k-1)] \quad \text{or} \quad \sum_{k=1}^{11} (2k+3)$$

53. $a_n = 12{,}000\left(1 + \dfrac{0.06}{4}\right)^n$

a. $n = 1$

$$a_1 = 12{,}000\left(1 + \frac{0.06}{4}\right)^1$$
$$= 12{,}000(1.015)$$
$$= 12{,}180$$

After 1 quarter, the account will have a balance of \$12,180.

b. $n = 4$ (4 quarters in each year)

$$a_4 = 12{,}000\left(1 + \frac{0.06}{4}\right)^4$$
$$= 12{,}000(1.015)^4$$
$$\approx 12{,}000(1.06136)$$
$$\approx 12{,}736.36$$

After 1 year (i.e. 4 quarters), the account will have a balance of \$12,736.36.

c. $n = 40$

$$a_{40} = 12{,}000\left(1 + \frac{0.06}{4}\right)^{40}$$
$$= 12{,}000(1.015)^{40}$$
$$\approx 12{,}000(1.8140184)$$
$$\approx 21{,}768.22$$

After 10 years (40 quarters), the account will have a balance of \$21,768.22.

55. $p_n = 313(1.011)^n$

a. In 2015, we have $n = 3$.
$$p_3 = 313(1.011)^3 \approx 323$$
In 2015, the population of the United States will be an estimated 32.3 million.

b. In 2050, we have $n = 38$.
$$p_{38} = 313(1.011)^{38} \approx 474$$
In 2050, the population of the United States will be an estimated 474 million.

57. $u_n = \dfrac{\left(1 + \sqrt{5}\right)^n - \left(1 - \sqrt{5}\right)^n}{2^n \cdot \sqrt{5}}$

$$u_1 = \frac{\left(1 + \sqrt{5}\right) - \left(1 - \sqrt{5}\right)}{2 \cdot \sqrt{5}} = \frac{2\sqrt{5}}{2 \cdot \sqrt{5}} = 1$$

$$u_2 = \frac{\left(1 + \sqrt{5}\right)^2 - \left(1 - \sqrt{5}\right)^2}{2^2 \cdot \sqrt{5}} = \frac{4\sqrt{5}}{4 \cdot \sqrt{5}} = 1$$

$$u_3 = \frac{\left(1 + \sqrt{5}\right)^3 - \left(1 - \sqrt{5}\right)^3}{2^3 \cdot \sqrt{5}}$$
$$= \frac{\left(16 + 8\sqrt{5}\right) - \left(16 - 8\sqrt{5}\right)}{8 \cdot \sqrt{5}} = \frac{16\sqrt{5}}{8\sqrt{5}} = 2$$

$$u_4 = \frac{\left(1 + \sqrt{5}\right)^4 - \left(1 - \sqrt{5}\right)^4}{2^4 \cdot \sqrt{5}}$$
$$= \frac{\left(56 + 24\sqrt{5}\right) - \left(56 - 24\sqrt{5}\right)}{16 \cdot \sqrt{5}} = \frac{48\sqrt{5}}{16\sqrt{5}} = 3$$

$$u_5 = \frac{\left(1 + \sqrt{5}\right)^5 - \left(1 - \sqrt{5}\right)^5}{2^5 \cdot \sqrt{5}}$$
$$= \frac{\left(176 + 80\sqrt{5}\right) - \left(176 - 80\sqrt{5}\right)}{32 \cdot \sqrt{5}} = \frac{160\sqrt{5}}{32\sqrt{5}} = 5$$

$$u_6 = \frac{\left(1+\sqrt{5}\right)^6 - \left(1-\sqrt{5}\right)^6}{2^6 \cdot \sqrt{5}}$$

$$= \frac{\left(576 + 256\sqrt{5}\right) - \left(576 - 256\sqrt{5}\right)}{64 \cdot \sqrt{5}} = \frac{512\sqrt{5}}{64\sqrt{5}}$$

$$= 8$$

$$u_7 = \frac{\left(1+\sqrt{5}\right)^7 - \left(1-\sqrt{5}\right)^7}{2^7 \cdot \sqrt{5}}$$

$$= \frac{\left(1856 + 832\sqrt{5}\right) - \left(1856 - 832\sqrt{5}\right)}{128 \cdot \sqrt{5}}$$

$$= \frac{1664\sqrt{5}}{128\sqrt{5}} = 13$$

$$u_8 = \frac{\left(1+\sqrt{5}\right)^8 - \left(1-\sqrt{5}\right)^8}{2^8 \cdot \sqrt{5}}$$

$$= \frac{\left(6016 + 2688\sqrt{5}\right) - \left(6016 - 2688\sqrt{5}\right)}{256 \cdot \sqrt{5}}$$

$$= \frac{5376\sqrt{5}}{256\sqrt{5}} = 21$$

$$u_9 = \frac{\left(1+\sqrt{5}\right)^9 - \left(1-\sqrt{5}\right)^9}{2^9 \cdot \sqrt{5}}$$

$$= \frac{\left(19456 + 8704\sqrt{5}\right) - \left(19456 - 8704\sqrt{5}\right)}{512 \cdot \sqrt{5}}$$

$$= \frac{17408\sqrt{5}}{512\sqrt{5}} = 34$$

$$u_{10} = \frac{\left(1+\sqrt{5}\right)^{10} - \left(1-\sqrt{5}\right)^{10}}{2^{10} \cdot \sqrt{5}}$$

$$= \frac{\left(62976 + 28160\sqrt{5}\right) - \left(62976 - 28160\sqrt{5}\right)}{1024 \cdot \sqrt{5}}$$

$$= \frac{56320\sqrt{5}}{1024\sqrt{5}} = 55$$

The first 10 terms of the sequence are 1, 1, 2, 3, 5, 8, 13, 21, 34, and 55.
Notice that, beginning with the third term, each term in the sequence can be obtained by adding the two previous terms together. For example, $2 = 1+1$, $3 = 1+2$, $5 = 2+3$, etc.

59. $a_1 = 10$
$a_2 = 1.05a_1 = 1.05(10) = 10.5$
$a_3 = 1.05a_2 = 1.05(10.5) = 11.025$
$a_4 = 1.05a_3 = 1.05(11.025) = 11.57625$
$a_5 = 1.05a_4 = 1.05(11.57625) = 12.1550625$
The first five terms of the sequence are 10, 10.5, 11.025, 11.57625, and 12.1550625.

61. $b_1 = 8$
$b_2 = 2 + b_1 = 2 + 8 = 10$
$b_3 = 3 + b_2 = 3 + 10 = 13$
$b_4 = 4 + b_3 = 4 + 13 = 17$
$b_5 = 5 + b_4 = 5 + 17 = 22$
The first five terms of the sequence are 8, 10, 13, 17, and 22.

63. a. $\dfrac{u_2}{u_1} = \dfrac{1}{1} = 1$; $\dfrac{u_3}{u_2} = \dfrac{2}{1} = 2$; $\dfrac{u_4}{u_3} = \dfrac{3}{2} = 1.5$;

$\dfrac{u_5}{u_4} = \dfrac{5}{3} = 1.\overline{6}$; $\dfrac{u_6}{u_5} = \dfrac{8}{5} = 1.6$;

$\dfrac{u_7}{u_6} = \dfrac{13}{8} = 1.625$; $\dfrac{u_8}{u_7} = \dfrac{21}{13} \approx 1.615385$;

$\dfrac{u_9}{u_8} = \dfrac{34}{21} \approx 1.619048$; $\dfrac{u_{10}}{u_9} = \dfrac{55}{34} \approx 1.617647$;

$\dfrac{u_{11}}{u_{10}} = \dfrac{89}{55} \approx 1.618182$

b. The ratio approaches a value around 1.618. Note: As n goes off to infinity, the ratio approaches the exact value $\dfrac{1+\sqrt{5}}{2}$.

c. $\dfrac{u_1}{u_2} = \dfrac{1}{1} = 1$; $\dfrac{u_2}{u_3} = \dfrac{1}{2} = 0.5$; $\dfrac{u_3}{u_4} = \dfrac{2}{3} = 0.\overline{6}$;

$\dfrac{u_4}{u_5} = \dfrac{3}{5} = 0.6$; $\dfrac{u_5}{u_6} = \dfrac{5}{8} = 0.625$;

$\dfrac{u_6}{u_7} = \dfrac{8}{13} \approx 0.615385$; $\dfrac{u_7}{u_8} = \dfrac{13}{21} \approx 0.619048$;

$\dfrac{u_8}{u_9} = \dfrac{21}{34} \approx 0.617647$;

$\dfrac{u_9}{u_{10}} = \dfrac{34}{55} \approx 0.618182$;

$\dfrac{u_{10}}{u_{11}} = \dfrac{55}{89} \approx 0.617978$

d. The ratio approaches a value around 0.618. Note: As *n* goes to infinity, the ratio approaches the exact value $\frac{\sqrt{5}-1}{2}$.

65. Answers may vary. A function is any relation that assigns each element of the domain to exactly one element of the range. A sequence is a function whose domain is the set of positive integers.

67. Answers may vary. The symbol Σ represents summation and indicates that we will need to add up the values of several terms.

69. a. The function is in slope-intercept form so the slope is $m=4$.

b. $f(x)=4x-6$
$f(1)=4(1)-6=-2$
$f(2)=4(2)-6=2$
$f(3)=4(3)-6=6$
$f(4)=4(4)-6=10$

71. a. The function is in slope-intercept form so the slope is $m=-5$.

b. $f(x)=-5x+8$
$f(1)=-5(1)+8=3$
$f(2)=-5(2)+8=-2$
$f(3)=-5(3)+8=-7$
$f(4)=-5(4)+8=-12$

73. $\{3n+5\}$

The first five terms of the sequence are 8, 11, 14, 17, and 20.

75. $\left\{\frac{n}{n+2}\right\}$

The first five terms of the sequence are $\frac{1}{3}$, $\frac{1}{2}$, $\frac{3}{5}$, $\frac{2}{3}$, and $\frac{5}{7}$.

77. $\{(-1)^n n\}$

The first five terms of the sequence are -1, 2, -3, 4, and -5.

79. $\{2^n+1\}$

The first five terms of the sequence are 3, 5, 9, 17, and 33.

81. $\sum_{i=1}^{4}(5i+1)$
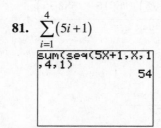
The sum is 54.

83. $\sum_{i=1}^{5}\frac{i^2}{2}$
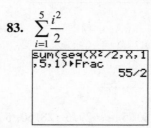
The sum is $\frac{55}{2}$.

85. $\sum_{k=1}^{3}2^k$
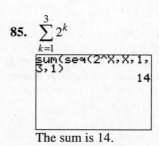
The sum is 14.

87. $\sum_{k=1}^{5}\left[(-1)^{k+1}\cdot 2k\right]$

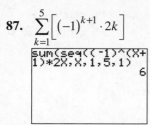

The sum is 6.

Section 10.2

Are You Ready for This Section?

R1. $y = -3x + 1$ is in the (slope-intercept) form $y = mx + b$. Therefore, the slope is $m = -3$.

R2. $g(x) = 5x + 2$
$g(3) = 5(3) + 2$
$\quad = 15 + 2$
$\quad = 17$

R3. $\begin{cases} x - 3y = -17 & (1) \\ 2x + y = 1 & (2) \end{cases}$
Multiply both sides of equation (1) by -2.
$\begin{cases} -2x + 6y = 34 & (1) \\ 2x + y = 1 & (2) \end{cases}$
Add equation (1) and equation (2).
$\begin{array}{r} \begin{cases} -2x + 6y = 34 \ \ (1) \\ \ \ 2x + y = 1 \ \ (2) \end{cases} \\ \hline 7y = 35 \\ y = 5 \end{array}$
+

Substitute this result into the original equation (1) and solve for x.
$x - 3y = -17$
$x - 3(5) = -17$
$x - 15 = -17$
$\quad x = -2$
The solution is the ordered pair $(-2, 5)$.

Section 10.2 Quick Checks

1. In an <u>arithmetic</u> sequence, the difference between consecutive terms is constant.

2. $-1 - (-3) = -1 + 3 = 2$
$1 - (-1) = 1 + 1 = 2$
$3 - 1 = 2$
$5 - 3 = 2$
The sequence $-3, -1, 1, 3, 5, \ldots$ is arithmetic because the difference between consecutive terms is constant. The first term is $a_1 = -3$ and the common difference is $d = 2$.

3. $9 - 3 = 6, \ 27 - 9 = 18, \ 81 - 27 = 54$
The sequence is not arithmetic because the difference between consecutive terms is not constant.

4. $a_n - a_{n-1} = [3n - 8] - [3(n-1) - 8]$
$\quad\quad\quad\quad = [3n - 8] - [3n - 11]$
$\quad\quad\quad\quad = 3n - 8 - 3n + 11$
$\quad\quad\quad\quad = 3$
The sequence is arithmetic because the difference between consecutive terms is constant. The first term is $a_1 = 3(1) - 8 = -5$ and the common difference is $d = 3$.

5. $b_n - b_{n-1} = \left[n^2 - 1\right] - \left[(n-1)^2 - 1\right]$
$\quad\quad\quad\quad = \left[n^2 - 1\right] - \left[n^2 - 2n + 1 - 1\right]$
$\quad\quad\quad\quad = \left[n^2 - 1\right] - \left[n^2 - 2n\right]$
$\quad\quad\quad\quad = n^2 - 1 - n^2 + 2n$
$\quad\quad\quad\quad = 2n - 1$
The sequence is not arithmetic because the difference between consecutive terms is not constant.

6. $c_n - c_{n-1} = [5 - 2n] - [5 - 2(n-1)]$
$\quad\quad\quad\quad = [5 - 2n] - [5 - 2n + 2]$
$\quad\quad\quad\quad = [5 - 2n] - [7 - 2n]$
$\quad\quad\quad\quad = 5 - 2n - 7 + 2n$
$\quad\quad\quad\quad = -2$
The sequence is arithmetic because the difference between consecutive terms is constant. The first term is $c_1 = 5 - 2(1) = 3$ and the common difference is $d = -2$.

7. For an arithmetic sequence $\{a_n\}$ whose first term is a_1 and whose common difference is d, the nth term is determined by the formula $\underline{a_n = a_1 + (n-1)d}$.

8. a. We have that $a_5 = 25$ and $d = 6$.

$$a_n = a_1 + (n-1)d$$
$$a_5 = a_1 + (5-1)d$$
$$25 = a_1 + 4(6)$$
$$25 = a_1 + 24$$
$$1 = a_1$$

Therefore, the nth term of the sequence is given by

$$a_n = 1 + (n-1)(6) = 1 + 6n - 6$$
$$a_n = 6n - 5$$

b. $a_{14} = 6(14) - 5 = 84 - 5 = 79$

9. a. We are given $a_5 = 7$ and $a_{13} = 31$. The nth term of an arithmetic sequence is given by $a_n = a_1 + (n-1)d$. Therefore, we have

$$\begin{cases} a_5 = a_1 + (5-1)d & \text{or} \\ a_{13} = a_1 + (13-1)d \end{cases}$$

$$\begin{cases} 7 = a_1 + 4d & (1) \\ 31 = a_1 + 12d & (2) \end{cases}$$

This is a system of linear equations in a_1 and d. We can solve the system by using elimination. Subtract equation (2) from equation (1) to obtain

$$-24 = -8d$$
$$3 = d$$

Let $d = 3$ in equation (1) and solve for a_1.

$$7 = a_1 + 4(3)$$
$$7 = a_1 + 12$$
$$-5 = a_1$$

The first term is $a = -5$ and the common difference is $d = 3$.

b. A formula for the nth term is

$$a_n = a_1 + (n-1)d = -5 + (n-1)(3)$$
$$= -5 + 3n - 3$$
$$a_n = 3n - 8$$

10. We have $a = 5$ and $d = 2$, and wish to find S_{100}.

$$S_n = \frac{n}{2}\left[2a_1 + (n-1)d\right]$$
$$S_{100} = \frac{100}{2}\left[2(5) + (100-1)(2)\right] = 50\left[10 + 198\right]$$
$$= 50(208) = 10,400$$

11. The first term is $a_1 = 1$ and the common difference is $d = 5 - 1 = 4$. We wish to find S_{70}.

$$S_n = \frac{n}{2}\left[2a_1 + (n-1)d\right]$$
$$S_{70} = \frac{70}{2}\left[2(1) + (70-1)(4)\right]$$
$$= 35\left[2 + 276\right]$$
$$= 35(278)$$
$$= 9730$$

12. We have $a_1 = 4$ and $a_{50} = 298$, and wish to find the sum of the first 50 terms, S_{50}.

$$S_n = \frac{n}{2}\left[a_1 + a_n\right]$$
$$S_{50} = \frac{50}{2}\left[a_1 + a_{50}\right]$$
$$= \frac{50}{2}\left[4 + 298\right]$$
$$= 25(302)$$
$$= 7550$$

13.
$$a_n = -3n + 100$$
$$a_1 = -3(1) + 100 = 97$$
$$a_{75} = -3(75) + 100 = -125$$

$$S_n = \frac{n}{2}\left[a_1 + a_n\right]$$
$$S_{75} = \frac{75}{2}\left[a_1 + a_{75}\right]$$
$$S_{75} = \frac{75}{2}\left[97 + (-125)\right]$$
$$= \frac{75}{2}(-28)$$
$$= -1050$$

14. If we let a_n = the number of seats in the nth row, then we have an arithmetic sequence with first term $a_1 = 20$ and common difference $d = 2$. Thus, the total number of seats in the 30 rows is given by S_{30}.

$$S_n = \frac{n}{2}\left[2a_1 + (n-1)d\right]$$
$$S_{30} = \frac{30}{2}\left[2a_1 + (30-1)d\right] = 15\left[2(20) + 29(2)\right]$$
$$= 15(98) = 1470$$

There are 1470 seats in the 30 rows.

10.2 Exercises

15. $\{n+5\}$

$a_{n-1} = (n-1)+5 = n+4$

$a_n = n+5$

$a_n - a_{n-1} = (n+5)-(n+4)$
$\qquad = n+5-n-4$
$\qquad = 1$

The difference between *any* consecutive terms is 1, so the sequence is arithmetic with common difference $d = 1$. To find the first term, we evaluate a_1 and find $a_1 = 1+5 = 6$.

17. $\{7n+2\}$

$a_{n-1} = 7(n-1)+2 = 7n-5$

$a_n = 7n+2$

$a_n - a_{n-1} = (7n+2)-(7n-5)$
$\qquad = 7n+2-7n+5$
$\qquad = 7$

The difference between *any* consecutive terms is 7, so the sequence is arithmetic with common difference $d = 7$. To find the first term, we evaluate a_1 and find $a_1 = 7(1)+2 = 9$.

19. $\{7-3n\}$

$a_{n-1} = 7-3(n-1) = 7-3n+3 = 10-3n$

$a_n = 7-3n$

$a_n - a_{n-1} = (7-3n)-(10-3n)$
$\qquad = 7-3n-10+3n$
$\qquad = -3$

The difference between *any* consecutive terms is -3, so the sequence is arithmetic with common difference $d = -3$. To find the first term, we evaluate a_1 and find $a_1 = 7-3(1) = 4$.

21. $\left\{\dfrac{1}{2}n+5\right\}$

$a_{n-1} = \dfrac{1}{2}(n-1)+5 = \dfrac{1}{2}n+\dfrac{9}{2}$

$a_n = \dfrac{1}{2}n+5$

$a_n - a_{n-1} = \left(\dfrac{1}{2}n+5\right)-\left(\dfrac{1}{2}n+\dfrac{9}{2}\right)$
$\qquad = \dfrac{1}{2}n+5-\dfrac{1}{2}n-\dfrac{9}{2}$
$\qquad = \dfrac{1}{2}$

The difference between *any* consecutive terms is

$\dfrac{1}{2}$, so the sequence is arithmetic with common difference $d = \dfrac{1}{2}$. To find the first term, we evaluate a_1 and find $a_1 = \dfrac{1}{2}(1)+5 = \dfrac{11}{2}$.

23. $a_1 = 4,\ d = 3$

$a_n = a_1 + (n-1)d$
$\qquad = 4+(n-1)3$
$\qquad = 4+3n-3$
$\qquad = 3n+1$

To find the fifth term we let $n = 5$.

$a_5 = 3(5)+1 = 15+1 = 16$

The fifth term of the sequence is 16.

25. $a_1 = 10,\ d = -5$

$a_n = a_1 + (n-1)d$
$\qquad = 10+(n-1)(-5)$
$\qquad = 10-5n+5$
$\qquad = -5n+15$

To find the fifth term we let $n = 5$.

$a_5 = -5(5)+15 = -25+15 = -10$

The fifth term of the sequence is -10.

27. $a_1 = 2,\ d = \dfrac{1}{3}$

$a_n = a_1 + (n-1)d = 2+(n-1)\left(\dfrac{1}{3}\right)$
$\qquad = 2+\dfrac{1}{3}n-\dfrac{1}{3}$
$\qquad = \dfrac{1}{3}n+\dfrac{5}{3}$

To find the fifth term we let $n = 5$.

$a_5 = \dfrac{1}{3}(5)+\dfrac{5}{3} = \dfrac{5}{3}+\dfrac{5}{3} = \dfrac{10}{3}$

The fifth term of the sequence is $\dfrac{10}{3}$.

29. $a_1 = 5,\ d = -\dfrac{1}{5}$

$a_n = a_1 + (n-1)d = 5+(n-1)\left(-\dfrac{1}{5}\right)$
$\qquad = 5-\dfrac{1}{5}n+\dfrac{1}{5}$
$\qquad = -\dfrac{1}{5}n+\dfrac{26}{5}$

To find the fifth term we let $n=5$.
$$a_5 = -\frac{1}{5}(5)+\frac{26}{5} = -1+\frac{26}{5} = \frac{21}{5}$$
The fifth term of the sequence is $\frac{21}{5}$.

31. $2,7,12,17,\ldots$
Notice that the difference between consecutive terms is $d=5$. Since the first term is $a=2$, the nth term can be written as
$$\begin{aligned} a_n &= a_1+(n-1)d \\ &= 2+(n-1)(5) \\ &= 2+5n-5 \\ &= 5n-3 \end{aligned}$$
To find the twentieth term, we let $n=20$.
$$a_{20}=5(20)-3=100-3=97$$
The twentieth term of the sequence is 97.

33. $12,9,6,3,\ldots$
Notice that the difference between consecutive terms is $d=-3$. Since the first term is $a_1=12$, the nth term can be written as
$$\begin{aligned} a_n &= a_1+(n-1)d \\ &= 12+(n-1)(-3) \\ &= 12-3n+3 \\ &= -3n+15 \end{aligned}$$
To find the twentieth term, we let $n=20$.
$$a_{20}=-3(20)+15=-60+15=-45$$
The twentieth term of the sequence is -45.

35. $1,\frac{5}{4},\frac{3}{2},\frac{7}{4},\ldots$
Notice that the difference between consecutive terms is $d=\frac{1}{4}$. Since the first term is $a_1=1$, the nth term can be written as
$$\begin{aligned} a_n &= a_1+(n-1)d \\ &= 1+(n-1)\left(\frac{1}{4}\right) \\ &= 1+\frac{1}{4}n-\frac{1}{4} \\ &= \frac{1}{4}n+\frac{3}{4} \end{aligned}$$
To find the twentieth term, we let $n=20$.
$$a_{20}=\frac{1}{4}(20)+\frac{3}{4}=\frac{20}{4}+\frac{3}{4}=\frac{23}{4}$$
The twentieth term of the sequence is $\frac{23}{4}$.

37. We know that the nth term of an arithmetic sequence is given by $a_n=a_1+(n-1)d$ where a_1 is the first term and d is the common difference. Since $a_3=17$ and $a_7=37$, we have
$$\begin{cases} a_3=a_1+(3-1)d \\ a_7=a_1+(7-1)d \end{cases} \text{ or } \begin{cases} 17=a_1+2d & (1) \\ 37=a_1+6d & (2) \end{cases}$$
This is a system of linear equations in two variables, a_1 and d. We can solve the system by elimination. If we subtract equation (2) from equation (1), we obtain
$$-20=-4d$$
$$5=d$$
Let $d=5$ in equation (1) to find a_1.
$$17=a_1+2(5)$$
$$17=a_1+10$$
$$7=a_1$$
The first term is $a_1=7$ and the common difference is $d=5$. Thus, a formula for the nth term is
$$\begin{aligned} a_n &= 7+(n-1)(5) \\ &= 7+5n-5 \\ &= 5n+2 \end{aligned}$$

39. We know that the nth term of an arithmetic sequence is given by $a_n=a_1+(n-1)d$ where a_1 is the first term and d is the common difference. Since $a_4=-2$ and $a_8=26$, we have
$$\begin{cases} a_4=a_1+(4-1)d \\ a_8=a_1+(8-1)d \end{cases} \text{ or } \begin{cases} -2=a_1+3d & (1) \\ 26=a_1+7d & (2) \end{cases}$$
This is a system of linear equations in two variables, a_1 and d. We can solve the system by elimination. If we subtract equation (2) from equation (1), we obtain
$$-28=-4d$$
$$7=d$$
Let $d=7$ in equation (1) to find a.
$$-2=a_1+3(7)$$
$$-2=a_1+21$$
$$-23=a_1$$
The first term is $a_1=-23$ and the common difference is $d=7$. Thus, a formula for the nth term is
$$\begin{aligned} a_n &= -23+(n-1)(7) \\ &= -23+7n-7 \\ &= 7n-30 \end{aligned}$$

41. We know that the *n*th term of an arithmetic
sequence is given by $a_n = a_1 + (n-1)d$ where
a_1 is the first term and *d* is the common
difference. Since $a_5 = -1$ and $a_{12} = -22$, we
have
$$\begin{cases} a_5 = a_1 + (5-1)d \\ a_{12} = a_1 + (12-1)d \end{cases} \text{ or } \begin{cases} -1 = a_1 + 4d \quad (1) \\ -22 = a_1 + 11d \quad (2) \end{cases}$$
This is a system of linear equations in two
variables, a_1 and *d*. We can solve the system by
elimination. If we subtract equation (2) from
equation (1), we obtain
$$21 = -7d$$
$$-3 = d$$
Let $d = -3$ in equation (1) to find a_1.
$$-1 = a_1 + 4(-3)$$
$$-1 = a_1 - 12$$
$$11 = a_1$$
The first term is $a_1 = 11$ and the common
difference is $d = -3$. Thus, a formula for the *n*th
term is
$$a_n = 11 + (n-1)(-3)$$
$$= 11 - 3n + 3$$
$$= -3n + 14$$

43. We know that the *n*th term of an arithmetic
sequence is given by $a_n = a_1 + (n-1)d$ where *a*
is the first term and *d* is the common difference.
Since $a_3 = 3$ and $a_9 = 0$, we have
$$\begin{cases} a_3 = a_1 + (3-1)d \\ a_9 = a_1 + (9-1)d \end{cases} \text{ or } \begin{cases} 3 = a_1 + 2d \quad (1) \\ 0 = a_1 + 8d \quad (2) \end{cases}$$
This is a system of linear equations in two
variables, a_1 and *d*. We can solve the system by
elimination. If we subtract equation (2) from
equation (1), we obtain
$$3 = -6d$$
$$-\frac{1}{2} = d$$
Let $d = -\frac{1}{2}$ in equation (1) to find *a*.
$$3 = a_1 + 2\left(-\frac{1}{2}\right)$$
$$3 = a_1 - 1$$
$$4 = a_1$$
The first term is $a = 4$ and the common
difference is $d = -\frac{1}{2}$. Thus, a formula for the

*n*th term is
$$a_n = 4 + (n-1)\left(-\frac{1}{2}\right)$$
$$= 4 - \frac{1}{2}n + \frac{1}{2} = -\frac{1}{2}n + \frac{9}{2}$$

45. We know that the first term is $a_1 = 2$ and the
common difference is $d = 8 - 2 = 6$. The sum of
the first $n = 30$ terms of this arithmetic sequence
is given by
$$S_{30} = \frac{30}{2}\left[2a_1 + (30-1)d\right]$$
$$= 15\left[2(2) + 29(6)\right]$$
$$= 15(178)$$
$$= 2670$$

47. We know that the first term is $a_1 = -8$ and the
common difference is $d = -5 - (-8) = 3$. The
sum of the first $n = 25$ terms of this arithmetic
sequence is given by
$$S_{25} = \frac{25}{2}\left[2a_1 + (25-1)d\right]$$
$$= \frac{25}{2}\left[2(-8) + 24(3)\right]$$
$$= \frac{25}{2}(56)$$
$$= 700$$

49. We know that the first term is $a_1 = 10$ and the
common difference is $d = 3 - 10 = -7$. The sum
of the first $n = 40$ terms of this arithmetic
sequence is given by
$$S_{40} = \frac{40}{2}\left[2a_1 + (40-1)d\right]$$
$$= 20\left[2(10) + 39(-7)\right]$$
$$= 20(20 - 273)$$
$$= 20(-253)$$
$$= -5060$$

51. The first term of the sequence is given by
$a_1 = 4(1) - 3 = 1$ and the 40th term is given by
$a_{40} = 4(40) - 3 = 157$. We can now use the
formula $S_n = \frac{n}{2}\left[a_1 + a_n\right]$ to find the sum.
$$S_{40} = \frac{40}{2}\left[1 + 157\right]$$
$$= 20(158)$$
$$= 3160$$
The sum of the first 40 terms is 3160.

53. The first term of the sequence is given by
$a_1 = -5(1)+70 = 65$ and the 75th term is given
by $a_{75} = -5(75)+70 = -305$. We can now use

the formula $S_n = \dfrac{n}{2}[a_1 + a_n]$ to find the sum.

$S_{75} = \dfrac{75}{2}\big[65+(-305)\big]$

$\quad = \dfrac{75}{2}(-240)$

$\quad = -9000$

The sum of the first 75 terms is -9000.

55. The first term of the sequence is given by
$a_1 = 5+\dfrac{2}{3}(1) = \dfrac{17}{3}$ and the 30th term is given by

$a_{30} = 5+\dfrac{2}{3}(30) = 25$. We can now use the

formula $S_n = \dfrac{n}{2}[a_1 + a_n]$ to find the sum.

$S_{30} = \dfrac{30}{2}\left[\dfrac{17}{3}+25\right]$

$\quad = 15\left(\dfrac{92}{3}\right)$

$\quad = 460$

The sum of the first 30 terms is 460.

57. To be an arithmetic sequence, the difference
between successive terms must be constant.
Therefore, we start by finding the differences
between the terms.

$d_1 = (2x+1)-(x+3)$
$\quad = 2x+1-x-3$
$\quad = x-2$
$d_2 = (5x+2)-(2x+1)$
$\quad = 5x+2-2x-1$
$\quad = 3x+1$

Now we set the two differences equal to each
other and solve the resulting equation for *x*.

$d_1 = d_2$
$x-2 = 3x+1$
$-2x = 3$
$x = -\dfrac{3}{2}$

59. To determine the total number of cans, we first
need to determine how many rows are in the
stack. Letting the bottom row be the first row,
we have $a_1 = 35$. Since each row decreases by 1
can, we have $d = -1$.

$a_n = a_1 + (n-1)d$
$1 = 35 + (n-1)(-1)$
$1 = 35 - n + 1$
$-35 = -n$
$35 = n$
The total number of cans in the 35 rows is
$S_{35} = \dfrac{35}{2}(35+1) = \dfrac{35}{2}(36) = 630$
There are 630 cans in the stack.

61. Since the first row has 40 seats, we know that
$a_1 = 40$. Each of the $n = 25$ successive rows has
2 more seats, so we also know that $d = 2$. To
determine the total number of seats, we first need
to know how many seats are in the last row.
$a_{25} = a_1 + (25-1)d$
$\quad = 40 + 24(2)$
$\quad = 40 + 48$
$\quad = 88$
The last row has 88 seats so the total number of
seats is given by
$S_{25} = \dfrac{25}{2}[40+88] = \dfrac{25}{2}(128) = 1600$
There are 1600 seats in the auditorium.

63. From the terms listed, we see that the first term
is $a_1 = -5$, the last term is $a_n = 244$, and the
common difference is $d = -2-(-5) = 3$. We can
determine the number of terms in the sequence
by using the formula for the *n*th term.
$a_n = a_1 + (n-1)d$
$244 = -5 + (n-1)(3)$
$244 = -5 + 3n - 3$
$252 = 3n$
$84 = n$
There are 84 terms in the sequence.

65. From the terms listed, we see that the first term
is $a_1 = 108$, the last term is $a_n = -326$, and the
common difference is $d = 101-108 = -7$. We
can determine the number of terms in the
sequence by using the formula for the *n*th term.
$a_n = a_1 + (n-1)d$
$-326 = 108 + (n-1)(-7)$
$-326 = 108 - 7n + 7$
$-441 = -7n$
$63 = n$
There are 63 terms in the sequence.

67. Since your starting salary is $32,000 and you get a $2500 raise each year, your salary each year will form an arithmetic sequence with $a_1 = 32,000$ and $d = 2500$. Your aggregate salary at any point is the sum of what you have earned up to that point. To find how long it would take to have an aggregate of $757,500 we can use the formula

$$S_n = \frac{n}{2}\left[a_1 + a_n\right] = \frac{n}{2}\left[2a_1 + (n-1)d\right]$$

$$757,500 = \frac{n}{2}\left[2(32,000) + (n-1)2500\right]$$

$$757,500 = \frac{n}{2}\left[64,000 + 2500n - 2500\right]$$

$$1,515,000 = n(61,500 + 2500n)$$

$$0 = 2500n^2 + 61,500n - 1,515,000$$

$$0 = n^2 + 24.6n - 606$$

We can solve this quadratic equation by using the quadratic formula.

$$n = \frac{-24.6 \pm \sqrt{(24.6)^2 - 4(1)(-606)}}{2(1)}$$

$$= \frac{-24.6 \pm \sqrt{3029.16}}{2}$$

Since n must be positive, we only need to consider the positive solution.

$$n = \frac{-24.6 + \sqrt{3029.16}}{2} \approx 15.22$$

It will take about 15.22 years for you to have an aggregate salary of $757,500.

69. Answers may vary. A sequence is arithmetic only if the difference between consecutive terms is constant. That is, $a_n - a_{n-1} = d$ where d is a constant.

71. a. $f(x) = 3^x$ is in the form $y = a^x$ so the base is 3.

    b. $f(1) = 3^1 = 3$
$$f(2) = 3^2 = 9$$
$$f(3) = 3^3 = 27$$
$$f(4) = 3^4 = 81$$

73. a. $f(x) = 10\left(\dfrac{1}{2}\right)^x$ is in the form $y = k \cdot a^x$ so the base is $\dfrac{1}{2}$.

    b. $f(1) = 10\left(\dfrac{1}{2}\right)^1 = 5$
$$f(2) = 10\left(\frac{1}{2}\right)^2 = \frac{5}{2}$$
$$f(3) = 10\left(\frac{1}{2}\right)^3 = \frac{5}{4}$$
$$f(4) = 10\left(\frac{1}{2}\right)^4 = \frac{5}{8}$$

75. $\{3.45n + 4.12\}$; $n = 20$

```
sum(seq(3.45n+4.
12,n,1,20))
            806.9
■
```

The sum of the first 20 terms of this arithmetic sequence is 806.9.

77. $a_1 = 85.9$; $d = 83.5 - 85.9 = -2.4$
$$a_n = 85.9 + (n-1)(-2.4)$$
$$= 85.9 - 2.4n + 2.4$$
$$= 88.3 - 2.4n$$
$$\{88.3 - 2.4n\}; \ n = 25$$

```
sum(seq(88.3-2.4
n,n,1,25))
            1427.5
■
```

The sum of the first 25 terms of this arithmetic sequence is 1427.5.

Section 10.3

Are You Ready for This Section?

R1. $g(x) = 4^x$
$$g(1) = 4^1 = 4$$
$$g(2) = 4^2 = 16$$
$$g(3) = 4^3 = 64$$

R2. $\dfrac{x^4}{x^3} = x^{4-3} = x$

Section 10.3 Quick Checks

1. In a <u>geometric</u> sequence the ratio of consecutive terms is constant.

2. $\frac{8}{4} = 2$, $\frac{16}{8} = 2$, $\frac{32}{16} = 2$, $\frac{64}{32} = 2$

The sequence is geometric because the ratio of consecutive terms is constant. The first term is $a_1 = 4$ and the common ratio is $r = 2$.

3. $\frac{10}{5} = 2$, $\frac{16}{10} = \frac{8}{5}$

The sequence is not geometric because the ratio of consecutive terms is not constant.

4. $\frac{3}{9} = \frac{1}{3}$, $\frac{1}{3} = \frac{\frac{1}{3}}{1} = \frac{1}{3}$, $\frac{\frac{1}{9}}{\frac{1}{3}} = \frac{1}{9} \cdot 3 = \frac{1}{3}$

The sequence is geometric because the ratio of consecutive terms is constant. The first term is $a_1 = 9$ and the common ratio is $r = \frac{3}{9} = \frac{1}{3}$.

5. $\frac{a_n}{a_{n-1}} = \frac{5^n}{5^{n-1}} = \frac{5 \cdot 5^{n-1}}{5^{n-1}} = 5$

The sequence is geometric because the ratio of consecutive terms is constant. The first term is $a_1 = 5^1 = 5$ and the common ratio is $r = 5$.

6. $\frac{b_n}{b_{n-1}} = \frac{n^2}{(n-1)^2} = \frac{n^2}{n^2 - 2n + 1}$

The sequence is not geometric because the ratio of consecutive terms is not constant.

7. $\frac{c_n}{c_{n-1}} = \frac{5\left(\frac{2}{3}\right)^n}{5\left(\frac{2}{3}\right)^{n-1}} = \frac{5 \cdot \frac{2}{3} \cdot \left(\frac{2}{3}\right)^{n-1}}{5\left(\frac{2}{3}\right)^{n-1}} = \frac{2}{3}$

The sequence is geometric because the ratio of consecutive terms is constant. The first term is $c_1 = 5\left(\frac{2}{3}\right)^1 = \frac{10}{3}$ and the common ratio is $r = \frac{2}{3}$.

8. The nth term of a geometric sequence is given by $a_n = a_1 \cdot r^{n-1}$. We are given $a_1 = 5$ and $r = 2$. Therefore, a formula for the nth term of the sequence is $a_n = 5 \cdot 2^{n-1}$.

$a_9 = 5 \cdot 2^{9-1} = 5 \cdot 2^8 = 5(256) = 1280$

9. The nth term of a geometric sequence is given by $a_n = a_1 \cdot r^{n-1}$. We are given that the first term is $a_1 = 50$ and the common ratio is $r = \frac{25}{50} = \frac{1}{2}$. Therefore, a formula for the nth term of the sequence is $a_n = 50 \cdot \left(\frac{1}{2}\right)^{n-1}$.

$a_9 = 50 \cdot \left(\frac{1}{2}\right)^{9-1} = 50 \cdot \left(\frac{1}{2}\right)^8 = \frac{50}{256}$

$= \frac{25}{128} = 0.1953125$

10. For a geometric sequence with first term a_1 and common ratio r, where $r \neq 0$, $r \neq 1$, the sum of the first n terms is $S_n = a_1 \cdot \frac{1 - r^n}{1 - r}$.

11. We wish to find the sum of the first 13 terms of a geometric sequence with first term $a_1 = 3$ and common ratio $r = \frac{6}{3} = 2$.

$S_n = a_1 \cdot \frac{1 - r^n}{1 - r}$

$S_{13} = 3 \cdot \frac{1 - 2^{13}}{1 - 2} = 3(8191) = 24{,}573$

12. We wish to find the sum of the first 10 terms of a geometric sequence with first term $a_1 = 8\left(\frac{1}{2}\right)^1 = 4$ and common ratio $r = \frac{1}{2}$.

$S_n = a_1 \cdot \frac{1 - r^n}{1 - r}$

$S_{10} = 4 \cdot \frac{1 - \left(\frac{1}{2}\right)^{10}}{1 - \frac{1}{2}} = 4\left(\frac{\frac{1023}{1024}}{\frac{1}{2}}\right) = 4\left(\frac{1023}{512}\right)$

$= \frac{1023}{128} = 7.9921875$

13. If $-1 < r < 1$, then the sum of the geometric series $\sum_{n-1}^{\infty} a_1 r^{n-1} = \frac{a_1}{1 - r}$.

14. This is an infinite geometric series with $a_1 = 10$

and $r = \dfrac{\frac{5}{2}}{10} = \dfrac{5}{2} \cdot \dfrac{1}{10} = \dfrac{1}{4}$. Since the common ratio

is between -1 and 1, we can use the formula for the sum of an infinite geometric series to find

$$S_\infty = \frac{a_1}{1-r}$$

$$S_\infty = \frac{10}{1 - \frac{1}{4}} = \frac{10}{\frac{3}{4}} = 10 \cdot \frac{4}{3} = \frac{40}{3}$$

15. This is an infinite geometric series with

$a_1 = \left(\dfrac{1}{3}\right)^1 = \dfrac{1}{3}$ and $r = \dfrac{1}{3}$. Since the common

ratio is between -1 and 1, we can use the formula for the sum of an infinite geometric series to find

$$S_\infty = \frac{a_1}{1-r}$$

$$S_\infty = \frac{\frac{1}{3}}{1 - \frac{1}{3}} = \frac{\frac{1}{3}}{\frac{2}{3}} = \frac{1}{3} \cdot \frac{3}{2} = \frac{1}{2}$$

16. The line over the 2 indicates that the 2 repeats indefinitely. That is, we can write

$0.\overline{2} = 0.2 + 0.02 + 0.002 + 0.0002 + \ldots$

This is an infinite geometric series with $a_1 = 0.2$ and common ratio $r = 0.1$. Since the common ratio is between -1 and 1, we can use the formula for the sum of a geometric series to find that

$$0.\overline{2} = 0.2 + 0.02 + 0.002 + 0.0002 + \cdots$$

$$= \frac{0.2}{1 - 0.1}$$

$$= \frac{0.2}{0.9}$$

$$= \frac{2}{9}$$

17. The total impact of the $500 tax rebate on the U.S. economy is

$\$500 + \$500(0.95) + \$500(0.95)^2 + \$500(0.95)^3 + \ldots$

This is an infinite geometric series with first term $a = 500$ and common ratio $r = 0.95$. The sum of this series is

$\$500 + \$500(0.95) + \$500(0.95)^2 + \cdots$

$$= \frac{\$500}{1 - 0.95}$$

$$= \frac{\$500}{0.05}$$

$$= \$10,000$$

The U.S. economy will grow by $10,000 because of the child tax credit to Roberta.

18. This is an ordinary annuity with $n = 4 \cdot 30 = 120$ payments with deposits of $P = \$500$. The interest rate per payment period is

$i = \dfrac{0.08}{4} = 0.02$. This gives

$$A = 500 \left[\frac{(1 + 0.02)^{120} - 1}{0.02} \right]$$

$$= 500(488.258152)$$

$$= 244,129.08$$

After 30 years, the IRA will be worth $244,129.08.

10.3 Exercises

19. $\left\{ 4^n \right\}$

$a_{n-1} = 4^{n-1}$

$a_n = 4^n$

$$\frac{a_n}{a_{n-1}} = \frac{4^n}{4^{n-1}} = \frac{4^n}{4^n \cdot 4^{-1}} = \frac{1}{4^{-1}}$$

The ratio of *any* two consecutive terms is 4, so the sequence is geometric, with common ratio $r = 4$. To find the first term, we evaluate a_1 and find $a_1 = 4^1 = 4$.

21. $\left\{ \left(\dfrac{2}{3} \right)^n \right\}$

$a_{n-1} = \left(\dfrac{2}{3} \right)^{n-1}$

$a_n = \left(\dfrac{2}{3} \right)^n$

$$\frac{a_n}{a_n-1} = \frac{\left(\frac{2}{3}\right)^n}{\left(\frac{2}{3}\right)^{n-1}}$$

$$= \frac{\left(\frac{2}{3}\right)^n}{\left(\frac{2}{3}\right)^n \cdot \left(\frac{2}{3}\right)^{-1}}$$

$$= \frac{1}{\left(\frac{2}{3}\right)^{-1}}$$

$$= \frac{2}{3}$$

The ratio of *any* two consecutive terms is $\frac{2}{3}$, so the sequence is geometric, with common ratio $r = \frac{2}{3}$. To find the first term, we evaluate a_1 and find $a_1 = \left(\frac{2}{3}\right)^1 = \frac{2}{3}$.

23. $\{3 \cdot 2^{-n}\}$

$$a_{n-1} = 3 \cdot 2^{-(n-1)} = 3 \cdot 2^{-n+1}$$

$$a_n = 3 \cdot 2^{-n}$$

$$\frac{a_n}{a_{n-1}} = \frac{3 \cdot 2^{-n}}{3 \cdot 2^{-n+1}}$$

$$= \frac{3 \cdot 2^{-n}}{3 \cdot 2^{-n} \cdot 2}$$

$$= \frac{1}{2}$$

The ratio of *any* two consecutive terms is $\frac{1}{2}$, so the sequence is geometric, with common ratio $r = \frac{1}{2}$. To find the first term, we evaluate a_1 and find $a_1 = 3 \cdot 2^{-1} = \frac{3}{2}$.

25. $\left\{\dfrac{5^{n-1}}{2^n}\right\}$

$$a_{n-1} = \frac{5^{(n-1)-1}}{2^{n-1}} = \frac{5^{n-2}}{2^{n-1}}$$

$$a_n = \frac{5^{n-1}}{2^n}$$

$$\frac{a_n}{a_{n-1}} = \frac{\dfrac{5^{n-1}}{2^n}}{\dfrac{5^{n-2}}{2^{n-1}}} = \frac{5^{n-1}}{2^n} \cdot \frac{2^{n-1}}{5^{n-2}} = \frac{5}{2}$$

The ratio of *any* two consecutive terms is $\frac{5}{2}$, so the sequence is geometric, with common ratio $r = \frac{5}{2}$. To find the first term, we evaluate a_1 and find $a_1 = \frac{5^{1-1}}{2^1} = \frac{5^0}{2} = \frac{1}{2}$.

27. a. $a_n = a_1 \cdot r^{n-1} = 10 \cdot 2^{n-1}$

   b. $a_8 = 10 \cdot 2^{8-1} = 10 \cdot 2^7 = 10 \cdot 128 = 1280$

29. a. $a_n = a_1 \cdot r^{n-1} = 100 \cdot \left(\frac{1}{2}\right)^{n-1}$

   b. $a_8 = 100 \cdot \left(\frac{1}{2}\right)^{8-1} = 100 \cdot \left(\frac{1}{2}\right)^7 = \frac{100}{128} = \frac{25}{32}$

31. a. $a_n = a_1 \cdot r^{n-1} = 1 \cdot (-3)^{n-1} = (-3)^{n-1}$

   b. $a_8 = (-3)^{8-1} = (-3)^7 = -2187$

33. a. $a_n = a_1 \cdot r^{n-1} = 100 \cdot (1.05)^{n-1}$

   b. $a_8 = 100 \cdot (1.05)^{8-1} = 100 \cdot (1.05)^7 \approx 140.71$

35. $r = \frac{6}{3} = 2$; $a_1 = 3$

$$a_n = a_1 \cdot r^{n-1}$$

$$a_{10} = 3 \cdot 2^{10-1} = 3 \cdot 2^9 = 3 \cdot 512 = 1536$$

37. $r = \dfrac{-2}{4} = -\dfrac{1}{2}$; $a_1 = 4$

$a_n = a_1 \cdot r^{n-1}$

$a_{15} = 4 \cdot \left(-\dfrac{1}{2}\right)^{15-1}$

$\quad\;\; = 4 \cdot \left(-\dfrac{1}{2}\right)^{14}$

$\quad\;\; = \dfrac{4}{16,384}$

$\quad\;\; = \dfrac{1}{4096}$

39. $r = \dfrac{0.05}{0.5} = \dfrac{1}{10}$; $a_1 = 0.5$

$a_n = a_1 \cdot r^{n-1}$

$a_9 = 0.5 \cdot (0.1)^{9-1} = 0.5 \cdot (0.1)^8 = 0.000000005$

41. $2 + 4 + 8 + \cdots + 2^{12}$

This is a geometric sequence with $a_1 = 2$ and common ratio $r = 2$. We wish to find the sum of the first 12 terms, S_{12}.

$S_n = a_1 \cdot \dfrac{1-r^n}{1-r}$

$S_{12} = 2 \cdot \dfrac{1-2^{12}}{1-2} = 2 \cdot \dfrac{-4095}{-1} = 8190$

$2 + 4 + 8 + \cdots + 2^{12} = 8190$

43. $50 + 20 + 8 + \dfrac{16}{5} + \cdots + 50\left(\dfrac{2}{5}\right)^{10-1}$

This is a geometric sequence with $a_1 = 50$ and common ratio $r = \dfrac{2}{5}$. We wish to find the sum of the first 10 terms, S_{10}.

$S_n = a_1 \cdot \dfrac{1-r^n}{1-r}$

$S_{10} = 50 \cdot \dfrac{1 - \left(\dfrac{2}{5}\right)^{10}}{1 - \dfrac{2}{5}} = 50 \cdot \dfrac{\dfrac{9,764,601}{9,765,625}}{\dfrac{3}{5}} \approx 83.3245952$

$50 + 20 + 8 + \dfrac{16}{5} + \cdots + 50\left(\dfrac{2}{5}\right)^{10-1} \approx 83.3245952$

45. $\displaystyle\sum_{n=1}^{10}\left[3 \cdot 2^n\right] = \sum_{n=1}^{10}\left[3 \cdot 2 \cdot 2^{n-1}\right] = \sum_{n=1}^{10}\left[6 \cdot 2^{n-1}\right]$

Here we want to find the sum of the first 10 terms of a geometric sequence with $a_1 = 6$ and common ratio $r = 2$.

$S_n = a_1 \cdot \dfrac{1-r^n}{1-r}$

$S_{10} = 6 \cdot \dfrac{1 - 2^{10}}{1-2} = 6 \cdot \dfrac{-1023}{-1} = 6138$

$\displaystyle\sum_{n=1}^{10}\left[3 \cdot 2^n\right] = 6138$

47. $\displaystyle\sum_{n=1}^{8}\left[\dfrac{4}{2^{n-1}}\right] = \sum_{n=1}^{8}\left[4 \cdot \left(\dfrac{1}{2}\right)^{n-1}\right]$

Here we want the sum of the first eight terms of a geometric sequence with $a_1 = 4$ and common ratio $r = \dfrac{1}{2}$.

$S_n = a_1 \cdot \dfrac{1-r^n}{1-r}$

$S_8 = 4 \cdot \dfrac{1 - \left(\dfrac{1}{2}\right)^8}{1 - \dfrac{1}{2}} = 4 \cdot \dfrac{\dfrac{255}{256}}{\dfrac{1}{2}} = 8 \cdot \dfrac{255}{256} = 7.96875$

$\displaystyle\sum_{n=1}^{8}\left[\dfrac{4}{2^{n-1}}\right] = 7.96875$

49. This is an infinite geometric series with $a_1 = 1$ and common ratio $r = \dfrac{\dfrac{1}{2}}{1} = \dfrac{1}{2}$. Since the common ratio, r, is between -1 and 1, we can use the formula for the sum of an infinite geometric series.

$1 + \dfrac{1}{2} + \dfrac{1}{4} + \cdots = \dfrac{a}{1-r} = \dfrac{1}{1 - \dfrac{1}{2}} = \dfrac{1}{\dfrac{1}{2}} = 2$

51. This is an infinite geometric series with $a_1 = 10$ and common ratio $r = \dfrac{\dfrac{10}{3}}{10} = \dfrac{1}{3}$. Since the common ratio, r, is between -1 and 1, we can use the formula for the sum of an infinite geometric series.

$10 + \dfrac{10}{3} + \dfrac{10}{9} + \cdots = \dfrac{10}{1 - \dfrac{1}{3}} = \dfrac{10}{\dfrac{2}{3}} = 15$

53. This is an infinite geometric series with $a_1 = 6$

and common ratio $r = \dfrac{-2}{6} = -\dfrac{1}{3}$. Since the

common ratio, r, is between -1 and 1, we can
use the formula for the sum of an infinite
geometric series.

$$6 - 2 + \frac{2}{3} - \frac{2}{9} + \cdots = \frac{6}{1 - \left(-\dfrac{1}{3}\right)} = \frac{6}{\dfrac{4}{3}} = 6 \cdot \frac{3}{4} = \frac{9}{2}$$

55. This is an infinite geometric series with

$a_1 = 5 \cdot \left(\dfrac{1}{5}\right)^1 = 1$ and common ratio $r = \dfrac{1}{5}$. Since

the common ratio, r, is between -1 and 1, we
can use the formula for the sum of an infinite
geometric series.

$$\sum_{n=1}^{\infty}\left(5 \cdot \left(\frac{1}{5}\right)^n\right) = \frac{1}{1 - \dfrac{1}{5}} = \frac{1}{\dfrac{4}{5}} = \frac{5}{4}$$

57. This is an infinite geometric series with

$a_1 = 12 \cdot \left(-\dfrac{1}{3}\right)^{1-1} = 12$ and common ratio

$r = -\dfrac{1}{3}$. Since the common ratio, r, is between

-1 and 1, we can use the formula for the sum of
an infinite geometric series.

$$\sum_{n=1}^{\infty}\left(12 \cdot \left(-\frac{1}{3}\right)^{n-1}\right) = \frac{12}{1 - \left(-\dfrac{1}{3}\right)} = \frac{12}{\dfrac{4}{3}}$$

$$= 12 \cdot \frac{3}{4}$$

$$= 9$$

59. $0.\overline{5} = 0.5 + 0.05 + 0.005 + 0.0005 + \cdots$

$$= \sum_{n=1}^{\infty}\left[0.5 \cdot \left(\frac{1}{10}\right)^{n-1}\right]$$

This is an infinite geometric series with $a = 0.5$

and common ratio $r = \dfrac{1}{10}$. Since the common

ratio, r, is between -1 and 1, we can use the
formula for the sum of an infinite geometric
series.

$$0.\overline{5} = \frac{0.5}{1 - \dfrac{1}{10}} = \frac{\dfrac{1}{2}}{\dfrac{9}{10}} = \frac{1}{2} \cdot \frac{10}{9} = \frac{5}{9}$$

61. $0.\overline{89} = 0.89 + 0.0089 + 0.000089 + \cdots$

$$= \sum_{n=1}^{\infty}\left[0.89 \cdot \left(\frac{1}{100}\right)^{n-1}\right]$$

This is an infinite geometric series with $a = 0.89$

and common ratio $r = \dfrac{1}{100}$. Since the common

ratio, r, is between -1 and 1, we can use the
formula for the sum of an infinite geometric
series.

$$0.\overline{89} = \frac{0.89}{1 - \dfrac{1}{100}} = \frac{0.89}{0.99} = \frac{89}{99}$$

63. $\{5n + 1\}$

$$\begin{aligned} a_n - a_{n-1} &= [5n + 1] - [5(n - 1) + 1] \\ &= [5n + 1] - [5n - 5 + 1] \\ &= 5n + 1 - 5n + 4 \\ &= 5 \end{aligned}$$

Since the difference between successive terms is
constant, this is an arithmetic sequence. The
common difference is $d = 5$.

65. $\{2n^2\}$

$$\begin{aligned} a_n - a_{n-1} &= \left[2n^2\right] - \left[2(n - 1)^2\right] \\ &= \left[2n^2\right] - \left[2n^2 - 4n + 2\right] \\ &= 2n^2 - 2n^2 + 4n - 2 \\ &= 4n - 2 \end{aligned}$$

Since the difference between consecutive terms
is not constant, the sequence is not arithmetic.

$$\frac{a_n}{a_{n-1}} = \frac{2n^2}{2(n - 1)^2} = \frac{n^2}{n^2 - 2n + 1}$$

Since the ratio of consecutive terms is not
constant, the sequence is not geometric.

67. $\left\{\dfrac{2^{-n}}{5}\right\}$

$$\begin{aligned} a_n - a_{n-1} &= \frac{2^{-n}}{5} - \frac{2^{-(n-1)}}{5} \\ &= \frac{2^{-n}}{5} - \frac{2^{-n+1}}{5} \\ &= \frac{2^{-n}}{5} - \frac{2 \cdot 2^{-n}}{5} \\ &= -\frac{2^{-n}}{5} \end{aligned}$$

Since the difference between consecutive terms is not constant, the sequence is not arithmetic.

$$\frac{a_n}{a_{n-1}} = \frac{\left(\frac{2^{-n}}{5}\right)}{\left(\frac{2^{-(n-1)}}{5}\right)} = \frac{2^{-n}}{5} \cdot \frac{5}{2^{-n+1}}$$

$$= 2^{-n-(-n+1)}$$

$$= 2^{-1}$$

$$= \frac{1}{2}$$

Since the ratio of consecutive terms is constant, the sequence is geometric. The common ratio is $r = \frac{1}{2}$.

69. $a_2 - a_1 = 36 - 54 = -18$

$a_3 - a_2 = 24 - 36 = -12$

Consecutive differences are not the same so the sequence is not arithmetic.

$$\frac{a_2}{a_1} = \frac{36}{54} = \frac{2}{3}; \quad \frac{a_3}{a_2} = \frac{24}{36} = \frac{2}{3}; \quad \frac{a_4}{a_3} = \frac{16}{24} = \frac{2}{3}$$

The ratio of consecutive terms is constant so the sequence is geometric. The common ratio is $r = \frac{2}{3}$.

71. $a_2 - a_1 = 6 - 2 = 4$

$a_3 - a_2 = 10 - 6 = 4$

$a_4 - a_3 = 14 - 10 = 4$

The difference between consecutive terms is constant so the sequence is arithmetic. The common difference is $d = 4$.

73. $a_2 - a_1 = 2 - 1 = 1$

$a_3 - a_2 = 3 - 2 = 1$

$a_4 - a_3 = 5 - 3 = 2$

$a_5 - a_4 = 8 - 5 = 3$

The difference between consecutive terms is not constant so the sequence is not arithmetic.

$$\frac{a_2}{a_1} = \frac{2}{1} = 2$$

$$\frac{a_3}{a_2} = \frac{3}{2} = 1.5$$

$$\frac{a_4}{a_3} = \frac{5}{3} \approx 1.67$$

$$\frac{a_5}{a_4} = \frac{8}{5} = 1.6$$

The ratio of consecutive terms is not constant so the sequence is not geometric.

75. To be a geometric sequence, the ratio of consecutive terms must be the same. Therefore, we need to solve the equation

$$\frac{x+2}{x} = \frac{x+3}{x+2}$$

$$x(x+2) \cdot \frac{x+2}{x} = x(x+2) \cdot \frac{x+3}{x+2}$$

$$(x+2)^2 = x(x+3)$$

$$x^2 + 4x + 4 = x^2 + 3x$$

$$4x + 4 = 3x$$

$$x + 4 = 0$$

$$x = -4$$

77. Your annual salaries will form a geometric sequence with $a_1 = 40,000$ and common ratio $r = 1.05$.

a. Your salary at the beginning of the second year is the value of the second term in the sequence.

$$a_2 = a_1 \cdot r^{2-1} = 40,000 \cdot (1.05)^1 = 42,000$$

Your salary at the beginning of the second year will be \$42,000.

b. Your salary at the beginning of the tenth year is the value of the tenth term in the sequence.

$$a_{10} = a_1 \cdot r^{10-1} = 40,000 \cdot (1.05)^9 \approx 62,053$$

Your salary at the beginning of the tenth year will be about \$62,053.

c. Your cumulative earnings after completing your tenth year is the sum of the first ten terms of the sequence.

$$S_{10} = a_1 \cdot \frac{1 - r^{10}}{1 - r}$$

$$= 40,000 \cdot \frac{1 - (1.05)^{10}}{1 - 1.05}$$

$$\approx 40,000 \cdot (12.577893)$$

$$\approx 503,116$$

Your cumulative earnings after finishing your tenth year will be about \$503,116.

79. The value of the car at the beginning of the year forms a geometric sequence with $a_1 = 20,000$ and common ratio $r = 0.92$. The value of the car after you have owned it for five years will be the sixth term of the sequence (the beginning of the sixth year of ownership).

$$a_6 = 20,000 \cdot (0.92)^{6-1}$$
$$= 20,000 \cdot (0.92)^5 \approx 13,182$$

After five years of ownership, the car will be worth about $13,182.

81. The lengths of the arc of the pendulum swings form a geometric sequence with $a_1 = 3$ and common ratio $r = 0.95$.

a.
$$a_{10} = a_1 \cdot r^{10-1}$$
$$= 3 \cdot (0.95)^9$$
$$\approx 1.891$$

The length of the arc in the 10th swing is about 1.891 feet.

b. Here we need to determine the term number. We start with the following equation:
$$a_n = 1$$
$$a_1 \cdot r^{n-1} = 1$$
$$3 \cdot (0.95)^{n-1} = 1$$
$$(0.95)^{n-1} = \frac{1}{3}$$
$$\ln\left[(0.95)^{n-1}\right] = \ln\left(\tfrac{1}{3}\right)$$
$$(n-1) \cdot \ln(0.95) = \ln\left(\tfrac{1}{3}\right)$$
$$n-1 = \frac{\ln\left(\tfrac{1}{3}\right)}{\ln(0.95)}$$
$$n = 1 + \frac{\ln\left(\tfrac{1}{3}\right)}{\ln(0.95)}$$
$$n \approx 22.42$$

Since the terms in the sequence are decreasing, the length of the arc will be less than 1 foot on the 23rd swing.

c.
$$S_{10} = a_1 \cdot \frac{1 - r^{10}}{1 - r}$$
$$= 3 \cdot \frac{1 - (0.95)^{10}}{1 - 0.95} \approx 24.08$$

After 10 swings, the pendulum will have swung a total of about 24.08 feet.

d. To find the total length the pendulum will swing requires an infinite geometric series.
$$S_\infty = \frac{a_1}{1 - r}$$
$$= \frac{3}{1 - 0.95}$$
$$= \frac{3}{0.05}$$
$$= 60$$

The pendulum will swing a total of 60 feet.

83. The annual salaries under option A form a geometric sequence with $a_1 = 30,000$ and common ratio $r = 1.05$. The annual salaries under option B form a geometric sequence with $a_1 = 31,000$ and common ratio $r = 1.04$.

To determine which option gives the larger annual salary in the final year of the contract, we need to find the 5th term of each sequence.

Option A: $a_5 = 30,000 \cdot (1.05)^{5-1} \approx 36,465.19$

Option B: $a_5 = 31,000 \cdot (1.04)^{5-1} \approx 36,265.62$

To determine which option gives the largest cumulative salary for five years, we need the sum of the first five terms of each sequence.

Option A: $S_5 = 30,000 \cdot \dfrac{1 - (1.05)^5}{1 - 1.05} \approx 165,768.94$

Option B: $S_5 = 31,000 \cdot \dfrac{1 - (1.04)^5}{1 - 1.04} \approx 167,906.00$

Therefore, option A will yield the larger annual salary in the final year of the contract and option B will yield the larger cumulative salary over the life of the contract.

85. The multiplier in this case is the geometric series with first term $a_1 = 1$ and common ratio $r = 0.98$.
$$1 + 0.98 + (0.98)^2 + \ldots = \frac{1}{1 - 0.98} = 50$$

The multiplier is 50.

87. $\dfrac{1+i}{1+r} = \dfrac{1+.02}{1+.09} = \dfrac{1.02}{1.09} \approx 0.93578$

The price is an infinite geometric series with $a_1 = 2$ and a common ratio of $\dfrac{1.02}{1.09}$. The maximum price is given by

$$\text{Price} = \frac{2}{1 - \frac{1.02}{1.09}} \approx 31.14$$

The maximum price you should pay for the stock is $31.14 per share.

89. $A = P \cdot \dfrac{(1+i)^n - 1}{i}$

In this case we have $P = 100$, $n = 30 \cdot 12 = 360$,

and $i = \dfrac{0.08}{12}$.

$$A = 100 \cdot \left[\frac{\left(1 + \frac{0.08}{12}\right)^{360} - 1}{\frac{0.08}{12}} \right]$$

$$= 100 \cdot [1490.359449]$$

$$= \$149,035.94$$

After 30 years, Christina's 401(k) will be worth $149,035.94.

91. $A = P \cdot \dfrac{(1+i)^n - 1}{i}$

In this case we have $P = 500$, $n = 25 \cdot 4 = 100$,

and $i = \dfrac{0.06}{4} = 0.015$.

$$A = 500 \cdot \left[\frac{(1 + 0.015)^{100} - 1}{0.015} \right]$$

$$= 500 \cdot [228.8030433]$$

$$= \$114,401.52$$

After 25 years, Jackson's IRA will be worth $114,401.52.

93. $A = P \cdot \dfrac{(1+i)^n - 1}{i}$

In this case we have $A = 1,500,000$,

$n = 35 \cdot 12 = 420$, and $i = \dfrac{0.10}{12}$.

$$1,500,000 = P \cdot \left[\frac{\left(1 + \frac{0.10}{12}\right)^{420} - 1}{\frac{0.10}{12}} \right]$$

$$1,500,000 = P \cdot [3796.638052]$$

$$P = \$395.09$$

Aaliyah will need to contribute $395.09, or about $395, each month in order to achieve her goal.

95. $0.4\overline{9} = 0.4 + 0.09 + 0.009 + 0.0009 + \cdots$

After the 0.4, the rest of the sum forms an infinite geometric series with $a_1 = 0.09$ and common ratio $r = 0.1$. We can find the sum of

this part as $S_\infty = \dfrac{a_1}{1-r} = \dfrac{0.09}{1-0.1} = \dfrac{0.09}{0.9} = 0.1$

Thus, $0.4\overline{9} = 0.4 + 0.1 = 0.5 = \dfrac{1}{2}$.

97. The sum is a geometric series with $a_1 = 2$ and common ratio $r = 2$. We can find the sum by using the summation formula, after we determine how many terms are in the sum.

$$a_n = a_1 \cdot r^{n-1}$$
$$1,073,741,824 = 2 \cdot 2^{n-1}$$
$$1,073,741,824 = 2^n$$
$$\ln 1,073,741,824 = \ln 2^n$$
$$\ln 1,073,741,824 = n \cdot \ln 2$$
$$\frac{\ln 1,073,741,824}{\ln 2} = n$$
$$30 = n$$
$$S_{30} = a_1 \cdot \frac{1 - r^{30}}{1 - r} = 2 \cdot \frac{1 - 2^{30}}{1 - 2} = 2,147,483,646$$

99. Answers may vary. The comparison of growth rates really depends on the common difference and the common ratio. It is possible for an arithmetic sequence to grow at a faster rate initially. However, the geometric sequence will always end up growing at a faster rate if given enough time. For example, consider the following sequences:

arithmetic: $\{20 + (n-1) \cdot 2\}$

geometric: $\{20 \cdot 1.05^{n-1}\}$

These two sequences begin at the same value, but then the arithmetic sequence begins growing at a faster rate. After about 27 years, the geometric sequence will take over.

Intersection
X=27.583449 Y=73.166897

If Malthus' conjecture is correct, it would mean that we would eventually be unable to produce enough food to feed everyone on the planet, even if current supply exceeds demand.

101. A geometric series has a finite sum if $-1 < r < 1$.

103. $\dfrac{1}{3} = 0.333333\ldots$

$\dfrac{2}{3} = 0.666666\ldots$

$\dfrac{1}{3} + \dfrac{2}{3} = (0.333333\ldots) + (0.666666\ldots)$

$\qquad = 0.999999\ldots$

Since $\dfrac{1}{3} + \dfrac{2}{3} = \dfrac{3}{3} = 1$, we would conjecture that $0.999999\ldots = 1$.

105. This is a geometric series with $a_1 = 4$ and common ratio $r = 1.2$. Based on the form of the last term, it appears that there are $n = 15$ terms in the series.

The sum is approximately 288.1404315.

107. $\displaystyle\sum_{n=1}^{20}\left[1.2(1.05)^n\right]$

This is a geometric series with $a_1 = 1.26$ (that is, $1.2(1.05)^1$), common ratio $r = 1.05$, and $n = 20$ terms.

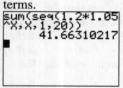

The sum is approximately 41.66310217.

Putting the Concepts Together (Sections 10.1–10.3)

1. $\dfrac{3/16}{3/4} = \dfrac{3}{16} \cdot \dfrac{4}{3} = \dfrac{1}{4}$

$\dfrac{3/64}{3/16} = \dfrac{3}{64} \cdot \dfrac{16}{3} = \dfrac{1}{4}$

$\dfrac{3/256}{3/64} = \dfrac{3}{256} \cdot \dfrac{64}{3} = \dfrac{1}{4}$

The ratio of consecutive terms is a constant so the sequence is geometric with $a_1 = \dfrac{3}{4}$ and common ratio $r = \dfrac{1}{4}$.

2. $a_n - a_{n-1} = \left[2(n+3)\right] - \left[2(n-1+3)\right]$

$\qquad = 2(n+3) - 2(n+2)$

$\qquad = 2n + 6 - 2n - 4$

$\qquad = 2$

The difference between consecutive terms is a constant so the sequence is arithmetic with $a_1 = 2(1+3) = 8$ and common difference $d = 2$.

3. $a_n - a_{n-1} = \left[\dfrac{7n+2}{9}\right] - \left[\dfrac{7(n-1)+2}{9}\right]$

$\qquad = \left(\dfrac{7n+2}{9}\right) - \left(\dfrac{7n-7+2}{9}\right)$

$\qquad = \dfrac{7n+2-7n+5}{9}$

$\qquad = \dfrac{7}{9}$

The difference between consecutive terms is a constant so the sequence is arithmetic with $a_1 = \dfrac{7(1)+2}{9} = 1$ and common difference $d = \dfrac{7}{9}$.

4. $-4 - 1 = -5$

$9 - (-4) = 13$

$\dfrac{-4}{1} = -4$

$\dfrac{9}{-4} = -\dfrac{9}{4}$

The sequence is neither arithmetic nor geometric. The difference between consecutive terms is not constant so the sequence is not arithmetic. Likewise, the ratio of consecutive terms is not constant so the sequence is not geometric.

5. $\dfrac{a_n}{a_{n-1}} = \dfrac{3 \cdot 2^{n+1}}{3 \cdot 2^{(n-1)+1}} = \dfrac{3 \cdot 2^n \cdot 2}{3 \cdot 2^n} = 2$

The ratio of consecutive terms is a constant so the sequence is geometric with $a_1 = 3 \cdot 2^{1+1} = 12$ and common ratio $r = 2$.

6. $a_n - a_{n-1} = \left[n^2 - 5\right] - \left[(n-1)^2 - 5\right]$

$\qquad\qquad\quad = n^2 - 5 - \left(n^2 - 2n + 1 - 5\right)$

$\qquad\qquad\quad = n^2 - 5 - n^2 + 2n + 4$

$\qquad\qquad\quad = 2n - 1$

$\dfrac{a_n}{a_{n-1}} = \dfrac{n^2 - 5}{(n-1)^2 - 5} = \dfrac{n^2 - 5}{n^2 - 2n - 4}$

The sequence is neither arithmetic nor geometric. The difference between consecutive terms is not constant so the sequence is not arithmetic. Likewise, the ratio of consecutive terms is not constant so the sequence is not geometric.

7. $\displaystyle\sum_{k=1}^{6}\left[3k + 4\right]$

$= (3 \cdot 1 + 4) + (3 \cdot 2 + 4) + (3 \cdot 3 + 4)$
$\quad + (3 \cdot 4 + 4) + (3 \cdot 5 + 4) + (3 \cdot 6 + 4)$

$= 7 + 10 + 13 + 16 + 19 + 22$

$= 87$

8. Each term is written in the form $\dfrac{1}{2(6+i)}$, where

i is the term number, and there are $n = 12$ terms.

$\dfrac{1}{2(6+1)} + \dfrac{1}{2(6+2)} + \cdots + \dfrac{1}{2(6+12)}$

$= \displaystyle\sum_{i=1}^{12} \dfrac{1}{2(6+i)}$

9. $a_n = a_1 + (n-1)d$

$\quad = 25 + (n-1)(-2)$

$\quad = 25 - 2n + 2$

$\quad = 27 - 2n$

$a_1 = 27 - 2(1) = 25$

$a_2 = 27 - 2(2) = 23$

$a_3 = 27 - 2(3) = 21$

$a_4 = 27 - 2(4) = 19$

$a_5 = 27 - 2(5) = 17$

The first five terms of the sequence are 25, 23, 21, 19, and 17.

10. $\qquad a_n = a_1 + (n-1)d$

$\qquad a_4 = a_1 + (4-1)(11) = 9$

$a_1 + 33 = 9$

$\quad a_1 = -24$

$a_n = -24 + (n-1)(11)$

$\quad\, = -24 + 11n - 11$

$\quad\, = 11n - 35$

$a_1 = 11(1) - 35 = -24$

$a_2 = 11(2) - 35 = -13$

$a_3 = 11(3) - 35 = -2$

$a_4 = 11(4) - 35 = 9$

$a_5 = 11(5) - 35 = 20$

The first five terms of the sequence are -24, -13, -2, 9, and 20.

11. $\qquad a_n = a_1 \cdot r^{n-1}$

$\qquad a_4 = a_1 \cdot \left(\dfrac{1}{5}\right)^{4-1} = \dfrac{9}{25}$

$a_1 \cdot \dfrac{1}{125} = \dfrac{9}{25} \qquad \rightarrow \qquad a_1 = 125 \cdot \dfrac{9}{25} = 45$

$a_n = 45 \cdot \left(\dfrac{1}{5}\right)^{n-1}$

$a_1 = 45 \cdot \left(\dfrac{1}{5}\right)^{1-1} = 45$

$a_2 = 45 \cdot \left(\dfrac{1}{5}\right)^{2-1} = 9$

$a_3 = 45 \cdot \left(\dfrac{1}{5}\right)^{3-1} = \dfrac{9}{5}$

$a_4 = 45 \cdot \left(\dfrac{1}{5}\right)^{4-1} = \dfrac{9}{25}$

$a_5 = 45 \cdot \left(\dfrac{1}{5}\right)^{5-1} = \dfrac{9}{125}$

The first five terms of the sequence are 45, 9, $\dfrac{9}{5}$, $\dfrac{9}{25}$, and $\dfrac{9}{125}$.

12. $a_n = a_1 \cdot r^{n-1}$

$\quad\quad = 150 \cdot (1.04)^{n-1}$

$\quad a_1 = 150 \cdot (1.04)^{1-1} = 150$

$\quad a_2 = 150 \cdot (1.04)^{2-1} = 156$

$\quad a_3 = 150 \cdot (1.04)^{3-1} = 162.24$

$\quad a_4 = 150 \cdot (1.04)^{4-1} = 168.7296$

$\quad a_5 = 150 \cdot (1.04)^{5-1} = 175.478784$

The first five terms of the sequence are 150, 156, 162.24, 168.7296, and 175.478784.

13. The terms form a geometric sequence with first term $a_1 = 2$, common ratio $r = 3$, and $n = 11$ terms.

$$S_{11} = 2 \cdot \frac{1 - 3^{11}}{1 - 3} = 177{,}146$$

14. The terms form an arithmetic sequence with first term $a_1 = 2$, common difference $d = 5$, and $n = 20$ terms.

$$a_{20} = 2 + (20 - 1) \cdot 5 = 97$$

$$S_{20} = \frac{20}{2}[2 + 97] = 990$$

15. This is an infinite geometric series with $a_1 = 1000$ and $r = \dfrac{1}{10}$.

$$S_\infty = \frac{a_1}{1 - r} = \frac{1000}{1 - \frac{1}{10}} = \frac{1000}{\frac{9}{10}} = \frac{10{,}000}{9} \text{ or } 1111\frac{1}{9}$$

16. The number of people who can be seated forms an arithmetic sequence with first term $a = 4$ and common difference $d = 2$. The number of tables required is the same as the term number, n.

To find the number of tables required to seat a party of 24 people, we solve the following for n:

$a_n = a_1 + (n-1)d$

$24 = 4 + (n-1)(2)$

$24 = 4 + 2n - 2$

$24 = 2n + 2$

$22 = 2n$

$11 = n$

A party of 24 people would require 11 tables.

Section 10.4

Are You Ready for This Section?

R1. $(x-5)^2 = x^2 - 2 \cdot x \cdot 5 + 5^2$

$\quad\quad\quad = x^2 - 10x + 25$

R2. $(2x+3)^2 = (2x+3)(2x+3)$

$\quad\quad\quad = \left((2x)^2 + 2 \cdot 2x \cdot 3 + 3^2 \right)$

$\quad\quad\quad = 4x^2 + 12x + 9$

Section 10.4 Quick Checks

1. If $n \geq 2$ is an integer, then $n! = \underline{n(n-1)(n-2) \cdot \cdots \cdot 3 \cdot 2 \cdot 1}$.

2. $0! = \underline{1}$; $1! = \underline{1}$.

3. $5! = 5 \cdot 4 \cdot 3 \cdot 2 \cdot 1 = 120$

4. $\dfrac{7!}{3!} = \dfrac{7 \cdot 6 \cdot 5 \cdot 4 \cdot 3!}{3!} = 7 \cdot 6 \cdot 5 \cdot 4 = 840$

5. False. The numerator should be $n!$ and the denominator should be $j! \, (n-j)!$.

6. True.

7. $\dbinom{7}{1} = \dfrac{7!}{1! \cdot (7-1)!} = \dfrac{7!}{6!} = \dfrac{7 \cdot 6!}{6!} = 7$

8. $\dbinom{6}{3} = \dfrac{6!}{3! \cdot (6-3)!} = \dfrac{6 \cdot 5 \cdot 4 \cdot 3!}{3! \cdot 3!} = \dfrac{6 \cdot 5 \cdot 4}{3 \cdot 2 \cdot 1} = 20$

9. $(x+2)^4 = \dbinom{4}{0} x^4 + \dbinom{4}{1} 2^1 \cdot x^{4-1} + \dbinom{4}{2} 2^2 \cdot x^{4-2} + \dbinom{4}{3} 2^3 \cdot x^{4-3} + \dbinom{4}{4} 2^4$

    $= 1 \cdot x^4 + 4 \cdot 2 \cdot x^3 + 6 \cdot 4 \cdot x^2 + 4 \cdot 8 \cdot x + 1 \cdot 16$

    $= x^4 + 8x^3 + 24x^2 + 32x + 16$

10. $(2p-1)^5$

    $= \left(2p + (-1)\right)^5$

    $= \dbinom{5}{0}(2p)^5 + \dbinom{5}{1}(-1)^1 \cdot (2p)^{5-1} + \dbinom{5}{2}(-1)^2 \cdot (2p)^{5-2} + \dbinom{5}{3}(-1)^3 \cdot (2p)^{5-3} + \dbinom{5}{4}(-1)^4 \cdot (2p)^{5-4} + \dbinom{5}{5}(-1)^5$

    $= 1 \cdot (2p)^5 + 5 \cdot (-1) \cdot (2p)^4 + 10 \cdot (-1)^2 \cdot (2p)^3 + 10 \cdot (-1)^3 \cdot (2p)^2 + 5 \cdot (-1)^4 \cdot (2p) + 1 \cdot (-1)^5$

    $= 32p^5 + 5(-1)\left(16p^4\right) + 10(1)\left(8p^3\right) + 10(-1)\left(4p^2\right) + 5(1)(2p) + 1(-1)$

    $= 32p^5 - 80p^4 + 80p^3 - 40p^2 + 10p - 1$

10.4 Exercises

11. $3! = 3 \cdot 2 \cdot 1 = 6$

13. $8! = 8 \cdot 7 \cdot 6 \cdot 5 \cdot 4 \cdot 3 \cdot 2 \cdot 1 = 40,320$

15. $\dfrac{10!}{8!} = \dfrac{10 \cdot 9 \cdot 8!}{8!} = 10 \cdot 9 = 90$

17. $\dfrac{8!}{5!} = \dfrac{8 \cdot 7 \cdot 6 \cdot 5!}{5!} = 8 \cdot 7 \cdot 6 = 336$

19. $\dbinom{7}{2} = \dfrac{7!}{2!(7-2)!} = \dfrac{7!}{2! \cdot 5!} = \dfrac{7 \cdot 6 \cdot 5!}{2 \cdot 5!} = \dfrac{7 \cdot 6}{2} = 21$

21. $\dbinom{10}{4} = \dfrac{10!}{4!(10-4)!} = \dfrac{10!}{4! \cdot 6!}$

$\qquad = \dfrac{10 \cdot 9 \cdot 8 \cdot 7 \cdot 6!}{4 \cdot 3 \cdot 2 \cdot 1 \cdot 6!}$

$\qquad = \dfrac{10 \cdot 9 \cdot 8 \cdot 7}{4 \cdot 3 \cdot 2 \cdot 1}$

$\qquad = 210$

23. $(x+1)^5 = \dbinom{5}{0}x^5 + \dbinom{5}{1}1^1 \cdot x^{5-1} + \dbinom{5}{2}1^2 \cdot x^{5-2} + \dbinom{5}{3}1^3 \cdot x^{5-3} + \dbinom{5}{4}1^4 \cdot x^{5-4} + \dbinom{5}{5}1^5$

$\qquad = x^5 + 5x^4 + 10x^3 + 10x^2 + 5x + 1$

25. $(x-4)^4 = (x+(-4))^4$

$\qquad = \dbinom{4}{0}x^4 + \dbinom{4}{1}(-4)^1 \cdot x^{4-1} + \dbinom{4}{2}(-4)^2 \cdot x^{4-2} + \dbinom{4}{3}(-4)^3 \cdot x^{4-3} + \dbinom{4}{4}(-4)^4$

$\qquad = x^4 + 4(-4)x^3 + 6(16)x^2 + 4(-64)x + 256$

$\qquad = x^4 - 16x^3 + 96x^2 - 256x + 256$

27. $(3p+2)^4 = ((3p)+2)^4$

$\qquad = \dbinom{4}{0}(3p)^4 + \dbinom{4}{1}2^1 \cdot (3p)^{4-1} + \dbinom{4}{2}2^2 \cdot (3p)^{4-2} + \dbinom{4}{3}2^3 \cdot (3p)^{4-3} + \dbinom{4}{4}2^4$

$\qquad = 81p^4 + 4 \cdot 2 \cdot 27p^3 + 6 \cdot 4 \cdot 9p^2 + 4 \cdot 8 \cdot 3p + 16$

$\qquad = 81p^4 + 216p^3 + 216p^2 + 96p + 16$

29. $(2z-3)^5$

$\qquad = ((2z)+(-3))^5$

$\qquad = \dbinom{5}{0}(2z)^5 + \dbinom{5}{1}(-3)^1 \cdot (2z)^{5-1} + \dbinom{5}{2}(-3)^2 \cdot (2z)^{5-2} + \dbinom{5}{3}(-3)^3 \cdot (2z)^{5-3} + \dbinom{5}{4}(-3)^4 \cdot (2z)^{5-4} + \dbinom{5}{5}(-3)^5$

$\qquad = 32z^5 + 5(-3) \cdot 16z^4 + 10(9) \cdot 8z^3 + 10(-27) \cdot 4z^2 + 5(81) \cdot 2z + (-243)$

$\qquad = 32z^5 - 240z^4 + 720z^3 - 1080z^2 + 810z - 243$

31. $\left(x^2+2\right)^4 = \dbinom{4}{0}\left(x^2\right)^4 + \dbinom{4}{1}2^1 \cdot \left(x^2\right)^{4-1} + \dbinom{4}{2}2^2 \cdot \left(x^2\right)^{4-2} + \dbinom{4}{3}2^3 \cdot \left(x^2\right)^{4-3} + \dbinom{4}{4}2^4$

$\qquad = x^8 + 4 \cdot 2x^6 + 6 \cdot 4x^4 + 4 \cdot 8x^2 + 16$

$\qquad = x^8 + 8x^6 + 24x^4 + 32x^2 + 16$

33. $\left(2p^3+1\right)^5 = \left(\left(2p^3\right)+1\right)^5$

$\qquad = \dbinom{5}{0}\left(2p^3\right)^5 + \dbinom{5}{1}1^1 \cdot \left(2p^3\right)^{5-1} + \dbinom{5}{2}1^2 \cdot \left(2p^3\right)^{5-2} + \dbinom{5}{3}1^3 \cdot \left(2p^3\right)^{5-3} + \dbinom{5}{4}1^4 \cdot \left(2p^3\right)^{5-4} + \dbinom{5}{5}1^5$

$\qquad = 32p^{15} + 5 \cdot 16p^{12} + 10 \cdot 8p^9 + 10 \cdot 4p^6 + 5 \cdot 2p^3 + 1$

$\qquad = 32p^{15} + 80p^{12} + 80p^9 + 40p^6 + 10p^3 + 1$

35. $(x+2)^6 = \binom{6}{0}x^6 + \binom{6}{1}2^1 \cdot x^{6-1} + \binom{6}{2}2^2 \cdot x^{6-2} + \binom{6}{3}2^3 \cdot x^{6-3} + \binom{6}{4}2^4 \cdot x^{6-4} + \binom{6}{5}2^5 \cdot x^{6-5} + \binom{6}{6}2^6$

$ = x^6 + 6 \cdot 2x^5 + 15 \cdot 4x^4 + 20 \cdot 8x^3 + 15 \cdot 16x^2 + 6 \cdot 32x + 64$

$ = x^6 + 12x^5 + 60x^4 + 160x^3 + 240x^2 + 192x + 64$

37. $\left(2p^2 - q^2\right)^4 = \left(\left(2p^2\right) + \left(-q^2\right)\right)^4$

$ = \binom{4}{0}\left(2p^2\right)^4 + \binom{4}{1}\left(2p^2\right)^{4-1} \cdot \left(-q^2\right)^1 + \binom{4}{2}\left(2p^2\right)^{4-2} \cdot \left(-q^2\right)^2 + \binom{4}{3}\left(2p^2\right)^{4-3} \cdot \left(-q^2\right)^3 + \binom{4}{4}\left(-q^2\right)^4$

$ = 16p^8 + 4 \cdot 8p^6 \cdot \left(-q^2\right) + 6 \cdot 4p^4 \cdot \left(q^4\right) + 4 \cdot 2p^2 \cdot \left(-q^6\right) + \left(q^8\right)$

$ = 16p^8 - 32p^6q^2 + 24p^4q^4 - 8p^2q^6 + q^8$

39. $(1.001)^4 = (1 + 0.001)^4$

$ = \binom{4}{0}1^4 + \binom{4}{1}1^{4-1} \cdot (0.001)^1 + \binom{4}{2}1^{4-2} \cdot (0.001)^2 + \binom{4}{3}1^{4-3} \cdot (0.001)^3 + \binom{4}{4}(0.001)^4$

$ = 1 + 4(0.001) + 6(0.000001) + 4(0.000000001) + (0.000000000001)$

$ = 1 + 0.004 + 0.000006 + 0.000000004 + 0.000000000001$

$ = 1.004006004001$

$ \approx 1.00401$

41. $(0.998)^5 = (1 - 0.002)^5$

$ = \left(1 + (-0.002)\right)^5$

$ = \binom{5}{0}1^5 + \binom{5}{1}1^{5-1}(-0.002)^1 + \binom{5}{2}1^{5-2}(-0.002)^2 + \binom{5}{3}1^{5-3}(-0.002)^3 + \binom{5}{4}1^{5-4}(-0.002)^4 + \binom{5}{5}(-0.002)^5$

$ = 1 + 5(-0.002) + 10(0.000004) + 10(-0.000000008) + 5(0.000000000016) + (-0.000000000000032)$

$ = 1 - 0.01 + 0.00004 - 0.00000008 + 0.00000000008 - 0.000000000000032$

$ = 1.000040000080000 - 0.010000080000032$

$ = 0.990039920079968$

$ \approx 0.99004$

43. The 3rd term of the expansion of $(x+2)^7$ is

$\binom{7}{2}2^2 \cdot x^{7-2} = 21 \cdot 4x^5 = 84x^5$

45. The 6th term of the expansion of $(2p-3)^8 = \left((2p) + (-3)\right)^8$ is

$\binom{8}{5}(-3)^5 \cdot (2p)^{8-5} = 56(-243) \cdot 8p^3$

$\phantom{\binom{8}{5}(-3)^5 \cdot (2p)^{8-5}} = -108,864p^3$

47. $\binom{n}{n-1} = \dfrac{n!}{(n-1)!\left(n - (n-1)\right)!} = \dfrac{n!}{(n-1)!1!} = \dfrac{n \cdot (n-1)!}{(n-1)!} = n$

$\binom{n}{n} = \dfrac{n!}{n!(n-n)!} = \dfrac{n!}{n!0!} = \dfrac{n!}{n!} = 1$

49.

$$\begin{array}{ccccccc} & & & 1 & & & \\ & & 1 & & 1 & & \\ & 1 & & 2 & & 1 & \\ 1 & & 3 & & 3 & & 1 \end{array}$$

51. The degree of each monomial (sum of the exponents) in the expansion of $(x+y)^n$ is equal to n.

53. $f(x) = x^4$

$$f(a-2) = (a-2)^4$$
$$= \binom{4}{0}a^4 + \binom{4}{1}(-2)^1 \cdot a^{4-1} + \binom{4}{2}(-2)^2 \cdot a^{4-2} + \binom{4}{3}(-2)^3 \cdot a^{4-3} + \binom{4}{4}(-2)^4$$
$$= a^4 + 4(-2)a^3 + 6(4)a^2 + 4(-8)a + (16)$$
$$= a^4 - 8a^3 + 24a^2 - 32a + 16$$

55. $H(x) = x^5 - 4x^4$

$$H(p+1) = (p+1)^5 - 4(p+1)^4$$
$$= \left[\binom{5}{0}p^5 + \binom{5}{1}1^1 \cdot p^{5-1} + \binom{5}{2}1^2 \cdot p^{5-2} + \binom{5}{3}1^3 \cdot p^{5-3} + \binom{5}{4}1^4 \cdot p^{5-4} + \binom{5}{5}1^5 \right]$$
$$\quad -4\left[\binom{4}{0}p^4 + \binom{4}{1}1^1 \cdot p^{4-1} + \binom{4}{2}1^2 \cdot p^{4-2} + \binom{4}{3}1^3 \cdot p^{4-3} + \binom{4}{4}1^4 \right]$$
$$= \left[p^5 + 5p^4 + 10p^3 + 10p^2 + 5p + 1 \right]$$
$$\quad -4\left[p^4 + 4p^3 + 6p^2 + 4p + 1 \right]$$
$$= p^5 + 5p^4 + 10p^3 + 10p^2 + 5p + 1 - 4p^4 - 16p^3 - 24p^2 - 16p - 4$$
$$= p^5 + p^4 - 6p^3 - 14p^2 - 11p - 3$$

Chapter 10 Review

1. $a_n = -3n + 2$

$$a_1 = -3(1) + 2 = -1$$
$$a_2 = -3(2) + 2 = -4$$
$$a_3 = -3(3) + 2 = -7$$
$$a_4 = -3(4) + 2 = -10$$
$$a_5 = -3(5) + 2 = -13$$

The first five terms of the sequence are $-1, -4, -7, -10,$ and -13.

2. $a_n = \dfrac{n-2}{n+4}$

$a_1 = \dfrac{1-2}{1+4} = \dfrac{-1}{5} = -\dfrac{1}{5}; \quad a_2 = \dfrac{2-2}{2+4} = \dfrac{0}{6} = 0;$

$a_3 = \dfrac{3-2}{3+4} = \dfrac{1}{7}; \quad a_4 = \dfrac{4-2}{4+4} = \dfrac{2}{8} = \dfrac{1}{4};$

$a_5 = \dfrac{5-2}{5+4} = \dfrac{3}{9} = \dfrac{1}{3}$

The first five terms of the sequence are $-\dfrac{1}{5}$, 0,

$\dfrac{1}{7}, \dfrac{1}{4}$, and $\dfrac{1}{3}$.

3. $a_n = 5^n + 1$

$a_1 = 5^1 + 1 = 5 + 1 = 6$

$a_2 = 5^2 + 1 = 25 + 1 = 26$

$a_3 = 5^3 + 1 = 125 + 1 = 126$

$a_4 = 5^4 + 1 = 625 + 1 = 626$

$a_5 = 5^5 + 1 = 3125 + 1 = 3126$

The first five terms of the sequence are 6, 26, 126, 626, and 3126.

4. $a_n = (-1)^{n-1} \cdot 3n$

$a_1 = (-1)^{1-1} \cdot 3(1) = (-1)^0 \cdot 3 = 3$

$a_2 = (-1)^{2-1} \cdot 3(2) = (-1)^1 \cdot 6 = -6$

$a_3 = (-1)^{3-1} \cdot 3(3) = (-1)^2 \cdot 9 = 9$

$a_4 = (-1)^{4-1} \cdot 3(4) = (-1)^3 \cdot 12 = -12$

$a_5 = (-1)^{5-1} \cdot 3(5) = (-1)^4 \cdot 15 = 15$

The first five terms of the sequence are 3, -6, 9, -12, and 15.

5. $a_n = \dfrac{n^2}{n+1}$

$a_1 = \dfrac{1^2}{1+1} = \dfrac{1}{2}, \qquad a_2 = \dfrac{2^2}{2+1} = \dfrac{4}{3},$

$a_3 = \dfrac{3^2}{3+1} = \dfrac{9}{4}, \qquad a_4 = \dfrac{4^2}{4+1} = \dfrac{16}{5},$

$a_5 = \dfrac{5^2}{5+1} = \dfrac{25}{6}$

The first five terms of the sequence are $\dfrac{1}{2}, \dfrac{4}{3}$,

$\dfrac{9}{4}, \dfrac{16}{5}$, and $\dfrac{25}{6}$.

6. $a_n = \dfrac{\pi^n}{n}$

$a_1 = \dfrac{\pi^1}{1} = \pi; \quad a_2 = \dfrac{\pi^2}{2};$

$a_3 = \dfrac{\pi^3}{3}; \qquad a_4 = \dfrac{\pi^4}{4};$

$a_5 = \dfrac{\pi^5}{5}$

The first five terms of the sequence are π, $\dfrac{\pi^2}{2}$,

$\dfrac{\pi^3}{3}, \dfrac{\pi^4}{4}$, and $\dfrac{\pi^5}{5}$.

7. The terms can be expressed as multiples of -3.

$-3 = -3 \cdot 1$

$-6 = -3 \cdot 2$

$-9 = -3 \cdot 3$

$-12 = -3 \cdot 4$

$-15 = -3 \cdot 5$

The nth term of the sequence is given by $a_n = -3n$.

8. The terms are rational numbers with a denominator of 3. The numerators are consecutive integers beginning with 1. The nth term of the sequence is given by $a_n = \dfrac{n}{3}$.

9. The terms can be expressed as the product of 5 and a power of 2.

$5 = 5 \cdot 2^0$

$10 = 5 \cdot 2^1$

$20 = 5 \cdot 2^2$

$40 = 5 \cdot 2^3$

$80 = 5 \cdot 2^4$

The exponent is one less than the term number. Therefore, the nth term of the sequence is given by $a_n = 5 \cdot 2^{n-1}$.

10. Rewrite as $-\dfrac{1}{2}, \dfrac{2}{2}, -\dfrac{3}{2}, \dfrac{4}{2}, \dots$

The terms are rational numbers with alternating signs. The denominator is always 2, and the numerators are consecutive integers beginning with 1. The nth term of the sequence is given by

$a_n = (-1)^n \cdot \dfrac{n}{2}$.

11. The terms can be expressed as 5 more than the square of the term number.

$$6 = 1^2 + 5$$
$$9 = 2^2 + 5$$
$$14 = 3^2 + 5$$
$$21 = 4^2 + 5$$
$$30 = 5^2 + 5$$

The nth term of the sequence is given by
$$a_n = n^2 + 5.$$

12. Rewrite the terms as $\dfrac{0}{2}, \dfrac{1}{3}, \dfrac{2}{4}, \dfrac{3}{5}, \ldots$

The terms are rational numbers whose numerators are consecutive integers beginning with 0 and whose denominators are consecutive integers beginning with 2. The nth term of the sequence is given by $a_n = \dfrac{n-1}{n+1}$.

13. $\displaystyle\sum_{k=1}^{5} (5k - 2)$
$$= (5(1) - 2) + (5(2) - 2) + (5(3) - 2)$$
$$+ (5(4) - 2) + (5(5) - 2)$$
$$= 3 + 8 + 13 + 18 + 23$$
$$= 65$$

14. $\displaystyle\sum_{k=1}^{6} \left(\dfrac{k+2}{2} \right)$
$$= \dfrac{1+2}{2} + \dfrac{2+2}{2} + \dfrac{3+2}{2} + \dfrac{4+2}{2} + \dfrac{5+2}{2} + \dfrac{6+2}{2}$$
$$= \dfrac{3}{2} + \dfrac{4}{2} + \dfrac{5}{2} + \dfrac{6}{2} + \dfrac{7}{2} + \dfrac{8}{2}$$
$$= \dfrac{33}{2}$$

15. $\displaystyle\sum_{i=1}^{5} (-2i)$
$$= (-2 \cdot 1) + (-2 \cdot 2) + (-2 \cdot 3) + (-2 \cdot 4) + (-2 \cdot 5)$$
$$= -2 - 4 - 6 - 8 - 10$$
$$= -30$$

16. $\displaystyle\sum_{i=1}^{4} \dfrac{i^2 - 1}{3} = \dfrac{1^2 - 1}{3} + \dfrac{2^2 - 1}{3} + \dfrac{3^2 - 1}{3} + \dfrac{4^2 - 1}{3}$
$$= \dfrac{0}{3} + \dfrac{3}{3} + \dfrac{8}{3} + \dfrac{15}{3}$$
$$= \dfrac{26}{3}$$

17. Each term takes on the form $4 + 3i$ and there are $n = 15$ terms.
$$(4 + 3 \cdot 1) + (4 + 3 \cdot 2) + \ldots + (4 + 3 \cdot 15) = \sum_{i=1}^{15} (4 + 3i)$$

18. Each term takes on the form $\dfrac{1}{3^i}$ and there are $n = 8$ terms.
$$\dfrac{1}{3^1} + \dfrac{1}{3^2} + \ldots + \dfrac{1}{3^8} = \sum_{i=1}^{8} \dfrac{1}{3^i}$$

19. Each term takes on the form $\dfrac{i^3 + 1}{i + 1}$ and there are $n = 10$ terms.
$$\dfrac{1^3 + 1}{1 + 1} + \dfrac{2^3 + 1}{2 + 1} + \ldots + \dfrac{10^3 + 1}{10 + 1} = \sum_{i=1}^{10} \dfrac{i^3 + 1}{i + 1}$$

20. This is an alternating series. Each term takes on the form $(-1)^{i-1} \cdot i^2$ and there are $n = 7$ terms.
$$(-1)^{1-1} \cdot 1^2 + (-1)^{2-1} \cdot 2^2 + \ldots + (-1)^{7-1} \cdot 7^2$$
$$= \sum_{i=1}^{7} \left[(-1)^{i-1} \cdot i^2 \right]$$

21. $10 - 4 = 6$
$16 - 10 = 6$
$22 - 16 = 6$
The sequence is arithmetic with a common difference of $d = 6$.

22. $\dfrac{1}{2} - (-1) = \dfrac{3}{2}$
$$2 - \dfrac{1}{2} = \dfrac{3}{2}$$
$$\dfrac{7}{2} - 2 = \dfrac{3}{2}$$
The sequence is arithmetic with a common difference of $d = \dfrac{3}{2}$.

23. $-5 - (-2) = -3$
$-9 - (-5) = -4$
$-14 - (-9) = -5$
The difference between consecutive terms is not constant. Therefore, the sequence is not arithmetic.

24. $3-(-1)=4$

$-5-3=-8$

$7-(-5)=12$

The difference between consecutive terms is not constant. Therefore, the sequence is not arithmetic.

25. $a_n - a_{n-1} = [4n+7] - [4(n-1)+7]$

$= 4n+7-(4n-4+7)$

$= 4n+7-4n+4-7$

$= 4$

The sequence is arithmetic with a common difference of $d = 4$.

26. $a_n - a_{n-1} = \left[\dfrac{n+1}{2n}\right] - \left[\dfrac{(n-1)+1}{2(n-1)}\right]$

$= \dfrac{n+1}{2n} - \dfrac{n}{2(n-1)}$

$= \dfrac{(n+1)(n-1)}{2n(n-1)} - \dfrac{n(n)}{2n(n-1)}$

$= \dfrac{n^2-1-n^2}{2n(n-1)}$

$= \dfrac{-1}{2n(n-1)}$

The difference between consecutive terms is not constant. Therefore, the sequence is not arithmetic.

27. $a_n = a_1 + (n-1)d$

$= 3 + (n-1)(8)$

$= 3 + 8n - 8$

$= 8n - 5$

$a_{25} = 8(25) - 5 = 200 - 5 = 195$

28. $a_n = a_1 + (n-1)d$

$= -4 + (n-1)(-3)$

$= -4 - 3n + 3$

$= -3n - 1$

$a_{25} = -3(25) - 1 = -75 - 1 = -76$

29. $d = \dfrac{20}{3} - 7 = -\dfrac{1}{3}$; $a_1 = 7$

$a_n = a_1 + (n-1)d$

$= 7 + (n-1)\left(-\dfrac{1}{3}\right)$

$= 7 - \dfrac{1}{3}n + \dfrac{1}{3}$

$= -\dfrac{1}{3}n + \dfrac{22}{3}$

$a_{25} = -\dfrac{1}{3}(25) + \dfrac{22}{3} = -\dfrac{3}{3} = -1$

30. $d = 17 - 11 = 6$; $a_1 = 11$

$a_n = a_1 + (n-1)d$

$= 11 + (n-1)(6)$

$= 11 + 6n - 6$

$= 6n + 5$

$a_{25} = 6(25) + 5 = 150 + 5 = 155$

31. Since $a_3 = 7$ and $a_8 = 25$, we have

$\begin{cases} 7 = a_1 + (3-1)d \\ 25 = a_1 + (8-1)d \end{cases}$ or $\begin{cases} 7 = a_1 + 2d \\ 25 = a_1 + 7d \end{cases}$

Subtract the second equation from the first equation to obtain

$-18 = -5d$

$\dfrac{18}{5} = d$

Let $d = \dfrac{18}{5}$ in the first equation to find a_1.

$7 = a_1 + 2\left(\dfrac{18}{5}\right)$

$7 = a_1 + \dfrac{36}{5}$

$-\dfrac{1}{5} = a_1$

The nth term of the sequence is given by

$a_n = a_1 + (n-1)d$

$= -\dfrac{1}{5} + (n-1)\left(\dfrac{18}{5}\right)$

$= -\dfrac{1}{5} + \dfrac{18}{5}n - \dfrac{18}{5}$

$= \dfrac{18}{5}n - \dfrac{19}{5}$

$a_{25} = \dfrac{18}{5}(25) - \dfrac{19}{5} = \dfrac{450}{5} - \dfrac{19}{5} = \dfrac{431}{5}$

32. Since $a_4 = -20$ and $a_7 = -32$, we have

$$\begin{cases} -20 = a_1 + (4-1)d \\ -32 = a_1 + (7-1)d \end{cases} \text{ or } \begin{cases} -20 = a_1 + 3d \\ -32 = a_1 + 6d \end{cases}$$

Subtract the second equation from the first equation to obtain

$$12 = -3d$$
$$-4 = d$$

Let $d = -4$ in the first equation to find a_1.

$$-20 = a_1 + 3(-4)$$
$$-20 = a_1 - 12$$
$$-8 = a_1$$

The nth term of the sequence is given by

$$\begin{aligned} a_n &= a_1 + (n-1)d \\ &= -8 + (n-1)(-4) \\ &= -8 - 4n + 4 \\ &= -4n - 4 \end{aligned}$$

$$a_{25} = -4(25) - 4 = -100 - 4 = -104$$

33. We know the first term is $a_1 = -1$ and the common difference is $d = 9 - (-1) = 10$.

$$\begin{aligned} S_n &= \frac{n}{2}\left[2a_1 + (n-1)d\right] \\ S_{30} &= \frac{30}{2}\left[2(-1) + (30-1)(10)\right] \\ &= 15\left[-2 + 29(10)\right] \\ &= 15(-2 + 290) \\ &= 15(288) \\ &= 4320 \end{aligned}$$

34. We know the first term is $a_1 = 5$ and the common difference is $d = 2 - 5 = -3$.

$$\begin{aligned} S_n &= \frac{n}{2}\left[2a_1 + (n-1)d\right] \\ S_{40} &= \frac{40}{2}\left[2(5) + (40-1)(-3)\right] \\ &= 20\left[10 + 39(-3)\right] \\ &= 20(10 - 117) \\ &= 20(-107) \\ &= -2140 \end{aligned}$$

35. $a_n = -2n - 7$

$$a_1 = -2(1) - 7 = -9$$
$$a_{60} = -2(60) - 7 = -127$$

$$S_n = \frac{n}{2}\left[a_1 + a_n\right]$$

$$\begin{aligned} S_{60} &= \frac{60}{2}\left[-9 + (-127)\right] \\ &= 30(-136) \\ &= -4080 \end{aligned}$$

36. $a_n = \frac{1}{4}n + 3$

$$a_1 = \frac{1}{4}(1) + 3 = \frac{13}{4}$$
$$a_{50} = \frac{1}{4}(50) + 3 = \frac{62}{4}$$

$$S_n = \frac{n}{2}\left[a_1 + a_n\right]$$

$$\begin{aligned} S_{50} &= \frac{50}{2}\left[\frac{13}{4} + \frac{62}{4}\right] \\ &= 25\left(\frac{75}{4}\right) \\ &= \frac{1875}{4} \quad \text{or} \quad 468.75 \end{aligned}$$

37. We could list terms of the sequence formed by the years when the Brood X cicada returns, or we could use the general formula for the nth term. We have $a_1 = 2004$ and $d = 17$. We wish to know the first time when $a_n \geq 2101$.

$$\begin{aligned} a_n &\geq 2101 \\ a_1 + (n-1)d &\geq 2101 \\ 2004 + (n-1)(17) &\geq 2101 \\ 2004 + 17n - 17 &\geq 2101 \\ 17n + 1987 &\geq 2101 \\ 17n &\geq 114 \\ n &\geq \frac{114}{17} \approx 6.71 \end{aligned}$$

The Brood X cicada will first appear in the 22nd century when $n = 7$.

$$a_7 = 2004 + (7-1)(17) = 2106$$

The Brood X cicada will first appear in the 22nd century in the year 2106.

38. Since the distance to each line is the same as the distance back to the goal line, we can consider the sequence of distances run away from the goal line and double our result. The sequence of distances is 10, 20, 30, 40, and 50. This is an arithmetic sequence with first term $a_1 = 10$ and fifth term $a_5 = 50$.

$$S_5 = \frac{5}{2}[10+50] = \frac{5}{2}(60) = 150$$

Doubling this result gives a total distance of 300 yards run by each player during wind sprints.

39. $\dfrac{2}{\frac{1}{3}} = 2 \cdot \dfrac{3}{1} = 6$

$\dfrac{12}{2} = 6$

$\dfrac{72}{12} = 6$

The ratio of consecutive terms is constant so the sequence is geometric with common ratio $r = 6$.

40. $\dfrac{3}{-1} = -3$

$\dfrac{-9}{3} = -3$

$\dfrac{27}{-9} = -3$

The ratio of consecutive terms is constant so the sequence is geometric with common ratio $r = -3$.

41. $\dfrac{1}{1} = 1$

$\dfrac{2}{1} = 2$

$\dfrac{6}{2} = 3$

The ratio of consecutive terms is not constant so the sequence is not geometric.

42. $\dfrac{4}{6} = \dfrac{2}{3}$

$\dfrac{\frac{8}{3}}{4} = \dfrac{8}{3} \cdot \dfrac{1}{4} = \dfrac{2}{3}$

$\dfrac{\frac{16}{9}}{\frac{8}{3}} = \dfrac{16}{9} \cdot \dfrac{3}{8} = \dfrac{2}{3}$

The ratio of consecutive terms is constant so the sequence is geometric with common ratio $r = \dfrac{2}{3}$.

43. $\dfrac{a_n}{a_{n-1}} = \dfrac{5 \cdot (-2)^n}{5 \cdot (-2)^{n-1}} = \dfrac{(-2)^{n-1} \cdot (-2)}{(-2)^{n-1}} = -2$

The ratio of consecutive terms is constant so the sequence is geometric with common ratio $r = -2$.

44. $\dfrac{a_n}{a_{n-1}} = \dfrac{3n-14}{3(n-1)-14} = \dfrac{3n-14}{3n-17}$

The ratio of consecutive terms is not constant so the sequence is not geometric.

45. $a_n = a_1 \cdot r^{n-1}$

$a_n = 4 \cdot 3^{n-1}$

$a_{10} = 4 \cdot 3^{10-1} = 4 \cdot 3^9 = 4(19,683) = 78,732$

46. $a_n = a_1 \cdot r^{n-1}$

$a_n = 8 \cdot \left(\dfrac{1}{4}\right)^{n-1}$

$a_{10} = 8 \cdot \left(\dfrac{1}{4}\right)^{10-1}$

$= 8 \cdot \left(\dfrac{1}{4}\right)^9$

$= 8 \cdot \left(\dfrac{1}{262,144}\right)$

$= \dfrac{1}{32,768}$

47. $a_n = a_1 \cdot r^{n-1}$

$a_n = 5 \cdot (-2)^{n-1}$

$a_{10} = 5 \cdot (-2)^{10-1} = 5 \cdot (-2)^9 = 5(-512) = -2560$

48. $a_n = a_1 \cdot r^{n-1}$

$a_n = 1000 \cdot (1.08)^{n-1}$

$a_{10} = 1000 \cdot (1.08)^{10-1}$

$= 1000 \cdot (1.08)^9$

$\approx 1000(1.999005)$

$= 1999.005$

49. This is a geometric series with first term $a_1 = 2$, common ratio $r = \dfrac{4}{2} = 2$, and $n = 15$ terms.

$$S_n = a_1 \cdot \frac{1 - r^n}{1 - r}$$

$$S_{15} = 2 \cdot \frac{1 - 2^{15}}{1 - 2}$$

$$= 2 \cdot \frac{-32,767}{-1}$$

$$= 65,534$$

50. This is a geometric series with first term $a_1 = 40$, common ratio $r = \dfrac{1}{8}$, and $n = 13$ terms.

$$S_n = a_1 \cdot \frac{1 - r^n}{1 - r}$$

$$S_{13} = 40 \cdot \frac{1 - \left(\dfrac{1}{8}\right)^{13}}{1 - \dfrac{1}{8}}$$

$$\approx 45.71428571$$

51. This is a geometric series with $n = 12$ terms. The first term is $a_1 = \dfrac{3}{4} \cdot 2^{1-1} = \dfrac{3}{4}$ and the common ratio is $r = 2$.

$$S_n = a_1 \cdot \frac{1 - r^n}{1 - r}$$

$$S_{12} = \frac{3}{4} \cdot \frac{1 - 2^{12}}{1 - 2}$$

$$= \frac{3}{4} \cdot \frac{-4095}{-1}$$

$$= \frac{12,285}{4} \quad \text{or} \quad 3071.25$$

52. This is a geometric series with $n = 16$ terms. The first term is $a_1 = -4\left(3^1\right) = -12$ and the common ratio is $r = 3$.

$$S_n = a_1 \cdot \frac{1 - r^n}{1 - r}$$

$$S_{16} = -12 \cdot \frac{1 - 3^{16}}{1 - 3}$$

$$= -12(21,523,360)$$

$$= -258,280,320$$

53. This is an infinite geometric series with first term $a_1 = 20 \cdot \left(\dfrac{1}{4}\right)^1 = 5$ and common ratio $r = \dfrac{1}{4}$.

Since the ratio is between -1 and 1, we can use the formula for the sum of an infinite geometric series.

$$S_\infty = \frac{a_1}{1 - r} = \frac{5}{1 - \dfrac{1}{4}} = \frac{5}{\dfrac{3}{4}} = 5 \cdot \frac{4}{3} = \frac{20}{3}$$

54. This is an infinite geometric series with first term $a_1 = 50 \cdot \left(-\dfrac{1}{2}\right)^{1-1} = 50$ and common ratio $r = -\dfrac{1}{2}$. Since the ratio is between -1 and 1, we can use the formula for the sum of an infinite geometric series.

$$S_\infty = \frac{a_1}{1 - r} = \frac{50}{1 - \left(-\dfrac{1}{2}\right)} = \frac{50}{\dfrac{3}{2}} = 50 \cdot \frac{2}{3} = \frac{100}{3}$$

55. This is an infinite geometric series with first term $a_1 = 1$ and common ratio $r = \dfrac{\dfrac{1}{5}}{1} = \dfrac{1}{5}$. The ratio is between -1 and 1, so we can use the formula for the sum of an infinite geometric series.

$$S_\infty = \frac{a_1}{1 - r} = \frac{1}{1 - \dfrac{1}{5}} = \frac{1}{\dfrac{4}{5}} = \frac{5}{4}$$

56. This is an infinite geometric series with first term $a_1 = 0.8$ and common ratio $r = \dfrac{0.08}{0.8} = 0.1$.

Since the ratio is between -1 and 1, we can use the formula for the sum of an infinite geometric series.

$$S_\infty = \frac{a_1}{1 - r} = \frac{0.8}{1 - 0.1} = \frac{0.8}{0.9} = \frac{8}{9}$$

57. Here we consider a geometric sequence with first term $a_1 = 200$ and common ratio $r = \dfrac{1}{2}$. Each term in the sequence represents the amount after each half-life, or the amount remaining after every 12 years. Since $\dfrac{72}{12} = 6$, we are interested in the 7$^\text{th}$ term of the sequence (that is, after 6 common ratios).

$a_n = a_1 \cdot r^{n-1}$

$a_7 = 200 \cdot \left(\dfrac{1}{2}\right)^{7-1}$

$\quad = 200 \cdot \left(\dfrac{1}{2}\right)^{6}$

$\quad = \dfrac{200}{64}$

$\quad = 3.125$

After 72 years, there will be 3.125 grams of the Tritium remaining.

58. The number of emails in each cycle forms a geometric sequence with first term $a_1 = 5$ and common ratio $r = 5$. To find the total e-mails sent after 15 minutes, we need to find the sum of the first 15 terms of this sequence.

$S_n = a_1 \cdot \dfrac{1 - r^n}{1 - r}$

$S_{15} = 5 \cdot \dfrac{1 - 5^{15}}{1 - 5}$

$\quad\ \approx 3.815 \times 10^{10}$

After 15 minutes a total of about 38.15 billion e-mails will have been sent.

59. $A = P \cdot \left[\dfrac{(1+i)^n - 1}{i}\right]$

Note that the total contribution each quarter is the sum of Scott's contribution and the matching contribution from his employer.
In this case we have $P = 900 + 450 = 1350$,

$n = 25 \cdot 4 = 100$, and $i = \dfrac{0.07}{4} = 0.0175$.

$A = 1350 \cdot \left[\dfrac{(1+0.0175)^{100} - 1}{0.0175}\right]$

$\quad = 1350 \cdot [266.7517679]$

$\quad = \$360,114.89$

After 25 years, Scott's 403(b) will be worth $\$360,114.89$.

60. Lump sum option:
For this option, we use the compound interest formula.

$A = P(1+i)^n$

In this case we have $P = \$28,000,000$,

$n = 26 \cdot 1 = 26$, and $i = \dfrac{0.065}{1} = 0.065$.

$A = 28,000,000 \cdot (1 + 0.065)^{26} = \$143,961,987.40$

Annuity option:
For this option, we use the annuity formula.

$A = P \cdot \left[\dfrac{(1+i)^n - 1}{i}\right]$

In this case we have $P = 2,000,000$,

$n = 26 \cdot 1 = 26$, and $i = \dfrac{0.065}{1} = 0.065$.

$A = 2,000,000 \cdot \left[\dfrac{(1 + 0.065)^{26} - 1}{0.065}\right]$

$\quad = 2,000,000 \cdot [63.71537769]$

$\quad = \$127,430,755.40$

The lump sum option would yield more money after 26 years.

61. $A = P \cdot \left[\dfrac{(1+i)^n - 1}{i}\right]$

In this case we have $A = 2,500,000$,

$n = 40 \cdot 12 = 480$, and $i = \dfrac{0.09}{12} = 0.0075$.

$2,500,000 = P \cdot \left[\dfrac{(1 + 0.0075)^{480} - 1}{0.0075}\right]$

$2,500,000 = P \cdot [4681.320273]$

$\qquad\quad P = \$534.04$

Sheri would need to contribute $\$534.04$, or about $\$534$, each month to reach her goal.

62. $A = P \cdot \left[\dfrac{(1+i)^n - 1}{i}\right]$

In this case we have $P = 400$, $n = 10 \cdot 12 = 120$,

and $i = \dfrac{0.0525}{12} = 0.004375$.

$A = 400 \cdot \left[\dfrac{(1 + 0.004375)^{120} - 1}{0.004375}\right]$

$\quad = 400 \cdot [157.3769632]$

$\quad = \$62,950.79$

$\dfrac{62,950.79}{340} \approx 185.15$

When Samantha turns 18, the plan will be worth $\$62,950.79$ and will cover about 185 credit hours.

63. $5! = 5 \cdot 4 \cdot 3 \cdot 2 \cdot 1 = 120$

64. $\dfrac{11!}{7!} = \dfrac{11 \cdot 10 \cdot 9 \cdot 8 \cdot 7!}{7!} = 11 \cdot 10 \cdot 9 \cdot 8 = 7920$

65. $\dfrac{10!}{6!} = \dfrac{10 \cdot 9 \cdot 8 \cdot 7 \cdot 6!}{6!} = 10 \cdot 9 \cdot 8 \cdot 7 = 5040$

66. $\dfrac{13!}{6!7!} = \dfrac{13 \cdot 12 \cdot 11 \cdot 10 \cdot 9 \cdot 8 \cdot 7!}{6 \cdot 5 \cdot 4 \cdot 3 \cdot 2 \cdot 1 \cdot 7!}$

$= \dfrac{13 \cdot 12 \cdot 11 \cdot 10 \cdot 9 \cdot 8}{6 \cdot 5 \cdot 4 \cdot 3 \cdot 2 \cdot 1}$

$= \dfrac{13 \cdot \cancel{12} \cdot 11 \cdot \cancel{10} \cdot \cancel{9}^{3} \cdot \cancel{8}^{4}}{\cancel{\cancel{2}} \cancel{6} \cdot \cancel{5} \cdot \cancel{4} \cancel{3} \cdot \cancel{2} \cdot 1}$

$= 13 \cdot 11 \cdot 3 \cdot 4$

$= 1716$

67. $\dbinom{7}{3} = \dfrac{7!}{3!4!}$

$= \dfrac{7 \cdot 6 \cdot 5 \cdot 4!}{3 \cdot 2 \cdot 1 \cdot 4!}$

$= \dfrac{7 \cdot 6 \cdot 5}{3 \cdot 2 \cdot 1}$

$= 35$

68. $\dbinom{10}{5} = \dfrac{10!}{5!5!}$

$= \dfrac{10 \cdot 9 \cdot 8 \cdot 7 \cdot 6 \cdot 5!}{5! \, 5 \cdot 4 \cdot 3 \cdot 2 \cdot 1}$

$= \dfrac{10 \cdot 9 \cdot 8 \cdot 7 \cdot 6}{5 \cdot 4 \cdot 3 \cdot 2 \cdot 1}$

$= 252$

69. $\dbinom{8}{8} = \dfrac{8!}{8!0!} = 1$

70. $\dbinom{6}{0} = \dfrac{6!}{0!6!} = 1$

71. $(z+1)^4 = \dbinom{4}{0}z^4 + \dbinom{4}{1}z^3 + \dbinom{4}{2}z^2 + \dbinom{4}{3}z + \dbinom{4}{4}$

$= 1 \cdot z^4 + 4 \cdot z^3 + 6 \cdot z^2 + 4 \cdot z + 1$

$= z^4 + 4z^3 + 6z^2 + 4z + 1$

72. $(y-3)^5 = \dbinom{5}{0}y^5 + \dbinom{5}{1}y^4 \cdot (-3) + \dbinom{5}{2}y^3 \cdot (-3)^2 + \dbinom{5}{3}y^2 \cdot (-3)^3 + \dbinom{5}{4}y \cdot (-3)^4 + \dbinom{5}{5} \cdot (-3)^5$

$= 1 \cdot y^5 + 5 \cdot y^4 \cdot (-3) + 10 \cdot y^3 \cdot 9 + 10 \cdot y^2 \cdot (-27) + 5 \cdot y \cdot 81 + (-243)$

$= y^5 - 15y^4 + 90y^3 - 270y^2 + 405y - 243$

73. $(3y+4)^6$

$$= \binom{6}{0}(3y)^6 + \binom{6}{1}(3y)^5(4) + \binom{6}{2}(3y)^4(4)^2 + \binom{6}{3}(3y)^3(4)^3 + \binom{6}{4}(3y)^2(4)^4 + \binom{6}{5}(3y)(4)^5 + \binom{6}{6}(4)^6$$

$$= 1 \cdot 729y^6 + 6 \cdot 243y^5 \cdot 4 + 15 \cdot 81y^4 \cdot 16 + 20 \cdot 27y^3 \cdot 64 + 15 \cdot 9y^2 \cdot 256 + 6 \cdot 3y \cdot 1024 + 4096$$

$$= 729y^6 + 5832y^5 + 19,440y^4 + 34,560y^3 + 34,560y^2 + 18,432y + 4096$$

74. $\left(2x^2 - 3\right)^4 = \binom{4}{0}\left(2x^2\right)^4 + \binom{4}{1}\left(2x^2\right)^3(-3) + \binom{4}{2}\left(2x^2\right)^2(-3)^2 + \binom{4}{3}\left(2x^2\right)(-3)^3 + \binom{4}{4}(-3)^4$

$$= 16x^8 + 4 \cdot 8x^6 \cdot (-3) + 6 \cdot 4x^4 \cdot 9 + 4 \cdot 2x^2 \cdot (-27) + 81$$

$$= 16x^8 - 96x^6 + 216x^4 - 216x^2 + 81$$

75. $(3p - 2q)^4 = \binom{4}{0}(3p)^4 + \binom{4}{1}(3p)^3(-2q) + \binom{4}{2}(3p)^2(-2q)^2 + \binom{4}{3}(3p)(-2q)^3 + \binom{4}{4}(-2q)^4$

$$= 81p^4 + 4 \cdot 27p^3 \cdot (-2q) + 6 \cdot 9p^2 \cdot \left(4q^2\right) + 4 \cdot 3p \cdot \left(-8q^3\right) + 16q^4$$

$$= 81p^4 - 216p^3q + 216p^2q^2 - 96pq^3 + 16q^4$$

76. $\left(a^3 + 3b\right)^5 = \binom{5}{0}\left(a^3\right)^5 + \binom{5}{1}\left(a^3\right)^4(3b) + \binom{5}{2}\left(a^3\right)^3(3b)^2 + \binom{5}{3}\left(a^3\right)^2(3b)^3 + \binom{5}{4}\left(a^3\right)(3b)^4 + \binom{5}{5}(3b)^5$

$$= a^{15} + 5a^{12} \cdot 3b + 10a^9 \cdot 9b^2 + 10a^6 \cdot 27b^3 + 5a^3 \cdot 81b^4 + 243b^5$$

$$= a^{15} + 15a^{12}b + 90a^9b^2 + 270a^6b^3 + 405a^3b^4 + 243b^5$$

77. The fourth term of the expansion of $(x-2)^8 = (x + (-2))^8$ is

$$\binom{8}{3}(-2)^3 x^5 = 56x^5 \cdot (-8) = -448x^5$$

78. The seventh term of the expansion of $(2x+1)^{11} = ((2x)+1)^{11}$ is

$$\binom{11}{6}1^6(2x)^5 = 462 \cdot 32x^5 = 14,784x^5$$

Chapter 10 Test

1. $-7 - (-15) = 8$

$\quad 1 - (-7) = 8$

$\quad 9 - 1 = 8$

The difference between consecutive terms is a constant so the sequence is arithmetic with $a_1 = -15$ and common difference $d = 8$.

2. $\dfrac{a_n}{a_{n-1}} = \dfrac{(-4)^n}{(-4)^{n-1}} = \dfrac{-4 \cdot (-4)^{n-1}}{(-4)^{n-1}} = -4$

The ratio of consecutive terms is a constant so the sequence is geometric with $a_1 = (-4)^1 = -4$ and common ratio $r = -4$.

3. $a_n = \dfrac{4}{n!}$; $a_{n-1} = \dfrac{4}{(n-1)!}$

   $a_n - a_{n-1} = \dfrac{4}{n!} - \dfrac{4}{(n-1)!} = \dfrac{4}{n!} - \dfrac{4n}{n(n-1)!}$

   $\qquad = \dfrac{4-4n}{n!}$

   $\dfrac{a_n}{a_{n-1}} = \dfrac{\frac{4}{n!}}{\frac{4}{(n-1)!}} = \dfrac{4}{n!} \cdot \dfrac{(n-1)!}{4} = \dfrac{1}{n}$

   The sequence is neither arithmetic nor geometric. The difference between consecutive terms is not constant so the sequence is not arithmetic. Likewise, the ratio of consecutive terms is not constant so the sequence is not geometric.

4. $a_n - a_{n-1} = \left[\dfrac{2n-3}{5}\right] - \left[\dfrac{2(n-1)-3}{5}\right]$

   $\qquad = \dfrac{2n-3-(2n-2-3)}{5}$

   $\qquad = \dfrac{2n-3-2n+5}{5} = \dfrac{2}{5}$

   The difference between consecutive terms is constant so the sequence is arithmetic with

   $a_1 = \dfrac{2(1)-3}{5} = -\dfrac{1}{5}$ and common difference

   $d = \dfrac{2}{5}$.

5. $2-(-3) = 5$

   $0-2 = 2$

   $\dfrac{2}{-3} = -\dfrac{2}{3}$

   $\dfrac{0}{2} = 0$

   The sequence is neither arithmetic nor geometric. The difference between consecutive terms is not constant so the sequence is not arithmetic. Likewise, the ratio of consecutive terms is not constant so the sequence is not geometric.

6. $\dfrac{a_n}{a_{n-1}} = \dfrac{7 \cdot 3^n}{7 \cdot 3^{n-1}} = \dfrac{7 \cdot 3 \cdot 3^{n-1}}{7 \cdot 3^{n-1}} = 3$

   The ratio of consecutive terms is a constant so the sequence is geometric with $a_1 = 7 \cdot 3^1 = 21$ and common ratio $r = 3$.

7. $\displaystyle\sum_{i=1}^{5}\left[\dfrac{3}{i^2}+2\right]$

   $= \left(\dfrac{3}{1^2}+2\right) + \left(\dfrac{3}{2^2}+2\right) + \left(\dfrac{3}{3^2}+2\right) + \left(\dfrac{3}{4^2}+2\right) + \left(\dfrac{3}{5^2}+2\right)$

   $= 3 + 2 + \dfrac{3}{4} + 2 + \dfrac{1}{3} + 2 + \dfrac{3}{16} + 2 + \dfrac{3}{25} + 2$

   $= 13 + \dfrac{900+400+225+144}{1200} = 13 + \dfrac{1669}{1200}$

   $= 14\dfrac{469}{1200}$ or $\dfrac{17269}{1200}$

8. Note that $\dfrac{2}{3} = \dfrac{4}{6}$, $\dfrac{3}{4} = \dfrac{6}{8}$, and $\dfrac{5}{6} = \dfrac{10}{12}$.

   Each term can be written as $\dfrac{i+2}{i+4}$, where i is the term number, and there are $n = 8$ terms.

   $\dfrac{3}{5} + \dfrac{2}{3} + \dfrac{5}{7} + \dfrac{3}{4} + ... + \dfrac{5}{6} = \displaystyle\sum_{i=1}^{8}\dfrac{i+2}{i+4}$

9. $a_n = a_1 + (n-1)d$

   $\qquad = 6 + (n-1)(10)$

   $\qquad = 6 + 10n - 10$

   $\qquad = 10n - 4$

   $a_1 = 10 \cdot 1 - 4 = 6$

   $a_2 = 10 \cdot 2 - 4 = 16$

   $a_3 = 10 \cdot 3 - 4 = 26$

   $a_4 = 10 \cdot 4 - 4 = 36$

   $a_5 = 10 \cdot 5 - 4 = 46$

   The first five terms of the sequence are 6, 16, 26, 36, and 46.

10. $a_n = a_1 + (n-1)d$

    $\qquad = 0 + (n-1)(-4)$

    $\qquad = -4n + 4$

    $\qquad = 4 - 4n$

    $a_1 = 4 - 4 \cdot 1 = 0$

    $a_2 = 4 - 4 \cdot 2 = -4$

    $a_3 = 4 - 4 \cdot 3 = -8$

    $a_4 = 4 - 4 \cdot 4 = -12$

    $a_5 = 4 - 4 \cdot 5 = -16$

    The first five terms of the sequence are 0, -4, -8, -12, and -16.

11. $a_n = a_1 \cdot r^{n-1}$
$\quad\;\; = 10 \cdot 2^{n-1}$
$\quad a_1 = 10 \cdot 2^{1-1} = 10$
$\quad a_2 = 10 \cdot 2^{2-1} = 20$
$\quad a_3 = 10 \cdot 2^{3-1} = 40$
$\quad a_4 = 10 \cdot 2^{4-1} = 80$
$\quad a_5 = 10 \cdot 2^{5-1} = 160$
The first five terms of the sequence are 10, 20, 40, 80, and 160.

12. $\quad\;\; a_n = a_1 \cdot r^{n-1}$
$\qquad\; a_3 = a \cdot (-3)^{3-1} = 9$
$\quad a(-3)^2 = 9$
$\qquad 9a = 9$
$\qquad\;\; a = 1$
$\quad a_n = 1 \cdot (-3)^{n-1}$
$\qquad\; = (-3)^{n-1}$
$\quad a_1 = (-3)^{1-1} = 1$
$\quad a_2 = (-3)^{2-1} = -3$
$\quad a_3 = (-3)^{3-1} = 9$
$\quad a_4 = (-3)^{4-1} = -27$
$\quad a_5 = (-3)^{5-1} = 81$
The first five terms of the sequence are 1, -3, 9, -27, and 81.

13. The terms form an arithmetic sequence with $n = 20$ terms, common difference $d = 4$, and first term $a_1 = -2$.

$$S_n = \frac{n}{2}\big[2a_1 + (n-1)d\big]$$
$$S_{20} = \frac{20}{2}\big[2(-2) + (20-1)(4)\big]$$
$$= 10(72)$$
$$= 720$$

14. The terms form a geometric sequence with $n = 12$ terms, common ratio $r = -3$, and first term $a_1 = \dfrac{1}{9}$.

$$S_n = a_1 \cdot \frac{1 - r^n}{1 - r}$$
$$S_{12} = \frac{1}{9} \cdot \frac{1 - (-3)^{12}}{1 - (-3)}$$
$$= \frac{1}{9} \cdot \frac{1 - 3^{12}}{1 + 3}$$
$$= -\frac{132{,}860}{9}$$

15. The terms form an infinite geometric series with first term $a_1 = 216$ and common ratio $r = \dfrac{1}{3}$.

$$S_\infty = \frac{a_1}{1-r} = \frac{216}{1-\left(\dfrac{1}{3}\right)} = \frac{216}{2/3} = 216 \cdot \frac{3}{2} = 324$$

16. $\dfrac{15!}{8!7!} = \dfrac{15 \cdot 14 \cdot 13 \cdot 12 \cdot 11 \cdot 10 \cdot 9 \cdot 8!}{8! \cdot 7 \cdot 6 \cdot 5 \cdot 4 \cdot 3 \cdot 2 \cdot 1}$

$\qquad = \dfrac{15 \cdot 14 \cdot 13 \cdot 12 \cdot 11 \cdot 10 \cdot 9}{7 \cdot 6 \cdot 5 \cdot 4 \cdot 3 \cdot 2 \cdot 1}$

$\qquad = 6435$

17. $\dbinom{12}{5} = \dfrac{12!}{5!7!} = \dfrac{12 \cdot 11 \cdot 10 \cdot 9 \cdot 8 \cdot 7!}{5 \cdot 4 \cdot 3 \cdot 2 \cdot 1 \cdot 7!}$

$\qquad = \dfrac{12 \cdot 11 \cdot 10 \cdot 9 \cdot 8}{5 \cdot 4 \cdot 3 \cdot 2 \cdot 1}$

$\qquad = 792$

18. $(5m - 2)^4 = \dbinom{4}{0}(5m)^4 + \dbinom{4}{1}(5m)^3(-2) + \dbinom{4}{2}(5m)^2(-2)^2 + \dbinom{4}{3}(5m)(-2)^3 + \dbinom{4}{4}(-2)^4$

$\qquad = 625m^4 + 4 \cdot 125m^3 \cdot (-2) + 6 \cdot 25m^2 \cdot 4 + 4 \cdot 5m \cdot (-8) + 16$

$\qquad = 625m^4 - 1000m^3 + 600m^2 - 160m + 16$

19. The values of the car form a geometric sequence with the first term $a_1 = 31,000$ and a common ratio of $r = 0.85$. To determine the value after 10 years, find a_{11}.

$a_{11} = a_1 \cdot r^{11-1}$

$a_{11} = 31,000 \cdot (0.85)^{11-1}$

$a_{11} = 6103.11$

The value of the car is \$6103.11 after 10 years.

20. Find the sum of the arithmetic sequence.

Sets	1	2	3	4	5
Pounds	100	130	160	190	220

$$\sum \frac{n}{2}(a_1 + a_n) = \sum \frac{5}{2}(100 + 220)$$
$$= 800$$

If he does 10 reps in each set, the total weight is $800 \times 10 = 8000$ lb.